International Exhibition & Conference for Power Electronics, Intelligent Motion, Renewable Energy and Energy Management (PCIM Europe 2024)

Nuremberg, Germany
11 – 13 June 2024

Volume 1 of 5

ISBN: 978-1-7138-9966-2

Printed from e-media with permission by:

Curran Associates, Inc.
57 Morehouse Lane
Red Hook, NY 12571

Some format issues inherent in the e-media version may also appear in this print version.

Copyright© (2024) by Mesago Messe Frankfurt GmbH
All rights reserved.

Printed with permission by Curran Associates, Inc. (2024)

For permission requests, please contact VDE VERLAG GMBH
at the address below.

VDE VERLAG GMBH
Bismarckstr. 33
P.O.B. 12 01 43
10625 Berlin, Germany

Phone: +49 30 34 80 01 - 0
Fax: +49 30 34 80 01 - 9088

kundenservice@vde-verlag.de

Additional copies of this publication are available from:

Curran Associates, Inc.
57 Morehouse Lane
Red Hook, NY 12571 USA
Phone: 845-758-0400
Fax: 845-758-2634
Email: curran@proceedings.com
Web: www.proceedings.com

TABLE OF CONTENTS

VOLUME 1

KEYNOTE

K01   AI BETWEEN HYPE AND INDUSTRIAL-GRADE - THE IMPACT OF AI ON THE ENTIRE POWER ELECTRONICS LIFECYCLE.. 1
  Rolf Hellinger

K02   INFRASTRUCTURE REQUIREMENTS FOR ELECTRIFIED HEAVY GOODS TRANSPORT IN GERMANY AND THE EU.. 7
  Martin Wietschel

K03   CHALLENGES AND SOLUTIONS TO POWER LATEST PROCESSOR GENERATIONS FOR HYPER SCALE DATACENTERS 15
  Gerald Deboy

GAN RUGGEDNESS

OP001   AN IMPROVED ULTRAFAST DESATURATION-BASED PROTECTION SCHEME FOR GAN HEMT 19
  Juncheng Lu

OP002   THE PERFORMANCE OF A GAN EMODE HEMT IN SURGE CURRENT SCENARIOS SUCH AS THE ACTIVE SHORT CIRCUIT.. 24
  Dominik Nehmer

OP003   GATE RESISTANCE EFFECT ON SHORT-CIRCUIT ROBUSTNESS OF P-GAN HEMTS 34
  Mohamed Lemine Dedew

ADVANCED PACKAGING TECHNOLOGIES

OP004   NEURAL NETWORK ASSISTED NUMERICAL SIMULATION BENCHMARKING FOR ELECTRIC VEHICLE THERMAL MANAGEMENT SYSTEM 40
  Ekin Alp Bicer

OP005   RELATIONSHIP BETWEEN POROSITY IN CU SINTERED BONDING AND BONDING RELIABILITY.. 49
  Hideo Nakako

OP006   HIGH THERMAL DURABILITY OF THIN COPPER DIE-ATTACH LAYERS AND FINITE ELEMENT MODEL SIMULATION.. 56
  Takaaki Eyama

THERMAL CYCLING RELIABILITY

OP007   THERMAL SHOCK TEST LIFETIME IMPROVEMENT WITH OPTIMIZED ADHESIVE STRENGTH BETWEEN EPOXY RESIN AND COPPER 62
  He Kangjia

OP008 POWER CYCLING RELIABILITY AND FAILURE MODE ANALYSIS OF POL 67
Kenichi Koi

OP009 ACCELERATED POWER CYCLING OF GAN HEMTS USING SWITCHING LOSS
AND FAST TEMPERATURE MEASUREMENT 74
Wing Tai Leung

HIGH POWER CONVERTERS

OP010 CONTROL OF AN MMC-BASED HYBRID TRANSFORMER WITH STAR-POINT
VOLTAGE INJECTION 84
Rui Wang

OP011 PROTECTION AND CONTROL OF A DUAL MMC MEDIUM VOLTAGE SUPPLY 93
Max Dupont

OP012 STATION POWER ELECTRONICS CONVERTER WITH HIGH THERMAL
ENDURANCE TO POLE-TO-POLE SHORT CIRCUITS FOR LVDC DISTRIBUTION GRID 103
Frédéric Reymond-Laruina

GATE DRIVERS

OP013 SUPPRESSION OF OSCILLATIONS IN A SIC BRIDGE-LEG USING A CUSTOM
SINGLE-CHIP DIGITAL ACTIVE GATE DRIVER WITH 2×255 STRENGTH LEVELS 113
Qilei Wang

OP014 SIC MOSFET SHORT-CIRCUIT PROTECTION: A FASTER SOFT SHUT DOWN
METHOD FOR GATE DRIVERS 121
Julien Weckbrodt

OP015 PARAMETER IDENTIFICATION: GATE SENSOR FOR POWER TRANSISTOR
TOLERANCE COMPENSATION IN ADVANCED GATE DRIVER ICS 128
Christopher Wille

ADVANCED CONTROL TECHNIQUES ON ELECTRICAL DRIVES I

OP016 AN INNOVATIVE HIGH-SPEED TRACK RANGE RESTART STRATEGY FOR
PERMANENT MAGNET SYNCHRONOUS MOTOR 135
Anna Corbitt

OP017 STEADY-STATE ERROR REDUCTION OF REINFORCEMENT LEARNING BASED
INDIRECT CURRENT CONTROL OF PERMANENT MAGNET SYNCHRONOUS
MACHINES 140
Tobias Schindler

OP018 PERFORMANCE COMPARISON OF USING SHUNT-BASED AND INTEGRATED
CURRENT SENSING FOR SENSORLESS FIELD-ORIENTED CONTROL 150
John Emmanuel Tan

GAN CONVERTERS

OP019 DESIGN OF HIGH-POWER INVERTER WITH 12 PARALLEL GAN DEVICES 161
Takashi Sawada

OP020 OVER 99.7% EFFICIENT GAN-BASED 6-LEVEL CAPACITIVE-LOAD POWER CONVERTER 167
Stefan Mönch

OP021 CASCADED PRIMARY-SIDE-ONLY CONTROL OF A COMPACT 2 MHZ 500 W WIRELESS POWER TRANSFER SYSTEM 174
Tim Krigar

ADVANCED MATERIALS AND TECHNOLOGIES

OP022 POWER MODULE EVALUATION USING ULTRA HIGH HEAT DISSIPATION AND HIGH HEAT RESISTANCE RESIN SHEET CONTAINING CARD HOUSE TYPE BORON NITRIDE FILLER 180
Ayano Imai

OP023 INVESTIGATING TEMPERATURE DEPENDENT WARPAGE IN METAL CERAMIC SUBSTRATES FOR POWER ELECTRONICS DEVICES 190
Benjamin Fabian

OP024 DEGRADATION MODE ANALYSIS OF DIFFERENT BONDING TECHNOLOGIES OF SIC POWER SEMICONDUCTORS STRESSED BY ACTIVE POWER CYCLING 197
Rasched Sankari

CHARGING STATION TECHNOLOGY

OP025 IMPLEMENTATION AND VERIFICATION OF A 50KW OPPORTUNITY WIRELESS CHARGER DESIGN 205
Carlos Costas Sos

OP026 PERFORMANCE EVALUATION OF SILICON-BASED 3-LEVEL VIENNA RECTIFIER IN ISOPLUS SMPD PACKAGE 214
Karsten Haehre

OP027 PERFORMANCE ANALYSIS OF A 25-KW SIC-BASED DUAL ACTIVE BRIDGE CONVERTER BASED ON PARALLEL-CONNECTED DEVICES 222
Francesco Porpora

MODELLING AND MONITORING

OP028 SEMICONDUCTOR CHIP MODELS ARE THE KEY FOR ENABLING VIRTUAL DESIGN AND OPTIMIZATION WORKFLOWS OF POWER ELECTRONIC SYSTEMS 230
Stefan Haensel

OP029 IMPROVED RESONANT FREQUENCY-BASED PARASITIC INDUCTANCE ESTIMATION METHOD FOR SIC MOSFET HALF-BRIDGE CIRCUIT 238
Hongpeng Zhang

OP030 FAST SIMULATOR WITH INVERTER TEMPERATURE ESTIMATION FOR TRACTION EDRIVES IN VEHICLES SUBJECTED TO DRIVING CYCLES 248
Simone Giuffrida

SOLID STATE TRANSFORMERS

OP031 A NEW FAMILY OF THREE-PHASE-UNFOLDER-BASED MVAC-LVDC SOLID-STATE TRANSFORMERS.. 254
Jonas Huber

OP032 VOLTAGE BALANCING OF A SPLIT-CAPACITOR IGCT 3L-NPC LEG FOR THE RESONANT DC TRANSFORMER... 264
Renan Pillon Barcelos

OP033 COMPARATIVE ANALYSIS OF UNIDIRECTIONAL HIGH STEP-UP CONVERTERS FOR MEDIUM VOLTAGE APPLICATIONS 274
Stefan Subotic

ADVANCED CONTROL TECHNIQUES ON ELECTRICAL DRIVES II

OP034 STARTUP BEHAVIOR OF HARMONIC SUPPRESSION IN ELECTRICAL MACHINES USING ITERATIVE LEARNING CONTROL AND NEURAL NETWORKS......................... 284
Annette Mai

OP035 ANALYTICAL APPROACH OF THE VECTOR CURRENT CONTROL FLUX-WEAKENING STRATEGY FOR PERMANENT MAGNET SYNCHRONOUS MACHINES 290
Oriol Subirats Rillo

POWER ELECTRONICS FOR E-MOBILITY

OP036 INVESTIGATION ON DIRECT LIQUID COOLING DESIGN OF POWER MODULES WITH FLAT BASEPLATE FOR AUTOMOTIVE APPLICATION.. 298
Nobuhide Arai

OP037 A NOVEL APPROACH FOR AFFORDABLE ELECTRIC VEHICLES BASED ON DUAL 48V BATTERY SYSTEM WITH MULTI-FUNCTIONAL 3-LEVEL CONVERTER........................ 305
Radovan Vuletic

OP038 AN INNOVATIVE 3-LEVEL SOLUTION FOR AUTOMOTIVE APPLICATIONS: EMPACK.. 315
Pranav Panchal

OP039 GATED RECURRENT UNITS-ASSISTED STATE-SPACE MODELING FOR ELECTRIC VEHICLE TEMPERATURE PREDICTION 322
Xinyuan Liao

OP040 NOVEL BIDIRECTIONAL SINGLE-STAGE ISOLATED 600-V GAN M-BDSBASED SINGLE/THREE-PHASE-OPERABLE EV ON-BOARD CHARGER 330
Sven Weihe

ENCAPSULATION MATERIALS

OP041 APPLICATION-SPECIFIC INVESTIGATION OF INORGANIC POTTING MATERIAL IN DRIVE TRAINS 338
Soenke Fleck

OP042   THE INFLUENCE OF THE GLASS TRANSITION TEMPERATURE OF EPOXY MOLD COMPOUNDS ON THE RELIABILITY OF A SEMICONDUCTOR DEVICE 343
   Stefan Schwab

OP043   CORROSION RESISTANT PACKAGING FOR POWER SEMICONDUCTOR MODULES - MODIFIED INSULATION MATERIALS FOR CONTAMINATED ENVIRONMENTS 351
   Michael Hanf

OP044   INVESTIGATION OF INORGANIC ENCAPSULATION MATERIALS IN POWER ELECTRONIC SYSTEMS FOR HIGH POWER DENSITY APPLICATIONS 361
   Stefan Behrendt

OP045   CHARACTERIZATION OF THERMALLY AGED SILICONE GELS FOR POWER SEMICONDUCTOR MODULES.. 369
   Sonja Madloch

POWER QUALITY

OP046   A COORDINATED CONTROL OF HYBRID SINGLE-PHASE AC/DC MICROGRIDS BASED ON THE NATURAL HARMONIC INJECTION CONCEPT 378
   Mehdi Baharizadeh

OP047   A HIGH-POWER DENSITY SIC BASED TP PFC WITH HIGH-FREQUENCY RIPPLE CANCELLATION LEG.. 383
   Serkan Dusmez

OP048   HIGH FREQUENCY ACTIVE FILTER FOR AC-DC HIGH POWER CONVERTERS 390
   Sarah Sifoune

OP049   LABORATORY SETUP FOR ACCURACY INVESTIGATION OF ELECTRICITY METERS AND MONITORS UNDER INDUSTRY-TYPICAL OPERATING CONDITIONS...................... 397
   Matthias Schmidt

GRID CONNECTED CONVERTERS

OP050   REAL-TIME EVALUATION OF WEIGHTING FACTORLESS PREDICTIVE CONTROL OF LCL FILTER EQUIPPED GRID-SIDE CONVERTERS USING SORTING NETWORKS.. 403
   Kristóf Bándy

OP051   RELAXED ROBUST CONTROL WITH PRAGMATIC SHORTAGE OF PASSIVITY FOR WIND, STORAGE AND PV POWER CONVERTERS 411
   Sergio De Lopez Diz

OP052   AN EFFECTIVE DC VOLTAGE REGULATION OF ACTIVE FRONT-END RECTIFIER THROUGH MODEL PREDICTIVE CONTROL.. 419
   Mobina Pouresmaeil

OP053   BI-DIRECTIONAL 11KW MULTI-LEVEL ACTIVE-NEUTRAL-POINT-CLAMPED AC-DC CONVERTER USING 600V/750V SI SUPER-JUNCTION AND SIC MOSFETS FOR HIGH-EFFICIENCY AND HIGH-DENSITY APPLICATIONS... 424
   Mengxing Chen

OP054   A STUDY OF GRID-FORMING INVERTER CONTROL STRATEGY FOR FAULT-RIDE-THROUGH CAPABILITY.. 433
Hirofumi Uemura

PASSIVE COMPONENTS

OP055   FILM CAPACITORS FOR HIGH TEMPERATURE AC-DC INVERTER APPLICATIONS... 440
Adel Bastawros

OP056   LOSS REDUCTION IN HF-TRANSFORMERS USING LAMINATED FERRITE E-CORES... 447
Lukas Reißenweber

OP057   MULTIGAP TOROIDAL TRANSFORMER AND INDUCTORS FOR OVERCOMING FRINGING LOSSES IN HIGH FREQUENCY CONVERTERS.................................... 456
Pau Colomer

OP058   STUDY ON SAMPLE GEOMETRIES FOR FERRITE CHARACTERISATION IN THE MHZ RANGE.. 463
Till Piepenbrock

OP059   FEM-SUPPORTED AND NON-DESTRUCTIVE MAGNETIC CHARACTERIZATION METHOD FOR NON-LAMINATED STEEL 472
Stefan Tobler

DRIVES FOR HIGH DEMANDING APPLICATIONS

OP060   HIGHLY-COMPACT BEARINGLESS AXIAL-FLUX MOTOR FOR A PEDIATRIC IMPLANTABLE FONTAN BLOOD PUMP 480
Andreas Horat

OP061   A NOVEL PERMANENT MAGNET SYNCHRONOUS MOTOR DRIVE FOR REACTION WHEELS IN SATELLITES.. 490
Baris Colak

OP062   EXPLORING HIGH FREQUENCY OPERATION OF MOTOR DRIVES: PRACTICAL INSIGHTS ON EFFICIENCY AND LOSS 497
Asantha Kempitiya

OP063   HIGH POWER DENSITY SYSTEM DESIGN FOR GAN-BASED LV MOTOR DRIVES 502
Marco Cannone

OP064   DESIGN OF GAN TRANSISTOR BASED VARIABLE SPEED DRIVE INVERTER WITH OUTPUT VOLTAGE FILTERING.. 510
Kaspars Kroics

IGBT

OP065   THE 8TH GENERATION LV100 IGBT MODULE WITH HIGHER CURRENT RATING 518
Daichi Otori

LOSS REDUCTION BY LAMINATIGN FERRITE E CORES.. 525
Lukas Reißenweber

OP066   NEW PLANAR 4.5 KV SPLIT-GATE (SG) SI-IGBT DEVICE FOR IMPROVED
SWITCHING CHARACTERISTICS AND HIGH FREQUENCY OPERATION 534
Gaurav Gupta

OP067   4.5 KV DOUBLE-GATE REVERSE-CONDUCTING PRESS-PACK IEGT 543
Satoshi Yoshida

DEVICE CONCEPTS

OP068   EVALUATION OF A 3 KV POLARIZATION SUPERJUNCTION GAN HEMT.......................... 549
Alireza Sheikhan

OP069   MORE THAN 1200 V BREAKDOWN AND LOW AREA-SPECIFIC ON STATE
RESISTANCES BY PROGRESS IN LATERAL GAN-ON-SI AND GAN-ON-INSULATOR
TECHNOLOGIES.. 557
Richard Reiner

OP070   NOVEL 200 V MOSFET TECHNOLOGY PUSHES MOTOR DRIVE INVERTER
EFFICIENCY TO AN UNPRECEDENTED LEVEL 564
Mark Thomas

DEGRADATION MECHANISMS

OP071   MOISTURE ROBUST CHIP DESIGN - IMPROVED EDGE-TERMINATIONS FOR
HIGH LIFETIME UNDER HIGH HUMID CONDITIONS.. 571
Michael Hanf

OP072   METHOD FOR MEASURING THE INITIAL STATE OF A SOLDER JOINT
DELAMINATION IN A 3D PCB INTEGRATION ASSEMBLY OF SIC MOSFETS................. 581
Souhila Bouzerd

OP073   GENERIC LIFETIME MODEL FOR WIRE BONDS DEGRADATION IN IGBT
MODULES BASED ON A FRACTURE MECHANICS PARAMETER...................................... 589
Merouane Ouhab

ADVANCED CONVERSION CONCEPTS

OP074   MODULAR COAXIAL POWER CONVERTER FOR HIGH-DENSITY INTEGRATION
INTO MEDIUM-VOLTAGE CABLES 599
Mark Cairnie

OP075   CONTROLLED INDUCTOR BASED BCM BUCK CONVERTERS 608
Ziv Gellman

OP076   INFLUENCE OF VARYING COMMON MODE CHOKE SIZES ON THE
PERFORMANCE AND STABILITY OF AN ACTIVE EMI FILTER....................................... 615
Patrick Körner

PHOTOVOLTAIC SYSTEMS

OP077   A HIGH EFFICIENCY BATTERY CHARGER WITH MAXIMUM POWER POINT TRACKING FOR MAGNETIC ENERGY HARVESTERS 625
Antonio Miguel Munoz Gomez

OP078   SYMMETRIC FLYING-CAPACITOR BOOST CONVERTER FOR MEDIUM-VOLTAGE PHOTOVOLTAIC APPLICATIONS 635
Luis Alves Rodrigues

OP079   COMPARISON OF SI IGBT, SIC MOSFET AND ADJUSTABLE HYBRID SWITCH PV INVERTERS FOR DIFFERENT GEOGRAPHICAL LOCATIONS 645
Tanya Thekemuriyil

MODEL BASED SYSTEM ANALYSIS

OP080   OPTIMISING A POWER MODULE FOR ELECTRICAL AND THERMAL PERFORMANCE AND SYMMETRY USING EDA TOOLS 655
Wilfried Wessel

OP081   CONDUCTOR-BASED MODELING OF VOLTAGE DISTRIBUTION ALONG A SINGLE-TOOTH WINDING OF ELECTRICAL MACHINES 665
Hujun Peng

OP082   REDUCTION OF PWM HARMONICS WITH CARRIER PHASE SHIFTING IN A DUAL-STATOR PMSM WITH MAGNETIC COUPLED WINDINGS 672
Bünyamin Tekir

VOLUME 2

SIC DEVICES

OP083   THE NEW COOLSIC MOSFET 1200 V G2: ELECTRICAL PERFORMANCE AND COMPACT MODELLING 681
Andreas Huerner

OP084   PARALLELING SIC-POWER-MOSFET BODY DIODES UNDER HARSH SWITCHING CONDITIONS 690
Michael Rauh

OP085   3.3KV SBD-EMBEDDED SIC-MOSFET MODULE FOR TRACTION USE 699
Yoichi Hironaka

OP086   DEAD TIME OPTIMIZATION FOR HIGH POWER SIC MOSFET MODULE IN CONSIDERATION OF PARASITIC COMPONENTS 707
Pham Ha Trieu To

WBG RELIABILITY

OP087   PERFORMANCE INSTABILITY OF 650 V P-GAN GATE HEMT DEVICE UNDER TEMPERATURE-RELATED POSITIVE GATE BIAS STRESSES 717
Renze Yu

OP088 GATE OXIDE RELIABILITY OF CURRENT GENERATION 1.2 KV SIC MOSFETS UNDER STEP-WISE INCREASED GATE VOLTAGE... 723
Roman Boldyrjew-Mast

OP089 AN ACCELERATED DYNAMIC GATE SWITCHING STRESS TEST CONCEPT OF SIC MOSFETS AT HIGH DRAIN-SOURCE VOLTAGE (HV-GSS) 731
Clemens Herrmann

OP090 SILICON CARBIDE POWER DEVICE USE IN SPACECRAFT AND AIRCRAFT 739
Akin Akturk

POWER ELECTRONICS FOR E-MOBILITY/ CONTROL

OP091 CURRENT RIPPLE REDUCTION BY COMBINATION OF SI IGBT AND SIC MOSFETS IN HEAVY DUTY FUEL CELL TRUCKS.. 745
Yavuz Gürlek

OP092 EVALUATION OF ACTIVE GATE DRIVERS WITH SWITCHABLE GATE RESISTORS AND INTERMEDIATE VOLTAGE LEVELS FOR SIC MOSFETS IN WLTC 754
Michael Frank

OP093 PERFORMANCE EVALUATION OF TCM-BASED, ZERO-VOLTAGE SWITCHING (ZVS) THREE-PHASE INVERTER FOR ELECTRIC VEHICLE DRIVE SYSTEMS 764
Khizra Abbas

OP094 A PARTIAL LOAD THREE-PHASE TRIANGULAR CURRENT MODE MODULATION CONCEPT WITH AN OPTIMIZED FILTER INDUCTOR FOR HIGH EFFICIENCY TRACTION DRIVES 774
Bhaskar Chatterjee

DC-DC CONVERTERS I

OP095 GAN VS SI SYNCHRONOUS RECTIFIER FOR LLC CONVERTER 784
Gokhan Sen

OP096 CO-SIMULATION DESIGN OF A GAN-BASED THREE-PHASE LLC CONVERTER WITH INTEGRATED THREE-PHASE MAGNETICS 791
Jhih-Cheng Hu

OP097 SWITCHING ASSISTING CIRCUIT IMPROVING THE EFFICIENCY OF DC-DC CONVERTERS BASED ON PIEZOELECTRIC RESONATORS... 797
Ghislain Despesse

OP098 TRANSFORMER-BASED FIXED-RATIO RESONANT DC-DC CONVERTERS FOR 48V DATA CENTERS 803
Xufu Ren

PFC CONVERTERS

OP099 HIGH-DENSITY 3.3 KW GAN RECTIFIER FOR SERVER APPLICATIONS COMPRISING A 130 KHZ TOTEM-POLE PFC AND A 500 KHZ LLC....................................... 812
Manuel Escudero Rodriguez

OP100 ADDRESSING POWER SWITCH TECHNOLOGY SELECTION SI/SIC/GAN IN HIGH EFFICIENCY ZVS-PFC RESONANT CONVERTERS 822
Marco Torrisi

OP101 BUCK-TYPE CURRENT UNFOLDING CONVERTER WITH DISCONTINUOUS CONDUCTION MODE IN ULTRA-LOW POWER-FACTOR OPERATION 831
Tomoyuki Mannen

OP102 GAN BASED BI-DIRECTIONAL 6.6KW INTERLEAVED TOTEM-POLE PFC WITH 13KW/L POWER DENSITY AND HIGH EFFICIENCY 837
Juncheng Lu

SIC MODULES

OP103 THE DESIGN OF A 2KV 1700A SIC MOSFET DUAL MODULE 843
Jorge Mari

OP104 TECHNOLOGICAL APPROACHES TO HIGH-POWER DENSITY SIC POWER MODULE FOR AUTOMOTIVE 849
Takeshi Tokorozuki

OP105 EXTREMELY COMPACT SIC POWER MODULE FOR EV TRACTION INVERTERS IN THE 250 KW CLASS 855
Raffael Schnell

OP106 BENEFITS OF .XT INTERCONNECTION TECHNOLOGY FOR 3.3 KV XHP 2 MODULE WITH 3.3 KV COOLSIC MOSFET 863
Matthias Bürger

ADVANCED COOLING

OP107 LARGE-AREA BONDING WITH LMEE: SUPPRESSION OF THE DEGRADATION OF THE JUNCTION-TO-WATER THERMAL RESISTANCE IN POWER MODULES 870
Yo Mochizuki

OP108 ACTIVE THERMAL CONTROL OF SIC MOSFETS UTILIZING TRANSIENT THERMAL CHARACTERIZATION 875
Varaha Satya Bharath Kurukuru

OP109 THERMAL MANAGEMENT SOLUTIONS BY ADDITIVE MANUFACTURING – POWDER BED FUSION AND DIFFUSION BONDING 883
Simon Jahn

OP110 ADVANCED PUMPED TWO-PHASE COLD PLATE FOR COOLING POWER ELECTRONICS 888
Elizabeth Seber

DC-DC CONVERTERS II

OP111 FEASIBILITY STUDY OF HIGH-POWER DENSITY ISOLATED CLLC DC-DC INTERFACE WITH WIDE RANGE OF VOLTAGE/CURRENT REGULATION 893
Oleksandr Husev

OP112   DC-BIAS REDUCTION IN HIGH-FREQUENCY DUAL ACTIVE BRIDGE DC-DC
CONVERTERS THROUGH SLOW DC MEASUREMENTS 903
Patrick Lenzen

OP113   OPTIMIZED CURRENT SHARING TECHNIQUE FOR INTERLEAVED CLLC
CONVERTERS FOR MINIMAL OUTPUT CURRENT DISTORTION 909
Martin Gendrin

OP114   PRIMARY-SIDE OUTPUT REGULATION PRINCIPLES IN DYNAMIC MULTI-MHZ
INDUCTIVE POWER TRANSFER SYSTEMS AND ISOLATED DC/DC CONVERTERS 916
Ioannis Nikiforidis

SMART GRID

OP115   LOW VOLTAGE DC-GRIDS WITH GALVANIC ISOLATION: SYSTEM
DISCUSSION, EFFICIENCY AND PERFORMANCE COMPARISON TO AC-FEEDING 926
Lukas Fräger

OP116   IMPLEMENTATION AND EXPERIMENTAL EVALUATION OF AN ADAPTIVE DC
GRID CONTROLLER FOR DECENTRALISED GRID CONTROL 933
Steffen Menzel

OP117   DEMONSTRATING THE EFFECTIVENESS OF A DC SOLID-STATE CIRCUIT
BREAKER'S FAST RESPONSE TIME 942
Ehab Tarmoom

OP118   MODELLING AND SIZING SENSITIVITY ANALYSIS OF A FULLY RENEWABLE
ENERGY-BASED ELECTRIC VEHICLE CHARGING STATION MICROGRID 949
David A. Stone

MEASUREMENT TECHNIQUES AND METHODS

OP119   LED POWERED ROTOR TELEMETRY SYSTEM 958
Raphael Beyerle

OP120   'INFINITY GATE SENSOR': A DIFFERENTIAL MAGNETIC FIELD SENSOR FOR
MEASURING GATE CURRENT OF SIC POWER TRANSISTORS 966
Yushi Wang

OP121   CHARACTERISING WIDE BANDGAP POWER MODULES: VALIDATING THE M-
SHUNT CONCEPT FOR HIGH-POWER APPLICATIONS IN THE KILOAMPERE RANGE 976
Hauke Lutzen

OP122   CHARACTERIZATION OF POWER-MODULE PARASITICS: SUB-NANOSECOND
LARGE SIGNAL PULSING VS. DOUBLE-PULSE TESTING 986
Gerhard Groos

STATISTICAL VARIATIONS IN THE PARASITIC CAPACITANCE OF A COIL 997
Kevin Talits

HIGH VOLTAGE SWITCHES

PP001   A 4.5 KV FAST RECOVERY DIODE PLATFORM FOR HIGH-CURRENT IGBTS 1002
Jan Vobecky

PP002 6.5 KV INNOVATIVE SILICON POWER DEVICE (I-SI) MODULE WITH HIGH POWER DENSITY AND LOW LOSS BY STORED CARRIER CONTROL 1007
Takashi Hirao

PP003 HIGH CURRENT DENSITY 4.5KV PRESSPACK IGBTS PUSH SOA LIMITS 1013
Hossein Davoodi

PP004 2.5KV IGBT MODULE WITH HIGH RELIABILITY FOR RENEWABLE APPLICATIONS 1018
Akiyoshi Masuda

PP005 NEW GENERATION 4.5KV IGCT AND FAST RECOVERY DIODE FOR RAILWAY POWER SUPPLY APPLICATIONS 1025
Umamaheswara Reddy Vemulapati

PP006 NEXT GENERATION 4.5 KV IGBT-ONLY STAKPAK MODULE WITH REDUCED LOSSES AND HIGH TEMPERATURE CAPABILITY 1031
Jeremy Jones

THERMAL MODELLING AND SIMULATIONS

PP007 FINITE ELEMENT ANALYSIS OF THE UPSCALING OF WARPAGE AND BIFURCATION HYSTERESIS LOOPS: FROM CU/SI DIE TO LARGE WAFERS 1039
Vincenzo Vinciguerra

PP009 MAXIMUM JUNCTION TEMPERATURE SIMULATION AND VALIDATION FOR THE HOT SPOT IN MULTI-CHIP SIC POWER MODULE 1046
Wonjin Dylan Cho

PP010 INTEGRATION OF CFD-SIMULATION RESULTS IN PLECS USING LOOKUP TABLES 1051
Simon Cepin

PP011 PCB ONLY THERMAL MANAGEMENT TECHNIQUES FOR EGAN FETS IN A HALF-BRIDGE CONFIGURATION 1057
Adolfo Herrera

HIGH POWER DENSITY DESIGNS

PP013 FROM 4X TO 3X STPAK – OPTIMIZATION FOR A MORE COMPACT EV TRACTION INVERTER SOLUTION 1065
Vittorio Giuffrida

PP014 A MULTI-OBJECTIVE STRUCTURAL OPTIMIZATION METHOD BASED ON MULTI-PHYSICS SIMULATIONS FOR POWER MODULE 1072
Baihan Liu

PP015 HOLISTIC APPROACH TO MAXIMIZE LIFETIME AND POWER DENSITY IN HIGH POWER SEMICONDUCTOR MODULES 1077
Martin Schulz

PP016 REGULATED HIGH DENSITY SWITCH CAPACITOR TOPOLOGY 1082
Pierrick Ausseresse

PP017   SILICON INTERPOSER AS A SUBSTRATE FOR POWER MODULES WITH HIGH
POWER DENSITY AND SUPERIOR THERMAL PERFORMANCE 1087
    Ahmed Ammar

SPECIAL CONVERTER APPLICATIONS

PP018   ANALYTICAL MODELING AND STABILITY CHARACTERIZATION OF A
DAMPED VSCC CM ACTIVE EMI FILTER FOR SINGLE- AND THREE-PHASE AC-DC
APPLICATIONS 1092
    Timothy Hegarty

PP020   A REPETITIVE HIGH VOLTAGE NANOSECOND PULSE GENERATOR: FIRST
PROTOTYPE DESIGN AND TEST RESULTS 1101
    Serge Gavin

PP021   FREQUENCY SHIFT KEYED DUAL SIDE CONTROL OF INDUCTIVE POWER
TRANSFER: AN APPLICATION OF TALKATIVE POWER CONVERSION 1105
    Hamzeh Beiranvand

PP022   STUDY OF A MULTI-ACTIVE BRIDGE CONVERTER FOR A DOMESTIC
ELECTRICAL GRID 1113
    Abdennour Merrouche

INTEGRATION TECHNOLOGIES AND RELIABILITY DESIGN

PP023   FABRICATION DEVELOPMENT FOR GATE DRIVER EMBEDDED DOUBLE-
SIDED COOLING SIC POWER MODULE FOR ELECTRIC VEHICLE APPLICATION 1123
    Anna Corbitt

PP024   PRINTED CIRCUIT EMBEDDING OF PREPACKAGED 150V POWER MOSFETS IN
A PORTABLE WELDING APPLICATION 1128
    Thomas Gebhard

PP025   PROCESS CHALLENGES AND PROGRESS TOWARDS DIRECT CONNECTION OF
AUTOMOTIVE POWER MODULES (TMM) TO HEATSINK..1133
    Indrajit Paul

PP026   OPTIMIZING PCB STACKUPS FOR ENHANCED GAN TRANSISTOR
PERFORMANCE IN HIGH-POWER APPLICATIONS..1139
    Philipp Czerwenka

PP027   NEW GENERATION CERAMIC SUBSTRATES – KEY COMPONENTS FOR POWER
ELECTRONIC APPLICATIONS: PROCESSING AND CHARACTERIZATION 1147
    Stefanie Schindler

PP028   AI-ENHANCED VACUUM REFLOW OVEN: PRECISION CONTROL FOR
RELIABLE LARGE-AREA SOLDERING 1152
    Chih Hui Lee

PP030   CORROSION-COMPATIBLE DRIVE ELECTRONICS FOR ELECTRIC VEHICLES
AND INDUSTRIAL POWER MODULES 1158
    Tom Petzold

PP031 EVALUATING THE SAFETY ISOLATION OF THE PACKAGE IN AN INTEGRATED POWER DEVICE 1168
Thomas Anthony Capobianco

CONTROL METHODS I

PP032 FLEXIBLE CONTROL SYSTEM FOR MODULAR ONE-PHASE INTERLEAVED GAN-BASED TOTEM POLE PFC USING REAL-TIME HARDWARE 1174
Oleksandr Solomakha

PP033 A PEAK CURRENT MODE CONTROL METHOD FOR PFC 1180
Sean Yu

PP034 ADAPTIVE RESONANT CONTROLLER FOR A THREE-PHASE PFC CONVERTER FOR AN ON-BOARD CHARGE APPLICATION 1185
Rami Troudi

PP035 SYNTHESIS OF A FIELD ORIENTED CONTROL ALGORITHM BY USING TWO DIFFERENT POLE-ZERO COMPENSATION APPROACHES...1192
Marco Denk

PP037 AVERAGE CURRENT MODE CONTROL AND ITS LOOP DESIGN 1200
Niklas Schwarz

PP038 NOVEL POWER FEED-FORWARD REGULATION FOR DUAL STAGE PFC+DCDC CONVERTERS 1207
Alfredo Medina-Garcia

HIGH POWER AC-DC AND DC-AC CONVERTER

PP039 22 KW BI-DIRECTIONAL WALL-BOX CHARGER WITH 1200 V SIC MOSFET................... 1212
Sanbao Shi

PP040 DYNAMIC SWITCHING FREQUENCY SELECTION FOR EFFICIENCY OPTIMIZATION IN ON-BOARD CHARGER PFC STAGE BASED ON NOVEL SIC MOSFET POWER MODULE.. 1217
Giuseppe Aiello

PP041 DESIGN AND OPTIMIZATION OF SIC-BASED 11KW MOTOR DRIVE WITH HIGH EFFICIENCY 1222
Iris Liu

PP042 MODEL DESIGN DEVELOPMENT FOR FALSE TURN-ON CHARACTERIZATION IN SIC-BASED ACTIVE T-TYPE CONVERTER CONSIDERING ALL PARASITICS 1227
Amir Babaki

PP043 EFFICIENCY INVESTIGATIONS OF AN AUXILIARY RESONANT COMMUTATED POLE INVERTER.. 1233
Markus Zocher

PP044 A NOVEL HYBRID TWO-STAGE AC-DC CONVERTER WITH SOFT-SWITCHED CCM PFC STAGE FOR EVS CHARGING APPLICATIONS.. 1242
Lei Wang

PP045   A METHOD FOR TUNING LEAKAGE INDUCTANCE IN TRANSFORMERS 1249
Rosemary O'Keeffe

PP046   LOW COST HIGH DENSITY 300W/20V AC-DC CONVERTER ENABLED BY GAN
POWER ICS.. 1254
Tom Ribarich

PP047   25KVA GRID-TIED BI-DIRECTIONAL T-TYPE INVERTER WITH HIGH-
EFFICIENCY AND HIGH-POWER DENSITY USING SIC MOSFETS....................................... 1259
Tamanna Bhatia

PP048   COST-EFFECTIVE EFFICIENCY ENHANCEMENT IN AC-DC CONVERTERS: A
STUDY ACROSS THE FULL LOAD CYCLE 1264
Sebastian Gick

E-MOBILITY TRACTION I

PP049   NEXT GENERATION POWER MODULE WITH PARALLEL CONNECTED SIC
MOSFETS FOR BEV TRACTION INVERTERS.. 1272
Kohei Tanikawa

PP051   INVESTIGATION OF COMMON SOURCE FEEDBACK IN SIC POWER MODULES
REGARDING PERFORMANCE AND SHORT CIRCUIT ROBUSTNESS..................................... 1277
Dominik Ruoff

PP052   HYBRIDPACK DRIVE POWER MODULES WITH SIC-MOSFET'S AND
MONOLITHIC RC- SNUBBER CHIPS FOR OPTIMIZED POWER DENSITY.............................. 1283
Andre Uhlemann

PP053   ROBUST AUXILIARY POWER SUPPLY FOR EVS BASED ON INNOVATIVE
STI2GAN 650V IC... 1289
Federica Cammarata

PP054   IMPACT OF VARIOUS SILICON DIODES ON THE HYBRID SWITCH INVERTER 1297
Michael Walter

PP055   ADVANCED PULSE SEQUENCE FOR SALIENCY-BASED HIGH-ACCURATE
ROTOR POSITION ESTIMATION OF RAILWAY TRACTION LOCOMOTIVE MOTORS 1307
Markus Vogelsberger

CONTROL TECHNIQUES

PP056   OPTIMIZED HALF-BRIDGE GATE-DRIVE WITH LOW TIME-SKEW FOR RC-
IGBTS AND SIC-MOSFET DEAD-TIME CONTROL 1315
Jan Fuhrmann

PP057   DESIGN OF A TRACTION INVERTER BASED ON PCB-EMBEDDED GAN
DEVICES 1322
Maurizio Tranchero

PP058   OPTIMIZING ELECTRIC VEHICLE PERFORMANCE WITH GAN DESIGN......................... 1330
Andrew Patterson

PP059 FAST ANALYTICAL CALCULATION OF THE MAGNETIC FIELD IN PERMANENT MAGNET SYNCHRONOUS MACHINES WITH FLUX BARRIERS INCLUDING SATURATION 1336
   Martin Ackermann

PP060 MODELING AND CONTROL OF LCL FILTERED 3L-VSCS IN INTERLEAVED TOPOLOGY 1346
   Adeel Jamal

PP062 ENHANCING SAFETY AND EFFICIENCY FOR ISOLATED PLC I/O DESIGNS WITH SPI DAISY CHAIN 1352
   Travis Lenz

VOLUME 3

PP063 COST-EFFECTIVE METHOD TO DISCHARGE DC LINK CAPACITORS WITH SIC POWER MODULES 1361
   Paul Kanatzar

POWER QUALITY

PP064 A STUDY ON CIRCULATION CURRENT IN PARALLEL OPERATION OF TRANSFORMER LESS UPS 1368
   Koji Kato

PP065 DESIGN CHALLENGES AND CONSIDERATIONS FOR GATE DRIVERS OF SIC MOSFETS AND THEIR TESTING 1374
   Niranjan Hegde

PP066 A PORTABLE EFFICIENCY CHARACTERIZATION SETUP FOR TECHNOLOGY DEMONSTRATION OF POWER MODULES 1380
   Sebastian Tengvall

PP067 FAST EME CHARACTERIZATION OF BARE-DIE SIC MOSFETS 1385
   Robert Kragl

PP068 THEORETICAL COMPARISON OF COMPONENT-RELATED MEASUREMENT METHODS OF PHOTOVOLTAIC INVERTERS FOR LONG-TERM TESTING 1393
   Niclas Reitz

DYNAMIC TRANSIENTS AND RELIABILITY OF HIGH-VOLTAGE SILICON & 4H-SIC BIPOLAR JUNCTION TRANSISTORS UNDER AVALANCE AND SHORT-CIRCUITS 1402
   Mana Hosseinzadehlish

PP069 POWER CYCLING TEST OPTIMIZATION TOWARD RELIABILITY ASSESSMENT OF SINTERED POWER MODULES 1410
   Robert Graham

PP070 REAL-TIME ESTIMATION AND SENSITIVITY ANALYSIS OF PARASITIC CAPACITANCES IN ELECTRIC DRIVE SYSTEMS 1418
   Mohammadreza Bagheribavaryani

MODELLING AND TESTING

PP071 PARASITIC COMPONENT EFFECTS OF INTERNAL AND EXTERNAL PACKAGE
LEVEL ON SWITCHING PERFORMANCE OF SIC POWER MODULE 1428
Nguyen Nghia Do

PP072 A MULTI-PHYSICS ITERATIVE APPROACH FOR TEMPERATURE ESTIMATION
IN SIC POWER MODULE FOR ELECTRIC VEHICLE 1434
Stefano Orlando

PP073 VOLTAGE BALANCING METHOD FOR SERIES CONNECTION OF 50 SIC
MOSFETS 1441
Antoine Philippe

VOLTAGE BALANCING METHOD FOR SERIES-CONNECTION OF 50 SIC MOSFETS 1449
Antoine Philippe

PP074 A LABORATORY-SCALE MMC-BASED DC SYSTEM WITH RCP AND PHIL
SIMULATION CAPABILITIES 1457
Marc René Lotz

PP075 FILM CAPACITOR STANDARD SERIES DIGITALIZATION: ELECTROMAGNETIC
& THERMAL MODELLING IMPLEMENTATION IN CLARA WEB TOOL 1467
Fernando Aunon

PP076 ACCURACY EVALUATION AND PROPOSED DYNAMIC TUNING PROCEDURE
OF A COMPACT SIC SPICE MODEL 1475
Austin Curbow

PP077 INVESTIGATION OF USE-CASE-DEPENDENT MODELING APPROACH FOR
SWITCHED-MODE POWER CONVERTER FOR LVDC GRID EVALUATION 1485
Melanie Lavery

PP078 AVERAGED MODEL WITH BLOCKING CAPABILITY FOR SOLID-STATE
TRANSFORMERS 1495
Ahmed Meligy

ADVANCED COMPONENTS

PP080 SURFACANT-MODIFIED NANOCOMPOSITE THIN-FILM CAPACITORS 1504
Bartosz Gackowski

PP081 INCREASING ENERGY STORAGE CAPABILITIES OF POWDER CORES BY
ADAPTING THE WINDING AND THE USE OF FRINGING FLUX 1511
Paul Winkler

PP082 PEEC-BASED THERMAL MODELING OF PASSIVE COMPONENTS 1516
Sascha Langfermann

PP083 GALVANICALLY ISOLATED POWER SUPPLY FOR GATE DRIVERS IN HIGH
VOLTAGE APPLICATIONS 1523
Priyanka Ghosh

PP084 FABRICATION TECHNIQUE FOR NOVEL NANOCRYSTALLINE CORES WITH HIGH SATURATION POLARIZATION AND LOW LOSSES 1532
Merlin Thamm

PP085 EXCITATION-DEPENDENT TEMPERATURE BEHAVIOR OF THE QUASI-STATIC HYSTERESIS LOSS ENERGY DENSITY OF N87 FERRITE MATERIAL...... 1538
Jeremias Kaiser

PP087 PASSIVE METHODS LIMITING LEAKAGE CURRENT IN METAL-OXIDE VARISTOR AS VOLTAGE CLAMPING DEVICE USED DC LOW VOLTAGE POWER ELECTRONICS-BASED CIRCUIT BREAKERS 1545
Kenan Askan

GAN DEVICES AND APPLICATIONS

PP088 ESD SOLUTIONS FOR 650V NORMALLY-OFF ALGAN/GAN HEMTS 1555
Thanh Hai Phung

PP089 A SIMULATIVE STUDY OF MEASUREMENT ERRORS DURING DOUBLE PULSE TESTING OF GAN DEVICES...... 1561
Severin Klever

PP090 PARALLEL CONNECTION OF GAN FETS: AN EXPERIMENTAL INVESTIGATION APPROACH...... 1568
Marco Palma

PP091 REPETITIVE SHORT CIRCUITS ON 650 V GAN 1574
Adrien Lambert

PP092 COMPARISON OF SWITCHING LOSSES AND DYNAMIC ON RESISTANCE OF 600 V-CLASS GAN HEMTS...... 1584
André Thönnessen

PP093 PERFORMANCE EVALUATION OF DEADTIME AND GATE RESISTANCE FOR PARALLEL CONNECTED GAN HEMTS 1590
Junhyeok Jegal

PP094 REACHING BEYOND 1200V: LATERAL GAN HEMTS FOR HIGH-RELIABILITY EV AND INDUSTRIAL APPLICATIONS 1598
Kamal Varadarajan

SIC DEVICES AND TECHNOLOGIES

PP095 SMARTSIC 150 & 200MM ENGINEERED SUBSTRATE: INCREASING SIC POWER DEVICE CURRENT DENSITY UP TO 30%...... 1604
Eric Guiot

PP096 DYNAMIC TRANSIENTS IN HIGH-VOLTAGE SILICON AND 4H-SIC NPN BIPOLAR JUNCTION TRANSISTORS 1610
Mana Hosseinzadehlish

PP097 AN ADVANCED MULTI-ASPECT PERFORMANCE ANALYSIS OF PLANAR-GATE 1.2 KV SIC POWER MOSFETS 1613
Anja Katerina Brandl

PP098   SIC MOSFET DIE SORTING AND PARALLEL FOR OPTIMAL MODULE DESIGN 1621
Zhong Ye

PP099   SIMULATION APPROACH FOR RADIATED ELECTRO-MAGNETIC FIELDS
ESTIMATION ON ACEPACK DRIVE SIC POWER MODULE 1627
Andrea Cusumano

CONTROL METHODS II

PP100   EXACT ANALYSIS OF CONTROL-TO-OUTPUT TRANSFER FUNCTIONS OF
PWM-CONVERTERS - A COMPARISON OF TWO METHODS.................................... 1634
Daniel Breidenstein

PP101   3-LEVEL FLYING CAPACITOR MULTILEVEL TOPOLOGY WITH DELTA-SIGMA
MODULATION 1642
Jannik Maier

PP102   MODEL BASED CONTROLLED POWER CONVERTER TEST PLATFORM......................... 1651
Dawid Koczy

PP103   EDUCATIONAL HARDWARE TRAINER FOR TEACHING THE DUAL ACTIVE
BRIDGE IN A DC GRID 1658
Peter Van Duijsen

PP104   STUDY OF THE OPERATING PERFORMANCE OF A FCS-MPC-CONTROLLED
MATRIX-CONVERTER FOR PMSM AT DIFFERENT FREQUENCY RATIOS 1664
Robert Zipprich

PP105   ENHANCING REACTIVE POWER CAPACITY IN BATTERY-FED POWER
CONDITIONING SYSTEMS.. 1673
Lucas Araujo

PP106   PULSE SHARING: ACHIEVING HIGH EFFICIENCY AND EXCELLENT REGULA-
TION IN MULTI-OUTPUT FLYBACK POWER SUPPLIES 1680
Xingda Yan

PP107   RELIABILITY-OPTIMIZED SPACE VECTOR MODULATION (RO-SVM) FOR
SEMICONDUCTORS LIFETIME ENHANCEMENT 1686
Amin Rezaeizadeh

INTELLIGENT POWER MODULES

PP108   ANALYSIS AND OPTIMIZATION OF INTERNAL COUPLING INTERFERENCE IN
INTEGRATED SIC POWER MODULE BASED ON DBC 1693
Chenhang Zeng

PP109   MULTISPECTRAL ELECTROLUMINESCENCE SENSING OF SIC MOSFETS FOR
JUNCTION TEMPERATURE AND CURRENT EXTRACTION... 1703
Lukas Ruppert

PP110   SIC-IPM FOR COMPACT AND ENERGY EFFICIENT LOW-POWER MOTOR
DRIVES 1712
Jongmu Lee

PP111 CONCEPT FOR A GAN-BASED INTELLIGENT MOTOR CONTROLLER WITH
INTEGRATED FAILURE PREDICTION FOR THE INVERTER AND THE DRIVE 1717
Christoph Blechinger

PP112 INTRODUCING THE NEW 1200 V CIPOS MAXI IM817 INTELLIGENT POWER
MODULE FOR MOTOR DRIVE APPLICATIONS 1724
Kihyun Lee

PP113 THERMAL PERFORMANCE OF INFINEON'S NEW 600 V CIPOSTM MICRO IM241
IPM FOR LOW POWER MOTOR DRIVE SYSTEMS WITHOUT HEATSINK 1732
David Jo

INTRODUCING THE NEEW 1200 V CIPOSTM MAXI IM12BXXXC1 INTELLIGENT POWER
MODULE FOR MOTOR DRIVE APPLICATIONS 1737
Kihyun Lee

INTELLIGENT GATE DRIVE UNITS

PP114 AN ADAPTIVE DEAD TIME CONTROL BASED ON SWITCH NODE VOLTAGE
DERIVATIVE 1745
Lukas Knappstein

PP115 COUPLING COIL DESIGN AND POSITIONING OPTIMIZATION ON NEW HIGH
POWER SEMICONDUCTOR MODULE FOR FAST SHORT CIRCUIT DETECTION 1751
Yannick Dumollard

PP116 ENABLING ACTIVE THERMAL CONTROL VIA AN ADAPTIVE MULTI-VOLTAGE
GATE DRIVER 1759
Tianlong Albert

PP117 INNOVATIVE GATE DRIVE METHOD TRIC3 FOR MOTOR 1765
Hisashi Sugie

PP118 A NEW CLASS OF SOLID STATE ISOLATORS ENHANCES THE RELIABILITY OF
SOLID STATE RELAYS 1770
Wolfgang Frank

PP119 A SELF-DRIVING 3-LEVEL ACTIVE GATE DRIVER NETWORK TO CONTROL
THE SWITCHING SLEW RATE FOR SIC MOSFETS 1775
Vin Loong Choo

E-MOBILITY TRACTION II

PP121 ANALYSIS OF LONG-TERM RELIABILITY OF SIC IN TRACTION INVERTER
CONSIDERING VTH INSTABILITY 1781
Chi Zhang

PP122 EFFICIENT MAPPING OF ON-DEMAND DRIVE LOAD PROFILES ON INVERTER
STRESS 1788
Zlatko Bosnjic

PP123 EV TRACTION INVERTER OPTIMAL DESIGN IS DOMINATED BY 3-LEVEL
ANPC 1797
Timothé Delaforge

PP124   INTRODUCTION OF POWER SEMICONDUCTOR OPTIONS FOR AN EXCITER OF
ELECTRICALLY EXCITED SYNCHRONOUS MOTOR 1804
Yeriel Bai

PP125   A NOVEL HIGH POWER DENSITY THREE PHASE TRACTION INVERTER
ARCHITECTURE FOR ELECTRIC VEHICLE (EV) APPLICATIONS....................................... 1809
Yiyang Yan

PP126   A MODULAR DC-LINK CAPACITOR SOLUTION FOR THE MAIN POWERTRAIN
INVERTER OF XEV 1814
David Olalla

PP127   FAULT IDENTIFICATION TESTING METHODS FOR A COMMERCIAL TRACTION
INVERTER 1821
Anna Corbitt

PP128   SHORT CIRCUIT ROBUSTNESS FOR TRACTION INVERTERS FROM AN
APPLICATION POINT OF VIEW 1828
Karl Oberdieck

INVESTIGATIONS OF PARTICULAR SIC DEVICE PHENOMENON

PP129   THE IMPACT OF THE DEADTIME ON THE STABILITY OF 1.2KV SIC MOSFET
BODY DIODE UNDER HARD SWITCHING WITH SYNCHRONOUS RECTIFICATION.................... 1835
Mohammed Amer Karout

PP130   RC-DC SNUBBER IMPLEMENTATION FOR SUPPRESSION OF DIODE VOLTAGE
PEAK AND RINGING IN A FULL SIC HALF-BRIDGE POWER MODULE 1844
Emanuela Alfonzetti

PP131   SUB-5 SECOND WIDE-BANDGAP POWER DEVICE CALORIMETRIC
MEASUREMENTS UTILZIING OPTICAL SENSORS AND PELTIER ELEMENTS 1851
Ruben Schnitzler

PP132   SIC TRENCH MOSFETS IN AVALANCHE MODE WITH RC SNUBBER CIRCUIT.............. 1858
Sebnem Tuncay

PP133   HIGH-FREQUENCY OSCILLATIONS IN SIC MOSFET POWER MODULES
DURING TURN-ON SWITCHING TRANSIENT – ANALYSIS BASED ON SIMULATIONS
AND MITIGATION METHODS... 1865
Rajani Kumar Thirukoluri

PP134   A DYNAMIC CURRENT BALANCING METHOD USING FULL-COUPLED
INDUCTORS IN PARALLELED GATE BRANCHES.. 1872
Jianwei Lv

PP135   QUANTITATIVE PERFORMANCE COMPARISON OF LARGE-FORMAT SIC
MOSFET AND SI IGBT MODULES 1878
Arthur Boutry

THERMAL MANAGEMENT AND ADVANCED COOLING

PP136   SOLDER PREFORM TECHNOLOGY FOR IMPROVED THERMOMECHANICAL
PERFORMANCE IN MOLDED POWER MODULE PACKAGE-ATTACH 1886
Joseph Hertline

PP138   EFFECT OF FLIP-CHIP DIE-ATTACH ON THE THERMAL BEHAVIOR OF POWER GAAS DIODES 1891
Felix Steiner

PP139   INFLUENCES OF SOLDER DELAMINATION ON THE THERMAL PERFORMANCE IN AUTOMOTIVE TRACTION MODULE 1896
Hansol Seo

PP141   DEVELOPMENT OF A PASSIVE CAPILLARY-PUMPED COOLING SYSTEM FOR HIGH-PERFORMANCE ELECTRONICS 1902
Justin Fey

PP143   ADVANCED COOLING OF POWER ELECTRONICS WITH COPPER COLD SPRAYED ALUMINIUM HEATSINKS & BUSBARS 1907
Michael Dasch

PP144   COLD PLATE DESIGN FOR COOLING LV100 SILICON CARBIDE POWER MODULE PACKAGING 1910
Wahid Cherief

PP145   AN IMPROVED DOUBLE-LAYER SPACER IN DOUBLE-SIDED COOLING POWER MODULE 1917
Linhao Ren

RELIABILITY TESTING

PP146   POWER CYCLING OF 1.7KV MULTI-CHIP POWER MODULES – SIC MOSFETS VS SILICON IGBTS 1923
Nick Baker

PP147   POWER CYCLING CAPABILITY OF DISCRETE SIC MOSFET DEVICES WITH DIFFERENT DESIGNS 1930
Luhong Xie

PP148   MODEL-BASED PARAMETER TUNING OF SEMICONDUCTOR DEVICES IN DC POWER CYCLING TEST 1936
Yi Zhang

PP149   INFLUENCE OF TRANSFER MOLDING ON THE RELIABILITY OF DCM SIC POW-ER MODULES 1942
Jacek Rudzki

PP150   DAMP HEAT BEHAVIOR OF HIGH HEAT CAPACITORS FOR APPLICATIONS IN ELECTRIC VEHICLES 1951
Adel Bastawros

PP151   INFLUENCE OF THE GATE VOLTAGE DURING ON-TIME ON THE POWER CYCLING CAPABILITY OF SIC MOSFETS 1955
Patrick Heimler

PP152   INVESTIGATION OF THE TEMPERATURE MEASUREMENT VIA VSD(T)-METHOD APPLIED TO PARALLELED SIC MOSFET CHIPS DURING POWER CYCLING 1964
Kevin Ladentin

PP153   APPROACHES OF TSEP MEASUREMENTS FOR POWER SEMICONDUCTORS 1969
Philipp Hauenschild

PP154　REALTIME JUNCTION TEMPERATURE ESTIMATION IN SIC POWER MODULES
BASED ON MULTIPLE TSEP ACQUISITION 1978
　　Kevin Muñoz Barón

HIGH VOLTAGE WBG DEVICES

PP155　ENHANCED CURRENT MEASUREMENT APPROACH FOR NON-ISOLATED 6.5
KV SILICON CARBIDE MOSFETS 1987
　　Xinyuan Du

PP156　NEW 2KV SIC-MOS TECHNOLOGY FOR APPLICATION FIELDS IN THE
INDUSTRIAL LANDSCAPE 1991
　　Igor Kasko

PP157　HIGH TEMPERATURE EXPERIMENTAL CHARACTERIZATIONS OF COSS OF 3.3
KV SIC MOSFET FOR MEDIUM VOLTAGE PV APPLICATIONS 1999
　　Paul Schmidt

PP158　IMPACT OF GATE CONTROL ON THE SWITCHING PERFORMANCE OF 3.3KV
SBD-EMBEDDED SIC-MOSFET 2006
　　Junya Sakai

PP159　COMPARATIVE ASSESSMENT OF OVERLOADABILITY POTENTIAL OF 3.3 KV
SI-IGBTS AND SIC-MOSFET POWER MODULES 2013
　　Muhammad Nawaz

PP160　IMPROVED RELIABILITY OF A 2200 V SIC MOSFET MODULE WITH AN EPOXY-
ENCAPSULATED INSULATED METAL SUBSTRATE 2022
　　Hiroshi Kono

PP161　PARALLELING 3.3-KV/800-A RATED SIC-MOSFET MODULES – AN
OPTIMIZATION METHOD 2028
　　Hiroyuki Irifune

PP162　PERFORMANCE ASSESSMENT OF 10 KV SIC MOSFET AND PIN DIODE IN 3L-
NPC CONVERTER TOPOLOGY 2036
　　Renato Amaral Minamisawa

VOLUME 4

PP163　PERFORMANCE EVALUATION OF COOLSIC 2 KV SIC MOSFET DISCRETE IN
1500 V DC LINK SYSTEMS 2041
　　Ajith Kumar Sekar

PP164　A NEW 2.3 KV RATED SIC MOSFET MODULE WITH LOW-INDUCTANCE HIGH-
POWER PACKAGE HPNC FOR 1500 VDC APPLICATIONS 2049
　　Junya Kawabata

PACKAGING AND INTERCONNECTION MATERIALS

PP166　MECHANISM FOR IMPROVING THE HEAT-RESISTANCE OF ADHESIVE
INTERFACE IN FLEXIBLE PRINTED CIRCUITS 2053
　　Keita Suzuki

PP167 A SYSTEMATIC COMPARISON STUDY OF DIFFERENT BONDING
TECHNOLOGIES FOR SUBSTRATE ATTACHMENT OF POWER ELECTRONICS............................ 2060
    Lisheng Wang

PP168 STABILITY OF PRESSURE SINTERED INTERCONNECTS AS A FUNCTION OF
TEMPERATURE AND ENVIRONMENTAL CONDITIONS.. 2067
    Kentaro Yoshioka

PP169 THE EFFECT OF NANO-CU INTERCONNECTION MATERIALS ON THE
THERMOMECHANICAL PROPERTIES OF SIC DOUBLE-SIDED POWER MODULES 2074
    Suhang Wei

PP170 ALL-IN-ONE-SINTERING: DIE-ATTACH AND SUBSTRATE-ATTACH ON BARE
COPPER IN A PRESSURE ASSISTED SINTERING ONE-STEP PROCESS.................................... 2082
    Battist Rabay

PP171 SEQUENTIAL MANUFACTURING OF HIGHLY FUNCTIONALIZED THREE-
DIMENSIONAL CERAMIC COMPONENTS FOR POWER ELECTRONICS 2088
    Lars Rebenklau

PP173 PARAMETRIC STUDY OF DAMAGE EVOLUTION IN SILVER SINTERED
LAYERS OF DOUBLE SIDED POWER ELECTRONICS MODULES OF ELECTRICAL
VEHICLES.. 2094
    Saeed Akbari

DC-DC CONVERTER I

PP174 TRISTATE MODIFIED BOOST CONVERTER.. 2104
    Johannes Gragger

PP175 COMPARATIVE EVALUATION OF THE CENTER TAPPED BOOST CONVERTER
TOPOLOGY 2112
    Bryan Radix

PP176 COMPARISON OF MULTI-LEVEL TOPOLOGIES TO REDUCE THE
COMPONENTS VOLTAGE STRESSES WHEN POWERED FROM INDUSTRIAL DC GRIDS..............2119
    Katharina Machtinger

PP177 HARD-SWITCHING HIGH-FREQUENCY GAN-BASED DC-DC CONVERTERS
WITH CONCOMITANT DATA TRANSMISSION FUNCTIONALITY 2128
    Abdelmoumin Allioua

PP178 EFFICIENT DESIGN OF HIGH-CURRENT, LOW-OUTPUT VOLTAGE DC-DC
CONVERTERS USING ARTIFICIAL INTELLIGENCE-BASED TOPOLOGY SELECTION
AND OPTIMIZATION 2138
    Thomas Harmand

HIGH POWER DC-DC CONVERTER I

PP180 A SIC BASED 60KW LLC CONVERTER WITH NOVEL TRANSFORMER DESIGN
FOR IMPROVING VOLTAGE BALANCE 2146
    Frank Wei

PP181 ANALYSIS OF INVERTER OPERATION MODES OF AN IGBT-BASED ZCS LLC CONVERTER FOR A 2 KW AUTOMOTIVE ON-BOARD DC-DC 2152
Daniel Urbaneck

PP182 DUAL OUTPUT HYBRID CONVERTER FOR 48 V DATA CENTERS: M-HSC 2162
Simone Mazzer

PP183 3.6KW HIGH EFFICIENCY SIC-BASED HV/LV DC-DC CONVERTER FOR EVS 2167
Veera Bharath Chandra Reddy Gandluru

PP184 BIDIRECTIONAL DC-DC TOPOLOGIES COMPARISON FOR 800 V AUTOMOTIVE APPLICATIONS INTEGRATING 650 V GAN-ON-SI DEVICES 2175
Ilias Chorfi

PP185 ANALYSIS OF PHASE SHIELDING METHOD BASED ON ?-CR-Y THREE-PHASE INTERLEAVED LLC CONVERTER 2182
Jin Wen

PP186 22KW IMS-BASED BIDIRECTIONAL DC-DC CONVERTER USING SURFACE MOUNT SIC MOSFETS FOR OBCS 2185
Hamlin Wang

PP187 COMPARATIVE ANALYSIS OF DC-DC CONVERTERS FOR ELECTROLYZERS USING GEOMETRIC PROGRAMMING 2190
Tim McRae

PP188 DESIGN CONSIDERATION OF BI-DIRECTIONAL CLLLC RESONANT CONVERTER IN ENERGY STORAGE SYSTEMS 2200
Sheng-Yang Yu

SMART-GRID TECHNOLOGIES

PP189 ADAPTIVE FAST CHARGING SYSTEM WITH SECOND LIFE BATTERIES - AN OVERVIEW OF A RESEARCH PROJECT 2208
Lukas Böhning

PP190 PARALLEL OPERATION AND SYNCHRONIZATION OF MICROGRIDS BY USING THE THEVENIN THEOREM 2217
Marius Block

PP192 21 KA SOLID STATE DC BREAKER FOR SUPERGRID INSTITUTE'S HIGH POWER TEST FACILITY 2227
Christophe Conilh

PP193 DESIGN AND ANALYSIS OF A 50KW SIC-BASED ACTIVE FRONT END WITH A VERY SMALL LINE CHOKE FOR DC-GRIDS 2234
Raphael Otte

PP194 INVESTIGATION OF LOAD TRANSITIONS BETWEEN LOADED AND LOAD FREE CONDUCTOR SEGMENTS IN INDUSTRIAL CONDUCTOR SYSTEMS 2240
Jan-Niklas Koch

PP195 A METHOD TO CONTROL VOLTAGE AND POWER FLOW IN A DC GRID 2248
Peter Van Duijsen

ENERGY STORAGE SYSTEMS

PP196   CONSIDERATIONS ON A HIGH-CELL-COUNT CONVERTER-BASED BATTERY STORAGE SYSTEM WITH REDUCED COMMUNICATION EFFORT 2258
Paul Aspalter

PP197   STUDYING CONVERTORS FOR VOLTAGE EQUALIZATION IN ENERGY STORAGE SYSTEM WITH ACTIVE BMS 2268
Dimitar Arnaudov

PP198   CHALLENGES OF HIGH SIDE GATE DRIVER AND DISCONNECT MOSFET FOR BATTERY PROTECTION UNIT DURING START-UP, TURN-OFF AND OVER CURRENT EVENTS 2273
Niranjan Suravarapu Reddy

PP199   ELECTRIC INSULATION COORDINATION TO PREVENT ELECTRIC ARCS IN LITHIUMION BATTERIES 2278
Daniel Chatroux

PP201   BATTERY CHARGER WITH IMPEDANCE SPECTROSCOPY CAPABILITY FOR LI-ION CELLS 2286
Christian Branas

EMC

PP202   EFFICIENCY, VOLUME AND CO2 EMISSIONS IMPACT IN A PFC CONVERTER WITH AN ACTIVE FILTER SOLUTION FOR OBC APPLICATION 2294
Kelly Ribeiro

PP203   ANALYTICAL AND EXPERIMENTAL VALIDATION COMMON MODE FEEDBACK LOOP FOR A THREE-PHASE_LEVEL VIENNA RECTIFIER 2303
Daniel San Laureano Igartuburu

PP204   ROBUSTNESS OF FREQUENCY-DOMAIN TERMINAL MODELING OF ELECTROMAGNETIC INTERFERENCES IN STATIC CONVERTERS 2309
Mehyeddine Singer

PP205   STUDY OF EMI BEHAVIOR OF A 2-LEVEL GAN-INVERTER – SIMULATION AND MEASUREMENT 2316
Benedikt Kohlhepp

COMMON MODE CURRENTS IN RESONANT CIRCUITS GENERATED WITH A DELTA-SIGMA MODULATED VOLTAGE SOURCE INVERTER 2326
Tobias Haas

PP206   ANALYSIS OF COMMON-MODE NOISE GENERATED DUE TO FAST-SWITCHING GAN DEVICES IN TOTEM-POLE PFCS 2334
Serkan Dusmez

PP207   CONDUCTED EMI FROM GAN-BASED 48V TO 12V DC-DC-CONVERTERS FOR AUTOMOTIVE APPLICATIONS 2342
Erik Kampert

ADVANCED DESIGN

PP208   APPLIED DESIGN AUTOMATION FOR FINDING FEASIBLE DESIGNS FOR HIGH-FREQUENCY PLANAR TRANSFORMERS 2350
Rando Raßmann

PP209   FREQUENCY DEPENDENT AREA PRODUCT METHOD 2359
Alfonso Martínez

HIGH RESOLUTION MIXED-SIGNAL PULSE WIDTH MODULATOR FOR HIGH-FREQUENCY DC-DC CONVERTERS 2364
Tim McRae

PP210   DESIGNING A CONTROL LIBRARY FOR GRID-FOLLOWING AND GRID-FORMING POWER INVERTERS 2370
Lars Lindner

PP211   INTELLIGENT OPTIMISATION OF A WIND TURBINE DIGITAL TWIN MODEL 2377
René Reimann

PP212   THERMAL TRANSIENT DIGITAL TWIN MODELLING FOR POWER CONVERTERS 2386
Xianghao Mo

PP213   A DIGITAL TWIN APPROACH TOWARD LIFETIME ANALYSIS AND PREDICTIVE MAINTENANCE OF POWER SEMICONDUCTORS FOR RAILWAY APPLICATION 2394
Emmanuel Batista

INDUCTORS

PP214   SATURABLE FERRITE CORE INDUCTORS IN LCL FILTERS OF THREE-PHASE VOLTAGE SOURCE INVERTERS 2400
Marius Kaufmann-Bühler

PP215   2D COPPER LOSS ANALYTICAL MODEL FOR PLANAR INDUCTOR COMBINING HIGH AND LOW PERMEABILITY MATERIALS 2408
Idriss Nachete

PP216   CNC-MANUFACTURED POWER INDUCTORS WITH EXCELLENT BANDWIDTH FOR MULTI-MEGAWATT CONVERTERS 2416
Thomas Kreppel

PP217   ANALYTICAL EVALUATION OF DIFFERENTIAL MODEL DC EMI FILTER INDUCTORS USING MATERIAL SATURATION COEFFICIENT 2425
Lukas Mueller

PP218   DESIGN AND PERFORMANCE EVALUATION OF AIR CORE INDUCTORS FOR VERY HIGH FREQUENCY POWER CONVERSION 2431
Florentin Salomez

PP220   IMPROVING MULTI-PHASE FERRITE MAGNETICS BY COUPLING FOR MV AND UPS CONVERTERS 2438
Michael Schmidhuber

E-MOBILITY CHARGING

PP221   22-KW BIDIRECTIONAL SINGLE-STAGE DIRECT-AC-AC POWER CONVERSION ON-BOARD CHARGER WITH HIGH-POWER-DENSITY IMPLEMENTATION.................................... 2448
Oscar Lucia

PP222   BENCHMARKING DC FAST CHARGERS: A COMPARATIVE ANALYSIS OF POWER CONVERTER STRUCTURES FOR WIDE VOLTAGE RANGE 2453
Sadik Cinik

PP223   PERFORMANCE OPTIMIZATION OF SINGLE-PHASE ON-BOARD CHARGERS WITH RIPPLE PORT 2461
Davide Gottardo

PP224   A REDUCED-SENSOR MODULAR DUAL ACTIVE BRIDGE-BASED BATTERY CHARGING SYSTEM FOR ELECTRIC VEHICLES USING AN IMPROVED LINEAR EXTENDED STATE OBSERVER.. 2469
Armel Asongu Nkembi

PP225   BIDIRECTIONAL NON-ISOLATED THREE-PHASE ONBOARD CHARGER WITH A LOW-VOLTAGE LOWER-PHASE OPERATION MODE.................................... 2478
Steffen Frei

PP226   CONTROL OF A THREE-PHASE INDUCTIVE POWER TRANSFER SYSTEM BASED ON DD²Q COIL TOPOLOGY 2488
Nikola Mirkovic

PP227   COMPARISON OF TWO BIDIRECTIONAL 11KW 400V CLLC AND CLLLC RESONANT CONVERTERS FOR EV APPLICATIONS 2494
Hasan Mousavi Somarin

PP228   DYNAMIC WIRELESS CHARGING SYSTEM DESIGN FOR EXTRA-URBAN AREAS BASED ON RESONANT INDUCTIVE POWER TRANSFER 2503
Irene Maria Torres Alfonso

PP229   BIDIRECTIONAL ISOLATED 400-12V DC-DC CONVERTER WITH IMPROVED POWER DENSITY AND FULL-RANGE OPERAION FOR EV APPLICATIONS 2513
Oscar Lucia

HIGH POWER DC-DC CONVERTER II

PP230   GAIN OPTIMIZATION CONTROL METHOD FOR CLLLC RESONANT CONVERTERS UNDER PHASE SHIFT MODE 2518
Sean Yu

PP231   ANALYSIS OF COMMON AND SPLIT DC-BUS INTERLEAVED H-BRIDGE CONVERTERS FOR HIGH-CURRENT LOW-RIPPLE APPLICATIONS................................. 2524
Bhavana Gudala

PP232   OPTIMAL FREQUENCY OPERATING POINTS FOR HYBRID SWITCHED CAPACITOR CONVERTERS AND LOSSLESS CURRENT SENSE METHOD 2532
Simone Mazzer

PP233 DESIGN AND TESTING OF A 250 KW 50 KHZ SIC-BASED HALF-BRIDGE-SERIES-RESONANT-CONVERTER 2538
Daniel Haake

PP234 30KW - 97% EFFICIENCY ISOLATED DC-DC CONVERTER WITH LARGE INPUT VOLTAGE RANGE BASED ON A BOOST DAB ASSOCIATION 2547
Jean-Jacques Huselstein

PP235 ANALYSIS OF A FULL-BRIDGE PUSH-PULL FORWARD DUAL ACTIVE BRIDGE DC-DC CONVERTER 2557
Gean Sousa

DC-DC CONVERTER II

PP236 SYMMETRICAL OPERATION OF FOUR CHANNEL RESONANT BOOST DC-DC CONVERTERS IN CONTINUOUS CONDUCTION MODE 2566
Kristóf Bándy

PP237 IMPACT OF MAGNETICS TOLERANCE ON THE POWER SHARING OF PARALLEL DUAL-OUTPUT PHASE-SHIFT FULL-BRIDGE CONVERTERS..................... 2576
Riccardo Mandrioli

PP238 A BALANCING CONVERTER WITH SERIES CONNECTED MOSFETS FOR +/- 700V BIPOLAR DC GRIDS... 2583
Sachin Yadav

PP239 OPTIMIZATION AND DESIGN OF LOW-VOLTAGE AND HIGH-CURRENT POINT-OF-LOAD CONVERTER UNDER 48V BUS ARCHITECTURE.................................... 2591
Jiajia Guan

PP240 INTERLEAVED BOOST CONVERTER EFFICIENCY AND POWER DENSITY MODEL FOR ACTIVE AND PASSIVE COMPONENT DESIGN................................... 2596
Damien Lemaitre

NOVEL AND ADVANCED SEMICONDUCTOR DEVICES

PP241 EVALUATION OF A HYBRID POWER SWITCH BASED ON TRENCH CLUSTERED IGBT AND SIC MOSFET 2606
Alireza Sheikhan

PP242 CONTRIBUTIONS FOR BUILDING BLOCKS FOR NORMALLY-OFF 650V GAN-ON-SI POWER INTEGRATED CIRCUITS.. 2612
Thanh Hai Phung

PP243 NEW BIDIRECTIONAL ASYMMETRIC HIGH VOLTAGE TVS (TRANSIENT VOLTAGE SUPPRESSOR) DIODE... 2620
Boris Rosensaft

PP244 ISO247: HIGH PERFORMANCE CERAMIC BASED ADVANCED ISOLATED DISCRETE PACKAGE TO FULLY EXPLOIT THE ADVANTAGES OF SIC MOSFET 2627
Sachin Shridhar Paradkar

PP245 IMPACT OF CURRENT RIPPLE REDUCTION USING HIGH SWITCHING FREQUENCIES ON PMSM EFFICIENCY 2632
Jannik Fuchs-Gade

PP246 MAXIMIZING COST-EFFICIENCY IN ELECTRIC DRIVETRAINS: A SIC/SI FUSION SWITCH APPROACH 2638
Matthias Ippisch

ADVANCED CONTROL

PP247 CONCISE AND RELIABLE SIC MOSFET DRIVER CIRCUITS 2646
Zhong Ye

PP248 ARTIFICIAL INTELLIGENCE ENHANCED RESOLVER SYSTEM FOR AUTOMOTIVE TRACTION INVERTER APPLICATIONS BASED ON AURIX TC4X 2651
David Zipperstein

PP250 MULTIFUNCTIONAL GRID MANAGER TOPOLOGY WITH CONFIGURABLE OUTPUT 2657
Peter Van Duijsen

PP252 CO2 FOOTPRINT OF MEDIUM VOLTAGE DC SOLID STATE TRANSFORMER 2663
Adriana Campos

SIC MOSFET

PP253 THERMO-ELECTRICAL ANALYSIS AND PERFORMANCE: A COMPARATIVE STUDY BETWEEN MODULAR AND DISCRETE APPROACHES 2673
Stefano Orlando

PP254 IMPACT OF PARAMETER SPREAD IN PARALLEL-OPERATED SIC MOSFETS FOR HARD-SWITCHING CONVERSION 2680
Andrea Piccioni

PP255 ASSESSMENT OF THE RDS,ON OF SIC MOSFET DIES THROUGH KELVIN WIRE CONNECTION 2686
Philipp Rehlaender

PP256 CHALLENGES IN SCALING SIC SINGLE-CHIP MEASUREMENTS TO CORRESPONDING POWER MODULES 2693
Hao Wang

PP257 SWITCHING PERFORMANCE EVALUATION OF HIGH-POWER 1.7 KV SIC MOSFET MODULES USING A COMMON BUSBAR DESIGN 2700
Sebastian Neira

PP258 CHARACTERIZING THE SWITCHING BEHAVIOR OF A 1.2 KV MIXED SIC JFET AND MOSFET HALF BRIDGE 2708
Tim Ringelmann

VOLUME 5

WBG HIGH FREQUENCY APPLICATION

PP259 PERFORMANCE EVALUATION OF THE PACKAGING OF SIC DIODES IN A 6.78 MHZ WIRELESS POWER TRANSFER SYSTEM 2718
Ioannis Nikiforidis

PP260   VOLTAGE WAVEFORM GENERATION FOR SAWYER-TOWER COSS LOSS
MEASUREMENTS USING A HYBRID POWER CONVERTER 2724
Malachi Hornbuckle

PP261   EVALUATION OF SIC DEVICES FOR OVER 500KHZ APPLICATION BASED ON
BUCK CIRCUIT 2730
Minli Jia

PP262   LINEARIZATION OF DRAIN-SOURCE CAPACITANCES FOR ANTISERIAL
CONFIGURATED SIC MOSFETS IN HIGH FREQUENCY SOLID STATE SWITCHES 2737
Lars Dresel

SIC RUGGEDNESS

PP263   EFFECTS OF NON-KILLER DEFECTS ON SIC MOSFET SHORT-CIRCUIT
RUGGEDNESS AND RELIABILITY 2745
Sara Kuzmanoska

PP264   DYNAMIC REVERSE BIAS TEST: ELECTRO-THERMAL CHARACTERIZATION
OF SIC MOSFETS 2751
Giuseppe Mauromicale

PP266   RADIATION HARDNESS OF SIC BASED INVERTERS BASED ON AN EV
MISSION PROFILE 2758
Hadiuzzaman Syed

PP267   RAPID SHORT CIRCUIT PROTECTION USING DIDT DETECTION FOR SIC
POWER MODULES 2764
Koki Samura

PP268   COMPARISON OF DYNAMIC GATE STRESS TEST RESULTS OF SIC MOSFETS 2769
Mathias Gebhardt

PP279   EXTENDING SIC MOSFET SHORT-CIRCUIT WITHSTANDING TIME BY TWO-
LEVEL TURN-OFF GATE DRIVING 2778
Kwokwai Ma

PP270   EXPERIMENTAL INVESTIGATIONS ON PARASITIC TURN-ON OF 1.2KV SIC
MOSFET DISCRETE DEVICES 2786
Thanh-Toan Pham

PP271   BEHAVIOR MODELLING THE SHORT CIRCUIT CHARACTERISTICS OF SIC
MOSFETS USING COMPACT MODELS 2791
Qing Sun

THERMAL CHARACTERIZATION

PP273   THERMAL ANALYSIS AND MODELLING OF CHARGING STATIONS FOR
ELECTRIC VEHICLES 2796
Ruben Kopischke

PP274   JUNCTION TEMPERATURE MEASUREMENT OF A 3.3 KV SILICON CARBIDE
MOSFET POWER MODULE 2803
Michael Gleissner

PP275 INNOVATIVE 3D POWER MODULE DEFAULTS DETECTION VIA THERMAL IMPEDANCE ANALYSIS AND SIMULATIONS...2811
Louis Alauzet

PP276 THERMAL CHARACTERIZATION OF AN AIR-COOLED PEBB BASED ON SIC MOSFET POWER MODULES 2819
Alexandre Marie

PP277 THERMAL BEHAVIOUR OF SIC MOSFET WITH PLANAR PACKAGING TECHNOLOGY 2826
Yijun Ye

RELIABILITY AND AVAILABILITY

PP279 IMPLEMENTING MODULE HEALTH MONITORING IN EV TRACTION INVERTERS 2831
Karol Rendek

PP280 RELIABILITY TESTS OF COPPER THICK-FILM SUBSTRATES FOR POWER ELECTRONIC APPLICATIONS... 2838
Henry Barth

PP281 POWER MODULE SOLUTIONS WITH IMPROVED RELIABILITY FOR ELEVATOR DRIVE APPLICATIONS 2843
Tiago Jappe

PP282 FAIL-OPERATIONAL LLC TOPOLOGIES WITH FAULT-TOLERANCE INTEGRATED REDUNDANT CAPABILITIES 2850
Aswathy M. Prince

PP283 THERMAL AND RELIABILITY OPTIMIZATION OF CLIPS IN SIC MOSFET POWER MODULES.. 2860
Zexiang Zheng

PP284 CONDITION MONITORING OF A GAN FULL-BRIDGE BY MEANS OF FORWARD VOLTAGE IN CONTINUOUS OPERATION.. 2866
Michael Vogt

PP285 A SIMPLE AND LOW COST OVERCURRENT PROTECTION SYSTEM BASED ON COMMERCIAL SHUNT FOR WIDE-BANDGAP DEVICES.................................... 2874
Emanuele Martano

PP286 SVM-BASED FAULT-TOLERANT CONTROL FOR A CASCADED H-BRIDGE MULTILEVEL CONVERTER UNDER MULTIPLE OPEN-CIRCUIT SWITCH FAULTS........................ 2880
Dong Xie

PP287 REVOLUTIONIZING MOBILITY: THE SECOND LIFE OF ONBOARD CHARGING SYSTEMS IN COMMERCIAL VEHICLES 2886
Ajay Krishna Voppu Muralikrishna

LOW VOLTAGE SWITCHES

PP288 A BEHAVIORAL TRANSIENT MODEL FOR IGBT DEVICE WITH ANTI PARALLEL FREEWHEELING DIODE.. 2893
Shiwu Zhu

PP289   PARAMETER EXTRACTION FOR AN ANN-ASSISTED IGBT MODEL IN TRANSIENT SIMULATIONS 2901
Huaiyuan Zhang

PP290   FABRICATION OF 600V RC-IGBT USING 300MM WAFER 2909
Masaki Ueno

PP291   NEXT LEVEL OF POWER MODULE SOLUTION FOR PV C&I STRING INVERTER WITH 1200V H7 TECHNOLOGY IN EASY3B PACKAGE 2914
Tilo Poller

PP292   ANALYSIS OF MOSFET SWITCHING LOSSES IN RESONANT CONVERTERS USING ELECTRICAL AND THERMAL MEASUREMENTS AND LOSS TRENDS WITH MOSFET SIZE VARIATION 2921
Alfio Scuto

PP293   OPTIMOS 6 135V FOR HIGH POWER MOTOR DRIVES 2930
Kunal Jha

PP294   AUTO POWER-SOI: SHAPING THE FUTURE OF BATTERY MONITORING TECHNOLOGY 2937
Alex Lim

LIFETIME MODELLING AND CONDITION MONITORING

PP295   UNDERSTANDING THE IMPACT OF IEC60747-17 ON CAPACITIVE AND MAGNETIC COUPLERS 2942
Shu Ee Ong

PP296   PARIS LAW APPLIED TO WIRE BONDS DEGRADATION USING CRACK GROWTH MEASUREMENT 2948
Merouane Ouhab

PP297   CONDITION MONITORING TECHNIQUE OF POWER ELECTRONIC MODULES VIA SQUARE-WAVE GATE SIGNAL EXCITATION 2956
Isabel Austrup

PP298   STATISTICS-BASED LIFETIME SIMULATION ENVIRONMENT FOR POWER MODULES INCORPORATING DEGRADATION MODELS 2963
Karthik Debbadi

PP299   POWER CYCLING RESULTS FOR RELIABILITY STUDIES OF SIC-INVERTERS 2972
Robert Keilmann

PP300   GAN CASCODE IN HIGH SPEED DRIVEN AIR COMPRESSORS FOR AUTOMOTIVE FUEL CELLS 2981
Florian Lippold

PP301   PROGNOSTIC ANALYSIS OF IGBT HEALTH: REAL-TIME ON-STATE VOLTAGE PREDICTION THROUGH MACHINE LEARNING 2986
Tanya Thekemuriyil

PP302   ROBUSTNESS ANALYSIS OF TEMPERATURE-SENSITIVE ELECTRICAL PARAMETERS OF IGBTS 2995
Laurids Schmitz

PP303 OBSERVATION OF THERMAL-RESISTANCE INCREASE OF DEGRADED IGBT MODULES BY VCE (SAT) MEASUREMENT IN A CHOPPER CIRCUIT 3002
Kazunori Hasegawa

PULSE WITH MODULATION METHODS

PP304 MODULATION TECHNIQUE FOR REDUCED AC CONTENT OF THE DC LINK CURRENT IN THREE-PHASE TWO-LEVEL INVERTERS 3007
Steffen Frei

PP305 COMMON MODE CURRENTS IN RESONANT CIRCUITS GENERATED WITH A DELTA-SIGMA MODULATED VOLTAGE SOURCE INVERTER 3017
Tobias Haas

PP306 EVALUATION OF NEW MODULATION SCHEME FOR 3L-ANPC USING BOTH CURRENT PATHS IN ZERO STATE 3020
Felix Eichler

PP307 AN INNOVATIVE SYNCHRONOUS RECTIFICATION METHOD FOR 11KW CLLC CONVERTER 3029
Sanbao Shi

PP308 INTERLEAVED ASYNCHRONOUS DELTA-SIGMA MODULATION CONCEPT FOR DYNAMIC POWER CONVERTERS 3034
Philipp Czerwenka

PP309 HIGH RESOLUTION MIXED-SIGNAL PULSE WIDTH MODULATOR FOR HIGH-FREQUENCY DC-DC CONVERTERS 3042
Tim McRae

PP310 IMPLEMENTATION AND CONTROL OF OPTIMIZED PULSE PATTERNS FOR SALIENT PERMANENT MAGNET SYNCHRONOUS MACHINES IN ELECTRIC VEHICLES........... 3045
Maximilian Hepp

PP311 A 3-LEG INTERLEAVED TP PFC WITH A 90° PHASE-SHIFTED ASYMMETRIC LEG FOR REDUCED MAGNETICS... 3060
Serkan Dusmez

PP312 FAULT-TOLERANT OPERATION ANALYSIS OF A FIVE-PHASE THREE-LEVEL TNPC INVERTER FOR ELECTRIC AIRCRAFT PROPULSION SYSTEMS 3067
Chanuch Chaisakdanugull

AC-DC AND DC-AC CONVERTER

PP313 CCM TOTEM-POLE PFC FOR ULTRA-HIGH POWER DENSITY USB-PD CHARGERS... 3077
Manuel Escudero Rodruigez

PP314 COMPARISON OF HYBRID SI/SIC AND SIC TWO-LEVEL AND THREE-LEVEL CONVERTERS FOR LOW-VOLTAGE LOW-POWER APPLICATIONS.. 3086
Tim Augustin

PP315 ANALYSIS OF ANALOGUE CURRENT AND FLUX BALANCING FOR THE DUAL-ACTIVE-BRIDGE CONVERTER.. 3096
Christophe Basso

PP316   DESIGN AND OPTIMIZATION OF A SINGLE-STAGE PHOTOVOLTAIC
MICROINVERTER WITH INTEGRATED MAGNETICS 3103
    Jin Wen

PP317   EXPERIMENTAL INVESTIGATION OF CLASS F INVERTER UNDER VARIOUS
LOAD CONDITIONS..3110
    Baptiste Daire

PP318   ANALYSIS, MODELING, DESIGN, AND LIMITATIONS OF CURRENT INJECTION
BASED UPF RECTIFIER WITH SMALL DC-LINK CAPACITOR..3118
    Ramkrishan Maheshwari

PP319   HIGH-EFFICIENT ISOLATED AC-DC CONVERTER WITH CIRCULATING
CURRENT REDUCTION FOR AC ADAPTERS 3125
    Hiroki Watanabe

PP320   A PHASE-LOCKED LOOP (PLL) BASED STRATEGY FOR ACCURATE BLANKING
TIMES IN BRIDGELESS TOTEM-POLE PFCS... 3130
    Sandu Tigira Tigira

PP321   CIRCULATING CURRENTS IN COUPLED MULTI-TERMINAL HYBRID AC-DC
GRIDS.. 3136
    Fabian Herzog

ADVANCED CONVERTER TOPOLOGIES

PP322   COMPARISON OF 4500V STATE-OF-THE-ART XHP3 IGBT AND CONVENTIONAL
IHV IGBT FOR 3300V 3-LEVEL ANPC MEDIUM VOLTAGE DRIVES 3142
    Martin Knecht

PP323   GENERALIZED SWITCHING SEQUENCE FOR VOLTAGE BALANCING IN A
FLYING CAPACITOR DC-DC CONVERTER WITH QUASI-2-LEVEL MODULATION......................... 3150
    Jose Andres Aguilar Croston

PP324   OPTIMIZATION-BASED SIZING OF A MODULAR MULTILEVEL CONVERTER
BASED ON 650 V GAN MODULES FOR NEW LVDC/MVDC GRIDS...................................... 3160
    Gregoire Le Goff

PP325   A NOVEL THREE-PHASE LOW-SWITCH-COUNT AC-DC GRID CONVERTER
TOPOLOGY WITH GALVANIC ISOLATION 3169
    Liska Steenbock

PP326   SINGLE-STAGE LED DRIVER BASED ON COUPLED INDUCTOR POWER
FACTOR CORRECTION AND LLC CONVERTER.. 3175
    Alireza Ramezan Ghanbari

PP327   A INVERSE COUPLED DC-DC BOOST INDUCTOR WITH 2-KV SIC MOSFET
MODULE FOR 1500V SOLAR INVERTER MPPT... 3181
    Yusi Liu

PP328   ENVIRONMENTAL IMPACT OF MODULAR POWER ELECTRONICS SYSTEMS
CONSIDERING DIAGNOSTIC-DRIVEN UNIT REPLACEMENT 3187
    Briac Baudais

POWER ELECTRONICS FOR RAILWAY APPLICATIONS

PP329 SWITCHING PERFORMANCE COMPARISON OF 3.3 KV SIC MOSFET AND SI IGBT POWER MODULES FOR RAILWAY TRACTION SYSTEMS 3197
Yue Zhao

PP330 COMPARISON OF THREE-LEVEL INVERTER TOPOLOGIES FOR MVDC REVERSIBLE RAILWAY SUBSTATIONS 3206
Luc Bimmel

PP331 CONTROL OF BIDIRECTIONAL POWER FLOW IN RAILWAY CATENARY OVERHEAD LINES 3213
Peter Van Duijsen

PP332 A RAIL TRACTION CONVERTER PLATFORM BASED ON POWER MODULE IMPLEMENTATIONS WITH 450 A, 600 A AND 800 A 3.3 KV IGBT MODULES 3221
Ekrem R. Gunes

PP333 COMPARISON OF SELECTED MEGAWATT-LEVEL TRACTION CONVERTER POWER MODULE IMPLEMENTATIONS IN TERMS OF COMMUTATION INDUCTANCE AND PRACTICALITY 3229
Abdulkerim Ugur

CURRENT RELATED TESTING

PP334 PITFALLS AND THEIR AVOIDABILITY IN THE DOUBLE-PULSE TEST 3237
Nikolas Förster

PP335 MODELING AND SIMULATION OF FLUXGATE BASED CURRENT SENSOR 3247
Yunus Çay

PP336 SIGMA-DELTA BASED CURRENT ACQUISITION WITH REDUCED SETTLING TIME 3256
Joschka Randerath

PP337 CHARACTERISATION OF WIDE-BANDGAP SEMICONDUCTORS IN DOUBLE PULSE TESTING USING OPTICALLY ISOLATED PROBES 3264
Lennart Hoffmann

PP338 NON-INVASIVE BATTERY CONDITION TESTING USING ELECTRICAL SIGNALS AND OSCILLOSCOPES 3269
Srikrishna N. H

PP339 INSTRUMENTATION REQUIREMENTS FOR FAST 130 V/NS SWITCHING OF 1700 V, 35 M? SIC MOSFETS 3276
Matthew Appleby

POWER ELECTRONICS FOR AEROSPACE APPLICATIONS

PP340 CONCEPTUALIZATION AND EXPERIMENTAL ASSESSMENT OF DESIGN ASPECTS FOR 3-LEVEL ANPC INVERTERS 3286
Lukas Radomsky

PP341   DESIGN OF A HIGH POWER DENSITY INVERTER AND FOC IMPLEMENTATION FOR UAVS 3296
  Matthias Neuner

PP342   HIGHLY-INTEGRATED, FLEXIBLE POWER SOLUTION FOR AEROSPACE 5KVA – 20 KVA MOTOR DRIVE APPLICATIONS 3305
  Alain Calmels

PP343   DATABASE-SUPPORTED PRELIMINARY DESIGN, SIMULATION AND EVALUATION OF POWER CONVERTERS IN ELECTRIC AIRCRAFT PROPULSION SYSTEMS 3315
  Jeff Kugener

PP344   DESIGN AND ANALYSIS OF GATE-DRIVER FOR SIC-BASED INVERTER FOR MEGAWATT SCALE ALL ELECTRIC AIRCRAFT 3318
  Jeff Kugener

MEASUREMENT TECHNIQUES AND METHODS

PP345   ADDRESSING TESTING CHALLENGES FOR POWER MODULES AND THREE-LEVEL INVERTERS 3328
  Oleg Fotteler

PP346   CHARACTERIZATION OF THE BONDING QUALITY OF SILVER SINTERED COMPOUNDS BY MEANS OF LASER-INDUCED BREAKDOWN SPECTROSCOPY 3334
  Yannick Bockholt

PP347   INVERTER-INTEGRATED MEASUREMENT OF THE FREQUENCY-DEPENDENT WINDING IMPEDANCE OF ELECTRIC MACHINES 3340
  Christian Mühlfeld

PP348   COMPENSATION TECHNIQUES FOR BANDWIDTH-DISTORTED MEASUREMENTS OF FAST TRANSIENTS IN DOUBLE PULSE TESTS 3347
  Christian Lottis

PP349   AN AERODYNAMIC LOAD MEASUREMENT TECHNIQUE FOR AUTONOMOUS AERIAL VEHICLES 3353
  Mehmet Oguz Girgin

COMPENSATION TECHNIQUES FOR BANDWIDTH-DISTORTED MEASUREMENTS OF FAST TRANSIENTS IN DOUBLE PULSE TESTS 3358
  Christian Lottis

PP350   A HIGH-BANDWIDTH MULTILEVEL COUNTER CIRCUIT FOR BEARING CURRENT EVALUATION 3364
  Felix Schulte

TRANSFORMERS

PP351   CORE LOSS MODEL FOR CONSIDERING ANISOTROPY AND TEMPERATURE EFFECTS ON ELECTRICAL STEEL UNDER POWER ELECTRONIC CONDITIONS 3371
  Michael Owzareck

PP353   CIRCULAR ECONOMY ORIENTED AND RECONFIGURABLE PLANAR
TRANSFORMER DESIGN FOR ISOLATED DC-DC CONVERTERS 3380
  Fabian Groon

PP354   CONTROLLABLE MAGNETICS: VARIABLE TRANSFORMERS AND VARIABLE
INDUCTORS, THEORY – PRODUCTION – APPLICATION.. 3390
  Florian Fenske

PP355   A THREE-PHASE INTERLEAVED LLC INTEGRATED TRANSFORMER USING
PCB WINDINGS FOR FUEL CELL DCDC CONVERTERS 3395
  Jiajia Guan

PP356   TESTING THE PRIMARY-SECONDARY COIL COUPLING OF HIGH-FREQUENCY
TRANSFORMER IMPLEMENTED ON ETD AND TOROIDAL CORES 3400
  Alexis Gioda

Author Index

PCIM Europe 2024, 11– 13 June 2024, Nuremberg   DOI: 10.30420/566262480

Artificial Intelligence – between hype and industrial grade
The Impact of AI on the Entire Power Electronics Lifecycle

Prof. Dr. Rolf Hellinger[1], Dr. Martin Bischoff[2]

[1] Siemens AG, Germany

[2] Siemens AG, Germany

Corresponding author: Dr. Martin Bischoff, martin.bischoff@siemens.com
Speaker: Prof. Rolf Hellinger, rolf.hellinger@siemens.com

Abstract

Artificial Intelligence (AI) is experiencing a surge in recent breakthroughs. Advances in generative AI enable machines to create texts, pictures, videos, and other digital assets that were unimaginable just few years ago. Its potential to transform the global economy is already evident. For industry, in particular for the power electronics sector, these innovations offer huge potential.

We give an overview on the state of power electronics and where AI can help us to overcome the present main challenges. Considering the power electronics lifecycle, we highlight selected references where artificial intelligence is already generating added value. We discuss present trends showing how AI will help us to innovate, design, develop, manufacture, and operate our systems.

1   State of Power Electronics

1.1   Why is Power Electronics so super important?

The current societal emphasis on environmental awareness and the pursuit of ecologically sustainable practices underscores a significant shift from the combustion era to an all-electric era. This transition encompasses not only environmental preservation and sustainability but also entails technological advancements, societal transformations, and the emergence of novel business models. At the core of this transition stands Power Electronics, serving as a vital technology driving this shift. Consequently, Power Electronics is undergoing a renaissance, and is poised to play a central role in shaping our electrified and digitalized future. This renaissance also requires a completely new understanding of power electronics in the context of control, swarm, and new methodology such as AI.

1.2   What are the hurdles and needs?

The primary obstacles in navigating this rapid and expansive transition phase are twofold: keeping abreast of the transition's momentum and addressing the sustainability requirements through-

out the entire product lifecycle of power electronics. How does AI assist us? This entails optimizing the system across multiple fronts simultaneously:

- Ensuring **sustainability** throughout the entire product lifecycle, from the selection and extraction of raw materials to a robust eco-design for repair, refurbish, re-use, re-manufacturing, disassembly, and recycling as well as achieving high energy efficiency during operation.
- Enhancing **speed** throughout the Product Lifecycle Management (PLM) process, encompassing design, manufacturing, operation, service, and end-of-life considerations.
- Integrating **smart**, intelligent functionalities and performance enhancements to streamline commissioning, operation, and service processes.
- Using a **network** of power electronic devices to support infrastructure such as smart grids.

1.3   How can we translate these demands?

- Establishing **closely interconnected R&D processes** among various partners, including academia, research institutes, suppliers, and customers, leveraging digital twins and accurate simulations.
- Achieving **scalability and flexibility** by employing software-defined re-usable power electronic building blocks.

- Encouraging the **exchange of information** both vertically across system stacks and horizontally throughout the entire Product Lifecycle Management (PLM) process.

We firmly believe that AI represents a potent methodology for expediting these approaches. Subsequent sections will delve into the strengths and limitations of AI for power electronics systems.

2 AI for Power Electronics

The data flow for power electronics systems consists of the following three key dimensions:

1. The overall lifecycle, ranging from first concept developments to refurbishment and recycling .

2. The system levels, ranging from semiconductor material to complete Power Electronics applications, such as photovoltaic power plants, charging stations, electric cars, production machines or wind turbines.

3. The actual task or work package for which the data is generated, collected, processed, and/or provided (use cases).

In this three-dimensional space, a wide variety of AI application possibilities exist, most of which still need to be explored.

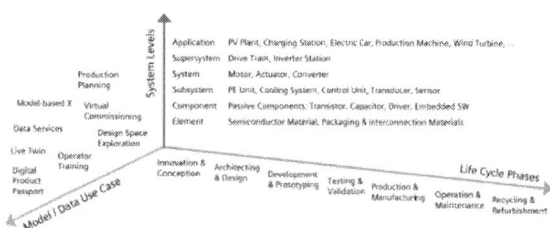

Fig. 1 AI application space for power electronics

In this section we present a small selection of AI application examples along the product lifecycle which have proven to generate added value.

2.1 Inverter certification

Traditionally, product certification demanded exhaustive physical tests, often subjecting a single product to a battery of assessments. Simulation based on the digital twin is the answer to the special challenges and cost factor of test units and certified laboratories when high power electronics components are involved.

Siemens and UL Solutions, however, have introduced a paradigm shift. A product now may not need to undergo every test physically; instead, the process can integrate the power of Siemens' digital twin technology. UL Solutions spearheads this transformative approach, navigating the certification process with unmatched expertise.

By harnessing the power of digital simulations, Siemens is ushering in an era in which traditional testing boundaries are being transcended and product insights can be gained with unprecedented speed and accuracy. The future of certification will be characterized by reduced costs, accelerated time-to-market, and seamless integration of digital modeling tools. Comprehensive digital twins are becoming the cornerstone for not just helping to meet compliance requirements, but also for driving innovation at high speed without compromising safety, performance, or quality.

Fig. 5 Simulation replacing physical certification tests.

In Siemens' pioneering venture into the industrial metaverse, the demarcation between physical and digital spaces is becoming increasingly indistinct. The resulting combination is fostering a dynamic environment conducive to innovation.

The first certification of an industrial product with simulations was for the "SINAMICS G220", a sophisticated system that enables the precise control and efficient operation of electrical drives in various applications. Leveraging Siemens' comprehensive digital twin for such a task is a compelling demonstration of the industrial metaverse's potential to usher in unprecedented possibilities for product development and other engineering tasks.

2.2 Inverter production

The Siemens Electronics Factory in Erlangen (GWE) is often denoted as "customer zero" since it also serves, along with producing inverters and controls, as an in-house test field for industrial metaverse innovations. New technologies are incubated and evaluated here. Only when they have proven of value are they offered externally to our customers.

2.2.1 Through-Hole-Technology (THT) – wave soldering

In THT production systems, an AI-based inspection system was applied at GWE which reduced false calls. With conventional methods, 5.0% to 25% of Printed Circuit Board Assemblies (PCBAs) must be controlled manually, whereas 98% are false calls. The AI-based inspection system reduces the number of necessary controls to only 2.5% to 12.5%.

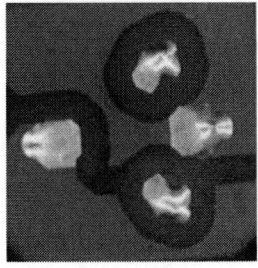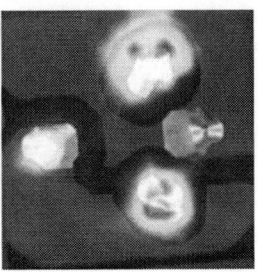

Fig. 2 AI localizes anomalies in THT wave soldering

At the end of the wave soldering line, an automatic optical inspection system (AOI) continuously evaluates PCB camera image data for possible solder bridges or open solder joints to either accept or reject the corresponding PCBs. Enhancing this AOI with AI technology helped improving its decision quality.

At GWE, this AI-based AOI system increases the overall efficiency while reducing human labor effort. Currently, there is one THT wave soldering line in operation; a rollout to further lines is planned.

2.2.2 Robot object picking

Typical robotic handling applications require the manual sorting of randomly supplied parts.

Fig. 3 Left: Pre-sorted material, Right: Enhanced robot bin picking station

SmartGrasp is an innovation which avoids this monotonous, time- and cost-intensive work. Here, an AI-trained robot performs bin picking, picking up objects with random poses. The AI is trained on image data to identify graspable objects which are not covered or jammed.

Whereas the acquisition and labeling of real image data is time and cost intensive, synthetic data, i.e. computer-generated images, is automatically generated from CAD data of the objects to be grasped. Often, this data is already available, such as from the digital twin of virtual commissioning. The benefits of synthetic image data are a reduction of the commissioning time, improved model performance, and more streamlined adjustments during production. Training on synthetic image data helps scale this AI technology.

At GWE, three SmartGrasp bin picking stations are currently in operation and three more are planned for the upcoming months. So far, they have performed more than 1.5 million picks. Bin picking closes the last gap for end-to-end production automation.

2.3 Inverter operation

In the coming years, inverters, drives and industry systems will increasingly be operated and controlled by AI. Enhanced controls that optimize the overall systems performance, availability, and economic efficiency generally also entail higher complexity. It must be ensured that transparency and manageability are not compromised.

2.3.1 Data services

Growing numbers of devices have interfaces for communicating actual operational data. AI services help process this flood of data to obtain unprecedented insights and efficiency gains. IoT, connectivity and AI are three powerful technologies that amplify and drive each other.

For low-voltage motors that drive such devices as pumps, fans or compressors, Siemens offers SIMOTICS CONNECT 400 together with Insights Hub analytics app SIDRIVE IQ Fleet for condition monitoring and predictive maintenance. Analytics here are based on comparing actual operational data with motor-specific digital twins based on motor-specific electrical and mechanical models (including electric equivalent circuit diagrams).

For Siemens motors, advanced motor models with instance-specific motor data are automatically provided from a database during commissioning. For non-Siemens motors, these models are being created during commissioning using rating plate information. In both cases drive train health can be accurately determined without the need for collecting additional data or training application-specific AI.

The benefits for operators, service providers and Original Equipment Manufacturers (OEMs), are manifold:

- Extended Product Lifespan: Defining optimum maintenance levels based on actual condition state prolongs product life by up to 30%.
- Enhanced Performance and Efficiency: A continuous monitoring and condition state analysis reduces maintenance costs and efforts by up to 30%. In combination with the reduction of downtimes, plant productivity can be increased by 8% to 12%.
- Enhanced Energy Efficiency and Sustainability: By using artificial intelligence (AI) and data analytics, digital motors identify inefficiencies in complex processes. Optimizing those processes can lower energy consumption and CO_2 emissions by up to 10%. This saves costs and reduces the environmental footprint.
- Remote Services: Service providers can quickly support operators from remote by accessing the most recent condition states. Avoiding physical inspections and on-site interventions reduces outages and saves time and effort.
- Improved Maintainability: Deep insights into component operation enhance maintainability, simplify troubleshooting and reduce the complexity of maintenance tasks.
- Real-time Fault Reporting: Immediate reporting of faults and errors ensures swift remedial action, preventing potential damage or safety hazards. This proactive approach to maintenance enhances operational reliability and safety across industrial environments.

In summary, AI-based data services already provide huge benefits for maintenance. Thinking further, continuous dataflows from shopfloor to cloud, together with the ability to process component data from an entire production plant with AI, will provide unparalleled performance optimization potential on the factory level.

2.3.2 Copilots

Chatbots have gained tremendous popularity in recent years. It took ChatGPT only five days to reach one million users and there has been a significant increase in its application since then. Text-based development environments are currently enhanced with chatbots to provide continuous support to users. A copilot can be considered as a chatbot which acts like a pair programmer and continuously offers autocomplete-style suggestions. It processes both existing code and natural language input and offers suggestions from within the text editor. Even though copilots generate text by themselves and can be considered as generative AI, the developer remains in full control of the text.

This integration proves particularly advantageous for code development by assisting with various tasks, including error correction, documentation, generation of unit tests, and code completion. The benefits: A drastic reduction of development costs and effort, and higher-quality code in shorter time. Copilots also help alleviate skilled worker shortage.

The Siemens Industrial Copilot exemplifies the advances in chatbot capabilities in two applications:

Firstly, it helps create PLC code by translating code from different programming languages and optimizing its efficiency. Furthermore, this functionality extends to human language as well, enabling application developers to express their automation requirements in natural language. The Copilot then seamlessly translates these instructions into PLC code.

Secondly, it helps analyze error states in operation and machine failures. In addition to identifying the reason for the error, it provides suggestions on how it can be resolved. Machine operators access this technology via a simple chat program and communicate in natural language to obtain the required information.

The Siemens Industrial Copilot is currently being tested and evaluated internally and with selected external partners.

2.3.3 AI-generated automation applications

The efficient management of discrete product flows is one key challenge in today's automation applications:

- merging products from multiple lines on one single line
- feeding products from one conveyor line in a compartment belt at the outlet
- aligning distances between products
- adjusting speeds (of e.g. unwinding packaging foil) to the amount of incoming products

These frequent and standard applications in factory intralogistics and packaging offer huge optimization potential since they typically involve buffer zones, such as intermediate conveyor belts for parking products or dancer tension control systems for packaging foils. By leveraging these buffer zones, accelerations can be avoided, and a maximized homogeneous product flow throughput is ensured.

This potential is currently often not utilized due to time and effort limitations in the development of automation applications and the complexity of conventional logic-based control algorithms.

Instead, simple controls are applied that switch between full-stop and full-speed modes, depending on whether products are currently available or not.

At Siemens, we are currently investigating reinforcement learning for identifying optimized controls for complex motion control tasks.

In contrast to supervised or unsupervised learning, reinforcement learning does not require any data. An agent learns an optimal policy by interacting with its environment.

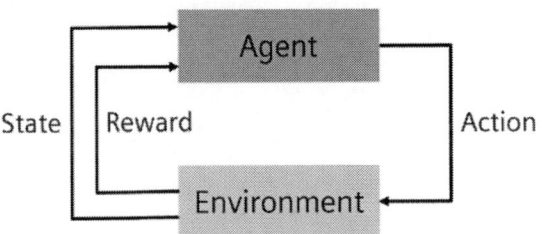

Fig. 6   Basic reinforcement learning process

In detail, a neural network is trained via reinforcement learning on a physical simulation model of the machine. The resulting AI-optimized controls, i.e., trained neural networks, are integrated into TIA projects as conventional function blocks, where they are (virtually) commissioned and subsequently deployed on Siemens SIMATIC Control units.

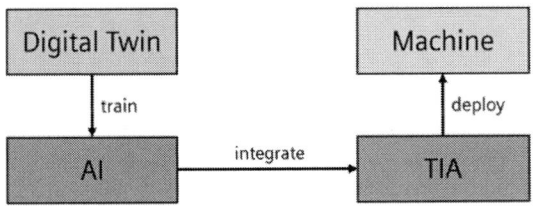

Fig. 6   AI training & deployment workflow

This drastically reduces engineering effort and commissioning time.

It enables the development of optimized machine designs and increases reconfiguration flexibility on the shopfloor since an optimized control can be generated for every design variant.

These smarter, forward-looking controls increase production efficiency and throughput in operations. And by maintaining steady velocities over rapid accelerations, energy consumption and mechanical stress are minimized. This has a positive impact on the overall lifetime and maintenance costs of all affected components. As for the overall machine layout, smart controls even allow for using lower powered motors without compromising

throughput, resulting in more cost-efficient production systems. The first co-creation projects with machine builders have proven the benefits of this novel approach [4].

This technology effectively automates the development of automation applications and drastically reduces engineering effort and commissioning time. We consider this technology to be a fundamental milestone for industrial AI. In connection with generative AI, this technology has the potential for creating completely novel systems exceeding today's design and control limitations.

3   Outlook

Considering the recent advances in AI as well as the challenges and digitalization innovations in industry, certain development trends for the coming years and decades can already be surmised.

3.1   Generative AI

Generative AI comprises all AI technologies focused on creating new contents, typically in the form of text, images, audio or even video. It mimics or resembles the input data it was trained on and often exhibits creativity and produces outputs that are indistinguishable from those created by humans. Large Language Models as Chatbots are one form of generative AI. Starting with copilots as assist functions for text-based developments, generative AI has huge potential to be unlocked in the upcoming years.

One very promising application field is model-based design space exploration. Here, generative AI is applied to create a huge set of candidate designs and evaluate them by simulation. The results are fed back to the generative AI to identify even better designs in subsequent iterations. In short, generative AI has the potential to bring the design of experiments (DOE) to a new level.

3.2   Industrial metaverse

The industrial metaverse includes all industry technologies focused on merging the digital and the real worlds and reducing their distinctiveness. AI and the industrial metaverse foster and leverage one another. Synthetic data, especially photorealistic images, is already of great value today for image-processing AI. The same holds true for physically accurate simulation models, which enable reinforcement learning on a new scale.

The key advantage of simulation models / digital twins is that they are spatially and temporarily independent and instantiated multiple times to simulate a variety of what-if-scenarios in a safe and risk-free manner. This fact has proven especially useful for model-based development over decades and has helped identify defects and optimization potential early as well as reduce development time and efforts for our systems.

Siemens, in close collaboration with partners like Nvidia, is now lifting the Digital Twin to the next level and bringing the industrial metaverse to life.

3.3 Data management

As already highlighted in Figure 1 above, data is generated along the entire product lifecycle across all system levels and for a variety of different use cases.

The data is further connected to other domains, such as the production process. It consists of multiple variants of past, present, and future products. It is collected, processed, and owned by different entities, i.e. suppliers and customers. At present, there is still considerable room for improvement in managing this data for re-use in other applications along these multiple dimensions.

Considering the advances in digitalization, such as data services, connectivity, and the industrial metaverse, increasingly more data will be generated in the future.

One current challenge for streamlining this dataflow is that data is stored in different formats. Moreover, often only a small share of data gathered in one application is relevant for another, and extracting this share in an automated manner requires a forbiddingly high programming effort. Here Large Language Models (LLMs) will come to play in the future. One key feature of LLMs is that they can translate texts between different languages or data formats.

Looking ahead, future LLMs will enable us to abstract information from its actual representation. That is, the user of the data will not need to know how the data is originally stored, since an AI translates it into the required format.

LLM technology will be a key enabler for streamlining data flows in industrial product development.

3.4 Collaboration & partnership

AI holds immense potential for various industries and sectors. Its capabilities in data analysis, automation, and decision-making have the power to revolutionize the way we work and live.

As engineers and researchers, it is our responsibility to collaborate with customers, partners, and academia to shape the future with our innovations, including AI. Through collaboration, we can realize and evaluate our innovations most efficiently and minimize time to industrial readiness.

The rapidly increasing pace of innovation is being further boosted by AI technologies. To keep at the forefront here, it is crucial for us to work as team. By fostering an open, collaborative environment that fosters continuous learning and adapting, we can embrace the changes and harness the full potential of AI. We can do more with less.

3.5 References

[1] Industrial AI and AIoT Market Report 2021-2026, IoT Analytics GmbH, 11/2021

[2] Current and future activities in energy-efficient industrial drives, Koellensperger, Peter; Tsotoulidis, Savvas, 2023, ETG-Fachtagung Baulemente der Leistungselektronik und ihre Anwendungen 2023

[2] AI-Based Motion Control with System Dynamics Flexibility, Körwer, Niklas, Bischoff, Martin, De Doncker, Rik W., Laumen Michael, 2024, ECCE Europe 2024, submitted

[3] "Siemens and UL Solutions redefine future of certification process with ground-breaking digital-twin technology", Jil Huber, Steven Breewster, Press Release, Siemens AG, 01/2024

[4] "Siemens en Sollas zetten AI in op de shopfloor", Siemens Stories, March 2024

PCIM Europe 2024, 11– 13 June 2024, Nuremberg   DOI: 10.30420/566262481

Infrastructure Requirements for Electrified Heavy Road Transport in Germany and the EU

Daniel Speth[1], Steffen Link[1], Martin Wietschel[1]
[1] Fraunhofer Institute for Systems and Innovation Research, Germany

Corresponding author: Martin Wietschel, wietschel@isi.fraunhofer.de
Speaker: Martin Wietschel, wietschel@isi.fraunhofer.de

Abstract

Heavy road freight transport is responsible for about 7% of energy-related greenhouse gas emissions in Germany and Europe. The electrification of trucks via batteries, overhead lines or fuel cells are promising options to meet the European climate targets for heavy road freight. The development of the necessary infrastructures and the development of the market ramp-up of the alternative truck technologies will be presented.

1 Introduction

The fast decarbonization of heavy road freight transport is crucial in limiting global warming in line with the Paris Climate Agreement and eventually reaching climate neutrality by mid-century [1]. This follows since heavy-duty vehicles (HDVs) account for approximately 7-8% of annual energy-related greenhouse gas (GHG) emissions in both Germany and Europe, despite comprising relatively small fractions of the total vehicle stock (cf. **Fig.1**). Plus, the ongoing increase in transport performance will aggravate GHG emissions if the heavy reliance on fossil fuels remains unabated.

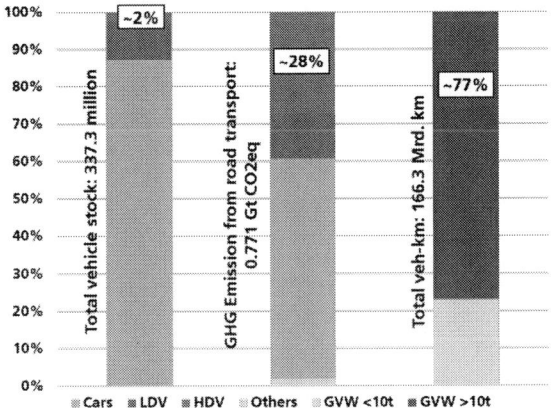

Fig. 1 Comparison of vehicle stock, road transport GHG emissions, and annual vehicle kilometers travelled (road freight only) for the EU. HDVs over 3.5t of gross vehicle weight (GVW). Reference year: 2021. Stock figures for EU-27, EFTA, and the UK. GHG emission and veh-km for EU-27 and EFTA. Own illustration based on [2, 3].

Accordingly, global key markets have instituted specific CO_2 reduction targets for newly sold HDVs. Notably, the European Union has ratified ambitious tailpipe emission reduction targets of -43% by 2030, -65% by 2035, and -90% by 2040 in comparison to 2019/20 levels [4]. Similarly, California has mandated the phasing-out of conventional combustion trucks by 2036, while the general US emission standards (Phase 3) likely require around 30% zero-emission truck (ZET) sales by 2032 [5]. Last, China may soon tighten its tailpipe emission reduction targets to align with its near-zero emission target by 2060.

To decarbonize HDVs, two technological pathways may be distinguished [6]: the direct use of electric energy in battery-electric trucks (BETs) and the adoption of carbon-neutral fuels such as hydrogen and synthetic liquid/gaseous fuels produced from renewable energy sources or biomass. Hydrogen can be deployed in fuel cell trucks (FCETs) and in trucks equipped with modified internal combustion engines (H2-ICETs). BETs can be designed as plug-ins or catenary electric trucks (CETs), with the latter being dynamically charged via overhead lines. However, each of these technologies relies on associated alternative refueling/charging infrastructures that still need to be built. In contrast, sustainable liquid fuels (BtL, PtL) may serve as a drop-in replacement for current fossil fuels, supporting the existing vehicle stock and ongoing utilization of existing refueling stations.

This paper reviews potential market diffusion scenarios of ZETs in Europe and examines the associated infrastructure requirements. Specifically, we focus on charging networks for BETs and hydrogen refueling station (HRS) networks for hydrogen-powered trucks, analyzing the necessary infrastructure dimensions, locations, and power requirements while discussing potential bottlenecks. We conclude with implications and recommendations for policymakers and industry stakeholders.

2 Market diffusion of zero emission truck drivetrains

Today, less than 2% of all newly registered HDVs (class N2-N3) in the European Union (EU) are ZETs [7]. However, all major truck manufacturers already started or announced the production of ZETs, especially BETs. **Fig.2** visualizes the expected annual sales across manufacturers in Germany, as stated by themselves during so-called cleanroom talks hosted by the German Federal Ministry of Digital and Transport in 2023 [8]. Accordingly, more than half of the sales are expected to be BETs while FCETs may account for about 15% by 2030.

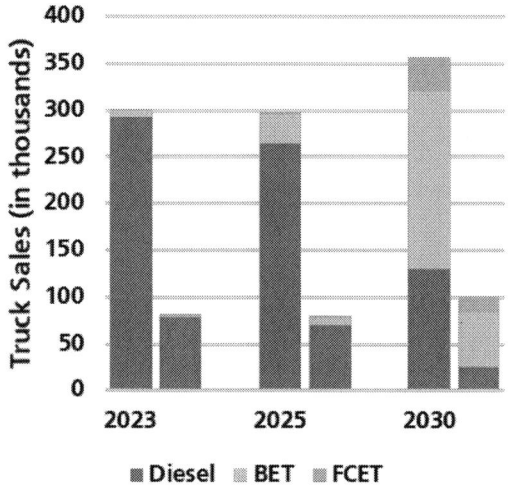

Fig. 2 Expected truck sales figures in Europe (large bars) and Germany (small bars) based on the cleanroom talks with truck manufactures (class N2-N3) (data source [2])

To reach German climate targets by 2030, approximately one quarter of the HDV fleet should be BETs, while another 15% may be hybrids. With respect to fleet turnover and a gradual diffusion within the 2020s, this requires that more than half of all HDV sales must be battery-electric [9, 10], which is well in line with stated manufacturers' announcements.

Looking ahead to 2045/50, the potential role of different technological alternatives - from niche to mass market - is subject to ongoing discussions, largely depending on the overall energy system and the respective role of different energy carriers [4]. Nonetheless, increasing evidence suggests that BETs may comprise a substantial relevance for sustainable road freight transport. Herein, synergies from battery-electric passenger cars, such as economies of scale and technological battery advances, drive cost-effectiveness, availability, and technical competitiveness with diesel trucks [11, 12]. Conversely, alternatives relying on green hydrogen and sustainable fuels are unlikely to play a major role, experiencing high costs due to scarcity and implying that these shall rather be used in other hard-to-abate sectors [10, 13].

Herein, we highlight [14], who performed a Monte-Carlo simulation on the techno-economic feasibility of diesel trucks, BETs and FCETs to estimate likely future market shares. By varying 16 parameters in 1,000 iterations, the authors found "that zero-emission vehicles should generally become cost-competitive with diesel-propelled trucks between 2030 and 2040 [...]", while stating that "FCEVs are cost competitive in only a small number of marginal cases that assume ambitiously low hydrogen fuel costs and very conservative assumptions for BETs."

3 Alternative recharging/refueling infrastructure requirements

3.1 European regulation

The Alternative Fuel Infrastructure Regulation (AFIR) [15], as part of the EU Fit for 55 package, defines the deployment of publically accessible alternative infrastructures based on annualized plans within distinct corridors. Specifically, targets for 2025, 2027, and 2030 focus on infrastructure deployment alongside the EU's main transport corridors (TEN-T network), with the latter further split into TEN-T Core and TEN-T Comprehensive and exemptions due to required minimum truck traffic volumes.

To specify the TEN-T network [16], the Comprehensive network has a total length of approximately 136,700 km, of which the Core network makes up 49,700 km. Plus, there are more than 420 urban nodes on the network, typically defined as cities with at least 100,000 inhabitants.

For recharging stations, a minimum single output of 350 kW is required every 60 km along the Core network and every 100 km on the larger Comprehensive network from 2025 onwards (15% coverage), with 50% coverage by 2027 and complete network coverage by 2030. Each recharging pool shall be located on the TEN-T road network or within 3 km driving distance from the nearest exit. Likewise, the minimum available power per recharging pool must increase from 1,400 kW by 2025 to 1,500 kW (Comprehensive) and 3,600 kW (Core) by 2030. Further targets apply to safe and secure truck parking areas (SSTPAs) and urban nodes along the TEN-T network, with targets for the latter requiring at least 150 kW (single) and 1,800 kW (per pool) by 2030 [15].

For HRS, stations with at least 1 t per day (cumulative capacity) and at least one 700 bar dispenser (compressed hydrogen) must be deployed from 2030 onwards in all urban nodes and every 200 km along the TEN-T Core network. Each HRS shall be located on the TEN-T road network or within 10 km driving distance from the nearest exit [15].

Other than that, the latest AFIR version mandates reporting of market development, technical preferences, and technical progress for several technologies yet without explicit deployment schedules, such as high-power recharging standards including the Megawatt Charging System (MCS), electric road systems (ERS) such as CETs, and the use of liquid hydrogen in HRS [15].

To illustrate and quantifiy the AFIR requirements for charging infrastructure, **Table1** provides an assessment of number of locations and total installed power in MW along the TEN-T network:

Region	Time	Number of locations	Total power in MW
Germany	2025	32	66
	2030	314	918
EU-27	2025	304	589
	2030	2,805	7,494

Table 1 Number of locations and installed power along the TEN-T network, as required by the AFIR.

3.2 HRS networks for Germany and Europe

Scientific literature on comprehensive HRS networks in Europe is limited. [17] and [18] evaluated potential HRS networks for Germany, both assuming full adoption of FCETs by 2050. Using an optimization model (flow refueling location model - FRLM), they found that 100 to 140 stations, depending on capacity restrictions (up to 30 t per day), could meet the national hydrogen refuelling demand. This approach was later extended to the European network [19], suggesting that 920 HRS stations could cover about 80% of the European truck traffic on major roads.

3.3 Electric road systems for Germany and Europe

The role of ERS remains uncertain, as it was initially excluded from the AFIR and without definitive model announcements from manufacturers. [20] has outlined three potential scenarios for ERS, ranging from niche application to widespread deployment, either ad-hoc or delayed until the 2030s. Their proposed ERS network spans approximately 12,000 km, prioritizing Europe's busiest highways. Similarly, [21] have proposed a collaborative infrastructure network for Germany, integrating overhead lines and high-power charging stations with the selective electrification of Germany's busiest highways (2,000 km) to ease long-distance transports, downsize large charging stations, relief local power grids, and reduce operational constraints.

3.4 Charging networks for Germany and Europe

Today, the debate often focusses on the deployment of public fast charging (FC) infrastructure, as for example in the AFIR, where the mandatory driving break after no more than 4.5 hours is used for recharging. However, alongside public charging, private charging at own depots and semi-public charging at customer depots, logistics centers, or industrial hubs are key, particularly regarding slow overnight charging (ONC). Here, scientific literature indicates that private charging will probably serve most of the charging events [22, 23]. Therefore, we show some key results on the general driving and charging behavior in 3.4.1. Afterwards, we focus on the potential demand for public charging infrastructure in 3.4.2. Finally, we identify potential challenges for the fast deployment of charging infrastructure in 3.4.3.

3.4.1 Particularities of public and private charging locations

Simulating 2,410 German truck driving profiles from diesel HDVs (class N2-N3) as BET shows that approximately 70% to 80% of all trucks - depending on battery sizes and reference years - will probably use exclusively private or semi-private charging infrastructure at depots. Additionally, about two third of all trucks can purely rely on private or public slow charging (SC) with up to 44 kW [22]. **Fig.3** shows the simulated fleet over one day, assuming a "as slow as possible" charging strategy, a vehicle range of 450 km, and an average maximum charging power of 1 MW.

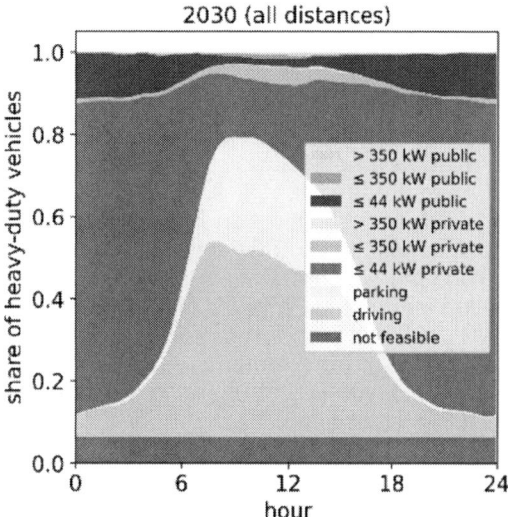

Fig. 3 Driving and charging behavior of German battery-electric HDVs (> 12 t GVW) fleet, simulated for 2030. Adopted from [22].

Herein, private charging is shown in shades of green, public charging in shades of blue. While electric vehicles are on the road during the day, the nighttime presents an opportune time to charge them at lower power levels. However, particularly during midday hours, the demand for FC capabilities becomes essential. Today's typical performance allows for up to 350 kW using the Combined Charging System (CCS), but future advancements enabled by MCS may enable even higher power outputs.

Fig.4 shows the corresponding simulated relative load curve. Again, it is evident that private SC will be responsible for the majority of the energy demand, in the simulation about half of the energy demand. However, the load curve also shows that approximately 20% (7% private, 13% public) of the energy demand stems from FC with more than 350 kW. Additionally, FC as intermediate charging

causes a load peak at midday hours. Analyses also show that the midday load peak decreases and charging events will be shifted to nighttime hours if battery sizes increase.

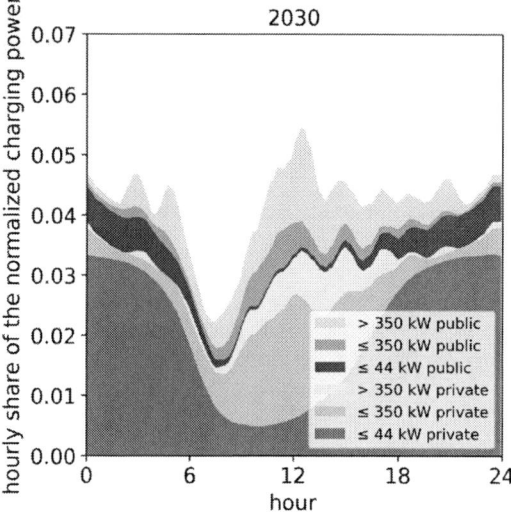

Fig. 4 Load curve of a German battery electric HDV (class N2-N3) fleet, simulated for 2030. Adopted from [22].

In conclusion, it is evident that SC, especially when implemented at private depots, will form the backbone of future charging infrastructure networks. Thus, any delay in expanding this private infrastructure will substantially delay the market diffusion of BETs. Additionally, public FC, exceeding 44 kW and up to MCS, plays a crucial role in facilitating long-distance transportation. Initial case studies, such as [24], emphasize this importance. Furthermore, note that public FC infrastructure accounts for approximately 20% of the BET fleet energy demand [22].

3.4.2 Potential number of charging stations

From a scientific perspective, there are three major approaches to calculate the demand for charging locations [25]. Node-based approaches position a charging location at one node so that the demand from neighboring nodes can be fulfilled. Path-based approaches consider not only the demand for charging events at a given node, but are based on origin-destination flows. Charging locations are positioned so that they can serve as many passing origin-destination flows as possible. Finally, tour-based approaches consider not only single trips (flows), but all tours of the vehicles. However, path- or tour-based models need more information about the driving behavior than node-based models. In principle, all models can be defined as simulation or optimization models and can be combined with queuing theory. For a more detailed introduction, see [26].

For Germany, there are three scientific publications, calculating a potential public fast charging infrastructure for trucks. Using a node-based approach with charging locations at regular intervals combined with queuing theory, [27] calculated approximately 1,200 MCS points at 267 charging locations (serving both directions), to serve 15% of the total truck traffic. Single locations consist of 2 to 8 charging points. Assuming a similarity factor of 0.6, as usual for passenger car charging locations today, a total power of up to of approximately 5 MW will be necessary for large locations. **Fig.5** shows an exemplary map. It is evident that large locations are positioned at highly trafficked routes, for example the highway A2 from Dortmund via Hanover to Berlin, while smaller locations ensure the coverage of the whole country.

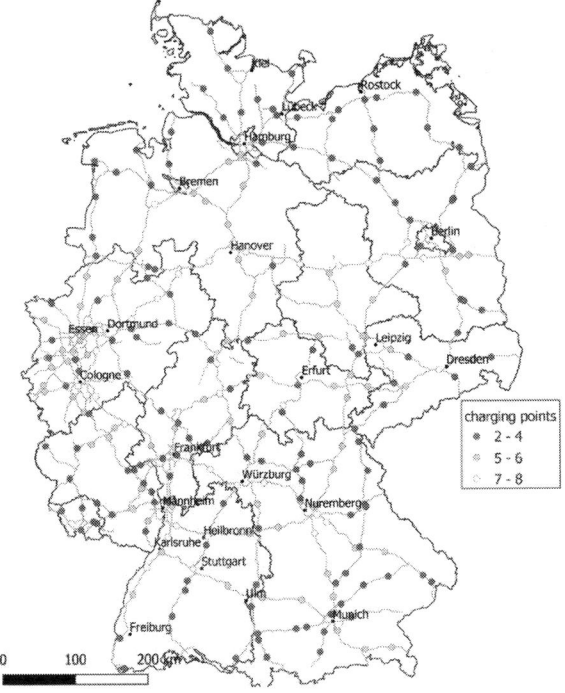

Fig. 5 Location of 267 fast-charging locations at regular intervals of 50 km in Germany. Adopted from [27], background map from GeoBasis-DE / BKG 2020

Similar results are shown in a tour-based simulation that calculates 1,150 FC points at 456 locations (serving one direction), also for 15% of the fleet. Single locations consist of 1 to 15 charging points [28]. As shown in a third publication [29], doing a path-based optimization approach on the green field, 42 charging locations (serving both directions) could serve Germany. However, when focusing on existing parking areas and converting the full truck fleet to BET, more than 10,000 FC charging points at 124 charging locations (serving both directions) will be necessary [29]. It should be mentioned that this is an optimization results with conservative values on the vehicle range (max. 300 km), so that a future network will likely have more locations but possibly less charging points. It can be concluded that for a first electrification in Germany until 2030, slightly more than 1,000 FC charging points at several hundred locations will be needed. For a full electrification, - depending on the vehicle range development - several thousand CPs will be necessary.

Similar calculations can be found for a European public charging infrastructure. Using a node-based approach, less than 5,000 FC points at approximately 1,500 locations (serving both directions) can serve 15% of the truck fleet in the EU27+3 [26]. Again, large locations with up to 11 FC points cover highly trafficked routes, while small locations cover less trafficked road sections. On average, a location is equipped with 3 FC points. Using a path-based simulation approach, [30] calculate a demand for 9,000 MCS points to serve 15% of the European truck fleet. They also point out that for each MCS point, probably approximately 4 CPs with less than 100 kW for ONC will be needed. Differences are particularly due to the expected arrival rates and the resulting utilization of the charging points as well as the expected proportion of private charging.

All in all, the mentioned publications can give first insights in the demand for public charging infrastructure, but need to be validated with real-world data when battery electric trucking becomes reality.

3.4.3 Challenges for the fast deployment of charging infrastructure

Deploying charging infrastructure for BETs faces several challenges.

First, grid extensions to accommodate the increased demand, in terms of energy and power, from European-wide charging infrastructure networks, especially considering the required medium- or even high-voltage connections, involve significant costs and require long time scales (up to ten years), as indicated by [31, 32]. Herein, the fragmented logistic market with numerous individual depots might require tailored solutions, complicating the standardization and scalability of large-scale deployment efforts. Here, battery-backed charging solutions, using buffer batteries for peak load balancing, may provide at least a short-term solution to grid expansion. Last, acceptance from logistics and fleet operators and addressing potential concerns is crucial for the widespread adoption of BETs and their charging infrastructure [33].

Second, selecting suitable locations for charging infrastructure presents challenges, involving green- and brownfield installations and securing adequate area. Best potentials layouts for large-scale charging hubs, involving slow and fast charging spots yet without losing too much space, are still under discussion [34]. To identify relevant locations, this dataset containing real-world truck parking locations may be helpful [24]. Note that parking spaces are already scarce along the highways.

4 Conclusions

This paper reviewed potential market diffusion scenarios of ZETs in Germany and Europe and examined associated infrastructure requirements. We draw three conclusions from our analysis.

First, battery-electric trucks (BETs) likely constitute the backbone of sustainable road freight transport in Europe, while scarcity of green hydrogen and sustainable fuels likely decelerate or even prevent their widespread adoption.

Second, BETs will heavily rely on private charging infrastructure below 50 kW at depots, usually for slow overnight charging. In contrast, public fast charging infrastructure along the highways, particularly above 350 kW and over 1 MW with the new MCS standard, will be the enabler for long-haul transportation and crucial for safeguarding tour completion.

Third, grid extensions and new connections to the medium- or even high-voltage grid will involve significant costs and require long time scales, potentially decelerating the fast BET market diffusion.

Future research might focus on peak shaving and load shifting possibilities for charging hubs, combined charging hubs with the synergetic set-up for cars and trucks, and embedding BET batteries in the energy system (vehicle-to-grid).

5 Acknowledgements

The Federal Ministry for Digital and Transport (BMDV) in Germany funded this research within the project HoLa under grant agreement No 03EMF0404A. The authors received funding from KAMO - High Performance Center Profilregion as a national high performance center funded by the Fraunhofer Gesellschaft.

6 References

[1] P. Jaramillo *et al.*, "Climate Change 2022: Mitigation of Climate Change.: Contribution of Working Group III to the Sixth Assessment Report of the Intergovernmental Panel on Climate Change. Chapter 10: Transport," Cambridge University Press, Cambridge; New York.

[2] European Commission (EC) - Directorate-General for Mobility and Transport (DG-Move), *EU transport in figures – Statistical pocketbook 2023.* [Online]. Available: https://data.europa.eu/doi/10.2832/319371

[3] Eurostat, *Dataset road_go_ta_mplw and road_go_ta_tott.* [Online]. Available: https://ec.europa.eu/euros-tat/web/main/data/database

[4] European Council (EC), *Heavy-duty vehicles: Council and Parliament reach a deal to lower CO2 emissions from trucks, buses and trailers.* Brussels, 18th 2024. [Online]. Available: https://www.consilium.europa.eu/en/press/press-releases/2024/01/18/heavy-duty-vehicles-council-and-parliament-reach-a-deal-to-lower-co2-emissions-from-trucks-buses-and-trailers/

[5] U.S. Environmental Protection Agency (EPA), *Greenhouse Gas Emissions Standards for Heavy-Duty Vehicles - Phase 3: Final Rule.* [Online]. Available: https://www.epa.gov/system/files/documents/2024-03/hd-phase3-veh-standrds-ghg-emission-frm-2024-03.pdf

[6] P. Plötz *et al.*, "Greenhouse gas emission budgets and policies for zero-Carbon road transport in Europe," *Climate Policy*, vol. 23, no. 3, pp. 343–354, 2023, doi: 10.1080/14693062.2023.2185585.

[7] ACEA, *New commercial vehicle registrations: vans +14.6%, trucks +16.3%, buses +19.4% in 2023.* [Online]. Available: https://www.acea.auto/cv-registrations/new-commercial-vehicle-registrations-vans-14-6-trucks-16-3-buses-19-4-in-2023/ (accessed: Apr. 2 2024).

[8] NOW, "Market development of climate-friendly technologies in heavy-duty road freight transport in Germany and Europe: Evaluation of the 2022 cleanroom talks with truck manufacturers," Im Auftrag des Bundesministeriums für Digitales und Verkehr (BMDV), NOW GmbH, Berlin, 2023.

[9] T. Gnann, D. Speth, M. Krail, and M. Wietschel, "Langfristszenarien für die Transformation des Energiesystems in Deutschland 3," 2024.

[10] T. Gnann, D. Speth, M. Krail, M. Wietschel, and S. Oberle, "Pathways to Carbon-Free Transport in Germany until 2050," *WEVJ*, vol. 13, no. 8, p. 136, 2022, doi: 10.3390/wevj13080136.

[11] M. Muratori, B. Borlaug, C. Ledna, P. Jadun, and A. Kailas, "Road to zero: Research and industry perspectives on zero-emission commercial vehicles," *iScience*, vol. 26, no. 5, p. 106751, 2023, doi: 10.1016/j.isci.2023.106751.

[12] B. Nykvist and O. Olsson, "The feasibility of heavy battery electric trucks," *Joule*, vol. 5, no. 4, pp. 901–913, 2021, doi: 10.1016/j.joule.2021.03.007.

[13] P. Plötz, "Hydrogen technology is unlikely to play a major role in sustainable road transport," *Nature Electronics*, vol. 5, no. 1, pp. 8–10, 2022, doi: 10.1038/s41928-021-00706-6.

[14] M. Craglia, "Decarbonising Europe's Trucks: How to Minimise Cost Uncertainty," International Transport Forum (ITF), Paris, International Transport Forum Policy Papers 107, 2022. Accessed: Mar. 23 2023.

[15] *On the deployment of alternative fuels infrastructure, and repealing Directive 2014/94/EU*, 2023. [Online]. Available: https://data.consilium.europa.eu/doc/document/PE-25-2023-INIT/en/pdf

[16] European Court of Auditors, *The EU core road network: shorter travel times but network not yet fully functional*. [Online]. Available: https://www.eca.europa.eu/Lists/ECA-Documents/SR20_09/SR_Road_network_EN.pdf

[17] P. K. Rose and F. Neumann, "Hydrogen refueling station networks for heavy-duty vehicles in future power systems," *Transportation Research Part D: Transport and Environment*, vol. 83, p. 102358, 2020, doi: 10.1016/j.trd.2020.102358.

[18] R. Nugroho, P. K. Rose, T. Gnann, and M. Wei, "Cost of a potential hydrogen-refueling network for heavy-duty vehicles with long-haul application in Germany 2050," *International Journal of Hydrogen Energy*, vol. 46, no. 71, pp. 35459–35478, 2021, doi: 10.1016/j.ijhydene.2021.08.088.

[19] J. Neuhausen, C. Foltz, P. Rose, and F. Andre, *Making zero-emission trucking a reality: Truck Study 2020: Routes to decarbonizing commercial vehicles*. [Online]. Available:

https://www.strategyand.pwc.com/de/en/industries/transport/green-trucking-2020/truck-study-2020.pdf

[20] P. Plötz, M. Andersson, A. Scherrer, and E. Johansson, "The possible future of electric road systems in Europe—time to decide and act," *Environ. Res.: Infrastruct. Sustain.*, vol. 4, no. 1, p. 13001, 2024, doi: 10.1088/2634-4505/ad3576.

[21] P. Plötz et al., *Infrastruktur für Elektro-Lkw im Fernverkehr: Hochleistungsschnelllader und Oberleitung im Vergleich – ein Diskussionspapier*. [Online]. Available: https://www.isi.fraunhofer.de/content/dam/isi/dokumente/cce/2021/BOLD_Truck_charging_discussion%20paper.pdf

[22] D. Speth and P. Plötz, "Depot slow charging is sufficient for most electric trucks in Germany," *Transportation Research Part D: Transport and Environment*, vol. 128, p. 104078, 2024, doi: 10.1016/j.trd.2024.104078.

[23] B. Borlaug et al., "Heavy-duty truck electrification and the impacts of depot charging on electricity distribution systems," *Nat Energy*, vol. 6, no. 6, pp. 673–682, 2021, doi: 10.1038/s41560-021-00855-0.

[24] S. Link, P. Plötz, J. Griener, and C. Moll, *Lieferverkehr mit Batterie-Lkw. Machbarkeit 2021*. [Online]. Available: https://doi.org/10.24406/publica-fhg-301266

[25] M. O. Metais, O. Jouini, Y. Perez, J. Berrada, and E. Suomalainen, "Too much or not enough? Planning electric vehicle charging infrastructure: A review of modeling options," *Renewable and Sustainable Energy Reviews*, vol. 153, p. 111719, 2022, doi: 10.1016/j.rser.2021.111719.

[26] D. Speth, V. Sauter, and P. Plötz, "Where to Charge Electric Trucks in Europe—Modelling a Charging Infrastructure Network," *WEVJ*, vol. 13, no. 9, p. 162, 2022, doi: 10.3390/wevj13090162.

[27] D. Speth, P. Plötz, S. Funke, and E. Vallarella, "Public fast charging infrastructure for battery electric trucks – a model-based network for Germany," *Environ. Res.: Infrastruct. Sustain.*, vol. 2, no. 2, p. 25004, 2022, doi: 10.1088/2634-4505/ac6442.

[28] J. Menter, T.-A. Fay, A. Grahle, and D. Göhlich, "Long-Distance Electric Truck Traffic: Analysis, Modeling and Designing a Demand-Oriented Charging Network for Germany," *WEVJ*, vol. 14, no. 8, p. 205, 2023, doi: 10.3390/wevj14080205.

[29] D. Speth, P. Plötz, and M. Wietschel, "Modelling a capacity-constrained public charging

infrastructure network for electric trucks in Germany," 2023.

[30] W. Shoman, S. Yeh, F. Sprei, P. Plötz, and D. Speth, "Battery electric long-haul trucks in Europe: Public charging, energy, and power requirements," *Transportation Research Part D: Transport and Environment*, vol. 121, p. 103825, 2023, doi: 10.1016/j.trd.2023.103825.

[31] K. Burges and S. Kippelt, "Grid-related challenges of high-power and megawatt charging stations for battery-electric long-haul trucks: Study on behalf of Transport & Environment," ef.Ruhr GmbH, 2021.

[32] S. Kippelt, F. Probst, M. Greve, and K. Burges, "Einfach laden an Rastanlagen: Auslegung des Netzanschlusses für E-Lkw-Lade-Hubs," Gefördert durch das Bundesministerium für Digitales und Verkehr (BMDV), 2022. Accessed: Mar. 13 2023.

[33] A. Scherrer, M. Helferich, D. Speth, and S. Link, *Requirements of German logistics companies for charging battery-electric trucks.* [Online]. Available: https://doi.org/10.24406/publica-2615

[34] P. Plötz, D. Speth, L. Kappler, F. Klausmann, and B. Satvat, *Megawatt charging in long-haul trucking: First findings on challenges and solutions.* [Online]. Available: https://hochleistungsladen-lkw.de/hola-wAssets/docs/publikationen/HoLa_LessonsLearnt-en.pdf

PCIM Europe 2024, 11– 13 June 2024, Nuremberg   DOI: 10.30420/566262482

Challenges and Solutions to Power Latest Processor Generations for Hyper Scale Datacenters

Gerald Deboy, Matthias Kasper, Martin Wattenberg and Roberto Rizzolatti
Infineon Technologies Austria AG, Austria

Speaker: Gerald Deboy, gerald.deboy@infineon.com

Abstract

This paper deals with the challenges created by the extreme power demand by latest processor generations as being used to train large models for Artificial Intelligence and for other use cases in Hyper Scale Datacenters. Very high load currents combined with very fast transients at processor level, lead to changes along the entire power conversion chain from AC/DC power supplies down to the Point-of-load power stages.

The article is organized along the power flow from the processor backward to the AC/DC power supply and gives insights into underlying concepts and architectures to cope with the challenges to power modern AI processors.

1   Introduction

Generative AI, which means creation of data based on pre-configured models, has led to a tremendous growth of Server installations. New applications such as Chat GPT, Microsofts Copilot or large language models with its trillions of parameters require ever larger and ever powerful processors to train the models.

AI training no longer uses serial data processing as in the classic von-Neumann architecture and its underlying x86 processors. Instead, it is based on massive parallel computing and uses different processor architectures such as Graphic processing units (GPU) and Tensor processing units (TPU). These processors have hundreds of cores with high bandwidth memory being either monolithically integrated or arranged laterally as stacked dies in the same package using a technology called Chips-on-Wafer-on-Substrate (CoWoS).

Chiplets and techniques to connect reticle-limited dies into bigger Super-chips drive compute capability of these Systems at an exponential scale.

As power lines impede further scaling of logic transistors, power will be fed to buried power lines at the backside of the processor and be connected to the frontside by nano through-silicon-vias.

The power requirements of modern processors have increased massively both in terms of load currents as well as transient response. The power intake has surpassed 1 kW per processor at supply voltages of around 0.7V, which translates to load currents in the range of 1500A.

Multi-phase buck converters being arranged laterally on the motherboard around the processor can no longer cope with the insatiable power hungry of GPUs and TPUs due to very high losses along lateral power distribution within the motherboard. Vertical power delivery with DC modules being located below the processor are required to shorten the distance to the point-of-load.

With several GPUs and co-packaged CPUs on one board, power levels have reached a couple of kWs. Feeding the boards with 12V is hence no longer an option. The transition to a 48V eco-system is currently in full swing at large cloud service providers.

As floor space in datacenters is always precious, the trend to maximize compute pay load per installation has led to power levels of more than 100 kW per rack. At this power level forced air cooling no longer works. Liquid cooling with heat exchangers at the backside of the rack are needed to cope with these tremendous power levels.

With limited rack space for power, the power rating of Server power supplies continuously moves up from 3 kW into 5 kW and in future to more than 10 kW driving a massive increase of power density at very high efficiencies of more than 97.5%.

2 Powering modern processors

2.1 Point-of-load converter

With load currents of more than 1500A per processor, lateral power delivery based on Discrete Power stages and separate inductors reaches limitations with respect to power distribution losses. As these losses scale with the square of the load current, the situation will get ever worse with every new generation of modern processors being designed for ever higher dissipated power. Consequently, power delivery needs to be arranged vertically as shown in Fig. 1.

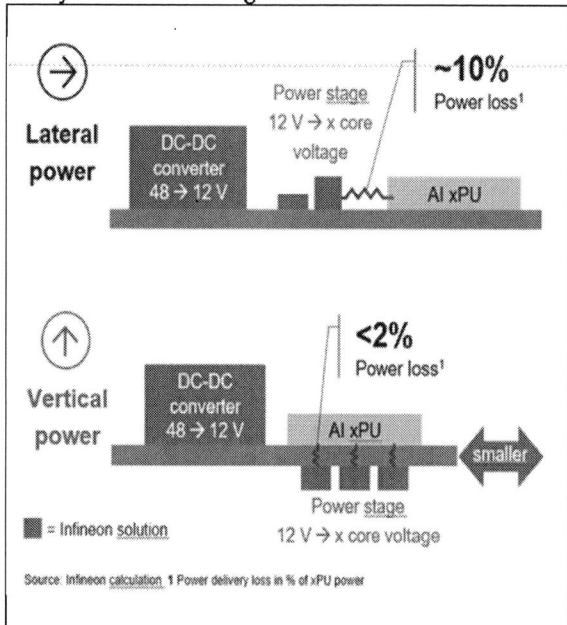

Fig. 1 Vertical power flow helps significantly to reduce power losses on 12V to core power conversion.

Due to limited space, a heterogenous 3D integration of power stage, gate driver and inductor into a compact DC module is required. Usually, these modules are constructed in such a way as to cool the power stage directly; this requires the routing of all input signals from the motherboard across the module, which makes the 3D setup fairly complex. Infineon Technologies took a different route and cools the power stage through the inductor. Furthermore, we use a new magnetic material with ultra-low core losses and a soft-saturation behavior yielding an up to 2% better efficiency than competing solutions on the market.

Fig. 2 Vertical backside modules in a compact 9x10 mm² footprint delivering up to 160A peak current.

Figure 2 shows the recently released dual-phase modules, TDM22544D and TDM22545D delivering up to 160A. The module is available in two different heights, 5 mm and 8 mm.

2.2 Intermediate Bus Converter

The intermediate bus converter (IBC) is the link between 48V provided at the backplane of the rack and the voltage level feeding the Point-of-load stages (PoL) discussed in the previous section. Input voltages may vary between 40 to 60V or may be tightly regulated to 50V as in the current release of the Open Compute Spec V3.0. The voltage level chosen for the output voltage needs to balance power distribution losses between the IBC and the power stages feeding the processor versus switching losses in the Point-of-load stages. At a lower intermediate bus voltage, PoLs can switch at a higher switching frequency, e.g. 1 MHz or above, and can take benefits from lower rated power MOSFETs such as 15V MOSFETs. A recent study by Intel and Google published at APEC 2024 sheds some light on this trade-off: with a resistance of the Power distribution Network between 3 and 6 mOhm a 6:1 divider with a respective intermediate bus voltage level of 8V proves to be the best[1]. As accelerator cards for GPUs or TPUs have typically a Power distribution resistance below 1 mOhm, even an 8:1 divider is effective, resulting in an intermediate bus voltage between 5 and 7.5V at wide voltage range input of 40 to 60V.

At a divider ratio of 8:1 pure capacitive dividers turn out to be too costly and consume too much space. Transformer based solutions such as an LLC power stage is an alternative. The best option in our eyes is a combination of capacitive and transformer-based power transfer, which we call Hybrid Switched Capacitance Converter. Fig. 3

shows a hardware demonstrator and Fig. 4 the underlying topology combining an auto transformer with a capacitive divider.

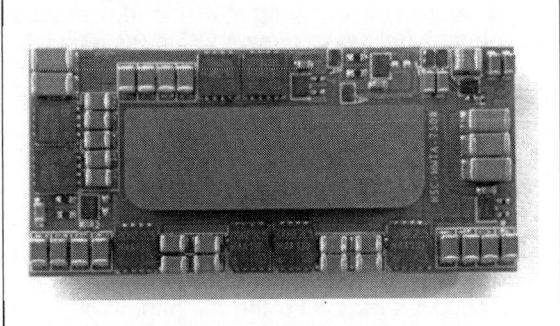

Fig. 3 Hardware demonstration of a Hybrid Switched Capacitor Converter providing a 8:1 divider ratio.

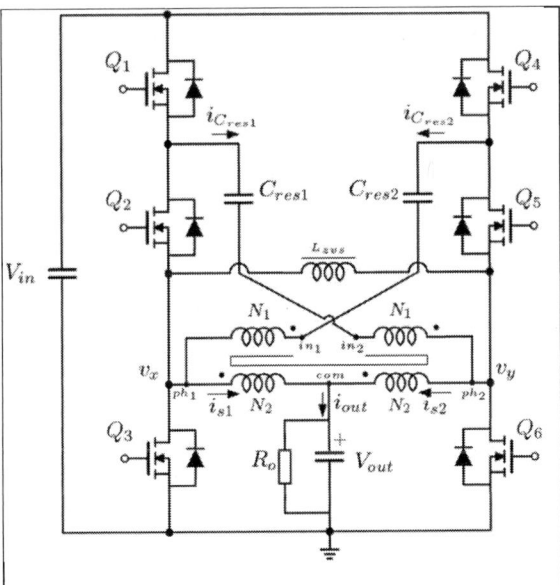

Fig. 4 Hybrid Switch Capacitor Converter using an auto-transformer and a dual capacitive divider setup.

2.3 Back-up battery unit

Large Cloud Service providers have abandoned the use of Uninteruptible Power Supplies (UPS), which were in series with the main power distribution. With its double AC/DC and DC/AC conversion UPSs cause an efficiency loss of the overall power flow of datacenter between 4 to 6 % of the total energy consumption and drive hence Operational Expenditure Cost significantly up. Backup power is now provided in a parallel path at the 48V DC level by Lithium-Ion batteries commissioned within the same rack or in a separate rack close-by to the compute trays. In case of the Open Com-

pute Spec the output voltage of the Back-up Battery Unit (BBU) is regulated to 48V. A bi-directional DC/DC converter is located between the Li-ion cells and the 48V rails being connected to the Intermediate bus converter.

Rather than processing the full power of the battery stack, the battery voltage range of the Li-ion cells can be chosen in such a way that the resulting voltage level of series and parallel connected cells is centered around 48V. In this case only a fraction of the power needs to be processed to provide a full regulated 48V output.

We recently demonstrated such a system in hardware achieving an outstanding peak efficiency of 99.6% at the mid-point of the battery voltage range. Fig. 5 shows the measured efficiency curve and Fig. 6 the employed partial power concept.

We use a combination of an LLC running on resonance frequency and a buck converter for regulation. As positive or negative voltages need to be added to the voltage of the Li-ion battery stack to deliver 48V at the output, a circuit changing the polarity of the output is included as well.

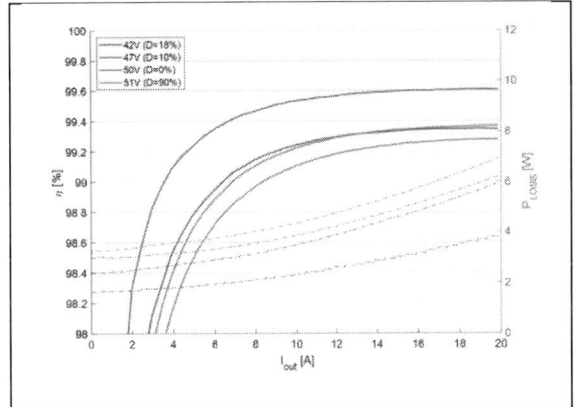

Fig. 5 Measured efficiency of a partial power converter at different battery voltages.

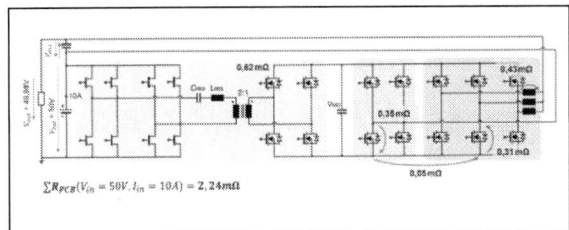

Fig. 6 Topology of the partial power converter.

2.4 AC/DC Power supply

With power levels of compute racks continuously moving up, either the number of power shelfs or the rating of Server power supplies need to go up accordingly. For training of Artificial Intelligence models the established 3 kW AC/DC power supply units (PSU), will soon be replaced by 5.5 kW rated PSUs. In the future power levels may move up to 8 kW and potentially beyond 10 kW per single-phase power supply. As efficiency remains a key concern due to its direct impact on electricity cost and consumption, these power supplies need to master very high power density of up to 100W/in³ and very high efficiency of up to 97.5% at the same time.

The use of Wide bandgap power devices based on SiC or GaN in this context is an obvious choice. Totem pole Power factor correction circuits employing SiC MOSFETs and DC/DC stages running on high switching frequency and taking full benefit of GaN HEMTs with their favorable FoM RDSon*Qoss and zero reverse recovery charge are the preferred choices for many designers.

Another interesting alternative is the use of multilevel topologies for the power factor correction. We have recently demonstrated a 3-level flying capacitor PFC stage achieving an outstanding efficiency of 99.2% at a power density of more than 150W/in³. The concept is based on a 400V rated SiC MOSFET, which will be launched soon. We furthermore use a proprietary circuit to utilize nearly the full energy content of the electrolytic capacitor cells while providing a constant input voltage to the DC/DC converter.

Figure 7 shows an implementation for a 3 kW Server power supply at a height of only 1/2U, hence allowing to push 6 kW of power into a standard 1U-high power supply.

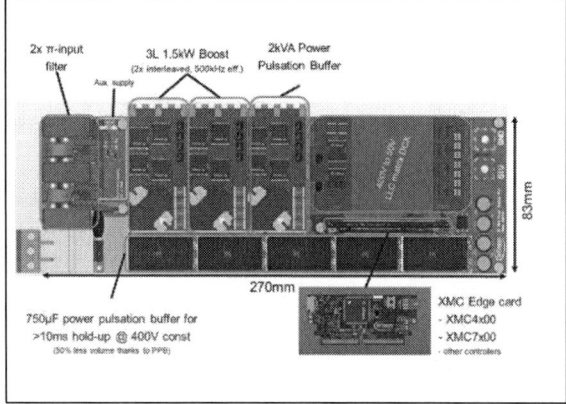

Fig. 7 Hardware demonstrator for a 3 kW Server power supply unit at a height of only 22 mm.

The concept can be naturally extended into Server power supplies being rated for 8 kW or more output power.

3 Summary

The insatiable power hunger of modern processors requires new solutions on all stages along the power conversion chain: vertical backside power modules to drive ever large processor currents and intermediate bus converters with an output voltage level between 6 to 8V to balance carefully power distribution losses versus switching losses at Point-of-load.

Partial power converters help to provide back-up power in case of loss of AC power while AC/DC power supplies may take the full benefit of multilevel architectures combined with latest Wide bandgap power devices based on SiC and GaN to cope with the challenge of very high power density and efficiency.

Infineon Technologies is supporting all these trends with a rich portfolio of products and solutions.

References

[1] Patt Chang, Chi Hsu, Shuai Jiang, Zichao Ye, "The Challenges of 12V Input for Dual Entry Power Design and the Future of 48V HVDC in Data Center", Proc. APEC 2024.

PCIM Europe 2024, 11– 13 June 2024, Nuremberg   DOI: 10.30420/566262001

An improved ultrafast Desaturation-based Protection scheme for GaN HEMT

Hossein Khoun Jahan[1,2], Hamidreza Esmaeilian[1,2], Lei Kou[1], Roy Hou[1], Lucas Lu[1], and Xiaoyu Wang[2]

[1] Infineon Technologies, Ottawa, Ontario, Canada

[2] Carleton University, Ottawa, Ontario, Canada

Corresponding author and speaker: Hossein Khoun Jahan, Khounjahan.external@infineon.com

Abstract

In this paper, an improved ultra-fast desaturation-based protection scheme for GaN HEMTs (Gallium Nitride High Electron Mobility Transistors) is proposed. The conventional desaturation-based overcurrent protection (OCP) scheme encounters performance challenges when a negative gate voltage is employed to turn off GaN devices. The proposed scheme addresses this issue. To verify the performance of the proposed scheme, experiments were conducted using a high-power prototype consisting of three paralleled 150-A and six paralleled 30-A GaN switches, and the experimental results are presented.

Introduction

Due to the high switching capability of GaN devices, an ultra-fast overcurrent (OC) and short circuit (SC) protection circuit is required [1-4]. However, conventional protection schemes face challenges when applied to GaN devices due to various limitations. Since the threshold gate voltage of these devices is low, they require elaborate gating circuits. To enhance their performance, a negative turn-off voltage is typically applied to the gate-source. The existing OC and SC protection schemes are not readily applicable when a negative turn-off gate voltage is needed for GaN HEMTs.

To address this challenge, this paper proposes an improved ultra-fast desaturation-based OC and SC protection scheme. Notably, this proposed scheme incorporates all the advantages of existing ultra-fast desaturation based OCP schemes and is compatible with GaN applications requiring negative turn-off gate voltages. The performance of the proposed scheme is evaluated using a high-power prototype featuring three paralleled 150-A and six paralleled 30-A GaN-based switches to provide experimental verification.

Conventional Desaturation Based OPC for GaN HEMT

For ease of reference, the conventional desaturation based OCP is depicted in Fig. 1. The operational principle of this OCP is detailed in [1], and for brevity, it is not reiterated here. However, to simplify matters, the equivalent circuits for the turn-on and turn-off stages of the OCP from [1] are presented in Fig. 2(a) and (b), respectively. As shown in Fig. 2(b), the voltage across C_1 (the sensing capacitor) discharges during the turn-off stage. The discharging current during this stage is constrained only by the parasitic impedances of D_3. This implies that if a negative voltage is applied to turn off the GaN device, a short circuit path will be established between D_4, D_3, and the negative gate voltage, as depicted in Fig. 3. Consequently, this OCP is not suitable when a safe turn-off with a negative voltage is required.

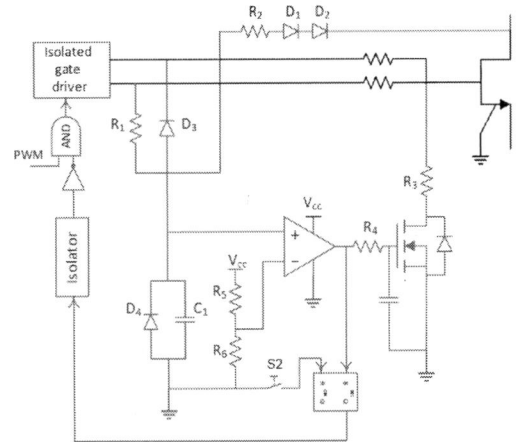

Fig. 1. Conventional ultra-fast desaturation based OCP for GaN (suggested in [1]).

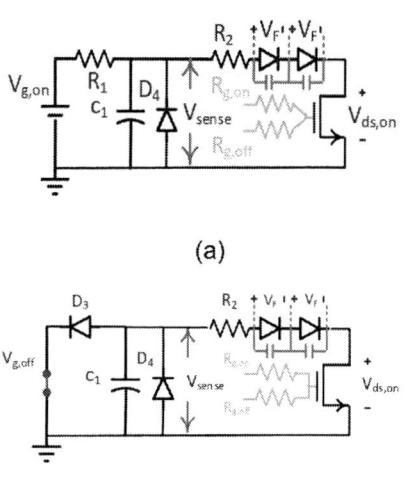

(a)

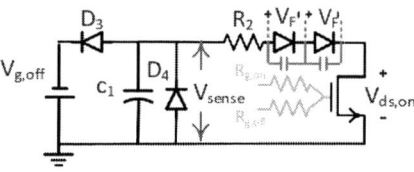

(b)

Fig. 2. Equivalent circuit during (a) turn on stage, and (b) turn off stage.

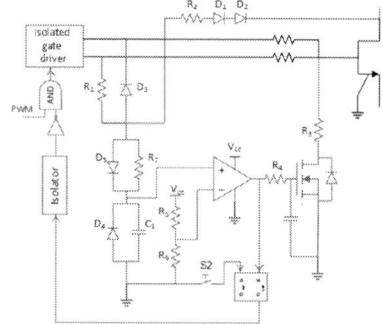

Fig. 3. Equivalent circuit during turn off stage if negative gate voltage is used.

Proposed Improved Desaturation-based OPC for GaN HEMT

The issue of a short circuit is addressed in the proposed OCP scheme by introducing a resistor (R_7) into the discharge path of C_1 and enabling the bypass of this resistor through diode D_5 during the charging interval. The overall configuration of the proposed OCP is illustrated in Fig. 4, and the equivalent circuits for the turn-on and turn-off stages are presented in Fig. 5(a) and (b), respectively.

Fig. 4. The proposed OCP circuit.

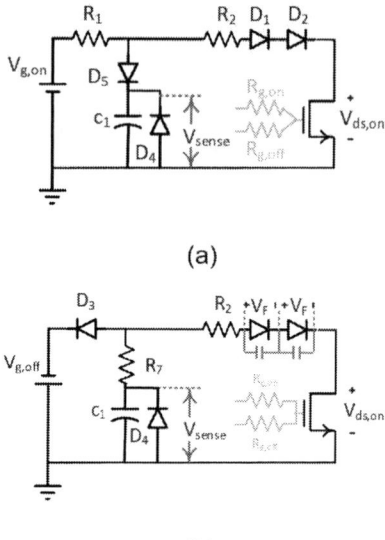

(a)

(b)

Fig. 5. Equivalent circuit of the proposed OCP during (a) turn on stage, and (b) turn off stage.

As it is seen in Fig. 5(a) the C_1 is charged through R_1 and D_5 and, as it is seen in Fig. 5(b), the discharging current is limited by R_7 during turn off stage.

The sensing voltage value can be ascertained by the following equation.

$$V_{sense} = \frac{R_1}{R_1+R_2}\left(V_{g,ON} - V_{FD5} - \frac{R_1}{R_2}\left(V_{ds,ON} + V_{FD1} + V_{FD2}\right)\right) \quad (1)$$

Where $V_{g,ON}$, $V_{ds,ON}$, V_{FD1}, V_{FD2} and V_{FD5} represent the gate voltage during the turn-on stage, on-state voltage of the GaN device, and the forward voltages of diodes D_1, D_2, and D_5, respectively. The values of resistors R_1 and R_2 are determined based on the specified overcurrent value. Referring to the Id vs. V_{ds} curve in the GaN device datasheet allows us to obtain the V_{ds} value for a given I_d.

The blanking time in the proposed OCP can be obtained as follows:

$$T_{blk} = R_1 C_1 \ln\frac{(V_{g,ON}-V_{FD5})}{(V_{g,ON}-V_{FD5}-V_{ds,ON})} \quad (2)$$

Experimental results

To evaluate the performance of the proposed OCP circuit, it is tested using a half-bridge configuration. To achieve a high current of 510 A, three 150-A GaN switches are paralleled on both the high and low sides, furthermore, to assess the performance of the proposed OCP/SCP it is employed in six paralleled 30-A (total current of 180A) GaN switches. A photo of the experimental prototype are shown in Figs. 6, and the characteristics are listed in Table I. moreover, the driver configuration for the parallel switches is shown in Fig. 7.

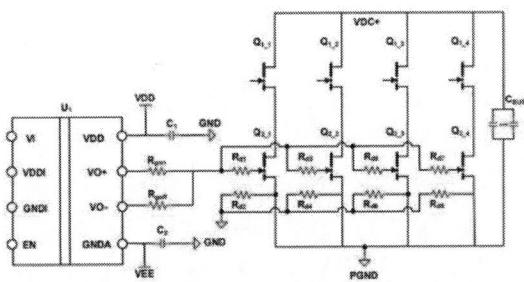

Fig. 7. Driver configuration for paralleled GaN switches

TABLE. 1. Setup characteristics

Device	Characteristics and Values
Power switch for four-parallel board	GS-065-150-1-D2
Power switches for six-parallel board	GS66508B
Gate driver	IEDB7275FXUMA1
Ron, Roff, R1,R2,R7	10, 2, 1, 0.7, 3.3k Ω
C1	10 pF
DC link Voltage	400 V
Load current	510 A
inductor	20 µH

The overcurrent threshold that triggers the protection is set at 510 A. Fig. 8 displays the performance of the prototype before reaching the threshold current. As shown in this figure, the sensing voltage remains below the reference voltage, indicating that the protection has not been activated. It's worth mentioning that since the on-state resistance of GaN switches varies with temperature, current can be distributed evenly among the paralleled GaN switches, as depicted in Fig. 8(a). This figure demonstrates the capability of paralleled 150-A GaN switches to share and handle 510 A under a 400 V DC input.

Fig. 8(b) illustrates the load current, voltage across the paralleled switches, reference and sensing voltages. As the current approaches the set point, the sensing voltage crosses the reference voltage, activating the protection circuit.

(a)

(b)

Fig. 6. GaN based high power prototype, (a) four-parallel GS-065-150-1-D2 (die) and (b) Six-parallel GS66508B.

PCIM Europe 2024, 11– 13 June 2024, Nuremberg   DOI: 10.30420/566262001

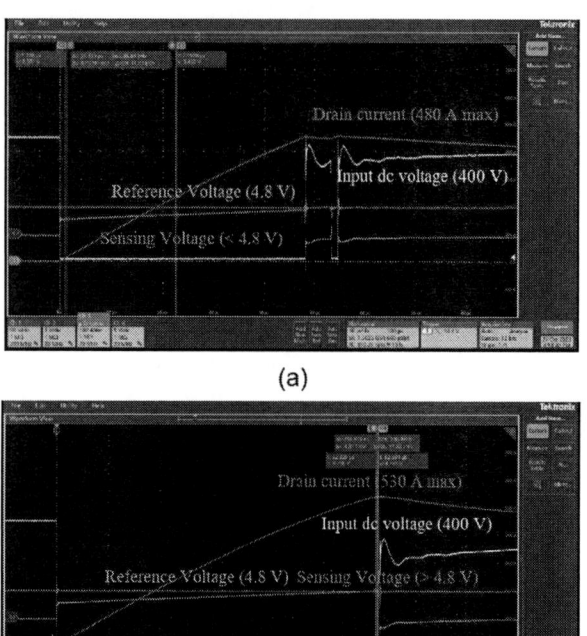

(a)

(b)

Fig. 8. Voltage across GaN switch, load current, sensing and reference voltages, (a) before triggering the protection, and (b) after triggering the protection.

The proposed scheme also was tested by using a six-in-parallel GaN (GS66508B) based half bridge. The six paralleled GS66508B can reach to a peak current of 180A. The double pulse test result is shown in Fig. 9(a). As shown in this figure, the current reaches to 180A. whereas the OCP is set to 200A, the OCP is not triggered in this current. the switching speed, which is shown in fig. 9 (b), is 25.4 nV/s. in order to trigger the OCP, the current is increased to above 200A. As shown in Fig. 10(a) the OCP, is triggered in this current turns off the switches. Furthermore, the SCP is tested by shorting the high-side switches and applying the dc link voltage directly to the low-side switches. The result is shown in Fig.10(b), as seen in this figure the SCP is trigged around 300 nS.

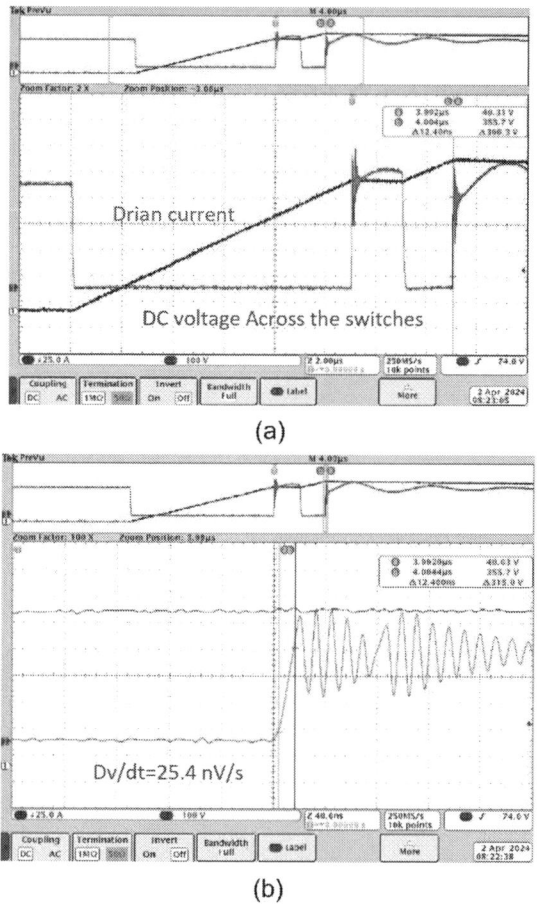

(a)

(b)

Fig. 9. The DPT result, (a) voltage across the switches and load current, and (b) switching speed.

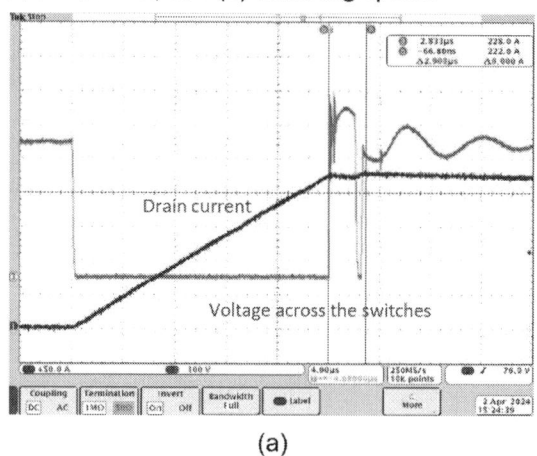

(a)

22

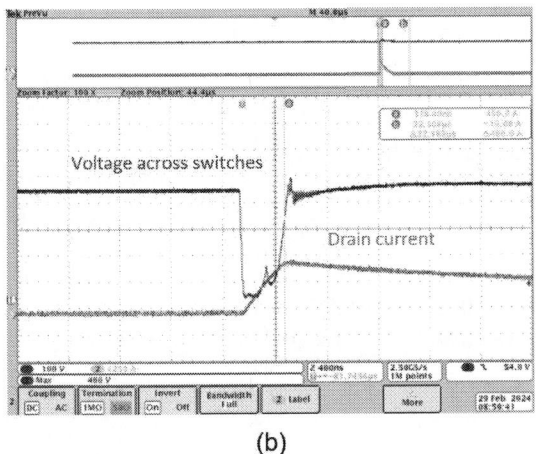

(b)

Fig. 10. SCP/OCP performance, (a) OCP, and (b) SCP.

Conclusion

In this article, an enhanced desaturation based OCP is introduced. The proposed OCP effectively addresses the challenge of safeguarding GaN switches when utilizing a negative gate voltage for turn-off. The performance of the proposed OCP was validated using a high-power half-bridge configuration, which was constructed with three 150-A and six paralleled 30-A GaN switches in parallel. The experimental results demonstrate the outstanding performance of the proposed OCP and the paralleled GaN switches in high-power applications.

References

[1] R. Hou, J. Lu, Z. Quan and Y. W. Li, "A Simple Desaturation-Based Protection Circuit for GaN HEMT With Ultrafast Response," in *IEEE Transactions on Power Electronics*, vol. 36, no. 6, pp. 6978-6987, June 2021, doi: 10.1109/TPEL.2020.3040727.

[2] J. Lu et al., "Applying variable-switching-frequency variable-phase-shift control and E-mode GaN HEMTs to an indirect matrix converter-based EV battery charger", *IEEE Trans. Transp. Electrific.*, vol. 3, no. 3, pp. 554-464, Sep. 2017.

[3] T. Modeer, N. Pallo, T. Foulkes, C. Barth and R. Pilawa-Podgurski, "Design of a GaN-based interleaved 9-level flying capacitor multilevel inverter for electric aircraft applications", *IEEE Trans. Power Electron*, vol. 35, no. 11, pp. 12153-12165, Nov. 2020.

[4] H. Li et al., "Robustness of 650-V enhancement-mode GaN HEMTs under various short-circuit conditions", *IEEE Trans. Ind Appl.*, vol. 55, no. 2, pp. 1807-1816, Mar./Apr. 2019.

PCIM Europe 2024, 11– 13 June 2024, Nuremberg   DOI: 10.30420/566262002

The Performance of a GaN eMode HEMT in Surge Current Scenarios such as the Active Short Circuit

Dominik Nehmer [1], Maximilian Hepp [2], Wolfgang Wondrak[2], Mark-M. Bakran [1]

[1] University Bayreuth, Germany
[2] Mercedes-Benz Group AG, Germany

Corresponding author: Dominik Nehmer, dominik.nehmer@uni-bayreuth.de
Speaker: Dominik Nehmer, dominik.nehmer@uni-bayreuth.de

Abstract

In addition to normal operation, the inverter of an electric vehicle (EV) has to provide certain fault operations. One of these is active short circuit (ASC). This paper presents surge current measurements. The defect is initiated by current saturation. To prevent saturation of the eMode GaN during ASC and thus a defect, a safe operating area (SOA) is specified. Therefore, a simple loss model and a saturation voltage are defined. Finally, the derating of a virtual module is presented for two representative machines. The derating is up to about 65.4 %.

1   Indroduction

In addition to normal operation, an electric vehicle (EV) inverter must provide certain fault modes. One of the fault modes is Active Short Circuit (ASC). ASC is used as a fail-safe state [1] when control of the drive is lost, such as during a watchdog timer reset or loss of high and low power supply to the drive control. To prevent defects in the electrical machine and the battery the ASC will be applied. Therefore either all high-side switches (HSS) or all low-side switches (LSS) are turned on [2].

In this paper surge current measurement will be presented for a GaN eMode HEMT [3] with turned-on switches. The defect mechanism will be analyzed and a criteria for the safe operation area (SOA) under ASC conditions will be given. Finally, the influence of the virtual module design, considering the ASC robustness, will be evaluated based on the virtual module calculation of [4], [5].

In [2], [6], [7], ASC measurements have been performed for SiC. In [2], IGBT modules were tested under ASC waveforms. For eMode GaNs there are some surge current evaluations in the third quadrant e.g. [8], [9]. The ASC robustness of GaN HEMTs has not been evaluated yet.

In [2], [7], the ASC current waveform will be imitated in the measurement setup by passive components. So, in fact only one operating point is tested. In this paper, the defect mechanism will be analyzed, and a criteria for the safe operating area under ASC will be defined. The compliance with the criteria for all operating points will be ensured by simulations similar to [6].

2   Basis of the evaluation

2.1   Concept

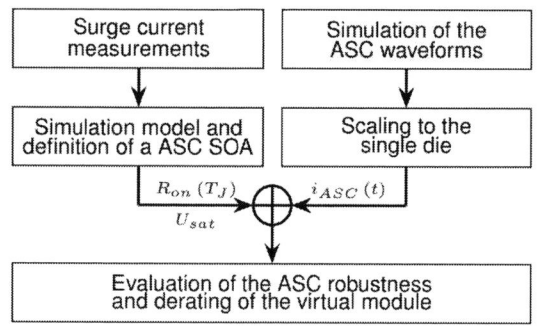

Fig. 1: Concept of the evaluation

In Figure 1, the ASC evaluation concept is shown. The evaluation is divided into three parts. The first part is the evaluation of the surge current robustness of the semiconductor. The second part is the calculation of the ASC waveforms for the electrical machine. The third part of the evaluation is the ASC simulation.

The first part includes 7 surge current series of measurements for different pulse durations and ambient temperatures. All measurements will be analyzed.

Therefore the junction temperature is calculated using a Foster thermal network. Furthermore, the output characteristics are modeled to forecast the losses during any surge current waveform with an on-state resistance model. Additionally, the saturation voltage will be defined as a criterion for the SOA under ASC. If the voltage calculated with the on-state resistance model is below the saturation voltage, the device will not saturate and hence no defect will occur under any surge current event.

For the second part the ASC waveforms for two typical permanent magnet synchronous machines (PMSMs) are calculated. The waveforms will be scaled to the the single die in the ratio of the nominal currents accordingly:

$$i_{ASC,single\,die}(t) = i_{ASC,machine}(t)\frac{I_{N,single\,die}}{I_{N,machine}} \quad (1)$$

For the final conclusion and virtual module design considering the ASC robustness, simulations of the single die with the scaled waveforms are performed. Therefore, the output characteristics of the die will be utilized. The thermal model is now based on the conditions of the virtual module. The ASC SOA criteria will be used to calculate the virtual module design and compared to the nominal current. The nominal current is therefore given for normal operation without and with switching losses.

All calculations and simulations were performed in MATLAB and Simulink.

2.2 Thermal model

The following two thermal models will be presented. The first model represents the thermal environment of the single die in the GaNPx package and is used for the estimation of the junction temperature of the surge current measurements. The second model represents the bare die and a scaled thermal model of a reference power module, which is used for the virtual module design.

2.2.1 Surge current measurements

Figure 2 illustrates the two prevalent classes of thermal networks. For computational purposes, a four-part Foster network is employed. The thermal network is presented in [3] as a Cauer model. Consequently, it must be transformed into a Foster model in accordance with [10].

In [3], the thermal network is depicted with the physical layers of the GaN die in the GaNPx package. According to [11], the first link represents the GaN

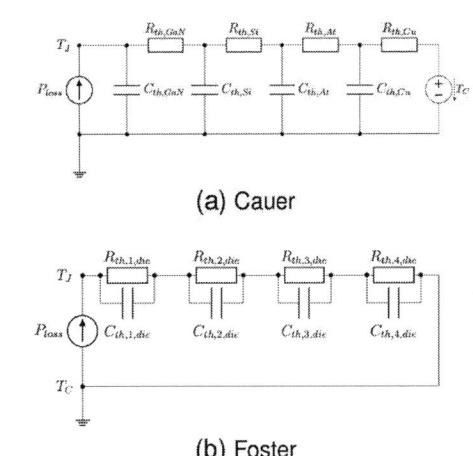

(a) Cauer

(b) Foster

Fig. 2: Thermal model of the single die

layer itself, the second link the Si substrate of the GaN HEMT, the third link the attachment, and the fourth link the copper baseplate.

	Cauer		Foster
$R_{th,GaN}$	8 mK/W	$R_{th,1,die}$	6.14 mK/W
$R_{th,Si}$	80 mK/W	$R_{th,1,die}$	57.9 mK/W
$R_{th,At}$	222 mK/W	$R_{th,1,die}$	2.80 mK/W
$R_{th,Cu}$	40 mK/W	$R_{th,1,die}$	283 mK/W
$C_{th,GaN}$	0.18 mJ/K	$C_{th,1,die}$	0.255 mJ/K
$C_{th,Si}$	1.3 mJ/K	$C_{th,1,die}$	1.75 mJ/K
$C_{th,At}$	9.5 mJ/K	$C_{th,1,die}$	44.7 mJ/K
$C_{th,Cu}$	3.7 mJ/K	$C_{th,1,die}$	10.3 mJ/K

Tab. 1: Single die thermal model

Table 1 presents the parameters of the Cauer and Foster network of the single die within the GaNPx package.

2.2.2 Virtual module

Figure 3 shows the thermal network of the virtual module. The bare die is placed on the structure of the reference module [12]. The Foster model of the reference module is fitted from the thermal impedance and is given in [12]. As previously stated, the first two links of the Cauer network of the single die refer to the bare GaN HEMT. These two links are added to the scaled module thermal network. For the initial step, the Foster network of the module must be scaled. The scaled thermal resistance can be calculated using the following equation:

$$R_{th,x,sF} = R_{th,x,module}\frac{A_{module}}{A_{bare\,die}} \quad (2)$$

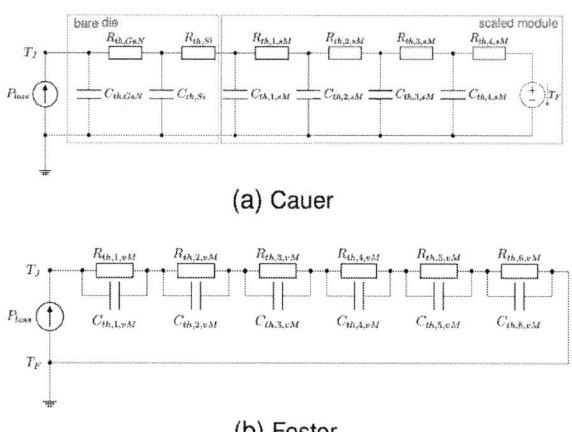

(a) Cauer

(b) Foster

Fig. 3: Thermal model of the scaled virtual module

	Cauer		Foster
$R_{th,GaN}$	8 mK/W	$R_{th,1,vM}$	6.14 mK/W
$R_{th,Si}$	80 mK/W	$R_{th,2,vM}$	75.8 mK/W
$R_{th,1,sM}$	205 mK/W	$R_{th,3,vM}$	90.3 mK/W
$R_{th,2,sM}$	327 mK/W	$R_{th,4,vM}$	332 mK/W
$R_{th,3,sM}$	276 mK/W	$R_{th,5,vM}$	320 mK/W
$R_{th,4,sM}$	78.4 mK/W	$R_{th,6,vM}$	151 mK/W
$C_{th,1,GaN}$	0.18 mJ/K	$C_{th,1,vM}$	0.206 mJ/K
$C_{th,2,Si}$	1.3 mJ/K	$C_{th,2,vM}$	1.51 mJ/K
$C_{th,3,sM}$	38.0 mJ/K	$C_{th,3,vM}$	60.2 mJ/K
$C_{th,4,sM}$	97.6 mJ/K	$C_{th,4,vM}$	122 mJ/K
$C_{th,5,sM}$	936 mJ/K	$C_{th,5,vM}$	853 mJ/K
$C_{th,6,sM}$	13.3 mJ/K	$C_{th,6,vM}$	7.69 J/K

Tab. 3: Thermal network of the scaled virtual module

Accordingly, the thermal capacitance must be scaled. The scaled thermal capacitance can be calculated by the following equation:

$$C_{th,x,sF} = C_{th,x,module} \frac{A_{bare\,die}}{A_{module}} \quad (3)$$

Following the scaling process, the Foster network must be transformed into a Cauer network.

	scal. Foster		scal. Cauer
$R_{th,1,sF}$	89.2 mK/W	$R_{th,1,sM}$	205 mK/W
$R_{th,2,sF}$	329 mK/W	$R_{th,2,sM}$	327 mK/W
$R_{th,3,sF}$	318 mK/W	$R_{th,3,sM}$	276 mK/W
$R_{th,4,sF}$	150 mK/W	$R_{th,4,sM}$	78.4 mK/W
$C_{th,1,sF}$	59.4 mJ/K	$C_{th,1,sM}$	38.0 mJ/K
$C_{th,2,sF}$	122 mJ/K	$C_{th,2,sM}$	97.6 mJ/K
$C_{th,3,sF}$	856 mJ/K	$C_{th,3,sM}$	936 mJ/K
$C_{th,4,sF}$	7.70 J/K	$C_{th,4,sM}$	13.3 J/K

Tab. 2: Thermal network of the scaled reference module

Table 2 presents the parameters of the scaled Foster network of [12], and the scaled Cauer network. The Cauer model of the bare die can be added by the Cauer network of the scaled module.

Table 3 presents the parameters of the virtual module as a Cauer network. Additionally, the parameters of the Foster model for the virtual module are provided. These are calculated from the Cauer network and utilized for the simulation of the ASC of the virtual module.

2.2.3 DGL of the Foster network

As previously stated, the Foster network is employed for the junction temperature calculation of the measurement as well as the ASC simulation of the virtual module. The temperature difference of each Foster link T_x can be described by the following differential equation:

$$\frac{dT_x}{dt} = \frac{1}{R_{th,x}C_{th,x}} \left(-T_x + R_{th,x} P_{loss} \right) \quad (4)$$

The temperature of the case T_C in the measurement evaluation as well as the fluid temperature T_F of the virtual module is considered as constant. The junction temperature of the Foster network with n links is then defined as follows:

$$T_J = T_{C/F} + \sum_{x=1}^{n} T_x \quad (5)$$

For the fluid temperature, a maximum value of 65 °C is utilized in the simulations.

In case of the measurements, the junction temperature is equivalent to the case temperature prior to the surge current pulse being applied. For the worst case conditions of the virtual module, the temperature of the junction is at its maximum operating temperature. As described in [4], [5], the maximum permitted operating temperature is set to 125 °C. For the simulation of the ASC, an initial temperature must be considered. The initial temperature of each link can be calculated as follows:

$$T_{x,0} = (T_{J,0} - T_F) \frac{R_{th,x}}{\sum_{x=1}^{n} R_{th,x}} \quad (6)$$

The junction temperature is also simulated in Simulink.

2.3 Machine model

$$
\begin{bmatrix} u_d \\ u_q \end{bmatrix} = \begin{bmatrix} R_s & 0 \\ 0 & R_s \end{bmatrix} \begin{bmatrix} i_d \\ i_q \end{bmatrix} + \begin{bmatrix} L_d & 0 \\ 0 & L_q \end{bmatrix} \frac{d}{dt} \begin{bmatrix} i_d \\ i_q \end{bmatrix}
$$
$$
+ \omega_{el} \begin{bmatrix} 0 & -L_q \\ L_d & 0 \end{bmatrix} \begin{bmatrix} i_d \\ i_q \end{bmatrix} + \omega_{el} \begin{bmatrix} 0 \\ \Psi_{PM} \end{bmatrix} \quad (7)
$$

Equation 7 is the matrix voltage equation of a PMSM in the rotor with fixed dq coordinates according to [13]. Therefore, the inductances are modeled as a constant value individually for each axis.

$$
\frac{d}{dt} \begin{bmatrix} i_d \\ i_q \end{bmatrix} = \begin{bmatrix} \frac{1}{L_d} & 0 \\ 0 & \frac{1}{L_q} \end{bmatrix} \begin{bmatrix} u_d \\ u_q \end{bmatrix} - \begin{bmatrix} \frac{R_s}{L_d} & 0 \\ 0 & \frac{R_s}{L_d} \end{bmatrix} \begin{bmatrix} i_d \\ i_q \end{bmatrix}
$$
$$
- \omega_{el} \begin{bmatrix} 0 & -\frac{L_q}{L_d} \\ \frac{L_d}{L_q} & 0 \end{bmatrix} \begin{bmatrix} i_d \\ i_q \end{bmatrix} - \omega_{el} \begin{bmatrix} 0 \\ \frac{\Psi_{PM}}{L_q} \end{bmatrix} \quad (8)
$$

In equation 8, the system of differential equations is derived from the voltage equation 7. The equation can be solved analytically as described in [7], [14] and [15] or simulated. MATLAB and Simulink are used to calculate the waveforms.

$$
\frac{d}{dt} \gamma = \omega_{el} \quad (9)
$$

Equation 9 is the differential equation of the rotor position γ. Therefore, the rotor speed is considered to be constant during the active short-circuit event.

	Machine 1	Machine 2
I_N	350 A	850 A
U_N	480 V	480 V
f_N	188 Hz	232 Hz
f_{max}	600 Hz	518 Hz
R_s	12.3 $m\Omega$	12 $m\Omega$
L_d	385 μH	200 μH
L_q	784 μH	200 μH
Ψ_{PM}	155.8 mVs	120 mVs
p	4	3
M_N	656 Nm	649 Nm

Tab. 4: Machine parameter values

Table 4 shows the relevant parameters of the two PMSMs considered. As already mentioned, machine 1 has a reluctance torque, while machine 2 does not.

For the ASC simulations of the virtual module the initial current in dq-coordinates is given by the maximum torque per ampere (MTPA) curve. Additionally, the rotor angle is varied to take every possible failure rotor position into account.

3 Measurement setup

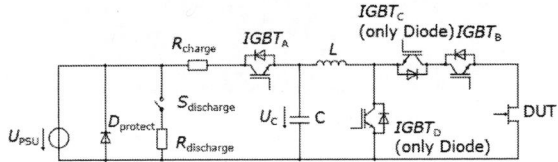

Fig. 4: Circuit diagram of the measurement setup

Figure 4 shows the circuit diagram of the surge current measurement setup. The capacitor C is precharged by the power supply unit (PSU). Then the IGBT A is turned off. IGBT B is then turned on for the required pulse duration. During the pulse, the current through the DUT and the inductor is nearly sinusoidal. The device under test is constantly turned on with the recommended gate-source voltage of 6 V. The pulse duration can be calculated from:

$$
T_P = \pi \sqrt{LC} \quad (10)
$$

With a voltage step of 1 V, the minimum resolution R of the peak current values can be calculated as follows:

$$
R = \frac{\Delta I_{max}}{\Delta U_C} = \sqrt{\frac{C}{L}} \quad (11)
$$

For each series of measurements, the voltage on capacitor C is increased in small increments until the failure occurs.

conf.	C	L	T_P	R
1	12.5 mF	690 μH	9.2 ms	4.26 A/V
2	1 mF	87 μH	0.93 ms	3.39 A/V
3	25 mF	7.15 mH	42 ms	1.87 A/V

Tab. 5: Overview of the measurement setup configurations

Table 5 shows the different configurations for the three pulse durations tested. The calculated and measured pulse durations are almost identical.
Figure 5 shows the measurement setup for the surge current measurements. Two different ambient temperatures were tested for each pulse duration. To achieve this, a copper bar was placed on

Fig. 5: Measurement setup

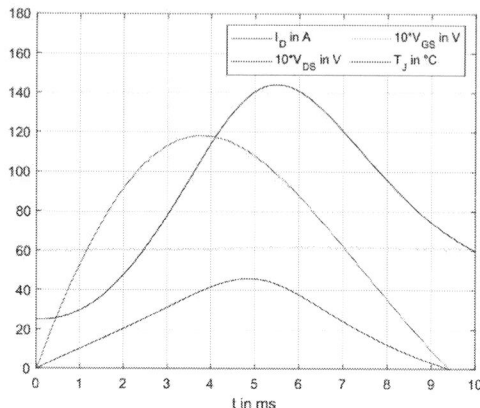

Fig. 6: Example of a LGP waveform

the die. The copper was heated to the required ambient temperature.

Passive probes are used to measure the gate-source and drain-source voltages. The current is measured with a current transducer based on the Hall effect.

Device number	T_A	direction	T_P
Device 1	25 ℃	forward	9.5 ms
Device 2	125 ℃	forward	9.5 ms
Device 3	25 ℃	forward	1 ms
Device 4	125 ℃	forward	1 ms
Device 5	25 ℃	forward	45 ms
Device 6	125 ℃	forward	45 ms
Device 7	25 ℃	reverse	9.5 ms

Tab. 6: Overview of the measurement series

Table 6 shows the conditions for the different measurements. Since the on-resistance in reverse direction is always slightly lower than in forward direction, the measurements are mostly in the worst-case forward direction. Only the measurement series of device 7 is in reverse direction to confirm the assertion.

4 Measurement results and defect analysis

The surge current measurements are presented below. The focus is on two pulses of the measurement series that are characteristic of the surge current robustness. One is the last good pulse (LGP) and the other is the first bad pulse (FBP) in which the failure occurs. The LGP value is the common benchmark of the surge current robustness. The analysis of the FBP pulse provides information about the failure mechanism and must therefore be considered.

Figure 6 shows the LGP measured waveforms as

well as the simulated junction temperature for device 1. It can be seen that the current shape is almost sinusoidal. The voltage reaches its maximum after the current reaches its maximum. The junction temperature increases even after the maximum voltage occurs.

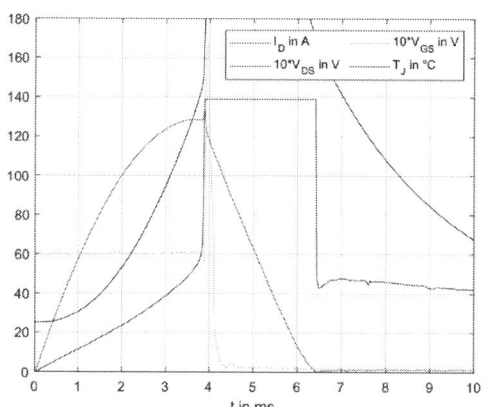

Fig. 7: Example of a FBP waveform

Figure 7 shows the FBP waveforms and simulated junction temperature for device 1. It can be seen that the drain-source voltage rises rapidly at about 3.8 ms. Since a gate resistance of 20 Ω was used and the capacitance coupling of the gate-drain and gate-source capacitance is high [5], the gate-source voltage also rises rapidly. The high drain-source voltage also leads to high losses and a significant rise in junction temperature. The device finally fails with a defect in the gate-source and drain-source distance.

Table 7 shows the LGP values. The results show that the robustness is lower in the forward direction

Device number	I_{max}	I^2t	T_{max}
Device 1	118.9 A	63.0 A^2s	145.2 ℃
Device 2	80.4 A	28.7 A^2s	211.3 ℃
Device 3	191.0 A	17.8 A^2s	122.9 ℃
Device 4	105.9 A	5.6 A^2s	174.5℃
Device 5	86.1 A	155.8 A^2s	134.1 ℃
Device 6	59.5 A	74.8 A^2s	193.6 ℃
Device 7	130.2 A	75.0 A^2s	178.0 ℃

Tab. 7: Evaluation of the LGPs

than in the reverse direction. In addition, the surge robustness decreases with increasing pulse duration and ambient temperature. There is no constant junction temperature that leads to a failure, so a different criterion for the SOA has to be defined.

Device number	Gate-Source	Drain-Source
Device 1	4.6 Ω	5.8 Ω
Device 2	3.5 Ω	11.1 Ω
Device 3	2.0 Ω	5.1 Ω
Device 4	2.2 Ω	441 Ω
Device 5	2.2 Ω	14.0 Ω
Device 6	0.3 Ω	4.2 Ω
Device 7	3.1 Ω	0.7 Ω

Tab. 8: Resistence value after the defect

Table 8 shows the gate-source and drain-source distance resistances for all devices after the FBP. All values show a defect of the device in the gate-source as well as in the drain-source distance. As suggested by the decapsulation in [9], the entire channel of the GaN device burns out step by step, leading to a complete failure of the device.

5 Modeling and ASC SOA

5.1 Static measurements and loss model

A simple approach to modeling the output characteristic is a temperature-dependent on-state resistance. Therefore, static measurements are made. Figure 8 shows the on-state resistance values calculated from the static measurement as well as its fitting. For temperature modeling, the values of each temperature were evaluated at 0 A. The equation for fitting and modeling is given by:

$$R_{DS,on} = R_0 \left(1 + \alpha\, T_J + \beta\, T_J^2\right) \quad (12)$$

The equation models the on-state resistance as a second order polynomial.

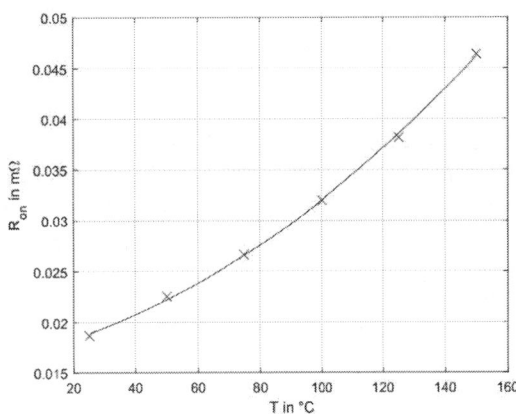

Fig. 8: Measured on state resistance

	R_0 in $m\Omega$	α in $\frac{1}{K}$	β in $\frac{1}{K^2}$
Device 1	16.2	5.88e-3	3.75e-5

Tab. 9: R_{on} fitting of the static measurements

Table 9 shows the parameters for the on-state resistance model for Device 1.

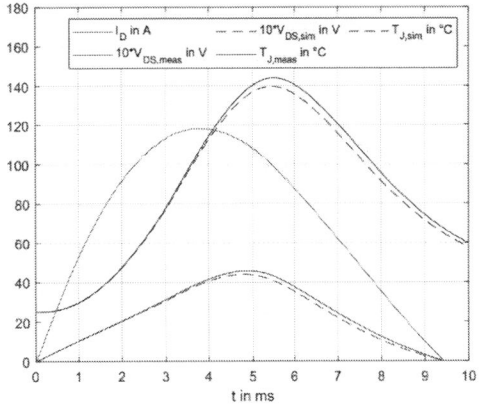

Fig. 9: Example of a simulated LGP waveform

Figure 9 shows an example of LGP waveforms and the waveforms of the on-state resistance simulation model. The voltage and junction temperature of the measurement and simulation match well.

The on-state resistance model used for ASC simulation, which is also employed in [6], [16], is sufficient. A more complex model also considering the current saturation based on [17] can also model the voltage run away, which can be observed in figure 7. However, this model requires a large number of parameters, making fitting complicated and unsuitable for all devices due to high parameter deviation.

5.2 ASC SOA criteria

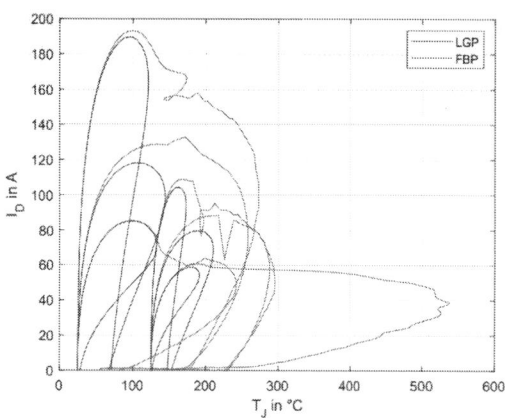

Fig. 10: Current-temperature trajectory

Figure 10 shows the current-temperature trajectories for the LGPs and the FBPs. Even if the LGP, e.g. in Figure 7, implicitly assumes that there is a saturation current, the trajectory does not show a clear dependence on the saturation current and the junction temperature for all pulse durations. A linear dependence can only be assumed for the same pulse duration. Furthermore, there is no clear breakdown temperature for all devices.

Another approach is to consider the saturation voltage. The saturation voltage is defined as:

$$U_{DS,sat} = U_{Drive,+} - U_{th} \qquad (13)$$

For this eMode GaN, the threshold voltage is 1.7 V with a range of 1.1 V to 2.6 V. With a drive voltage of 6 V, the saturation voltage is 4.3 V with a range of 3.4 V to 4.9 V.

In Figure 11 the voltage-temperature trajectories for the LGPs and the FBPs are shown. One can observe, that during the FBP all trajectories bend off at a certain voltage. For all trajectories this point is above the minimum saturation voltage of 3.4 V (marked with the lower dashed black line). The criterion is valid for all devices and is therefore defined as an SOA criterion.

Figure 12 shows the voltage-current characteristics for the measurements and the simulation. Therefore, the maximum voltage and current during each pulse of a measurement series was evaluated. The simulated values agree quite well with the measured values, especially up to the lower saturation limit. Therefore, the lower saturation value together

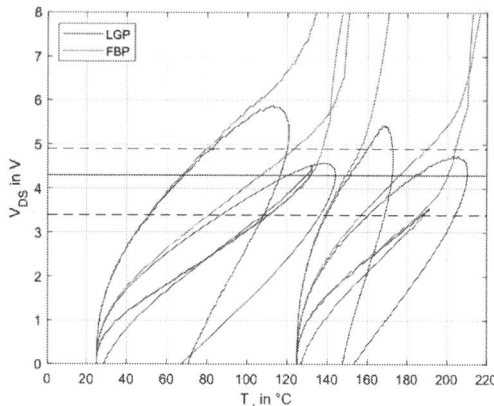

Fig. 11: Voltage-temperature trajectory

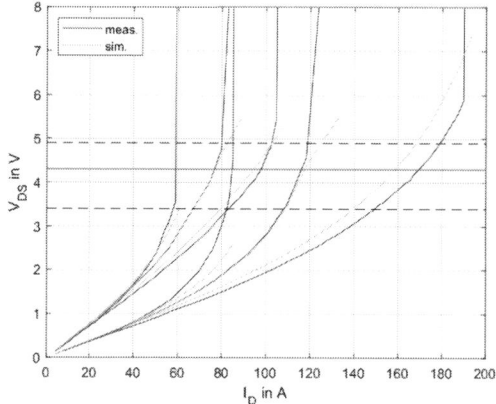

Fig. 12: Voltage current characteristic

with the one-state resistance value is a good SOA criterion for the ASC simulations.

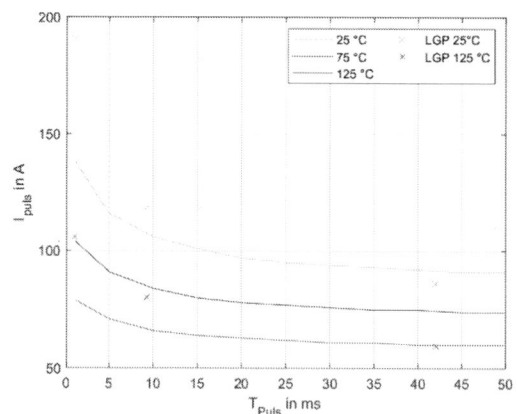

Fig. 13: Simulated maximum surge surrent robustness

Figure 13 shows the maximum surge current for

simulated ideal pulses and for the LGP measurements. For the simulation, an ideal pulse is generated and the current is ramped up until the saturation criteria is reached. It can be seen that for long pulse durations the measurement result and the simulation with ideal pulses are in good agreement. For short pulses, the simulation underestimates the surge current robustness.

6 Design regarding ASC

6.1 Loss model

As already described, a simple on-state resistance model in combination with the saturation voltage defines the SOA under ASC. For the virtual device, the on-state resistance model is based on the data sheet values from [3], since these are the only values guaranteed by the manufacturer.

R_0 in $m\Omega$	α in $\frac{1}{K}$	β in $\frac{1}{K^2}$
20.5	7.64e-3	4.55e-5

Tab. 10: R_{on} fitting with datasheet values

Table 10 shows the fitting parameters of the on-state resistor values in the data sheet.

6.2 Transient ASC

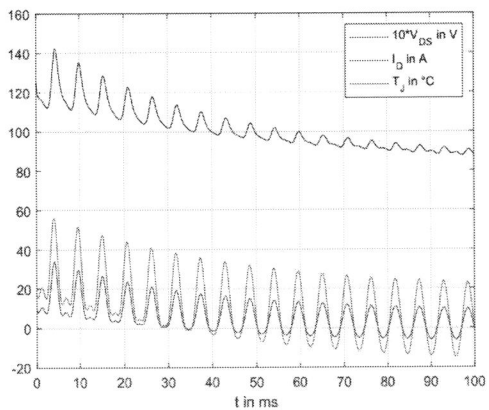

Fig. 14: Worst case ASC waveforms for machine 1

Figure 14 shows the voltage, current and junction temperature during the worst case conditions of machine 1. In order to remain within the SOA, the nominal current was scaled down. The frequency for the worst case conditions of machine 1 is 175 Hz. Figure 15 shows the ASC waveforms for the worst case conditions of Machine 2. For machine 2, the worst case frequency is 200 Hz.

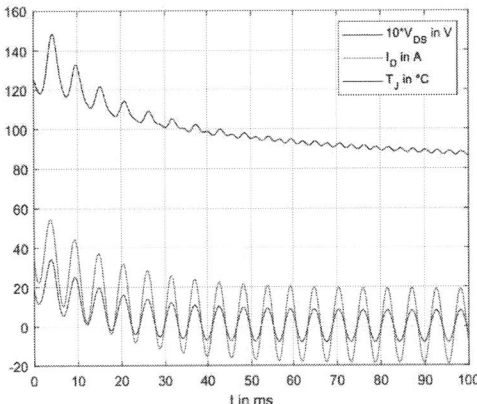

Fig. 15: Worst case ASC waveforms for machine 2

6.3 Steady state ASC

In addition to transient ASC, the semiconductor must also withstand steady state ASC. During ASC, a switch conducts during the positive and negative current wave. The nominal current under steady state ASC can be calculated with:

$$I_{N,ASC} = \sqrt{\frac{T_{J,max} - T_F}{R_{on} R_{th}}} \tag{14}$$

The steady state ASC in dq coordinates can be calculated by setting the voltage in equation 7 to zero and solving the equations according to the current. The d component of the steady-state ASC current is given by:

$$I_d = \frac{-\omega_{el}^2 L_q \Psi_{PM}}{R_s^2 + \omega_{el}^2 L_d L_q} \tag{15}$$

The q component of the steady-state ASC is then given by:

$$I_q = \frac{R_s \omega_{el} \Psi_{PM}}{R_s^2 + \omega_{el}^2 L_d L_q} \tag{16}$$

The RMS value of the absolute current is given by the following equation:

$$I_{ASC} = \sqrt{\frac{I_d^2 + I_q^2}{2}} \approx \frac{\Psi_{PM}}{\sqrt{2} L_d} \tag{17}$$

For high frequencies, the effective value of the ASC current can be approximated by the right side of Equation 17. The steady-state ASC current for machine 1 is 286 A and 424 A for machine 2.

	abs. I_N	rel. w/o $P_{loss,sw}$	rel. w/ $P_{loss,sw}$
w/o $P_{loss,sw}$	47.7 A	-	+ 31.9 %
w/ $P_{loss,sw}$	36.2 A	- 24.2 %	-
M1 transient	16.5 A	- 65.4 %	- 54.4 %
M1 steady	41.3 A	- 13.4 %	+ 14.1 %
M2 transient	27.0 A	- 43.3 %	- 25.4 %
M2 steady	67.6 A	+ 41.7 %	+ 86.7 %

Tab. 11: Nominal current, absolute and relative derating

6.4 Virtual module design

Table 11 shows the nominal currents considering the steady state and transient ASC as well as the relative derating for machine 1 (M1) and machine 2 (M2). It can be seen that the steady-state values always lead to a lower derating than the transient, which is particularly high for machine 1 (65.4 %).

7 Comparison of surge current robustness

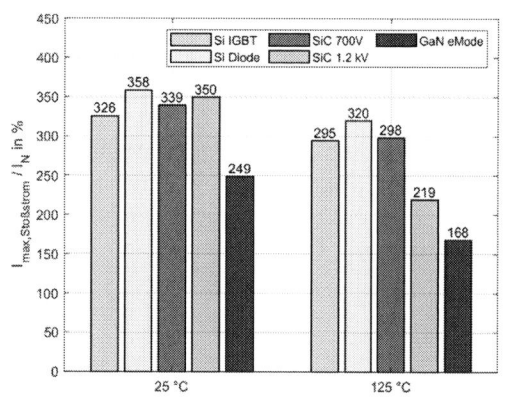

Fig. 16: Comparison of the surge current robustness of Si, SiC and GaN for 9.5 ms pulse duration

Figure 16 shows the LGP values of the surge current measurements for Si IGBT with Si diode [18], SiC MOSFET with two different voltage ratings [19], [20] and eMode GaN. The pulse duration is 9.5 ms in all cases. For all devices, the values are given for different ambient temperatures. The values are scaled to the nominal current under normal operation without switching losses. The eMode GaN has a significantly lower surge current robustness than the other devices, especially at high ambient temperatures. This is mainly due to the high temperature coefficient of the eMode GaN one-state

resistance and the low saturation voltage.

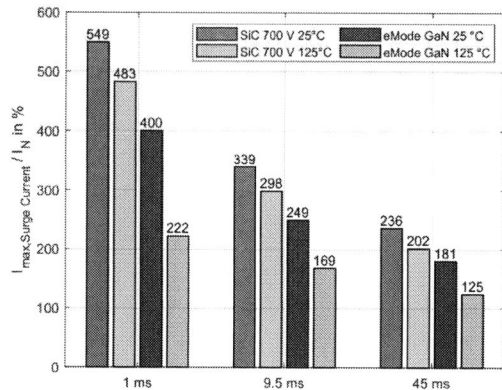

Fig. 17: Comparison of the surge current robustness of GaN and SiC for different pulse durations

Figure 17 shows the LGP values for the SiC 700V and the eMode GaN for different ambient temperatures and pulse durations. The eMode GaN has a lower surge current than the SiC MOSFET. Especially for short pulse durations and high ambient temperatures, the surge current robustness is significantly lower for the eMode GaN.

8 Conclusion and outlook

Surge current measurements were performed in this paper. The eMode GaN shows a saturation current initiated defect under surge current events. A simple loss model and saturation criteria were defined for the safe operating region during the active short circuit to avoid the saturation-initiated fault. For two permanent magnet synchronous machines, the derating of the virtual module was calculated, showing a derating of up to 65.4 %.

In future investigations, the surge current robustness will be evaluated for other GaN HEMTs. The LGP values will also be reported and the derating calculated for all devices.

References

[1] B. Welchko, T. Jahns, W. Soong, and J. Nagashima, "IPM synchronous machine drive response to symmetrical and asymmetrical short circuit faults," *IEEE Transactions on Energy Conversion*, vol. 18, no. 2, pp. 291–298, 2003. DOI: 10.1109/TEC.2003.811746.

[2] T. Appel and A. Bieler, "Novel Method for Active Short Circuit (ASC) Tests of Power Module in Automotive Traction Application," in *2022 24th European Conference on Power Electronics and Applications (EPE'22 ECCE Europe)*, 2022, P.1–P.7.

[3] GaNSystems, *GS-065-060-5-T-A, Automotive 650 V GaN E-mode transistor, Datasheet*, online, https://gansystems.com/wp-content/uploads/2022/01/GS-065-060-5-T-A-DS-Rev-220127.pdf, Jan. 2022.

[4] D. Nehmer, M. Hepp, W. Wondrak, and M.-M. Bakran, "A Comparison of GaN HEMT and SiC MOSFET Power Inverter Modules for Electric Vehicles (EV)," in *PCIM Europe 2023; International Exhibition and Conference for Power Electronics, Intelligent Motion, Renewable Energy and Energy Management*, 2023, pp. 1–9. DOI: 10.30420/566091139.

[5] D. Nehmer, M. Hepp, W. Wondrak, and M.-M. Bakran, "Switching a emode gan hemt under conditions of an inverter module for electrical vehicles (ev)," in *2023 25th European Conference on Power Electronics and Applications (EPE'23 ECCE Europe)*, 2023, pp. 1–8. DOI: 10.23919/EPE23ECCEEurope58414.2023.10264634.

[6] A. Ruiz, A. Meseamanolis, L. Santolaria, M. Maleki, and A. Baschnagel, "Active Short Circuit Capability of Half-Bridge Power Modules Towards E-Mobility Applications," in *PCIM Europe digital days 2021; International Exhibition and Conference for Power Electronics, Intelligent Motion, Renewable Energy and Energy Management*, 2021, pp. 1–7.

[7] A. Lanzafame, L. D. Tornello, G. Scelba, E. Venuti, A. Raffa, *et al.*, "Experimental Test Setup for Thermal Stress Analysis of SiC Devices under Active Short Circuits," in *2022 IEEE Energy Conversion Congress and Exposition (ECCE)*, 2022, pp. 1–6. DOI: 10.1109/ECCE50734.2022.9947533.

[8] X. Wang, W. Chen, R. Sun, C. Liu, Y. Xia, *et al.*, "Degradation Behavior and Mechanism of GaN HEMTs With P-Type Gate in the Third Quadrant Under Repetitive Surge Current Stress," *IEEE Transactions on Electron Devices*, vol. 69, no. 10, pp. 5733–5741, 2022. DOI: 10.1109/TED.2022.3200928.

[9] X. Wang, W. Chen, R. Sun, C. Liu, X. Chen, *et al.*, "Failure Behavior and Mechanism of p-GaN Gate AlGaN/GaN HEMTs in the Third Quadrant Under Repetitive Surge Current Stress," *IEEE Transactions on Electron Devices*, vol. 71, no. 3, pp. 1694–1701, 2024. DOI: 10.1109/TED.2023.3345258.

[10] K. Murthy and R. Bedford, "Transformation between Foster and Cauer equivalent networks," *IEEE Transactions on Circuits and Systems*, vol. 25, no. 4, pp. 238–239, 1978. DOI: 10.1109/TCS.1978.1084459.

[11] GaNSystems, *GN007 Application Note - Modeling the Thermal Behavior of GaNPX and PDFN packages Using RC Thermal SPICE Models*, online, https://gansystems.com/wp-content/uploads/2018/02/GN007-%E2%80%93-Modelling-Thermal-Behavior-of-GaNPX-packages-Using-RC-Thermal-SPICE-Models-Rev-180216.pdf, Mar. 2022.

[12] Infineon Technologies AG, *FS03MR12A6MA1LB: HybridPACK™ Drive module - datasheet*, online, https://www.infineon.com/cms/en/product/power/mosfet/silicon-carbide/modules/fs03mr12a6ma1lb/, Sep. 2022.

[13] U.-P. D.-I. D. Schröder, *Elektrische Antriebe - Regelung von Antriebssystemen -*. Berlin Heidelberg New York: Springer-Verlag, 2013.

[14] G. Choi and T. M. Jahns, "Investigation of Key Factors Influencing the Response of Permanent Magnet Synchronous Machines to Three-Phase Symmetrical Short-Circuit Faults," *IEEE Transactions on Energy Conversion*, vol. 31, no. 4, pp. 1488–1497, 2016. DOI: 10.1109/TEC.2016.2594223.

[15] C. Poulain, B. de Metz-Noblat, and F. Dumas, *Calculation of short-circuit currents*, online, https://www.researchgate.net/publication/273381153_Calculation_of_short-circuit_currents, Nov. 2005.

[16] A. Raffa, P. P. Veneziano, A. Manzitto, and G. Bazzano, "A new analog behavioral SPICE macro model with self-heating effects and 3rd quadrant behavior for Silicon Carbide Power MOSFETs," in *PCIM Europe digital days 2020; International Exhibition and Conference for Power Electronics, Intelligent Motion, Renewable Energy and Energy Management*, 2020, pp. 1–8.

[17] A. Endruschat, "Simulationsmodell für einen GaN-HEMT mit Schottky p-GaN-Gate," doctoralthesis, Friedrich-Alexander-Universität Erlangen-Nürnberg (FAU), 2021.

[18] Infineon Technologies AG, *AIKQ120N75CP2 - Datasheet*, online, https://www.infineon.com/dgdl/Infineon-AIKQ120N75CP2-DataSheet-v01_10-EN.pdf?fileId=8ac78c8c7e7124d1017f0672206b14ee.

[19] Microsemi, *MSC015SMA070B Silicon Carbide N-Channel Power MOSFET*, online, https://www.microsemi.com/document-portal/doc_view/1244454-msc015sma070b-datasheet, Sep. 2020.

[20] Wolfspeed Inc., *C3M0016120D - Datasheet*, online, https://assets.wolfspeed.com/uploads/2024/01/Wolfspeed_C3M0016120D_data_sheet.pdf.

PCIM Europe 2024, 11– 13 June 2024, Nuremberg   DOI: 10.30420/566262003

Gate Resistance Effect on Short-circuit Robustness of p-GaN HEMTs

Mohamed Lemine Dedew[1,2,*], Tien Anh Nguyen[2], Thanh Long Le[3], Matthieu Landel[2], Valeria Rustichelli[1], Joao Oliveira[1], Maroun Alam[1], Fabio Coccetti[1], Stéphane Lefebvre[2]

[1] IRT Saint-Exupéry, France

[2] SATIE, Cnam, CNRS, ENS Paris-Saclay, Université Paris-Saclay, France

[3] SAFRAN TECH, France

*Corresponding author: mhamed.dedew@irt-saintexupery.com

Abstract

This paper depicts the effect of Gate Resistance (R_G) on the robustness of 650 V normally-off Gallium Nitride (GaN) High Electron Mobility Transistors (HEMTs) under Short-Circuit (SC) conditions through qualitative and experimental analysis. Tests were performed on Devices Under Test (DUTs) at various R_G values, while maintaining the Drain-Source Voltage (V_{DS}) and the Gate-Source Voltage (V_{GS}) constant. Overall, the DUTs demonstrated good robustness under single SC conditions. By increasing R_G, a significant increase in the energy dissipated during the SC stress (E_{SC}) before failure was observed. Besides, the initial peak of the Drain current (I_D) appears to decrease as R_G increases. However, by conducting repetitive SCs of very short durations, E_{SC} does not seem to be affected by the variation in R_G and the DUTs appear to be able to withstand more SC cycles by increasing R_G. The results suggest that two failure mechanisms may occur. The first one, under long SC stresses, where failure appears to be primarily caused by dissipated energy and temperature increase. The second one, under repetitive SCs of short duration, which remains unclear, especially as in this situation, energy does not seem to be the main cause of failure.

1   Introduction

Compared to silicon, wide bandgap semiconductors such as GaN are more promising for power electronics applications due to their intrinsic properties such as wide bandgap energy, high-switching frequency, high charge carrier mobility, high breakdown critical field, and good thermal conductivity [1]. Despite GaN HEMTs are emerging as a potential choice for the automotive and aerospace industries [2], questions remain regarding their SC robustness, a factor that may limit their diffusion in the market.

The root cause of GaN HEMTs failures during SC conditions is still not fully understood, leading to conflicting explanations. Some suggest that the cause is the increased temperature due to power dissipation, which is inversely proportional to the SC Withstanding Time (SCWT) [3]. On the contrary, [4] attributes the failures to a high electric field due to hot electron trapping in the Aluminum-Gallium Nitride (AlGaN) layer. In [5], it is suggested that at lower V_{DS}, failures with SCWT (>1 µs) are due to the device temperature rise, while

the physical mechanism of the failure with shorter SCWT is still under investigation.

In the literature, few articles have adequately discussed the effect of R_G on SC robustness of GaN HEMTs. For instance, in [6], it is suggested that there might be a trade-off between the value of R_G and the SC capability of GaN HEMTs. However, the authors do not provide an explanation for the premature failure of devices with low R_G, nor for the good robustness as R_G increases. Nevertheless, it should be noted that increasing R_G leads to a decrease in the switching speed of the device, which in turn affects its performance [1]. Thus, the variation in SCWT observed in the literature could be attributed to R_G variation [7].

The objective of this paper is to present an experimental study on the effect of gate resistance R_G on the robustness of GaN HEMT under SC. The DUTs used in this study are commercial normally-off 650 V-30 A Schottky p-GaN HEMTs.

2   Experimental setup

To carry out SC tests, an experimental test bench was developed, as illustrated in **Fig. 1** and in **Fig. 2**. The bench allows to conduct type I SC tests (meaning that the DUT is subjected to a SC stress while it was in the OFF-state). The DUT is inserted with its driver on a daughterboard, as shown in **Fig. 3**.

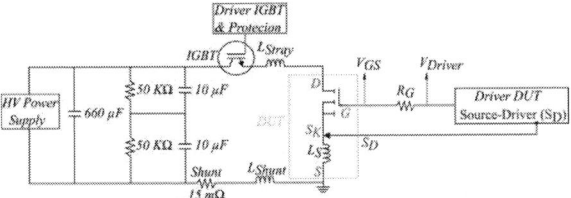

Fig. 1 Electric schematic of the test bench.

Fig. 2 Photograph of the test bench.

Fig. 3 Photograph of the daughterboard.

The large capacitive ballast of the bench guarantees, at a voltage of 500 V, a maximum voltage drop ≤ 5% for a current pulse of 300 A during 55 µs, as shown in Eq. (1).

$$I_D = C \frac{dV_{DS}}{dt} \qquad \text{Eq. (1)}$$

The DUT is driven by a specific GaN driver (UCC27511). The driver is controlled by a pulse generator.

To protect the capacitive ballast when a SC fault occurs, the Insulated Gate Bipolar Transistor (IGBT) in **Fig. 1** is turned OFF as soon as the current I_D exceeds 450 A. The protection circuit at the IGBT level can react within 2 µs after the DUT failure. However, the driver of the DUT does not integrate an overcurrent detection circuit. Thus, during the failure, only the IGBT is blocked, while the DUT remains ON for the rest of the programmed pulse.

The DUT is switched to the ON-state by applying a positive voltage between its gate and its Kelvin source, as shown in **Fig. 1**. Throughout this paper, V_{GS} is actually V_{GSK}. Indeed, the V_{GS} voltage is measured by measuring the potential difference between the gate and the Kelvin source of the DUT. Similarly, the output voltage of the driver is measured by measuring the potential difference between the driver output and the Kelvin source of the DUT. These measurements were carried out using two voltage probes (LeCroy PP018 500 MHz 10:1). The gate leakage current (I_G) is measured by measuring the voltage drop across R_G. On the power loop side, the measurement of V_{DS} voltage is performed by measuring the potential difference between the drain and the power source of the DUT, using a Tektronix P5100A 500 MHz 100:1 voltage probe. The drain current I_D is measured by using a current viewing resistor (15 mΩ - 1200 MHz) of T&M Research Products.

3   Design of experiments

This paper presents a qualitative and experimental analysis of the effect of R_G on the robustness of GaN HEMTs under SC conditions. The DUTs were tested at different values of R_G, namely R_G=10 Ω; 22 Ω; 47 Ω. V_{DS} was maintained at 400 V and V_{GS} was maintained at 5 V in the ON-state. All tests were conducted at room temperature. The protective IGBT was always in ON-state, and the DUT was initially in OFF-state (SC type I), and is/was driven in ON-state for triggering SC fault.

3.1   Single SC event capability

In this case, long duration destructive SC tests were performed, at a pulse width (t_p) set at 2 ms (sufficient for failure occurrence). The SCWT is determined from the start of the SC until the moment just before failure. The E_{SC} is therefore calculated by integrating over the SCWT, as shown in Eq. (2).

$$E_{SC} = \int_0^{SCWT} V_{DS}.I_D.dt \qquad \text{Eq. (2)}$$

3.2   Robustness after a repetitive SCs of short duration

In this case, repetitive SC tests were performed, at a pulse width (t_p) set at 500 ns. The E_{SC} is calculated only for the first cycle, by integrating over t_p, as shown in Eq. (3). The waiting time between consecutive cycles is not controlled and varies around one minute. For clarity, only waveforms of the first and the last cycles are plotted.

$$E_{SC} = \int_0^{t_p} V_{DS}.I_D.dt \qquad \text{Eq. (3)}$$

4 Results and analysis

4.1 Single SC event

In this section, the behavior of the DUT under destructive SC conditions was investigated by varying R_G. Firstly, two devices were tested at R_G=10 Ω. The DUTs exhibited very homogeneous behaviors. They sustained a SCWT of 629 μs and 630 μs, as shown in **Fig. 4** and **Fig. 5**, respectively. It can also be seen that the maximum of Drain saturation current ($I_{Dsat-max}$) is almost the same in both cases, typically $I_{Dsat-max}$=117.6 A in **Fig. 4**, and $I_{Dsat-max}$=120 A in **Fig. 5**. The energy dissipated is also very similar, approximately 1.38 J as in the case of **Fig. 4**, or 1.47 J as in the case of **Fig. 5**.

By zooming-in on **Fig. 4**, a transient peak of the gate current I_G ($I_{Gmax-trans}$) can be oberved (see **Fig. 6**). This transient peak, which typically occurs within 100 ns, corresponds to the charging of the parasitic gate capacitances. Once these capacitances are fully charged (meaning that V_{GS} has reached its set value), I_G returns to 0 A.

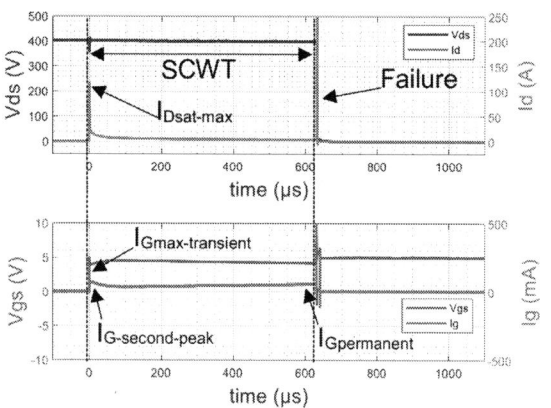

Fig. 4 DUT 1. Test conditions: V_{DS}=400 V; V_{GS}=5 V; R_G=10 Ω; tp=2 ms; SCWT=629 μs.

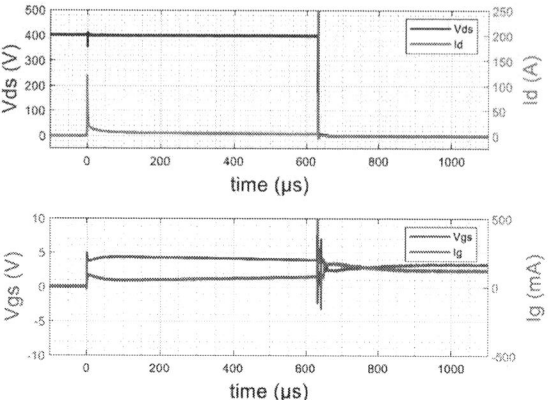

Fig. 5 DUT 2. Test conditions: V_{DS}=400 V; V_{GS}=5 V; R_G=10 Ω; t_p=2 ms; SCWT=630 μs.

Then, I_G begins to increase again until it reaches a second peak after a few microseconds. This new rise in I_G is believed due to the rise in temperature inside the chip [8]. Afterward, I_G starts decreasing again, then rises slightly before stabilizing at a permanent value inferior to its second peak until failure occurs, as illustrated in **Fig. 4**. It is not yet clear why I_G decreases a second time as the temperature rises.

At the drain level, it can be observed that I_D drops drastically from its maximum value of 117.6 A to 19.1 A after 5 μs, which represents a reduction of 83.7 %, as shown in **Fig. 6**. This current collapse is beneficial for the robustness of the DUT and it has several potential causes. The most apparent cause is the decrease in electron mobility due to component heating [9]. But also, the combined effect of the device self-heating and the increase in I_G (thus leading to a decrease in V_{GS}), can contribute to such reduction [6], [10].

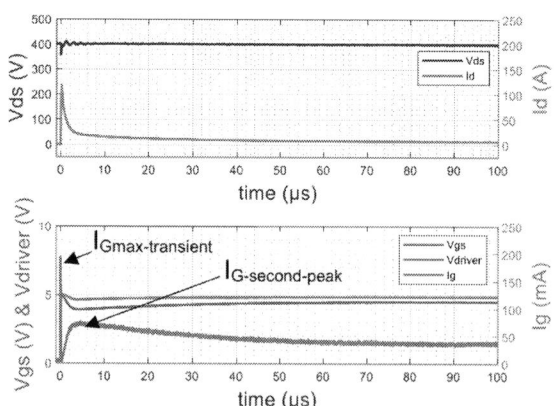

Fig. 6 Zoom-in on the beginning of the SC event in **Fig. 4** (DUT1).

PCIM Europe 2024, 11– 13 June 2024, Nuremberg   DOI: 10.30420/566262003

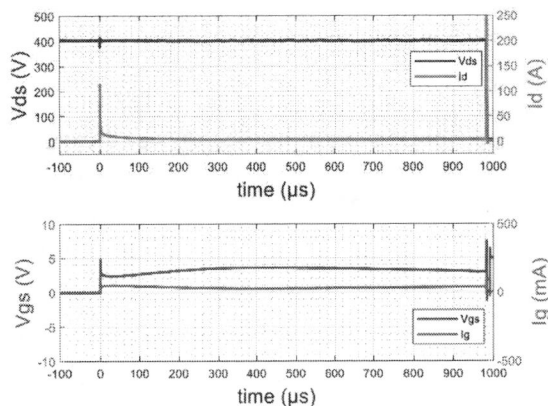

Fig. 9 DUT4. Test conditions: V_{DS}=400 V; V_{GS}=5 V; R_G=47 Ω; t_p=2 ms.

Two other components were tested under destructive SC conditions at R_G=22 Ω and at R_G=47 Ω. At R_G=22 Ω, the DUT withstood a SCWT of 884 µs, with an $I_{Dsat-max}$ of 113.04 A and and an E_{SC} of 1.74 J, as shown in **Fig. 8**. At R_G=47 Ω, the DUT withstood a SCWT of 982 µs, with an $I_{Dsat-max}$ of 114 A and an E_{SC} of 1.81 J, as shown in **Fig. 9**. The same I_G behavior seen above at R_G=10 Ω can also be observed with these two latter cases, but with much lower levels.

Table 1 summarizes the above discussed results with destructive SCs at different R_G values. An increase in E_{SC} was observed with increasing R_G. This implies that increasing R_G enhances the component's robustness by constraining one or more failure origins. Assuming that failure is primarily due to a critical temperature being reached, it can be seen that the DUTs failed at different temperatures (different dissipated energies). Nonetheless, it is yet to be determined if this energy difference corresponds to a similar order of magnitude difference in temperature. In other words, if the cause of failure is the increase in chip temperature, one might wonder why do the DUTs do not all break at the same temperature. This indicates that in this scenario, there may be one or more causes of failures, besides temperature.

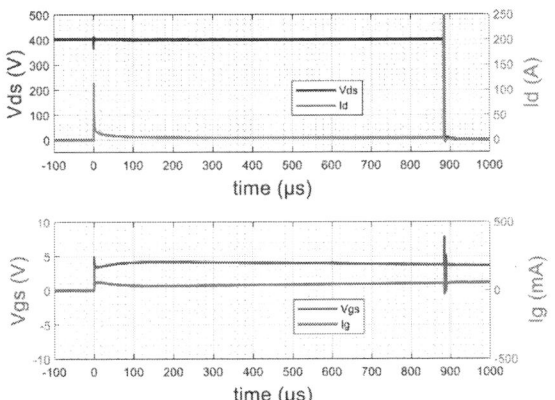

Fig. 7 Zoom in on the end of the SC event in **Fig. 4** (DUT1).

After failure, the IGBT is driving OFF within 2 µs. Consequently, V_{DS} decreases towards 0 V, as shown in **Fig. 7**. The gate and source are short-circuited, and the driver signal continues to be applied for the rest of the programed pulse. The driver cannot maintain the set voltage, leading to a decrease from 5 V to 3 V. A high leakage current (high current density), only limited by R_G, flows through the gate until it breaks and becomes an open-circuit, as shown in **Fig. 7**. However, with some DUTs, the gate leakage current is relatively low after the failure. Consequently, the gate is not completely damaged. Thus, it continues to conduct this leakage current until the end of the programmed pulse, as shown in **Fig. 5**.

Fig. 8 DUT3. Test conditions: V_{DS}=400 V; V_{GS}=5 V; R_G=22 Ω; t_p=2 ms.

Table 1 Summary of destructive single event SC results at various R_G values

R_G / Param	10 Ω (DUT1)	10 Ω (DUT2)	22 Ω (DUT3)	47 Ω (DUT4)
SCWT (µs)	629	630	884	982
$I_{Dsat-max}$ (A)	117.6	120	113.04	114
$I_{Gmax-trans}$ (mA)	193.9	196.3	103.3	73.6
E_{SC} (J)	1.38	1.47	1.74	1.81

37

4.2 Robustness after a repetitive SCs of short duration

In this section, the robustness of the DUT under repetitive SC conditions was investigated by varying R_G. By conducting repetitive SC at R_G=10 Ω and R_G=22 Ω, both devices were broken at the second cycle, as shown in **Fig. 10** and in **Fig. 11**, respectively. However, with R_G=47 Ω, the DUT breaks at the seventh cycle, as shown in **Fig. 12**. In these three cases, the energy dissipated during the first cycle remains almost constant (20.2 mJ at R_G=10 Ω; 19.7 mJ at R_G=22 Ω; 18.6 mJ at R_G=47 Ω). $I_{Dsat\text{-}max}$ is found to decrease as R_G increases.

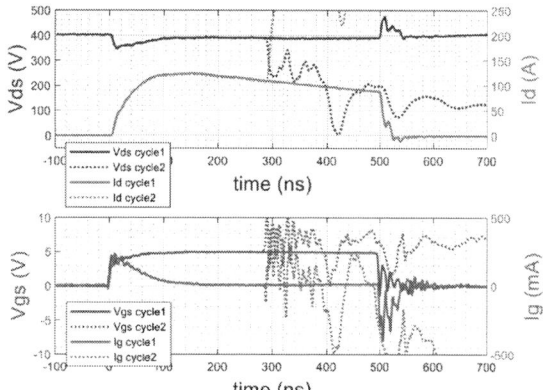

Fig. 10 DUT5. Test conditions: V_{DS}=400 V; V_{GS}=5 V; R_G=10 Ω; t_p=500 ns.

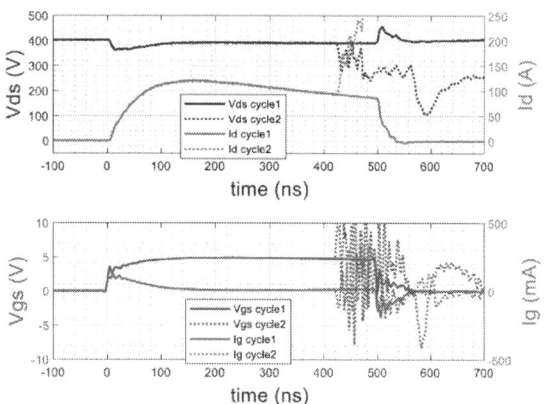

Fig. 11 DUT6. Test conditions: V_{DS}=400 V; V_{GS}=5 V; R_G=22 Ω; t_p=500 ns.

In parallel, it can be observed that as R_G increases, $I_{Gmax\text{-}trans}$ decreases, and therefore the gate parasitic capacitances charge more slowly. Consequently, V_{GS} reaches its set value more slowly, decreasing in turn the switching speed of

the DUT. The results of this section are summarized in **Table 2**.

It can be observed that in the case of DUT8, DUT9, and DUT10, there is no gate leakage current except for the initial peak for charging the gate capacitances. This could imply that not only there is almost no variation in E_{SC}, but also that the gate leakage current plays a minor role in the failure. Additionally, compared to the previous section (long destructive SCs) where the Esc energy was in the range of 1 to 2 J, in short repetitive SCs, the Esc energy is in the range of a few tens of mJ. This excludes the energy (temperature) as a possible cause of failure in this case.

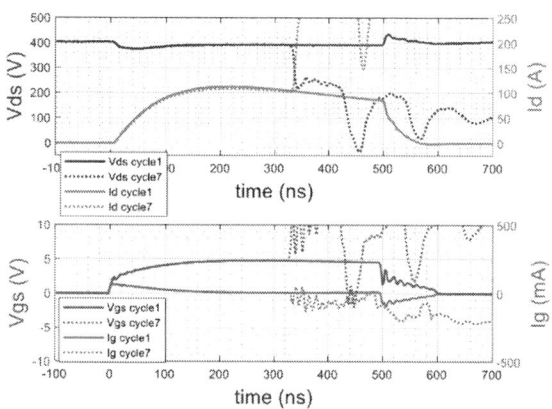

Fig. 12 DUT7. Test conditions: V_{DS}=400 V; V_{GS}=5 V; R_G=47 Ω; t_p=500 ns.

Table 2 Summary of repetitive SC results at various R_G values

R_G Param	10 Ω (DUT5)	22 Ω (DUT6)	47 Ω (DUT7)
Number of SC cycles until failure occurs	2	2	7
$I_{Dsat\_max\_cycle1}$ (A)	125.6	121.04	112.8
$I_{Gmax\text{-}trans\_cycle1}$ (mA)	276	121.3	83.4
E_{SC_cycle1} (mJ)	20.2	19.7	18.6

5 Conclusions and perspectives

This study investigates the impact of gate resistance R_G on the robustness of available 650 V normally-off p-GaN HEMTs under SC conditions through qualitative and experimental analysis. In the experiments, R_G values were varied while V_{DS} and V_{GS} were kept constant. Two scenarios were explored: single SC events of long duration and repetitive SC events of short duration. In single SC event tests, DUTs demonstrated varying degrees of robustness depending on the R_G value. Lower R_G values led to shorter SCWT and higher $I_{Dsat-max}$. Conversely, higher R_G values result in improved robustness with longer SCWT, lower $I_{Dsat-max}$, and increased energy dissipation. It was also observed that increasing R_G improves robustness by limiting potential failure sources; however, the exact cause of failure remains multifaceted.

Repetitive SC tests conducted at a very short pulse width reveal a decrease in $I_{Dsat-max}$ as R_G increases, along with slower charging of gate parasitic capacitances, which affects the switching speed of the device. Notably, gate leakage current appears to play a minimal role in failure under these conditions.

The results highlight the complex interplay between R_G, E_{SC} and failure mechanisms. Thus, further investigations are necessary to fully understand GaN HEMT robustness under SC scenarios.

Moreover, it will be necessary to test more components to ensure the homogeneity of their behaviors.

Acknowledgements

This work was carried out in the framework of the IRT Saint Exupery's project GANRET (GaN Reliability Evaluation for Transport). We acknowledge the financial support from the GANRET's Industrial and Academic Members and the financial support from the French National Research Agency (ANR).

References

[1] J. Sun, J. Wei, Z. Zheng, et K. J. Chen, « Short Circuit Capability Characterization and Analysis of p-GaN Gate High-Electron-Mobility Transistors Under Single and Repetitive Tests », *IEEE Transactions on Industrial Electronics*, vol. 68, n° 9, p. 8798-8807, sept. 2021, doi: 10.1109/TIE.2020.3009603.

[2] N. Keshmiri, D. Wang, B. Agrawal, R. Hou, et A. Emadi, « Current Status and Future Trends of GaN HEMTs in Electrified Transportation », *IEEE Access*, vol. 8, p. 70553-70571, 2020, doi: 10.1109/ACCESS.2020.2986972.

[3] T. Nagahisa, H. Ichijoh, T. Suzuki, A. Yudin, A. O. Adan, et M. Kubo, « Robust 600 V GaN high electron mobility transistor technology on GaN-on-Si with 400 V, 5 µs load-short-circuit withstand capability », *Jpn. J. Appl. Phys.*, vol. 55, n° 4S, p. 04EG01, févr. 2016, doi: 10.7567/JJAP.55.04EG01.

[4] K. Tanaka *et al.*, « Reliability of hybrid-drain-embedded gate injection transistor », in *2017 IEEE International Reliability Physics Symposium (IRPS)*, avr. 2017, p. 4B-2.1-4B-2.10. doi: 10.1109/IRPS.2017.7936308.

[5] D. Wieland *et al.*, « A common hard-failure mechanism in GaN HEMTs in accelerated switching and single-pulse short-circuit tests », in *2023 IEEE International Reliability Physics Symposium (IRPS)*, mars 2023, p. 1-6. doi: 10.1109/IRPS48203.2023.10117943.

[6] M. Riccio, G. Romano, L. Maresca, G. Breglio, A. Irace, et G. Longobardi, « Short circuit robustness analysis of new generation Enhancement-mode p-GaN power HEMTs », in *2018 IEEE 30th International Symposium on Power Semiconductor Devices and ICs (ISPSD)*, mai 2018, p. 104-107. doi: 10.1109/ISPSD.2018.8393613.

[7] J. P. Kozak *et al.*, « Stability, Reliability, and Robustness of GaN Power Devices: A Review », *IEEE Transactions on Power Electronics*, vol. 38, n° 7, p. 8442-8471, juill. 2023, doi: 10.1109/TPEL.2023.3266365.

[8] C. Abbate, G. Busatto, A. Sanseverino, D. Tedesco, et F. Velardi, « Experimental study of the instabilities observed in 650V enhancement mode GaN HEMT during short circuit », *Microelectronics Reliability*, vol. 76-77, p. 314-320, sept. 2017, doi: 10.1016/j.microrel.2017.07.020.

[9] H. Li *et al.*, « Robustness of 650-V Enhancement-Mode GaN HEMTs Under Various Short-Circuit Conditions », *IEEE Transactions on Industry Applications*, vol. 55, n° 2, p. 1807-1816, mars 2019, doi: 10.1109/TIA.2018.2879289.

[10] C. Abbate, F. Iannuzzo, et G. Busatto, « Thermal instability during short circuit of normally-off AlGaN/GaN HFETs », *Microelectronics Reliability*, vol. 53, p. 1481-1485, sept. 2013, doi: 10.1016/j.microrel.2013.07.119.

PCIM Europe 2024, 11– 13 June 2024, Nuremberg   DOI: 10.30420/566262004

Neural Network Assisted Numerical Simulation Benchmarking for Electric Vehicle Thermal Management System

Ekin Alp Bicer[1,2], Pascal Schirmer[1], Peter Schreivogel[1], Gabriele Schrag[2]

[1] BMW Group, Germany
[2] Technical University of Munich, Germany

Corresponding author: Ekin Alp Bicer, ekin-alp.bicer@bmw.de
Speaker: Ekin Alp Bicer, ekin-alp.bicer@bmw.de

Abstract

Thermal Management System (TMS) in Electric Vehicles (EVs) is tasked with providing optimal thermal conditions for the EV components while keeping the passengers comfortable. An accurate TMS model prevents overengineered components during the early design phase, but high-fidelity models like CFD or FEM become computationally infeasible when simulating the whole system. Neural Networks (NNs) provide accuracy without heavy computational loads, however, their extrapolation capabilities can be limited when predicting coolant temperatures for EVs in the design phase. To solve this, the authors introduce an NN-based TMS simulation approach using analytical equations and dedicated look-up tables. The results show that the proposed approach outperforms the baseline approach only utilizing neural networks up to 11.5% during dynamic driving.

1   Introduction

The transition towards Electric Vehicles (EVs) is driven by a global imperative to reduce carbon emissions and mitigate the effects of climate change [1]. EVs represent a pivotal shift in the automotive industry, offering a sustainable alternative to internal combustion engine vehicles. As EVs become more widespread, the focus on their performance and reliability intensifies. Efficiency and reliability of EVs and their components depend on the effectiveness of the Thermal Management System (TMS), which has the purpose of maintaining the optimal operating conditions for vehicle parts and ensuring a comfortable cabin environment for passengers [2]. During the early development of EVs, having an accurate and fast TMS model plays an important role[3] since the design is not constrained based on the worst-case scenario of the coolant temperature and the heat exchange with the TMS. Most power electronic components' reliability analyses depend on their operating temperatures [4]; thus, avoiding worst-case scenario designs reducing the likelihood of over-engineered components, and consequently, the cost.

In order to model the specific temperatures of the components in the TMS, high-fidelity methods such as Finite Element Method (FEM) or Computational Fluid Dynamics (CFD) can be utilized. However, when it comes to modeling and simulating the entire TMS, these methods suffer from computational intensity due to long mission profiles and size of the model. On the other hand, Reduced Order Models (ROMs), such as State-Space Models (SSMs) or Lumped Parameter Thermal Networks (LPTNs) can be utilized with a lower requirement of calculations [5]. However, as the model complexity reduces, the model becomes less accurate. In order to reduce the amount of calculations, data driven methods can be utilized. However, data driven methods require data to be already present, and their interpretability and extrapolation capabilities are open to debate [6].

As the TMS models are most useful to have during the early development phase, using NNs to predict the coolant temperatures create a problem of extrapolation, where the data from existing vehicles is used to model the non existing vehicle. At that point, there is no guarantee that the predictions are accurate. In order to mitigate this, physics informed neural networks can be employed [7].

To the best of authors' knowledge, there are not many attempts to model the complete TMS in previous works. In a previous work from the authors of this paper [8], different feature generation methods were tested for coolant temperature prediction. In [9] a TMS simulation framework for the long range tests is proposed. In [10], a Simulink model for the main coolant loop of the TMS is proposed and in [11], the model is extended to the refrigerant loops. In [12], coolant temperatures are predicted by separating active and passive vehicle states. In some of the previous works, some parts or components of the TMS are modeled. In [13], NNs are utilized together with cross-domain data to predict battery temperatures. In [14], [15], electric motor temperatures are modeled using neural networks, and in [16], authors use physics informed NNs for the same task.

In this paper, an iterative simulation that leverages neural networks is developed and benchmarked. In Section 2, TMS architecture is introduced. In Section 3, a simulation architecture that models the inverter heat contribution is proposed and detailed. Section 4 details the experimental setup to benchmark the simulation versus generic NNs. Section 5 presents the results for different scenarios. The paper is concluded in Section 6.

2 Thermal Management System

The TMS plays a crucial role in an electric vehicle's efficiency, and therefore driving range, as well as safety and reliability. A simplified schematic of the TMS is depicted in Fig. 1

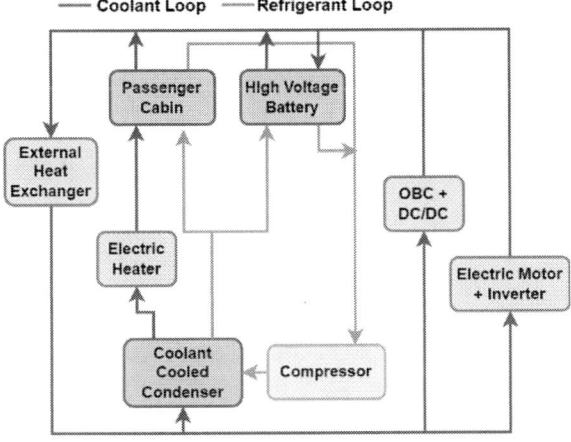

Fig. 1: Simplified block diagram of the TMS. Components depicted in blue only interact with the coolant, yellow ones with the refrigerant, and red ones with both.

As illustrated in Fig. 1, the TMS consists of multiple loops, namely the coolant loop and refrigerant loop. Operation of different sections of each loop can depend on the operating mode of the TMS. Therefore, coolant temperature in the TMS can be modeled as a nonlinear function of heat contributions of each component that is connected to the TMS and can be written as:

$$\hat{T}_c(t) = f_\theta(P_{cont}(t)) \tag{1}$$

where $\hat{T}_c$ is the predicted coolant temperature, $f_\theta(\cdot)$ is an non-linear regression function parameterized by θ representing the architecture of the TMS, P_{cont} is the heat contribution to the coolant of each component. It is important to differentiate between the heat contribution to the coolant and the heat generation of the TMS components. Heat generated at a component's hot-spot must traverse material with thermal resistance and capacitance to reach the coolant, acting as a low-pass filter for the heat generation when assessing the heat injected into the coolant. Additionally, not all generated heat is transferred to the coolant; some is also dissipated into the ambient environment through convection and radiation. Consequently, the heat contribution of a component to the coolant can be written as:

$$P_{cont} = g(P_{loss}, \dot{Q}_{amb}, C_{th}, R_{th}) \tag{2}$$

where P_{cont} is the heat contribution of each component to the coolant , $g(\cdot)$ is a component-dependent function, P_{loss} are the power losses that occur in the component, $\dot{Q}_{amb}$ is the thermal energy dissipated into the environment that does not reach the coolant, C_{th} and R_{th} are the thermal capacitance and resistance respectively.

Accurately calculating the heat contribution is also critical for ensuring transferabilty when compared to using only power losses of the components. The path taken by the heat is dependent on component geometry and using the heat contribution makes the model indifferent to the geometry of a component. This implies that replacing the component with one of a completely different geometry does not diminish the accuracy of the NN predictions. Furthermore, most of the TMS component's losses are a function of their hot-spot temperature (e.g. copper losses of stator-windings or losses of lithium ion cells). The hot-spot temperature is directly affected by the coolant temperature since the coolant acts as reference temperature and thermal sink.

Since the complete TMS is computationally infeasible to model on high-fidelity methods, NNs can be used to dramatically increase the calculation speed. The coolant temperature prediction is a temporal problem, meaning that the current state of the coolant temperature is dependent on the previous states of the system. However, as the coolant temperature is not uniform over the TMS, each heat contribution will have a different effect, depending on where the coolant temperature is measured. This complexity can be overcome by using NN architectures that can capture the temporal dependencies. The TMS model, that utilizes NNs to capture the nonlinear relation between the coolant temperature and the design inputs is shown in Fig. 2.

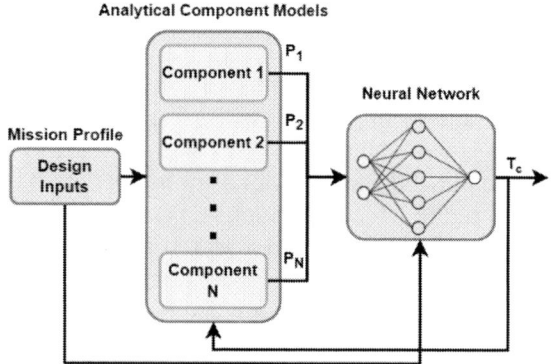

Fig. 2: Block diagram of the complete TMS model. $P_1, P_2, ...P_N$ are heat contributions of each TMS component, T_c is the coolant temperature.

3 Proposed Architecture

In this paper, a NN-assisted simulation is developed to solve the interdependent problem of the heat loss-coolant temperature with one of the major heat contributors to the TMS: The inverter. A single component is chosen to establish a framework for the simulation as a proof of concept. A detailed block diagram of the simulation that includes the inverter model can be seen in Fig. 3.

3.1 Inverter Loss Model

In order to be able to accurately model the heat contribution of the inverter to the coolant two aspects must be considered: heat generation and heat propagation. By using the input features, the heat generation can be calculated using the heat generation model. After calculation of the heat loss, heat contribution to the coolant can be calculated by using the heat propagation model. Inverter losses can be described as a sum of power module

losses, ohmic losses from AC and DC bus bars, DC-link capacitors and EMC filters. The losses are dominated by the power module losses, therefore, only power module losses are considered. Power module losses can be described by:

$$P_{PM} = P_T + P_D \tag{3}$$

where P_{PM} are the power module losses, P_T are the IGBT losses and P_D are the free-wheeling diode losses. These losses can further be split into conduction and switching losses:

$$P_T = P_{T,s} + P_{T,c} \tag{4}$$

$$P_D = P_{D,s} + P_{D,c} \tag{5}$$

where $P_{X,s}$ and $P_{X,c}$ are the switching and conduction losses. Analytical description of each loss component can be found in Eqs. (6) to (9) for space vector PWM, in terms of phase currents (I_p), power factor (ϕ), switching frequency (f_{sw}), DC-link voltage (V_{dc}), modulation index (M_i), as well as datasheet parameters.

$$P_{T,s} = (E_{T,on} + E_{T,off}) \cdot f_{sw} \tag{6}$$

$$P_{D,s} = E_{rec} \cdot f_{sw} \tag{7}$$

In Eqs. (6) to (9) T_j is the junction temperature, $E_{T,on/off}$ are the temperature dependent turn on/off energies, E_{rec} is the reverse recovery energy, and R_{ce} and V_{ce} are the temperature dependent collector-emitter resistance and voltage, R_f and V_f is temperature dependant resistance and voltage of the freewheeling diode [17]. M_i is dependant on the phase voltage requirement of the electric engine and the DC link voltage:

$$M_i = \frac{V_{12}/\sqrt{3}}{V_{DC}/2} \tag{10}$$

where V_{12} is the line to line voltage requirement of the electric engine and V_{DC} is the DC-link voltage of the battery. Since φ, I_p, V_{12} depend only on the torque and Rotations Per Minute (RPM) requirements of the electric motor, they can be mapped to the torque and RPM range of the motor. Since most of the parameters are temperature dependent, the junction temperature is used to update the parameters iterative for each operating point.

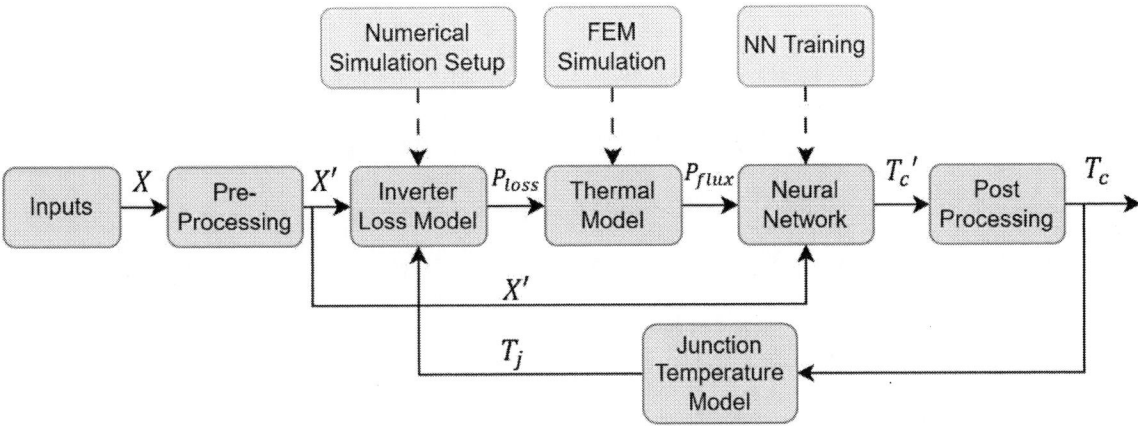

Fig. 3: Block diagram of the proposed architecture with X representing the raw inputs, X' representing the pre-processed inputs, P_{loss} representing power losses in the inverter, and P_{flux} representing the heat flux into the coolant. Further, $\hat{T}_c'$ and $\hat{T}_c$ are the normalized and absolute coolant temperature, and T_j is the junction temperature of the power module.

$$P_{T,c} = \frac{R_{ce}(T_j)I_p^2}{8}\left(1 + \frac{8 \cdot M_i \cdot \cos\varphi}{3\pi} - \frac{8\sqrt{3} \cdot M_i \cdot \cos(3\varphi)}{40\pi^2}\right) + \frac{V_{ce}(T_j) \cdot I_p}{2\pi}\left(1 + \frac{2\pi \cdot M_i \cdot \cos\varphi}{8}\right) \quad (8)$$

$$P_{D,c} = \frac{R_f(T_j)I_p^2}{8}\left(1 - \frac{8 \cdot M_i \cdot \cos\varphi}{3\pi} + \frac{8\sqrt{3} \cdot M_i \cdot \cos(3\varphi)}{40\pi^2}\right) + \frac{V_f(T_j) \cdot I_p}{2\pi}\left(1 - \frac{2\pi \cdot M_i \cdot \cos\varphi}{8}\right) \quad (9)$$

3.2 Inverter Thermal Model

Heat propagation modeling of the inverter can be done by conducting FEM or CFD simulations. The heat flux into the coolant and the junction temperature can be measured with respect to the heat input to the junction. For the inverter the heat transfer properties are dependant on the coolant flow rate describing a single input single output discrete time linear time invariant system that can be described as:

$$\hat{G}(\Phi) : \begin{cases} x_{t+1} = A(\Phi)x_t + B(\Phi)u_t \\ y_t = C(\Phi)x_t \end{cases} \quad (11)$$

where $A(\Phi) \in R^{n \times n}$, $B(\Phi) \in R^{n \times 1}$, $C(\Phi) \in R^{1 \times n}$ are flow rate dependent matrices with the scaler parameter Φ, which is the flow rate. With $\hat{G}$ being stable for every flow rate and having no repeated poles and zeros a transfer function $\hat{H}_\Phi$ can be described as:

$$\hat{H}_\Phi(z) = C(\Phi)(zI - A(\Phi))^{-1}B(\Phi) \quad (12)$$

In the high fidelity simulation, a constant power input can be given as a unit step input to observe the heat flux into the coolant over time. Obtained step responses then can be used to estimate a transfer function at various flow rates. After some transfer functions are estimated, the transfer function at any flow rate can be calculated by firstly pairing the poles and zeros of two transfer functions. After matching the poles and zeros, a hyperbolic line is placed on both of the poles, and the interpolated pole can be described with:

$$p_i(\rho) = \frac{p_{i1} - \rho w_i}{1 - \rho w_i \overline{p_{i1}}} \quad (13)$$

where $\rho \in [0,1]$ is the scaling value between the poles that is determined by the flow rate in question, $p_i(\rho)$ is the hyperbolically interpolated pole, p_{i1} is the first pole that the hyperbolic line passes through, and w_i is an appropriate constant, that is determined by:

$$w_i = \frac{p_{i1} - p_{i2}}{1 - p_{i2}\overline{p_{i1}}} \quad (14)$$

where p_{i2} is the second pole that the hyperbolic line passes. After all poles are interpolated, zeros can be calculated by the same method. Detailed explanation of the method can be found in [18]. During the operation, the flow rate of the coolant changes, and the obtained heat flux transfer functions must be interpolated according to the flow rate of the operation. At each time step, the interpolated transfer function can be multiplied in the frequency domain with the heat input as a discrete impulse. As the system is linear, the heat impulse responses can be summed up to obtain the accurate heat flux over the mission profile. In order to estimate the losses, the junction temperature of the power module needs to be calculated. Similar to the heat propagation approach, the junction temperature can be modeled using a transfer function that is dependent on the flow rate.

3.3 Neural Network Model

The NN part of the simulation is used to model the non-linear relationship between the input features and the coolant temperature. As explained earlier, the coolant temperature is dependant on previous states of the system, therefore neural networks that are suited for time series data, such as LSTMs and 1-D CNN are applicable. As found in earlier research [8], 1-D CNNs perform slightly better when it comes to the prediction accuracy in the coolant temperature prediction problem.

4 Experimental Setup

In this section, details about the experiments are explained. The experiments are used to benchmark the iterative closed loop approach, versus a plain NN that uses the basic design inputs.

4.1 Data

The data used in this paper are obtained from a test vehicle of a large automotive manufacturer (BMW i7 xDrive60). The vehicle is driven for about 140,000 kilometers, with over 4300 hours of recorded raw data that includes stand-by and charging. The data is sampled at 1 Hz and the data recording is stopped when the vehicle is turned off. Features that are used for the experiments can be found on Tab. 1, ambient and coolant temperature distributions of the data is illustrated in Fig. 4.

4.2 Data Processing

The following pre-processing steps have been applied to the data: The data is separated into driving sessions. Then, sessions where non-driving time exceeded 20% were excluded. Since driving is

Signal Name	Unit	Min	Max
Motor RPM	min^{-1}	-3000	16500
Motor Torque	Nm	-330,	390
Ambient Temp.	°C	-10	50
Coolant Temp.	°C	-2	55
Battery Voltage	V	290	440
Battery Current	A	-1000	575
Volume Flow	l/hour	0	840

Tab. 1: Names, ranges and units of the mission profile variables and the target value

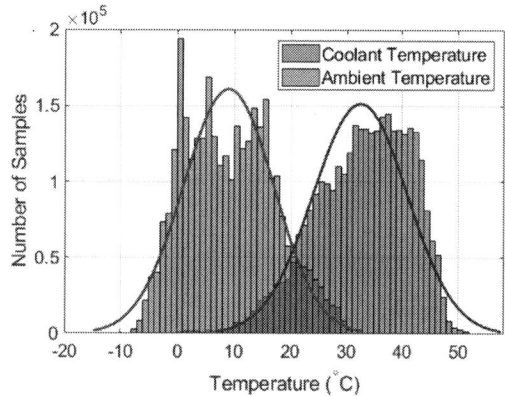

Fig. 4: Temperature distribution of coolant and ambient temperature for the dataset

the focus of this paper, the charging parts were removed, remaining parts of the session are labeled with IDs and kept in the dataset. At the end, the data is refined into 890 hours of dynamic driving, with a total of 410 separate driving sessions. Since the simulation needs the initial condition of the coolant temperature it is added as an additional feature for each driving session. Furthermore, the coolant temperature is sampled at 1K intervals, the data has been filtered using zero-phase delay moving average filter with a window length of 60 sec to preserve the physical nature of the signal.

Since the data is made from 410 individual driving sessions, train-validation-test splits are done per session. It is important for the models to be able to capture the highly dynamic behaviour of the coolant temperature, therefore the first 15 sessions with the highest range of coolant temperatures are selected. The test dataset is then built by randomly selecting two highly dynamic sessions and three random sessions from the remaining set. Similarly, the validation set is created using three random sessions from the highly dynamic set and seven sessions from the remaining data. The rest of the data is used for NN training. Statistics of the chosen test set can be seen in Tab. 2.

Session ID	1	2	3	4	5
min (°C)	0.0	-1.9	19.4	19.7	15.6
max (°C)	47.1	35.7	40.0	41.0	29.0
mean (°C)	25.4	20.3	27.9	36.0	20.4
median (°C)	30.0	25.6	28.0	37.8	19.5
std (K)	14.4	11.9	4.5	5.0	3.5

Tab. 2: Statistical properties for test sessions.

4.3 Inverter Model

As discussed in Section 3.2 the inverter must be modeled considering both its junction temperature of the power module as well as the heat contribution to the coolant. To obtain the transfer functions five transient heat-up curves have been simulated using a unit step response of power losses using different flow rates. The response plots of both junction temperature and heat fluxes for each flow rate are illustrated in Fig. 5.

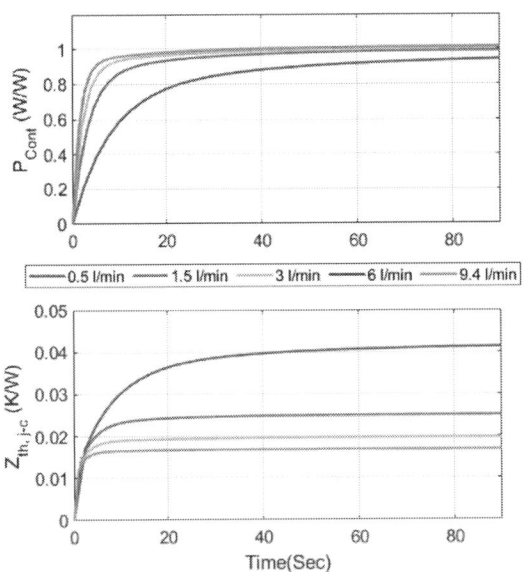

Fig. 5: Unit step response of the inverter contribution, and junction temperature, that is obtained and fitted from the CFD simulation

Observation of Fig. 5 shows that as the flow rate increases, thermal convection properties from the coolant side improve until it saturates between 3 l/min and 6 l/min. Moreover, it can also be noted that as the flow-rate goes down, thermal resistance between the junction and coolant increases, allowing more heat to flow to the ambient instead of going into the coolant.

4.4 Naive Predictor

In order to be able to check whether the trained models have meaningful accuracy, a naive predictor is proposed. Due to the setup of the TMS, the coolant temperature is always above or equal to the ambient. This can also be seen from the difference between the peaks of the envelopes of ambient and coolant temperature in Fig. 4. Therefore, a naive predictor that minimizes the error between the coolant temperature and a constant offset to the ambient temperature is proposed:

$$\hat{T}_c(t) = T_a(t) + \delta \qquad (15)$$

where δ is the constant offset that the naive predictor adds to the ambient temperature to predict the coolant temperature. To find the optimal δ, we solve the following optimization problem over the training data:

$$\delta^* = \arg \min_{\delta} e(T_c(t), \hat{T}_c(t)) \qquad (16)$$

where $e(\cdot)$ is an error function. In this work, $e(\cdot)$ is chosen to be the root mean squared error.

4.5 Model Parameters and Experimental Protocols

Models in both baseline and inverter heat contribution model have the same neural network structure that is given in Tab. 3. This network structure has given results that are comparable to state of the art models [19] in electric vehicle temperature predictions, thus the same network structure is utilized.

In order to have a closed loop simulation, NNs are trained on the mission profile variables as well as the heat flux contribution. In order to generate the dataset to train the neural networks, firstly, the aforementioned inverter model is used with the ground truth of the coolant temperature to generate the actual heat contribution of the inverter. Afterwards four experimental protocols are created. The first one considering the naive prediction model (Naive) described in Section 4.4, the second one being the baseline model (Base) using only the features from the mission profile, the third one utilizing the actual heat contribution of the inverter (Heat), and the

Layers	Training Configs
Conv1D(f=30, k=10) Conv1D(f=30, k=8) Conv1D(f=40, k=6) Conv1D(f=50, k=5) Conv1D(f=50, k=5) MaxPool1D(p.s.=5, str=5) Flatten() Dense(u=32) Dense(u=32) Dense(u=32) Dense(u=1)	Max Epochs = 40 Batch Size = 128 Plat. Multiplier = 0.3 Plat. Patience = 3 Plat. Min = 0.00005 Early Stop Pat. = 4 Window Length = 90 L. Rate = 0.0005 Optimizer = "SGD" Loss = "mse"

Tab. 3: NN structure using ReLU activation and other parameters. (f: filters, k: kernel size, p.s.: pooling size, str: strides, u: units, Plat: Plateau)

fourth one where the best performing Heat model is taken and tested in the closed loop simulation (CL Sim). In detail, for the NN based approaches ten models have been trained to consider the statistical variation of NN based approaches. As seen from Tab. 2, session #1 has the highest standard deviation and the coolant temperature range. Therefore this session is most difficult to make predictions on and will be discussed in more detail. The normal-ized torque and RPM profiles of session #1 can be seen in Fig. 6 (a).

5 Results

The architecture presented in Section 3 was evaluated according to the experimental setup described in Section 4. For the performance metric, Mean Absolute Error (MAE) was chosen.

$$\text{MAE} = \frac{1}{N} \sum_{t=1}^{N} |T_c(t) - \hat{T}_c(t)| \qquad (17)$$

where N is the number of samples, and $T_c(t)$ or $\hat{T}_c(t)$ are the measured and predicted coolant temperature at time step t. The average results for the five testing sessions are tabulated in Tab. 4.

For the naive predictor, upon running the algorithm given in Eq. (16) over the training data, δ is found to be 23.51 K, with a MAE of 5.01 K over the training dataset and 8.08 K over the test dataset. Any NNs trained should result in a better MAE than these values and will be shown as the minimum acceptable accuracy for the results. Performance of the best performing one of the ten Base models, Heat models and the CL Sim on each of the test IDs can be found on Tab. 4. For the ten NN

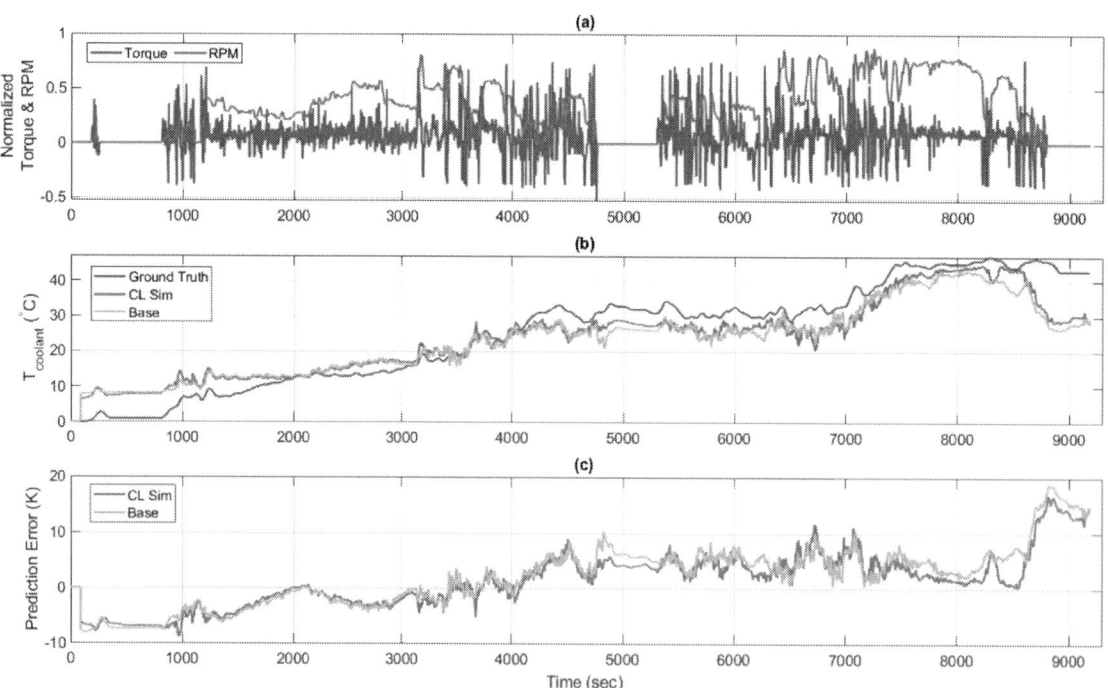

Fig. 6: (a) Normalized torque and RPM values of session #1, (b) ground truth coolant temperature and model predictions, (c) prediction errors

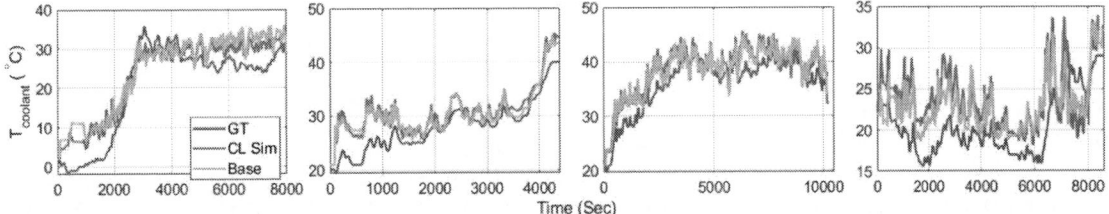

Fig. 7: Ground Truth (GT), simulation results and baseline model results for the sessions 2 (a), 3 (b), 4 (c), 5 (d)

ID	Naive	Base	Heat	CL Sim
1	12.7 K	5.55 K	**4.89 K**	4.91 K
2	8.92 K	4.50 K	**3.41 K**	3.42 K
3	6.94 K	1.77 K	**1.71 K**	1.71 K
4	6.47 K	**1.60 K**	1.65 K	1.65 K
5	4.76 K	**2.05 K**	2.07 K	2.07 K
Avg	8.07 K	3.01 K	**2.75 K**	2.75 K

Tab. 4: MAE results for the best Base model, best Heat model and CL Sim on different testing sessions.

miscalculation of the junction temperature and therefore a slight miscalculation of the heat flux values, as during the early design phase, exact heat propagation or heat loss properties might not be well known, there will be an error within the heat contribution calculation. To quantify the effect a sensitivity study has been done adding noise to the heat contribution for session #1. Statistics of the results can be seen in Tab. 5.

SNR(W/W)	inf	40	30	20	10	5
MAE (K)	4.91	5.08	5.11	5.28	5.59	6.13

Tab. 5: Results of the tests done for the noisy inputs.

models explained in Section 4.5, p-score evaluation of the ten Base versus ten Heat MAE values result in 0.0539. The best of the ten Base models reached 3.01 K average MAE, while the best Heat model reached an average of 2.75 K average MAE over the test set. On sessions #4 and #5 Base model has slightly outperformed the Heat model, but, as can be observed from Tab.2, sessions #3, #4, #5 have smaller ranges and smaller standard deviations, therefore not very dynamic.

The Heat model and CL Sim results are very similar due to the relatively high accuracy of the coolant temperature predictions. Low coolant temperature prediction error leads to a low error in the junction temperature and heat contribution calculation. Since the Heat model is used as the NN for the simulation, a slight error in the heat contribution leads to a slightly worse prediction of the coolant temperature. On the most dynamic session (#1), Base model managed to reach 5.55 K MAE, while the CL Sim reached 4.91 K leading to a 11.5% increase in the prediction accuracy. Plots of the coolant temperature ground truth versus the predictions on session #1 for both the CL Sim and the Base model is given in Fig. 6. The prediction profiles for the remaining test sessions for both models are given in Fig. 7

The coolant temperature prediction errors cause

From the Tab. 5, it can be seen that the simulation is not very sensitive to the noise. The prediction quality becomes comparable to the baseline model at a SNR value of 10.

6 Conclusion & Outlook

In this paper, a heat contribution model that uses neural networks for the TMS is developed and benchmarked against a naive predictor, and NNs that only use the mission profile inputs. The iterative simulation models the inverter's losses and heat transfer properties to accurately calculate the amount of heat transferred from inverter to coolant at every time step, thus abstracting the inverter's type and geometry from the model. The calculated heat input to the coolant is then fed to a NN to enhance the predictions. The simulation achieved 11.5% better results than a NN that only uses the mission profile inputs. The simulation is also tested against noise sensitivity, and found to be outperforming the baseline model until an SNR of ten. In the future, more components will be added to the simulation setup, broader scenarios (such as charging) will be explored, the TMS topology information will be conveyed to the NN, and physics informed neural network approaches will be tested.

References

[1] C. Li, Y. Cao, M. Zhang, J. Wang, J. Liu, *et al.*, "Hidden benefits of electric vehicles for addressing climate change," *Scientific reports*, vol. 5, p. 9213, Mar. 2015. DOI: 10.1038/srep09213.

[2] L. He, H. Jing, Y. Zhang, P. Li, and Z. Gu, "Review of thermal management system for battery electric vehicle," *Journal of Energy Storage*, vol. 59, p. 106 443, 2023. DOI: https://doi.org/10.1016/j.est.2022.106443.

[3] I. M. Sofi, T. Q. Dinh, A. Mohanadass, J. Jeffs, Q. T. Truong, and N. M. Truong Bui, "Advanced simulation tool to develop efficient thermal management systems for electric vehicles," in *2021 24th International Conference on Mechatronics Technology (ICMT)*, 2021, pp. 1–6. DOI: 10.1109/ICMT53429.2021.9687213.

[4] H. Wang and F. Blaabjerg, "Power electronics reliability: State of the art and outlook," *IEEE Journal of Emerging and Selected Topics in Power Electronics*, vol. 9, no. 6, pp. 6476–6493, 2021. DOI: 10.1109/JESTPE.2020.3037161.

[5] O. Wallscheid, "Thermal monitoring of electric motors: State-of-the-art review and future challenges," *IEEE Open Journal of Industry Applications*, vol. 2, pp. 204–223, 2021. DOI: 10.1109/OJIA.2021.3091870.

[6] K. Xu, J. Li, M. Zhang, S. Du, K.-i. Kawarabayashi, and S. Jegelka, "How neural networks extrapolate: From feedforward to graph neural networks," Sep. 2020.

[7] D. Kim and J. Lee, "A Review of Physics Informed Neural Networks for Multiscale Analysis and Inverse Problems," *Multiscale Science and Engineering*, Feb. 2024. DOI: 10.1007/s42493-024-00106-w.

[8] E. Bicer, P. Schirmer, P. Schreivogel, and G. Schrag, "Electric vehicle thermal management system modeling with informed neural networks," Sep. 2023, pp. 1–8. DOI: 10.23919/EPE23ECCEEurope58414.2023.10264482.

[9] T. J. Shelly, J. A. Weibel, D. Ziviani, and E. A. Groll, "A dynamic simulation framework for the analysis of battery electric vehicle thermal management systems," in *2020 19th IEEE Intersociety Conference on Thermal and Thermomechanical Phenomena in Electronic Systems (ITherm)*, 2020, pp. 538–546. DOI: 10.1109/ITherm45881.2020.9190543.

[10] T. Kiss, J. Lustbader, and D. Leighton, "Modeling of an electric vehicle thermal management system in matlab/simulink," *SAE Technical Papers*, vol. 2015, Apr. 2015. DOI: 10.4271/2015-01-1708.

[11] G. Titov and J. Lustbader, "Modeling control strategies and range impacts for electric vehicle integrated thermal management systems with matlab/simulink," Mar. 2017. DOI: 10.4271/2017-01-0191.

[12] M. S. Padrós, P. A. Schirmer, and I. Mporas, "Estimation of cooling circuits' temperature in battery electric vehicles using karhunen loeve expansion and lstm," in *2022 30th European Signal Processing Conference (EUSIPCO)*, 2022, pp. 1546–1550. DOI: 10.23919/EUSIPCO55093.2022.9909690.

[13] A. M. Billert, M. Frey, and F. Gauterin, "A method of developing quantile convolutional neural networks for electric vehicle battery temperature prediction trained on cross-domain data," *IEEE Open Journal of Intelligent Transportation Systems*, vol. 3, pp. 411–425, 2022. DOI: 10.1109/OJITS.2022.3177007.

[14] W. Kirchgässner, O. Wallscheid, and J. Böcker, "Data-driven permanent magnet temperature estimation in synchronous motors with supervised machine learning: A benchmark," *IEEE Transactions on Energy Conversion*, vol. 36, no. 3, pp. 2059–2067, 2021. DOI: 10.1109/tec.2021.3052546.

[15] W. Kirchgässner, O. Wallscheid, and J. Böcker, "Estimating electric motor temperatures with deep residual machine learning," *IEEE Transactions on Power Electronics*, vol. 36, no. 7, pp. 7480–7488, 2020. DOI: 10.1109/tpel.2020.3045596.

[16] W. Kirchgässner, O. Wallscheid, and J. Böcker, "Thermal neural networks: Lumped-parameter thermal modeling with state-space machine learning," *Engineering Applications of Artificial Intelligence*, vol. 117, no. 105537, 2022. DOI: 10.1016/j.engappai.2022.105537.

[17] Y. Zhu, M. Xiao, X. Su, G. Yang, K. Lu, and Z. Wu, "Modeling of conduction and switching losses for igbt and fwd based on svpwm in automobile electric drives," *Applied Sciences*, vol. 10, no. 13, 2020. DOI: 10.3390/app10134539.

[18] I. Gőze and A. Soumelidis, "A parametric lti interpolation for minimum phase systems with guaranteed stability and bounds," in *2018 Annual American Control Conference (ACC)*, 2018, pp. 1224–1229. DOI: 10.23919/ACC.2018.8431875.

[19] P. A. Schirmer and I. Mporas, "Pydts: A python toolkit for deep learning time series modelling," *Entropy*, vol. 26, no. 4, 2024. DOI: 10.3390/e26040311.

PCIM Europe 2024, 11– 13 June 2024, Nuremberg   DOI: 10.30420/566262005

Relationship Between the Porosity in Cu Sintered Bonding and Thermal Cycle Reliability

Hideo Nakako, Michiko Natori, Dai Ishikawa, Minami Kazuhiko, and Seiji Matsushima

[1] Resonac Corporation, Japan

Corresponding author and Speaker: Hideo Nakako, nakako.hideo.xikgo@resonac.com

Abstract

The bonding reliability of a Cu sintered die-bond is strongly influenced by the porosity of the sintered Cu. Therefore, the relationship between the bonding reliability and porosity is important to determine the sintering conditions and estimate a product's life. Accordingly, thermal cycle tests (TCTs) were performed on test vehicles die-bonded using a Cu sintered layer with different porosities. Cu sintered die-bonds with different porosities were obtained at different sintering pressures. When the sintering pressure increased from 4 to 20 MPa, the porosity of the Cu sintered layer decreased from 26 % to 15 %. The TCT was performed up to 3000 cycles with the conditions of 200 °C / 15 min ⇔ - 40 °C / 15 min. The test vehicles with Cu sintered die-bonds with porosities of 26 % and 21 % failed till 500 and 2500 cycles, respectively, while those with lesser porosities of 20 %, 16 %, and 15 % did not fail up to 3000 cycles. Therefore, the less porous Cu sintered die-bond had higher thermal cycle reliability. Additionally, for the same sintering pressure or porosity, the Cu sintered die-bond had superior thermal cycle reliability than the Ag sintered die-bond. After 3000 cycles of TCT, the Cu sintered bond underwent fatigue damage at the chip edge area and exhibited small point-like fatigue in all areas. In the sintered Cu layer with porosities as high as 26 % and 21 %, the small point-like fatigue expanded with the TCT and acted as the starting point of fatigue failure. The small point-like fatigue originated from the depression caused by grain boundary deformation owing to thermal stress in the underlying Cu electrode of a SiN-active metal brazing substrate (SiN-AMB).

1   Introduction

Silicon carbide (SiC) power devices have many advantages such as reduced power loss, high-temperature operation, high-speed switching operation, and high heat dissipation. However, SiC has a very high Young's modulus, which increases its thermal strain in a die-bonding layer, thereby lowering the bonding reliability. The high-temperature operation also lowers the bonding reliability. To improve the bonding reliability and heat dissipation, Ag sintered materials have been applied to die-bonding materials.

Sintered materials consist of small metal particles and are obtained by sintering at lower temperatures (200–300 °C) relative to their melting points [1]. Metallic surface is unstable cause of a termination of the metallic bonding therefore the contacting between metallic surfaces can easily make new metallic bonding each other. In the case of small metallic particles, the many number of contacting between particles' surfaces and the large ratio of unstable metallic atoms on particles' surfaces can accelerate a sintering and allow a low temperature sintering.

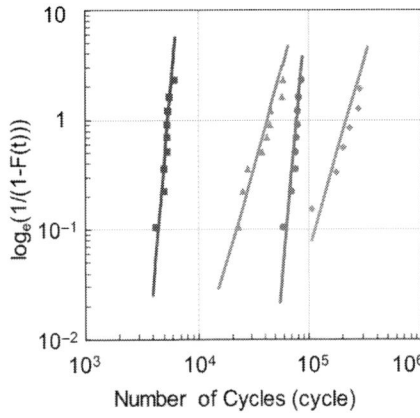

Fig. 1: Weibull plot obtained from the power cycle test ($T_{j,max}$ = 175 °C) with samples die bonded with ◆ Cu sinter at a pressure of 10 MPa, ● Cu sinter bonded without pressure, ▲ Ag sinter at a pressure of 20 MPa, and ■ high-Pb solder.

Consequently, a porous sintered metal is formed. The mechanical properties of sintered metals originate not only from the metal but also from the porous structure [2]. A previous study [3] shows that a difference in the porous structures of Cu sintered die-bonds can affect the power cycle life. The pressure-assisted Cu sintered die-bond with 6 % porosity exhibits a 2.8-fold better characteristic life compared to the pressure-free Cu sintered die-bond with 27 % porosity. The two Cu sintering pastes have different material composition, because of which it is likely that the power cycle life will be affected by other factors. However, the porosity can also be an important factor for determining the power cycle life.

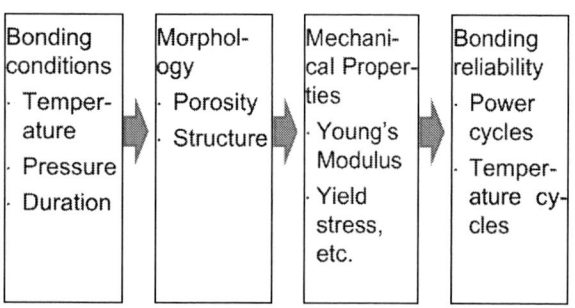

Fig. 1 Relationship from bonding conditions to bonding reliability for Cu sintering paste.

In the case of a Cu sintering paste, porosity of the sintered layer is strongly influenced by the sintering conditions such as the sintering temperature, pressure, and duration. This also implies that the sintering conditions affect the bonding reliability. This is the different point from soldering and causes difficulty in choosing the bonding conditions and controlling the bonding quality.

The relationships can be described as shown in Figure 1. The bonding conditions of a Cu sintering paste affects the morphology of the Cu sintered bonding. Morphological features such as the porosity affect the mechanical properties, which determine the bonding reliability of a Cu sintered die-

bond. The final goal is to establish the bonding conditions for estimating the bonding reliability or to establish a range of bonding conditions to achieve a target bonding reliability. In this study, the relationship between the morphology (porosity) and bonding reliability was established based on thermal cycle tests (TCTs). The relationship between the bonding conditions and morphology and the mechanical properties of the Cu sintered strip with different porosities are currently under investigation and will be reported in due course.

2 Results and Discussion

2.1 Initial morphology and porosity of the Cu sintering paste

To perform the TCT using the sintered die-bonding layers with different porosities, pressure-assisted sintering was carried out at different sintering pressures.

The test vehicle structure comprised a Si-insulated gate bipolar transistor (IGBT) die-bonded to a SiN ceramic active metal brazing (SiN-AMB) circuit board using sintering pastes. The dimension of the IGBT was 13 mm × 11 mm, with a thickness of 0.12 mm. The dimension of SiN-AMB was 34 mm × 29 mm, with the thickness of Cu, Si_3N_4, and Cu being 0.8, 0.3, and 0.8 mm, respectively. The

Items	Sintering pressure (MPa)		
	7	12	17
Cross section ▬ 5 µm	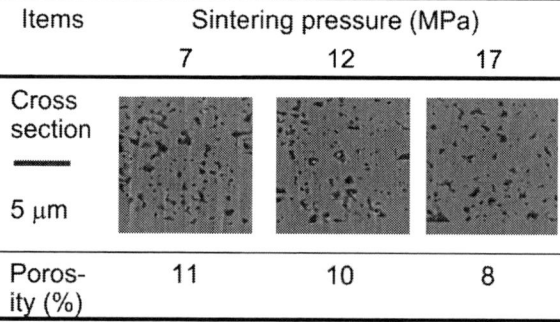		
Porosity (%)	11	10	8

Table 2 The morphology and porosities of Ag-sintered bonding from various sintering pressures.

Property	Sintering pressure (MPa)				
	4	7	12	17	20
Cross section ▬ 5 µm	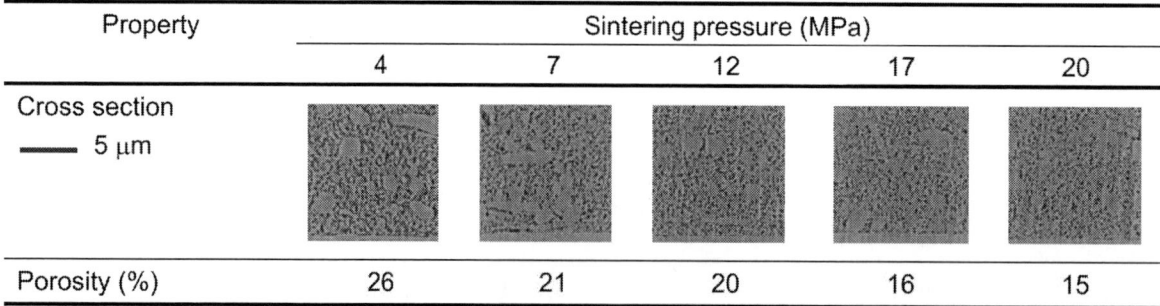				
Porosity (%)	26	21	20	16	15

Table 1 Morphology and porosity of the Cu-sintered bonding at various sintering pressures.

bonding conditions of the sintered Cu were 260 °C and 5 min, with varying sintering pressures of 4, 7, 12, 17, and 20 MPa.

The morphology and porosity of the Cu-sintered bonding are shown in Table 1. The porosity was influenced by the sintering pressure, and as the sintering pressure increased from 4 to 20 MPa, the porosity decreased from 26 to 15 % in the Cu-sintered die-bond.

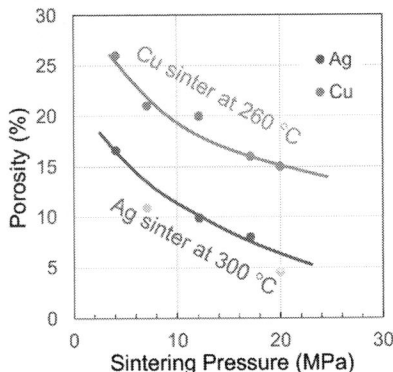

Fig. 2 Relationship between sintering pressure and porosity in sintered bonding. Cu and Ag sintering pastes were bonded at 260 and 300 °C for 5 min, respectively.

Ag sintered bonding was previously developed by us and used as the reference (Table 2). The sintered bonding was performed at 300 °C for 5 min at 7, 12, and 17 MPa, achieving porosities of 11, 10, and 8 %, respectively. The relationship between the sintering pressure and porosity is plotted in Figure 2. As evident, smooth non-linear curves were obtained in the porosity vs sintering pressure plots. The extent of decrease in the porosity with increasing sintering pressure was less at high sintering pressures. A similar curve was obtained for Ag sintered bonding. At the same sintering pressure, the soft Ag particles and the higher sintering temperature can lower the porosity by ~8 % compared to the porosity of the Cu sintering paste.

2.2 TCTs for sintered die-bond with varying porosities

The test vehicles die-bonded using the Cu sintered layer with various porosities were applied to the gas phase TCT (Table 1). The test vehicle was placed in an aluminum cup and potted by silicone gel. Five test vehicles for each sintering pressure were subjected

to the TCT. The TCT was carried out between 200 °C / 15 min and −40 °C / 15 min. After 1000 and 2000 cycles, one test vehicle was chosen from the test and its cross-sectional scanning electron microscopy (SEM) image was obtained. Three test vehicles were subjected to TCT for up to 3000 cycles, following which some test vehicles were remarkably damaged through peeling-off in the SiN-AMB.

Scanning acoustic tomography (SAT) images of the Cu sintered die-bonding were obtained after 0, 500, 1000, 1500, 2000, 2500, and 3000 cycles for each test vehicle to examine the damage in the die-bonding. Up to 1500 cycles, the SAT images were acquired from the side of the substrate, whereas from 2000 cycles onward, the image was acquired from the side of the device because the defects were clearer. The change in thermal resistance was also examined after 500 cycles; however, only solder, 4 MPa-sintered Cu sintering, and 4 MPa-sintered Ag sintering samples were detected with increasing thermal resistances. Other Cu and Ag sintering samples obtained at higher pressures did not show any clear change in the thermal resistance.

2.2.1 TCTs for Cu sintering paste

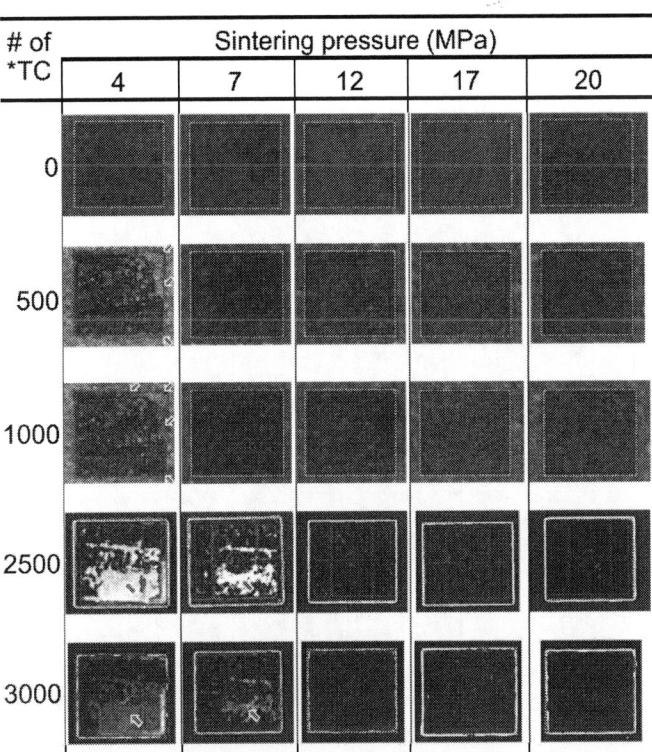

# of *TC	Sintering pressure (MPa)				
	4	7	12	17	20
0					
500					
1000					
2500					
3000					

Table 3 SAT images obtained through TCT for Cu sintering bonded test vehicle (*TC denotes thermal cycle)

Table 3 shows the SAT images of the test vehicles die-bonded using the Cu sintering paste through TCT. For the 4 MPa-sintered Cu sintering layer (26 % porosity), the SAT image at 500 cycles shows the damage at a corner, as evident from the light color.

Moreover, the fatigue area ratio exceeded 10 %, because of which it was concluded to be a failure. In the case of the 7 MPa-sintered sample (21 % porosity), the fatigue damages occurred in the chip edge area and internal area. The damage in the chip edge area occurred at ~2000 cycles but did not grow through the TCT. In contrast, in the internal area, color variation was observed at 500 cycles, while the small white spots were generated at ~2000 cycles and grew through the TCT. The 7 MPa-sintered sample (21 % porosity) underwent failure at 2500 cycles because of the growth of fatigue damage from the small white spots in the internal area.

The 12 MPa-sintered (20 % porosity), 17 MPa-sintered (16 % porosity), and 20 MPa-sintered (15 % porosity) test vehicles did not fail till 3000 cycles in the TCT. Most edges of the bonding area and the white spots in the internal area underwent damage, although it did not grow through the TCT. The number of spots increased through the TCT. The sintering pressure and the number of white spots and its increase were correlated to the porosity— higher pressure-sintered Cu sintering exhibited smaller number of white spots on the SAT images.

These results suggest that Cu sintered die-bonding achieved at higher sintering pressures had lower porosities inside the Cu sintered layer, with less fatigue damage through TCT. Particularly, the Cu sintered die-bonds obtained at a pressure of 12, 17, and 20 MPa did not break up to 3000 cycles.

2.2.2 TCTs for Ag sintering paste and sol der

Table 4 shows the SAT images of the test vehicles die-bonded using the Ag sintering paste and solder through TCT. The failure point for the test vehicle die-bonded using Ag sintering at a sintering pressure of 4 MPa was determined based on the thermal resistance, which increased by 50 % after 500 thermal cycles. The SAT images at 500 and 1000 cycles were not clear. Since the SAT images were acquired from the chip side after 2000 cycles, the damage in all the areas of the Ag sintered die-bond was revealed. In the case of the Ag sintered die-bond obtained at 12 MPa, the damage occurred at the edge area from 500 cycles. The damaged area grew from the edge and exceeded 10% at 1000 cycles. The thermal cycle reliability was superior at a higher sintering pressure of 17 MPa, which resulted in lower porosity in the Ag sintered layer. The SAT images of the test vehicle bonded at 17 MPa using the Ag sintering paste showed small damages at the corner from 1000 cycles; the damage expanded over 10 % of the area at 3000 cycles.

The SAT images of the test vehicles die-bonded by soldering through TCT were not clear. However, the thermal resistance increased by the thermal cycles and excessed 10 % of the thermal resistance change at 1000 cycles. Therefore, the thermal cycle life of the test vehicles with solder was 1000 cycles.

2.2.3 Comparison between Cu sintering paste and Ag sintering paste

The failure cycles through TCT with Cu and Ag sintering pastes were compared with the failure cycle of the solder for 1000 cycles. The bonding reliabilities of the 4 MPa-sintered Cu sintering (26 % porosity), 4 MPa-sintered Ag sintering (11 % porosity), and 12 MPa-sintered Ag sintering (10 % porosity) were less or equivalent to that of the solder. Therefore, for Cu s

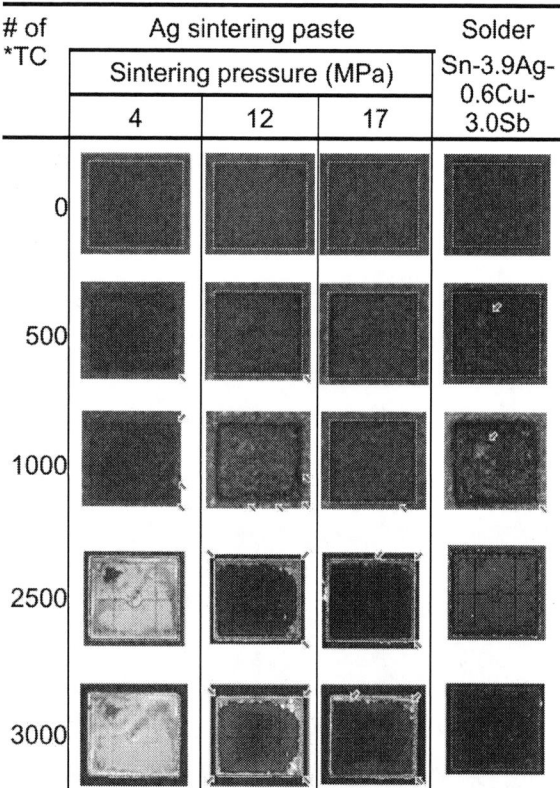

# of *TC	Ag sintering paste			Solder Sn-3.9Ag-0.6Cu-3.0Sb
	Sintering pressure (MPa)			
	4	12	17	
0				
500				
1000				
2500				
3000				

Table 4 SAT images obtained through TCT for Ag sintering or solder die-bonded test vehicle (TC denotes thermal cycle).

intering, a sintering pressure of 7 MPa or higher, which can result in 21 % or lesser porosity, can achieve a higher bonding reliability than that of the solder. In contrast, Ag sintering required 17 MPa pressure, resulting in 8 % porosity, to achieve a higher bonding reliability than that of the solder.

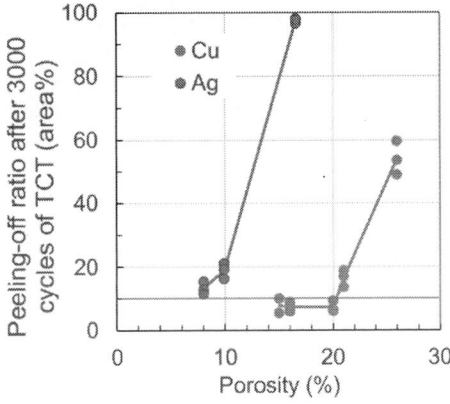

Fig. 3 Relationship between the peeling-off ratio after 3000 cycles of TCT and porosity in the sintered die-bonding.

In figure 3, the peeling-off area ratio after 3000 cycles of TCT was plotted against the porosity in the Cu and Ag sintered die-bonding. For Cu sintering, the peeling-off area ratio was not related to the porosity up to 5–10 % porosity. However, peeling off occurred at very edge area of die-bonding when the porosity in the Cu sintering was between 15 % and 20 %. In most of the edge areas, the Cu sintered layer exhibited incomplete sintering because of a printing defect and paste push-out. Therefore, although this weak sintered layer at the chip edge could be damaged, the cracking did not easily progress onto the well-sintered layer. When the porosity in the Cu sintered layer was more than 20 %, the peeling-off ratio increased with increasing porosity. Sufficient data to determine the linear relationship between the porosity and peeling-off ratio were not available.

A similar relationship between the peeling-off ratio and porosity was observed in the case of Ag sintering. However, to obtain a same peeling-off ratio of 20 % or 50 %, the sintered Ag must be ~12% denser than the sintered Cu.

These results reveal that Cu sintering has a superior bonding reliability against thermal cycles than Ag sintering when die-bonded at the same sintering pressure or for the same porosity ratio in the sintered bonding. It is convenient that a low sintering pressure such as 7 and 12 MPa for the Cu sintering paste is acceptable because of sufficient

thermal cycle reliability. Recently, the top electrode of a SiC power device was also bonded using Ag sintering [4]. In this case, when the upper electrode was bonded by pressure-assisted sintering, the pressure was applied over a smaller area compared to that of a device chip, causing chip cracking owing to the uneven pressure on the device chip. Therefore, a lower sintering pressure such as 10 MPa or less is preferred. The Cu sintering paste is especially suitable for this application because of the low sintering pressure and good bonding reliability.

2.2.4 Failure analysis after TCT by cross-sectional SEM

After 3000 cycles of TCT, the test vehicle was cut and observed by SEM to determine the failure mode. Figure 4 shows the cross-sectional SEM images for the Cu-sintered die-bond achieved at a sintering pressure of 17 MPa. At the chip edge, a 100–200 μm long crack was observed. The crack was initiated at the interface between the IGBT and Cu sintered bond, and it progressed downward toward the interface between the Cu sintered bond and the Cu electrode of SiN-AMB. A healthy bonding, without failure, was observed in the central area of the die-bonding. The porosity in the Cu sintered bond was not significantly different between the initial state and after the thermal cycles or between the chip edge area and central area.

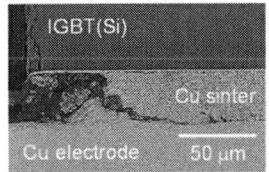 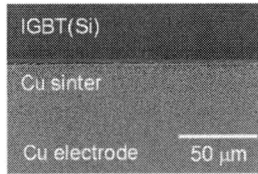

(a) chip edge area   (b) central area

Fig. 4 Cross sectional SEM images for the test vehicle die-bonded using Cu sintering paste at a sintering pressure of 17 MPa.

The Ag sintered layer sintered at 17 MPa had fatigue cracks from the chip edge area, and these cracks were observed both on the interfaces between the IGBT and Ag sintered bond and between the Ag sintered bond and Cu electrode of SiN-AMB. Through the TCT, the porosity in the Ag sintered bond changed from its initial state. Particularly, after TCT, the porosity in the central area was smaller than that in the chip edge area. The soft mechanical properties of Ag allowed easy deformation upon thermal stress, facilitating any change in its porosity.

 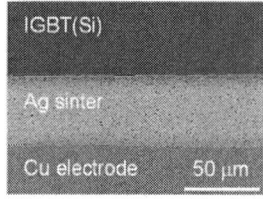

(a) chip edge area (b) central area

Fig. 5 Cross sectional SEM images for the test vehicle die-bonded using Ag sintering paste at a sintering pressure of 17 MPa.

Through the TCT, white spots were observed in the SAT images and were analyzed by cross sectional SEM (Figure 6). These white spots were observed for both Cu and Ag sintered bonds. In the cross-sectional SEM images, cracking was observed at the interface between the sintered bond and the Cu electrode and was generated by the depressing of the Cu electrode. The Cu electrode on the AMB substrate was deformed and exhibited increased roughness after the TCT. This deformation can be caused by the grain boundary deformation owing to thermal stress through the TCT. This was not the direct fatigue in the sintering bond from the thermal stress; however, in the case of Cu sintered bonding, the small point-like fatigue acted as the starting point and enlarged the cracking in the Cu sintered layer.

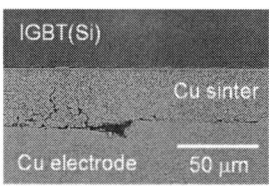 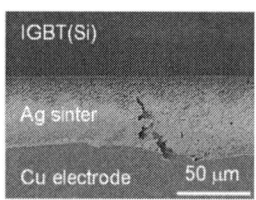

(a) Cu sintered bond (b) Ag sintered bond

Fig. 6 Cross-sectional SEM images at the white spotted area in the SAT image after the TCT for the Cu and Ag sintered bond.

Conclusions

To clarify the relationship between the bonding reliability and porosity, TCTs were carried out for the test vehicles die-bonded using the Cu sintered bonds with various porosities that originated because of different sintering pressures. The porosity in the Cu sintered bond was strongly influenced by the sintering pressure; as the sintering pressure increased from 12 to 20 MPa, denser sintered layers were obtained, with porosities decreasing from 26% to 15%.

TCTs with the condition of 200 °C / 15 min ⇔ −40 °C / 15 min revealed that the Cu sintered bond with

a porosity of 26 % failed after 500 cycles, while that with a porosity of 21 % failed after 2500 cycles. In contrast, those with porosities less than 20 % (i.e., those prepared at sintering pressures of 12, 17, and 20 MPa) could stand up to 3000 cycles. Ag sintered bond obtained at sintering pressures of 7, 12, and 17 MPa (porosities of 11 %, 10 %, and 8 %, respectively) failed after 500, 1000, and 3000 cycles, respectively. Therefore, Cu sintering pastes have superior bonding reliability compared to Ag sintering pastes for the same porosity or same sintering pressure. Cu sintered layers can allow a lower sintering pressure such as 12 MPa with sufficient bonding reliability. This is convenient for die top sintering because the small pressure can suppress chip cracking due to uneven pressure.

After the TCT, a fatigue damage in the Cu sintered bond was observed at the edge area and small point-like fatigue was observed in all the bonding areas. At the edge area, the cracking area was limited to few hundreds of micrometer in length from the chip edge, and the cracking was initiated from the interface between the IGBT and Cu sintered bond, progressing downward to the interface between the Cu sintered bond and the Cu electrode of the SiN-AMB. Damage like the small point-like fatigue originated from the peeling off at the interface between the Cu sintered bond and depressed Cu electrode of the SiN-AMB and was caused by depressed deformation at the grain boundary in the Cu electrode.

Our final goal is to clarify the relationship among the sintering conditions, porosity, physical properties, and bonding reliability. At present, we obtained the physical properties of Cu sintered strips that had different porosities. In the future, we shall apply a simulation method to establish these relationships.

References

[1] U. Scheuermann, P. Wiedl: "Low Temperature Joining Technology – A High Reliability Alternative to Solder Contacts", Workshop on Metal Ceramic Composites for Functional Applications, 4-6 June 1997, Wien, Austria, 181-192.

[2] T. Suzuki, Y. Yasuda, T. Terasaki, M. Morita, Y. Kawana, D. Ishikawa, M. Nishimura, H. Nakako, K. Kurafuchi: "Macro- and Micro-Deformation Behavior of Sintered Copper Die-attach Material", IEEE Transactions on Device and Materials Reliability, Vol. 18, Issue 1, pp. 54-63 2018.

[3] H. Nakako, M. Natori, D. Ishikawa, T. Tanaka, Y. Ejiri: Copper Sintering Pastes for Die Bonding, PCIM Europe 2021, 3–7 May 2021, Nuremberg, Germany, 453–458.

[4] J. Rudzki, F. Osterwald, M. Becker, R. Eisele: Novel Cu-bond Contacts on Sintered Metal Buffer for Power Module with Extended Capabilities, PCIM Europe 2012, 8–10 May 2012, Nuremberg, Germany, 784–791.

PCIM Europe 2024, 11– 13 June 2024, Nuremerg   DOI: 10.30420/566262006

High Thermal Durability of Thin Copper Die-attach Layers and Finite Element Model Simulation

Takaaki Eyama[1], Shuichi Inaya [1], Ukyo Suzuki[1], Taiki Fukuda[1], Takumi Miyamoto[1], and Masafumi Takesue[1]

[1] Kao Corporation, Japan

Corresponding author: Takaaki Eyama, eyama.takaaki@kao.com

Abstract

A thin but highly durable bonding layer is strongly desired for cost reduction and improving reliability in power device market. Dense Cu bonding layers having a thickness of 15 µm provided high thermal stability for a thermal cycle test (TCT) operated in 4000 cycles at between –55 and 200 °C under air, because was no significant delamination and no changed microstructure. Microstructure of the sintered Cu layers was homogenous and the porosities were 4.8%, originating in high dispersibility of the submicron Cu particles. The thermal reliability of the dense sintered Cu layers was subjected to the Coffin-Manson law, which the plastic strain amplitude calculated by a finite element model simulation was proportional to the number of cycles to failure (cycles of delamination ratio 10% over during TCT). The results can be of assistance to predict the life of the dense sintered Cu bonding layers being operated in power devices.

1 Introduction

Wide band gap (WBG) power devices such as silicon carbide (SiC), gallium nitride (GaN) provide higher current density, higher switching speed, lower switching loss, and capability to higher operating temperature than Si devices. It is said that power modules can simplify cooling systems and decrease the size and weight of WBG power devices. As WBG power devices are expected to operate at temperatures higher than 200 °C [1–6], die attach materials need to endure a harsh thermal environment for a long life. In recent two decades, the Ag sinter materials have gradually spread to the market instead of solders. However, the Ag sinter materials are expensive because of using a precious metal.

It is well known that Ag bonding layers degrade in microstructure and decrease bonding strength according to the exposure at high temperatures between 200–300 °C [7–9]. Therefore, Ag sinter materials is difficult to be widely applied to WBG power devices operated at the high temperatures. Alternatively, Cu is the most candidate for cost reduction and high durability. Recent studies say that Cu sinter materials showed better reliability than Ag sinter materials with power cycles test results and microstructure observation for Cu bonding layers [10, 11]. Unfortunately, thermal durability and thermal fatigue behavior of thin Cu bonding layers are not fully studied in previous literature.

We reported high thermal stability of a dense Cu bonding layer at PCIM Europe 2023, whereas it was in the case of a thicker bonding layer of about 80 µm [11]. Hattori et al. evaluated Cu bonding layers having a thickness of about 35 µm, but the thermal cycle tests were conducted in a narrow temperature range [12]. The dense Cu bonding layer might overcome an increased thermal stress with a thinner bonding layer. A thin Cu bonding layer also has advantages in cost reduction.

In this study, we evaluated the relationship between bonding layer thickness and thermal durability of the dense Cu bonding layers formed by the Cu sinter paste introduced at PCIM Europe 2023 [11] to verify that thin Cu bonding layers had high thermal durability. Plastic strain amplitude of the Cu layers was calculated by a finite element model simulation (FEM) to discuss a behaver of low cycles fatigue of the dense Cu bonding layers.

2 Experimental

2.1 Cu Sinter Paste

Submicron Cu particles were synthesized by a chemical method at Kao Corporation (Fig. 1a). The submicron Cu particles having an average diameter of 180 nm showed high dispersibility. Cu sinter pastes were prepared by mixing the submicron Cu particles with micro-sized Cu particles into a solvent mixture (Fig. 1b).

56

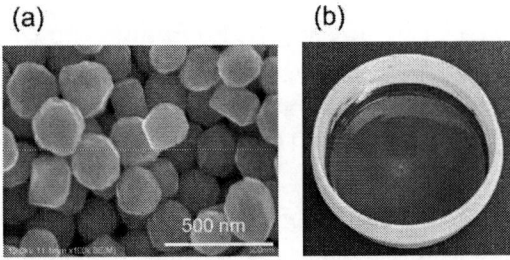

Fig. 1 (a) Submicron Cu particles, (b) Cu sinter

2.2 Cu Bonding Samples

Cu bonding samples prepared by a pressure sintering process are shown in Fig. 2. Au-metalized Si dummy dies of 5 × 5 mm (thickness, 0.4 mm) and oxygen-free Cu substrates were used for bonding tests. The oxygen-free Cu substrates of 30 × 30 mm (thickness, 1 mm) were cleaned with an aqueous solution of sulfuric acid (2 vol%) before printing. The Cu sinter paste was printed on the Cu substrate by stencil printing with a stencil mask having a hole size of 7.6 × 7.6 mm (thickness,

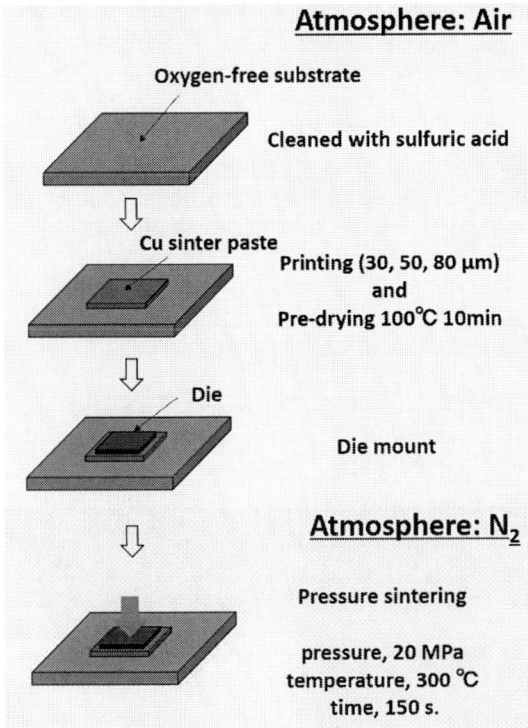

Fig. 2 Schematic of the pressure sintering process.

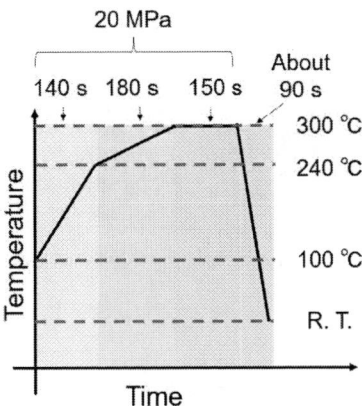

Fig. 3 Pressure sintering profile.

30, 50, and 80 µm). Pre-drying process was performed at 100 °C for 10 min under air on a hot plate. The Si die was mounted on the pre-dried Cu paste. Sintering was conducted in N_2 with a pressure sintering equipment. Sintering conditions were as follows: pressure, 20 MPa; temperature, 300 °C; time, 150 s. Pressure sintering profile is shown in Fig. 3. Bonding layer thickness and porosity after sintering were measured by a cross-section scanning electron microscopy (SEM).

2.3 Thermal Cycle Tests and Calculation of Delamination Ratio

TCT of the Cu bonding samples was performed at a temperature range of between –55 and 200 °C (each hold time, 15 min) under air for 4000 cycles. Delamination in the bonding area was examined by scanning acoustic tomography (SAT) using a 50 MHz probe. The delamination ratio was calculated from the delaminated area to the die area in the SAT image every 250 cycles during TCT. Porosities of the Cu bonding layers after 4000 cycles of TCT were measured with cross-sectional SEM images.

2.4 Finite Element Model Simulation of Cu Bonding Samples

FEM calculations were performed to evaluation the low cycle fatigue tendency under thermal cycle with the sintered Cu. The FEM solver used was a COMSOL Multiphysics ver. 6.2 software. The calculation model for the Cu bonding samples cut diagonally and view from the side is shown in Fig. 4. Parameters of the mechanical properties in the sintered Cu for the FEM calculations were obtained by stress-strain curves, whereas those of

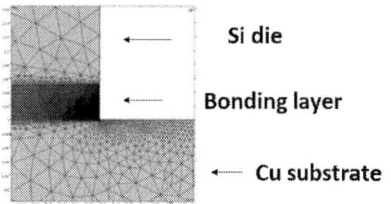

Fig. 4 Calculation model for FEM of Cu bonding samples.

the bulk materials were used for the Si die and Cu substrate. The plastic strain amplitude between to –55 to 200 °C were calculated with an assumption that the thermal stress was free at 200 °C (the maximum temperature of the TCT).

3   Results and Discussion

3.1   Cross-section SEM Evaluation of Cu Bonding Samples in different thicknesses

Cross-sectional SEM images of the Cu bonding samples in different thicknesses are shown in Fig. 5. The images showed that the Cu sinter layer was well-bonded without significant defects onto the Au back metal and the Cu substrate. Porosities of the bonding layers in thicknesses of 15, 30, and 50 µm (after sintering) were 4.8%, 5.6%, and 5.9%, respectively. It can be considered that the Cu sinter paste formed dense Cu bonding layers at a wide bonding thickness range of at least between 15 and 50 µm regardless of the same sintering conditions.

3.2   Thermal Cycle Tests and Calculation of Delamination Ratio Evaluation of Cu Bonding Samples

SAT images of the bonding samples in different thicknesses are shown in Fig. 6. The SAT images showed no significant delamination and no voids at initial states because the corners of the bonding areas clearly appeared. As an increase in cycle number, the bonding areas gradually decreased due to a little delamination from the four corners, but it showed no significant delamination and voids until 4000 cycles. Cross-sectional SEM images (Fig. 7) also showed that the Cu bonding layers after 4000 cycles had no significant delamination. The porosities in thicknesses of 15, 30, and 50 µm were 5.0%, 5.2%, and 5.4%, respectively, indicating that the microstructure showed no significant change and no degradation during TCT. Therefore, it can be considered that the thin dense Cu bonding layer endured the harsh thermal environments and showed high thermal durability. The delamination ratio verses the number of thicknesses, the delamination ratios of the Cu bonding layers gradually increased. Finally, the delamination ratios of the Cu layers in thicknesses of 15, 30, and 50 µm were 15.0%, 13.3%, and 10.5% after TCT, respectively. The delamination ratio tended to increase as thickness of bonding layer thermal cycles is shown in Fig. 8. Regardless of decreased.

3.3   Examine of Low Cycle Fatigue of Dense Cu Bonding Layer by FEM

It is important to examine the thermal fatigue behavior of the Cu bonding layer. In this study, the results of TCT were used for evaluation of thermal

Magnification	×700		
Initial			
Bonding layer Thickness (µm)	15	30	50
Porosity (%)	4.8	5.6	5.9

Fig. 5   Cross-sectional SEM images of the Cu bonding layers in different bonding layer thicknesses.

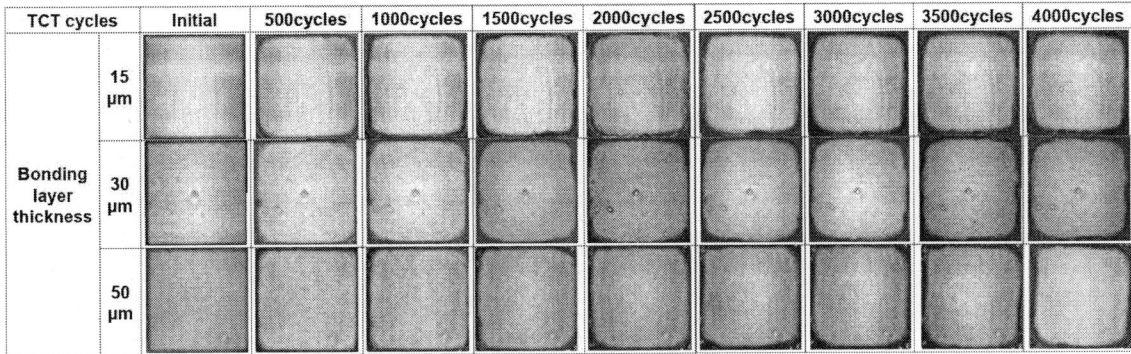

Fig. 6 SAT images of the Cu bonding samples in different bonding layer thicknesses during TCT.

Magnification	×700		
After TCT	Si / Cu bonding layer / Cu / 50μm	Si / Cu bonding layer / Cu / 50μm	Si / Cu bonding layer / Cu / 50μm
Bonding layer Thickness (μm)	15	30	50
Porosity (%)	5.0	5.2	5.4

Fig. 7 Cross-sectional SEM images of the Cu bonding layers in different bonding layer thicknesses after TCT.

fatigue of the dense Cu bonding layer. The strain-life low cycle fatigue equation is generally expressed as the Coffin-Manson law (Eq. 1) [14, 15],

$$\Delta \varepsilon_p \cdot N_f{}^a = C \qquad (1)$$

where $\Delta \varepsilon_p$ is plastic strain amplitude, N_f is the number of cycles to failure, α is an empirical constant known as the fatigue ductility exponent, and C is an empirical constant known as the fatigue ductility coefficient. The $\Delta \varepsilon_p$ values of the Cu bonding layers in different thicknesses were calculated by FEM (Table 2). The $\Delta \varepsilon_p$ increased as Cu bonding layer thickness decreased. In this study, N_f was defined as a cycle number reaching a delamination ratio of 10%. The N_f of the bonding layers in thicknesses of 15, 30, and 50 μm were 2500, 3250, and 3750, respectively. The $\Delta \varepsilon_p$ verses N_f is shown in Fig. 9. Low cycle fatigue of the dense Cu bonding layer was subjected to the Coffin-Manson law ($\alpha = 0.48$, $C = 0.80$). The results can be of

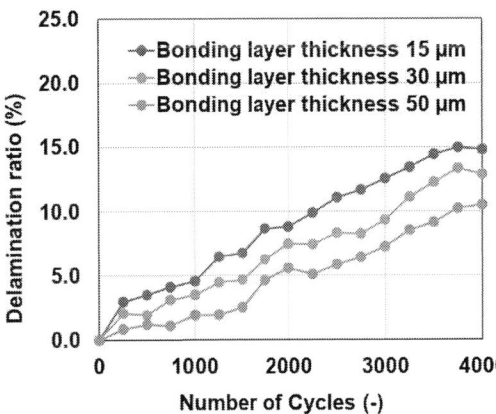

Fig. 8 Relationship between the number of cycles in TCT and delamination ratio of the Cu layers in different thicknesses.

Table 2 Plastic strain amplitude of Cu bonding layers.

Bonding layer thickness (µm)	$\Delta\varepsilon_p$
15	0.0188
30	0.0170
50	0.0154

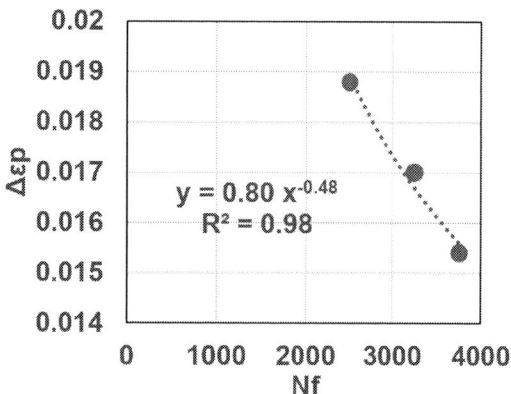

Fig. 9 $\Delta\varepsilon_p$ verses Nf of Cu bonding layers in different thicknesses.

assistance to predict the life of the dense sintered Cu bonding layers being operated in power devices.

4 Summary

We evaluated the thermal durability of the thin and dense Cu bonding layers formed with Cu submicron particles having high dispersibility. The dense Cu bonding layers had low porosities under 6%. The thickness of 15 µm in Cu bonding layer showed no significant delamination and no change in microstructure during TCT until 4000 cycles at between −55 and 200 °C in air. The thermal fatigue behavior of the dense Cu bonding layer was examined by plastic strain amplitude calculated by FEM. The low cycle fatigue of Cu bonding layer was subjected to the Coffin-Manson law (α = 0.48, C = 0.80). The high thermal durability of the thin Cu bonding layer having dense microstructure would be assistance to cost reduction and improving reliability for power devices.

References

[1] Y. Gao, S. Takata, C. Chen, S. Nagao, K. Sugamuma, A Sajjad Bahman, F. Iannuzzo, Reliability analysis of sintered Cu joints for SiC power devices under thermal shock condition, Microelectronics Reliability, 100-101, 2019, 113456.

[2] N. Murayama, K. Hirao, M. Sando, T. Tsuchiya, H. Yamaguchi, High-temperature electro-ceramics and their application to SiC power modules, 44(4), 2018, 3523-3530.

[3] M. J. Palmer, R. W. Johnson, T. Autry, R. Aguirre, V. Lee, J. D. Scofield, Silicon Carbide Power Modules for High-Temperature Applications, IEEE TRANSACTIONS ON COMPONENTS, PACKAGING AND MANUFACTURING TECHNOLOGY, 2(2), 2012, 208-216.

[4] C. Chen, F. Luo, Y. Kang, A Review of SiC Power Module Packaging: Layout, Material System and Integration, CPSS TRANSACTIONS ON POWER ELECTRONICS AND APPLICATIONS, 2 (3), 2017, 170-186.

[5] F. Roccaforte, P. Fiorenza, G. Greco, M. Vivona, R. Lo Nigro, F. giannazzo A. Patti, M. Saggio, Recent advances on dielectrics technology for SiC and GaN power devices, Applied Surface Science, 301, 2014, 9-18.

[6] K. Matocha, Challenges in SiC power MOSFET design, Solid-State Electronics, 52, 2008, 1631-1635.

[7] S. Yan Zhao, X. Li, Y. Hui Mei, G. Quan Lu, Study on high temperature bonding reliability of sintered nano-silver joint on bare copper plate, Microelectronics Reliability, 55, 2015, 2524-2531.

[8] T. Watanabe, M. Takesue, T. Matsuda, T. Sano, A. Hirose, Thermal stability and characteristic properties of pressureless sintered Ag layers formed with Ag nanoparticles for power device applications, Journal of Materials Science: Materials in Electronics, 31, 2020, 17173-17182.

[9] F. Yang, W. Zhu, W. Wu, H. Ji, C. Hang, M. Li, Microstructural evolution and degradation mechanism of SiC–Cu chip attachment using sintered nano-Ag paste during high-temperature ageing, Journal of Alloys and Compounds, 846(15), 2020, 156442.

[10] H. Nakako, M. Natori, D. Ichikawa, T. Tanaka, Y. Ejiri, Copper sintering pastes for die bonding, PCIM Europe 2021, 3-7 May 2021, Nuremberg, Germany, 453-458.

[11] T. Eyama, S. Inaya, U. Suzuki, M. Takesue, Crystallographic Examination of High Thermal Stability of Dense Sintered Copper Layer,

PCIM Europe 2023, 9-11 May 2023, Nuremberg, Germany.

[12] T. Hattori, S. Yamauchi, K. Anai, Bonding Properties and Reliability Evaluation of Cu Sinter Paste for Pressure Sintering, PCIM Europe 2023, 9-11 May 2023, Nuremberg, Germany.

[13] E. A. Brandes, G. B. Brook, Smithells Metals Reference Book, Butterworths Heinemann, England, 7, 1992, 22-28.

[14] L. F. Coffin Jr., A Study of the Effects of Cyclic Thermal Stresses on a Ductile Metal, Trans. ASME, 76(6), 1954, 931-949.

[15] Manson S. S., Behavior of Materials Under Conditions of Thermal Stress, NACA-TN-2933, 1953.

PCIM Europe 2024, 11– 13 June 2024, Nuremberg DOI: 10.30420/566262007

Thermal Shock Test Lifetime Improvement with Optimized Adhesion Strength between Epoxy Resin and Copper

Kangjia He[1], Yukihiro Kumagai[1], Seiichi Hayakawa[1], Kan Yasui[1], Osamu Ikeda[2], Kisho Ashida[2] and Takayuki Kushima[1]

[1] Hitachi Power Semiconductor Device, Ltd., Japan
[2] Hitachi Ltd., Japan

Corresponding author: Kangjia He, koka.ka.sr@hitachi.com
Speaker: Kangjia He, koka.ka.sr@hitachi.com

Abstract

In this paper, we examined and verified an improvement of thermal shock test lifetime in epoxy resin encapsulated power semiconductor modules by improving the adhesion reliability between the epoxy resin and the copper plates of the insulating substrate. When power modules encapsulated with epoxy resin are subjected to thermal shock test (TST), significant thermal stress is generated due to the difference in the thermal expansion coefficient of each component inside the power module. As a result, the epoxy resin delaminates from copper plates of a insulation substrate. Delamination of the epoxy resin from the copper gradually increases with the number of TST cycles, which adversely affects the reliability of the power module. This work confirmed that the adhesion strength with the epoxy resin can be doubled by forming an appropriate oxide film on the copper plates. Power modules with oxidized copper showed that the epoxy resin delamination occurred area after the TST target cycle was about 70% less than that of the module without the oxide film. Finally, the TST lifetime of the power module was enhanced to more than TST 1,000 cycles.

1 Introduction

With the development of electric vehicles (EVs), there is a growing interest in power module technology that can efficiently control power conversion. To protect the internal connections of power modules from mechanical or chemical stress, it is important to develop packaging technology for power modules. In recent years, epoxy resin has been used as encapsulants to meet the requirements of high temperature operation, high shock and vibration resistance, high current density, and high insulation reliability [1]. On the other hand, the epoxy resin is subject to more pronounced mechanical stress

concentration due to thermal history during the process and thermal expansion and contraction caused by temperature cycling than silicone gel, which was used in conventional power modules for EVs as encapsulant material. As power modules are required to be smaller, to have higher current capability, and to be cappable of operating at higher temperature in the future, it is necessary to develop the technology to ensure the reliability of the package.

With these requirements, we have developed a 6-in-1 IGBT module for EVs (Fig. 1) combined with epoxy resin encapsulation.

2 Delamination Issue between Epoxy Resin and Copper

2.1 Confirmed Issue in Thermal Shock Test

Compared to the silicone gel, the epoxy resin has advantages such as higher temperature stability, insulation durability, and moisture resistance. However, it is hard due to its large elastic modulus,

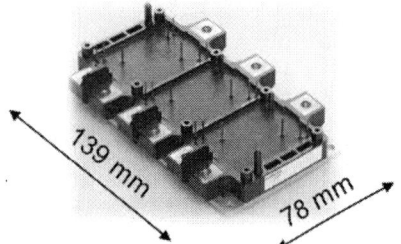

Fig. 1 Developed power module with epoxy resin encapsulation (6-in-1 IGBT module).

and stress concentration tends to occur at the interface of different materials. As mentioned above, package reliability of power modules is extremely important. Thermal shock test is one of the most important reliability tests for the package reliability of power modules. In this development, TST conditions in accordance with the AQG324 standard were applied (Table 1) [2]. Fig. 2 shows a cross-sectional images of the internal structure of the power module used in this verification. SAT (Scanning Acoustic Tomography) was also used as a means of checking the condition of each connection inside the power modules. Fig. 3 shows the SAT results of the interface between the epoxy resin and the copper plate on a portion of the power module's AMB substrate. The white-colored area shown in Fig. 3 is the area where delamination occurred. The epoxy resin delamination occurred from the edge of the copper of AMB substrate (Fig. 4), and the area where delamination occurred expanded with the number of TST cycle (Fig. 3). When the epoxy resin delamination occurs over a large area, electrical failures and solder cracks, etc. may happen incidentally, leading to reduced reliability of power modules. It is necessary to suppres the epoxy resin delamination to improve TST lifetime.

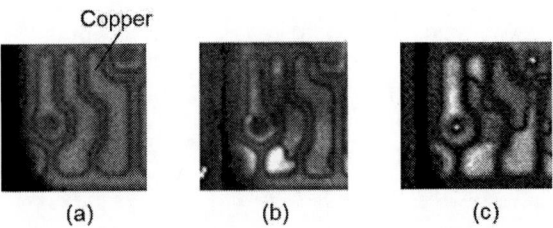

Fig. 3 SAT results of (a)TST 0 cycle, (b)TST 500 cycles, (c)TST 1,000 cycles from top side view of portion of AMB substrate.

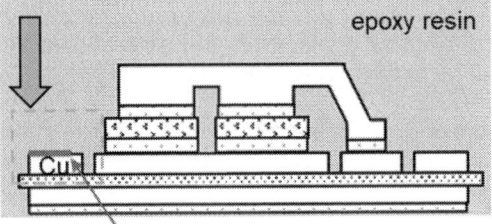

Fig. 4 Location of SAT results on AMB substrate from cross-sectional image.

Table 1 Parameters of TST (AQG324 standard).

Parameter	Symbol	Condition
Lowest value of the storage temperature	$T_{stg,min}$	-40 $°C^{0}_{-10}$
Highest value of the storage temperature	$T_{stg,max}$	+125 $°C^{+15}_{0}$
Minimum dwell time for highest/lowest temperature	t_{dwell}	> 15 min
Minimum number of cycles without failures	Nc	> 1,000

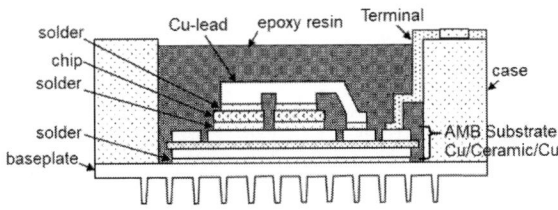

Fig. 2 Cross-sectional images of power module.

2.2 Study of Adhesion Strength Improvement

Methods to improve the adhesion reliability between the epoxy resin and the copper of AMB substrate were investigated. There has been a method to improve adhesion strength by roughening copper plating film for a high surface area ratio film on the copper surface to achieve a physical anchor [3]. In addition, an oxide layer formed on copper surface by 85℃/85% relative humidity test may improve adhesion strength due to the anchor caused by contact with the ferrules in epoxy resin [4]. On the other hand, there is a report that hydrogen bonding is the main factor for improving the adhesion strength of the metal-epoxy interface based on molecular theoretical analysis of the adhesion mechanism using quantum scientific calculations [5]. As described above, studies have been conducted to improve the adhesion strength between the copper and the epoxy resin, which is one of the most important issues in the encapsulation technology of power modules. In this paper, two methods were investigated to improve the adhesion strength between the epoxy resin and the copper plates: a formation of an appropriate oxide film on the copper surface of AMB substrate and an addition of an adhesion promoter to the epoxy resin. The intentional formation of an oxide film on the copper is expected to promote hydrogen bonding with the epoxy resin and to have an anchor effect by increasing the surface area ratio. Additives for the epoxy resin is expected to improve its interaction with copper, thereby discovering adhesion strength.

3 Approach

3.1 Shear Test by Pudding Cup Resin

To confirm the effect of the improvement, a pudding cup-shaped epoxy resin was formed on the copper plate to which the improvement was applied, and the shear strength was measured by a shear test, which was called the pudding cup test. An illustration of the test is shown in Fig. 5. The copper oxide film is formed by heat treatment. For the oxidation conditions used for verification, samples were prepared by applying oxidation treatment conditions with fixed temperature and divided implementation time (0.5h, 1h, 4h, 7h). Samples were also prepared using an existing

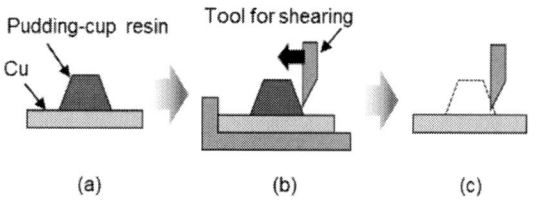

Fig. 5 Images of Pudding Cup Test: (a) Prepare sample, (b) Set sample for test, (c) Finished test.

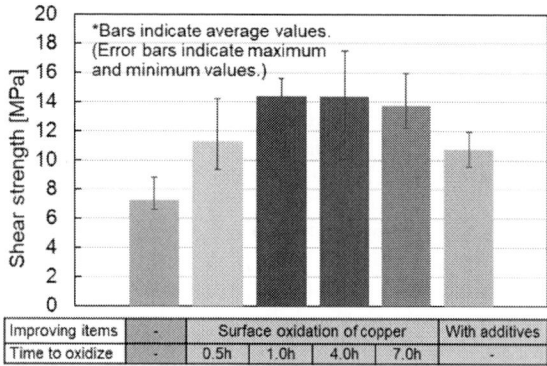

Fig. 6 Results of Pudding Cup Test.

epoxy resin filled with an adhesion strength promoter. The results of the pudding cup test are shown in Fig. 6, and the conditions of samples before and after test (representative samples) are shown in Table 2. All the improvement plans had higher adhesion strength with the epoxy resin than the reference sample. Also, the fracture mode was confirmed to be an interface fracture because the fracture interface was between the resin and the copper from the appearance after test. Of these, the adhesion strength of the oxidized samples stabilized after 1 hour oxidation treatment and was twice as strong as that before the countermeasure. The adhesion strength of the additives sample was inferior to that of the oxidized samples and was approximately 1.5 times stronger than that of the reference sample. Based on the results of the pudding cup test, we selected oxidation treatment as the improvement plan and proceeded to the next study.

Table 2 Sample before and after Pudding Cup Test (representative sample).

Improving item	Before Test	After Test
None		
Surface oxidation of copper		
With additives		

3.2 Investigation of Oxide Film on Copper Surface

It is assumed that the formation of an oxide film on the surface of copper contributes to the improvement of adhesion strength with the epoxy resin in this verification. To clarify the relationship between the formed oxide film and the adhesion strength, the thickness of the oxide film formed on the copper surface was measured. To measure the oxide film thickness, we used SERA (Sequential Electrochemical Reduction Analysis), which is a method for measuring specific metallic films. Fig. 7 shows the relationship between the thickness of the oxide film formed under each oxidation condition and the adhesion strength measured in the pudding cup test. Although the thickness of the oxide film formed on the copper surface grew with time of oxidation treatment, we estimate that the oxide film expected to improve adhesion strength was saturated in about 1 hour oxidation treatment time. Since the adhesion strength was at the same level within the range of 1 to 7 hours of oxidation treatment time confirmed, we consider that the minimum oxide film that can be expected to improve the adhesion strength between the epoxy resin and the copper surface is 16 nm. In addition, the adhesion strength does not change significantly as the thickness of oxide film, and from the appearance after pudding cup test shown in Table 2, it is inferred that the adhesion strength is contributed near the interface between the epoxy resin and the oxide film.

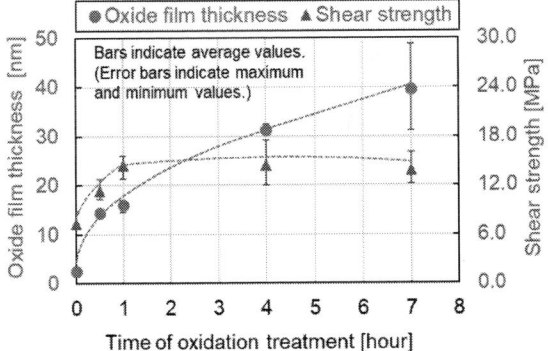

Fig. 7 Correlation between oxide thickness and shear strength.

3.3 Observation of bonding interface

Cross-sectional observation by SEM (Scanning Electron Microscope) was conducted to investigate the mechanism of improvement of adhesion strength with the epoxy resin by the oxide film formed on the copper surface. A module sample was used in which the oxide film was formed on the copper of the AMB substrates with the epoxy resin encapsulation. The results of the cross-sectional observation are shown in Fig. 8. At the interface between the epoxy resin and the copper, no mechanical irregularities that could be described as anchor were observed. Therefore, it is inferred that the dominant factor contributing to the improvement in adhesion strength with the epoxy resin is the hydrogen bonding between the OH groups in the epoxy resin and Cu-O in the copper oxide film. In addition, we believe that the intentional formation of the oxide film promoted hydrogen bonding.

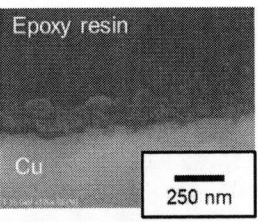

Fig. 8 Cross-sectional SEM images of interface between oxidized copper and epoxy resin.

3.4 Confirmation in Power Module

The oxidation treatment conditions that can improve the adhesion strength with the epoxy resin were applied to the power module samples, and the delamination status and progress of the epoxy resin and the copper were confirmed by TST (Test conditions are in Table 1). The delamination inside the module was confirmed by SAT. To obtain clearer SAT images, the epoxy resin on the top side of the power modules were thinned down to a thickness suitable for SAT by machining, which we called experimental sample. To evaluate the delamination phenomenon of the epoxy resin more quantitatively, a delamination area rate was defined. The delamination rate is a ratio of the area of delamination observed on the copper on the AMB substrates to the total area of the copper on the power module. The power modules used for the verification were those to which oxidation was applied, which are expected to improve adhesion strength, in addition to those to which without oxidation treatment, which were used as a reference. TST ⇔ SAT observation / calculation of delamination area rate was repeated from TST 0 cycle to TST 1,000 cycles (some samples were up to TST 2,000 cycles). The results of the epoxy resin delamination transition were shown in Fig. 9. The oxidation-applied samples in Fig. 9 include samples with oxidation treatment time conditions of 1 to 7 hours, which were confirmed to have the same level of adhesion strength. The delamination area rate of the oxidation-applied samples were found to be 70% lower than that of the reference sample. In fact, we also confirmed that the power module with oxidation treatment applied passed the TST without any defects (F.R.=0/6) after 1,000 TST cycles, which is a reliability test with condition specified in AQG324. SAT observation using the reliability test sample after TST 1,000 cycles confirmed that the epoxy resin delamination inside the power module was at the same level as the experimental samples, with a slight delamination of approximately 3%.

The above results show that, in the power module with the oxidation-applied, the adhesion strength between the epoxy resin and the copper of AMB substrate is improved, which suppressed the occurrence of delamination. This means that the occurrence of stress concentration due to delamination of epoxy resin is reduced. The verification above envices that the improvement in thermal shock reliability of power module due to copper surface oxidation treatment of AMB substrate has been verified.

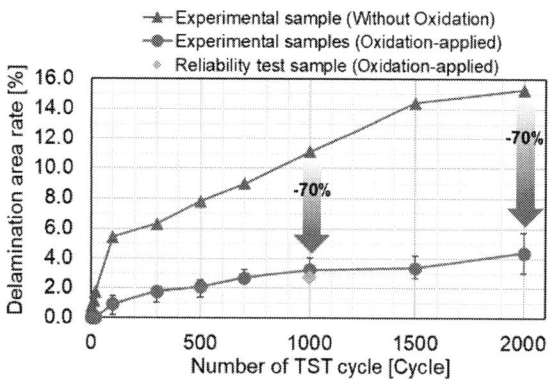

Fig. 9 Delamination condition after TST in power modules.

4 Conclusion

In this paper, we focused on the adhesion reliability between epoxy resin and copper of AMB substrate to improve the thermal shock durability of power module. We confirmed that forming oxide film on the copper surface before epoxy resin encapsulation is an effective method to improve the adhesion strength. The relationship between the oxide thickness formed on the copper surface and the adhesion strength of the epoxy resin was clarified. The formation of an oxide film of at least 16 nm is expected to improve the adhesion strength. The dominant factor for the improvement in adhesion strength was concluded to be hydrogen bonding rather than an anchor by the confirmed of SEM of the interface between the copper and the epoxy resin. Modules with an appropriate oxide film formed on the surface of copper plate showed that epoxy resin delamination was suppressed up to 2,000 TST cycles. The power module used in this verification passed TST, which satisfied the AQG324 standard.

References

[1] Y. Takeuchi, M. Kamikawa, Y. Kumagai, T. Wada, T. Tanimura, T. Ouchi, H. Tsuruoka, S. Hayakawa, T. Kushima, K. Hisada, Y. Kato, M. Shiraishi, T. Oda, "A New High Power Density 6-in-1 IGBT Module Enabling Acceleration of Vehicle Electrification", PCIM Europe 2022, pp.131-136.

[2] ECPE Working Group, ECPE Guideline AQG324, 2021

[3] R. Naka, J. Yoshikawa, T. Ishihara, K. Oono, "Effect of Roughened Copper Plating Film on Adhesion with Mold Resin", MES 2023, pp.155-158.

[4] S. Zhao, C. Chen, M. Ueshima, M. Haga, H. Suzuki, H. Takenaka, K. Suganuma, "An abnormal phenomenon observed for copper/epoxy bonding after 85℃/85% relative humidity test and its possibile mechanism", MES 2023, pp. 429-432.

[5] F. Ohsako, K. Yoshizawa, "Molecular Theory of Adhesion of Metal/Epoxy Resin Interface", J-STAGE, Kobunshi Ronbunshu 2011, Volume 68, Issue 2, pp. 72-80.

PCIM Europe 2024, 11– 13 June 2024, Nuremberg   DOI: 10.30420/566262008

Power Cycling Reliability and Failure Mode Analysis of POL

Kenichi Koi [1], Jumpei Tokutake [1], Koji Bando [1]

[1] SHINKO ELECTRIC INDUSTRIES CO., LTD., JAPAN

Corresponding author: Kenichi Koi, ke.koi@shinko.co.jp
Speaker: Kenichi Koi, ke.koi@shinko.co.jp

Abstract

Power Overlay (POL) is a power semiconductor package that utilizes advantages of wide bandgap semiconductors. POL has Copper (Cu) wiring formed on a polyimide substrate and Cu vias joining the top-side of devices. In this paper, we investigated the reliability results and failure mode analysis of POL in a power cycling test. Further, we studied the correlation between remaining joint area and reliability on power cycling test using finite element method (FEM) simulation. The failure mode of POL is crack between Cu vias and electrodes of top side of power devices. We have found that even if the Cu via joining area was less than 40%, the electrical and thermal resistance increase were minor. The FEM results also showed that the characteristics degradation in a power cycling test are minor even if the Cu via joining area is reduced, and that the solder joining area on the bottom side of the devices is dominant. These results demonstrate that Cu via joining, a feature of POL, have high power cycling reliability.

1   Introduction

To reduce the size of power converters and to increase their power conversion efficiency, wide bandgap semiconductors such as silicon carbide (SiC) and gallium nitride (GaN) are increasingly being applied to power converters. However, power modules with conventional structures such as aluminum wire bond modules cannot utilize full advantages of the characteristics of those devices. Furthermore, lift-off of aluminum wire bonds is a reliability concern [1]. POL is a power semiconductor package structure with low inductance and high reliability, with Cu wiring formed on a polyimide substrate and Cu vias joining the topside of devices. POLs with power converters can be expected to make power converters smaller and more efficient [2]. We had demonstrated that POL passed thermal cycling and high temperature storage tests [3]. POL passed 300,000 cycles in power cycling test with ΔT_j=90 deg.C, while aluminum wire bond power modules failed at less than 10,000 cycles [4].

We evaluated the power cycling reliability of POL up to 300,000 cycles with ΔT_j=90 deg.C. The POL sample mounts SiC diodes with Cu vias. The back side of SiC diodes are joined to an active metal brazing copper substrate (AMB substrate) using solder. Heat generated from SiC diodes dissipates to the cooling plate through the AMB substrate. All samples did not show characteristic degradation that would indicate failure even after 300,000 power cycling. After power cycling test, we evaluated   solder layer at bottom side of SiC diodes and Cu vias at top side of SiC diodes using scanning acoustic tomograph (SAT) images. In SAT images, we found solder joining areas on were reduced about 76-98 %, and the Cu via joining areas were reduced about 41-45 %. From these results, we have concluded that reducing the POL's Cu via joining area has the even minor effect on the power cycling reliability.

Further, we performed FEM simulation to verify the validity of degradations in characteristic values after the power cycling test. We used FEM simulation models that changed the solder joining area on the bottom side of SiC diodes and Cu vias joining area on the top side of SiC diodes and verified the effect of reducing the joining area. From FEM simulation results, we confirmed that reducing Cu via joining area to 20% has the even minor effect on power cycling reliability. On the other hand, solder jointing areas reduced to 60%, power cycling reliability is compromised due to reduction of heat dissipation path.

From these results, the power cycling failure mode of POL is the reduction of the Cu via joining area on the top side of the devices, but even if the Cu via joining area is significantly reduced, the effect of power cycling reliability is minor.

2 POL structure

Fig.1 shows Cu re-wiring side of POL and Fig.2 shows component side of POL. Multiple alumina spacer, SiC diodes, test chips and Cu spacer are mounted on a POL. Those chips are joined to Cu wiring formed on the polyimide (PI) substrate through Cu vias.

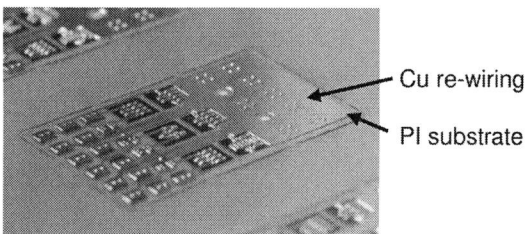

Fig. 1   Cu re-wiring side of POL

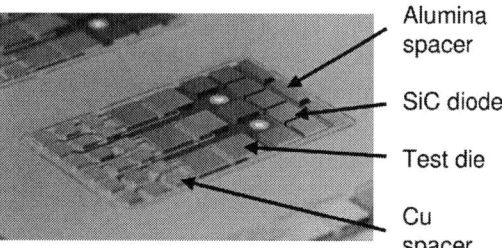

Fig. 2   Component side of POL

Fig.3 shows POL layer structure. A Cu re-wiring layer was formed on the 50μm thick polyimide substrate and was directly connected to dies through the polyimide film with Cu vias. 100μm thick Cu trace re-wiring layer was formed by subtractive process and the Ti (Titanium) /Cu seed metal layer was deposited by sputtering.

From these structures, POL achieves lower Inductance and lower impedance and high heat dissipation and withstand higher voltage.

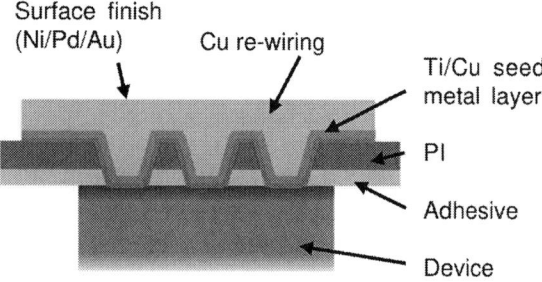

Fig. 3   POL layer structure

3 Power cycling test

3.1 Test sample structure

Fig.4. shows the power cycling sample structure. A POL mounts 3mm square 8 SiC diodes in parallel with 750μm diameter 4 cu vias. An AMB substrate is joined at bottom side of diodes with lead-free solder. Then, a POL is encapsulated with epoxy resin and joined busbars with lead-free solder. In this evaluation, the samples were cooled from the back side of the AMB substrate during the power cycling test.

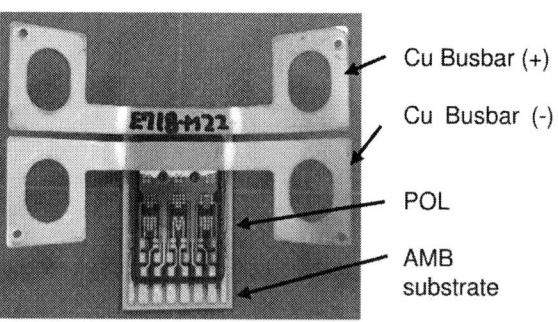

Fig. 4   Power cycling sample

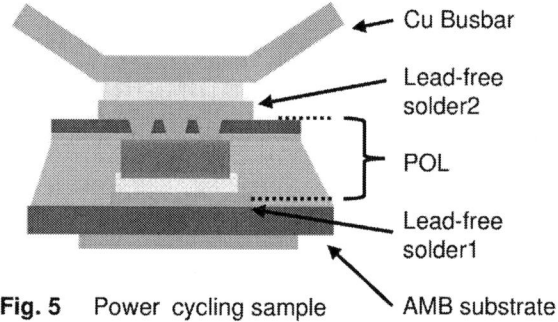

Fig. 5   Power cycling sample

3.2 Test condition

Table 1 shows power cycling test conditions in this study. We evaluated power cycling reliability up to 300,000 cycles with $\Delta Tj=90$ deg.c on 3 same samples. To evaluate the reliability of the POL that join the SiC diode, the ON/OFF time is 1 second ON and 2 seconds OFF. The Tj values for the evaluated samples were applied in accordance with JEDEC51-14 [5], corrected by Square Root Fitting. For each sample, the current load values were adjusted so that $\Delta Tj= 90$ deg.c from the 1-101 cycles, and a constant current load values were applied after the 101 cycles.

As shown in Fig. 5, evaluation samples are fixed on the cooling plate by the pressure mechanism of the power cycling device.

Table 1 Power cycling test conditions

Conditions	Unit	Value
ΔTj	deg.C	90
Tj min.	deg.C	40
Tj max.	deg.C	130
Load current	A	142-147 (constant)
On/Off	s	2/1
Cycling number	-	300,000

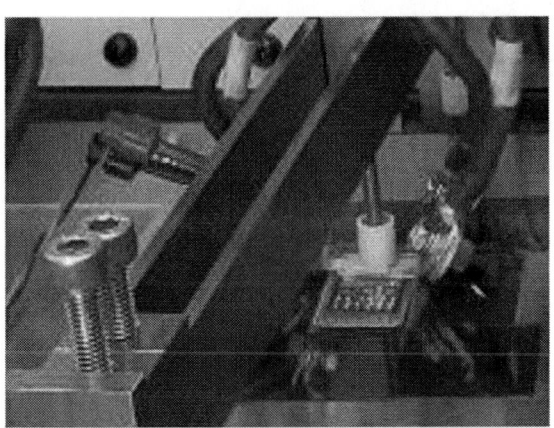

Fig. 5 Power cycling test view

3.3 Test results

Fig.6-7 show results of ΔTj and Von by power cycling test. All samples showed small ΔTj increase (2.47- 4.28 deg.C) and Von increase (0.004-0.024 V) after 300,000 cycles.

After power cycling test, we evaluated joining interfaces of the lead-free solder 1 and the POL Cu via joining interfaces using scanning acoustic tomograph (SAT) images. By comparing with the virgin sample, white areas in the SAT images were observed in the power cycling samples, which are presumed to be delamination or cracks (Fig. 8-9).

To confirm that the white parts in the SAT images represent delamination or crack, we compared cross sectional images of copper via joining with SAT images (Fig.9-12). From cross-sectional images, crack and delamination were observed at

the interface between adhesive & aluminum electrode and Cu via & aluminum electrode, corresponding to the white areas in the SAT image.

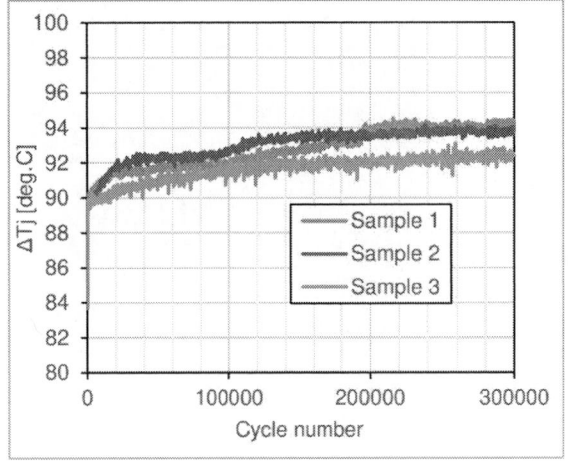

Fig. 6 Result of ΔTj by power cycling test

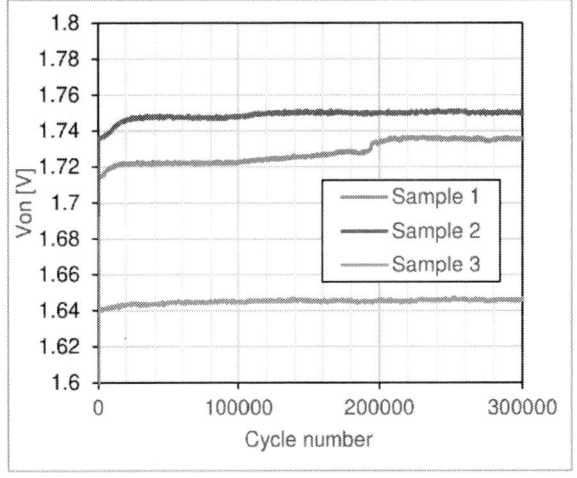

Fig. 7 Result of Von by power cycling test

From these results, correlation between the white part in the SAT images and the cracks and delamination in the cross-sectional images, it shows possible to evaluate via joints from SAT observations.

To evaluate the joining area, we used image processing using binarization processing from the SAT images. The joining areas remaining of Lead-free solder 1 were 76.3 to 98.5 % and the joining area remining of Cu via was 41.4 to 45.4 % with after power cycling samples (Table 2).

These results show that even if the POL's Cu via joining areas remaining are by about 40 %, the effect on electrical and thermal resistance are small and does not affect the reliability of power cycling.

Sample	SAT images
Sample 1	
Sample 2	
Sample 3	
Untested Sample	

Fig. 8 SAT images of lead-free solder 1

Sample	SAT images
Sample 1	
Sample 2	
Sample 3	
Untested sample	

Fig. 9 SAT images of Cu Vias

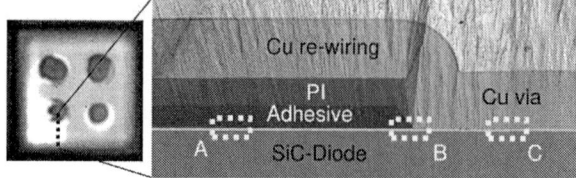

Fig. 9 Cross-sectional image of the white part in the SAT image

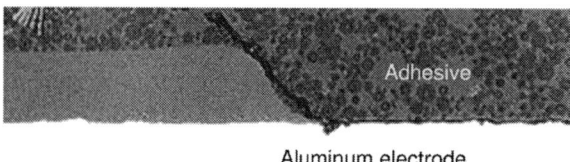

Fig. 10 Cross-sectional zoomed image of A in Fig. 9

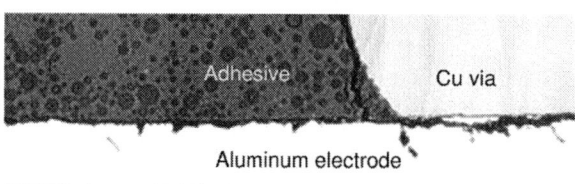

Fig. 11 Cross-sectional zoomed image of B in Fig. 9

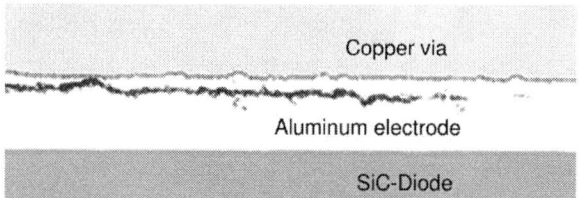

Fig. 12 Cross-sectional zoomed image of C in Fig. 9

Table 2 Joint area remaining and ΔVon, ΔTj increase value after power cycling

Sample No.	Lead-free solder1	Cu via	ΔVon [V]	ΔTj [deg.C]
1	76.3 %	43.9 %	0.024	4.28
2	84.6 %	45.5 %	0.019	3.61
3	98.5 %	41.4 %	0.004	2.47

4 FEM simulation

4.1 Simulation model

To assess the validity of characteristic degradations values after the power cycling test using FEM simulation, electrical-thermal-stress coupled simulation was performed using ABAQUS 2022.HF6. Electrical characteristics of SiC diodes were input into 3D model of devices to model the heat generated by the device due to load current. Delamination and cracks of vias and solder in actual samples were reflected in the simulation model by creating a void layer at each boundary. Cu busbars in the actual sample were omitted from the 3D model in order to simply consider the simulation results in this study.

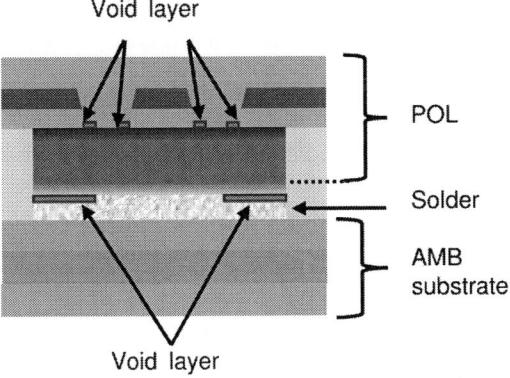

Fig. 13 Simulation model with void layer

4.2 Simulation results

To evaluate the Von and ΔTj characteristic transition, the current load of 140 A was imputed to the simulation model. The initial temperature was 40 deg.C, which is the same as Tjmin in the power cycling test condition. Fig.13-14 show the simulation results of Von and ΔTj using 3D models in which the via and solder joining area has been changed. These results showed that if only the Cu via joining areas were reduced, the characteristics degradation was minor, but if only the solder joint areas were reduced, the characteristics degradation was major. When the solder joining areas and Cu via joining areas decreased, the characteristics degraded to the same extent as when only the solder joint area was reduced.

From these results, we confirmed the correlation between the power cycling test result and the simulation result that the characteristics degradation in the actual sample was due to the reduce of the solder joining area.

Model	Joint area remaining [%]	Von results [V]
A	Via:100 Solder:100	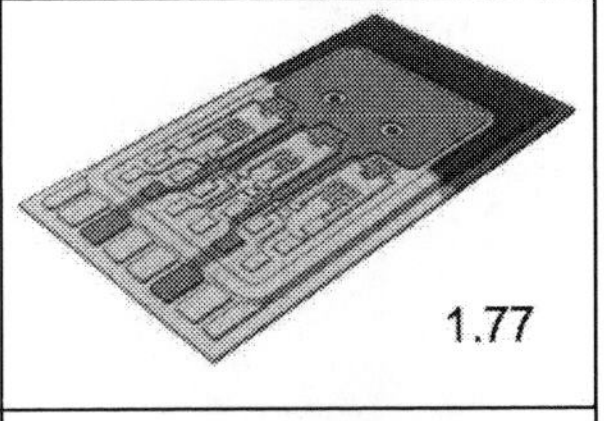 1.77
B	Via:20 Solder:100	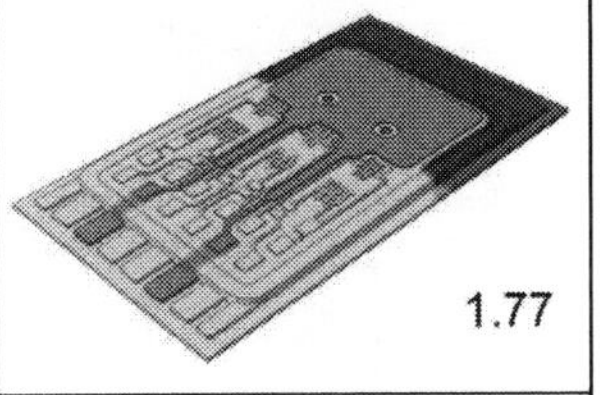 1.77
C	Via:60 Solder:100	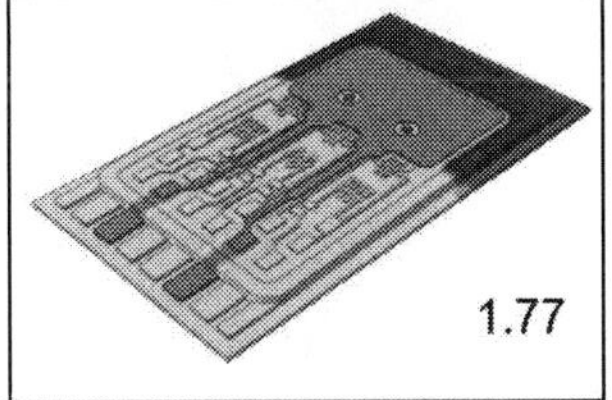 1.77
D	Via:100 Solder:20	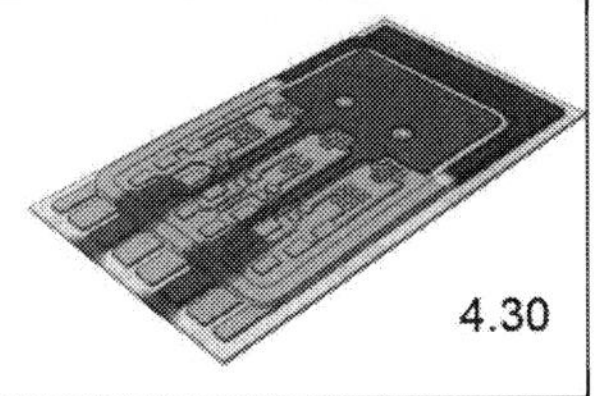 4.30
E	Via:100 Solder:60	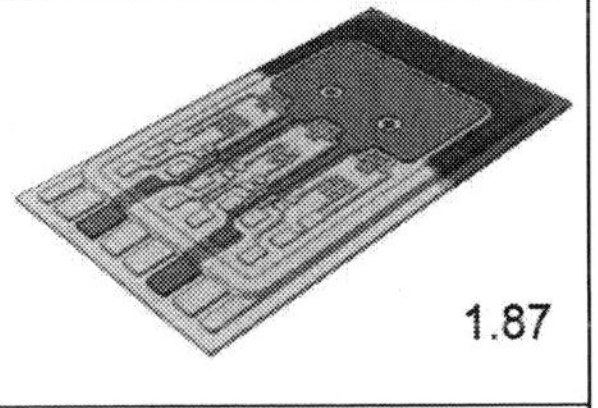 1.87
F	Via:20 Solder:20	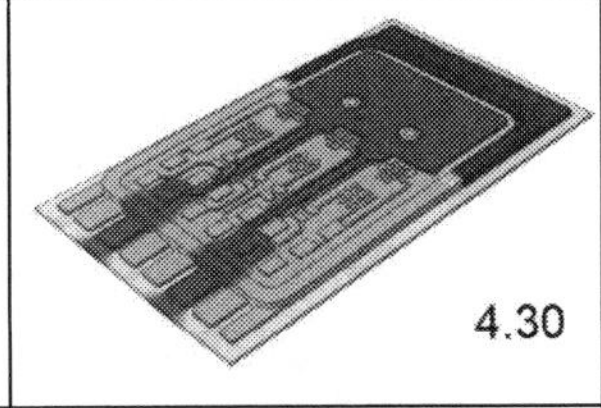 4.30
G	Via:60 Solder:60	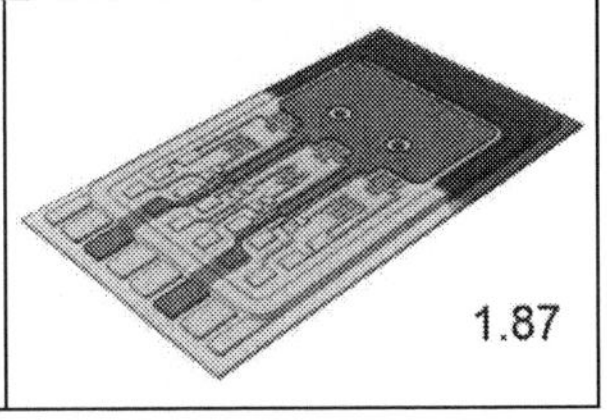 1.87

Fig. 14 Simulation results of Von

Model	Joint area remaining [%]	ΔTj results [deg.C]
A	Via:100 Solder:100	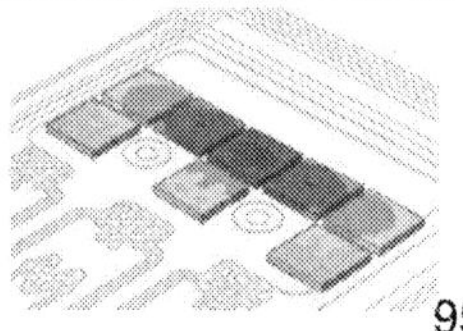 95.4
B	Via:20 Solder:100	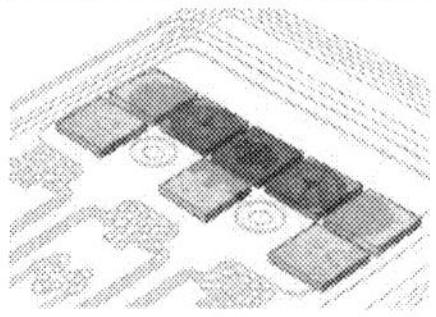 95.5
C	Via:60 Solder:100	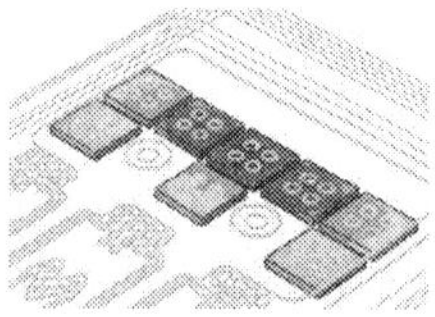 95.4
D	Via:100 Solder:20	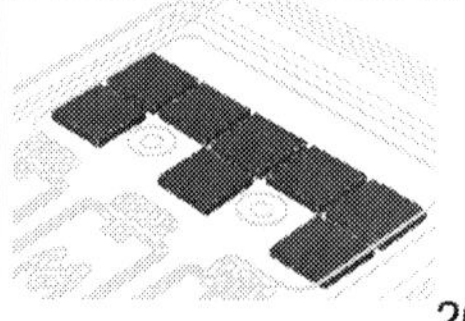 266.4
E	Via:100 Solder:60	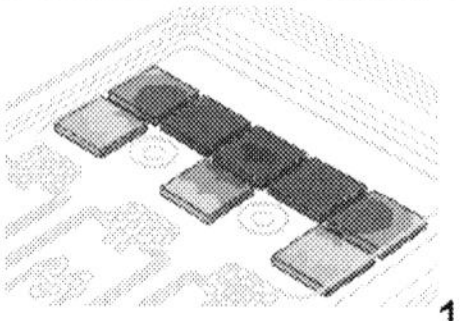 102.5
F	Via:20 Solder:20	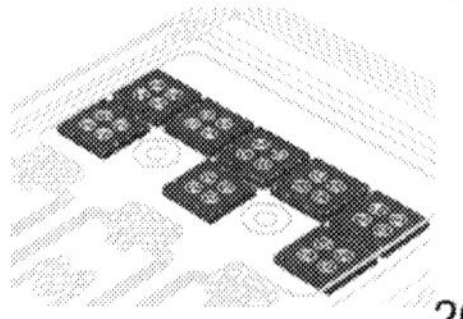 266.7
G	Via:60 Solder:60	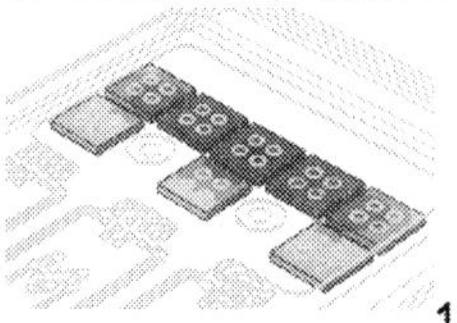 102.5

Fig. 15 Simulation results of ΔT

5 Conclusion

To evaluate the power cycling reliability and failure modes of POL, we performed power cycling tests on actual samples and analyzed the failure modes. And we evaluated the characteristics in models with reduced joining areas using FEM simulation to confirm the influence of the reduction in the Cu via joining area on the POL device joint interfaces after the power cycle test on the characteristic degradation.

The power cycle results of the actual sample and the simulation results generally match and are well correlated. The slight increase in Von and Tj shown in the power cycling test results are considered to be caused by the decrease in the solder joint area on the back side of the SiC diode. The heat dissipation path of the POL sample is on the solder joining side, and as the solder joining area decreases, the heat dissipation path decreases and Von increases as Tj increases due to the temperature characteristics of the SiC diode. The sample structure in this study, the Cu via joining side was not the cooling path, so the heat dissipation performance did not deteriorate even if the Cu via joining area decreased.

These results of will help in considering the POL'S Cu via joining structure has the reliability required for power semiconductor packages. POL can be densely joined with large diameter 300-750um Cu vias on the top side of device. The large joining area on the top side of device provides the high joining reliability required for WBG power semiconductor packages.

Furthermore, we are planning power cycling tests with Tjmax 175 deg.C-200 deg.C to verify the high temperature resistance of POL.

References

[1] J. Lutz, T. Herrmann, M. Feller, R. Bayerer, T. Licht, Raed Amro," Power Cycling Induced Failure Mechanisms in the Viewpoint of Rough Temperature Environment", Proceedings of CIPS 2008, Nuremberg, Germany, pp.55–58.

[2] Shingo Hayashibe, Kei Murayama, Koji Bando, Hitoshi Ito, Hirotsugu Suzuki, Takanori Sugita, "Evaluation of Half-Bridge Power Module with POL-kW", Proceedings of ICSJ2021, November 2021, Kyoto, Japan, pp.130-133.

[3] Liang Yin, Kaustubh Nagarkar, Arun Gowda, Christopher Kapusta, Risto Tuominen, PaulGillespie, Donna Sherman, Tammy Johnson, Shingo Hayashibe, Hitoshi Ito, Tadashi

Arai," Reliability of POL-kw power modules", Proceedings of ICEP2017, April 2017, Yamagata,Japan, pp.106-111.

[4] Youichi Nishihara, Koji Bando, Shingo Hayashibe, Takumi Yumoto, Taturo Yoshida, Hiroko Ota, "Thermal stress reliability comparison of WB and POL-kW structure by Power cycling and simulation", Proceedings of Applied Power Electronics Conference and Exposition (APEC)2023, Orlando, United States, pp.1150-1154

[5] JEDEC standard, JESD51-14, "Transient Dual Interface Test Method for the Measurement of the Thermal Resistance Junction to Case of Semiconductor Devices with Heat Flow Through a Single Path," JEDEC Solid State Technology Association (2010).

[6] Tomoaki Hara, Yoshitaka Aoki and Tsuyoshi Funaki, "Thermal fluid simulation modeling based on thermal transient test and fatigue analysis of asymmetric structural double-sided cooling power module", Journal of The Japan Institute of Electronics Packaging, vol. 24, no. 1, pp. 130-142, 2021.

[7] H. Notsu, H. Michikoshi, D. Ishikawa, Y. Ejiri, S. Hatsukawa and K. Fukuda: "A full SiC module operational at 200°C junction realized by a new fatigue-free structure", Proceedings of PCIM Europe 2017, pp. 556-560

[8] S. Butow and M. Spang, "Investigation of Stability and Oscillations at Power Modules with Low Stray Inductance", 2022 IEEE 34th International Symposium on Power Semiconductor Devices and ICs (ISPSD), May 2022, pp. 17-20.

[9] L. Wang, W. Wang, R.J.E. Hueting, G. Rietveld, J.A. Ferreira, Review of topside interconnections for wide bandgap power semiconductor packaging, IEEE TRANSACTIONS ON POWER ELECTRONICS, VOL. 38, NO. 1, JANUARY 2023, pp. 472-490.

PCIM Europe 2024, 11– 13 June 2024, Nuremberg   DOI: 10.30420/566262009

Accelerated Power Cycling of GaN HEMTs using Switching Loss and Fast Temperature Measurement

Wing Tai Leung[1] , Mehdi Niroomand[1] , Saeed Jahdi[1] , Bernard H. Stark[1]
[1] University of Bristol, U.K.

Corresponding author: Wing Tai Leung, trevor.leung@bristol.ac.uk
Speaker: Wing Tai Leung, trevor.leung@bristol.ac.uk

Abstract

Power cycling is typically performed by periodically self-heating a power device using a DC current. This paper demonstrates a technique to boost the heating power to shorten the heating phase, by the addition of switching loss. This power cycling technique is demonstrated on 190 mΩ, 600 V Gallium Nitride (GaN) discrete devices, where it achieves 240,000 thermal cycles per week with a junction temperature swing ΔT_J of 100°C, and where the device remains integrated in a switching converter. The device under test is heated rapidly from 30°C to 130°C in 0.5 s, by hard-switching at 1 MHz, at rated current and 400 V. An optimized thermal path cools the junction back to 30°C in 2 s. The junction temperature is closed-loop controlled to maintain an approximately constant temperature swing. This requires junction temperature sensing with low ms-scale latency, implemented here using peak turn-on di/dt as the junction temperature indicator. The inferred temperature is fed into a control system that governs the heating and cooling durations. The resulting closed-loop-controlled heating time is shown to be a sensitive real-time indicator of device change. The paper discusses the practicality of temperature calibration methods, in light of temperature-sensitive electrical parameters' known drift and sensitivity to bias temperature instability, and the problem of self-heating during calibration. Experimental results show one GaN device surviving 400,000 cycles with a ΔT_J of 102°C with no apparent thermal degradation, and another GaN device cycling at a ΔT_J of 136°C, whose heating duration reduces from 500 to 10 ms over the course of 30,000 cycles, indicating an apparent degradation of the device's thermal properties.

1   Introduction

Power cycling is a method of investigating the reliability of semiconductor devices. Typically, a DC current self-heats the device, and a cooling period with no device current returns the device to its original temperature. This cycling strains bonds between different materials due to their different temperature coefficients of expansion. Generally, 100s of thousands of cycles are performed, which can take months. The maximum achievable cycling frequency or minimum cycling duration is limited by the maximum allowed device heating current, the thermal performance of the cooling system, and the time required for heat to soak from the heat-generating device junction to the material bonds of interest. Typical cycle times vary between 3 – 9 s for discrete GaN HEMTs and Si MOSFETs [1], [2], [3], to 10 – 30 s for >100 A rated IGBT

modules [4], [5], with typical junction temperature swings ΔT_J of 70 – 110°C. Failures are typically due to bond wire liftoff or solder delamination inside the package [6], [7].

GaN power semiconductor packages, with their low thermal resistances and mass, have the potential for much faster cycling compared to their silicon counterparts. For instance, the thermal capacity of the 600 V, 190 mΩ GaN device used in this paper is 14 mJ/K, around 2 orders lower than an equivalent IGBT. Power cycling of GaN power devices with cycle times of 6 s has been reported using two temperature-sensitive electrical parameters (TSEPs) of drain-source voltage and gate-source voltage [2], to observe the temperature between heating and cooling. These electrical temperature indicators remove the latency of thermocouples or infrared imaging. Reported temperature sensitive electrical parameters include drain-source on-resistance, turn-on rise time, gate threshold voltage, peak value of turn-on di/dt and more [8], [9]. Downsides of using them for GaN devices are parameter drifts [10] and that they are affected by bias temperature instability [11], [12]. As a result of these effects, continuously switching applications and hard-switched double pulse tests show slightly different switching waveforms, in particular, differences are seen in turn-on dv/dt transients and total energy loss [13].

The contributions of this paper are as follows:

Section 2 presents four concepts that when applied together, allow an increase in power cycling frequency over that of cycling with DC current heating. These are the addition of switching loss by hard-switching the device under test, the increase in cooling performance by direct soldering of the thermal pad to a copper interface piece, the use of a closed-loop control system to maintain the required junction temperature swing ΔT_J and peak junction temperature T_{HOT}, and the use of the temperature-sensitive electrical parameter 'peak turn-on di/dt' to determine junction temperature.

Section 3 presents how these concepts are implemented, including the bridge-leg circuit that is used for in-circuit power cycling, the implementation of a forced air cooling system with a low thermal resistance of 6 K/W, and the closed-loop control system architecture that regulates the junction temperature swing. In addition, a method is proposed of mapping the temperature sensitive electrical parameter values onto junction temperature.

Section 4 shows the mapping results, and results of a finite element (FE) simulation to estimate error due to self-heating. The system is shown to achieve cycle times of around 2.5 s, or 240,000 power cycles per week, and the closed-loop controlled heating duration is shown to be a function of device age and temperature swing, and therefore a potential real-time indicator of device degradation.

2 Accelerated power cycling

2.1 Enhanced self-heating using switching loss

The goal of this work is to reduce power cycling times by exploring the use of switching loss in addition to DC current. The heating power of DC current is limited by on-resistance and rated current. Therefore, in order to increase heating power using DC current only, the gate voltage could be lowered to increase the forward voltage drop, or the rated current could be exceeded. However, neither of these are in the spirit of establishing the reliability of a device when operated correctly. Hard-switching the device, however, is part of normal operation in a converter. This provides additional heating power of a similar order of magnitude, which is a function of switching frequency and gate resistor. It may also activate additional, realistic failure or aging modes, due to the presence of high electric field strengths during blocking and voltage overshoots, and due to the temperature excursions in the junction during ns-scale switching events. Fig. 1 illustrates the difference between DC-only on the left, and the switched power cycling demonstrated in this paper on the right. In this work, a power cycle consists of a heating phase in excess of 10,000 switching cycles, and a cooling cycle when the device is off. This increases both average and peak heating power, reducing the required heating duration, and increasing the power cycling frequency.

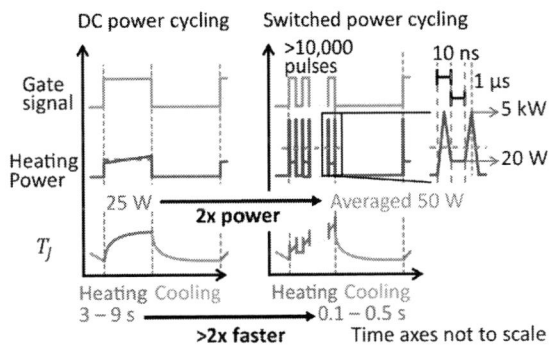

Fig. 1 DC-only (left) versus faster switched power cycling (right) used in this work.

2.2 Enhanced cooling, and the interdependency of heating and cooling durations

The cooling phase of power cycling is usually longer than the heating phase, and therefore the potential for increased heating to increasing the cycling frequency is limited. Improving the cooling system's performance reduces the length of the cooling period, however it increases the heating duration, since heating power is being drained into the cooling system, again providing diminishing returns. Fig. 2 illustrates the fact that the increased heating that is available due to hard-switching of a device permits the use of enhanced cooling to a) overcome the additional heating requirement due to having better cooling, and b) to shorten the heating duration.

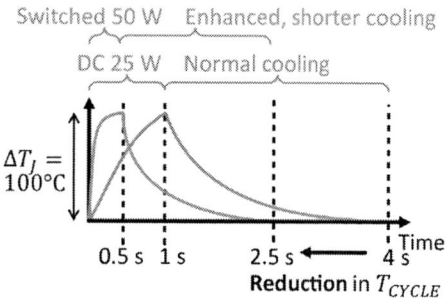

Fig. 2 Reduction in cycle duration T_{CYCLE} due to use of switching to increase heating power, and use of enhanced cooling system.

Enhanced cooling is implemented here, by soldering the thermal pad of the device to a copper interface piece, to establish a thermally efficient connection to a fan-assisted heatsink. This avoids the use of thermal vias for the bottom-cooled devices used here. A range of thermal via configurations were explored for this work, however these achieved thermal resistances of 2 – 5 K/W, which is similar to the 2 K/W junction to case thermal resistance of the GaN device used in this work. As a result, thermal vias would significantly lower the cooling performance, and increase the cooling duration, and therefore their use is avoided in this work, though this may not always be viable.

2.3 Control of the changeover times between heating and cooling

The concept of controlling the heating and cooling periods during power cycling to keep the ΔT_J constant between the threshold temperatures T_{HOT} and T_{COLD} is shown in Fig. 3. Cycle durations are continuously updated based on the previous cycle's temperature values. This method always ties the peak temperature, when heating is stopped, to a preset threshold temperature T_{HOT}.

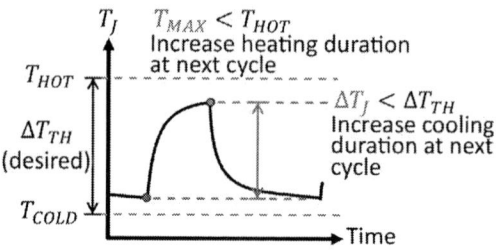

Fig. 3 Concept of setting heating and cooling durations by sensing junction temperature T_J at the end of heating phase to regulate peak temperature T_{HOT} and temperature swing ΔT_J.

The constant T_{HOT} and ΔT_J control method, see also [14], [15] for similar approaches, is kinder to the device than reported methods that maintain constant cycle periods, since it holds the device in its intended operating area by shortening the heating duration rather than promoting thermal runaway as the device's thermal resistance increases with age.

2.4 Calibration of junction temperature measurement via peak di/dt

To maintain the desired junction temperature swing, the GaN device's junction temperature is monitored, by monitoring the temperature-sensitive electrical parameter 'peak di/dt'. Alternative methods such as thermal cameras and thermocouples, whilst typically being more stable and accurate, only have access to the outer surface of the case, and they typically have readout latencies of >10 ms. Using these methods would likely require support from a model, and therefore in this work, real-time inference from a temperature indicator is adopted.

A hot device will incur a drop in transconductance, and thus also in peak turn-on di/dt [16]. This temperature indicator is relatively simple to measure using a magnetic di/dt sensor and peak detection, and therefore it is adopted here. To accommodate device-to-device variation, prior to each power cycling test, the mapping of T_J onto peak di/dt is established at the two temperatures of interest, T_{HOT} and T_{COLD}.

Peak turn-on di/dt is also a function of other switching parameters, such as gate current and drain-source voltage. Therefore, the calibration method needs to mimic the switching conditions of power cycling as much as possible. In the literature, a popular calibration method is the double pulse test, with the benefits of shorter test time and simpler experimental setup compared to a longer continuous test.

However, due to self-heating and bias temperature instability, the peak di/dt measurement values of both double-pulse and continuous switching should be compared to ensure that the final calibration method produces the correct temperature swing.

3 Hardware and control system implementation

3.1 In-circuit power cycling setup

The system architecture is shown in Fig. 4. An oscilloscope captures the GaN device's switching waveforms, and passes these to a PC running MATLAB, to extract peak di/dt and thus the temperature, and to monitor evolving device characteristics. A Bristol 1 GHz Infinity Current Sensor, available at infinitysensor.com [17] measures di/dt, whose peak value, the temperature indicator, is fed into the top-level controller. This controller sets new heating and cooling durations as illustrated in Fig. 3. These are passed to a slave controller that controls a function generator that triggers a gate driver.

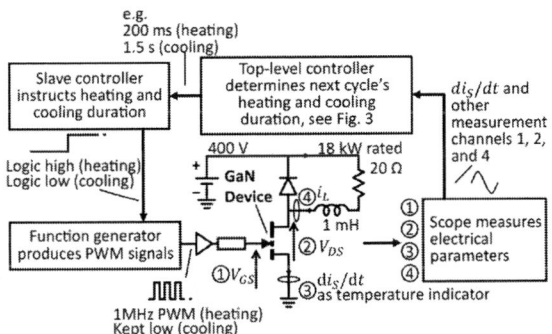

Fig. 4 System diagram of the power cycling setup.

During the heating phase, the GaN device is hard-switched at 1 MHz, 0.5 duty ratio, 400 V and 10 A. Fig. 5 shows the cycling facility mounted on a forced-air cooling system. The GaN device's thermal case is soldered directly onto a copper block through a cutout on the PCB, avoiding the need for thermal vias and their higher thermal resistances.

The thermal characteristics of this cooling system have been determined from a steady-state DC heating experiment that yields a total thermal resistance of 6 K/W. The thermal capacitance of

the GaN device is obtained from the commercial Spice model, and the other capacitances are estimated from reported material weights and known specific heat capacities.

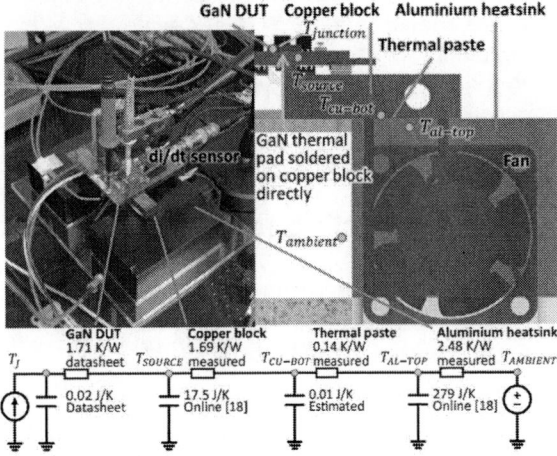

Fig. 5 Power cycling setup with enhanced cooling, and a thermal model of the cooling path. Specific thermal specific capacities are obtained from [18].

3.2 Test procedures

The calibration sequence that is run prior to power cycling is shown in Fig. 6.

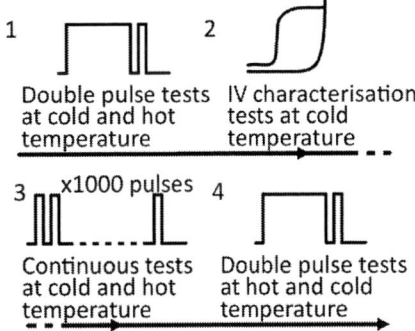

Fig. 6 Calibration tests performed before power cycling.

Four tests are carried out in this sequence:

1. Double pulse test at cold and hot temperatures.
2. Device IV characterisation at cold temperature.
3. 1 ms continuous switching at cold and hot temperatures.
4. Repeat of cold and hot double pulse tests.

Note that Step 4 starts from a hot temperature first since Step 3 ends with a hot temperature.

Both double pulse and continuous tests are used to obtain the peak turn-on di/dt for the two threshold temperatures T_{HOT} and T_{COLD}, since GaN devices are known to have different waveforms for double-pulse and continuous switching [13]. The respective results are shown in Section 4.1. The device characterisation step is used to check that the GaN device is within specification of the datasheet, and to allow comparison after power cycling. The test device remains soldered into the switching converter throughout.

4 Power cycling results

4.1 Simulated device temperature rise ΔT_{SH} due to self-heating, for double pulse and continuous switching

Both double pulse and continuous switching tests are performed to obtain the peak di/dt values for the hot and cold switch-over temperatures T_{HOT} and T_{COLD}. Whist both tests start in thermal equilibrium at either T_{HOT} or T_{COLD}, by the time that the peak di/dt values are measured, the junction has undergone self-heating and increased by ΔT_{SH}. To investigate this rise, a 3D finite element model of the chip and thermal path to ambient is developed, using published CT scans, photos of etched packages [19], and the manufacturer's material content sheet to model the GaN device. The simulation results are shown in Fig. 7.

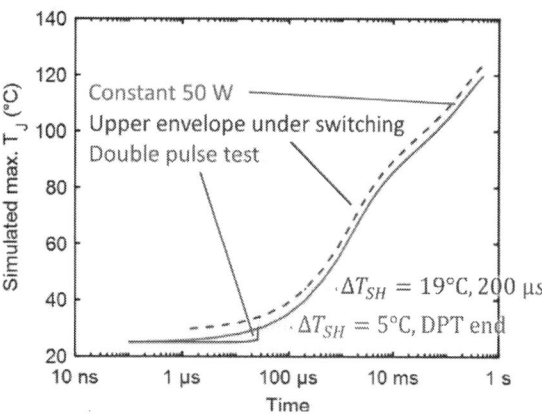

Fig. 7 Simulated temperature rise ΔT_{SH} due to self-heating of double pulse and steady 50 W heating, and inferred temperature (dashed) at the switching instant of continuous switching.

The orange line is the simulated junction temperature for a constant 50 W of heat generation at the junction, spread evenly over the active area. The power dissipation value of 50 W

has been determined by sweeping the heat generation in the junction in successive simulations, until the simulated outer package temperature matches the temperature observed by a thermal camera, sampling at 80 Hz of a real-life 1 s heating event using 1 MHz switching at rated 400 V and 10 A. The blue line represents the simulated double pulse test, showing a 4°C uptick at the end due to the turn-on event of the second pulse, where the simulation is stopped. The input power of the first pulse corresponds to a current ramp from 0 – 10 A over 26 μs, assuming a constant on-resistance of 200 mΩ as quoted on the datasheet. The total temperature rise due to self-heating is 5°C.

The simulations show that at the point of measuring peak turn-on di/dt, the junction temperature exhibits are sharp rise of around 4°C above the short-term average, which decays with a time constant of around 30 ns. This is indicated by the purple dashed line in Fig. 7, which represents the upper envelope of the junction temperature under switching with an average power dissipation of 50 W. For example, the total temperature rise due to self-heating of a continuous switching test that lasts 200 μs and ends in a turn-on event is 15°C + 4°C = 19°C. The results show that self-heating under continuous switching becomes significant after just a few tens of switching cycles.

4.2 Temperature calibration of peak turn-on di/dt as a temperature indicator

To permit the temperature regulation of the power cycle according to Fig. 3, measured peak di/dt values are required for $T_J = T_{HOT}$ and $T_J = T_{COLD}$. This is achieved here by pre-heating a GaN HEMT to either 30°C or 100°C, and for each starting temperature, running continuous and double pulse switching tests. Measured peak di/dt for these four tests are plotted against inferred junction temperature, which equals the pre-heated temperature plus a self-heating delta derived from Fig. 7. The results are plotted in Fig. 8.

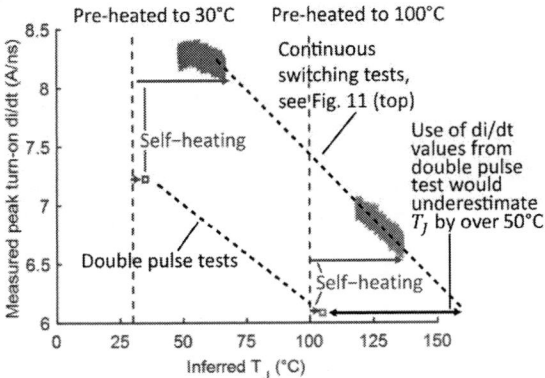

Fig. 8 Measured turn-on peak di/dt of double pulse and continuous switching against junction temperature that has been inferred from the preheat temperature and the self-heating data of Fig. 7.

The double pulse test results consist of a single point, or (x,y) pair, whose first entry is the preheat temperature plus the self-heating temperature rise up to when the di/dt value is measured (see Fig. 7), and whose second entry is the measured peak di/dt value at the turn-on of the second pulse. The colder double pulse test shows a higher turn-on di/dt as expected. The continuous tests' peak di/dt measurements are presented as a cluster of points, starting at 200 µs into switching when the load current has settled to 10 A, running up to 1 ms when around 20°C of self-heating has occurred. Since the switching frequency is 1 MHz, there are 800 turn-on peak di/dt points in the cluster. The time values of each point are converted into a self-heating value for each using the purple envelope of ΔT_{SH} from Fig. 7, and added to the pre-heating temperature to infer T_J. It is apparent that, again, the colder switching has a higher di/dt. However, it is also apparent, that continuous switching provides higher di/dt values for the same junction temperature than double pulse switching does. Given that power cycling is a continuous switching test, we adopt the di/dt values for the threshold temperatures from the continuous switching tests. From Fig. 8 it is apparent that the use of the di/dt values from the double-pulse switching tests would underread the temperature by over 50°C.

There are noticeable peak di/dt variations in the continuous tests' cluster for a given temperature. When zoomed in, a 50 kHz oscillation is seen to be riding on top of the measurements. This is assumed to be noise, since it is a function of the width of the oscilloscope window used, and therefore the actual di/dt is taken to lie between the top and bottom envelopes of the cluster.

The upper dashed line in Fig. 8 is used to determine the two peak di/dt values that delineate the heating and cooling phases for the desired threshold temperatures T_{COLD} and T_{HOT}, as illustrated in Fig. 9. We refer to these di/dt threshold values as di/dt_{COLD} and di/dt_{HOT}.

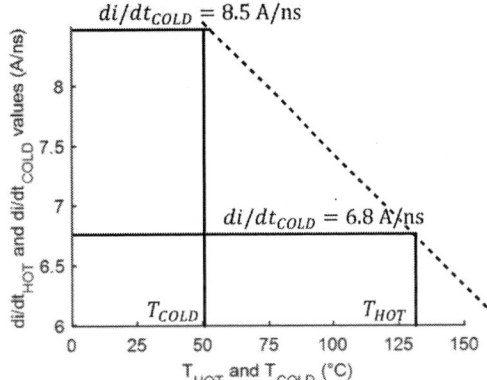

Fig. 9 Determining the threshold values for di/dt_{HOT} and di/dt_{COLD} from Fig. 8.

4.3 Regulation of junction temperature swing ΔT_J during power cycling

During the heating phase of each power cycle, the controller determines the actual junction temperature swing by sampling peak di/dt at two points. First, 700 µs into the cycle, and second at the end of the heating phase, as illustrated in Fig. 10. On each cycle, di/dt at the end of heating is compared to di/dt_{HOT}, and the heating duration is stepped by 10 ms in the direction to reduce the error and therefore maintain the desired peak T_J. Also in each cycle, the difference of the two measured di/dt values is calculated and compared to the desired difference ΔT_{TH}. A correction is made, according to the self-heating graph Fig. 7, to allow for the fact that the first measurement is made 700 µs into the cycle and not at the start. The cooling phase of the next cycle is stepped by 0.1 s in the direction to reduce this error, thereby regulating the temperature swing to the desired value.

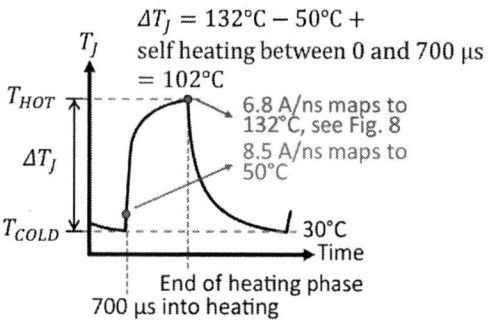

Fig. 10 Concept of regulating ΔT_J.

In the following, this method will be used to power cycle two GaN HEMTs with different cycling parameters, see Table 1.

	GaN HEMT A	GaN HEMT B
T_{COLD}	30°C	30°C
T_{HOT}	132°C	166°C
ΔT_J	102°C	136°C

Table 1 Cycling parameters for GaN HEMTs A and B.

GaN HEMT A will remain within the datasheet's stated safe operating area, whereas GaN HEMT B's junction temperature will be set to exceed the datasheet's 150°C limit by 16°C.

4.4 Validation of use of turn-on peak di/dt as a temperature indicator

In this section, an attempt is made to validate the chosen temperature sensitive electrical parameter against a 3D finite element simulation, and thermal camera measurements. The results are plotted in Fig. 11.

To obtain the measurements in the top half of the figure, a single heating phase is run using 1 MHz switching at 400 V and 10 A. The first peak di/dt value is captured at 300 µs as the load current has built up to 10 A, and the subsequent peaks are measured at 50 ms intervals. In the bottom half, there are three plots. The first is simulated junction temperature T_J, using an averaged power input of 50 W. The second is the simulated temperature averaged over the top surface of the package, and the third is the temperature measured on the top of the package using a TIM 40 thermal camera from MicroEpsilon with a special resolution of 0.27 mm and a sampling rate of 80 Hz. Polyimide tape is applied to the package surface to control the emissivity, and an emissivity value of 0.95 is used.

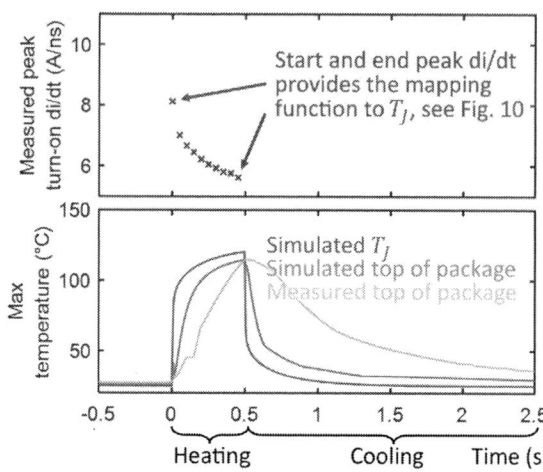

Fig. 11 Temperature and peak turn-on di/dt course over one heating and cooling phase. Device starts from 25°C.

As expected, peak di/dt follows an inverse relationship with temperature, which appears to broadly correlate with the simulated junction temperature.

The thermal camera records the surface of the package heating up in 0.5 s to 120°C and cooling back down in just over 2 s. Relative to the simulated package temperature, it appears to have a higher time constant. This may be due to the use of polyimide tape, or latency in the measurement, or overly simplifying assumptions in the model or simulation inputs, and should be further investigated. Nevertheless, the absolute junction temperature peak and swing appear to correlate well between the different methods.

4.5 Switching waveforms during power cycling at the threshold temperatures

Figure 12 shows two turn-on switching instances during power cycling, captured by an oscilloscope as illustrated in Fig. 4. The blue waveforms are measured at the start of the heating phase (225 µs from start) of power cycle number 2300, and the orange waveforms at the end (400 ms) of the same heating phase.

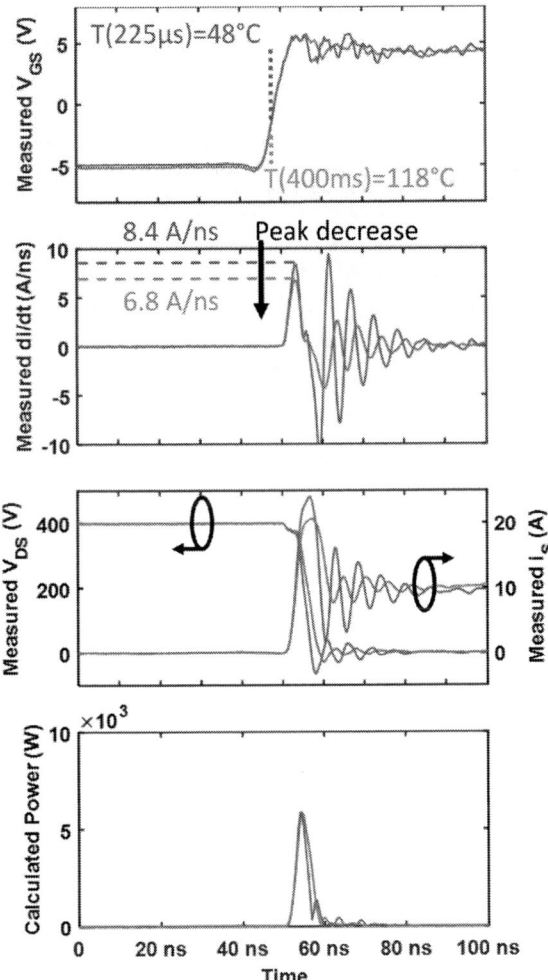

Fig. 12 Waveforms captured at start and end of heating cycle no. 2300 for GaN HEMT A. Reduction in turn-on peak turn-on di/dt due to self-heating.

The GaN device's source current i_S is measured using an Infinity Sensor [17] which provides a direct measurement of di/dt. This is integrated using the script available at infinitysensor.com.

The peak turn-on di/dt is seen to drop by 1.6 A/ns as a result of heating. The GaN device has a switching time of approximately 10 ns and the peak instantaneous power calculated from device current multiplied by device voltage reaches 5 kW.

4.6 Power cycling results and regulated heating duration as a real-time indicator of device change

The two GaN HEMTs A and B are put into the proposed accelerated power cycling scheme using the cycling parameters of Table 1. The

closed-loop controlled heating duration is plotted against time on a log scale in Fig. 13.

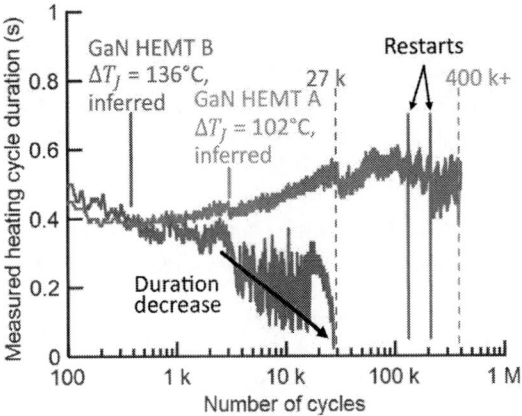

Fig. 13 Heating duration vs cycle number. GaN HEMT A ΔT_J=102°C, 400 k power cycles. GaN HEMT B, ΔT_J=136°C, gradual decay. ΔT_J inferred from Section 4.3.

GaN HEMT A which is cycled within the datasheet recommended operating range of 150°C undergoes over 400,000 power cycles in 12 days without showing a significant change in heating duration, apart from variation that is due to changes in the laboratory's ambient temperature. GaN HEMT B which is cycled up to 162°C shows a reduction in heating duration from the start of the cycling, and by 30,000 cycles the heating duration has dropped from 500 to only 10 ms. This drop in heating duration is likely a sign of a deterioration of the device's ability to conduct heat to its thermal pad, due to damaged material interfaces. There could also be changes in heating power and a drift of the temperature dependent electrical parameter used for the temperature regulation.

5 Conclusion

This paper successfully demonstrates that by the use of inductive switching instead of the injection of a constant current, power cycling can be accelerated. In the case of discrete 600 V, 10 A GaN HEMTs with a low thermal capacity, 240,000 cycles are achievable in a week. A GaN device is cycled 400,000 times with a ΔT_J of 102°C, and appears to undergo no deterioration, showing that the GaN device tested here is potentially very robust. However, multiple devices should be tested for conclusive results. Due to the closed-loop regulation of the heating duration, a real-time indicator of device health is available, since any voids or material separation increase the thermal resistance from the junction to the package's thermal pad, which, in turn, visibly shortens the heating duration. This is demonstrated by exceeding the safe operating range by a relatively small amount. It is conceivable that due to the use of switching, and therefore the presence of relatively high field strengths, relevant failure modes are introduced that would otherwise not be present in DC power cycling.

Further work should include in-detail validation of the temperature sensitive electrical parameter used here, over the course of power cycling, to determine if there is any drift or overheating of the junction. More devices should be tested at the edge of the datasheet recommended operating range. Finite element modelling should be used to verify that the material interfaces of interest are undergoing the desired temperature swings, since by shortening the power cycle, the heat has less time to spread through the volume of a device.

Acknowledgement – This work was supported by Toshiba's Bristol Research and Innovation Laboratory (BRIL).

6 References

[1] S. Song, S. Munk-Nielsen, C. Uhrenfeldt, and K. Pedersen, "Power cycling test of a 650 V discrete GaN-on-Si power device with a laminated packaging embedding technology," in *2017 IEEE Energy Conversion Congress and Exposition (ECCE)*, 2017: IEEE, pp. 2540-2545.

[2] M. Goller, J. Franke, J. Lutz, and T. Basler, "Power Cycling Results of Discrete Gallium Nitride Gate Injection Transistors," in *2021 23rd European Conference on Power Electronics and Applications (EPE'21 ECCE Europe)*, 2021: IEEE, pp. 1-7.

[3] F. Hoffmann and N. Kaminski, "Impact of device design on the power cycling capability of discrete SiC MOSFETs at different temperature swings," in *2020 32nd International Symposium on Power Semiconductor Devices and ICs (ISPSD)*, 2020: IEEE, pp. 533-536.

[4] A. Hensler, J. Lutz, M. Thoben, and J. Zachariae, "Power cycling tests at high temperatures with IGBT power modules for hybrid electrical vehicle applications," in *3rd Electronics System Integration Technology Conference ESTC*, 2010: IEEE, pp. 1-6.

[5] V. Smet, F. Forest, J.-J. Huselstein, A. Rashed, and F. Richardeau, "Evaluation of $V_{\rm ce}$ Monitoring as a Real-Time Method to Estimate Aging of Bond Wire-IGBT Modules Stressed by Power Cycling," *IEEE Transactions on Industrial Electronics,* vol. 60, no. 7, pp. 2760-2770, 2012.

[6] L. R. GopiReddy, L. M. Tolbert, and B. Ozpineci, "Power cycle testing of power switches: A literature survey," *IEEE Transactions on Power Electronics,* vol. 30, no. 5, pp. 2465-2473, 2014.

[7] C. Durand, M. Klingler, D. Coutellier, and H. Naceur, "Power cycling reliability of power module: A survey," *IEEE Transactions on Device and Materials Reliability,* vol. 16, no. 1, pp. 80-97, 2016.

[8] H. Sayed, G. S. Kulothungan, and H. S. Krishnamoorthy, "Characterization of GaN HEMTs' Aging Precursors and Activation Energy under a Wide-Range of Thermal Cycling Tests," *IEEE Open Journal of the Industrial Electronics Society,* 2023.

[9] R. Boldyrjew-Mast and J. Lutz, "Reliability of GaN GIT devices in power cycling tests with RDS (on)(T) and VGS (T) for junction temperature calculation," in *PCIM Europe digital days 2020; International Exhibition and Conference for Power Electronics, Intelligent Motion, Renewable Energy and Energy Management*, 2020: VDE, pp. 1-8.

[10] J. Franke, C. Bäumler, D. Kretzschmar, and J. Lutz, "Advanced temperature estimation in low Rds, on p-GaN HEMT devices for performing power cycling tests," in *PCIM Europe 2019; International Exhibition and Conference for Power Electronics, Intelligent Motion, Renewable Energy and Energy Management*, 2019: VDE, pp. 1-6.

[11] J. O. Gonzalez, B. Etoz, and O. Alatise, "Characterizing threshold voltage shifts and recovery in Schottky gate and ohmic gate GaN HEMTs," in *2020 IEEE Energy Conversion Congress and Exposition (ECCE)*, 2020: IEEE, pp. 217-224.

[12] J. O. Gonzalez, B. Etoz, and O. Alatise, "Gate stresses and threshold voltage instability in normally-OFF GaN HEMTs," in *2020 22nd European Conference on*

Power Electronics and Applications (EPE'20 ECCE Europe), 2020: IEEE, pp. P. 1-P. 10.

[13] M. H. Hedayati, H. C. Dymond, R. Goswami, and B. H. Stark, "Investigating GaN power device double-pulse testing efficacy in the face of V TH-shift, dynamic R dson, and temperature variations," in *2021 IEEE Applied Power Electronics Conference and Exposition (APEC)*, 2021: IEEE, pp. 2291-2298.

[14] S. Schuler and U. Scheuermann, "Impact of test control strategy on power cycling lifetime," in *Proc. PCIM*, 2010, vol. 57, pp. 355-360.

[15] G. Zeng, F. Wenisch-Kober, and J. Lutz, "Study on power cycling test with different control strategies," *Microelectronics Reliability,* vol. 88, pp. 756-761, 2018.

[16] E. A. Jones *et al.*, "Characterization of an enhancement-mode 650-V GaN HFET," in *2015 IEEE Energy Conversion Congress and Exposition (ECCE)*, 2015: IEEE, pp. 400-407.

[17] M. H. Hedayati, H. C. Dymond, D. Liu, and B. H. Stark, "Fast temperature sensing for GaN power devices using E-field probes," in *2020 IEEE 21st Workshop on Control and Modeling for Power Electronics (COMPEL)*, 2020: IEEE, pp. 1-7.

[18] S. Milczanowski, "Changing temperatures." Florida State College at Jacksonville. Accessed: Apr. 7, 2024. [Online]. Available: https://web.fscj.edu/Milczanowski/psc/lect/Ch4/slide5.htm

[19] A. Schletz, A. Endruschat, and T. Heckel, State of the Art Packaging: Fraunhofer-Gesellschaft, (2021). [Online]. Available: https://publica.fraunhofer.de/handle/publica/411241.

PCIM Europe 2024, 11– 13 June 2024, Nuremberg   DOI: 10.30420/566262010

Control of an MMC-based Hybrid Transformer with Star-point Voltage Injection

Rui Wang[1], Maurice G. L. Roes[1], Henk Huisman[1], C.G.E. (Korneel) Wijnands[1]

[1] Eindhoven University of Technology

Corresponding author: Rui Wang, r.wang.1@tue.nl
Speaker: Rui Wang, r.wang.1@tue.nl

Abstract

Hybrid transformers (HTs) can provide both voltage and current compensation in the utility grid. HTs include a line-frequency transformer and a converter rated for partial transformer power. This paper presents a HT using a modular multilevel converter (MMC) connected at the medium-voltage (MV) side of the transformer, which allows scaling to a wide range of power and voltage ratings. The application in a HT requires an unconventional control approach for the MMC, as its output voltages have equal phase angles, but different amplitudes.

1   Introduction

With the rapid growth of the modern energy market, residential loads are becoming increasingly diverse. The installation of heat pumps, solar panels, and electric vehicles contributes to the transformation towards sustainability. However, due to these increasing loads, voltage fluctuations at the grid edge have become larger and more frequent.

To maintain a well-regulated low voltage at the customer side, transformer tap changers are used for voltage regulation: the turns ratio of transformers can be altered using mechanical switches. This method of voltage control is step-wise and slow. Previous research has proposed fully electronic tap changers by inserting semiconductor switches between multiple tap positions, to reduce the wear in the mechanical switches that are traditionally used [1], [2]. The transformer voltage can be controlled continuously over the full length of the tap winding. However, those designs need bidirectional switches and require multiple tap positions. Recently, the concept of hybrid transformers was proposed. A hybrid transformer comprises a power electronic converter connected to a partial transformer winding [3]–[6]. Such HTs are capable of compensating both voltage and current because of the series-shunt-connected converters. Common HTs use 2-level voltage source converters. The voltage and power ratings of semiconductor devices often limit

the design of such HTs [4]. Beside this, bulky filters are required to eliminate switching harmonics.

HTs at the low-voltage (LV) side need to be able to withstand short-circuit currents that can typically reach more than 20 times the nominal current. For example, in the 100 kVA HT prototype in [7], the converter is protected by a pair of bypass thyristors with current ratings above 3 kA. This solution is difficult to scale for transformers with higher power ratings. In some studies [8], [9], one side of the back-to-back converter is parallel-connected to the LV grid, and the other side is series-connected to the MV grid via coupling transformers. The short-circuit protection is implemented at the MV side, where the required current rating is reduced significantly. However, the coupling transformers must withstand high common-mode voltages due to the MV grid, which was not discussed in [8], [9].

Alternatively, the switch-mode power converters can be series-shunt-connected to the MV-side winding without additional coupling transformers. In this situation, the voltage stress is challenging for most semiconductor devices currently available in the market. For instance, the minimum blocking voltage can reach 6.5kV considering 20% of the 23 kV MV grid voltage, which makes it impossible to use 2-level topologies directly.

The high voltage stress can be reduced by series operation of semiconductor switches for 2-level topologies. To achieve proper voltage sharing during switching transients, accurate timing is essential

PCIM Europe 2024, 11– 13 June 2024, Nuremberg   DOI: 10.30420/566262010

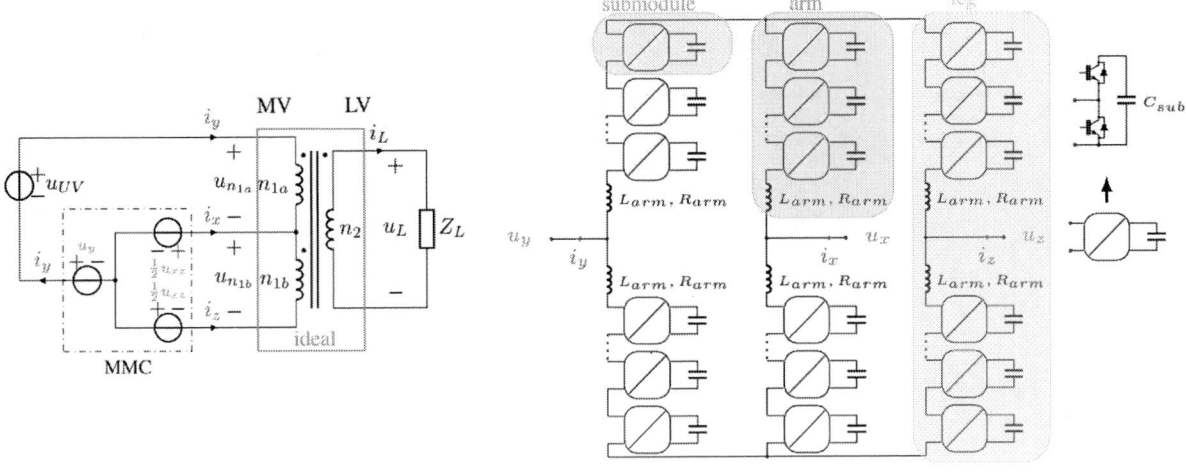

(a) Simplified circuit

(b) Converter topology of the MMC-HT.

Fig. 1: Single-phase representation of an MMC-HT

but difficult to guarantee. Another solution can be to use a multilevel topology. A neutral-point-clamped (NPC) 3-level converter was proposed for this application in [3]. However, since it requires access to the neutral point, additional tap positions on the transformer are still necessary. A recent study presented an MMC-HT [10]. This topology is derived from the 2-level HT in [4]. All three phases are controlled by one integrated back-to-back MMC. This topology can only be used when the converter is placed at the Y-winding side of the transformer.

In this paper, a different MMC-based HT will be investigated. The topology is derived from the 2-level solutions in [3], [5], [6]. The proposed MMC-HT has the advantage of straightforward control for every phase regardless of the configuration of the transformer windings.

2 MMC-based hybrid transformer

The simplified per-phase circuit diagram of a hybrid transformer is shown in Fig. 1a. The distribution transformer is presented as a three-winding transformer with turns ratio of $n_{1a} : n_{1b} : n_2$. The tap winding is located at the MV side with n_{1b} turns. The converter presentation is simplified using controlled voltage sources. Voltage compensation can be achieved by injecting a voltage u_y in series to the MV grid. Reactive current compensation is realized by controlling the voltage u_{xz} across the tap winding.

2.1 Realization of HT functionalities

Ideally, the converter draws zero energy from the grid in steady state; therefore, the power flow should be controlled. As a result, the voltage and current compensation are not decoupled. For convenience, a control parameter K is introduced in [6] and refers to different operation points:

$$u_y = \frac{K}{2}u_{xz}, \tag{1}$$

$$i^{\circlearrowleft} = i_x - i_z = -Ki_y + i_\perp^{\circlearrowleft}, \tag{2}$$

$$n_{1e} = n_{1a} + \frac{1}{2}(1 - K)n_{1b}, \tag{3}$$

where $i^{\circlearrowleft}$ is the circulating current between the converter and the tap winding, and is controlled by u_{xz}; $i_\perp^{\circlearrowleft}$ injects reactive current with a phase angle 90-degree shifted from the phase angle of u_{xz}. The MV grid is indirectly connected to the transformer winding due to the power converter, where u_y is controlled equal to a voltage potential at one point of the tap winding. As a result, the grid voltage u_{UV} is effectively coupled to n_{1e} turns at the MV side. In the nominal scenario, K is zero and n_{1e} is equal to $(n_{1a} + 0.5n_{1b})$; the MV grid voltage is effectively coupled to the main winding combined with half of the tap winding.

As shown in Fig. 1b, an MMC-HT is derived from the 2-level topology in [3], [5], [6]. The MMC-HT comprises three legs, modulating u_x, u_z and u_y respectively; the voltage difference between u_x and u_z controls the circulating current $i^{\circlearrowleft}$. One MMC leg

85

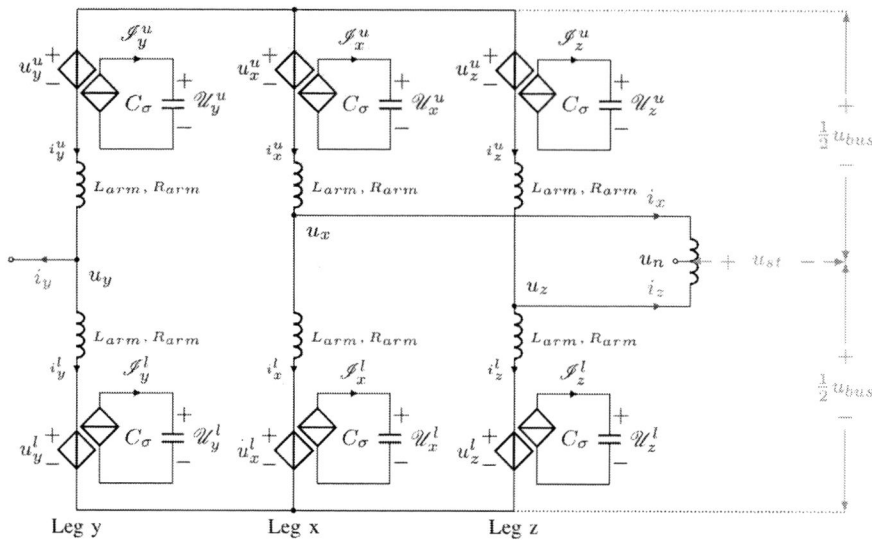

Fig. 2: Averaged equivalent circuit of the MMC used in a single-phase HT; the MMC outputs x and z are connected to the tap winding. u_n denotes the voltage potential at the center of the tap winding. This point is not accessible. The MMC is connected to the MV grid as shown Fig. 1a.

consists of two symmetric arms. Each arm includes N series-connected half-bridge submodules and an inductor L_{arm} to reduce the current ripple due to switching events; R_{arm} represents the resistance in the arm.

The 2-level HT requires a bus capacitor to maintain a stable dc voltage, which is not necessary in the MMC-HT because each submodule contains a bus capacitor C_{sub}.

2.2 MMC basics

An averaged model of the MMC is shown in Fig. 2 [11]. The N series-connected submodules can be represented using controlled voltage and current sources, and the equivalent capacitance C_σ in one arm is C_{sub}/N. In the MMC, variables u, i denote the arm voltage and current; $\mathscr{U}$ and $\mathscr{I}$ represent the equivalent current flowing to and voltage across an arm capacitor C_σ. The subscript $a \in \{x, z, y\}$ and the superscript $b \in \{u, l\}$ of a variable indicate the leg and arm (upper, lower) respectively. In the averaged model, the power flowing into an arm capacitor is equal to the power absorbed by its corresponding arm:

$$p_a^b = u_a^b i_a^b = \mathscr{U}_a^b \mathscr{I}_a^b. \tag{4}$$

The voltage reference u_n is chosen equal to the middle-point voltage of the tap winding; and u_{st} is defined as the voltage from u_n to the center of the floating bus voltage u_{bus}.

As an example, the following equations can be obtained using Kirchhoff's voltage law for leg a

$$u_{an} + u_{st} + u_a^u + L_{arm}\frac{\mathrm{d}i_a^u}{\mathrm{d}t} + R_{arm}i_a^u = \frac{1}{2}u_{bus}, \tag{5}$$

$$-u_{an} - u_{st} + u_a^l + L_{arm}\frac{\mathrm{d}i_a^l}{\mathrm{d}t} + R_{arm}i_a^l = \frac{1}{2}u_{bus}. \tag{6}$$

These two equations can be rewritten using common-mode (Σ) and differential-mode (Δ) components in order to simplify the analysis

$$u_a^\Sigma + L_{arm}\frac{\mathrm{d}i_a^\Sigma}{\mathrm{d}t} + R_{arm}i_a^\Sigma = \frac{1}{2}u_{bus}, \tag{7}$$

$$u_a^\Delta + L_{arm}\frac{\mathrm{d}i_a^\Delta}{\mathrm{d}t} + R_{arm}i_a^\Delta = -u_{an} - u_{st}, \tag{8}$$

where

$$i_a^\Sigma = \frac{1}{2}(i_a^u + i_a^l), \tag{9}$$

$$i_a^\Delta = \frac{1}{2}(i_a^u - i_a^l), \tag{10}$$

$$u_a^\Sigma = \frac{1}{2}(u_a^u + u_a^l), \tag{11}$$

$$u_a^\Delta = \frac{1}{2}(u_a^u - u_a^l). \tag{12}$$

As shown in Eqs. (7) and (8), the MMC arm quantities are decoupled into control variables at the grid side ($f = f_1$) and the dc bus side ($f = f_2 = 0$)

respectively. Consequently, the MMC can be regarded as an ac/dc converter, where the dc side is open and draws zero power. Each MMC leg separately is able to exchange power with the grid, but the sum of the powers is zero.

3 Control strategy

The circuit shown in Fig. 1b looks similar to a 3-phase grid-connected MMC; however, the operating principle of the MMC-HT is very different. The characteristics of grid-side voltages lead to concerns regarding the inner energy control of the MMC. Figure 3 shows typical voltage and current waveforms at the grid side of the MMC-HT. The amplitudes of voltages u_{xn}, u_{zn} and u_{yn} are different, and they vary at different operating points of K according to Eq. (1). These voltages have the same phase angle; therefore using the Park transformation does not simplify the control. The currents shown in Fig. 3 also have the same phase angles; however, in case of reactive current injection, the phase angles can be different.

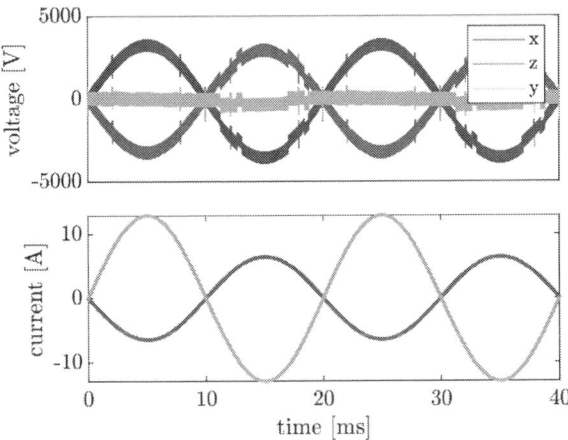

Fig. 3: An example of the MMC-HT output voltages $\{u_{xn}, u_{zn}, u_{yn}\}$ and currents $\{i_x, i_z, i_y\}$ when $K = 0$. The distortions in voltage waveforms are due to the switching events of submodules.

3.1 MMC arm balancing control

For stable operation, the MMC arm capacitor voltages should be controlled to a constant dc value. Due to the coupling described in Eq. (4), the capacitor voltages can be influenced by the power absorbed by the arms. For convenience, the common-mode and differential-mode arm power are introduced:

$$p_a^\Sigma = \frac{1}{2}(p_a^u + p_a^l) = u_a^\Sigma i_a^\Sigma + u_a^\Delta i_a^\Delta, \qquad (13)$$

$$p_a^\Delta = \frac{1}{2}(p_a^u - p_a^l) = u_a^\Delta i_a^\Sigma + u_a^\Sigma i_a^\Delta. \qquad (14)$$

If the voltage drops across L_{arm} and R_{arm} are ignored in Eqs. (7) and (8), then the common-mode and differential-mode power can be written as

$$p_a^\Sigma = \frac{1}{2}u_{bus} i_a^\Sigma - \frac{1}{2}(u_{an} + u_{st})\, i_a, \qquad (15)$$

$$p_a^\Delta = \frac{1}{4}u_{bus}\, i_a - (u_{an} + u_{st})i_a^\Sigma. \qquad (16)$$

The averaged power flowing in individual arms should be zero in steady state; however, if the capacitor voltages are unbalanced, p_a^Σ and p_a^Δ should be non-zero to change the common-mode and differential-mode capacitor voltages ($\mathscr{U}_a^\Sigma$ and $\mathscr{U}_a^\Delta$) respectively. In classic MMC control, different frequency components in the common-mode current, i_{a,f_1}^Σ at the frequency of f_1 and i_{a,f_2}^Σ at the frequency of f_2, control $\mathscr{U}_a^\Delta$ and $\mathscr{U}_a^\Sigma$ respectively [12]. The control diagrams for such a control scheme are shown in Fig. 4. Due to Kirchhoff's current law, the sum of the common-mode arm currents should be zero, which reduces the number of control degrees of freedom. Therefore, the three common-mode current references are transformed into a vector with two elements

$$\begin{bmatrix} i_1^{\Sigma*} \\ i_2^{\Sigma*} \end{bmatrix} = T_{3\to2} \begin{bmatrix} i_x^\Sigma \\ i_z^\Sigma \\ i_y^\Sigma \end{bmatrix} \qquad (17)$$

where the transform matrix

$$T_{3\to2} = \begin{bmatrix} 1 & -1 & 0 \\ -1 & -1 & 1 \end{bmatrix}. \qquad (18)$$

The transform matrix $T_{3\to2}$ is not unique, a generalized discussion about $T_{3\to2}$ can be found in [13].

As shown in Fig. 4b, the outputs of the voltage controller H_Δ are multiplied with the normalized voltages

$$v_a^* = \frac{u_{an}^*}{\mathscr{U}^{\Sigma*}}, \qquad (19)$$

in order to create ac current references at the grid frequency f_1. However, the amplitude of the output voltage in leg a can change the response of

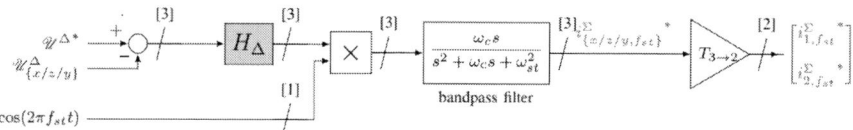

(a) Common-mode arm capacitor voltage control

(b) Differential-mode arm capacitor voltage control.

Fig. 4: Classic control methods for MMC inner energy balance.

Fig. 5: Control differential-mode arm voltage with the star-point voltage injection.

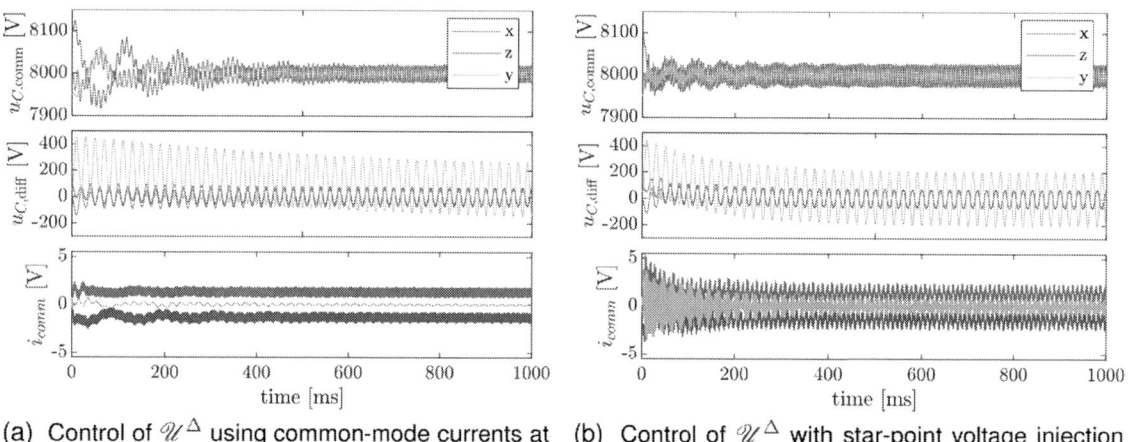

(a) Control of $\mathscr{U}^{\Delta}$ using common-mode currents at $f_1 = 50$ Hz (*Method I*).

(b) Control of $\mathscr{U}^{\Delta}$ with star-point voltage injection (*Method II*), $u_{st} = 0.04\mathscr{U}^{\Sigma*}\cos(800\pi t)$

Fig. 6: Simulation results using different control methods and $K = 0$. The arm capacitor voltages are simulated with unbalanced initial conditions.

the control loop of $\mathscr{U}_a^{\Delta}$ [14]. In the MMC-HT, the amplitude of u_{yn} depends on the LV side voltage reference and is different from the amplitudes of u_{xn} and u_{zn}. In the example shown in Fig. 3, the amplitude of u_{yn} is zero; as a result, controlling $\mathscr{U}_y^{\Delta}$ is much slower compared to the control of $\mathscr{U}_x^{\Delta}$ and $\mathscr{U}_z^{\Delta}$.

Alternatively, the differential-mode capacitor voltage can be controlled using star-point voltage injection, which has been investigated in [15]–[17] in order to achieve a broad range of frequency operation in ac/ac MMC applications. In the MMC-HT, this method can help to eliminate the changing loop

gain due to the amplitude of u_{yn}. Equation (16) indicates that the product of u_{st} and i_a^{Σ} can also create a dc component, which is independent of u_{an}. The control diagram is shown in Fig. 5. A higher frequency star-point voltage is injected in all three legs

$$u_{st} = \epsilon\mathscr{U}^{\Sigma*}\cos(2\pi f_{st}t) \qquad (20)$$

where $\mathscr{U}^{\Sigma*}$ is the voltage reference for the arm capacitors, and ϵ determines the amplitude of u_{st}. As shown in Fig. 5, the outputs of controller H_{Δ} are multiplied with a sinusoidal signal synchronous to u_{st}, which creates current references at the fre-

PCIM Europe 2024, 11–13 June 2024, Nuremberg   DOI: 10.30420/566262010

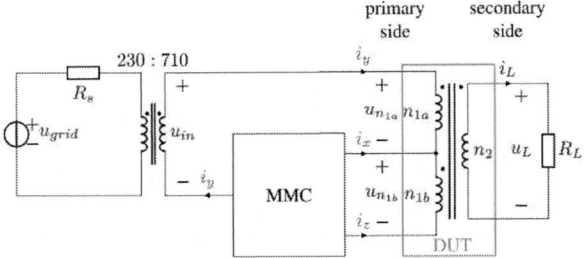

Fig. 7: Schematic of the test setup

quency of f_{st}.

Figure 6 shows simulation results at the operating point $K = 0$ for the desired output voltages and currents shown in Fig. 3. In Fig. 6a, the conventional control scheme (*Method I*) is implemented, the control of the differential-mode capacitor voltage $\mathscr{U}^\Delta$ in leg y is much slower compared to the control in the other two legs; and the error of $\mathscr{U}_y^\Delta$ does not converge to zero at the end of the simulated time. In Fig. 6b, $\mathscr{U}^\Delta$ is controlled with star-point voltage injection (*Method II*). In the beginning of the simulation, the common-mode currents are much higher compared to the currents in *Method I*; and the error of u_y^Δ is removed after 0.5s. However, the common-mode currents at f_{st} result in higher RMS arm currents; *Method II* can therefore lead to higher loss compared to *Method I*.

4 Experimental results

A scaled-down test was conducted using a single-phase transformer as shown in Fig. 7. The transformer under test (DUT) has turn ratios of $n_{1a} : n_{1b} : n_2 = 2.80 : 0.66 : 1$, where n_{1b} refers to the tap winding. The MMC has 4 submodules per arm; the arm inductance L_{arm} is 2.36 mH and the equivalent arm capacitance C_σ is 1.25 mF. The submodules switch at 20 kHz and phase-shifted carriers are used for the modulation. During the test, a $31\,\Omega$ resistance R_L is used as load. Another transformer scales the 230 Vrms grid voltage to 710 Vrms, which provides the source voltage u_{in} for the transformer under test. However, resistances of cables, protection circuit and the transformer winding at the grid side together give rise to an equivalent resistance $R_s = 2.6\Omega$. This resistance is not negligible compared to the effective load impedance seen at the grid side. As a result, R_s can cause 6-9% voltage drop before the first transformer.

The arm capacitor voltages are controlled to 300 V for *Method I*. In *Method II*, the injected star-point

voltage has a frequency $f_{st} = 200$ Hz and the amplitude parameter ϵ is 4%. The arm capacitor voltages are increased to 320 V for *Method II* in order to achieve the same voltage regulation range as *Method I*.

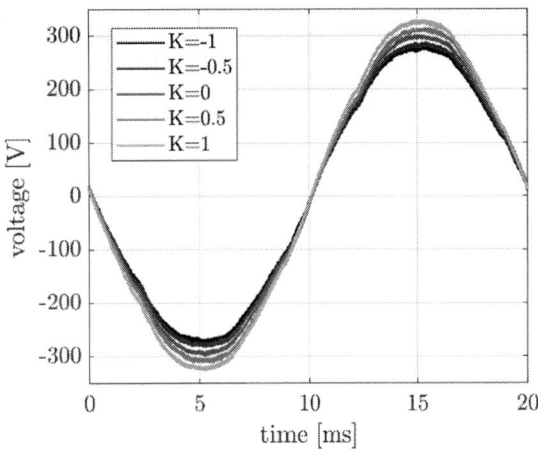

(a) Load voltage wrt. K (*Method II*)

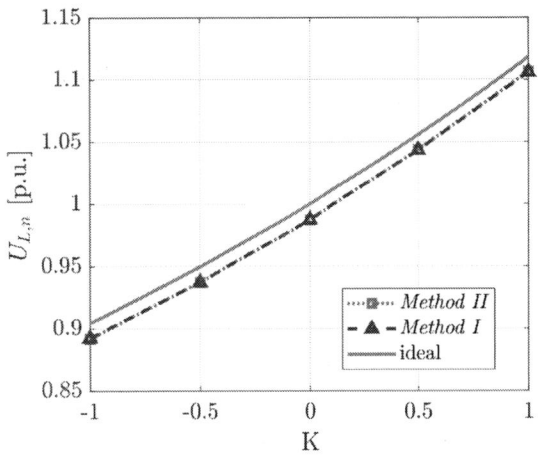

(b) Normalized load voltage.

Fig. 8: Load voltage regulation

Figure 8a shows the load voltage for different operating points; the amplitude of the load voltage changes by varying K. Figure 8b shows the load voltage as a function of K. To eliminate the influence due to the variation of the test voltage u_{in} (as defined in Fig. 7), the voltage regulation performance is evaluated by the normalized load voltage

$$U_{L,n} = \frac{u_{L,RMS}}{u_{in,RMS}} \cdot 710\,\text{V} \tag{21}$$

and the ideal load voltage is calculated using the

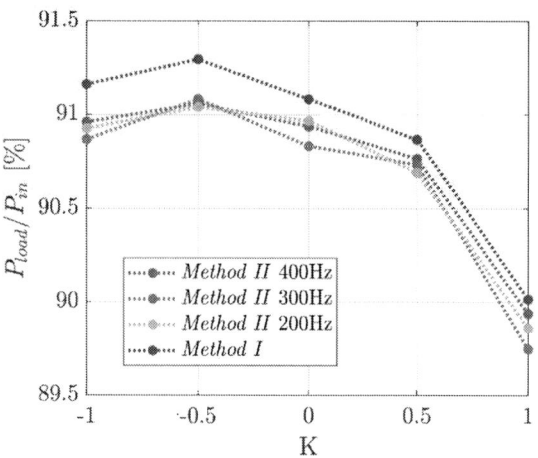

Fig. 9: Efficiency of the MMC-HT under different test conditions

expected turns ratio:

$$U_{L,ideal} = \frac{n_2}{n_{1e}} \cdot 710\,\text{V} \qquad (22)$$

where the ideal load voltage at the nominal scenario ($K = 0$) is defined as 1 pu (234 Vrms). *Method I* and *II* perform similarly in terms of voltage regulation; the methods used for MMC inner energy balance do not have a significant impact on the load voltage of the HT.

The efficiency of the MMC-HT was measured and the results are shown in Fig. 9. *Method I* results in a higher efficiency compared to *Method II*. The efficiency was measured for three different frequencies f_{st} of the injected star-point voltage. Changing f_{st} does not significantly influence the measured efficiency.

Figure 10 shows an example of what happens when the operating point K changes suddenly. In both test cases, the amplitude of u_{Load} decreases immediately after the step. Similar to the simulation results shown in Fig. 6, the differential-mode arm capacitor voltage of leg y converges faster using *Method II*.

5 Conclusions

An MMC-HT was developed and tested in a scaled-down single-phase transformer. The proposed HT is a scalable solution, which can be used at the MV side of the transformer. To realize voltage regulation, the desired output voltages of the MMC have the same phase angles but very different amplitudes, which brings two challenges into control.

Due to the equal phase angles, a per-leg control strategy is implemented instead of the more common approaches using the Park transformation. The unequal amplitudes of the voltages can lead to different responses in controlling the differential-mode arm capacitor voltages. The impact of the unequal amplitudes on the loop gains can be avoided by injecting a star-point voltage at a different frequency.

Compared to the classic approach to controlling the MMC capacitor voltages, using star-point voltage injection requires extra submodules to provide higher capacitor voltages to avoid overmodulation. The amplitude of the injected voltage is restricted by the number of extra submodules. However, the amplitude of u_{st} also has an impact on the amplitudes of the arm currents at the frequency of f_{st}, which influence the conduction loss of the MMC. In the experiments, changing the frequency f_{st} does not have a significant impact on the efficiency of the MMC-HT. Therefore, the amplitude of the star-point voltage should be optimized considering the efficiency of the MMC and cost.

References

[1] P. Bauer and S. de Haan, "Electronic tap changer for 500 kVA/10 kV distribution transformers: Design, experimental results and impact in distribution networks," in *Conference Record of 1998 IEEE Industry Applications Conference. Thirty-Third IAS Annual Meeting (Cat. No.98CH36242)*, vol. 2, 1998, 1530–1537 vol.2. DOI: 10.1109/IAS.1998.730344.

[2] J. de Oliveira Quevedo, F. E. Cazakevicius, R. C. Beltrame, T. B. Marchesan, L. Michels, *et al.*, "Analysis and design of an electronic on-load tap changer distribution transformer for automatic voltage regulation," *IEEE Transactions on Industrial Electronics*, vol. 64, no. 1, pp. 883–894, 2017. DOI: 10.1109/TIE.2016.2592463.

[3] R. P. Kandula, A. Iyer, R. Moghe, J. E. Hernandez, and D. Divan, "Power router for meshed systems based on a fractionally rated back-to-back converter," *IEEE Transactions on Power Electronics*, vol. 29, no. 10, pp. 5172–5180, 2014. DOI: 10.1109/TPEL.2013.2292854.

[4] J. Burkard and J. Biela, "Evaluation of topologies and optimal design of a hybrid distribution transformer," pp. 1–10, 2015. DOI: 10.1109/EPE.2015.7309097.

[5] H.-J. Lee, S. W. Yoon, and Y.-D. Yoon, "Hybrid distribution transformer based on an existing distribution transformer and a series-connected power

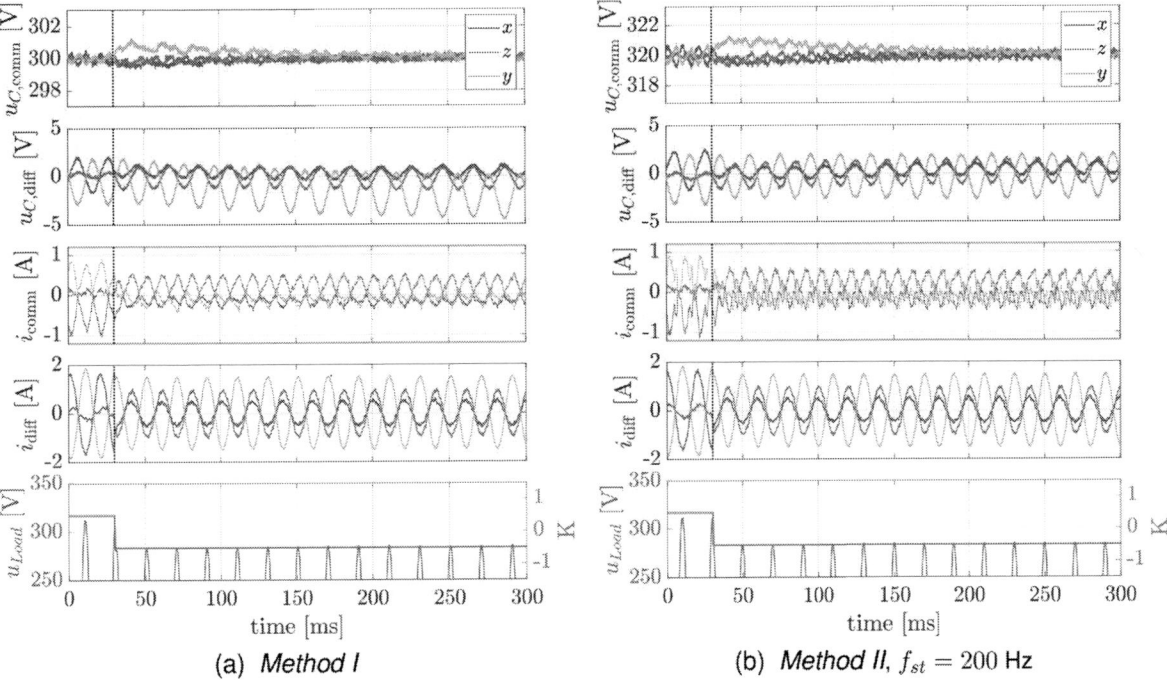

(a) *Method I*   (b) *Method II*, $f_{st} = 200$ Hz

Fig. 10: Step response at $t = 30$ ms, K changes from 0.5 to -0.5.

converter," *IEEE Transactions on Power Delivery*, vol. 37, no. 5, pp. 4202–4211, 2022. DOI: 10.1109/TPWRD.2022.3147820.

[6] R. Wang, H. Huisman, and K. Wijnands, "Distribution transformer voltage control using a single-phase matrix converter," in *2022 24th European Conference on Power Electronics and Applications (EPE'22 ECCE Europe)*, IEEE, 2022, pp. 1–8.

[7] J. Burkard and J. Biela, "Design of a protection concept for a 100-kVA hybrid transformer," *IEEE Transactions on Power Electronics*, vol. 35, no. 4, pp. 3543–3557, 2020. DOI: 10.1109/TPEL.2019.2936888.

[8] Y. Liu, D. Liang, P. Kou, M. Zhang, S. Cai, *et al.*, "Compound control system of hybrid distribution transformer," *IEEE Transactions on Industry Applications*, vol. 56, no. 6, pp. 6360–6373, 2020. DOI: 10.1109/TIA.2020.3014058.

[9] A. Carreno, M. A. Perez, and M. Malinowski, "State-feedback control of a hybrid distribution transformer for power quality improvement of a distribution grid," *IEEE Transactions on Industrial Electronics*, vol. 71, no. 2, pp. 1147–1157, 2024. DOI: 10.1109/TIE.2023.3262872.

[10] Y. Shi, L. Qi, X. Zhang, L. Chen, G. Zhao, and G. Qiao, "A novel voltage-clamped submodule scheme for enhancing overvoltage withstand capability of high voltage hybrid transformers," *IEEE*

Transactions on Industrial Electronics, vol. 71, no. 6, pp. 5453–5462, 2024. DOI: 10.1109/TIE.2023.3292852.

[11] D. C. Ludois and G. Venkataramanan, "Simplified terminal behavioral model for a modular multilevel converter," *IEEE Transactions on Power Electronics*, vol. 29, no. 4, pp. 1622–1631, 2014. DOI: 10.1109/TPEL.2013.2268856.

[12] K. Sharifabadi, L. Harnefors, H.-P. Nee, S. Norrga, and R. Teodorescu, *Design, control, and application of modular multilevel converters for HVDC transmission systems*. John Wiley & Sons, 2016. DOI: 10.1002/9781118851555.

[13] S. Milovanović and D. Dujić, "Comprehensive comparison of modular multilevel converter internal energy balancing methods," *IEEE Transactions on Power Electronics*, vol. 36, no. 8, pp. 8962–8977, 2021. DOI: 10.1109/TPEL.2021.3052607.

[14] A. J. Korn, M. Winkelnkemper, P. Steimer, and J. W. Kolar, "Capacitor voltage balancing in modular multilevel converters," in *6th IET International Conference on Power Electronics, Machines and Drives (PEMD 2012)*, 2012, pp. 1–5. DOI: 10.1049/cp.2012.0158.

[15] W. Kawamura, M. Hagiwara, and H. Akagi, "Control and experiment of a modular multilevel cascade converter based on triple-star bridge cells," *IEEE Transactions on Industry Applications*,

vol. 50, no. 5, pp. 3536–3548, 2014. DOI: 10.1109/TIA.2014.2311759.

[16] F. Kammerer, M. Gommeringer, J. Kolb, and M. Braun, "Energy balancing of the modular multilevel matrix converter based on a new transformed arm power analysis," in *2014 16th European Conference on Power Electronics and Applications*, 2014, pp. 1–10. DOI: 10.1109/EPE.2014.6910939.

[17] D. Karwatzki and A. Mertens, "Generalized control approach for a class of modular multilevel converter topologies," *IEEE Transactions on Power Electronics*, vol. 33, no. 4, pp. 2888–2900, 2018. DOI: 10.1109/TPEL.2017.2703917.

PCIM Europe 2024, 11– 13 June 2024, Nuremberg   DOI: 10.30420/566262011

Protection and Control of a Dual MMC Medium Voltage Supply

Max Dupont, Drazen Dujic

École Polytechnique Fédérale de Lausanne (EPFL), Power Electronics Laboratory, Switzerland

Corresponding author/speaker: Max Dupont, max.dupont@epfl.ch

Abstract

The Modular Multilevel Converter (MMC) is a mature technology for high and medium voltage AC to DC power conversion. In the majority of applications, the MMC is interfacing well-defined voltage levels on either side. Contrary to that, this paper presents the use of the MMC as a bipolar DC power supply, operating both as a voltage or current source. Control methods ensuring correct operation while respecting operational voltage and current limits on the DC side are presented. Furthermore, results are validated on a single MMC, and extended to two MMC units connected in series/parallel on the DC side, using RT-HIL simulations.

1   Introduction

Due to its realization based on the series connection of low-voltage submodules, the Modular Multilevel Converter (MMC), initially introduced in [1], [2], is characterized by exceptional voltage scalability, high efficiency, and good output voltage quality. These features led to the adoption of this technology in applications such as High Voltage Direct Current transmission (HVDC) [3], Medium Voltage (MV) drives [4], railway interties [5], and more. Additionally, the MMC intrinsically features di/dt limiting in case of DC short circuit thanks to its branch inductors, which is especially advantageous for MV or HV DC applications. Moreover, when full-bridge submodules are used, the short circuit currents can be further limited and even completely canceled. A demonstration of this can be found in [6], [7], where the authors provide experimental results gathered on a medium voltage MMC rated for 1.25 MW.

While this feature was used for the particular case of DC short circuits, the same principles apply to the limitation of the MMC output current for any kind of overload conditions. This allows the MMC to provide the same current limiting behavior as many other laboratory power supplies.

Identifying this, the dual MMC based converter of Fig. 1 was realized as a flexible 4 quadrant (4Q) medium voltage supply. Its two MMCs can be configured in series or parallel at the DC terminals, thanks to galvanic isolation on the AC side. In these configurations, illustrated in Fig. 2, the supply shall be usable as both a voltage source or a current source with current/voltage limiting, respectively. Due to various disturbances and parameter variations in a real system (detailed in section 4), the series current sources and parallel voltage sources operation require additional control to ensure equal power-sharing.

In the existing literature, very little can be found regarding such configurations. In bipolar HVDC systems, series connection of MMCs is common, but a neutral conductor is generally included [8], [9]. This midpoint connection absorbs the small current mismatches which are the cause of power unbalances seen in the true series connection of current sources. In truly series connected systems, the MMCs are treated as open loop controlled voltage sources, and the problem is solved by implementing the DC current controller centrally [9]. However, this solution introduces communication delays, which need to be minimized when fast current limiting is desired, as in the case of the laboratory supply. Similarly, the use of parallel connected MMCs is seldom discussed, and existing works address the current control of parallel connected MMCs, MMCs with parallel legs, or branches [10]–[12], but in all cases, the DC bus is assumed to be controlled externally. Consequently, the power-sharing challenge appearing in the parallel connection of voltage controlling MMCs is

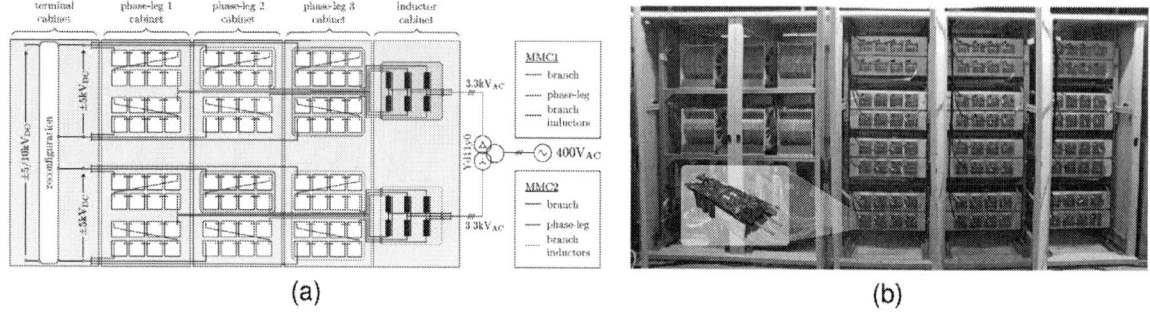

Fig. 1: (a) Layout of the dual MMC based 4Q MV supply. (b) Actual realization with zoomed in view of a submodule.

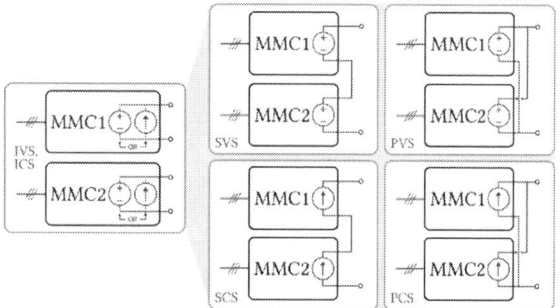

Fig. 2: MV supply internal configurations consisting of individual MMCs or series/parallel connected MMCs operated as voltage or current sources.

not addressed. Only studies on the 5 MW supply, based on four MMCs, which inspired this work, have proposed a master controller based solution, allowing arbitrary power-sharing in series/parallel configurations [13], [14].

Hence, this paper addresses the control and control based protection of a dual MMC four quadrant-supply rated for 500 kW, ±10 kV (Fig. 1). Its operation as a voltage or current source with user selected voltage and current limits is first detailed in the case of a single MMC, and extended to two units.

2 MV Supply Specifications

The realized MV supply (Fig. 1) is composed of two identical MMCs of which the key parameters are summarized in Table 1. At their DC terminals, the MMCs can be configured in series for ratings of ±10kV, 50A, or in parallel for ±5kV, 100A. This flexibility allows for covering a wide range of use cases without oversizing the converter's components. Since full-bridge submodules are used, the supply can operate in all four quadrants, and the

Tab. 1: Parameters of a single MMC.

MMC		Submodule	
P_{nom} [kW]	250	Type	Full bridge
$V_{\mathrm{DC,nom}}$ [kV]	±5	Nb/branch	8
$V_{\mathrm{AC,nom}}$ [kV]	±3.3	PWM	Unipolar
$f_{\mathrm{sw,eq}}$ [kHz]	16	f_{sw} [kHz]	1
L_{br} [mH]	1.75	V_{nom} [V]	650
R_{br} [mΩ]	66.4	C_{SM} [mF]	2.25

full operating range is illustrated by the green areas in Fig. 3. Furthermore, by manipulating voltage and current limits, the operation of the power supply can easily be confined to a smaller operational area than the maximum allowed one, and examples of possible user-defined limits are illustrated by the orange boxes in Fig. 3. This functionality is valuable for protecting the supply and the device being supplied. For instance, when supplying a unidirectional converter, the voltage should be restricted to positive values, and negative currents prevented. The realization of such constraints using a control-based implementation is explained in the next section.

3 Control of a Single MMC

3.1 Control structure overview

The MMC control structure used in this work, and depicted in Fig. 4, is adapted from [15], and leverages techniques proposed earlier in [16], [17]. It uses a cascaded implementation with two layers, and the control is divided according to three different objectives: the control of the MMC's total energy content, the balancing of this energy across branches, and control of the DC voltage or current.

To realize the first objective ① the outer loop uses a PI based controller to maintain the total energy

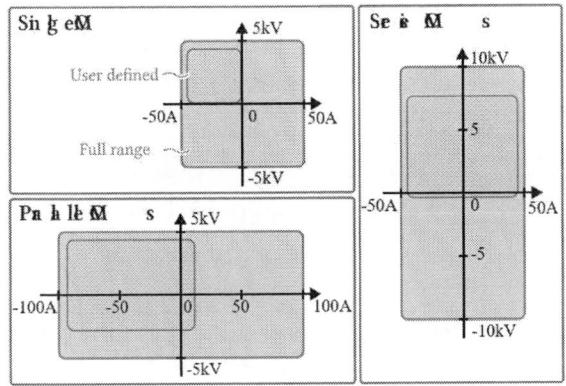

Fig. 3: Converter allowed operating range for all DC side configurations (green areas), and example of user-defined limits (orange areas).

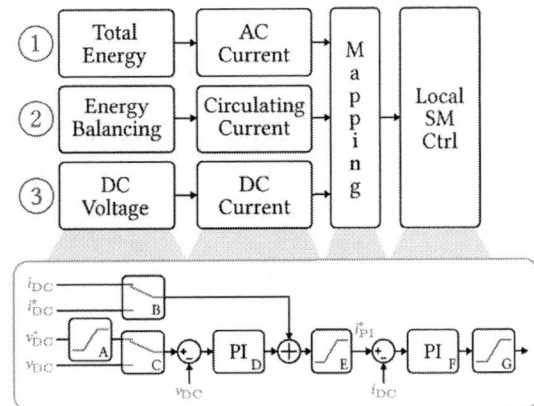

Fig. 4: MMC control structure, with details for the DC side control, which is the most relevant part for the features discussed in this work.

content of the converter to its desired value. Since the MMCs are used as rectifiers here, the output of this controller is a set of AC side current references used to exchange active power with the grid. An inner loop consisting of proportional resonant (PR) controllers at the fundamental, and 5th harmonics is used to track these references.

Energy balancing among the six branches of the MMC ② is ensured following the method of [17] and the output of this stage is a set of circulating current references. Another set of PR controllers tuned for the 1st and 2nd harmonics is used to track these references. The third set of controllers ③ is illustrated with more details as it is the most relevant part for the features discussed in this work. It consists of cascaded PI controllers to control the DC side current and voltage. Naturally, the outer loop voltage control is disabled when current source operation is desired.

All three objectives are achieved simultaneously, and the outputs of each current controller consist of voltage components that are mapped to form appropriate branch voltage references. These references are communicated to the submodules as modulation indices, where they are realized by phase-shifted carriers PWM, at the level of the branch. With modulation performed by the submodule controller, the balancing of submodules within a branch is also realized locally with a P controller as proposed in [16]. In addition, branch current references are also communicated to the submodules, where current control is enhanced by a local P controller.

3.2 DC side control

The DC side control structure, illustrated in the bottom part of Fig. 4, is realized to serve for current and voltage source operation, and selection is done by configuration of the signal selectors labeled "B" and "C".

During operation as a voltage source, the signal selectors are configured in the upper position as in Fig. 4. External current references are ignored by replacing them with the current measurements, while voltage references are accepted, and the PI controller labeled "D" is actively controlling the output voltage. In this instance, voltage limits are ensured by acting on the reference signal (saturation block "A"). Moreover, the saturation block "E" limits the reference entering the inner loop, effectively switching over to current source operation when a limit is reached. As a result, current limiting takes priority over voltage limiting in voltage source operation. The saturation limits of block "G" are set to the maximum values allowed by the MMC and are only used for anti-windup of the PI labeled "F".

When operating the MMC as a current source, both selectors are in the bottom position. Hence, the voltage reference is replaced by the voltage measurement, and the error signal entering the PI controller "D" is equal to zero, effectively switching it off. Only the inner current controller remains active, resulting in current source operation. In this configuration, current limiting is ensured by the saturation block "E" which prevents the reference from exceed-

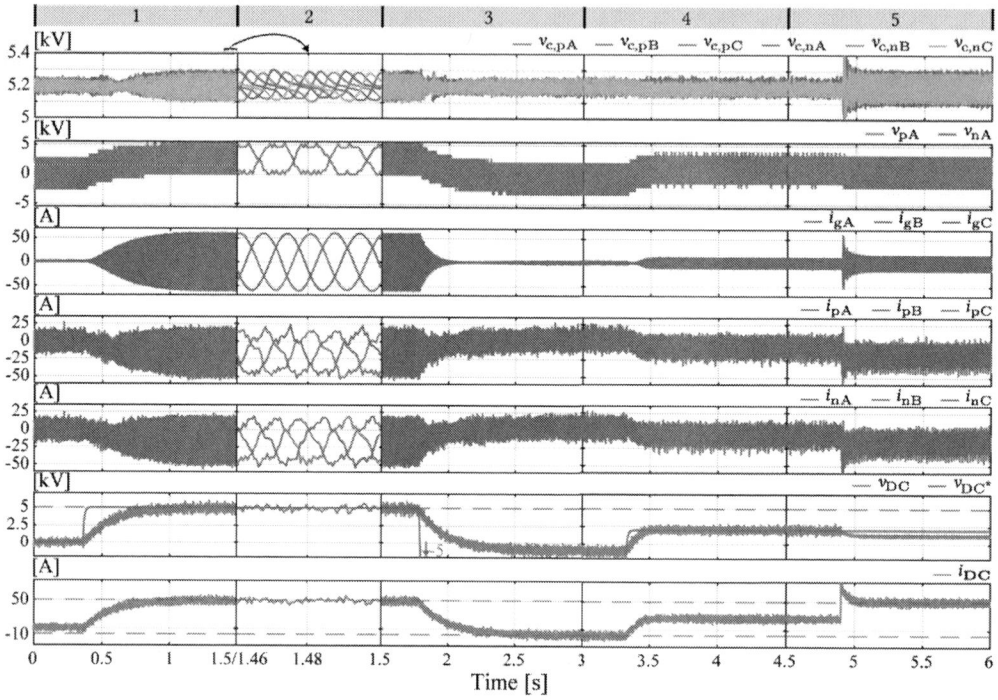

Fig. 5: RT-CHIL simulation results for a single MMC used as a voltage source. From top to bottom, the graphs show the sum of capacitor voltages per branches, the output voltage of the top and bottom branch of one phase leg, the AC terminal currents, the top branch currents, the bottom branch currents, the DC side output voltage, and DC side output current.

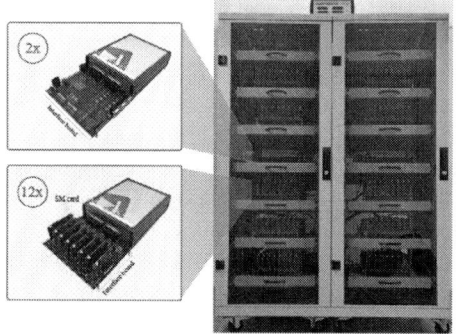

Fig. 6: RT-HIL test setup, detailed in [18].

Tab. 2: DC side control parameters.

Inner loop		Outer loop		Local	
K_p	1.11	K_p	0.07	$K_\mathrm{p,eq}$	1.33
K_i	511.8	K_i	0.001		
$f_\mathrm{s,central}$		8kHz		$f_\mathrm{s,SM}$	40kHz

each capable of simulating one MMC. Each MMC has 48 submodules and is simulated using seven PLEXIM RT Boxes, which interface with the central and submodule controllers. For dual MMC configurations, a 15th RT Box simulates the connection between the two MMCs and their load.

For the simulation of Fig. 5, the user-defined limits are set to ± 5kV, $[+50\mathrm{A}, -10\mathrm{A}]$, and the DC side control parameters used are listed in table 2 (note that $K_\mathrm{p,eq}$ is the equivalent contribution of the submodule's branch current controller to the DC side current control, acting in parallel to the central controller but with a faster sampling rate).

Initially, the MMC is turned ON and its DC voltage is regulated to zero. During the first segment, indi-

ing previously defined limits, and voltage limiting is ensured by the saturation block "G" which acts directly on the DC contribution of the branch voltage reference. Hence, in current source operation, voltage limits have priority over current limits.

3.3 Single MMC results

Figure 5 shows real time control hardware in the loop (RT-CHIL) simulation results for a single MMC operating as a voltage source. The waveforms are captured with the RT-CHIL setup shown in Fig. 6. This setup is divided into two cabinets,

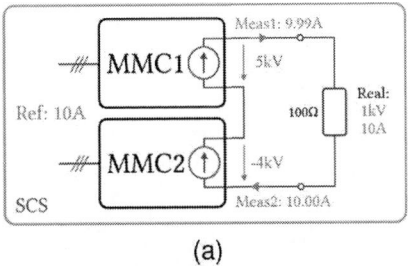 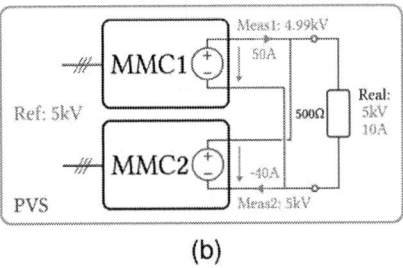

(a)   (b)

Fig. 7: Power unbalances in (a) series current source operation, and (b) parallel voltage source operation when no control action is taken to ensure power sharing.

cated on top, its voltage reference (second graph from the bottom) is stepped up to $+5kV$ at nominal load. The zoomed-in steady-state waveforms shown in the second segment confirm the proper control of the MMC. The third segment shows a voltage reference step to $-5kV$, however, the MMC does not reach this set point because the $-10A$ current limit is reached before (bottom plot), and the MMC seamlessly switches to current control to satisfy this limit rather than track the voltage reference. In the fourth segment, the voltage reference is set to $+3kV$, and the MMC switches back to voltage control and operates away from any limits. Finally, the last segment shows a sudden overload. After a brief overcurrent, the MMC manages to limit the current to $50A$ within approximately $100ms$.

4 Dual MMC Control

4.1 Series/parallel connection challenges

Among the possible dual MMC configurations (Fig. 2), series connection as voltage source (SVS) is relatively easy to achieve. The output voltage of each MMC can be set to half of the desired voltage, the current is defined by the load, and power is equally shared between the MMCs. Analogous considerations apply to the parallel connection of MMCs controlling their output current (PCS). In contrast, the series connection of current sources (SCS) and parallel connection of voltage sources (SVS) require more attention from the control point of view.

Problems arise because of the unavoidable measurement errors found in real systems, and offset errors are particularly problematic for dual MMC configurations. This issue is illustrated in the case of series connected current mode MMCs in Fig. 7a. In this example, MMC1 has a small negative offset in its current measurement, whereas MMC2's

current measurement is correct. Assuming that the actual current flowing through both MMCs is initially equal to the desired value, which is $10A$ in this example, the small current measurement error in MMC1 causes its controller to increase its output voltage. This increase leads to a current rise, and MMC2 detects a deviation from its setpoint and tries to counteract this change by decreasing its voltage. Eventually, MMC1 reaches its maximum output voltage, which is $5kV$, and MMC2, which has not reached its negative voltage saturation limit, takes over the current control. Similarly, when paralleling MMCs as voltage sources (Fig. 7b), voltage measurement errors lead to uneven current sharing.

These problems are further illustrated by the RT-CHIL simulation results of Fig. 8. In particular, Fig. 8a shows the instant when all controllers of parallel connected voltage mode MMCs are enabled. At first, the branch sum voltages (two upper graphs) are upregulated to reach the desired total internal energy. This requires active power to be taken from the AC side, hence the large AC currents (two middle graphs) during this phase. During this initial transient, small disturbances are observed on the DC side, but immediately after, the MMCs' DC terminal currents diverge and saturate to their limits.

In contrast, unbalances in series connected current mode MMCs (Fig. 8b occur much slower. This can be explained by the fact that current measurements are more accurate than the output voltage measurements in this system; therefore, the parallel voltage source configuration is more affected. In Fig. 8b, a change of current reference was even necessary to observe uneven voltage sharing.

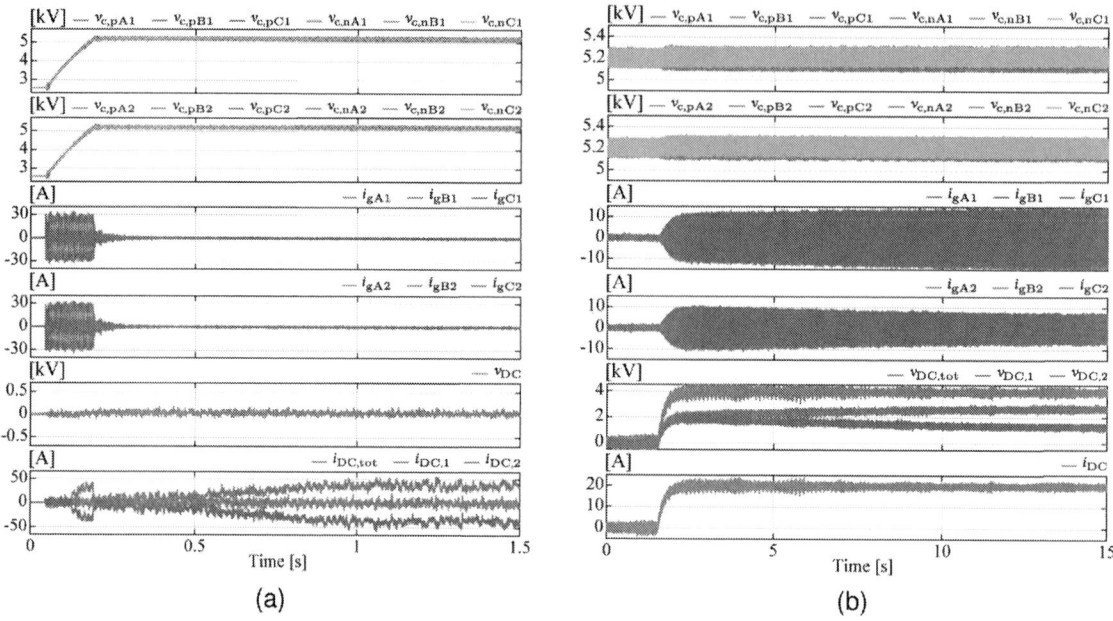

Fig. 8: RT-CHIL results showing power unbalances in (a) parallel voltage source operation, and (b) series current source operation when no control action is taken to ensure power sharing.

4.2 Master controller based solution

With an additional master controller monitoring the two MMCs, the DC side control structure discussed in section 3.2 can be leveraged to ensure equal power sharing. In particular, the signal selectors of Fig. 4 (blocks "B" and "C") are configured to accept the current and voltage references sent by the master controller. Figure 9 illustrates how the master computes the references and how the MMCs use them.

The top part of Fig. 9 shows the calculations done in the master controller and the local DC side control structure configuration corresponding to the case of SCS. In this configuration, the user decides the current setpoint that is transmitted, as is, to both local MMC controllers. The outer voltage control loop guarantees voltage sharing among series connected MMCs by using their average output voltage as a reference. Essentially, the output of PI controller "A" compensates the measurement offset in $i_{\text{DC,x}}$. The rest of the control remains identical to the case of a single current source MMC.

The bottom part of Fig. 9 illustrates the controller configuration used for PVS. The voltage setpoint, requested by the user, is transmitted

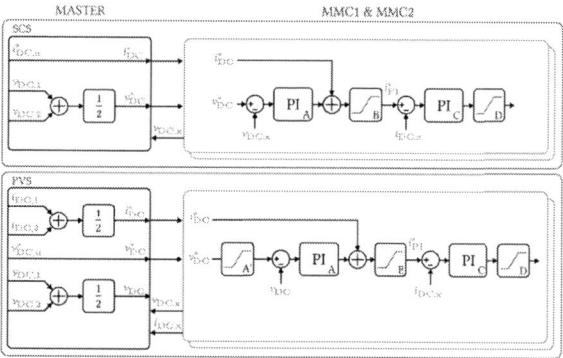

Fig. 9: Master controller based solution for power sharing in SCS and PVS.

to both MMCs. In the MMC controller, the local voltage feedback ($v_{\text{DC,x}}$) is replaced by the average of voltage measurements provided by the master controller ($(v_{\text{DC,1}} + v_{\text{DC,2}})/2$). This ensures that both MMC controllers use the exact same voltage measurement, and the outputs of the PI controllers labeled "A" in each MMC will not diverge. Additionally, the average output current is used to enforce equal current sharing.

For the SVS, the master controller communicates half of the voltage reference to each MMCs, and current is naturally determined by the load. Similarly, half of the current reference is sent to each MMC in the case of PCS.

5 RT-CHIL Simulation Results

The solutions discussed in the previous section are demonstrated through extensive RT-CHIL simulations of the complete system. While the power balancing issues mainly affect the SCS and PVS configurations, SVS and PCS are also simulated to ensure that a reasonable power balance is maintained even during current and voltage limiting.

5.1 Series current source MMCs

Starting with the case of series current mode MMCs, Fig. 10 presents simulated cases divided into four segments indicated above the graphs. Each segment presents a different transient and involves different limits. Throughout this RT-CHIL simulation, the control parameters of Table 2 are reused, and the user-defined limits are arbitrarily set to $[+8kV, -3kV]$, $[+40A, -10A]$.

Please note that the second graph from the bottom shows the output voltage of the supply and of each MMC. The plots of the MMC's outputs overlap, indicating excellent voltage sharing.

Initially, the MV supply is turned ON with a zero current reference. In the first segment, its reference (orange line in the bottom graph) is stepped to $+50A$. However, the voltage and current limits are reached, and the converter only provides $+40A$. In the second segment, its reference is changed to

$-40A$; this time, only the current limit constraints the output to $-10A$. Then, in the third segment, the setpoint is changed again to $20A$ and the supply operates away from any limit. Finally, in the fourth segment, an external voltage source is inserted in series with the load, disturbing the current control. To bring the current back toward the setpoint, the supply raises its output voltage, but the $+8kV$ limit is reached. Like with a single MMC in current source mode, the voltage limit takes priority over current limits, and the converter settles with an output current violating the $-10A$ limit. The new steady-state current is outside the user-defined limits but within the power supply's capability. For this example, the operation is maintained, but in a normal scenario, the supply would trip if the operating point is outside the user-defined limits for more than a predefined time limit.

In summary, this sequence shows that the master controller based solution maintains the same current and voltage limiting features as for a single MMC, while solving the issue of voltage sharing.

5.2 Series voltage source MMCs

A similar sequence is simulated for the case of series connected voltage source MMCs. The user-defined limits are arbitrarily set to $[+10kV, 0kV]$, $[+40A, -40A]$, and RT-CHIL results are shown in Fig. 11.

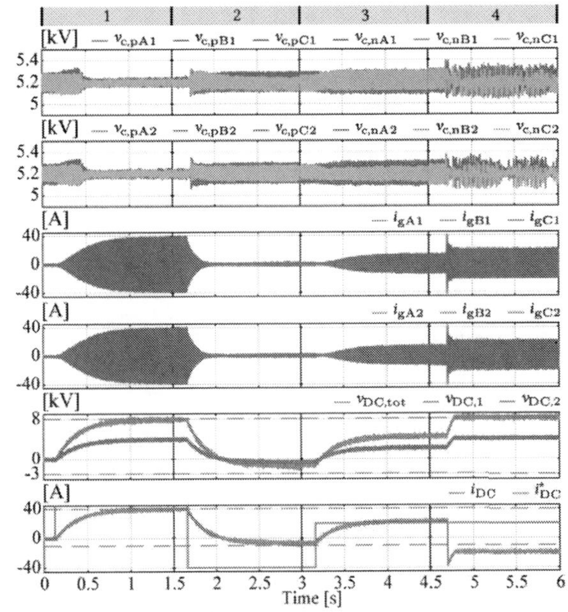

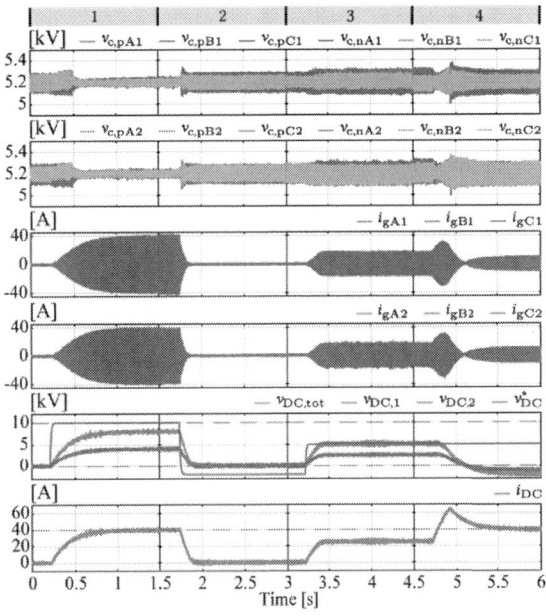

Fig. 10: RT-CHIL simulation results for SCS.

Fig. 11: RT-CHIL simulation results for SVS.

The sequence simulated is the following: The converter is initially turned ON with a voltage reference of $0V$. In the first segment, the voltage reference (pink line in the second graph from the bottom) is changed to $+10kV$. However, the output voltage of the converter (green line) does not reach this setpoint because the current limit is reached first (bottom graph). In the second segment, the voltage reference is changed to $-2kV$. However, the previously defined voltage limit prevents the converter from following it, and the output settles at $0V$. In the third segment, the setpoint is changed to $+5kV$. This time, the converter can reach the setpoint and operates away from the current and voltage limits. In the last segment, a load change pushes the converter into current limiting mode, and the $0V$ voltage limit is violated to allow the current limiting to work properly.

Throughout this simulation, the output voltages of the two MMCs are perfectly balanced (second graph from the bottom). This is natural for SVS, but this simulation also highlights that this balance is maintained during current limiting.

5.3 Parallel voltage source MMCs

Fig. 12 shows RT-CHIL simulation results for the case of parallel voltage source MMCs, when the user-defined limits are set to $[+5kV, 0kV]$, $[+50A, -100A]$.

The first segment illustrates a voltage reference step from $0V$ to $+5kV$. Constrained by the current limit of $+50A$, the voltage settles at $2.5kV$. In the second segment, due to a load change, the MV supply, which was delivering power, starts absorbing it. In these new conditions, the current is now far from the limit and the supply reaches its $5kV$ setpoint. In the third segment, this setpoint is then changed to $0V$. This time, the converter is prevented from reaching its voltage reference due to the negative current limit. Finally, the third segment shows a new reference step to $3kV$, bringing the converter's operating point away from the limits, and into normal operation.

The effectiveness of the presented method is once again demonstrated as the MMC's output current remained balanced during the whole simulation.

5.4 Parallel current source MMCs

Finally, the last set of RT-CHIL simulation results is shown in Fig. 13 for the case of parallel current source MMCs. The user-defined limits are set to $[0kV, -5kV]$, $[+50A, -100A]$.

The first segment illustrates a current reference step from $0A$ to $-100A$; no limit is encountered, and the output current reaches its reference. In the second segment, a voltage source is inserted in series with the load, which reduces

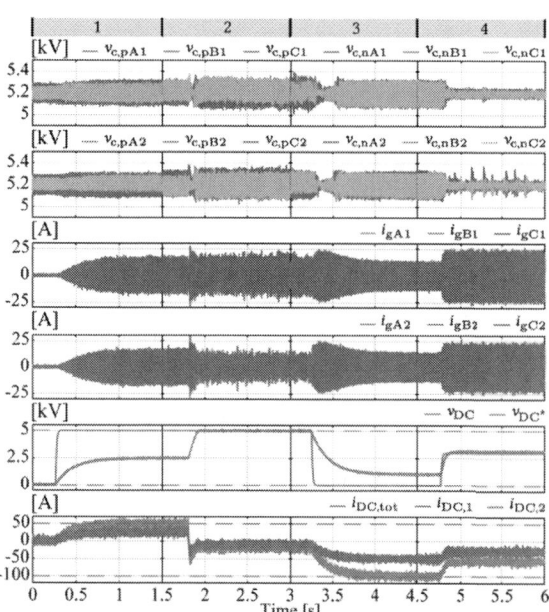

Fig. 12: RT-CHIL simulation results for PVS.

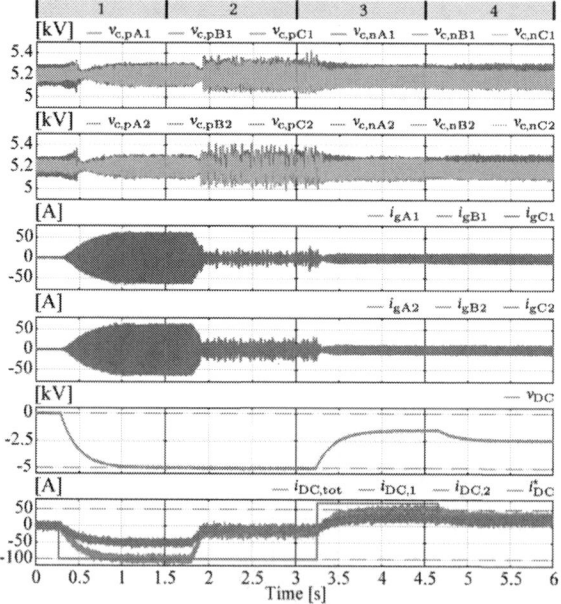

Fig. 13: RT-CHIL simulation results for PCS.

the output current. Since the MV supply was already operating at its maximum negative voltage, it could not bring the output current back toward the setpoint. In the third part of the sequence, the current reference is stepped to $+70A$. As a result, the output voltage leaves its $-5kV$ limit. Nevertheless, the new setpoint is not reached because it is outside of the user-defined current limits. Finally, the reference is changed to $+30A$ and the converter settles about this new setpoint.

Similarly to the case of PVS, the output current of each MMC remained balanced, but this outcome was not guaranteed during voltage limiting. Indeed, when voltage limiting is in effect, the arrangement becomes one of parallel voltage sources, and the MMCs are essentially operated in open loop on the DC side. As a result, current imbalances are caused by differences in internal resistances between the two MMCs. Since these differences are negligible in this system, the current remained balanced.

6 Conclusion

This work presented a dual MMC-based MV converter system intended for use as a general-purpose laboratory supply. Its control-based implementation of current and voltage limiting was discussed and demonstrated for the case of a single MMC and extended to multi-MMC configurations. Challenges of voltage sharing in series connected current mode MMCs, and current sharing in parallel connected voltage mode MMCs were discussed. A master controller-based solution was presented and its efficacy demonstrated by extensive RT-CHIL simulations for all intended configurations of the MV supply. The current and voltage limiting features were thoroughly tested for each of these configurations, further demonstrating the seamless switchover ensured by this control-based implementation.

References

[1] A. Lesnicar and R. Marquardt, "An innovative modular multilevel converter topology suitable for a wide power range," in *2003 IEEE Bologna Power Tech Conference Proceedings*, vol. 3, 2003, p. 6. DOI: 10.1109/PTC.2003.1304403.

[2] S. Allebrod, R. Hamerski, and R. Marquardt, "New transformerless, scalable Modular Multilevel Converters for HVDC-transmission," in *2008 IEEE Power Electronics Specialists Conference*, 2008, pp. 174–179. DOI: 10.1109/PESC.2008.4591920.

[3] SIEMENS. "HVDC PLUS," SIEMENS. (2012), [Online]. Available: https://p3.aprimocdn.net/siemensenergy/87cdd3cd-be98-4a79-b9c3-b03b00d3e37a/2022-03-11-HVDC-PLUS-pdf_Original%20file.pdf (visited on 12/22/2023).

[4] SIEMENS. "SINAMICS Perfect Harmony GH150," SIEMENS. (2022), [Online]. Available: https://assets.new.siemens.com/siemens/assets/api/uuid:18aa6fa6-7bfa-450a-8328-2cfe787e8844/gh150acbrochurejune2019.pdf (visited on 03/14/2024).

[5] M. Winkelnkemper, A. Korn, and P. Steimer, "A modular direct converter for transformerless rail interties," in *2010 IEEE International Symposium on Industrial Electronics*, 2010, pp. 562–567. DOI: 10.1109/ISIE.2010.5637826.

[6] M. Winkelnkemper, L. Schwager, P. Blaszczyk, M. Steurer, and D. Soto, "Short circuit output protection of MMC in voltage source control mode," in *2016 IEEE Energy Conversion Congress and Exposition (ECCE)*, 2016, pp. 1–6. DOI: 10.1109/ECCE.2016.7855427.

[7] K. Sun, D. Soto, M. Steurer, and M. Faruque, "Experimental verification of limiting fault currents in MVDC systems by using modular multilevel converters," in *2015 IEEE Electric Ship Technologies Symposium (ESTS)*, 2015, pp. 27–33. DOI: 10.1109/ESTS.2015.7157857.

[8] W. Guoqiang, W. Haiqing, Y. Rong, J. Lei, H. Ting, and L. Yu, "Dc resonance suppression for hybrid double-ended hvdc transmission systems," in *12th IET International Conference on AC and DC Power Transmission (ACDC 2016)*, 2016, pp. 1–5. DOI: 10.1049/cp.2016.0392.

[9] C. Hirsching, M. Goertz, S. Wenig, S. Beckler, M. Suriyah, and T. Leibfried, "On control and balancing of mmc-hvdc links in rigid bipolar configuration," in *15th IET International Conference on AC and DC Power Transmission (ACDC 2019)*, 2019, pp. 1–6. DOI: 10.1049/cp.2019.0026.

[10] F. Gao, D. Niu, H. Tian, C. Jia, N. Li, and Y. Zhao, "Control of Parallel-Connected Modular Multilevel Converters," *IEEE Transactions on Power Electronics*, vol. 30, no. 1, pp. 372–386, 2015. DOI: 10.1109/TPEL.2014.2313333.

[11] J. Pou, S. Ceballos, G. Konstantinou, G. J. Capella, and V. G. Agelidis, "Control strategy to balance operation of parallel connected legs of modular multilevel converters," in *2013 IEEE International Symposium on Industrial Electronics*, 2013, pp. 1–7. DOI: 10.1109/ISIE.2013.6563685.

[12] S. Milovanovic and D. Dujic, "On Power Scalability of Modular Multilevel Converters: Increasing Current Ratings Through Branch Paralleling," *IEEE Power Electronics Magazine*, vol. 7, no. 2, pp. 53–63, 2020. DOI: 10.1109/MPEL.2020.2984350.

[13] P. Blaszczyk, M. Steurer, D. Soto, M. Bosworth, and M. Winkelnkemper, "Power balancing in multi-converter systems composed of modular multilevel converters (MMCs)," in *2016 18th European Conference on Power Electronics and Applications (EPE'16 ECCE Europe)*, 2016, pp. 1–10. DOI: 10.1109/EPE.2016.7695321.

[14] P. Blaszczyk, M. Winkelnkemper, and L. Schwager, "Converter energy balancing in MMC system energy sharing using master controller," in *2015 International Conference on Electrical Drives and Power Electronics (EDPE)*, 2015, pp. 30–37. DOI: 10.1109/EDPE.2015.7325265.

[15] P. Blaszczyk, M. Steurer, D. Soto, F. Bogdan, J. Hauer, *et al.*, "Modular multilevel converter based test bed for MVDC applications — A case study

with a 12 kV, 5 MW setup," in *2016 IEEE International Power Electronics and Motion Control Conference (PEMC)*, 2016, pp. 139–145. DOI: 10.1109/EPEPEMC.2016.7751988.

[16] M. Hagiwara and H. Akagi, "Control and Experiment of Pulsewidth-Modulated Modular Multilevel Converters," *IEEE Transactions on Power Electronics*, vol. 24, no. 7, pp. 1737–1746, 2009. DOI: 10.1109/TPEL.2009.2014236.

[17] J. Kolb, F. Kammerer, M. Gommeringer, and M. Braun, "Cascaded Control System of the Modular Multilevel Converter for Feeding Variable-Speed Drives," *IEEE Transactions on Power Electronics*, vol. 30, no. 1, pp. 349–357, 2015. DOI: 10.1109/TPEL.2014.2299894.

[18] S. Milovanovic, I. Polanco, M. Utvic, and D. Dujic, "Flexible and efficient mmc digital twin realized with small-scale real-time simulators," *IEEE Power Electronics Magazine*, vol. 8, no. 2, pp. 24–33, 2021. DOI: 10.1109/MPEL.2021.3075803.

PCIM Europe 2024, 11– 13 June 2024, Nuremberg   DOI: 10.30420/566262012

A Station Power Electronics Converter with High Thermal Endurance to Pole-To-Pole Short Circuits for LVDC Distribution Grid.

Frédéric Reymond-Laruina[1,2,3], Loïc Quéval[2,3], Gustavo Alves de Lima Henn[2,3,4], Damien Huchet[2,3], Djamel Hadbi[1], Philippe Egrot[1], Stéphane Mercier[5] and Marc Petit[2,3]

[1] Laboratoire des Matériels Electriques, EDF R&D, France

[2] Laboratoire de Génie Electrique et Electrotechnique de Paris, CentraleSupélec, CNRS, Université Paris-Saclay, France

[3] Laboratoire de Génie Electrique et Electrotechnique de Paris, CNRS, Sorbonne Université, France

[4] Instituto de Engenharias e Desenvolvimento Sustentável, Universidade da Integração Internacional da Lusofonia Afro-Brasileira, Brazil

[5] Socomec Group, France

Corresponding author: Frédéric Reymond-Laruina, frederic.reymond-laruina@edf.fr
Speaker: Frédéric Reymond-Laruina, frederic.reymond-laruina@edf.fr

Abstract

Operating a LVDC distribution grid, to increase the insertion of native DC users, implies to respect the selectivity of the grid. Such a constraint has direct consequences on the sizing of the components and available converter topologies. In this paper, a novel topology is proposed. Based on high current parallel diode, this topology allows to maintain components within their operating range by adjusting the diode parameters. Experimental observations validate the simulation results, while a first technical-economic analysis highlights the interest of this solution compared to a traditional oversizing approach.

1   Introduction

Over the past few decades, the development of distributed electricity generation and energy storage systems, as well as the increase in the number of native direct current (DC) consumers, has led to an increasing focus on low voltage direct current (LVDC) distribution grids or at least dedicated DC feeders. Such a grid would allow better interaction between applications [1], but also, and more importantly, greater efficiency. Firstly, by suppressing skin effect in DC cables, as well as reactive power exchange, and secondly because, future standards could allow for voltages up to 1500 V, resulting in lower transmission losses [2]. Finally, a DC distribution grid would allow for centralization and standardization of the AC/DC conversion step, which is currently performed sub optimally by each user. Due to the large number of transformer stations, the use of an electronic transformer must be avoided, due to its cost, size, and high losses. In this context, the preferred solution is to keep the existing LVAC transformer and to implement a centralized AC/DC converter, called here station converter, on its secondary [3].

Despite a large number of mature and commercially available converters, electrical safety remains an obstacle to the deployment of these solutions. This is because the behavior of a DC short circuit differs largely from that of an AC short circuit, due to the presence of capacitors on the DC link. In the event of a fault, these capacitors discharge almost instantaneously, before the AC grid current flows through the converter [4]. The absence of zero-crossing of the current requires the generation of an opposite voltage to force the decrease of the DC current and interrupt the fault [5]. The cut-off time for traditional protection is then extended from at least tens of ms to several hundreds of ms. Combined with the limited overcurrent capabilities of semiconductor components, these factors have encouraged manufacturers to look for fast protective devices. Many different proposals can be found in the literature, such as hybrid or electronic circuit breakers [6], which can achieve response times of 2 to 0.1 ms. Another solution is to use fuses, already used in distribution grids, which are cheaper but offer longer response time [7], [8].

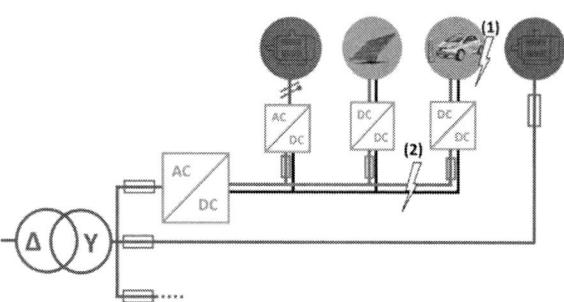

Fig. 1 Hybrid distribution grid with the substation converter interfacing AC and DC grids, the existing fuse protection and possible short-circuit locations.

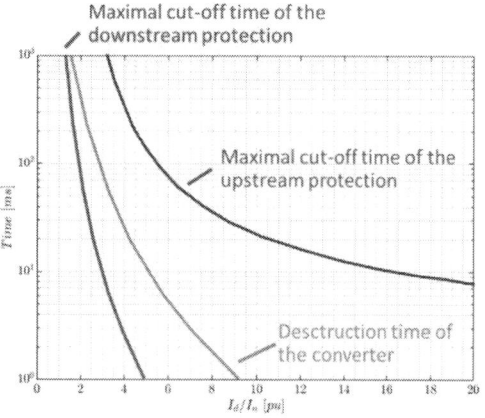

Fig. 2 Example of protection not coordinated with the admissible module thermal stress of the converter.

However, a fast protection is not sufficient for LVDC distribution grid: the interruption principle must also comply with the Distribution System Operator's (DSO) grid code. Among these constraints, a significant one is the "selectivity". During a fault event, the closest upstream protection to the fault must react first. Therefore, the further away a protection is from a consumer, the later the interruption should be. In the situation of **Fig. 1**, whatever the fault location (1) or (2), the AC/DC converter must withstand the fault up to the upstream blue protection react. It must have a high thermal endurance. To check the AC/DC converter sizing, time-current curves can be used, as illustrated in **Fig. 2**, For any short-circuit at location (1), the red protection will cut-off the fault in an acceptable range of time for the station converter. However, for short-circuit at location (2), the blue protection cutting off time is to slow: the AC/DC converter would have reached its maximal thermal constraint and could be damaged.

This paper addresses this issue by proposing a new topology of power electronics converter with high thermal endurance allowing it to withstand a pole-to-pole short circuit and respect the selectivity principle. The sizing method is detailed, as well as the operating constraints of the distribution grid. The results are compared to a traditional oversizing approach. Finally, a prototype is built to validate the operating principle of the proposed topology.

2 A High Thermal Endurance Topology

In the literacy, the simplest way to achieve this goal is to oversize the converter. This results in an additional cost representing at least half the price of a normal converter [9]. It also leads to a partial loading of the converter and therefore reduced efficiency.

De Olieira *et al.* [10] describe the use of by-pass thyristors in parallel of each arms, while Li *et al.* [11] propose to implement star-connected thyristors. Both solutions aim to deviate the short-circuit current towards components having high thermal endurance. Although promising, such active protections are not suitable for distribution grids: they require the correct operation of a fault detection circuit and a reliable control unit.

A passive solution would be the Implementation of a "sacrificial diode" in parallel of the DC-link [12]. In case of a short-circuit with a voltage reversal, the short circuit current is naturally deviated in the "sacrificial diode". The converter is protected, but only during a small amount of time in regards of the duration of the short-circuit.

Because of the limited applicability of the existing solutions, we propose to improve existing voltage source converter topology by adding diodes, called protection diodes, in parallel of every valve or arm. A valve is the association of a freewheeling diode with an IGBT or a MOSFET.

Three main associations can be distinguished depending on the converter topology and the DC voltage level considered, as illustrated in **Fig. 3** for a 2L-VSC. They are named respectively protection:

- **Per valve:** each valve has N protection diode in parallel.
- **Per arm:** each arm has N protection diodes in parallel; suitable for multilevel topologies.
- **Per arm with serial assembly:** each arm has N protection branch in parallel and M diodes in series; suitable to withstand high voltage.

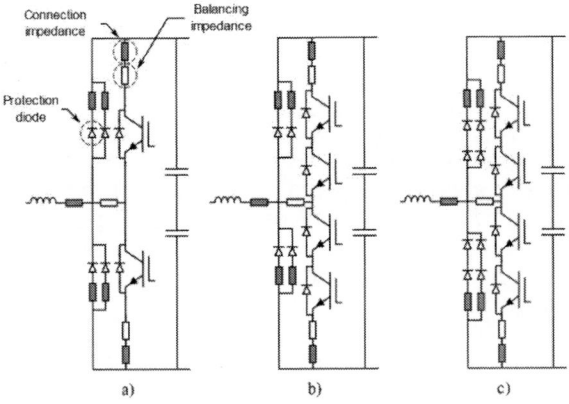

Fig. 3 2L-VSC with protection diodes: a) per valve ($N = 2, M = 1$), b) per arm ($N = 2, M = 2$), c) per arm with serial assembly (($N = 2, M = 2, L = 2$).

Whatever the association, the idea is to deviate as much current as possible in case of a DC short-circuit so that the valves remain within their acceptable operating range. This can be achieved by selecting protection diodes with relevant parameters and adjusting their number in parallel. If necessary, one can insert balancing impedances to offset the consequences of the connection impedance and protect the valve diodes.

Consider the 2L-VSC shown in **Fig. 3 c)**. Expressions for the protection branch currents $i_{P,tot}(t)$ and for the studied converter arm $i_{D,tot}(t)$ can be derived,

$$i_{P,tot}(t) = \frac{V_{D,eq} - V_{P,eq} + (R_{D,eq} + R_S).i_d(t)}{R_{P,eq} + R_{D,eq} + R_S} \quad (1)$$

$$i_{D,tot}(t) = \frac{V_{P,eq} - V_{D,eq} + R_{P,eq}.i_d(t)}{R_{P,eq} + R_{D,eq} + R_S} \quad (2)$$

Where

- $V_{D,eq} = \sum_{i=1}^{M} V_{D,i}$ is the equivalent threshold voltage of the valve diodes in the converter arm [V]

- $V_{P,eq} = R_{P,eq} \sum_{j=1}^{N} \frac{\sum_{i=1}^{L} V_{P,i,j}}{\sum_{i=1}^{L} R_{P,i,j}}$ is the equivalent threshold voltage of the protection diodes in the parallel protection branches [V]

- $R_{D,eq} = \sum_{i=1}^{M} R_{D,i}$ is the equivalent conduction resistance of the valve diodes in the converter arm [Ω]

- $R_{P,eq} = \sum_{j=1}^{N} \frac{1}{\sum_{i=1}^{L} R_{P,i,j}}$ is the equivalent conduction resistance of the protection diodes in the parallel protection branch [Ω]

- R_S is the balancing impedance connected in series with the converter arm [Ω]

- i_d is the short-circuit current in the converter arm [A]

- L is the number of protection diodes in series for each parallel protection branch [ ]

- M is the number of valves for the converter arm [ ]

- N is the number of protection branch in parallel of the converter arm [ ]

- $R_{P,i,j}$ is the conduction resistance of the ith protection diode in the jth protection branch [Ω]

- $R_{D,i}$ is the conduction resistance of the ith valve diode in the converter arm [Ω]

- $V_{P,i,j}$ is the threshold voltage of the ith protection diode in the jth protection branch [V]

- $V_{D,i}$ is the threshold voltage of the ith valve diode in the converter arm [V]

From these expressions, we can deduce that the fault current deviation is maximal by selecting protection diodes in such a way as to have $V_{P,eq} \leq V_{D,eq}$ and $R_{P,eq} \leq R_{D,eq} + R_S$. As a rule of thumb, the protection diode chosen should have a conduction resistance and threshold voltage lower than the minimum values of the diodes integrated in the valves of the studied arm.

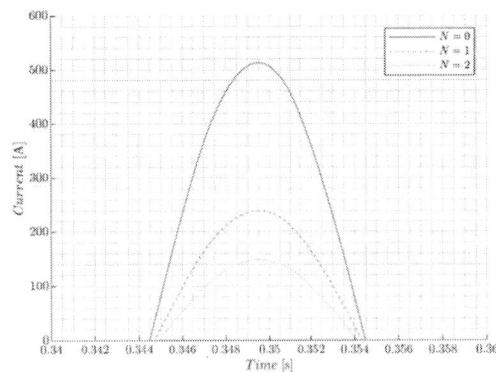

Fig. 4 Impact of the number of protection diodes N_D on the integrated diode current for a 2L-VSC

It should be noted that by selecting diodes with different threshold voltage, a delay in conduction occurs between them, in ascending order of their threshold voltage. Provided previous conditions are meet, the valve diodes of the converter arm should not conduct while the condition is not meet,

$$|i_d(t)| \leq \frac{V_{D,eq} - V_{P,eq}}{R_{P,eq}} \qquad (3)$$

An illustration of this phenomenon, in the case of a 2L-VSC with a protection per valve, is given in **Fig. 4**: protection diode having a lower conduction resistance and voltage threshold, the more protection diodes there are, the less current there is in the valve diodes and the longer the delay in valve diode conduction.

3 Case study

In order to validate the proposed solution, we compare it here with the traditional oversizing approach.

3.1 Constraints and assumptions

3.1.1 Grid requirements

Harmonics	Odd	Even
h < 11	4 %	1 %
11 ≤ h < 17	2 %	0.5 %
17 ≤ h < 23	1.5 %	0.375 %
23 ≤ h < 35	0.6 %	0.15 %
35 ≤ h	0.3 %	0.075 %

Table 1 Harmonic current constraints in relation to the magnitude of the fundamental current at rated power

3 constraints must be meet by a grid-connected converter in a LV distribution grid: power quality, efficiency, and selectivity.

The *power quality* on the DC side of the substation converter is still debated in the literature. The only consensus parameter is the voltage ripple, but its exact value is still undefined. Similarly to the accepted voltage deviation on the LVAC grid [13], we propose to select a ripple whose peak-to-peak value is less than 5% of the nominal DC voltage. For the AC side, an overall harmonic content $THD_i \leq 8\%$ at rated power, with the constraints per current harmonic specified in Table 1, is recommended [14].

An *efficiency* η_{min} of at least 97 % is required in order to limit the efficiency decrease of the transformer substation to 99% [3]. This value must include semiconductors and passive components losses. The efficiency constraint is bi-directional and therefore calculated for 100 %, 75 %, 50 % and 25 % of the rated power.

For *selectivity*, existing fuses in substations have a response time of a few seconds to hundreds of milliseconds [6] which is too slow for their use with power electronics. Therefore, we assume here

that a dedicated protection device is implemented, with a maximum cut-off time given by,

$$t_{trip,max} = e^{15.38\left(\frac{I_d}{I_n}-1\right)^{-0.5309}} \qquad (4)$$

where $I_d\,[A]$ is the estimated fault current in steady state and $I_n[A]$ the nominal current. Due to the specific behavior of a VSC during a short circuit [4], [12], the steady state fault current is almost perfectly sinusoidal. Its RMS value I_d can be estimated by,

$$I_d = \frac{V}{\sqrt{\left(\frac{2}{3}(2r_{DC}l_{fault}+R_d)+R_L\right)^2 + \left(\frac{4}{3}x_{DC}l_{fault}+X\right)^2}} \qquad (5)$$

where V is the phase to neutral RMS voltage, r_{DC} is the cable resistance per unit length, l_{fault} is the fault distance from the feeder, R_d is the fault resistance, R_L is the tie reactor resistance, x_{DC} is the cable reactance per unit length, and X is the tie reactor reactance.

3.1.2 Grid definition

The parameters of the studied distribution grid are summarized in Table 2. From a LVAC point of view, it is assumed that the substation converter is inserted in an existing MV/LV substation. Therefore, cables allowing the connection of the transformer to the converter are neglected in front of the transformer windings and converter filter. The short-circuit current of the converter being limited in front of the transformer capability, it is assumed that the transformer won't saturate.

From an LVDC point of view, it is assumed that existing LVAC cables are reused [15] in a 750 V DC monopolar symmetrical configuration. Due to space limitations in the substation, the DC-link mid-point is grounded at the same point as the MV/LV transformer. The Protective Earth conductor is distributed in order to implement a TN-S configuration and achieve a protection against indirect contact.

MV/LV Transformer	
Nominal power	400 KVA
Primary voltage L-L RMS	20 kV
Secondary voltage L-L RMS	400 V
Short-circuit voltage	4 %
Coupling	Dy11
DC Cable [22]	
Resistance per unit length r_{DC}	0.124 Ω/km
Reactance per unit length x_{DC}	0.071 Ω/km
Length	700 m
Fault	
Location l_{fault}	0 − 700 m

Starting time t_{fault}	0 – 20 ms
Impedance	1 mΩ
Resistive Load	
Power	20 kW

Table 2 LVAC and LVDC grid parameters

3.1.3 Station Converter definition

Because of its low number of semiconductors and simplicity, a 2L-VSC is considered. It is interfaced to the LVAC grid with a tie reactor, as illustrated in Figure 2. As regards the control of the converter, a PI controller is implemented, according to the method presented by S. Akkari [16]. Due to the grounding, only a Sinusoidal-PWM is considered.

The converter being added in a MV/LV transformer station, the maximal ambient operating temperature is 40°C [17] and cooling is by natural convection, based on the example of transformers.

To ensure compatibility with the various transformer stations, we are assuming that the station converter is composed of several 20 kVA elementary 2L-VSC converter which are parallelized. In the next paragraphs, we focus on this elementary converter.

3.2 Sizing Methodology

The following paragraphs detail the different steps of the converter sizing process.

3.2.1 Module selection

Due to the large number of available products, we focus on integrated 2L-VSCs, also referred to as "sixpack" and "dual" configuration. For each reference, we extract the module cost and its datasheet parameters: diode and IGBT switching losses as well as conduction losses, junction thermal model,

diode voltage threshold, diode conduction resistance, diode thermal stress $i^2 t_{max}$, maximum steady-state junction temperature $T_{j,max}$, dimensions and maximal repetitive current I_{RRM}. The commutation losses depend on the driver gate resistance. For modules under 150 A, there is usually no integrated driver. In that case, the gate resistance is assumed to be equal to 10 Ω. For modules over 150 A, we use the resistance value of the integrated driver. The selection of the switching frequency is left to the user's discretion but must stay within the range of the driver and module capabilities.

As for the protection diode, considering the need for a high thermal endurance, only the reference DD89N12K is considered. For this reference, all previous parameters are extracted from the manufacturer datasheet.

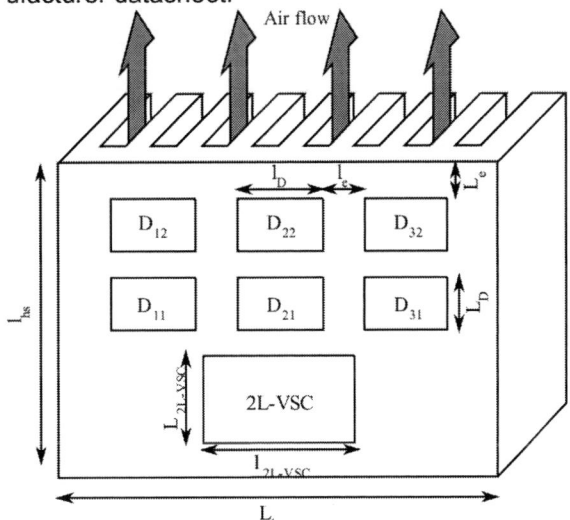

Fig. 6 Arrangement of the modules on the heatsink with 2 protection diodes per valves.

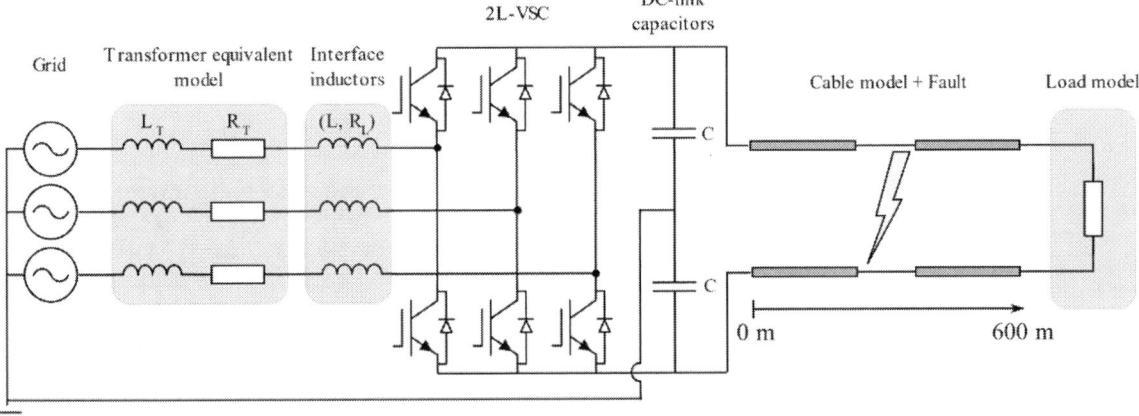

Fig 5 Circuit under study: grid, step-down transformer of the substation, 2-level voltage source converter with its tie reactor and DC-link capacitors and the DC grid.

3.2.2 Heat sink selection

Because diodes have a negative temperature coefficient [18], their layout is significant in our topology. For diodes in parallel, the higher the junction temperature of one diode, the greater the current it carries compared with the others. In extreme situation it can lead to the destruction of the diode. The solution is to homogenize junction temperatures by using the same heatsink for the components. During short-circuit, the layout in is adopted. The air flow ensures a rising temperature profile and a bigger deviation in the protection diodes.

A ABL177AB heatsink is considered. Its thermal parameters in natural convection can be determined from the datasheet as a function of the considered components.

3.2.3 Capacitors selection

During a pole-to-pole short circuit, the voltage can become negative [4], [19] and damage electrolytic capacitors, which can only accept a voltage reversal of 1 to 1.5 V [20]. To overcome this issue, film capacitors should be used.

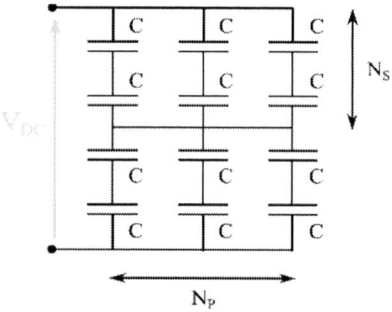

Fig. 7 Series-parallel association of capacitors for $N_P = 3$ and $N_S = 2$ with a middle point

For each C44U-M capacitor reference, the cost $C_{tot,C}$ [€] and volume $V_{tot,C}$ [L] of the series-parallel association of capacitors to reach a given value C_{obj} [F] given by,

$$C_{tot,C} = 2N_S N_P C_C \tag{1}$$

$$V_{tot,C} = 2N_S N_P V \tag{2}$$

$$N_S = \frac{V_{DC}}{2V_C} \tag{1}$$

$$N_P = \frac{2N_S C_{obj}}{C} \tag{2}$$

Where N_S is the number of capacitors in series, N_P is the number of capacitors in parallel, C_C is the capacitor reference cost [€], V is the capacitor

reference volume [L], V_C is the capacitor nominal voltage [V], V_{DC} is the DC-link nominal voltage [V], C is the reference capacitance [F].

In a representation of volume as a function of cost and among the non-dominated solutions, we will retain the one with the smallest distance to the origin.

3.2.4 Tie reactor selection

During a pole-to-pole short circuit, the magnetic core of the coil could saturate, decreasing its inductance. This behavior leads to a further increase in the fault current from the grid, as illustrated by equation (5). To avoid this phenomena, air-core coils are used.

Inductance sizing is a single-objective optimization problem, solved using the MOPSO (Multiple Objectives Particle Swarm Optimization) algorithm [21], [22], developed by J. Aubry [23]. It aims to to minimize losses in the coil and therefore its resistance. This performance indicator is an output of the semi-analytical model based on C Paul [24] and R. Medeiros [25] publications.

The optimization variables and their range are,

- Number of layers $N_l \in [\![1; 25]\!]$
- Number of turns $N_t \in [\![1; 25]\!]$
- Coil inside diameter [mm] $d_L \in [\![100; 200]\!]$
- Conductor diameter [mm] d_w based on the American Wire Gauge (AWG) values.

To validate the selection of the conductor, the following equations must be validated [26],

$$I_d \leq S\sqrt{\frac{0.0297}{\Delta t} log\left(\frac{T_2 + 234}{T_1 + 234}\right)} \tag{5}$$

$$I_{adm}(d_w) \geq I_n \tag{5}$$

$$L_{tot}(N_l, N_t, d_L, d_w) \geq L_{obj} \tag{5}$$

where T_1 [°C] is the initial temperature of the conductor, T_2 [°C] is the maximal temperature acceptable by the insulation, Δt [s] is the duration of the fault and S [m^2] is the section of the conductor, I_{adm} is the conductor rated RMS current and I_n is the converter RMS current. These parameters are set to 70 °C, 150 °C, 1 s, 10 mm², depends on Equations (5) and 30 A, respectively. The maximal temperature is due to the use of varnished conductors [26].

Air-core coils are usually custom-made, so there is no cost data available. Based on our experience, we assume that the cost C_L is expressed as,

$$C_L = 2{,}11\pi f_n L_{tot} I_n^2 \tag{5}$$

Where f_n is the grid frequency [Hz].

3.2.5 Performance Estimation

To validate the chosen sizing according to the presented constraints, two PLECS simulation sequences are performed. The first one simulates the steady state operation of the converter. It aims to validate the power quality and efficiency criteria. The temperature of each diode and each IGBT is monitored and must remain below $0.8T_{j,max}$.

The second sequence simulates the transient operation of the converter, during a pole-to-pole short circuit. It aims to validate the selectivity criterion. For each diode, the thermal stress $i_D{}^2t$ is calculated for different fault locations l_{fault}, starting times t_{fault}, and compared to its datasheet value $i_D^2 t_{max}$. The sizing is deemed correct when,

$$\int_0^{t_{trip,max}} i_D(t, l_{fault}, t_{fault})^2 dt \leq i_D^2 t_{max} \tag{5}$$

$$\max\left(i_D\left(t, l_{fault}, t_{fault}\right)\right) \leq I_{RRM} \tag{5}$$

And that the maximum current is inferior to the maximal

3.3 Simulation Results

Using the methodology described above, a first sizing of the converter is carried out considering only the power quality and efficiency criteria, without protection diode. The resulting sizing #A is summarized in the first column of Table 2. Then sizing #B is carried out, considering oversized module with selectivity criteria. Finally, sizing #C is carried out on the proposed topology.

The simulations results point out the economic interest of the proposed topology compared to an oversized converter. Indeed, taking into account the selectivity constraint leads to an extra cost of 37.5 % compared to 45,6 % with an oversizing approach.

Note that for sizing #A an air-core tie reactor was not necessary. As the converter cannot handle large currents, an iron-core tie reactor could be used instead, resulting in an even less expensive alternative [27]. In order to reach the efficiency constraints, a tradeoff is required between the switching frequency of the converter and tie reac-

tor sizing. A higher inductance limits the fault current, filter harmonics but leads to higher losses and a higher. This difference appears between sizing #B and #C.

Finally, it must be noted that the sizing cost are underestimated, insofar as drivers, calculators, metal frames, PCB, connectors and manufacturers' margins are not taken into account.

	#A	#B	#C
Topology	2L-VSC	2L-VSC	2L-VSC + D
Power quality	✓	✓	✓
Efficiency	✓	✓	✓
Selectivity	✗	✓	✓
I_n [A]	50	600	50 + 2×75
f_{sw} [kHz]	12	12	10
L [mH]	1,47	1,48	1,75
R_L [mΩ]	54	42	60
THD_i [%]	7,52	7,46	7,54
η [%]	97,09	97,43	98,2
Capacitor	2×C44UOGT7450M34K		
V_{tot}[L]	18,8	19,7	19,2
C_{tot}[€]	1385	2548	1905

Table 3 Summary of the 2L-VSC sizing for LVDC grids

4 Experimental validation

Based on the previous sizing, a substation converter prototype has been built (**Fig. 8**). Taking into account available components and a safety margin, an extra protection diode is added, provided inductance value is 2.22 mH / 59 mΩ and IGBT nominal current is 75 A.

Because of the laboratory's fuse protection limits the current to 40 A and the available voltage is 127 V L-L RMS, we characterized the converter behavior during short-circuit at reduced voltage by carrying out three experimental tests:

1. Test 1: in rectifier without tie reactors.
2. Test 2: in rectifier with tie reactors.
3. Test 3: connected to the grid in (P, Q) mode.

Short-circuits are generated using the system presented in Fig. 8. This system is characterized by an equivalent resistance of 11,8 mΩ and a 0.749 V threshold voltage. Measurements are performed using ST 1000 differential probe, TA 167 Pico Technology current probe and A622 Tektronix current probe.

PCIM Europe 2024, 11– 13 June 2024, Nuremberg   DOI: 10.30420/566262012

Fig. 8 2L-VSC+D prototype with, from left to right, the laboratory's 127 V three-phase grid, an autotransformer, a step-down transformer, air core coils, the IGBT module and protection diodes on the same heat sink, capacitors and a DC voltage source.

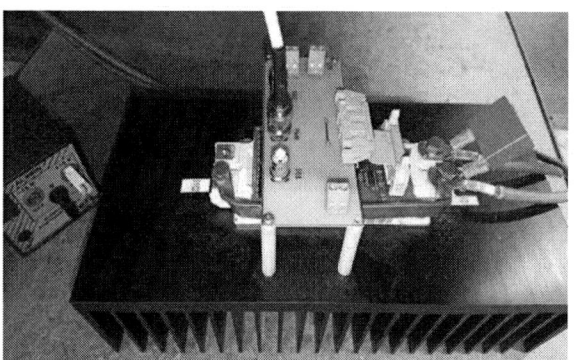

Fig. 9 Prototype short-circuit generator controlled by a PM 5716 pulse generator for short-circuit up to 100 ms, 1200 A and 600 V using FF600R06ME3 IGBT

The first test allows us to validate the principle of the topology: the thermal endurance of the converter is increased. The thermal constraint generated in the 2L-VSC diode associated to the first phase values 1350 A²s which is 12 % higher than the datasheet value (**Fig. 10**): without protection diode the converter may have been destroyed. By comparing the total current on the DC-link (**Fig. 11**) and the current outputted by 2L-VSC converter's diode (**Fig. 12**), a ratio of 10 appears between the two maximum values, validating the principle of current deviation.

Note that measurement have been verified with simulation and a deviation appears in **Fig. 12**. This discrepancy is largely explained by the model adopted. As long as the voltage across the diode is less than the threshold voltage, the current is zero, which is not the case in reality. This phenomenon is amplified by the parallelisation, as illustrated in **Fig. 4**.

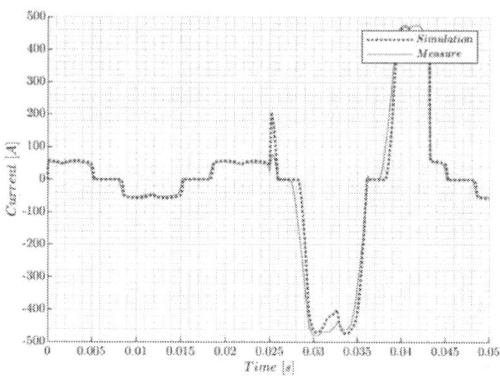

Fig. 10 AC grid current measured (orange) and simulated (dotted blue). The fault appears at t=25ms, until t=43ms

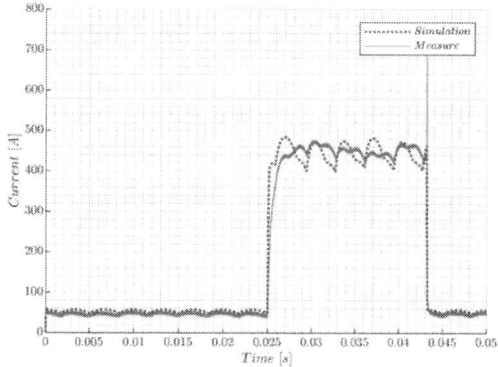

Fig. 11 DC current outputted by the converter measured (orange) and simulated (dotted blue).

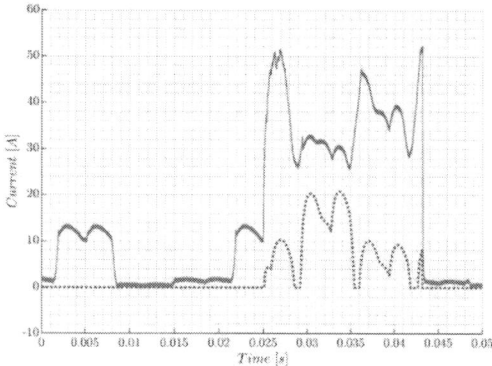

Fig. 12 DC current outputted by the 2L-VSC converter's diode measured (orange) and simulated (dotted blue).

The second test highlights the diode overlapping effect during the short-circuit (**Fig. 13**). Because of the tie reactors, the current cannot vary discontinuously. Therefore, diode switching cannot be instantaneous, resulting is simultaneous conduction of three diodes in the event of a short circuit on the DC bus. This phenomenon is called "overlapping". From AC grid point of view, due to the high current and inductance value, the overlapping is constant and leads to sinusoidal current whose value is limited only by the air-core coil inductances [12], [28].

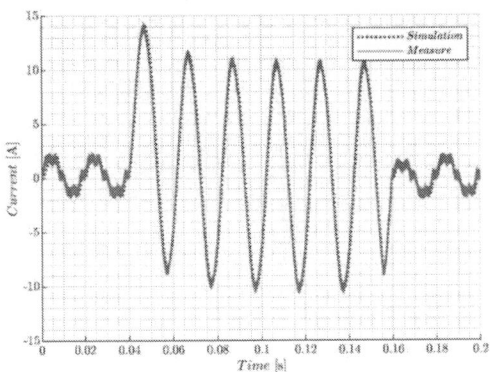

Fig. 13 AC grid current measured (orange) and simulated (dotted blue) in presence of the tie reactors.

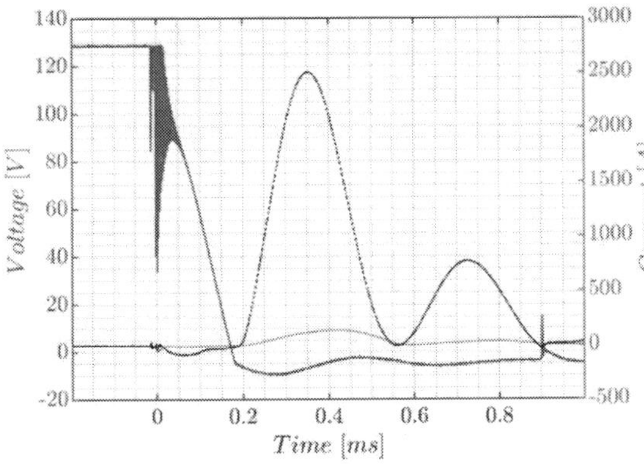

The last test points out the voltage reversal effect that could happening in a converter. During the test, for a 128 V DC-link, a voltage inversion of 7 V occurs. This voltage inversion means that diode conduction is forced and that the fault current corresponding to capacitor discharge also flows through the diodes, resulting in a current peak of 2.5 kA in the converter. By comparing current in the 2L-VSC (orange) and outputted by the converter (black), the role of protection diodes is highlighted. They deviated most of the current, allowing the 2L-VSC converter's diodes to remain within their operating range from both thermal constraint

and maximal repetitive current point of view. Because the reversal is obtained with parasitic inductance, the measure has not been compared with a simulation. Nonetheless, it is close to the phenomenon described in the literacy [4], [12], [28]. Further tests are planned with Socomec to validate this solution on an industrial converter.

5 Conclusion

In order to facilitate the development of hybrid distribution grid, combining AC and DC, a new converter topology has been proposed. Adaptable to several converter structures, this topology naturally deviates the fault current to diodes with a high thermal resistance. This deviation is possible by adjusting the diodes parameters, so that converter components remain within their operating range. This solution has been applied to a 2L-VSC converter and compared with a traditional oversizing. The sizing method of the different components has been detailed and allows to highlight the economic interest of the proposed topology. Finally, to demonstrate the technical feasibility a converter prototype has been built and tested in short-circuit condition. Given the limited short-circuit power available at the laboratory, tests have been performed at lower voltage. The 3 tests allowed to validate the increased thermal endurance of the converter, the deviation principle as well as the converter behavior.

References

[1] J. J. Justo, F. Mwasilu, J. Lee, and J.-W. Jung, 'AC-microgrids versus DC-microgrids with distributed energy resources: A review', *Renewable and Sustainable Energy Reviews*, Aug. 2013

[2] K. Smith, D. Wang, A. Emhemed, S. Galloway, and G. Burt, 'Overview paper on: low voltage direct current (LVDC) distribution system standards', *Int. J. Power Electronics*, 2018

[3] J. E. Huber and J. W. Kolar, 'Applicability of Solid-State Transformers in Today's and Future Distribution Grids', *IEEE Trans. Smart Grid.*, 2019.

[4] S.-M. Xue and C. Liu, 'Line-to-Line Fault Analysis and Location in a VSC-Based Low-Voltage DC Distribution Network', *Energies*, Mar. 2018

[5] Y. Pelenc, 'Interruption des circuits alimentés en courant continu', Technique de l'Ingénieur, 1998.

[6] R. Rodrigues, Y. Du, A. Antoniazzi, and P.

Fig. 14 DC-link voltage (blue), current outputted by the converter (black) and current in the IGBT (orange) with a fault appearing at t = 0 s

Cairoli, 'A Review of Solid-State Circuit Breakers', *IEEE Trans. Power Electron.*, Jan. 2021.

[7] D. Hadbi, L. C. Delgado, and F. Reymond-Laruina, 'DC short-circuit behaviour of LVAC fuses', 27th International Conference on Electricity Distribution (CIRED), Rome, Italy, Jun. 2023.

[8] EATON, 'Fusible Overcurrent Protection for DC Applications'. 2020.

[9] F. Reymond-Laruina, L. Quéval, D. Hadbi, P. Egrot, M. Cordonnier, and S. Mercier, 'Impacts of low voltage distribution grid resilience constraints on AC/DC converter sizing', 27th International Conference on Electricity Distribution (CIRED), Rome, Italy, Jun. 2023.

[10] A. L. P. de Oliveira, C. E. Tiburcio, M. N. Lemes, and D. Retzmann, 'Prospects of Voltage-Sourced Converters (VSC) applications in power transmission systems', *IEEE ANDESCON*, Bogota, Sep. 2010.

[11] X. Li, Q. Song, W. Liu, H. Rao, S. Xu, and L. Li, 'Protection of Nonpermanent Faults on DC Overhead Lines in MMC-Based HVDC Systems', *IEEE Trans. Power Delivery*, Jan. 2013.

[12] J. Yang, J. E. Fletcher, and J. O'Reilly, 'Short-Circuit and Ground Fault Analyses and Location in VSC-Based DC Network Cables', *IEEE Trans. Ind. Electron.*, Oct. 2012.

[13] Direction Technique d'Enedis, 'Principes d'étude et de développement du réseau pour le raccordement des clients', Aug. 2019.

[14] IEEE Power and Energy Society, 'IEEE Recommended Practice and Requirements for Harmonic Control in Electric Power Systems', 2014.

[15] S. Rupp, S. Krahmer, R. Adam, K. Backhaus, C. Hildmann, and M. Nilges, 'Conversion of Existing AC into DC Cable Links in Distribution Grids Benefits and Challenges', ETG Kongress, Berlin, 2021.

[16] M. Akkari, 'Control of a multi-terminal HVDC (MTDC) system and study of its interactions with the AC grids', Thèse de Doctorat, Université Paris-Saclay, 2016.

[17] D. Neyers, C. Guillaume, and P. Le Roux, 'SPECIFICATION TECHNIQUE EDF HN 62-S-83', 1999.

[18] STMicroelectronics, 'Current sharing in parallel diodes', Application note, AN4381, 2014.

[19] S. Ravyts, G. V. den Broeck, L. Hallemans, M. D. Vecchia, and J. Driesen, 'Fuse-Based Short-Circuit Protection of Converter Controlled Low-Voltage DC Grids', *IEEE Trans. Power Electron.*, Nov. 2020.

[20] G. Mouriès, 'Condensateurs utilisés en électronique de puissance', *Conversion de l'énergie électrique*, Aug. 2015.

[21] X. Jun and H. Chang, 'The Discrete Binary Version of the Improved Particle Swarm Optimization Algorithm', *International Conference on Management and Service Science*, Beijing, China, Sep. 2009.

[22] H. Nezamabadi-pour, M. Rostami-sharbabaki and M. Maghfoori-Farsangi, 'Binary Particle Swarm Optimization: Challenges and New Solutions', Apr. 2008.

[23] J. Aubry, 'Optimisation du dimensionnement d'une chaîne de conversion électrique directe incluant un système de lissage de production par supercondensateurs: application au houlogénérateur SEAREV', Thèse de Doctorat, École Normale Supérieure de Cachan, 2011.

[24] C. R. Paul, *Inductance: Loop and Partial*. John Wiley & Sons, 2010.

[25] R. Coelho-Medeiros, 'Cryo-MMC : a Modular Multilevel Converter with Superconducting Coupled Arm Coils', Thèse de Docotorat, Université Paris-Saclay, 2022.

[26] EATON, 'Component Protection - Wire & Cable'. 2005.

[27] R. Burkart and J. W. Kolar, 'Component cost models for multi-objective optimizations of switched-mode power converters', in *2013 IEEE Energy Conversion Congress and Exposition*, Denver, CO, USA, Sep. 2013.

[28] F. Reymond-Laruina, M. Petit, L. Quéval, T.-D. Le, D. Hadbi, and P. Egrot, 'Prise en compte de la résilience aux défauts entre pôles lors du dimensionnement d'un convertisseur pour un réseau LVDC', Conférence des Jeunes Chercheurs en Génie Electrique (JCGE), Jun. 2022.

PCIM Europe 2024, 11– 13 June 2024, Nuremberg   DOI: 10.30420/566262013

Suppression of Oscillations in a SiC Bridge-Leg using a Custom Single-Chip Digital Active Gate Driver with 2×255 Strength Levels

Qilei Wang[1], Harry C. P. Dymond[1], Dawei Liu[2], Yushi Wang[1], Saeed Jahdi[1], Bernard H. Stark[1]

[1] University of Bristol, UK

[2] Institute of Semiconductors, Guangdong Academy of Sciences

Corresponding author: Qilei Wang, qilei.wang@bristol.ac.uk
Speaker: Qilei Wang, qilei.wang@bristol.ac.uk

Abstract

Digital active gate driving application-specific integrated circuits (ASICs), with a 100 ps waveform update resolution, have been demonstrated in GaN-based power converters to improve switching waveform quality. However, this high time resolution was achieved using fast, low-voltage core transistors, leaving too little output voltage swing to drive SiC power device gates. This paper introduces a new single-chip digital driver capable of driving a SiC gate with a voltage swing in excess of 36 V and a maximum update resolution of 1.2 ns. The ASIC contains internal memory for two gate sequences, and provides a choice of 255 current levels in both polarities. The paper demonstrates the use of high-bandwidth gate current measurement to program the driver to emulate available gate drivers, and to create new gate current profiles that are not currently available. This is demonstrated on a 1200 V SiC MOSFET with a 15 V gate voltage requirement. This device is switched in a bridge-leg with a DC voltage of 800 V and load current of 10 A. A gate current profile is demonstrated that reduces the power of high-frequency spectral components of current by 4.5 dB without affecting switching loss.

1   Introduction

Silicon carbide (SiC) devices can achieve switching speeds beyond 50 V/ns, providing medium voltage power converters with lower loss, higher control bandwidth, and faster dynamic response. However, the high dv/dt and di/dt during switching transients introduces problematic switching features with high-frequency content, which can cause undesirable oscillations and EMI [1], potentially causing problems with radiated emissions compliance. There is also concern over near-field coupling and self-interference, which may increase the cost of filtering and shielding in practical converters.

Active gate driving has been shown to reduce ringing and overshoots in the power loop by precisely shaping the gate driving signals. Some publications show this being done whilst maintaining high efficiency [2], at the expense of higher control complexity. The challenge of active gate driving increases with switching speed, since the reduced switching duration requires faster control. There are a number of reports of active gate drivers with suitable output voltage ranges for SiC. For example, circuit-based modular multilevel

drivers in [4] use five discrete GaN drivers in series to achieve voltage swings from −5 V to 20 V in 5 V steps with a 2.5 ns resolution. Ref. [5] proposes a binary-weighted topology, using four submodules to achieve 16 voltage levels from 0 V to 17 V with a 5 ns resolution. Single-chip drivers can provide 2^n levels by n-bit totem-poles; i.e. [7],[8] contain 64-level pMOS-nMOS pairs, that shape the gate profiles with a time resolution in the region of 20-40 ns. The driver in [6] produces 8 current levels in both polarities, with a 1 ns resolution, and is shown to switch discrete SiC devices at 16 V/ns.

Previous work by the authors of [3] has demonstrated the first active gate driver with a 10 GHz update rate, which is fast enough to shape GaN FETs' nanosecond-scale switching edges. However, the use of high-speed transistors in this design limits the maximum output voltage range to 5 V, which is not enough to drive SiC gates. Therefore, new solutions are needed for the high-resolution, digitally programmable active gate driving of SiC MOSFETs.

This paper demonstrates use of a custom single-chip active gate driver for SiC, fabricated on a TSMC 180 nm 70 V HV BCD Gen2 process, occupying 7 mm². Its profiling flexibility exceeds that of the aforementioned drivers: It has the ability to set any of 255 strength levels of either polarity at each of its 100 waypoints, and output these with an update resolution of up to 1.2 ns. The driver uses sub-circuits of the 100 ps resolution GaN device driver of [3], but has a higher output voltage swing to 36 V to accommodate SiC devices with higher enhancement voltage or negative off-state bias requirements.

The significant contributions of this paper are:

- Presentation, in Section 2, of an active gate driver architecture with a propagation delay of only 2 clock cycles (2.4 ns at maximum clock speed), 2×2^8 drive strengths, and an update rate of up to 833 MHz, that provides almost arbitrarily designable gate current profiles. The dual profile memory can be overwritten during converter operation, e.g. to allow gate drive profile changes over a low-frequency AC cycle in AC/DC converter systems.

- In Section 3, details of a method, using observation of high-bandwidth gate current measurements, to determine profile settings required to achieve a desired output current profile, either to that measured on an available driver to emulate, or to validate simulated drivers.

- In Section 4, deployment of the gate driver chip into an 800 V, 10 A bridge-leg employing 1200 V 160 mΩ SiC MOSFETs, with a duplicate board with an identical power-loop layout and similar gate-loop dimensions, but with an off-the-shelf driver, for comparison.

- An experimental demonstration of the new driver chip in Section 5, showing two aspects: 1) the near-perfect emulation of an Analogue Devices, Inc. (ADI) ADuM4146 driver, and 2) the output of a gate current waveform that has been profiled to reduce ringing, with a measured 4.5 dB reduction in spectral content relative to a fast step-driven switching event of the same switching loss.

2 Active gate driver with programmable gate sequence for SiC devices

A custom SiC-device gate driver integrated circuit has been designed, whose architecture (Fig. 1) permits the fast recalling of strength-setting sequences from memory.

The driver integrates three main blocks: high-speed memory, a digitally-controlled gate drive output stage and logic circuits.

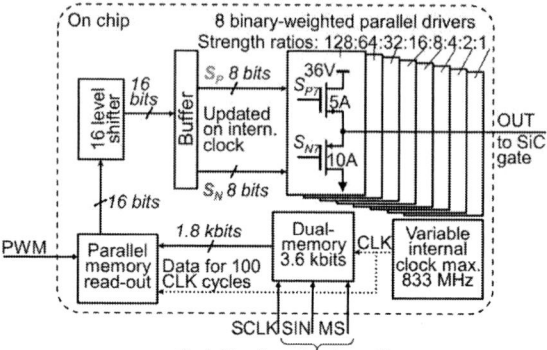

Fig. 1 Bristol Active SiC Gate Driver's functional architecture.

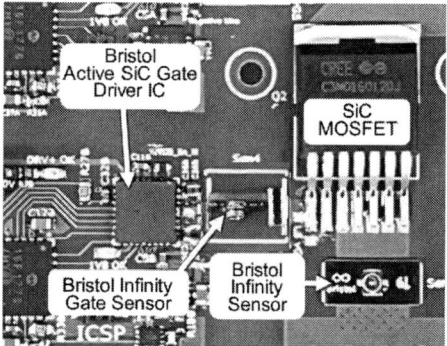

Fig. 2 PCB with Bristol active driver IC and Bristol current sensors.

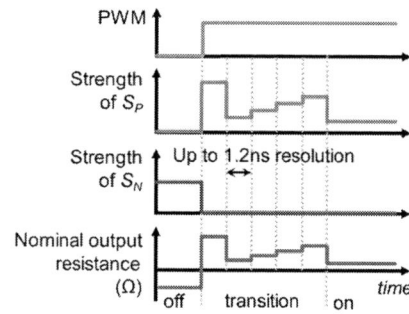

Fig. 3 Driver concept waveforms.

The driver output stage can be connected directly to the gate of the SiC device (Fig. 2), operating with up to 36 V supply voltage, ample for driving SiC devices. The driver stores sequences of 'nominal output resistance' (Fig. 3), in internal memory; these sequences are "played back" whenever a drive output transition is commanded at the logic input pin. The 'nominal output resistance' is a strength setting that is proportional to the combined pull-up or pull-down driver output transistor widths selected at any one time. These transistors operate in the saturation region for most of the gate voltage transition and thus behave like voltage-controlled current sources, e.g. see Fig. 4 of [11]. Therefore, for this duration, the gate current is approximately proportional to the strength setting, which simplifies parameter selection for a given desired gate current profile. The post-layout simulated steady-state output of a single pull-up transistor is 37.1 mA at 15 V, and for the pull-down transistors it is 77.9 mA. This corresponds to a maximum gate current, at 15 V and using all 255 transistors, of around 9.5 A for turn-on and 19 A for turn-off.

Two high-speed memories are integrated on the chip that hold two independent gate sequences. Each of them stores a turn-on and a turn-off sequence. The two internal memories can be switched between during live operation, which allows programming of a new sequence even when the power stage is switching. The total control interval is 100 time slots. On arrival of a PWM edge, the driver output activates a programmable sequence, read from memory. A nominal "pull-up" sequence is output when the input PWM control signal is logic high, and a nominal "pull-down" sequence is used when PWM is low. However, in either case, the driver can be instructed to pull up or to pull down in each of the 100 time slots, with any strength setting from 0 to 255.

An internal voltage-controlled oscillator determines the time resolution of the gate sequences, with a maximum resolution of 1.2 ns (833 MHz), to permit use for different turn-on and turn-off transients. It is worth noting that the gate driver output needs to be modulated during the entire switching transient, including the delay time before the switching performance becomes visible. For a SiC MOSFET with a 20 ns switching transient [22], the maximum update rate provides 16 waypoints, during this transient, at which the gate current can be set to a new value. The driver has the capability of setting 100 different waypoints per transient. The final setting is held until the next PWM transition.

3 Method for determining gate control sequences

The high flexibility of a digital active gate driver comes with the downside of having to find suitable values for all required drive sequence parameters. Therefore, viable methods are required to find these. Active gate drivers are generally aimed at closed-loop operation, falling into one of two categories: use of instantaneous feedback from the power or gate waveforms during the switching transient, and use of feedback from past transients [14]. These past transients could be from the

directly preceding switching cycle, as shown in [15], or from previous trials during the development or post-fabrication testing of the converter. Iterative optimisation trials have been reported, e.g. in [16] and [17], that use different degrees of switching measurements and modelling to determine optimised gate profiles for specific operating conditions. The aim of these trials is to predetermine gate profiles, so that in operation, a closed loop system senses the operating point and recalls the appropriate, previously determined gate profile. Each of these concepts have their merits in different scenarios. In terms of the output stage of the gate driver, there exist resistive, voltage, and current sources, for example those in [21].

Methods of finding appropriate gate profiles have been reported using various forms of machine learning [18]-[20], where relatively little system data are required. Reported methods generally perform Steps 1-3 of Fig. 4. In Step 1, the problem to be addressed is established, usually by inspecting problematic switching features in a measured switching waveform. As Step 2, a desired goal is defined, for example to reduce ringing or overshoot. Step 3 entails the actual computational search algorithm to find a gate voltage profile that would achieve the desired switching waveforms. In this work, we focus on Steps 4 and onwards.

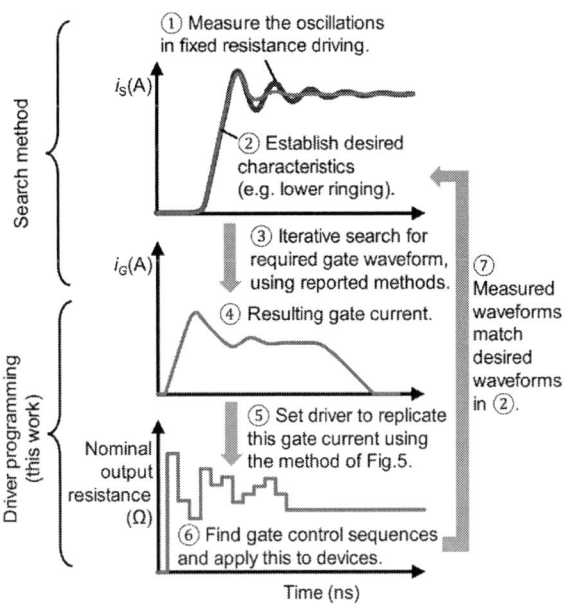

Fig. 4 A process of finding gate control sequences.

In Step 4, the gate current corresponding to the desirable power switching waveform is obtained. In Step 5, the gate driver is iteratively programmed

115

to approximate this gate driver current, without the DC link being energised. This results in having a complete set of driver parameters. Step 6 applies these settings to an energised double-pulse rig. Finally, in Step 7, the obtained measured waveforms are checked against the desired waveforms of Step 2.

For the purpose of demonstrating the driver of this work, two desirable gate current waveforms (Steps 4) are found. The first is measured on a standard commercial gate driver, allowing the active driver to emulate this commercial driver. This should demonstrate that the active gate driver can be reprogrammed to emulate a variety of existing, emerging, and future gate drivers. The second is found experimentally, by manual, iterative searching for a gate profile that minimises ringing. This gate current profile is used as if it had been found by known Step 3 methods, and the procedure of carrying out Steps 4-7 is demonstrated.

With the gate current known, Step 5 proceeds as illustrated in Fig. 5. A high-bandwidth Bristol Infinity Gate Sensor [12] is used to measure injected gate current. Starting at the waypoint t = 0, an initial drive strength is manually set to a relatively low value. The driver is activated with the power device's gate connected to the driver output, but without the DC link energised. The resulting gate current at the next waypoint t = 2 ns is observed. The strength setting for t = 0 is gradually increased to achieve the correct current at the next waypoint. Then, the same process is carried out for this next waypoint. The process continues until the drive strengths have been set at all waypoints where the gate driver has a noticeable influence over the switching transient. The time resolution of the gate driver can be reduced from 1.2 ns to simplify the resulting gate drive parameter sequences, and similarly, it is possible to skip waypoints. In this work, it is found that a sequence resolution of 2 ns is sufficient to accurately emulate the commercial driver.

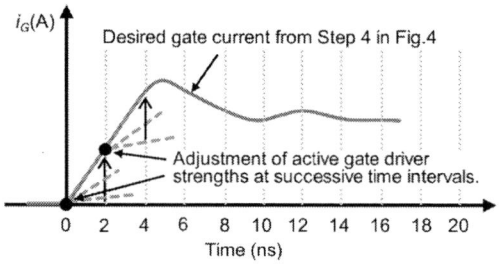

Fig. 5　Method of setting the active gate driver strengths to replicate a desired gate current profile.

4　Hardware implementation

A SiC converter circuit has been designed and operated in double-pulse mode at 800 V and 10 A, a simplified schematic of which is provided in Fig. 6(a).

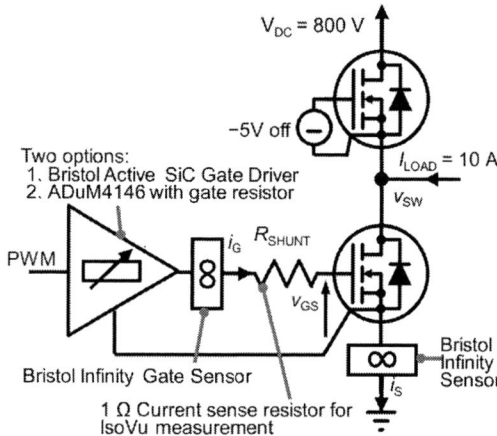

(a)

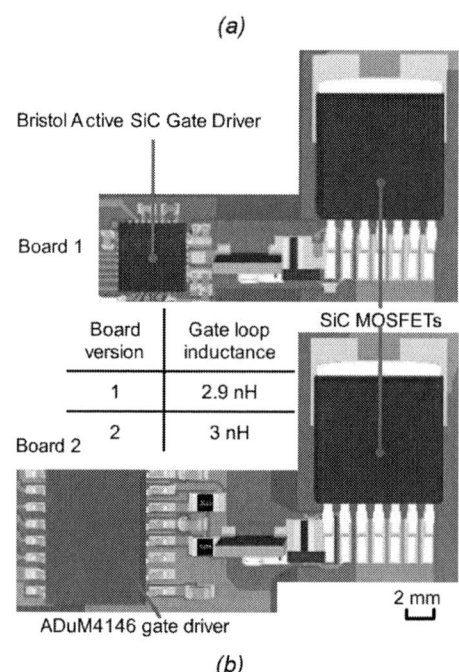

Board version	Gate loop inductance
1	2.9 nH
2	3 nH

(b)

Fig. 6　*(a)* Circuit schematic of double pulse test. *(b)* Close-up of gate circuit layout. Top: board with active driver. Bottom: board with commercial fixed-strength driver + external gate resistors.

To test active driving capability, two boards with identical power loop layouts have been constructed, as illustrated in Fig. 6(b). The only difference is the control method via the gate, where one board uses the active gate driver and another uses fixed turn-on and turn-off gate resistance driving by a commercial gate driver ADuM4146 from ADI. The respective gate loop inductances measured using an R&S ZVL Vector Network Analyzer are 2.9 nH and 3 nH.

The entire experimental facility is illustrated in Fig. 7. A 'Master+Slave' microcontroller system provides a user interface to configure the gate drivers and program them. Prior to testing, the host computer sends the gate control sequences via USB to the chipKit development board, which subsequently instructs the slave microcontroller to perform the relevant task. Communication is via a standard serial peripheral interface (SPI). During operation, a Keysight 33500B function generator provides the double-pulse gate-driver control signal, which is configured from MATLAB running on the host PC. This control signal is passed into and through the 'Master+Slave' microcontroller system to the gate drivers.

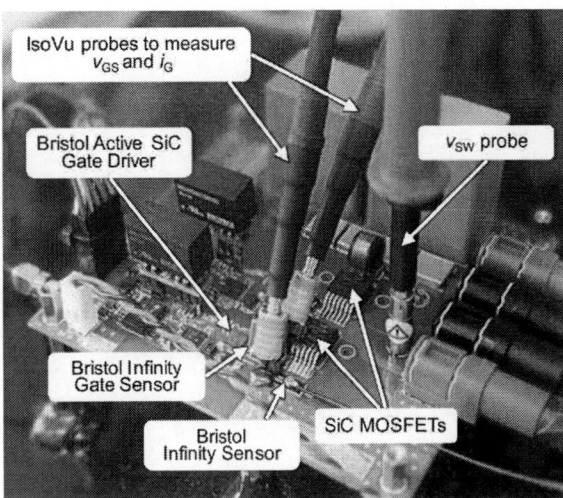

Fig. 8 Detailed view of the power board and measurement probes.

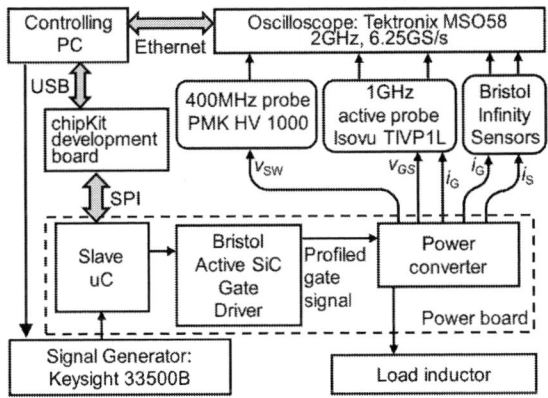

Fig. 7 Test facility with driver programming and waveform measurement.

The power board of this work, shown in Fig. 8, contains the active gate driver and equipment connections. Measurements are de-skewed, and 64 acquisitions of each double pulse switching waveform are captured on a Tektronix MSO58B 2 GHz 6.25 GSa/s oscilloscope in order to lower the measurement noise [10]. As shown in Fig. 8, the low side gate voltage v_{GS} is measured with a 1 GHz Tektronix IsoVu TIVH08 high common-mode rejection differential probe. The switch-node voltage v_{SW} is measured using a PMK HV1000 100:1 400 MHz passive voltage probe connecting through a PCB-mounted coaxial adaptor. The source current i_S is obtained using an Infinity Sensor [9], and integrating using the script provided at InfinitySensor.com. Both a Bristol Infinity Gate Sensor [12] and a Tektronix IsoVu optically-isolated voltage probe across a 1 Ω shunt measure gate current i_G.

5 Experimental results

5.1 Emulation of a commercial driver

When the active gate driver of this work is set to provide a constant output strength, it does not emulate a standard step-driver with a gate resistor. This is because the output source impedance is set by the active driver's output transistors which are in saturation, rather than being dominated by a gate resistor. Therefore, to provide a baseline switching waveform with the active driver, a commercial ADI driver with a gate resistance of 11 Ω is emulated, using the strategy described in Section 3. To this end, the gate current of the ADI driver is captured. Then the active gate driver's current is set to follow the same trajectory using the method of Fig. 5, using a time resolution of 2 ns. The measured waveforms are plotted in Fig. 9. The resulting strength sequence in the bottom graph shows that only five drive strength changes are required for a near perfect replication of the ADI's gate and power loop switching waveforms. These changes are all applied over the first 14 ns of the turn-on transient, well before the switch node's voltage transient. The driver is using only a small proportion of the available 2×255 strength levels.

The switching loss in the transient for each scenario is calculated using the method described in [13], and is provided in the line labels in the source current i_S graph. The switching loss of both waveforms lie within 2% of each other.

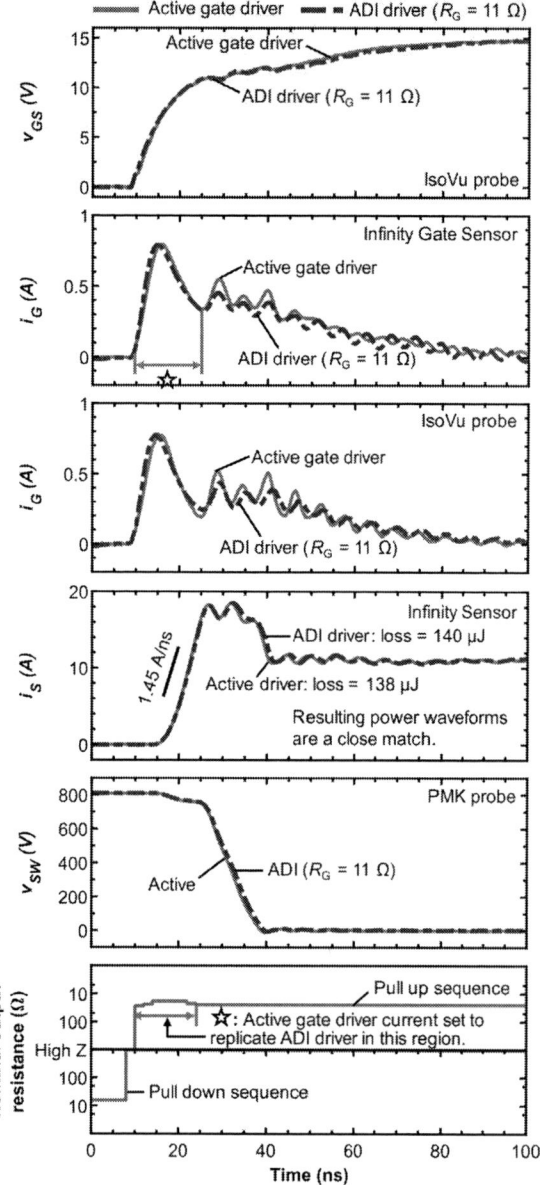

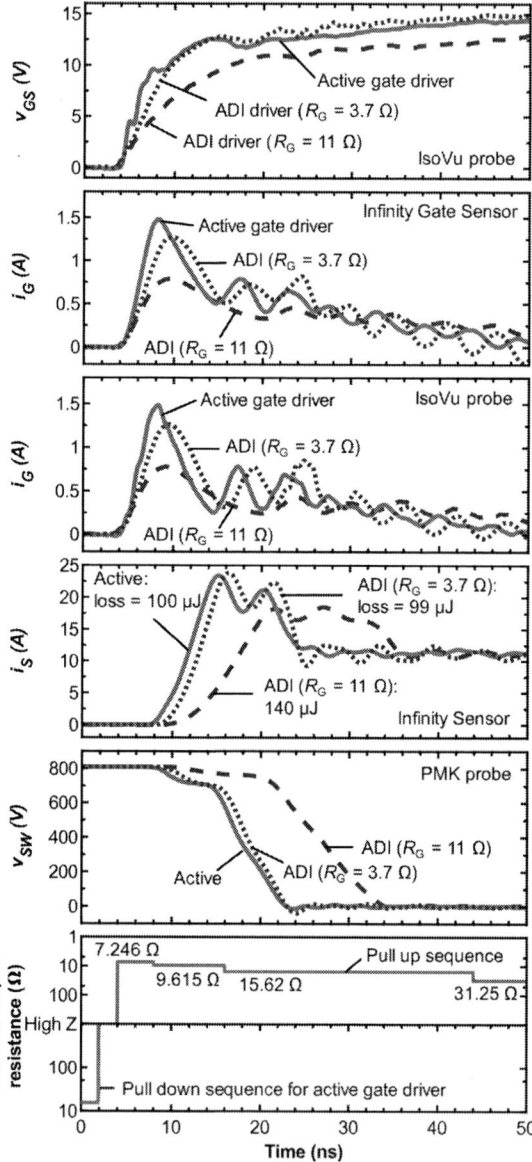

manual, iterative search. The driver settings shown in the bottom plot of Fig. 10 are derived using only knowledge of the optimised gate current profile, using the programming method of Step 5 in Section 3. This aims to demonstrate that if a suitable gate profile search method exists (Steps 1-3 in Fig. 4), and the active driver's current ratings are sufficient, then this profile can be programmed and applied to a real switching circuit.

Fig. 9 Emulation of an ADI ADuM4146 gate driver with the Bristol Active SiC Gate Driver, by replicating its gate current in the first 14 ns of a turn-on transient.

5.2 Damping of ringing: targeting specific spectral components

In this experiment, the active gate driver is set to suppress current oscillation in the power circuit at turn on, with the aim of reducing high-frequency spectral components and resulting EMI.

The measured turn-on switching waveforms in Fig. 10 show three scenarios: The ADI driver with 3.7 Ω and 11 Ω gate resistors, and the active gate driver, whose gate current profile was found by a

Fig. 10 Demonstration of the Bristol active driver using a 4-step sequence, to combine the fast switching speed and low loss of the ADI driver at R_G = 3.7 Ω, and the reduced ringing amplitude of the ADI driver at R_G = 11 Ω.

The turn-on switching loss of the control device for each scenario is provided in the line labels in the source current i_S graph. Current oscillation is observed at the capacitive discharge where i_S overshoots and v_{SW} begins to fall. Compared with 3.7 Ω driving, 11 Ω driving reduces current overshoot and oscillation in the source current i_S, however, the switching loss is increased by 41.4%.

In contrast to the results with the lower gate resistance value of 3.7 Ω, the relatively simple 4 step active gate driving sequence is seen to reduce the oscillation in the source current i_S whilst incurring a similar switching loss.

The spectral envelopes of the three measured source current i_S plots of Fig. 10, are shown in Fig. 11. With 3.7 Ω fixed-resistance driving, two spectral peaks are present at approximately 180 MHz and 300 MHz. Changing to the active gate driving, these peaks are reduced by 4.5 dB and 2 dB respectively. The switching loss has not significantly changed. Comparing to 11 Ω fixed-resistance driving, the active gate sequence has the similar damping effect at the 180 MHz ringing frequency but the switching loss is reduced by 40%.

Active driving, therefore, successfully clips both spectral peaks, with a negligible increase in switching loss. This adds to the reported examples of active gate driving allowing users to break the trade-off between switching loss and high frequency oscillations that are a potential cause of EMI in the radiation band of many EMC standards.

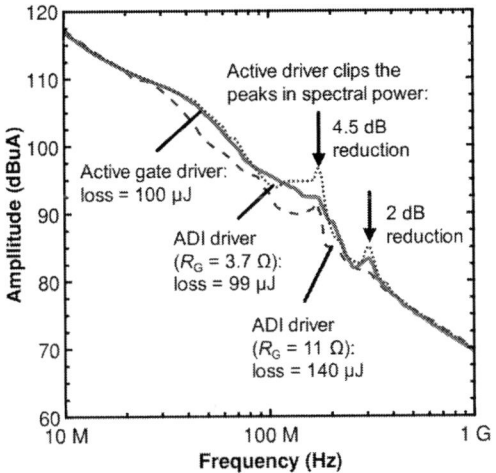

Fig. 11 Source current spectral envelopes, created by applying an FFT to the measured time-domain waveforms of Fig. 10.

6 Conclusion

This paper presents a custom single-chip active gate driver with a 1.2 ns resolution and 2×255 output current levels for driving SiC gates. It is shown, on an 800 V SiC MOSFET bridge leg switching at 10 A, that the driver can emulate an existing voltage-source gate driver with an 11 Ω gate resistor, and nearly perfectly recreate its switching waveforms, with the same switching loss. This is done by programming the active driver to have the same output current profile as the commercial driver with this gate resistor. Only 5 strength changes are required to achieve this base-line measurement. This illustrates how active drivers can be used to compare gate drivers, and potentially develop new ones with less complexity than the arbitrary waveform driver reported here. It also shows the importance of measuring gate current, which is carried out using both IsoVu probe on a current sense resistor, and an Infinity Gate Sensor, since the switching waveforms directly correlate with gate current.

The active driver is reprogrammed to improve on this baseline, showing that it maintains a similarly low spectral envelope at switching speeds and losses that are equivalent to driving with a 3.7 Ω gate resistor. The switching loss of a low gate resistor has been combined with the lower EMI of a larger gate resistor, in a relatively simple 4-step gate profile. Therefore, we anticipate that emerging gate drivers with lower numbers of strength changes per switching transient could have a similar positive impact on switching performance. The reported University of Bristol Active SiC Gate Driver can therefore be used as a research tool to help develop new gate drivers or to establish how to program them as a function of changes in operating conditions, in applications such as automotive drives.

The driver as presented here, would likely require a significant additional application-specific feature-set to allow use in commercial products. The cost of the silicon and package is low; the cost of such a chip, if used commercially, would likely be dominated by the chosen test procedure.

Future work will include developing search algorithms for optimal gate sequences, and to create control systems that automatically adapt to operating conditions.

References

[1] B. Zhang and S. Wang, "A Survey of EMI Research in Power Electronics Systems With Wide-Bandgap Semi-

conductor Devices," IEEE Journal of Emerging and Selected Topics in Power Electronics, vol. 8, no. 1, pp. 626-643, March 2020.

[2] J. Henn, C. Lüdecke, M. Laumen, S. Beushausen, S. Kalker, C. H. v. d. Broeck, G. Engelmann, R. W. d. Doncker, "Intelligent Gate Drivers for Future Power Converters," in IEEE Transactions on Power Electronics, vol. 37, no. 3, pp. 3484-3503, March 2022.

[3] D. Liu et al., "Full Custom Design of an Arbitrary Waveform Gate Driver With 10-GHz Waypoint Rates for GaN FETs," in IEEE Transactions on Power Electronics, vol. 36, no. 7, pp. 8267-8279, July 2021.

[4] M. Parker, I. Sahin, R. Mathieson, S. Finney and P. Judge, "Investigation Into Active Gate-Driving Timing Resolution & Complexity Re-quirements for a 1200 V 400 A SiC Half Bridge Module," IEEE Open Journal of Power Electronics, vol. 4, pp. 161-175, 2023.

[5] H. Takayama, S. Fukunaga and T. Hikihara, "Binary-weighted Modular Multi-level Digital Active Gate Driver," ECCE Europe, 2023.

[6] S. Kawai, T. Ueno, H. Ishihara, S. Takaya, K. Miyazaki, K. Onizuka, H. Ishikuro, "A Load Adaptive Digital Gate Driver IC With Integrated 500 ksps ADC for Drive Pattern Selection and Functional Safety Targeting Dependable SiC Application," IEEE Transactions on Power Electronics, vol. 38, no. 6, pp. 7079-7091, June 2023.

[7] D. Zhang, K. Horii, K. Hata and M. Takamiya, "Digital Gate Driver IC with Real-Time Gate Current Change by Sensing Drain Current to Cope with Operating Condition Variations of SiC MOSFET," ECCE Asia, 2023.

[8] D. Yamaguchi et al., "An Optimization Method of a Digital Active Gate Driver Under Continuous Switching Operation Being Capable of Suppressing Surge Voltage and Power Loss in PWM Inverters," in IEEE Transactions on Industry Applications, vol. 58, no. 1, pp. 481-493, Jan.-Feb. 2022.

[9] H. C. P. Dymond, Y. Wang, S. Jahdi and B. H. Stark, "Probing Techniques for GaN Power Electronics: How to Obtain 400+ MHz Voltage and Current Measurement Bandwidths without Compromising PCB Layout," PCIM Europe 2022; International Exhibition and Conference for Power Electronics, Intelligent Motion, Renewable Energy and Energy Management, Nuremberg, Germany, 2022, pp. 1-10.

[10] N. Oswald, B. H. Stark, N. McNeill and D. Holliday, "High-bandwidth, high-fidelity in-circuit measurement of power electronic switching waveforms for EMI generation analysis," 2011 IEEE Energy Conversion Congress and Exposition, Phoenix, AZ, USA, 2011, pp. 3886-3893, doi: 10.1109/ECCE.2011.6064297.

[11] H. C. P. Dymond et al., "A 6.7-GHz Active Gate Driver for GaN FETs to Combat Overshoot, Ringing, and EMI," in IEEE Transactions on Power Electronics, vol. 33, no. 1, pp. 581-594, Jan. 2018.

[12] Y. Wang, Q. Wang, M. Appleby, J. Yan, H. C. P. Dymond, S. Jahdi, and B. H. Stark, "'Infinity ate Sensor': a Differential Magnetic Field Sensor for Measuring ate Current of SiC Power Transistors," in PCIM , Nuremberg, 2024.

[13] J. J. O. Dalton et al., "Shaping switching waveforms in a 650 V GaN FET bridge-leg using 6.7 GHz active gate drivers," 2017 IEEE Applied Power Electronics Conference and Exposition (APEC), Tampa, FL, USA, 2017, pp. 1983-1989.

[14] W. T. Cui et al., "A Dynamic Gate Driver IC with Automated Pattern Optimization for SiC Power MOSFETs," 2022 IEEE 34th International Symposium on Power Semiconductor Devices and ICs (ISPSD), Vancouver, BC, Canada, 2022, pp. 33-36.

[15] E. Raviola and F. Fiori, "An Adaptive Method to Reduce Undershoots and Overshoots in Power Switching Transistors Through a Low-Complexity Active Gate Driver," in IEEE Transactions on Power Electronics, vol. 38, no. 3, pp. 3235-3245, March 2023.

[16] K. Miyazaki et al., "General-Purpose Clocked Gate Driver IC With Programmable 63-Level Drivability to Optimize Overshoot and Energy Loss in Switching by a Simulated Annealing Algorithm," in IEEE Transactions on Industry Applications, vol. 53, no. 3, pp. 2350-2357, May-June 2017.

[17] Yang L, Liu Y, Yu W, et al. Sequence Prediction for SiC MOSFET Active Gate Driving With a Recurrent Neural Network[J]. IEEE Open Journal of Industry Applications, 2023.

[18] H. Takayama, S. Fukunaga and T. Hikihara, "Digital-Twin-Compatible Optimization of Switching Characteristics for SiC MOSFETs Using Genetic Algorithm," in IEEE Journal of Emerging and Selected Topics in Industrial Electronics, vol. 4, no. 4, pp. 1024-1033, Oct. 2023.

[19] Y. S. Cheng et al., "High-Speed Searching of Optimum Switching Pattern for Digital Active Gate Drive to Adapt to Various Load Conditions," in IEEE Transactions on Industrial Electronics, vol. 69, no. 5, pp. 5185-5194, May 2022, doi: 10.1109/TIE.2021.3084169.

[20] S. Leonovs, S. Jahdi, H. C. P. Dymond and B. H. Stark, "Use of an NSGA-II Genetic Algorithm and Active Gate Driving to Improve Simulated GaN Power Electronic Switching Waveforms," PCIM Europe 2022; International Exhibition and Conference for Power Electronics, Intelligent Motion, Renewable Energy and Energy Management, Nuremberg, Germany, 2022, pp. 1-10.

[21] Y. Sukhatme, V. K. Miryala, P. Ganesan and K. Hatua, "Digitally Controlled Gate Current Source-Based Active Gate Driver for Silicon Carbide MOSFETs," in IEEE Transactions on Industrial Electronics, vol. 67, no. 12, pp. 10121-10133, Dec. 2020.

[22] Wolfspeed, "Silicon Carbide Power MOSFET C3MTM MOSFET Technology N-Channel Enhancement Mode," Rev. 02 December 2023, wolfspeed.com: https://assets.wolfspeed.com/uploads/2024/01/Wolfspeed_C3M0160120J_data_sheet.pdf

PCIM Europe 2024, 11– 13 June 2024, Nuremberg   DOI: 10.30420/566262014

SiC MOSFET Short-Circuit Protection: a Faster Soft Shut Down Method for Gate Drivers

Julien Weckbrodt[1], Thanh Long Le[1], Nicolas Ginot[2], Christophe Batard[2], Louison Gouy[2]

[1] Safran Tech, Safran group, Châteaufort, France
[2] Nantes Université, CNRS, IETR, UMR 6164, F-44000 Nantes, France

Corresponding author: Julien Weckbrodt, julien.weckbrodt@safrangroup.com
Speaker: Julien Weckbrodt, julien.weckbrodt@safrangroup.com

Abstract

Standard short-circuit protection features implemented in gate drivers are often calibrated for IGBT technology. However, the increase in switching slope of the wide-band Gap semiconductors and their lower reliability/maturity is a challenge for securing these new devices against SC events. Decreasing the delays in protection circuits is hence required to secure the operation of high power SiC MOSFETs. This paper presents an Advanced Soft Shut Down method with a reaction delay of 400ns after detection of a 650nH SC. The principle was demonstrated on a 1.2kV SiC module with a bipolar buffer stage based gate driver.

1   Introduction

Short-circuit (SC) events can have various origins in a power conversion system: cables manipulation, load failure, insulation material aging, component breakdown, design error… In order to avoid cascading failures, different protection features are implemented in high added value power inverters or critical systems to improve global reliability. Highly inductive SC events coming from the load are generally treated by current limitations and threshold voltages implemented in the software while lowly inductive SC must be treated by hardware protection embedded in the gate driver boards. Indeed, the dynamic of the regulation loops implemented in the control-command systems are unable to react at the μs scale. The protection features implemented in gate drivers are designed to avoid any destruction of the protected device in case of a fast increase of the drain current. The device is then locked in open circuit and a Fault signal is generally sent to the control unit. This article deals with the hardware implementation of SC protection features on the gate driver and optimizations dedicated to SiC power devices safety.

New SiC devices tend to reduce their $R_{DS(on)}$ to improve the conduction performances. E. Wiesner et al. [1] linked the low on-state resistance of the Silicon Carbide Metal-Oxide-Semiconductor Field Effect Transistors (SiC MOSFET) to their SC energy capabilities, which are lower than their IGBT competitors. An effort is therefore expected on the safety of the SiC devices during SC events.

2   Gate driver buffer conventional architectures

The gate driver buffer stage enables high peak current delivery to the gate input capacitances ($C_{gs} + C_{gd} = C_{iss}$) during the switching events. This transient current is required to achieve a fast switching of the power device leading to lower losses on power switches, which is particularly important for SiC power devices.

In voltage driven gate command, two topologies are commonly used to realize this function: the emitter-follower buffer based on bipolar junction transistors (see Fig.1) and the rail-to-rail push-pull buffer based on MOSFET technology (see Fig.2). Of course, each topology has its own advantages and weaknesses. The bipolar-based structure is easier and safer to use but the gate current injected to the power device highly depends on the current gain (β) of the bipolar transistors used which cannot be precisely characterized in most cases. The MOSFET-based structure can realize a rail-to-rail operation and deliver a maximal peak current but it is sensible to unintentional short-circuit of the gate driver supply voltage (V_{ISO_REG}) or high-impedance driven gate during the NMOS and PMOS transient transitions.

More complex topologies exist for the gate buffer stage, such as H-bridge gate drivers, resonant gate drivers or current-source controlled gate drivers… These circuits are rarely used in industrial applications and will not be discussed further in this paper.

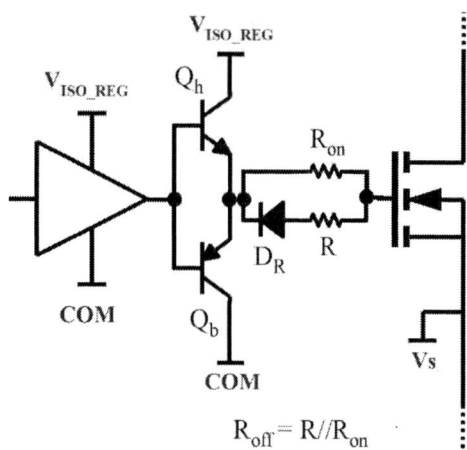

Fig. 1 Emitter-follower buffer stage using bipolar transistors

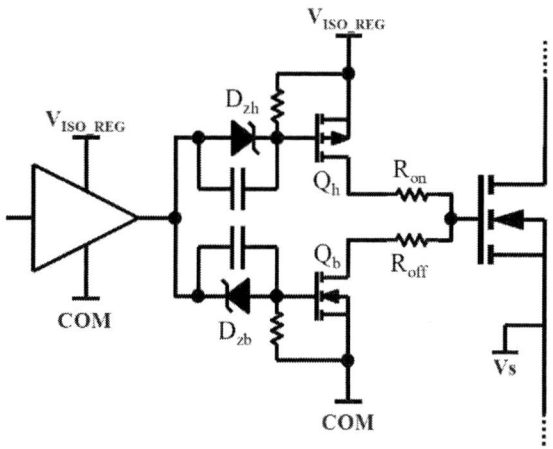

Fig. 2 Rail-to-rail push-pull buffer stage using MOSFETs

3 Conventional SC protection features and typical latencies

3.1 Definitions

Typical Over-Current Protection (OCP) features implemented in gate drivers can be divided in two categories:

- SC detection (e.g. DESAT function)
- SC reaction (e.g. SSD or TLTO)

The detection circuit can identify an abnormal operation characterized by a fast increase of the drain current during the conduction phase. Once a SC event is identified, an appropriate reaction method is required to extinguish the drain current before the breaking such as Soft Shut Down (SSD) / Safe Turn-Off (STO) or Two-level Turn Off (TLTO) [2], [3]. A reaction circuit is necessary because a regular switching applied on a very high drain current can lead to catastrophic Drain-Source overvoltage due to parasitic stray inductance. This overvoltage can be minimized by an exceptional slower turn-off during a SC event in order to keep V_{DS} under the Drain-Source breakdown voltage $V_{(BR)DSS}$.

Of course, both SC detection and reaction circuits have latencies due to their operating principles and the choice of components. For that reason, we can define 3 times in the SC protection process:

- The blanking time: configurable parameter (about 0.5-5µs);
- The detection delay: latency due to the capacitance charge and propagation delays in components on the SC detection circuitry (about 200-500ns);
- The reaction time: the time required to turn-off and reset the drain current without damage after a SC event detection in the gate driver (variable considering the SC impedance, power device, gate buffer topology, OCP method, components and settings).

The time required to extinguish a SC in a semiconductor power device is therefore the sum of the blanking time, the detection delay and the reaction delay required to reset the drain current to zero. As mentioned above, decreasing the delays in protection circuits is a key point for the safe operation of power SiC MOSFETs.

SC events can also be classified in several categories :

- Lowly inductive SC at turn-on (type I, worst case: requires a blanking time)
- Lowly inductive SC during the conduction (type II, no blanking time = faster detection)
- Highly inductive SC / overload (most of the time due to manipulation or load failures).

Highly inductive SC are generally treated at control level with current limitations implemented in the software since that kind of protection does not require a microsecond reaction; while lowly inductive SC requires the fastest hardware reaction without any control loop delay. It is the responsibility of the designer to check that all probable SC cases are covered by one of these safety features. In this paper, we focus on the protection circuits against lowly inductive SC which are located in the gate driver boards.

3.2 SC detection in gate drivers

Regarding SC detection, the well-known desaturation method (DESAT) is now commonly used in gate drivers. The DESAT circuit can identify an abnormal operation characterized by a fast increase of the drain current during the conduction phase. A desaturation diode is used to protect the sensing circuitry against drain potential during the turn-off. If a SC event occurs while the device conducts, the DESAT capacitor voltage increases from the clamped on-state drain-source voltage $V_{DS(on)}$ of the power device until reaching the pre-set threshold value (typically several Volts). Indeed, during a SC event the on-state drain-source voltage quickly increases due to the saturation behaviour of the MOSFETs under very high drain current. The $L \cdot di_D/dt$ also contributes to increase the drain-source voltage seen by the gate driver board.

In order to prevent false SC detection during normal operation due to a chaotic transient state after the turn-on, a blanking time is generally implemented – about 2-5µs for IGBTs or around 1µs for SiC MOSFETs, depending on the device. The detection circuit is therefore disabled during the blanking time. Sometimes, a template curve using a RC discharge is used at the end of the blanking time to enable the detection feature smoothly. Ideally, both threshold voltage and blanking time should be parametrized by successive SC tests in order to find the best compromise in between false error event immunity (robustness) and safety.

Others detections techniques are investigated in the literature such as Rogowski coil sensors [4], [5], [6] embedded in order to reduce the detection delay to less than 200ns [5]. A Rogowski coil measures the voltage induced by a varying magnetic field, which is proportional to the di/dt. However, an additional integrator is required and the coil is sensitive to the external magnetic field, which could impede accuracy. A Rogowski coil concept made by printed circuit board (PCB) could be a low-cost solution for SC detection.

3.3 Over current reaction

3.3.1 OCP using Soft-Shut-Down (SSD) / Soft-Turn-Off (STO)

A conventional SSD / STO circuit applied to gate buffer stages are presented in Fig.4 and Fig.5. In both case, the SC reaction is based on an exceptional slower turn-off realized by a gate impedance modification.

In the case of the emitter-follower buffer stage (Fig. 4), the SSD reaction is realized by the R_{ssd} resistance (470Ω in this setup) combined with a logic gate with an embedded high impedance (HiZ) feature. This feature is activated only if the detection circuitry detects a SC event (ERROR signal). In this circuit, the current gain of the bipolar current amplifier (β) contributes to the SC reaction but it enables the use of a lower power resistance and transistor Q_{ssd} since the peak current in the SSD branch is divided by the factor β.

In the case of the MOSFET-based buffer stage (Fig. 5), the SSD reaction is realized by an R_{ssd} resistance exceptionally connected in series with $R_{g,off}$ when the ERROR signal occurs. The chosen resistance has to support the peak current even if shorted most of the time.

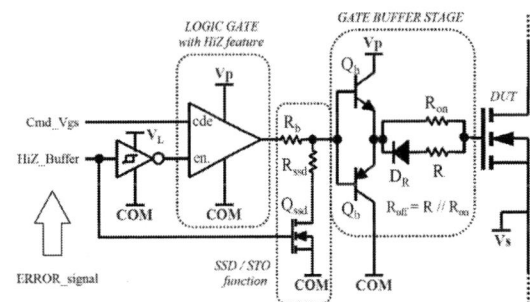

Fig. 4 SSD / STO protection feature implemented on a bipolar follower gate buffer stage

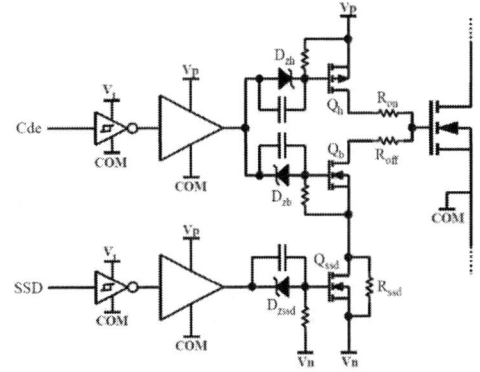

Fig. 5 SSD / STO function implemented on a rail-to-rail MOSFET-based gate buffer stage

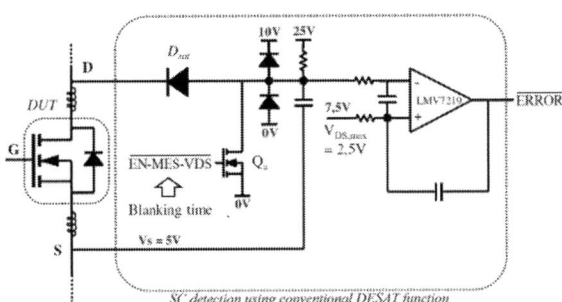

Fig. 3 Typical DESAT circuit for SC detection

In this setup, the emitter-follower buffer stage was used with a SSD protection feature and a blanking time of 1µs implemented in the DESAT circuit.

3.3.2 OCP using Two-level-Turn-Off (TLTO)

The TLTO feature [2], [7], [8] is more and more used for SiC MOSFET protection. The principle consists in reducing the overvoltage using a multi-step gate voltage at the turn-off. The lower gate voltage decreases the value of the saturation current which limits the SC current during the reaction time. Once again, the objective is to keep V_{DS} under the intrinsic breaking voltage (e.g. 1200V). More than two steps can be used if necessary but this would also increase the complexity of the gate driver.

3.3.3 Other OCP methods

Although conventional protection features are still applicable to SiC MOSFET as explained in [9], it is more and more difficult to guarantee a safe operation in all SC cases due to the faster and less robust nature of SiC MOSFETs compared to their IGBT competitors. Accordingly, it is highly recommended to proceed carefully when applying conventional protection methods to SiC MOSFET.
In order to reduce the delay in the SC detection path, a quasi-flying gate concept was introduced in 2023 by M. Picot-Digoix et al. [10]. The principle consist here to put the gate in high impedance (about 10kΩ) while the device conducts. In this case, the gate-source voltage is used to trigger the error signal. Of course, the high impedance mode is disabled before and after the switching transients but more investigations has to be done on the reliability of applying a high gate impedance during normal operation.

4 Proposed optimization for a faster SSD reaction: ASSD

The Advanced Soft Shut-Down (ASSD) principle consists in the implementation of a very short delay (e.g. t_{delay}=20ns) after the SC detection in between the turn-off request and the SSD activation (HiZ_Buffer signal) when a SC is detected, as described in figure 7. The principle consists in a quick pre-discharge of the gate capacitance C_{gs} of the SiC device before the real turn-off occurs at the Miller plateau. This optimization consists in connecting the R_{ssd} at the best moment to obtain both overvoltage reduction and a fast SC reaction. During the delay time, the regular turn-off gate resistance is still connected leading to a fast discharge of the gate-source capacitance C_{gs}. Characterization has to be done to quantify precisely the optimum delay considering that the SSD mode must be enabled when approaching the turn-off threshold voltage V_{th}- otherwise, an extra overvoltage appears.

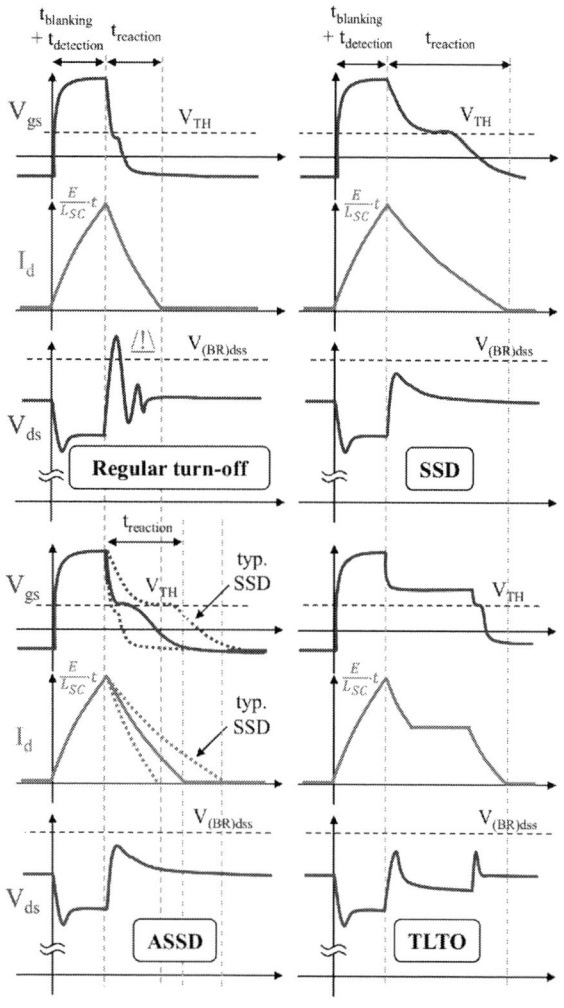

Fig. 6 Theoretical approach on OCP features: SSD, TLTO and proposed ASSD

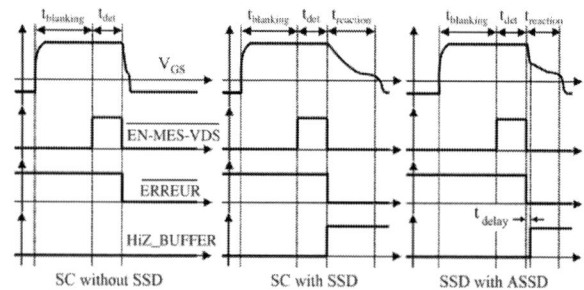

Fig. 7 Chronograms and command signals for the proposed ASSD method

If the introduced delay is too short then the result will be similar to the conventional SSD curve, if it is too long then a maximal overvoltage will occur. The simulations also show that this result is relatively independent from the stray inductance of the SC.

5 Simulations based on a SiC MOSFET SPICE model

5.1 Model building

Simulations were realized using Wolspeed's SPICE model of the 3rd generation SiC power module CAS350M12BM3. The bipolar-based gate buffer was modelled by SPICE NPN and PNP model class with respectively current gains of 140 and 90, corresponding to the components used STN851 and STN951. The short circuit inductance L_{sc} can be fixed to simulate different cases of SC event such as a half-bridge SC (lowest impedance – about 10nH) or load SC (highest impedance).

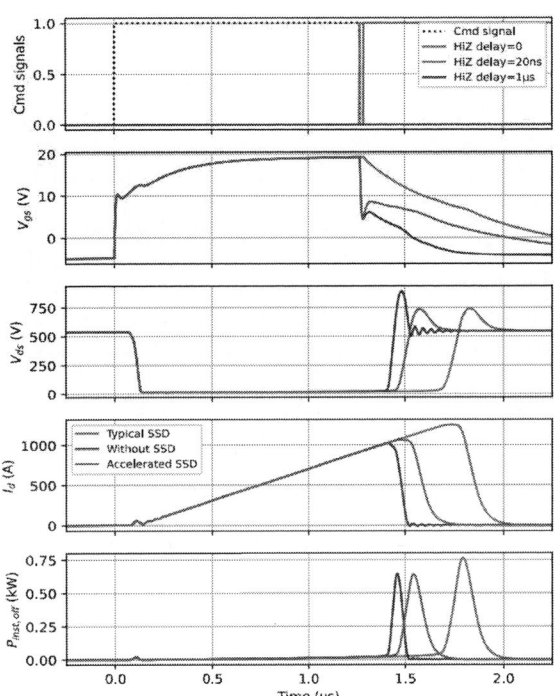

Fig. 9 SC simulation based on CAS350M12BM3 Wofspeed's SPICE model with $L_{SC} = 650nH$

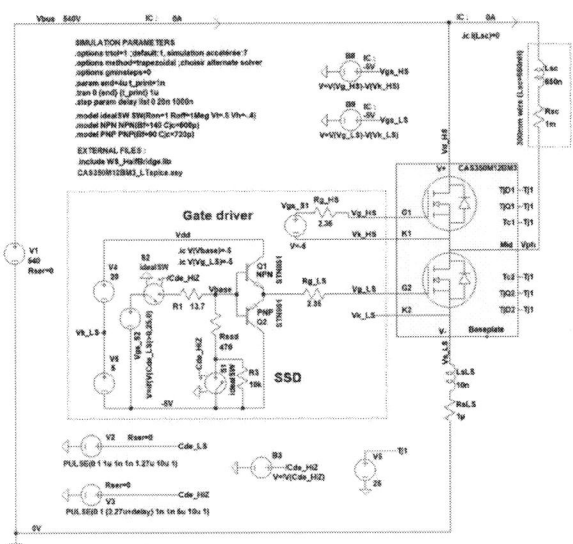

Fig. 8 LTspice model used for parametric analysis

5.2 Simulation results

The simulations based on SPICE models demonstrate the existence of an optimal delay time for a given SiC device and its associated gate driver, in the selected settings this optimal time is about 10-20ns. According to theses simulations, the STO energy during a SC event can be reduced by -20% while the overvoltage is reduced by -4% using the proposed ASSD method.

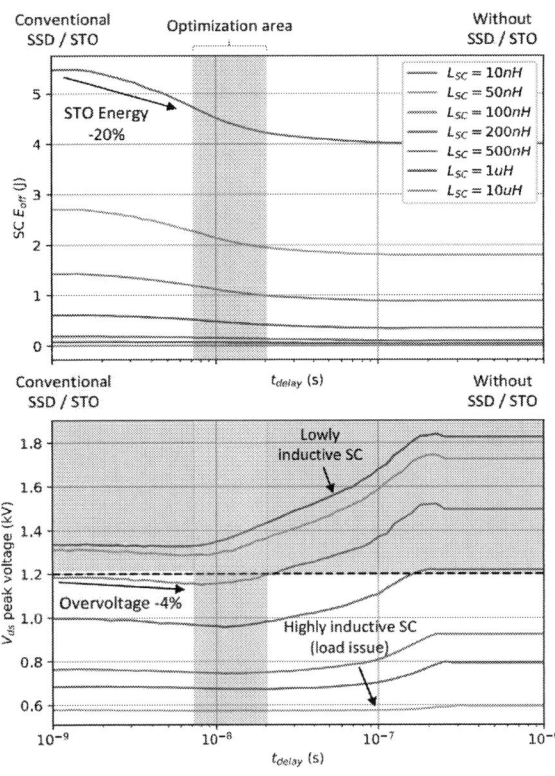

Fig. 10 Parametric simulation results with inductive SC from 10nH to 10µH: STO energy and peak drain-source voltage at turn-off during SC events

6 Experimental results

The method proposed in this article demonstrates a significant reduction of the reaction time of -52% using the same R_{ssd} value and switching slope. In figure 11, experimental waveforms are shown to highlight the impact of the turn-off under SC condition on the transient low-side (LS) voltages and current. The short-circuit impedance of the 30cm wire used to short the high-side device is here estimated at 650nH considering the current increasing slope. In these conditions, a regular turn-off without any reaction circuitry causes a V_{DS} overvoltage of 850V while the SSD feature limits it at 720V (ΔV_{DS} -42%) for a 540VDC operation. SC with lower impedance will result in a higher overshoot which can cause the voltage breakdown of the SiC device. The resulting overvoltage can be limited by the tuning of R_{ssd} but it can also increase the turn-off energy during the SC extinguishing E_{SC}. In addition, the red curve of the figure 13 shows that the drain current continues its increasing during the SSD process until the gate voltage reaches the threshold voltage V_{TH}. The proposed solution (green curve) takes advantages of the fast discharge of a regular turn-off and the R_{ssd} resistance connected to the gate buffer during the switching. This method enable both drain-source overshoot limitation and minimization of the soft turn-off energy E_{SC} dissipated in the SiC power device.

Both the blanking time and the R_{ssd} value are tunable as well as in the conventional SSD circuit. The blanking time reflects the compromise between robustness and safety while the R_{ssd} value is the balance between overvoltage tolerance and STO energy/reaction time.

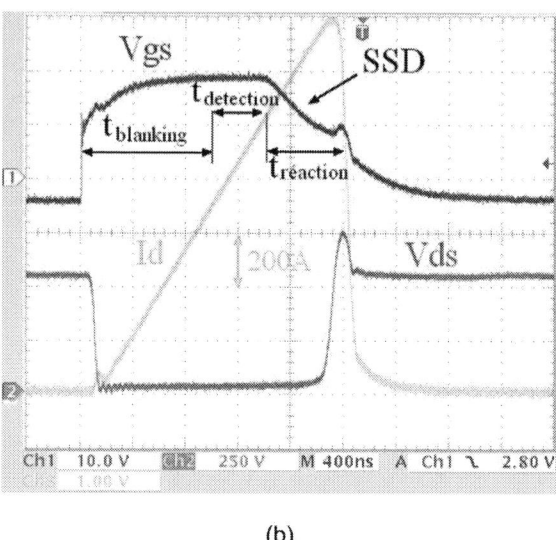

(b)

Fig. 11 Experimental waveforms for SC tests using a Rogowski sensor at V_{DC}=540V with $L_{SC} \approx$ 650nH realized with a CAS325M12HM2 SiC MOSFET : (a) turn-off without SSD; (b) turn-off with SSD (R_{ssd} = 470Ω)

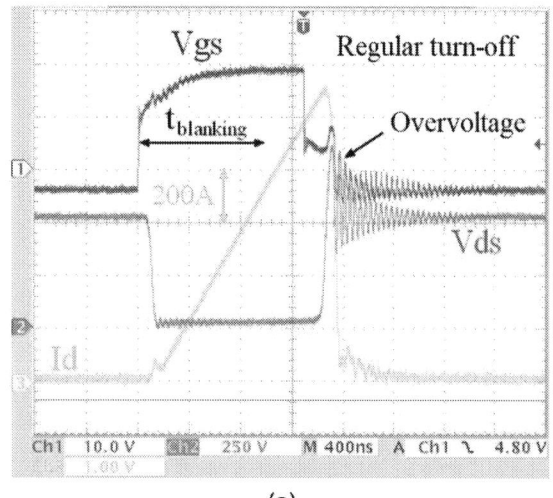

(a)

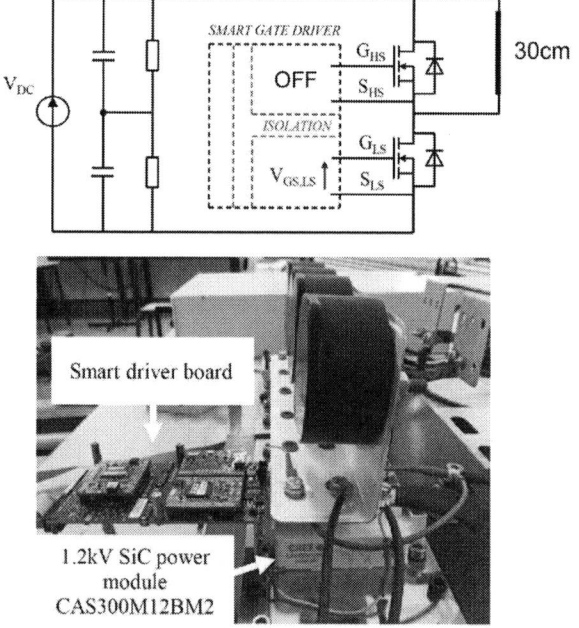

Fig. 12 Experimental setup for SC tests

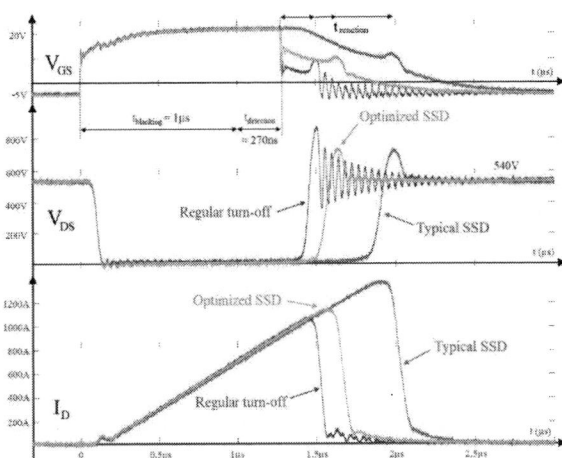

Fig. 13 Experimental comparison of SC test results using the ASSD method at V_{DC}=540V with $L_{SC} \approx$ 650nH and 1µs blanking time

7 Conclusion

The safe operating of SiC power devices requires a faster SC protection. Conventional protection methods are still applicable but require some optimizations for SiC MOSFETs. The process of interruption of a SC event can be decomposed into 3 steps: the blanking time, the detection time and the reaction time. The proposed ASSD solution consists in an optimization of the reaction time after SC detection by adding a few nanoseconds delay (e.g. 20ns) in the SSD process to achieve both V_{DS} overvoltage limitation and fastest reaction. Simulations based on SPICE models have shown the existence of an optimum point for a given SiC device and its associated gate driver. Results have proven a reduction of the reaction time of -300ns, a SC turn-off energy of -20% and an overvoltage reduction of -4% compared to the conventional SSD method using the same R_{ssd} value. The experimental results were obtained with a SC impedance of 650nH on a CAS325M12HM2 SiC module at V_{DC}=540V. The ASSD proposed method can be easily realized using a basic RC circuit at the appropriate location on a conventional gate driver.

This optimization was implemented on a bipolar gate driver architecture but the principle can be easily extended to all types of gate drivers and combined with fast SC detection to improve the robustness of SiC devices. Although the optimum delay value can be predicted by an appropriate SPICE simulations, it is highly recommended to set it by several SC tests for a given design. Fur-

ther investigations has to be done to check the repeatability of these SC capabilities considering chips disparities and various devices.

References

[1] E. Wiesner, E. Thai, A. Volke, and K. Fink, "Advanced protection for large current full SiC-modules," presented at the PCIM Europe, 2016.

[2] A. Volke, M. Hornkamp, and J. Wendt, *IGBT modules. Technologies, Driver and application*, Infineon Technologies. 2017.

[3] N. Ginot, C. Batard, and P. Lahaye, "MOSFET et IGBT : circuits de commande, sécurisation et protection du composant à semi-conducteur," in *Techniques de l'ingénieur*, vol. V1, 2017.

[4] T. L. Le, J.-S. Ngoua-teu, and T. Youssef, "High power module GaN with integrated current sensor for fast short circuit protection," presented at the EuroSimE, Apr. 2022. doi: 10.1109/EuroSimE54907.2022.9758909.

[5] J. Walter, J. Acuna, and I. Kallfass, "Design and Implementation of an Integrated Current Sensor for a Gallium Nitride Half-Bridge," presented at the PCIM Europe, Nüremberg, 2018.

[6] J. Fu, Z. Zhang, Y. F. Liu, P. C. Sen, and L. Ge, "A New High Efficiency Current Source Driver With Bipolar Gate Voltage," *IEEE Trans Power Electron*, vol. 27, no. 2, pp. 987–997, Feb. 2012, doi: 10.1109/TPEL.2010.2077741.

[7] M. Shim, K. Lee, J. Kim, and K. Kim, "Multistep Soft Turn-Off Time Control to Suppress the Overvoltage of SiC MOSFETs in Short-Circuit State," *IEEE Access*, vol. 10, 2022, doi: 10.1109/ACCESS.2022.3169764.

[8] Y. Shi, R. Xie, L. Wang, Y. Shi, and H. Li, "Short-circuit protection of 1200V SiC MOSFET T-type module in PV inverter application," presented at the ECCE, Sep. 2016. doi: 10.1109/ECCE.2016.7855428.

[9] V. Thayumanasamy, C. Fuentes, K. Lenz, I. Rabl, and J. Engstler, "Short Circuit Protection of a Power Module with Trench-SiC-MOSFET. Can DESAT be Fast Enough ?," presented at the PCIM Europe, May 2022. doi: 10.30420/565822111.

[10] M. Picot-Digoix, F. Richardeau, S. Azzopardi, and T.-L. Le, "Quasi-flying gate concept used for short-circuit detection on SiC power MOSFETs based on a dual-port gate driver," *IEEE Trans. Power Electron.*, 2023.

PCIM Europe 2024, 11– 13 June 2024, Nuremberg   DOI: 10.30420/566262015

Parameter Identification: Gate Sensor for Power Transistor Tolerance Compensation in Advanced Gate Driver ICs

Rakshith Satheesh [1], Christopher Wille [1], Pushkar Kulkarni [1], Andreas Menzel [1], Thoralf Rosahl [1]

[1]Robert Bosch GmbH, Mobility Electronics, Reutlingen, Germany

Corresponding Author: Andreas Menzel, andreas.menzel@de.bosch.com

Speaker   : Christopher Wille, christopher.wille@de.bosch.com

Abstract

This paper introduces an innovative approach to optimize power module performance by implementing a Parameter Identification (PI) mode in advanced high voltage gate drivers, which can enhance both the efficiency and lifespan of power electronic systems. The paper details the development and integration of the PI mode within the gate driver architecture. We present a comprehensive analysis of the test conditions and experimental results, demonstrating several options PI offers to enhance system performance. The findings underscore the potential of integrating parameter identification capabilities in gate drivers to achieve optimal power module operation, offering substantial benefits in terms of energy efficiency and system robustness.

1 Introduction

Silicon Carbide (SiC) power modules present a compelling alternative to traditional silicon (Si) devices for high voltage applications. With their high breakdown voltage and ability to operate at high switching frequencies, SiC material facilitate a substantial increase in power density. This improvement not only enhances the overall efficiency of power electronic systems but also simplifies their design by reducing the complexity involved [1]. The technique of gate driving is crucial for improving the high-speed switching capabilities of power devices. In current power electronic system, it is required to connect several power transistors in parallel. However, the load unbalancing in power electronic system becomes a crucial effect. A current-source gate driver (CSGD), which allows for the adjustment of gate shapes, offers the flexibility to modify the gate current during various phases of a switching event for every single power transistor. This capability is particularly beneficial and represents a significant advantage over voltage source gate drivers (VSGD) [2]. Especially, over lifetime power transistors will see aging effects which may cause a shift in its electrical characteristics and switching behaviour causing negative effects in

its performance. In addition, variations in its operation conditions also change the device characteristics which can be hardly controlled by common VSGD devices.

A unique gate driver test chip was developed to facilitate the implementation of specific gate current profiles. This chip, distinguished by its open-loop slope-shaping capability, marks a departure from conventional designs. Unlike state-of-the-art standard gate drivers, this device can modify its driving strength mid-commutation, guided by three programmable intervals and gate current levels. This feature grants the ability to independently fine-tune the switching speeds for both current and voltage commutation [3].

1.1 Gate Charging Mechanism

A MOSFET's switching process can be divided into four phases, which will be described in Figure 2. From time t0 to t1, the gate drive circuit fills both the gate-source (C_{GS}) and gate-drain (C_{GD}) capacitances until the gate voltage hits the threshold voltage (V_{th}). Between t1 and t2, as the gate to source voltage (V_{GS}) surpasses V_{th}, it initiates a drain current flow, eventually becoming the primary current, while continuing to charge C_{GS} and C_{GD}. The gate voltage's rise boosts the drain current until the gate voltage aligns with the Miller voltage, $V_{GS(pl)}$, at t2.

From t2 to t3, the gate voltage stabilizes at V_{GS} (pl) due to the Miller effect, maintaining a steady state. During this phase, the main gate current flows through the MOSFET, and the drain voltage hits its turn-on threshold. The constant gate voltage during this interval directs the drive current towards C_{GD} rather than C_{GS}. The charge gathered in $C_{GD}(Q_{GD})$ in this time frame equals the current to the gate circuit multiplied by the voltage drop duration (t3-t2).

Finally, from t3 to t4, the gate voltage is driven to an oversaturated state, charging both C_{GS} and C_{GD} until V_{GS} matches the gate supply voltage. Since the turn-on transient has completed, the MOSFET experiences no switching loss in this last period [4].

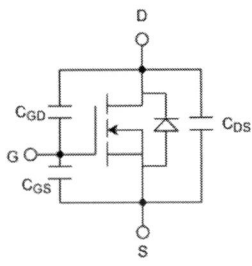

Fig1.: Capacities of a MOSFET

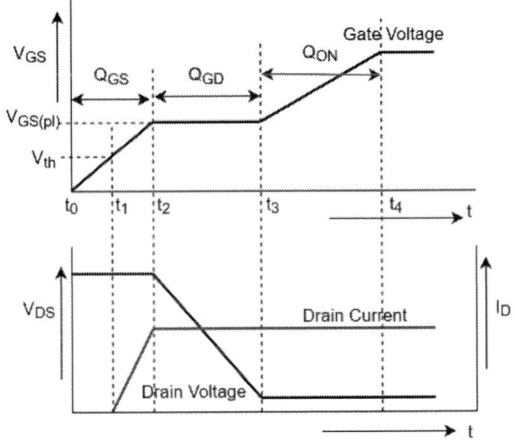

Fig2.: Gate Charge Waveform [4]

1.2 Parameter Identification Mode

Gate control strategies for the CSGD dynamically modify the gate current using gate current profiles, aiming to minimize switching losses and mitigate the adverse impacts associated with voltage and current slew rates. The flexibility offered by these techniques has attracted significant research interest in recent years, Parameter Identification is one the topic which is significantly useful in tuning the current profile and minimizing the switching losses during operation.

In the latest CSGD IC from Bosch PI has been implemented as a dedicated operation mode ("PI mode") (see chapter 2) that enables the IC to ascertain the values of various parameters of the power semiconductor that affect the switching process. This mode has specifically been designed to analyse each gate channel individually among each of the IC's gate outputs, allowing a precise determination of the characteristics of each single power semiconductor chip.

The activation of PI mode is initiated through a command sent via the Low-Voltage (LV) communication interface, which sets the state of the Application-Specific Integrated Circuit (ASIC) to a designated PI state. Once activated, the PI process utilizes a configurable constant gate current source for the measurement, and it includes an option to set an Analog-to-Digital Converter (ADC) offset. This offset determines the voltage range within which the ADC will sample the gate voltage, ensuring that values outside this range are not considered in the measurement. This flexibility allows the PI to be conducted either on a single gate output or on a combination of outputs, depending on the specific requirements of the analysis.

The outcome of the PI process is captured and then transmitted to an MCU, where undergoes further data processing to extract the necessary parameters. The ability to accurately identify these parameters is crucial for optimizing the performance of the power semiconductor within the system, ensuring efficient and reliable operation of the gate driving.

Within the MCU, the raw data received from the ASIC during the PI mode is transformed into crucial parameters that are essential for optimizing the efficiency and lifespan of the system. These parameters include the Gate to Source Charge (Q_{GS}), the Gate to Drain Charge (Q_{GD}), and the Turn-On Charge (Q_{ON}), see Figure 2. These key outputs from the PI raw data processing mode play a significant role in enhancing the performance of the power semiconductor devices [5].

Understanding and accurately determining these parameters allows for a more precise control over the switching behaviour of the semiconductors. For instance, by knowing the exact values of Q_{GS}, Q_{GD}, and Q_{ON}, fine tuning of gate current profiles can be performed which is critical for minimizing switching losses, reducing stress on the components, and thereby increasing the overall efficiency of the power electronic system. Additionally, by optimizing these parameters, the system can achieve better performance, which directly contributes to extending the operational lifespan of the devices. This meticulous approach to parameter adjustment and system optimization is a cornerstone in the development of high-performance, reliable, and efficient power electronic systems.

1.3 Alternative method to sense power transistor gate characteristics

In order to discuss limitations to VSGD a brief overview of the state of the art of known available gate driver ICs need to be given, which have a similar function as the parameter identification described in this paper: Some VSGD devices have a gate threshold monitor function designed to assess the gate turn-on voltage of the power transistor during the startup phase. This action initiates the charging of the power transistor's gate capacitance by a constant current source, causing the gate voltage to increase steadily. As the transistor begins to conduct, its gate voltage stabilizes at the threshold level, mimicking a diode's behaviour. Following a predetermined blanking period, the built-in voltage sensing takes a sample of the stabilized gate voltage and records this data in the register. The recorded measurement represents a scaled-down value of the actual gate voltage. This threshold voltage can be used as a critical parameter for

the Microcontroller Unit (MCU) to evaluate the actual condition of the power transistor.

Comparison to PI mode from chapter 1.2: Depending on the resolution of the stabilized gate voltage sensing this concept could be used to identify changes or tolerances in the gate turn-on voltage. But the method in VSGD devices is not capable to determine changes in the gate charges required to get through the single segments of a switching event according to Figure 2. Additionally, changes in the Miller-Plateau region, highly dependent on drain voltage, drain current and temperature (see also Figures 6,7,8 in chapter 3) cannot be identified. As a temporary conclusion, this concept might be capable to identify bigger changes or drifts of the power transistor turn-on voltage, but its output might not be accurate enough to adjust the gate drive strength (when VSGD is used) or gate drive current profile (when CSGD is used).

1.4 Optimizing through parameter identification

Parameter Identification (PI) in power electronics can serve as a fundamental tool for several critical functions that enhance the efficiency and lifespan of power electronic systems. In the following is an overview of where PI can contribute:

Tuning Current Profiles for CSGD gate drivers: As described before (section1.2), one of the primary uses of PI is in the tuning of current profiles. By accurately identifying the gate charge characteristics (Q_{GS}, Q_{GD}, Q_{ON}), tailoring the gate driving signals to minimize switching losses, reduce electromagnetic interference (EMI), and optimize the switching speed, directly contributing to the efficiency and durability of the system. When combining Q_{GS}, Q_{GD}, Q_{ON} results from the ASICs inline measurement with detailed measurement data and modelling from the power-transistors characterisation, the MCU can calculate the actual turn-on voltage V_{th} and use it for the current profile adjustment.

Changing gate drive strength for VSGD gate drivers: When a VSGD supports different gate drive strengths the PI output can be taken as decision criteria for the actual working point. The gate drive strength selection is than more orientated towards the actual condition in the embedded system than to the conditions during the design phase even the strength resolution is much smaller than for CSGD.

Thermal Management: When knowing V_{th} PI helps in predicting the thermal behaviour of power devices under various operating conditions, especially during the turn-on event. This information can be used to optimize thermal management strategies, especially balancing the load between different parallel switching transistors, reducing the risk of overheating, and improving the system's reliability and lifespan.

Fault Detection and Predictive Maintenance: Parameters obtained through PI can be monitored over time to detect early signs of wear or failure in power devices. For example, a gradual increase in the turn-on threshold voltage or changes in the gate charge could indicate device degradation. This capability enables predictive maintenance, power degradation strategies and therefore preventing unexpected failures and extending the lifespan of the system.

Adaptive Control Systems: In advanced power electronic systems, PI data can be used to implement adaptive control algorithms that adjust the operation of the system in real-time based on the identified parameters.

2 PI Analysis: Core Concepts

The Data Acquisition Mode is a comprehensive process designed for capturing and analysing raw data or parameters from a system. This mode is structured into several key phases, each critical for preparing, executing, and finalizing the data capture process. These phases must be followed in a specific sequence to ensure the system is correctly set up for data acquisition, allowing for the collection of detailed and accurate data under various conditions.

Initialization Phase:
This initial phase is focused on configuring the system with essential settings based on user-defined criteria. These settings might include parameters like measurement sensitivity, ADC input offset, and specific operational configurations tailored to the data capture objectives.

Users can customize the setup to match their specific requirements, choosing from different operational modes or configurations. This flexibility ensures that the system can be adapted to capture data under a wide range of conditions, optimizing the relevance and quality of the collected data.

Proper configuration during this phase is crucial for ensuring that the system is accurately tuned to capture the desired data, directly influencing the effectiveness of the subsequent data acquisition process.

Data Acquisition Phase:
This phase is subdivided into two distinct segments, each designed to capture data under specific conditions or intervals.

First Segment: The first segment of data collection begins with a command to start, accompanied by a signal defining its duration. This phase is critical for determining how data is sampled and stored, focusing on capturing an initial set of data points based on predefined triggers or conditions. These triggers could include specific events or thresholds indicating the start of relevant data collection. The main goal is to establish a baseline dataset for further analysis, providing insights into system behaviour or performance. This data is relied upon to pinpoint potential concerns, evaluate functionality, and ensure that subsequent data collection aligns with the goals of the experiment.

Second Segment: Continues the data collection process, often under extended or slightly modified conditions to gather a more comprehensive dataset. This segment is essential for building upon the initial data set, providing a deeper and more detailed insight into the system's behaviour or performance.

Data Transmission Phase:
The final phase involves transmitting the collected data to a processing or storage unit. This step is critical for making the data accessible for analysis, review, or archival purposes.

Data is sent in structured batches, ensuring that all collected information is efficiently and accurately relayed for further processing. This systematic approach to data transmission is vital for maintaining the integrity and usability of the data for subsequent analysis or decision-making processes [3].**Raw-data processing phase:**

The MCU analyses the ADC raw data to derive critical parameters, ensuring a thorough evaluation of the power module's functionality via the ASIC's PI mode. Initially, upon the first activation, it extracts a trigger voltage value. ASIC detects the trigger at the end of the Miller-plateau-region, see Figure 3. Subsequently, during the second activation, it determines additional parameters including the gate to source charge (Q_{GS}), drain to source charge (Q_{DS}), turn-on charge (Q_{ON}), and the total charge (Q_{TOTAL}), see Figure 4. The calculation requires the definition of application specific thresholds for "Above $V_{GS(pl)}$" and "Below $V_{GS(pl)}$". This methodical process facilitates a detailed assessment of the power module's operational characteristics.

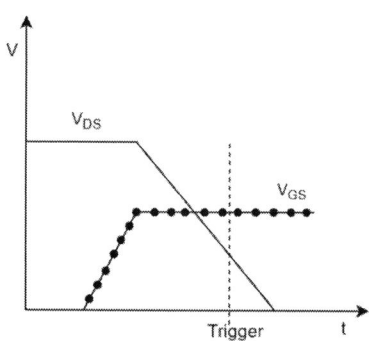

Fig 3. PI first turn-on

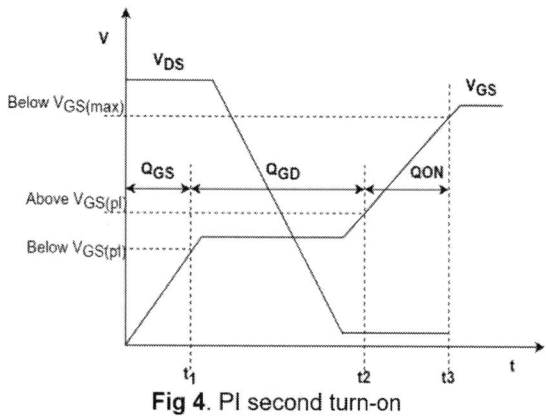

Fig 4. PI second turn-on

3 PI Mode Evaluation

PI mode evaluation was conducted, examining five distinct parameters: ambient temperature variation, different drain-to-source voltage V_{DS} levels, PI charge current (I_G) configurations, variations in drain current performed by different constant resistive loads. The measurements have been carried out on a typical B2 power module for an automotive traction inverter application. Detailed results of these assessments are outlined in the subsequent sections. The plots below illustrate the comparison of a working point across the different dimensions.

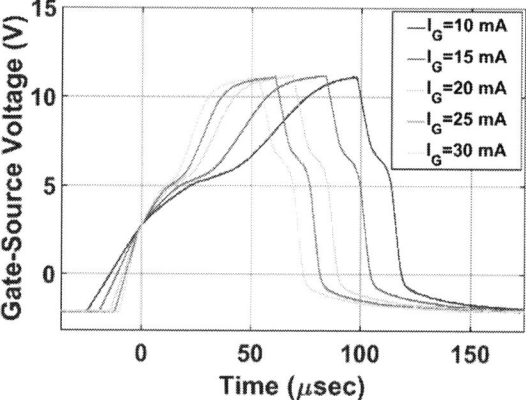

Fig 5. V_{GS} vs Time for variable gate current during PI

Figure 5 illustrates that an increase in gate current results in a faster completion of the turn-on phase of the gate-source voltage curve, reducing the required time but the impact on the PI output accuracy has not been evaluated in this work yet.

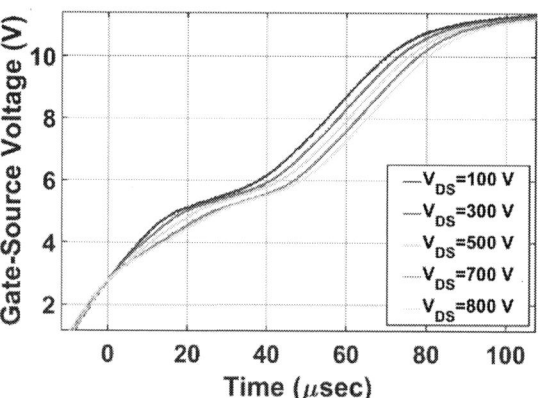

Fig 6. V_{GS} vs Time Miller Plateau region for variable V_{DS} during PI

PCIM Europe 2024, 11– 13 June 2024, Nuremberg   DOI: 10.30420/566262015

Temperature (C°)	V_{DS} (V)	I_G (mA)	I_D (A)	Q_{GS} (nC)	Q_{GD} (nC)	Q_{ON} (nC)	Q_{TOTAL} (nC)
21	500	10	40	606	129	542	1277
21	600	10	40	619	129	542	1290
21	700	10	36	619	155	529	1303
21	800	10	40	645	155	516	1316
21	600	10	40	619	129	542	1290
21	600	10	200	710	116	451	1277
21	600	10	300	735	103	439	1277
21	600	10	400	761	103	413	1277
21	500	10	40	606	129	542	1277
90	500	10	40	555	129	593	1277
110	500	10	40	542	142	606	1290

Table 1: Measured charge values from PI for different ambient conditions on the same power module

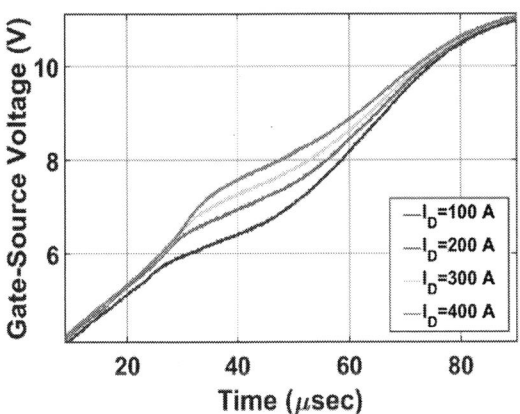
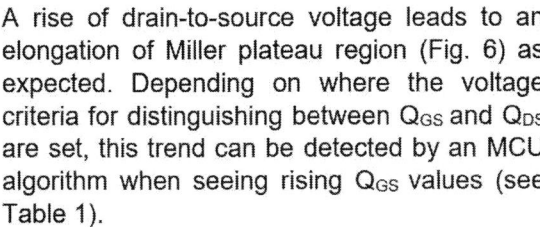

Fig 7. V_{GS} vs. Time Miller Plateau region for variable I_D during PI

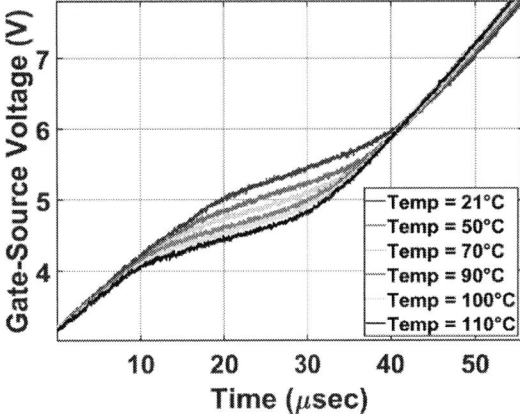

Fig 8. V_{GS} vs. Time Miller Plateau for variable Temperatures during PI

A rise of drain-to-source voltage leads to an elongation of Miller plateau region (Fig. 6) as expected. Depending on where the voltage criteria for distinguishing between Q_{GS} and Q_{DS} are set, this trend can be detected by an MCU algorithm when seeing rising Q_{GS} values (see Table 1).

Figure 7 shows the effect of constant drain current (I_D) variation (by a resistive load), which causes a vertical drift in the Miller plateau. It is possible to identify a rising Q_{GS} trend together with rising I_D within this steady state laboratory test-condition. But within a traction inverter application with inductive loads where I_D is

changing during the PI execution, it is expected to be very hard to derive analysis results from the gate drivers raw-data output perspective.

Furthermore, rising transistor temperatures lead to lower Miller-plateau voltage levels (see Fig 8) without elongation resulting in falling Q_{GS} values (see Table 1).

Conclusion

In conclusion, the implementation of a PI mode in a gate driver IC represents a significant advancement in the field of semiconductor testing and evaluation in the real traction inverter application. Through its comprehensive approach to parameter inspection and analysis, PI mode offers a versatile fundamental base for understanding and enhancing power module performance. This is not only possible – like other known methods – during the product development phase with accurate external measurement equipment. It offers the unique chance to analyse and identify trends or changes in the power module characteristic during the products service life in various operation conditions. To ensure that future products meet both current technological demands and are equipped to tackle future challenges, PI mode can play a crucial role in advancing the efficiency and reliability of switching applications requiring gate driver ICs.

The detailed data provided in this paper explain the test conditions and the PI output results during the evaluation, including temperature variations, different V_{DS} levels, PI charge current configurations and I_D variations.

Next planned steps after this work are a detailed characterisation of the PI output accuracy compared to an external reference measurement and the evaluation with different power transistors and power modules. Additionally, the influence of inductive loads on PI output results should be investigated. Instead of extracting gate-charges out of the PI raw-data output, a differential capacity could also be used as criteria for further analysis.

As an outlook it is highly recommended to investigate the potential of using the PI output data to adjust the gate drive strength current profiles. This would show how efficient a tuning of the switching behaviour based on the actual detected power transistor condition can be. When having access to more than one gate pin of a power module PI can play an even more important role when trying to minimize load imbalance between parallel switched transistors.

Acknowledgments

The IPCEI ME/CT project is supported by the Federal Ministry for Economic Affairs and Climate Action on the basis of a decision by the German Parliament, by the Ministry for Economic Affairs, Labor and Tourism of Baden-Württemberg based on a decision of the State Parliament of Baden-Württemberg, the Free State of Saxony on the basis of the budget adopted by the Saxon State Parliament, the Bavarian State Ministry for Economic Affairs, Regional Development and Energy and financed by the European Union - NextGenerationEU.

References

[1] Xiaoling Li, Yuxiang Chen, Yuheng Wu,"High Voltage SiC Power Module Optimized for Low Parasitics and Compatible System Interfaces" 2022 IEEE Applied Power Electronics Conference and Exposition, pg.1-5.

[2] M. Riefer, J. Winkler, S. Strache and I. Kallfass, "Implementation of Current-Source Gate Driver with Open-Loop Slope Shaping for SiC-MOSFETs," PCIM Europe digital days 2021; International Exhibition and Conference for Power Electronics, Intelligent Motion, Renewable Energy and Energy Management, Online, 2021, pp. 1-8.

[3] T. Rosahl, A. Barner, "On-Chip and Inline Characterization of SiC-MOSFET physical Parameters for Optimized Gate Drive Control", APEC 2023 Industry Session IS25.3, March 2023, pp. 1-18.

[4] J. Brown, "Power MOSFET Basics: Understanding Gate Charge and Using It To Assess Switching Performance,"Vishay Intertechnology, Inc., Application Note,2004.

[5] Samar K. Saha "Compact models for Integrated Circuit Design", in 2016 MOSFETs Capacitance Model, pp.226-228

PCIM Europe 2024, 11– 13 June 2024, Nuremberg   DOI: 10.30420/566262016

An Innovative High-speed Track Range Restart Strategy for Permanent Magnet Synchronous Motor

Weiping Fu[1], Hao Chen, Anna Corbitt, Cheng Tang, H. Alan Mantooth[1]

[1]University of Arkansas, USA

Corresponding author: Weiping Fu, weipingf@uark.edu

Speaker: Anna Corbitt, amcorbit@uark.edu

Abstract

In speed sensorless motor drive systems, restarting a free-running motor requires knowledge of the instantaneous rotor position and motor speed. Those restart methods based on zero voltage pulses have been widely used. However, conventional restart methods are not suitable for those conditions with high ratational frequency and long current sample time. When motor operates at high speed the phase error between two consecutive short current may over 180 degrees, conventional zero voltage sequence will cause incorrect rotor speed and position estimation. The novelty of the proposed method in this paper is that it is suitable for both high rotaional frequency and low rotaional frequency and needs no machine paramters identification.

1   Introduction

Permanent Magnet Synchronous Motors are widely used in aerospace, electric vehicle, and industrial control fields due to their advantages such as high torque inertia ratio, high energy density, and high efficiency. However, conventional control systems often utilize mechanical sensors like photoelectric encoder and resolver to obtain speed signals. These mechanical sensors have issues such as high cost, complex installation, and susceptibility to the environment. To overcome these drawbacks, speed sensorless control technology has drawn great attention on industries [1][2][3].

For PMSMs, the initial rotor position and rotor speed are needed for closed loop control. Several restart methods have been proposed to get those signals [4]-[10]. The method in [7] uses external voltage sensor to detect stator voltage for sensorless flux observer. To eliminate the external voltage sensor, some methods based on high frequency signal injection is proposed [4][5][6]. But those methods can be only applied in internal mounted PMSMs. [8] proposes a restarting strategy based on two zero voltage and its suitable for VF and space vector control, but only suitable for those condition of motor frequency is low. [9] and [10] use three voltage pulses to achieve high rotor frequency range estimation, but its drawback is low speed estimation precision. The proposed method in this paper is suitable for both high frequency and low frequency compared to the conventional method like [10], and maintain good speed estimated precision.

The mathematical model of a synchronous motor can be expressed by the following formula:

$$\begin{bmatrix} V_d \\ V_q \end{bmatrix} = \begin{bmatrix} R_s + sL_d & -\omega_r L_q \\ \omega_r L_q & R_s + sL_d \end{bmatrix} \begin{bmatrix} I_d \\ I_q \end{bmatrix} + \begin{bmatrix} 0 \\ \omega_r \phi_M \end{bmatrix} \quad (1)$$

At higher speeds, the voltage drop across the stator resistance can be ignored. equation (1) can be rewritten as follows:

$$\begin{bmatrix} V_d \\ V_q \end{bmatrix} = \begin{bmatrix} sL_d & -\omega_r L_q \\ \omega_r L_q & sL_d \end{bmatrix} \begin{bmatrix} I_d \\ I_q \end{bmatrix} + \begin{bmatrix} 0 \\ \omega_r \phi_M \end{bmatrix} \quad (2)$$

When the stator is shorted with zero voltage, equation (2) can be rewritten as

$$\begin{bmatrix} 0 \\ 0 \end{bmatrix} = \begin{bmatrix} sL_d & -\omega_r L_q \\ \omega_r L_q & sL_d \end{bmatrix} \begin{bmatrix} I_d \\ I_q \end{bmatrix} + \begin{bmatrix} 0 \\ \omega_r \phi_M \end{bmatrix} \quad (3)$$

Assume that the initial currents are zeros,

$$\begin{bmatrix} I_d(t_{pulse}) \\ I_q(t_{pulse}) \end{bmatrix} = \begin{bmatrix} -\dfrac{\phi_M}{L_d}(1 - cos(\omega_r t_{pulse})) \\ -\dfrac{\phi_M}{L_q} sin(\omega_r t_{pulse}) \end{bmatrix} \quad (4)$$

t_{pulse} in equation (4) is the short circuit time. The equation reveals that in permanent magnet synchronous motors with fixed parameters, the short-circuit current amplitudes I_d and I_q depend solely on time and motor speed, independent of the motor's spatial position.

From Equation (5), it can be seen that the angle between I_s and the D axis is only related to time and motor speed and motor parameters.

135

$$\theta_0(\omega_r t_{pulse}) = arctan\left[\frac{L_d \sin \omega_r t_{pulse}}{L_q(1 - \cos \omega_r t_{pulse})}\right] + \pi \quad (5)$$

Equation (5) can be depicted in Fig.1.

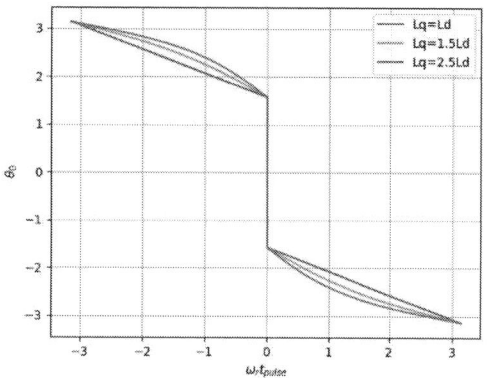

Fig. 1   Relative position between D-axis and short current vector

When the short time is very small, and if $\omega_r > 0$, $\theta_0(\omega_r t_{pulse}) \approx -0.5\pi$. if $\omega_r < 0$, $\theta_0(\omega_r t_{pulse}) \approx 0.5\pi$.

And the rotor speed and be calculated from two short pulses current. The theory can be described as follows: $I_{\alpha 1}$ and $I_{\beta 1}$ are the stator current for first pulse, $I_{\alpha 2}$ and $I_{\beta 2}$ are the stator current for the second pulse.

θ_1 and θ_2 can be calculated from inverse trigonometric functions.

$$\theta_1 = arctan\left(\frac{I_{\beta 1}}{I_{\alpha 1}}\right)$$

$$\theta_2 = arctan\left(\frac{I_{\beta 2}}{I_{\alpha 2}}\right) \quad (6)$$

The estimated rotational speed ω_{est} can be expressed as

$$\omega_{est} = \frac{\theta_2 - \theta_1 + 2k\pi}{t_2 - t_1} = \frac{2k\pi + \Delta\theta_{12}}{\tau} \quad (7)$$

where k is the number of rotations between two pulses, τ is the time interval between two pulses. To make sure we get the right rotate direction and speed. τ should satisfy the following equation.

$$\tau < \frac{\pi}{|\omega_r|} \quad (8)$$

Where ω_r is the practical rotational speed. When motor runs in a very high frequency, τ need to be very small. On that condition, equation (8) is hard to satisfied.

In Fig. 2, the black arrows are the current vectors for zero voltage pulses. Both counterclockwise rotation and clockwise rotation may share the same current vector position. Conventional restart method would think the absolute angular difference is less than 180°. Therefore, the conventional restart method may erroneously determine the rotational direction

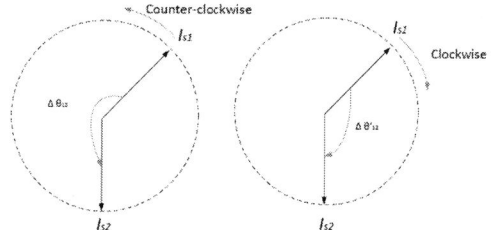

Fig. 2   short circuit current vector with conventional restart method

2   Proposed restart method

The proposed method for the restart procedure of PMSM can be depicted in Fig. 3.

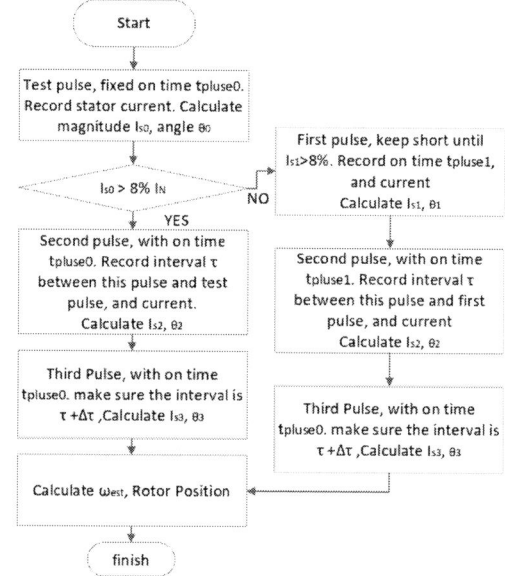

Fig. 3 Flow chart of proposed restart method

The detailed conceptual diagram of the proposed restart method is depicted in Fig. 4. In Fig. 4, I_s is the state current vector magnitude. The interval between t_1 and between t_2 and the interval between t_2 and t_3 are different, their difference is $\Delta\tau$. And the coarse value of estimated speed can be described as the following equation:

$$\omega_{est\_coarse} = \frac{\Delta\theta_{23} - \Delta\theta_{12}}{\Delta\tau} \quad (9)$$

Where $\Delta\theta_{23}$ is the angular difference between the 2nd pulse θ_2 and the 3rd pulse θ_3, $\Delta\theta_{12}$ is the angular difference between the 1st pulse θ_1 and the 2nd pulse θ_2.

From the coarse speed we can know that the coarse angular difference between I_{s2} and I_{s3} can be described as follows:

$$\Delta\theta_{23\_coarse} = \omega_{est\_coarse} \cdot (\tau + \Delta\tau) \qquad (10)$$

Calculate the five values described in the following equation:

$$\Delta\theta_{23\_k} = [arctan(\frac{I_{\beta2}}{I_{\alpha2}}) - arctan(\frac{I_{\beta1}}{I_{\alpha1}})] \\ + 2\pi \cdot k, (k-2,-1,0,1,2) \qquad (11)$$

Choose the value of k when $\Delta\theta_{23\_k}$ is closest to $\Delta\theta_{23\_coarse}$ than the other four values, then that value is the accurate angle difference $\Delta\theta_{23}$ between pulse 2 and pulse 3. The accurate rotor speed can be calculated by the equation:

$$\omega_{est} = \Delta\theta_{23}/(\tau + \Delta\tau) \qquad (12)$$

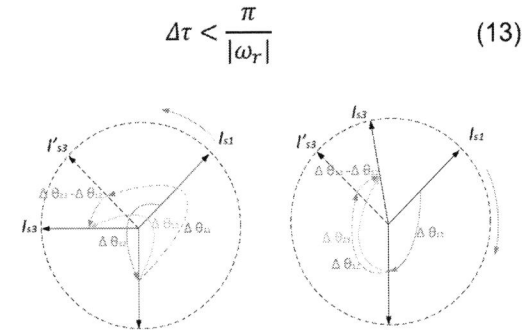

Fig. 4 conceptual diagram of proposed method

The short-circuit current vectors' positions are depicted in Fig. 5. If the three zero voltage pulses are executed in a short time, the motor speed can be regarded as constant. If the interval between the third pulse and the second pulse are equal $(t_3 - t_2 = t_2 - t_1)$, the angular difference between the third pulse and the sec

ond pulse should be equal to that between the second pulse and the first pulse as well $(\Delta\theta_{23} = \Delta\theta_{12})$. In that condition, the position of the third short circuit current vector I'_{s3} is Indicated with a dashed arrow. Now $(t_3 - t_2) - (t_2 - t_1) = \Delta\tau$, the actual position of the third current vector is indicated with a solid arrow.

If the motor is rotating counter-clockwise, then $\Delta\theta_{23} - \Delta\theta_{12} > 0$. If the motor is rotating clockwise, then $\Delta\theta_{23} - \Delta\theta_{12} < 0$. Actually, $\Delta\theta_{23} - \Delta\theta_{12}$, which should be in the range of -180°~180°, is the angle that motor rotates in the time of $\Delta\tau$. That means equation (13) needs to be satisfied. For digital controller, if the control cycle is less than $\Delta\tau$, equation (13) can be satisfied.

For example, if the maximum tracking frequency is 500Hz, $\Delta\tau$ should be less than 1ms. If the control frquency is 5kHz, set $\Delta\tau$ to 1, 2, 3 or 4 cycles will readily fulfill equation (13).

$$\Delta\tau < \frac{\pi}{|\omega_r|} \qquad (13)$$

Fig. 5 Short-circuit current vector with proposed restart method (a)counter-clockwise rotation, (b)counter-clockwise rotation

3   Simulation results

To validate the effectiveness of the proposed restart method, a simulation model was built with PLECS. The PMSM motor parameter is listed below:

Rated Voltage	240V
Rated Current	160A
Rated RPM	4245

Rated Frequency	283Hz
Back EMF	180V
Ld	0.393 mH
Lq	0.625 mH
Rs	60mΩ

Table 1 Parameters of the PMSM

Fig .2 shows the simulation results when motor rotor at frequency of 50Hz and 500Hz respectively. In upside of Fig. 6 are the three phase currents and the middle and downside of Fig .6 are rotor position and rotor sped respectively, with conventional method marked with yellow line and proposed method marked with blue line.

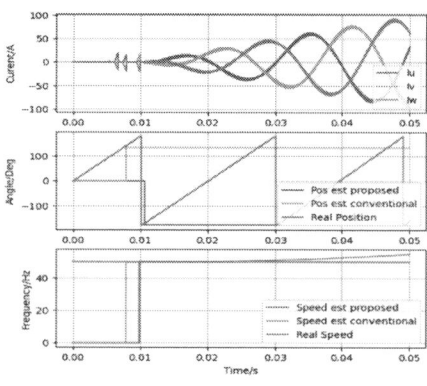

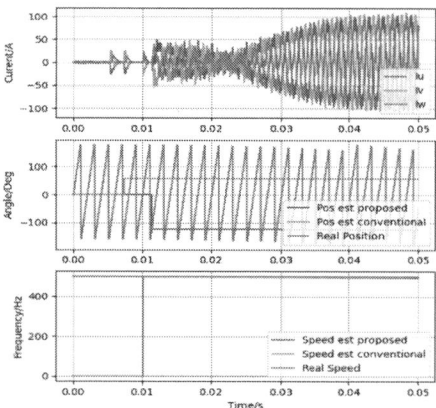

Fig. 6 (a)speed track when motor speed is 50Hz, (b)speed track when motor speed is 500Hz

when rotational frequency is 50Hz both conventional method and proposed method can get the right speed. But when frequency is 500Hz, only proposed methods can get the right speed. In Fig .6(b), the angle of I_{s1} and I_{s2} and I_{s3} are 148.3°, 148.1° and -77.8 ° respectively. τ is 0.002s, $\Delta\tau$ is 0.000746s.

using conventional methods, the speed is $-1.745 rad/s$. it is incorrect.

$$\omega_{est} = \frac{(148.1 - 148.3) \cdot \pi}{180 \cdot 0.002s} \quad (13)$$

$$= -1.745 rad/s$$

With proposed method $\Delta\theta_{23}$ is 134.1 °, $\Delta\theta_{12}$ is -0.2 °,

$$\omega_{est\_coarse} = \frac{[134.1 - (-0.2)] \cdot \pi}{180 \cdot 0.000746s} \quad (14)$$

$$= 3142 rad/s$$

Then $\Delta\theta_{23\_coarse}$ is 494.3°, that is approximately 360°+134.1 ° when k equals 1. The correct and accurate $\Delta\theta_{23}$ is 494.1°.

$$\omega_{est\_coarse} = \frac{494.1 \cdot \pi}{180 \cdot 0.002746s} \quad (14)$$

$$= 3140.4 rad/s$$

3140.4rad/s equals to 499.8Hz, much close to 500Hz.

Based on the simulation results, it is evident that the proposed method is well-suited for a higher speed range when compared to conventional restart methods.

4 Conclusion

This paper introduces a restart strategy for Permanent Magnet Synchronous Motors (PMSMs). The approach involves three consecutive zero voltage pulses, each separated by two different intervals. Notably, the simulation results demonstrate that this method is well-suited for a higher speed range compared to conventional approaches while maintaining good precision.

5 Reference

[1] J. Lee, J. Hong, K. Nam, R. Ortega, L. Praly, and A. Astolfi, "Sensorless control of surface-mount permanent-magnet synchronous motors based on a nonlinear observer," IEEE Trans. Power Electron., vol. 25, no. 2,pp. 290–297, Feb. 2010.

[2] H. Kim, J. Son, and J. Lee "A high-speed sliding-mode observer for the sensorless speed control of a PMSM," IEEE Trans. Ind. Electron., vol. 58, no. 9, pp. 4069–4077, Sep. 2011.

[3] P. Niedermayr, L. Alberti, S. Bolognani and R. Abl, "Implementation and Experimental Validation of Ultrahigh-Speed PMSM Sensorless Control by Means of Extended Kalman Filter," IEEE Journal of Emerging and Selected Topics in Power Electronics, vol. 10, no. 3, pp. 3337-3344, June 2022.

[4] Z. Q. Zhu and L. M. Gong, "Investigation of effectiveness of sensorless operation in carrier-signal-injection-based sensorless-control

methods," IEEE Trans. Ind. Electron., vol. 58, no. 8, pp. 3431–3439, Aug. 2011.

[5] Lin, T.C.; Zhu, Z.Q., "Sensorless operation capability of surface mounted permanent magnet machine based on high-frequency signal injection methods," IEEE Transactions On Industry Applications, vol. 51, pp. 2161-2171, 2015.

[6] G. W. Zhang, H. Wang, D. Xiao, L. Li, and D. G. Xu, "Pseudo random frequency sinusoidal injection based sensorless IPMSM drives with tolerance for system delays," IEEE Trans. Power Electron., vol. 34, no. 4, pp. 3623–3632, Apr. 2019.

[7] K. Yasui, Y. Nakazawa, O. Yamazaki, and I. Yasuoka, "Development of rotor position sensorless control for PRM applied to railway traction drive," in Proc. IPEC, Niigata, Japan, Apr. 2005, pp. 1671–1675. S50-4.

[8] K. Lee, S. Ahmed, and S. M. Lukic, "Universal restart strategy for high-inertia scalar-controlled PMSM drives," IEEE Transactions On Industry Applications, vol. 52, no. 5, pp. 4001–4009, Sep./Oct. 2016.

[9] T. Yamakawa, S. Wakao, K. Kondo, T. Yoneyama, S. Taniguchi, and S. Mochiduki, "Starting procedure of rotation sensorless PMSM at coasting condition for railway vehicle traction," IEEE Transactions On Industry Applications, vol. 127, no. 7, pp. 700–706, 2007.

[10] S. Taniguchi, S. Mochiduki, T. Yamakawa, S. Wakao, K. Kondo and T. Yoneyama, "Starting Procedure of Rotational Sensorless PMSM in the Rotating Condition," IEEE Transactions On Industry Applications, vol. 45, pp. 194-202, 2009.

PCIM Europe 2024, 11– 13 June 2024, Nuremberg   DOI: 10.30420/566262017

Steady-State Error Reduction of Reinforcement Learning based Indirect Current Control of Permanent Magnet Synchronous Machines

Tobias Schindler[1,2], Lara Broghammer[1], Dennis Hufnagel[1], Nina Diringer[1], Benedikt Hofmann[1], Armin Dietz[1], Petros Karamanakos[3], Ralph Kennel[2]

[1] Technische Hochschule Nuremberg, Germany
[2] Technical University of Munich, Germany
[3] Tampere University, Finland

Corresponding author: Tobias Schindler, tobias.schindler@th-nuernberg.de
Speaker: Tobias Schindler, tobias.schindler@th-nuernberg.de

Abstract

Deep reinforcement learning (DRL) can achieve favorable dynamic performance compared to conventional control methods. However, steady-state errors are often present. This paper investigates the reduction of steady-state error in DRL-based current control of permanent magnet synchronous machines (PMSMs) by augmenting the integrated tracking error to the observation vector. More specifically, this paper assesses the performance of a DRL-based method under nominal and adverse operating conditions by considering PMSMs with linear and nonlinear magnetic circuits, which exhibit saturation, cross-coupling, and spatial harmonics. The latter include parameter mismatches between the training model and the physical system and misalignment of the dq-frame with respect to the identified position of the d-axis. As shown with the presented experimental results, the DRL-based control method can successfully operate the drive system under all operating conditions, with the steady-state and dynamic performance being similar to that of field-oriented control.

1 Introduction

Field-oriented control (FOC) is the state-of-the-art control scheme for permanent magnet synchronous machines (PMSMs). While FOC performs reasonably well in many applications, tuning the controller can be cumbersome. This is particularly true for drive systems with PMSMs that exhibit a non-linear magnetic circuit, e.g., highly utilized PMSMs for traction applications [1], more-electric aircraft, or machines without rare-earth magnets [2]. Controlling these more challenging machine types requires additional effort, e.g., implementing gain scheduling and decoupling based on flux maps [1], [3].

Deep reinforcement learning (DRL) control algorithms offer an alternative to FOC, potentially improving transient performance and reducing the tuning effort. On the one hand, several publications indicate that deep deterministic policy gradient (DDPG) [4] is a promising DRL algorithm for current control of PMSMs [5], [6], [7]. On the other hand, a persistent steady-state error (SSE) is fre-

quently present in DRL-based control concepts and also in applications other than electrical machine control. In [8], SSE is present in some operating points when controlling plasma in a tokamak. Using recurrent neural networks in the actor to compensate for the SSE is suggested [8]. In [7], a DDPG-based current controller is trained in a simulation environment using a linear machine model. Training continues on the test bench, where the used PMSM exhibits significant saturation. [7] reports an average normalized SSE of $\approx 1\%$ and concurs with [8] on the use of recurrent neural networks. A method to deal with SSE in DRL-based controllers is to add a dedicated integral action, as shown in [9]. The proposed concept features an additional control output, i.e., one action per control signal is fed to an integrator and added to the existing action. Previous work of the authors, presented in [6], found a persisting SSE in DDPG-based indirect current control (DDPG-CC) of a PMSM with linear magnetic circuit. The SSE is compensated by augmenting the observation of the DDPG agent with a discrete-time integrator for the tracking error.

This paper examines whether DDPG-CC can effectively address SSE when controlling more challenging PMSMs. Furthermore, adverse operating conditions are considered in which the assumptions of the training model are not applicable.

Two scenarios for adverse operating conditions are carried out. In the first scenario, additional resistors are added to the machine. The resistance varies within production batches [10], and motor cables of different lengths may be used in the installation. These effects are typically not accounted for when tuning the controller, and FOC is robust regarding these slight deviations. The effective stator resistance also depends on the frequency of the stator currents (skin & proximity effect) and the rotor temperature. Both effects can be modeled but are often neglected. Therefore, the ability of DDPG-CC to control PMSM when the stator resistance does not match the model value is investigated.

FOC and DDPG-CC control the stator currents in the rotating dq-frame, which has to be aligned with the α-axis of the machine. Sophisticated algorithms to perform the alignment exist [11], but the alignment is always imperfect due to measurement uncertainty. Additionally, the precision and bandwidth of the rotor position measurement are limited during operation. Therefore, some degree of misalignment, i.e., a difference between the measured position of the d-axis used by the controller and the d-axis of the machine, is always present in the control system. Thus, the second scenario investigates the effect of an intentionally induced misalignment of the dq-frame used by DDPG-CC.

A case study is carried out on two machines. One machine has a linear magnetic circuit, and the other shows significant saturation and spatial harmonics. DDPG-CC agents are trained on machine models with different degrees of fidelity to account for these effects. Control performance is evaluated for adverse scenarios and nominal operating conditions, i.e., the absence of adverse scenarios.

2 Model

The stator currents of the three-phase PMSM in the abc-frame are given by $\boldsymbol{i}_{\mathrm{s}}^{abc} = [i_{\mathrm{s}}^a\ i_{\mathrm{s}}^b\ i_{\mathrm{s}}^c]^\top$. The reduced, amplitude-invariant Clarke transformation

$$\boldsymbol{T}_{\mathrm{c}} = \frac{2}{3}\begin{bmatrix} 1 & -\frac{1}{2} & -\frac{1}{2} \\ 0 & \frac{\sqrt{3}}{2} & -\frac{\sqrt{3}}{2} \end{bmatrix} \quad (1)$$

is used to transform the phase quantities of the stator variables into the stator-fixed $\alpha\beta$-frame, e.g., the stator currents $\boldsymbol{i}_{\mathrm{s}}^{\alpha\beta} = \boldsymbol{T}_{\mathrm{c}}\boldsymbol{i}_{\mathrm{s}}^{abc}$. Note that the system is assumed to be balanced. Therefore, the zero component i_{s}^0 is neglected. Applying the Park transformation

$$\boldsymbol{T}_{\mathrm{P}}(\varphi) = \begin{bmatrix} \cos(\varphi) & \sin(\varphi) \\ -\sin(\varphi) & \cos(\varphi) \end{bmatrix} \quad (2)$$

yields the quantities in the rotating dq-frame, e.g., the stator currents $\boldsymbol{i}_{\mathrm{s}}^{dq} = [i_{\mathrm{s}}^d\ i_{\mathrm{s}}^q]^\top$ using $\boldsymbol{i}_{\mathrm{s}}^{dq} = \boldsymbol{T}_{\mathrm{P}}(\varphi)\boldsymbol{i}_{\mathrm{s}}^{\alpha\beta}$. Herein, the rotor angle φ is defined between the α- and the d-axis of the machine. With (1) and (2), the generic PMSM model in the rotating dq-frame is given by

$$\boldsymbol{v}_{\mathrm{s}}^{dq} = R_{\mathrm{s}}\boldsymbol{i}_{\mathrm{s}}^{dq} + \omega_{\mathrm{el}}\boldsymbol{J}\boldsymbol{\psi}_{\mathrm{s}}^{dq}(\boldsymbol{i}_{\mathrm{s}}^{dq},\varphi) + \frac{\mathrm{d}}{\mathrm{d}t}\boldsymbol{\psi}_{\mathrm{s}}^{dq}(\boldsymbol{i}_{\mathrm{s}}^{dq},\varphi), \quad (3)$$

with the stator voltages $\boldsymbol{v}_{\mathrm{s}}^{dq}$, the stator currents $\boldsymbol{i}_{\mathrm{s}}^{dq}$, the per-phase stator resistance R_{s}, the flux-linkage $\boldsymbol{\psi}_{\mathrm{s}}^{dq}$, and $\boldsymbol{J} = \begin{bmatrix} 0 & -1 \\ 1 & 0 \end{bmatrix}$ [3], [10], [12]. The electrical angular speed $\omega_{\mathrm{el}} = \omega_{\mathrm{m}}p$ is given by the rotational angular speed ω_{m} and the number of pole pairs p.

2.1 Linear Case

Assuming a linear magnetic circuit, the stator flux-linkages in d- and q-axis are given by

$$\psi_{\mathrm{s}}^d = \psi_{\mathrm{PM}}^d + L_{\mathrm{s}}^d i_{\mathrm{s}}^d, \quad (4)$$
$$\psi_{\mathrm{s}}^q = L_{\mathrm{s}}^q i_{\mathrm{s}}^q, \quad (5)$$

and the machine model is described using

$$v_{\mathrm{s}}^d = R_{\mathrm{s}}i_{\mathrm{s}}^d + L_{\mathrm{s}}^d\frac{\mathrm{d}i_{\mathrm{s}}^d}{\mathrm{d}t} - \omega_{\mathrm{el}}\psi_{\mathrm{s}}^q, \quad (6a)$$

$$v_{\mathrm{s}}^q = R_{\mathrm{s}}i_{\mathrm{s}}^q + L_{\mathrm{s}}^q\frac{\mathrm{d}i_{\mathrm{s}}^q}{\mathrm{d}t} + \omega_{\mathrm{el}}\psi_{\mathrm{s}}^d, \quad (6b)$$

where ψ_{PM}^d is the flux-linkage of the permanent magnets, L_{s}^d is the absolute stator inductance in the direct axis, and L_{s}^q is the absolute stator inductance in the quadrature axis.

2.2 Linear Case including Iron Losses

The iron losses are modeled by an additional parallel resistance in the equivalent circuit as depicted in Fig. 1 [11], [14], [13]. The machine model, including

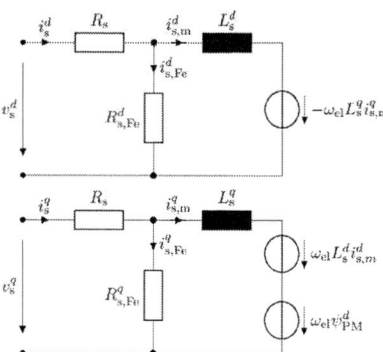

Fig. 1: Equivalent circuit of fundamental wave model including fixed iron loss resistance [11], [13].

iron losses [14] is given by

$$v_s^d = i_s^d \left(R_s + \frac{L_s^d L_s^q \omega_{el}^2}{R_{s,Fe}^q} \right) - \omega_{el} L_s^q i_s^q + L_s^d \frac{di_s^d}{dt}$$
$$+ \frac{L_s^q \omega_{el}^2 \psi_{PM}^d}{R_{s,Fe}^q}, \tag{7a}$$

$$v_s^q = i_s^q \left(R_s + \frac{L_s^d L_s^q \omega_{el}^2}{R_{s,Fe}^d} \right) + \omega_{el} L_s^d i_s^d + L_s^q \frac{di_s^q}{dt}$$
$$+ \omega_{el} \psi_{PM}^d, \tag{7b}$$

with the equivalent iron loss resistances $R_{s,Fe}^d$ and $R_{s,Fe}^q$ in the d- and q-axis, respectively. The currents $i_{s,Fe}^d$ and $i_{s,Fe}^q$ account for the power loss due to iron losses [11]

$$P_{Fe} = \frac{3}{2} \left(R_{s,Fe}^d (i_{s,Fe}^d)^2 + R_{s,Fe}^q (i_{s,Fe}^q)^2 \right). \tag{8}$$

2.3 Saturation, Cross-coupling, and Spatial Harmonics

A model that accounts for saturation, cross-coupling, and the dependency on the rotor angle is presented in [3], [12], [15]. The stator flux in d- and q-axis is assumed to be a function of the stator current i_s^{dq} and the rotor angle φ [3] [12] described by

$$\psi_s^d = f_1(i_s^{dq}, \varphi), \tag{9a}$$
$$\psi_s^q = g_1(i_s^{dq}, \varphi). \tag{9b}$$

The differential inductances

$$\frac{\partial \psi_s^d(i_s^{dq}, \varphi)}{\partial i_s^d} = L_s^{dd}(i_s^{dq}, \varphi), \tag{10a}$$

$$\frac{\partial \psi_s^q(i_s^{dq}, \varphi)}{\partial i_s^q} = L_s^{qq}(i_s^{dq}, \varphi), \tag{10b}$$

$$\frac{\partial \psi_s^d(i_s^{dq}, \varphi)}{\partial i_s^q} = L_s^{dq}(i_s^{dq}, \varphi), \tag{10c}$$

$$\frac{\partial \psi_s^q(i_s^{dq}, \varphi)}{\partial i_s^d} = L_s^{qd}(i_s^{dq}, \varphi) \tag{10d}$$

are defined by the partial derivative of the stator flux $\psi_s^{dq}(i_s^{dq}, \varphi)$ with respect to the stator currents i_s^{dq}. In addition, the partial derivative of the stator flux-linkage $\psi_s^{dq}(i_s^{dq}, \varphi)$ with respect to the rotor angle φ is given by

$$\frac{\partial \psi_s^d(i_s^{dq}, \varphi)}{\partial \varphi} = \Lambda_s^d(i_s^{dq}, \varphi), \tag{11a}$$

$$\frac{\partial \psi_s^q(i_s^{dq}, \varphi)}{\partial \varphi} = \Lambda_s^q(i_s^{dq}, \varphi). \tag{11b}$$

Note that the dependency of the stator flux $\psi_s^{dq}(i_s^{dq}, \varphi)$ (and other quantities such as differential inductances) on the stator current and rotor angle is omitted in the following notation for readability. Rewriting (3) using (10) and (11) yields

$$v_s^d = R_s i_s^d + L_s^{dd} \frac{di_s^d}{dt} + L_s^{dq} \frac{di_s^q}{dt} + \omega_{el}(\Lambda_s^d - \psi_s^q), \tag{12a}$$

$$v_s^q = R_s i_s^q + L_s^{qq} \frac{di_s^q}{dt} + L_s^{qd} \frac{di_s^d}{dt} + \omega_{el}(\Lambda_s^q + \psi_s^d). \tag{12b}$$

2.4 Saturation and Cross-coupling

If the spatial harmonics are neglected, i.e., if the flux-linkage $\psi_s^{dq}(i_s^{dq})$ is assumed to be independent of the rotor angle ($\Lambda_s^{dq} = 0$), the model described in (12) is simplified to

$$v_s^d = R_s i_s^d + L_s^{dd} \frac{di_s^d}{dt} + L_s^{dq} \frac{di_s^q}{dt} - \omega_{el} \psi_s^q, \tag{13a}$$

$$v_s^q = R_s i_s^q + L_s^{qq} \frac{di_s^q}{dt} + L_s^{qd} \frac{di_s^d}{dt} + \omega_{el} \psi_s^d. \tag{13b}$$

3 Control Algorithm

This section briefly summarizes the control concept of DDPG-CC, as introduced in [6], for drive systems consisting of a three-phase PMSM and a two-level voltage source inverter (VSI). Additionally, the used reference FOC is described.

3.1 DDPG-CC

The stator currents i_{s}^{dq} of the PMSM are controlled in the rotating dq-plane to track the reference currents $i_{\mathrm{s,ref}}^{dq} = [i_{\mathrm{s,ref}}^{d} \; i_{\mathrm{s,ref}}^{q}]^{\top}$ and minimize the tracking error $e_{\mathrm{s}}^{dq} = i_{\mathrm{s,ref}}^{dq} - i_{\mathrm{s}}^{dq}$. The control output (action) of DDPG-CC is defined as $v_{\mathrm{s,ref}}^{dq} = [v_{\mathrm{s,ref}}^{d} \; v_{\mathrm{s,ref}}^{q}]^{\top}$. Since the available output voltage of the VSI is limited by the available dc-link V_{dc}, the applied action is limited to the linear region of a space vector modulation (SVM) $v_{\mathrm{s,lim}}^{dq} = [v_{\mathrm{s,lim}}^{d} \; v_{\mathrm{s,lim}}^{q}]^{\top}$. Therefore, the modulation index is $m \leq 2/\sqrt{3}$ and $\|v_{dq}^{*}\|_2 \leq V_{\mathrm{dc}}/\sqrt{3}$ applies. The limitation is implemented according to [16].

Similar to [6], two different observation vectors are considered. The observation

$$o_1(k) = \begin{bmatrix} e_{\mathrm{s}}^{dq}(k) \\ \int e_{\mathrm{s}}^{dq}(k) \cdot f_{\mathrm{c}} \\ i_{\mathrm{s}}^{dq}(k) \\ v_{\mathrm{s,lim}}^{dq}(k-1) \\ n(k) \end{bmatrix} \quad (14)$$

features the integral of the tracking error using Euler-forward discretization, scaled by the control frequency f_{c} and the rotational speed n. The observation

$$o_2(k) = \begin{bmatrix} e_{\mathrm{s}}^{dq}(k) \\ i_{\mathrm{s}}^{dq}(k) \\ v_{\mathrm{s,lim}}^{dq}(k-1) \\ n(k) \end{bmatrix} \quad (15)$$

does not feature the integral over the tracking error. Note that all quantities are normalized by their rated values.

The investigated agents are trained using the reward

$$r_1(k) = \begin{cases} -\|e_{\mathrm{s}}^{dq}(k)\|_1, & \text{for } i_1(k) \leq I_{\max} \\ -\|e_{\mathrm{s}}^{dq}(k)\|_1 - i_1(k), & \text{for } i_1(k) > I_{\max} \end{cases} \quad (16)$$

where an additional penalty is given to the agent if the amplitude of the current $i_1(k) = \|i_{\mathrm{s}}^{dq}\|_2$ exceeds the maximum allowed current $I_{\max}$.

3.2 Field-oriented control

The reference FOC, including a decoupling network, is tuned according to the modulus optimum method [1]. The small time constants of the system are lumped into the time constant $\tau_\sigma \approx 1.5 f_{\mathrm{c}}^{-1}$ with

Tab. 1: Drive parameters for M1 (Heidrive HMD06-005) and M2 (prototype).

Parameter	Symbol	Unit	M1	M2
d-axis inductance	L_{s}^{d}	mH	1.13	0.44
q-axis inductance	L_{s}^{q}	mH	1.42	2.45
Stator resistance	R_{s}	mΩ	543	249
PM flux linkage	ψ_{PM}^{d}	mVs	16.9	20
Pole pairs	p		3	4
Rated dc-link	$V_{\mathrm{dc,r}}$	V	48	48
Rated speed	n_{r}	min⁻¹	3000	1000
Rated current	I_{r}	A	4.2	15
Maximum current	$I_{\max}$	A	10.8	20
Control frequency	f_{c}	kHz	10	10

Tab. 2: Trained agents, the used observation vector, the training model, and the machine.

Parameter	A1	A2	A3	A4	A5
Observation	(14)	(15)	(14)	(15)	(14)
Integrator	✓	✗	✓	✗	✓
Model	(6)	(6)	(13)	(13)	(12)
Spatial harmonics	✗	✗	✗	✗	✓
Saturation	✗	✗	✓	✓	✓
Cross-coupling	✗	✗	✓	✓	✓
Machine	M1	M1	M2	M2	M2

the control frequency f_{c}. For machines with a linear magnetic circuit, the controller parameters

$$K_{\mathrm{I}}^{d} = \frac{R_{\mathrm{s}}}{2\tau_\sigma}, \qquad K_{\mathrm{I}}^{q} = \frac{R_{\mathrm{s}}}{2\tau_\sigma}, \qquad (17a)$$

$$K_{\mathrm{P}}^{d} = \frac{L_{\mathrm{s}}^{d}}{2\tau_\sigma}, \qquad K_{\mathrm{P}}^{q} = \frac{L_{\mathrm{s}}^{q}}{2\tau_\sigma} \qquad (17b)$$

are used. For non-linear machines, i.e., machines that exhibit saturation behavior, a gain scheduling approach [1] is used for the parallel PI controllers with

$$K_{\mathrm{P}}^{d} = \frac{\psi_{\mathrm{s}}^{d}(i_{\mathrm{s,ref}}^{d}, i_{\mathrm{s}}^{q}) - \psi_{\mathrm{s}}^{d}(i_{\mathrm{s}}^{d}, i_{\mathrm{s}}^{q})}{2\tau_\sigma e_{\mathrm{s}}^{d}}, \qquad (18a)$$

$$K_{\mathrm{P}}^{q} = \frac{\psi_{\mathrm{s}}^{q}(i_{\mathrm{s}}^{d}, i_{\mathrm{s,ref}}^{q}) - \psi_{\mathrm{s}}^{q}(i_{\mathrm{s}}^{d}, i_{\mathrm{s}}^{q})}{2\tau_\sigma e_{\mathrm{s}}^{q}}. \qquad (18b)$$

Decoupling is achieved by adding

$$v_{\mathrm{s,decoup}}^{d} = -\omega_{\mathrm{el}}\psi_{\mathrm{s}}^{q}(i_{\mathrm{s}}^{dq}), \qquad (19a)$$

$$v_{\mathrm{s,decoup}}^{q} = \omega_{\mathrm{el}}\psi_{\mathrm{s}}^{d}(i_{\mathrm{s}}^{dq}) \qquad (19b)$$

to the outputs of the PI controllers.

4 Experimental Setup

Two different machines in three operating conditions, see Sec. 4.1, are used to evaluate the performance of DDPG-CC. Table 1 lists the nominal

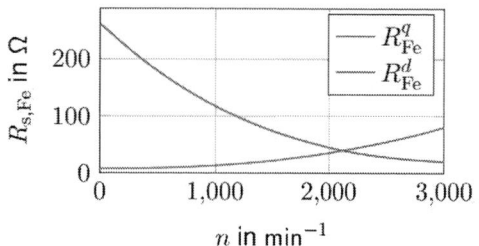

Fig. 2: Identified equivalent iron loss resistance of M1.

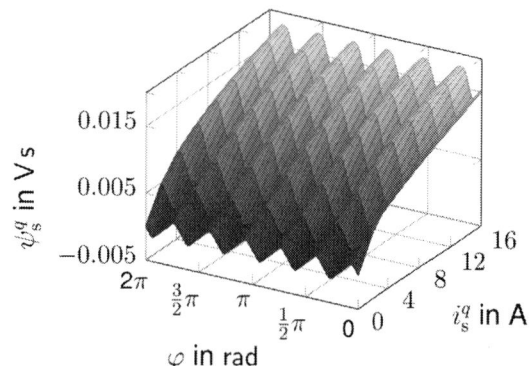

Fig. 3: Flux map ψ_s^q at $i_s^d = 0$ A with spatial harmonics of M2 (FEA data).

values of the machines. Machine M1 is a commercial off-the-shelf PMSM with low anisotropy. It is also used in [6].

The linear model (6) for M1 is parameterized based on the values presented in Table 1. To consider iron losses as in (7), the equivalent iron loss resistance $\boldsymbol{R}_{s,Fe}^{dq}$ is identified according to [11]. For simulation, the equivalent iron loss resistance is assumed to be only dependent on the rotational speed n, i.e., the dependency on $\boldsymbol{i}_s^{dq}$ is neglected. The identification result is depicted in Fig. 2.

The prototype machine M2, which serves as a benchmark, is intentionally designed to exhibit significant saturation and spatial harmonics, see Fig. 3. Thus, this machine is modeled based on flux maps accounting for spatial harmonics by (12) using $\psi_s^{dq}(i_s^{dq}, \varphi)$ based on FEA data. The model neglecting spatial harmonics in (13) is parametrized by $\psi_s^{dq}(i_s^{dq})$ using FEA data as well. A detailed description of M2 is presented in [10].

Table 2 shows the evaluated agents, along with their corresponding training models and observation vectors. Agents A1 and A2 control machine M1 and are taken from [6] without modification. Agents A3 - A5 are trained to control M2 according to Sec. 4.2. The agents A1, A3, and A5 use the observation

vector with the augmented integral of the tracking error.

DDPG-CC is compared to FOC in Sec. 5. FOC is tuned according to (17) for M1 based on the machine parameters of Table 1. To control M2, gain scheduling and decoupling according to (18) and (19) is employed. For this, the controller requires the flux maps of $\psi_s^{dq}(i_s^{dq})$ during runtime. The flux maps are implemented using the analytical approximation approach of [2].

DDPG-CC and FOC are implemented on the R5 processor of the UltraZohm, see [6], [17].

4.1 Operating conditions

The performance of all controllers is investigated in three different operating conditions:

1. Nominal operating condition. The d-axis of the controller is aligned to the α-axis of the machine according to the principle outlined in [11], which accounts for iron losses in the machine. No changes to the drive system.

2. Increased stator resistance by adding $100\,\text{m}\Omega$ resistors to each phase of the machines. This corresponds to an increase of $20\,\%$ and $40\,\%$ for M1 and M2, respectively, compared to Tab. 1.

3. Induced misalignment of the dq-frame by $5°$ compared to nominal operating conditions. Effectively, the Park transformation $\boldsymbol{T}_{\mathrm{P}}\big(\varphi(k) + 5°\big)$ and the inverse Park transformation $\boldsymbol{T}_{\mathrm{P}}^{-1}\big(\varphi(k) + 5°\big)$ are used by the controller.

Note that all reported measurement and simulation data are the sampled data of the controller, which is misaligned in operating condition 3.

4.2 Training

The DDPG-CC agents A3 - A5 are trained using the Matlab/Simulink Reinforcement Learning Toolbox 2023a [18]. The machine models (12) and (13) are used for M2 as per Table 2. The VSI is assumed to be ideal in the training environment. For each agent, a set of 60 hyperparameter combinations is trained using random search to find appropriate values for the discount factor, critic and actor learn rates and exploration standard deviation. Table 3 provides a summary of the hyperparameters.

The reference currents $i_{s,\mathrm{ref}}^{dq}$ and rotational speed n are set to uniformly distributed random values within the machine's operating range during each

Tab. 3: Hyperparameter of agents A3 - A5.

Parameter	Value
Training samples	1,8 Mio.
Minibatch size	64
Experience buffer length	900,000
L2 regularization actor & critic	0.01
Target network update frequency actor & critic	1
Target smooth factor	$1 \cdot 10^{-3}$
Discount factor	$0.95 - 0.999$
Learn rate critic	$1 \cdot 10^{-6} - 1 \cdot 10^{-4}$
Learn rate actor	$1 \cdot 10^{-6} - 1 \cdot 10^{-4}$
Exploration standard deviation	$0.1\% - 1\% V_{\max}$
Exploration decay rate	10 % of samples
Actor neurons in hidden layer	64
Actor hidden layer	1
Actor hidden layer activation function	ReLU
Actor output layer activation function	tanh
Critic neurons in hidden layer	256
Critic hidden layer	5
Critic hidden layer activation function	ReLU
Critic output layer activation function	linear

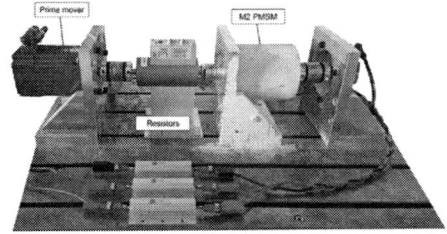

Fig. 4: Test bench with M2 and additional resistors.

training episode. These values are then kept constant for the duration of the episode. The duration of each training episode is set to $7 \cdot L_s^q / R_s$. Training 60 hyperparameter combinations for 1,800,000 samples in parallel takes 14 h on a workstation with an AMD Ryzen Threadripper Pro 3995WX.

5 Performance evaluation

Performance evaluation of the trained agents is done on the test bench in nominal and adverse operating conditions according to Sec. 4.1. Measurements are conducted at different rotating speeds set by a speed-controlled prime mover. Fig. 4 shows the test bench setup with M2, including the additional phase resistance. The test bench setup for M1 can be found in [6].

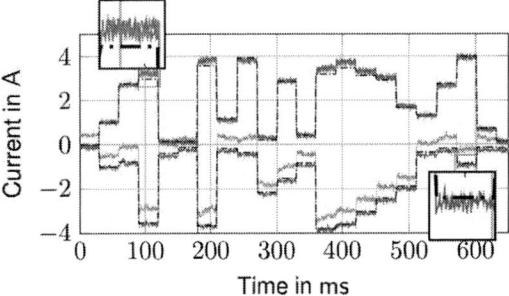

Fig. 5: Measured currents of i_s^d (purple) and i_s^q (blue) with sampling delay compensation for φ and without compensation i_s^d (green) and i_s^q (red) of A2 controlling M1 at $n = 3000\,\mathrm{min}^{-1}$. Reference values are black.

An evaluation profile with arbitrarily chosen reference values for $i_{s,\mathrm{ref}}^{dq}$ within the machine's operating range is used. For machine M1, the evaluation profile is depicted in Fig. 5 and matches [6]. For machine M2, the profile is scaled by the rated current of M2, and the duration between reference changes is increased from 30 ms to 1.1 s. This allows for the calculation of the Fast Fourier Transform (FFT) in Sec. 5.4.

The steady-state performance is quantified by the mean tracking-error

$$\bar{Q}_{\mathrm{SSE}} = \frac{1}{22} \sum_{m=1}^{22} \frac{100\%}{2 I_{\mathrm{r}}} \|\bar{e}_s^{dq}(m)\|_2^2 \qquad (20)$$

over all $m = 22$ reference values $i_{s,\mathrm{ref}}^{dq}$ in the evaluation profile. The mean tracking-error $\bar{e}_s^{dq}$ is averaged over 20 ms at each operating point.

Transient performance is evaluated using the average of the integral absolute error (IAE) criterion

$$\bar{Q}_{\mathrm{IAE}} = \frac{1}{22} \sum_{m=1}^{22} \int_{t_{\mathrm{start}}}^{t_{\mathrm{end}}} \|\bar{e}_s^{dq}(m)\|_2^2 \, \mathrm{d}t . \qquad (21)$$

Here, t_{start} is set to each reference change in the validation profile and $t_{\mathrm{end}} = t_{\mathrm{start}} + 5\,\mathrm{ms}$ for M1 and $t_{\mathrm{end}} = t_{\mathrm{start}} + 10\,\mathrm{ms}$ for M2. Similar to $\bar{Q}_{\mathrm{SSE}}$, the mean IAE $\bar{Q}_{\mathrm{IAE}}$ over all operating points is reported.

5.1 Sample Delay Compensation

The SSE of agent A2, which does not use the augmented integral of the tracking error, increases with the rotational speed of the machine in [6]. The measured SSE is much more significant compared to simulation results. Due to faulty implementation in [6], the sample delay of acquiring $\varphi(k)$ is

PCIM Europe 2024, 11– 13 June 2024, Nuremberg DOI: 10.30420/566262017

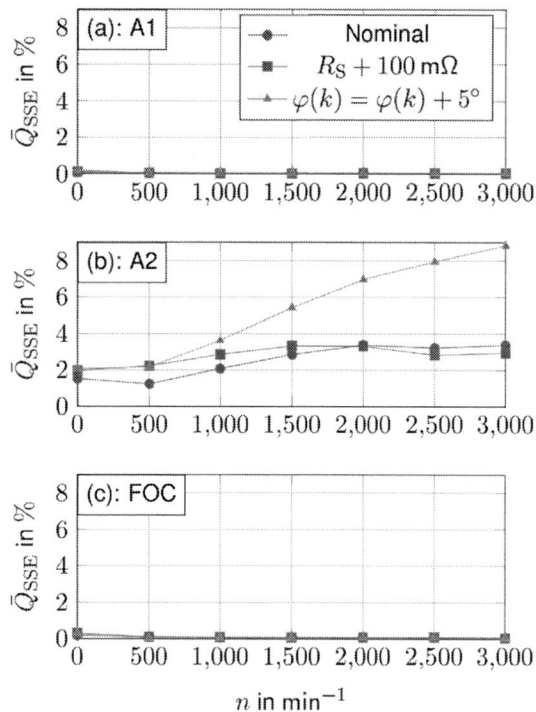

Fig. 6: Measured average SSE of controllers for M1 in different operation conditions.

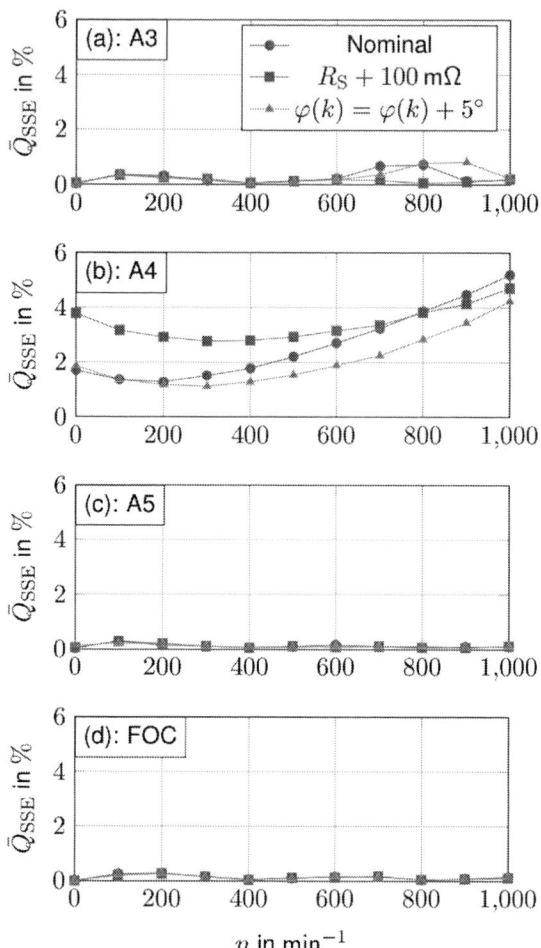

Fig. 7: Measured average SSE of controllers for M2 in different operation conditions.

not compensated in the inverse dq-transformation $T_{\mathrm{P}}^{-1}(\varphi(k))$. Compensation is required since the rotor position of the machine changes between the discrete sample instance and the application of the voltage by the VSI using SVM. The angle φ changes linearly between two concurrent sampling instances, assuming that the machine speed only changes marginally compared to the sampling time [19]. Therefore, the delay of φ is compensated by adding

$$\Delta\varphi(k) = \omega_{\mathrm{el}}(k)\tau_\sigma \tag{22}$$

to the inverse dq-transformation $T_{\mathrm{P}}^{-1}(\varphi(k)+\Delta\varphi(k))$ [19]. Fig. 5 shows the measurement results of A2 with and without proper compensation of the sample delay. The SSE, especially in the d-axis, is significantly reduced when using (22). Note that this effect is distinct from the intentionally induced misalignment of Sec. 4.1. The compensation of the sample delay (22) is used in all subsequent experiments.

5.2 Steady-state Performance

The steady-state performance of all controllers for machine M1 and M2 are shown in Fig. 6 and Fig. 7, respectively, in nominal and adverse operating conditions. Agents A1, A3, and A5 using the obser-

vation (14) with integrator compensate the SSE. Agents A1 and A5 match the performance of FOC in all nominal and adverse operating conditions. Agent A3 shows increased SSE from $n = 600\,\mathrm{min}^{-1}$ to $900\,\mathrm{min}^{-1}$. The adverse operating conditions do not significantly degrade the steady-state performance of DDPG-CC with augmented integration state in the measurements.

The agent A2 does not compensate the SSE in nominal operation (Fig. 6-(b)). Additionally, misalignment of the d-axis further deteriorates the steady-state performance. The increase in stator resistance results in additional SSE at slow rotational speeds but reduces the SSE slightly above $n = 2000\,\mathrm{min}^{-1}$. A possible explanation for the improvement is that agent A2 does show significant SSE, and a mismatch of the parameters compared to the training environment potentially moves the

146

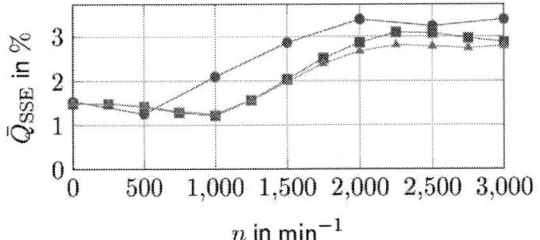

Fig. 8: Measured SSE of A2 (red) compared to to simulation based on (7) (including iron losses) (blue) and without iron losses modelled based on (6) (green).

control action in the right direction due to effects not accounted for by the model. This behavior is consistent for agent A4. However, contrary to A2, the SSE of A4 improves when misalignment of the dq-frame is present.

Fig. 8 compares the measured SSE of agent A2 to simulation results obtained using the linear machine model that includes iron losses (7). As shown, the SSE of agent A2 is generally increased in the measurement compared to simulation. Significant SSE is observed even when the agent controls the machine in ideal conditions, i.e., in simulation where the evaluation plant matches the training plant. The mismatch between simulation and measurement is attributed to neglected effects in the simulation, e.g., the inverter is assumed to be ideal in simulation, and parameter deviations.

5.3 Transient Performance

Fig. 9 and Fig. 10 depict the transient performance of M1 and M2, respectively, in nominal and adverse operating conditions. FOC (M1 & M2) as well as A1 is relatively robust regarding the adverse operating conditions. The transient response of FOC is decreased when the additional resistors are added to M1 and M2. The decrease is more severe for A4 and A5, especially at low rotational speed. On the contrary, the transient performance of agent A1 improves slightly when additional resistors are added. [6] indicates that agent A1 is prone to overshoot, which might be reduced in this operating condition. All control algorithms with integrator show no significant impact when misalignment of the dq-frame is induced.

Similar to the behavior in steady-state, the transient response of agent A2 deteriorates noticeably at higher speeds when misalignment is present. Unexpectedly, $\bar{Q}_{\mathrm{IAE}}$ of agent A4 is not affected by misalignment in the conducted experiments.

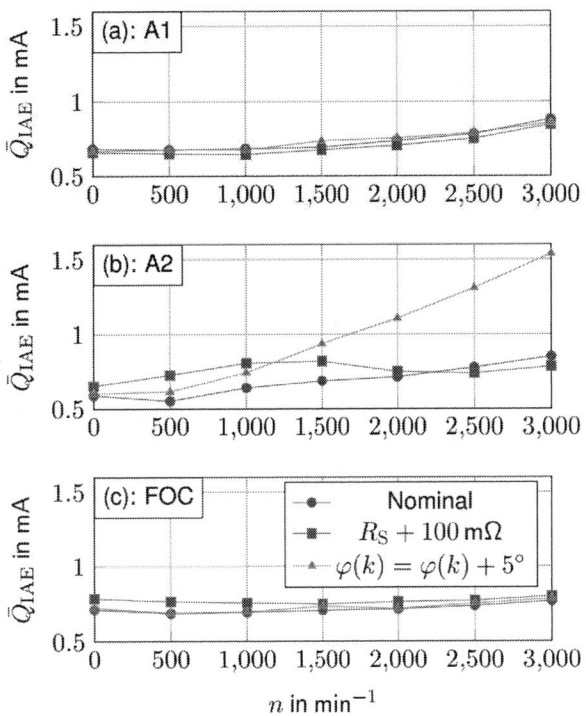

Fig. 9: Measured transient performance of controllers for M1 in different operating conditions.

The behavior is consistent for the machines M1 and M2 except A4. In general, the transient performance of the investigated agents is comparable to FOC at higher rotational speeds. However, Agent A5 exhibits subpar transient performance compared to the other controllers at low rotational speed. Furthermore, the additional resistors degrade the transient performance of A4 and A5 stronger compared to the other algorithms.

5.4 Current Harmonics

Fig. 11 depicts the phase currents of M2 for the first reference change in the evaluation profile. Noticeable distortion of the phase currents is visible due to the spatial harmonics of the machine. The harmonic components of the phase currents, based on FFT over 20 periods, is displayed in Fig. 12 at one operating point. The agent A5, trained on the model accounting for spatial harmonics (12), can compensate for the fourth and seventh-order harmonics. This results in a decrease of the total harmonic distortion (THD) to $11.6\,\%$ using A5 compared to $16.9\,\%$ using FOC.

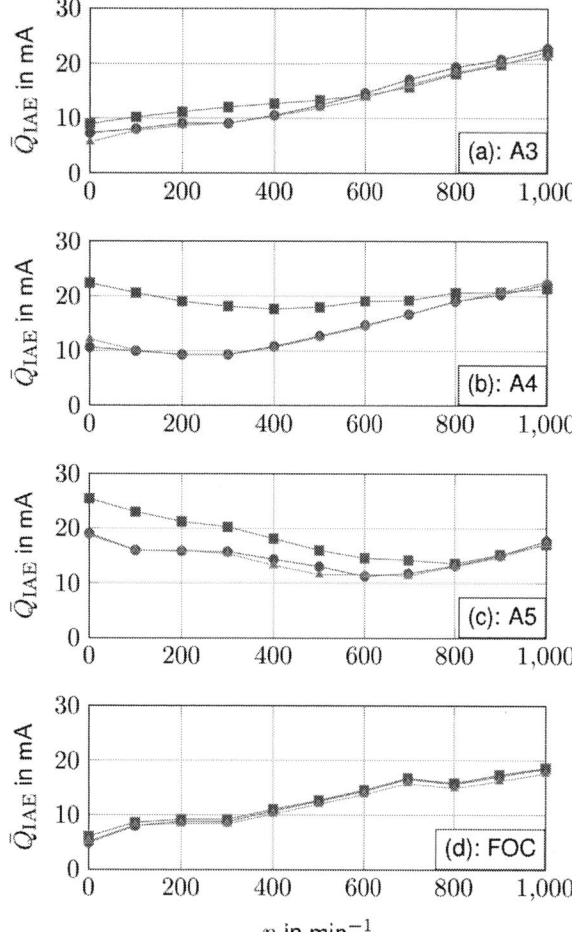

Fig. 10: Measured transient performance of controllers for M2 in nominal operation (red), additional resistors (blue), and misalignment (red).

6 Conclusion

This paper investigated the ability of of DDPG-CC to address the SSE that is present when controlling PMSMs with linear and nonlinear magnetic circuit. Nominal and adverse operating conditions were considered. A high-fidelity model was used in the training to control a PMSM that exhibits saturation, cross-coupling, and spatial harmonics. As demonstrated by the presented experimental results, when an integrator is not included in the observation vector, DDPG-CC is sensitive to parameter mismatches, and it becomes less effective when operating under adverse conditions. On the other hand, adding an integrating element to the observation vector significantly improves the steady-state performance of DDPG-CC as it removes any SSE under all operating conditions. Unfortunately, this results in a reduced dynamic performance. The latter could be improved by increasing the size of the neural network of the actor but at the expense of a higher computational burden. Finally, as shown, accounting for spatial harmonics allows DDPG-CC to produce currents of higher quality compared to FOC, albeit at the cost of slower dynamics.

Acknowledgment

The authors thank Philipp Dölger for his work on the flux-map approximation used in the FOC.

References

[1] T. Gemaßmer, "Effiziente und dynamische Drehmomenteinprägung in hoch ausgenutzten Synchronmaschinen mit eingebetteten Magneten," German, Ph.D. dissertation, 2015. DOI: 10.5445/KSP/1000046666.

[2] S.-W. Su, C. M. Hackl, and R. Kennel, "Analytical prototype functions for flux linkage approximation in synchronous machines," *IEEE Open Journal of the Industrial Electronics Society*, vol. 3, pp. 265–282, 2022. DOI: 10.1109/OJIES.2022.3162336.

[3] J. Richter, *Modellbildung, Parameteridentifikation und Regelung hoch ausgenutzter Synchronmaschinen.* Karlsruhe: KIT Scientific Publishing, Oct. 2016. DOI: 10.5445/KSP/1000057097.

[4] T. P. Lillicrap, J. J. Hunt, A. Pritzel, N. Heess, T. Erez, *et al.*, *Continuous control with deep reinforcement learning*, 2015.

[5] S. Bhattacharjee, S. Halder, Y. Yan, A. Balamurali, L. V. Iyer, and N. C. Kar, "Real-time SIL validation of a novel PMSM control based on deep deterministic policy gradient scheme for electrified vehicles," *IEEE Transactions on Power Electronics*, vol. 37, no. 8, pp. 9000–9011, 2022. DOI: 10.1109/TPEL.2022.3153845.

[6] T. Schindler, L. Broghammer, P. Karamanakos, A. Dietz, and R. Kennel, "Deep reinforcement learning current control of permanent magnet synchronous machines," in *2023 IEEE International Electric Machines & Drives Conference (IEMDC)*, 2023, pp. 1–7. DOI: 10.1109/IEMDC55163.2023. 10238988.

[7] G. Book, A. Traue, P. Balakrishna, A. Brosch, M. Schenke, *et al.*, "Transferring online reinforcement learning for electric motor control from simulation to real-world experiments," *IEEE Open Journal of Power Electronics*, vol. 2, pp. 187–201, 2021. DOI: 10.1109/OJPEL.2021.3065877.

[8] J. Degrave, F. Felici, J. Buchli, M. Neunert, B. Tracey, *et al.*, *Magnetic control of tokamak plasmas through deep reinforcement learning*, 2022. DOI: 10.1038/s41586-021-04301-9.

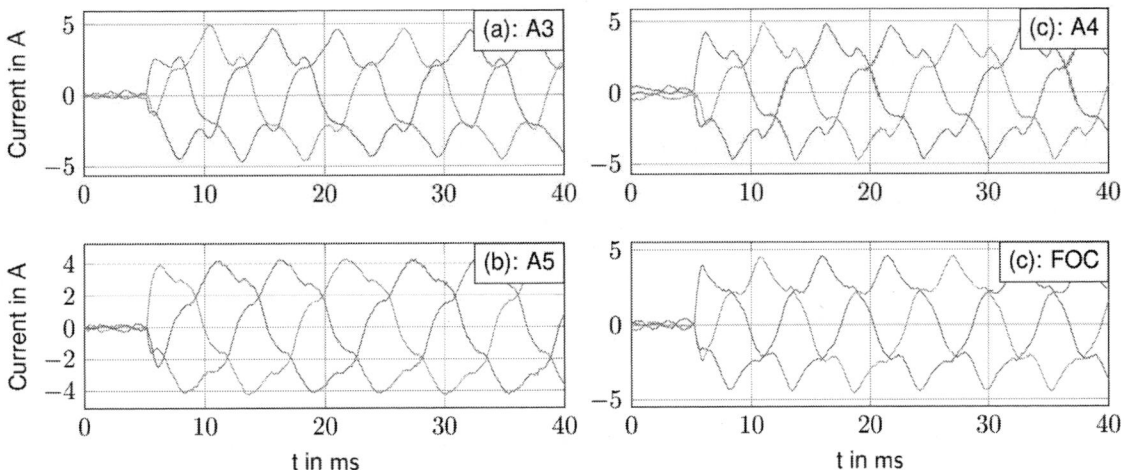

Fig. 11: Measured phase currents i_s^a (red), i_s^b (blue), and i_s^c (green) of M2 controlled by different algorithms at $n = 900\,\text{min}^{-1}$.

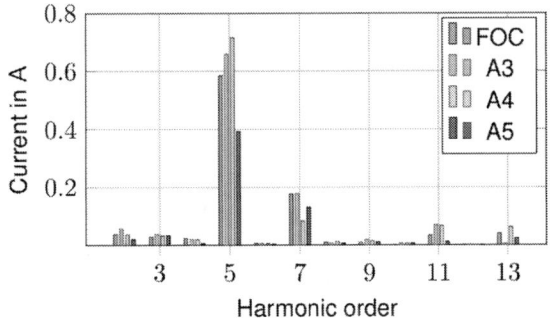

Fig. 12: Experimental results of the harmonic components of the phase currents of M2 calculated by FFT at $n = 900\,\text{min}^{-1}$, $i_s^q = 3.57\,\text{A}$, and $i_s^q = -1.0\,\text{A}$.

[9] D. Weber, M. Schenke, and O. Wallscheid, "Steady-state error compensation for reinforcement learning-based control of power electronic systems," *IEEE Access*, vol. 11, pp. 76 524–76 536, 2023. DOI: 10.1109/ACCESS.2023.3297274.

[10] S. Wendel, "Model predictive control of small mechatronic drive systems with pm synchronous machines," Ph.D. dissertation, Technische Universität München, 2022.

[11] J. Richter, A. Dollinger, and M. Doppelbauer, "Iron loss and parameter measurement of permanent magnet synchronous machines," in *2014 International Conference on Electrical Machines (ICEM)*, 2014, pp. 1635–1641. DOI: 10.1109/ICELMACH.2014.6960401.

[12] X. Chen, J. Wang, B. Sen, P. Lazari, and T. Sun, "A high-fidelity and computationally efficient model for interior permanent-magnet machines considering the magnetic saturation, spatial harmonics, and iron loss effect," *IEEE Transactions on Industrial Electronics*, vol. 62, no. 7, pp. 4044–4055, 2015. DOI: 10.1109/TIE.2014.2388200.

[13] D. Schroeder, *Elektrische Antriebe - Regelung von Antriebssystemen*. Springer, 2015.

[14] S. Kellner, "Parameteridentifikation bei Permanenterregten Synchronmaschinen.," Ph.D. dissertation, 2012.

[15] S. Li, D. Han, and B. Sarlioglu, "Modeling of interior permanent magnet machine considering saturation, cross coupling, spatial harmonics, and temperature effects," *IEEE Transactions on Transportation Electrification*, vol. 3, no. 3, pp. 682–693, 2017. DOI: 10.1109/TTE.2017.2679212.

[16] P. Q. Nguyen and D. Jörg-Andreas, *Vector control of three-phase AC Machines*. Springer, 2015.

[17] S. Wendel, A. Geiger, E. Liegmann, D. Arancibia, E. Durán, *et al.*, "UltraZohm— a powerful real-time computation platform for MPC and multi-level inverters," Quanzhou, China, 2019, pp. 1–6.

[18] *Matlab documentation: Deep deterministic policy gradient (DDPG) agents*, https://de.mathworks.com/help/reinforcement-learning/ug/ddpg-agents.html, Accessed: 2023-04-12.

[19] N. U and R. A, *Benötigt ein pulsweitenmoduliert betriebener Drehstromantrieb einen Stromzustandsregler?* Technischer Bericht - IAF-Report, Nr. 8, 2012.

PCIM Europe 2024, 11– 13 June 2024, Nuremberg   DOI: 10.30420/566262018

Performance Comparison of Using Shunt-Based and Integrated Current Sensing for Sensorless Field-Oriented Control

John Emmanuel Tan[1], Juan Paolo Quismundo[1], Jhaebhee Mark Calderon[1]

[1] Power Integrations, Philippines

Corresponding author: John Emmanuel Tan, JohnEmmanuel.Tan@power.com
Speaker: John Emmanuel Tan, JohnEmmanuel.Tan@power.com

Abstract

Sensorless Field-Oriented Control (FOC) for 3-phase Permanent Magnet Synchronous Motors (PMSM) requires motor phase current information. This paper compares the performance of shunt-based and low-side switch integrated current sensing in terms of system efficiency and ease of use for predictive maintenance. The use of integrated current sensing in the low-side switch along with a current reconstruction algorithm doubles the information resolution and reduces the number of required components. Predictive maintenance techniques use sensed information to infer the health of the system. Integrated current sensing provides more information and better noise immunity, resulting in enhanced system performance and easier use for predictive maintenance analysis.

1 Introduction

Sensorless Field-Oriented Control (FOC) of three-phase motors makes use of phase current information [1,2]. This paper examines the performance of two current sensing methods: shunt-based and low-side switch-integrated sensing [3,4]. For shunt-based sensing, signal conditioning circuits are required to add offset and amplification to the sense signal. In contrast, integrated current sensing reduces the number of discrete components, including the current shunt resistors, which dissipate energy and are more expensive than the two small-signal resistors used in an integrated approach. However, with integrated current sensing, only the negative motor phase current is sensed, necessitating the implementation of a reconstruction algorithm to create the phase-current waveform for an entire cycle [5].

The current information can also be used for protection and to gain insights into early motor degradation, which may then be addressed through predictive maintenance [6,7]. Two approaches for predictive maintenance are explored in this paper: Park Circle [6,8], and Motor Current Signal Analysis (MCSA) [9]. The Park Circle is performed by plotting the Alpha and Beta phase-currents on an XY plot. Ideally, the plot should form a perfect circle. Deviations from this shape may suggest abnormal conditions or asymmetric operation. MCSA is performed using Fast Fourier Transform (FFT) to compare the signature of the motor phase-current with and without fault conditions. The features extracted from the current signature are then used to develop a Random Forest (RF) classifier model to determine if the motor is "healthy" or "faulty."

This paper will describe the FOC algorithm in Section 2 to determine where the current information is used. Section 3 will discuss how the current information is sensed. Section 4 explores how predictive maintenance can be implemented using the sensed current information. Finally, the experimental results of the study are discussed in Section 5.

2 Field-oriented control

FOC is performed by controlling the stator current vector components, namely the flux- and torque-producing components [1,2]. For Permanent Magnet Synchronous Motors (PMSM), the flux-producing component of the stator current vector is normally adjusted to zero since the flux is provided by the permanent magnets in the rotor. This maximizes the torque produced by the motor.

This method of control requires knowledge of rotor position and stator current measurement. The FOC block diagram is illustrated in Fig. 1.

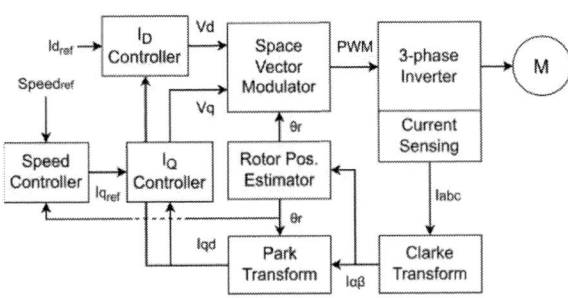

Fig. 1 Field-oriented control block diagram.

2.1 Rotor position

The position of the rotor during operation is used by the control system to provide the appropriate control signals. Sensors, such as rotary encoders, can easily be used to determine this rotor position. However, integrating such sensors adds extra cost and introduces reliability concerns.

2.1.1 Rotor/PM flux estimation

Sensorless control schemes, such as the rotor flux observer method, are used in place of the position sensor. In the rotor flux observer method, the motor model equation in Eq. (1) is used.

$$V = iR + L\frac{di}{dt} + e \tag{1}$$

$$e = V - iR - L\frac{di}{dt} = \frac{d\Psi_{PM}}{dt}$$

$$\Psi_{PM} = \int e\,dt = \int (V - iR)dt - Li \tag{2}$$

$$\Psi_S = \int \left(V - iR - V_{comp}\right)dt$$

$$V_{comp} = \left(k_p + \frac{k_i}{s}\right)\cdot\Psi_S$$

$$\Psi_{PM} = \int \left(V - iR - V_{comp}\right)dt - Li \tag{3}$$

where: V is the applied voltage across the winding, i is the winding current, R is the winding resistance, L is the winding inductance, e is the Motor back-electromotive force or back-EMF, Ψ_{PM} is the flux of the rotor permanent magnet, Ψ_S is the stator flux, V_{COMP} is the DC offset compensator, while the k_p and k_i are the PI correction feedback gains.

Equation (2) describes the rotor or permanent magnet (PM) flux estimation. A DC offset drift is observed when ideal integrators are used as parameters and measurement inaccuracies are accumulated by the integrator. To mitigate these effects, Proportional-Integral (PI) correction feedback, V_{comp}, is used for compensation [1,2].

Figure 2 shows the structure used for estimating the rotor/PM flux as described by Eq. (3). Motor parameters such as Stator Resistance, R, and Stator Inductance, L, are used. The Applied Voltage, V, and Motor Phase Current, i, serve as inputs to the estimator block.

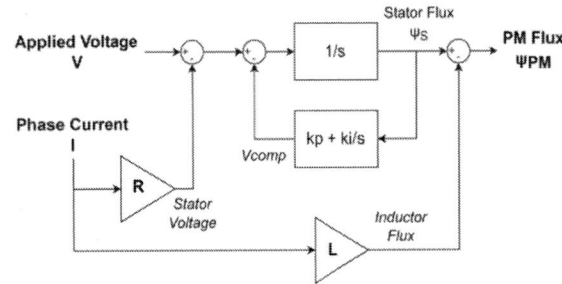

Fig. 2 Rotor/PM flux estimation.

2.1.2 Rotor position estimation

The estimation for the rotor position (θ) can be simplified by transforming the 3-phase reference frame (U-, V-, and W- axes) into the stationary 2-phase reference frame (α-, β- axes), see Fig. 3.

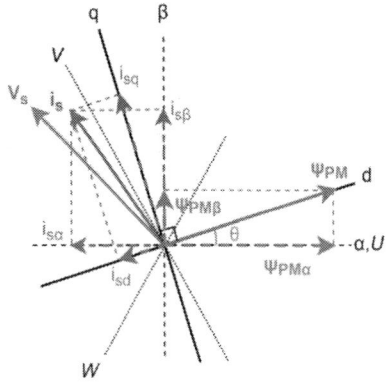

Fig. 3 Motor control space vector diagram

Equation (4) provides the translation of a 3- to 2-phase reference frame, known as the Clarke Transform. After transforming the stator current information into the stationary reference frame components $i_{s\alpha}$ and $i_{s\beta}$, they are used as inputs to the rotor flux estimator shown in Fig. 2.

$$\begin{bmatrix} i_\alpha \\ i_\beta \end{bmatrix} = \begin{bmatrix} 1 & 0 & 0 \\ \dfrac{\sqrt{3}}{3} & \dfrac{2\sqrt{3}}{3} & 0 \end{bmatrix} \begin{bmatrix} i_U \\ i_V \\ i_W \end{bmatrix} \quad (4)$$

$$\begin{bmatrix} i_q \\ i_d \end{bmatrix} = \begin{bmatrix} -sin\,\theta & cos\,\theta \\ cos\,\theta & sin\,\theta \end{bmatrix} \begin{bmatrix} i_\alpha \\ i_\beta \end{bmatrix} \quad (5)$$

where: i_α and i_β are the stator current components in the stationary reference frame, i_U, i_V, and i_W are the stator current components in the 3-phase windings, while i_q and i_d are the stator current components in the synchronous reference frame.

The rotor flux components estimated in the stationary reference frame, $\Psi_{PM\alpha}$, and $\Psi_{PM\beta}$, are then used to estimate the rotor position [2]. The examination confirms that the trigonometric function arctangent is usable, albeit with additional filtering in the form of a Phase-Locked Loop (PLL) structure. Similarly, the Quadrature PLL or Q-PLL shown in Fig. 4 can also be used to achieve the same angle calculation and noise filtering functionality.

2.2 Current control

The stator current can further be analyzed and broken down into its flux- and torque- producing components using the Park Transform described in Eq. (5). By using the estimated rotor position (θ) as the input to the transformation, the translated horizontal axis (Direct or d- axis) is parallel to the rotor position, while the translated vertical axis (Quadrature or q- axis) is perpendicular to the rotor position. This is also known as the synchronous or rotating reference frame since it rotates with the rotor.

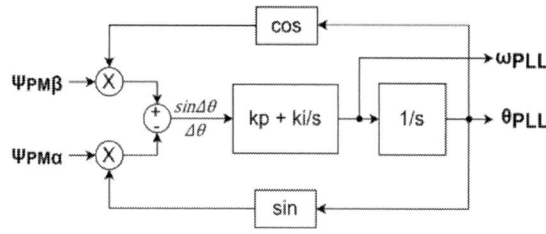

Fig. 4 Quadrature phase-locked loop structure.

The stator current component parallel to the rotor, or the Direct Current (I_D) is referred to as the flux-producing current; while the component perpendicular to the rotor, or the Quadrature Current (I_Q) is known as the torque-producing current.

I_Q and I_D control are both implemented in FOC, as shown in Fig. 1. Typically, I_D is reduced to zero since the flux is already provided by the rotor permanent magnets. A special control technique called Flux Weakening can reduce I_D to a negative value to increase the motor operating speed at the expense of reduced torque. I_Q is used to indirectly regulate the output torque, such that for Speed Control operation, the target I_Q is provided by the Speed Controller to reach the target speed.

The Quadrature and Direct Currents are DC values since they are in the synchronous reference frame. Proportional-Integral (PI) controllers are now suitable for regulating these components.

The Space Vector Modulator applies the appropriate stator voltage vector by providing the Pulse-Width Modulated signals to the inverter. The modulator takes the target Quadrature and Direct Voltages (V_Q and V_D) as inputs, see Fig. 1. The magnitude and angle of the applied vector vary depending on the output of the current controllers which is finite until the torque is maximized (when I_D becomes zero).

3 Current sensing

In FOC, the stator current information is used for rotor position estimation and stator current component control.

Various current sensing methods are available, such as different current sensors [3, 4]. In this paper, current-shunt based sensing and low-side switch integrated current sensing is scrutinized.

Current shunt-based sensing requires a resistive element placed on the path of the current to be sensed. The voltage across the resistive element is measured to determine the shunt current. This sensed voltage is processed by a signal pre-processing circuit then fed into an Analog-to-Digital Converter (ADC) used by the motor controller.

3.1 Shunt-based current sensing schemes

Figure 5 illustrates the different locations in the inverter and motor where the phase current information can be sensed [3]: 1) High-side DC-bus current sensing, 2) Motor in-line current sensing, 3) Low-side inverter leg current Sensing, and 4) Low-side DC-bus current sensing.

High-side current sensing techniques, such as high-side DC-bus sensing and motor in-line sensing, require differential sensing since it is riding on a high-voltage signal. The additional challenge of resolving a millivolt level signal combined with the high-voltage signal containing fast dv/dt needs to be resolved.

Conversely, low-side current sensing techniques, such as low-side DC-bus sensing and low-side inverter leg current sensing, are already referenced to the ground, allowing simpler sensing methods. In low-side inverter leg current sensing, the sensed information is a negative representation of the actual current since sensing is performed adjacent to the low-side switch.

For DC-bus current sensing — both high-side and low-side — an additional reconstruction algorithm is required to determine the 3-phase motor currents. This current reconstruction requires more complex ADC sampling.

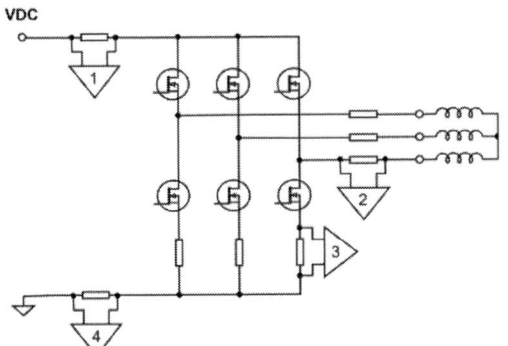

Fig. 5 Inverter current sensing schemes.

3.1.1 Shunt-based sensing in the low-side inverter leg

Low-side shunt-based sensing increases the power dissipation of the system as it is placed along the high-current path. This reduces the advantage of using a low-channel resistance switch. To limit the dissipation, the shunt resistance is minimized.

Before feeding the sense current signal to the controller ADC, a signal processing circuit is necessary. Amplification is required to increase the magnitude of the sense signal which is very low due to the need for minimum sense resistance. Additionally, a DC offset is required to account for the negative current passing through the shunt. Figure 6 shows the signal processing circuit required for all three phases.

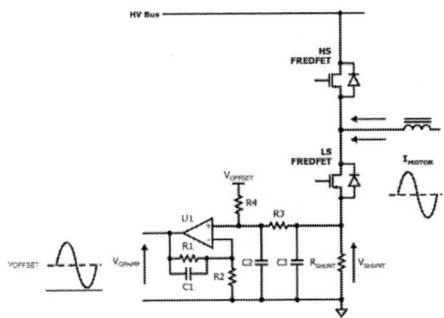

Fig. 6 Low-side switch shunt-based sensing.

The sampled ADC signal is processed by the controller by removing the offset and reversing the amplification gain to calculate the actual motor phase current.

Both positive and negative current information must be represented in the sensed signal to be sampled by the ADC.

3.2 Integrated current sensing in the low-side switch

Current sensing can also be integrated into the power switches to reduce the need for external resistive sensing elements. Figure 6 shows lossless current sensing integrated using a parallel sensing transistor. This sensing transistor outputs a current proportional to the current passing through the main power transistor (see Fig. 7).

An amplifier is also integrated to increase the magnitude of this sense current signal. An inexpensive small-signal resistor is used to convert this sense current to a voltage signal captured by the ADC.

Only half of the current information is sensed since the current only passes through the sensing transistor when the power switch is on. An additional current reconstruction algorithm [5] is required to reproduce the missing information for correct FOC operation.

As the current vector rotates, the available sensed current information alternates between highlighted and non-highlighted sectors in Fig. 8. In the highlighted sectors current information is available for one phase; in the non-highlighted phases, current from 2 phases is known.

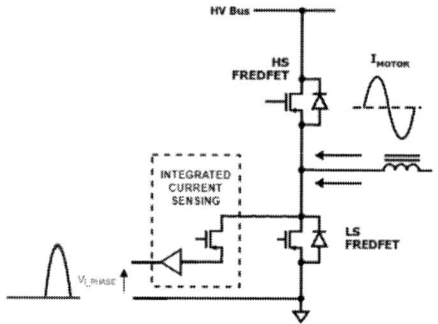

Fig. 7 Low-side switch integrated sensing.

3.2.1 Current reconstruction

As outlined at an IAS conference [5], the feasibility of using the integrated current sensor and the processing required to use the current information must be determined.

If current information is available for two phases (non-highlighted sectors, e.g., i_b and i_c), the third phase current (e.g., i_a) is reconstructed using Eq. (6). If current information is only available for one phase (highlighted sectors, e.g., i_c), the second and third phase currents (e.g., $i_a{}^*$ and $i_b{}^*$) are reconstructed using Eq. (7), (8), and (9) depending on which two are applicable. The current vector angle, $\theta_{\alpha\beta}$, is required for the reconstruction.

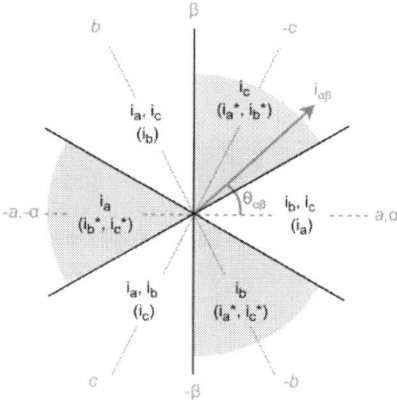

Fig. 8 Current reconstruction diagram.

$$i_a + i_b + i_c = 0 \qquad (6)$$

$$i_a = i_{mag} \cos \theta_{\alpha\beta} \qquad (7)$$

$$i_b = i_{mag} \cos(\theta_{\alpha\beta} - 120°) \qquad (8)$$

$$i_c = i_{mag} \cos(\theta_{\alpha\beta} + 120°) \qquad (9)$$

where: i_a, i_b, and i_c are the stator current components in the 3-phase windings (to be used interchangeably with i_U, i_V, and i_W), i_{mag} is the stator current magnitude, while $\theta_{\alpha\beta}$ is the current vector angle.

Since only negative current information needs to be represented in the sensed signal to be sampled by the ADC, and the positive current information is reconstructed based on the algorithm, the available information resolution is twice that of shunt-based sensing.

4 Predictive maintenance

Predictive maintenance techniques used in inferring the health of induction motors have been extensively explored [6,7]. These methods are also applicable for PMSM.

Table 9 describes common methods for performing predictive maintenance on different motor control subsystems and their significance for fault detection. The following predictive maintenance techniques have been listed: Extended Park Vector Approach (EPVA), Motor Current Signature Analysis (MCSA), and Instantaneous Power Signature Analysis (IPSA). These techniques are also known as Electrical Signature Analysis (ESA). ESA is a non-invasive method since no added sensors are required. Instead, it is performed on the sensed voltage and current information, which may already be used for control.

This paper focuses on techniques that utilize sensed current information. Mechanical imbalance and misalignment are simulated and compared with normal operation to provide comparative performance data.

4.1 Park vector approach

For steady-state FOC operation with constant torque, the phase current vector rotates at a constant magnitude. The representation of the phase current vector varies depending on the reference frames discussed in the FOC section.

Subsystem	Method	Significance
Supply	Power Quality	High
	EPVA (Voltage)	Medium
Mechanical	MCSA	High

Imbalance or Misalignment	Vibration	High
	EPVA	Medium
Insulation Faults	Partial Discharge	High
	EPVA	Low
Stator Electrical Imbalance	EPVA	High
	MCSA	Medium
	Power Quality	Medium
Bearing Faults	Vibration	High
	Wavelet on Current	Medium
	MCSA, EPVA and IPSA	Low
Coupling and Load Mechanical Failures	Vibration	High
	MCSA, EPVA and IPSA	Medium

Table 1 Predictive maintenance techniques for different motor subsystems.

In the stationary reference frame, the phase current components, $i_{s\alpha}$ and $i_{s\beta}$, are sinusoidal in nature. Plotting these components in an XY plot yields a circular graph for a healthy system [8].

In contrast, in the synchronous reference frame, the phase current components, i_{sq} and i_{sd}, are DC in nature. Plotting these components in a polar graph using the estimated rotor position as the angle also yields a circular graph for a healthy system.

Deviation from the ideal circular shape denotes issues in the balance of the 3-phase current which can be identified by analyzing the images.

4.2 Motor current signature analysis

Further analysis can be performed with the stator current information by using transformation tools to extract the signal signature or features [6, 9].

Fast-Fourier Transform (FFT) is a technique used to extract signal features. FFT extracts the magnitude of the signal components across the frequency spectrum. These features are further processed to determine the health of the system.

5 Experimental results

For this paper, sensorless FOC is used to drive a 300 W 3-phase motor capable of running up to 5000 RPM at a control-loop frequency of 8 kHz. For the inverter, the Power Integrations Reference Design Kit RDK-853 is used. This inverter makes use of the integrated half-bridge driver BridgeSwitch™ BRD1265C, which features an integrated lossless current mirror (IPH) to provide motor current information. The controller used is the 32-bit Arm® Cortex®-M0 XMC1400, a low-cost microcontroller from Infineon. The control code library utilized is the MotorXpert™ Suite from Power Integrations, which supports both shunt-based and integrated current sensing operation.

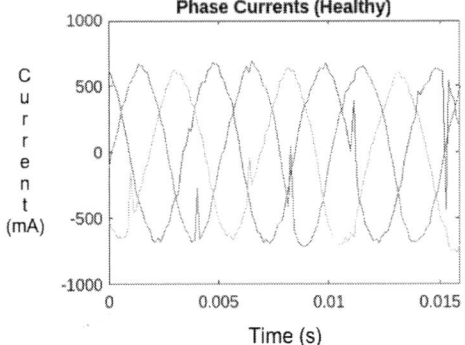

Fig. 9 Current information with shunt-based sensing.

5.1 Current information

The motor was operated at 3000 RPM for both shunt-based and integrated current sensing schemes. Figures 9 and 10 show the current information used by the FOC operation for both schemes. Noticeably, the current information in Fig. 9 had more noise compared to Fig. 10.

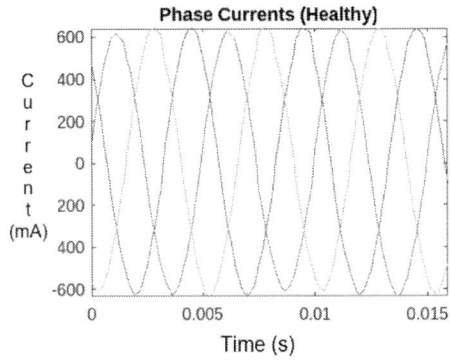

Fig. 10 Current information with integrated sensing.

5.2 Motor drive system efficiency

Figure 10 shows the visualization for how the motor drive system efficiency was calculated.

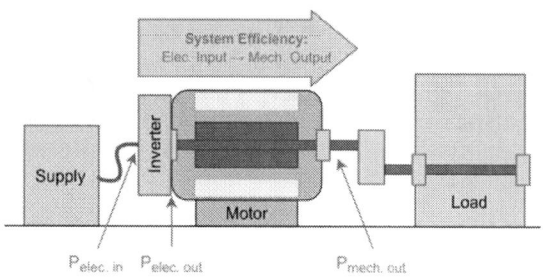

Fig. 11 Motor drive system efficiency.

For the measurement of this performance, the motor output torque was regulated at 0.45 Nm and the speed maintained at 3000, 4000, and 4500 RPM. The input electrical power was measured at the input of the inverter, while the mechanical output power was calculated using Eq. (10). The system efficiency was calculated using Eq. (11). The results are shown in Tables 2 and 3.

$$P_{mech\_out} = \tau \cdot \omega \qquad (10)$$

$$\eta_{system} = \frac{P_{mech\_out}}{P_{elec\_in}} \times 100\% \qquad (11)$$

where: P_{mech_out} is the output mechanical power, τ is the torque, ω is the angular speed, P_{elec_in} is the input electric power, and η_{system} is the system efficiency.

Table 4 shows the improvement in system performance using the integrated current sensing scheme. Across the three conditions recorded, the average reduction in wasted power was ~1.03 W when integrated current sensing was used.

The measured improvement was higher than just the loss dissipated by the shunt resistors. Better signal quality of the current information due to the increased resolution and lower noise contributed to the system efficiency improvement by optimizing the FOC operation such that lower electrical input power was required to provide the same mechanical power output.

Speed (RPM)	Mechanical Output Power (W)	Input Electrical Power (W)	System Efficiency (%)
3000	142.2	176.8	80.43
4000	189.7	230.4	82.34
4500	211.9	255.5	82.94

Table 2 System efficiency for shunt-based sensing.

Speed (RPM)	Mechanical Output Power (W)	Input Electrical Power (W)	System Efficiency (%)
3000	142.2	175.3	81.18
4000	189.7	229.6	82.62
4500	211.9	254.7	83.20

Table 3 System efficiency for integrated sensing.

Speed (RPM)	Input Electrical Power Delta (W)	System Efficiency Delta (%)
3000	-1.5	+0.75
4000	-0.8	+0.28
4500	-0.8	+0.26

Table 4 System efficiency improvement between shunt-based and integrated sensing.

5.3 Predictive maintenance — park vector approach

The current information in Figs. 9 and 10 is from the operation of a properly mounted motor. To simulate a mechanical imbalance or misalignment fault, mounting supports were intentionally loosened. Based on Table 1, EPVA had a Medium Significance while MCSA had a High Significance in detecting this fault condition.

Figures 12 and 13 show the Park Circle visualization for healthy and faulty operation, respectively, for a motor with shunt-based sensing. It is difficult to determine which condition is faulty due to the spikes in the Park Circle caused by the noise in the sampled current information.

PCIM Europe 2024, 11– 13 June 2024, Nuremberg   DOI: 10.30420/566262018

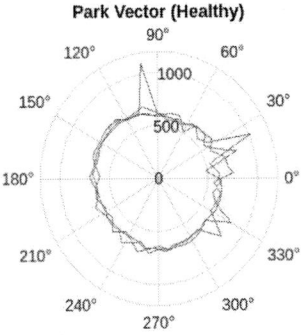

Fig. 12 Park Circle with shunt-based current sensing for a healthy system.

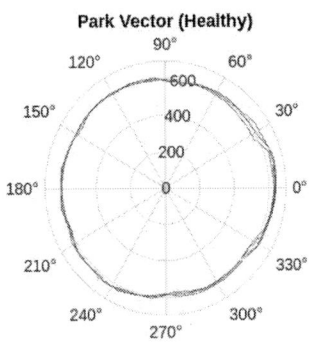

Fig. 14 Park Circle with integrated current sensing for a healthy system.

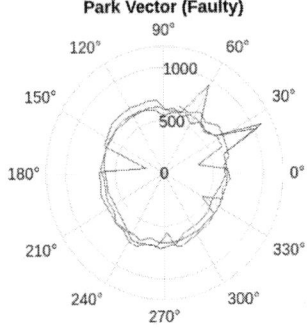

Fig. 13 Park Circle with shunt-based current sensing for a faulty system simulation.

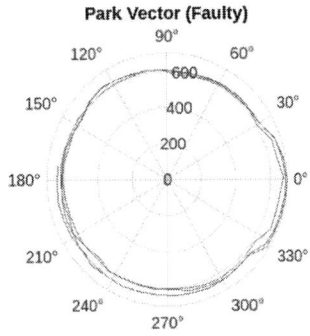

Fig. 15 Park Circle with integrated current sensing for a faulty system simulation.

Figures 14 and 15 show the Park Circle visualization for healthy and faulty operation respectively for a motor employing Integrated current sensing. The waveform in Fig. 15 has a more noticeable bulge from the 330° to 30° part of the circle.

The integrated current sensing scheme enables easier analysis of Park Circle due to the better signal quality.

5.4 Predictive maintenance – motor current signal analysis

MCSA is applied to the current information by using FFT to extract the features.

Figures 16 and 17 show the current signature from shunt-based sensing during healthy and faulty operation, respectively. The magnitude of the 0 Hz component can be interpreted as the magnitude of the current vector. The signature during faulty condition shows more peaks compared to the signature for healthy operation.

Figures 18 and 19 show the current information signature from integrated sensing during healthy and faulty operations, respectively. Similarly, the magnitude of the 0 Hz component can be interpreted as the magnitude of the current vector. The magnitude during the faulty condition is higher (626.2) compared to that of the healthy condition (607.9). This means that a higher current is required to drive the motor when supporting the same load and speed.

Furthermore, the feature at 93 Hz which is closest to the motor speed in terms of rotation frequency, is also highlighted by these signatures. For integrated current sensing, a higher magnitude is observed for the faulty operation (21.20) compared to the healthy operation (7.07) and suggests that the signal has 3x the magnitude at that frequency.

157

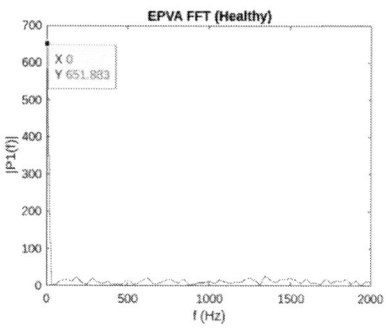

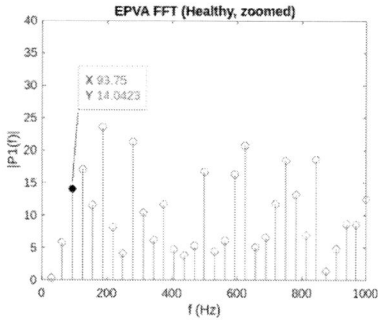

Fig. 16 Motor current signature analysis with shunt-based current sensing for a healthy system.

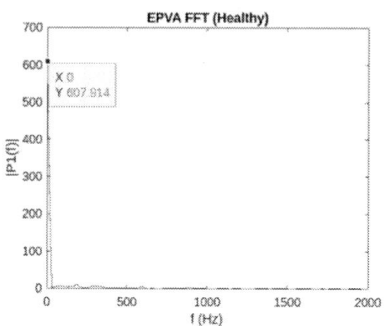

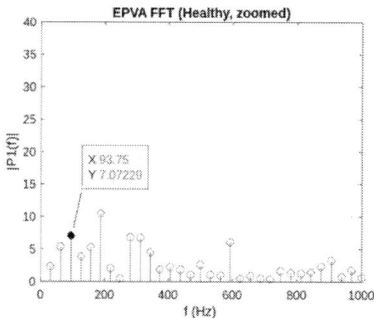

Fig. 18 Motor current signature analysis with integrated current sensing for a healthy system.

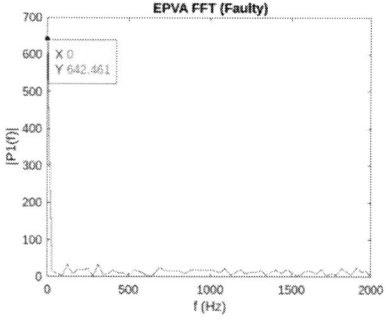

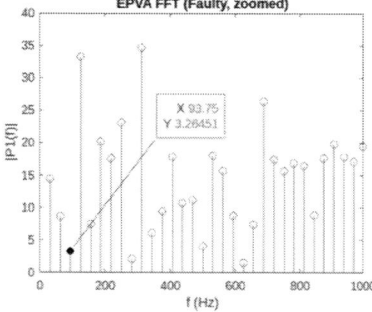

Fig. 17 Motor current signature analysis with shunt-based current sensing for a faulty system simulation.

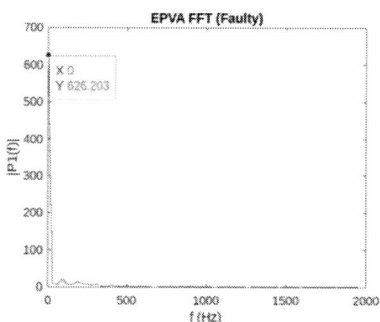

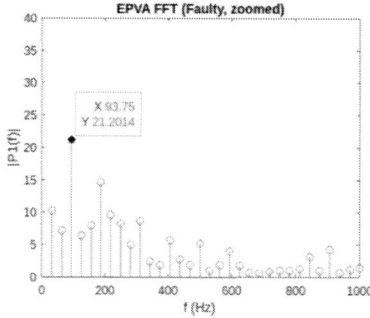

Fig. 19 MCSA with integrated current sensing for a faulty system simulation.

A method of classification is required to take advantage of these extracted signatures and determine the health of the system.

5.4.1 Random forest classification

The motor was run at 3000 RPM for both healthy and faulty conditions. Data was captured and split into training and test data. MCSA was applied to the current data, and the extracted features used to train a Random Forest (RF) classification model that was then verified using the test data [10, 11].

A trained RF classifier is composed of decision trees, which are also called estimators. The final classification (i.e., "Healthy" or "Faulty") is determined by a majority vote from the estimators.

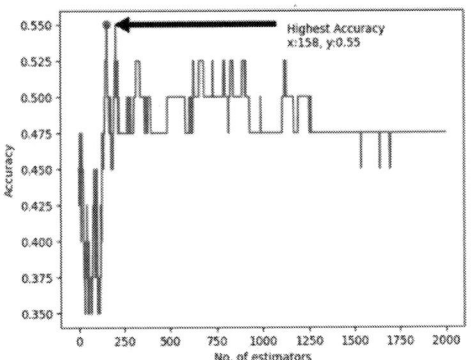

Fig. 20 RF classifier testing with current information from shunt-based sensing.

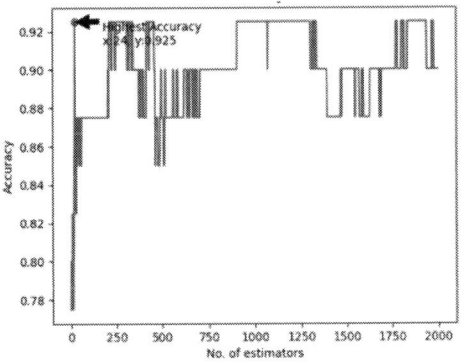

Fig. 21 RF classifier testing with current information from integrated sensing.

The model has been developed and tested after iterating with the number of estimators (up to 2000). The accuracy compared with the number of estimators was graphed and shown in Figs. 20 and 21. For shunt-based sensing, the maximum accuracy achieved was 55% with 158 estimators, while for integrated sensing, the maximum accuracy was 92.5% with 24 estimators.

This result showed that due to the better quality of the current information, developing more accurate predictive maintenance models is easier with integrated sensing compared to shunt-based sensing.

6 Conclusion

In this paper, the performance of shunt-based and integrated current sensing was compared in terms of system efficiency and compatibility with predictive maintenance techniques.

FOC was explored to describe where the current information is mainly used - rotor position estimation and current control. The different current sensing techniques were also discussed to give an overview of shunt-based and integrated sensing.

An improvement in the system efficiency was observed due to the increased resolution and better quality of the current information. Further study is recommended to determine if the removal of the resistive element between the low-side switch and the ground has more impact on metrics of performance beyond system efficiency.

The better quality of current information also helps in the development and implementation of predictive maintenance systems. Two methods of analysis were explored: Park Vector approach, and MCSA, both confirming easier analysis when integrated sensing was used compared to shunt-based sensing. The features extracted with MCSA were also used to develop an RF classifier. The model developed with integrated sensing was more accurate with fewer estimators compared to when the model was developed with shunt-based sensing. Further study was recommended to confirm the findings with different predictive maintenance techniques.

References

[1] Shen J, Hao H, Wang C, Jin M. 2013. Sensorless control of IPMSM using rotor flux observer. COMPEL - The international journal for computation and mathematics in electrical and electronic engineering. Vol. 32 No. 1, pp. 166-181.

[2] Sreepriya R, Rajagopal R. Sensorless control of three phase BLDC motor drive with improved flux observer. 2013 International Conference on Control Communication and Computing (ICCC). India, 2013. pp. 292-297.

[3] Persson E. Motor current measurement using time-modulated signals. Proceedings of the Power Conversion Conference-Osaka 2002 (Cat. No.02TH8579). Japan. 2002. pp. 716-720 vol.2.

[4] Patel A, Ferdowsi M. Current Sensing for Automotive Electronics—A Survey. IEEE Transactions on Vehicular Technology. vol. 58, no. 8, pp. 4108-4119. 2009.

[5] Chakrabarti S, Jahns T M, Lorenz R D. A current reconstruction algorithm for three-phase inverters using integrated current sensors in the low-side switches. 38th IAS Annual Meeting on Conference Record of the Industry Applications Conference. 2003. USA. 2003. pp. 925-932 vol.2.

[6] Bonaldi E L, de Oliveira L E L, Borges da Silva J G, Lambert-Torres G, Borges da Silva L E. 2012. Predictive Maintenance by Electrical Signature Analysis to Induction Motors. In: R. Araujo, Ed., Induction Motor Modelling and Control. InTech, Croacia, 487-520.

[7] Siddiqui K, Sahay K, Giri V, Scholar P. Health monitoring and fault diagnosis in induction motor-a review. International Journal of Advanced Research in Electrical, Electronics and Instrumentation Engineering. 2014. 3: 6549–65.

[8] Boudiaf M, Cherroun L, Benbrika M. 2018. Real-time diagnosis of three-phase induction machine using Arduino-Uno card based on park's circle method. Diagnostyka. pp. 19(3), 63-71.

[9] Bouslimani S, Drid S, Chrifi-Alaoui L, Bussy P, Ouriagli M, Delahoche L. An extended Park's vector approach to detect broken bars faults in Induction Motor. 2014 15th International Conference on Sciences and Techniques of Automatic Control and Computer Engineering (STA), Tunisia. 2014. pp. 411-416.

[10] Prihatno A T, Nurcahyanto H, Jang Y M, Predictive Maintenance of Relative Humidity Using Random Forest Method. 2021 International Conference on Artificial Intelligence in Information and Communication (ICAIIC). Korea (South). 2021. pp. 497-499.

[11] Taşcı B, Omar A, Ayvaz S. 2023. Remaining useful lifetime prediction for predictive maintenance in manufacturing. Computers & Industrial Engineering. Volume 184, Article 109566.

PCIM Europe 2024, 11– 13 June 2024, Nuremberg   DOI: 10.30420/566262019

Design of High-Power Inverter with 12 Parallel GaN Devices

Takashi Sawada[2,3], Shunsuke Takuma[1], Yoshiya Onuma[1], Kenichi Takagi[2], Hiroshi Tadano[2], and Koji Shiozaki[2]

[1] Nagaoka Power Electronics Co., Ltd., Japan

[2] Nagoya University, Japan

[3] Naturanix Co., Ltd., Japan

Corresponding author: Takashi Sawada, sawada.takashi.d7@f.mail.nagoya-u.ac.jp

Speaker: Takashi Sawada, sawada.takashi.d7@f.mail.nagoya-u.ac.jp

Abstract

Parallel placement and current imbalance are widely discussed in the application of GaN devices to high-power inverters. This paper describes a circuit layout that reduces the parasitic component of the printed circuit board by mounting GaN devices in 12 parallel in a three-phase inverter circuit. The current imbalance in the multi-pulse test is less than 20% at 600 A current output.

1   Introduction

The wide band-gap (WBG) power semiconductor devices such as Silicon Carbide (SiC) and Gallium Nitride (GaN) power devices are expected to be applicated for various power conversion circuits and improve their performances [1], owing to their high efficiency and fast switching characteristics [2][3]. However, the GaN wafer manufacturing and device processing technology limit the size and current rating of GaN power devices. Therefore, the GaN power devices are advantageously applied to power conversion circuits only for several hundred watts. The high-power converters and inverters require a number of WBG power devices to be paralleled to improve their rated current, and previous studies have attempted to connected the WBG devices in parallel [4]-[11].

Modularization is an effective method of parallelization; however, it lacks flexibility in circuit design due to structural constraints such as package size. Placing parallel devices directly on the printed circuit board (PCB) allows efficient heat dissipation, reduced structural constraints and scalable designs, since the appropriate number of parallel devices can be determined for the output power [4]-[6]. For example, PCB can mount the DC link capacitors next to the half-bridge circuit to reduce their stray inductance. GaN devices have input capacitance (Ciss) small enough to be charged by a single gate drive buffer and a positive on-resistance thermal coefficient, making them easy to drive in parallel [6][7].

However, designing circuit layouts with multi-parallel devices is a major challenge in WBG power applications [8]-[10]. GaN device has low threshold voltage and not small enough for Miller capacitance (Crss). The increased device mounting area causes an additional parasitic inductance and capacitance between the drain and gate of the WBG device, which raises the possibility of false turn-on and gate oscillation [11]-[13]. The individual gate drivers for each parallel device can suppress stray inductance in the gate drive circuit, but it is difficult to reduce the difference in switching delay time of each device. In addition, the larger parasitic inductance of the inverter circuit can lead to a significant current imbalance. Therefore, most of the previous researches have limited the number of GaN devices in parallel to four or less.

This research is appropriate for the stray capacitance between drain and gate PCB patterns (C_{DG_PCB}) and the individual gate resistance of GaN devices in the half-bridge circuit. Then, an inverter circuit board with 12 parallel GaN devices is constructed and verified to operate while suppressing false turn-on, gate oscillation and current imbalance of GaN devices.

2 Structure Consideration of 12 Parallel GaN Inverter

2.1 Arrangement of Inverter Main Circuit and Gate Drive Circuit Board

Increasing the number of power devices in parallel expands the area of the gate drive circuit in addition to the main inverter board. If 12 GaN Systems GS66516Ts are placed in parallel, the area for mounting them needs to be 120 mm wide, which causes an increase in the gate stray inductance (L_G). Figure 1 shows examples of gate voltage (V_{GS}) waveforms from simulated multi-pulse operation with 4 parallel and 12 parallel circuits. The input voltage is 350 V, and the output current is 200 A and 600 A, respectively. Q_2 in 4 parallel and Q_6 in 12 parallel are the center devices of each half-bridge circuit. The large L_G and complex LC ladder of 12 parallel circuit cause V_{GS} oscillation after the deadtime, the timing when V_{GS} often glitches.

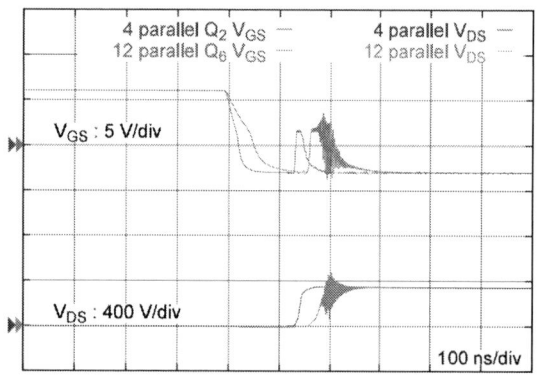

Fig. 1 Simulation results of V_{GS} for 4 and 12 parallel

In this paper, the 12 parallel inverter circuit is considered with smaller C_{DG_PCB} instead of decreasing L_G. When the main power circuit and the gate drive circuit are built on a single PCB, drain and gate patterns in close proximity have large C_{DG_PCB}.

Figure 2(a) shows the cross-section of drain and gate circuit. The Drain pattern, shown in blue, must be wide in order to reduce the inductance and the resistance in the main circuit and to connect with the DC link capacitors. Therefore, it extends below the gate circuit pattern shown in dark brown. The gate buffers driving the 12 GaN elements are located in the center of the gate drive PCB, indicated by yellow circles, to make the circuit as symmet-

rical as possible. When the gate drive circuit is integrated with the main circuit board, the gate pattern is 70 µm above the drain pattern (h_1 = 70 µm) and C_{DG_PCB} is 8.2 pF, which is comparable to the Crss of GS66516T, 5.9 pF. This is a factor of device self-turn-on, and adversely affects the stable operation of the inverter circuit. Separated gate circuit floats 1.6 mm from the drain pattern (h_2 = 1.6 mm), and C_{DG_PCB} is 2.1 pF. Increasing the distance between two PCBs can reduce C_{DG_PCB} by a factor of four. Figure 2(b) compares the differences in waveforms due to these circuit placements. The red and brown lines are the same condition. In this way, the gate drive board is separated from the main inverter board in this paper.

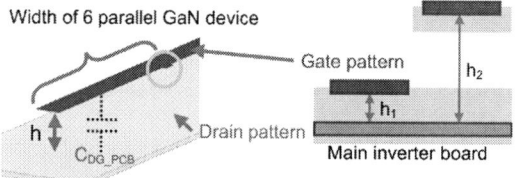

(a) Gate drive circuit placement

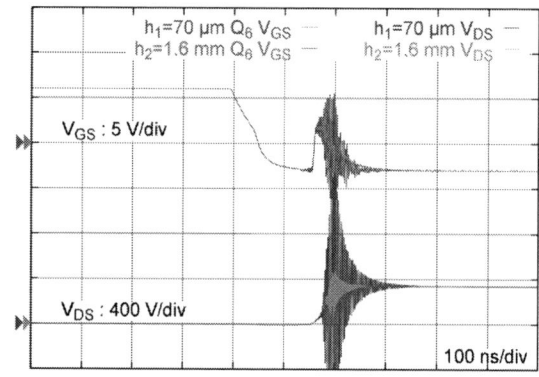

(b) Comparison of waveforms by gate circuit placement

Fig. 2 C_{DG_PCB} model and circuit simulation result

2.2 Avoid the Parallel Oscillation

2.2.1 Gate Drive Circuit Analysis setup

The GaN devices have small Ciss, and a single buffer such as a push-pull amplifier circuit can drive their gates. When parallel devices are aligned in a straight line, it is better to place the gate drive buffer in the center of the line, which makes the shape of the gate circuit symmetrical as shown in Fig. 3(a). Therefore, the gate drive circuit can be analyzed with a six-parallel circuit, which is half of the total number of parallels. The buffer circuit is placed in the right or left corner.

Figure 3(b) shows the analytical schematic of the gate drive circuit network of six-parallel GaN devices simulated on LTspice. The stray inductance and resistance of the provisionally designed PCB patterns are calculated with ANSYS Q3D Extractor and set. This half-bridge has gate resistances for turn-on and turn-off, six GS66516T spice models in each arm, individual resistors to each gate, stray components of PCB, and gate noise source.

Figure 3 is an example of a noise source installed in the gate drive circuit of Q_1, and similarly, the gate drive circuits with the noise source in the gates of Q_2 to Q_6 are also analyzed.

Although components such as gate mutual inductance, gate-to-source resistance, DC link capacitors, and output voltage feedback for gate signals are omitted in Fig. 3, they are taken into account in the actual simulation.

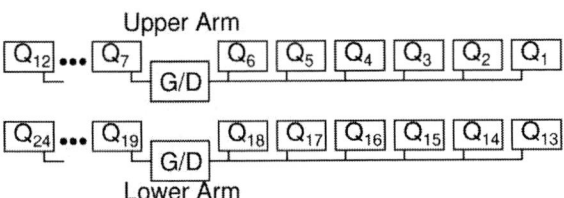

(a) Numbering of GaN devices and Gate drivers in upper and lower arms.

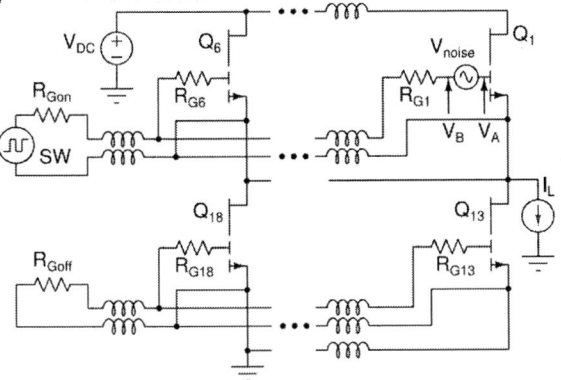

(b) Gate circuit for 6 parallel GaN devices with noise source at Q_1 (excerpt).

Fig. 3 Analytical schematic of 12 parallel inverter gate circuit.

The noise source simulates the displacement current flowing through Crss, which causes gate self-turn-on. In addition, the noise source causes oscillation of the gate drive circuits. This is because the stray capacitance of the six parallel GaN devices and the stray inductance of each gate

drive circuit constitute an LC ladder circuit, which can lead to more complicated oscillations than when the devices are used singly or a few in parallel.

2.2.2 Loop Gain of Gate Drive Circuit

The AC analysis of this circuit is used to calculate the gain of the noise source.

Figure 4 shows the Bode diagram of the gate drive circuit noise for each device, which is the result of simulations using the parameters listed in Table 1. Two simulations were conducted with different gate resistance conditions. The individual gate resistance connected to each device is important not only to suppress gate impedance variations, but also to suppress parasitic oscillation of the gate.

Gate resistance condition	(a)	(b)
DC voltage V_{DC}	350 V	
Output current I_L	100 A	
Output voltage V_{OUT}	50 V	
Frequency of noise source	1 MHz to 1 GHz	
Switching arm	Upper arm	
Concentrated gate resistance R_{Gon} / R_{Goff}	30 Ω / 6.2 Ω	1 mΩ / 1 mΩ
Individual gate resistance R_{Gi}	1 Ω	10 Ω

Table 1 Circuit simulation parameters

Figure 4(a) shows the simulation results for the first condition. Individual gate resistances are minimal, and the common gate resistance primarily determines the switching performance of the GaN device. While this circuit pattern is effective in adjusting the switching noise and surge voltage of GaN devices, Fig. 4(a) warns of the oscillation: the phase shifts around the gate drive loops of Q_1 and Q_2 exceed 360° at about 60 MHz, and the loop gains for that frequency is greater than 0 dB.

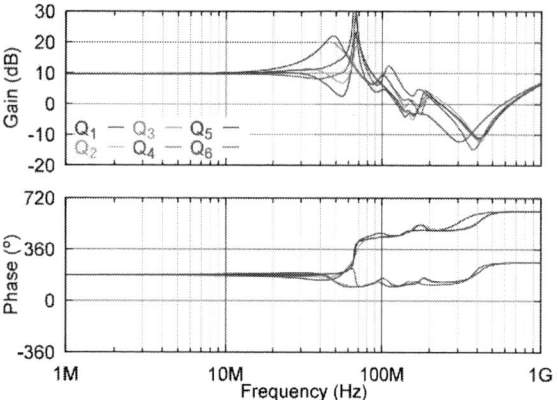

(a) Gate IC output resistance 10 Ω, individual resistance 2.2 mΩ

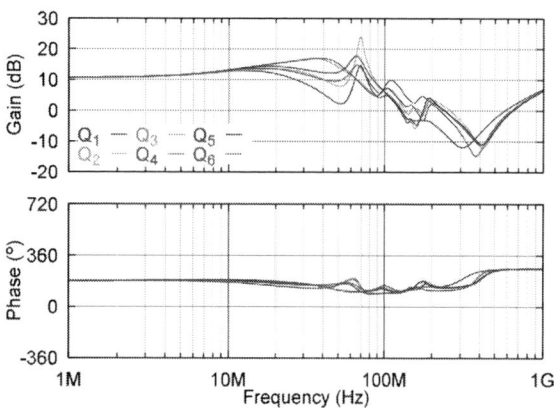

(b) Gate IC output resistance 1 mΩ, individual resistance 10 Ω

Fig. 4 Loop gain analysis of gate circuit for 6 parallel GaN devices.

This satisfies Barkhausen's criteria. The switching speed of GaN devices is much faster, potentially causing gate voltages to oscillate at frequencies above 60 MHz. Figure 4(b) shows the simulation results for the second condition, minimal common gate resistance and optimal individual gate resistances. The phases do not cross 360°×n, and it can be seen that the gate drive circuit is stable without parallel oscillation. Figure 4 shows the model only for an output voltage of 50 V. Simulation results for other output voltages are similar; the lower the common gate resistance and the higher the individual gate resistance achieves more stable gate circuit. Every transistor in the inverter circuit operates as an inverting amplifier simultaneously and constitutes a number of closed loops in Fig. 3, therefore, the phase diagram of feedback loop is not shifted by 180° in Fig. 4, and the oscillation criteria is also set to 360°×n.

3 Structure Consideration and Experimental Measurement of 12 Parallel GaN Inverter

3.1.1 Appearance of GaN Inverter Board

The topside of the inverter board has a number of ceramic capacitors, DC link film capacitors and three phase output terminals as shown in Fig. 5(a). The ceramic capacitors suppress the stray inductances of the GaN devices to reduce their switching surge voltages. Each phase has six output terminals which conduct the current of two adjacent GaN devices. The 12 parallel three-phase GaN inverter has a total of 72 GaN devices on the back of main circuit board as shown in Fig. 5(c).

(a) Top view of inverter main board.

(b) Gate driver and output bus-bar placed on the main board phase W.

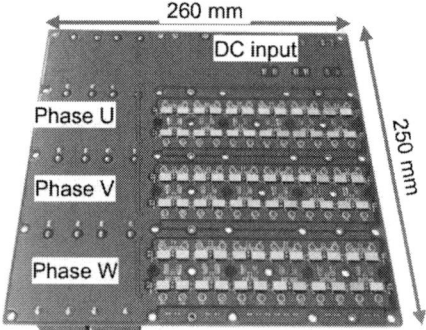

(c) Back view of inverter main board.

Fig. 5 Appearance of GaN inverter circuit board.

(a) Measurement setup

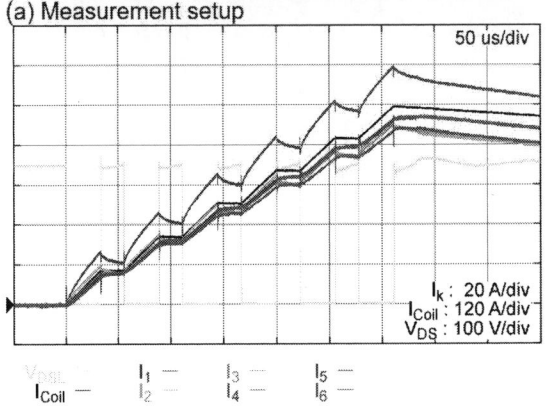

(b) Current sharing of six output poles

Fig. 6 Multi-pulse test with Rogowski coil probe measuring the current of output terminal.

3.1.2 Current Imbalance Measurement of Multi-Pulse Test

The gate drive board and output busbars are placed separately from the main board and connected to it, as shown in Fig. 5(c). Each of the six output terminals is independent, allowing current measurement with the Rogowski coil current probe, which is useful for checking the current balance of the devices without adding extra current paths and increasing inductance.

Figure 6 presents a multi-pulse switching test performed on phase W with 350 VDC applied to the inverter board. The six output terminals were called No. 1 to No. 6 in order of proximity to the inductive load (air core coil 107 µH), and their currents were measured repeatedly, and all waveforms were overlaid in Fig. 6(b). The total current I_{Coil}, indicated by the black line, reaches 600 A without operational defects such as self-turn-on. The current at output terminal No. 1, represented

by the red line, is the largest among the currents at the six output terminals. The current at other five terminals are almost the same. The current imbalance at the output pins is kept below 20 % due to optimal gate drive layout.

3.1.3 Continuous Operation Experiment of Three-Phase Inverter

The continuous three-phase inverter operation of the GaN inverter circuit was verified.

Figure 7 shows the waveforms of continuous inverter operation. The gate voltage spikes are acceptable and there is no oscillation in gate and drain voltage.

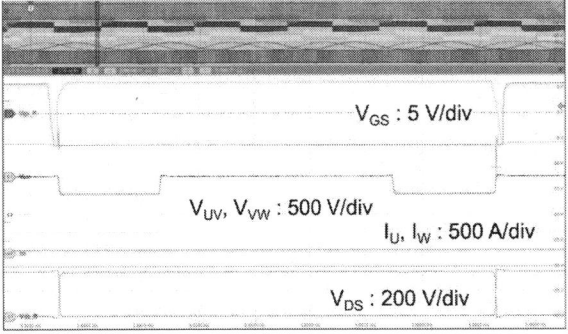

Fig. 7 Continuous inverter operation waveforms.

4 Conclusion

A three-phase inverter with 12 parallel GaN devices in each arm of each phase was designed and evaluated. The layout of the main circuit board and gate drive circuit was optimized to suppress drain current imbalance. The output current reached 600 A in the multi-pulse test without gate voltage spike and oscillation. The inverter output current is up to 200 Arms, which demonstrates the potential of GaN devices for high-power applications.

5 Acknowledgement

This research was partly supported by the Ministry of the Environment, Japan, Project of GaN technology innovation for enabling decarbonized society and lifestyle.

References

[1] J. Millán, P. Godignon, X. Perpiñà, A. Pérez-Tomás and J. Rebollo, "A Survey of Wide

Bandgap Power Semiconductor Devices," *IEEE Transactions on Power Electronics*, vol. 29, no. 5, pp. 2155-2163, 2014.

[2] S. Tiwari, T. Undeland, S. Basu, and W. Robbins, "Silicon carbide power transistors, characterization for smart grid applications," in *15th International Power Electronics and Motion Control Conference (EPE/PEMC)*, Novi Sad, Serbia, Sep. 2012, p. LS6d.2-1 - LS6d.2-8.

[3] T. Ueda, "GaN power devices: current status and future challenges," *Japanese Journal of Applied Physics*, vol. 58, no. SC, p. SC0804, Jun. 2019.

[4] S. Cheng and Po-Chien Chou, "Investigation on the parallel operation of All-GaN power module and thermal performance evaluation," in *2014 International Power Electronics Conference (IPEC-Hiroshima 2014 - ECCE ASIA)*, Hiroshima, Japan, May 2014, pp. 3425-3431.

[5] J. Burkard and J. Biela, "Paralleling GaN switches for low voltage high current half-bridges," in *IEEE Energy Conversion Congress and Exposition (ECCE)*, Baltimore, MD, USA, Sep. 2019, pp. 3245-3252.

[6] J. L. Lu and D. Chen, "Paralleling GaN E-HEMTs in 10kW-100kW systems," in *IEEE Applied Power Electronics Conference and Exposition (APEC)*, Tampa, FL, USA, Mar. 2017, pp. 3049-3056.

[7] P. P. Das, S. Satpathy, S. S. Shah, S. Bhattacharya, and V. Veliadis, "Paralleling of Four 650V/60A GaN HEMTs for High Power Traction Drive Applications," in *IEEE Energy Conversion Congress and Exposition (ECCE)*, Vancouver, BC, Canada, Oct. 2021, pp. 5269–5276.

[8] H. Li, S. Munk-Nielsen, C. Pham, and S. Beczkowski, "Circuit mismatch influence on performance of paralleling silicon carbide MOSFETs," in *16th European Conference on Power Electronics and Applications*, Lappeenranta, Finland, Aug. 2014, pp. 1-8.

[9] X. Long, Z. Jun, L. Pu, D. Chen, and W. Liang, "Analysis and Suppression of High Speed Dv/Dt Induced False Turn-on in GaN HEMT Phase-Leg Topology," *IEEE Access*, vol. 9, pp. 45259-45269, 2021.

[10] Q. Li and G. Schröder, "Minimization of parasitic capacitance for proper function of 3 level ANPC with GaN switches," in *25th European Conference on Power Electronics and Applications (EPE'23 ECCE Europe)*, Aalborg, Denmark, 2023, pp. 1-7.

[11] B. F. Kjærsgaard, A. B. Jørgensen, T. S. Aunsborg, J. K. Jørgensen, G. Liu, H. Zhao, S. Munk-Nielsen and B. Rannestad, "Analysis of Miller Region Sustained Oscillations during Turn-on of High-Side 10kV SiC MOSFET," in *25th European Conference on Power Electronics and Applications (EPE'23 ECCE Europe)*, Aalborg, Denmark, 2023, pp. 1-8.

[12] R. Matsumoto, K. Umetani, and E. Hiraki, "Optimization of the balance between the gate-drain capacitance and the common source inductance for preventing the oscillatory false triggering of fast switching GaN-FETs," in *2017 IEEE Energy Conversion Congress and Exposition (ECCE)*, Cincinnati, OH, Oct. 2017, pp. 405-412.

[13] K. Umetani, R. Matsumoto, and E. Hiraki, "Prevention of Oscillatory False Triggering of GaN-FETs by Balancing Gate-Drain Capacitance and Common-Source Inductance," *IEEE Transactions on Industry applications*, vol. 55, no. 1, pp. 610-619, Jan. 2019.

PCIM Europe 2024, 11– 13 June 2024, Nuremberg   DOI: 10.30420/566262020

Over 99.7% Efficient GaN-based 6-Level Capacitive-Load Power Converter

Stefan Mönch [1,2], Richard Reiner [1], Michael Basler [1], Kilian Bartholomé [3], Patrick Waltereit [1], Rüdiger Quay [1,4]

[1] Fraunhofer Institute for Applied Solid State Physics IAF, Germany
[2] Institute of Electrical Energy Conversion IEW, University of Stuttgart, Germany
[3] Fraunhofer Institute for Physical Measurement Techniques IPM, Germany
[4] Department for Sustainable Systems Engineering (INATECH), Albert Ludwigs University of Freiburg, Germany

Corresponding author: Stefan Mönch, stefan.moench@iaf.fraunhofer.de
Speaker: Stefan Mönch, stefan.moench@iaf.fraunhofer.de

Abstract

This work presents a highly-efficient 6-level multilevel converter, which uses two back-to-back connected 650 V GaN HEMTs as bidirectional blocking switch for each of the four inner voltage levels and one unidirectional HEMT for each of the the highest and lowest voltage level. The converter is applied to cyclical charging and discharging of a 40 μF capacitive load between 0 and 435 V at 5.5 Hz. External power is provided only to the highest voltage level; the inner levels are operated sufficiently self-balanced for the low-loss, linear and almost ideal capacitive load. A power-stage efficiency as high as 99.75% is experimentally achieved by zero-voltage-switching with an analog hysteretic current control after optimization of the dead times and the peak and valley current setpoints. A low-loss gate signal and power isolation is realized, which reduced the system efficiency only slightly to 99.71%. The high number of 10 discrete unidirectional GaN HEMTs used can be replaced with just 4 monolithic bidirectional switches and 2 unidirectional switches in future. Such highly efficient converters for capacitive loads enable a high system coefficient of performance for emerging emission-free, solid-state (electrocaloric) heat-pumps.

1   Introduction

Wide bandgap power semiconductors enabled high converter efficiencies above 99%. Significantly higher efficiencies are possible with advanced topologies such as partial [1] or differential power processing. An emerging application, which requires highly-efficient charging of capacitive loads are so-called electrocaloric [2] heat pumps, where an electroactive dielectric (capacitive load) has to be cyclically charged and discharged by electronics. While there were recent significant advancements in the power density of electrocaloric systems [3], temperature span [4], adiabatic temperature change [5] and entropy change [6], this work focuses on the efficiency of such systems. The low loss and high permittivity of some ceramic dielectrics (e.g. electrocaloric PMN [7], [8], PST [9]–[11], or lead-free material such as BSSnT [12]),

typically as part of multilayer ceramic capacitors [2], [13]–[16], results in a ultra electrical efficiency requirement to achieve a high heat pump system performance. The importance of efficient electric-energy recovery was explored in [7], [17], and addressed in [18] by an electrical resonant circuit ($\approx$ 80% efficiency), and adapted in [19]. For higher operation voltages where one transistor is not capable of blocking the full voltage, advanced circuits such as as a Marx-generator were presented [20]. In [21] the authors proposed to use switched-mode power converters, and achieved around 99.2% charging efficiency. However, it was also calculated that to surpass 50% of the Carnot-efficiency (which then would be competitive [22] to today's vapor compression heat pumps), over 99.71% electrical charging efficiency is required (for the considered ceramic PMN material and with an ideal thermal system approach without thermal heat regeneration). Several heat pump system approaches were

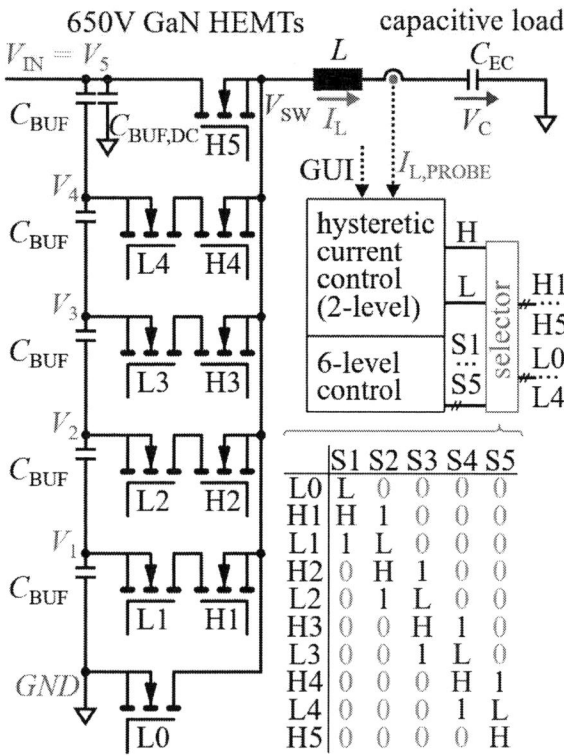

Fig. 1: 6-level π-type multilevel inverter, and hysteretic current control logic.

	S1	S2	S3	S4	S5
L0	L	0	0	0	0
H1	H	1	0	0	0
L1	1	L	0	0	0
H2	0	H	1	0	0
L2	0	1	L	0	0
H3	0	0	H	1	0
L3	0	0	1	L	0
H4	0	0	0	H	1
L4	0	0	0	1	L
H5	0	0	0	0	H

proposed and reviewed in [23]–[25]. In [26], the authors then achieved 99.74% power-stage efficiency by combining a 2-level low-voltage GaN half-bridge with a high-voltage Si-based 5-level multiplexer by partial power processing. The functionality of that converter system can be combined into one fully GaN-based multilevel converter. This work now presents such a capacitive-load 6-level multilevel inverter, and compares it to the previously presented GaN/Si hybrid converter. Furthermore, this work now presents a low-loss gate driver solution which enables a converter efficiency (power-stage and gate-drive) of 99.71% (control board and current sensor not included).

Fig. 1 shows a schematic of the topology, which is discussed in Sec. 2. Efficiency optimization and measurement results are presented in Sec. 3, and compared to a previous work.

2 Methods

2.1 6-Level π-Type Multilevel Topology

Fig. 1 shows the converter topology, where the maximum input voltage $V_{IN} = V_5$ is externally provided, and the inner voltages ($V_1...V_4$) from self

balancing buffer capacitors C_{BUF} (300 µF in this work). The input is additionally buffered by another C_{BUF} (600 µF). The 6-level converter is of π-type (also called E-type), a generalization of the 3-level T-type. All buffer voltages are statically referenced to a common ground, in contrast to flying capacitor or other (modular) multilevel converters [27]. The π-type topology uses parallel branches of bidirectional blocking high voltage where only one is conducting at any time, compared to flying capacitor converters [28] which use series branches of many unidirectional low voltage transistors where half of all transistors are conducting at any time. While the utilization of each transistor in this work consequently is reduced, the fact that only one bidirectional transistor is conducting seems still beneficial for high efficiencies, especially at low load currents. The target application of this work is cyclic charging and discharging of an unipolar capacitive load C_{EC}. A power inductor L filters the switch-node voltage to a continuous current, which changes the load voltage V_C.

2.2 Multilevel Hysteretic Current Control

Hysteretic current control of the inductor current I_L is used to achieve high efficiency by resonant switch-node voltage V_{SW} transitions and zero-voltage-switching (ZVS). The isolated current probe signal $I_{L,PROBE}$ is compared to software-adjustable peak and valley current setpoints ($I_{PEAK/VALLEY}$) provided by digital-to-analog converters on a programmable system on chip (PSoC), which then toggle a flip-flop. From this signal, two high-side (H) and low-side (L) half-bridge signals are derived, and two individual software-adjustable dead times (peak and valley dead times $t_{DT,PEAK/VALLEY}$) added. This quasi-square-wave zero-voltage-switching (QSW-ZVS) 2-level half-bridge control is then extended by a multilevel logic: A selector logic combines the H/L signals with the state of a cyclic up-down counter (...,S1,S2,...,S4,S5,S5,S4,...,S1,...) to the 10 switch signals. The logic table provided in Fig. 1 shows that the simple 2-level half-bridge control signals H/L are output to the desired levels, and the nearest two other switches of the inner branches are turned constantly on (1), while all other switches are off (0). The selector thus simply selects two buffer voltages for an inner conventional 2-level control. The two-level hysteretic current control and implementation is explained in detail in [29], and a similar multilevel extension is published in [26]. This work's control

PCIM Europe 2024, 11– 13 June 2024, Nuremberg   DOI: 10.30420/566262020

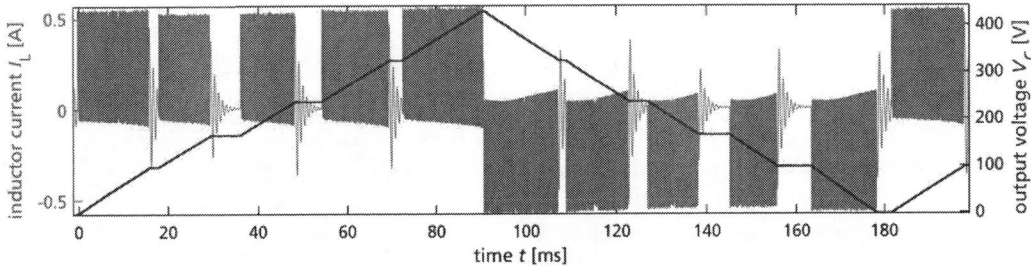

Fig. 2: Measured load (output) voltage V_C during one complete cycle with step-wise (multilevel) charging and discharging (40 µF load). The inductor current I_L is controlled by hysteretic current control with slightly negative valley currents for efficient zero-voltage-switching.

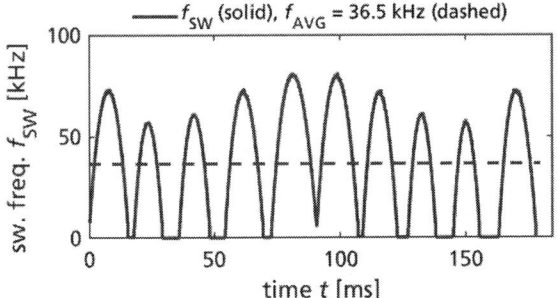

Fig. 3: Measured instantaneous and average switching frequencies.

differs from [26] only by the additional selector.

Fig. 2 shows the capacitor voltage V_C and inductor current I_L during a complete charging and discharging cycle. Fig. 3 shows the extracted instantaneous (solid) and average switching frequency (dashed, here: 36.5 kHz) f_{SW}. The switching frequency is output voltage dependent due to the hysteretic current control method. The measurement in Fig. 2 is also the later discussed efficiency-optimal operation point. Even though an almost ideal capacitive load was used for the characterization, the voltage levels still are not equally balanced. This results partly from the fact that the maximum input voltage is buffered by an additional dc-link capacitance (see Fig. 1), higher than the buffer capacitances of the inner voltage levels, and partly because the voltage probes connected to the load cause unsymmetrical power loss to GND, pulling the inner voltage levels slightly towards GND. For the future target application where non-linear and lossy capacitive (electrocaloric) load will be used instead of an almost ideal load capacitor, a more significant voltage level unbalance is expected. Possible countermeasures are for example to provide input power not only to the highest voltage level, but also

to the inner voltage levels and to actively control each voltage level.

There are four short voltage plateau (standby) phases visible in Fig. 2 during charging and discharging. This is because in this work a very simple control strategy is implemented: The cycle time is divided in equal timesteps, which then trigger the 6-level control of the control system. A continuous charging is possible with advanced control strategies. In [30] a voltage-sensorless control is published which solves this issue by detecting when the next voltage level is just reached to immediately trigger the next level of the multi-level control.

2.3 Efficiency Measurement

Since the accurate measurement of ultra-efficient converters is challenging even with precise power analyzers, this work uses a modified method, where the power loss is directly measured. The control ensures that the load is always fully charged between 0 and the input voltage V_{IN}. The load capacitance is a linear, almost ideal capacitor (MKP film type) with a known and voltage-independent value (40 µF). Only the input power P_{IN} to the converter is measured with the power analyzer. The input power is furthermore provided through a $1.1\,\text{k}\Omega$ resistor, which forms a low-pass such that the power analyzer can measures this continuous low-frequency (dc-like) signals with high accuracy. As byproduct of the input power (loss) measurement, the power analyzer also provides the RMS, and a minimum input (dc-link) voltage. The input voltage slightly varies during each system cycle, because energy is bidirectionally transferred (within the system) between the input buffer capacitor and the load capacitance. If the load capacitance is at its maximum (final) value, (end of state S5 during charging), it is connected to the input buffer capacitor, which

169

at that times is at its lowest value $V_{\text{IN,MIN}}$. The system cycle frequency f_{SYS} is precisely set by the controller. The average (low-pass filtered and rectified) load power (charging power) is calculated as $P_{\text{CHARGE}} = C_{\text{EC}}V_{\text{IN,MIN}}^2 f_{\text{SYS}}$, with no factor of $\frac{1}{2}$, because the energy stored in the capacitor is moved twice (during charging and discharging) in each cycle. The losses of the voltage probes of the power analyzer at the input and output (load) voltage and of the output (load) voltage probe from the oscilloscope $P_{\text{PROBE,IN/OUT}} = \dfrac{V_{\text{IN/OUT,RMS}}^2}{R_{\text{PROBE,IN/OUT}}}$ are calculated from the measured RMS input voltage from the power analyzer and the known probe resistances ($2.69\,\text{M}\Omega$ from the power analyzer and $66.7\,\text{M}\Omega$ from the oscilloscope).

The power-stage efficiency (without gate drive losses) is extracted as $\eta = 1 - \dfrac{P_{\text{LOSS}}}{P_{\text{CHARGE}}}$, where $P_{\text{LOSS}} = P_{\text{IN}} - P_{\text{PROBE,IN}} - P_{\text{PROBE,OUT}}$ is the power stage loss.

The converter efficiency includes the gate drive losses (isolated power supply and signal isolation), but exlucdes the control board power, and is extracted as $\eta = 1 - \dfrac{P_{\text{LOSS}} + P_{\text{LOSS,DRV}}}{P_{\text{CHARGE}}}$, where $P_{\text{LOSS,DRV}}$ is the total measured gate driver supply and signal isolation power during switching.

3 Results

3.1 Converter Prototype

Fig. 4 shows the 6-level converter prototype built in this work, and a zoom on one gate driver power supply and signal isolation. Similar components as reported in [21] are used, and main differences and results listed in Tab. 1. The power stage consists of 650 V normally-off (e-mode) GaN HEMTs (Tab. 1). The large dc-link buffer capacitors (B25631B0307K600) are not visible in Fig. 4 but directly connected under the power board.

The gate driver power supply is transformer-based (TPS60402DBV and 750314839) as in a reference design (TIDA-00349), which has high efficiency at milliwatts power. A low-power digital isolator (MAX22421BASA+) and a low-power gate driver (FAN3111ESX) is used. Operating the complete converter at higher charging powers might reduce the relative contribution of the gate driver power loss further. Optical receivers (AFBR-2624) before the digital isolators are not required for the converter operation and are only used for lab safety reasons. The optical receivers' losses are thus not included in the results. A controller is con-

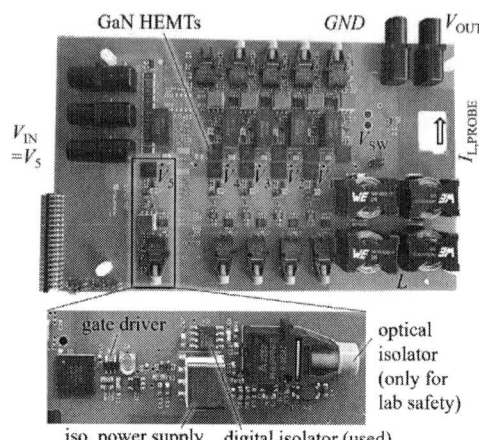

Fig. 4: Photo of the 6-level multilevel inverter and detail on gate driver.

nected through the optical isolators, but could be directly connected to the digital isolators through a pinheader (left in Fig. 4).

3.2 Efficiency Measurement and Discussion

The control parameters (peak and valley current setpoints and peak and valley dead times) were optimized by parameter sweeps during continuous operation. At the efficiency optimal operation point, the control parameters were: Peak and valley currents $I_{\text{PEAK/VALLEY}} = 0.485/-0.048\,\text{A}$, dead times $t_{\text{DT,PEAK/VALLEY}} = 85/1200\,\text{ns}$. The system cycle frequency is $5.5\,\text{Hz}$.

Significantly faster cycle frequencies are not possible because then the charging within each level is not finished any more (vanishing standby phases). Slower cycle frequencies would increase the standby phases between the cycles, and the static (leakage) losses reduce the efficiency.

For the optimized operation point, the efficiency and accuracy (based on the power analyser's datasheet, ZES LMG671) is calculated after several minutes continuous operation to stabilize all voltages: From the measured total input power $192.5\,\text{mW}$, the $70.82\,\text{mW}$ loss in the input probe is subtracted based on the measured RMS input voltage ($436.46\,\text{V}$) and the $21.905\,\text{mW}$ loss in the output probes are also subtracted based on the measured RMS output voltage ($237.99\,\text{V}$). The power stage loss (after subtracting both probe losses) is $99.78\,\text{mW}$. A charging power of $40.82\,\text{W}$ is calculated for the $40\,\mu\text{F}$ load and from the minimum input voltage ($430.75\,\text{V}$, which is also the peak-to-peak

Tab. 1: Comparison of Capacitor Charging Multilevel Converters.

	this work	Ref. [26]
power	40.82 W	19.947 W
power-stage loss	99.78 mW	51.854 mW
gate-drive loss	16 mW	n/a
power-stage effic.[1]	99.75%	99.74%
converter effic.[2]	99.71%	n/a
topology	multilevel (π-type)	half-bridge + multiplexer
levels	6	6 (2+4)
switches	10× GaN	10× (2× GaN + 8× Si)
load, freq.	40 µF, 435 V, 5.5 Hz	10 µF, 317 V, 19 Hz
dead times	85/1200 ns	50/250 ns
peak/valley	0.485/-0.048 A	0.3/-0.05 A
GaN HEMTs	190 mΩ, 650 V	190 mΩ, 600 V
	Nexperia GAN190-650EBEZ	Infineon IGLD60R190D1

[1] Power-stage (without gate drivers, control board and current sensor).

[2] Power-stage and gate drivers (without control board and current sensor).

charging voltage of the load). The efficiency in that operation point is 99.7556%, accurate within a range of 99.7435%...99.7661% after considered uncertainties. The efficiency accuracy of and measurement method is discussed in detail in [26].

The power loss of all gate drivers and signal isolators was measured during the continuous operation and was just 16 mW. Considering the additional losses of the gate driver, the power-stage efficiency is reduced to the converter efficiency of 99.71%. Due to the low-power design of the gate driver, the efficiency is only slightly reduced. This is a significant improvement to the previous work [26], where off-the-shelf dc-dc converters were used for power isolation and only the high power-stage, but not the additional gate driver power losses were not reported. The control is realized by a current probe and programmable system on chip on an evaluation board, and could be realized with a very-low power controller. The power loss of the control and current sensor is not included in this work's efficiency discussion. This work uses an PSoC evaluation board and conventional isolated current probe, both with high power consumption. Future works will focus on an ultra-low power design of the control and current sensing, which then will allow to also report complete power system efficiencies.

The required bidirectional blocking capability of the four inner levels is realized by two back-to-back connected (common drain) unipolar blocking GaN HEMTs. The on-state resistance of each such switch circuit thus is twice the on-resistance of one GaN HEMT. These bidirectional blocking switch circuits can be replaced in future by monolithic bidirectional blocking transistors [31] with just one shared drain channel region, which will reduce the required semiconductor area to around 1/4 for the same on-resistance. Despite their obvious advantages, such high-voltage switches are not yet commercially available, but part of ongoing research and product roadmaps. In future works, the high number of 10 discrete unidirectional GaN HEMTs used can then be replaced with just 4 monolithic bidirectional switches (of also lower on-resistance or lower semiconductor area) and 2 unidirectional switches. Tab. 1 compares this work's GaN multilevel inverter and results to a previous partial power processing Si/GaN hybrid converter. One main difference is that in the multilevel topology the sum of the output capacitances of seven GaN HEMTs reduce the (resonant) switch node voltage transition slew-rate, while in the previous converter the only two half-bridge transistors slow down the switching significantly less. Longer dead times were thus required in this work for optimal ZVS. Also, six GaN transistors are effectively paralleled and thus charged and discharged with the switching frequency even though only two contribute to the actual power flow, such that output capacitance hysteresis loss (if any) might be increased [32]. On the other hand, in the previous converter the Si FETs were hard-switched between the levels (increasing switching loss), which is not required in this work any more, because the transitions between the levels in this work do not require additional hard-switching because in the multilevel converter the lower voltage of the next level is already the higher voltage of the previous level, and such a commutation between levels thus is also ZVS.

4 Conclusion

GaN-based multilevel converters enable ultra-efficient capacitive load charging. This work demonstrates that beside a highly-efficient power stage (99.75% efficiency), a highly-efficient (low-power) gate driver supply and signal isolation is required and can be realized, which then reduces the overall converter efficiency just slightly to 99.71%.

In future works, the advantage of a replacement of the back-to-back connected transistors with monolithic bidirectional ones should be investigated. Furthermore, the inverter should be applied to actual electrocaloric heat pump systems to demonstrate the efficiency advantage also for applications. Then, also balancing multilevel voltage balancing has to be considered which was not required with this work's almost ideal capacitive load.

Acknowledgment

This work was partially supported by the Fraunhofer Society in the Fraunhofer lighthouse project "ElKaWe - Electrocaloric Heat Pumps" (www.ElKaWe.org). This work was partially supported by the Federal Ministry for Economic Affairs and Climate Action (BMWK) in the project "GaN4EmoBiL" (grant ID 01MV23003A).

References

[1] K. A. Kim, P. S. Shenoy, and P. T. Krein, "Converter Rating Analysis for Photovoltaic Differential Power Processing Systems," *IEEE Transactions on Power Electronics*, vol. 30, no. 4, pp. 1987–1997, 2015. DOI: 10.1109/TPEL.2014.2326045.

[2] B. Nair, T. Usui, S. Crossley, S. Kurdi, G. G. Guzmán-Verri, *et al.*, "Large electrocaloric effects in oxide multilayer capacitors over a wide temperature range," *Nature*, vol. 575, no. 7783, pp. 468–472, 2019. DOI: 10.1038/s41586-019-1634-0.

[3] J. Metzdorf, P. Corhan, D. Bach, S. Hirose, D. Lellinger, *et al.*, "Electrocaloric cooling system utilizing latent heat transfer for high power density," *Communications Engineering*, vol. 3, no. 1, p. 55, 2024. DOI: 10.1038/s44172-024-00199-z.

[4] J. Li, A. Torelló, V. Kovacova, U. Prah, A. Aravindhan, *et al.*, "High cooling performance in a double-loop electrocaloric heat pump," *Science*, vol. 382, no. 6672, pp. 801–805, 2023. DOI: 10.1126/science.adi5477.

[5] X. Qian, D. Han, L. Zheng, J. Chen, M. Tyagi, *et al.*, "High-entropy polymer produces a giant electrocaloric effect at low fields," *Nature*, vol. 600, no. 7890, pp. 664–669, 2021. DOI: 10.1038/s41586-021-04189-5.

[6] S. Zheng, F. Du, L. Zheng, D. Han, Q. Li, *et al.*, "Colossal electrocaloric effect in an interface-augmented ferroelectric polymer," *Science*, vol. 382, no. 6674, pp. 1020–1026, 2023. DOI: 10.1126/science.adi7812.

[7] U. Plaznik, M. Vrabelj, Z. Kutnjak, B. Malič, A. Poredoš, and A. Kitanovski, "Electrocaloric cooling: The importance of electric-energy recovery and heat regeneration," *EPL (Europhysics Letters)*, vol. 111, no. 5, p. 57009, 2015. DOI: 10.1209/0295-5075/111/57009.

[8] C. Molin and S. Gebhardt, "PMN-8PT device structures for electrocaloric cooling applications," *Ferroelectrics*, vol. 498, no. 1, pp. 111–119, 2016. DOI: 10.1080/00150193.2016.1169062.

[9] S. Crossley, B. Nair, R. W. Whatmore, X. Moya, and N. D. Mathur, "Electrocaloric Cooling Cycles in Lead Scandium Tantalate with True Regeneration via Field Variation," *Physical Review X*, vol. 9, no. 4, 2019. DOI: 10.1103/PhysRevX.9.041002.

[10] Y. Nouchokgwe, P. Lheritier, C.-H. Hong, A. Torelló, R. Faye, *et al.*, "Giant electrocaloric materials energy efficiency in highly ordered lead scandium tantalate," *Nature Communications*, vol. 12, no. 1, p. 3298, 2021. DOI: 10.1038/s41467-021-23354-y.

[11] T M Correia, S Kar-Narayan, J S Young, J F Scott, N D Mathur, *et al.*, "PST thin films for electrocaloric coolers," *Journal of Physics D: Applied Physics*, vol. 44, no. 16, p. 165407, 2011. DOI: 10.1088/0022-3727/44/16/165407.

[12] Z. Li, C. Molin, and S. E. Gebhardt, "Influence of Grain-Growth Inhibitors on Modified (Ba,Sr)(Sn,Ti)O3 for Electrocaloric Application," *Materials*, vol. 17, no. 5, p. 1036, 2024. DOI: 10.3390/ma17051036.

[13] C. Molin, P. Neumeister, H. Neubert, and S. E. Gebhardt, "Multilayer Ceramics for Electrocaloric Cooling Applications," *Energy Technology*, vol. 6, no. 8, pp. 1543–1552, 2018. DOI: 10.1002/ente.201800127.

[14] S. Crossley, J. R. McGinnigle, S. Kar-Narayan, and N. D. Mathur, "Finite-element optimisation of electrocaloric multilayer capacitors," *Applied Physics Letters*, vol. 104, no. 8, p. 082909, 2014. DOI: 10.1063/1.4866256.

[15] R. Faye, H. Strozyk, B. Dkhil, and E. Defay, "Large heat flux in electrocaloric multilayer capacitors," *Journal of Physics D: Applied Physics*, vol. 50, no. 46, p. 464002, 2017. DOI: 10.1088/1361-6463/aa8d0f.

[16] J. Li, A. Torelló, Y. Nouchokgwe, T. Granzow, V. Kovacova, *et al.*, "Electrocaloric effect in BaTiO 3 multilayer capacitors with first-order phase transitions," *Journal of Physics: Energy*, vol. 5, no. 2, p. 024017, 2023. DOI: 10.1088/2515-7655/acc972.

[17] Y. Meng, J. Pu, and Q. Pei, "Electrocaloric cooling over high device temperature span," *Joule*, vol. 5, no. 4, pp. 780–793, 2021. DOI: 10.1016/j.joule.2020.12.018.

[18] E. Defay, R. Faye, G. Despesse, H. Strozyk, D. Sette, *et al.*, "Enhanced electrocaloric efficiency via energy recovery," *Nature Communications*, vol. 9, no. 1, 2018. DOI: 10.1038/s41467-018-04027-9.

[19] Y. Meng, Z. Zhang, H. Wu, R. Wu, J. Wu, *et al.*, "A cascade electrocaloric cooling device for large temperature lift," *Nature Energy*, vol. 5, no. 12, pp. 996–1002, 2020. DOI: 10.1038/s41560-020-00715-3.

[20] M. Almanza, T. Martinez, M. Petit, Y. Civet, Y. Perriard, and M. LoBue, "Adaptation of a Solid-State Marx Modulator for Electroactive Polymer," *IEEE Transactions on Power Electronics*, vol. 37, no. 11, pp. 13014–13021, 2022. DOI: 10.1109/TPEL.2022.3183437.

[21] S. Moench, R. Reiner, P. Waltereit, C. Molin, S. Gebhardt, *et al.*, "Enhancing Electrocaloric Heat Pump Performance by Over 99% Efficient Power Converters and Offset Fields," *IEEE Access*, vol. 10, pp. 46571–46588, 2022. DOI: 10.1109/ACCESS.2022.3170451.

[22] D. E. Schwartz, "Thermodynamic Cycles and Electrical Charge Recovery in High-Efficiency Electrocaloric Cooling Systems," *International Journal of Refrigeration*, 2021. DOI: 10.1016/j.ijrefrig.2021.02.003.

[23] A. Torelló and E. Defay, "Electrocaloric Coolers: A Review," *Advanced Electronic Materials*, p. 2101031, 2022. DOI: 10.1002/aelm.202101031.

[24] A. Greco and C. Masselli, "Electrocaloric Cooling: A Review of the Thermodynamic Cycles, Materials, Models, and Devices," *Magnetochemistry*, vol. 6, no. 4, p. 67, 2020. DOI: 10.3390/magnetochemistry6040067.

[25] Hicham Johra, "Performance overview of caloric heat pumps: magnetocaloric, elastocaloric, electrocaloric and barocaloric systems," 2022. DOI: 10.54337/aau467469997.

[26] S. Mönch, R. Reiner, K. Mansour, P. Waltereit, M. Basler, *et al.*, "A 99.74% Efficient Capacitor-Charging Converter using Partial Power Processing for Electrocalorics," *IEEE Journal of Emerging and Selected Topics in Power Electronics*, vol. 11, no. 4, pp. 4491–4507, 2023. DOI: 10.1109/JESTPE.2023.3270375.

[27] F. Z. Peng, W. Qian, and D. Cao, "Recent advances in multilevel converter/inverter topologies and applications," in *The 2010 International Power Electronics Conference - ECCE ASIA -*, 2010, pp. 492–501. DOI: 10.1109/IPEC.2010.5544625.

[28] D. Chou, Y. Lei, and R. C. N. Pilawa-Podgurski, "A Zero-Voltage-Switching, Physically Flexible Multilevel GaN DC–DC Converter," *IEEE Transactions on Power Electronics*, vol. 35, no. 1, pp. 1064–1073, 2020. DOI: 10.1109/TPEL.2019.2914213.

[29] S. Moench, K. Mansour, R. Reiner, M. Basler, P. Waltereit, *et al.*, "A GaN-based DC-DC Converter with Zero Voltage Switching and Hysteretic Current Control for 99% Efficient Bidirectional Charging of Electrocaloric Capacitive Loads," in *PCIM Europe 2022; International Exhibition and Conference for Power Electronics, Intelligent Motion, Renewable Energy and Energy Management*. DOI: 10.30420/565822251.

[30] S. Mönch, R. Reiner, M. Basler, P. Waltereit, and R. Quay, "Voltage-Sensorless Control and GaN Multilevel Converter for Charging Non-Linear and Lossy Electrocaloric Capacitors," *13th International Conference on Integrated Power Electronics Systems (CIPS 2024)*, pp. 202–207, 2024.

[31] T. Morita, M. Yanagihara, H. Ishida, M. Hikita, K. Kaibara, *et al.*, "650 V 3.1 mΩcm² GaN-based monolithic bidirectional switch using normally-off gate injection transistor," in *IEEE International Electron Devices Meeting, 2007*, Piscataway, NJ: IEEE Service Center, 2007, pp. 865–868. DOI: 10.1109/IEDM.2007.4419086.

[32] N. Perera, A. Jafari, L. Nela, G. Kampitsis, M. S. Nikoo, and E. Matioli, "Output-Capacitance Hysteresis Losses of Field-Effect Transistors," in *2020 IEEE 21st Workshop on Control and Modeling for Power Electronics (COMPEL)*, 2020, pp. 1–8. DOI: 10.1109/COMPEL49091.2020.9265823.

PCIM Europe 2024, 11– 13 June 2024, Nuremberg   DOI: 10.30420/566262021

Cascaded Primary-Side-Only Control of a compact 2 MHz 500 W Wireless Power Transfer System

Tim Krigar[1], Tim Egener[1], Martin Pfost[1]

[1] Chair of Energy Conversion, TU Dortmund University, Germany

Corresponding author: Tim Krigar, tim.krigar@tu-dortmund.de
Speaker: Tim Krigar, tim.krigar@tu-dortmund.de

Abstract

In this paper, we present the integration of an innovative control strategy implemented solely on the primary side of a 2 MHz Wireless Power Transfer (WPT) system. This system transfers 500 W with a power density of 25 W/cm^2 and has a peak efficiency of 92.8 %. The control technology uses a cascaded design where the fast inner control loop is for safety functions and the outer control loop provides a constant output voltage. The control operates without communication between the primary and secondary sides, which has the advantage of eliminating the need for a controller on the compact secondary side, saving space and cost. This method handles not only high-frequency power transfer, but also offers promising controlling performance.

1 Introduction

Modern industrial applications are raising substantial interest in Wireless Power Transfer (WPT) systems with compact, low-cost receivers. These systems eliminate common problems associated with traditional power transfer methods, such as brush contact wear and cable drag. To achieve the high power density required in compact WPT systems, switching frequencies in the MHz range are required. This is made possible by the implementation of resonant power converters.

In our research, we focus on a WPT system using an LLC topology that operates at 2 MHz, transmits 500 W, and achieves a power density of 25 W/cm^2, all while maintaining an efficiency of 92.8 %. Notably, our design differs from conventional systems, cf. [1–3], which typically address stray inductance compensation on both the transmitter and receiver sides. Instead, our approach only compensates for stray inductance on the primary side, a concept that has been demonstrated in previous work such as [4].

To ensure a constant output voltage with current limiting, we use a cascaded primary-side-only control system, eliminating the need for communication between the transmitter and receiver. This approach saves cost and space by eliminating the need for a controller on the receiver side. In the inner control loop, we use the resonant current amplitude as direct feedback, which increases system robustness. In the outer control loop, the lack of direct primary-secondary communication presents a control challenge that we address by using a model that estimates the output voltage based on primary-side measurements.

The challenge in this model-based approach is to accurately determine the primary-side measurements of this high-frequency system, since traditional measurements of phase information, as shown in [5–8], are not applicable. Therefore, high-frequency analog signals are converted to DC quantities before being fed to the microcontroller, allowing us to effectively control higher-frequency systems. This control strategy is critical to accurately regulate the output voltage, especially in the presence of parameter variations due to high frequency operation and load variations on the secondary side.

2 Proposed System

The proposed system is shown in Fig. 1 (a). It uses a half-bridge circuit with GaN HEMTs to generate a square wave voltage for the wireless power link, which is formed by a resonant capacitor and a coil

174

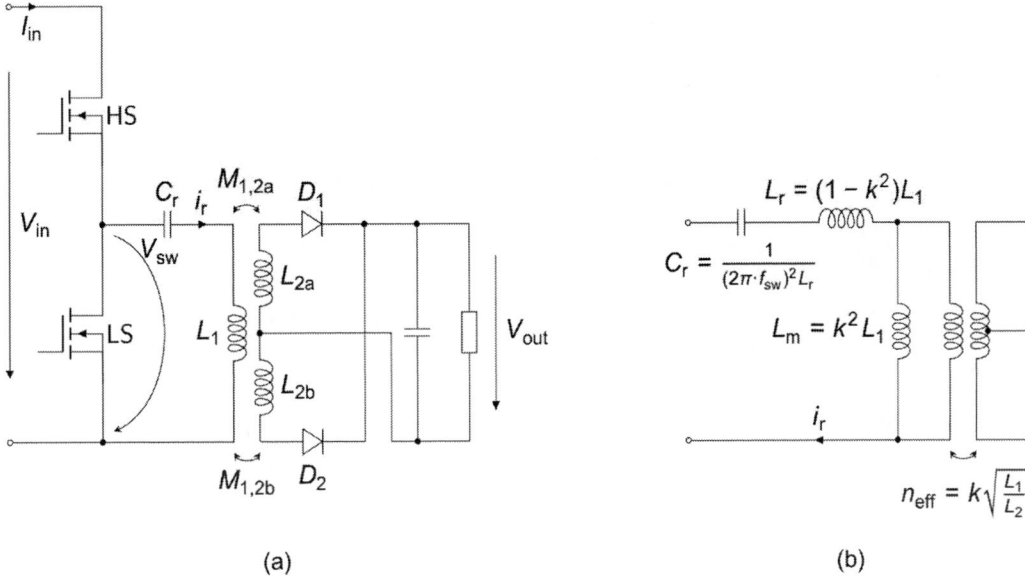

(a)   (b)

Fig. 1: Subfigure (a) shows the equivalent circuit of the proposed system with the transmitter circuit on the left, consisting of the half bridge C_r and L_1. The receiver circuit includes L_{2a}, L_{2b}, the rectifier and the load. Subfigure (b) shows the All-Primary-Referred equivalent circuit of the wireless power link, showing an LLC topology formed by C_r, L_m, and L_r.

assembly. This consists of one transmitter coil and two receiver coils, which are designed as PCB-integrated center-tapped coils to enable full-wave rectification using only two diodes. This also increases the compactness and cost efficiency of the receiver unit, simplifies the design, and achieves symmetry in the center-tapped transformer. The system has an input voltage of 400 V, a desired output voltage of 48 V, and a switching frequency of 2 MHz. The voltage transformation is achieved by careful coil design and coil turn ratio.

Due to the loosely coupled coils, it is necessary to compensate for stray inductance. An approach that differs from the traditional method of providing compensation on both the transmitter and receiver sides is to implement compensation only on the primary side. This method uses the All-Primary-Referred-Equivalent circuit, see Fig. 1(b), to simplify the LLC topology consisting of L_r, L_m, and C_r.

In Fig. 2, a photo of the prototype shows the transmitter on the left from an overhead perspective, highlighting different functional areas. These include a galvanically isolated voltage measurement, an input filter with a common mode choke and differential mode capacitors, and the placement of switching GaN HEMTs on a separate GaN module below. In addition, C_r and the resonant current

measurement, realized by a shunt measurement and a peak detection circuit, are located in this area. The microcontroller is also galvanically isolated from the power circuit. The input current measurement is not visible in these images because it is located on the back of the board. To the right of the transmitter circuit is the coil assembly area. The secondary rectifier is visible, but only in side view. Since this configuration is only a prototype, it is important to emphasize that a commercial product could be designed to be much more compact.

3 Controller Design

In theory, the LLC converter has a load-independent output voltage at its resonant frequency. However, due to parameter variations and parasitic characteristics of the electronic components, the system presented in the previous chapter must be controlled. Communication between the transmitter and receiver electronics should be avoided in the control loop to keep the receiver electronics low-cost and compact. For this purpose, a cascaded control loop was designed, as shown in Fig. 3. The controller is divided into an inner and an outer control loop, each with different control objectives.

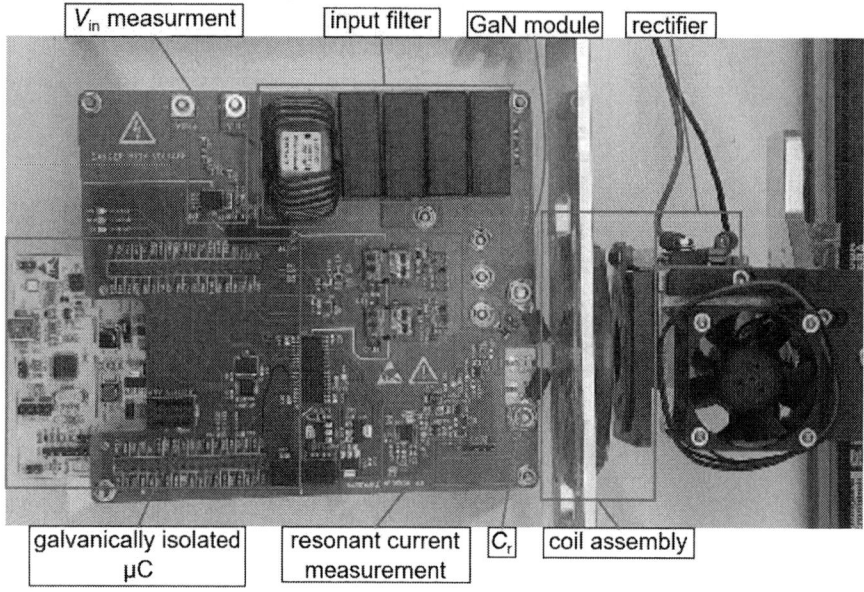

Fig. 2: Foto of the system hardware.

The inner control loop uses the switching frequency f_{sw} as control variable and the peak value of the resonance current $I_{r,ref}$ as reference variable. In the outer control loop, the output voltage and $I_{r,ref}$ are the reference and control variables, respectively. Due to the design constraint that prohibits direct communication between the transmitter and receiver, the actual output voltage cannot be fed back directly for processing by the outer control loop. Therefore, V_{out} is estimated by a model using only the primary side measurements $\hat{I}_r$, I_{in}, V_{in}, and f_{sw}.

The estimation model consists of two stages: the load model and the voltage model. The load model, expressed as

$$R_{load,mod} = x_1 \frac{V_{in}}{I_{in}} + \frac{x_2}{P_{in}}(f_{sw} - f_{r,0}) + \frac{x_3}{P_{in}}(f_{sw} - f_{r,0})^2 - x_4,$$

uses the input dimensions to calculate $R_{load,mod}$. Here, x_{1-4} are fitting parameters, and $f_{r,0}$ is the resonant frequency of the LLC converter.

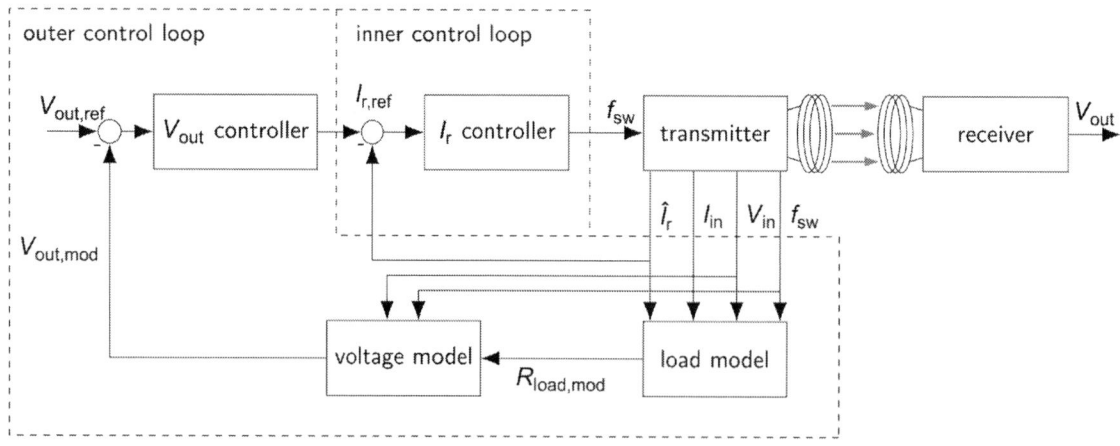

Fig. 3: Schematic of the control loop

The voltage model then calculates $V_{out,mod}$ as follows

$$V_{out,mod} = V_{in} H(f_{sw}, R_{load,mod}) \frac{R_{load,mod}}{R_{load,mod} + R_{sec}} - V_{sec},$$

where V_{sec} and R_{sec} account for offsets caused by rectifier forward voltage and parasitic ohmic components. $H(f_{sw}, R_{load,mod})$ is based on the equivalent circuit modeling described in [9].

The model parameters were determined using particle swarm optimization based on measured results of the uncontrolled system at various operating points. The results, shown in Fig. 4 (a) and (b), illustrate the accuracy of the model in estimating load and voltage highlighting the effectiveness of the cascaded control loop in managing system dynamics without the need for direct transmitter-receiver communication.

Note that both control loops are implemented as proportional-integral (PI) controllers to pro-vide steady-state error correction and dynamic response. The design includes a frequency constraint that limits f_{sw} to a range between 2 MHz and 2.6 MHz to ensure safe operation. In addition, an anti-windup mechanism was incorporated into the controller design to mitigate integral windup and ensure stable and responsive control under varying operating conditions.

Importantly, the inner control loop is designed to respond faster than the outer control loop. This fast response is particularly useful for safety features such as soft start and overload protection. The soft start function of the inner control loop ramps f_{sw} down from the starting frequency to its nominal operating frequency.

By allowing the inner control loop to focus on fast, protective actions and the outer control loop to maintain the correct output voltage, the system is both safe and effective. This setup ensures that the system is well protected against unexpected problems that could arise due to changes in component behavior or overload scenarios. This approach makes our system robust in handling a wide range of conditions, such as charging supercapacitors or on-board batteries.

4 Experimental Results

The performance of the proposed cascaded controller is demonstrated in Figs. 5 and 6.

Fig. 5 shows the behavior of the inner control loop in the soft start scenario. It can be seen that the specified reference value of $I_{r,ref} = 5$ A is successfully regulated. The system is switched on with a frequency of 2.6 MHz and then settles down to its load depending final value of approx. 2.1 MHz - 2.2 MHz after about 150 ms.

In Fig. 6 (a) and (b), the output voltage is considered without and with the outer control loop (voltage regulation) activated. While in Fig. 6 (a) the output voltage shows a load dependent behavior without voltage regulation enabled, it settles to the desired reference value when regulation is enabled, see Fig. 6 (b).

These figures clearly illustrate the effectiveness of the control system in regulating the output voltage and ensuring that it stays within the desired range of ±5%.

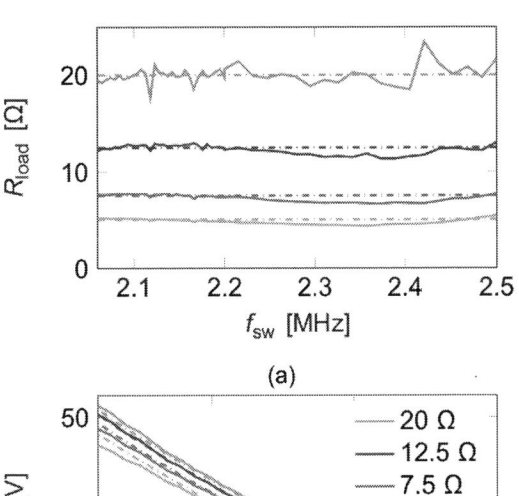

(a)

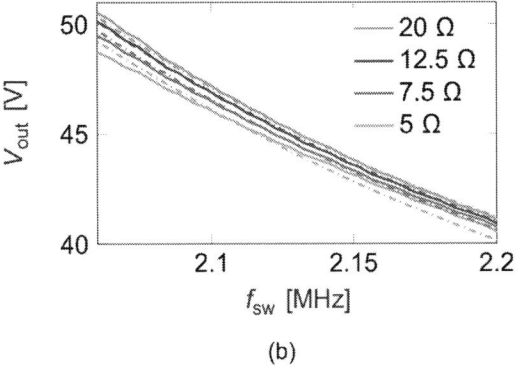

(b)

Fig. 4: Results of (a) the load model and (b) the voltage model. The solid lines represent the calculated model values compared to the dashed lines representing the reference values derived from measurements.

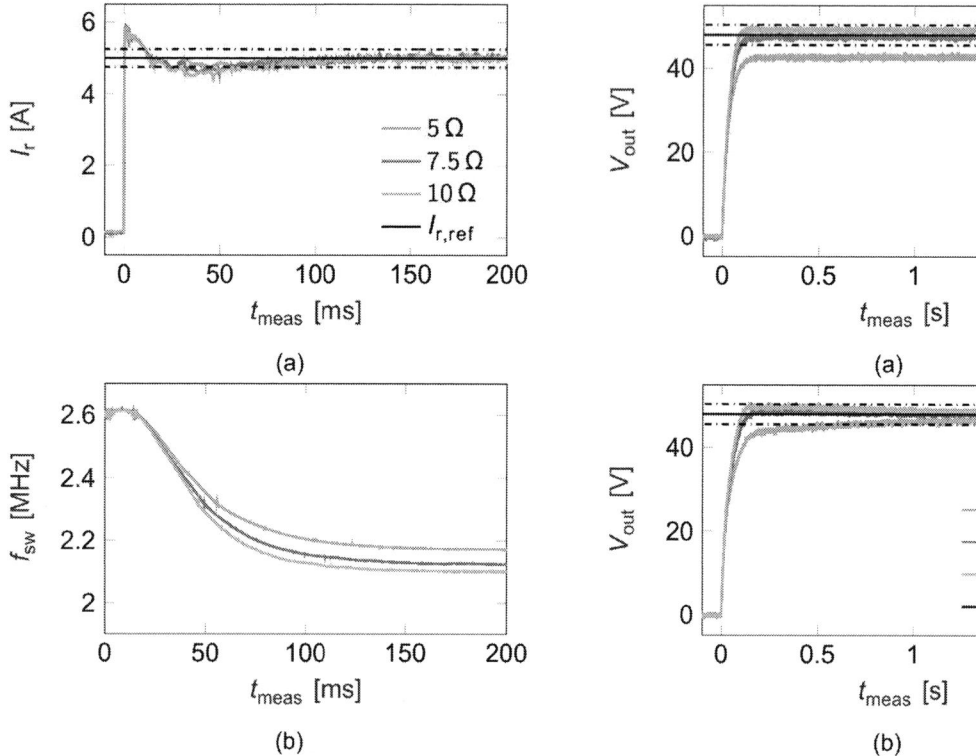

Fig. 5: The start-up performance of the inner control loop, operating independently of the outer control loop, is shown. (a) The amplitude of the resonant current I_r, along with its reference value and a 5% tolerance band, and (b) the resulting frequency are displayed.

Fig. 6: Output voltage results: (a) without control and (b) with control system. The active control system successfully maintains the output voltage within a 5% tolerance band around the reference value of 48 V.

Figs. 7 and 8 show a comparison of the proposed system under conditions with and without active control in a power sweep test. The uncontrolled system with fixed f_{sw} is shown as a dashed red line and the system with cascaded control enabled is shown as a solid blue line. In Fig. 7 it can be seen that the implemented control does not significantly affect the operating efficiency of the system. This indicates that the control can be well implemented without any loss of efficiency.

From Fig. 8 (a) it can be seen that the system with cascaded control enabled adjusts the switching frequency and responds to load changes by changing the frequency accordingly, while the uncontrolled system has a constant switching frequency.

Fig. 8 (b) shows the effect of load changes on the output voltage of both systems. It can be seen that the output voltage of the uncontrolled system is sensitive to load variations. In contrast, the controlled system shows the ability to maintain a con-

stant output voltage regardless of load variations, highlighting the benefits of the implemented control.

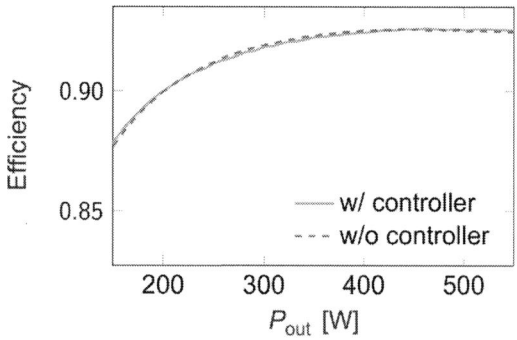

Fig. 7: Comparing the efficiency of the system with and without the control enabled in the solid blue and dashed orange lines, respectively. It can be seen that the system efficiency is not significantly affected by the presented control loop.

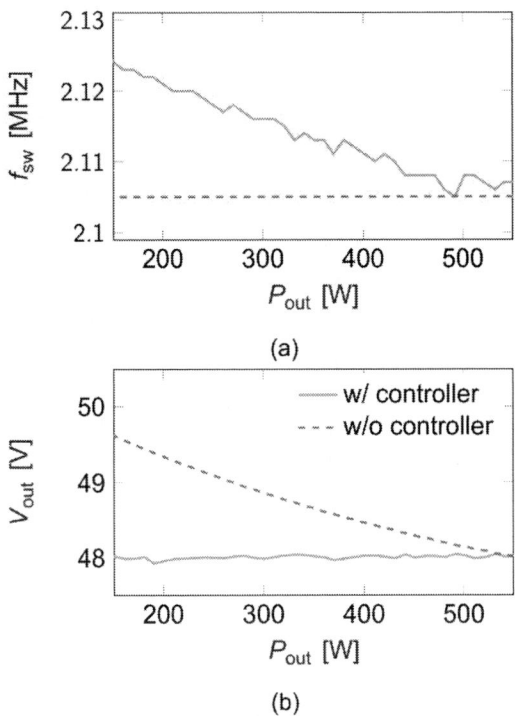

(a)

(b)

Fig. 8: Comparative analysis of system with and without control during the power sweep test. (a) Frequency remains constant without control and adjusts with load under control. (b) Output voltage fluctuates with load without control, but remains stable with control. The effect of the control mechanism on system efficiency.

5 Conclusion

This paper successfully demonstrates an innovative approach to primary-side-only control in high-frequency WPT systems. Using a model-based strategy, the controller can efficiently predict the output voltage, allowing constant power while providing soft start and overload protection. It is also shown that the system efficiency is not affected by the control. This method opens up new possibilities for efficient, compact and reliable WPT system design in various applications.

References

[1] W. Zhang, S.-C. Wong, C. K. Tse, and Q. Chen, "Analysis and comparison of secondary series- and parallel-compensated inductive power transfer systems operating for optimal efficiency and load-independent voltage-transfer ratio," *IEEE Transactions*

on Power Electronics, vol. 29, no. 6, pp. 2979–2990, 2014.

[2] Y. H. Sohn, B. H. Choi, E. S. Lee, G. C. Lim, G. Cho, and C. T. Rim, "General unified analyses of two-capacitor inductive power transfer systems: Equivalence of current-source SS and SP compensations," *IEEE Transactions on Power Electronics*, vol. 30, no. 11, pp. 6030–6045, Nov 2015.

[3] L. Gu, G. Zulauf, A. Stein, P. A. Kyaw, T. Chen, and J. M. R. Davila, "6.78-MHz wireless power transfer with self-resonant coils at 95% DC–DC efficiency," *IEEE Transactions on Power Electronics*, vol. 36, no. 3, pp. 2456–2460, Mar. 2021.

[4] T. Krigar and M. Pfost, "Optimization of a 2 MHz 500 W compact wireless power transfer system with a large voltage conversion ratio," in *PCIM Europe 2022*, 2022, pp. 1–6.

[5] K. Song, Z. Li, J. Jiang, and C. Zhu, "Constant current/voltage charging operation for series–series and series–parallel compensated wireless power transfer systems employing primary-side controller," *IEEE Transactions on Power Electronics*, vol. 33, no. 9, pp. 8065–8080, 2018.

[6] E. Chung, G. C. Lim, J.-I. Ha, and K. Y. Kim, "Output voltage control for series-series compensated wireless power transfer system without direct feedback from measurement or communication," in *2017 IEEE Energy Conversion Congress and Exposition (ECCE)*, 2017, pp. 4035–4040.

[7] P. Zheng, W. Lei, F. Liu, R. Li, and C. Lv, "Primary control strategy of magnetic resonant wireless power transfer based on steady-state load identification method," in *2018 IEEE International Power Electronics and Application Conference and Exposition (PEAC)*, 2018, pp. 1–5.

[8] F. Sadeque and F. Fateh, "Voltage control by transmitter-side measurements for onboard wireless EV battery charger," in *2021 IEEE Kansas Power and Energy Conference (KPEC)*, 2021, pp. 1–6.

[9] S. Tian, F. C. Lee, and Q. Li, "Equivalent circuit modeling of LLC resonant converter," in *2016 IEEE Applied Power Electronics Conference and Exposition (APEC)*, 2016, pp. 1608–1615.

Power Module Evaluation Using Ultra High Heat Dissipation and High Heat Resistance Resin Sheet Containing Boron Nitride Filler

Katsuhiko Hidaka[1], Ayano Imai[1], Shuji Suzuki[1], Jun Matsui[1], Toshiyuki Sawamura[1], Yuya Koga[1], Yasushi Yamada[2], Shinichi Yasaka[3], Hitoshi Habuka[4]

[1] Mitsubishi Chemical Corporation, Japan

[2] Daido University, Japan

[3] Kanagawa Institute of Industrial Science and Technology, Japan

[4] Yokohama Jisso Consortium, Japan

Corresponding author: Yuya Koga, yuuya.koga.ma@mcgc.com
Speaker: Ayano Imai, ayano.imai.mb@mcgc.com

Abstract

SiC has the advantage of high-temperature operation at above 200°C, but existing peripheral materials have not been sufficient in terms of heat resistance and durability. This research succeeded in developing a high heat-resistant module that enables continuous high-temperature operation of SiC by using a resin substrate with high heat dissipation of 20 W/(m·K), high heat resistance, and high insulation, which can withstand high temperature and high pressure. It achieved durability of 50,000 cycles at Tj_{max} 225°C and 20,000 cycles at Tj_{max} 250°C. It was confirmed that the module using the heat-dissipating resin substrate has almost reached the stage of practical application for the 200°C high-temperature operation required for SiC.

1 Introduction

Recently, SiC, a wide bandgap semiconductor, has been applied to power modules, and since SiC can operate at temperatures above 200°C unlike Si, development of modules with high heat dissipation and high heat resistance is underway. We have previously developed an ultra-high heat dissipation insulation metal baseplate (IMB) with both high thermal conductivity and insulation properties [1]. For this IMB, card house structure BN originally developed by Mitsubishi Chemical Corporation was used to achieve high heat dissipation. This filler has a unique structure in which each tip of BN flakes with anisotropic thermal conductivity is sintered to form a card house structure. By introducing this structure, surfaces with high thermal conductivity are connected to each other, improving thermal conductivity in all directions (Fig. 1).

The development of high heat-resistant modules requires not only heat resistance against high chip temperature during high temperature operation, but also resistance to high temperature and high-pressure conditions such as Ag sintering conditions (300°C, 10 MPa). The Kanagawa Advanced MOdule for Material Evaluation (KAMOME) project, the Yokohama Jisso Consortium (YJC) has promoted material development by evaluating and providing feedback on mounting technologies and materials for SiC devices and other high-temperature applications [2]. Therefore, in this study, by selecting a thermoplastic resin with ultra-high heat resistance for the resin layer of the IMB, heat resistance was greatly improved, and power modules were mounted and evaluated in KAMOME-project using the developed resin substrate.

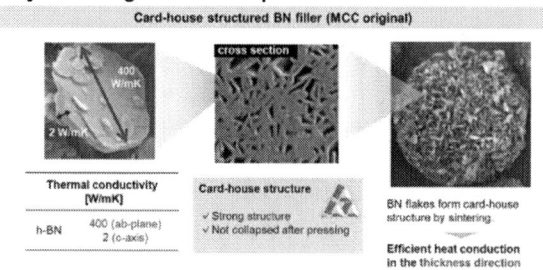

Fig.1 Card house structure BN

2 Evaluation of Insulated Metal Baseplate

2.1 Physical properties of sheet and high heat resistant Insulated Metal Baseplate

The physical properties of the developed sheets and their breakdown voltage performance during IMB formation are listed in Table 1. The measurement method was the same as previously reported, and the thermal conductivity was measured by the steady-state method [1]. 5 different film thicknesses were prepared, including 150 µm, and the Simcenter DynTIM was used to measure the film thickness and thermal resistance at each film thickness. Thermal conductivity was calculated from the thermal resistance and film thickness slope obtained by the steady-state method. The breakdown voltage measurement sample was prepared by laminating resin sheet in between 0.5 mm oxygen-free copper and 2.0 mm oxygen-free copper of 40 mm x 80 mm, heat pressing under pressure. The 0.5 mm thick side copper of the prepared substrate was etched to leave a φ25 coin pattern. For breakdown voltage measurement, jig was connected to the coin-shaped pattern copper and voltage was applied to the resin sheet in insulating oil. When measuring, 0.5 kV was applied for 1 minute, and if it does not breakdown, another 0.5kV higher voltage was applied for 1 minute. The elastic modulus at 225°C was measured by Dynamic Mechanical Analysis (DMA) (DMS6100, Hitachi High-Tech Science Corporation). To ensure stable measurements, samples with the thickness of 500 µm of resin sheets were cut into approximately 10 mm width and 50 to 60 mm length for evaluation. The measurement conditions were as follows: frequency: 1 Hz, chuck-to-chuck distance: 20 mm, strain amplitude: 5 µm, and measurement temperature: -110°C to 320°C. Coefficient of thermal expansion (CTE) was measured by TMA. The equipment was TMA SS6100 (Hitachi High-Tech Science, Inc.), measuring with tensile mode, and sample with the thickness of 500 µm of resin sheets were cut into approximately 3 mm width and 30 mm length for evaluation. The measurement conditions were as follows: chuck-to-chuck distance: 20 mm, tension: 49 mN, and temperature was increased at a rate of 5°C/min between -50 and 230°C. The high temperature modulus of elasticity and CTE of the resin sheet layer in substrate were significantly improved compared with the conventional product [1]. It also exhibited a high thermal conductivity of 20 W/(m·K) and a high breakdown voltage of 10 kV. As previously reported, this can be attributed to the effect of introducing card house structured BN fillers developed in MCC.

Physical properties	Form	Unit	Developed	Conventional
Thermal Conductivity	Sheet	W/(m·K)	20	16.5
		µm	125	155
Breakdown Voltage	IMB	kV	10	7
		µm	125	155
Modulus of elasticity 225 °C	Sheet	GPa	6.5	5
CTE α1	Sheet	ppm / K	14	18

Table 1 Physical properties of sheet and breakdown voltage of substrate

2.2 Comparison of IMB and Si_3N_4 by silver sintering

In power modules using SiC, silver sintering is used to improve durability and reduce thermal resistance. Since silver sintering is bonded under high temperature and high-pressure conditions, high durability is also required for the substrate. The durability was compared between ultra-high heat resistant resin substrate (MCC-IMB) and Si_3N_4. The configuration and cross-sectional images of the IMB and Si_3N_4 modules used in the study are shown in Table 2, Fig. 2 and Fig. 3.

Components	Si_3N_4	IMB
Upper copper	0.30 mm	0.8 mm
Insulator thickness	0.32 mm	0.12 mm
Insulator Thermal conductivity	90 W/(m·K)	20 W/(m·K)
Lower copper	0.30 mm	1.0 mm
Mo-Cu Al heat diffusion bonding	1.0 mm	-

Table 2 Composition of substrate materials

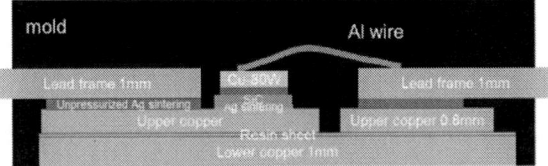

Fig. 2 Module configuration structure with IMB

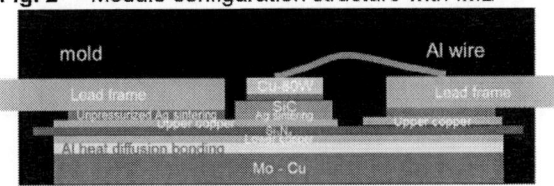

Fig. 3 Module configuration structure with Si_3N_4

There is an issue of large warpage of Si_3N_4 substrate during mounting process due to CTE difference between ceramic and copper. On the other hand, since IMB is softer and tougher than ceramic, warpage is much smaller. Therefore, to suppress the warpage of Si_3N_4 module during the mounting process and improve heat dissipation property, Mo-Cu with low CTE was applied for Si_3N_4 substrate. Also, thermal diffusion bonding was used instead of soldering to improve the durability between Si_3N_4 substrate and Mo-Cu. For both modules, silver sintering was used for SiC Schottky barrier diodes (SiC-SBD) bonding, and pressure sintering was applied. In addition, to extend the lifetime of wire bonding, Cu-80W, which acts as a CTE controlling material, was silver sintered on the SiC-SBD chip without pressure. The lead frame (LF) and Cu-80W on the chip were bonded with Al wire and encapsulated with transfer-molded resin with Tg higher than 270°C.

The bonding conditions for silver sintered materials are shown in Table 3. In the bonding process of SiC-SBD chip, which requires higher reliability, pressure sintering was selected. In other parts, Ag sintering material, which allows no-pressure sintering, was selected, considering the damage to the chip. The LF and the top surface of the chip were bonded with Cu-80W at 260°C for 30 minutes.

	Pressure	Non-pressure
Joint material	SiC - upper copper	SiC – Cu-80W
Temperature	300°C	260°C
Time	15 min	30 min
Pressure	10 MPa	-

Table 3 Sintering conditions

Two types of resin substrates were used as references. Table 4 shows the SAT image. There was no delamination regarding developed MCC resin substrate and Si_3N_4, but the resin substrates used as references showed delamination. Resin substrates are generally not resistant to high temperatures and pressures of silver sintering, but it was found that the high heat-resistant resin used in this study can withstand the high temperatures and pressures of silver sintering, such as 300°C and 10 MPa. This is thought to be owing to the high thermal decomposition temperature, high temperature modulus, and high adhesion at high temperatures of the introduced resin, which prevented delamination. The module was evaluated using these silver-sintered substrates.

Table 4 SAT images after silver sintering

3 Module Evaluation

The silver-sintered substrates were wired and encapsulated to create power modules. Thermal resistance measurement (R_{th}), power cycle test (PCT), and thermal cycle test (TCT) were conducted to evaluate heat dissipation and durability of the power module.

3.1 Comparison of thermal resistance between IMB and Si_3N_4

The SiC chip is a ROHM SiC-SBD with 1200 V withstand voltage, S6305 or equivalent was selected. Thermal resistance measurements were performed under the following conditions: heating power of 50 W, indirect water cooling, cooling water temperature of 25°C, heating time of 100 s, and cooling time of 100 s. The structure function of each module was calculated using these measurements according to JEDEC standard (JESD51-14) [3]. For each module, the heat capacity was estimated from the area, thickness, density, and specific heat capacity of each material, and the cumulative heat capacity to the lower copper plate was calculated. The cumulative thermal resistance corresponding to this cumulative heat capacity in the structure function was defined as the thermal resistance of each substrate and used for comparison. The thermal resistance was compared as the thermal resistance of the module. T_{jmax} was estimated using the temperature conversion method (I_f = 20 mA) based on microcurrent V_f measurements. Figure 4 shows a graph of thermal resistance. The thermal resistance of the Si_3N_4 substrate was 0.84 K / W, while that of the resin substrate was 0.66 K / W. The total thermal resistance of the Si_3N_4 module was 1.6 K / W and that of the resin substrate was 1.4 K / W, including the TIM used to attach the module to the water-cooled heat sink. Furthermore, the thermal resistance of the Si_3N_4 substrate varied with module mounting due to warpage. On the other hand, the resin substrate had a smaller difference in thermal resistance when mounted on the sample due to its smaller warpage. Therefore, MCC-IMB showed 21 % reduction in thermal resistance compared to Si_3N_4 in the form of evaluated module.

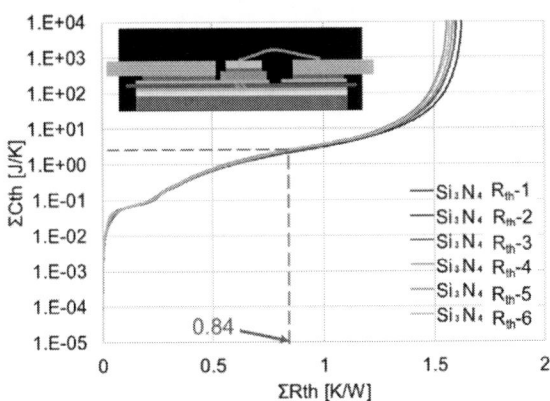

(a) structure function of Si_3N_4

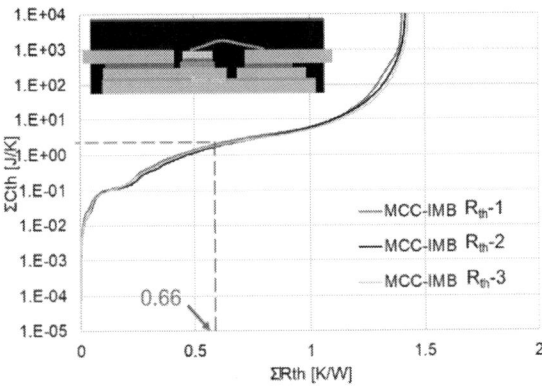

(b) structure function of MCC-IMB

Fig. 4 Comparison of structure functions of Si_3N_4 and MCC-IMB modules

The power required to reach 225°C for Tj was 162 W for the Si_3N_4 module, while 191 W for the MCC-

IMB module. This is because the MCC IMB module has lower thermal resistance than the Si_3N_4 module. Therefore, 17 % higher power can be applied when the modules are driven at the same temperature. The ability to laminate thicker copper and lower warpage allows MCC-IMB to have fewer components and significantly lower thermal resistance compared to Si_3N_4.

3.2 Thermal cycle test

TCT was performed on the mounted modules at setpoints from -40°C to 225°C for 30 minutes each. After 500 cycles, no delamination was observed for MCC-IMB module, but chip delamination was observed with Si_3N_4 module (Table 5).

Insula-tor	Cy-cles	SAT image
Si_3N_4	0	No delamination
	500	delamination Ag layer delamination
MCC-IMB	0	No delamination
	500	No delamination

Table 5 SAT images before and after TCT at -40°C / -225°C conditions

The structure function was compared before and after TCT. There was no significant difference in the structure function before and after TCT for MCC-IMB, but the thermal resistance increased after TCT for Si_3N_4. The SAT image and structure function suggest that the reason lies on the delamination of the silver sintering layer of the chip. The reason for the delamination of the chip is thought to be that the IMB substrate with lower modulus and CTE matched to copper has less warpage due to heat, whereas the Si_3N_4 substrate has more stress on the chip due to thermal warpage caused by the CTE difference between copper and Si_3N_4.

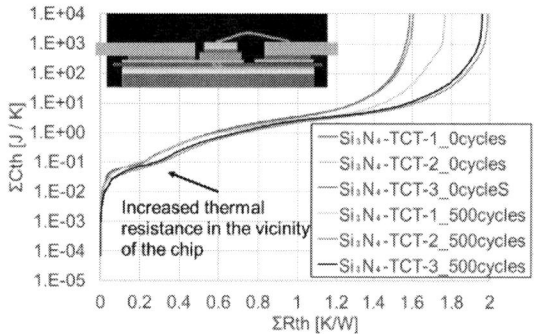

(a) Structure function of Si_3N_4 after TCT 500cycles

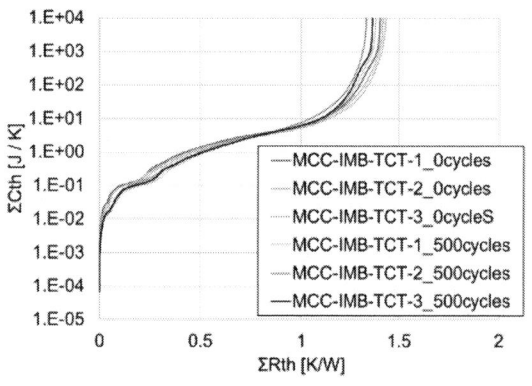

(b) Structure function of MCC-IMB after TCT 500cycles

Fig. 5 Comparison of thermal resistance of MCC-IMB and Si_3N_4 modules before and after TCT

3.3 Power Cycle Test

PCT was performed. 50% ethylene glycol solution was used as the cooling solution, and power was applied for 2 seconds at a cooling solution temperature of 65°C to bring T_j to 225°C. Cooling was 18 seconds. T_j was estimated from the V_f of the SiC-SBD when 20 mA was applied to the chip. V_{fmax} and R_{th} were checked after each cycle of the thermal cycling test, and the test was stopped when either value increased by 10%. As a result, the test was completed after 12,000 cycles for the ceramic module and 33,000 cycles for the resin module (Fig. 6).

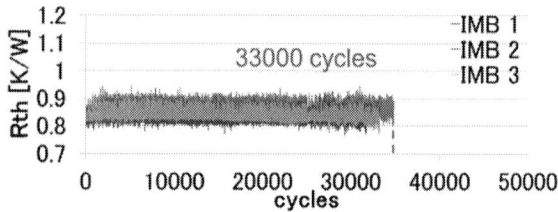

(a) MCC-IMB thermal resistance

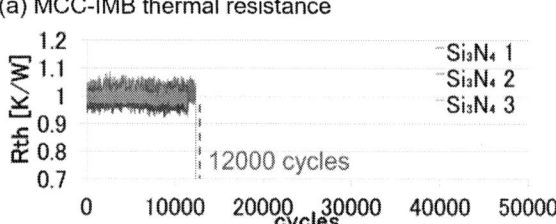

(b) Si_3N_4 thermal resistance

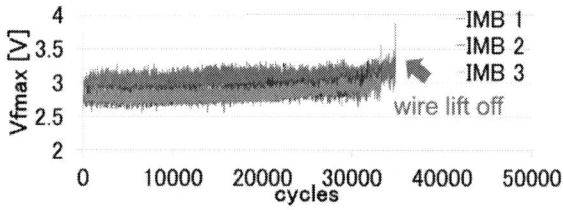

(c) MCC-IMB V_{fmax}

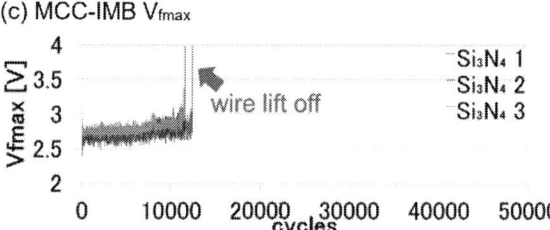

(d) Si_3N_4 V_{fmax}

Fig. 6 Power cycle test results (a) Thermal resistance of IMB, (b) Thermal resistance of Si_3N_4 (c) V_{fmax} of IMB, (d) V_{fmax} of Si_3N_4

To confirm the failure mode, samples were analyzed after PCT. Figure 7 shows an image of the Si_3N_4 sample. The left side is the chip side and the right side is the non-chip side. Figure 8 shows an example of the cross-sectional image.

Continuity was checked between the wire and LF as shown in Fig. 7. The chip side was measured in the diode measurement mode of the tester and the LF side was measured in the resistance measurement mode. Table 6 compares the open numbers between the wires and LF of each module. The wire on the chip side dropped off in all modules. On the other hand, only Si_3N_4 wires on the non-chip side dropped off. Based on the number of Al wires on the non-chip side that dropped off, it is thought that they are due to warpage, not temperature factors.

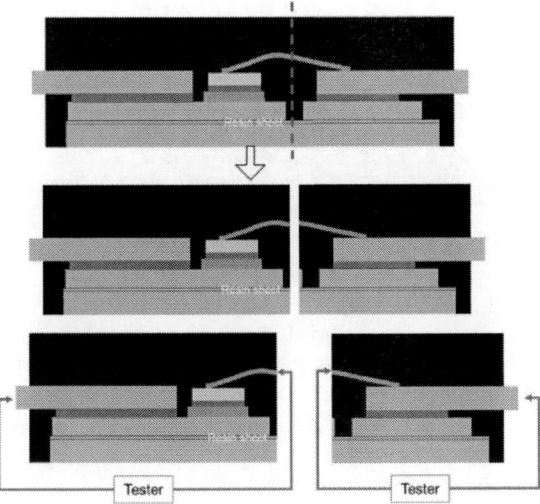

Fig. 7 Cross-sectional image of module and example of image diagram of continuity check using a tester

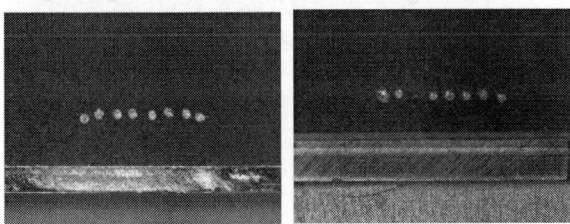

Chip side of IMB   non-chip side of Si₃N₄

Fig. 8 Example of appearance of cross section after cutting

	Si₃N₄	IMB
Current (Power) T$_j$ 225°C	60 A (162 W)	68 A (191 W)
PCT cycles	12000	33000
Chip side	5/8 open	5/8 open
Non-chip side	2/8 open	0/8 open

Table 6 Checking the conductivity of the chip side and non-chip side at each module

The warpage of the Sl₃N₄ substrate also shortens the lifetime of the wires on the chip side and breaks the wires on the non-chip side. Generally, resin sheets warp less than ceramics in response to thermal changes because CTE of resin is close to copper. Regarding ceramic substrate, wires are pulled by the warpage. Therefore, it is thought that the wire on the non-chip sides still did not break. Thus, it was found that IMB with less warpage greatly extends the life of the wire and dramatically improves durability in power cycling tests at very high temperature (T$_{jmax}$=225°C).

3.4 Clip structure

The Al wire, which was a factor in power cycle test lifetime, was replaced with a structure using Mo-Cu clips to match the CTE of the SiC chip for further improvement with the IMB module. Cross-sectional images are shown in Fig. 9 and Fig. 10.

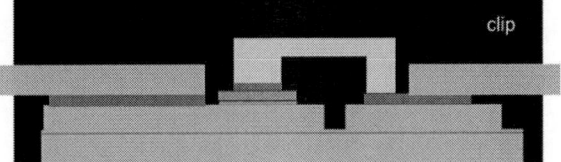

Fig. 9 MCC-IMB Module configuration structure with Mo-Cu Clip

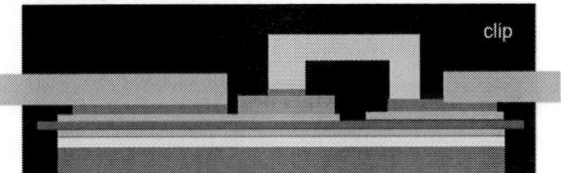

Fig. 10 Si₃N₄ module configuration structure with Mo-Cu Clip

3.4.1 Comparison of thermal resistance be tween Si₃N₄ and IMB

Figure 11 shows the result of thermal resistance measurements of Si₃N₄ and IMB in the clip structure.

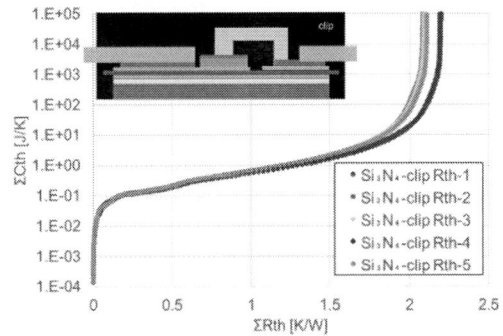

(a)Si₃N₄

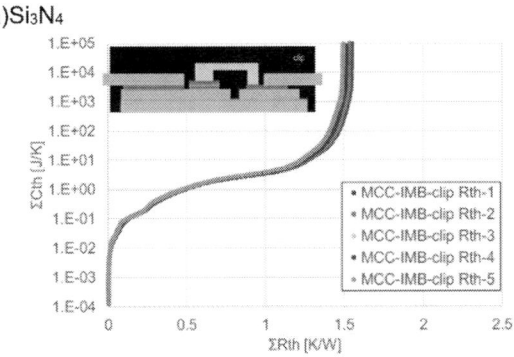

(b)MCC-IMB

Fig. 11 Thermal resistance in clip structure (a) Si₃N₄, (b) MCC-IMB

A similar trend was observed with the wires: the thermal resistance was 1.35 K / W for the Si_3N_4 substrate and 0.65 K / W for the resin substrate. The total thermal resistance of the Si_3N_4 module was 2.2 K / W and that of the resin substrate was 1.55 K / W. The total thermal resistance of the Si_3N_4 substrate varied with the module mounting due to warpage. On the other hand, the resin substrate had a smaller difference in thermal resistance when mounted on the sample due to its smaller warpage. Therefore, compared to Si_3N_4, MCC-IMB showed 52 % reduction in thermal resistance in the form of evaluated module.

3.4.2 Power cycle test at $T_{jmax}225℃$

The power cycle test was performed with IMB only. Test conditions were the same as described above; when the wiring of the IMB module was changed to a Mo-Cu clip structure, durability was improved at T_{jmax} 225°C, and 4/5 of the samples achieved 50,000 PCT cycles (Fig. 12).

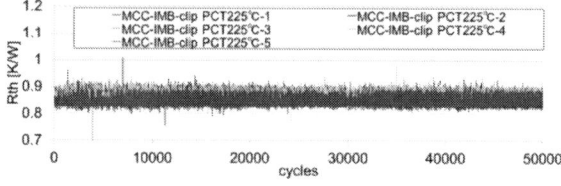

(a) Thermal Resistance

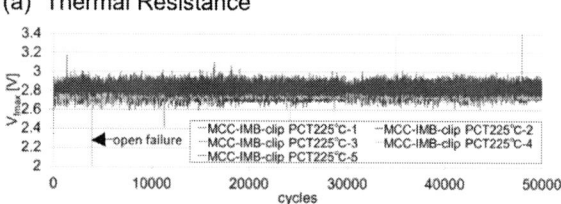

(b) V_{fmax}

Fig. 12 PCT results for IMB module with clip structure, (a) thermal resistance, (b) V_{fmax}

The failed module had a chip open at 3000 cycles, and the fact that the SAT image showed no delamination on the resin substrate suggests that the problem occurred during the mounting process (Fig. 13).

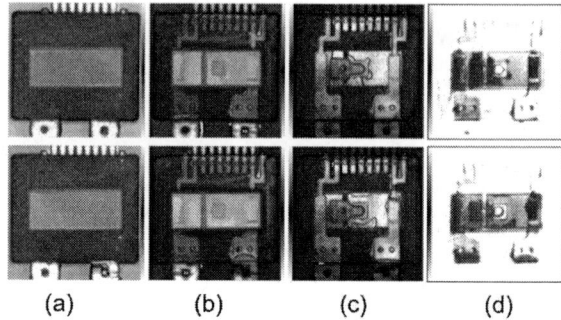

(a)　　　(b)　　　(c)　　　(d)

Fig. 13 SAT images of MCC-IMB-clip PCT225°C-3 before and after PCT. Top: before, bottom: after. (a) Observation of sheet bonding from the lower copper, (b) Observation of silver sinter junction from lower copper, (c) Observation from mold side to upper copper, (d) Transmission image from mold side

3.4.3 Power cycle test at $T_{jmax}250℃$

Next, PCT at T_{jmax} 250°C was performed to obtain further high temperature durability. The test conditions were the same as described above except that T_{jmax} was 250°C. The IMB module structure with clips passed 20,000 cycles (Fig. 14).

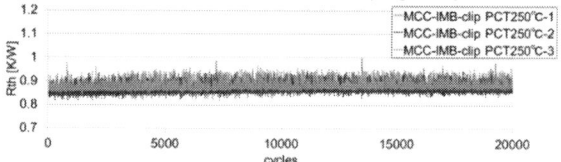

(a) Thermal resistance

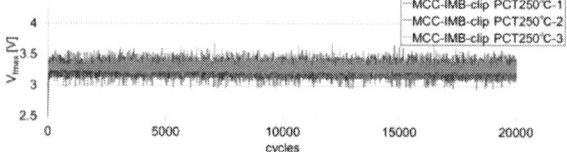

(b) V_{fmax}

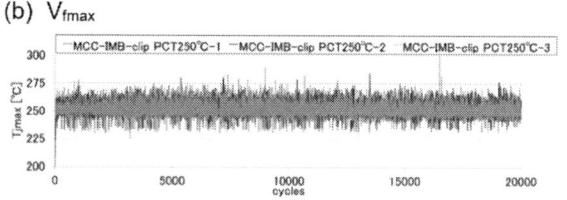

(c) T_{jmax}

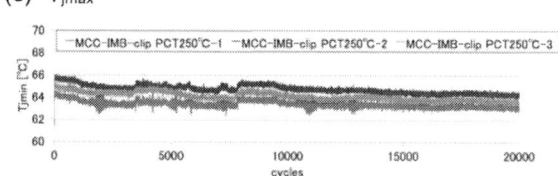

(d) T_{jmin}

Fig. 14 Power cycle test of MCC-IMB with clip

Comparison of thermal resistance before and after the test showed no difference (Fig. 15 and Fig. 16). However, there was some peeling of the mold resin (Fig. 17). The ability to bond with patterned copper needs to be improved in the future.

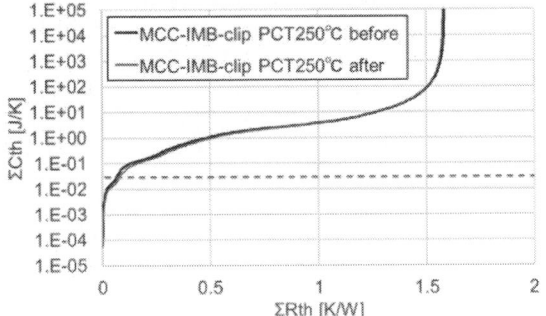

Fig. 15 Structure function comparison before and after power cycle test at Tjmax250°C

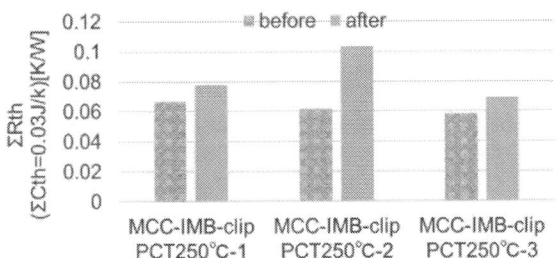

Fig. 16 Comparison of thermal resistance before and after power cycle test at $T_{jmax}250°C$

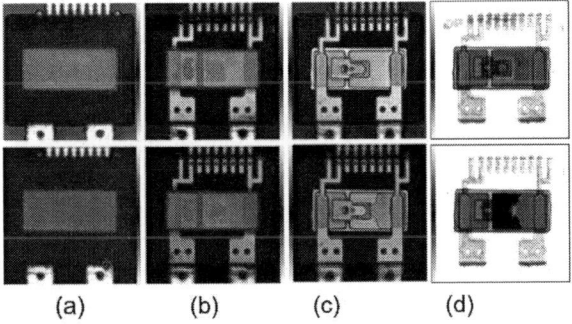

Fig. 17 SAT images before and after power cycle test at T_{jmax} 250°C. Top: before, bottom: after. (a) Observation of sheet bonding from the lower copper, (b) Observation of silver sinter junction from lower copper, (c) Observation from mold side to upper copper, (d) Transmission image from mold side

3.4.4 TCT

Since the clip structure improved resistance to PCT, TCT was performed at setpoints from -40°C to 200°C for 30 minutes each. Evaluated modules were taken out after 100 cycles, 200 cycles, and 500 cycles, and thermal resistance and SAT images were compared. Details are presented by

Yamada et al [4]. Comparison of thermal resistance and structure function showed no degradation up to 200 cycles, but degradation was observed at 500 cycles (Fig. 18 and Fig. 19)

Comparison of SAT images shows no degradation of the resin sheet even after 500 cycles (Fig. 20). Also, no noticeable degradation was observed at the die bonding area. However, delamination was observed at the joint with the encapsulant from 200 cycles. The degradation of thermal resistance was probably due to this delamination. Therefore, it is necessary to improve the bonding with the encapsulant in the future.

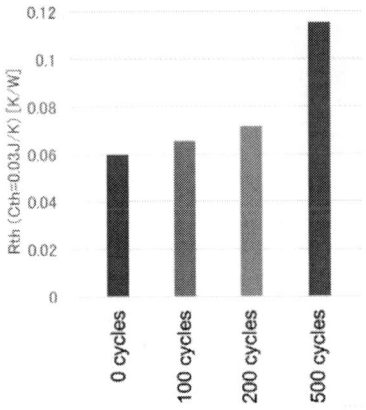

Fig. 18 Comparison of thermal resistance for each cycle of TCT at -40°C/200°C

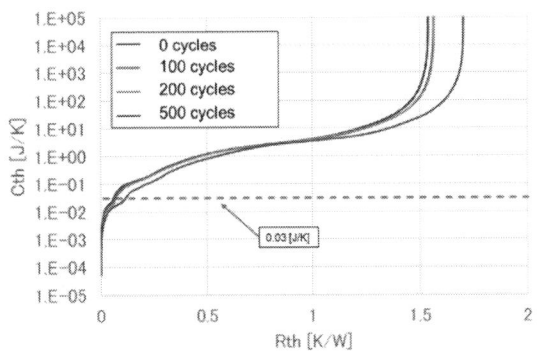

Fig. 19 Comparison of structure functions for each cycle of TCT at -40°C/200°C

(a)   (b)   (c)   (d)

Fig. 20 SAT images at 0, 100, 200 and 500 cycles of TCT under -40°C / 200°C conditions. Top: before, 2nd line: 100cycles, 3rd line: 200cycles bottom: 500cycles. (a) Observation of sheet bonding from the lower copper, (b) Observation of silver sinter junction from lower copper, (c) Observation from mold side to upper copper, (d) Transmission image from mold side

Conclusion

The IMB exhibited high heat dissipation, high breakdown voltage, high elastic modulus at high temperatures, and CTE close to that of copper. Table 7 summarizes the results for the evaluated module. The MCC-IMB analysis showed that the resin substrate had less warpage than the Si_3N_4 substrate, which resulted in less wire breakage and longer life. When the wire part, which was a weak point in the PCT, was changed to a clip structure, the durability of the resin substrate module was improved. Therefore, when T_{jmax} was raised to 250°C and evaluated, it passed 20,000 cycles. However, while the wire structure passed 500 cycles at -40°C / 225°C TCT, the clip structure passed 200 cycles at -40°C / 200°C TCT, but at 500 cycles, delamination of the encapsulating resin and increase in thermal resistance were observed. It leaves further improvements regarding TCT durability. From all the evaluated results in this paper, it can be said that modules using heat-dissipating resin substrates have almost reached the stage of practical application for the 200°C high-temperature operation required for SiC.

Bonding	Spec	Unit	Condition	Si_3N_4	IMB
Structure					
Ag sinter				No delamination	No delamination
Wire	R_{th}	K / W		0.84	0.66
	Input power	A W		60 162	68 191
	PCT	Cycles	Tjmax225°C Δ160°C	12000	33000
	TCT	Cycles	-40°C ~ 225°C	<500	>500
Clip	R_{th}	K / W		1.35	0.65
	PCT	Cycles	Tjmax225°C Δ160°C	-	>50000
	PCT	Cycles	Tjmax250°C Δ185°C	-	>20000
	TCT	Cycles	-40°C~200°C	-	200 OK 500 delamination

Table 7 Summary of module evaluation

Acknowledgments

We would like to thank everyone involved in KA-MOME PJ, the companies involved in mounting materials, and the advisors.

References

[1] K. Hidaka, et al., "Super High Heat Dissipation Resin Sheet by Card House Type BN Filler", PCIM Europe 2023, ISBN 978-3-8007-6091-6

[2] A. Takahashi et al., "Result of 12-year KA-MOME PJ and New Development ", Proceedings of MES2023, pp.11-15(Sep.2023).

[3] JESD51-14, "Transient dual interface test method for the measurement of the thermal resistance junction-to-case of semiconductor devices with heat flow through a single path", JEDEC (2010).

[4] Y. Yamada, et al., "Reliability Study of Packaging Structure of Power Semiconductor Device for High Temperature Operation", Proceedings of MES2023, pp.279-282(Sep.2023).

PCIM Europe 2024, 11– 13 June 2024, Nuremberg   DOI: 10.30420/566262023

Investigating Temperature Dependent Warpage in Metal Ceramic Substrates for Power Electronics Devices

Benjamin Fabian[1], Felix Koser[2], Daniel Schnee[1], Peter Prenosil[1], Marco Müller[1], Sebastian Fritzsche[1]

[1] Heraeus Electronics Deutschland GmbH, Germany

[2] Technische Hochschule Aschaffenburg, Germany

Corresponding author: Dr. Benjamin Fabian, benjamin.fabian@heraeus.com
Speaker: Dr. Benjamin Fabian, benjamin.fabian@heraeus.com

Abstract

This research investigates the temperature-dependent warpage behavior of metal ceramic substrates used in power electronics devices. It examines Active Metal Brazed (AMB) substrates with Silicon Nitride (Si_3N_4) ceramics and silver-based brazing material (Condura®.prime), comparing them to innovative AMB 2.0 substrates (Condura®.ultra). The AMB 2.0 technology addresses silver migration issues by using a Ag-free brazing paste. The study analyzes warpage outcomes of master cards and single units with different copper thicknesses, utilizing a comprehensive statistical method and Finite Element Method (FEM) simulations. It explores the impact of ceramic material variation and processing temperature on AMB substrate warpage. These findings contribute to better component selection and process parameter determinations for power electronics device assembly.

1   Introduction

In the dynamic landscape of power electronics, metal-ceramic substrates play a critical role in the performance and reliability of advanced devices. These substrates form the foundation upon which semiconductor power components are mounted, thus serving as a vital link in the thermal management and electrical insulation of the system. However, during the operation and manufacturing processes, substrates are exposed to elevated temperatures which can induce warpage—a deviation from desired flatness. Warpage compromises the mechanical integrity and thermal performance of power electronic modules by affecting the uniformity of bond layers, potentially leading to increased thermal resistances and stress concentrations.

Additionally, this work will address the emergence of a newly developed silver-free active metal brazed (AMB) substrate, which offers a cost-effective alternative to traditional silver-containing AMB substrates. The comparison of the warpage behavior of these two substrate types under elevated temperatures will be a focal point of this publication. Understanding the warpage characteristics of the silver-free AMB substrate in comparison with its silver-containing counterpart is instrumental in

elucidating its viability for use in power electronic applications.

Predicting the warpage of metal-ceramic substrates is of paramount importance, as it provides a proactive means of ensuring product quality and reliability. By understanding and anticipating this phenomenon, engineers can design and manufacture substrates that maintain their shape and functionality even under harsh operating conditions, during manufacturing and operation. This foresight aids in mitigating the risks of early component failures, which are not only costly but also detrimental to the reputation of product manufacturers.

As power electronics move towards higher power densities and stricter performance specifications, the need to accurately predict and control warpage becomes increasingly critical. Traditional passive inspection approaches are insufficient to capture the complex, transient interactions that occur at elevated temperatures during processes such as soldering, sintering, or power cycling. Thus, there is a strong impetus to develop robust predictive models and simulation tools that can account for the material properties, geometrical designs, and process parameters influencing substrate warpage.

Understanding the warpage behavior under these elevated temperatures is crucial, as it allows for the optimization of the substrate design and processing techniques, enhancing thermal contact and reducing thermomechanical stresses. It also facilitates the selection of appropriate materials and the design of layered structures which exhibit minimal deformation, thereby extending the service life of power electronic components.

Furthermore, the development of the silver-free AMB substrate presents a significant advancement, particularly in cost-sensitive industries where the cost of silver has been a limiting factor in adopting advanced substrates. Additionally, the environmental benefits of reducing silver usage further underscore the importance of evaluating and understanding the performance of this alternative material under elevated process temperatures.

In this paper, we explore the significance of warpage prediction for metal-ceramic substrates in power electronics, while addressing the emerging silver-free active metal brazed (AMB) substrate. We discuss the challenges associated with high-temperature processes and the ramifications of substrate deformation. Additionally, we present the latest advancements in predictive modeling techniques and elucidate their impact on the design, manufacture, and operational stability of power electronic modules. By emphasizing the critical nature of warpage control and the comparison of warpage behavior between silver-containing and silver-free AMB substrates, this publication aims to contribute to the continual improvement of power electronic systems and pave the way for more resilient and efficient technological solutions.

2 Experiment

The characterization of substrate warpage involved a comprehensive exploration of sample layouts and material combinations. Although the substrate's ceramic material and copper remained constant, variations in the thickness of these materials, as well as the dimensions and configurations of the front and back copper sides, were investigated. Specifically, two different thicknesses of ceramic material (250 µm and 320 µm) and three different thicknesses of copper (300 µm, 500 µm, and 800 µm) were studied at both the mastercard (MC) and single unit (SU) levels (refer to Fig. 1). It is important to note that the study was limited to substrates with equal copper thickness on each side, aligning with the common practice among most customers, thereby increasing its relevance.

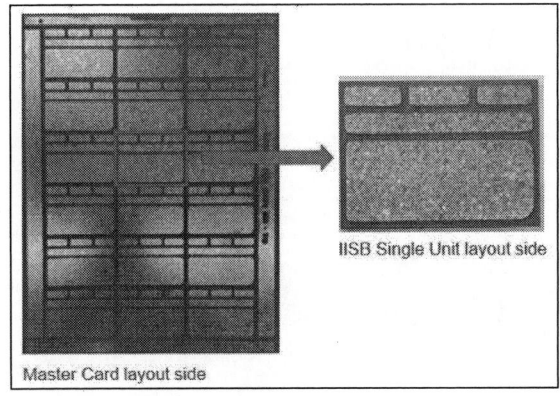

Fig. 1   example for a mastercard and a single unit.

In the assessment of warpage relative to substrate temperature, the AXP 2.0 testing device from Akrometrix, based on the shadow Moire effect, was utilized. The specimens (SU or MC) were positioned on a glass plate within the machine, with heat generated by heated rods beneath the plate. A temperature probe was placed in close proximity to the substrates on the glass plate, allowing for the establishment of specific temperature profiles through controlled heating power (see Fig. 2). The accuracy of temperature measurements and the positioning of the temperature probe were scrutinized and refined in previous work.

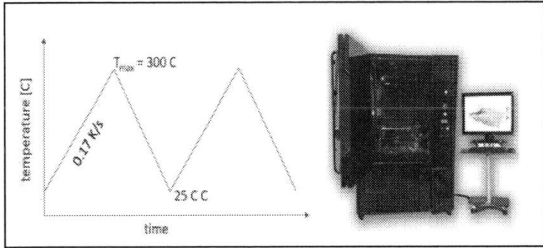

Fig. 2   picture of the AXP 2.0 and the used temperature profile for hotwarpage measurement

The temperature was intentionally cycled from 25°C to 300°C and back to 25°C, repeated twice to simulate typical temperatures encountered during soldering or sintering processes. The chosen temperature range encompasses the conditions prevalent in die attach and baseplate attach procedures. The temperature ramps were executed linearly with a slope of 0.17 K/min, balancing measurement time and sample temperature homogeneity. Throughout the temperature excursions, the machine recorded and stored surface deformation plots (as depicted in Fig. 3) and calculated warpage by determining the difference between maximum and minimum deformations for each sample. Additionally, the shape of each plot at various temperatures was categorized using specific descriptors. Notably, for single units, multiple cards (up to 18) could be measured simultaneously, while for MC, only one card was assessed.

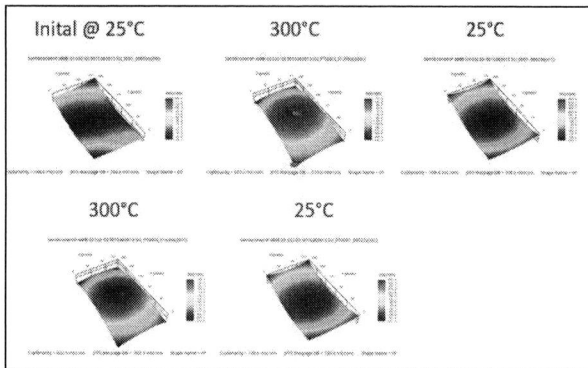

Fig. 3 Typical development of warpage surfaces over temperature while the layout is oriented towards the bottom

3 Data analysis and results

The typical results obtained from the study consist of diagrams illustrating the relationship between warpage and temperature. To derive meaningful insights and correlations between input parameters (such as layout, material thicknesses, etc.) and output parameters (specifically, warpage as a function of temperature), the data must be prepared for subsequent data analysis.

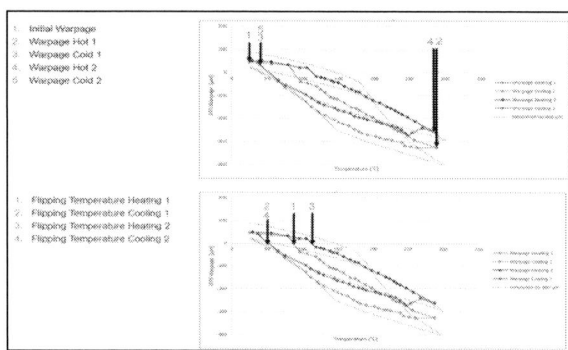

Fig. 4 Hotwarpage curve (Warpage in dependence of temperature) and definition of extraction parameters for statistical data analysis

Given that the typical representation of layout is in the form of a drawing and warpage in relation to temperature is expressed as curves, both forms

are not readily amenable to statistical analysis for correlations. Consequently, the layout and material thickness were distilled into condensed values, namely the total area and volume of the copper on the front and back sides, along with the ratio of these values. Similarly, the warpage curves were deconstructed to facilitate access to specific parameters. This involved segmenting the curves into heating 1 and 2, as well as cooling 1 and 2, to cover each heating or cooling process for the first and second time. Moreover, key metrics such as the maximum and minimum warpage of each branch, and the intersection point with the x-axis (warpge = 0) — referred to as the flipping temperature—were also extracted (refer to Fig. 4). This extraction process was essential to parameterize the curves, rendering them suitable for subsequent statistical analysis.

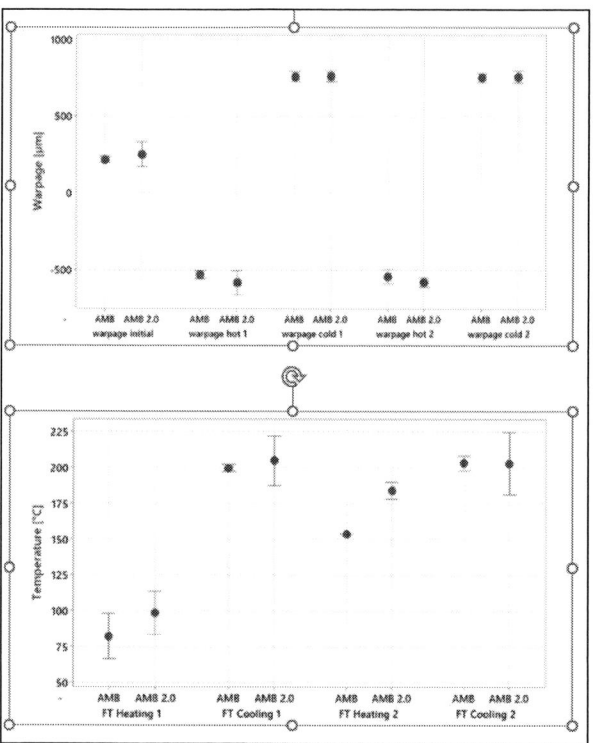

Fig. 5 development of the warpage and flipping temperatures of AMB and AMB2.0 SU in dependence of the temperature profile

A comprehensive analysis of the parameters revealed that the initial warpage is generally lower than the warpage at room temperature after heating, a trend observed for both SU and MC (as evident in Fig. 5, Fig. 6, and Fig. 7). Furthermore, a comparative evaluation of layouts prepared by the DCB process with a zirconia toughened alumina ceramic (ZTA) (Condura®. Extra) and AMB show-cased higher warpage at room temperature and 300°C for AMB compared to DCB (see Fig. 6). This observation led to a comparative study with thermo-mechanical FEM simulation, suggesting a potential correlation with the different coefficients of thermal expansion (CTE) of the ceramics.

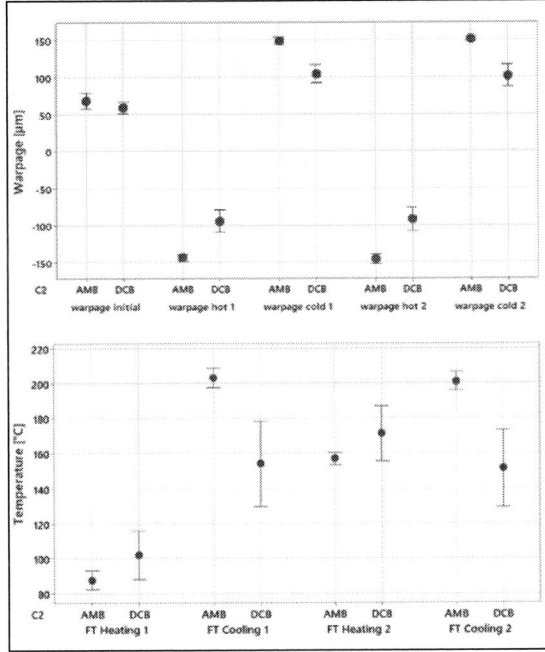

Fig. 6 development of the warpage and flipping temperatures of AMB and ZTA SU in dependence of the temperature profile

Additionally, a direct comparison of AMB and AMB2.0 substrates with identical layout and material combination was conducted on two different layouts. The comparison revealed no significant difference in hotwarpage behavior within the error bars (refer to Fig. 7).

Finally, an investigation into the hotwarpage behavior for varying process temperatures from 150 to 300°C demonstrated that the flipping temperatures during cooling are not inherent material properties, as they change in response to the process temperature (See Fig. 8). This effect was further examined through FEM simulation, suggesting that higher temperatures result in increased substrate warpage and strain, leading to a hardening of the copper. Consequently, the copper exhibits effectively different plastic properties during cooling, resulting in different warpage and flipping temperatures.

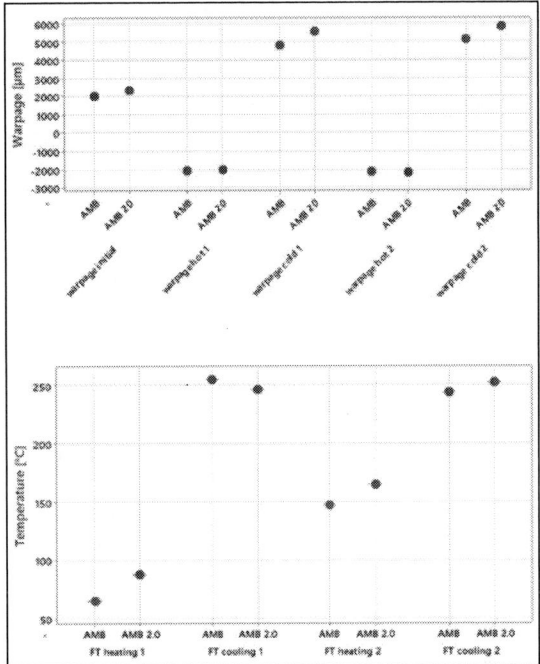

Fig. 7 development of the warpage and flipping temperatures of AMB and AMB2.0 MC in dependence of the temperature profile

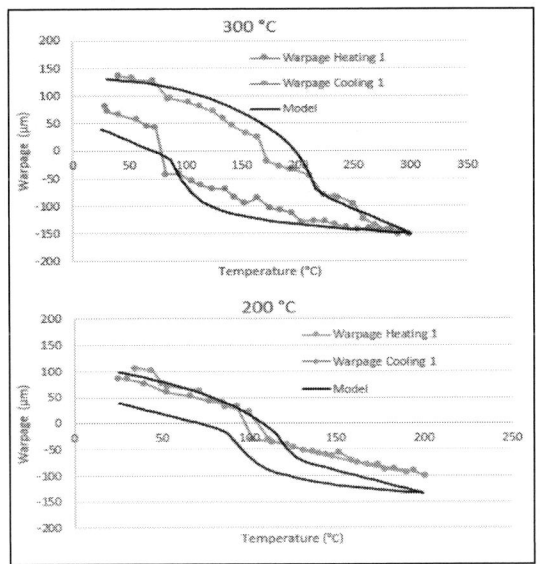

Fig. 8 influence of maximum heating temperature on the hotwarpage behaviour

4 Simulation

To gain further insight, thermo-mechanical simulations were conducted using the finite element method in ANSYS Mechanical to simulate the non-linear deformation of substrates (SU and MC).

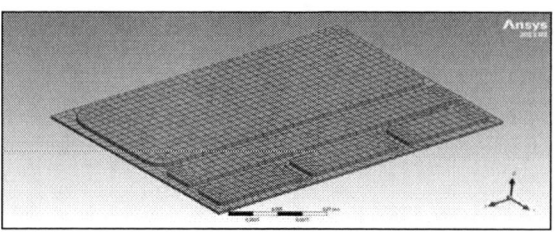

Fig. 9 example of a meshed geometry of a substrate

The geometry was constructed in Spaceclaim and subsequently imported into Mechanical, where it underwent meshing with a granularity of approximately 1 mm (refer to Fig. 9). To facilitate stable convergence and unrestricted substrate movement, a three-point support was implemented. Although explicit heat transport within the materials was not simulated, a homogeneous temperature variation over time was assumed. The experimental temperature profile, encompassing heating up to 300°C and cooling down to 25°C, was utilized. Additionally, the simulation process integrated the cooling of the bonding process from 1000°C to room temperature to account for isotropic hardening of the copper at the interface, arising from disparate coefficients of thermal expansion between the copper and the ceramic, giving rise to internal stresses. Subsequently, the stiffness of the copper grooves was significantly reduced using a kill-command to simulate the etching process of the copper grooves.

Material	E [GPa]	CTE [ppm/K]	Yield stress [MPa]	Plasticity model	SFT [C]
Cu Model 1	[2] [a]	17	15 [b]	MIH	1000
Cu Model 2	[2] [a]	17	15 [b]	MIH	700
Cu Model 3	120	17	10 [c]	MIH	1000
Cu Model 4	120	17	26 [d]	MIH	1000
Cu Model 5	[2] [a]	[2] [a]	[2] [a]	VOCE	1000
Cu Model 6	120	17	15 [b]	MIH	1000
Si3N4	300	3.4	-	-	-

Table 1 overview of the material data used for copper and ceramic simulation, SFT as stress free temperature & MIH as multilinear isotropic hardening, [a]: T-Dependent, [b]: PNA404, [c]: Cu-SE, [d]: DCB-Cu

The ceramic material, specifically Si_3N_4, was characterized as linear elastic, utilizing data from Toshiba [1]. Conversely, the copper was modeled using nonlinear approaches, incorporating different forms of isotropic hardening models such as multilinear descriptions or the Voce model. These models of copper also accounted for potential variations in coefficients of thermal expansion (CTE) or Young's modulus (E), including their temperature dependencies. Consequently, six distinct copper models were employed and cross-validated with the experimental results [refer to Table 1]. Importantly, all the copper models were rooted in empirical measurements, capturing the annealed state at 1000°C before the cooling process.

5 Model extraction and comparison with experimental results

The experimental results underwent comprehensive analysis using statistical methods within the Minitab and Optislang software programs. Directly accessible substrate parameters, including the total area of the ceramic, copper thickness, area of the top and bottom sides, and the ratio of the top and bottom side areas or volumes, were scrutinized for correlations with different warpage outcomes (initial, hot, and cold states) and flipping temperatures. Correlations were modeled in quadratic form and could include cross-correlations.

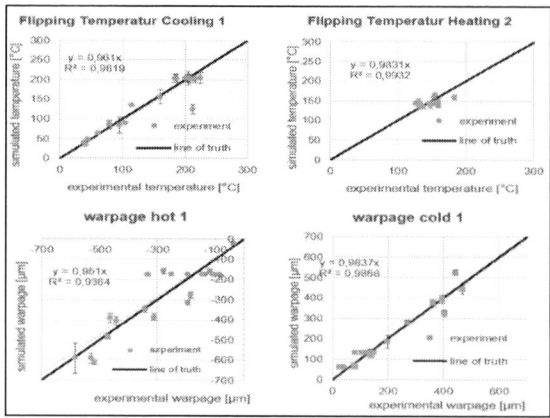

Fig. 10 comparison of measured and calculated parameters using a reduced order model from Minitab for SU

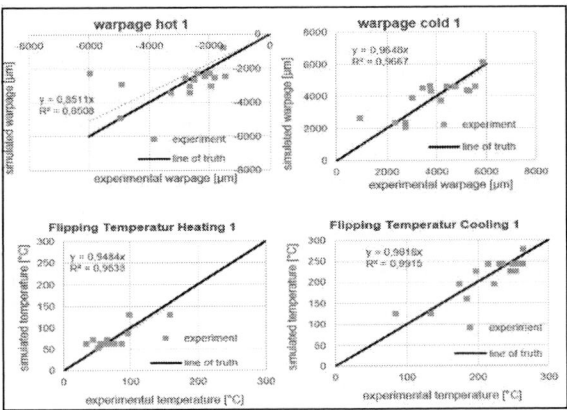

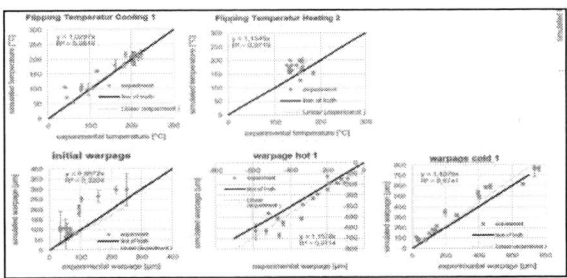

Fig. 11 comparison of measured and calculated parameters using a reduced order model from Minitab for MC

For the SU data, clear correlations were observed for most variables, enabling the calculation of specific warpage and flipping temperatures for a given layout and material combination. Notably, the initial warpage and the first flipping temperature (heating 1) did not exhibit a correlation. Comparing the model predictions to the experimental results for the remaining parameters revealed good prediction quality (refer to Fig. 10). Upon examining the correlation parameters, it was found that the warpage and flipping temperatures for SU were primarily influenced by the copper area ratio of the layout to the backside and the copper thickness.

Similarly, the analysis of the MC results followed the same approach as for the SU. While no correlation was found for the initial warpage, the prediction was deemed good to reasonable for all other parameters (see Fig. 11). The most influential parameters affecting the warpage and flipping temperature for MC were identified as the copper volume ratios of the layout and backside, the area of the copper backside, and the total volume of the MC.

The results from the finite element simulations were also compared with the experimental results, showcasing the prediction quality of the six different copper models when examined against the standard deviation of each warpage value and flipping temperature. These multiple standard deviations were then averaged for each copper model to yield a single quality parameter for comparison.

Fig. 12 comparison of measured and simulated parameters using the FEM method and Model 5 for SU

Notably, five out of six models exhibited similar prediction quality, with only model 4 demonstrating significantly poorer performance (refer to Fig. 14). While some models excelled in predicting flipping temperatures but showed slightly reduced accuracy in warpage prediction, others demonstrated the opposite pattern. Despite this, all five models accurately predicted the substrate's shape and yielded reasonable predictions for warpage and flipping temperatures.

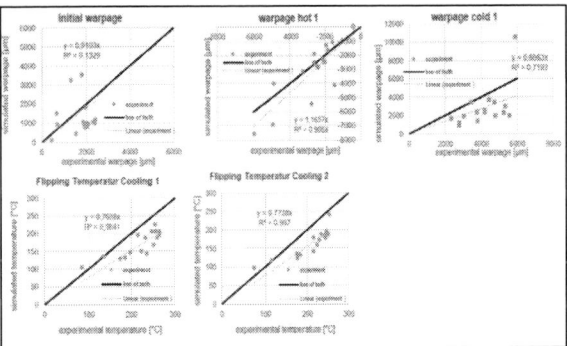

Fig. 13 comparison of measured and simulated parameters using the FEM method and model 1 for MC

Comparison of the simulation data with the MC results indicated that the overall precision of the simulation was not as strong as for SU, suggesting the potential existence of systematic differences for some parameters. This divergence may be explained by factors such as the limited statistics of the experimental results, given that only one MC was measured per layout and material combination, or the presence of physical effects such as time-dependent phenomena like creep, the influence of prewarped ceramics or residual stress, and the impact of inhomogeneous temperature distribution during cooling down. Notably, these effects might be more pronounced in MC due to its larger area and may not be as evident after singularization into SU.

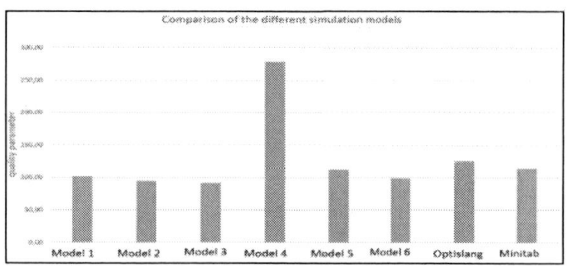

Fig. 14 Overview of the prediction quality for different copper models and reduced order models for SU

6 Conclusion

The analysis of the hotwarpage results revealed that it is possible to predict warpage and flipping temperatures for both SU and MC based on fundamental substrate properties using a reduced order model. However, an accurate prediction was challenging for the initial warpage and the first flipping temperature. Nonetheless, the calculation could be completed within seconds with good accuracy, although it should be noted that the reduced order model can only predict the warpage at a hot state of 300°C.

The finite element simulation also demonstrated a good correlation with the experimental results and even exhibited higher accuracy in predicting the initial warpage and first flipping temperature compared to the reduced order model. Moreover, the simulation was capable of predicting the substrate shapes. It is worth noting that the simulation model could be adapted to specific process temperature profiles with varying temperatures from 300°C, as well as incorporate changes in material properties and mechanical boundaries (e.g., fixation). However, the simulation time is longer, typically taking up to one hour including geometry preparation.

Both methods offer their own benefits and can be applied based on specific needs. These methods can help to determine predicting the warpage behaviour in application processes like large area sintering and soldering where the warpage can play a crucial role. The reduced order model is particularly valuable in enabling customers to evaluate the warpage and flipping temperatures for new layout designs quickly. If more detailed information is required, an additional FEM simulation can provide further insights.

In conclusion, no significant difference in hotwarpage behavior was observed between AMB and AMB2.0 substrates on the same layout. Even through statistical analysis, no distinct trend indicating different behavior between the two products could be discerned.

7 outlook

In the next phase of the study, the model can be expanded by Heraeus Electronics GmbH Deutschland to encompass substrates with different copper thicknesses on each side as well as AlN-based ceramics. Additionally, if there is demand from customers, the focus can be directed towards extending the model to DBC substrates with Al_2O_3 (Condura®.classic) and ZTA ceramics. The feasibility of implementing the reduced order model as a quality benchmark tool for new layouts will be explored, as it has the potential to assist in achieving the desired warpage behavior in response to customer requirements.

References

[1] https://www.toshiba-tmat.co.jp/en/product/ce_sin_plain.htm

[2] Patrick Gaiser, Markus Klingler, Jürgen Wilde, The influence of strain hardening of copper on the crack path in Cu/Al2O3/Cu direct bonded copper substrates, International Journal of Fatigue, Volume 140, 2020,105821, ISSN 0142-1123.

PCIM Europe 2024, 11– 13 June 2024, Nuremberg

DOI: 10.30420/566262024

Degradation Mode Analysis of Different Bonding Technologies of SiC Power Semiconductors Stressed by Active Power Cycling

Rasched Sankari[1,2], Ulrich Keßler[1], Martin Rittner[1], Manfred Reinold[1], Thomas Kaden[1], Emilia Schwindt[1], Martin Schneider-Ramelow[2,3]

[1] Robert Bosch GmbH, Corporate Sector Research and Advance Engineering, Germany
[2] Technical University Berlin, Germany
[3] Fraunhofer-Institute for Reliability and Microintegration IZM, Germany

Corresponding author: Rasched Sankari, Rasched.Sankari@de.bosch.com
Speaker: Rasched Sankari, Rasched.Sankari@de.bosch.com

Abstract

SiC is being used as a new generation of power semiconductor to meet the increasing demands of modern electrified automotive powertrains. The standard AIT technology is Al-plated Si semiconductor wirebonded with Al wire bonding as the topside contact. With the establishment of highly reliable Ag-sinter die-attach technology, the Al bond wire has come into focus as the weakest component in the assembly. The typical and well-studied failure mode is a crack located near the interface between the Al bond wire and the Al metallization. A Cu-plated SiC with Cu bond wires is used to improve the robustness and reliability of the topside contact. This study compares the degradation mode of the Al-plated SiC with Al bond wire to a Cu-plated SiC with Cu bond wire in terms of degradation modes. The SiC MOSFETs investigated are plated with 4 μm Al or 30 μm Cu metallization and wirebonded on top with 300 μm Al or Cu bond wires. To evaluate the reliability of the devices, they are subjected to various stress configurations in active power cycling tests. Investigations of the Cu system showed that the Cu topside is more reliable than the Al topside contact and that the degree of oxidation affects the degradation mode of the Cu-AIT.

1   Introduction

An important subsystem of electric vehicle technology is the electrified powertrain. The inverter, with its power semiconductors, is a key element in the performance of the powertrain [1].

The growing performance demands of electrified vehicles require higher power density, high thermal performance, and high voltage operation. While silicon (Si) has been the traditional power semiconductor material, it can no longer meet these increasing demands. Silicon carbide (SiC), on the other hand, has the ability to meet these requirements and enable high voltage and high temperature applications due to its higher critical field strength and thermal conductivity compared to Si. However, the higher Young's modulus of SiC, which is three times that of Si, requires a focus on robust assembly and interconnection technology (AIT) [2].

Before a new SiC power module is released for production, the reliability of the power electronic components must be verified. Reliability is assessed by the active power cycling (APC) test, which determines the resulting number of cycles [1].

In the past, power modules were typically designed using the following AIT: Si power semiconductors are soldered to a ceramic power substrate (DBC or AMB) and wirebonded at the top by Al bond wires. The assembly is then soldered or contacted by a thermal interface material (TIM) to a baseplate or cooler. The entire module is either hard cast or transfermolded beforehand or soft encapsulated afterwards. The components in this design concept that limit reliability are the lead-free solder joints and the Al bond wires. The Al bond wire has been identified as the weakest part in terms of failure when using the more reliable Ag-sinter die-attach technology. This failure has been investigated by experiments and simulations [3]–[5]. Another factor influencing the reliability of thick Al bond wire is the bond wire geometry, as shown in [6].

An alternative to Al wire bonding to improve reliability is Cu wire bonding technology. Among other applications, Cu bond wires have been investigated as topside contacts in the Cu Danfoss Bond Buffer (DBB). The Cu DBB is sintered onto an Al-plated chip with a Ni-Au finish, and the Cu bond wires are wirebonded to the Cu pad. Although this structure is an improvement over the Al wire bonding structure, it provides only a factor of two improvement in reliability. The observed failure mode is the formation of a crack in the Al metallization [7]. It is clear that an Al-free system is required to significantly increase reliability. This study compares the reliability of the design using Al-plated SiC with Al bond wire technology to the Al-free Cu-plated SiC with Cu bond wire technology. The full Cu-AIT should provide a significant increase in reliability compared to the Al-AIT. This paper discusses these variants in terms of their degradation modes.

2 Device Under Test (DUT)

In the schematic, Fig. 1, a 270 µm thick SiC MOSFET is connected to the AMB substrate (Cu-Si$_3$N$_4$-Cu) by a sintered Ag layer. The SiC die is plated with 30 µm Cu and wirebonded with four 300 µm Cu bond wires. In contrast to the Al system, the Cu system is additionally soldered to a Cu baseplate using SAC solder and connected to the heatsink via a TIM. This structure is unsuitable for cycling the topside contact in the APC until the end of its life, as the TIM connection to the cooler degrades much faster than the chip-AIT. However, since the degradation mode is of primary importance for this study, the devices are not cycled to End of Life (EoL), but are removed and analyzed by electrical tests and cross sections according to defined cycle numbers.

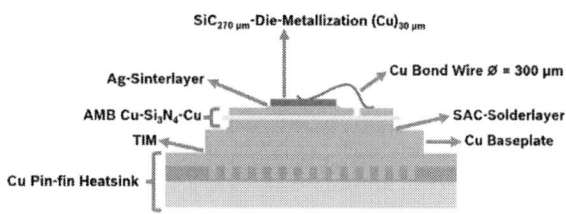

Fig. 1: Heatsink setup with 300 µm Cu bond wires on a 30 µm thick Cu-plated SiC die.

The schematic Fig. 2 shows an 180 µm thick SiC MOSFET plated with 4 µm Al metallization and wirebonded with seven 300 µm Al bond wires. The Al-plated SiC has a 24% larger chip area compared to the Cu-plated SiC. Based on the previously described cooler setup with TIM and the unstable thermal path of the Cu-AIT, the Al-AIT is cycled

with a direct cooler concept. The direct cooling of the AMB allows for the design without a TIM, Cu baseplate, or SAC-solderlayer. This results in a more stable thermal and electrical behavior during APC, allowing the Al-AIT to be cycled until EoL. The shorter thermal path with less components enables a shorter cycle time of $5\ s$ instead of $14\ s$ for the Cu system.

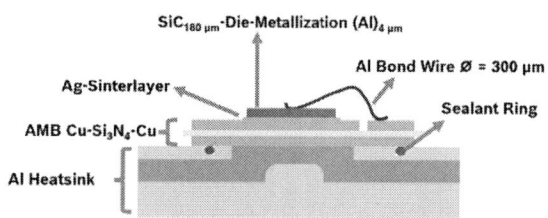

Fig. 2: Heatsink setup with 300 µm Al bond wires on a 4 µm thick Al-plated SiC die.

Fig. 3 shows the setups, the APC boundary conditions, and their analyses.

3 Measurement Setup

3.1 Active Power Cycling (APC)

During the APC, the devices are subjected to a temperature swing, turning the current on and off, resulting in repetitive heating of the device. The temperature is determined by measuring the voltage of the inverse diode, which acts as a temperature-sensitive electrical parameter during the cooling cycle.

The temperature swing ΔT causes thermomechanical stress that leads to degradation and failure of the device. Module degradation is always accompanied by changes in electrical parameters. According to [8], [9], failure during APC is defined as an increase of 5% V_{DS} or 20% R_{th}. The devices are subjected to temperature variations within the range of $T_{j,\ min}$ = 45, 55, and 65°C and $T_{j,\ max}$ = 165, 175, and 195°C, with a temperature swing of $\Delta T = 110, 130$, and $150\ K$. To verify the temperature distribution, an APC with $\Delta T = 80K$ and a thermal camera was carried out. As illustrated in Fig. 4, the infrared measurement reveals that the SiC chip surface exhibits a slightly inhomogeneous temperature distribution. At an average temperature of $T_{j,\ max} = 160°C$, there is a temperature difference of $5\ K$ between the bond wire feet #2 and #6. Bond wire #2 experiences the highest temperature on the chip surface and is subjected to the

	Boundary Conditions	**Analyses**	**Setup**
Al-System	t_{on} = 2 s, t_{off} = 3 s ΔT = 110 K, 130 K, 150 K	Lifetime modelling considering EoL Criteria [8,9]	Al Heatsink – AMB $SiC_{180 \mu m}$-Die-Metallization $(Al)_{4 \mu m}$ Al Bond Wire $\varnothing$ = 300 µm
Cu-System	t_{on} = 3 s, t_{off} = 11 s ΔT = 130 K, 150 K	Degradation mode analysis by defined number of cycles without consideration of EoL	Cu Pin-fin Heatsink – TIM Cu Baseplate – AMB $SiC_{270 \mu m}$-Die-Metallization $(Cu)_{30 \mu m}$ Cu Bond Wire $\varnothing$ = 300 µm

Fig. 3: Setups were tested with their APC boundary conditions.

highest thermomechanical stress compared to the other bond wire feet. However, bond wire #2 has a small difference of $160.2°C$ from $T_{j, max} = 160.0°C$, so the temperatures are considered to be in the acceptable range as measured with the inverse diode.

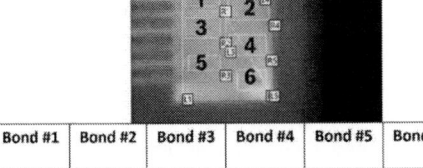

Bond #1	Bond #2	Bond #3	Bond #4	Bond #5	Bond #6	$T_{j, max}$
159.2°C	160.2°C	158.0°C	157.8°C	155.2°C	155.0°C	160.0°C

Fig. 4: Temperature distribution in APC measured by infrared camera at $80\ K$ temperature swing $(T_{j, min} = 80°C,\ T_{j, max} = 160°C)$ from the bond wire feet compared to the averaged T_j.

3.2 Encapsulation

A module is often cast or transfermolded to protect it from the environment [4], [5]. To study the degradation mode for the Cu and Al systems, a non-encapsulated variant is selected, as shown in Fig. 5A. A silicone soft potting is also used as a reference for a realistic encapsulation of a potential future product, as shown in Fig. 5B. To exclude oxidation, a N_2 flooded cap is used Fig. 5C. The N_2 flow ensures that the elements of the Cu-AIT do not visibly oxidize. The Al system is also equipped with a cap that allows the test area to be flooded with N_2.

3.3 Degradation Mode Al

The degradation of the Al-plated SiC MOSFET results in changes in semiconductor characteristics, which in turn probably influence the functionality of

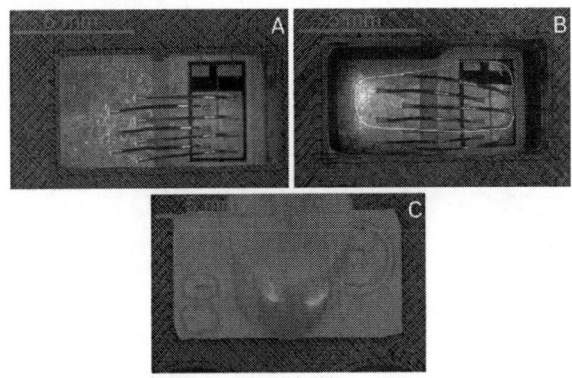

Fig. 5: Non-encapsulated DUT (A), with frame and soft potting (B), and with N_2 flooded cap (C).

the device. The Al system at all three temperature swings, regardless of the encapsulation method, $\Delta T = 110, 130, 150\ K$, shows a sudden increase in $V_{DS} > 5\%$, which is defined by [8], [9] as an EoL criterion. An increase in V_{DS} indicates damage to the topside contact, such as degradation of the weld zone between the Al bond wire foot and the Al metallization. Fig. 6 illustrates the normalized V_{DS} increase of the Al DUTs at $\Delta T = 130\ K$. The curve of DUT A1 clearly demonstrates that the EoL criterion of $V_{DS} > 5\%$ is reached within a very short number of cycles, indicating a bond wire lift-off.

Six Al DUTs are tested non-encapsulated at a ΔT of $110, 130,$ and $150\ K$. The failure mode is consistently a bond wire lift-off at bond position 2, shown in red in Fig. 7. The electrical parameter change described above correlates with the bond wire lift-off at position 2 for all temperature swings. A correlation can be observed between the first bond wire to fail, and the temperature distribution measured by infrared on the chip surface (see Fig. 4). Thus, the hottest bond wire is found to fail at bond wire position 2.

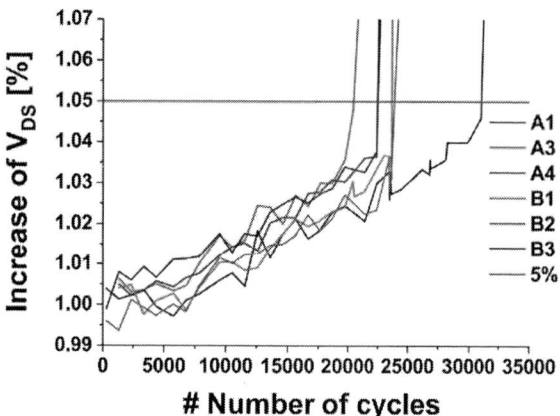

Fig. 6: Normalized V_{DS} increase of non-encapsulated Al DUTs at $\Delta T = 130\ K$.

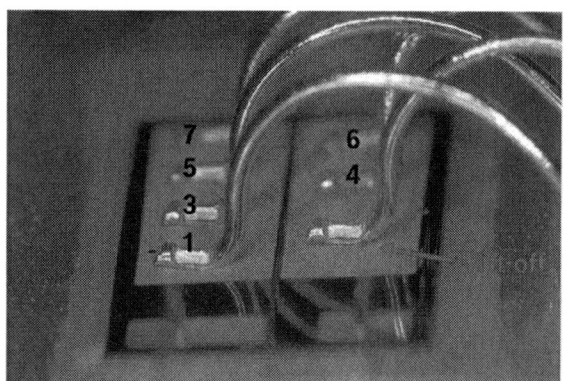

Fig. 7: Non-encapsulated Al bond wire lift-off after APC $30,000$ cycles at $\Delta T = 130\ K$.

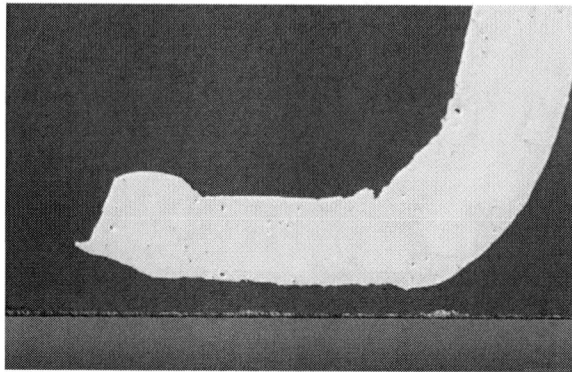

Fig. 8: Cross section of a non-encapsulated Al DUT at bond wire position 2 with a lift-off as a failure after $32,000$ cycles at $\Delta T = 130\ K$.

Fig. 8 shows a cross section of a non-encapsulated Al DUT after $32,000$ cycles at $\Delta T = 130\ K$. The bond wire lift-off that occurs corresponds to the failure mode described in the literature. Thermomechanical stresses and strains due to the mismatch of the coefficient of thermal expansion (CTE) occur in the bond area. According to [10], [11], the periodic plastic strains in metals cause the crack to propagate along the lower end of the bond foot as well as partially along the interface, resulting in a bond wire lift-off. The lift-off results in a higher current density, which leads to a higher temperature load in the remaining bond wires, thus accelerating the degradation process. This behavior can be measured in the electrical measurements during the APC. The crack starts at the complete wedge of the weld in the bond at the outline of the foot and then propagates inward along the interface between the Al bond wire and the Al metallization [12].

Fig. 9A depicts the center section of an EBSD analysis of a 300 µm Al bond wire and a 4 µm thick Al metallization. In this measurement, the Al bond wire exhibits a weak (111) texture on the y-axis. At the transition to the Al metallization, the measurement shows only a weak texture between the (110) and (111) textures on the x-axis. The sample displays a relatively coarse, partially globular grain to the left and right of the bond wire. There are few coarse grains in the bond area; most of the grains are very fine and globular (recrystallized). The greatest recrystallization and grain disorientation are observed in the interface region of the Al metallization, as illustrated in Fig. 9B. This is where the microstructure is more influenced by the bonding process, resulting in the formation of a polycrystalline microstructure. The results of this study are consistent with the findings of [3]. As described in reference [3], the crack is identified as propagating along the area close to the interface between the Al wire bonding and the AlSiCu metallization.

With the bond wire lift-off and the correlating EoL criterion, the lifetime analysis of Al DUTs is summarized in Fig. 10. For the non-encapsulated variant, a connecting line as a guide for the eye between data points is created by analyzing the Weibull distribution according to the Mean Time to Failure (MTTF) value, which reflects the characteristic lifetime. The lifetime of the non-encapsulated Al system decreases by approximately $10,000$ cycles as ΔT increases by $20\ K$. An APC run was performed on three DUTs each at $\Delta T = 130\ K$ with soft encapsulation, non-encapsulation, and with a N_2 flooded cap. The failure mode for all

PCIM Europe 2024, 11–13 June 2024, Nuremberg DOI: 10.30420/566262024

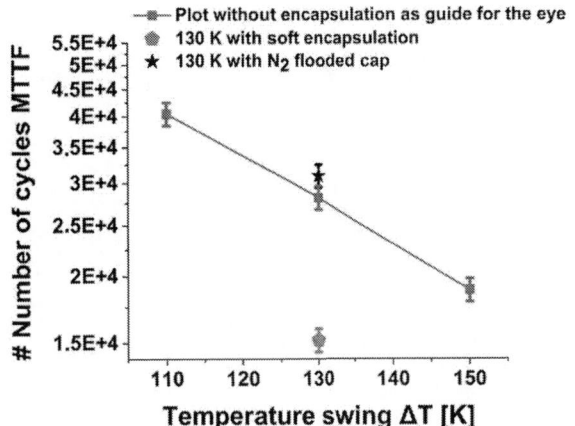

Fig. 10: MTTF vs. temperature swing of the Al setup connecting line as a guide for the eye between data points without encapsulation and data points of Al DUTs at $\Delta T = 130 \ K$ with soft encapsulation and with N_2 flooded cap.

lifetime between the soft encapsulation variant and the variant without encapsulation. It is assumed at this point that no suitable silicone gel was identified that met the specified boundary conditions of our APC tests. Another possible issue could be the hardening of the gel during the APC test, which has a thermomechanical impact on the topside contacts, resulting in higher stress and potentially reducing the lifetime.

3.4 Degradation Mode Cu

After a temperature swing of $150 \ K$ ($T_{j, \ min} = 45 \ °C$, $T_{j, \ max} = 195°C$), the Cu-AlT shows significant differences in the degree of oxidation, as shown in Fig. 11, depending on their encapsulation variants. The non-encapsulated method (left) is subject to heavy oxidation on the chip surface and on the bond wires. On the other hand, the soft potting variant (center) shows light oxidation of the chip surface and the bond wires. The N_2 cap variant (right) shows no visible oxidation on the chip surface or on the bond wires.

Fig. 12 compares the crack propagation of three variants after APC at a temperature swing of $150 \ K$ and $35,000$ cycles. Fig. 12A shows the non-encapsulated variant. A darker zone around the crack is clearly visible in the SEM image (Fig. 12B), indicating CuO. This variant shows 30 μm vertical crack propagation. For the die with soft encapsulation (Fig. 12C), a comparable type of vertical crack propagation with a length of 19 μm is observed. It seems that the degree of oxidation affects how fast

Fig. 9: Microstructure after bond wire process in the center section (inverse pole figure (IPF) map of bond wire (A) and interface near area (B)).

variants is a bond wire lift-off at bond position 2. From Fig. 10, it is clear that the N_2 flooded cap variant is close to the lifetime curve for the Al DUTs. The small discrepancy between the connecting line and the two data points is probably due to the statistical distribution of the three DUTs. In contrast, the soft encapsulation variant exhibits a lifetime of $15,175$ cycles when analyzed according to MTTF. This represents a 46% reduction in

201

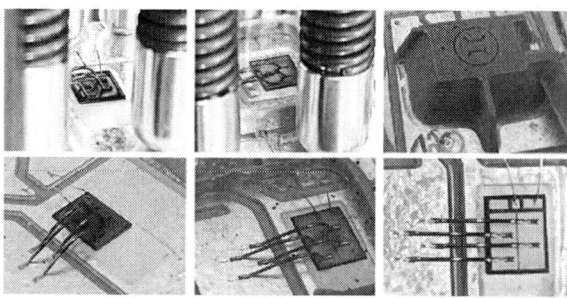

Fig. 11: Cu-AIT DUTs after $35,000$ cycles at $150\ K$ APC, without encapsulation (left), with frame and soft potting (center), and with a N_2 flooded cap (right).

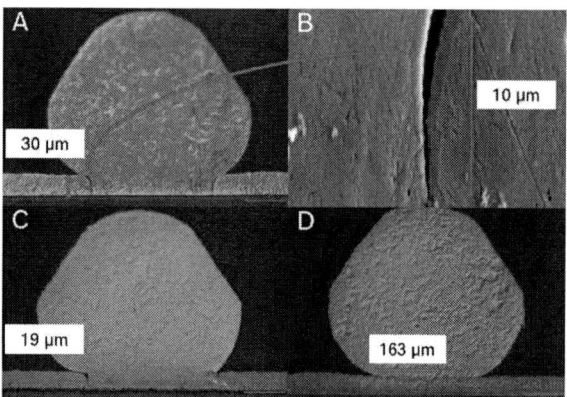

Fig. 12: Cross section of Cu DUTs after $35,000$ cycles at $\Delta T = 150\ K$, without encapsulation (A), SED-SEM image (B), with soft potting (C), and with N_2 flooded cap (D).

Fig. 13: Cross section of Cu DUTs after $100,000$ cycles at $\Delta T = 150\ K$, with N_2 flooded cap (A) and non-encapsulated (B).

the crack propagates. For the N_2-flooded variant, a different mode of degradation can be observed in the form of a horizontal crack, as illustrated in Fig. 12D. By suppressing the oxidation of the copper, horizontal crack propagation is recognized.

As outlined by Kuźnicka and Junik [13], crack propagation follows the oxide cracking mechanism, whereby local oxidation of the copper (stress corrosion cracking, SCC) occurs. The stress level and distribution determine the crack path and propagation, and thus the fracture of the oxide layer and crack opening.

A bond wire lift-off can be observed for APC in an N_2 atmosphere after $100,000$ cycles (Fig. 13A). It is assumed that when oxidation is excluded, the Cu metallization beneath the bond is more robust and allows greater plastic deformation than the interface area with a polycrystalline microstructure. As a result, the crack propagates horizontally through the interface between the 243 µm thick bond foot

and the 30 µm thick Cu metallization, with vertical branches into the metallization. After bond wire lift-off, the MOSFET is considered non-functional.

The results of long-term non-encapsulated APC testing after $100,000$ cycles, as shown in Fig. 13B, indicate that the vertical crack stops at the barrier, with the formation of horizontal microcracks at the transition to the SiC material. However, these cracks have not yet affected the functionality of the MOSFET. The same degradation mode after $100,000$ cycles at $\Delta T = 150\ K$ is seen in the soft encapsulation variant.

The same non-encapsulated Cu-AIT is tested at $\Delta T = 130\ K$. The DUTs are not cycled to EoL due to the degradation of the heatsink connection (TIM). The test had to be stopped after $370,000$ cycles, with all samples still being electrically functional. In more detail, the sample exhibits vertical cracks with darker zones surrounding the crack, indicative of CuO that has been stopped at the barrier (see Fig. 13B).

Fig. 14 shows the plot of the number of power cycles of non-encapsulated Al DUTs, with consideration of MTTF with a bond wire lift-off. It also presents the data points from the Cu setup at temperature swings of $\Delta T = 130$ ($370,000$ cycles) and $150\ K$ ($100,000$ cycles). The Cu setup failed due to the degradation of the heatsink connection (TIM), so the actual number of cycles to EoL is probably considerably higher. The highly reliable topside contacting of the Cu-AIT leads to a failure in the

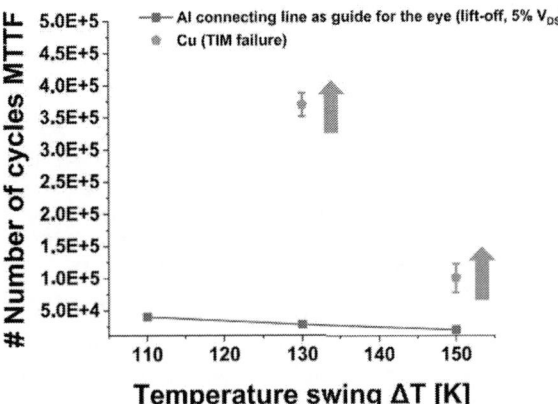

Fig. 14: MTTF vs. temperature swing of non-encapsulated samples. Al setup connecting line as guide for the eye between data points. The Cu setup failed due to the degradation of the TIM foil.

thermal path. The unstable thermal path in this setup does not allow cycling with consideration of EoL criteria. After the failure, the DUTs are still electrically functional, despite the clear cracks. From Fig. 14 a preliminary reliability assessment can be recognized. At $\Delta T = 130$ and $150\ K$, the Cu-AIT shows at least 13 and 5 times higher APC-reliability than Al-AIT.

4 Conclusion

The present study examines the degradation phenomenon of Al 300 µm and Cu 300 µm bond wires on SiC power semiconductors with 4 µm Al and 30 µm Cu metallization in APC. The bond wire with the highest temperature, as measured via infrared measurements, is also the first bond wire to fail in the APC. The results of the degradation of the Al-AIT are consistent with those reported in the literature. As a failure mode at temperature swings of $\Delta T = 110, 130$ and $150\ K$, a bond wire lift-off always occurred on the same bond wire, regardless of the loading conditions.

In conclusion, a novel degradation mode with non-encapsulated complete Cu-AIT has been identified on SiC with 30 µm Cu metallization and 300 µm Cu bond wire, which is vertical cracking. This same degradation mode can be observed in the variant with soft encapsulation. This phenomenon is likely a result of a reduction in hardness due to oxide formation, which creates a stress situation in the metallization, leading to vertical cracking. Under a N_2 atmosphere, the metallization is more robust and allows for greater plastic deformation compared to the interface. This results in a horizontal near-interface crack, which can lead to lift-off. Long-term non-encapsulated APC results after $100,000$ cycles indicate that the crack stops at the barrier, with the formation of horizontal microcracks at the transition to the SiC material. However, these cracks have not yet affected the functionality of the SiC MOSFET. A preliminary reliability assessment can be conducted. At a temperature swing of $\Delta T = 130$ and $150\ K$, the Cu-AIT exhibits at least a 13 and 5 times higher APC-reliability than the Al-AIT. The full topside Cu-AIT has a robust topside contact, which suggests that other AIT components such as die attach or heatsink connection are more likely to fail. To date, the topside contact has not been provoked to an EoL failure. The topside Cu-AIT lifetime data reported in this study demonstrate a strong upward trend, which cannot yet be quantified with EoL criteria. Current measurements of the Cu-AIT run with the direct cooling concept, which was described in Chapter 2.

Acknowledgement

The supply of chip samples for this paper was supported by the H2020 KDT JU program of the European Union under the grant of the TRANSFORM project 'Trusted European SiC Value Chain for a greener Economy' - KDT Grant No. 101007237 -, and the German Federal Ministry of Education and Research under grant number 16MEE0127K. The authors want to say 'Thanks' to all supporting institutions and all Transform project partners.

References

[1] M. Ciappa, "Lifetime prediction on the base of mission profiles," *Microelectronics Reliability*, vol. 45, no. 9, pp. 1293–1298, 2005, Proceedings of the 16th European Symposium on Reliability of Electron Devices, Failure Physics and Analysis. DOI: https://doi.org/10.1016/j.microel.2005.07.060.

[2] R. Brown, A. Al-Khalidi, D. Macfarlane, S. Taking, G. Ternent, *et al.*, "Novel high performance algan/gan based enhancement-mode metal-oxide semiconductor high electron mobility transistor," *physica status solidi (c)*, vol. 11, Feb. 2014. DOI: 10.1002/pssc.201300179.

[3] M. Broll, U. Geissler, J. Höfer, S. Schmitz, O. Wittler, *et al.*, "Correlation between mechanical properties and microstructure of different aluminum wire qualities after ultrasonic bonding," *Microelectronics Reliability*, vol. 55, no. 9, pp. 1855–1860,

2015, Proceedings of the 26th European Symposium on Reliability of Electron Devices, Failure Physics and Analysis. DOI: https://doi.org/10.1016/j.microrel.2015.06.061.

[4] U. Geissler, M. Schneider-Ramelow, and H. Reichl, "Hardening and softening in alsi1 bond contacts during ultrasonic wire bonding," *Components and Packaging Technologies, IEEE Transactions on*, vol. 32, pp. 794–799, Jan. 2010. DOI: 10.1109/TCAPT.2008.2009930.

[5] M. Schneider-Ramelow and C. Ehrhardt, "The reliability of wire bonding using ag and al," *Microelectronics Reliability*, vol. 63, pp. 336–341, 2016. DOI: https://doi.org/10.1016/j.microrel.2016.05.009.

[6] B. Kilian, Y. Maniar, J. Gleichauf, O. Wittler, and M. Schneider-Ramelow, "Influence of Geometry Effects on Thermo-Mechanical Reliability of Aluminum Bond Wires in Discrete SiC MOSFETs," in *2024 25th International Conference on Thermal, Mechanical and Multi-Physics Simulation and Experiments in Microelectronics and Microsystems (EuroSimE)*, Catania, Italy: IEEE, Apr. 2024, pp. 1–10. DOI: 10.1109/EuroSimE60745.2024.10491463.

[7] A. Streibel, M. Becker, O. Muehlfeld, B. Hull, S. Sabri, *et al.*, "Reliability of sic mosfet with danfoss bond buffer technology in automotive traction power modules," in *PCIM Europe 2019; International Exhibition and Conference for Power Electronics, Intelligent Motion, Renewable Energy and Energy Management*, 2019, pp. 1–7.

[8] J. Lutz, T. Herrmann, M. Feller, R. Bayerer, T. Licht, and R. Amro, "Power cycling induced failure mechanisms in the viewpoint of rough temperature environment," Jan. 2008, pp. 1–4.

[9] M. Held, P. Jacob, G. Nicoletti, P. Scacco, and M.-H. Poech, "Fast power cycling test of igbt modules in traction application," in *Proceedings of Second International Conference on Power Electronics and Drive Systems*, vol. 1, 1997, 425–430 vol.1. DOI: 10.1109/PEDS.1997.618742.

[10] Puqi Ning, T. G. Lei, Fei Wang, Guo-Quan Lu, K. D. T. Ngo, and K. Rajashekara, "A Novel High-Temperature Planar Package for SiC Multichip Phase-Leg Power Module," *IEEE Transactions on Power Electronics*, vol. 25, no. 8, pp. 2059–2067, Aug. 2010. DOI: 10.1109/TPEL.2010.2046498.

[11] Manson SS., "Behavior of materials under conditions of thermal stress," *NACA TN*, vol. 2933, 1953.

[12] J. Goehre, M. Schneider-Ramelow, U. Geissler, and K.-D. Lang, "Interface degradation of al heavy wire bonds on power semiconductors during active power cycling measured by the shear test," Apr. 2010, pp. 1–6.

[13] B. Kuźnicka and K. Junik, "Intergranular stress corrosion cracking of copper – a case study," *Corrosion Science*, vol. 49, no. 10, pp. 3905–3916, 2007. DOI: https://doi.org/10.1016/j.corsci.2007.05.014.

PCIM Europe 2024, 11– 13 June 2024, Nuremberg   DOI: 10.30420/566262025

Implementation and Verification of a 50kW Opportunity Wireless Charger Design

Carlos Costas-Sos[1], Irene-Maria Torres-Alfonso[1], Juan M. Perie-Buil[1], Antonio-Miguel Munoz-Gomez[1]

[1] CIRCE Foundation, Spain

Corresponding author & speaker: Carlos Costas Sos, ccostas@fcirce.es

Abstract

This paper presents the hardware and control implementation of the primary side of an LCC-C SP-S (Series Parallel-Series) Inductive Power Transfer (IPT) system. The system is designed to provide variable power transfer up to 50kW without the need for communication by controlling a fixed current in the primary coil. The advantages of this approach include its simplicity in design and implementation, optimization and adaptability for new designs. Validation of the system's performance is conducted through a combination of simulations and experimental testing. Key aspects related to design and modeling, addressing critical considerations in the development process are also discussed.

Keywords: Inductive power transfer (IPT), wireless power transfer (WPT), SPS (Series-Parallel-Series), LCC Resonant inverters.

Nomenclature:

α	Phase shift between voltage and current	Q2	Secondary side reactive power
C_s	Series capacitor	SiC	Silicon Carbide
C_p	Parallel capacitor	V1	Primary inverter output voltage
C2	Secondary compensation network series capacitor	Vm1	Voltage induced from secondary to primary inductor
EMC	Electromagnetic Compatibility	Vm2	Voltage induced from primary to secondary inductor
EMI	Electromagnetic Interference	$V1_{DC}$	Primary side DC bus voltage
EV	Electric Vehicle	VC_s	Voltage across primary side series capacitor
f_{sw}	Switching frequency	VC_p	Voltage across primary side parallel capacitor
I1	Primary inverter output current	Vp	Voltage across parallel capacitor
Im1	Primary inductor current	WPT	Wireless Power Transfer
Im2	Secondary inductor current	w_{sw}	Angular switching frequency
Ip	Parallel capacitor current	Z2	Secondary compensation network impedance
IPT	Inductive Power Transfer	Zs	Primary side series impedance
L_s	Compensation Series coil	Zp	Primary side parallel impedance
L1	Primary inductor leakage inductance	ZVS	Zero voltage switching
L2	Secondary inductor leakage inductance	Z_{L1}	Primary coil leakage inductance impedance.
M	IPT mutual inductance	Z_{L1}	Primary coil leakage inductance impedance.
Po	Output power	SP-S	Series parallel-series compensation network
P2	Secondary side active power		

1 Introduction

Electric vehicles play an increasingly important role in the context of global decarbonization. The projected growth of the electric vehicle fleet over the current decade will require the development of a large charging infrastructure [1]. The expected large-scale integration into the transport sector poses major technical challenges, some of which are related to the development of new charging technologies. Nowadays, the use of conductive charging systems is widespread compared to inductive charging or battery swapping [2]. Despite the maturity of the conductive charging technology, it suffers from potential drawbacks, such as the need for the user to manipulate the cables, which can be considerably unergonomic for charging stations with high power transfer, and the present a notable risk of electrical damage related to the high DC output voltages [3], [4].

Even though it has a lower level of development and implantation, IPT systems are becoming one of the most advantageous options. Its advantages of isolation, convenience and safety features, have attracted considerable attention in recent years, especially for high-power applications such as battery charging of electric vehicles (EVs) [5].

Based on electromagnetic field power transfer, IPT is a wireless charging method that eliminate the need for a physical connection between the charging station and EV. This technology is especially useful for public transport and delivery vehicles, allowing them to avoid service interruptions, taking advantage of stop periods and decreasing the size of batteries required [6], [7]. Currently, main efforts are focused on improving efficiency, as well as the reduction of associated costs [2], [4].

The main principle of IPT systems is the magnetic coupling between a primary or charging station inductor and the secondary or EV on-board inductor. The primary inductor must be energized by a high frequency current, which requires the use of primary side power converters and a compensation network [8].

Systems based on inductive coupling have a much lower efficiency than those based on resonance. In IPT frameworks, the leakage inductance is higher than in conventional transformers due to the large air gap between the two coils. To compensate the increased reactive power on the primary and secondary sides, additional compensation capacitors are required.

There are four basic topologies for compensation circuits depending on how the compensation capacitors are added to the transmitting and receiving coils: series-series (SS), series-parallel (SP), parallel-series (PS), and parallel-parallel (PP) topologies [9]. Nevertheless, primary side's series-based resonant topologies, such as the commonly used series-series (S-S) or series-parallel (S-P) networks, present a notable limitation in relation to misalignment due to their intrinsic impedance decrement and the consequent source current increasement. Primary side parallel-based topologies overcome that limitation, but misalignment implies a substantial power transfer drop [10].

On the basis of traditional compensation topologies, second-order compensation methods such as LCL and LCC are derived. They are used in applications where tolerance to high loads and misalignment is desirable [11]. That is the case of opportunity charging, where additionally, the necessity of a careful alignment of the EV with the charging area is particularly inconvenient due to the limited available time.

There are also significant challenges in terms of control implementation and communication between the primary and secondary sides when developing IPT charging systems. Establishing wireless communication between EV and charging station has been suggested by [12], [13], [14], [15], [16]. In these articles, a power control shared between the primary and secondary sides is developed. Despite offering high flexibility in the control mode, the low latency of wireless communication results in lower reliability and increased cost and complexity, making it a less attractive alternative. Other proposals seek to control the power transferred from the primary side instead of relying on communication [17]. Nevertheless, their disadvantage is that the secondary side cannot control the charging power, making it hard to implement for private EV chargers.

The aim of this paper is to define a simultaneous primary and secondary control without communication. The primary control method presented in this paper establishes primary inductor current as a controlled variable acting on the secondary side via secondary induced voltage, enabling secondary power control. As a result, interoperable and communication-free control is achieved. For the compensation network, an LCC-C, SP-S is chosen due to its high efficiency and reliability [9], [18], [19].

This work continues from the previous study conducted in [20], which presented an initial system design. In this work, a new calculation method is proposed, the control is improved and the final simulation and experimental results are presented and evaluated.

Summarizing, this paper outlines a design proposal of a 50kW IPT system, and is structured as follows: after the introduction, section 2 is focused on the system overview, where hardware is analyzed and modelled, and firmware is explained. In section 3, the novel design methodology will be presented and in section 4 the simulation and experimental results are shown. Finally, in section 5, conclusions from the results are extracted.

2 System overview

As previously stated, the system to be used is described within the framework established in the article [20], where the design decisions were investigated and the initial design was developed.

2.1 Hardware

The core of the system comprises primary (ground-based) and secondary (onboard the vehicle) coils integrated in a resonant circuit. Alternating current conduction through the primary coil enables wireless power transfer between the charger and the vehicle. The chosen topology for the global resonant circuit is an SP-S with an LCC arrangement on the primary side (Fig 1), which allows the circulation of a high current in the primary coil without requiring excessively high current in the inverter. For simplicity and to accommodate a great number of inductive on-board chargers, a series topology is employed in the secondary coil. The remaining hardware comprises an AC/DC converter that supplies a DC bus that can be set on a voltage between 700 and 800V, followed by a silicon carbide (SiC) MOSFET H-bridge that allows the generation of high frequency AC current. To design the primary compensation network parameters and study the effect of its variations in the primary inverter, resonant circuits on both sides of the system are first decoupled.

2.1.1 Secondary side modelling

The secondary side equivalent model is presented in Fig. 2, where power control is assumed. The secondary current (Im2) is set as the phase reference in the developed equations. To attain maximum secondary power, the switching frequency will be set as a fixed value close to the secondary resonant frequency.

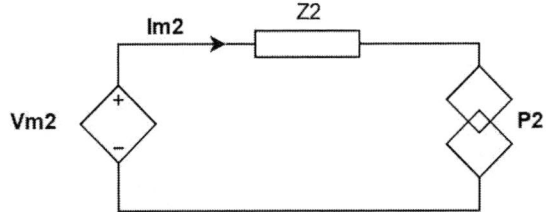

Fig. 1 Secondary side equivalent model diagram [20]

The equations that govern the behavior of the secondary are presented in Eq.3-9.

$$Vm2 = Im1 * M * w_{sw} \tag{1}$$

$$P2 = Vm2_x * Im2 \tag{2}$$

$$Im2 = \frac{Vm2_y}{Z2} \tag{3}$$

$$P2 = Vm2_x * \frac{Vm2_y}{Z2} \tag{4}$$

$$P2 = \frac{|Vm2|^2}{Z2} * sen(\alpha) * \cos(\alpha) \tag{5}$$

$$P2_{max}(Vm2, Z2) = \frac{|Vm2|^2}{2 * Z2} \tag{6}$$

$$Q2 = Im2^2 * Z2 \tag{7}$$

$$(Vm2 * Im2)^2 = Q2^2 + P2^2 \tag{8}$$

$$Q2^2 - \frac{Vm2^2}{Z2} * Q2 + P2^2 = 0 \tag{9}$$

2.1.2 Primary side modelling

Once the secondary power and current values have been outlined, the primary circuit can be analyzed. Fig 2 shows the simplified model for the primary side, where the relationship between primary and secondary is represented in the dependent current source Vm1 by Eq.11. The controlled variable in this approach is the primary inductor current (Im1), assuring a maximum induced voltage in the secondary side for a given coupling. The Im1 current is considered as phasor reference.

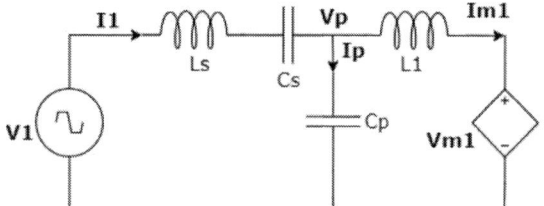

Fig. 2 Primary side equivalent model diagram [20]

The secondary side formulas are developed in Eq.11-24.

$$Vm1 = Im2 * M * w_{sw} \tag{11}$$

$$Zs = w_{sw} * Ls - \frac{1}{w_{sw} * Cs} \tag{12}$$

$$Zp = \frac{-1}{w_{sw} * Cp} \tag{13}$$

$$Vm1_y = \frac{Q_2}{Im1} \tag{14}$$

$$Vm1_x = Vp_x = \frac{P_2}{Im1} \tag{15}$$

$$Vp_y = Vm1_y + Im1 * Z_{L1} \tag{16}$$

$$Z_p = \frac{|Vp|}{|Ip|} \tag{17}$$

$$\vec{Ip} = \vec{I1} - \vec{Im1} \tag{18}$$

$$I1_{min} = \frac{Po_{max}}{V1_{x,max}} \tag{19}$$

$$V1_{y,ref_{I1}} = (Vp_{y,ref_{I1}} + I1 * Zs) \tag{20}$$

$$V1_{ref_{I1}} > \frac{2 * \sqrt{2} * V1_{DC}}{\pi * 4} \tag{21}$$

$$V1_{ref_{I1}} < \frac{2 * \sqrt{2} * V1_{DC}}{\pi} \tag{22}$$

$$Vp_{y,ref_{I1}} = Vp_{y,ref_{Im1}} * \cos(\alpha) - Vp_{x,ref_{Im1}} * \sin(\alpha) \tag{23}$$

$$Vp_{x,ref_{I1}} = Vp_{y,ref_{Im1}} * \sin(\alpha) + Vp_{x,ref_{Im1}} * \cos(\alpha) \tag{24}$$

2.1.3 Primary and secondary coils

The primary and secondary coils design is defined and described in [20]. For the primary coil, the objective of simplifying the mechanical assembly and consequently the installation was the main constraint considered. Therefore, a single turn coil with a ferrite plate was designed. In the case of the secondary coil, the unit created in the framework of the project has been taken as reference for the implementation presented here. The design consists of a multi-turn winding ferrite-based coil installed on EV's underbody, which establishes the shielding. On the other hand, the secondary coil takes part of a series resonant network topology.

2.2 Control

The implemented control is based on the primary coil current, which is maintained at a fixed value to ensure the maximum power transfer to the secondary coil with optimal coupling between the coils. Primary inverter output voltage amplitude, phase cancellation of the voltage between legs of the inverter and switching frequency can be defined as regulated variables to act over the primary inductor current. Nevertheless, for robustness, the switching frequency will remain fixed and close to the secondary resonant frequency. Therefore, the control regulation is performed via DC bus amplitude modulation and voltage phase cancellation.

This control is achieved through phase cancellation of the H-bridge output voltage and by modifying the DC bus voltage at the input. It utilizes a proportional-integral (PI) control strategy with current feedback in the primary coil (Fig 3). It also establishes communication with the AC/DC stage to act over the bus voltage, thus changing the voltage amplitude.

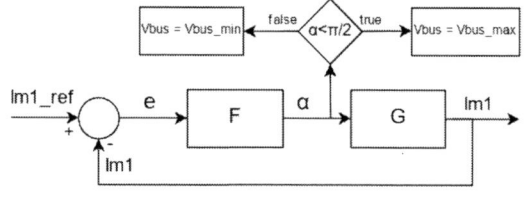

Fig. 3 . Control diagram of the primary side

The modulation technique used is Asymmetric Voltage Cancellation (AVC), which enables phase cancellation and facilitates Zero Voltage Switching (ZVS) by advancing the V1 voltage wave. ZVS is a condition where the voltage across transistors is zero when the current begins to flow, thereby achieving a significant reduction of switching losses and electromagnetic interference [21], [22].

3 Design methodology

The charger's design is based on the developments presented in [20], where primary and secondary coils and corresponding compensation circuits were defined. In the current study, improvements have been made in the calculation of the resonant circuit. The proposed design methodology involves introducing the contour conditions of the system in an iterative model that uses the equations presented in the previous section (see Figure 4a). The circuit's unknown values are filled with an array of values to cover all possible combinations. Meanwhile, the known values, which are tolerance, position and output power dependent, are randomly assigned in each iteration to cover all possible component combinations (see Figure 4b). Therefore, this method is based on a Monte Carlo algorithm.

4 Results

The development of the OWPT charger is part of a pilot demonstrator of the INCIT EV project. For the experimental setup of the charger, the reference secondary system described in section 2.1.3 is employed. The values of L2 and C2, along with other given values are presented in table 1. The circuit parameters presented form the basis for the calculating the resonant tank, primary current and system switching frequency. Even though the charger is designed to be capable of transferring 50kW, the secondary system is developed to demand a maximum power of 30 kW, which constitutes a limit for the experimental setup. The output values of the system design, as well as the corresponding measurements on the system implementation are presented in table 2, where the divergence between them is also included.

Input data	Parameters
L1	[2.74-2.94] µH
L2	[140-144] µH
M	[0-4.6] µH
C2	33 nF
Vm2$_{max}$	800 V
Po$_{max}$	30 kW
V1$_{DC}$	800 V
I1_max	160A
[X;Y;Z]	[-1,1;0.1,-0.1;0.2,0.25] m

Table 1 Values of the proposed system input data in the frame of the INCIT EV project

Output data	Calculated values	Measured values	Variation (%)
L1[uH]	2.74-2.94	2.62	4.37-10.88
L2 [uH]	140-144	137.77	1.59-4.32
C2 [nF]	33	33.51	1.57
Cs [nF]	565	560.65	0.77
Ls [uH]	16.5	17.59	6.25
Cp [nF]	1230	1254	1.91
Im1[A]	230A	-	-
f$_{sw}$ [kHz]	73.4	-	-

Table 2 Obtained values by the iterative calculation method for the given secondary and measurements in the implemented setup.

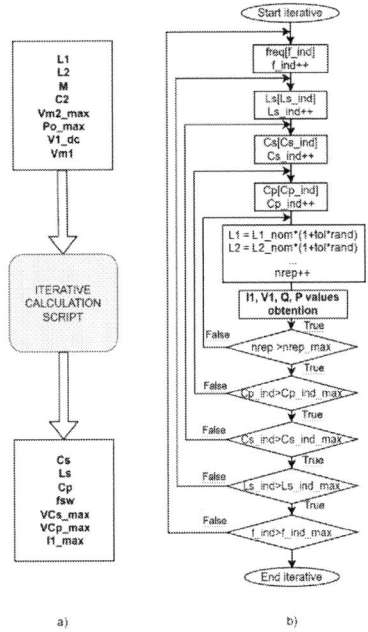

Fig. 4 Iterative calculation method a) Blackbox model b) Flowchart

The following sections present the results of the simulation and field tests. Two setups are considered in order to analyze the system behavior. The first scheme includes the designed IPT system, and the onboard charger is simplified by a diode bridge with a resistive load and a capacitor connected to its output. This circuit is designed for experimental tests to model the secondary power stages and the battery without implementing a secondary control (Fig. 5). Secondary side resistive load is set with a value of 6.8 Ω with the aim of achieving a power transfer

209

close to 30 kW. The second model reflects the proposed secondary circuit, where a buck-boost DC/DC converter and the EV battery are designed and implemented (Fig. 6).

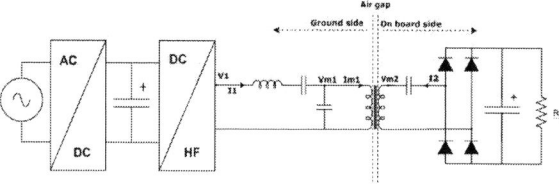

Fig. 5. OWPT charger model based on secondary side resistive load.

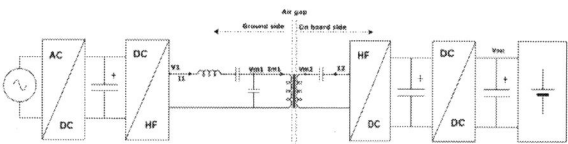

Fig. 6 OWPT charger model based on secondary side complete scheme, including a DC/DC converter and a battery (final setup).

4.1 Simulations

The simulations were conducted in MATLAB Simulink®.

4.1.1 Resistive load

Prior to the implantation of the system in laboratory, a simulation of the first defined scheme was carried out. Results of the coils perfectly aligned are illustrated in Fig. 7. The first graph displays inverter and ground primary coil RMS currents. The inverter current does not exceed the 160A, as limited in the design stage. In the second graph, circuit relevant signals are represented. They show the expected behavior, since the MOSFETs' turn-on is made when the antiparallel diodes are conducting, thus assuring the ZVS condition. The output power, displayed in the lower panel of Figure 7, is slightly over 30kW. The estimated system efficiency is around 89.7%.

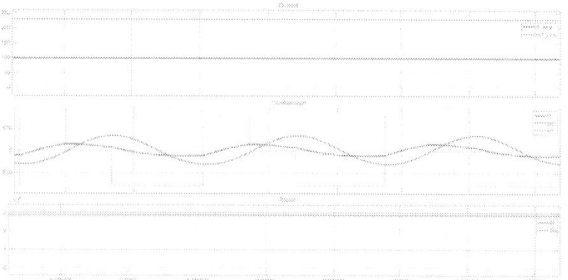

Fig. 7 Simulation results of the resistive load model with both coils perfectly aligned and minimum height distance on stationary. From top to bottom, Im1 and I1

rms currents, main waveforms and Pin and Pout are displayed.

The case of maximum misalignment in axis y is presented in Figure 8. In this scenario the results are very similar, but with a lower power delivery to the secondary side, only managing to charge 22.5kW due to the lower mutual impedance. The efficiency of this case is 88.5%.

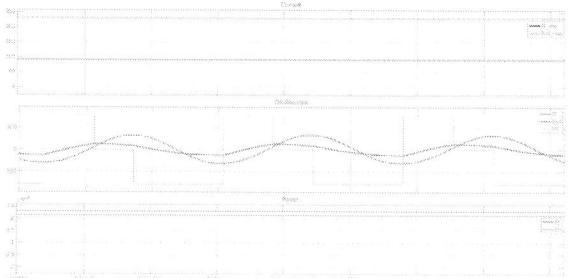

Fig. 8 Simulation results of the resistive load model with both coils with maximum misalignment at y axis and minimum height distance on stationary. From top to bottom, Im1 and I1 rms currents, main waveforms and Pin and Pout are displayed.

4.1.2 DC/DC power control

In order to simulate the final setup of the OWPT charger, a DC/DC buck-boost power control has been designed and implemented (Fig.6). The chosen setpoint for the secondary is the design power, 50kW.

The results included in Fig. 9 show that output power is obtained with a voltage phase similar to the 30kW previous case, but with an increase in the inverter output current. Nevertheless, I1 does not exceed the current limitation. ZVS condition is also achieved and there is a margin to reduce the current phase to be more capacitive without leaving this switching mode. The efficiency showed in this case is around 91.2%.

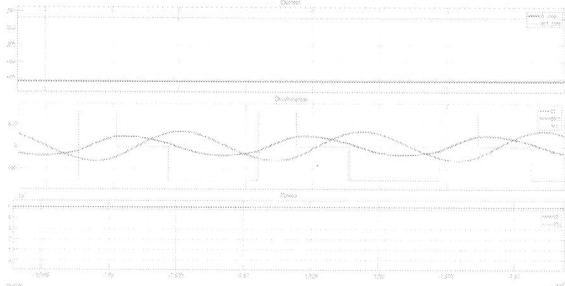

Fig. 9 Simulation results of the DC/DC converter-based model with both coils perfectly aligned and minimum height distance on stationary. From top to bottom, Im1 and I1 rms currents, main waveforms variables and Pin and Pout are displayed.

4.2 Experimental testing

Once the validation methods and simulations were performed and demonstrated a correct behavior of the calculated system, the experimental setup was developed. The laboratory layout can be found in Fig. 10, where the resistive load model has been assembled. To avoid electromagnetic emissions to the surrounding, the IPT system is located into a Faraday cage.

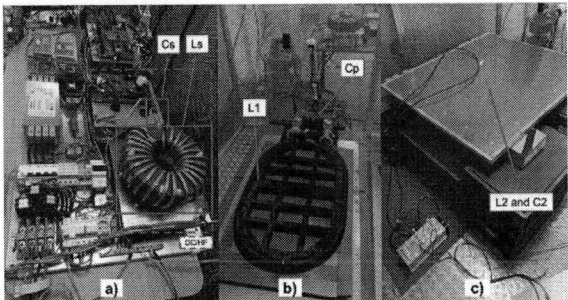

Fig. 10 Experimental laboratory setup of the OWPT charger. a) AC/DC, DC/AC and a part of the compensation network, b) primary coil and parallel compensation capacitor and c) primary and secondary coils coupled with the minimum z distance.

The electrical results are presented in Fig.11. It can be observed that results are similar to those of the simulation. The I1 is slightly higher in the experimental setup, presenting a 122 A value. Achieved AVC phase cancellation is similar, and inverter keeps working in ZVS condition. The load current (I_load) is established in 63.34A, what follows that transferred power is 27.28 kW, with an estimated efficiency of 85.1%.

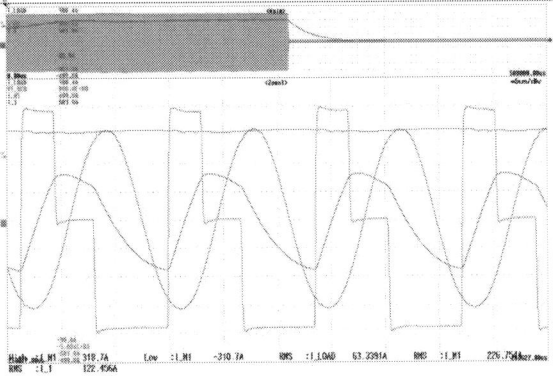

Fig. 11 Waveforms of the experimental test, with coils perfectly aligned and a 6.8 Ω resistive load. Red:I_load; Blue: V1; Orange: Im1; Purple: I1.

4.3 Demonstrator

The final setup of the implemented OWPT charger is installed as part of the Mobility City in in Zaragoza, constituting one of the demonstrations of wireless charging of Horizon 2020 INCIT-EV project (Fig. 12).

The primary coil is located below the ground level in a concrete manhole, establishing the charging area. Composite manhole cover and frames have been specially designed considering their exposure to magnetic field.

The use case also includes two main cabinets. AC connection, protection and filtering stages, AC/DC and DC/AC converters and part of the primary resonant network are installed on the electronics' cabinet. A second cabinet contains an oil-based cooling system to meet the refrigeration needs identified during the design phase. Cooling is required for primary coil and parallel branch for full-charging tests.

Fig. 12 OWPT charger installed at INCIT-EV Zaragoza demo site. a) Final result and b) During tests.

5 Conclusions

This work presents the design process of an LCC-S compensated OWPT system for electric vehicle charging, maximising the transferred power under misalignment conditions while reducing the inverter output current to maximise system efficiency. Analysis and experimental results suggest that ZVS can always be achieved, resulting in significantly reduced switching losses and a simplified thermal design.

Further work is required to enhance peak power transfer and system efficiency. This can be achieved by increasing the mutual inductance between the coils and the switching frequency (thus increasing Zm), or by increasing the DC bus voltage amplitude. This would result in lower currents in the primary coil and in the inverter, respectively.

The drawback found in the design method is its dependence on an accurate previous modeling, study and simulation. Slight variations in the final setup on unexpected elements (such as parasitic resistances or mutual inductance) can cause a moderate divergence between calculated and experimental results. However, its simplicity makes it easy to execute a recalculation based on laboratory measurements.

6 Acknowledgement

The authors gratefully acknowledge the support from Vedecom (https://www.vedecom.fr/) and Tria Ingenieria (https://www.triarail.com/) and its contribution in the pilot project. The research has been carried out in the framework of the INCIT-EV project (https://www.incit-ev.eu/), funded by the European Union under the Horizon 2020 Programme (H2020-LC-GV-2019).

7 References

[1] "Global EV Sales by Scenario, 2020–2030—Charts—Data & Statistics—IEA. Accessed: Apr. 05, 2024. [Online]."

[2] S. Aghajan-Eshkevari, S. Azad, M. Nazari-Heris, M. T. Ameli, and S. Asadi, "Charging and Discharging of Electric Vehicles in Power Systems: An Updated and Detailed Review of Methods, Control Structures, Objectives, and Optimization Methodologies," *Sustainability*, vol. 14, no. 4, p. 2137, Feb. 2022, doi: 10.3390/su14042137.

[3] B. J. Varghese, A. Kamineni, and R. A. Zane, "Investigation of a DD^2Q Pad Structure for High Power Inductive Power Transfer," in *2019 IEEE PELS Workshop on Emerging Technologies: Wireless Power Transfer (WoW)*, IEEE, Jun. 2019, pp. 129–133. doi: 10.1109/WoW45936.2019.9030644.

[4] M. Longo, D. Zaninelli, G. Cipriani, V. Di Dio, and R. Miceli, "Economic analysis on the use of wired and wireless recharging systems," in *2017 IEEE International Conference on Environment and Electrical Engineering and 2017 IEEE Industrial and Commercial Power Systems Europe (EEEIC / I&CPS Europe)*, IEEE, Jun. 2017, pp. 1–6. doi: 10.1109/EEEIC.2017.7977704.

[5] J. Yang, X. Zhang, K. Zhang, X. Cui, C. Jiao, and X. Yang, "An LCC-SP Compensated Inductive Power Transfer System and Design Considerations for Enhancing Misalignment Tolerance," *IEEE Access*, vol. 8, pp. 193285–193296, 2020, doi: 10.1109/ACCESS.2020.3032793.

[6] M. Amjad, M. Farooq-i-Azam, Q. Ni, M. Dong, and E. A. Ansari, "Wireless charging systems for electric vehicles," *Renewable and Sustainable Energy Reviews*, vol. 167, p. 112730, Oct. 2022, doi: 10.1016/j.rser.2022.112730.

[7] S. R. Khutwad and S. Gaur, "Wireless charging system for electric vehicle," in *2016 International Conference on Signal Processing, Communication, Power and Embedded System (SCOPES)*, IEEE, Oct. 2016, pp. 441–445. doi: 10.1109/SCOPES.2016.7955869.

[8] Q. Zhu, L. Wang, Y. Guo, C. Liao, and F. Li, "Applying *LCC* Compensation Network to Dynamic Wireless EV Charging System," *IEEE Transactions on Industrial Electronics*, vol. 63, no. 10, pp. 6557–6567, Oct. 2016, doi: 10.1109/TIE.2016.2529561.

[9] X. Liu, L. Clare, X. Yuan, C. Wang, and J. Liu, "A Design Method for Making an LCC Compensation Two-Coil Wireless Power Transfer System More Energy Efficient Than an SS Counterpart," *Energies (Basel)*, vol. 10, no. 9, p. 1346, Sep. 2017, doi: 10.3390/en10091346.

[10] J. L. Villa, J. Sallan, J. F. Sanz Osorio, and A. Llombart, "High-Misalignment Tolerant Compensation Topology For ICPT Systems," *IEEE Transactions on Industrial Electronics*, vol. 59, no. 2, pp. 945–951, Feb. 2012, doi: 10.1109/TIE.2011.2161055.

[11] A. Ridge, K. K. Ahamad, R. McMahon, and J. Miles, "Development of a 50 kW Wireless Power Transfer System," in *2019 IEEE PELS Workshop on Emerging Technologies: Wireless Power Transfer (WoW)*, IEEE, Jun. 2019, pp. 406–409. doi: 10.1109/WoW45936.2019.9030672.

[12] L. H. Chan, Y. Yang, and K.-W. E. Cheng, "Comparative Studies on the Primary-Side Frequency and Phase Shift Control for Series-Series Compensated Inductive Power Transfer," in *2020 8th International Conference on Power Electronics Systems and Applications (PESA)*, IEEE, Dec. 2020, pp. 1–5. doi: 10.1109/PESA50370.2020.9343994.

[13] A. Mostafa, Y. Wang, H. Zhang, and F. Lu, "Output Power Control of an S-S IPT System Based on Voltage and Frequency Tuning for EV Charging," in *2021 IEEE PELS Workshop on Emerging Technologies: Wireless Power Transfer (WoW)*, IEEE, Jun. 2021, pp. 1–5. doi: 10.1109/WoW51332.2021.9462886.

[14] Tiefu Zhao, B. Pahl, Jun Xu, B. Wu, P. Nirantare, and M. Kothekar, "Optimal operation point tracking control for inductive power transfer system," in *2015 IEEE Wireless Power Transfer Conference (WPTC)*, IEEE, May 2015, pp. 1–4. doi: 10.1109/WPT.2015.7139134.

[15] A. Vulfovich, M. M. Peretz, and A. Kuperman, "Output Voltage Feedback Control Method for Series-Series Compensated Inductive Wireless Power Transfer Link with Varying Primary Capacitor," in *2021 IEEE International Conference on Microwaves, Antennas, Communications and Electronic Systems (COMCAS)*, IEEE, Nov. 2021, pp. 89–92. doi: 10.1109/COMCAS52219.2021.9629002.

[16] Y. Chen, H. Zhang, S.-J. Park, and D.-H. Kim, "A Switching Hybrid LCC-S Compensation Topology for Constant Current/Voltage EV Wireless Charging," *IEEE Access*, vol. 7, pp. 133924–133935, 2019, doi: 10.1109/ACCESS.2019.2941652.

[17] S. Mukherjee *et al.*, "Control of Output Power in Primary Side LCC and Secondary Series Tuned Wireless Power Transfer System without Secondary Side Sensors," in *2020 IEEE Energy Conversion Congress and Exposition (ECCE)*, IEEE, Oct. 2020, pp. 5532–5536. doi: 10.1109/ECCE44975.2020.9236098.

[18] L. Zhang, H. Li, Q. Guo, S. Xie, and Y. Yang, "Research on Constant Voltage/Current Output of LCC–S Envelope Modulation Wireless Power Transfer System," *Energies (Basel)*, vol. 15, no. 4, p. 1562, Feb. 2022, doi: 10.3390/en15041562.

[19] B. Chen, G. Hu, X. Zhang, and X. Ma, "Modeling and Efficiency Analysis of LCC-S Inductive Power Transmission (IPT) system," *J Phys Conf Ser*, vol. 1754, no. 1, p. 012066, Feb. 2021, doi: 10.1088/1742-6596/1754/1/012066.

[20] I. -M. Torres-Alfonso *et al.*, "An Interoperable 50 kW Inductive Power Transfer Design for Opportunity Wireless Vehicle Charging," PCIM Europe 2023; International Exhibition and Conference for Power Electronics, Intelligent Motion, Renewable Energy and Energy Management, May 2023. doi: 10.30420/566091102.

[21] S. N. Manias, "DC–DC Converters. Resonant Switch Converters.," in *Power Electronics and Motor Drive Systems*, Elsevier, 2017, pp. 501–611. doi: 10.1016/B978-0-12-811798-9.00007-X.

[22] T. Saha, H. Wang, B. Riar, and R. Zane, "Analysis of zero voltage switching requirements and passive auxiliary circuit design for DC-DC series resonant converters with constant input current," in *2016 IEEE 2nd Annual Southern Power Electronics Conference (SPEC)*, IEEE, Dec. 2016, pp. 1–6. doi: 10.1109/SPEC.2016.7846177.

PCIM Europe 2024, 11– 13 June 2024, Nuremberg   DOI: 10.30420/566262026

Performance Evaluation of Silicon-Based 3-Level Vienna Rectifier in ISOPLUS®SMPD Package

Karsten Haehre[1], Muhammad Yassof[1]

[1] Littelfuse, Germany

Corresponding author: Karsten Haehre, khaehre@littelfuse.com
Speaker: Karsten Haehre

Abstract

Public DC-charging infrastructure is connected to the public three-phase low-voltage grid. Therefore, it must comply to international standards and the corresponding grid codes. The widely adopted Vienna rectifier allows fulfilling this requirement and offers very high efficiency at low circuit complexity. It allows the use of fast 650 V silicon-based switches and 1200 V diodes. In this paper, measurement results are presented how the Vienna rectifier in the compact ISOPLUS® surface mounted power device (SMPD) performs and what power- and efficiency-levels can be achieved based on silicon technology.

1   Introduction

Latest developments in electric vehicle (EV) DC-chargers show power levels up to 400 kW and are normally built in a modular manner based on power subunits with up to 100 kW output power each [1, 2]. This approach is well-known in the design of server power supplies in data centers, which comply with requirements for high efficiency, high reliability, and simple maintainability. A block diagram of such a power subunit is illustrated in Fig. 1.

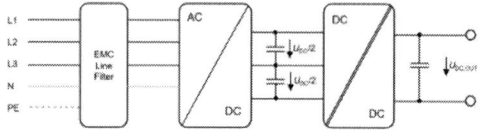

Fig. 1   Block diagram of a power supply unit of a DC-charger.

The power electronics inside the AC-DC subcircuit converts the line voltages into DC-voltage while maintaining sinusoidal input currents, as well as performing a power factor correction (PFC). A common approach is to use the Vienna rectifier topology, also known as symmetric boost PFC or 3-level T-type neutral boost PFC.

1.1   The 3-level Vienna rectifier

Figure 2 depicts the circuit diagram of the Vienna rectifier. It is a three-phase boost converter, which can draw sinusoidal grid currents, correct the power factor of the line currents, and control the DC-output voltage U_{DC}. When using silicon devices, like 650 V Power MOSFETs or fast 650 V IGBTs, it can offer low cost and low filtering efforts but operate at high efficiency.

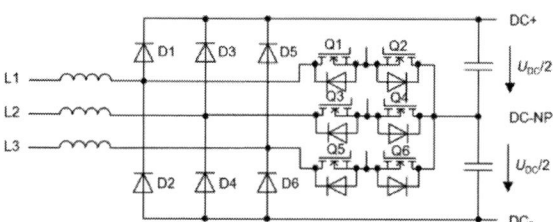

Fig. 2   Circuit diagram of the 3-level Vienna rectifier.

The diodes D1 to D6 need to block the DC-link voltage U_{DC} and need to be rated for a blocking voltage of 1200 V. Thus, to achieve high switching speeds, 1200 V Si fast recovery diodes (FREDs) or SiC schottky barrier diodes (SBD) are often used. Alternatively, series-connected 600 V Si FREDs can be used. The boost switches Q1 to Q6 only need to block half the DC-link voltage $U_{DC}/2$, which allows the selection of 650 V MOSFETs.

To achieve the effect of PFC, the DC-link voltage needs to be boosted to a higher voltage than the rectified line voltage, for example to U_{DC}=750 V. The switch then needs to block $U_{DC}/2$=375 V. Keeping a margin of 25 %, a MOSFET with a breakdown voltage of $U_{DS,max} \geq 375/0.75 \geq 500$ V would be suitable.

Consequently, two 650 V devices are suitable for DC-link voltages up to $U_{DC} \leq 975$ V keeping the 25 % margin into account. To achieve the output voltage required by the IEC61851-23 standard for DC-charging systems of 250 V$\leq U_{charge} \leq$950 V, either the use of a transformer with a voltage-ratio $U_{in}/U_{out} \geq 1$ in the DC-DC subcircuit or boosting the DC-link voltage to the required level is suitable.

1.2 ISOPLUS® surface mounted power device (SMPD) package

The ISOPLUS® surface mounted power device (SMPD) package combines the advantages of isolated power modules with surface mounted device (SMD) package technology. The SMD power package allows automatic assembly compared to through-hole technology (THT) devices and additionally soldering in the same reflow process as other passive components. Therefore, process steps and thus cost can be cut. The ISOPLUS® SMPD package is based on a direct copper bonding (DCB) ceramic substrate, which eases insulation coordination and thermal management. The package is illustrated in Fig. 3.

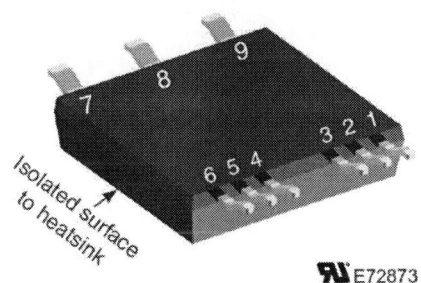

Fig. 3 Rendering of the Surface Mounted Power Device (SMPD) package, variant B with three main terminals 7-9 and six auxiliary terminals 1-6.

Thanks to the DCB substrate, different circuits can be integrated. It covers a PCB footprint area of 25×32.7 mm² and has a height of 5.5 mm. The low weight of approximately 8.5 grams allows automatic assembly. The total DCB area is 21×23 mm²=483 mm². Due to the lead-frame, an area of approximately 330 mm² is available for die placing and bonding. Additionally, the lead-frame provides a limited number of pins. The main terminals 7-9 can handle currents up to 100 A and terminals 1-6 up to 50 A, each.

1.3 Fitting the 3-level Vienna rectifier into the ISOPLUS® SMPD package

The 3-level Vienna rectifier requires four power terminals and three gate drive terminals. To fit one phase-leg, as sketched in Fig. 4, into the ISOPLUS® SMPD package, the MOSFETs Q1 and Q2 need to be mounted in common-source configuration.

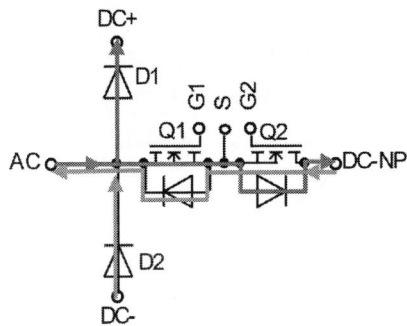

Fig. 4 One phase-leg of the 3-level Vienna rectifier showing the current paths during positive (red) and negative (blue) half-wave of the AC grid-voltage.

The main current paths during positive and negative half-wave of the AC grid-voltage are indicated with red and blue arrows in Fig. 4, as well.

Consequently, the power terminals 7-9 of the ISOPLUS® SMPD package can be used for the DC-link connection and three of the auxiliary terminals for the AC line connection. The resulting terminal configuration is sketched in Fig. 5, which was previously presented in [3] as a concept.

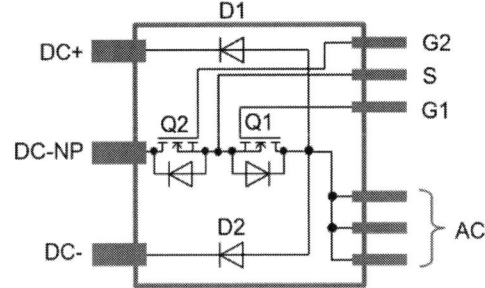

Fig. 5 Circuit schematic of the 3-level Vienna rectifier phase-leg fit into the SMPD footprint.

The terminal configuration was optimized to achieve the lowest possible stray inductances within the commutation loops, when considering the PCB layout. As the AC terminals normally are connected to a line inductor, compare to Fig. 2, the stray inductance of this path is less critical than the commutation loop from the DC-neutral-point to the positive or negative DC-rail.

215

1.4 DCB layout for the 3-level Vienna rectifier in ISOPLUS® SMPD package

Based on the terminal configuration and the results presented in [3], a DCB design was created as sketched in Fig. 6, excluding gate wire bonding.

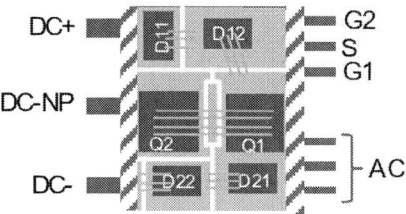

Fig. 6 DCB layout concept for estimation if the available DCB area is sufficient for the maximum selected die sizes. Shaded areas indicate the placing of the lead frame.

The layout concept allows for other combinations of MOSFETs, and series-connected tandem diodes with lower blocking voltage. This helps to address different application specific operating conditions or the use of SiC devices for higher switching frequencies.

First engineering samples of the 3-level Vienna rectifier were built based on the results in [3]. For the first batch, an IXFH22N65X2 MOSFET chip with a chip area of 25 mm² was selected [4]. On the Al_2O_3 DCB ceramic substrate it achieves an approximate thermal resistance of R_{thJC}=0.8 K/W. This will lead to a maximum drain-current of $I_{D,max}$=15 A at a junction temperature of T_{vJ}=150°C and a given heatsink temperature of T_{HS}=80°C. Littelfuse diodes DSEP12-12A [5] rated for 1200 V, 12 A were utilized as D1 and D2.

2 Measurement setup

In order to evaluate the switching performance of the 3-level Vienna rectifier prototype devices, double pulse tests were conducted. Therefore, a PCB was designed and manufactured, which allows for configuring different test cases. The PCB is shown in Fig. 7 and Fig. 8.

On the top side, the gate drive power supply and circuitry are assembled. For this test, a 9 A-rated Littelfuse gate driver IC IXD_609 was used and the possibility to employ separate gate resistances for turn-on and turn-off was implemented. The gate driver power supply was built using a DC-DC converter with output voltages of $U_{GS,on}$=+15 V and $U_{GS,off}$=-3 V.

To connect the test board to the DC-link and the load inductor, screw terminal blocks are available.

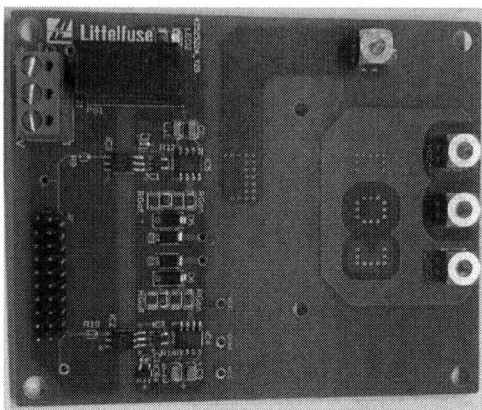

Fig. 7 Top view of the test PCB having gate driver power supply and gate driver ciruitry, as well as the power connection terminal blocks.

Fig. 8 Bottom view of the test PCB holding the SMPD package with integrated 3-level Vienna rectifier.

Both MOSFET and diode combinations Q1+D1 and Q2+D2 were tested. For this, an air-cored coil of 330 µH was connected between the terminals DC+ and AC, as well as DC- and AC, respectively.

The test conditions and parameters for the double pulse testing are listed in Table 1. A wide range of test conditions using different gate resistance values from $R_{G,ext}$=6.7 Ω to $R_{G,ext}$=47 Ω were applied at different load currents from I_D=10 A to I_D=40 A.

Parameter	Abbrev.	Value
DC-link voltage	U_{DC}	400 V
Gate resistors	R_G	6.7...47 Ω
Drain Current	I_D	10...40 A
Gate Voltages	V_{GS}	-3/+15 V

Table 1 Test conditions for double pulse testing of the Vienna rectifier SMPD sample.

The DC-link voltage for all test points was set to $U_{DC}=400$ V. All tests were conducted at room temperature $T_{amb}=25°C$.

Voltages were measured using Teledyne LeCroy HVD3106 high-voltage differential probes and the drain currents using a CWT Ultra-mini Rogowski coil.

From the measurements it is also possible to calculate the stray inductances of the test setup during turn on of the MOSFETs Q1 and Q2, respectively, according to Eq. (1).

$$L_\sigma = \Delta u/(di/dt), \qquad (1)$$

where Δu indicates the voltage drop of the drain-to-source voltage during the transient of the drain current di/dt of the MOSFET.

3 Measurement results

The characterization of the first prototypes includes measurement of the stray inductances, the static output characteristics, the dynamic switching performance, and the thermal resistance. The following section will present and discuss the measurement results.

3.1 Determination of stray inductances

From the measured waveforms in Fig. 9 for the loop of Q1 & D1, the stray inductances of the PCB can be calculated according to Eq. (1):

$$L_\sigma=90 \text{ V}/(2.96 \text{ A/ns})=\textbf{30 nH} \text{ for Q1 \& D1} \quad (2)$$
$$L_\sigma=70 \text{ V}/(3.08 \text{ A/ns})=\textbf{23 nH} \text{ for Q2 \& D2} \quad (3)$$

These are very low values, considering the DC-link connection via screw terminal blocks. The difference of the two stray inductances most likely originates from the internal DCB layout combined with the PCB layout.

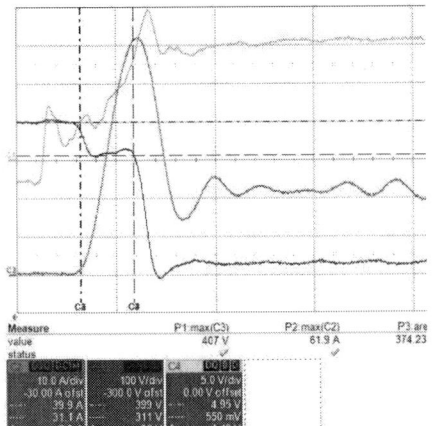

Fig. 9 Turn-on waveforms of Q1 and D1 for determination of the stray inductance at $U_{DC}=400$ V, $I_D=20$ A, with C2 (red): i_D, C3 (blue): u_{DS}, C4 (green): u_{GS}.

3.2 Static device characteristic

The static output characteristics of Q1 at a junction temperature of $T_J=25°C$ are represented in Fig. 10 and at $T_J=125°C$ in Fig. 11.

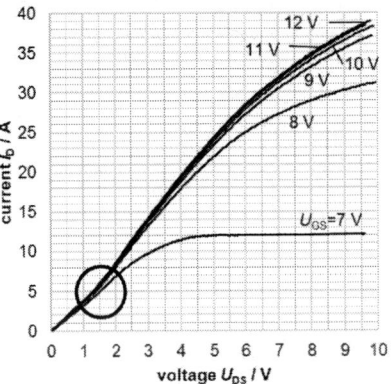

Fig. 10 Static output characteristics of Q1, $T_J=25°C$.

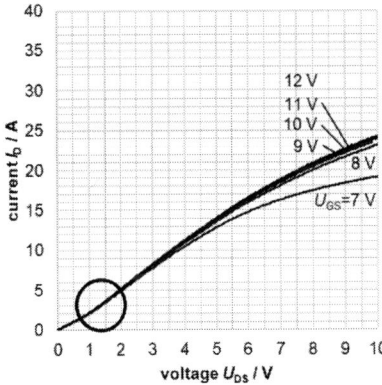

Fig. 11 Static output characteristics of Q1, $T_J=125°C$.

Since the forward voltage is measured on both Q1 and Q2 the voltage includes the combined body diode voltage drop and channel resistance of Q2. This effect is represented by the inverting slope of the curve marked by the circles in Fig. 10 and Fig. 11. Additionally, it is possible to derive that the voltage U_{DS}, at which the slope changes, reduces with higher temperature from $U_{DS}\approx1.3$ V at $T_J=25°C$ to $U_{DS}\approx1.1$ V at $T_J=125°C$. This originates from the negative temperature coefficient of the bipolar body diode of Q2.

Furthermore, the conductivity modulation starts saturating at gate voltages $U_{GS}\geq10$ V.

3.3 Switching performance of the MOSFET

Besides the stray inductances of the setup and the static output characteristics, the dynamic double pulse test was utilized to determine the switching losses of the MOSFETs, as well as reverse recovery charge of the 1200 V diode in use. For this both the combinations Q1-D1, as well as Q2-D2 were tested.

The turn-off waveforms of Q1 at I_D=40 A are depicted exemplarily in Fig. 12 for a gate resistance of $R_{g,ext}$=47 Ω.

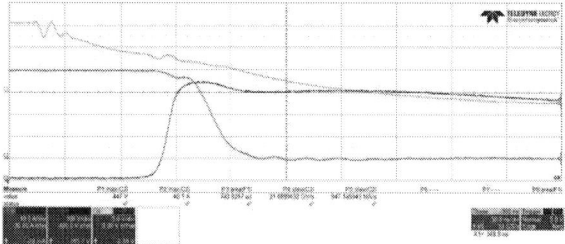

Fig. 12 Turn-off switching waveforms of Q1 at I_D=40 A and with $R_{G,ext}$=47 Ω. Green: u_{GS} (5 V/div), red: i_D (10 A/div), blue: u_{DS} (100 V/div).

The turn-on waveforms are illustrated in Fig. 13 for the same test conditions.

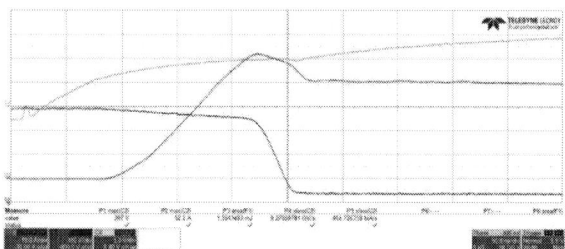

Fig. 13 Turn-on switching waveforms of Q1 at I_D=40 A and with $R_{G,ext}$=47 Ω. Green: u_{GS} (5 V/div), red: i_D (10 A/div), blue: u_{DS} (100 V/div).

With an external gate resistance of $R_{G,ext}$=47 Ω, the switching losses are comparably high, namely E_{off}=744 µJ and E_{on}=1.55 mJ respectively.

Therefore, it is of interest what performance can be achieved with lower external gate resistance. The resulting switching waveforms for $R_{G,ext}$=6.7 Ω are shown in Fig. 14 for turn-off and in Fig. 15 for turn-on transition.

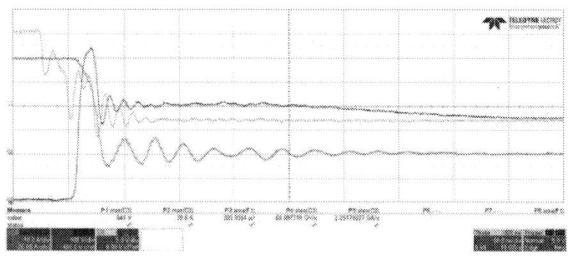

Fig. 14 Turn-off switching waveforms of Q1 at I_D=40 A and with $R_{G,ext}$=6.7 Ω. Green: u_{GS} (5 V/div), red: i_D (10 A/div), blue: u_{DS} (100 V/div).

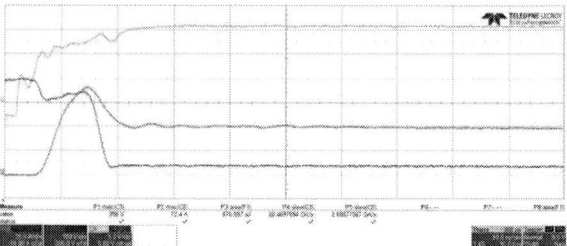

Fig. 15 Turn-on switching waveforms of Q1 at I_D=40 A and $R_{G,ext}$=6.7 Ω. Green: u_{GS} (5 V/div), red: i_D (20 A/div), blue: u_{DS} (100 V/div).

The resulting switching losses then are E_{off}=326 µJ and E_{on}=871 µJ respectively.

The turn-off waveforms in Fig. 14 reveal that some oscillations occur and the high dv/dt causes some feedback to the gate voltage. However, it did not cause parasitic turn-on.

On the other hand, the turn-on waveforms in Fig. 15 do not show oscillations of the gate voltage.

3.4 Reverse recovery of the Diode

The reverse recovery waveforms of the Diode D1 at I_D=40 A are depicted exemplarily in Fig. 16 for a gate resistance of $R_{g,ext}$=47 Ω. The resulting di/dt is 0.49 kA/µs and the corresponding reverse recovery charge can be calculated to be E_{rec}=111 µJ. The maximum reverse recovery current is I_{rrm}=12 A.

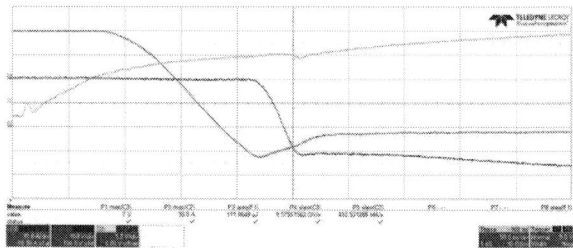

Fig. 16 Diode reverse recovery waveforms for D1 at I_D=40 A and with $R_{G,ext}$=47 Ω. Green: $u_{GS,Q1}$ (5 V/div), red: $i_{F,D1}$ (10 A/div), blue: $u_{R,D1}$ (100 V/div).

With lower gate resistance the switching speed increases accordingly as illustrated in Fig. 17 when switching with $R_{G,ext}$=6.7 Ω. There, the reverse recovery charge reduces slightly to E_{rec}=97 µJ at a di/dt=2.3 kA/µs. However, the maximum reverse recovery current peaks to I_{rrm}=16 A. Temperature dependance of reverse recovery was not investigated.

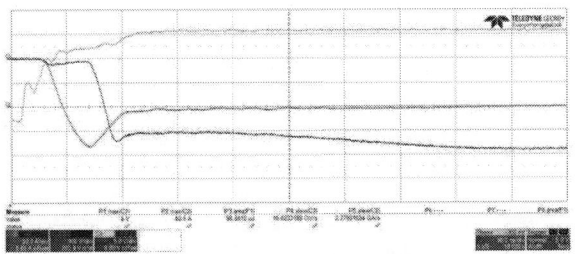

Fig. 17 Diode reverse recovery waveforms for D1 at I_D=40 A and with $R_{G,ext}$=6.7 Ω. Green: $u_{GS,Q1}$ (5 V/div), red: $i_{F,D1}$ (10 A/div), blue: $u_{R,D1}$ (100 V/div).

3.5 Switching loss comparison

The total switching losses of Q1 and D1 are illustrated in Fig. 18.

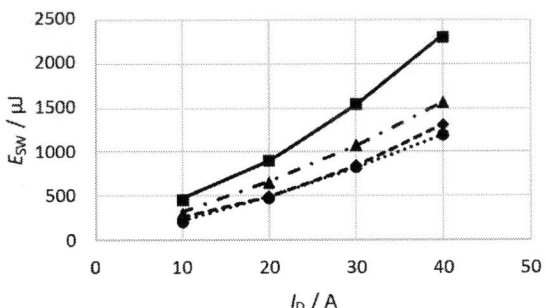

Fig. 18 Switching loss energy E_{sw} of Q1 with $R_{G,ext}$=47 Ω (■), 22 Ω (▲), 10 Ω (♦) and 6.7 Ω (●).

The measurements revealed only very slight differences between the combinations of Q1-D1 and Q2-D2 as illustrated in Fig. 19 for $R_{G,ext}$=22 Ω. The deviations result from the PCB and DCB layout.

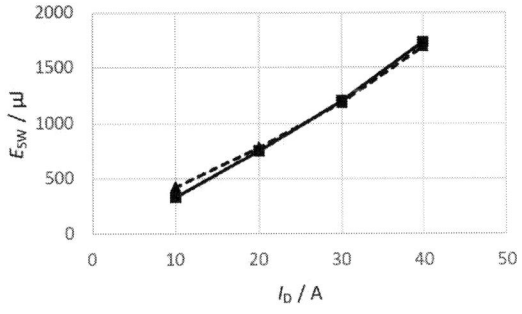

Fig. 19 Comparison of total switching loss energy E_{sw} between Q1-D1 (▲) and Q2-D2 (■) combination, when using $R_{G,ext}$=22 Ω.

3.6 Switching speed comparison

The switching speed curves dv/dt and di/dt for Q1-D1 during turn-on are illustrated in Fig. 20 and Fig. 21, respectively.

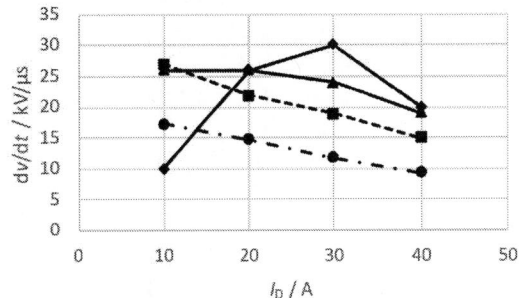

Fig. 20 Turn-on switching speed dv/dt of Q1-D1 with different $R_{G,ext}$=47 Ω (●), 22 Ω (■), 10 Ω (▲) and 6.7 Ω (♦).

From the dv/dt values given in Fig. 20 it can be concluded that the switching speed can be tuned with the external gate resistance as long as the value is selected in the range of $R_{G,ext}$=10...47 Ω.

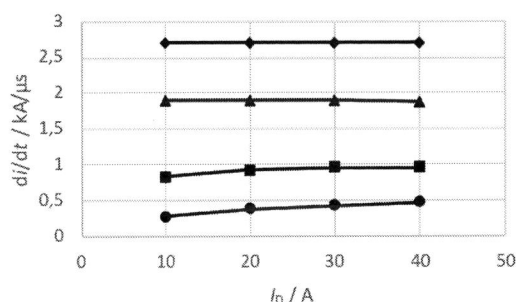

Fig. 21 Turn-on switching speed di/dt of Q1-D1 with different $R_{G,ext}$=47 Ω (●), 22 Ω (■), 10 Ω (▲) and 6.7 Ω (♦).

The same can be concluded for the current slope di/dt. Here, the difference from $R_{G,ext}$=6.7 Ω to 10 Ω is still significant. Since the dv/dt is not influenced at $R_{G,ext}$=6.7 Ω in a proportional way to di/dt, it can be concluded that the parasitic capacitances of the test setup and of the devices limit the dv/dt.

The switching speed curves dv/dt and di/dt for Q1-D1 during turn-off are illustrated in Fig. 22 and Fig. 23.

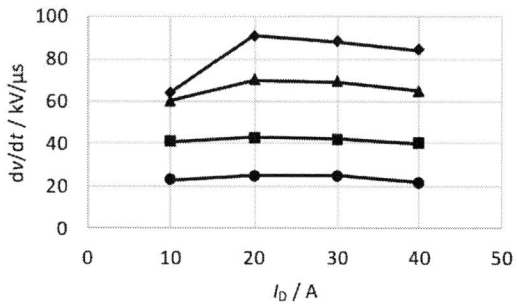

Fig. 22 Turn-off switching speed dv/dt of Q1-D1 with different $R_{G,ext}$=47 Ω (●), 22 Ω (■), 10 Ω (▲) and 6.7 Ω (♦).

The turn-off waveforms exhibit contrary behavior with dependance on $R_{G,ext}$. From dv/dt curves in Fig. 22, it can be concluded that by finetuning $R_{G,ext}$ the voltage transient can be controlled. Only at small drain currents, there is a limit to it due to the parasitic capacitances.

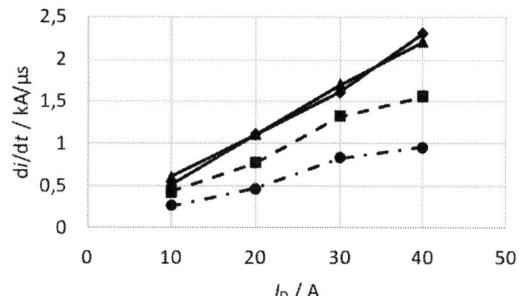

Fig. 23 Turn-off switching speed di/dt of Q1-D1 with different $R_{G,ext}$=47 Ω (●), 22 Ω (■), 10 Ω (▲) and 6.7 Ω (♦).

The current slope in Fig. 23 is controllable in the range of $R_{G,ext}$=10…47 Ω. There is no significant difference between $R_{G,ext}$=6.7 Ω and 10 Ω.

The test results presented above can be used to support the application specific design.

3.7 Thermal performance of the 3-level Vienna SMPD

The thermal resistance R_{thJH} of each chip in the engineering samples was measured using a Littelfuse specific test fixture. It was measured using a thermal grease thickness of 40 μm rolled on the backside of the SMPD package. This represents a practical approach used by customers and still leaves room for optimization. The results of R_{thJH} are listed in Table 2.

Device	Abbrev.	R_{thJH}
Diode 1	D1	2.2 K/W
Diode 2	D2	2.1 K/W
MOSFET 1	Q1	1.0 K/W
MOSFET 2	Q2	1.0 K/W

Table 2 Measured thermal resistance values of SMPD engineering sample.

The measured values match closely to the estimation given in [3].

4 Application specific design

Based on the presented measurement results above, the application specific losses can be estimated. The calculation method can be adopted from [3] and modified to the new target values.

Hence, the target grid current will be I_{AC}=16 A at an input voltage of U_{AC}=400 V and switching frequency of f_{sw}=48 kHz. The switching duty cycle as a function of the grid current is illustrated in Fig. 24.

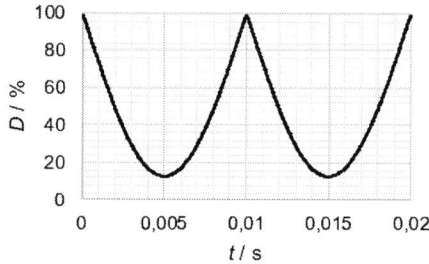

Fig. 24 Duty cycle as a function of time for a DC-link voltage of U_{DC}=750 V at a line voltage of $U_{AC,rms}$=400 V and an input current of IAC=16 A.

The switching losses during the grid cycle can be linearized from the measured switching losses by the regression functions (4)-(6) for different gate resistances $R_{G,ext}$.

$R_{G,ext}$=47 Ω:
$$E_{SW}(i_D)=0.825 \cdot i_D^2+19.99 \cdot i_D+190.5 \qquad (4)$$
$R_{G,ext}$=22 Ω:
$$E_{SW}(i_D)=0.3275 \cdot i_D^2+25.195 \cdot i_D+31.25 \qquad (5)$$
$R_{G,ext}$=10 Ω:
$$E_{SW}(i_D)=0.525 \cdot i_D^2+8.85 \cdot i_D+117 \qquad (6)$$

Based on these equations the overall losses can be estimated as listed in Table 3. The values include the doubled switching losses of the MOSFET.

Parameter	@R_g = 10 Ω	@R_g = 22 Ω	@R_g = 47 Ω
Switching losses per MOSFET	4.6 W	5.8 W	8.3 W
Conduction losses per MOSFET		22.3 W	
Conduction losses per MOS Bodydiode		3.5 W	
Conduction losses per Diode D1&D2		18.2 W	

Table 3 Calculated semiconductor losses of the SMPD prototype in a Vienna PFC at f_{sw}=48 kHz and input current of I_{AC}=16 A at T_{vJ}=125°C.

At the target operating point in Table 3, total semiconductor losses of 145.8...156.9 W for the complete input stage can be expected. This results in a semiconductor efficiency of η=98.58...98.68 %, which matches very well with the expected values in [3].

The resulting increase of the junction temperatures ΔT_J can be estimated as given in Table 4, based on the given thermal resistances of the chips in Table 2,. Consequently, a heatsink temperature of T_{HS}=80°C will keep the power semiconductors below their thermal limits at these the targeted operating conditions.

Device	R_{thJH}	ΔT_J
Diode 1	2.2 K/W	40.0 K
Diode 2	2.1 K/W	38.2 K
MOSFET 1	1.0 K/W	34.1 K
MOSFET 2	1.0 K/W	34.1 K

Table 4 Increase of junction temperatures for the given losses of Table 3 for $R_{G,ext}$=47 Ω.

5 Summary

In this paper, the design proposal for fitting a Vienna rectifier phase-leg into the Littelfuse ISOPLUS® SMPD package is presented. The article presents the concept of the prototype DCB layout as well as the background of the pin layout for stray inductance optimization. Additionally, a test setup to conduct dynamic characterization of the SMPD samples with integrated Littelfuse Ultrajunction silicon MOSFET IXFH22N65X2 and Littelfuse DSEP12-12A silicon FRED is presented. The thorough analysis of dynamic switching performance as well as static characterization provides a good basis to further extent the product range. The application related design process to estimate losses in a practical approach showcases that the presented solution perfectly combines the cost-ef-

fective silicon-based technology with high efficiency without the ultimate need for wide band-gap devices. It might be obvious that SiC or GaN devices will dissipate lower losses. However, depending on the target operating conditions highest efficiency is not always the highest priority. For design engineers, system designers, and for keeping a stable supply chain, a full silicon solution is still a considerable choice, especially in trending topics as for example the EV-charging infrastructure or other systems connected to the public grid.

References

[1] Hypercharger by alpitronic, https://www.hypercharger.it/hyc400/, access on 2023-10-17

[2] ABB Terra 360 HPC, https://new.abb.com/ev-charging/terra-360, access on 2023-10-17

[3] Haehre et. al., Evaluation of Silicon-Based 3-Level T-Type Neutral Boost Rectifier Integrated into SMPD Package for EV-Charger Application, PCIM Europe 2023 conference, Nuremberg, Germany

[4] Littelfuse IXFH22N65X2, https://www.littelfuse.com/media?resource-type=datasheets&itemid=768f9992-b376-4042-9534-504d4728e1df&filename=littelfuse-discrete-mosfets-n-channel-ultra-junction-ixf-22n65x2-datasheet, access on 2023-10-17

[5] Littelfuse DSEP12-12A, https://www.littelfuse.com/media?resource-type=datasheets&itemid=2bd25575-2e36-4173-905b-8067e799d3f1&filename=littelfuse-power-semiconductors-dsep12-12a-datasheet, access on 2023-10-17

PCIM Europe 2024, 11– 13 June 2024, Nuremberg   DOI: 10.30420/566262027

Performance Analysis of a 25-kW SiC-based Dual Active Bridge Converter based on Parallel-connected Devices

Francesco Porpora [1], Daniele Marciano [2], Emanuele Di Fazio [2], Mauro Di Monaco [1], Vito Nardi[1], Giuseppe Tomasso [1], Eric Benedict[4], Ryan Schnell[4], Maurizio Granato[3]

[1] University of Cassino and Southern Lazio, Italy
[2] E-Lectra s.r.l, Cassino (FR), Italy
[3] Analog Devices s.r.l., Vimercate (MB), Italy
[4] Analog Devices, USA

Corresponding author: Francesco Porpora, francesco.porpora@unicas.it

Abstract

This paper presents the design of a 25-kW Dual Active Bridge (DAB) converter based on parallel-connected SiC MOSFETs with the aim of demonstrating the applicability of discrete solutions for medium-power applications. A symmetric layout for the related gate-driving paths has been implemented and an experimental characterization of the DAB converter has been carried out considering different operating conditions in terms of input/output voltages and output power. The performances achieved in terms of electrical and thermal behavior of the paralleled SiC devices validate the proposed symmetric design.

1   Introduction

Silicon Carbide (SiC) MOSFETs have become predominant in different application fields, including electric vehicle chargers and photovoltaic inverters, thanks to their capability of covering a wide power range due to their best-in-class features in terms of switching speed, efficiency and thermal properties [1], [2], which require an accurate and reliable losses modeling to be exploited [3]. However, considering the limited maximum current rating of SiC-based commercial MOSFETs (usually below 120 A), the adoption of parallel-connected discrete devices or power module solutions is definitely required to ensure the high performances requested by high power/high current applications. Currently, with reference to the specific power rating of the desired application, the definition of the most suitable solution between discrete and module architectures is still a challenge due to performances, costs, sizes and reliability constraints. Indeed, power modules guarantee an optimal layout (low stray inductance), high insulation and thermal properties due to their optimal level of integration at the expense of increasing the cost and size of the power converter. On the other hand, the adoption of parallel-connected discrete devices allows

for optimizing the overall cost and the size of the power converter due to the possibility of easily displacing the SiC MOSFETs as well as the greater availability of different current rating solutions within the portfolio of the semiconductor manufacturers, which further contributes to avoid excessive oversizing that could negatively impact on the performances. However, the paralleling of discrete devices requires precise symmetric design of both power and driving layouts in order to minimize potential imbalance conditions for the current sharing between the parallel-connected devices [4], which is currently limiting their adoption for high-power applications. As widely discussed in literature, this undesired imbalance condition can occur mainly due to manufacturing tolerances for the SiC devices that cause differences in terms of threshold voltage and on-state resistance [5], [6] and their sensitivity to the junction temperature [7]. Specifically, the mismatch between the on-state resistance of parallel-connected discrete devices produce a current imbalance that affects the conduction losses distribution among them. Instead, the mismatch between the threshold provides an imbalance during the switching transients, which affects the switching losses distribution and is strongly influenced by the driving layout design. Consequently, designs based on parallel connections are crucial to minimize the effects of asymmetric layout and current imbalances on the static and dynamic performances of the SiC devices [8], [9]. A different approach for reducing this potential imbalance condition in discrete-based applications also includes the Active Gate Driver (AGD), which performs an active control of the driving voltage to each parallel device to equalize the current sharing [10], [11]. In particular, the rising and falling edges of the driving voltages are adjusted with specific delays to achieve an optimal synchronization, needed to compensate the current imbalance.

With the aim of increasing the knowledge of paralleling solutions and demonstrating their applicability for medium-power applications, this paper presents a performance analysis of a 25-kW SiC-based Dual Active Bridge (DAB) converter, properly designed with an optimal symmetry of the gate-driving sections to the parallel-connected devices. A flexible design that includes up to three SiC MOSFETs in parallel is considered and different operating conditions are implemented for evaluating the correct functionalities of the proposed DAB converter. In particular, experimental tests at low and high power are carried out to verify both the electrical and thermal behaviors of the parallel-connected SiC MOSFETs, focusing on the symmetry of the gate-source voltages and the thermal gradients.

2 Proposed DAB Architecture

This section presents the details of the proposed 25-kW DAB converter, based on parallel-connected SiC MOSFETs and designed with a symmetric layout for the gate-driving sections. In particular, the architecture can be divided in three main sections: power, driving/measurement and control. It is important to highlight that the design of the proposed DAB converter has been carried out with the aim of ensuring modularity and flexibility to the architecture, without taking into account any specific requirement in terms of power density or efficiency.

2.1 Power Section

The proposed DAB converter has been designed for a rated power of 25 kW, an input/output voltage range of 300-600 V and an operating switching frequency of 100 kHz. Three parallel-connected SiC MOSFETs manufactured by Infineon (model IMZ120R030M1H) have been considered for each switch that composes both full-bridge converters at the input and output ports of the DAB architecture, thus enabling the possibility of evaluating the performances achievable with a different number of parallel connections. In order to minimize the potential occurrence of unbalanced current sharing among the SiC MOSFETs, symmetric paths have been adequately implemented between the gate driver boards and the three parallel devices. Indeed, as shown in Fig. 1(a), the parallel-connected SiC devices have been displaced along the circumference centered in the gate trace coming from the driver board to correctly ensure the equidistance for the three gate-driving paths. Moreover, all the copper trace lengths of the gate-driving paths have been equally designed to avoid different parasitic parameters for the driving layout, which would further contribute to increase the severity of potential imbalance conditions for the parallel connection. Likewise, the selection of the components included in the gate driving paths and the related tolerances have been properly selected.

Figure 1(b) shows the experimental layout of one full-bridge converter, including the related power, gate-driving and measurement sections. Note that a modular and configurable design has been considered for the gate-driving section, which can be easily disconnected from the power section, thus allowing the evaluation of different driving solutions as well. Moreover, DC-link capacitors based on a mixed technology, including both Film and Multilayer Ceramic (MLC) solutions, have been adopted to fulfill the high current rating of the converter and improve the switching performances of SiC devices through the minimization of the power-loop layout, respectively. Indeed, due to their compact leadless design, the MLC capacitors have been displaced closer to the SiC devices, allowing a strong reduction of the stray inductances of the layout. A pre-charge circuit has been also added to the power section for safely charging the DC-link capacitors and thus avoiding high inrush currents. This circuit presents an Insulated-gate Bipolar Transistor (IGBT), model IGW60T120 by Infineon without anti-parallel diode, in series with a 47Ω high-power resistor, model AY470KE by Ohmite, which limits the current through the DC-link capacitors during the pre-charge operation. Then, once completed, the IGBT is turned off during the normal operation of the converter and the current flows through two parallel-connected SiC MOSFETs, model C3M0021120D by Wolfspeed, resulting in lower power losses thanks to the re-

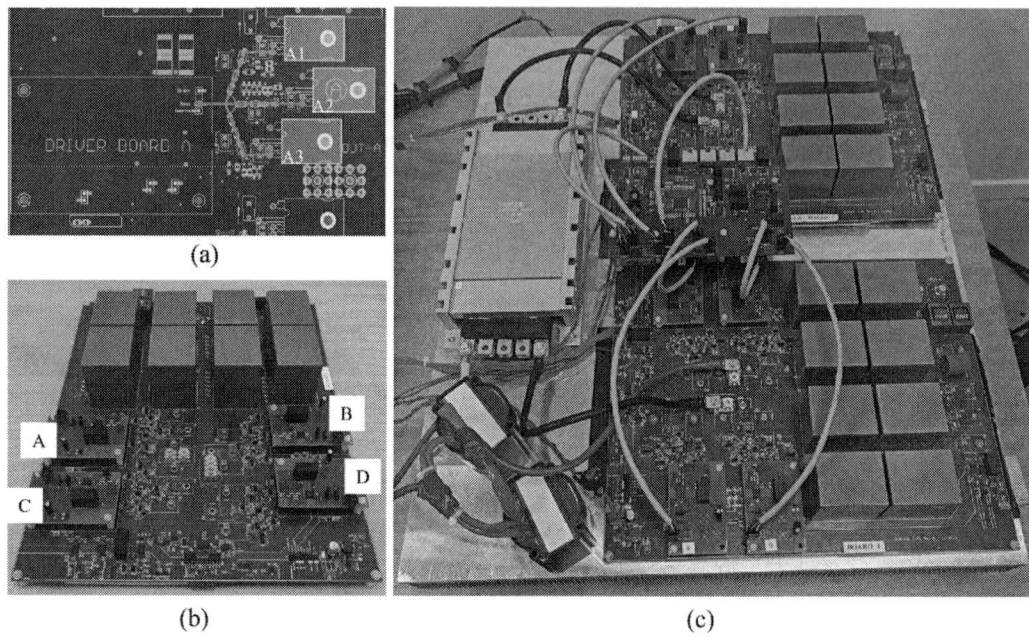

Fig. 1: (a) Concept of the symmetric layout of the gate driving paths to the three parallel-connected SiC devices; (b) full-bridge converter with the related gate driver sections; (c) overall architecture for the proposed 25-kW DAB converter.

duced on-resistance of SiC technology. Figure 1(c) illustrates the overall architecture of the proposed 25-kW DAB converter, including the power sections of both full-bridges at the input and output ports, respectively named as *Board 1* and *Board 2*, which are connected through a high-frequency transformer (planar technology by Payton) with an external series inductor, as well as the gate-driving, measurement and control sections.

2.2 Driving and Measurement Sections

As previously discussed, a single gate driver has been implemented for all the three parallel-connected SiC MOSFETs. Considering the large number of switches for *Board 1* and *Board 2*, they have been named depending on the specific gate-driving board connected. In particular, with reference to Fig. 1, the four gate-driving boards that compose each full-bridge have been named as A, B, C and D, while the specific SiC MOSFETs controlled as $X1$, $X2$ and $X3$ with $X = A, B, C, D$.

For each gate driver, evaluation boards developed by Analog Devices Inc. (ADI) have been taken into account as reference, which implement different solutions for isolated gate drivers with a large variety of isolated power supplies, able to fulfill the voltage driving requested by SiC devices from different manufacturers. Therefore, a flexible inter-connection between the gate-driving and power sections has been considered to enable the possibility of adopting different driving solutions from ADI while adopting the same power section. To ensure this flexibility, the form factor of the evaluation boards and the position of the main connectors (gate-source, logic inputs and power supply) have been shared between the layouts of both the power and the gate-driving sections. Note that, despite the possibility of adopting different gate-driving solutions from ADI, the evaluation board based on the MAX22701E has been selected for carrying out the performance analysis of the proposed DAB architecture, which is featured by ultra-high common-mode transient immunity (CMTI) and Miller Clamp feature as well as an integrated digital galvanic isolation using proprietary process technology.

On the other hand, the measurement section includes all the components devoted to the monitoring of the correct operating conditions of the DAB converter in terms of voltage, current and temperature as well as to the protection in case of undesired operation. Most of the components have been selected from the ADI portfolio and summarized in Table 1, highlighting the specific model and function for each one. In detail, voltage measurements have been considered upstream and downstream of the

Tab. 1: Details of all the ADI components adopted for the measurement section of the proposed DAB converter.

Model	Function
ADUM4195-1	Voltage measurement
AD7402	Current measurement
AD8210	Current measurement
ADG1208	Temperature measurement
ADA4891-2	
ADUM4120	Pre-charge Gate driver
AD2302	5V global supply
ADP165AUJZ-3.3-R7	3.3V sensor supply
ADA4891-2	Protection against short-circuit
	Hardware control pre-charge

DC-link capacitors for correctly performing the pre-charge operation, whereas current measurements have been implemented to either monitor the leg current of each full-bridge or ensure protection for the DAB converter against shoot-through events. Then, measurement components for monitoring the temperatures of both SiC devices and the board have been implemented, resulting in a complete management of the DAB converter during all the operating conditions.

2.3 Control Section

A custom design has been carried out for the control section, based on a microcontroller by STMicroelectronics, model STM32G484, and located between the two full-bridges that compose the DAB converter, as shown in Fig. 1(c). To improve the gate-driving paths and minimize any issues due to electromagnetic interferences, twisted and shielded cables have been adopted for transferring the control signals and the power supply from the control section to all the gate driver boards. Then, the proposed DAB converter has been managed through an open-loop control based on a traditional phase-shift modulation [12]. In detail, a custom software tool has been implemented in MATLAB as a graphical interface for communicating with the microcontroller and setting the control parameters, including switching frequency, the power flow direction (Grid-to-Vehicle or Vehicle-to-Grid), the dead time and the phase-shift. In this way, the operating conditions of the DAB converter in terms of input/output power can be directly adjusted by varying the phase shifting between the two full-bridges.

3 Performance Analysis

This section presents a performance analysis of the proposed DAB converter. In particular, the experimental setup is accurately described, highlighting all the power and measurement instrumentation adopted. Then, low-power (up to 5kW) and high-power (around 25kW) testing results are illustrated to demonstrate the correct functionalities of the DAB converter in different operating conditions and the effectiveness of the symmetric layout implemented for the gate driving paths to the parallel-connected SiC MOSFETs. Note that, despite the design of the full-bridges has been carried out considering the possibility of adopting up to three parallel-connected SiC devices for each power switch, a configuration based on two devices in parallel only has been taken into account. Indeed, this configuration represents the worst condition for the parallel-connected SiC devices since they experience a higher current and thus a higher thermal stress, which could strongly contribute to emphasize the effect of imbalance conditions among the switches due to an incorrect gate-driving layout.

3.1 Experimental Setup

Figure 2 illustrates all the instrumentation and tools within the experimental setup implemented for carrying out the performance analysis of the proposed DAB converter and gate-driving layout. In particular, a diversified instrumentation is considered for correctly monitoring both the electric and thermal behaviors of the SiC MOSFETs. More details regarding their characteristics are reported as follows.

- Two programmable DC bidirectional power supplies by Regatron (model G5.54) for powering the input and output ports of the proposed DAB converter.

- A 7-channel power analyzer with the related voltage and current sensors by Zes Zimmer for accurately computing the input/output powers processed, thus the efficiency of the proposed DAB converter.

- 1-GHz oscilloscope by Tektronix (model MSO58B) for voltage and current measurements by means of high-accuracy isolated probes (gate-source voltages), model IsoVu TIVP1, differential voltage probes (transformer voltages), model P5205A, and a current probe (inductor current), model TCP303.

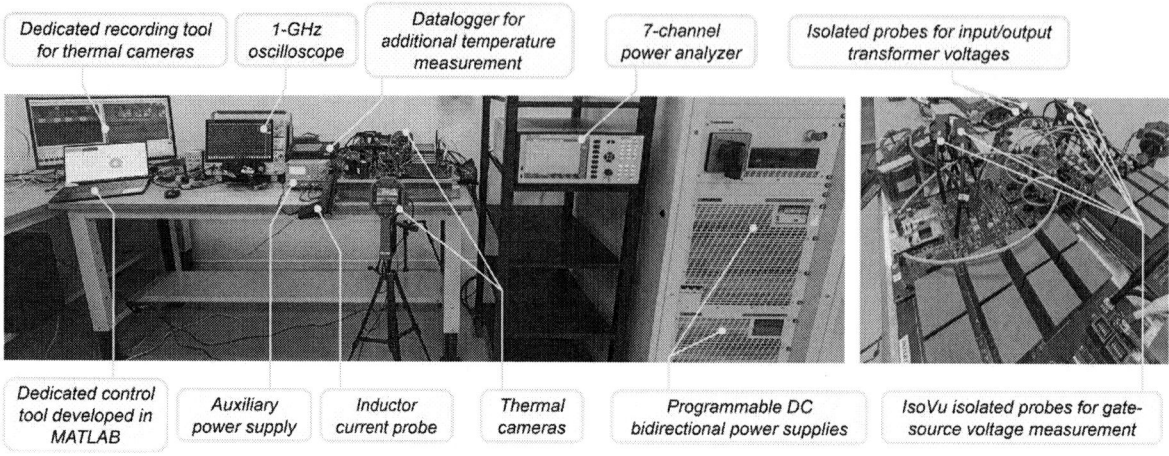

Fig. 2: Pictures of the experimental setup implemented for carrying out the performance analysis of the proposed DAB converter and gate-driving layout.

– Auxiliary power supply (model MX100TP by Aim-TTi Instruments) for powering the components within the control and power sections of the proposed DAB converter.

– Two thermal cameras, model 345A by Fotric, one for each full-bridge of the proposed DAB converter, to accurately evaluate the thermal gradients among the parallel-connected SiC MOSFETs over the heat sink at both input and output ports. Then, a dedicated software is used for continuously recording the variations of the thermal gradients over time from both thermal cameras.

– An additional datalogger (models LR8450 and U8552 by Hioki) for measuring the temperature of all the SiC devices of the entire DAB architecture by using Negative Temperature Coefficient (NTC) thermistors as sensor technology, directly placed between the power section and the top side of the MOSFETs' package.

This experimental setup has been adopted for testing the proposed DAB converter in different operating conditions in terms of input/output voltage and output power and consequently evaluating its thermal and electrical behaviors. Moreover, these experimental tests have been carried out to verify the temperature distributions between the parallel-connected SiC MOSFETs, which were expected to be balanced due to the symmetric layout of the gate-driving paths implemented for the proposed DAB converter.

3.2 Low Power Testing

With the aim of evaluating the thermal distribution among the parallel-connected SiC MOSFETs over time, the performance analysis has started by operating the DAB converter at a lower power with respect to the rated one. Indeed, high powers would have requested the adoption of a forced convection cooling system attached to the heat sink, leading to the unfeasibility of accurately monitoring the thermal gradients through the two thermal cameras. Therefore, an output power of around 5kW has been selected as an optimal operating condition for testing the thermal behavior of the SiC devices over time interval without encountering hazardous temperatures. Moreover, different operating conditions in terms of input and output voltages have been taken into account to verify the severity of potential temperature imbalances among the parallel-connected SiC MOSFETs depending on the voltage level at the input and output ports of the proposed DAB converter. In detail, two voltage levels have been considered, equal to 450V and 500V, properly combined between *Board 1* and *Board 2* for reproducing three different case scenarios. A fixed dead time equal to 500ns has been also implemented between the control signals of a single full-bridge leg for all the operating conditions selected.

Figure 3 shows the electrical and thermal behaviors of the proposed DAB converter in the three case scenarios. With reference to the electrical characteristics, different waveforms are reported, including the transformer voltages at the primary

PCIM Europe 2024, 11– 13 June 2024, Nuremberg   DOI: 10.30420/566262027

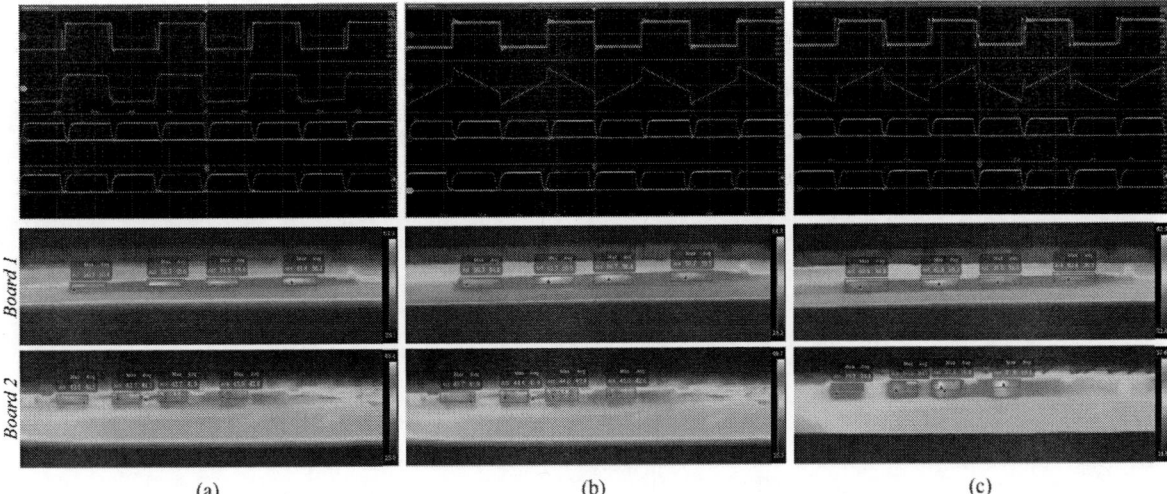

Fig. 3: Experimental results achieved at low power for the three different case scenarios: (a) $V_{in} = V_{out} = 500V$; (b) $V_{in} = 450V$, $V_{out} = 500V$; (c) $V_{in} = 500V$, $V_{out} = 450V$.

and secondary sides (*yellow* and *blue*), the inductor current (*red*) and multiple gate-source voltages (*orange* and *cyan* for *Board 1*, *green*, *white* and *magenta* for *Board 2*). The latter have been measured for highlighting the symmetry of the gate-driving signals for the parallel-connected SiC devices and the phase-shifting between the control signals of both full-bridges. These results validate the correct functionalities of the proposed DAB converter since the waveforms observed for the transformer voltages and the inductor current are in line with the ones expected in three different case scenarios. Moreover, the gate-source voltages of two parallel-connected SiC MOSFETs (*white* and *magenta*) result completely overlapped in all the test conditions, confirming the goodness of the symmetric layout implemented for the gate-driving paths of the proposed DAB converter.

On the other hand, the thermal images of a single leg at *Board 1* and *Board 2* are depicted to illustrate the thermal gradients achieved after a time interval of 20 minutes. Therefore, the temperature imbalances between the parallel-connected SiC MOSFETs ($B1$-$B3$ and $D1$-$D3$ for *Board 1*, $A1$-$A3$ and $C1$-$C3$ for *Board 2*) have been observed, highlighting their maximum and average values. Note that the initial conditions have not been considered equal for all the case scenarios since the main target of the performance analysis at low power was oriented to verify the temperature imbalance reached at the end of the test only, regardless of the overtemperature with respect to the initial thermal

distribution. However, as possible to notice from the thermal images of Fig. 3, $D3$ and $A1$ devices exhibit the lowest temperature for *Board 1* and *Board 2* in all the case scenarios. This is only due to the displacement of the SiC MOSFETs with respect to the heatsink. Indeed, $D3$ and $A1$ devices exchange heat with the coolest part of the heatsink, whereas the SiC MOSFETs displaced in the center of the power sections are surrounded by other devices, which result as additional thermal sources.

Table 2 summarizes the experimental results achieved at low power, highlighting the input/output voltages and powers for each case scenario considered (respectively V_{in} and P_{in} for *Board 1*, V_{out} and P_{out} for *Board 2*), the overall efficiency of the proposed DAB converter and the maximum thermal gradients observed after 20 minutes (ΔT_{20}) among each pair of parallel-connected SiC MOSFETs of *Board 1* and *Board 2*.

As possible to notice, a ΔT_{20} lower than or equal to $2°C$ has been achieved in all the case scenarios. Despite the manufacturing tolerances of the SiC MOSFETs, this value results compatible with the resolution of the thermal cameras adopted within the experimental setup and thus confirms the correct current sharing between the parallel-connected devices due to the proposed symmetric layout of the gate-driving paths. Moreover, according to the average temperature values shown in Fig. 3, note that the thermal gradients among the SiC MOSFETs result totally balanced, further confirming the symmetry of the parallel connections.

227

Tab. 2: Maximum thermal gradients achieved experimentally after 20 minutes (ΔT_{20}) at a fixed output power (P_{out}) of around 5kW considering the three different case scenarios.

V_{in} (V)	V_{out} (V)	P_{in} (W)	P_{out} (W)	η (%)	ΔT_{20} (°C)	
					Board 1	Board 2
500	500	5505.15	5181.07	94.113	2.0	0.7
450	500	5465.32	5003.20	91.544	1.0	0.7
500	450	5504.65	5165.74	93.843	1.7	1.0

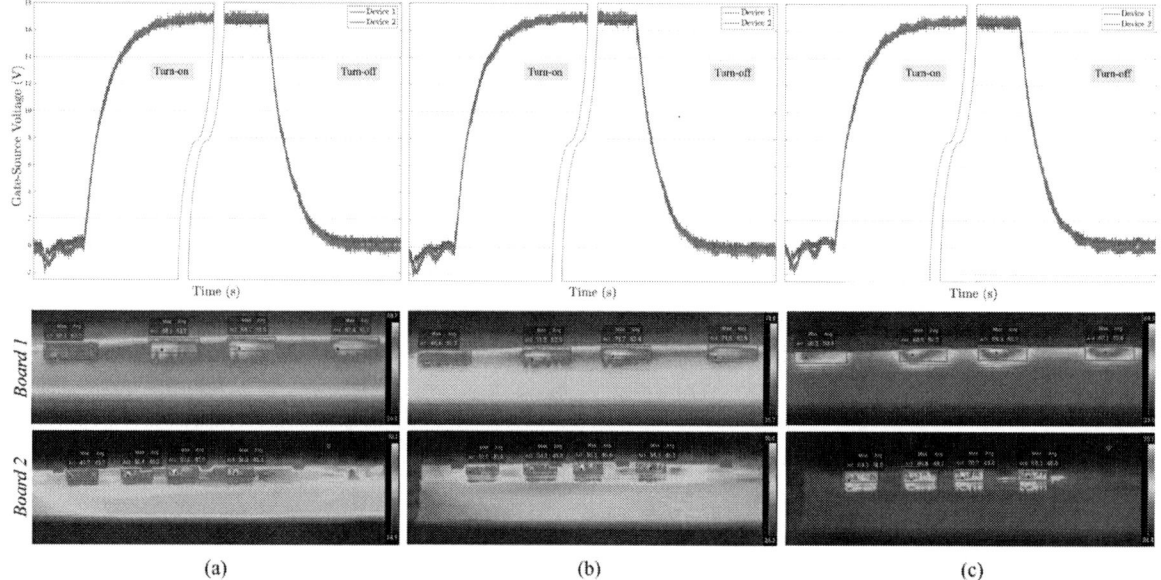

Fig. 4: Experimental results achieved at high power for three different operating conditions: (a) $P_{out} = 10kW$; (b) $P_{out} = 15kW$; (c) $P_{out} = 25kW$.

3.3 High Power Testing

Despite the unfeasibility of continuously operating the proposed DAB converter at powers greater than 5kW without using a forced cooling system, experimental tests have been also performed at higher output powers up to reach the rated one (25kW) by limiting the time interval of operation. In particular, considering the same input and output voltages ($V_{in} = V_{out} = 500V$) for *Board 1* and *Board 2*, three different operating conditions in terms of output power have been considered: around 10kW, 15kW and 25kW. In this case, each test has been terminated upon the detection of a maximum temperature for the SiC devices greater than $70°C$.

These experimental tests have been carried out to verify the presence of any thermal imbalance among the SiC MOSFETs or asymmetries in the gate-driving signals at higher powers processed by the proposed DAB converter. Figure 4 shows the gate-source voltages of two parallel-connected

SiC MOSFETs during the turn-on and turn-off operations as well as the thermal images of a single leg at *Board 1* and *Board 2* at the end of each high-power test. As possible to notice, the gate-source voltages result overlapped during the rise and fall times, further validating the proposed symmetric design of the gate-driving loop at high power as well. Moreover, according to the thermal gradients observed, the differences in the maximum temperature of two parallel-connected SiC devices increases as the output power processed by the DAB converter rises, resulting in a maximum value equal to $8.2°C$ achieved at 25kW. However, it is important to highlight that these results are strongly related to the different heat dissipation of the $D3$ and $A1$ devices, as discussed in section 3.2. Indeed, by comparing the thermal gradients between parallel-connected SiC devices that experience similar heat dissipation ($B1$-$B3$ for *Board 1*, $C1$-$C3$ for *Board 2*), the differences in the maximum temperature result lower than $2.5°C$ for all the three output powers processed. This validates the goodness of the proposed symmetric layout for the gate-driving paths also at high power.

4 Conclusion

In this paper, the design of a symmetric layout for the gate-driving section of a 25-kW DAB converter based on parallel-connected discrete SiC devices has been presented. Experimental tests have been carried out at low power (5kW) and high power (25kW) to illustrate a performance analysis of the electrical and thermal behaviors of the parallel-connected SiC MOSFETs in different operating conditions. The results have demonstrated the applicability of discrete solutions based on parallel connections for medium-power applications since symmetric gate-source voltages regardless the specific output power processed and thermal gradients compatible with the accuracy of the instrumentation adopted have been observed. Moreover, the performance analysis focused on the temperature imbalances among the parallel-connected SiC MOSFETs have also highlighted the importance of their displacement on the heatsink, which has to be optimized in order to ensure similar heat dissipation area for all the devices and thus further reduce potential thermal gradients.

References

[1] X. She, A. Q. Huang, Ó. Lucía, and B. Ozpineci, "Review of silicon carbide power devices and their applications," *IEEE Transactions on Industrial Electronics*, vol. 64, no. 10, pp. 8193–8205, 2017. DOI: 10.1109/TIE.2017.2652401.

[2] J. Biela, M. Schweizer, S. Waffler, and J. W. Kolar, "Sic versus si—evaluation of potentials for performance improvement of inverter and dc–dc converter systems by sic power semiconductors," *IEEE Transactions on Industrial Electronics*, vol. 58, no. 7, pp. 2872–2882, 2011. DOI: 10.1109/TIE.2010.2072896.

[3] F. Porpora, D. Marciano, F. P. Caruso, M. Di Monaco, and G. Tomasso, "Accurate data-driven losses modeling for sic-based converters," in *2024 IEEE Applied Power Electronics Conference and Exposition (APEC)*, 2024.

[4] C. Zhao, L. Wang, and F. Zhang, "Effect of asymmetric layout and unequal junction temperature on current sharing of paralleled sic mosfets with kelvin-source connection," *IEEE Transactions on*

Power Electronics, vol. 35, no. 7, pp. 7392–7404, 2020. DOI: 10.1109/TPEL.2019.2954716.

[5] H. Li, S. Munk-Nielsen, C. Pham, and S. Beczkowski, "Circuit mismatch influence on performance of paralleling silicon carbide mosfets," in *2014 16th European Conference on Power Electronics and Applications*, 2014, pp. 1–8. DOI: 10.1109/EPE.2014.6910835.

[6] H. Li, S. Munk-Nielsen, X. Wang, R. Maheshwari, S. Beczkowski, *et al.*, "Influences of device and circuit mismatches on paralleling silicon carbide mosfets," *IEEE Transactions on Power Electronics*, vol. 31, no. 1, pp. 621–634, 2016. DOI: 10.1109/TPEL.2015.2408054.

[7] G. Wang, J. Mookken, J. Rice, and M. Schupbach, "Dynamic and static behavior of packaged silicon carbide mosfets in paralleled applications," in *2014 IEEE Applied Power Electronics Conference and Exposition - APEC 2014*, 2014, pp. 1478–1483. DOI: 10.1109/APEC.2014.6803502.

[8] D.-P. Sadik, J. Colmenares, D. Peftitsis, J.-K. Lim, J. Rabkowski, and H.-P. Nee, "Experimental investigations of static and transient current sharing of parallel-connected silicon carbide mosfets," in *2013 15th European Conference on Power Electronics and Applications (EPE)*, 2013, pp. 1–10. DOI: 10.1109/EPE.2013.6634432.

[9] S. La Mantia, L. Abbatelli, C. Brusca, M. Melito, and M. Nania, "Design rules for paralleling of silicon carbide power mosfets," in *PCIM Europe 2017; International Exhibition and Conference for Power Electronics, Intelligent Motion, Renewable Energy and Energy Management*, 2017, pp. 1–6.

[10] Y. Xue, J. Lu, Z. Wang, L. M. Tolbert, B. J. Blalock, and F. Wang, "Active current balancing for parallel-connected silicon carbide mosfets," in *2013 IEEE Energy Conversion Congress and Exposition*, 2013, pp. 1563–1569. DOI: 10.1109/ECCE.2013.6646891.

[11] Y. Xue, J. Lu, Z. Wang, L. M. Tolbert, B. J. Blalock, and F. Wang, "Active compensation of current unbalance in paraleled silicon carbide mosfets," in *2014 IEEE Applied Power Electronics Conference and Exposition - APEC 2014*, 2014, pp. 1471–1477. DOI: 10.1109/APEC.2014.6803501.

[12] K. Zhang, Z. Shan, and J. Jatskevich, "Large- and small-signal average-value modeling of dual-active-bridge dc–dc converter considering power losses," *IEEE Transactions on Power Electronics*, vol. 32, no. 3, pp. 1964–1974, 2017. DOI: 10.1109/TPEL.2016.2555929.

PCIM Europe 2024, 11– 13 June 2024, Nuremberg   DOI: 10.30420/566262028

Semiconductor Chip Models are the Key for Enabling Virtual Design and Optimization Workflows of Power Electronic Systems

Stefan Haensel[1] , Sebastian Nielebock[1], Christian Radüge[1], Rolf Hellinger[1], Markus Pfeifer[1], Abby Shih[2]

[1] Siemens AG, Germany

[2] Keysight Technologies, Germany

Corresponding author: Stefan Haensel, Stefan.haensel@siemens.com
Speaker: Stefan Haensel, Stefan.haensel@siemens.com

Abstract

Virtual optimization of power electronic systems is based on accurate semiconductor models which exactly describe the switching behavior, current sharing and over-voltages of the converter system. A combined static & dynamic characterization and parameter fitting approach is described in this paper which enables semiconductor users to build and improve their models. To benchmark the overall workflow three different IGBTs are characterized and compared with predefined evaluation criteria. The results are promising and enable users to compare different designs but there is still an improvement required for a complete optimization of the entire system.

1 Introduction

Power electronic systems have become an integral part of our everyday lives, from chargers, power supplies in the consumer sector, electromobility up to the integration of renewable sources and energy distribution. The key element in all applications is the power semiconductor, which influences the overall system design in terms of electrical behavior, cooling requirements, gate driver circuits, filter components, and control effort. Virtual design of power modules or optimization of complete converter systems starts with a sufficiently accurate simulation of the semiconductor switching behavior which has to cover various influences, e.g., dynamic/static current sharing of parallel dies inside a power module or over-voltage in different operation points. A lot of experience and several design iterations are required to develop an adequate power module for a certain application. The development process takes years with various prototypes which are characterized by a broad range of operation conditions to identify the best-suited solution. Simulations in various domains accompany the development process today, although there is currently no closed-loop workflow from the die characterization to the power module model which is accurate enough to cover all design criteria of a power electronic converter system.

This paper provides an overview of available design workflows, challenges of the modeling and introduces a parameter extraction workflow to do a bare-die characterization. Based on the measurement results an equivalent circuit model is fitted to cover a broad range of operation points e.g., short-circuit behavior. Several evaluation criteria for the modeling accuracy are defined and applied during the workflow.

2 Overview of current virtual design workflows and challenges

Several virtual design workflows for this multi-objective-optimization of power electronic systems have been proposed, e.g. [1], [2]. All workflows have in common, that they rely on an accurate transient electrical simulation of the semiconductor switching behavior to extract e.g., current distribution among parallel semiconductor dies or the losses of devices. Once the transient simulation is built up, it can be used for an electrical and thermal optimization of the power module [3] or an entire converter. The question is how to achieve an accurate transient simulation model. First of all a semiconductor device model is needed, which may be supplied by the vendor itself, or can be generated by the user with tools from e.g. SABER [4] or Keysight [5]. Furthermore,

230

PCIM Europe 2024, 11– 13 June 2024, Nuremberg   DOI: 10.30420/566262028

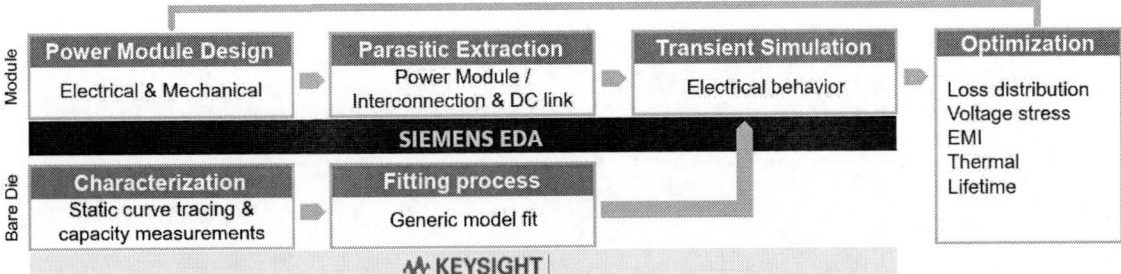

Fig. 1   Design workflow for the overall characterization, fitting, and modelling process

an exact representation of the surrounding network is needed, which includes parasitic elements from wire bonds and busbars, the DC link capacitor and the driver circuit [6]. The parasitic elements can be extracted by FEM simulation with well-known software like Q3D, HFSS, or PEPro [7], which achieve remarkable accuracy [8]. The DC link capacitor and the driver circuit can be implemented by simple equivalent circuit models, which can be built up based on measurements or datasheet values. All elements are merged into an electrical circuit simulator. A good accuracy can be achieved, if the circuit simulator supports S-parameter as an import file, so no translation from the frequency-based description of the parasitic elements to an equivalent circuit model is needed.

Therefore, from a user perspective, most of the challenges of this design workflow are already addressed by universities or well-known software companies but the semiconductor device model is still an open issue. Although some vendors of power semiconductors already supply compact models of their devices, either for free or for a fee, the accuracy of those models must be checked, every time a new simulation is set up or an operation point is changed. Semiconductor devices are complex devices, which can only be accurately simulated in each specific operation point using complex physical models. However, those models take too much computation time for an optimization workflow of a power module or a converter. Therefore, compact or behavior models were introduced, which simplify the complex 3D structure of a semiconductor drastically to a single-point device. In the past, a good alignment of simulation and measurement data of compact models was achieved by vendors like Infineon but not for the complete portfolio and the model accuracy is limited to certain operation points. Due to the encryption of the model users couldn't extend them for certain needs, e.g., improved accuracy in short-circuit operation.

To overcome this limitation a cooperation with Keysight was initiated to improve their toolset in terms of achievable accuracy and usability.

In this paper, the models of three different IGBTs and diodes from two different vendors are presented, which were extracted without any knowledge about the internal structure of the devices or any details about the production applying the described workflow from Fig 1.

2.1 Extraction Workflow

For the extraction workflow of IGBTs the transfer characteristic, the collector emitter voltage-dependent input, output and reverse transfer capacitance, and the gate emitter voltage-dependent input capacitance are needed. For the diode, the output characteristic and the voltage-dependent capacitance are needed. At least three different temperature operating points must be characterized to implement the temperature dependency of the modell. Since those values are not listed in a datasheet, the data was measured by a power device analyzer B1505A from Keysight. This data was imported to IC-CAP [9] to fit the generic device model provided by Keysight. In the first step, the static measurements were used for

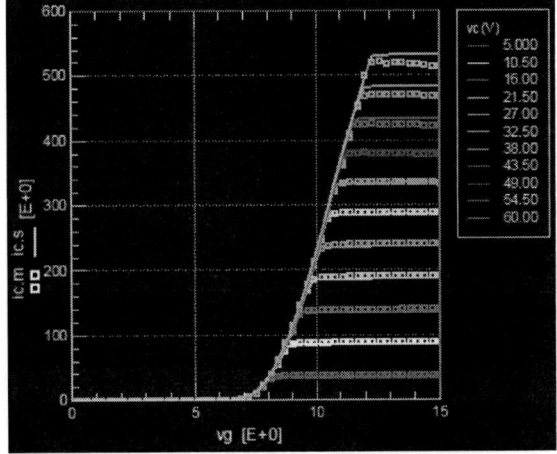

Fig. 2   Measured (dotted) and modelled (solid) output characteristic of IGBT

231

PCIM Europe 2024, 11– 13 June 2024, Nuremberg   DOI: 10.30420/566262028

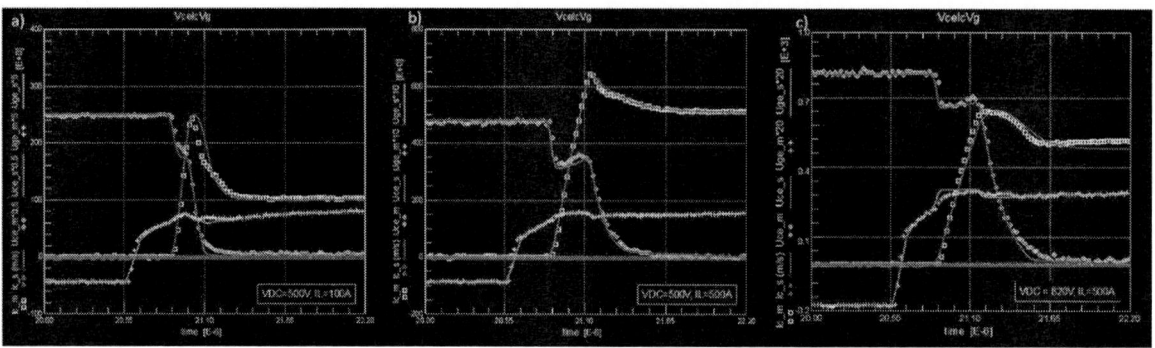

Fig. 4   Transient waveforms for a 1200 V / 250 A IGBT switch on Event at a) 500 V/100 A
b) 500 V/500 A c) 820 V/500 A; Dotted lines are measurements, solid lines simulations

the model extraction only. However due to the reverse recovery behavior of the diode [10], and due to the small values of the reverse transfer

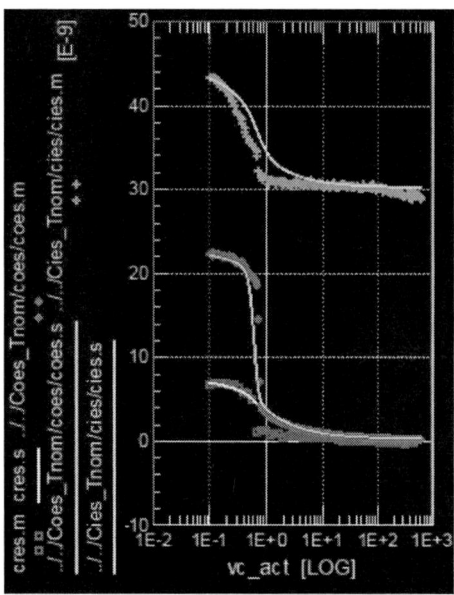

Fig. 3   Measurements (dotted) and modelled (solid) Voltage dependent capacitance of a 1200 V / 250 A IGBT

capacitance in the switch-off state of the IGBT, which is difficult to measure in a static measurement, a transient double pulse measurement was applied to extract the parameters. The measurements (dotted) and the modelled (solid line) output characteristic of the 250 A / 1200 V IGBT can be found in Fig. 2. The voltage dependent capacitances of the device are shown in Fig. 3.

After the modelling of the IGBT and the diode, the IGBT module and the double pulse measurement setup was replicated in simulation. The double pulse measurements at different voltages and currents were conducted in simulation and compared to the measurements. The results of

three operation points for the 250 A / 1200 V IGBT are shown in Fig. 4. The results for a 150 A / 650 V and a 75 A / 1200 V from Infineon [9] are shown in Fig. 5 and Fig. 6. It can be seen, that for all devices and all operation point a good accuracy was achieved. It can be concluded, that the underlaying generic chip model achieves a good accuracy for different power ratings and different vendors.

2.2   Definition of Evaluation Criteria

In most publications about semiconductor device modelling the achieved accuracy of the model is given as an error of the switching losses. But the losses describe only an overall accuracy of the model, and the simulated circuit (e.g. stray inductance, gate driver circuit, ...) and the error includes several different errors which may be present in the simulation. Since the value of the errors may be positive or negative, a good loss accuracy may be achievable although major errors may be present in the model or circuit. This becomes a problem, if the model shall be used in different environments, e.g. with variable gate

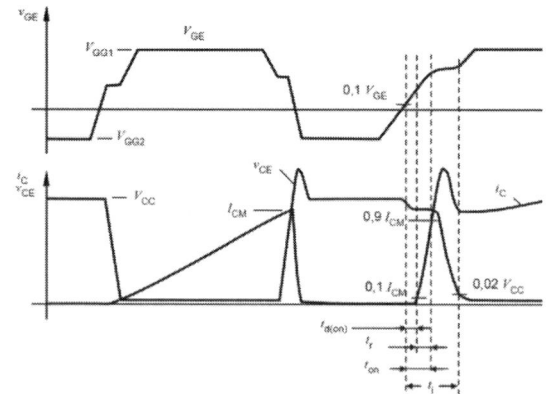

Fig. 7   Switch-on event of IGBT with corresponding levels and times defined by IEC 60747-9

232

PCIM Europe 2024, 11– 13 June 2024, Nuremberg   DOI: 10.30420/566262028

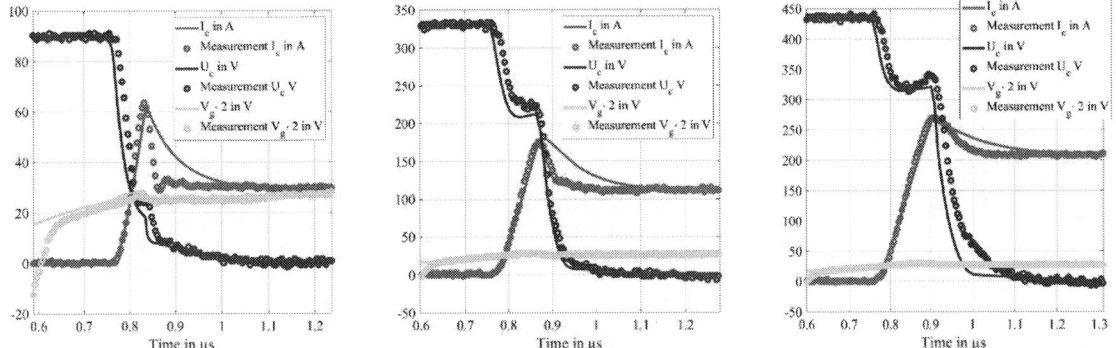

Fig. 5   Transient waveforms for 650 V / 150 A IGBT switch on Event at a) 90 V/30 A b) 330 V/100 A
c) 450 V/200 A; Dotted lines are measurements, solid lines simulations.

drive circuit or various values of parasitic elements of the module. A significant loss error might appear depending on the system setup. Furthermore, this can lead to a major error, if the losses shall not be investigated in the simulation but different aspects like current sharing, over voltages or the impact of complex driver circuits on the switching behavior.

To overcome these issues 13 criteria are proposed to evaluate the different aspects of the transient simulation, which have a significant impact on the switching losses.

2.2.1 Evaluation criteria for the switch-on event of IGBT

The first five evaluation criteria deal with the switch-on event of the IGBT. All criteria are based on the description of the waveform given by IEC 60747-9 [10], shown in Fig. 7. The first criteria describe the difference in % of the di/dt of the collector current of the IGBT. The definition is given by (1).

$$e_{\text{on didt}} = \frac{\frac{\Delta i_C}{\Delta t}_{\text{sim}} - \frac{\Delta i_C}{\Delta t}_{\text{meas}}}{\frac{\Delta i_C}{\Delta t}_{\text{meas}}} \quad (1)$$

$$= \frac{I_{\text{CM sim}} t_{\text{r meas}}}{I_{\text{CM meas}} \cdot t_{\text{r sim}}} - 1$$

With I_{CM} the collector's current and t_{r} the rise time. The second criterion is the peak collector current. Its error definition is given by (2). The error is mainly defined by the reverse recovery behavior of the diode. Nevertheless, it has a significant impact on the tail time (t_i) of the IGBT and therefore on the losses.

$$e_{\text{on Icmax}} = \frac{\max(i_{\text{C sim}})}{\max(i_{\text{C meas}})} \quad (2)$$

The third criterion is the error integral of the tail collector-emitter voltage of the IGBT. It describes

the difference in voltage tail behavior of the measurement and the simulation. Its definition is given by (3).

$$e_{\text{on VceInt}}$$
$$= \frac{\int_{t_0}^{t_0+t_i} |v_{\text{CE sim}}(t) - v_{\text{CE meas}}(t)| \, dt}{\int_{t_0}^{t_0+t_i} v_{\text{CE meas}}(t) \, dt} \quad (3)$$

For this integral it is important to define a common time basis of the simulation and the measurement. For the presented results, the simulation is shifted in time so that the time, where the simulation and measurement reaches half of the peak collector current is equal. The next criterion evaluates the voltage plateau during switch-on. The time to sample the voltage will be assumed to be $t_0 + t_{\text{d(on)}} + t_{\text{r}}/2$, which is half of the rise time. The criterion definition is given by (4).

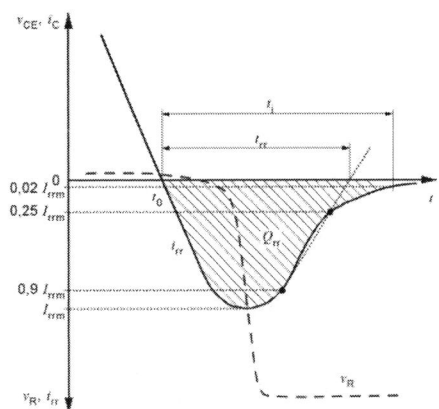

Fig. 8   Waveform of diode current and voltage according to IEC 60747-9.

233

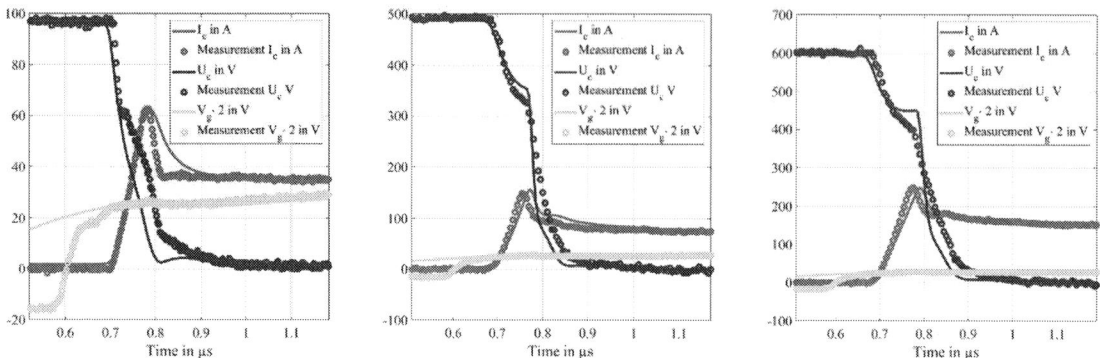

Fig. 6 Transient waveforms for 1200 V / 75 A IGBT switch on Event at a) 100 V/30 A b) 500 V/90 A c) 600 V/150 A; Dotted lines are measurements, solid lines simulations.

$$e_{\text{on Plateau}} = \frac{v_{\text{C sim}}\left(t_0 + t_{\text{d(on)}} + \frac{t_{\text{r}}}{2}\right)}{V_{\text{CC meas}}}$$
$$- \frac{v_{\text{C meas}}\left(t_0 + t_{\text{d(on)}} + \frac{t_{\text{r}}}{2}\right)}{V_{\text{CC meas}}} \quad (4)$$

The last criterion for the switch-on event is the difference in power losses. It is proposed to (3).

$$e_{\text{on loss}} = \frac{\int_{t_0}^{t_0+t_{\text{i}}} v_{\text{CE sim}}(t) \cdot i_{\text{CE sim}}(t) dt}{\int_{t_0}^{t_0+t_{\text{i}}} v_{\text{CE meas}}(t) \cdot i_{\text{CE meas}}(t) dt} \quad (5)$$

2.2.2 Evaluation criteria for switch-off event of diode

The reverse recovery behavior of the diode has a significant impact on the losses of the IGBT. Therefore, three different criteria are defined to describe the accuracy of the reverse recovery behavior. The typical reverse recovery behavior of a silicon diode is given in Fig. 8 [10]. The first criterion is the error in reverse recovery charge, which is defined by (6).

$$e_{\text{Qrr}} = \frac{\int_{t_0}^{t_0+t_{\text{i}}} i_{\text{rr sim}}(t)\,dt}{\int_{t_0}^{t_0+t_{\text{i}}} i_{\text{rr meas}}(t)\,dt} - 1 \quad (6)$$

The second criterion evaluates the error of the voltage waveform of the diode. It is given by (7).

$$e_{\text{off VcaInt}}$$
$$= \frac{\int_{t_0}^{t_0+t_{\text{i}}} |v_{\text{CA sim}}(t) - v_{\text{CA meas}}(t)|\,dt}{\int_{t_0}^{t_0+t_{\text{i}}} v_{\text{CA meas}}(t)\,dt} \quad (7)$$

The last criterion for the switch-off event of the diode is the difference in switching losses. It is similar to (3), but the voltage and current are replaced by the voltage and current of the diode.

The integration interval stays the same as for the IGBT.

2.2.3 Evaluation criteria for switch-off event of IGBT

For the switch-off event of the IGBT six criteria are proposed. The underlying definition of switch-off times is listed in IEC 60747-9, which is shown in Fig. 9. The first criterion defines the error of peak value of the switch-off overvoltage of the IGBT, which is given by (8).

$$e_{\text{off Vcemax}} = \frac{\max(V_{\text{CE sim}})}{\max(V_{\text{CE meas}})} \cdot \frac{V_{\text{CC meas}}}{V_{\text{CC sim}}} \quad (8)$$

The second criterion evaluates the di/dt of the collector current during the switch-off event. Its definition is the same as for the di/dt evaluation during switch-on and is given by (1).

The next criterion is the error in voltage rise. It is the integral between the collector-emitter voltage

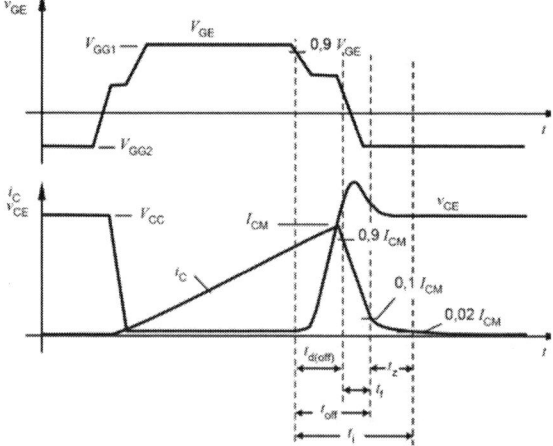

Fig. 9 Switch-off event of IGBT with corresponding levels and times defined by IEC 60747-9

PCIM Europe 2024, 11– 13 June 2024, Nuremberg   DOI: 10.30420/566262028

from the beginning of the switch-off event till the current decreases by 10 % of its maximum value. Its definition is given by (9).

$$e_{\text{off VceInt}} = \frac{\int_{t_0}^{t_0+t_{\text{d(off)}}} |v_{\text{CE sim}}(t) - v_{\text{CE meas}}(t)| dt}{\int_{t_0}^{t_0+t_{\text{d(off)}}} v_{\text{CE meas}}(t) dt} \quad (9)$$

Since there is a big difference in tail current observed in various simulations, the error in tail time (t_z) is proposed as the next criterion. Its definition is given by (10).

$$e_{IcTail} = t_{z\ \text{sim}} - t_{z\ \text{meas}} \quad (10)$$

The last criterion is the error in switching losses. Its definition for the switch-off event is similar to the switch-on event, given by (3).

3 Error investigation based on evaluation criteria

To evaluate the achieved accuracy of the model generated with the proposed workflow, the previous defined evaluation criteria are applied. During this work, 192 operating points (eight different currents, eight different voltages, and three different temperatures) of the three IGBTs were investigated. In this paper only the results of the 250 A / 1200 V IGBT model at 25°C are shown. The accuracy of the model can be plotted in various ways. In this paper, three possible ways shall be introduced. First, the error in switching losses of the IGBT for the switch-on event is investigated in dependency of the DC voltage and the switching current. The corresponding data is shown in Fig. 10.

It can be seen, that at small currents and voltages, the error is around -25 %, while at high-power regions this error becomes positive and is in the

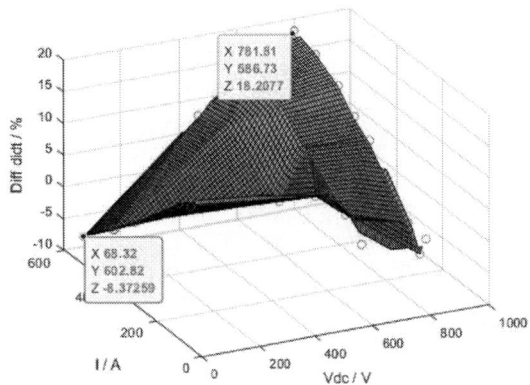

Fig. 11   Error in di/dt for different DC voltages and switching currents

range of 7 %. The di/dt is shown in Fig. 11. In the low voltage region the error is in the range of -8 %, while it increases to 18 % at high current and high voltage.

A second way of plotting the data is by the usage of a histogram. This gives a good overview about the frequency of occurrence in the simulation for

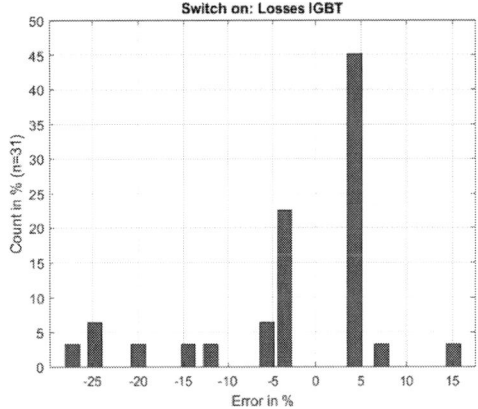

Fig. 12   Histogram of switch-on losses @25°C

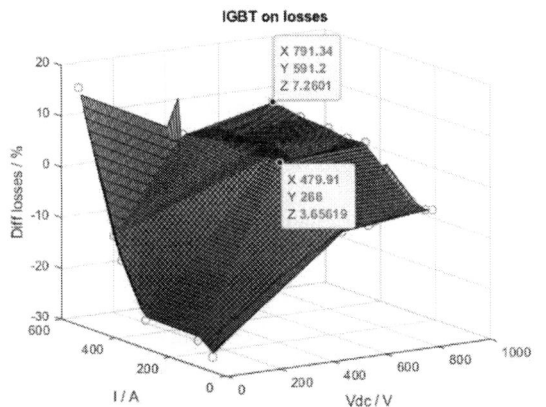

Fig. 10   Error in switching losses for different DC voltages and switching currents

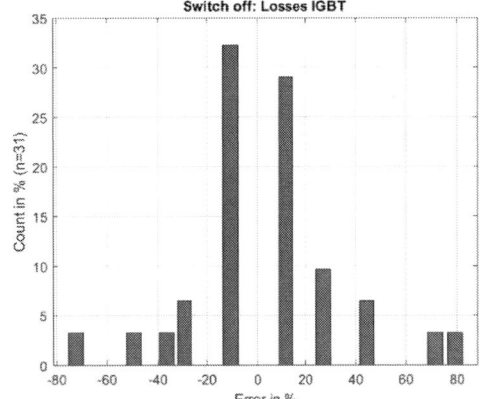

Fig. 13   Histogram of switch-off losses @25°C.

235

each error group. For the switch-on and switch-off losses these histograms are given in Fig. 12 and Fig. 13. It can be seen, that for both switching events the histogram is centered around zero. For the switch-on, almost 70 % of the total operation points are in the range of ± 5 %, while for the switch-off event, 60 % are in the range of ± 15 %. The third way of presenting the data is based on acceptance criteria. For this, a limit for each evaluation criterion is chosen and it is calculated how many operation points are within the limits. A detailed list of passed operation points with loos and tight limits is shown in Table I.

Table I: Criteria to evaluate the overall accuracy of the extracted model and accuracy results

	Loose Limit	Passed in %	Tight Limit	Passed in %
$e_{\text{on didt}}$	15 %	90	5 %	29
$e_{\text{on Icmax}}$	15 %	58	5 %	13
$e_{\text{on Vcelnt}}$	60 %	42	10 %	6
$e_{\text{on Plateau}}$	50%	77	10 %	0
$e_{\text{on loss}}$	50%	87	10 %	77
e_{Qrr}	50 %	68	10 %	0
$e_{\text{off Vcalnt}}$	100 %	71	10 %	0
$e_{\text{off d loss}}$	50 %	42	10 %	0
$e_{\text{off didt}}$	70 %	84	10 %	0
$e_{\text{off Vcemax}}$	15 %	94	5 %	26
$e_{\text{off Vcelnt}}$	40 %	84	10 %	0
e_{IcTail}	3 µs	87	0.1 µs	0
$e_{\text{off loss}}$	60 %	84	10 %	23

The column "passed" shows how many operation points fulfill the limit. It can be seen, that e.g., 90% of the operation points meet the $\pm 15\%$ limit for the slope of the current during the switch-on event, but only 29% meet the $\pm 5\%$ limit. For the reverse recovery, the accuracy is still worse and zero of the operation points meet the 10 % limit. Similar results can be achieved with the two other IGBT chips.

Although it may look like these models have insufficient accuracy based on the tight limits, this accuracy is sufficient to answer important questions during the design process and save hardware iterations already today. Compared to encrypted models supplied by semiconductor device manufacturers, these models can be adjusted if accuracy is insufficient in specific operation points. Therefore they are better suited for optimization process.

4   Conclusion and outlook

The investigation shows that quite good accuracy of the semiconductor model was achieved by combining transient and static measurements to fit the chip model. A basic comparison of different power module designs is possible with this model but there is still more accuracy needed for an overall optimization of converter systems, e.g. better match of switching losses, dynamic switching behavior and virtual tuning of different gate driver configurations. The paper shows that the characterization and generation of a physical model of semiconductors is also possible for end-users who don't have the necessary knowledge about manufacturing processes. Considering the overall virtual design and optimization trend, it would be helpful that semiconductor suppliers deliver detailed chip models in the future which could be easily integrated into several circuit simulators based on a generic model description language. Suppliers of power electronic systems can use these models to optimize their products and generate virtual models of the complete converter which can be used by system integrators or customers to analyze e.g., the plant behavior.

References

[1] J. Biela, J. W. Kolar, A. Stupar, U. Drofenik und A. Müsing, "Towards virtual prototyping and comprehensive multi-objective optimisation in power electronics" in *PCIM Europe 2010: International Exhibition & Conference for Power Electronics Intelligent Motion Power Quality*, 2010.

[2] N. Hingora, X. Liu, Y. Feng, B. McPherson und A. Mantooth, "Power-CAD: A novel methodology for design, analysis and optimization of Power Electronic Module layouts" in *2010 IEEE Energy Conversion Congress and Exposition (ECCE)*, Atlanta, GA, 2010, S. 2692–2699, doi: 10.1109/ECCE.2010.5618043.

[3] J. Wang, W. Chen, Y. Wu, J. Zhang, L. Wang und J. Liu, "Chip-Level Electrothermal Stress Calculation Method of High-Power IGBT Modules in System-Level Simulation", *IEEE Trans. Power Electron.*, Jg. 37, Nr. 9, S. 10546–10561, 2022, doi: 10.1109/TPEL.2022.3163199.

[4] M. Zhang, A. Courtay und Z. Yang, "An improved behavioral IGBT model and its characterization tool" in *2000 IEEE Hong Kong Electron Devices Meeting*, Hong Kong, China, 2000, S. 142–145, doi: 10.1109/HKEDM.2000.904235.

[5] Simon Muff, Abby Shih, Ludwig Eichinger, Bernhard Holzinger und Hiroaki Tanigawa, "From device modeling to characterization: a complete end to end design flow for SiC devices half bridge design" in *PCIM 2020*.

[6] M. Nagel, S. Race, I. Kovacevic-Badstuebner, T. Ziemann und U. Grossner, "Virtual PCB Layout Prototyping: Importance of Modeling Gate Driver and Parasitic Capacitances" in *2022 IEEE Design Methodologies Conference (DMC)*, Bath, United Kingdom, 2022, S. 1–5, doi: 10.1109/DMC55175.2022.9906542.

[7] M. Bucolo *et al.,* "A Comparative Analysis of Computer-Aided Design Tools for Complex Power Electronics Systems", *Energies*, Jg. 14, Nr. 22, S. 7729, 2021, doi: 10.3390/en14227729.

[8] D. Popescu und M. Treiber, "Broadband TCAD mixed-mode simulation framework for predictive modeling of fast dynamic switching events" in *2019 31st International Symposium on Power Semiconductor Devices and ICs (ISPSD)*, Shanghai, China, 2019, S. 327–330, doi: 10.1109/ISPSD.2019.8757578.

[9] Infineon, "FS75R12KT4: EconoPACK2 module withTrench/Fieldstop IGBT4 and Emitter Controlled 4 diode and NTC" [Online]. Verfügbar unter: https://www.infineon.com/dgdl/Infineon-FS75R12KT4-DataSheet-v03_00-EN.pdf?fileId=5546d4626d82c047016d8c09 290f43d9.

[10] *Semiconductor devices; Part 9: Discrete devices - Insulated-gate bipolar transistors (IGBTs)*, 60747-9, IEC, Nov. 2019.

Improved Resonant Frequency-based Parasitic Inductance Estimation Method for SiC MOSFET Half-bridge Circuit

Hongpeng Zhang[1], Felix Steiner[1], Horst Demattio[1], Thomas Blank[1]
[1] Karlsruhe Institute of Technology, Germany

Corresponding author: Hongpeng Zhang, zhang.hongpeng@kit.edu
Speaker: Hongpeng Zhang, zhang.hongpeng@kit.edu

Abstract

Estimating parasitic inductance by resonant frequency via double pulse tests is an easy-to-implement method. This method faces challenges in estimating the parasitic due to the short switching time of Wide-bandgap (WBG) semiconductor applications, causing measurement errors and poor resolution. This paper utilizes data processing methods such as Fast Fourier Transform, Continuous Wavelet Transform, and Variational Mode Decomposition to improve the accuracy and sensitivity of the conventional resonant frequency-based method. Besides, the finite element method (FEM) and experimental impedance measurement are carried out to validate the inductance estimation result. As a result, the proposed methods provide more accurate estimated parasitic inductance, reducing the estimation error by 40% to 90%.

1   Introduction

To fully release the potential advantage of the SiC MOSFETs, the switching frequency of the power converters has steadily increased. In modern power converter devices, the switching frequency of SiC MOSFETs is approaching 200 kHz and even higher. Within this frequency range, even a few nanohenry of stray inductance can lead to severe failures, such as commutation current overshoot, switching oscillation [1], and gate driver crosstalk. Thus, estimating the power commutation loop parasitic inductance is vital for power electronics robust design.

The stray inductance is typically measured by an impedance analyzer [2], such as a vector network analyzer (VNA). The VNA directly measures the circuit RLC parameters, including the total stray inductance. During an impedance measurement, the semiconductors in the circuit under test are either excluded or shorted to form a simple two-port network. However, this action, at the same time, excludes the parasitic effects on the semiconductor packaging. Another measurement of these packaging parasitic elements is required to fully reveal the stray inductance of the commutation loop.

The FEM is also popular in estimating the parasitic inductance values [3]. FEM provides both self- and mutual inductance simulation results under different frequencies in the electromagnetic field. Since the multilayer PCBs usually have thin copper layers and thin dielectric layers, their 3D models consist of many tiny geometry structures, which significantly increase the computational time.

The resonant frequency-based method utilizing the double pulse test (DPT) is an alternative method that does not require a professional impedance meter [4] or hours of computation. This method converts the commutation circuit model into an equivalent LC resonant circuit according to Kirchhoff's Circuit Laws. Its resonant frequency will only depend on the value of its equivalent inductance and capacitance. Consequently, the equivalent inductance can be estimated by the resonant frequency extracted via DPT and the equivalent capacitance value provided in the product datasheets.

The conventional approach to extracting the resonant frequency counts the period length of the ringing in the voltage signal by detecting zero-crossing points and then taking the reciprocal of the period length to derive the resonant frequency. However, two factors affect zero-crossing detection in practical measurement. Firstly, the resonant ringing is usually damped within 50 ns, resulting in insufficient sampling points considering the oscilloscopes' fixed sampling rate of, e.g., 2 GHz. The sparse sampling points limit the resolution of the extracted resonant frequency [2]. Secondly, the resonant ringing usually overlaps with low-frequency oscillations caused by long probes or common-mode noise. This phenomenon shifts the zero-crossing points away from its actual value.

This paper introduces the resonant frequency-based method and investigates data processing methods based on the Fast Fourier Transform [5], Continuous Wavelet Transform [6], and Variational Mode Decomposition [5] in Section 2. We explore how these data processing methods enhance the accuracy and sensitivity of frequency extraction, both theoretically and experimentally. Section 3 illustrates and compares the results of inductance estimation obtained from voltage - current differential method, the impedance measurements, the FEM simulation, the conventional resonant frequency-based method, and the proposed method. Finally, Section 4 concludes this paper.

2 Ringing Frequency-based Parasitic Inductance Estimation Method

2.1 Clamped Inductive Switching Circuit for the Double-Pulse Test

This paper utilizes a clamped inductive switching circuit as an example to introduce the concept of the conventional resonant frequency-based inductance estimation method.

The clamped inductive switching circuit without a snubber circuit consists of parasitic inductance and parasitic capacitance. The parasitic elements cause significant resonant behaviors on the MOSFET drain-to-source voltage during the switching transient in the DPT. The resonant frequency-based method indirectly measures the stray inductance by detecting the circuit's resonant frequency. The clamped inductive switching circuit for DPT is shown in Fig. 1.

The clamped inductive switching circuit comprises one half-bridge circuit and a load inductor. A power supply U_{dc} and a bulk capacitor C_{bulk} are on the DC side. The equivalent series inductance of the C_{bulk} is illustrated as L_{esl}. The inductance of the bus bar is modeled as L_{DC+} and L_{DC-}.

The half-bridge circuit contains two discrete SiC MOSFETs, low-side MOSFET Q1 and high-side

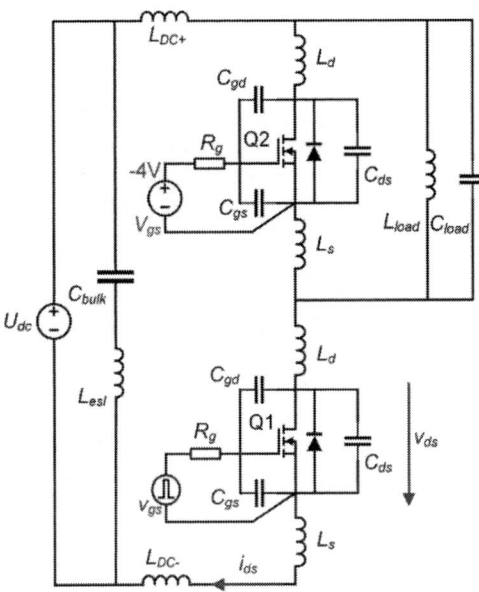

Fig. 1 Clamped inductive switching circuit model and the DPT waveform.

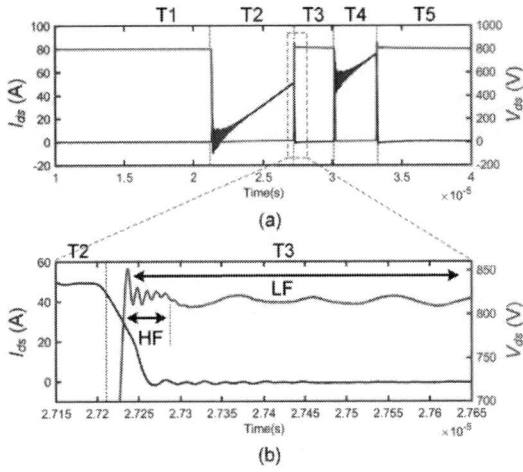

Fig. 2 The experimental double pulse test result on MOSFET Q1 at U_{dc}=800 V and I_{load}=50 A. (a) The Q1 drain-to-source voltage and current waveforms. (b) The zoomed-in waveforms during the Q1 turn-off transient. The high-frequency ringing (HF) and the low-frequency ringing (LF) are indicated.

MOSFET Q2. Their parasitic capacitors are shown besides Q1 and Q2, including the drain-source capacitance C_{ds}, the gate-source capacitance C_{gs}, and the gate-drain capacitance C_{gd}. The parasitic inductance on their leads is also modeled as two pairs of L_D and L_S. The body diodes are illustrated as a freewheeling loop. Besides, the isolated gate driver loops of Q1 and Q2 are shown with their gate resistors.

During the DPT, the high-side gate driver is kept in a negative off state, and the low-side gate driver outputs the double-pulse pattern signal.

A load inductor L_{load} is clamped to Q1 in the commutation loop. The parasitic resistors of these components mentioned above are ignored because the circuit resonant frequency is independent of them.

The proposed parasitic inductance estimation method is based on the double pulse test characterization results. The double pulse test is done on the MOSFET Q1 and Q2 (C2M0025120D) at U_{dc}=800 V and I_{load}=50 A. The bulk film capacitors have a total capacitance of 240 µF. The load inductance is a 90 µH air coil. A pair of active differential voltage probes, PMK BumbleBee® ±2000 V with a bandwidth of 400 MHz, measures V_{ds} on Q1. A Rogowski coil current probe, PEM CWTUM/1/R with a bandwidth of 30 MHz, measures I_{ds} on Q1. The sampling rate of the 12-bit oscilloscope, LeCroy HRO 66Zi, is 2 GS/s. The gate driver output voltage is +20 V/-5 V, and the external gate resistor is 10 Ω.

The experimental result is shown in Fig. 2(a). The double pulse test consists of five periods. During period T1, the DC power supply turns on and charges C_{bulk}. The period T2 represents the first pulse during which Q1 turns on, and C_{bulk} charges I_{load} to 50 A. At the beginning of period T3, Q1 turns off, and the body diode of Q2 conducts the freewheeling current from L_{load}. The period T4 represents the second pulse during which Q1 turns on again. Finally, in the period T5, Q1 turns off. The pulse width of T2 is 6 µs, the dead time of T3 is 2.75 µs, and the pulse width of T4 is 3 µs.

It can be seen from Fig. 2(a) that the turn-on/-off behaviors of Q1 cause significant resonant ringing on V_{ds} and I_{ds}. Considering that the voltage probe has a higher bandwidth than the Rogowski coil, this paper focuses on the resonant ringing of the V_{ds} waveform at the beginning of period T3, as shown in Fig. 2(b). During the Q1 turn-off transient, V_{ds} approaches bus voltage, overshoots, and starts resonant ringing. It is worth noting that different frequency ringings exist in the turn-off transient. As depicted in Fig. 2(b), the voltage probe recorded a high-frequency ringing (HF) lasting about 50 ns and a low-frequency ringing (LF) lasting about 400 ns. The parasitic elements usually cause high-frequency ringing, and the common-mode noise in the half-bridge circuit usually causes low-frequency ringing.

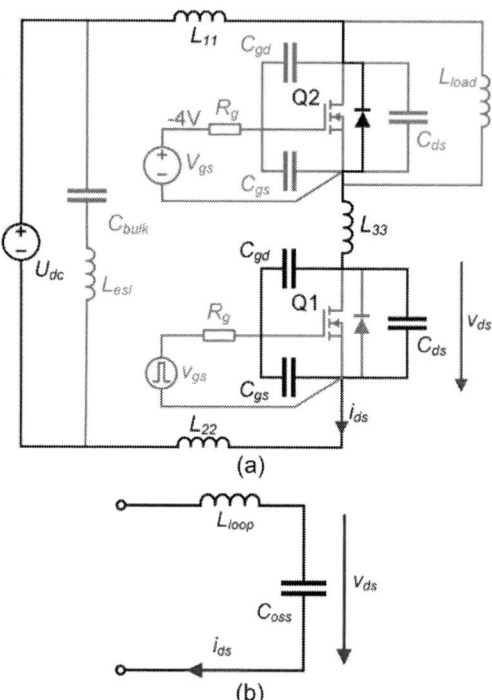

Fig. 3 Small signal equivalent circuit model for turn-off transient. (a) The commutation loop during turn-off transient. (b) The equivalent LC resonant circuit.

2.2 Parasitic Inductance Estimation Method

To simplify the resonant circuit analysis, the small signal equivalent model for the clamped inductive switching circuit turn-off transient is shown in Fig. 3(a). The stray inductance, the parasitic capacitance of Q1, and the body-diode of Q2 form the LC resonant circuit. The L_{DC+} and L_D of Q2 are equivalently combined as L_{11}; the L_{DC-} and L_S of Q1 are equivalently combined as L_{22}; the L_S of Q2 and L_D of Q1 are equivalently combined as L_{33}. The mutual inductance between L_{11}, L_{22}, and L_{33} are defined as M_{xy}, where $x, y = 1,2,3$, $x \neq y$. The loop inductance matrix $\boldsymbol{L}$ is shown in Eq. (1).

$$\boldsymbol{L} = \begin{bmatrix} L_{11} & M_{12} & M_{13} \\ M_{21} & L_{22} & M_{23} \\ M_{31} & M_{31} & L_{33} \end{bmatrix} \quad (1)$$

The small signal equivalent LC circuit of the turn-off commutation loop is shown in Fig. 3(b). The loop total stray inductance is defined as L_{loop}. The equivalent capacitance C_{oss} is the output capacitance of the MOSFET Q1. L_{loop} and C_{oss} are defined in Eq. (2).

$$\begin{cases} L_{loop} = \sum L = \sum_{i=1}^{3}\sum_{j=1}^{3} L_{ij} \\ C_{oss} = C_{ds} + C_{gd} \end{cases} \quad (2)$$

Thus, the function of the resonant frequency f_r inductance is defined by Eq. (3).

$$f_r = \frac{1}{2\pi\sqrt{L_{loop}C_{oss}}} \quad (3)$$

The parasitic inductance estimation method derives L_{loop} out of C_{oss} from the datasheet and f_r extracted from the double pulse test. The C_{oss} of Q1 is illustrated in Fig. 4. Based on Eq. (3), the parasitic inductance estimation equation considering frequency measurement resolution Δf is defined as Eq. (4).

$$L_{loop} \pm \Delta L_{loop} = \frac{1}{4\pi^2(f_r \pm \Delta f)^2 C_{oss}} \quad (4)$$

The derived loop inductance is sensitive to C_{oss}. Considering that the output capacitance of the MOSFET varies dramatically when V_{ds} is low, this test configures the double pulse test power supply as 800 V to reduce the C_{oss} error in calculation. The C_{oss} of Q1 is 227.9 pF when V_{ds} is 800 V.

2.3 Data Processing Methods

As shown in Fig. 2, the V_{ds} signal performs a step response to the turn-off behavior of Q1. V_{ds} consists of a high dv/dt step, an overshoot, high-frequency, and low-frequency ringing. The low-frequency signal makes the conventional method of counting the period length of the high-frequency signal unreliable. This is because the overlapping of the two ringings causes the zero-crossing points of the high-frequency signal to shift, which increases the error in the period length counting result.

Another factor adversely affecting the resonant frequency measurement is the voltage sampling frequency. Even with a voltage probe whose bandwidth is up to 500 MHz, the V_{ds} resonant signal is often sampled at less than 30 points per period. Consequently, this results in diminished accuracy in determining the resonant frequency.

This paper applies time-frequency domain methods to the V_{ds} resonant signal to avoid the above drawbacks. Three well-known data processing methods are utilized in this section.

2.3.1 Short-Time Fourier Transform

Applying the Fast Fourier Transform to capture the high-frequency characteristics of the double pulse pattern driven V_{ds} signal becomes

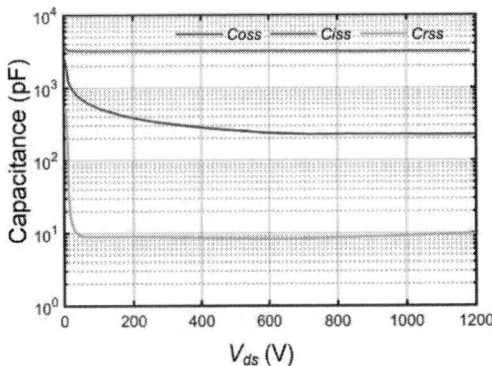

Fig. 4 The voltage dependent C_{oss} of the Q1 MOSFET from datasheet.

problematic. This is due to the Fast Fourier Transform having no time resolution. The Short-Time Fourier Transform (STFT) effectively extracts the frequency characteristics of the transient signals, which involves segmenting the original signal into shorter segments and subsequently subjecting these segmented signals to Fast Fourier Transform analysis. The STFT algorithm applies a longer window to short-term data to generate a localized moving spectrum in time history. The transient ringing behavior can be captured with proper window length and window type. The discrete format STFT definition is Eq. (5).

$$X_k = \sum_{n=0}^{N-1} x[n]w[n-kL]e^{-i2\pi n\frac{k}{N}} \quad (5)$$

Equation (5) segments the N length signal sampling data $\{x[n]\} := x[n0], x[1], \dots, x[n-1]$ by window function $w[n-kL]$ where L is the segments overlap. Then Eq. (5) transforms the segments into the STFT $\{X_k\} := X_0, X_1, \dots, X_{N-1}$.

STFT algorithm is based on the Fast Fourier Transform. The theoretical frequency resolution of STFT can be defined similarly, as shown in Eq. (6).

$$\Delta f = \frac{f_s}{N} \quad (6)$$

where Δf is the frequency resolution, and f_s is the sampling rate.

The power spectral density spectrogram of STFT applied to the V_{ds} resonant signal is illustrated in Fig. 5 as a waterfall plot. The STFT algorithm is done with $N=2048$, $L=1004$, and a 1024-length Kaiser window with a beta of 2. The input data starts from the overshoot behavior in V_{ds} prior to the resonant ringing. There are two magnitude peaks on the waterfall plot's first spectrogram marked by labels 1 and 2. The first

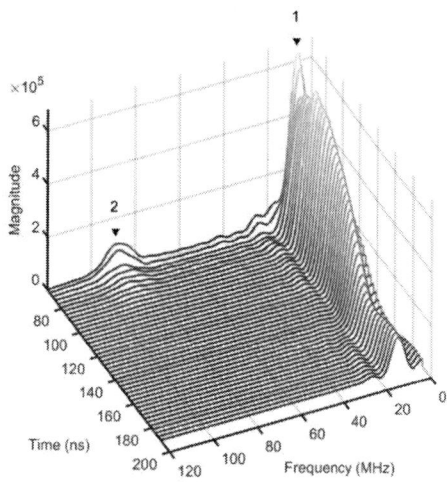

Fig. 5 The waterfall spectrogram of STFT applied to the V_{ds} resonant signal. Two magnitude peaks are marked by labels 1 and 2.

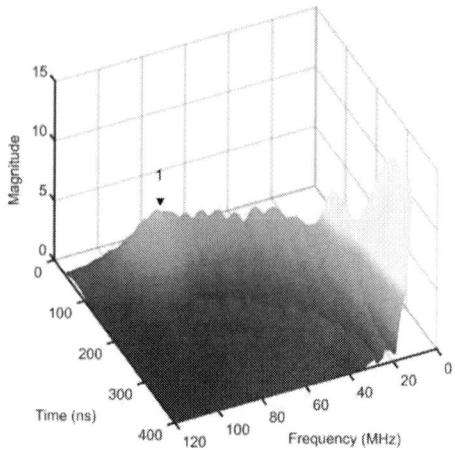

Fig. 6 The magnitude spectrogram of CWT applied to the V_{ds} resonant signal. One magnitude peak is marked by label 1. The CWT outside the cone of influence is ignored.

peak represents the low-frequency ringing at 6.8360 MHz, and the second peak represents the high-frequency ringing at 87.8906 MHz. The frequency resolution of the spectrum Δf is ± 1.9531 MHz.

2.3.2 Continuous Wavelet Transform

The Continuous Wavelet Transform (CWT) algorithm uses a wavelet function to highlight the transient ringing. Unlike traditional Fourier analysis, where sinusoidal functions are used, wavelets are localized in both time and frequency. Compared to the STFT, the CWT algorithm performs a more robust time resolution for analyzing non-stationary signals. When analyzing the frequency characteristics of V_{ds} signals, the CWT algorithm is more likely to identify transient behaviors.

In this study, we employ the analytic bump wavelet function for the CWT algorithm. The bump wavelet is a band-limited function defined in the frequency domain. This wavelet function offers notable advantages in frequency localization, although it exhibits relatively less efficient time localization than alternative wavelets. According to [7], the bump wavelet is defined as Eq. (7).

$$\widehat{\Psi}(s\omega) = 2e^{1 - \frac{1}{1 - \frac{(s\omega - \mu)^2}{\sigma^2}}} \mathbb{1}(s\omega) \qquad (7)$$

where s is the scaling factor, $\mathbb{1}$ is indicator function for the interval $[\mu - \sigma(1 - \varepsilon), \mu + \sigma(1 - \varepsilon)]$, the center frequency $\mu = 5$, the bandwidth $\sigma = 0.6$, and scaling factor ε is 2.2204×10^{-16}.

The discrete format CWT algorithm is defined as Eq. (8).

$$W_f^{\widehat{\Psi}}(s, \tau) = \frac{1}{\sqrt{s}} \sum_{n=0}^{N-1} x[n] \overline{\Psi\left[\frac{n - \tau}{s}\right]} \frac{1}{f_s} \qquad (8)$$

where τ is the time shift factor, $\Psi[n]$ is the discrete format inverse Fourier transform of $\widehat{\Psi}(\omega)$.

The bandwidth of the bump wavelet defines the frequency resolution of the CWT algorithm as Eq. (9).

$$\Delta f(s) \approx \frac{\sigma(1 - \varepsilon)}{2\pi s} \qquad (9)$$

The frequency resolution of the CWT algorithm is variable with the scaling factor s. The $\Delta f(s)$ decreases when the CWT frequency is higher.

The magnitude spectrogram of the CWT applied to the V_{ds} resonant signal is illustrated in Fig. 6. The CWT algorithm is done with $N=2048$ and 48 scaling factors per octave. The input data starts from the step behavior in V_{ds} prior to the overshoot and the resonant ringing. There is one magnitude peak in the spectrogram marked by label 1. The peak represents the high-frequency ringing at 83.8954 MHz. The ringing frequency has the frequency resolution Δf as -1.2028 MHz to below and +1.2203 MHz above.

2.3.3 Variational Mode Decomposition – Hilbert Huang Transform

The Variational Mode Decomposition - Hilbert-Huang Transform (VMD-HHT) method donates an advanced adaptive decomposition and signal analysis approach. Like the other decomposition method, VMD can decompose an input signal into a set of modes, such as offset, oscillation, and transient behaviors. The decomposing of the input signal is defined as

$$x(t) = \sum_{k=1}^{K} u_k(t) + r_k(t) \qquad (10)$$

where K is the total number of IMFs, $u_k(t)$ is the time domain IMFs, and $r_k(t)$ is the residual.

The common expression of IMFs is defined as

$$\mathrm{IMF}_k = u_k(t) = A_k(t) \cos(\phi_k(t)) \qquad (11)$$

where $A_k(t)$ is the instant amplitude and $\phi_k(t)$ is the instant phase.

As for the V_{ds} signal in this paper, these IMFs represent various modes of the step response decomposed by time and frequency features. IMFs consist of the continuous background noise, the transient resonant ringing, the high dv/dt step, and the residual.

The optimization goal of VMD is to obtain a series of smooth Intrinsic Mode Functions (IMFs) whose frequencies on the time span are limited to their center frequency and thereby optimally reconstruct the original signal. The VMD optimization is accomplished by solving a proposed constrained variational problem. The VMD applies the Hilbert transform to the IMFs to obtain analytic signals. Then, VMD modulates the analytic IMFs to a specified baseband by mixing with an exponential tuned to the respective estimated center frequency. Thus, VMD constructs a Gaussian smoothness optimization model of the analytic IMFs using a constrained variational problem, as in Eq. (12) and Eq. (13).

$$\min_{\{u_k\},\{\omega_k\}} \left\{ \sum_{k=1}^{K} \left\| \partial_t \left[\left(\delta(t) + \frac{j}{\pi t} \right) * u_k(t) \right] e^{-j\omega_k t} \right\|_2^2 \right\} \qquad (12)$$

$$\sum_{k=1}^{K} u_k(t) = x(t) \qquad (13)$$

where $\{\omega_k\}$ is the center frequency, $\delta(t)$ is the Dirac distribution, and $\| \ \|_2$ donates the L^2-norm.

[5] suggests rendering the problem unconstrained by using a quadratic penalty and Lagrangian multipliers in an iterative algorithm. Therefore, the IMFs with minimum bandwidth and stable center frequency are reconstructed.

The most significant feature of VMD is its ability to set the number of IMF components according to practical needs. Subsequently, VMD dynamically matches the optimal center frequency and finite bandwidth for each IMF component during the searching and solving process in an adaptive approach.

With the IMFs information, their instantaneous frequencies can be extracted by the HHT. The Hilbert transform in this paper is based on discrete Fourier transform approximation. Hilbert transform algorithm computes the discrete Fourier transform of the IMFs. Then, the Hilbert transform algorithm removes the negative frequency components and executes the inverse Fourier transform to return the Hilbert transform results. HHT algorithm constructs an analytic signal $z_k(t)$ by the input signal and the Hilbert transform results as shown in Eq. (14).

$$z_k(t) = u_k(t) + iH\{u_k(t)\} \qquad (14)$$

Where $H\{u_k(t)\}$ is the Hilbert transform of the IMFs.

HHT algorithm rebuilds the Eq. (14) as the instantaneous amplitude $A_k(t)$ and the instantaneous phase $\phi_k(t)$ as shown in Eq. (15).

$$z_k(t) = A_k(t) e^{i\phi_k(t)} \qquad (15)$$

Therefore, the instantaneous frequencies of the IMFs can be defined as the derivative of $\phi_k(t)$. The frequency resolution of the VMD-HHT method can be defined as Eq. (16).

$$\Delta f(t) = \frac{df(t)}{dt} = \frac{1}{4\pi^2} \frac{d^2\phi_k(t)}{dt^2} \qquad (16)$$

Researchers have different opinions about the frequency resolution. [8] defined the VMD-HHT frequency resolution by the resolution of the discrete Fourier transform algorithm, which is $n_{min}m_{min}f_s/N$ where n_{min} is the number of samples to define the maximum analysis frequency and m_{min} is the number of samples for the derivative calculation. While [9] proposed that the frequency resolution of VMD-HHT is not explicitly the same as that of the discrete Fourier transform because the HHT algorithm utilizing the gradient of phase information provides more detailed frequency information than the integral method in the discrete Fourier transform. The frequency resolution of VMD-HHT is able to achieve f_s/N with a proper filter.

This paper proposes that the HHT algorithm using the gradient method is able to surpass the limitation defined by the Nyquist–Shannon sampling theorem. The Nyquist–Shannon sampling theorem is the relationship between two integral functions and ignores the localized time information. In contrast, the instantaneous frequency given by the HHT algorithm is a differential function based on time [10]. Therefore, the frequency resolution of the HHT algorithm should be defined as that of the differentiation algorithm implementing the derivative calculation of the discrete input signal.

The IMFs and their instantaneous frequencies of VMD-HHT applied to the V_{ds} resonant signal are illustrated in Fig. 7. The VMD-HHT algorithm is done with $N=2048$ and $K=13$. The input data starts before the step behavior in V_{ds}. Fig. 7(a) illustrates that the IMF1 to the IMF5 are white noise continuous at all time, the IMF7 to the IMF13 represent the step behavior harmonics and the low-frequency ringing, and the IMF6 is the decomposed LC resonant ringing.

Fig. 7(b) illustrates the time-frequency dynamic of the IMF6 to the IMF13. The color of each IMF is proportional to its instantaneous amplitude. A circle marks the amplitude peak in the IMF6. The peak represents the high-frequency ringing at 85.4999 MHz. The VMD-HHT algorithm achieved a frequency resolution Δf of ±0.1000 MHz by IMFs interpolation.

3 Parasitic Inductance Extraction Methods

This section illustrates the parasitic inductance measurement results implemented by traditional methods to validate our proposed improved resonant frequency-based parasitic inductance estimation method.

3.1 Voltage – Current Differential Method

During the Q1 turn-on transient, the high di/dt rising current in the commutation loop causes a voltage drop across the parasitic inductance, as shown in Fig. 8. By measuring the voltage drop according to the current rising slope, the loop total parasitic inductance can be extracted as given in Eq. (17).

$$L_{loop} = \frac{\Delta V_{ds}}{\Delta I_{ds}/\Delta t} \quad (17)$$

The turn-on transient is located at the beginning of T4. The measured ΔV_{ds} is 20.68 V and the measured current slope $\Delta I_{ds}/\Delta t$ is 1.15 kA/µs. The extracted parasitic inductance is 18.0377 nH according to Eq. (17).

The utilization of differential methods to extract parasitic inductance may lead to an

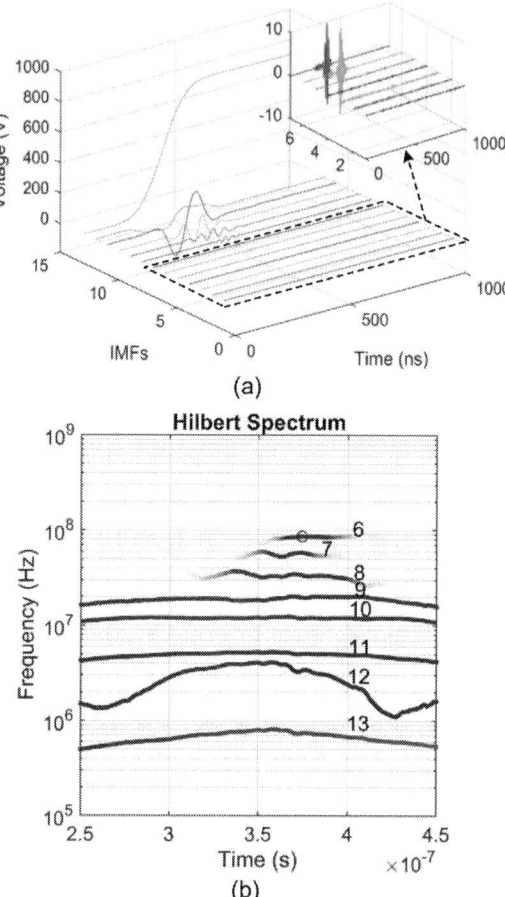

Fig. 7 The analysis results of the VMD-HHT method applied to the V_{ds} resonant signal. (a) The decomposed IMFs with a zoomed-in view on IMF1 to IMF7. (b) The HHT spectrum of IMF6 to IMF13 with a marker on IMF6 indicates the amplitude peak.

overestimation. This phenomenon can be attributed to several factors. Firstly, the issue of parasitic inductance estimation is influenced by the slew rate of the Rogowski coil. Despite the Rogowski coil's datasheet indicating a peak di/dt capability of 20 kA/µs, which surpasses the measured current slope rate, the measured current rise slope rate can be underestimated due to the attenuation of high-frequency harmonics. Another critical factor is the de-skewing problem. The current slope varies rapidly during the current rising transient. Hence, even a minor mismatch in de-skewing alignment could result in a notable error in the measured current slope rate.

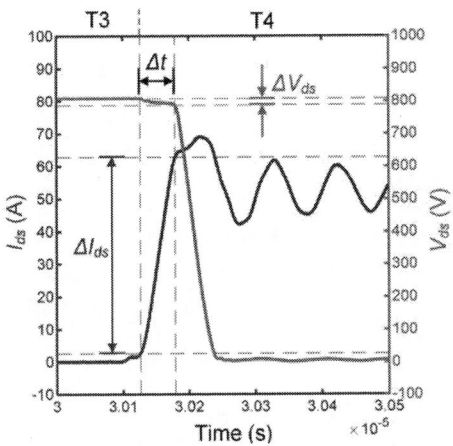

Fig. 8 The V_{ds} and I_{ds} waveform during Q1 turn-on transient at the beginning of T4.

3.2 COMSOL FEM Method

The loop inductance matrix of the PCB is estimated by COMSOL FEM simulation in Fig. 9. The PCB model consists of four copper layers and air as dielectric layers. The first and third layers are the positive power layers, and the second and fourth layers are power ground. This PCB model is simplified by removing tiny structures. The simulated loop inductance matrix L at 100 KHz is shown in Eq. (18). It can be seen that the mutual inductance of this laminated circuit is relatively high according to the self-inductance, which reduces the total loop inductance of this PCB.

$$L = \begin{bmatrix} 20.2794 & -18.7692 & 0.0461 \\ -20.7072 & 29.5270 & -2.4557 \\ 0.1030 & -2.5558 & 5.3489 \end{bmatrix} \quad (18)$$

It is worth noting that the DC bulk capacitance C_{bulk} as well as the Q1 and Q2 MOSFET pin inductance are missing in the FEM model due to the fact that only the copper plate of the PCB is simulated. Therefore, the FEM simulation results may be lower than the actual loop inductance.

A frequency-sweep simulation is implemented in COMSOL. The results are shown in Tab. 1. The simulated L_{loop} is a function of the model frequency. Considering the switching frequency in the DPT, the L_{loop} simulation result at 100 kHz better matches the PCB under test.

3.3 Vector Network Analyzer Measurement

The Vector Network Analyzer, Bode 100 from Omicron Lab, measures the turn-off loop

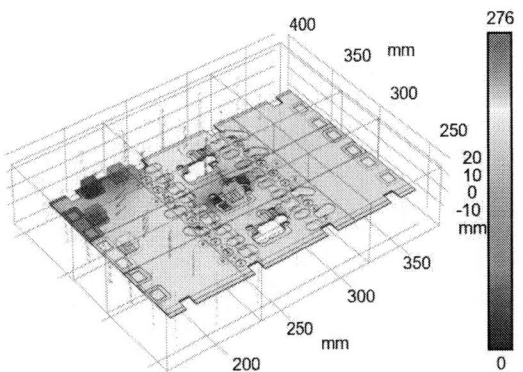

Fig. 9 The COMSOL FEM simulation result with the colour bar donating the electric potential (μV) and the arrow indicating the magnetic flux density.

f_{model} (kHz)	L_{loop} (nH)
0	20.5837
1	19.2381
10	13.9285
100	10.8165

Tab. 1 COMSOL frequency-sweep inductance simulation result

inductance. The probes are connected to the drain and source leads of the Q1, as in Fig. 10. The Q2 is shorted on the footprint by a copper bar, the bulk capacitors are soldered, and the load inductor is removed. The measurement report is shown in Fig. 10. The resonant loop is inductive between 100 kHz and 10 MHz, with a positive phase response. Two specific results are 18.0240 nH at 100 kHz and 15.4596 nH at 200 kHz.

3.4 Proposed Method Result

The extracted resonant frequency is more accurate and precise based on the proposed data processing methods. Thus, the resonant frequency-based parasitic inductance estimation method is improved. The extracted resonant frequencies f_r, the estimated parasitic inductance L_{loop}, and their resolutions Δf and ΔL_{loop} are shown in Tab. 2.

As can be seen from the table, the L_{loop} results obtained by all four methods are in line with the results obtained by the traditional methods in the previous sections. The maximum difference of

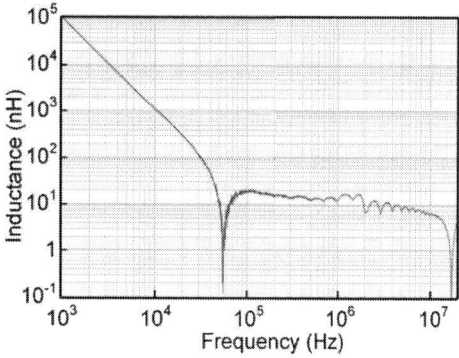

Fig. 10 Vector Network Analyzer measured PCB stray inductance.

their results with respect to each other does not exceed 1.5 nH. Among them, the inductance error ΔL_{loop} calculated without data processing is the largest, followed by the STFT method results. The result obtained using the CWT method has a smaller ΔL_{loop} compared to the first two, and lastly, the ΔL_{loop} of the results obtained using the VMD-HHT method is the smallest.

4 Conclusion

This paper demonstrates the difficulty of applying the traditional periodical counting method to high-frequency SiC MOSFET DPT. The data processing methods are utilized to improve the accuracy and the robustness of frequency extraction, improving the accuracy of the estimated loop inductance. The results from the proposed improved resonant frequency-based parasitic inductance estimation method is in line with that of the voltage – current differential method, the FEM simulation, and the VNA impedance measurement. The estimated inductance with proper data processing is more accurate and precise. The inductance estimation error is reduced by 40% to 90%.

References

[1] T. Liu, R. Ning, T. T. Y. Wong, and Z. J. Shen, "Modeling and Analysis of SiC MOSFET Switching Oscillations," in *IEEE Journal of Emerging and Selected Topics in Power Electronics*, vol. 4, no. 3, pp. 747-756, Sept. 2016, doi: 10.1109/JESTPE.2016.2587358.

[2] A. Lemmon, S. Banerjee, K. Matocha and L. Gant, "Analysis of Packaging Impedance on Performance of SiC MOSFETs," *PCIM Europe 2016; International Exhibition and Conference for Power Electronics, Intelligent Motion, Renewable Energy and Energy Management*, Nuremberg, Germany, 2016, pp. 1-8.

[3] B. Zhang and S. Wang, "Parasitic Inductance Modeling and Reduction for Wire-Bonded Half-Bridge SiC Multichip Power Modules," *in IEEE Transactions on Power Electronics*, vol. 36, no. 5, pp. 5892-5903, May 2021, doi: 10.1109/TPEL.2020.3032521

[4] S. Chen and H. S. Krishnamoorthy, "Analysis of Parasitic Inductance in Multi-MHz Frequency Switching Power Converters," *2022 IEEE International Conference on Power Electronics, Drives and Energy Systems (PEDES)*, Jaipur, India, 2022, pp. 1-5, doi: 10.1109/PEDES56012.2022.10079999

[5] K. Dragomiretskiy and D. Zosso, "Variational Mode Decomposition," in *IEEE Transactions on Signal Processing*, vol. 62, no. 3, pp. 531-544, Feb.1, 2014, doi: 10.1109/TSP.2013.2288675

[6] Mallat S, "A Wavelet Tour of Signal Processing: The Sparse Way (Third Edition)," *Academic Press*, 2009, pp. 89-153, doi: 10.1016/B978-0-12-374370-1.00008-2.

Data processing	None	STFT	CWT	VMD-HHT
f_r (MHz)	86.9565	87.8906	83.8954	85.4999
Δf (MHz)	(-3.6232, +3.9526)	(-1.9531, +1.9531)	(-1.2028, +1.2203)	(-0.1000, +0.1000)
L_{loop} (nH)	14.6991	14.3884	15.7914	15.204
ΔL_{loop} (nH)	(-1.3060, +1.2504)	(-0.6614, +0.6188)	(-0.4627, +0.4496)	(-0.0356, +0.0355)

Table 2 Proposed inductance estimation methods results

[7] A. Silik, M. Noori, W. Altabey A., R. Ghiasi, Z. Wu, "Comparative Analysis of Wavelet Transform for Time-Frequency Analysis and Transient Localization in Structural Health Monitoring," *Structural Durability & Health Monitoring*, vol. 15, no. 1, pp. 1–22, march 2021, doi: 10.32604/sdhm.2021.012751

[8] Huang, Norden E., et al. "The Empirical Mode Decomposition and the Hilbert Spectrum for Nonlinear and Non-Stationary Time Series Analysis," Proceedings: *Mathematical, Physical and Engineering Sciences*, vol. 454, no. 1971, 1998, pp. 903–95.

[9] Q. Gai, "Theoretical research and application of local wave time-frequency analysis method," Ph.D. dissertation, Dalian University of Technology，Dalian, 2001.

[10] Y. Zhong, "Research on the Local-instantaneous Signal Analysis Theory of the Hilbert-Huang Transform," Ph.D. dissertation, Chongqing University, Chongqing, 2002.

PCIM Europe 2024, 11– 13 June 2024, Nuremberg   DOI: 10.30420/566262030

Fast simulator with inverter temperature estimation for traction eDrives in vehicles subjected to driving cycles

Simone Giuffrida[1], Luisa Tolosano[1], Fabio Mandrile[1], Claudio Romano[2], Maurizio Tranchero[2], Radu Bojoi[1]

[1] Politecnico di Torino, Italy

[2] Ideas&Motion, Italy

Corresponding author: Radu Bojoi, radu.bojoi@polito.it
Speaker: Simone Giuffrida, simone.giuffrida@polito.it

Abstract

Accurate estimation of the losses and of the junction temperature of power devices is a well-known problem in the design of traction inverters. Very often the inverter design must be validated before the prototyping, using specific operating conditions that must be properly simulated using digital twins of the powertrain. This paper presents a fast simulation model of a traction eDrive providing an accurate inverter temperature estimation. The proposed model is implemented in Matlab/Simulink and uses inverter efficiency maps obtained with PLECS simulation environment. The model is extremely useful when the eDrive is subjected to long driving cycles. The eMotor model requires the motor flux maps, without being dependent on the specific eMotor torque control strategy. The proposed model has been simulated using the flux maps of a commercial traction motor and the loss data of a SiC three-phase inverter power module.

1   Introduction

The verification of losses and of the junction temperature is a well-known problem in the design of traction inverters. A common approach is to design the inverter for the worst case in terms of maximum allowed junction temperature, maximum input battery voltage and maximum output current, considering the maximum coolant temperature. Then, the inverter design must be verified according to a realistic driving cycle of the vehicle.

The eDrive (inverter and eMotor) can be tested together on a test rig. Usually, the target eMotor is torque-controlled, while its speed is imposed by a driving machine. The references for torque and speed are imposed by the test rig controller, according to a specific driving cycle that must be tested. Obviously, this testing approach requires a calibrated eMotor torque control that needs time to be properly implemented on the target inverter.

The next step is the verification of junction temperatures of the inverter power modules/switches. The junction temperature can be estimated in real-time using direct optical methods (infrared cameras) [1], physical contact methods (thermocouples applied on die surfaces) [1] or electrical methods, where the temperature is estimated via the measurement of thermosensitive parameters, such as the on-state voltage drop [2]. However, all these methods need special power modules or expensive additional hardware. A possible alternative is to obtain accurate inverter loss models [3]-[6] to be integrated in a fast simulator that must include the electrical machine, the vehicle model and the battery. Commercial simulators are available, such as IPOSIM (Infineon) [7], SemiSel (Semikron Danfoss) [8] or Semis (Hitachi) [9]. However, their focus is only on the power converter and not to the eDrive as a whole. Moreover, they are not intended to simulate long cycles and therefore the simulation time becomes excessively long.

Another problem for eDrives digital twins is the torque control of the electrical machine. Even if the traction motor model is available, a realistic simulation should consider the implemented torque control scheme.

Therefore, this paper proposes a fast simulation model of an eDrive (inverter and motor) subjected to long working cycles. With respect to past solutions that are based on analytical equations [4]-[6], the inverter loss model is based on efficiency maps calculated with PLECS software using the inverter modules datasheet. Moreover, the eMotor model is based on flux maps and does not depend on a specific torque control strategy, such as vector control or direct torque control.

The paper is organized as follows. The drive eDrive model is introduced in Section 1, then the inverter and motor models are described in detail in Section 2 and Section 3, respectively. Simulation results, obtained for an WLTC driving cycle, are provided in Section 4.

2   Traction eDrive model

The traction eDrive model is shown in Fig.1 and consists of a vehicle dynamic model, a motor model and the inverter model.

According to a specific driving cycle that imposes the vehicle speed and acceleration, the vehicle model provides to the eMotor the torque setpoint and the motor rotational speed.

Given the torque setpoint and speed, the motor model provides to the inverter model the rms current, the peak voltage request and the power factor. Moreover, the motor model may include the total losses and the internal temperatures (not considered in this paper).

Given the input DC link voltage and the motor voltage, current and power factor provided by the motor model, the inverter model provides the total losses and the estimated junction temperature.

The eDrive model can be expanded with a battery model that provides the inverter DC link supply voltage. The integration of the battery model is currently under development and subject of future work.

The inverter and motor models are built on multi-dimensional maps, as shown in detail in the next sections.

3   Inverter model

As well-known from the literature [3-6], the inverter conduction and switching losses depend on:

- RMS load current (I_{rms})
- DC-link voltage (V_{dc})
- Switching frequency (f_{sw})
- Load power factor ($cos\phi$)
- Modulation index (M)
- Junction temperature (T_j)
- Modulation technique

In case of SiC MOSFET inverters, the power factor does not significantly impact the inverter losses. Moreover, if the switching frequency is considered as constant and the modulation strategy does not change, the total inverter losses will depend only on four parameters, as in (1):

$$P_{loss} = f\left(I_{rms}, T_j, M, V_{dc}\right) \qquad (1)$$

The loss computation can be solved using analytical approaches, but the formulae are usually provided for Sinusoidal Pulse-Width Modulation (S-PWM) and for the inverter working in its linear region, while very often overmodulation is employed along with more advanced modulation techniques with respect to the S-PWM.

An alternative to estimate the junction temperature and the inverter losses is to map the losses at device level, as implemented by the PLECS simulation software [10].

The PLECS software can simulate properly the inverter seen as circuital model, using 3D thermal models of the power devices that are usually provided by the semiconductor manufacturers or it can be built from datasheet curves. The motor can be easily integrated in the simulation model, using available built-in models or custom designed ones.

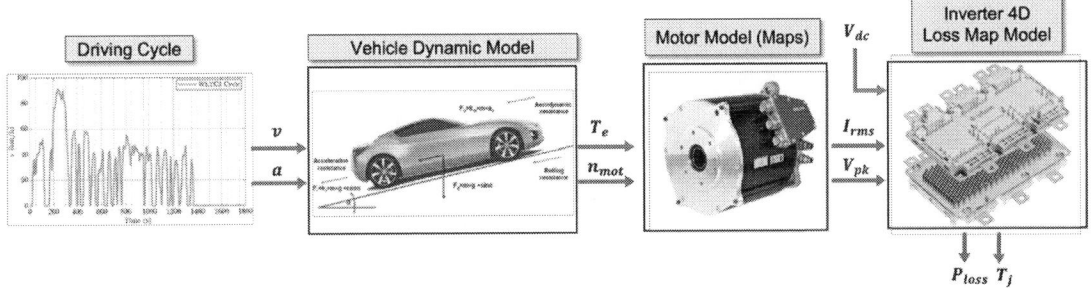

Fig. 1   Proposed eDrive model for a vehicle subjected to driving cycles.

However, the main issue is the simulation time when long driving cycles must be considered, as the losses are calculated at each switching period.

Therefore, the methodology proposed in this paper is to exploit the PLECS capability to easily get the switches loss maps, having as inputs the rms current, the temperature and the modulation index. This approach is very fast but the correct evaluation of the junction temperature ripple for low output frequency operation becomes problematic.

The proposed inverter loss model does not intend to simulate the junction temperature variation but the peak junction temperature. The adopted method is the one presented in [11] for IGBT inverters that was reconsidered for SiC inverters.

Considering the total power loss (conduction loss and switching loss) for one fundamental period, the loss is sum between an average value P_{avg} and a ripple whose maximum deviation respect to the average value is ΔP, as shown in Fig. 2.

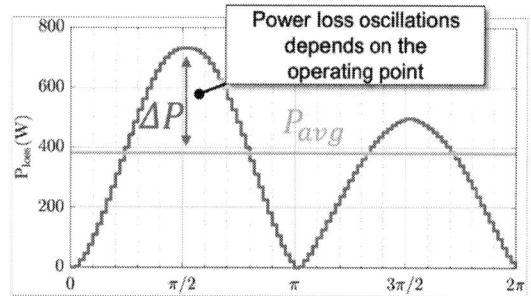

Fig. 2   Power loss for one fundamental period.

If the power loss ripple is approximated as sinusoidal having the amplitude equal with ΔP and the frequency equal with the double of fundamental frequency f_o, then the junction temperature can be estimated using the switch Foster thermal model, as shown in Fig. 3. The loss oscillation is corrected according to a corrective factor $G(f_o)$ depending on the Foster thermal model, as in (2).

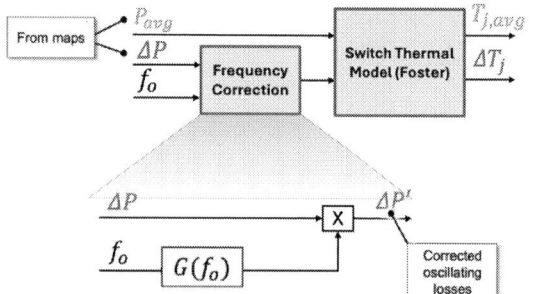

Fig. 3   Power loss for one fundamental period.

$$G(f_o) = \frac{1}{R_{jf}^{tot}} \sum_{i=1}^{N} \frac{R_{jf,i}}{\sqrt{1+(2\pi 2 f_o \tau_i)^2}} \qquad (2)$$

where: R_{jf}^{tot} is the total steady-state thermal resistance, $R_{jf,i}$ is the single thermal resistance from the Foster network, while τ_i is the single time constant of the Foster network.

The proposed inverter model is shown in Fig. 4.

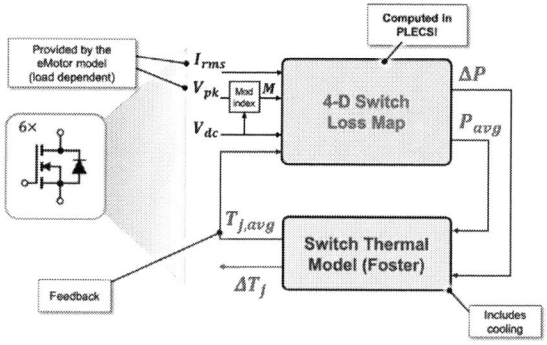

Fig. 4   Inverter loss model with 4D loss maps.

The 4D switch loss maps are generated in PLECS using multiple simulations for the schematic shown in Fig. 5.

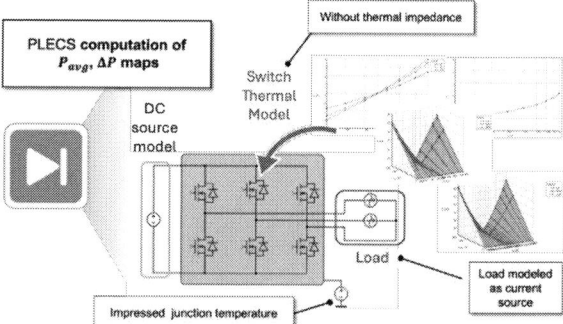

Fig. 5   Generation of inverter 4D loss maps.

The maps are generated at variable impressed junction temperature and without using any thermal impedance of the switches. The load is modelled as a simple current source with variable rms current. Obviously, the current ripple is neglected. The DC link is modelled as a simple ideal voltage source with variable voltage in the range of interest. The loss maps for the single switches are XML files that are usually provided by the semiconductor manufacturer, according to the selected power module.

The loss maps computation is very fast and can be done by launching PLECS with a Matlab script.

4 Motor model

The motor model is based on the flux maps (current-to-flux linkage relationship), defined in the rotor (dq) frame as (3)

$$\begin{cases} \lambda_d = f(i_d, i_q) \\ \lambda_q = f(i_d, i_q) \end{cases} \tag{3}$$

The flux maps can be generated with Finite Element Analysis (FEA) of the target eMotor, if the motor design is known. As alternative, the flux maps can be obtained experimentally, as in [12].

In model proposed in this paper uses the experimental flux maps of a commercially available traction motor (Brusa HSM1, rated 93 kW, 4900 rpm, 165 Nm). The flux maps are shown in Fig. 6.

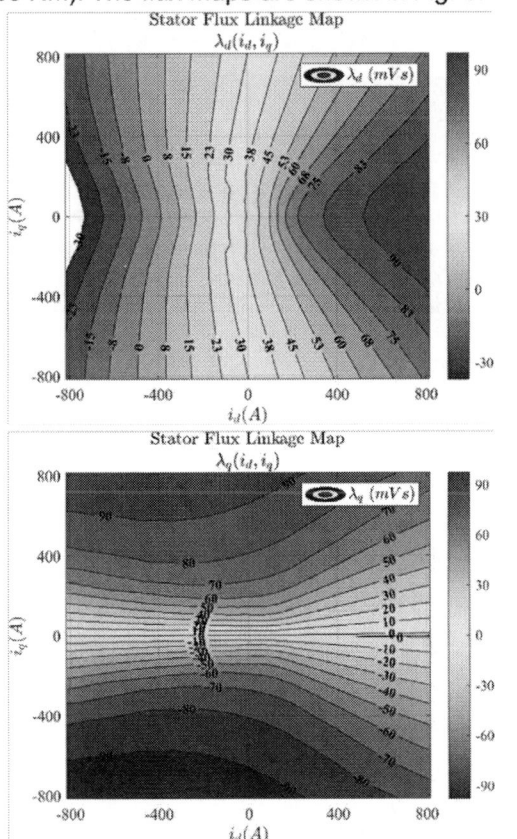

Fig. 6 Target motor flux maps.

The flux maps are mandatory to control well an electrical motor exhibiting heavy magnetic saturation, as happens for the traction motors. Indeed, the knowledge of the flux maps is equivalent to predict the motor performance for Maximum Torque per Ampere (MTPA) operation (required below the base speed), Maximum Torque per Volt operation (required above the base speed for heavy flux-weakening operation) and Maximum

Torque per Speed (MTPS). The MTPS represents the best torque production versus speed, given a maximum inverter current limit and the inverter DC link voltage. As example, the MTPA/MTPV and the MTPS curves for the target motor are shown in Fig. 7 and Fig. 8, respectively.

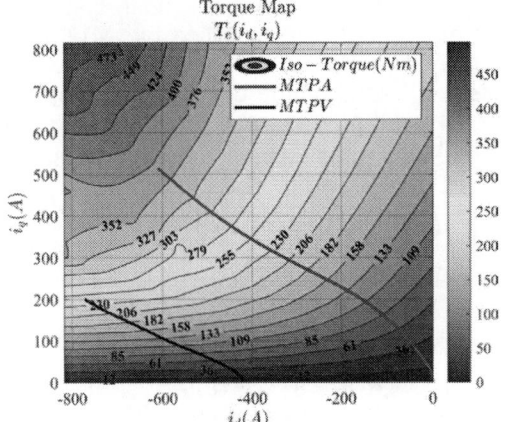

Fig. 7 MTPA/MTPV curves for the target motor.

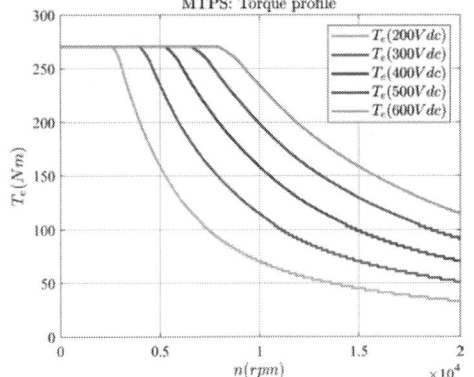

Fig. 8 MTPS curves for the target motor.

Given the MTPS, the motor model (Fig. 9) provides the voltage peak value and the rms phase current for any operating point in the torque-speed plane that is below the MTPS curve.

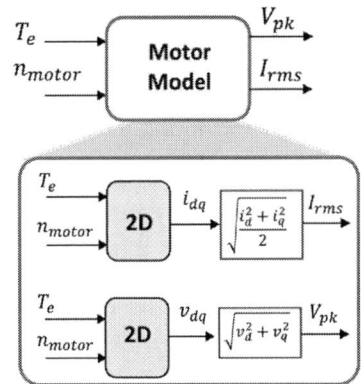

Fig. 9 Motor model.

As shown in Fig. 9, the motor model consists of multiple 2D maps having as input the motor torque and speed. If required, the model can be enhanced with loss estimation maps along a thermal model (not considered in this paper).

The motor model provides then the required voltage and current to get a specific operating point in the torque-speed plane fulfilling the MTPS. This approach is thus independent on the torque control as it is based on the real torque capability of the machine that is found through the flux maps processing.

5 Simulation results

The proposed model (Fig .1) has been simulated in Matlab/Simulink for a WLTC driving cycle (Fig. 10) of a vehicle with the following data: mass $m = 2000\,kg$, friction model $F_{road} = a + bv + c\dot{v} + mgsin(\theta)$, with $\theta = 0°$ $a = 196.2\,N$, $b = 0\,Ns/m$, $c = 0.536\,Ns^2/m^2$, gear ratio $\tau = 10$, gear efficiency $\eta = 94\%$.

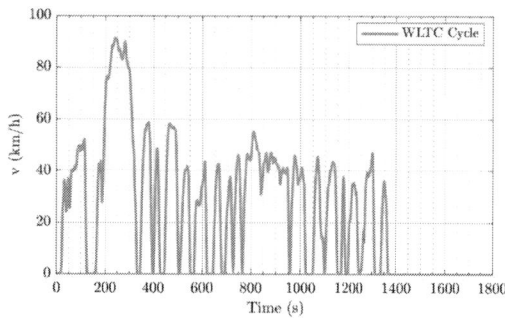

Fig. 10 Simulated WLTC for the proposed model.

The inverter DC link voltage has been considered as constant and equal with 600Vdc. The inverter power module is the 1200V/400A SiC Hybrid-PACK FS03MR12A6MA1B from Infineon. The generation of the inverter loss maps has been performed using the PLECS model provided by Infineon. The inverter loss maps have been obtained as $10 \times 10 \times 10$ 3D map with 1000 PLECS simulations using the methodology shown in Fig. 5.

As already mentioned in Section 4, the eMotor is an Interior Permanent Magnet (IPM) motor from Brusa Elektronik having the following parameters: rated power 93 kW, rated speed 4900 rpm, rated torque 165 Nm, peak power 185 kW.

The simulation results for the selected cycle are shown in Figs. 11-14. The required motor torque and speed, representing the inputs of the motor model, are shown in Fig. 11. The motor model provides to the inverter model the rms current and the peak voltage that is subsequently transformed into modulation index, as shown in Fig. 12.

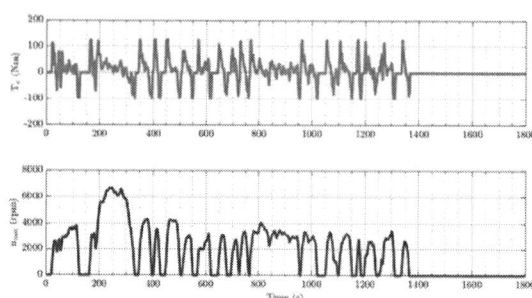

Fig. 11 Motor torque (Nm) and speed (rpm).

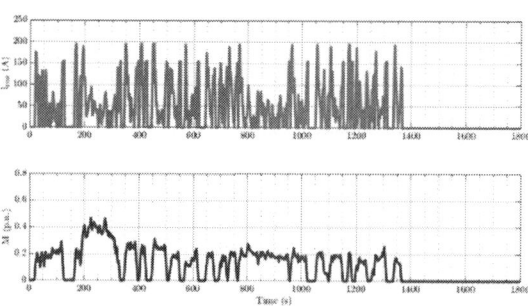

Fig. 12 Phase RMS current (A) and modulation index.

The estimated average junction temperature (corresponding to the average losses) and the temperature variation (due to the loss oscillations) are shown in Fig. 13, while the estimated inverter loss and the junction temperature are shown in Fig. 14.

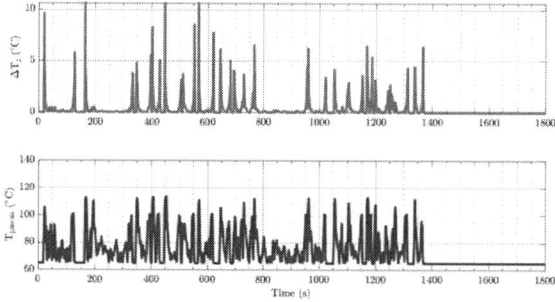

Fig. 13 Estimated ΔT_j and $T_{j,avg}$.

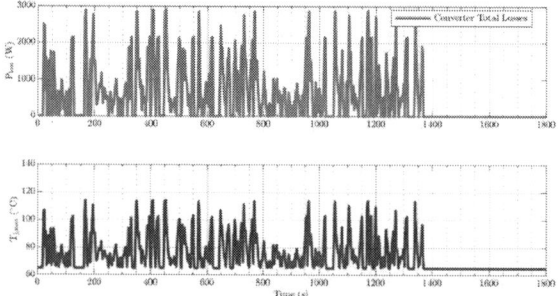

Fig. 14 Estimated inverter losses (W) and junction temperature (°C).

The estimated inverter junction temperature for the considered cycle does not exceed 120 °C, as shown in Fig. 14. The total simulation time of the proposed Matlab/Simulink model for the above-mentioned cycle has been performed in 6.17 seconds, including maps computation, on a computer equipped with Intel I9 processor.

6   Conclusions

This paper presented a methodology to obtain a fast simulator with inverter temperature estimation for traction eDrives that are subjected to long driving cycles.

The proposed model is implemented in Matlab/Simulink. The inverter model is based on efficiency maps that are calculated in advance using the PLECS simulation environment. The motor model is using 2D maps that are calculated from the flux maps and therefore does not depend on a particular torque control strategy.

The proposed model is very useful in digital twins for quick evaluation of an inverter design before the final prototyping and testing. The simulation time of the proposed eDrive model for 30 minutes driving cycle was about 7 seconds, demonstrating the feasibility of the proposed solution.

The proposed model can be enhanced with a battery model along with a vehicle cooling model, as part of future work.

7   Acknowledgement

This work was supported by HiEFFICIENT project. This project has received funding from the ECSEL Joint Undertaking (JU) under grant agreement no. 101007281. The JU receives support from the European Union's Horizon 2020 research and innovation programme and Austria, Germany, Slovenia, Netherlands, Belgium, Slovakia, France, Italy, and Turkey.

References

[1] M. H. M. Sathik, J. Pou, S. Prasanth, V. Muthu, R. Simanjorang, and A. K. Gupta, "Comparison of IGBT junction temperature measurement and estimation methods—A review," in Proc. Asian Conf. Energy, Power Transp. Electrif., 2017, pp. 1–8.

[2] Y. Avenas, L. Dupont, and Z. Khatir, "Temperature measurement of power semiconductor devices by thermo-sensitive electrical parameters—A review," IEEE Trans. Power Electron., vol. 27, no. 6, pp. 3081–3092, 2012.

[3] A. Allca-Pekarovic, Ph. J. Kollmeyer, J. Reimers, P. Mahvelatishamsabadi, T. Mirfakkhrai, P. Naghshtabrizi, and A. Emadi, "Loss Modeling and Testing of 800 V DC Bus IGBT and SiC Traction Inverter Modules," in IEEE Transactions on Transportation Electrification, early access.

[4] H. Ye, K. Yang, H. Ge, P. Magne and A. Emadi, "A drive cycle based electro-thermal analysis of traction inverters," 2014 IEEE Transportation Electrification Conference and Expo (ITEC), Dearborn, MI, USA, 2014, pp. 1-6, doi: 10.1109/ITEC.2014.6861761.

[5] J. Ye, K. Yang, H. Ye and A. Emadi, "A Fast Electro-Thermal Model of Traction Inverters for Electrified Vehicles," in IEEE Transactions on Power Electronics, vol. 32, no. 5, pp. 3920-3934, May 2017, doi: 10.1109/TPEL.2016.2585526.

[6] L. Giraudi, M. Tranchero, C. Romano, P. Santero, "A Fast and Reliable Simulator for the Evaluation of Losses in Power Devices Based on a Mixed Analytical and Empirical Model," 29th International Workshop on Thermal Investigations of ICs and Systems (THERMINIC), 2023.

[7] IPOSIM, Infineon Online Power Simulation Platform, https://www.infineon.com/cms/en/tools/landing/iposim-infineon-online-power-simulation-platform/

[8] SemiSel Semikron Danfoss simulator, https://www.semikron-danfoss.com/service-support/semisel-simulation.html.

[9] Hitachi Energy, SEMIS thermal simulation platform, https://www.hitachienergy.com/products-and-solutions/semiconductors/semislaunchpage

[10] PLECS Simulation Platform for Power Electronic Systems, User Manual version 4.8, https://www.plexim.com/download/documentation.

[11] V. Blasko, R. Lukaszewski and R. Sladky, "On line thermal model and thermal management strategy of a three phase voltage source inverter," Conference Record of the 1999 IEEE Industry Applications Conference. Thirty-Forth IAS Annual Meeting (Cat. No.99CH36370), Phoenix, AZ, USA, 1999, pp. 1423-1431 vol.2.

[12] E. Armando, R. I. Bojoi, P. Guglielmi, G. Pellegrino and M. Pastorelli, "Experimental Identification of the Magnetic Model of Synchronous Machines," in IEEE Transactions on Industry Applications, vol. 49, no. 5, pp. 2116-2125, Sept.-Oct. 2013.

PCIM Europe 2024, 11– 13 June 2024, Nuremberg   DOI: 10.30420/566262031

Comparative Evaluation of Three-Phase-Unfolder-Based MVAC-LVDC Solid-State Transformers

Jonas Huber [1], Uwe Drofenik[2], Francisco Canales[3], Johann W. Kolar [1]

[1] ETH Zurich, Switzerland
[2] TU Wien, Austria
[3] ABB Switzerland Ltd., Switzerland

Corresponding author: Jonas Huber, huber@lem.ee.ethz.ch
Speaker: Jonas Huber, huber@lem.ee.ethz.ch

Abstract

Solid-state transformers (SSTs) for MVac-LVdc conversion, e.g., in future high-power EV charging stations, should feature low conversion losses and low complexity. Therefore, topologies based on a three-phase unfolder, i.e., a six-pulse rectifier, which thus eliminate PFC rectifier input stages and instead rely on modular dc-dc isolation stages for sinusoidal input current shaping, are of high interest. This paper discusses state-of-the-art and new topologies in this category, explains the operating principles of the systems, and, finally, provides a comparative evaluation of the SST topologies concerning realization effort, component stresses, and general complexity.

1 Introduction

Solid-state transformers (SSTs) have been proposed for various applications like traction and future smart grids or dc distribution systems [1], [2]. In particular, SSTs are also considered for supplying low-voltage dc (LVdc) loads from the medium-voltage ac (MVac) three-phase mains, e.g., high-power EV charging stations [3]–[5], datacenters [6], or electrolyzers [7], where in all cases mostly unidirectional power flow is required.

To handle the MV input, SSTs typically employ input-series output-parallel (ISOP) arrangements of converter cells, e.g., in the per-phase branches of the well-known cascaded H-bridge (CHB) structure [8]. There, each cell contains an active ac-dc PFC rectifier stage and a downstream isolated dc-dc converter. Even if realized with only unidirectional power flow capability [9], the complexity is relatively high and each branch processes a single-phase power flow with a correspondingly large fluctuation at twice the mains frequency.

In contrast, this paper discusses SST topologies with a three-phase unfolder stage at the MVac input, i.e., a six-pulse (B6) rectifier [3], [4], [10]–[13]. Compared to fully phase-modular approaches like the CHB system, the unfolder stage improves the operating conditions of the converter branches that are arranged on the dc side and not directly connected

to the MVac grid.[1] The then unipolar (but pulsating) input voltages facilitate the direct application of isolated dc-dc converter cells without ac-dc stages and the processed power is roughly constant.

Thus, **Section 2** summarizes the known B6-bridge-based MVac-LVdc SST topologies shown in **Fig. 1** and **Section 3** introduces the new topologies from **Fig. 2**. **Section 4** provides a high-level comparative evaluation of all topologies and **Section 5** concludes the paper.

2 State-of-the-Art Topologies

For each considered topology, we introduce an idealized equivalent circuit, which is useful for explaining the operating principles and facilitates the derivation of straightforward yet meaningful indicators for realization efforts and component stresses. In the equivalent circuits (see, e.g., **Fig. 3**), controlled current sources model the output branches (OBs), i.e., the ISOP configuration of isolated dc-dc converters that deliver power to the LVdc load, and the injection branches (IBs), i.e., the series connections of half-bridge cells (HBCs) or full-bridge cells

---

[1]Note that there is a structural but not operational similarity to "controlled transition bridge" converters [14]. Further, there are topologies that integrate a B6 rectifier with a non-modular isolation stage, but these are limited in power rating due to discontinuous input currents [5].

PCIM Europe 2024, 11–13 June 2024, Nuremberg   DOI: 10.30420/566262031

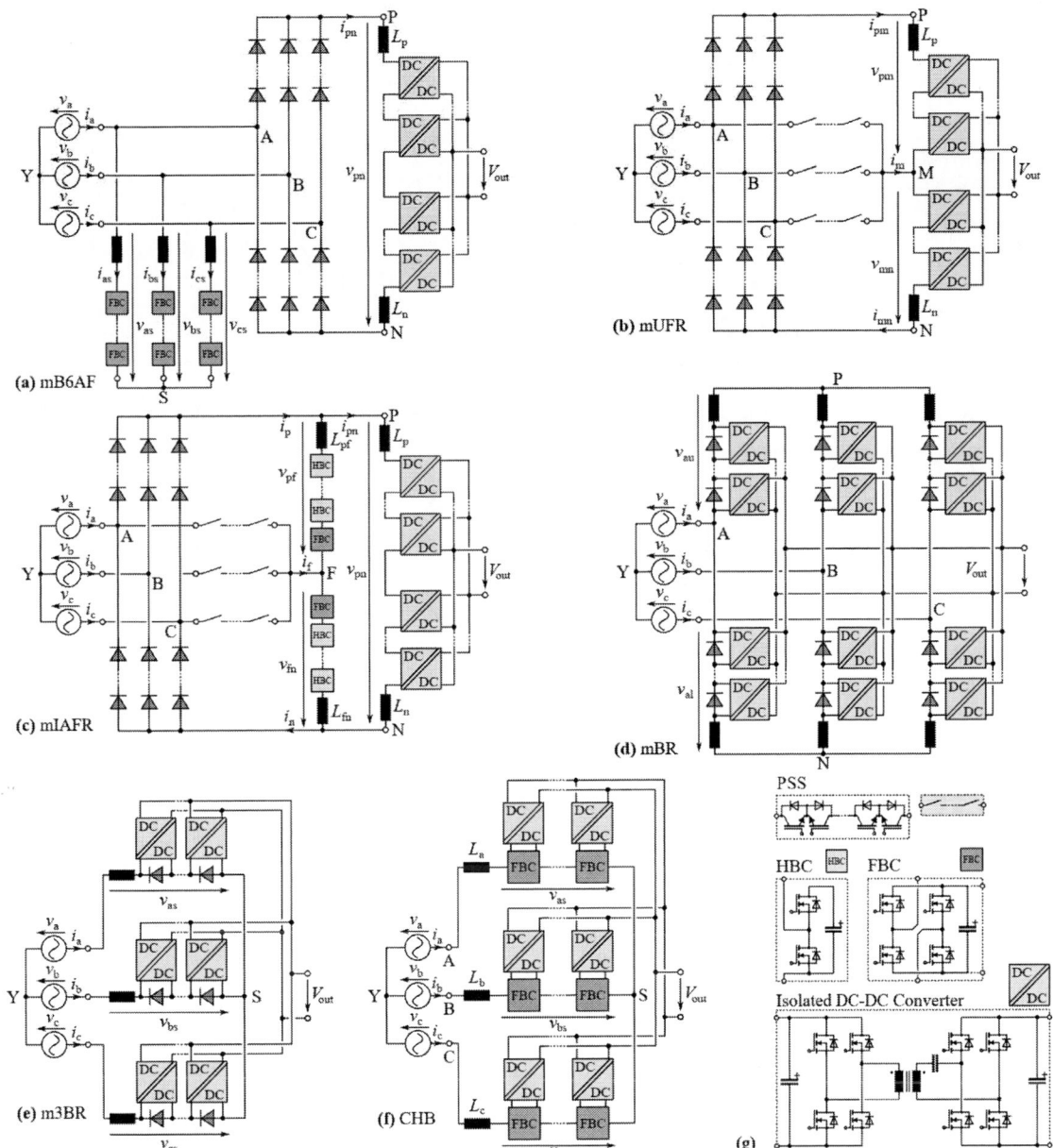

Fig. 1: State-of-the-art MVac-LVdc SST topologies discussed in **Section 2**. **(a)** mB6AF; **(b)** mUFR [15]; **(c)** mIAFR [3], [10]; **(d)** mBR [4], [11]; **(e)** mB3R [12], [13]. **(f)** CHB (for reference). **(g)** Legend.

(FBCs) that only process reactive power. Simple control block diagrams explain the generation of the OB and IB current or voltage references from an exemplary reference power flow, $P^* = 1\,\text{MW}$, and for ohmic mains behavior achieved by using the reference conductance $G^* = P^*/V_g^2$ ($V_g = 10\,\text{kV}$ is the MVac mains line-to-line rms voltage) to obtain phase current references that are proportional to the phase voltages. Further, the mains voltages define the B6 rectifier switching state such that always

the phase with the maximum voltage is connected to the dc-side terminal P and the phase with the minimum voltage to N. It is thus useful to define a mapping of the phase voltages (and currents) v_a, v_b, and v_c to v_{max}, v_{mid}, and v_{min} such that $v_{max} > v_{mid} > v_{min}$; the mapping changes in each 60°-wide sector of the mains period.

2.1 mB6AF: B6 Rectifier with Active Filter

The mB6AF (**Fig. 1a** and **Fig. 3**) features a B6 rectifier, one OB that operates from the characteristic

255

PCIM Europe 2024, 11– 13 June 2024, Nuremberg   DOI: 10.30420/566262031

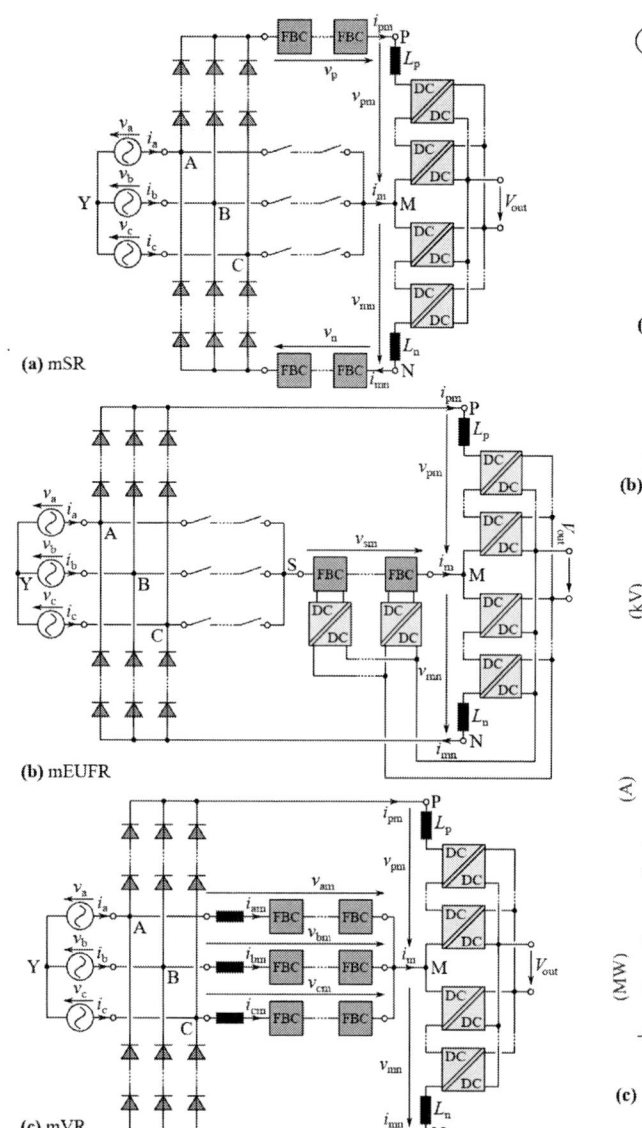

Fig. 2: Proposed MVac-LVdc SST topologies discussed in **Section 3. (a)** mSR; **(b)** mEFUR; **(c)** mVR.

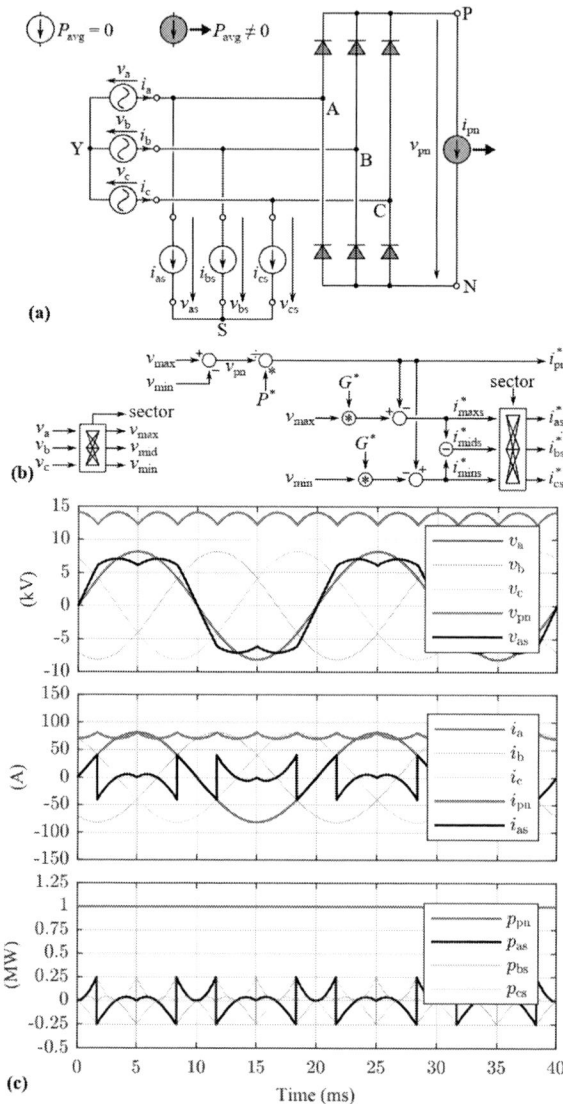

Fig. 3: mB6AF (see **Fig. 1a**). **(a)** Idealized equivalent circuit, **(b)** generation of OB and IB current references, and **(c)** simulated key waveforms.

six-pulse-shaped dc-side voltage $v_{pn} = v_{max} - v_{min}$, and an ac-side active filter (AF) consisting of three IBs with FBCs.[2] Controlling the OB input current to $i_{pn}^* = P^*/v_{pn}$ results in constant power flow P^*. The IB current references are calculated such that the AF compensates harmonic distortions (i.e., the average power processed by the IBs is zero) of the diode rectifier with impressed output current i_{pn}, resulting in sinusoidal phase currents and ohmic mains behavior.

---

[2]Note that a local, slow control loop regulates the average value of the floating FBC dc capacitor voltage, compensating losses.

2.2 mUFR: Modular Unfolder Rectifier

The mUFR shown in **Fig. 1b** and in **Fig. 4** is a direct extension of the non-modular LV variant proposed in [15] to MV applications. In contrast to the mB6AF, the mUFR features two OBs and a thus formed node M. There is no AF (no IBs), but three four-quadrant phase-selector switches (PSSs) can connect one phase terminal at a time to M. The OB currents i_{pm} and i_{mn} directly define the max. and min. phase currents, respectively, and the PSSs route the current $i_m = i_{mn} - i_{pm}$ to the mid phase. Whereas thus sinusoidal mains currents result, the two OB input voltages v_{pm} and v_{mn} vary widely be-

256

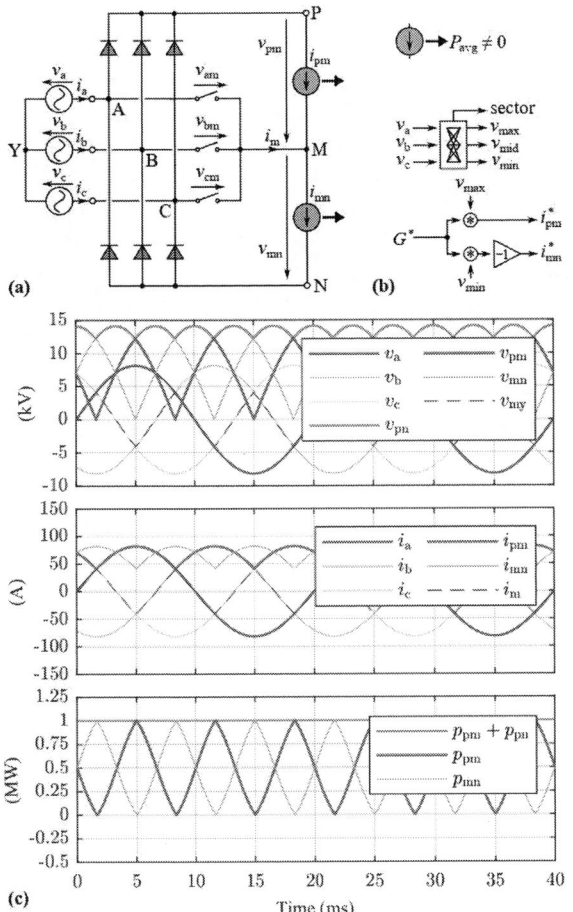

Fig. 4: mUFR (see **Fig. 1b**). **(a)** Idealized equivalent circuit, **(b)** generation of OB and IB current references, and **(c)** simulated key waveforms.

tween $0\,\mathrm{V}$ and $\sqrt{3/2}V_\mathrm{g}$, and each OB processes power fluctuating between zero and the total output power.

2.3 mIAFR: Modular Integrated Active Filter Rectifier

The mIAFR (see **Fig. 1** and **Fig. 5a**) has been proposed in [10] and then analyzed in [3]; it is a modularized MV version of the well-known LV IAF rectifier [16], [17]. In essence, the ac-side AF of the mB6AF is moved to the dc-side, leaving the operation of the OB with $i_\mathrm{pn}^* = P^*/v_\mathrm{pn}$ unchanged. However, three PSSs and two IBs suffice for implementing the AF functionality by injecting a current that is proportional to v_mid into the mid phase. The IB voltages v_pf and v_fn are always positive and therefore the IBs employ HBCs only.[3] The IB voltages and

---

[3]Note that practical realizations require one FBC in each IB to compensate inductive voltage drops [3], [10].

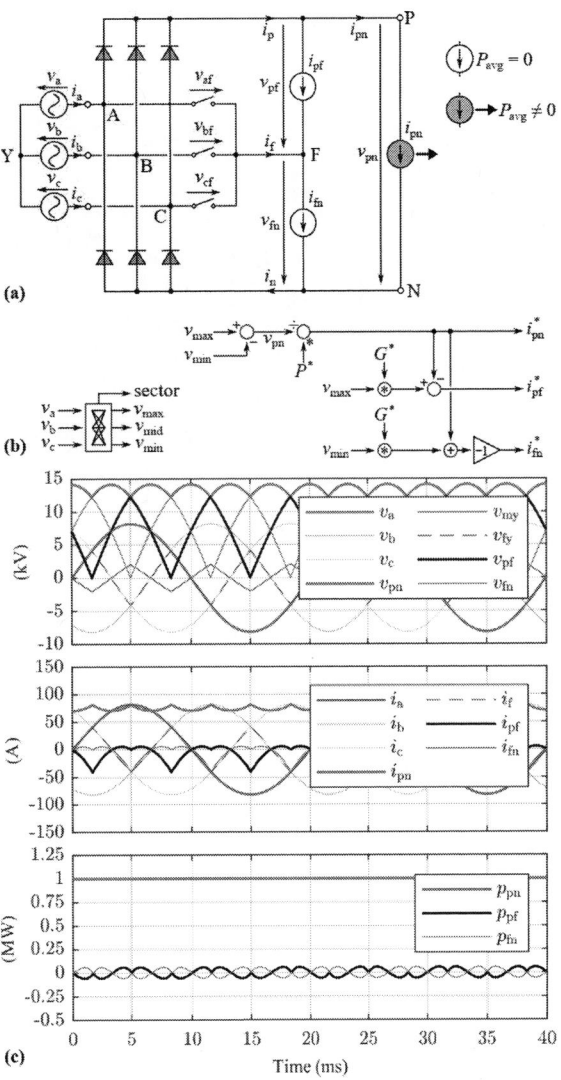

Fig. 5: mIAFR (see **Fig. 1c**). **(a)** Idealized equivalent circuit, **(b)** generation of OB and IB current references, and **(c)** simulated key waveforms.

the processed phase current i_mid, advantageously, result in relatively low IB peak power and low power fluctuation. On the other hand, v_pf and v_fn vary between zero and $\sqrt{3/2}V_\mathrm{g}$ (as the mUFR OB voltages discussed in **Section 2.2**) and hence *each* IB requires almost as many converter cells as the OB.

2.4 mBR: Modularized Bridge Rectifier

An alternative way of utilizing a B6 rectifier has been proposed in [4] and analyzed in [11]: The mBR shown in **Fig. 1d** and in **Fig. 6** uses the voltages across blocking rectifier diodes as unipolar supply voltages for isolated dc-dc converters connected in parallel to the diodes; i.e., the OBs are integrated into the B6 rectifier, which, advanta-

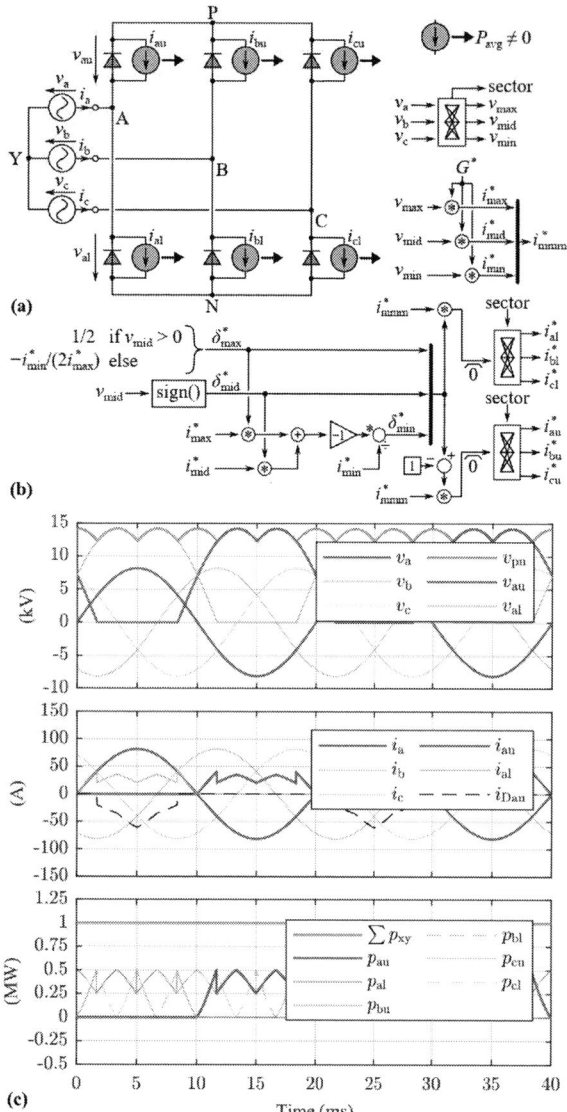

Fig. 6: mBR (see **Fig. 1d**). **(a)** Idealized equivalent circuit, **(b)** generation of OB and IB current references, and **(c)** simulated key waveforms.

geously, ensures defined voltage sharing among the series-connected diodes. Still, the mains voltages define the conduction state of the B6 bridge and thus which OBs have a non-zero input voltage needed for drawing power. Specifically, the lower diode of the max. phase, the upper diode of the min. phase, and both diodes of the mid phase are blocking. The direction of the mid phase current defines whether the upper or the lower OB should impress this current and thus deliver power to the output, i.e., $\delta_{\text{mid}}^* = \text{sign}(v_{\text{mid}}) = \pm 1$ with δ denoting the share of the respective phase current provided by the respective lower OB. In contrast,

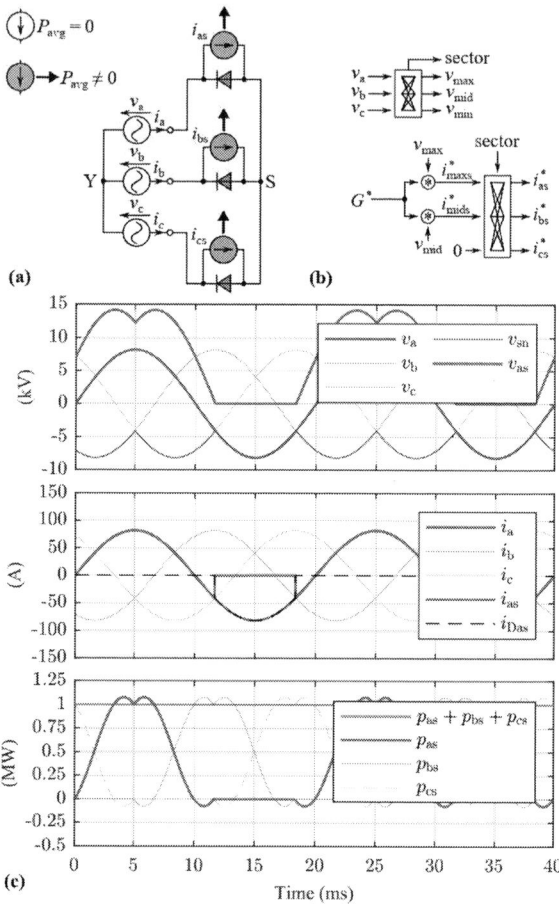

Fig. 7: mB3R (see **Fig. 1e**). **(a)** Idealized equivalent circuit, **(b)** generation of OB and IB current references, and **(c)** simulated key waveforms.

δ_{max}^* and δ_{min}^* are degrees of freedom (coupled by Kirchhoff's current law in the nodes P and N) that allow optimization of the OB stresses [11]; here, δ_{max}^* and δ_{min}^* are selected such that the two active OBs (lower OB of the max. phase and upper OB of the min. phase) process equal power. Note that the resulting currents in the conducting diodes are non-zero but lower than in normal B6 operation. However, all 6 OBs must be rated for an input voltage of $v_{\text{pn}} = v_{\text{max}} - v_{\text{min}}$, i.e., the line-to-line voltage magnitude, and at any given time, only 3 out of the 6 OBs deliver power to the load.

2.5 mB3R: Modularized B3 Rectifier

Similarly, the mB3R proposed in [12], [13] and shown in **Fig. 1e** and in **Fig. 7** employs a reduced B3 rectifier with isolated dc-dc converters connected in parallel to the diodes. Essentially, a single diode replaces the ac-dc converter stage used in conventional CHB-based SSTs (see **Sec-**

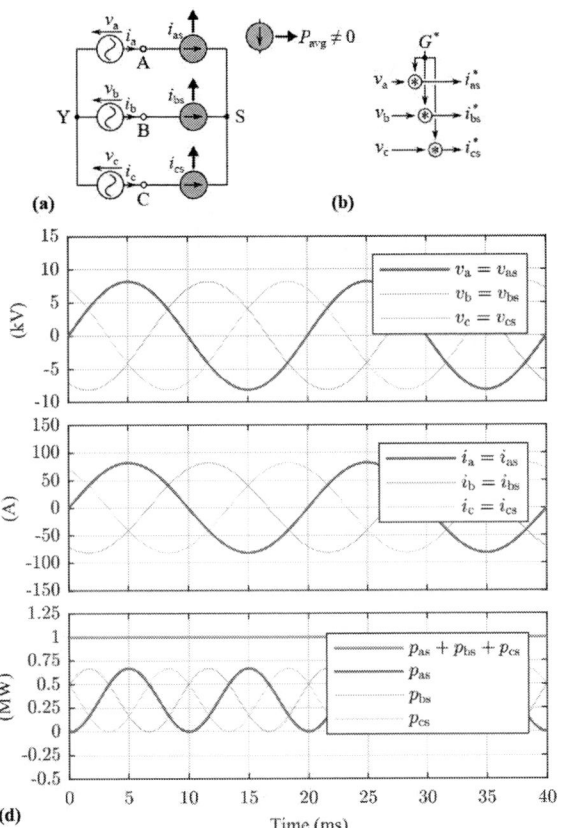

Fig. 8: CHB (see **Fig. 1f**). **(a)** Idealized equivalent circuit, **(b)** generation of OB and IB current references, and **(c)** simulated key waveforms.

tion 2.6), which comes at the price of large fluctuations of the voltage and power processed by the OBs.

2.6 CHB: Cascaded H-Bridge SST

For completeness, **Fig. 1f** and **Fig. 8** show the well-known CHB-based SST structure mentioned in the introduction. Unlike in the other topologies, the OBs process ac input voltages and currents, i.e., each cell consists of an ac-dc converter stage (i.e., an FBC) and an isolated dc-dc converter. Being a fully phase-modular system, each OB processes power that shows the characteristic twice-mains frequency fluctuation of single-phase systems.[4]

3 Proposed Topologies

This section introduces further MVac-LVdc SST topologies based on a B6 rectifier input stage.

---

[4]Alternatively, the power fluctuation could be buffered on the cells' MV side in correspondingly large capacitors.

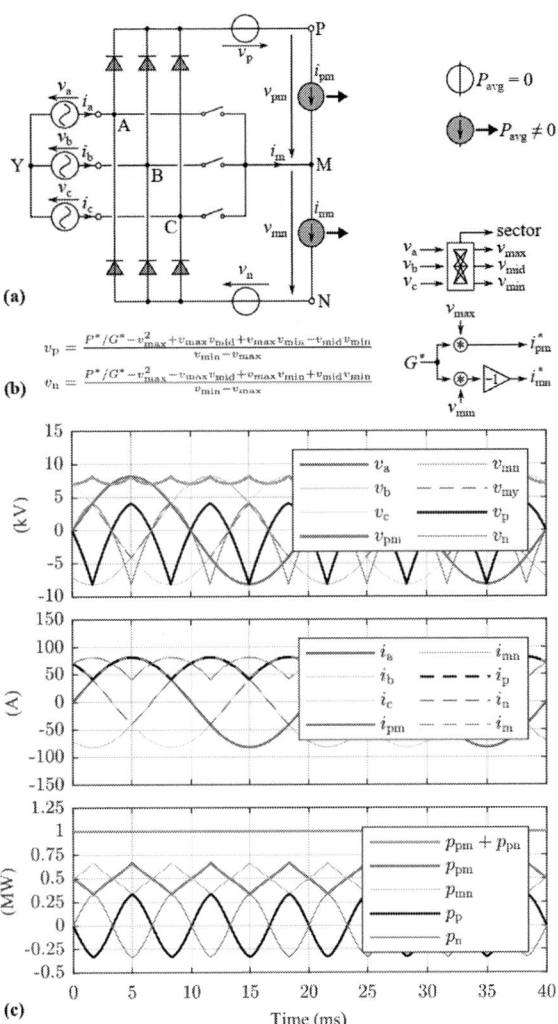

Fig. 9: mSR (see **Fig. 2a**). **(a)** Idealized equivalent circuit, **(b)** generation of OB and IB current references, and **(c)** simulated key waveforms.

3.1 mSR: Modular Swiss Rectifier

The proposed mSR shown in **Fig. 2a** and in **Fig. 9** follows from the mIAFR using duality considerations: whereas the mIAFR employs current-source-type IBs connected in parallel to the OB, the mSR uses voltage-source-type IBs connected in series to two OBs. The OBs form the dc midpoint M to which the PSS network connects. Structurally, the mSR is a modular extension of the LV Swiss rectifier [17]. As in the mUFR (see **Section 2.2**), the two OBs directly define the phase currents i_{max} and i_{min}. Connecting the mid phase via the PSSs to M implies large voltages between the B6 bridge's dc terminals and M, which the mSR IBs provide (in contrast to the mUFR, where the OBs are subject to these voltage variations). Specifically, the selected

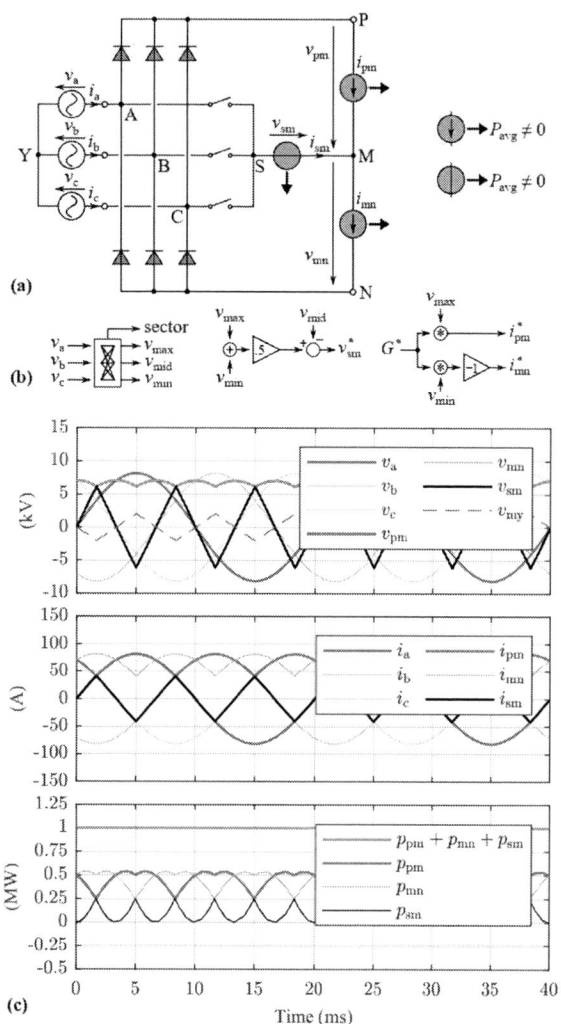

Fig. 10: mEUFR (see **Fig. 2b**). **(a)** Idealized equivalent circuit, **(b)** generation of OB and IB current references, and **(c)** simulated key waveforms.

IB voltage references v_p^* and v_n^* ensure equal and almost constant OB voltages ($v_{pn} = v_{mn}$).[5] However, v_p^* and v_n^* are ac voltages and hence the IBs must be realized with FBCs. As the IBs reside in the main dc current path (in series to the OBs), they process high power fluctuations, which implies relatively large energy buffering requirements.

3.2 mEUFR: Modular Extended Unfolder Rectifier

The mEUFR shown in **Fig. 2b** and in **Fig. 10** combines the mSR's two IBs into a single voltage-

---

[5]Alternatively, equal power processed by the two OBs (i.e., $p_{pm} = p_{mn}$) could be achieved at the price of increased voltage requirements for the OBs and the IBs.

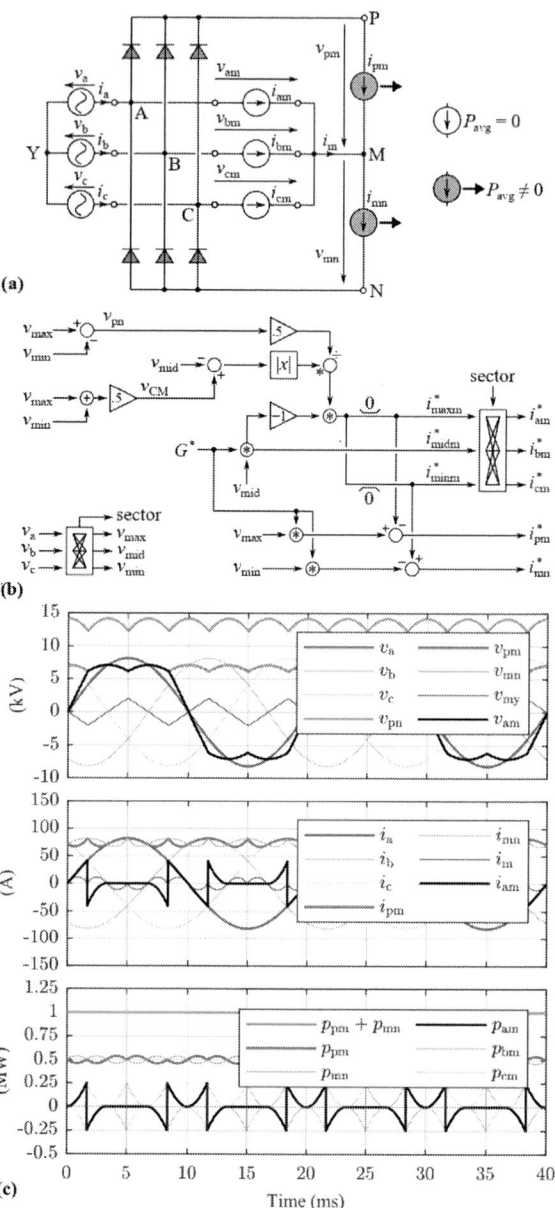

Fig. 11: mVR (see **Fig. 2c**). **(a)** Idealized equivalent circuit, **(b)** generation of OB and IB current references, and **(c)** simulated key waveforms.

source-type IB connected between the dc midpoint M and the star-point S of the PSS network. The two OBs again directly impress the currents of the max. and the min. phase. The IB voltage reference v_{sm}^* follows from the requirement $v_{pm} = v_{mn}$, which implies $v_{my} = 0.5(v_{max} + v_{min})$. However, the IB processes a non-zero average power, i.e., it's FBCs must be connected to isolated dc-dc converters, essentially forming a third OB that contributes to the total output power.

PCIM Europe 2024, 11– 13 June 2024, Nuremberg   DOI: 10.30420/566262031

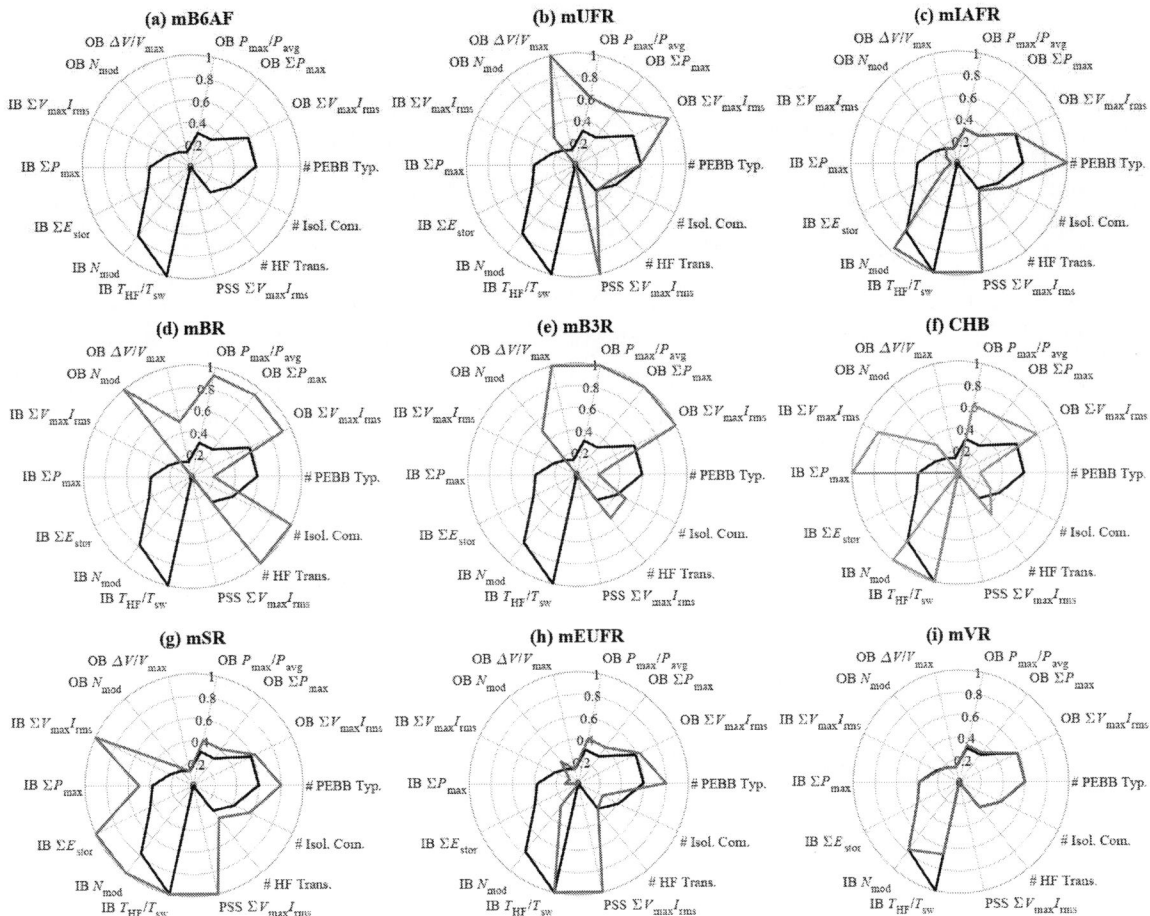

Fig. 12: Comparative evaluation. Smaller values indicated more favorable characteristics and the values on each axis are normalized to the respective maximum amongst all topologies; **Fig. 13** provides absolute values.

3.3 mVR: Modular Vienna Rectifier

Connecting the star-point S formed by the three IBs of the mB6AF (see **Section 2.1**) to a dc midpoint M formed by two OBs, the mVR topology shown in **Fig. 2c** and in **Fig. 11** results; the name follows from the close structural similarity to the LV Vienna Rectifier (VR) [18]. Inspired by an advantageous operating mode of LV VRs using pre-shaped dc-side voltages [19], the operating mode of the mVR ensures that only two out of the three IBs operate with high-frequency (HF) switching at any given time ("2/3-PWM"): **Fig. 11c** confirms that always one IB operates with zero current, i.e., all transistors of the FBCs can be turned off, avoiding switching losses. The price to pay is a minor fluctuation in the powers p_{pm} and p_{mn} processed by the OBs.

4   Comparative Evaluation

Based on the idealized equivalent circuits and considering exemplary specifications (10-kV line-to-line rms grid voltage, 1-MW output power, 1700-V transistors with 70% blocking voltage utilization, nominal maximum modulation index $M = 0.85$, 10% peak-to-peak IB capacitor LF voltage ripple), key indicators for realization efforts and component stresses of all discussed topologies are compared in **Fig. 12** and **Fig. 13**; the mB6AF is considered as a baseline: $\sum V_{max}I_{rms}$ represents the total installed converter power, i.e., isolated dc-dc converters for OBs, HBCs or FBCs (considering a factor 2 over HBCs) for the IBs. Similarly, $\sum P_{max}$ is the sum of the peak power processed by the OBs or IBs, P_{max}/P_{avg} quantifies the utilization of the OBs, and $\Delta V/V_{max}$ characterizes the input voltage range of the dc-dc converters. N_{mod} gives the number of converter modules; for OBs, this corresponds to the number of medium-frequency transformers (MFTs) providing the galvanic separation between the MV and the LV side. $\sum E_{stor}$ is the stored capacitive energy (not considering small capacitors

261

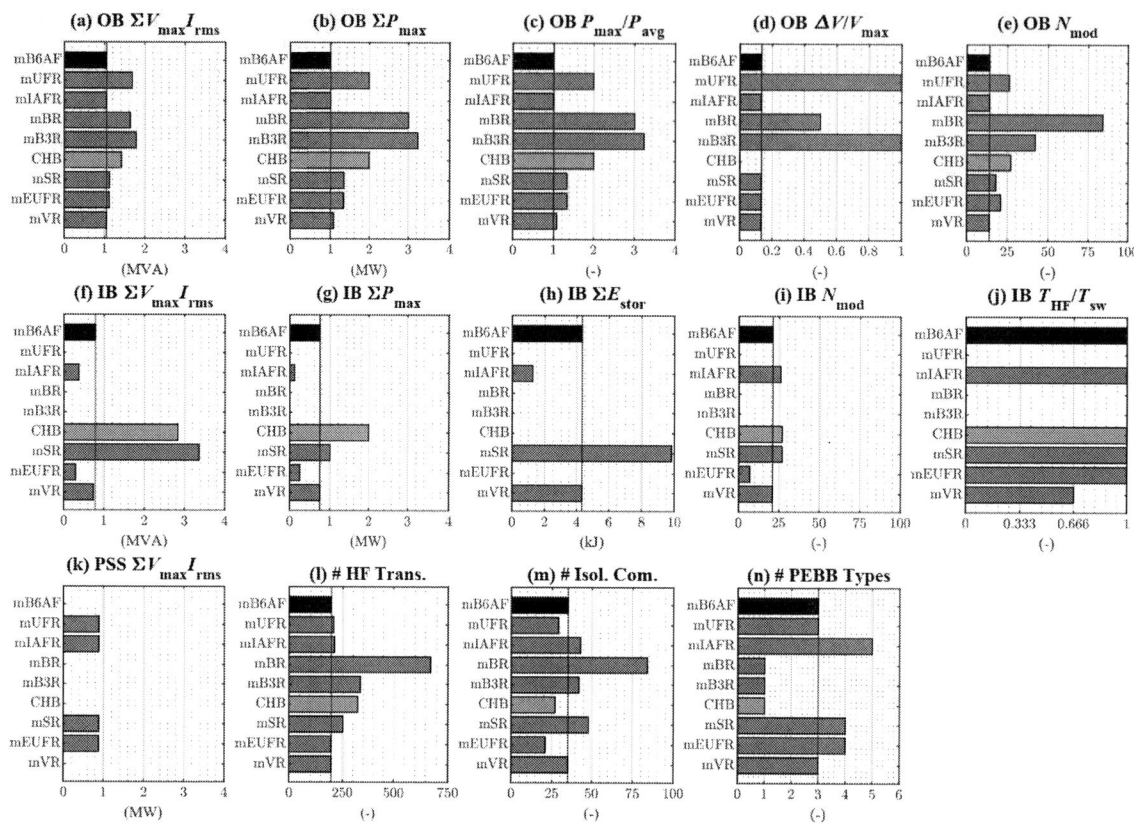

Fig. 13: Absolute values for the performance indicators used in the comparative evaluation from **Fig. 12** for 1-MW output power, 10-kV grid (line-to-line rms), 1700-V transistors (70% blocking voltage utilization), nominal maximum modulation index $M = 0.85$, and 10% peak-to-peak IB capacitor LF voltage ripple.

used solely for HF ripple filtering), and T_{HF}/T_{sw} is the time share during which the IB converter cells switch with HF. Finally, the number of HF-operated transistors (i.e., excluding the PSSs that could be realized with elements of higher blocking voltage and slow switching speeds like 6.5-kV IGBTs), the number of isolated communication links between a central controller and the converter cells, and the number of different power-electronic building block (PEBB) types are given.

Interestingly, the conventional/baseline mB6AF performs very favorably in most dimensions. If PSSs are accepted (despite the potential issues with static and transient voltage sharing among series-connected devices), the mIAFR with the AF moved to the dc side is an interesting alternative; the mEUFR features overall fewer converter modules but requires two types of dc-dc converters with different power ratings. The mBR, the mB3R, and also the standard CHB solution[6] are highly modular

(only one PEBB type), but suffer from relatively low utilization of the installed power electronics. The mSR does not show advantages, mostly because the IBs process the full dc-side current. Finally, the mVR shows very similar performance as the baseline mB6AF, but the "2/3-PWM" operating method reduces the switching losses of the IBs while introducing only a marginal fluctuation of the power processed by the OBs. Note that for many topologies, certain implementation variants and simplifications are conceivable; for example, replacing and/or complementing the diodes of the B6 bridge by anti-parallel switches (e.g., IGBTs, thyristors) facilitates bidirectional power flow.

5 Conclusion & Outlook

This paper provides an overview on existing and new three-phase-unfolder-based MVac-LVdc SST topologies. Based on idealized equivalent circuits, we explain the basic operating principles and provide a first comparative evaluation considering key

---

[6]The ac-dc stages in the converter cells are treated as IBs with FBCs.

indicators for the realization efforts, component stresses, and complexity. All in all, some (in particular, the mB6AF, the mIAFR, and the mVR) MVac-LVdc SSTs based on robust three-phase unfolder stages (B6 rectifiers) are interesting alternatives to fully modular systems, especially for high-power applications with (mostly) unidirectional power flow, e.g., for future high-power EV charging stations. Further research should extend the comparative evaluation to SST topologies with a single MFT like [20] and hybrid concepts based on low-frequency transformers as discussed in [6], e.g., employing 12-pulse rectifiers and active filters. Also, the cost aspect should be considered, whereby concepts requiring only one type of PEBB (like the mBR) might benefit from economies of scale.

References

[1] J. Fabre, P. Ladoux, H. Caron, A. Verdicchio, J.-M. Blaquière, *et al.*, "Characterization and implementation of resonant isolated DC/DC converters for future MVdc railway electrification systems," *IEEE Trans. Transport. Electrific.*, vol. 7, no. 2, pp. 854–869, Jun. 2021.

[2] G. Ulissi, S. Kim, and D. Dujic, "Solid-state technology for shipboard DC power distribution networks," *IEEE Trans. Ind. Electron.*, vol. 68, no. 12, pp. 12 100–12 108, Dec. 2021.

[3] M. Makoschitz, "Unidirectional medium voltage rectifier utilizing sinusoidal input currents," *Electron. Lett.*, vol. 58, no. 12, pp. 474–476, 2022.

[4] S. Götz, "Charging apparatus with a phase unit having multiple strands," U.S. Pat. 10 439 407B2, Oct. 2019.

[5] C. Zhang, R. Wang, Z. Shen, T. Sadilek, A. Anurag, and P. Barbosa, "A single-stage three-phase isolated AC-DC converter for medium voltage solid state transformer applications," in *Proc. Appl. Power Electron. Conf. Expo. (APEC)*, Orlando, FL, USA, Mar. 2023.

[6] J. Huber, P. Wallmeier, R. Pieper, F. Schafmeister, and J. W. Kolar, "Comparative evaluation of MVAC-LVDC SST and hybrid transformer concepts for future datacenters," in *Proc. Int. Power Electron. Conf. (IPEC/ECCE Asia)*, Himeji, Japan, May 2022, pp. 2027–2034.

[7] R. Unruh, F. Schafmeister, N. Froehleke, and J. Boecker, "1-MW full-bridge MMC for high-current low-voltage (100V-400V) DC-applications," in *Proc. Power Convers. Intelligent Motion Conf. (PCIM Europe)*, Nuremberg, Germany, Jul. 2020.

[8] C. Zhao, D. Dujic, A. Mester, J. K. Steinke, M. Weiss, *et al.*, "Power electronic traction transformer—Medium voltage prototype," *IEEE Trans.*

Ind. Electron., vol. 61, no. 7, pp. 3257–3268, Jul. 2014.

[9] W. van der Merwe and T. Mouton, "Solid-state transformer topology selection," in *Proc. IEEE Int. Ind. Techn. Conf. (ICIT)*, Gippsland, Australia, Feb. 2009.

[10] M. Hartmann, "Gleichrichterschaltung (in German)," Austria Pat. AT 516 643B1, Feb. 2018.

[11] G. Andrioli, S. Calligaro, R. Petrella, J. W. Kolar, and J. Huber, "Analysis and comparative evaluation of a modularized bridge rectifier MVAC-LVDC solid-state transformer," in *Proc. Int. Power Electron. Motion Control Conf. (IPEMC/ECCE Asia)*, Chengdu, China, May 2024.

[12] R. Raju, "AC solid-state transformer with DC-DC converters," U.S. Pat. 11 811 301B1, Nov. 2023.

[13] R. Raju and J. Leonard, "AC solid-state transformer using DC-DC converters and without added rectifier and inverter stages," in *Proc. IEEE Appl. Power Electron. Conf. Expo. (APEC)*, Long Beach, CA, USA, Feb. 2024, pp. 528–532.

[14] C. Oates and K. Dyke, "The controlled transition bridge," in *Proc. 17th Europ. Power Electron. Appl. Conf. (EPE/ECCE Europe)*, Geneva, Switzerland, Sep. 2015.

[15] W. W. Chen, R. Zane, and L. Corradini, "Isolated bidirectional grid-tied three-phase AC–DC power conversion using series-resonant converter modules and a three-phase unfolder," *IEEE Trans. Power Electron.*, vol. 32, no. 12, pp. 9001–9012, Dec. 2017.

[16] M. Jantsch and C. W. G. Verhoeve, "Inverters with three phase output and without electrolyte capacitor for improved lifetime, efficiency and costs of grid connected systems," in *Proc. 14th Europ. Photovoltaic Solar Energy Conf.*, Barcelona, Spain, Jun. 1997, pp. 2198–2200.

[17] J. W. Kolar and T. Friedli, "The essence of three-phase PFC rectifier systems—Part I," *IEEE Trans. Power Electron.*, vol. 28, no. 1, pp. 176–198, Jan. 2013.

[18] J. W. Kolar and F. C. Zach, "A novel three-phase utility interface minimizing line current harmonics of high-power telecommunications rectifier modules," *IEEE Trans. Ind. Electron.*, vol. 44, no. 4, pp. 456–467, Aug. 1997.

[19] J. A. Anderson, M. Haider, D. Bortis, J. W. Kolar, M. Kasper, and G. Deboy, "New synergetic control of a 20kW isolated VIENNA rectifier front-end EV battery charger," in *Proc. 20th IEEE Workshop Control Modeling Power Electron. (COMPEL)*, Toronto, Canada, Jun. 2019.

[20] M. Glinka and R. Marquardt, "A new AC/AC multilevel converter family," *IEEE Trans. Ind. Electron.*, vol. 52, no. 3, pp. 662–669, Jun. 2005.

PCIM Europe 2024, 11– 13 June 2024, Nuremberg   DOI: 10.30420/566262032

Voltage Balancing of a Split-Capacitor IGCT 3L-NPC Leg for the Resonant DC Transformer

Renan Pillon Barcelos, Drazen Dujic

École Polytechnique Fédérale de Lausanne (EPFL), Power Electronics Laboratory, Switzerland

Corresponding author: Renan Pillon Barcelos, renan.pillonbarcelos@epfl.ch
Speaker: Renan Pillon Barcelos, renan.pillonbarcelos@epfl.ch

Abstract

This work investigates the static voltage balancing challenges in an IGCT-based LLC resonant converter with split-capacitor 3L-NPC power stages. It focuses on its operation in a two-level mode with a 50% duty cycle, especially under conditions of medium frequency switching and extremely low turn-off currents. The paper conducts a comparative analysis of two static balancing strategies: one using the parallel balancing resistors for each IGCT, and another employing a single symmetrizing resistor across the two inner IGCTs. These methods are assessed in terms of performance and losses. Additionally, it investigates how the symmetrizing resistor influences the commutation process behavior under these specific conditions. Experimental results are provided to validate the effectiveness and trade-offs of each approach.

1 Introduction

Developing high-power, medium-voltage (MV) converters presents several challenges, such as ensuring their safe and reliable operation, integrating high-voltage devices, and operating at medium frequencies. Notably, the advancement of the MV DC-DC converter design has become a focus area due to significant academic and industrial efforts, positioning it as a key enabling technology for advanced MVDC systems.

In this context, several works have been focused on developing high-power MV DC-DC converters over the years. For instance, in [1], an IGCT-based 3 MW, 600 Hz, 10 kV : 10 kV, dual active bridge (DAB) was developed. In that work, the prototype was successfully demonstrated featuring soft commutation with a turn-off current of 92 A. Another work has developed an IGCT-based 5.6 MW, 1 kHz, 5 kV : 5 kV, 3-phase DAB, where the converter's operation was tested to refine control strategies [2]. These examples highlight the success of developing DC-DC converter prototypes at the MW and MV levels. However, aiming for higher switching frequencies and lower turn-off currents, enabled by resonant operation, brings additional challenges, including the design of snubber circuits.

Other works have been investigating the development of a high-power MV LLC resonant converter, taking advantage of its intrinsic load-independent behavior, and high conversion efficiency. This converter is often referred to as a DC transformer (DCT) as an analogy to the AC transformer when operating in an open loop. In [3], and [4], the authors explored and presented the design principles, and in [5] the operation characteristics were assessed. In summary, the challenges of building high-power MV DC-DC converters come down to the design using existing technologies and costs.

In this work, an IGCT-based 1 MW, 5 kHz, 10(5) kV : 5 kV DCT is investigated. Fig. 1 shows a simplified schematic of the converter which consists of a a split-capacitor three-level (3L) NPC LLC resonant converter. The 3L operation plays a crucial role in protecting the DCT, providing soft-start and current-limiting capabilities [6]. Nevertheless, under normal operating conditions, the power stage operates in a two-level (2L) mode, with a 50% duty cycle, as shown in Fig. 1. In this case, the two upper and two lower IGCTs are effectively in series. Consequently, it is crucial to ensure both dynamic and static balancing in this configuration.

Numerous works have previously investigated the

264

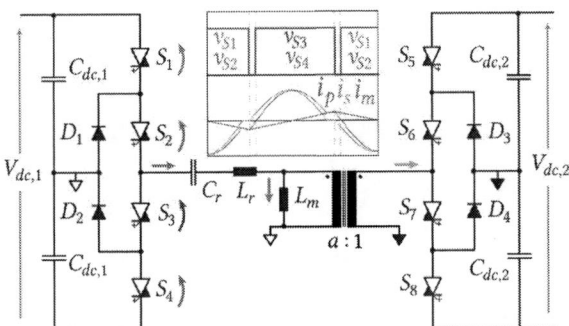

Fig. 1: 3L-NPC leg power stage of an LLC resonant converter and an illustration of the typical voltages and current waveforms for the 2L, 50% duty cycle operation.

static and dynamic voltage balancing in series-connected IGCTs [7]–[9]. These works mainly focus on hard-switched applications, which are typical in MV drives [7], [10]. Only recently there has been a shift towards investigating it in soft-switching applications, as explored in [5].

In [5], an evaluation was conducted on the resonant operation of series-connected IGCTs. That analysis highlighted the effectiveness of a low capacitance C-snubber connected in parallel - for the dynamic voltage balancing - in contrast to the conventional RCD snubber typically used in hard-switching scenarios. This simplification is primarily attributed to the soft-switching conditions (ZVS and QZCS), which reduce the energies involved during commutation. Moreover, that work examined the switching dynamics of the IGCTs under low turn-off currents scenarios (as low as 50 A) and used a parallel resistor approach for static voltage balancing.

Nevertheless, unlike the work described in [5], this study incorporates the IGCT stack within a 3L-NPC leg power stage. This setup is used in conjunction with the 1 MW medium frequency transformer (MFT) developed in [11]. The focus of the analysis is on evaluating the power stage performance with an ultra-low turn-off current of approximately 6 A, defined by magnetizing inductance of the MFT at 5 kHz operating frequency. The assessment includes tests to validate dynamic voltage balancing using the C-snubber, as well as examining the static voltage balancing.

The main contributions of this work are as follows: i) it presents a comparative assessment of two

static voltage balancing strategies - parallel resistors and symmetrizing resistor; ii) it demonstrates the no-load operation of a 3L-NPC in a 2L mode for a 5 kV, 5 kHz IGCT-based DCT prototype, featuring an ultra-low turn-off current condition; and iii) it demonstrates the effective use of the symmetrizing resistor in a soft-switched 2L-operated 3L-NPC, exploring its impact on switching transients in conjunction with a low C-snubber value is used for the dynamic voltage balancing.

The paper is organized as follows: In Section II, the developed MV DCT prototype is described; Section III delves into the designs of the static voltage balancing, focusing on the parallel resistor and the symmetrizing resistor strategies, with experimental verification; Section IV presents the evaluation of the dynamic impact of the symmetrizing impact of the switches transient; and Section V, concludes this article.

2 MV DCT prototype

The MV DCT prototype consists of a 10(5) kV : 5 kV split-capacitor 3L-NPC LLC resonant converter. In this work, only the 1:1 turns ratio configuration is investigated. The power stages operate with 4.5 kV RC-IGCTs (5SGX1445H0001), and the NP clamping diodes (5SDF0545F0001).

Fig. 2a shows the MV DCT prototype. The entire converter is fitted inside a cabinet (200 x 180 x 80 cm), excluding the deionized water cooling unit, which is externally situated and connected to the laboratory's water supply system. The integration inside the cabinet has not been optimized towards any power density-driven considerations, but only for easy of access to relevant parts. Table 1 shows the main parameters of the DCT.

The RC-IGCTs are controlled by the ABB's AC800PEC controller. Additionally, the system incorporates ABB's COMBI-IO for interfacing with different peripheral devices and ABB's PEC-MI to connect with voltage and current sensors. Fig. 2b shows the schematic of the prototype.

The MFT prototype was built with a nanocrystalline air-cooled core and hollow copper oil-insulated water-cooled windings [11]. Therefore, this prototype serves as a research platform for this study.

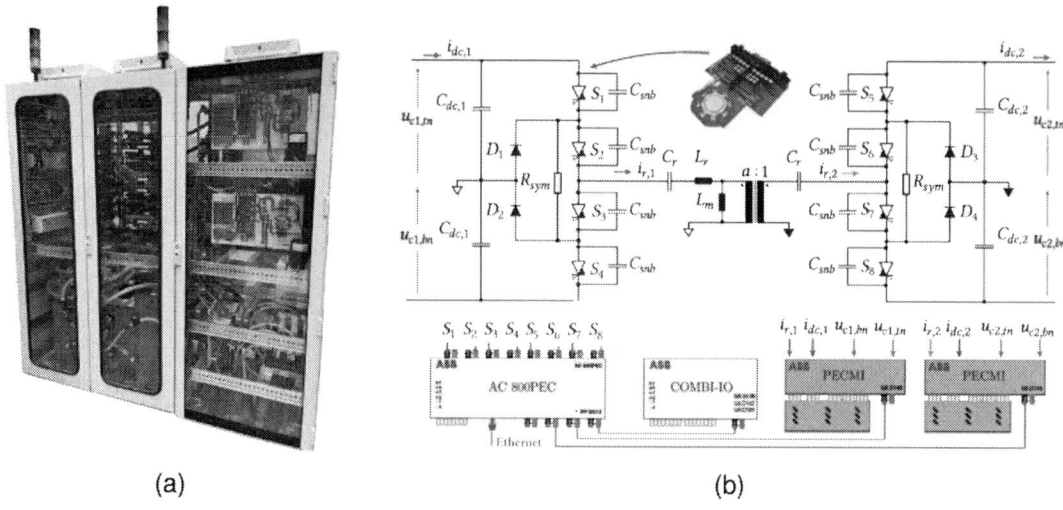

(a)   (b)

Fig. 2: (a) Photo of the MV DCT prototype. (b) Schematic of the MV DCT, illustrated using the symmetrizing resistor for static voltage and a single C-snubber for dynamic voltage balancing.

Tab. 1: General information about the MV DCT

Description	Symbol (Unit)	Value
Rated power	P_n (MW)	1
DC Voltage 1	$V_{dc,1}$ (kV)	10(5)
DC Voltage 2	$V_{dc,2}$ (kV)	5
DC-link capacitance 1	$C_{dc,1}$ (μF)	400
DC-link capacitance 2	$C_{dc,2}$ (mF)	2.6
Leakage ind. (2:1)	L_r (μH)	42.86
Leakage ind. (1:1)	L_r (μH)	11.1
Magnetizing ind.	L_m (mH)	10.7
Resonant capacitor	C_r (μF)	61
Operating frequency	f_s (kHz)	5

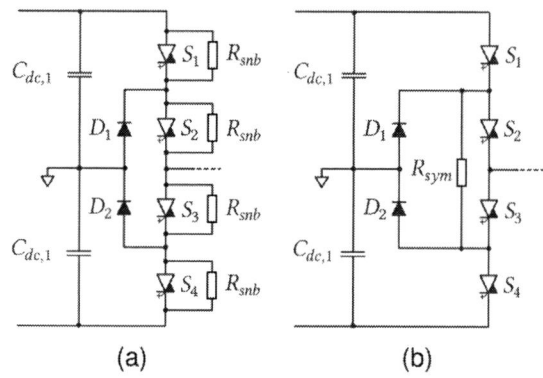

(a)   (b)

Fig. 3: (a) Schematic of the parallel resistor snubber. (b) Schematic of the 3L-NPC symmetrizing resistor snubber.

While the snubbers examined here are customized to meet these particular specifications, their design can be adapted for other applications.

3 Static Voltage Balancing

Unlike IGBTs that can benefit from active voltage balancing, series-connected RC-IGCTs depend on additional snubbers and balancing resistors for effective dynamic and static voltage balancing. Particularly, for static voltage balancing, the snubber is designed to ensure the switches operate safely during blocking periods.

Two known methods for static voltage balancing are evaluated, shown in Fig. 3. The first one, shown in Fig. 3a, uses the resistors in parallel with each power switch. This approach is both effective and straightforward for addressing the issue. However, it requires an individual resistor for each device, leading to increased power losses.

The second method, shown in Fig. 3b, uses a single resistor in parallel to the inner two IGCTs of the NPC-leg, solving the static voltage balancing with only one resistor, benefiting from the lower losses, as demonstrated next.

3.1 Parallel resistor

The static voltage balancing resistors ensure voltage balance across the series-connected devices

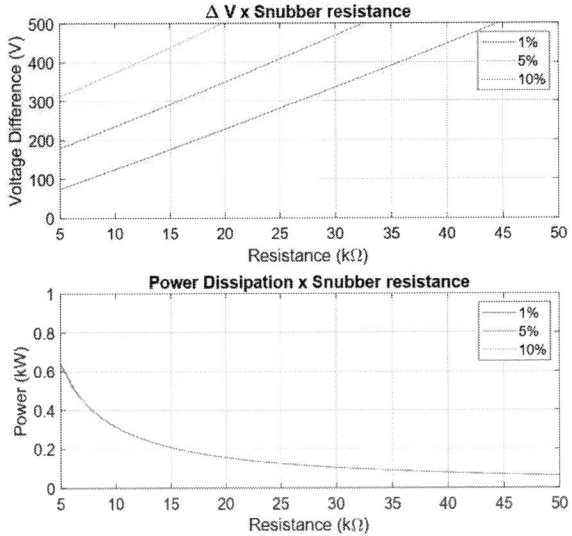

Fig. 4: Plots for the parallel resistor design. On top the expected/allowed voltage unbalance is shown and below the total corresponding power dissipation of the selected resistor is given.

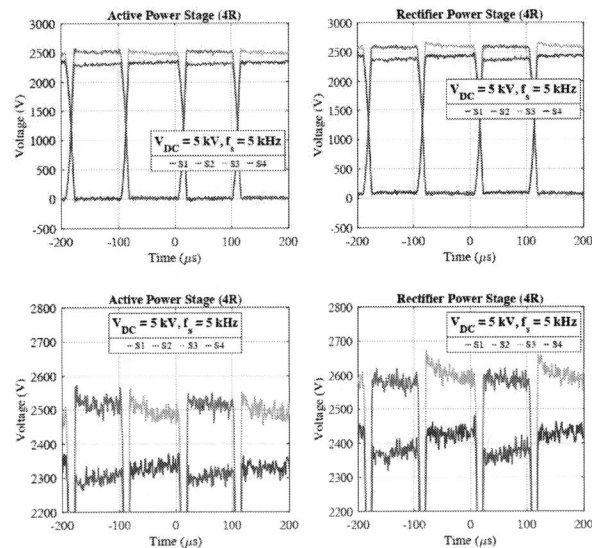

Fig. 5: Experimental waveform of the voltage across the IGCTs for the active and passive stack for the 5 kV and 5 kHz no-load operation with parallel resistor snubber.

by conducting a current greater than the maximum leakage current of these devices. Specifically, the resistors are required to carry a current that exceeds the leakage current of the RC-IGCTs when they are in the OFF state.

A sizing rule for these resistors was derived in [12], where the resistor value is sized considering the allowed voltage difference between the series connected devices, and the resistance tolerance of the resistors. This sizing rule leads to the following relationship:

$$\Delta V = \frac{V_{op} + \Delta V}{R_{snb} + \Delta R}\Delta R + I'_{leak,0}(R_{snb} + \Delta R) \quad (1)$$

where $I'_{leak,0}$ is the maximum leakage current of the devices at the operating voltage V_{op}, R_{snb} and ΔR are the values of the balancing resistor and its tolerance ($1\% \rightarrow \Delta R = R \times 0.01$), and ΔV is the maximum voltage deviation between the series connected devices. In order to correlate the leakage current with the actual operating voltage, the leakage current is,

$$I'_{leak,0} = \hat{I}_{leak,0}\sqrt{\frac{\frac{V_{op}}{n} + \frac{\Delta V}{n-1}}{V_{IGCT,0}}} \quad (2)$$

where $\hat{I}_{leak,0}$ is the maximum leakage current of the device, n is the number of series connected

devices, and $V_{IGCT,0}$ is the reference voltage of the IGCT test.

Thus, this equation can be solved and the relationship between the allowed voltage deviation and the resistance value is drawn. Fig. 4 shows the resistance value versus the voltage deviation for different tolerances. At the bottom of Fig. 4, the power dissipation is shown.

From this plot, a resistor can be selected to maintain the difference between the voltages of the IGCTs below 10% of the DC-link voltage. ($\Delta V = 500$ V for the $V_{DC} = 5$ kV DC-link). Thus, resistors of $R_{snb} = 10$ kΩ can be selected with a 5% tolerance, expecting a voltage difference of $\Delta V \approx 240$ V for the demonstration in this prototype. In total, this selection leads to $P_{snb} = 4 \times 320$ W $= 1280$ W in power losses per stack.

These resistors were tested with the MV DCT prototype. The experiment was carried out by operating the DCT with the 5 kV input voltage, in a $1:1$ turns ratio configuration. The primary power stage is active with a 50% duty cycle, secondary power stage is a passive rectifier with no load. The dead time was set to $\delta = 30$ μs (mainly due to a very low turn-off current) and a C-snubber of $C_{snb} = 20$ nF was used for dynamic voltage balancing.

Fig. 5 shows the experimental waveforms for the balancing resistors, with $R_{snb} = 10$ kΩ. With this selection, a total of $P_{snb,tot} = 2 \times 1280 = 2560$ W is dissipated for the static voltage balancing, considering both stacks. This power dissipation represents 0.256% of the nominal power (1 MW).

3.2 NPC symmetrizing resistor

Taking advantage of the NPC-leg, which has the NP clamping diodes between the devices and the neutral point, a single symmetrizing resistor can be used for static voltage balancing. The resistor is positioned in parallel to the inner two IGCTs of the NPC-leg, as shown in Fig. 3b. This solution is broadly used in industrial high-power 3L-NPC inverters [13]–[17].

Differently from the simple parallel resistor strategy, which is often designed for the maximum leakage current of the device, the symmetrizing resistor needs to compensate only for the difference in leakage currents of devices. Fig. 6a illustrates the role of the clamping diode on the static voltage balancing, and Fig. 6b illustrates the role of the symmetrizing resistor.

Firstly, Fig. 6a shows an illustration when S_1 and S_2 are blocking, and the device S_2 has a higher leakage current. In this case, the voltage on S_2 will decrease and forward bias the NP clamping diode D_1, preventing the voltage of S_1 from reaching a destructive level outside of the safe operating area (SOA). Consequently, the upper NP diode provides a current path to compensate for the leakage current, and voltage S_2 is clamped to the capacitor voltage $C_{dc,2}$, approximately $V_{dc}/2$. The symmetrizing resistor is not needed in this particular scenario.

On the other hand, Fig. 6b shows the case when the device S_1 has a higher leakage current. In this case, the voltage of S_1 will decrease, while the voltage of S_2 increases. Thus, the NP diode is reverse-biased and cannot offer protection as in the previous case. Yet, this time, the symmetrizing resistor provides the current path to compensate for the leakage current, preventing voltage across S_2 from reaching destructive voltage levels.

In other words, the symmetrizing resistor compen-

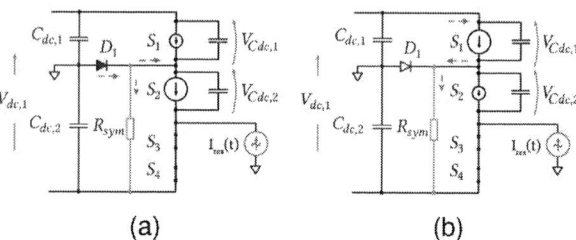

Fig. 6: (a) An illustrative example of the clamping diode acting when S_2 has a higher leakage current. (b) Example of the basic principles of the static voltage balance with an NPC symmetrizing resistor.

sates for the leakage current by draining extra current from S_2 to match the S_1 leakage current ($I_{leak,1} \approx I_{leak,2} + I_{Rsym}$). Hence, R_{sym} handles the difference between leakage current mismatch.

The current of the symmetrizing resistor should be big enough to properly compensate for the higher leakage current value of S_1. However, if the compensation is too big, the NP diode is forward biased again, and the scenario of Fig. 6a is repeated.

Consequently, following immediately this case, a simple design rule of this resistor can be derived considering the output capacitance rate of discharge and the current divider between the symmetrizing resistor and the equivalent resistance of the switch in parallel, resulting in:

$$\Delta V = \frac{R_{sym}}{R_{sym} + R_{leak,S_2}} I_{leak} \frac{T_{bk}}{C_{out}} \qquad (3)$$

where ΔV is the voltage difference between both series-connected IGCTs, R_{leak,S_2} is the equivalent resistance of the switch S_2 (which is the switch in parallel with the R_{sym} for the $-V_{dc}/2$ state - similar rule can be derived with $V_{dc}/2$ and S_3), I_{leak} is the leakage surplus current of the switch S_1, C_{out} is the output capacitance of the device, and T_{bk} is the actual blocking period considering dead time of the devices, which is half of the switching period minus dead time (considering the 2L operation of the resonant converter operating principles.)

Thus, (3) relates the voltage difference between the devices and the resistance value of the symmetrizing resistance, which is affected by the capacitance value in parallel with the device. This capacitance value impacts the rate of change to discharge the

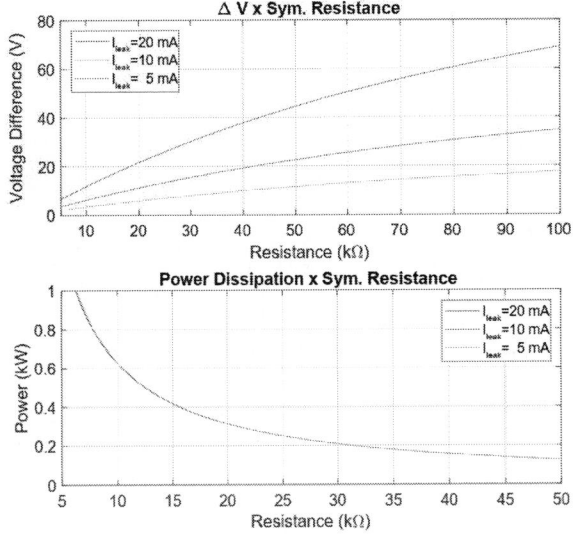

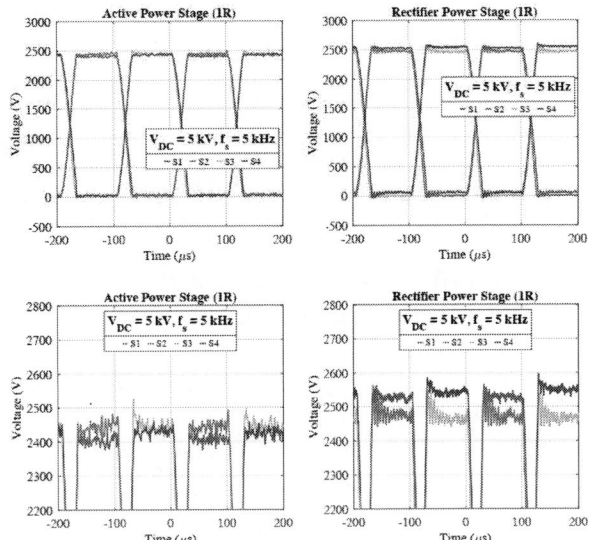

Fig. 7: Plots for the symmetrizing resistor design. On top, the maximum expected/allowed voltage unbalance during the blocking stage and below the total correspondent power dissipation of the selected resistor.

Fig. 8: Experimental waveform of the voltage across the IGCTs for the active and passive stack for the 5 kV and 5 kHz no-load operation with symmetrizing resistor snubber.

voltage with the leakage current. In (3), the capacitance is written as C_{out}, being the output capacitance of the device; however, if a snubber is included, it should be also considered for proper calculation.

Fig. 7 shows the design of the symmetrizing resistor using (3). At the bottom of Fig. 7, the power dissipation is shown for the symmetrizing resistor. This plot represents the maximum voltage deviation expected during the blocking state. The specifications of the MV DCT prototype were used for this simulation, where a switching frequency of $f_s = 5$ kHz has a switching period of $T_s = 200$ μs; hence, half a period is $T_{s,half} = 100$ μs, and with a dead time of $\delta = 30$ μs, the effective blocking period is $T_{bk} = 70$ μs. The considered C-snubber value is $C_{out} = 20$ nF. Thus, three scenarios of exceeding current of S_1, $I_{leak} = 5$ mA, $I_{leak} = 10$ mA, and the most extreme case $I_{leak} = 20$ mA, were simulated to evaluate their effect on the maximum voltage deviation.

Based on this plot, one should choose the considered worst case for designing the symmetrizing resistor. For instance, if a surplus current of $I_{leak} = 20$ mA is selected, this scenario assumes that only S_1 has leakage while S_2 is ideal. Consequently, the resistor can be chosen based on the

blue curve. Alternatively, if it is assumed that the switches have similar characteristics and the surplus current is lower, opting for a higher resistor can reduce losses. It's worth noting that the C-snubber value and the blocking time influence this curve, and similar conclusions can be drawn for different sets of parameters.

Therefore, using Fig. 7, a conservative resistance value is chosen to be $R_{sym} = 20$ kΩ, for the purpose of demonstration, resulting in a power consumption of $P_{Rsym} = 320$ W per stack.

Fig. 8 shows the experimental results using the symmetrizing resistor. The total power consumption, considering both stacks is $P_{snb,tot} = 2 \times 320 = 640$ W, which represents 0.064% of the nominal power. Although the power consumption of this snubber is already much lower than the other strategy, this experiment shows that the resistance value could increase even further to reduce the losses and allow a higher voltage difference between the IGCTs.

Purely analyzing the voltage static balancing, the symmetrizing resistor is a better solution regarding performance and losses. The two solutions are compared and Tab. 2 shows the side-by-side comparison. The slightly higher values for the passive stack result from a higher DC voltage at its terminals.

Tab. 2: Comparison between R_{sym} and R_{snb}.

	Sym-R	Parallel-R
Losses per stack	340 W	1260 W
ΔV of Active stack	41 V	182 V
ΔV of Passive stack	76 V	219 V

4 Dynamic impact of the R_{sym} on the switching transients

During the 2L operation, the symmetrizing resistor is always exposed to $V_{dc}/2$, which leads to a continuous current flowing through the resistor, including during transient moments. In this sense, the current flowing through the resistor impacts the switch's transition. In general, this current is much lower than the turn-off current, which makes this effect negligible. Nevertheless, for the DCT operation, the turn-off current is already low, and with IGCT switches, this effect can be observed.

Fig. 9a shows an illustration of the impact of the symmetrizing resistor on the voltage rise time. In this example the switches $S1$ and $S2$ were conducting current and a command to turn OFF was triggered. At this moment, the dynamic behavior of the voltage across the switches can be approximated by the switches's output capacitance (including snubber if present), the two well-balanced split-capacitor DC-link as voltage sources, and a constant turn-off current during this short period.

During the voltage rise of S_1 and S_2, the R_{sym} current contributes to the total current charging the capacitor of S_1, consequently, leading to a faster transient compared to S_2. More importantly, the resistor's current returns from S_4, acting against its discharge, leading to a slower transient compared to S_3. Thus, this contribution leads to an increase in voltage imbalance.

Fig. 9b illustrates the opposite effect of the symmetrizing resistor on switches S_1 and S_4, during the complementary transition, now slowing down the discharge of S_1, and speeding up S_4.

From the simplified circuit, shown in Fig. 9a, one can notice that the impact of the R_{sym} depends on its resistance value which defines the current, the switch's equivalent output capacitance, and the turn-off current.

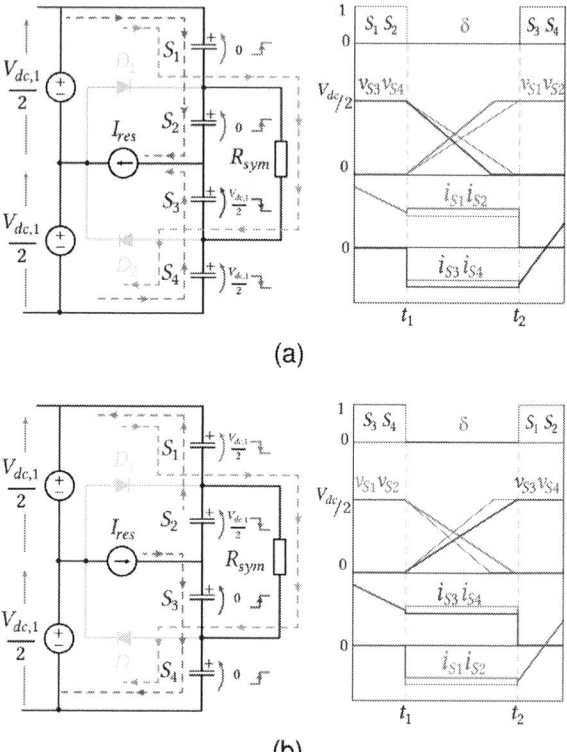

(a)

(b)

Fig. 9: Illustration with the impact of the NPC symmetrizing resistor on the switching transient, in (a) $V_{dc}/2 \rightarrow -V_{dc}/2$, and in (b) $-V_{dc}/2 \rightarrow V_{dc}/2$. The R_{sym} influences mainly the outer devices by reducing/increasing the turn-off current. During dead time, the full DC link voltage appears across $S_1 - R_{sym} - S_4$.

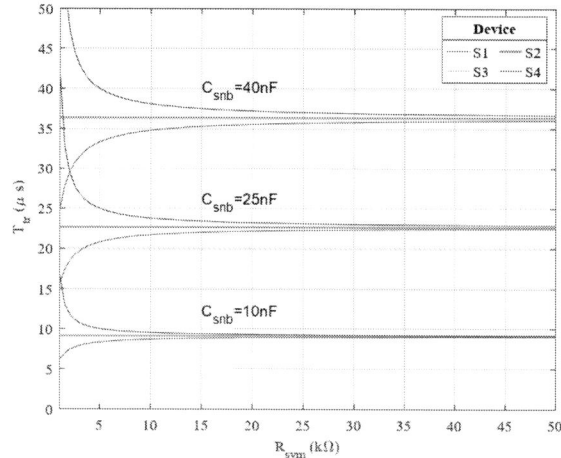

Fig. 10: Relationship between the transient response for the IGCTs and the symmetrizing resistor resistance for different snubber capacitances, considering $V_{dc} = 5$ kV, and $I_{off} = 5.5$ A, for the transition $V_{dc}/2 \rightarrow -V_{dc}/2$.

Thus, a simple equation can be derived to map its impact on the switches transients:

$$\begin{cases} i_{S1} = I_{off} \pm I_{Rsym} \\ i_{S2} = I_{off} \\ i_{S3} = I_{off} \\ i_{S4} = I_{off} \mp I_{Rsym} \end{cases} \quad (4)$$

where, $I_{Rsym} = V_{dc}/(2R_{sym})$. The first sign represents the case with transition from high to low state ($V_{dc}/2 \rightarrow -V_{dc}/2$). Consequently, the period of the voltage transition is given by:

$$t_{tr,Sx} = \frac{C \cdot V_{dc} \cdot n}{2 \cdot i_{Sx}}, \quad (5)$$

where n is the number of series-connected devices, and i_{Sx} is the IGCT's turn off current from (4).

In this way, (5) estimate the voltage rise time, taking into consideration the R_{sym} contribution. Fig. 10 shows the transient duration of each IGCT depending on the R_{sym} resistance value, for three different capacitor snubbers, for a turn-off current of $I_{off} = 5.5$ A.

It can be seen in this plot, that for a C-snubber value of $C_{snb} = 25$ nF, and a resistor of $R_{sym} = 10$ kΩ, the switches S_2 and S_3 would take (theoretically, with the presented assumptions), approximately $t_{tr} \approx 22.73$ μs for the transition, while the switch S_1 would take $t_{tr} \approx 21.74$ μs, and finally the switch S_4 would take $t_{tr} \approx 23.81$ μs for the transition from $V_{dc}/2 \rightarrow -V_{dc}/2$. It means that switch S_3 will discharge approximately 1 μs faster than its pair (S_4), creating a voltage imbalance, and impacting the dynamic voltage balancing.

As expected, this effect is only strongly affected by low resistor values. However, these values are in the range of the required resistance to have a good compensation with low C-snubbers - which are required to allow fast voltage rise transient due to the low turn-off current. Ultimately, this is a trade-off involving the symmetrizing resistor value, the C-snubber, and the turn-off current.

To evaluate the waveforms of both stacks and the symmetrizing resistor waveforms at the same time, two Yokogawa DLM4058, 8-channel oscilloscopes were synchronized to capture the 16 waveforms at the same time. Four Cal Test Electronics CT4079-NA differential probes were used to capture the

primary IGCTs voltage and the other voltages were recorded using GW Instek GDP-100 differential probe. The magnetizing current was recorded using the PEM CWT1 B/2.5/500 Rogowski coil, and the symmetrizing resistor current was recorded using the Keysight N2781B current probe.

Fig. 11 shows the complete no-load experiment showing the effect of the symmetrizing resistor on the voltage transient. The current on the symmetrizing resistor is around $I_{Rsym} \approx 0.2$ A, showing that the actual equivalent resistance value is $R_{sym} \approx 12.5$ kΩ, representing 3-5% of the turn-off current.

The total transition time is around $t_{tr} = 28$ μs. This extended duration can also be noted by visual inspection in Fig. 5 and Fig. 8, comparing the voltage shape of both strategies, where the total transition time for the parallel resistors is $t_{tr} < 20$ μs.

Furthermore, the voltage imbalance when the transient has finished while using the parallel resistor snubber is very similar to the imbalance present during the blocking state (approximately 200 V). This is a result of all the capacitors having the same current for discharging/charging. However, this is not the case for the symmetrizing resistor case, where the resistor's current contributes to increasing the voltage mismatch difference during this transient. The voltage imbalance when the first switch was completely discharged was around 290 V for the active power stage and approximately 380 V at the passive power stage.

Ultimately, this experiment showed that under these conditions, the symmetrizing resistor performed successfully the static voltage balancing, but also influenced the dynamic voltage balancing, slowing it down and increasing the voltage imbalance. Such conditions lead to the requirement for a dead time $\delta \geq 30$ μs to ensure a safe transition - representing 15% of the switching period, which is not ideal for the converter's operation.

5 Conclusion

This work detailed the challenges of operating an IGCT-based split capacitor 3L-NPC DCT at 5 kV and 5 kHz, under ultra-low turn-off current conditions. It focused on the snubber design for static

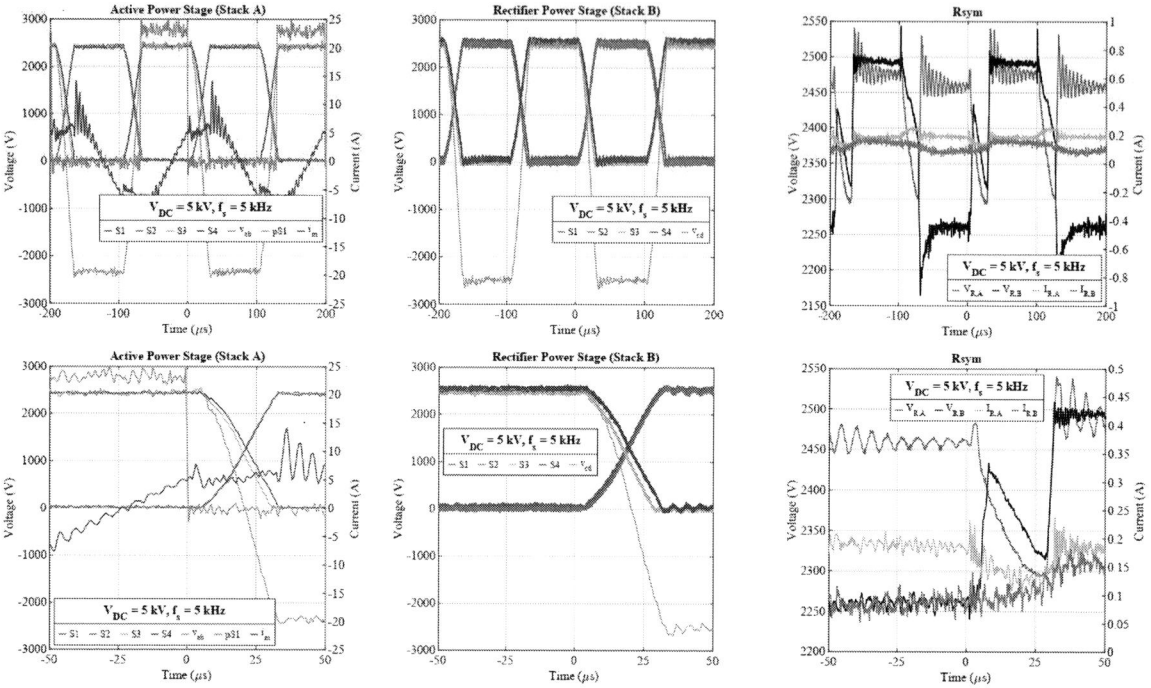

Fig. 11: Experimental waveforms for 5 kV and 5 kHz test with the voltage across the IGCTs, the voltages at the MFT terminals, the voltage across the symmetrizing resistors and their current, and the magnetizing current. The dead time was set to $30~\mu s$ and a C-snubber of 20 nF was used for the dynamic voltage balancing. The turn-off current is around $I_{off} = 5.5$ A. The total transition time is around $t_{tr} = 28~\mu s$.

voltage balancing, evaluating two strategies: the parallel resistor and the symmetrizing resistor.

The investigation has shown that the symmetrizing resistor snubber is as effective as the parallel resistors snubber, offering the advantage of using a single resistor compared to four in the alternative solution. Consequently, it successfully achieves static voltage balancing with lower losses.

On the downside, the symmetrizing resistor affects the voltage rise transient of the switches, which increases the dynamic voltage imbalance of the series-connected IGCTs. This effect is more critical when operating with ultra-low turn-off currents, as demonstrated with the MV DCT prototype. Nevertheless, this effect can be mitigated by using a higher resistance value and adjusting the dead time accordingly to allow safe commutation.

Consequently, additional trade-offs are required to allow the 5 kHz operation of the 3L-NPC DCT

with the symmetrizing resistor such as reducing even further the C-snubber or increasing the turn-off current by adjusting the magnetizing inductance. Also, further improvements on the dynamic impact evaluation of the symmetrizing resistor could be done by assessing its impact on the 3L operation and by calculating the actual turn-off current of the IGCTs including parasite resistances and device voltage drops.

Finally, due to ongoing work in integration and commissioning, we have only included preliminary results. Comprehensive tests at full power will be detailed in subsequent reports.

Acknowledgments

The results presented in this paper are a part of the EMPOWER project that has received funding from the European Research Council (ERC) under the European Union's Horizon 2020 research and innovation program (Grant agreement No. 818706.)

References

[1] Y. Ma, B. Zhao, B. Cui, X. Tang, L. Dong, *et al.*, "An IGCT-series-based DC transformer with quasi-zero switching loss modulation by minimum backflow power injection," *IEEE Transactions on Power Electronics*, vol. 38, no. 12, pp. 15 566–15 578, 2023. DOI: 10.1109/TPEL.2023.3306598.

[2] J. Voss, S. P. Engel, and R. W. De Doncker, "Control method for avoiding transformer saturation in high-power three-phase dual-active bridge DC–DC converters," *IEEE Transactions on Power Electronics*, vol. 35, no. 4, pp. 4332–4341, 2020. DOI: 10.1109/TPEL.2019.2938585.

[3] G. Ortiz, M. G. Leibl, J. E. Huber, and J. W. Kolar, "Design and experimental testing of a resonant DC–DC converter for solid-state transformers," *IEEE Transactions on Power Electronics*, vol. 32, no. 10, pp. 7534–7542, 2017. DOI: 10.1109/TPEL.2016.2637827.

[4] D. Dujic, G. Steinke, E. Bianda, S. Lewdeni-Schmid, C. Zhao, and J. K. Steinke, "Characterization of a 6.5kV IGBT for medium-voltage high-power resonant DC-DC converter," in *2013 Twenty-Eighth Annual IEEE Applied Power Electronics Conference and Exposition (APEC)*, 2013, pp. 1438–1444. DOI: 10.1109/APEC.2013.6520487.

[5] G. Ulissi, U. R. Vemulapati, T. Stiasny, and D. Dujic, "High-frequency operation of series-connected IGCTs for resonant converters," *IEEE Transactions on Power Electronics*, vol. 37, no. 5, pp. 5664–5674, 2022. DOI: 10.1109/TPEL.2021.3132200.

[6] J. Kucka and D. Dujic, "Current limiting in overload conditions of an LLC-converter-based DC transformer," *IEEE Transactions on Power Electronics*, vol. 36, no. 9, pp. 10 660–10 672, 2021. DOI: 10.1109/TPEL.2021.3060106.

[7] A. Nagel, S. Bernet, T. Bruckner, P. Steimer, and O. Apeldoorn, "Design of IGCT series connection for 6 kV medium voltage drives," in *IEE Seminar PWM Medium Voltage Drives (Ref. No. 2000/063)*, 2000, pp. 2/1–2/5. DOI: 10.1049/ic:20000334.

[8] H. Bai, Z. Zhao, M. Eltawil, and L. Yuan, "Optimization design of high-voltage-balancing circuit based on the functional model of IGCT," *IEEE Transactions on Industrial Electronics*, vol. 54, no. 6, pp. 3012–3021, 2007. DOI: 10.1109/TIE.2007.907002.

[9] S. Parashar, N. Kolli, R. Kumar Kokkonda, S. Bhattacharya, and A. Kumar, "Design optimization of the snubber circuit for three- level NPC converter pole for hard switching application," in *2020 IEEE Energy Conversion Congress and Exposition (ECCE)*, 2020, pp. 2459–2466. DOI: 10.1109/ECCE44975.2020.9235644.

[10] R. Bai, B. Zhao, T. Zhou, X. Tang, J. Li, *et al.*, "PWM-current source converter based on IGCT-in-series for DC buck and constant-current application: Topology, design, and experiment," *IEEE Transactions on Industrial Electronics*, vol. 70, no. 5, pp. 4865–4874, 2023. DOI: 10.1109/TIE.2022.3186385.

[11] N. Djekanovic and D. Dujic, "Design optimization of a MW-level medium frequency transformer," *PCIM Europe 2022; International Exhibition and Conference for Power Electronics, Intelligent Motion, Renewable Energy and Energy Management*, pp. 1–10, 2022. DOI: 10.30420/565822101.

[12] A. Nagel, S. Bernet, T. Bruckner, P. Steimer, and O. Apeldoorn, "Characterization of IGCTs for series connected operation," in *Conference Record of the 2000 IEEE Industry Applications Conference. Thirty-Fifth IAS Annual Meeting and World Conference on Industrial Applications of Electrical Energy (Cat. No.00CH37129)*, vol. 3, 2000, 1923–1929 vol.3. DOI: 10.1109/IAS.2000.882141.

[13] R. Jakob and P. Sadowski, *Method for operating an electric circuit*, EP2654190A2, April 2012.

[14] C. Li, X. Li, C. Zhu, Y. Li, C. Wang, *et al.*, *Integrated gate commutated thyristor (IGCT) three-level power module*, CN102064676A, 2011.

[15] B. Zhongwei, S. Honggui, M. Kai, and Z. Bingning, *Improved high-voltage high-power three-level NPC topological structure inversion parallel circuit*, CN214674943U, 2021.

[16] J. Jianguo, G. Ligang, W. Baoyu, Q. Gaowei, L. Yanliang, *et al.*, *Frequency converting circuit based on IGCT*, CN208143130U, 2011.

[17] Curou, W. Wubin, C. Jian, Rechau, Y. Zhanqing, *et al.*, *Three-level wind power converter inner tube overvoltage equalization circuit and method*, CN116155077A, 2023.

PCIM Europe 2024, 11– 13 June 2024, Nuremberg   DOI: 10.30420/566262033

Comparative Analysis of Unidirectional High Step-Up Converters for Medium-Voltage Applications

Stefan Subotic[1], Ralph Burkart[2], Thomas Gradinger[2], Drazen Dujic[1]

[1] École Polytechnique Fédérale de Lausanne (EPFL), Power Electronics Laboratory, Switzerland
[2] Hitachi Energy Research, Switzerland

Corresponding author/speaker: Stefan Subotic, stefan.subotic@epfl.ch

Abstract

Connection of renewable energy sources to the MVDC collection grid requires an efficient, reliable and economically viable high power DC/DC converter. With these requirements in mind, the paper presents a comparative analysis of two bulk-power processing converter topologies - the Single Active Bridge (SAB) and the Phase Shifted Full Bridge (PSFB) converter from the standpoint of suitability for the application, considering power hardware design and control requirements. The aim of the work is to provide a clear summary of trade-offs between the SAB and PSFB converter without their detailed design optimization. An overview of the converters' operating principles and steady-state models is presented first, laying the foundation for the comparison through converter characteristics. Based on carefully chosen exemplary designs the advantages and drawbacks of both topologies are identified and preliminary semiconductor losses are evaluated for each case. Finally, the input voltage regulation of the converters is addressed as well as additional considerations relevant for the applications of interest.

1 Introduction

The increased presence of renewable energy sources has brought forward numerous considerations toward their integration into the grid by means of Direct Current (DC) technologies. There is now strong indication that present Alternating Current (AC) systems will be complemented or replaced by their DC counterparts not only at High Voltage (HV) levels but at Medium Voltage (MV) levels as well [1], [2]. This is particularly true for the MVDC collection grids, which are being investigated for the large solar plants [3] and wind farms [4]. Depending on the considered power levels and required distances, the shift towards MVDC collection can lead to increased system efficiency [5]. The key factor still impeding this transition is the lack of a strong business case for all involved parties.

Recent advances in the field imply that, in the near future, single wind turbine power will be approaching $20\,\text{MW}$ [6]. Similar power levels can be processed by central inverters in large-scale solar plants. While the input voltage, depending on the source, is in the $(1.5-5)\,\text{kV}$ range, the MVDC collection grid voltage can reach several dozens of kV

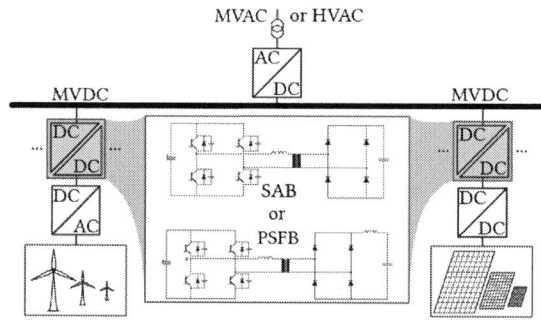

Fig. 1: Wind farm or solar plant connection to the MVDC collection grid

in both cases, with $20\,\text{kV}$ being taken as an example in this paper. This implies that the application requires a unidirectional high step-up high power DC/DC converter as the interfacing element. Fig. 1 illustrates the application and considered topological solutions on a conceptual level. The collection grid voltage is regulated by the DC/AC converter and it is assumed that maximum power point tracking and conversion from AC to DC (in case of wind turbines) is performed by the source-side converters. Consequently, it is required that the DC/DC converter regulates its input voltage.

274

PCIM Europe 2024, 11– 13 June 2024, Nuremberg   DOI: 10.30420/566262033

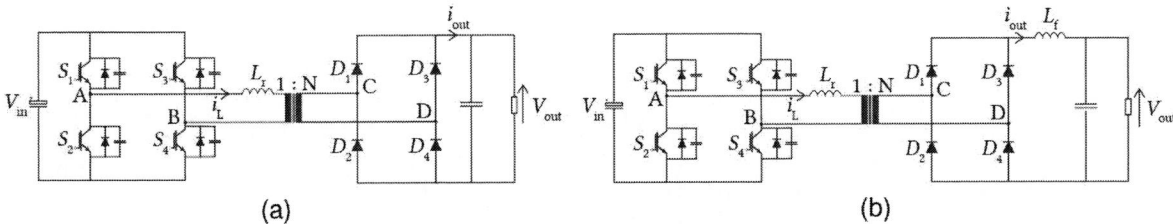

Fig. 2: Topologies analysed in this work: (a) SAB converter (b) PSFB converter

In contrast to the widely studied modular solid state transformer, the concept of monolithic (bulk-power processing) unidirectional DC/DC converters has not been extensively investigated for these applications, with few notable exceptions [7]–[15]. The lack of standardization and demand from the market has also led to a scarcity of pilot projects and demonstrators. Nonetheless, the monolithic DC/DC converter emerges as an attractive solution, potentially offering a reduction of isolation overhead and an overall more robust structure compared to the partial-power processing alternative.

In the scientific literature, bidirectional converters such as the dual active bridge and resonant converters are drawing significant attention. Although prototypes of these converters exist [16], [17], the resonant topologies present additional challenges for control in high power applications (usually requiring an additional converter to perform the control, otherwise leading to reduced efficiency), while an active rectifier introduces additional costs and complexity in the considered application, where, generally, only unidirectional power flow capabilities are required.

Among the monolithic unidirectional converters, two particularly attractive topologies, considering complexity and control requirements, are the SAB and PSFB converter. These converters have already been investigated to some extent as an interfacing element between wind [8], [9] and solar plants [15] and the power grid. Both converters offer high efficiency through soft switching as well as a comparatively simple architecture. These topologies, although well known [18], [19], have never been demonstrated nor comprehensively compared at MW-level ratings, indicating the need for further research and motivating this work.

With the presented considerations in mind, the

paper is focused on an analytical comparison of the SAB and PSFB converter, based on output power and current characteristics, and semiconductor losses. Additionally, other aspects relevant for the application such as input voltage control and specific hardware requirements are addressed. It should be emphasised that the presented methodology allows a juxtaposition of the two converters without their respective optimization, strictly through appropriately selected exemplary designs. This allows the evaluation of the converters' performance based on analytical models in steady state.

2 Specifications and Designs

In Tab. 1, the overall requirements for the converter are defined. The $\pm 10\%$ variation of the input/output voltage is taken as a representative design example, considering the input voltage control requirements under output voltage and input current disturbances. The comparison is carried out for the wind farm and power grid interconnection application, with similar conclusions possible for the solar plant case as well.

The two studied toplogies are shown in Fig. 2. It can be seen that the SAB converter features only an inductor at the primary side of the Medium Frequency Transformer (MFT), while the PSFB has an additional di/dt limiting filter inductor at the output (alternatively, split and installed in both DC lines). It should be mentioned that the primary-side inductance of the PSFB is typically much smaller than

Tab. 1: General converter specifications

Description	Symbol [Unit]	Value
Input voltage	V_{in} [kV]	$5 \pm 10\%$
Output voltage	V_{out} [kV]	$20 \pm 10\%$
Output power	P_{out} [MW]	20
Sw. frequency	f_s [Hz]	500

275

the filter inductance and is usually carried out as the leakage inductance of the MFT. For the SAB, the inductor is often realized as an additional discrete component. Both the SAB and the PSFB converter can operate in two possible modes of operation - Continuous Conduction Mode (CCM) and Discontinuous Conduction Mode (DCM).

In the analysis, three different designs, shown in Tab. 2, are considered for both converters. Each considered design conforms to the specifications from Tab. 1, enabling a fair comparison. The first and second design were oriented toward CCM and DCM operation respectively, in the entire input/output voltage range under nominal power. For the third design, the worst-case operational conditions are taken into account, without any operational mode constraints. Each design is elaborated in more detail in Section 4. It should be noted that none of the designs are optimized, and that their main purpose is to enable grounds for comparison between the SAB and the PSFB converter in terms of capability to serve the considered application.

3 Analytical Modeling

The phase-shift modulation, offering the possibility of input/output voltage or power-flow control, and being the most commonly proposed modulation strategy for both converters, is considered in the

Tab. 2: Exemplary converter designs

Design 1		
Parameter / Symbol [Unit]	**SAB1**	**PSFB1**
Leakage inductance / L_r [µH]	94.7	29.6
Filter ind. / L_f [mH]	n/a	2.7
Turns ratio / N	8	5.5
Nom. duty cycle / D_{nom}	0.62	0.75
Design 2		
Parameter / Symbol [Unit]	**SAB2**	**PSFB2**
Leakage inductance / L_r [µH]	10.74	1.79
Filter ind. / L_f [mH]	n/a	0.22
Turns ratio / N	5	5
Nom. duty cycle / D_{nom}	0.29	0.29
Design 3		
Parameter / Symbol [Unit]	**SAB3**	**PSFB3**
Leakage inductance / L_r [µH]	8.27	75.4
Filter ind. / L_f [mH]	n/a	9.6
Turns ratio / N	6.5	6.5
Nom. duty cycle / D_{nom}	0.59	0.82

analysis. In this work, it is considered that the leg containing switches S_1 and S_2 is the leading leg, while the other leg is the lagging leg. The value of phase shift (ϕ) translates into the duty cycle (D) of the primary side voltage V_{AB}. This relation can be described as:

$$D = 1 - \frac{\phi}{\pi} \tag{1}$$

Both the SAB and the PSFB converter can operate either in CCM or DCM depending on the value of the duty cycle and other, fixed, parameters (N, V_{in}, V_{out}). If a converter operates in CCM, by decreasing the duty cycle, at some point, it enters into DCM. This point corresponds to the boundary duty cycle, which is the same for both converters and equal to the DC voltage conversion ratio ($D_{BCM} = \frac{V_{out}}{NV_{in}}$).

In Fig. 3, the steady-state waveforms are shown for both converters operated in CCM/DCM along with conduction intervals for all semiconductor devices ("+" indicates the IGBT alone, while "-" corresponds to antiparallel diodes) during one switching cycle. It should be mentioned that the magnetizing inductance is considered to be infinite in this model which is a good approximation for the purposes of the analysis. The notation in the figure is self-explanatory. Therefore, some of the shown symbols are not explicitly explained in the text. Rather, a concise body of equations governing the operation of each converter in each operating mode is given, and comments are offered only where relevant.

3.1 SAB Converter

Referring to the left-hand side of Fig. 3a, the steady-state waveforms of the SAB converter operated in CCM can be described by the following equations:

$$\frac{V_{out}}{V_{in}} = N\frac{DT - 4t_\delta}{T} \tag{2}$$

$$I_1 = \frac{V_{in} - V_{out}/N}{L_r}(DT/2 - t_\delta) \tag{3}$$

$$I_2 = \frac{V_{in} - V_{out}/N}{L_r}(DT/2 - t_\delta) - \frac{V_{out}/N}{L_r}(1 - D)T/2. \tag{4}$$

The DCM mode operation is depicted on the right-hand side of Fig. 3a. The corresponding description can be easily derived:

$$\frac{V_{out}}{V_{in}} = N\frac{DT}{T - 2t_\alpha} \tag{5}$$

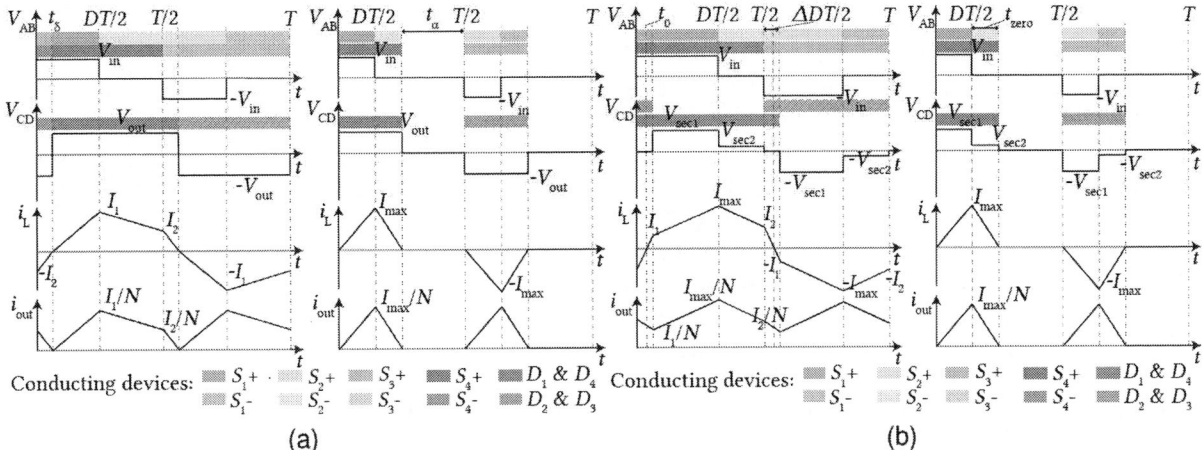

Fig. 3: Steady-state waveforms and conduction states: (a) SAB converter: CCM (left) and DCM (right) (b) PSFB converter: CCM (left) and DCM (right)

$$I_{max} = \frac{1}{L_r}(V_{in} - \frac{V_{out}}{NV_{in}})DT/2. \qquad (6)$$

3.2 PSFB Converter

CCM operation of the PSFB converter is illustrated on the left-hand side of Fig. 3b. The voltage drop across L_r is always taken into account, although it is usually neglected during the $(\frac{\Delta DT}{2}, \frac{T}{2})$ interval, which is only justified if L_r is much smaller than L_f (when reflected to the primary side of the transformer). Thus, the model contains no simplifications in this sense, appreciably enhancing precision when L_r is considerable, and can be described by:

$$V_{sec1} = \frac{NL_fV_{in} + N^2L_rV_{out}}{L_f + N^2L_r} \qquad (7)$$

$$V_{sec2} = \frac{N^2L_rV_{out}}{L_f + N^2L_r} \qquad (8)$$

$$D_{eff} = \frac{V_{out} - V_{sec2}(1-D)}{V_{sec1}} \qquad (9)$$

$$K = \frac{V_{in}\Delta D}{4L_rNf_s} + \frac{(V_{out} - V_{sec2})(1-D)}{4L_ff_s} \qquad (10)$$

$$\Delta I_{out} = \frac{(NV_{in} - V_{out})D_{eff}}{2f_s(L_f + N^2L_r)} \qquad (11)$$

$$I_1 = N(K - \frac{\Delta I_{out}}{2}) \qquad (12)$$

$$I_2 = N(K + \frac{\Delta I_{out}}{2} - \frac{(V_{out} - V_{sec2})(1-D)}{2L_ff_s}) \qquad (13)$$

$$I_{max} = N(K + \frac{\Delta I_{out}}{2}) \qquad (14)$$

$$t_0 = \frac{V_{sec2}(D-1) + NV_{in}D - V_{sec1}D_{eff}}{4NV_{in}f_s}. \qquad (15)$$

In the equations, f_s is the switching frequency. D_{eff} is equal to D-ΔD, and corresponds to the interval when V_{sec1} is applied at the MFT secondary. When L_f is much larger than L_r, this value becomes equal to $\frac{V_{out}}{NV_{in}}$ and is typically called the effective duty cycle of the secondary voltage. ΔI_{out} is the peak to peak ripple of the output current. While all values shown in Fig. 3b are accurately described, expression (10) also offers a good approximation of the average output current (which can be accurately expressed based on the presented model), by neglecting the change of output current slope during the duty cycle loss interval.

Equations (7) and (8) still hold true in DCM operation, shown on the right-hand side of Fig. 3b, as well as:

$$I_{max} = \frac{(NV_{in} - V_{sec1})D}{2N^2L_rf_s} \qquad (16)$$

$$t_{zero} = \frac{(NV_{in} - V_{sec1})D}{2V_{sec2}f_s}. \qquad (17)$$

Based on the presented steady-state analytical models and knowing the conduction states, through integration of linear segments, expressions for RMS and average currents can be derived for each semiconductor device of the converters. These expressions, although not shown, are used to obtain the converter characteristics in the following section as a basis for analysis.

4 Current Stress Comparison

In this section, the three pairs of SAB and PSFB converter designs are compared based on their output power and current characteristics. In the presented output power graphs, only the characteristics corresponding to extreme and nominal combinations of the input and output voltages are displayed, while all other characteristics fall between the shown ones. These cases are noted as: ① $(V_{in,min}, V_{out,max})$; ② $(V_{in,min}, V_{out,min})$; ③ $(V_{in,nom}, V_{out,nom})$; ④ $(V_{in,max}, V_{out,max})$; ⑤ $(V_{in,max}, V_{out,min})$. It should be noted that the devices within the same leg experience the same current stress, only in different half-cycles of switching. Therefore, only one device per leg is considered in the analysis. Only RMS currents are displayed, with similar conclusions being valid for average currents. Also, within current characteristics graphs, only nominal input/output voltage is shown with similar conclusions being applicable for any other input/output voltage combination.

4.1 CCM Operation

Converters SAB1 and PSFB1, defined in Tab. 2, are designed to operate in CCM at rated power in the entire input/output voltage range. It can be shown that the RMS currents on the primary side of the SAB and PSFB converters increase with the transformer turns ratio. Therefore, the turns ratio is kept as low as possible for the SAB1 and slightly above the theoretical minimum ($N_{min} = \frac{V_{out,max}}{NV_{in,min}}$) to avoid an excessively high slew rate of the output power in CCM and/or high secondary voltage, for the PSFB1. This, nonetheless, resulted in a considerably lower required turns ratio for PSFB1 than for SAB1. The output power characteristics are shown for both converters in Fig. 4.

In Fig. 5, the current stress of the primary and secondary side devices is shown for both converters. It

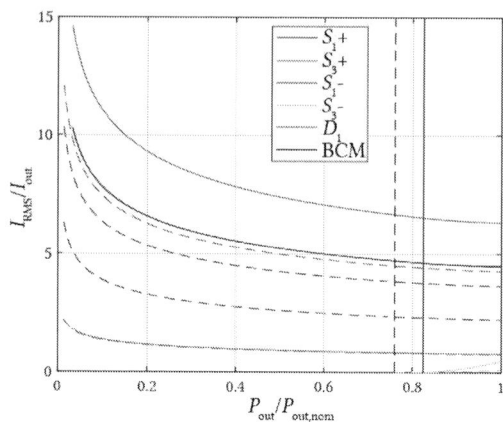

Fig. 5: Current characteristics: SAB1 - full line and PSFB1 - dashed line

can be seen that for all devices on the primary side, the PSFB1 achieves significantly lower RMS currents, while the rectifier diodes experience similar current stress.

4.2 DCM Operation

SAB2 and PSFB2, were designed to operate in DCM at rated power (hence, also at all partial power levels) for any input/output voltage combination. To make the comparison fair in this case, the same turns ratio (slightly higher than the theoretical minimum for nominal power transfer) is adopted for both converters. As a design rule, it was considered that the converters must operate in BCM (boundary mode) if the input voltage is minimal and output voltage is maximal. This rule helps to reduce the RMS currents which are drastically increased with deep DCM operation (low duty cycle for nominal power operation) and, again, provides leveled ground for the analysis. As obvious from the output power characteristics shown in Fig. 6, this resulted in identical behavior of both converters in DCM. It should be mentioned that the PSFB inductances can be selected in an infinite number of ways (as long as the total inductance reflected to one side of the

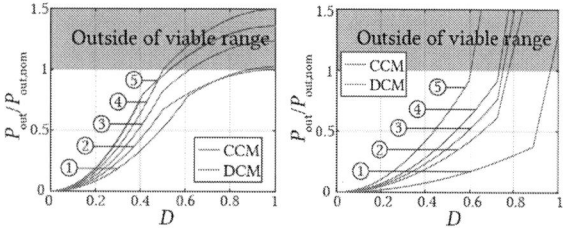

Fig. 4: Output power characteristics: SAB1 and PSFB1

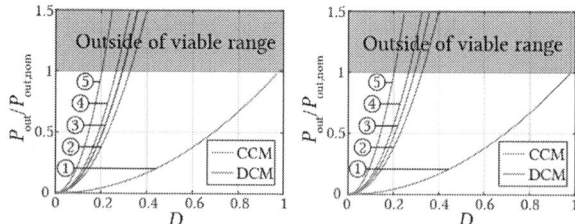

Fig. 6: Output power characteristics: SAB2 and PSFB2

PCIM Europe 2024, 11– 13 June 2024, Nuremberg   DOI: 10.30420/566262033

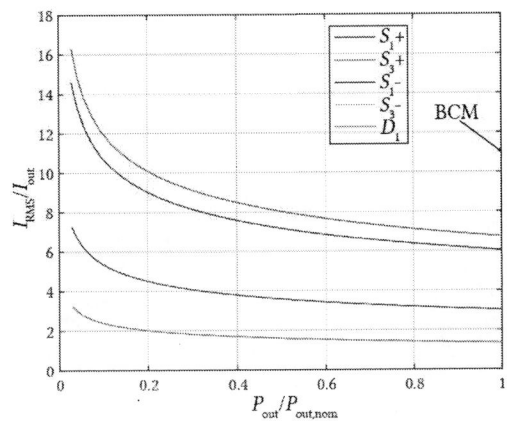

Fig. 7: Current characteristics: SAB2 - full line and PSFB2 - dashed line (coincides with the full line)

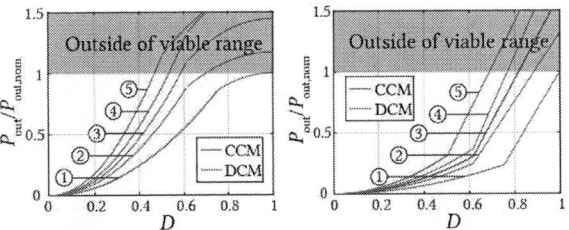

Fig. 8: Output power characteristics: SAB3 and PSFB3

transformer remains the same) to achieve exactly the same characteristics in DCM. However, the fact that a high output filter inductance increases the voltage across the rectifier should be kept in mind.

In Fig. 7, the device RMS current stress is shown for both converters. As expected, the current stress in DCM is exactly the same for SAB2 and PSFB2. Despite this fact, the SAB converter shows clear advantages compared to the PSFB when both are designed for DCM operation. Since the SAB does not feature an output inductor, the total required inductance can be reduced by a factor of almost N^2, and the rectifier voltage is clamped to V_{out} which reduces the rectifier voltage stress.

4.3 Operation Without Mode Constraints

Converters SAB3 and PSFB3 were designed without CCM/DCM operation requirements at rated power as, considering the entire input/output voltage range operation, an optimal design is likely the one allowing a combination of CCM and DCM operation, at least for the SAB. Again, it must be emphasised that SAB3 and PSFB3 are not optimized designs. As a basis for fair comparison, in this case, the same MFT turns ratio equal to 6.5 is selected for both converters. This value of N is selected as it provides a good trade-off between forcing the SAB converter toward DCM operation, which is generally the case with low turns ratios, and CCM operation, which is achievable with higher turns ratios. For the PSFB, the additional degree of freedom originating from the additional inductance, enables CCM operation in the entire voltage range

even with low turns ratios. The final requirement for both converters, ensuring a fair comparison, is that nominal power is achieved at $D = 1$ when input voltage is minimal and output voltage is maximal.

Output power characteristics are shown in Fig. 8. It can be seen that, for the SAB3 converter, nominal power is reached both in CCM and in DCM for various characteristics. On the other hand, the PSFB3 output power characteristics indicate continuous operation in CCM under nominal power. It should be mentioned that the PSFB3 could have been designed differently for this comparison. The shape of the output characteristics of a PSFB converter depends on the parameter r which can be defined as the ratio of the two inductance reflected on the same transformer side ($r = \frac{L_{\text{f}}}{N^2 L_{\text{r}}}$). Comparing two designs, one with a high value of r, and the other with the low value of r for the same L_{r}, it is possible to observe that the second design results in a higher filter inductance which reduces the current ripple of the output current. This results also in lower RMS currents of the primary. The downside of the second design is a large variation of output power with duty cycle change in CCM. Therefore, it

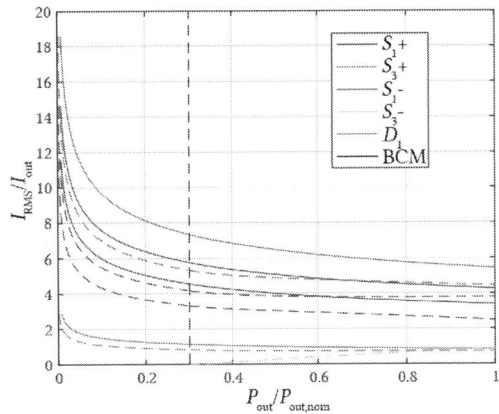

Fig. 9: Current characteristics: SAB3 - full line and PSFB3 - dashed line

279

PCIM Europe 2024, 11– 13 June 2024, Nuremberg   DOI: 10.30420/566262033

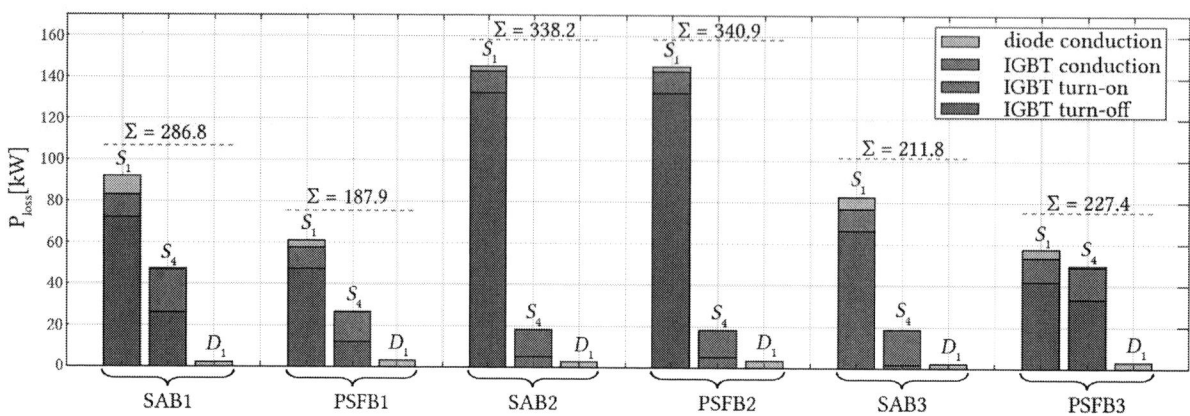

Fig. 10: Loss distribution of the 6 designs under nominal conditions: leading leg switch S_1, lagging leg switch S_4 and rectifier diode D_1. The sum represents the total losses of the converter ($\Sigma = 2P_{S1} + 2P_{S4} + 4P_{D1}$).

can be concluded that, by increasing the factor r, it is possible to achieve lower RMS current stress and smaller output current ripple, but at a price of loose control of the output power. A conservative value of $r = 3$ was adopted for PSFB3 as a good trade-off in this sense.

In Fig. 9, it can be observed that the additional degree of freedom of the PSFB resulted in lower RMS currents in all devices. Again, a different design of the PSFB3 could have resulted in an even more obvious difference at the cost of higher slew rate of the characteristics shown in Fig. 8 (right) in CCM. From Fig. 9, it can also be seen that the PSFB3 operates mainly in CCM, but at lower power levels (around $0.3P_{\text{nom}}$) enters into DCM. On the other hand, the SAB3 converter operates in DCM even at nominal power levels.

5 Semiconductor Losses

In this section, the converter designs from Tab. 2 are compared based on semiconductor losses. On the inverter side, conduction and switching losses of IGBT ($S+$), as well as conduction losses of the antiparallel diodes ($S-$) are taken into account. On the rectifier side, only conduction losses of the diodes are considered. Owing, to the soft switching properties of these converters, the reverse recovery losses of all diodes are considered negligible. The snubber circuitry related to voltage balancing and protection is not considered in this analysis, even though it is clear that addition of these elements would bring additional losses in practical design. The considered devices are 5SNA 1000G650300

(6.5 kV/ 1000 A) for inverter switches and 5SDD 75Y8500 (8.5 kV/ 6720 A) for the rectifier diodes. It should be mentioned that the selected devices serve only as an example for the comparison, and that the thermal and cooling aspects were not considered either.

In Fig. 10, the loss distribution is shown for each considered case. It should be pointed out that, in all cases, the devices in the same leg (leading or lagging) are subject to the same losses. Therefore, only the losses per single device position (e.g. upper switch) within the leading and lagging leg are shown. Likewise, only the losses per single rectifier diode position are displayed. It should be also mentioned that, although conduction losses of the antiparallel diodes of the lagging leg devices are low, and therefore hardly visible for SAB1 and PSFB1 converters, they still exist. The losses are, in all cases, shown for operation with nominal input and output voltages and output power. (Σ) represents the sum of all semiconductor losses for each converter.

Tab. 3 shows the required number of parallel connected IGBTs (N_p) and series connected rectifier diodes (N_s), determined based on the maximum current stress of the IGBTs (I_{max}) and maximum voltage across the rectifier (U_{max}) respectively, thus indicating the sizing rules. It is worth mentioning that the maximal voltage stress on the primary side corresponds to the input voltage, while the maximum current stress of the rectifier diodes is comparable with the peak output current which is in all cases well below the selected device ratings.

Tab. 3: The required number of used devices

Converter	Imax [kA]	Umax [kV]	Np	Ns
SAB1	14.8	20	10	4
PSFB1	9.7	25.6	7	6
SAB2	27.3	20	18	4
PSFB2	27.3	24.2	18	5
SAB3	13.6	20	9	4
PSFB3	8.6	29.4	6	6

Converters SAB1 (semiconductor efficiency $\eta = 98.59\%$) and PSFB1 ($\eta = 99.07\%$) feature a similar loss distribution with the turn-off losses (particularly of the leading leg) being dominant. As the converters are designed for CCM operation, turn-on losses do not appear, owing to the Zero Voltage Switching (ZVS) in this mode. Clearly, PSFB1 is superior to SAB1 in terms of losses and device count.

In DCM operation, converters SAB2 ($\eta = 98.34\%$) and PSFB2 ($\eta = 98.32\%$) are almost exactly the same, as explained in the previous section. The advantage of the SAB converter here comes to the fore again, this time, in the form of lower number of rectifier diodes due to lower secondary voltage stress, and slightly lower losses overall. Considering that the converters now operate in DCM, the turn-off losses of the lagging leg are zero as the current is zero at the corresponding instants owing to Zero Current Switching (ZCS) in this mode. However, the turn-on losses are present as the anti-parallel diodes of the lagging leg no longer conduct prior to IGBT turn-on. The main contributor in the losses are the leading leg turn-off losses, penalised by the high RMS current in DCM.

Without operational mode constraints, the design rules resulted in the SAB3 ($\eta = 98.95\%$) operating in DCM and PSFB3 ($\eta = 98.87\%$) in CCM for nominal input/output voltages. It can be seen that SAB3 has slightly lower losses than PSFB3. However, this is mainly a consequence of the higher required number of parallel connected devices in this converter, lowering the primary side losses. Therefore, it is again evident that the PSFB converter shows more advantages than drawbacks overall, in this comparison as well.

Comparing all six designs, it can be said that CCM operation is preferable from the standpoint of semiconductor efficiency and can lead to lower device count on the inverter side. The main disadvantage

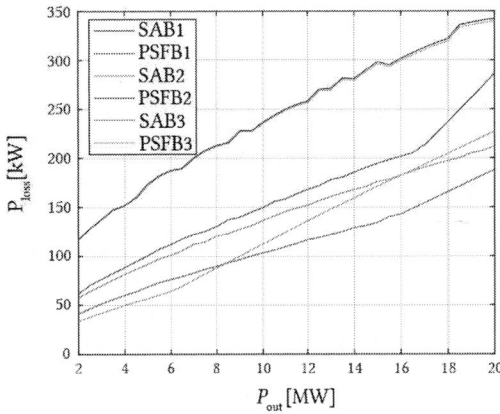

Fig. 11: Total semiconductor losses of the six converters under varying output power levels

of CCM operation of the PSFB, compared to the SAB is the higher secondary-side voltage, requiring more diodes for blocking. Therefore, it can be said that PSFB operated in CCM can lead to highest semiconductor efficiencies but at the expense of higher rectifier voltage stress.

In Fig. 11, the total losses of the six converters are shown for nominal voltages and varying output power levels, from 2MW up to 20MW. It can be concluded that, from the perspective of semiconductor losses, the PSFB converter is superior compared to the SAB converter. The only exception to this are designs for DCM operation over the entire operational range, where the SAB converter shows advantages.

6 Additional Considerations

In this section, additional considerations relevant for SAB and PSFB comparison at MW-levels are briefly addressed. The considerations are specific to the application and include feasibility of input voltage control and transient voltage suppression requirements.

6.1 Input Voltage Control

As mentioned, the DC/DC converters from Fig. 1 must regulate their input voltage under the disturbances caused by input current variation (dependant on the source power generation), and MVDC collection grid voltage fluctuations which are outside of its control.

To realize the input voltage control, only the low voltage at the input of the converter needs to be

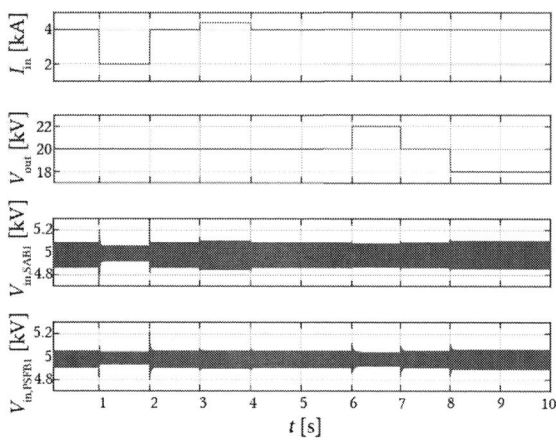

Fig. 12: Input voltage control under disturbances: SAB1 and PSFB1

measured and used as the feedback variable. The error between the wanted reference and measured value is fed to a simple PI regulator. The regulator directly controls the duty cycle of the primary side voltage. E.g. if generated power is increased, the input voltage has a tendency to increase, so in order to maintain the wanted reference, the duty cycle is increased and more current is drawn from the input capacitor. Conversely, under lower power production, the duty cycle is decreased, drawing less current from the input capacitor.

To demonstrate the feasibility of input voltage control, the above-explained controller is implemented for converters SAB1 and PSFB1. In Fig. 12, the results are shown. It can be seen that both converters can control the input voltage under varying profiles of input current and output voltage. Therefore, it is demonstrated that the control can be implemented in both cases. However, designing a PSFB converter to have a high output power variation with small duty cycle changes can lead to poor input voltage control in the described scenario as the resolution of the duty cycle variation is limited.

6.2 Additional Hardware Requirements

Unlike the SAB, the PSFB converter features an output filter inductance which can reduce the output current ripple. However, due to the existence of this inductance, in steady state operation in CCM, a voltage ringing occurs between the parasitic capacitors of the rectifier (and transformer) and the primary-side inductance. This phenomenon occurs after the duty cycle loss interval. While the maximum sec-

ondary voltage is defined by (7) in the ideal case, when the parasitic capacitance is considered it can be described by:

$$V_{\text{secMAX}} = \frac{2(L_{\text{f}} N V_{\text{in}} + L_{\text{r}} V_{\text{out}})}{(L_{\text{f}} + L_{\text{r}})}. \quad (18)$$

Assuming that $L_{\text{f}} >> L_{\text{r}}$, (7) can be approximated as $N V_{\text{in}}$, and (18) as $2 N V_{\text{in}}$. It becomes obvious that the ringing leads to an almost twofold increase of the secondary voltage, presenting itself as an inherent disadvantage of the PSFB topology. It should be emphasised that the rectifier voltage of the PSFB is inevitably higher that the rectifier voltage of the SAB (which is clamped to the grid voltage) even without considering the parasitics. The voltage ringing imposes additional requirements for the rectifier design in the case of the PSFB converter compared to the SAB converter. This issue can be addressed either by oversizing the rectifier, which may be impractical, considering the voltage levels, or by introducing additional clamping circuitry, imposing both losses and additional components.

7 Conclusion

In the paper, the SAB and PSFB converter were compared from the standpoint of current stress and semiconductor losses. It was shown that higher semiconductor efficiency can be achieved with the PSFB converter. By comparing several relevant cases, it was demonstrated that the advantages of the PSFB converter can be only utilized if the converter operates in CCM, while in DCM, the SAB is the superior topology. Several other considerations relevant for topology selection were also presented, such as the increased rectifier voltage stress of the PSFB converter and potential control issues. This makes the ultimate selection of the topology a trade-off between several presented factors.

References

[1] CIGRE WG C6.31, 'Medium Voltage Direct Current (MVDC) Grid Feasibility Study', https://e-cigre.org/publication/793-medium-voltage-direct-current-mvdc-grid-feasibility-study, Accessed: 2024-02-14.

[2] J. K. Steinke, P. Maibach, G. Ortiz, F. Canales, and P. Steimer, "MVDC Applications and Technology," in *PCIM Europe 2019; International Exhibition and Conference for Power Electronics, Intelligent Motion, Renewable Energy and Energy Management*, 2019, pp. 1–8.

[3] H. A. B. Siddique and R. W. De Doncker, "Evaluation of DC Collector-Grid Configurations for Large Photovoltaic Parks," *IEEE Transactions on Power Delivery*, vol. 33, no. 1, pp. 311–320, 2018. DOI: 10.1109/TPWRD.2017.2702018.

[4] P. K. Steimer and O. Apeldoorn, "Medium voltage power conversion technology for efficient wind-park power collection grids," in *The 2nd International Symposium on Power Electronics for Distributed Generation Systems*, 2010, pp. 12–18. DOI: 10.1109/PEDG.2010.5545750.

[5] P. Le Métayer, J. Paez, S. Touré, C. Buttay, D. Dujic, *et al.*, "Break-even distance for MVDC electricity networks according to power loss criteria," in *2021 23rd European Conference on Power Electronics and Applications (EPE'21 ECCE Europe)*, 2021, pp. 1–9. DOI: 10.23919/EPE21ECCEEurope50061.2021.9570416.

[6] GE Renewable Energy, 'World's Most Powerful Offshore Wind Platform: Haliade-X', https://www.ge.com/renewableenergy/wind-energy/offshore-wind/haliade-x-offshore-turbine, Accessed: 2023-09-29.

[7] C. Meyer and R. W. De Doncker, "Design of a Three-Phase Series Resonant Converter for Offshore DC Grids," in *2007 IEEE Industry Applications Annual Meeting*, 2007, pp. 216–223. DOI: 10.1109/07IAS.2007.40.

[8] L. Max and T. Thiringer, "Control method and snubber selection for a 5 MW wind turbine single active bridge DC/DC converter," in *2007 European Conference on Power Electronics and Applications*, 2007, pp. 1–10. DOI: 10.1109/EPE.2007.4417324.

[9] M. Mobarrez, M. Fazlali, M. A. Bahmani, and T. Thiringer, "Performance and loss evaluation of a hard and soft switched 2.4 MW, 4 kV to 6 kV isolated DC-DC converter for a wind energy application," in *IECON 2012 - 38th Annual Conference on IEEE Industrial Electronics Society*, 2012, pp. 5086–5091. DOI: 10.1109/IECON.2012.6389558.

[10] F. Deng and Z. Chen, "Control of Improved Full-Bridge Three-Level DC/DC Converter for Wind Turbines in a DC grid," *IEEE Transactions on Power Electronics*, vol. 28, no. 1, pp. 314–324, 2013. DOI: 10.1109/TPEL.2012.2198835.

[11] G. Ning, W. Chen, L. Shu, J. Zhao, W. Cao, *et al.*, "Hybrid Resonant ZVZCS PWM Full-Bridge Converter for Large Photovoltaic Parks Connecting to MVDC Grids," *IEEE Journal of Emerging and Selected Topics in Power Electronics*, vol. 5, no. 3, pp. 1078–1090, 2017. DOI: 10.1109/JESTPE.2017.2651020.

[12] J. Lam and P. K. Jain, "A new step-up DC/DC resonant converter with a capacitive output filter for medium voltage (MV) DC grid in wind energy power systems," in *IECON 2014 - 40th Annual Conference of the IEEE Industrial Electronics Society*, 2014, pp. 1084–1089. DOI: 10.1109/IECON.2014.7048637.

[13] C. Dincan, P. Kjaer, Y.-h. Chen, S. Munk-Nielsen, and C. L. Bak, "A High-Power, Medium-Voltage, Series-Resonant Converter for DC Wind Turbines," *IEEE Transactions on Power Electronics*, vol. 33, no. 9, pp. 7455–7465, 2018. DOI: 10.1109/TPEL.2017.2770220.

[14] X. Zhao, B. Li, B. Zhang, and D. Xu, "A High-Power Step-Up DC/DC Converter Dedicated to DC Offshore Wind Farms," *IEEE Transactions on Power Electronics*, vol. 37, no. 1, pp. 65–69, 2022. DOI: 10.1109/TPEL.2021.3102228.

[15] P. L. Métayer, Q. Loeuillet, F. Wallart, C. Buttay, D. Dujic, and P. Dworakowski, "Phase-Shifted Full Bridge DC–DC Converter for Photovoltaic MVDC Power Collection Networks," *IEEE Access*, vol. 11, pp. 19039–19048, 2023. DOI: 10.1109/ACCESS.2023.3247952.

[16] N. Soltau, H. Stagge, R. W. De Doncker, and O. Apeldoorn, "Development and demonstration of a medium-voltage high-power DC-DC converter for DC distribution systems," in *2014 IEEE 5th International Symposium on Power Electronics for Distributed Generation Systems (PEDG)*, 2014, pp. 1–8. DOI: 10.1109/PEDG.2014.6878696.

[17] M. S. Agamy, D. Dong, L. J. Garcés, Y. Zhang, M. E. Dame, *et al.*, "A High Power Medium Voltage Resonant Dual Active Bridge for MVDC Ship Power Networks," *IEEE Journal of Emerging and Selected Topics in Power Electronics*, vol. 5, no. 1, pp. 88–99, 2017. DOI: 10.1109/JESTPE.2016.2636365.

[18] A. Rodriguez, J. Sebastian, D. G. Lamar, M. M. Hernando, I. Ayarzaguena, *et al.*, "An Overall Analysis of the Static Characteristics of the Single Active Bridge Converter," *Electronics*, vol. 11, no. 4, 2022. DOI: 10.3390/electronics11040601.

[19] J. Sabate, V. Vlatkovic, R. Ridley, F. Lee, and B. Cho, "Design considerations for high-voltage high-power full-bridge zero-voltage-switched PWM converter," in *Fifth Annual Proceedings on Applied Power Electronics Conference and Exposition*, 1990, pp. 275–284. DOI: 10.1109/APEC.1990.66420.

PCIM Europe 2024, 11– 13 June 2024, Nuremberg   DOI: 10.30420/566262034

Startup Behavior of Harmonic Suppression in Electrical Machines Using Iterative Learning Control and Neural Networks

Annette Mai[1,2], Bernhard Wagner[1], Maximilian Hofmann[2]

[1] Technische Hochschule Nürnberg Georg Simon Ohm, Germany
[2] Fraunhofer Institute IISB, Germany

Corresponding author: Annette Mai, annette.mai@iisb.fraunhofer.de
Speaker: Annette Mai, annette.mai@iisb.fraunhofer.de

Abstract

Electrical machines generate unwanted flux and current harmonics. Harmonics can be suppressed by using various methods. In this paper, the harmonics are reduced by using iterative learning control (ILC) and neural networks (NNs). This paper focuses on the startup behavior of the control system. The ILC can compensate well for the harmonics in operation at constant speed and constant current reference values, but needs multiple rotations to learn. The NNs are trained with the data from the ILC and help to suppress the harmonics well even in transient operation and from the first rotation. The simulation model is based on flux and torque maps, depending on dq-currents and the electrical angle. The methods are also applied on the test bench and measurement results are presented.

1 Introduction

Current harmonics pose a significant challenge to the operation of permanent magnet synchronous machines (PMSMs) in both industrial and traction applications. The origin of these harmonics can be attributed to a variety of factors, including inverter behavior and inherent nonlinearities, as highlighted by [1]. Additionally, [2] emphasizes that flux harmonics within the motor itself contribute to the generation of current harmonics, which subsequently manifest as phase current harmonics.

Previous studies have explored the application of ILC in eliminating current harmonics, using methods such as a 2D-array for signal correction mapping at different speeds, as noted by [3], or a speed-independent mapping approach using a singular vector, as outlined by [4]. However, these approaches have limitations, particularly in scenarios involving rapid speed variations and the need to account for harmonics influenced by changes in torque and dq-currents.

Our recent work [5] introduces a novel methodology that combines the strengths of both iterative learning control (ILC) and neural networks (NNs) to address the issue of current harmonics in permanent magnet synchronous machines (PMSMs).

Building on our previous research, this paper expands its focus to analyze the startup behavior of the different control variations. To validate the effectiveness of our proposed approach, extensive simulations are performed and test bench measurement results are shown.

This paper provides a brief overview of the basic system model, various methods for reducing the harmonics are explained, and the simulation results are compared. The results of test bench measurements are shown and discussed.

2 Nonlinear drive system model

The simulation model of the motor is based on the linear flux equations 1 and 2. These equations are derived from the linear model according to [6].

$$\dot{\psi}_\mathsf{d}(t) = u_\mathsf{d}(t) - R_\mathsf{S}i_\mathsf{d}(t) + \omega_\mathsf{el} \cdot \psi_\mathsf{q}(t) \qquad (1)$$

$$\dot{\psi}_\mathsf{q}(t) = u_\mathsf{q}(t) - R_\mathsf{S}i_\mathsf{q}(t) - \omega_\mathsf{el} \cdot \psi_\mathsf{d}(t) \qquad (2)$$

In this system, $u_\mathsf{d,q}$ and $i_\mathsf{d,q}$ denote the dq-axes, the voltages and currents respectively. R_s represents the stator resistance and ω_el is the electric angular velocity, which can be calculated based on the mechanical speed and the number of pole pairs. $\psi_\mathsf{d,q}$ refers to the stator magnetic fluxes on the dq-axes, while ψ_PM represents the magnetic flux linkage of the permanent magnets.

284

To ensure accurate results, the model used in this research incorporates the effect of angle dependent harmonics. For the representation of flux and torque harmonics, both quantities are considered to be angle dependent, with the electrical rotor angle ϕ_{el} serving as the relevant parameter, and depending on the dq-currents. In addition, the non-linearity of the motor is influenced by cross-saturation, which affects the flux linkage.

The model used, shown in Figure 1, incorporates the angular dependence of the fluxes and the torque. In conjunction with the current-dependent fluxes, the torque is also represented non-linearly. The mappings of the functions for the dq-currents and the torque are implemented using look-up tables (LUTs). These LUTs are filled with data derived from a finite element method (FEM) simulation, see [5].

3 Control architecture

The primary control framework used in this analysis is based on field-oriented control (FOC). To mitigate current harmonics, correction signals are incorporated into the dq-current setpoints using current harmonic suppression controllers, such as the ILC and NNs, as shown in Figure 2. The ILC methodology used in this research is consistent with our previous publications [3], [5], [7].

The ILC can suppress current harmonics at steady state operating points. To achieve better results at transient operating points, NNs are used to replace or assist the ILC. The neural networks used in this study are multi-layer perceptron (MLP) networks with hyperbolic tangent sigmoid activation functions in the hidden layers. The NNs are trained with the values stored by the ILC at steady-state operating points.

The simplified block diagram of the ILC implemented as a current harmonic suppressor, with the current controller and the NNs, is shown in Figure 2.

The NNs were implemented in two different ways. The first variant maps the ILC directly and outputs the correction signals. The second variant outputs the phases and amplitudes of the Fourier analysis of the correction signal, from which the correction signal must then be calculated. The first variant requires fewer conversions, the second variant requires smaller networks.

A more detailed description of the two variants can be found in [5]. To compare the different variants,

torque and speed were varied and the root mean square error (RMSE) of the resulting dq-currents was calculated, see Figure 3.

Figure 3 shows that the NNs are more effective than the ILC for changing operating points, and the combination of ILC and NNs are even more effective.

The simulation results for the two variants of the neural networks are nearly identical. For this reason, only one of the two variants is considered further in this paper. Due to the smaller size of the networks, variant 2 was chosen.

Figure 4 shows the structure of variant 2 in more detail.

The correction signals are subjected to fast Fourier transformation (FFT) and the NNs are trained to map the frequency spectrum of these correction signals. The NNs use only the reference torque and speed as inputs, without requiring the angle. The outputs consist of amplitudes and phases for the corresponding harmonics of the correction signal. To obtain a continuous phase signal, the sine and cosine of the phases are used.

Due to the specific physical characteristics of PMSMs, only certain harmonics are significant. For efficiency reasons, the NNs map only relevant harmonics above a certain amplitude threshold. The NNs have h outputs, where h is the number of harmonics to map.

Separate neural networks are trained for the different output signal types, i.e., the amplitudes, the sine of the phases, and the cosine of the phases. The actual correction signals are then derived from these outputs.

Figure 4 illustrates how the NN inputs are determined from the reference torque and the mechanical rotor angle for variant 2. The correction signal is calculated from the NN outputs and can be combined with the ILC correction.

In this paper, the startup behavior of the different methods is considered in more detail in order to include it in the selection of the most suitable variant.

4 Simulation results

The control system is integrated into an FOC as shown in Figure 2. For simulation purposes, the inverter model is kept simple. The switching operations of the inverter are not simulated. The motor model is based on the structure shown in Figure 1. The dependencies between the fluxes, the dq-currents, the torque, and the rotor angle are

PCIM Europe 2024, 11– 13 June 2024, Nuremberg   DOI: 10.30420/566262034

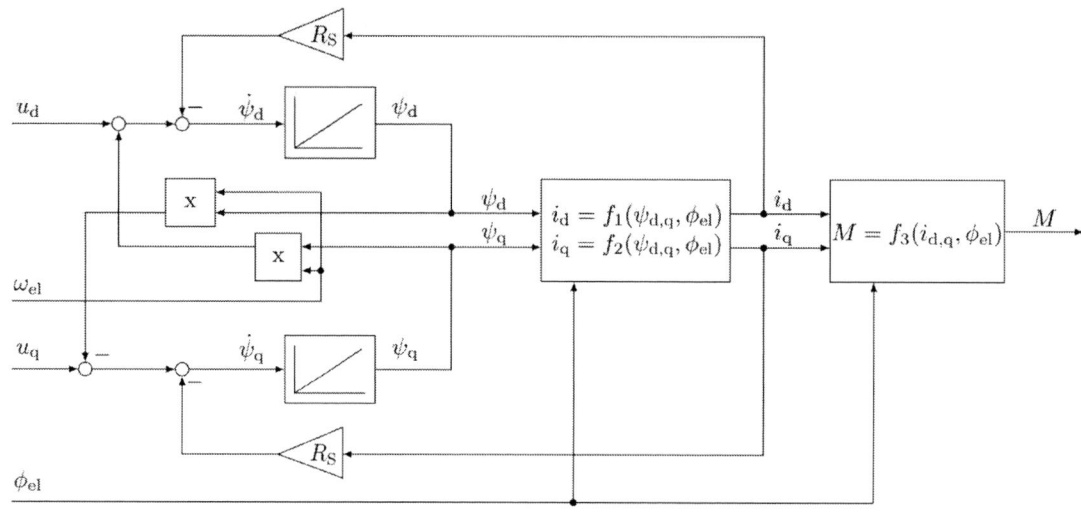

Fig. 1: Block diagram of the PMSM motor model used for simulations where the functions for the dq-currents and the torque are mapped from look-up tables of the benchmark motor, from [5]

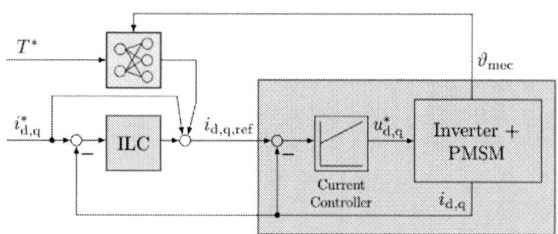

Fig. 2: Simplified block diagram of the ILC with the NNs in the FOC, from [5]

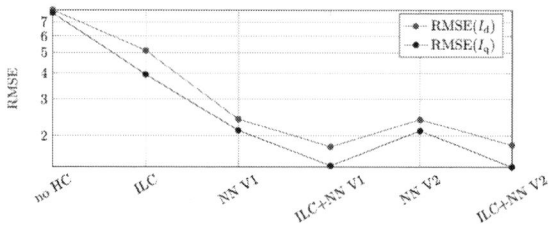

Fig. 3: RMSE for various controls: Variable speed n = 500 rpm...3000 rpm, variable torque T = -280 Nm...280 Nm, from [5]

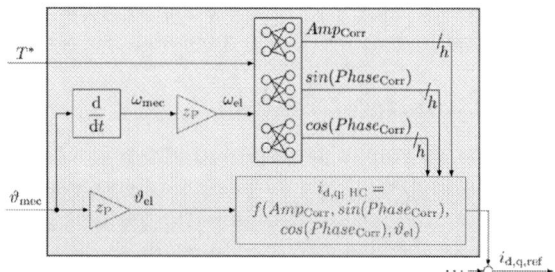

Fig. 4: Simplified block diagram of the NNs of Variant 2 in the FOC, from [5]

modeled with LUTs. These LUTs are filled with data from FEM simulations, see [5]. This allows this model to also represent the harmonics resulting from the fluxes. The linear motor parameters of the motor being modeled are listed in Table 2.

For a basic comparison in different operating modes, [5] shows the simulation results for the different variants. For the analysis of the startup behavior, constant speeds and constant torques are the most suitable. The total harmonic distortion (THD) from equation 3 is used as the comparison quantity.

$$\text{THD} = \frac{\sqrt{\sum_{k=5,7,11,13,17,19,...} I_k^2}}{I_1} \qquad (3)$$

I_1 is the fundamental component, and I_k^2 represents the kth harmonic of the phase current, calculated using fast Fourier transformation. Only odd harmonics that are not divisible by three are considered relevant for comparison due to the physical structure of the PMSM.

Figure 5 shows the THD of $I_{\text{ph,a}}$ plotted against the number of electrical revolutions for the different harmonic control (HC) variants on a logarithmic y-axis. The THD is calculated over one electrical rotation. Without harmonic control, the THD is approximately 1.67. With ILC, plotted in red, the THD initially decreases steadily and then approaches a final value of about 0.008. For NNs, plotted in dark blue, the THD is, as expected, constantly low from the beginning. At 0.15, the THD with the NNs is

286

PCIM Europe 2024, 11– 13 June 2024, Nuremberg   DOI: 10.30420/566262034

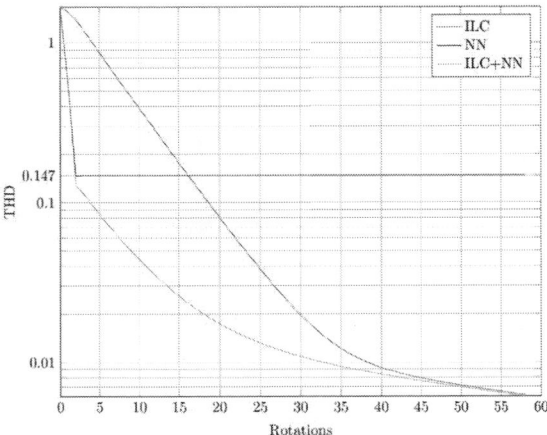

Fig. 5: THD from simulated $I_{ph,a}$ plotted over electrical rotations at constant speed and constant torque

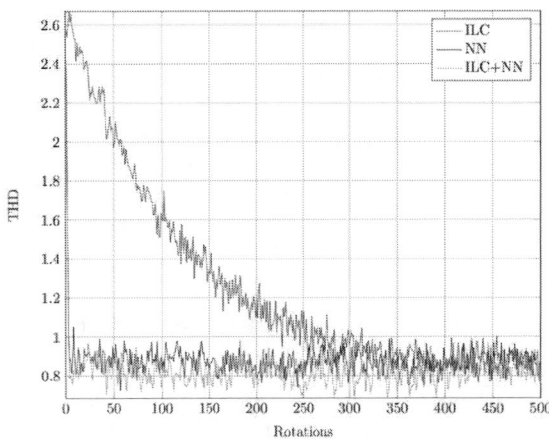

Fig. 6: THD from measured $I_{ph,a}$ calculated from three rotations plotted over electrical rotations at constant speed and constant torque

significantly higher than in the steady state of the ILC, but also significantly lower than without HC. In the combination of ILC and NN, shown in green, the THD also drops significantly from the beginning and then continues to drop with the same gradient as the ILC until it reaches the same value as the ILC without NNs.

5 Test bench results

The experiments on the test bench utilize a PMSM. Motor data can be found in Table 4. Data is collected for training the NNs by taking measurements along the MTPC curve for 17 operating points during motor operation at 13 different speed values. The ILC is utilized at each measurement point and the memory is recorded.

A Fourier analysis is performed on the collected data for the respective operating points using the ILC memory data, which serves as the correction signal. The phases and amplitudes of the correction signal are determined, and those of the relevant harmonics are stored. This data is then used to train the neural networks (NNs). The NNs that map the amplitudes have two hidden layers of 5 neurons each, while the NNs that map the sine and cosine of the phases have two hidden layers of 15 neurons each.

To save computation time, it is assumed that speed and torque remain constant within two sampling steps at a sampling frequency of 20 kHz. Therefore, it is sufficient to calculate the outputs of the NNs only in every other sampling step. The correction signal is calculated from the output phases and amplitudes in each sampling step.

To obtain a more precise measurement of THD, the FFT is performed averaging over three rotations. Figure 6 shows the THD calculated over these three rotations. The THD value without harmonic control is approximately 2.65. As anticipated, the THD values of the HC variants with NNs decrease rapidly, while the ILC alone requires multiple turns. All three variants can reduce the THD value to around 0.85 and thus more than halve it. On the test bench, the ILC cannot reduce the value much further. This is partly due to current measurement signals that are heavily affected by noise and the fast-switching power electronics.

6 Conclusion

This paper shows how the different current harmonic suppression variants differ in their startup behavior. All the variants shown are generally good at suppressing current harmonics, but they differ in their ability to do so at stationary or varying operating points in different scenarios. This paper demonstrates that the ILC requires a few initial turns to suppress harmonics. The neural networks and the iterative learning control combined with the NNs can directly suppress harmonics after activating the harmonic control.

The simulation results show that the harmonics can be reduced more quickly with NNs. It also shows that the NNs do not lead to an improvement in the final result compared to the ILC. The test bench results show that the improvement can also be achieved more quickly with the NNs. Measurements show that all three methods can reduce THD

Tab. 1: Simulation control parameters

Parameter	Value	Unit
ILC learn factor	0.3	[-]
simulated speed	666.67	rpm
reference d-current	-133.8	A
reference q-current	336.3	A
resulting torque	150	Nm
sample frequency	20	kHz

Tab. 3: Test bench control parameters

Parameter	Value	Unit
ILC learn factor	0.1	[-]
speed	1000	rpm
reference d-current	-8.6	A
reference q-current	28.8	A
resulting torque	34	Nm
sample frequency	20	kHz

Tab. 2: Linear PMSM motor parameters used for simulation

Parameter	Symbol	Value	Unit
number of pole pairs	z_P	8	[-]
stator resistance	R_S	0.02	Ω
induction of the d-axis	L_d	106.83	μH
induction of the q-axis	L_q	127.76	μH
permanent magnet flux	Ψ_{PM}	0.0468	Vs
DC link voltage	U_{DC}	330	V
max. torque	T_{max}	280	Nm
max. phase current	$I_{ph,max}$	360	A
base speed	n_{base}	3000	rpm
max. speed	n_{max}	7000	rpm

Tab. 4: MACHmotors motor parameters

Parameter	Symbol	Value	Unit
number of pole pairs	z_P	4	[-]
stator resistance	R_S	0.079	Ω
induction of the d-axis	L_d	1120	μH
induction of the q-axis	L_q	2750	μH
permanent magnet flux	Ψ_{PM}	0.1678	Vs
DC link voltage	V_{DC}	800	V
max. torque	T_{max}	115	Nm
max. phase current	$I_{ph,max}$	100	A
base speed	n_{base}	2500	rpm
max. speed	n_{max}	10000	rpm
sample frequency	f_s	20	kHz

by a third on the test bench. The ILC cannot further improve the results in this set up due to hardware influences. This result will be tested with other hardware in the future.

Funding

This research was funded by the German Federal Ministry for Economic Affairs and Climate Action (19I21030H - KIRA).

Appendix

Table 1 lists the control parameters used for the simulation.
Table 2 shows the linear motor data of the simulated motor.
Table 3 lists the control parameters used for the test bench control.
In Table 4, the linear parameters of the used test bench motor are displayed.

References

[1] J. Richter, "Modellbildung, Parameteridentifikation und Regelung hoch ausgenutzter Synchronmaschinen," Doctoral Dissertation, Karlsruhe, 2016. DOI: 10.5445/KSP/1000057097.

[2] S. I. Suriano-Sánchez, M. Ponce-Silva, V. H. Olivares-Peregrino, and S. E. De León-Aldaco, "A review of torque ripple reduction design methods for radial flux pm motors," *Eng*, vol. 3, no. 4, pp. 646–661, 2022. DOI: 10.3390/eng3040044.

[3] A. Mai, B. Wagner, and F. Streit, "Elimination of current harmonics in electrical machines with iterative learning control," in *2020 10th International Electric Drives Production Conference (EDPC)*, 2020, pp. 1–5. DOI: 10.1109/EDPC51184.2020.9388177.

[4] H. Zeng, R. D. Lorenz, C. H. van der Broeck, and R. W. De Doncker, "Spatial repetitive controller based harmonic mitigation methodology for wide varying base frequency range," in *2019 IEEE Energy Conversion Congress and Exposition (ECCE)*, 2019, pp. 1542–1549. DOI: 10.1109/ECCE.2019.8912281.

[5] A. Mai, X. Liu, B. Wagner, and M. Hofmann, "Current harmonics minimization of permanent magnet synchronous machine based on iterative learning control and neural networks," *Machines*, vol. 11, no. 8, 2023. DOI: 10.3390/machines11080784.

[6] N. P. Quang and J. Dittrich, *Vector Control of Three-Phase AC Machines*. Heidelberg: Springer Berlin, 2015.

[7] A. Mai, B. Wagner, and S. Arenz, "Comparison of control algorithms for the suppression of current harmonics in pmsms," in *PCIM Europe digital days 2021; International Exhibition and Conference for Power Electronics, Intelligent Motion, Renewable Energy and Energy Management*, 2021, pp. 1–8.

PCIM Europe 2024, 11– 13 June 2024, Nuremberg   DOI: 10.30420/566262035

Analytical Approach of the Vector Current Control Flux-Weakening Strategy for Permanent Magnet Synchronous Machines

Oriol Subirats Rillo[1], Carlos Miguel Espinar[1], Samuel Galceran Arellano[1], Daniel Montesinos i Miracle[1]

[1] Centre d'Innovació Tecnològica en Convertidors Estàtics i Accionaments (CITCEA-UPC)

Corresponding author: Oriol Subirats-Rillo, oriol.subirats.rillo@upc.edu
Speaker: Oriol Subirats-Rillo, oriol.subirats.rillo@upc.edu

Abstract

This paper proposes a Flux-Weakening (FW) controller based with an integrator controller for a Vector Current Control (VCC) strategy of Permanent Magnet Synchronous Machines (PMSMs) to operate above the rated speed. The controller will be defined throughout the small-signal model of the voltage control loop. Simulation results verify the worthiness of the method in a direct-drive e-motorbike.

1   Introduction

In cities, the use of mild-hybrid, full-electric, and micro-mobility vehicles is rapidly increasing, hence the importance of reducing Greenhouse Gas (GHG) emissions [1]. Transportation is one of the industrial sectors that contribute the most to GHG emissions, producing approximately 28% of the total amount [2], being road transportation the source of practically 75% of transport GHG emissions [3].

Their electric powertrain is based mainly on a battery, a traction inverter, and an electric machine (EM). High torque is demanded to accelerate quickly, hill climbing, engine auto-start, and reversing at high road gradients at lower speeds. Besides, a high torque should be delivered for city driving at a medium speed range. However, the EM should operate at a lower torque rate in the high-speed range. Permanent Magnet Synchronous Machines (PMSMs) are the most selected due to: i) high power density, ii) high overall efficiency in the range of nominal speed, and iii) easy heat dissipation [4]. Below the rated speed, the Maximum Torque per Ampere (MTPA) strategy maximises the efficiency of the EM because the Joule effect losses are minimised. Nevertheless, the back electromotive force increases when the mechanical speed increases and the rotor magnet flux must be reduced at a certain point. Then, the EM works at the denominated Flux Weakening (FW) zone.

To the authors' best knowledge, the literature shows several FW algorithms working at Constant Torque (CT), Current and Voltage Limits (CVL) and Maximum Torque per Voltage (MTPV) zones. These techniques have been widely classified and analysed in [5], citing the most important characteristics of each one: Analytical Direct Calculation [6], Direct Open Loop Algorithm with Experimental Look Up Tables (LUT) [7], [8], Single Current Regulator (SCR) [9], Torque and Flux Control Method with LUT [10], Unified Direct Flux Vector Control (UD-FVC) in the stator flux frame [11] and Vector Current Control (VCC) [12]–[14].

The most common FW transition method is the VCC through the voltage norm in closed loop [15], [16]. This methodology's main advantage is considering the resistor voltage drop and the saturation effects in the transition to/from the FW operation. In [13], Bolognani et. al propose an online gain adaptation based on the real-time compensation of the static gain showing a lower voltage margin that can be adopted with the proposed approach concerning standard solutions based on constant gains for the voltage controller. Nonetheless, they did not find the analytical equation to define the voltage control loop regulator from the time response.

This article proposes a static and dynamic analytical investigation of the voltage control loop for the FW operation of a PMSM. The objective is to define an analytical equation of the integral gain of the voltage regulator according to specific transient requirements. This equation seeks two different ob-

PCIM Europe 2024, 11– 13 June 2024, Nuremberg   DOI: 10.30420/566262035

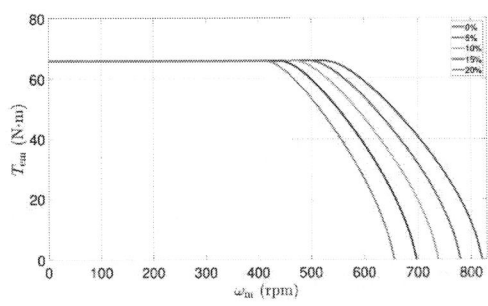

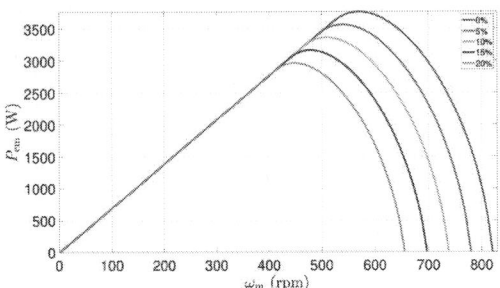

Fig. 1: Analysis of the voltage margin in the (a) torque-speed and (b) power-speed curves.

jectives: i) to reduce the voltage margin regarding the maximum voltage of the system at which the traction inverter starts to weaken the flux, and ii) to facilitate the integration of a new EM in a particular powertrain, knowing its characteristic parameters. Figure 1 shows the significance of the voltage margin in the FW performance of a PMSM. Different voltage margins from 0% to 20% regarding the maximum voltage are studied, and the EM's aftermath steady-state torque and speed curves are analysed. The higher the voltage margin, the lower the maximum speed and power, and the FW zone at a specific torque begins at a lower velocity. It is crucial to define the minimum voltage margin to exploit the full capabilities of a PMSM.

The content of the paper is organised into six sections. In Section 2, a time domain PMSM model is presented. In Section 3, the linearised PMSM model is explained. In Section 4, the static and dynamic analysis of the FW closed loop is analysed. Section 5 presents the simulation results for a real exterior-rotor Interior Permanent Magnet Synchronous Motor (IPMSM) direct-drive e-motorbike. Finally, Section 6 details the conclusion reached after this research was conducted.

2 Time Domain Model

Following [17], the PMSM electrical dynamic equations referred to the synchronous rotor frame with the d-axis aligned to the rotor flux, and by implementing the Park transformation, which maintains the current and voltage modulus invariant, are

$$v_d(t) = R_s i_d(t) + L_d \frac{di_d(t)}{dt} - \omega_e(t) L_q i_q(t) \quad \text{, and} \tag{1}$$

$$v_q(t) = R_s i_q(t) + L_q \frac{di_q(t)}{dt} + \omega_e(t) L_d i_d(t) + \omega_e(t)\lambda_m \quad , \tag{2}$$

where $v_{dq}(t)$ and $i_{dq}(t)$ are the time-dependent voltage and current dq components, L_{dq} are the dq-axes motor inductance, λ_m is the flux linkage due to the spinning of magnets, R_s is the winding phase resistance and $\omega_e(t)$ is the time-dependent electrical speed.

The relationship between mechanical and electrical speeds in a PMSM is expressed as

$$\omega_m(t) = \frac{\omega_e(t)}{pp} \quad , \tag{3}$$

where $\omega_m(t)$ is the time-dependent mechanical speed, and pp is the pair of poles.

The electromagnetic torque ($T_{em}(t)$) can be specified in terms of the dq-axes current distribution as

$$T_{em}(t) = \frac{3}{2} pp \left(\lambda_m i_q(t) + (L_d - L_q)\, i_d(t) i_q(t) \right) \quad . \tag{4}$$

The mechanical equation of an EM establishes the relationship between the mechanical variables (load torque and mechanical speed) with the electrical variables through the electromagnetic torque. This equation completes the system model with the relationship between the EM and its load, and it can be expressed as

$$T_{em}(t) = J \frac{d\omega_m(t)}{dt} + b\omega_m(t) + T_{load}(t) \quad , \tag{5}$$

where J is the inertia, b is the viscous friction of the EM and $T_{load}(t)$ is the torque applied by the load.

3 Linearised Model

Equations (1), (2), (4), and (5), define a complete model that describes the electrical and mechanical dynamic operation of a PMSM. These equations form a nonlinear system that cannot be directly analysed with the linear system theory, and in this section, they will be linearised. This linearised PMSM

291

model will be the tool for studying and finding the parameters of the FW regulator defined in the proposed control algorithm.

The linearisation of the nonlinear PMSM model is set forth by applying the Taylor series around an operating point. Hence, the resulting set of linear differential equations describes the dynamic behaviour during small displacement or small excursions around an operating point that is valid for operation with stator voltages of any frequency.

The nonlinear equations of a PMSM in its state variables have the following form

$$\frac{dx}{dt} = f(x, u, t) \quad , \tag{6}$$

where x is the vector of the state variables and f is the nonlinear function of the state variables x and inputs u.

The linearisation is based on expanding the Taylor series around a particular point and considering only the first-order terms [18]. The considered point is the operating point at the steady-state condition. Then, each state variable can be defined as

$$x_i = \overline{X}_i + \Delta x_i \quad , \tag{7}$$

where x_i is the ith state variable, $\overline{X}_i$ is its steady-state value, and Δx_i is a small perturbation around the steady-state.

Applying (6) and (7), then (1), (2), and (4) with (5) can be written in their matrix form as [18]

$$
\begin{bmatrix} \frac{d(\Delta i_d(t))}{dt} \\ \frac{d(\Delta i_q(t))}{dt} \\ \frac{d(\Delta \omega_e(t))}{dt} \end{bmatrix} =
$$

$$
\begin{bmatrix} -\frac{R_s}{L_d} & \frac{L_q}{L_d}\overline{\omega}_e & \frac{L_q}{L_d}\overline{I}_q \\ -\frac{L_d}{L_q}\overline{\omega}_e & -\frac{R_s}{L_q} & -\frac{L_d}{L_q}\overline{I}_d - \frac{\lambda_m}{L_q} \\ \frac{3pp^2}{2J}\left(L_d - L_q\right)\overline{I}_q & \frac{3pp^2}{2J}\left(\lambda_m + (L_d - L_q)\overline{I}_d\right) & -\frac{b}{J} \end{bmatrix}
\begin{bmatrix} \Delta i_d(t) \\ \Delta i_q(t) \\ \Delta \omega_e(t) \end{bmatrix}
$$

$$
+ \begin{bmatrix} \frac{1}{L_d} & 0 & 0 \\ 0 & \frac{1}{L_q} & 0 \\ 0 & 0 & -\frac{pp}{J} \end{bmatrix}
\begin{bmatrix} \Delta v_d(t) \\ \Delta v_q(t) \\ \Delta T_{load}(t) \end{bmatrix} , \tag{8}
$$

where $\overline{I}_{dq}$ are the dq-axes currents and $\overline{\omega}_e$ is the electrical speed at the operating point.

4 Voltage Control Loop Analysis

The linearised model of a PMSM obtained in Section 3 allows the dynamic analysis of the variation of the currents that circulate through it. The dq-axes voltage variation in the frequency domain is obtained by applying the Laplace transform to (8) that results in

$$\Delta V_d(s) = R_s \Delta I_d(s) + s L_d \Delta I_d(s)$$
$$- L_q \overline{\omega}_e \Delta I_q(s) - L_q \overline{I}_q \Delta \omega_e(s) \quad , \text{ and} \tag{9}$$

$$\Delta V_q(s) = R_s \Delta I_q(s) + s L_q \Delta I_q(s)$$
$$+ L_d \overline{\omega}_e \Delta I_d(s) + \left(\lambda_m + L_d \overline{I}_d\right) \Delta \omega_e(s) \quad , \tag{10}$$

where $\Delta V_{dq}(s)$ and $\Delta I_{dq}(s)$ are the Laplace transform of the small variation of the stator phase voltage and current vectors expressed in the rotating reference frame, $\Delta \omega_e(s)$ is the Laplace transform of the small variation of the electrical speed, and s is the complex Laplace variable.

In this section, a static and dynamic analytical investigation of the voltage control loop for the FW operation of a PMSM is carried out. The proposed scheme is based on the feedback of the voltage magnitude generated by the traction inverter, which intrinsically provides a smooth transition between the MTPA and FW regions. This method is the most studied in the literature, and its main advantage is the intrinsic consideration of the resistor voltage drop and the saturation effects during the transition.

This section pursues the objective of defining an analytical equation of the parameters of the voltage regulator according to specific transient requirements. This mathematical approach seeks two different goals: i) to reduce the voltage margin regarding the maximum voltage of the system at which the traction inverter starts to weaken the flux, and ii) to facilitate the integration of a new EM in a particular powertrain from its characteristic parameters.

The equivalent dynamical model is highly nonlinear due to the presence of speed-dependent terms (i.e., back-electromotive force) and other nonlinear terms (trigonometric and modulus functions), so small signal linearisation around the steady-state condition will be studied hereafter.

4.1 Static Analysis

The modulus of the voltage of a PMSM at steady-state in terms of the current modulus and angle (I_s and γ) can be obtained from (1) and (2) as

$$|V| = \sqrt{V_d^2 + V_q^2}$$
$$= \Big[\left(R_s I_s \cos(\gamma) - \omega_e L_q I_s \sin(\gamma) \right)^2$$
$$+ \left(R_s I_s \sin(\gamma) + \omega_e (L_d I_s \cos(\gamma) + \lambda_m) \right)^2 \Big]^{1/2} . \tag{11}$$

The current angle is unique for a limited voltage and a specific current module at constant mechanical speed and electromagnetic torque. At these conditions, the electrical speed of the EM considering the resistor voltage drop can be calculated from (11) as

$$\omega_e = \frac{-R_s I_s \sin(\gamma) \cdot (I_s \cos(\gamma)(L_d - L_q) + \lambda_m)}{L_q^2 I_s^2 \sin^2(\gamma) + (L_d I_s \cos(\gamma) + \lambda_m)^2}$$
$$+ \left[\frac{R_s^2 I_s^2 \sin^2(\gamma) \cdot (I_s \cos(\gamma)(L_d - L_q) + \lambda_m)^2}{\left(L_q^2 I_s^2 \sin^2(\gamma) + (L_d I_s \cos(\gamma) + \lambda_m)^2 \right)^2} \right.$$
$$\left. - \frac{\left(L_q^2 I_s^2 \sin^2(\gamma) + (L_d I_s \cos(\gamma) + \lambda_m)^2 \right) \left(R_s^2 I_s^2 - |V|^2 \right)}{\left(L_q^2 I_s^2 \sin^2(\gamma) + (L_d I_s \cos(\gamma) + \lambda_m)^2 \right)^2} \right]^{1/2} .$$

(12)

The small-signal model of the voltage magnitude will explain its deviation regarding a slight variation of the current angle. This small variation of the current angle near the steady-state working point will affect the current and voltage vectors. So, the voltage magnitude can be represented from a constant component plus a small variation as [13]

$$|V| = \sqrt{(\overline{V}_d + \widehat{V}_d)^2 + (\overline{V}_q + \widehat{V}_q)^2} \cong |\overline{V}| + |\widehat{V}| ,$$

(13)

where $\overline{V}_{dq}$ are the steady-state dq-axes voltages, and $\widehat{V}_{dq}$ are the small variation of dq-axes voltages. $|\overline{V}|$ is the limit of the voltage imposed by the traction inverter equal to $V_{s,max}$. $|\widehat{V}|$ is the small signal term that can be approximated by its First-Order Taylor (FOT) series approximation as

$$|V| \approx |\overline{V}| + \frac{\partial(|V|)}{\partial\gamma}\bigg|_{\substack{\gamma=\overline{\gamma} \\ I_s=\overline{I}_s}} \cdot (\gamma - \overline{\gamma})$$
$$= |\overline{V}| + \frac{\partial(|V|)}{\partial\gamma}\bigg|_{\substack{\gamma=\overline{\gamma} \\ I_s=\overline{I}_s}} \cdot \widehat{\gamma} ,$$

(14)

where $\widehat{\gamma}$ is the small variation of the current angle. The static gain of voltage magnitude regarding the current angle is estimated from the square of the voltage magnitude [13] as

$$\frac{\partial(|V|)}{\partial\gamma}\bigg|_{\substack{\gamma=\overline{\gamma} \\ I_s=\overline{I}_s}} = \frac{1}{2|V|} \cdot \frac{\partial\left(|V|^2\right)}{\partial\gamma}\bigg|_{\substack{\gamma=\overline{\gamma} \\ I_s=\overline{I}_s}}$$
$$= \frac{1}{2V_{s,max}} \cdot \frac{\partial\left(|V|^2\right)}{\partial\gamma}\bigg|_{\substack{\gamma=\overline{\gamma} \\ I_s=\overline{I}_s}} .$$

(15)

Substituting (11) in (15) results in

$$\frac{\partial(|V|)}{\partial\gamma}\bigg|_{\substack{\gamma=\overline{\gamma} \\ I_s=\overline{I}_s}} = \frac{1}{2V_{s,max}} \cdot \left(2\left(R_s \overline{I}_s \cos(\overline{\gamma}) - \overline{\omega}_e L_q \overline{I}_s \sin(\overline{\gamma})\right) \right.$$
$$\cdot \left(-R_s \overline{I}_s \sin(\overline{\gamma}) - \overline{\omega}_e L_q \overline{I}_s \cos(\overline{\gamma})\right)$$
$$+ 2\left(R_s \overline{I}_s \sin(\overline{\gamma}) + \overline{\omega}_e L_d \overline{I}_s \cos(\overline{\gamma}) + \overline{\omega}_e \lambda_m\right)$$
$$\left. \cdot \left(R_s \overline{I}_s \cos(\overline{\gamma}) - \overline{\omega}_e L_d \overline{I}_s \sin(\overline{\gamma})\right) \right) .$$

(16)

The static gains of the voltage magnitude regarding the dq-axes current can be obtained from the related derivatives as

$$\frac{\partial(|V|)}{\partial I_d}\bigg|_{\substack{\gamma=\overline{\gamma} \\ I_s=\overline{I}_s}} = \frac{\partial(|V|)}{\partial\gamma}\bigg|_{\substack{\gamma=\overline{\gamma} \\ I_s=\overline{I}_s}} \cdot \frac{\partial\gamma}{\partial I_d}\bigg|_{\substack{\gamma=\overline{\gamma} \\ I_s=\overline{I}_s}}$$
$$= -\frac{\partial(|V|)}{\partial\gamma}\bigg|_{\substack{\gamma=\overline{\gamma} \\ I_s=\overline{I}_s}} \cdot \frac{1}{\overline{I}_s \sin(\overline{\gamma})} , \text{ and}$$

(17)

$$\frac{\partial(|V|)}{\partial I_q}\bigg|_{\substack{\gamma=\overline{\gamma} \\ I_s=\overline{I}_s}} = \frac{\partial(|V|)}{\partial\gamma}\bigg|_{\substack{\gamma=\overline{\gamma} \\ I_s=\overline{I}_s}} \cdot \frac{\partial\gamma}{\partial I_q}\bigg|_{\substack{\gamma=\overline{\gamma} \\ I_s=\overline{I}_s}}$$
$$= \frac{\partial(|V|)}{\partial\gamma}\bigg|_{\substack{\gamma=\overline{\gamma} \\ I_s=\overline{I}_s}} \cdot \frac{1}{\overline{I}_s \cos(\overline{\gamma})} .$$

(18)

Figure 2 shows the normalised static gain of the stator voltage amplitude as a consequence of the variation of the current angle (blue), the d-axis current (black), and the q-axis current (red) for the IPMSM studied in this paper. In this analysis, the voltage limit is equal to the battery voltage without considering any margin, and the current limit equals the rated current of the EM specified in Table 1. The variation of the static gain in the FW region differs significantly among the three cases. Regarding the FW operation, the d-axis current scheme shows the lower static gain, and the current angle scheme offers the highest static gain. However, heavy gain variation occurs in the deep FW region using a method in which d-axis current is the voltage regulator output. This region's current circular trajectory (constant amplitude) has a high slope for angle values near π. The results of this analysis are valid for the actual motor considered in this paper.

4.2 Dynamic Analysis

In Section 4.1, the highly non-linear relationship between the voltage magnitude and the current angle was demonstrated. In this section, the authors will provide a dynamic model of the voltage magnitude regarding the current angle in the open-loop and closed-loop structures. The study of the dynamic behaviour in the Laplace domain will be done through the plant linearisation in the neighbourhood of a particular working point defined by a current magnitude setpoint ($\overline{I}_s^*$), a current angle ($\overline{\gamma}$), and an electrical speed ($\overline{\omega}_e$).

Figure 3 shows the block diagram of the voltage magnitude regarding the variation of the current angle in a closed loop. γ_{MTPA} is the current angle depending on the reference torque to work in

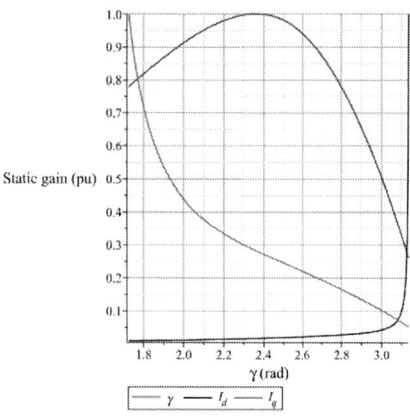

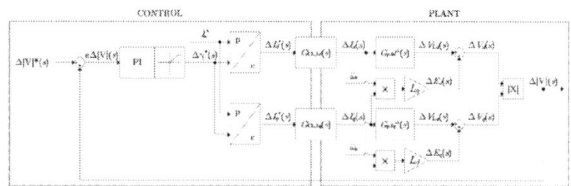

Fig. 2: Normalised static gains of the stator voltage amplitude regarding the current angle, d-axis component, and q-axis component at rated current.

Fig. 3: Block diagram of the voltage magnitude regarding the current angle in closed loop.

the MTPA region. $\overline{I}_s^*$ is the setpoint current magnitude that is estimated depending on the speed range. The current magnitude equals its nominal value when the EM works in the CVL region. Nevertheless, when the EM works in the CT or MTPV regions, the current magnitude varies as the current angle increases, perturbing the voltage control loop. Hence, during the FW operation, the inverter must change the current angle and/or modulus to work inside its voltage limits.

The open-loop transfer function between a variation of the voltage magnitude regarding a slight variation of the current angle in the working point can be defined as

$$
\begin{aligned}
G_{\mathrm{p},\Delta\gamma\to\Delta|V|}(s) &= \frac{\Delta|V|(s)}{\Delta\gamma^*(s)} \\
&\approx \frac{\overline{V}_d}{V_{\mathrm{s,max}}} \cdot \left(-G_{\mathrm{p,Id}}^{-1}(s)G_{\mathrm{CL,Id}}(s)\overline{I}_s^*\sin(\overline{\gamma}) - \overline{\omega}_e L_q G_{\mathrm{CL,Iq}}(s)\overline{I}_s^*\cos(\overline{\gamma}) \right) \\
&+ \frac{\overline{V}_q}{V_{\mathrm{s,max}}} \cdot \left(G_{\mathrm{p,Iq}}^{-1}(s)G_{\mathrm{CL,Iq}}(s)\overline{I}_s^*\cos(\overline{\gamma}) - \overline{\omega}_e L_d G_{\mathrm{CL,Id}}(s)\overline{I}_s^*\sin(\overline{\gamma}) \right),
\end{aligned}
\tag{19}
$$

where $G_{\mathrm{p,Id}q}^{-1}(s)$ are the inverse dq-axes currents transfer functions, and $G_{\mathrm{CL,I}dq}(s)$ are the current closed-loop transfer functions.

For the sake of defining the voltage magnitude reg-

ulator, this paper supposes a current closed-loop dynamics higher than the dynamic of the voltage magnitude closed-loop. This condition means that the transient response of the current closed loop is sufficiently high not to pay attention to its dynamics in the voltage magnitude closed loop, neither in time response nor frequency band interactions. So, equation (19) can be simplified as

$$
\begin{aligned}
G_{\mathrm{p},\Delta\gamma\to\Delta|V|}(s) &\approx \frac{\overline{V}_d}{V_{\mathrm{s,max}}} \cdot \left(-G_{\mathrm{p,Id}}^{-1}(s)\overline{I}_s^*\sin(\overline{\gamma}) - \overline{\omega}_e L_q \overline{I}_s^*\cos(\overline{\gamma}) \right) \\
&+ \frac{\overline{V}_q}{V_{\mathrm{s,max}}} \cdot \left(G_{\mathrm{p,Iq}}^{-1}(s)\overline{I}_s^*\cos(\overline{\gamma}) - \overline{\omega}_e L_d \overline{I}_s^*\sin(\overline{\gamma}) \right).
\end{aligned}
\tag{20}
$$

The controller of the voltage magnitude is defined as a pure integral as

$$
G_{\mathrm{c,V}}(s) = \frac{K_{\mathrm{i,V}}}{s},
\tag{21}
$$

where $K_{\mathrm{i,V}}$ is the integral parameter of the voltage PI controller. This condition reduces the order of the closed-loop system, simplifying the definition of the voltage controller.

The closed-loop transfer function between the voltage magnitude setpoint and the actual voltage magnitude with a positive feedback signal is defined as

$$
G_{\mathrm{CL},\Delta\gamma\to\Delta|V|}(s) = \frac{\Delta|V|(s)}{\Delta|V|^*(s)} = \frac{-G_{\mathrm{c,V}}(s)G_{\mathrm{p},\Delta\gamma\to\Delta|V|}(s)}{1-G_{\mathrm{c,V}}(s)G_{\mathrm{p},\Delta\gamma\to\Delta|V|}(s)}.
\tag{22}
$$

If the zero effect in (22) is considered negligible, the closed-loop dynamics can be adjusted as a first-order system through its time constant. The integral gain can be calculated as (23) if the settling time of the voltage control loop ($t_{\mathrm{s,V}}$) is imposed to be five times its time constant.

5 Simulation Results

Simulations from a MATLAB-Simulink model are carried out to analyse the performance of the voltage control loop in different scenarios and time responses. A direct-drive IPMSM, which runs as the rear wheel of an e-motorbike, is used. The most relevant parameters of the EM are listed in Table 1. Table 2 indicates the integral grain of the voltage control loop for different cases. These values have been obtained from (23) and considering the characteristic parameters of the PMSM used in this paper.

Figure 4 shows the dynamic behaviour of the PMSM state variables in the front of a square-wave voltage reference: a) voltage magnitude, b) current angle, and c) dq-axes currents. In blue, red, yellow,

$$K_{\text{i},V} = \cfrac{-5V_{\text{s,max}}}{\left[\begin{array}{l} 2\overline{I}_{\text{s}}^{*}\left(R_{\text{s}}\overline{\omega}_{\text{e}}t_{\text{s},V}\overline{I}_{\text{s}}^{*}\left(L_d - L_q \right)\cos^2(\overline{\gamma}^{*}) \right) \\[2mm] +\overline{I}_{\text{s}}^{*}\cos(\overline{\gamma}^{*})\left(5\left(L_d - L_q \right)\left(-\frac{t_{\text{s},V}(L_d + L_q)\overline{\omega}_{\text{e}}^2}{5} + R_{\text{s}} \right)\overline{I}_{\text{s}}^{*}\sin(\overline{\gamma}^{*}) + \lambda_{\text{m}}\overline{\omega}_{\text{e}}\left(R_{\text{s}}t_{\text{s},V} - 5L_q \right) \right) \\[2mm] -\overline{I}_{\text{s}}^{*}\overline{\omega}_{\text{e}}\left(\sin(\overline{\gamma}^{*})L_d\overline{\omega}_{\text{e}}t_{\text{s},V}\lambda_{\text{m}} + \left(R_{\text{s}}\left(L_d - L_q \right)t_{\text{s},V} + 5L_dL_q \right)\overline{I}_{\text{s}}^{*} \right) \end{array} \right]} \tag{23}$$

Tab. 1: Characteristic parameters of the IPMSM.

Symbol	Value	Unit
R_{s}	17	$m\Omega$
L_d	70	μH
L_q	79	μH
$V_{\text{ph-n,max}}$	27,7	V
$I_{\text{ph,max}}$	330	A
$\lambda_{\text{m,ph-n,max}}$	0,023	(V·s)/rad
pp	20	
$T_{\text{m,n}}$	66	N·m
$\omega_{\text{m,n}}$	510	min^{-1}
$\omega_{\text{m,max}}$	1000	min^{-1}

Tab. 2: Integral grain of the voltage control loop for different cases.

$t_{\text{s},V}$ (ms)	$K_{\text{i},V}$
100	5,1643
200	2,5645
300	1,7058
400	1,2779

and purple are drawn the dynamic response for different settling times specified in Table 2. In Figure 4a, the reference voltage is indicated in black, whereas the maximum synthesisable voltage is depicted in the discontinuous red line.

Initially, the reference voltage is fixed to 0,8 pu. At the time $t = 2$ s, a square-wave voltage reference of 1 s of period and 50% of duty cycle is applied between 0,75 pu and 0,85 pu. At the time $t = 4$ s, the minimum and maximum values of the square-wave voltage reference are changed to 0,70 pu and 0,90 pu. The reference voltage has been chosen to be lower than the maximum allowable voltage of the inverter to study the controller's behaviour in non-saturated conditions.

The main conclusion is that the dynamic behaviour of the voltage closed-loop follows the settling time for which the loop is designed. So, the approxima-

tion done in Section 4 is validated. This procedure streamlines the integration of a new EM into a powertrain, making it feasible, even for the FW zone. Figures 4c and 4d show that the d-axis current variation is significant, but the q-axis current maintains constant. This operation indicates that the working point lies in the constant torque region.

Figure 5 shows the dynamic behaviour of the PMSM state variables in front of a square-wave load torque: a) voltage magnitude, b) current angle, and c) dq-axes currents. In blue, red, yellow, and purple are drawn the dynamic response for different settling times specified in Table 2. The reference voltage is set to 0,8 pu, and the load torque is established at 25 N·m. At the time $t = 2$ s, a square-wave load torque with 2 s of period and 50% of duty cycle is applied between 37,5 N·m and 12,5 N·m.

The peak voltage reached in each case is lower than the maximum allowable voltage of the inverter. The higher the settling time of the closed-loop voltage controller, the higher the peak voltage during the load torque step because the regulator reacts slower. In the same way as the previous test, the working point lies in the constant torque region.

6 Conclusion

The voltage controller of the VCC control strategy under FW operation is examined by linearising the plant at a working point. In this study, the authors propose an integral gain to simplify the tuning of the voltage controller based on the characteristic parameters of the PMSM and the required temporal response. Simulation results prove the effectiveness of this methodology on a 2,5 kW IPMSM in a direct-drive e-motorbike powered by a 48-V system by simplifying the tuning of the FW controller.

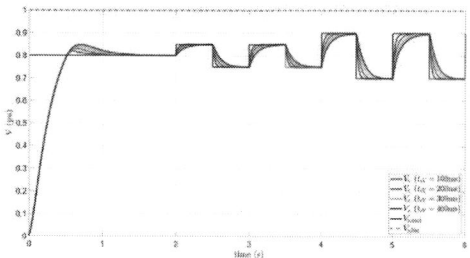

a) Voltage magnitude.

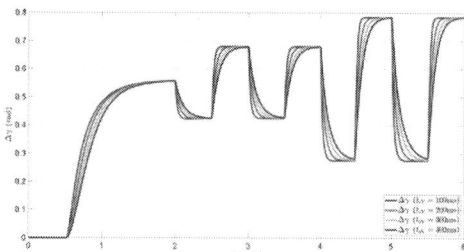

b) Variation of current angle.

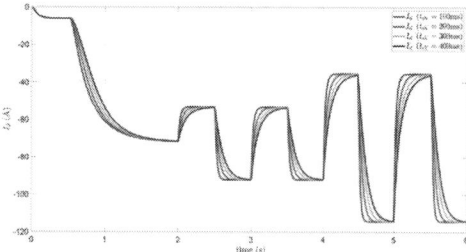

c) d-axis current.

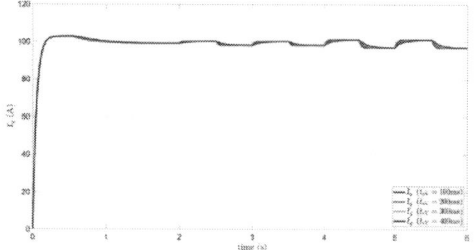

d) q-axis current.

Fig. 4: Dynamic response in the face of a square-wave voltage reference.

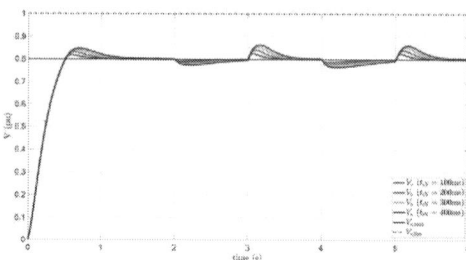

a) Voltage magnitude.

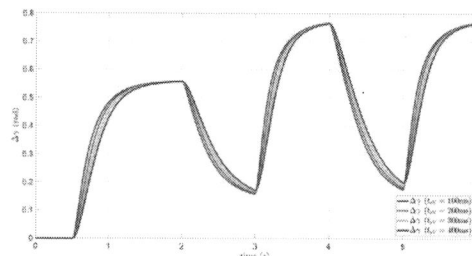

b) Variation of current angle.

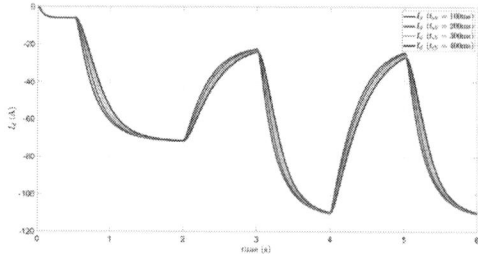

c) d-axis current.

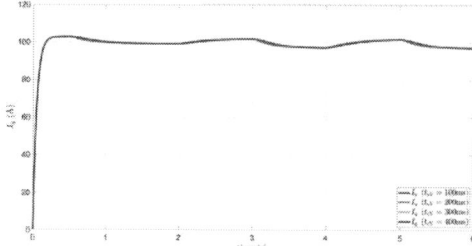

d) q-axis current.

Fig. 5: Dynamic response in the face of square-wave load torque in FW operation.

References

[1] T. H. Møller and J. Simlett, "Micromobility: moving cities into a sustainable future," EY Mobility Centre, Tech. Rep., 2020, pp. 1–36.

[2] Environmental Protection Agency, "Inventory of U.S. Greenhouse Gas Emissions and Sinks: 1990-2015," United States Environmental Protection Agency, Tech. Rep., 2017, pp. 1–633.

[3] S. Wang and M. Ge, *Everything You Need to Know About the Fastest-Growing Source of Global Emissions: Transport*, Oct. 2019.

[4] B.-G. Gu, J.-H. Choi, and I.-S. Jung, "Development and Analysis of Interturn Short Fault Model

of PMSMs With Series and Parallel Winding Connections," *IEEE Trans. Power Electron.*, vol. 29, no. 4, pp. 2016–2026, 2014. DOI: 10.1109/TPEL.2013.2265400.

[5] L. Sepulchre, M. Fadel, M. Pietrzak-David, and G. Porte, "Flux-Weakening Strategy for High Speed PMSM for Vehicle Application," in *2016 Int. Conf. Electr. Syst. Aircraft, Railw. Sh. Propuls. Road Veh. Int. Transp. Electrif. Conf.*, 2016, pp. 1–7. DOI: 10.1109/ESARS-ITEC.2016.7841413.

[6] S. Sudhoff, K. Corzine, and H. Hegner, "A Flux-Weakening Strategy for Current-Regulated Surface-Mounted Permanent-Magnet Machine Drives," *IEEE Trans. Energy Convers.*, vol. 10, no. 3, pp. 431–437, 1995. DOI: 10.1109/60.464865.

[7] W. Peters, T. Huber, and J. Böcker, "Control Realization for an Interior Permanent Magnet Synchronous Motor (IPMSM) in Automotive Drive Trains," in *PCIM Eur. Conf. Proc.*, 2011, pp. 1–6.

[8] T. Huber, W. Peters, and J. Böcker, "Voltage Controller for Flux Weakening Operation of Interior Permanent Magnet Synchronous Motor in Automotive Traction Applications," in *2015 IEEE Int. Electr. Mach. Drives Conf.*, IEEE, 2016, pp. 1078–1083. DOI: 10.1109/IEMDC.2015.7409195.

[9] T. Hu, F. Lin, L. Cui, Q. Yuan, and Z. Yang, "The Flux-Weakening Control of Interior Permanent Magnet Synchronous Traction Motors for High-Speed Train," in *Proc. 1st Int. Work. High-Speed Intercity Railw.*, vol. 1, Springer Link, 2012, pp. 451–461.

[10] P.-Y. Lin and Y.-S. Lai, "Voltage Control Technique for the Extension of DC-Link Voltage Utilization of Finite-Speed SPMSM Drives," *IEEE Trans. Ind. Electron.*, vol. 59, no. 9, pp. 3392–3402, 2012. DOI: 10.1109/TIE.2011.2173095.

[11] G. Pellegrino, R. I. Bojoi, and P. Guglielmi, "Unified Direct-Flux Vector Control for AC Motor Drives," *IEEE Trans. Ind. Appl.*, vol. 47, no. 5, pp. 2093–2102, 2011. DOI: 10.1109/TIA.2011.2161532.

[12] J. Li, Q. Wang, J. Yu, and J. Xiong, "Field-weakening Control Algorithm for Interior Permanent Magnet Synchronous Motor Based on Space-Vector Modulation Technique," *J. Converg. Inf. Technol.*, vol. 8, no. 3, pp. 167–175, 2013.

[13] S. Bolognani, S. Calligaro, and R. Petrella, "Adaptive Flux-Weakening Controller for Interior Permanent Magnet Synchronous Motor Drives," *IEEE J. Emerg. Sel. Top. Power Electron.*, vol. 2, no. 2, pp. 236–248, 2014. DOI: 10.1109/JESTPE.2014.2299153.

[14] S. Bozhko, M. Rashed, C. I. Hill, S. S. Yeoh, and T. Yang, "Flux-Weakening Control of Electric Starter-Generator Based on Permanent-Magnet Machine," *IEEE Trans. Transp. Electrif.*, vol. 3, no. 4, pp. 864–877, 2017. DOI: 10.1109/TTE.2017.2718221.

[15] Z. Lei, X. Shan, W. Xuhui, L. Yaohua, and K. Liang, "A New Deep Field-Weakening Strategy of IPM Machines Based on Single Current Regulator and Voltage Angle Control," in *2010 IEEE Energy Convers. Congr. Expo.*, 2010, pp. 1144–1149. DOI: 10.1109/ECCE.2010.5617844.

[16] L. Sepulchre, M. Fadel, M. Pietrzak-David, and G. Porte, "MTPV Flux-Weakening Strategy for PMSM High Speed Drive," *IEEE Trans. Ind. Appl.*, vol. 54, no. 6, pp. 6081–6089, 2018. DOI: 10.1109/TIA.2018.2856841.

[17] R. Krishnan, *Permanent Magnet Synchronous and Brushless DC Motor Drives*. Virgina: CRC Press, 2010, pp. 303–327. DOI: 10.1201/9781420014235.

[18] P. Krause, O. Wasynczuk, S. Sudhoff, and S. Pekarek, *Analysis of Electric Machinery and Drive Systems*, Third, I. Press, Ed. New Jersey: John Wiley and Sons, 2013, pp. 1–678.

PCIM Europe 2024, 11– 13 June 2024, Nuremberg   DOI: 10.30420/566262036

Investigation on Direct Liquid Cooling Design of Power Modules with Flat Baseplate for Automotive Application

Nobuhide Arai[1], Shinichiro Adachi[1], Kensuke Matsuzawa[1], Takahiro Koyama[1], Takanori Shintani[1], Steffen Ewald[2]

[1]Fuji Electric Co., Ltd, Japan
[2]Fuji Electric Europe GmbH., Germany

Corresponding author: Nobuhide Arai, arai-nobuhide@fujielectric.com
Speaker: Nobuhide Arai, arai-nobuhide@fujielectric.com

Abstract

Copper pin fins and aluminum integrated fins have generally been used in power modules for direct liquid cooling. The direct liquid cooling system using copper pin fins or aluminum integrated fins provides excellent heat dissipation, but it is expensive in terms of materials and their production costs. Copper flat baseplates, which are low cost and generally used for indirect liquid cooling or air cooling, have not been adequately evaluated as direct liquid coolers regarding to the cooling performance and reliability in automotive applications. This paper investigates suitable rib shape in direct liquid cooling for a flat baseplate power module from the viewpoints of heat dissipation and pressure loss. In addition, it demonstrates sufficient reliability in terms of coolant corrosion and pressure durability, which are essential for automotive applications.

1   Introduction

As electrification of automobiles progress, electrical components such as inverters for motor and generator need to be smaller to increase room space of vehicles. In addition, as conventional gasoline-powered vehicles are replaced with electrified vehicles, it is necessary to reduce the cost of power electronics components. The cost of automotive power modules can be reduced by increasing power density. Fuji Electric has developed direct liquid cooling technology with integrated fin, Reverse-Conducting Insulated Gate Bipolar Transistor (RC-IGBT), and lead frame wiring technology for high power density [1] [2]. Electrification is rapidly expanding not only to SUVs and luxury cars, but also to small and middle power range such as light and compact vehicles, so power range variations in power modules are also required. M682 series, which features a power range of 50-100 kW, have been developed with the concept of a compact package and low cost for the market of light and compact vehicles. The lineup of the M682 series is shown in Fig.1. The package footprint is the same, with a choice of capacity variants depending on the chip size and cooler combination. By using the module, it is possible to standardize peripheral components of the module while changing the capacity range of the inverter, thereby reducing costs, and improving development efficiency through standard-

ization of inverter components. A pin fin cooler is used for the 100 kW output, while a flat baseplate structure is applied to the coolers for the lower outputs of 50 and 75 kW. Flat baseplates have lower heat dissipation performance than copper pin fins. Baseplates have traditionally been used for indirect liquid or air cooling, but direct liquid cooling structures can improve heat dissipation. However, there are few research which the optimum liquid cooling channels for flat bases have been fully investigated for automotive applications. This paper describes the optimum liquid cooling channels for flat bases. In addition, the results of a reliability assessment of the direct liquid-cooled structure for automotive applications are presented. Finally, a comparison of the output performance of modules with flat-based structures in optimized liquid cooling channel is presented.

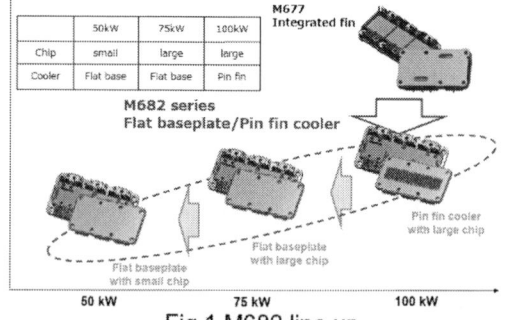

Fig.1 M682 line up

2 Direct liquid Cooling with a Flat Baseplate

2.1 Thermal resistance of flat baseplate power modules

Flat baseplate coolers are applied in industrial modules and are generally used with an indirect liquid cooling method. In this method, copper base type modules are installed in liquid-cooled heatsinks via grease, which have significantly high thermal resistance. Figure 2 shows the cross-sectional structures of modules and the analysis results of the thermal resistance breakdown. In the direct liquid cooling method, the use of coolers with heat dissipation fins cooled by a coolant circulation system achieves low thermal resistance [3], [4], [5], [6]. With the copper base type in this study, by eliminating grease and adopting a direct liquid cooling approach, a 36% improvement in thermal resistance can be achieved. Furthermore, applying a silicon nitride AMB (Si_3N_4, thermal conductivity 80 W/mK) from an alumina DCB (Al_2O_3, thermal conductivity 20 W/mK) which is widely used as insulated substrates, an additional 12% improvement in thermal resistance can be achieved. As the result, a total improvement of 48% compared to a conventional indirect liquid cooling configuration. The efficiency of heat dissipation performance in direct liquid cooling structures is a key point, thus a study of the flow paths was carried out.

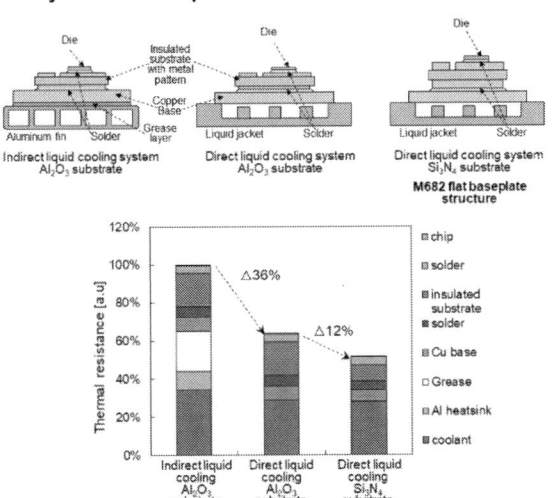

Fig. 2 Comparison of thermal resistance between conventional structure and M682 flat baseplate

2.2 Coolant Flow in Cooling Jacket

The flow of coolant in the liquid cooling jacket was analyzed using Computational Fluid Dynamics (CFD) simulations with a model of a 6 in 1 power module, flowing coolant serially shown in Fig. 3. The simulation condition includes a coolant mixture of long-life coolant and water at a ratio of 1:1, a flow rate of 10 L/min, and a water temperature of 70°C at the inlet. The simulation results of the velocity distribution in the flow channel of the model are shown in Fig. 4. In the flow channel, the coolant tends to flow uniformly, resulting in a uniform distribution of coolant velocity. Cooling of the device-generated heat involves transferring heat to the coolant from the surface of the flat baseplate, transporting it for cooling. Therefore, improving the coolant velocity near the surface of the flat baseplate is important for enhancing heat dissipation efficiency. However, increasing coolant velocity leads to increased pressure loss, which must be kept below the allowable pressure loss determined by an electric pump's capacity in automobiles. Hence, low pressure loss is required. Therefore, we investigated models to increase the flow velocity on the back surface of the cooler with lower pressure loss.

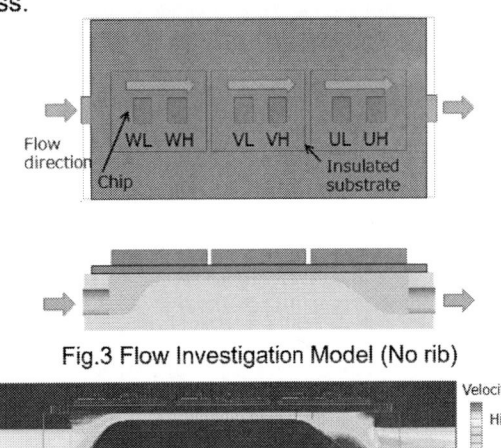

Fig.3 Flow Investigation Model (No rib)

Fig.4 Velocity in liquid cooling jacket (No rib)

2.3 The Flow Velocity Improvement Effect by Ribs

The rib geometries of the flow channel were studied on simple models that can be machined in the aluminum die-casting process. Table 1 shows three proposed models and the results of flow analysis at flow rate of 10 L/min and coolant temperature 70°C. Figure 5 shows the analysis of pressure loss and thermal resistance.

Table 1 Ribs layout and coolant velocity distribution comparison

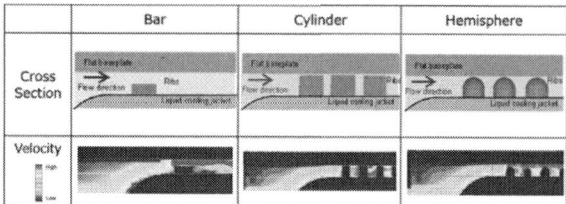

in flow velocity. As the result, hemisphere ribs are found to be more efficient rib shapes compared to cylinder ribs in terms of thermal resistance and pressure loss.

Table 2 Velocity distribution

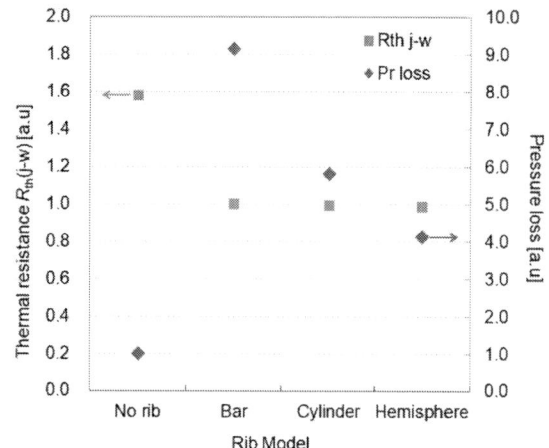

Fig.5 Thermal resistance and pressure loss at LLC50 %, 70 ℃, 10 L/min

By adding ribs, it is possible to increase the flow velocity beneath the surface of the flat baseplate. The bar structure, being arranged across the width of the flow channel, significantly increases pressure loss. On the other hand, cylinder ribs and hemisphere ribs can be selectively placed so that lower pressure loss is achieved. Moreover, since the cross-sectional area of the flow channel of the hemisphere is larger than that of the cylinder, the pressure loss of the hemisphere is smaller than the cylinder. Cylinder ribs and hemisphere ribs are compared with regards to coolant flow. Table 2 shows schematic diagrams of the cross-sectional structure and the distribution of coolant speed. hemisphere ribs, compared to cylinder ribs, facilitate the flow of coolant beneath the copper base plate due to the tapered shape of the ribs. In addition, since the coolant collides with the ribs in the direction of flow, it is difficult to flow the coolant in the back of the ribs. However, hemisphere ribs have a smaller effective rib diameter because of tapered shape, reducing the area where the flow is impeded in the back of the ribs. Consequently, hemisphere ribs can enhance flow velocity directly beneath the flat baseplate and promote uniform flow velocity. Analysis of the velocity distribution mapping plot confirms that hemispheres result in more uniform improvement

2.4 Optimization of Thermal Resistance and Pressure Loss by Rib Arrangement

High-density arrangement of ribs significantly increases pressure loss. Therefore, a study was conducted to reduce pressure loss by adjusting the number of ribs. In addition, experimental verification was carried out to validate the analysis. Experiments were conducted by making liquid cooling channel parts using a resin 3D printer, varying the density of the ribs, and investigating cooling performance.

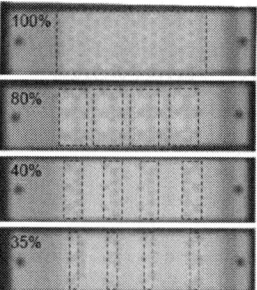

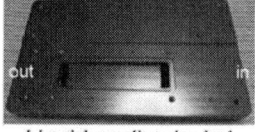

Liquid cooling jacket

3D prited pattern

Fig. 6 Evaluation liquid cooling jacket and 3D printed pattern

The evaluation liquid cooling jacket is made of aluminium and features several patterns of plastic coolant channel models for performance comparison. The testing environment involves a coolant circulation with LLC 50%, 70°C, and a flow rate of 10 L/min. The junction temperature of the chips was measured using on-chip sensor diodes. Pressure losses were measured at the inlet and

PCIM Europe 2024, 11– 13 June 2024, Nuremberg   DOI: 10.30420/566262036

outlet of the liquid jacket using a differential pressure measurement system. Figure 7 shows the impact of rib spacing on thermal resistance along with sample evaluations.

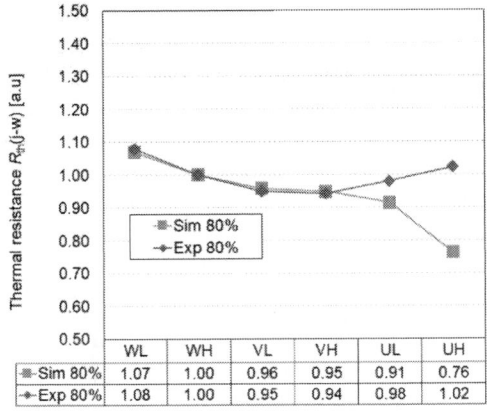

	WL	WH	VL	VH	UL	UH
Sim 80%	1.07	1.00	0.96	0.95	0.91	0.76
Exp 80%	1.08	1.00	0.95	0.94	0.98	1.02

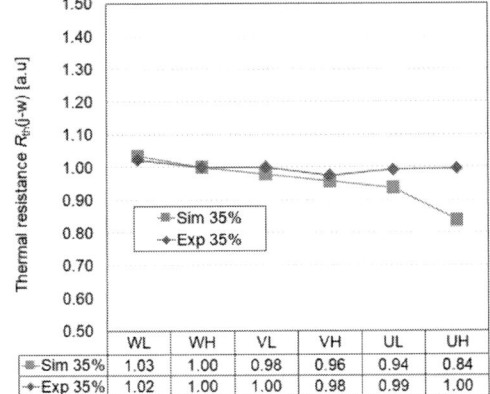

	WL	WH	VL	VH	UL	UH
Sim 35%	1.03	1.00	0.98	0.96	0.94	0.84
Exp 35%	1.02	1.00	1.00	0.98	0.99	1.00

Fig.7 Simulation and experiment results

Fig. 8 The coolant flow in hemisphere ribs

It can be found that the thermal resistance distribution is in good agreement between experiment and simulation, except for 1 arm (UH). The difference of thermal resistance UH could be caused by the generation of vortices causing discrepancies between experiment and simulation shown in Fig. 8. Countermeasures can be implemented in the outlet side flow path geometry to mitigate this effect. Since the thermal resistances of the chips placed in the middle of the 6 in 1 module such as WH, VL and VH have a good agreement between experiment and simulation, the relationship between thermal resistances, pressure losses, and the spacing ratio is analysed in VH shown in Fig. 9.

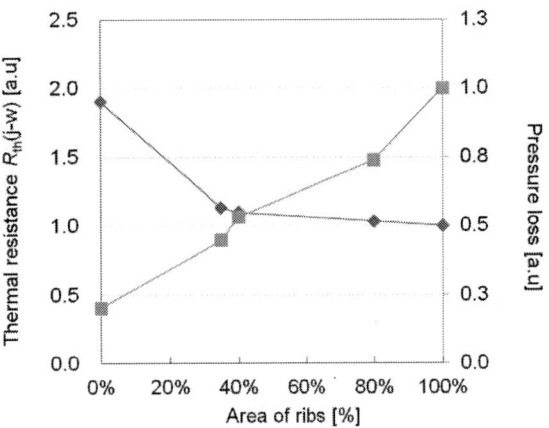

Fig. 9 Thermal resistance and Pressure loss Versus area of ribs percentage

Analysing the coolant velocity obtained through simulation, the relationship between velocity and spacing ratio is shown in Fig. 10. Plotting velocity against thermal resistance reveals that thermal resistance decreases gradually as the spacing ratio decreases from 100% but exhibits a sharp decline at 35%.

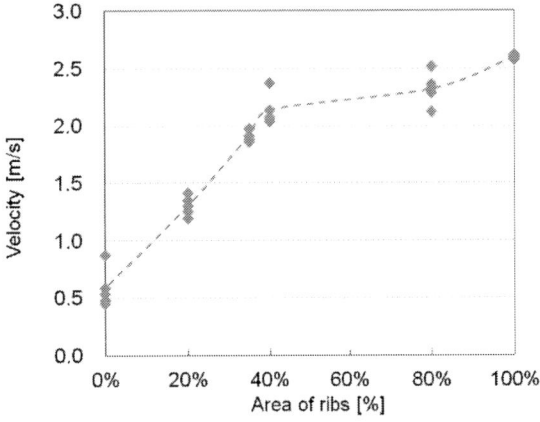

Fig. 10 Velocity versus area of ribs percentage

From these results, it is found that

· The thermal resistance and the pressure loss in direct liquid cooling with a flat baseplate are dependent on the flow velocity likewise fin structures. Flow velocity is influenced by the spacing ratio of the ribs, with a sharp decrease observed beyond a certain point, around 35%. This is because when the proportion of ribs in the flow channel cross-section is too low, the coolant does not flow directly beneath the base.

301

- By maintaining a spacing ratio of 40% or higher, where the flow velocity does not decrease significantly, it is possible to adjust thermal resistance and pressure distribution effectively by ensuring an appropriate proportion of ribs.

3 Reliability Test

3.1 Corrosion Test

In the performance validation of thermal resistance and pressure loss, evaluations were carried out using 3D printed plastic channel models for the sake of simplicity. However, in actual inverter housings, the liquid cooling channel is often formed together with ADC12 aluminum material through aluminum die-casting process. On the other hand, flat baseplate coolers are made of copper and are surface covered with nickel plating to prevent copper corrosion. However, under the worst-case scenario of rib tolerances, there is a possibility of the contact between the flat baseplate and the aluminum housing ribs of the flow channels. Concerns arise about corrosion due to the potential difference between the nickel plating and aluminum, which could lead to corrosion of the nickel plating and exposure of copper. This could accelerate copper corrosion and deteriorate thermal performance.

The evaluation of corrosion resistance using a corrosive liquid was conducted. The corrosion test was performed using a copper flat baseplate with Ni plating and an ADC12 inverter housing. As the corrosive liquid, OY water is applied. OY water is formulated to be even more corrosive than the most corrosive water found through global water quality survey [7]. The solutions of OY water and long-life coolant mixed in a 1:1 ratio is used. Table 3 shows the conditions for the corrosion test and Fig. 11 presents a schematic diagram of the sample and the test setup. The inverter test work which initially contact with the baseplate was prepared. An evaluation was conducted.

Table 3. Corrosion test condition

Corrosion test		
Test water	OY water : LLC 50% = 1: 1	
OY water	CL⁻	195 ppm
	SO_4^{2-}	60 ppm
	Cu^{2+}	1 ppm
	Fe^{3+}	30 ppm
LLC 50%	Long Life Coolant 50 %	
Temperature	65 ℃	
Flow rate	12 L/min	
Test time	1000 hr	

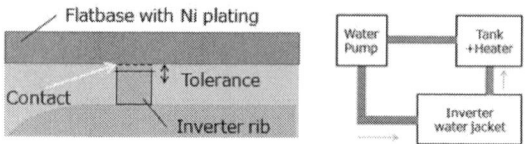

Fig. 11 Cross-section of sample and test set up

Fig. 12 shows the test results and analysis findings. Although there is some discoloration, no pitting corrosion was observed on the nickel plating of the cooler. Additionally, there was no discoloration or pitting corrosion observed on the liquid cooling jacket rib side. Surface analysis of the plating was conducted using Energy Dispersive X-ray Spectroscopy (EDX) for elemental analysis. EDX analysis at red circle area contact with the water revealed the presence of carbon (C) and oxygen (O), which are the main components of long-life coolant, and sulfur (S) and chlorine (Cl), elements of the corrosive fluid. It was confirmed that the plating remained intact without exposing copper. Fig. 13 describes the thermal resistance evaluation results at initial, 500 hours, and 1000 hours. There was no increase in thermal resistance on the cooler side after a 1000-hour corrosion test. Based on these results, it was confirmed that the cooler exhibited corrosion resistance in the corrosion cycling test.

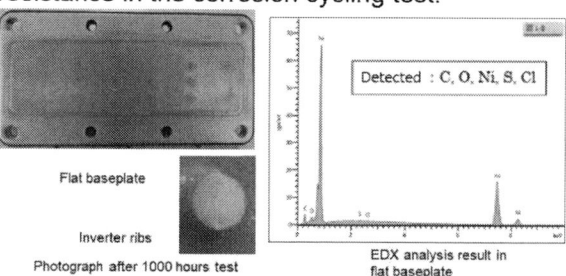

Fig. 12 Sample surface and EDX analysis result after the corrosion test

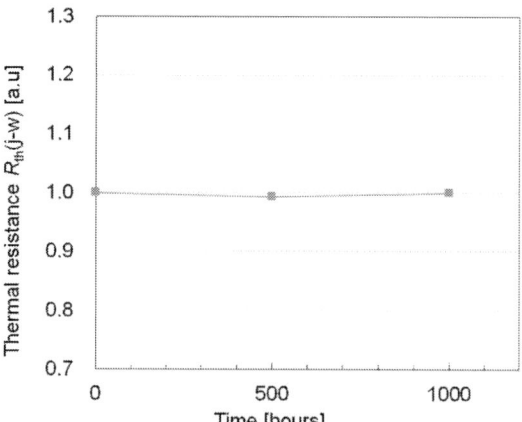

Fig. 13 Evaluation results of thermal resistance during the corrosion test

3.2 Intermittent Pressure Test

When an inverter system including a cooling system is started in vehicles, an electric pump starts to work, and high pressure applies to power modules from the cooling system. The seal of a cooling circuit is maintained by O-rings installed between an inverter housing and a power module, but inadequate Moreover, the degradation of the solder layer on a flat baseplate can be considered due to deformation of the flat baseplate.

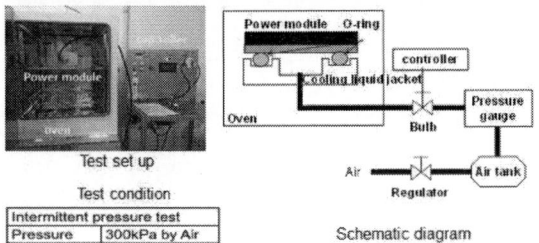

Fig. 14 Intermittent pressure test setup and condition

Therefore, an intermittent pressure test was conducted to verify the reliability of the sealing system under the pressure condition shown in Fig14. The test was carried out pressurizing the power module and fixture with air at 300 kPa. The results showed no increase in the thermal resistance in Fig.15. Moreover, ultrasonic inspection confirmed no solder cracks beneath the insulating substrate in Fig.16. Therefore, it was confirmed that the power module has the required pressure resistance for automotive applications.

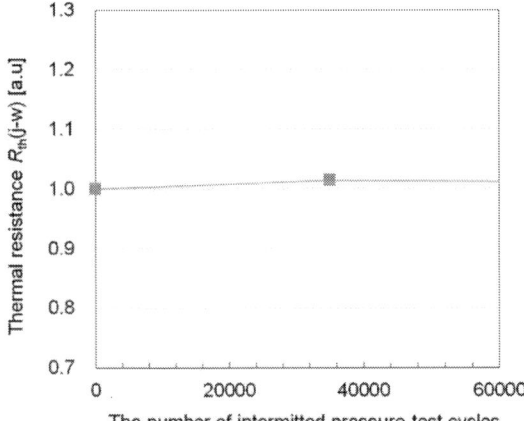

Fig. 15 Evaluation result in thermal resistance during the intermitted pressure test

	Initial	After the test
Scanning Acoustic Tomograph		

Fig. 16 Scanning acoustic tomograph of the solder layer under the insulated substrate

4 Output Power of M682 with a flat baseplate

The output current of the new power module M682 75 kW using a flat baseplate and silicon nitride insulating substrate was calculated compared with a conventional configuration power module having the same electrical performance of M682 75 kW with indirect liquid cooling and alumina insulating substrate. Figure 17 shows the junction temperature calculation results. It is found that the M682 75kW can achieve 1.5 times the output current compared to the conventional configuration power module.

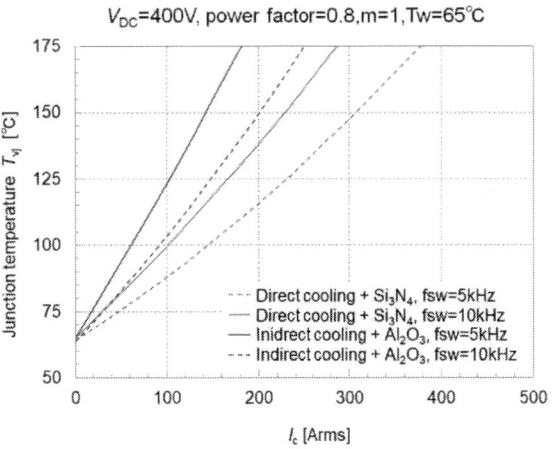

Fig. 17 Junction temperature calculation results

5 Conclusion

In this paper, effective ribs shapes that improves the cooling performance and pressure loss are discussed. Hemisphere rib structure is suitable for coolant flow beneath the flat baseplate and lower pressure loss. Moreover, it was clarified that the design of the cooler can adjust both thermal resistance and pressure loss by adjusting the arrangement of cooling channel ribs. Finally, the cooling system's reliability was confirmed through coolant corrosion test and intermittent pressure tests suitable for automotive applications. Finally, by using the optimal rib shape, the output performance can be improved by 50% percentage compared to the conventional indirect liquid-cooled configuration.

References

[1] S. Adachi, T. Obata, N. Arai, N. Higashi, Y. Tateishi, "Ultra-Compact Automotive Power Module for 100 kW xEV Application", PCIM Europe 2022, pp.1035-1038.

[2] Y. Sato, S. Adachi, H. Gohara, N. Higashi, S. Yoshida, "Advantage of Lead-Frame Wiring and High Reliable to Electromigration Package for High Power Density Automotive Power Module", PCIM Europe 2023, pp. 1674-1679.

[3] S. Adachi, F. Nagaune, H. Gohara, T. Hitachi, A. Morozumi, P. Dietrich, A. Nishiura, "High thermal conductivity technology to realize high power density IGBT modules for electric and hybrid vehicles, PCIM Europe 2012, pp.1378-1384.

[4] H. Gohara,Y. Nishimura, A. Morozumi, P. Dietrich, E. Mochizuki, Y. Takahashi, "Next-gen IGBT module structure for hybrid vehicle with high cooling performance and high temperature operation", PCIM Europe 2014, pp. 1187-1194.

[5] K. Higuchi, T. Koyama, A. Kitamura, S. Soyano, Y. Takamiya, H. Gohara, S. Yoshida, H. Kobayashi, Y. Nishimura, T. Heinzel, A. Nishiura, "New standard 800A/750V IGBT module technology for automotive application", PCIM Europe 2015, pp.1137-1144.

[6] Y. Tamai, S. Ewald, R. Kato, K. Yamauchi, H. Gohara, T. Yamazaki, "Fourth Generation Aluminum Direct Water Cooling Structure with High Reliability for Automotive Electric System", PCIM Asia 2020, pp.207-210.

[7] H. Tanaka, H. Ikeda, "Influence of Fe and Ni addition on corrosion resistance of aluminum alloy-clad sheet for automotive radiators in weakly alkaline LLC solution", Journal of The Japan Institute of Light Metals, Vol. 70, No. 10, 2020, pp. 451-458.

A Novel Approach to Affordable Electric Vehicles Based on Dual 48V Battery System with Multi-functional 3-level Converter

Akihiro Furukawa[1], Dušan Graovac[2], Tatsuya Arai[3], Radovan Vuletić[2]

[1]Mazda Motor Corporation, Hiroshima, Japan
[2]Infineon Technologies AG, Neubiberg, Germany
[3]Infineon Technologies Japan K.K., Tokyo, Japan

Corresponding author: Radovan Vuletić, radovan.vuletic@infineon.com
Speaker: Radovan Vuletić, radovan.vuletic@infineon.com

Abstract

Main building block of modern mobility are electric vehicles (EV). Mainstream power supplies of EV are 400 V or 800 V batteries and related to them inverter implementation based on wide-band gap semiconductors like SiC MOSFETs – however, efficiency and corresponding affordability (due to high voltage ratings and service of hazards) remain as a main challenges. Inverter efficiency is mainly influenced by losses in semiconductor power switches – for 400 V and higher battery voltages, high-voltage switching components must be used (650 V and higher). Based on experiences with 48 V mild hybrid vehicles, this paper gives an alternative approach for affordable smaller size EVs. 48 V as single voltage is not enough to cover the efficiency requirements of the whole driving cycle, therefore a dual power supply with two batteries (+/- 48 V, with a common ground connected to the chassis as shown on Fig. 1) was considered. In a 2-levele topology is difficult to find optimal power switches for 96 V DC bus, therefore the 3-level topology, Active Neutral Point Clamping (ANPC) inverter, based on 80 V MOSFET was implemented. Additionally, two-battery system requires active battery balancing, which is typically implemented as additional DC/DC converter – in this paper is presented a novel idea of battery balancing by simply selecting optimal voltage vectors of Space-Vector Modulation (SVM) and on that way avoiding need for additional hardware – with this approach system costs are reduced and overall system efficiency is increased.

1 Concept of 3-level Inverter

1.1 Selection of Topology

One of the main tasks during development of electric drives in automotive traction applications is to fulfill requirements on highest possible converter efficiency, and, related to limited space in passenger cars, smallest possible volume that translates into the highest possible power density. In the first part of this paper is presented a design concept that is developed by means of simulations, and later proven on the prototype. By usage of SPICE simulation tool, solution on reaching inverter highest possible efficiency is found – this includes selecting the right electrical power converter topology, selecting optimal semiconductors (power switches) and passive electronic components, and finally estimating the optimal operating conditions (like switching frequency or gate drivers strength). In

the second part of this paper is presented the design of algorithm for SVM on ANPC using Model-based Design (MBD) methodology by means of MathWorks' MATLAB/Simulink and automated code generation. Finally, in the last part of this paper is described utilization of battery balancing that is implemented by selection of switching voltage vectors instead of usage of additional DC/DC converter, that consequently lead to reduction of overall system complexity and increasing overall system efficiency.

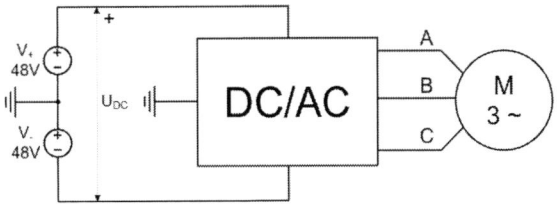

Fig. 1 Electrical drive train driven by 3-level inverter

Having on mind that Infineon Technologies has excellent low voltage (LV) MOSFET power switches (up to 80-100 V breakdown voltage) and having on mind that ANPC as a 3-level topology requires just 50 % of breakdown voltage rating comparing to B6 2-level topology, ANPC inverter, supplied with two batteries is designed, as shown on Fig. 2.

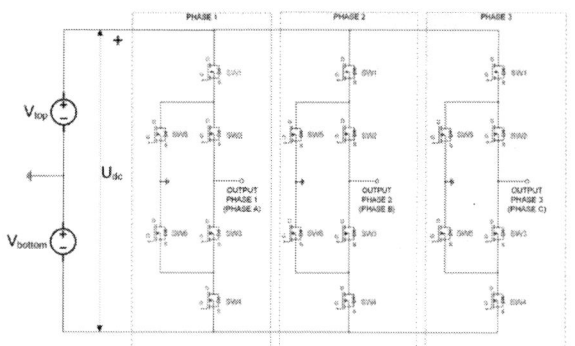

Fig. 2   ANPC inverter supplied by two batteries

1.2   Analog Simulations

SPICE simulations are done using the setup that is depicted on Fig. 3, where high fidelity electrical-

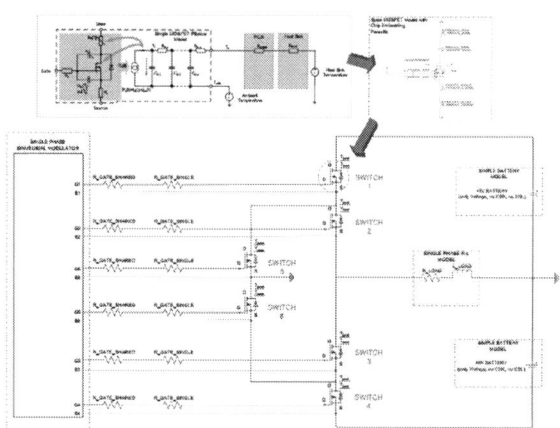

Fig. 3   Analog simulation setup

thermal models of OptiMOS™ power MOSFETs (models are explained in [1]) are used together with gate drivers model, assessed PCB parasitic (novel method of chip embedding) and assumed load (simplified model of electrical motor).

Electrical circuit parameters, like number of MOSFETs in parallel, used type of MOSFET switch, load and switching frequency are varied in order to find optimal operating conditions. Around 200 simulation setups (with 15 s simulation time,

up to 300k switching cycles) are sent to UNIX computing farm (Fig. 4), requiring average simulation time of around 48 h for one run.

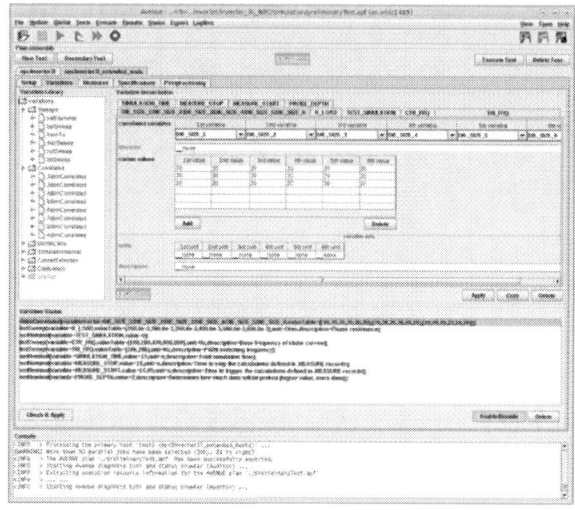

Fig. 4   Simulation setup parallelisation tool

Explanation for relatively long simulation times is that thermal time constant of the simulated circuit is much higher than the electrical time constant. Although electrically circuit comes relatively quickly to the steady state (already after a few hundreds of switching cycles), due to junction-to-case (Z_{JC}) thermal impedance of MOSFET and case-to-

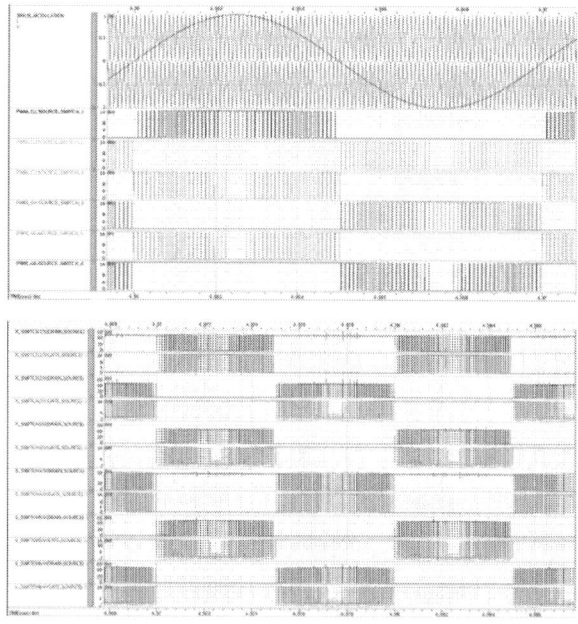

Fig. 5   Typical waveforms used for calculations of losses

ambient (Z_{CA}) thermal impedance of PCB and heatsink, reaching thermal equilibrium requires 100 to 1000 times more switching cycles comparing to "pure" electrical simulations. Possible workarounds to overcome above described limitations of electrical-thermal time limitations are presented in [2]. Based on SPICE simulation results (example waveforms on Fig. 5) and by simple integration of instantaneous values of voltage drops and currents, losses of every electronic component as well as overall system efficiency can be easily determined.

Based on above described concept, MAZDA developed a 3–level inverter prototype (Fig. 6) that was showing very good matching with simulations.

P2PACK board

Fig. 6 MAZDA prototype of 3-level inverter

2 Implementation of SVM for ANPC Inverter

2.1 Comparison between SVM for 2-level and 3-level Inverter

As it can be easily observed on Fig. 7, 3-level inverter utilizes much higher number of space vectors comparing to the 2-level inverter. 2-level inverter deploys 6 non-zero vectors (V_1 to V_6) and 2-zero vectors (V_0 and V_7). On the other side, 3-level inverter utilizes 18 non-zero vectors - however, if we consider the redundancy, or better said feature at 3-level inverter that zero and some of non-zero vectors can be expressed with different switch states (Fig. 9 and Fig. 10), selection and design of space vector modulation at 3-level inverter becomes much more complex and computing intensive operation comparing to the 2-level inverter!

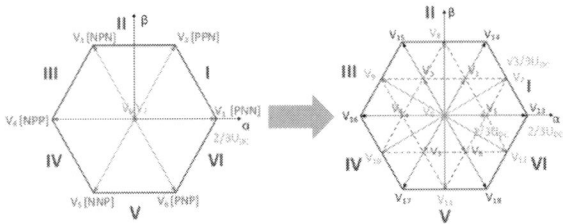

Fig. 7 Comparison between 2-level and 3-level inverter SVM

2.2 Algorithm for Generation of SVM for 3-level Inverter

Let's consider one arbitrary voltage vector at 3-level inverter, for example [PPU] as shown on Fig. 8 – as it can be observed, depending on the selected combination of the switches, either positive pole of battery ("P") or negative pole of battery ("N") or neutral point ("U" or "L") are connected to phase output.

- Example: Vector [PUU]

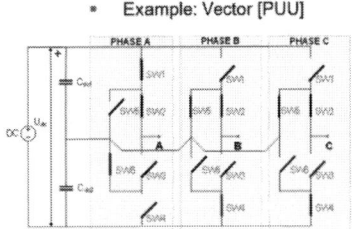

„P" means SW1, SW2, SW6 **ON**, SW3, SW4, SW5 **OFF**
„U" means SW2, SW5, SW4 **ON**, SW1, SW3, SW6 **OFF**
„L" means SW1, SW3, SW6 **ON**, SW2, SW4, SW5 **OFF**
„N" means SW3, SW4, SW5 **ON**, SW1, SW2, SW6 **OFF**

Complementary:
SW1 & SW4
SW2 & SW3
SW5 & SW6

Fig. 8 Example of vector for 3-level inverter

All possible voltage vectors at 3-level inverter are listed in Fig. 9 and Fig. 10. For the sake of the further analysis, as shown on Fig. 11, the whole SVM space is divided in 6 sectors (each comprising 60°), and each sector is divided 4 segment that are in the case of balanced system (balanced batteries) equilateral triangles. As it can be observed, zero vector (V0) has 10 different variants, each small vector (V1 – V6) is having 6 possible variants, each medium vector (V7 – V12) is having 2 variants and only large vectors (V13 – V18) have unique, single representation. This feature of 3-level inverter, that zero, small and medium vectors can be mapped on different ways, is used later to implement battery voltage balancing, without actually utilizing additional hardware for battery balancing system. Let us consider one example where the reference vector is in sector I, segment 2, as depicted on Fig. 12, and discuss dwell time calculation. Just as in the case of 2-level inverter, system of 3 equations with 3 unknown variables (times for each sub-vector) has to be solved. Difference comparing to the 2-level SVM, is that at 3-level SVM different sub-vectors for decomposing reference vector can be selected, in this case Nearest Three Vectors (NTV) method is selected (more details on NTV method can be found in [3]). Notice that a reference vector V_{ref} is more near to the vector V_2 (area "b") than to the vector V_1, therefore, the decomposition of reference vector will start with V_2.

SPACE VECTOR		SWITCHING STATE		VECTOR CLASSIFI-CATION	MAGNI-TUDE
$\vec{V_0}$		[PPP] [NNN] [UUU] [UUL] [ULL] [ULU] [LUU] [LUL] [LLU] [LLL]		Zero Vector	0
		P-type	N-type		
$\vec{V_1}$	$\vec{V_{1P}}$	[PUU] [PUL] [PLU] [PLL]			
	$\vec{V_{1N}}$		[LNN] [UNN]		
$\vec{V_2}$	$\vec{V_{2P}}$	[PPU] [PPL]			
	$\vec{V_{2N}}$		[UUN] [ULN] [LUN] [LLN]		
$\vec{V_3}$	$\vec{V_{3P}}$	[UPU] [UPL] [LPU] [LPL]			
	$\vec{V_{3N}}$		[NLN] [NUN]	Small Vector	$\frac{1}{3}U_{DC}$
$\vec{V_4}$	$\vec{V_{4P}}$	[UPP] [LPP]			
	$\vec{V_{4N}}$		[NUU] [NUL] [NLU] [NLL]		
$\vec{V_5}$	$\vec{V_{5P}}$	[UUP] [ULP] [LUP] [LLP]			
	$\vec{V_{5N}}$		[NNL] [NNU]		
$\vec{V_6}$	$\vec{V_{6P}}$	[PUP] [PLP]			
	$\vec{V_{6N}}$		[UNU] [UNL] [LNU] [LNL]		

Fig. 9 Space vectors and switching states (zero and small vectors)

SPACE VECTOR	SWITCHING STATE	VECTOR CLASSIFI-CATION	MAGNI-TUDE
$\vec{V_7}$	[PUN] [PLN]		
$\vec{V_8}$	[UPN] [LPN]		
$\vec{V_9}$	[NPU] [NPL]	Medium Vector	$\frac{\sqrt{3}}{3}U_{DC}$
$\vec{V_{10}}$	[NUP] [NLP]		
$\vec{V_{11}}$	[UNP] [LNP]		
$\vec{V_{12}}$	[PNU] [PNL]		
$\vec{V_{13}}$	[PNN]		
$\vec{V_{14}}$	[PPN]		
$\vec{V_{15}}$	[NPN]	Large Vector	$\frac{2}{3}U_{DC}$
$\vec{V_{16}}$	[NPP]		
$\vec{V_{17}}$	[NNP]		
$\vec{V_{18}}$	[PNP]		

Fig. 10 Space vectors and switching states (medium vector and large vector)

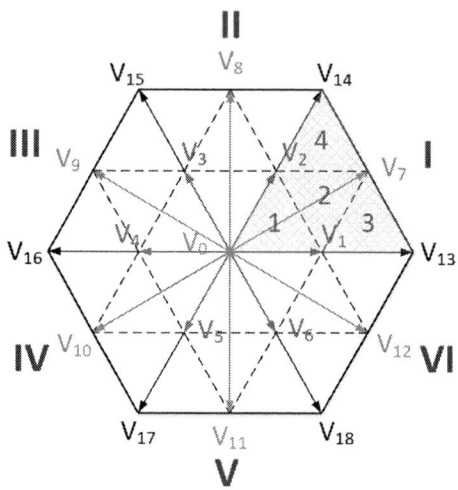

Fig. 11 SVM space with 6 sectors (I–VI) each having 4 segments

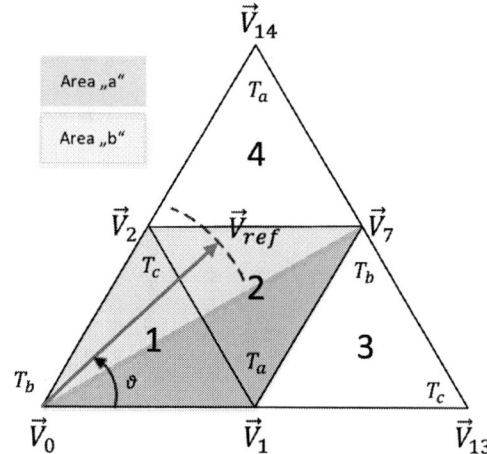

Fig. 12 Example of 3-level SVM with reference vector in sector I, segment 2b

Based on SVM theory, each reference vector can be decomposed on 3 sub-vectors (V_a, V_b and V_c) as shown in Eq. 1 with their corresponding dwelling times (T_a, T_b and T_c) that in sum must be equal to PWM switching period (T_s) as shown in Eq. 2.

$$\vec{V_a}T_a + \vec{V_b}T_b + \vec{V_c}T_c = \vec{V_{ref}}T_s \tag{1}$$

$$T_a + T_b + T_c = T_s \tag{2}$$

In this specific case (Fig. 12) vectors V_a, V_b and V_c correspond to vectors V_1, V_7 and V_2 respectively, on this way Eq.1 becomes Eq. 3.

$$\vec{V}_1 T_a + \vec{V}_7 T_b + \vec{V}_2 T_c = \vec{V}_{ref} T_s \qquad (3)$$

Vectors V_1, V_7 and V_2 can be written in a complex number notation (Eq. 4 to Eq. 6)

$$\vec{V}_1 = \frac{1}{3} U_{DC} \qquad (4)$$

$$\vec{V}_2 = \frac{1}{3} U_{DC} e^{j\pi/3} \qquad (5)$$

$$\vec{V}_7 = \frac{\sqrt{3}}{3} U_{DC} e^{j\pi/6} \qquad (6)$$

Reference vector itself can be written as complex number as well (Eq. 7):

$$\vec{V}_{ref} = V_{ref} e^{j\theta} \qquad (7)$$

When Euler's formula (Eq. 8) is taken into the account, the whole system can

$$e^{j\theta} = Re\{e^{j\theta}\} + jIm\{e^{j\theta}\} = \cos\theta + j\sin\theta \qquad (8)$$

be re-written and analytically solved as 3 equations with 3 unknown variables (Eq. 9 to Eq. 11)

$$Re\{\vec{V}_1 T_a + \vec{V}_7 T_b + \vec{V}_2 T_c\} = Re\{\vec{V}_{ref} T_s\} \qquad (9)$$

$$Im\{\vec{V}_1 T_a + \vec{V}_7 T_b + \vec{V}_2 T_c\} = Im\{\vec{V}_{ref} T_s\} \qquad (10)$$

$$T_a + T_b + T_c = T_s \qquad (11)$$

Finally, the calculated dwelling times has to be mapped to distinct switch states (7 segments sequence) – as shown on Fig. 13 and Fig. 14. As mentioned previously, since reference vector is in sector I, segment 2b (Fig. 12), switch state mapping will start with vector V_2 (PPU), since it is the nearest to the reference vector. V_2 (PPU) will be followed, in counter-clock direction by V_1 (PLU), followed by V_7 (PLN) and finally ending again with V_2, but now as LLN – after that the mapping is going in backwards direction using reverse sequence LLN -> PLN -> PLU -> PPU (see Fig. 14)

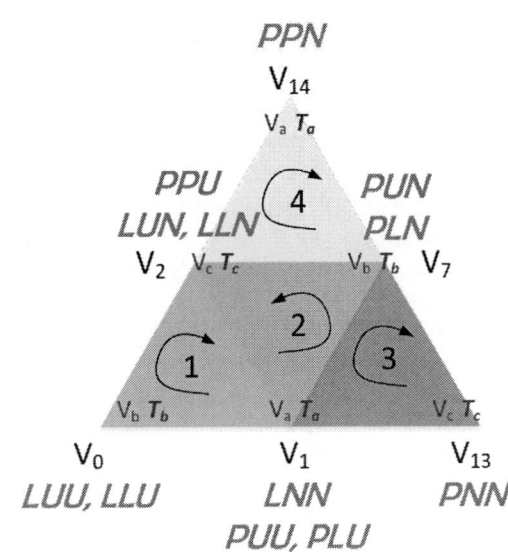

Fig. 13   States mapping for the sector I

Sector 1						
Segment	2b					

Vector	PPU	PLU	PLN	LLN	PLN	PLU	PPU
ON Time	Tc/4	Ta/2	Tb/2	Tc/2	Tb/2	Ta/2	Tc/4

Phase a	P	P	P	L	P	P	P
SW1a	1	1	1	1	1	1	1
SW2a	1	1	1	0	1	1	1
SW3a	0	0	0	1	0	0	0
SW4a	0	0	0	0	0	0	0
SW5a	0	0	0	0	0	0	0
SW6a	1	1	1	1	1	1	1

Phase b	P	L	L	L	L	L	P
SW1b	1	1	1	1	1	1	1
SW2b	1	0	0	0	0	0	1
SW3b	0	1	1	1	1	1	0
SW4b	0	0	0	0	0	0	0
SW5b	0	0	0	0	0	0	0
SW6b	1	1	1	1	1	1	1

Phase c	U	U	N	N	N	U	U
SW1c	0	0	0	0	0	0	0
SW2c	1	1	0	0	0	1	1
SW3c	0	0	1	1	1	0	0
SW4c	1	1	1	1	1	1	1
SW5c	1	1	1	1	1	1	1
SW6c	0	0	0	0	0	0	0

Fig. 14   States mapping sector I, segment 2b

Notice a few state mapping rules:

- only allowed transitions are between P and L, and between N and U;

- during each transition only one phase can changed the state;

- small vector used on the beginning (and end) of the sequence changed the polarity at the middle of sequence (PPU -> LLN -> PPU);

- always start with "P"- types of vectors (V_{1P}, V_{2P}, ... V_{6P}) as shown on Fig. 9;

Resulting PWM for each phase for previous example are shown on Fig. 15 (always have on mind that SW1 and SW4, SW2 and SW3, and SW5 and SW6 are complementary pairs)

$$V_{ref,max} = \frac{U_{dc}}{\sqrt{3}} \tag{13}$$

$$m_a = \frac{V_{ref}}{\left(\frac{U_{dc}}{\sqrt{3}}\right)} \tag{14}$$

$$m_n = \sqrt{3}m_a \tag{15}$$

$$0 \leq m_n \leq \sqrt{3} \tag{16}$$

$$0 \leq m_a \leq 1 \tag{17}$$

$$m_1 = m_n\left(\cos\theta - \frac{\sin\theta}{\sqrt{3}}\right) \tag{18}$$

$$m_2 = \frac{2}{\sqrt{3}}m_n \sin\theta \tag{19}$$

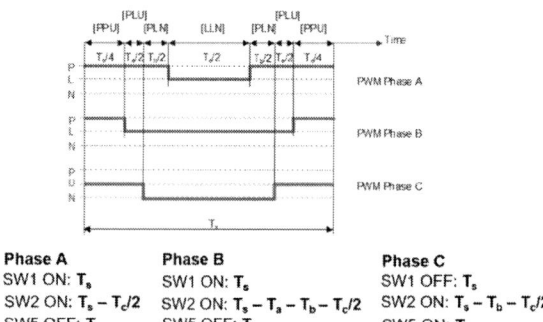

Fig. 15 Resulting PWM signals for sector I, segment 2b

Fig. 16 Dwell time calculations for per segments

2.3 Algorithm for Sector and Segment Determination and Dwell Time Calculation

Algorithm for determining sector and segment of reference vector, as well as formula for calculation of dwell time are explained in [4]. This procedure can be shortly described as following:

- Determine the sector and map reference vector angle to first sector ($0° \leq \theta \leq 60°$);

- Calculate auxiliary variables m_1 and m_2 (Eq. 12 to Eq. 19);

- Based on calculated auxiliary variables m_1 and m_2 calculate T_a, T_b and T_c timings as shown in Fig. 16;

- Once the dwell times are calculated, they can be easily re-mapped to every sector and segment, as shown on Fig. 17, Fig. 18 or Fig. 19.

$$m_a = \frac{V_{ref}}{V_{ref,max}} \tag{12}$$

Sector 1				Sector 2		
Segment	T_a	T_b	T_c	Segment	T_a	T_b	T_c
1	$\vec{V}_1$	$\vec{V}_0$	$\vec{V}_2$	1	$\vec{V}_2$	$\vec{V}_0$	$\vec{V}_3$
2	$\vec{V}_1$	$\vec{V}_7$	$\vec{V}_2$	2	$\vec{V}_2$	$\vec{V}_8$	$\vec{V}_3$
3	$\vec{V}_1$	$\vec{V}_7$	$\vec{V}_{13}$	3	$\vec{V}_2$	$\vec{V}_8$	$\vec{V}_{14}$
4	$\vec{V}_{14}$	$\vec{V}_7$	$\vec{V}_2$	4	$\vec{V}_{15}$	$\vec{V}_8$	$\vec{V}_3$

Fig. 17 Mapping of dwell times on 1st/2nd sector

Sector 3				Sector 4		
Segment	T_a	T_b	T_c	Segment	T_a	T_b	T_c
1	$\vec{V}_3$	$\vec{V}_0$	$\vec{V}_4$	1	$\vec{V}_4$	$\vec{V}_0$	$\vec{V}_5$
2	$\vec{V}_3$	$\vec{V}_9$	$\vec{V}_4$	2	$\vec{V}_4$	$\vec{V}_{10}$	$\vec{V}_5$
3	$\vec{V}_3$	$\vec{V}_9$	$\vec{V}_{15}$	3	$\vec{V}_4$	$\vec{V}_{10}$	$\vec{V}_{16}$
4	$\vec{V}_{16}$	$\vec{V}_9$	$\vec{V}_4$	4	$\vec{V}_{17}$	$\vec{V}_{10}$	$\vec{V}_5$

Fig. 18 Mapping of dwell times on 3rd/4th sector

Sector 5			
Segment	T_a	T_b	T_c
1	$\vec{V}_5$	$\vec{V}_0$	$\vec{V}_6$
2	$\vec{V}_5$	$\vec{V}_{11}$	$\vec{V}_6$
3	$\vec{V}_5$	$\vec{V}_{11}$	$\vec{V}_{17}$
4	$\vec{V}_{18}$	$\vec{V}_{11}$	$\vec{V}_6$

Sector 6			
Segment	T_a	T_b	T_c
1	$\vec{V}_6$	$\vec{V}_0$	$\vec{V}_1$
2	$\vec{V}_6$	$\vec{V}_{12}$	$\vec{V}_1$
3	$\vec{V}_6$	$\vec{V}_{12}$	$\vec{V}_{18}$
4	$\vec{V}_{13}$	$\vec{V}_{12}$	$\vec{V}_1$

Fig. 19 Mapping of dwell times on 5th/6th sector

Finally, the seven states decomposition starting with "P"-Type vector for sector 1 is given on Fig. 20. The same tables can be easily derived by using the rules described in previous section for all other sectors as well.

1a		1b		2a		2b		3		4	
Ta (V1)	PUU	Tc (V2)	PPU	Ta (V1)	PUU	Tc (V2)	PPU	Ta (V1)	PUU	Tc (V2)	PPU
Tb (V0)	LUU	Ta (V1)	PLU	Tb (V7)	PUN	Ta (V1)	PLU	Tb (V7)	PUN	Ta (V14)	PPN
Tc (V2)	LUN	Tb (V0)	LLU	Tc (V2)	LUN	Tb (V0)	PLN	Tc (V13)	PNN	Tb (V7)	PLN
Ta (V1)	LNN	Tc (V2)	LLN	Ta (V1)	LNN	Tc (V2)	LLN	Ta (V1)	LNN	Tc (V2)	LLN
Tc (V2)	LUN	Tb (V0)	LLU	Tc (V2)	LUN	Tb (V0)	PLN	Tc (V13)	PNN	Tb (V7)	PLN
Tb (V0)	LUU	Ta (V1)	PLU	Tb (V7)	PUN	Ta (V1)	PLU	Tb (V7)	PUN	Ta (V14)	PPN
Ta (V1)	PUU	Tc (V2)	PPU	Ta (V1)	PUU	Tc (V2)	PPU	Ta (V1)	PUU	Tc (V2)	PPU

Fig. 20 Seven states decomposition starting with "P"-type vector for all segments in sector I

3 Implementation of Battery Balancing with 3-level Inverter

3.1 Basic Idea for Battery Balancing

At the 3-level inverter, when being supplied with one battery (Fig. 7) depending on the type of small or medium vector ("P" or "N"-type) that is being used electrical current will be driven-in or drained-out of neutral point (V_{NP}) as shown on Fig. 21. As it can be seen (discussed in [5]), zero and large vectors are not influencing neutral point at all, whereas the small vectors and medium vectors (depending load) are influencing neutral point either by supplying it with current (V_{NP} voltage increases) or by draining the current out of neutral point (V_{NP} voltage decreases). In the case of 3-level inverter supplied by 2 batteries (Fig. 2), basically the same idea of using different types of small and medium vectors ("P"- and "N"-types) is being used to enforce more frequently usage of one or another battery (V_+ or V_-) depending on their state of the charge. In other words, if for example, battery V_+ is having higher state of the charge comparing to battery V_-, intention is to enforce longer usage of "P"-type vectors that will longer drain V_+ than V_-. Analog approach would be used when battery V_- is having higher state of charge compar-

ing to V_+ battery: in that case, longer usage of vectors of "N"-type would be imposed, to drain higher amount of current out of battery V_- than out of battery V_+.

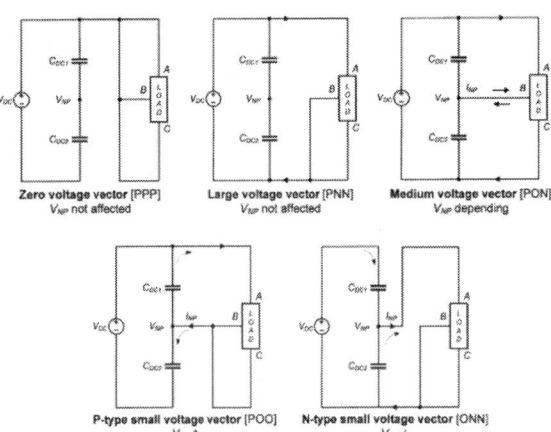

Fig. 21 Influence of different space vectors on neutral point voltage (state O can be either U or L)

3.2 Implementation of Basic Battery Balancing

Assume that the reference vector is in sector I, segment 1a (as shown on Fig. 22). Seven-segment sequence would be used: PPU -> LUU -> LUN -> LNN -> LUN -> LUU -> PPU. Notice that in balanced case (V_+ and V_- are having the same

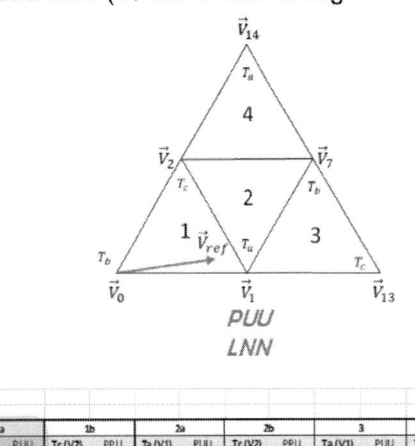

Fig. 22 Reference vector in sector I, segment 1a

state of charge) small vector V_1 is for 50% of dwell time (T_a) presented with it's "P"-type variant PUU and for 50% of dwell time (T_a) presented with it's "N"-type variant LNN. Let's assume now that V_+ battery has lower voltage (lower state of the charge) comparing to V_- battery: implementing basic idea of battery balancing means that "N"-type vectors (vectors that are draining V-) are longer used than "P"-type vectors. In this specific case ($V_- > V_+$) it would mean that dwell time of LNN is longer than 50%, whereas dwell time of PUU should be shorter than 50% (Fig. 23). Difference to the 50% of the dwell time is proportional to the amount of balancing that should be implemented.

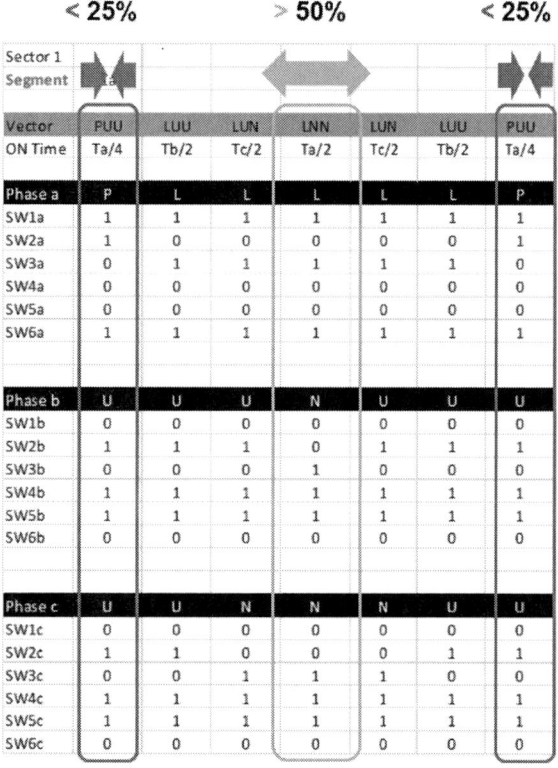

Fig. 23 Simple balancing for reference vector
in sector I, segment 1a when $V_- > V_+$

Conclusion is that for the price of the minor mismatch of resulting reference vector, simple balancing algorithm can be easily implemented.

3.3 Implementation of Advanced Battery Balancing

In the case of misbalance between V_+ and V_-, the "P"- and "N"-types variants of the same vectors are not matching anymore (they have different amplitudes), also the symmetricity and segmentation of

each sector into 4 equilateral triangles is not available anymore. Pou [6] provided a nice solution for this problem. In addition to Eq. 12 to Eq. 19, equations (Eq. 20 to Eq. 23) are introduced to describe misbalance:

$$m_{12} = 2 - m_1 - m_2 \tag{20}$$

$$\gamma_1 = \frac{V_-}{V_{DC}/2} \tag{21}$$

$$\gamma_2 = \frac{V_+}{V_{DC}/2} \tag{22}$$

$$\gamma_1 + \gamma_2 = \frac{V_+ + V_-}{V_{DC}/2} = \frac{V_{DC}}{V_{DC}/2} \tag{23}$$

With those additional equations, each sector now can be segmented into 12 segments (instead of "just" 4 segments as it was in the case of balanced batteries). Example for new segmentation for previous segment 4 is shown on Fig. 24. – it can be

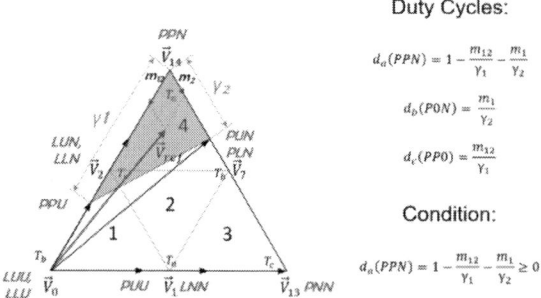

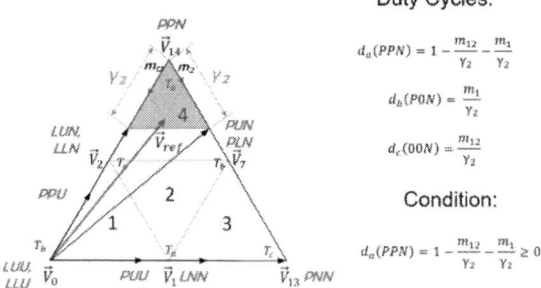

Fig. 24 Segmentation of Segment 4 in the case of battery misbalance

seen that previous segment 4 can be segmented on 2 different ways, depending on which small vector "P"-type (PPU) or "N"-type (LLN or LUN) is used. Calculations to determine if reference vector lies within one segment are done by calculating barycentric coordinates of each reference vector with respect to each segment. If sum of barycentric coordinates of reference vector equals to 1, reference vector is lying within segment – if sum of barycentric coordinates [7] of reference vector is greater than 1, reference vector is outside of given segment. Additional complexity on assigning reference vectors to the segments is due to fact that one reference vector might be assigned to more than one segment – in that case some additional criterions, like measure of misbalance should be introduced to assign reference vector to right segment. As a conclusion on implementation of advanced battery balancing, it is fair to say that for the price of high CPU calculation power (due to high complexity of the algorithm) very precise method to control draining/sinking of the battery with different state of the charge can be implemented.

3.4 Practical results and measurements

Both battery-balancing methods are implemented on physical hardware that was developed by MAZDA. SVM, together with Field-Oriented Control (FOC) which is driving Permanent Magnets Synchronous Motor (PMSM) are implemented in MathWorks' MATLAB/Simulink environment are being designed for automated code generation. Code is running on Infineon's microcontroller AURIX TC38A.

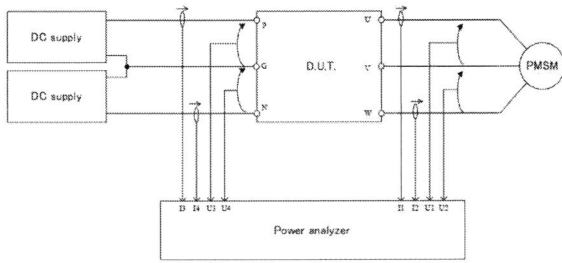

Fig. 26   Block diagram of measurements setup

The measurement setup is shown on Fig. 26 and measurement results are presented on Fig. 27 and 28. As it can be seen, due to high battery misbalance ration (56 V/ 40 V), in the case of non-balancing and basic balancing algorithm, the current waveform are showing, as expected, high distortion. With advanced control algorithm the current

waveform are showing a nice sinusoidal waveform. In cases of both, basic and advanced balancing algorithms, the currents are stronger pulled-out from battery that is having higher state of charge (higher voltage), supporting on this way balancing.

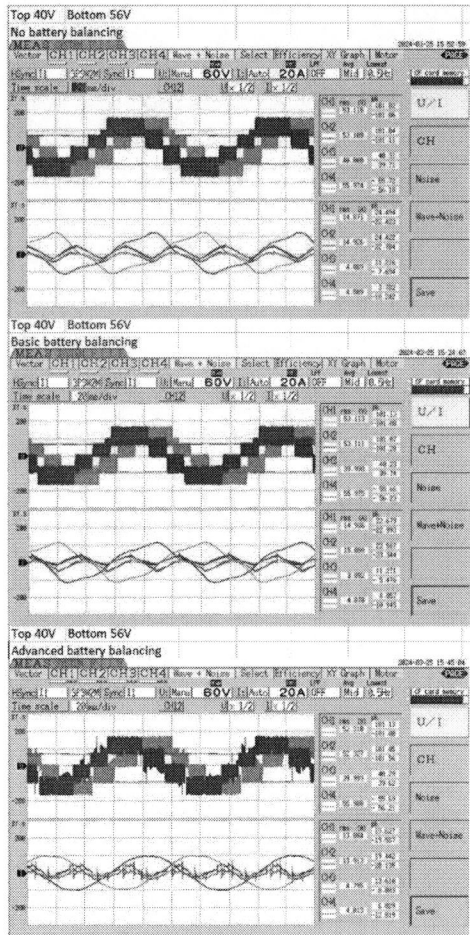

Fig. 27   Measurements for V_+ = 40 V, V_- = 56 V using non-balancing (top picture), basic balancing (middle picture) and advanced balancing (bottom picture) algorithm

3.5 Conclusions

3-level ANPC inverter supplied by two batteries (+/- 48 V) is implemented by MAZDA in both variants, first with discrete MOSFETs, using Infineon's AEC qualified 100 V IAUT300N10S5N015 in OptiMOS™-5 technology, and in another variant, with chip-embedding Smart p²Pack technology from

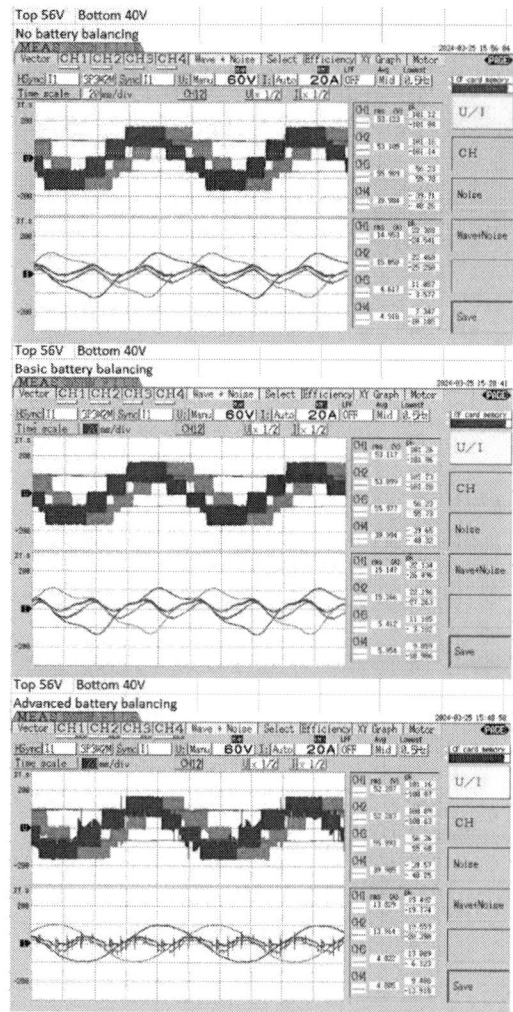

Fig. 28 Measurements for V_+ = 56 V, V_- = 40 V for non-balancing (top picture), basic balancing (middle picture) and advanced balancing (bottom picture) algorithm

Schweizer [8], using 80 V IAUE303N08S5N009 (OptiMOS™-5). Control algorithm for basic operating of 3-level inverter, as well as both (basic and advanced) variants of battery balancing are implemented on Infineon's microcontroller AURIX TC387 by usage of MathWorks' MATLAB/Simulink based MBD and automated code generation. Basic operation mode of 3-level inverter, as well as 2 different battery balancing operation modes are sucefully demonstrated and it is proven that additional hardware to perform battery balancing is not needed - on this way, overall system costs are reduced and system efficiency is increased.

References

[1] März, M. and Nance, P., "Thermal Modeling of Power-electronic Systems," February 2000. Available online at www.infineon.com/dgdl/Thermal+Modeling.pdf?fileId=db3a30431441fb5d011472fd33c70aa3..in 11 pt type size and 12 pt line spacing.

[2] Vuletic, R. and Hyde, R., John, D., "Infineon Buck Simscape Example," MathWorks File Exchange, February 2022. Available online at https://de.mathworks.com/matlabcentral/fileexchange/106925-infineon-buck-simscape-example.

[3] A. Bendre, S. Krstic, J. Vander Meer, G. Vekataramanan, "Comparative Evaluation of Modulation Algorithms for Neutral-Point-Clamped Converters," IEEE Transactions On Industry Applications, Vol. 41, No. 2, March/April 2005.

[4] B.Wu, "High-Power Converters and AC Drives," Hoboken, NJ, USA: Wiley, pp. 143-176, 2006.

[5] N. Celanovic and D. Boroyevich, "A comprehensive study of neutral-point voltage balancing problem in three-level neutral-point-clamped voltage source PWM inverters," IEEE Trans. Power Electron., vol. 15, no. 2, pp. 242–249, Mar. 2000.

[6] Pou, J.F., "Modulation and Control of Three phase Pwm Multilevel Converters," PhD thesis, Universitat Politècnica de Catalunya. Departament d'Enginyeria Electrònica, 2002. Available online at http://hdl.handle.net/10803/6327

[7] P. Szczepankowski and J. Nieznański, "Application of Barycentric Coordinates in Space Vector PWM Computations," in *IEEE Access*, vol. 7, pp. 91499-91508, 2019.

[8] D. J. Kearney, S. Kicin, E. Bianda and A. Krivda, "PCB Embedded Semiconductors for Low-Voltage Power Electronic Applications," in IEEE Transactions on Components, Packaging and Manufacturing Technology, vol. 7, no. 3, pp. 387-395, March 2017, doi: 10.1109/TCPMT.2017.2651646.

PCIM Europe 2024, 11– 13 June 2024, Nuremberg   DOI: 10.30420/566262038

An Innovative 3-level Solution for Automotive Applications: eMPack

Pranav Panchal[1], Arendt Wintrich[1], Andreas Wohlfart[1], Oliver Tamm[1], Ole Muehlfeld[1]

[1]Semikron-Danfoss, Nuremberg, Flensburg, Germany

Corresponding author: Pranav Panchal, pranav.panchal@semikron-danfoss.com
Speaker: Pranav Panchal, pranav.panchal@semikron-danfoss.com

Abstract

The automotive industry is undergoing a transformative shift towards electric and hybrid propulsion systems to meet stringent emissions regulations and enhance overall vehicle efficiency. Three-level inverters have emerged as a crucial component in these electrified drivetrains, offering advantages such as reduced harmonic distortion, improved system efficiency, and enhanced control capabilities. Three-level inverters offer distinct advantages over traditional two-level inverters in terms of reducing switching losses and harmonic distortion.

While three-level topologies offer significant advantages for electric vehicles, for a holistic comparison it's important to investigate their implementation with regards to system costs and complexity compared to traditional two-level inverters. This paper will focus on the solution which can use the best of both 2-level and 3-level topologies with a single power module to enhance the system efficiency while avoiding the higher cost of materials and implementation.

1   Introduction

Next-generation automotive drivetrains are expected to meet challenging requirements to address evolving driving experience, environmental protection, and regulatory standards. While specific requirements can vary depending on the type of drivetrain (electric, hybrid, internal combustion, etc.) and the intended application (e.g., passenger cars, trucks, or commercial vehicles), there are few important requirements which stands out to evolve the entire electric drivetrains system such as improving energy efficiency to maximize the use of available power, extended driving range and improved battery life durability including faster charging. This includes enhanced acceleration, torque, efficient power distribution for better handling and stability.

The drivetrains should be designed to withstand a wide range of driving profile and harsh environmental conditions. As technology continues to advance, the automotive industry will continue to refine and adapt drivetrain designs to meet these evolving requirements.

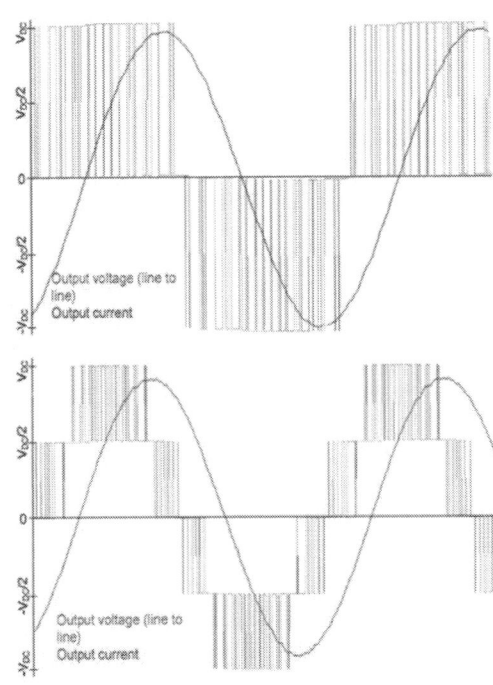

Fig. 1   2-lvl and 3-lvl voltage graph

One of the most important part of the electric drivetrain system is the inverter which has evolved over period of years from changing Si to SiC and IGBT to MOSFETs in most of the state-of-the-art hybrid/ electric passenger and commercial vehicles. SiC MOSFET semiconductor devices show significant lower static losses over the drive cycles due to their MOSFET forwarding characteristics and have lower switching losses compared to IGBT/FWD based semiconductors. This results in higher overall efficiency for the inverter. SiC devices also allow higher power density in the same form factor due to reduced losses and higher maximum junction temperatures. Reduced switching losses enables higher-frequency operation, which is beneficial for achieving smoother and more precise motor control.

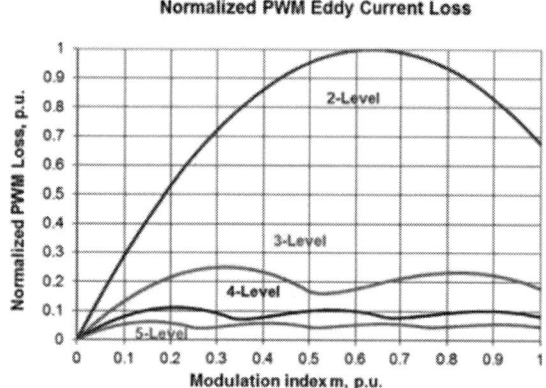

Fig. 2   Normalized integral PWM iron loss for 2, 3, 4, and 5 voltage levels [3]

Three level inverters gain popularity in automotive applications because of further increased motor efficiency and therefore extended driving range. Lower harmonic distortion due to the 3-level switching scheme (Fig. 1) leads to significant reduction of iron losses in the motor and THD which will improve the efficiency of the drivetrain system, resulting in improved vehicle performance and energy efficiency [1], [2]. The optimization criteria in automotive applications are different as for solar or energy storage application because the inverter operates with less than 20% of the rated power most of the time as shown in Fig. 10.

2   Losses in Electric Drivetrain

In an electric drivetrain, the most dominant loss typically varies depending on the specific components and the operating conditions. The inverter losses related to power electronics which include conduction, switching and reverse recovery losses are considered when designing the inverters and gate drive control board. There are other losses which are generated within the electric motor that contribute to the overall performance and efficiency of the electric drivetrain. These losses include mechanical losses, copper losses (I^2R losses), Iron core losses including THD induced losses.

To increase the overall efficiency and milage of the EVs, we must find solution to reduce the losses induced in the electric motor. One of the main contributors is the core losses that occur in the iron core of the motor's stator and rotor. These losses are primarily associated with the magnetic properties of the iron core and can be categorized into two main types i.e., hysteresis loss and eddy current loss.

2.1   Hysteresis Loss

Hysteresis loss is the result of the repeated magnetization and demagnetization of the iron core as the alternating current (AC) passes through the motor windings. The iron core has a certain level of retentivity, which means it retains some magnetization even after the magnetic field has reversed. As the magnetization continuously changes direction with the alternating current, energy is lost in the form of heat due to the hysteresis loop that represents this energy loss.

$$P_h = k_h f B_{max}^n \tau^2 v \qquad (1)$$

Where k_h is the coefficient of hysteresis of the ferromagnetic material, f is the frequency of power supply in hertz, B_{max} is the maximum flux density in weber/m^2 and volume of the ferromagnetic material in m^3.

2.2   Eddy Current Loss

Eddy current loss occurs due to the circulation of induced currents in the iron core because of the changing magnetic field. The iron core, being a good conductor, experiences eddy currents, which generate resistive losses in the form of heat. These currents can be minimized by laminating the iron core, with insulating material between the laminations, to reduce the closed-loop paths for the eddy currents.

$$P_e = k_e f^2 B_{max}^2 \tau^2 V \qquad (2)$$

Where, K_e is the eddy current constant, B_{max} is the maximum flux density, f is the frequency of the induced voltage including harmonics, V is the volume of the material.

2.3 Total Harmonic Distortion

Total Harmonic Distortion (THD) refers to the distortion of a waveform from its ideal sinusoidal shape due to the presence of harmonics. It is a critical parameter that effects the performance and reliability of electric motor in the drivetrain leading to various issues such as increased losses, reduced efficiency, and electromagnetic interference (EMI).

$$THD = \frac{V_{RMS\_without\_fundamental}}{V_{RMS\_fundamental}} \quad (3)$$

The major contributing factor for THD generation in motor current and voltage waveforms are from non-linearities in the motor's operation such as the saturation of magnetic materials, non-sinusoidal back electromagnetic force (EMF), and non-linear behavior of power modules used in the inverter of electric motors. This can be evaluated by following equations:

$$THD_I = \frac{1}{I_m} \sqrt{\sum_{n=2}^{\infty} I_n^2} \quad (4)$$

$$THD_V = \frac{1}{V_m} \sqrt{\sum_{n=2}^{\infty} V_n^2} \quad (5)$$

Where, I_m and I_n are the fundamental magnitude and the nth harmonic magnitude of phase current output and V_m and V_n are the fundamental magnitude and the nth harmonic magnitude of line-voltage output, respectively.

At low power, THD rises because the ripple current of PWM frequency stays almost constant at reduced current with fundamental frequency. The ripple current is determined by the applied voltage, inductance, and pulse length. It is possible to reduce V which leads to 3-lvl solution with $V_{dc}/2$ or to reduce dt with a higher switching frequency. Another way is by increasing L, which means more windings with a consequence of higher copper losses with thinner wire diameter or larger motor.

$$\Delta I = \frac{V}{L} dt \quad (6)$$

There are several ways to mitigate the THD in electric motor by implementing passive or active filters to suppress the harmonic components in the output waveforms, using advanced control algorithms such as predictive control or space vector modulation (SVPWM) and optimizing PWM control parameters, such as switching frequency and modulation index [5], [7].

3 Design of 3-level Power Module

Semikron-Danfoss has developed an advanced automotive module to which can live up to the harshest traction application environments, meeting the demands of the current and future drivetrain requirements. This state-of-the-art module has two variants showcasing different topologies i.e., traditional 2-level (Fig. 3a) and emerging 3-level (Fig. 3b) configuration in the same form factor.

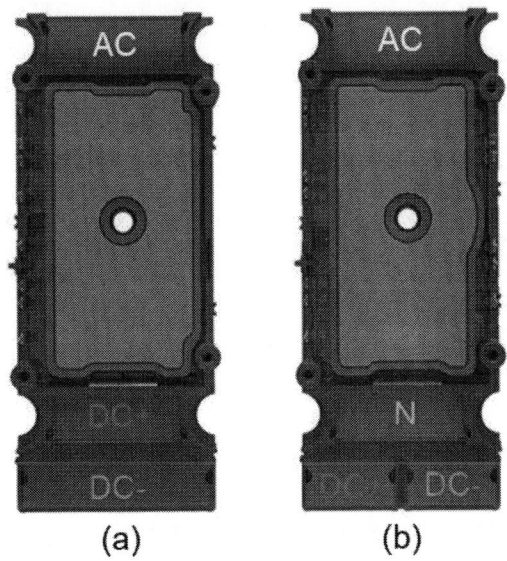

Fig. 3   (a) 2-lvl module (b) 3-lvl module

The outer dimensions are same for both the variants though there are marginal changes in the aux pins layout and the splitting of the main DC terminals as 3-level has additional neutral point voltage level. Many 3-level topologies are available with their respective advantages and disadvantages. TNPC (T-type Neutral Point Clamped) is a preferred topology for using MOSFET with regards to the implemented semiconductor area per Amp output current. The 3-level topology has 4 switches per half-bridge as compared to the 2 switches in 2-level (Fig. 4).

The 3-level automotive power module i.e., eMPack® uses TNPC topology as it has "short" commutation path compared to NPC configuration. A "longer" commutation path leads to higher inductance which produces very high voltage overshoot and hence contributing to reduced output power [4]. Another advantage TNPC topology offers is that it does not need additional diode as in the case with NPC configuration. This allows extra space to include more chips in the same package. Furthermore, the TNPC topology can be arranged using common Drain or common Source at the Neutral voltage level i.e., T2/T3 of Fig. 4(b).

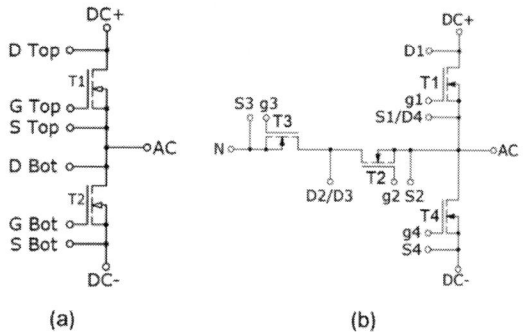

(a)   (b)

Fig. 4   (a) 2-level and (b) 3-level topology

In [6] and [8], studies have discussed on asymmetric T-type topology on synchronous reluctance machine to improve efficiency, reliability and cost during the low voltage operating point of the electric vehicle. The packaging technology and the flexible routing of the current path within the power module which is discussed below enables to realize eMPack® with asymmetric T-type topology.

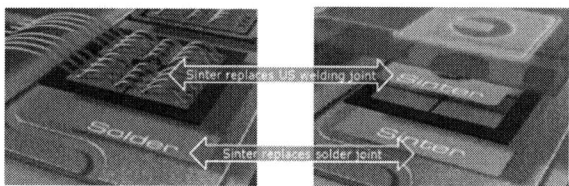

Fig. 5   Solder/ Wire-bond vs. Sinter/ Flex foil

The bottleneck of the automotive power modules is its bond-wire technology which uses ultrasonic welding process to solder current carrying Al wires in between of the chips. Lifetime reliability tests shows that the weakest part of the power module is the connection of Al wire-bond to the chip. This has been improved by replacing the wire-bonds with flex foil using DSS (Double Sided Sintering) process (Fig. 5). The chip will be sintered on the bottom side to the DCB as well as from the top side to the flex foil carrying the current. This gives us more routing possibilities as compared to wire-

bonds in terms of flexibility in current routing and chip positions.

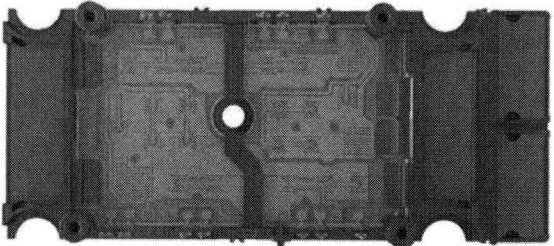

Fig. 6   eMPack® realization with Flex layer

Figure 6 shows the module with the flex-layer implemented using sintering process. The advantage of using the Flex foil technology allows us to design the internal chip layout with the same form factor as 2 level modules to design 3 level power modules. Furthermore, the flex-foil is providing the auxiliary connection, saving space on AMB substrate for adding more chips in the same form factor. The flex-foil based packaging technology is perfect for complex circuits. Three conductive layers i.e., substrate and both sides of flex-foil offer efficient current distribution. Cross-sectional schematic of the eMPack® can be seen in Fig. 7.

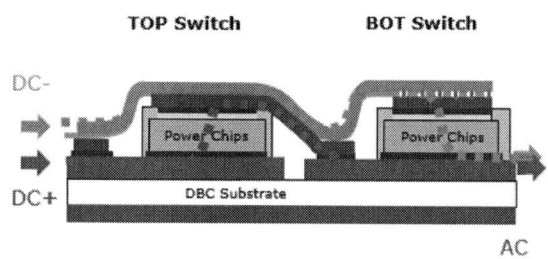

Fig. 7   Schematic cross-section of eMPack®

The flex layer allows to reuse the same packaging technology with very high package density which leads to 30% less rated current per switch in 3-level designs for the same inverter output power. Using flex-foil reduces the stray inductance which needs to be kept in mind during design phase.

The development of automotive power module improved further to increase the thermal performance of the thermal stack by introducing DPD (Direct Pressed Die). The DPD system presses directly on top of the chips with the help of the flex foil and the pressure element leading to the reduction in TIM layer thickness and smooths out underlying cavities to ensure low R_{th} and increased thermal efficiency. The DPD shows excellent thermal performance equivalent to that of a module sintered to the heatsink. Figure 8 shows the ceramic substrate (AMB) with sintered chips and a flex layer using DSS process. As the module is pressed down by DPD system, there is no rigid connection between substrate and heat-sink leading to lower thermo-mechanical stress on the substrate.

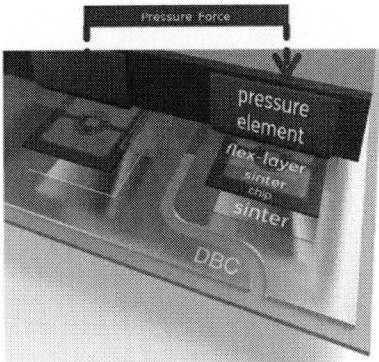

Fig. 8   DPD method to improve thermal efficiency

The complete mechanical assembly of the eMPack® is shown in Fig. 9. To ensure the homogeneous pressure distribution, the DPD system including plate and spring are screwed down using a single screw from top of the module to all way till heat-sink. Rivets are used to align and fix the 3-phase assembly to the heat-sink on the edges as shown in the Fig. 9. The rivets are also necessary for screwing the module to the customer interface in the inverter. The assembly is covered on top by the flame resistant, high temperature polyamides to provide mechanical stability and additionally to protect it from outer contamination.

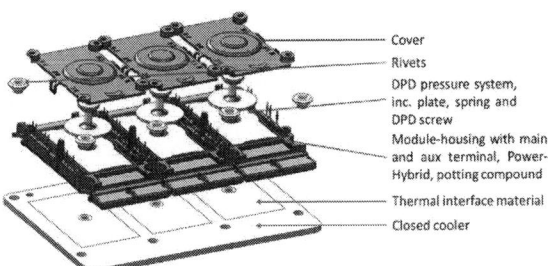

Fig. 9   Explosion view of eMPack®

The minimal change in the eMPack® package design of 3-level as compared to 2-level is an advantageous and economic solution for manufacturer because of least change in the production line. Since the form factor of 2 and 3-level module is same (Fig. 3), potential customers do not have to change the mechanical setup to incorporate 3-level module in their drivetrain system.

4 Simulation and Results

The simulation is carried out by the in-house developed simulation tool known as SemiSel™. This online tool uses a newly developed hybrid simulation engine which can simulate different mission profiles providing superior computation speed and simulation accuracy over a wide range of output frequencies including 0 Hz [9]. It is an optimum tool to estimate inverter losses and junction temperature of power semiconductors. The working principle of the hybrid simulation engine is described in detail in [9] and the tool is available to use at [10].

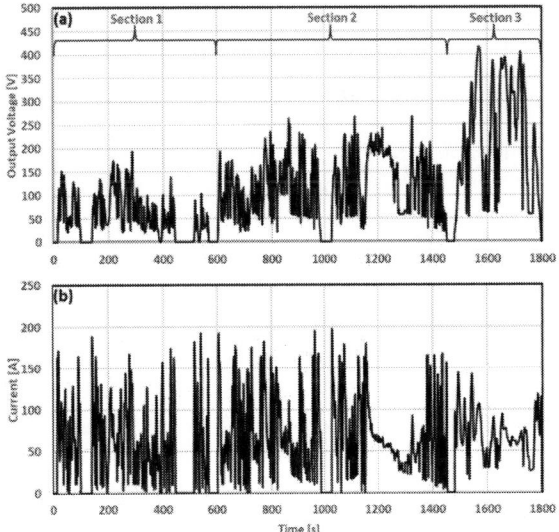

Fig. 10   A typical WLTP cycle; (a) Output voltage (with section 1, 2, 3), (b) Current.

The device model which is used in SemiSel™ to simulate the losses and junction temperature is eMPack® 1200V, 780A Gen3 SiC MOSFET (2-level) and 1200V-750V, 800A Gen3 SiC MOSFET (TNPC 3-level). The static, dynamic characteristics and thermal impedance were measured prior to the model generation in the SemiSel™. The thermal model consists of copper pin-fin heat-sink with water-glycol cooling medium at 65°C and 10 l/min with R_{thsa} = 0.016 K/W and the thermal resistances of the chips from datasheet ($R_{thjs,T1/4}$ = 0.0987 K/W and $R_{thjs,T2/3}$ = 0.129 K/W).

The approach used here to show the efficiency gain in 3-level configuration as compared to 2-level is to simulate a standard WLTP cycle shown in Fig. 10. The cycle is divided in 3 sections based on different average speeds i.e., low (section 1), medium (section 2) and high (section 3) as shown in Fig. 10(a). For the simulation, we have used a hypothetical motor with 160kW (300A) nominal power and 300kW (550A) peak power. The average power in the WLTP is only 15kW. The simulation results provide the conduction and switching losses over the entire mission profile. Figure 11 shows the loss distribution of each of the semiconductor chip per inverter. The underlined advantage of 3-lvl TNPC topology is the reduction in switching losses because the commutation voltage of the 1200V modules is only $V_{dc}/2$ as compared to V_{dc} for 2-level [11], [12]. The results shown in Fig. 11 verify that the switching losses are reduced to 80% with 3-level configuration as compared to 2-level.

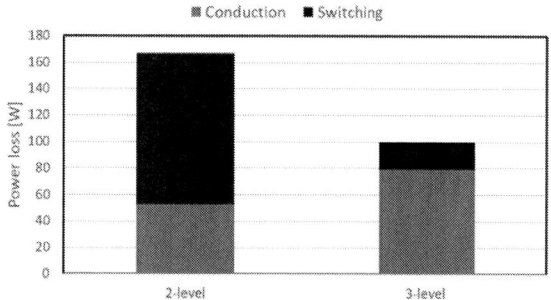

Fig. 11 Average loss distribution for individual chips during WLTP cycle at fs=8 kHz

The losses are computed into the total energy consumption of the inverter. The inverter efficiency is calculated by dividing the total inverter power by the inverter losses. The efficiency shown by 2 level topology is 98.9 % whereas for 3-level is 99.34%. The efficiency gain at the inverter level is at 0.44 %, which is a nice side-effect but not the main reason for the higher effort to change to 3-level topology.

Section	1	2	3
Time [s]	<600	600-1450	1450-1800
Avg. modulation index	0.25	0.38	0.75
PWM losses for 2-lvl	0.61	0.8	0.96
PWM losses for 3-lvl	0.23	0.25	0.22
Iron loss reduction	62%	69%	77%

Table 1 Loss reduction of WLTP cycle according to Fig. 2 and Fig. 10

In [13], the study shows that the ratio of motor loss in the total loss of the drivetrain system is higher than that of the inverter losses. At nominal power, the dominant loss in the motor losses is copper

losses followed by eddy-current losses and hysteresis losses. For lower output power range or higher switching frequencies (f_s > 10 kHz), eddy-current losses are major contributors to motor losses. In [11], it has been experimentally proven that the 3-level modulated output voltage reduces the harmonic losses by a factor of four, leading to lower heating and degradation in the electric drivetrain. The reduction in THD by using 3-level over 2-level topology is discussed and proved in detail in [11], [14], [15].

At this low power level in the WLTP cycle i.e., 15kW/160kW = 9.4% is the efficiency of the electric motor less than 80%. Means 3kW of the 15kW are dissipated in the motor. To show the effect of the 3-level inverter on the total losses, we simplify the WLTP cycle and make some general assumptions for the motor efficiency. For precise analysis, simulation with a motor model with complete drivetrain would be necessary. In our example, we cluster the WLTP in 3 sections. Low speed, from 0 -600 s, medium to high speed, from 600 -1500 s and extra-high speed, from 1500 -1800 s.

In Table 1, the averaged the modulation index for the sections 1-3 is shown together with the calculated reduction of iron losses for 3-lvl inverter topology compared to the 2-lvl according to the Fig. 2. For example, in section 1, the modulation index is 0.25 which means that for 2-lvl, the normalized losses are 61% and for 3-lvl are 23%. This means that the saved power dissipation is 1-(23%/61%) equals to 62%.

Section	1	2	3
Avg. power WLTP [kW]	7	16.3	25.8
Assumed motor efficiency	70%	75%	80%
Total motor losses [kW]	2.10	4.08	5.16
Iron losses [kW]	0.84	1.63	2.06
Saved losses by 3-lvl [kW]	0.52	1.12	1.59

Table 2 Efficiency calculation of 2- and 3-level drivetrain

For the loss calculation in Table 2, we have assumed a typical motor efficiency for different average power in the range of 70% at 8% nominal power (7kW) to 80% at 15.6% nominal power (25.8kW). Furthermore, we have assumed that the iron losses are 40% of the total motor losses. For example, considering section 2 with 16.3 kW in the WLTP and the motor efficiency of 75% results in 4.08 kW of total motor loss. Assuming iron losses to be 40% of motor losses, leading to 1.63 kW of iron losses. With a 3-level circuit, we can save 69% (Table 1) means about 1.1 kW.

The energy taken from the battery is 15kW for a drive cycle in half an hour is 7.5kWh. The saved energy in the 3 sections sums up to 0.51kWh which is 6.8% of the total energy. This leads to driving range extension of a given battery charge in the same range. This is only a rough calculation, but it should show the potential of the 3-level topology.

5   Conclusion

This paper discusses the advantages of 3-level configuration over 2-level in traction applications. The realization of the newly developed automotive power module package i.e., eMPack® is also described in detail, applying advanced packaging technologies of Semikron-Danfoss.

The package offers the advantages of compact geometry enabling high power density inverter designs while the modular design concept is usable for 2 and 3 level designs in the same enclosure. The paper investigated the t-type 3 level technology using 1200V SiC MOSFET as switches in a 300kWp application.

Simulations have shown significant efficiency gains in the WLTP cycle. For the given application energy savings of 6.8% were calculated in one cycle using the 3-level vs 2-level module design, mainly driven through the reduction in the motor losses. The gain in efficiency is very large at lower motor power range due to reduced ripple current and iron losses.

References

[1] L. Masisi, M. Ibrahim, P. Pillay, *The Effect of the Three Level Neutral Point Clamped (NPC) Inverter on the Core Loss of a Synchronous Reluctant Machine (SynRM)*, IEEE, 2015.

[2] P. Rasilo, A. Salem, A. Abdallh, F. De Belie, L. Dupre, *Effect of Multilevel Inverter Supply on Core Losses in Magnetic Materials and Electrical Machines*, IEEE Transactions on Energy Conversion, 2015.

[3] A. Ruderman, *Effect of Multilevel Inverter Supply on Core Losses in Magnetic Materials and Electrical Machines*, IEEE Transactions on Energy Conversion, 2015.

[4] I. Staudt, A. Wintrich, *Advantages and Disadvantages of 3L NPC and 3L TNPC Topology*, PCIM Asia, 2012.

[5] A. K. Iyer, M. Otten, J. Goudswaard, *Mitigation of PWM-Induced Losses in Electric Machines Using Multi-Level Converters*, PCIM Europe, 2023.

[6] N. X. Doan, N. V. Nguyen, *Virtual Space Vector Pulse Width Modulation for Asymmetric T-Type Neutral Point Clamped 3-Level Inverter*, Mathematical Problems in Engineering, Vol. 2021.

[7] A. Madan, E. Bostanci, *Comparison of Two-Level and Three-Level NPC Inverter Topologies for a PMSM Drive for Electric Vehicle Applications*, IEEE, 2019.

[8] P. Azer, J. Bauman, *An Asymmetric Three-Level T-Type Converter for Switched Reluctance Motor Drives in Hybrid Electric Vehicles*, IEEE, 2019.

[9] M. Roeblitz, C. Schmidt, A. Wintrich, *High Speed Hybrid Simulation Engine for Electrical Mission Profiles*, PCIM 2021.

[10] SemiSel5, available at https://semisel.semikron.com

[11] M. Schweizer, T. Friedli, J.W. Kolar, *Comparative Evaluation of Advanced Three-Phase Three-Level Inverter/Converter Topologies Against Two-Level Systems*, IEEE Transactions on Industrial Electronics, 2013.

[12] A. Choudhury, P. Pillay, M. Amar, S.S. Williamson, *Performance Comparison study of Two and Three Level Inverter for Electric Vehicle Application*, IEEE 2014.

[13] D. Sato, J. Itoh, *Total Loss Comparison of Inverter Circuit Topologies with Interior Permanent Magnet Synchronous Motor Drive System*, IEEE ECCE Asia, 2013.

[14] R. Mecke, *Energy efficiency of two-level and multilevel inverters - a drive system comparison*, EPE 2015.

[15] N. Soualhi, A. Makouf, N. Nait-Said, S. Hamada, *Comparison between a Two-Level and Three-Level Inverter fed Induction Motor including Losses and Efficiency*, IC-ASET 2020.

PCIM Europe 2024, 11– 13 June 2024, Nuremberg   DOI: 10.30420/566262039

Gated Recurrent Units-Assisted State-Space Modeling for Electric Vehicle Temperature Prediction

Xinyuan Liao [1], Shaowei Chen [1], Shuai Zhao [2]

[1] School of Electronics and Information, Northwestern Polytechnical University, China
[2] AAU Energy, Aalborg University, Denmark

Corresponding author: Shuai Zhao, szh@energy.aau.dk
Speaker: Xinyuan Liao, liaoxinyuan@mail.nwpu.edu.cn

Abstract

The electric system (electric motor, electric control, and battery pack) is the most critical component of e-mobility. Effective thermal management of the electric system is crucial for enhancing the range, reliability, and performance of e-mobility. Accurate prediction of the electric system temperature forms the foundation of effective thermal management. This paper presented a novel gated state-space model (GSSM) that utilized the gated recurrent unit (GRU) to assist the state-space model (SSM) in predicting electric system temperature. The gated state-space model is a physics-informed method, while it offers consistency with the efficiency of the data-driven method and dynamics system properties. The proposed method is experimentally verified with a real electric system temperature dataset provided by an electric vehicle (EV) company.

1 Introduction

The electric system (electric motor, electric control, and battery pack) as in Fig. 1 is the core component of electric vehicles (EVs) primarily affecting the range, reliability, and performance. However, the electric system is easily troubled by temperature rising under extreme operating conditions. For example, high temperatures can lead to permanent magnet synchronous motor damage (the stator winding insulation melting or permanent magnets irreversible demagnetization) [1], accelerated aging of the battery pack [2], etc. Model Predictive Control (MPC) serves as the primary framework for thermal management systems [3]. However, its effectiveness is primarily constrained by the accuracy of the EV thermal dynamics model. Balancing real-time performance with accuracy is a common challenge faced by thermal dynamics models. In this context, temperature prediction which is the foundation of the thermal management system has gained extensive attention [4, 5].

Temperature prediction methods can be divided into three groups including model-based methods, physics-informed methods, and data-driven methods [6, 7]. Model-based methods, like, computa-

Fig. 1: The electric system of EVs.

tional fluid dynamics (CFD) [8] and finite element analysis (FEA) [8], are challenging to deploy in practice industrial applications due to their complicated modeling processes. Although purely data-driven methods [1] offer feasibility and accuracy with simple modeling processes, the large model size and lack of interpretability are their inherent challenges, which is a crucial concern in the industrial area. Consequently, there is a critical need to develop a novel model that offers accurate predictions and consistency with dynamics system properties, i.e., physics-informed methods [9, 10]. Among the temperature-prediction-based methods,

322

the Lumped-parameter thermal network (LPTN) [11] is one of the notable exemplars of physics-informed methods, integrating physical models of heat transfer processes with empirical data-based model identification. LPTNs leverage heat conduction formulas, facilitating high prediction accuracy while preserving a compact model size. The state-space model (SSM) is a prevalent modeling approach in the industry [12], by disentangling system behavior into state transitions, control inputs, and system outputs. It serves as a robust backbone for physics-informed methods [7]. Kirchgässner et al. [13] further combined the SSM and heat conduction formula to achieve a thermal neural network with higher prediction accuracy and fewer parameters. Based on the advanced data-driven method and SSM, this paper proposed a novel gated state-space model (GSSM) to predict the temperature of the electric system. Specifically, GSSM leverages the robust sequence learning capabilities of the Gated Recurrent Unit (GRU) [14] to govern the state transition and control input in SSM, while approximating initial states by past measured temperature and maintaining smooth evolutionary characteristics of dynamics systems by adding an extra loss term. We verified the GSSM on an electric system temperature dataset, provided by an EV company, the average mean absolute error (MAE) of the prediction temperature of the electric system is 1.042 K, which holds significant practical importance.

The remainder of this paper is structured as follows. Section 2 mainly introduces the methodology and the details of the proposed method. Section 3 details the dataset and provides the result of the temperature prediction performance and the ablation study of the smooth regulation. Section 4 concludes this paper.

2 Methodology

2.1 State-Space Model

State-space models (SSMs) have been extensively applied in modeling physical systems owing to their simple structure and easy-controlled properties. The SSM is a modeling method that represents a physical system as a set of inputs, outputs, and states, and the relationship between inputs, outputs, and states can be described by many first-order differential equations. For a continuous time-invariant

system, its SSM is expressed as

$$
\begin{aligned}
\dot{x}_t &= Ax_t + Bu_t, \\
y_t &= Cx_t + Du_t.
\end{aligned}
\tag{1}
$$

where A is the state matrix, B is the input matrix, C is the output matrix, and D is the feedforward matrix. D denotes the residual connection linking the input to the output, often presumed to be 0 for the sake of computational simplicity. For a discrete time-invariant system, its SSM is

$$
\begin{aligned}
x_{t+1} &= Ax_t + Bu_t, \\
y_t &= Cx_t + Du_t.
\end{aligned}
\tag{2}
$$

The potential of SSM is still being explored. From the deep learning perspective, SSM is the combination of convolutional neural networks (CNNs) and recurrent neural networks (RNNs) [15], which has received wider attention in the field of natural language processing (NLP) [16, 17] and computer vision (CV) [18, 19].

2.2 Gated Recurrent Unit

The fundamental linear SSM raises performance concerns, necessitating the integration of modern artificial intelligence (AI) technologies to enhance its capabilities. GRU [14] is a variant of recurrent neural networks (RNNs) primarily employed for time series modeling. Numerous researchers prefer GRU due to its outstanding performance and small computational overhead. Its specific formula can be expressed as

$$
\begin{aligned}
z_t &= \sigma\left(W_z \cdot [h_{t-1}, x_t]\right), \\
r_t &= \sigma\left(W_r \cdot [h_{t-1}, x_t]\right), \\
\tilde{h}_t &= \tanh\left(W \cdot [r_t * h_{t-1}, x_t]\right), \\
h_t &= (1 - z_t) * h_{t-1} + z_t * \tilde{h}_t.
\end{aligned}
\tag{3}
$$

where h_{t-1} is the hidden state at the previous step, x_t is the input at current step, and σ is the gated activation function $sigmoid$. In contrast to vanilla RNNs, the distinctive update gate and reset gate mechanism equips GRU with enhanced learning and modeling capabilities. Furthermore, compared to the long short-term memory (LSTM) [20], GRU has similar performance but is computationally cheaper. Therefore, this paper employed the GRU to assist the SSM in predicting the electric system temperature.

PCIM Europe 2024, 11– 13 June 2024, Nuremberg   DOI: 10.30420/566262039

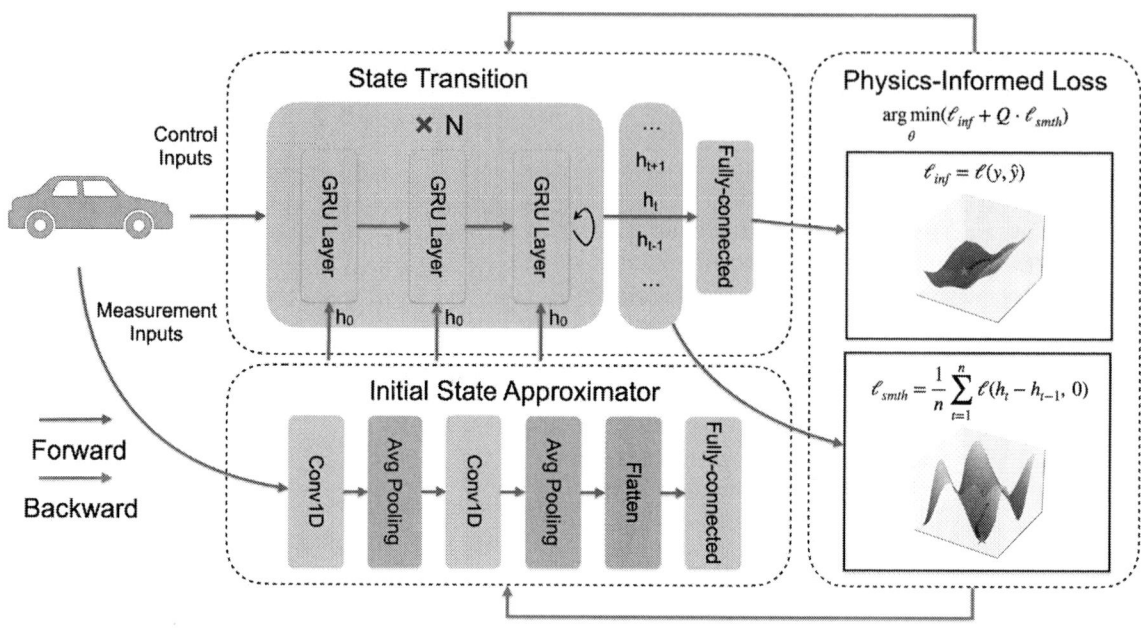

Fig. 2: Overview of the gated state-space model. f_0 is a one-dimensional convolutional neural network with a constant number of convolutional channels and each convolutional layer is followed by an average pooling layer. f_0 approximates the initial state h_0, which is cloned and passed to each GRU layer. The output of the last GRU layer gets the prediction outputs after passing through an output layer C. The physics-informed loss function contained an inference loss term and a smooth evolution regularization.

2.3 Gated State-Space Model

The backbone of GSSM is the state-space model, GRU is employed to process the interaction between the hidden state and the control input. It is well-known that the initial state is significant in a dynamics system [12]. In this context, we approximate the initial state of the GRU by past measured temperature, i.e., measurement inputs, instead of simply setting it to **0**. The specific formula can be expressed as

$$
\begin{aligned}
h_0 &= f_0\left([y_{1-n}, \ldots, y_0]\right), \\
h_t &= f_h(h_{t-1}, x_t), \\
y_t &= C\left(h_t\right).
\end{aligned} \tag{4}
$$

where f_0 is a neural network that predicts the initial state h_0 using the target measurement values of the first n moments of the system, f_h is the abstraction of GRUs, and C is the output layer.

Furthermore, it is widely recognized that the state of dynamics systems evolves smoothly over time. To address this, the paper incorporates a smooth loss term into the loss function of the GSSM. The loss function adding a smooth loss term can be expressed as

$$
Loss_{total} = Loss_{inf} + Q \times Loss_{smth}, \tag{5}
$$

where $\begin{cases} Loss_{inf} = \ell(y, \hat{y}), \\ Loss_{smth} = \frac{1}{n}\sum_{t=1}^{n} \ell(h_t, h_{t-1}), \end{cases}$ $\hat{y}$ is the prediction output, y is the measurement output, h_t is the hidden state of GSSM at time t, h_{t+1} is the hidden state of the model at time $t + 1$, ℓ is an arbitrary loss function, and Q is the weight coefficient of the smooth loss term. Fig. 2 shows the whole process of temperature monitoring of the electric system of an electric vehicle using the GSSM. Note that, f_0 is a flexible module, and different backbones can be selected for adapting different data, such as multi-layer perceptrons (MLP), temporal convolutional networks (TCN), RNNs, etc.

3 Experiment

3.1 Dataset

The dataset utilized to evaluate GSSM was collected by an EV company on real EVs. Due to confidential aspects, only limited information is shown here. The dataset contains 4 independent experiments, distinguished by $profile\_id$. This paper uses the data with $profile\_id = 1$ as the test set, and the other three sets of experimental data are divided into training sets and validation sets according to the setting 4:1. The dataset contains a total

PCIM Europe 2024, 11– 13 June 2024, Nuremberg   DOI: 10.30420/566262039

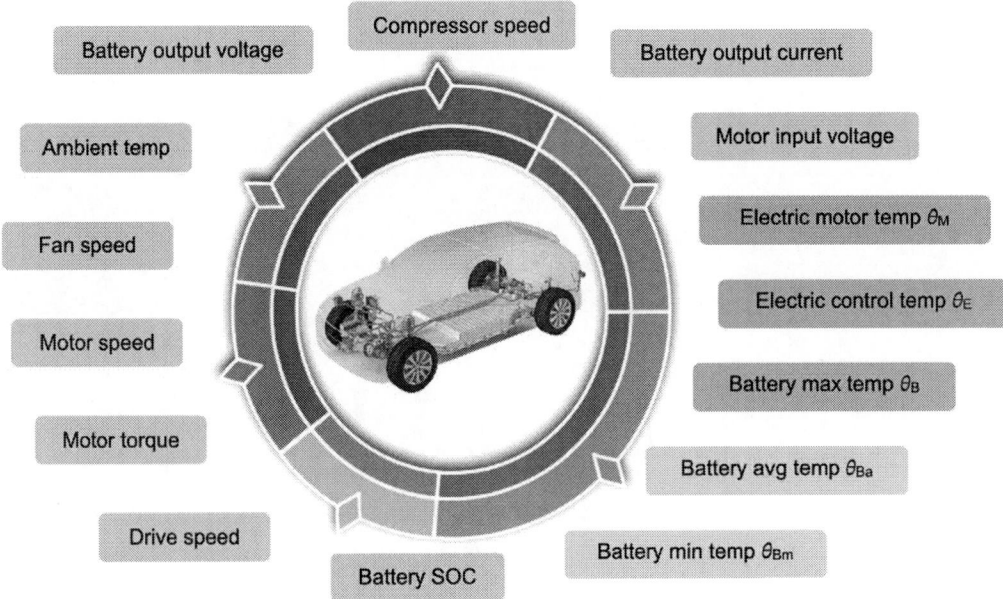

Fig. 3: The dataset details, where the yellow parts are the control inputs, the green parts are the target outputs, and the gray parts are the regular outputs.

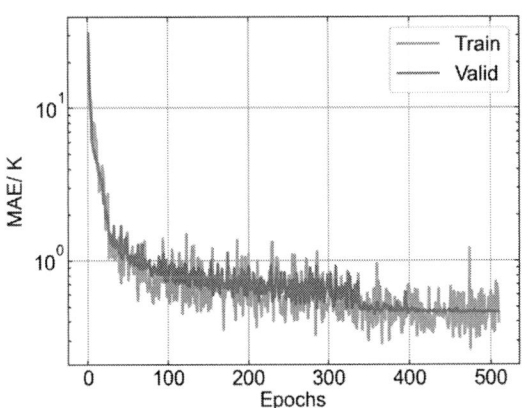

Fig. 4: The decline curve of the training set loss and the validation set loss.

Tab. 1: Hyperparameters of the GSSM

Structural	Value	Training	Value
Prediction len	128	Batch size	16
Estimation len	16	Q	0.1
Hidden size	16	Learning rate	1e-3
GRU layers	3	Dropout	0.5
Conv channels	5	Optimizer	Adam [21]
Conv kernel	3×1		
Pooling	Avg		

uniformly by 100 $^\circ C$ and divided other data by the maximum absolute value. The specific formula can be expressed as

$$\overline{\mathbf{S}} = \begin{cases} \mathbf{S}/100, & if\ \mathbf{S} \in Temperature, \\ \mathbf{S}/max(|\mathbf{S}|), & else. \end{cases} \quad (6)$$

where $\mathbf{S}$ is the raw sensor data, and $\overline{\mathbf{S}}$ is the normalized sensor data.

3.2 Prediction Result

In experiments, the hyperparameters were optimized by the grid search method, and the final selection is shown in Table 1. The estimation length is the length of the measurement input used to approximate the initial state, which is set to 16. The hidden state size is set to 16, comprising 3 hidden layers. The prediction length is 128, i.e., predict the

of 300 minutes of experimental data and the sampling frequency is 1 Hz. Fig. 3 shows the specific data details. Note that θ_{Ba} and θ_{Bm} are included in the target output only as regularizers and are not the temperatures we need to monitor. Furthermore, during the serialization of training data, to mitigate the similarity between sequence data within the training and validation sets, the step size of the sliding window is selected to be a quarter of the prediction length instead of 1. The step size of the sliding window for serializing test data is chosen to be the prediction length.

Normalizing the raw data is beneficial to the model training. This paper divided the temperature data

325

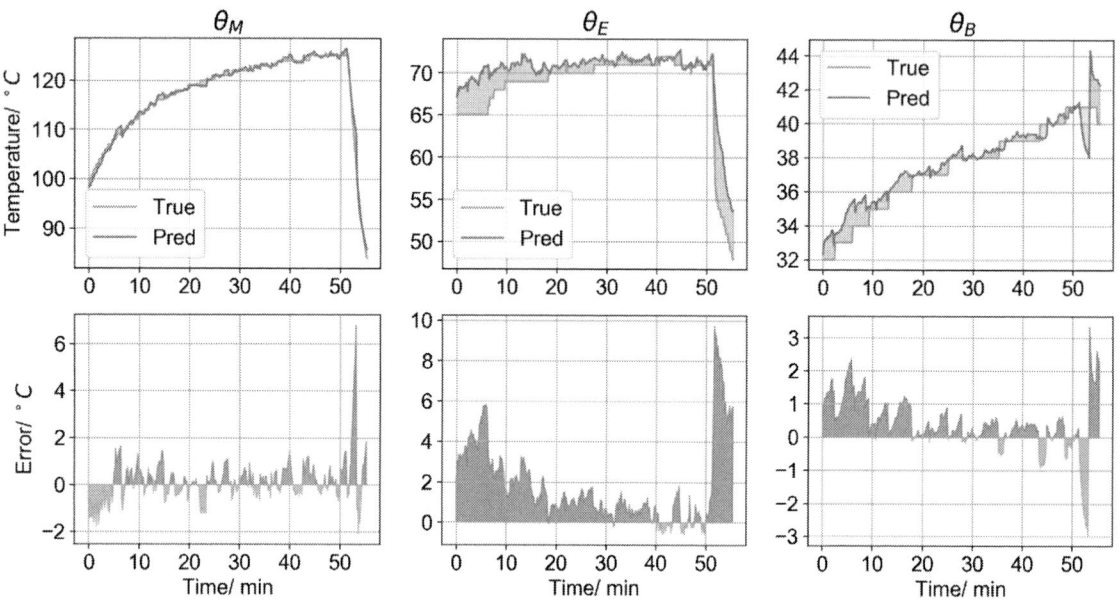

Fig. 5: The first row is the prediction results for the electric motor, electric control, and max battery temperatures, and the second row is the distribution of the error, which equals $pred - true$.

Tab. 2: Comparison of the model performance when utilizing different recurrent units to govern the state transition and control input of SSM under the same parameter amount or the same hidden size.

Method	Hidden Size	Model Params ↓	MAE/K ↓	ℓ_∞/K ↓	R^2 ↑	Inference Time/s ↓
RNN-SSM	26	5958	2.89	55.3	0.43	0.017
	16	**2928**	3.39	56.1	0.33	**0.015**
LSTM-SSM	14	6096	2.53	12.6	0.51	0.032
	16	7584	2.06	14.8	0.58	0.036
GRU-SSM	16	6032	**1.04**	**9.68**	**0.78**	0.040

temperature for the next 128 steps, and the dropout [22] is set to 0.5 to enhance its generalization performance. Moreover, the loss coefficient Q is set to 0.1, making the smooth loss term smaller than the inference loss term. This configuration allows the regularization term to play a regulatory role without perturbing the standard optimization trajectory. It is worth mentioning that, we employed the early-stopping scheme and the decay of the learning rate strategy to optimize the training process. The initial learning rate is set to 1e-3, and the learning rate is divided by 5 if the loss of the validation set does not decrease after 50 epochs. Stop training when the validation loss does not decrease for the fourth time. The mean absolute error (MAE) is used to train and evaluate the model, and its specific formula is

$$MAE(y, \hat{y}) = \frac{1}{m} \sum_{i=1}^{m} |y_i - \hat{y}_i| . \qquad (7)$$

The experimental platforms utilized are the Nvidia Geforce RTX3060 Laptop GPU and Ryzen 7 5800H CPU. The training framework is implemented in Python 3.7 and PyTorch 1.12. Given the hyperparameters outlined in Table 1, the model comprises 6032 parameters, and the average training time per epoch is 0.19 seconds. Meanwhile, the inference time of GSSM on the 60-minute test set is 0.04 seconds. Fig. 4 depicts the training progress of GSSM. It is evident that both the training set loss and the validation set loss exhibit a steady decline without significant oscillations, indicating the robustness of the training process.

The average MAE across multiple experimental results is 1.042 K. Due to the accuracy of the measured temperature being 1 K, which can be determined by Fig. 5, these experimental results can be considered sufficiently accurate. Specific prediction results are illustrated in Fig. 5, and the prediction outcomes of GSSM lean towards being pessimistic. Moreover, acknowledges that pessimistic predictions, i.e., the predicted temperature surpasses the actual temperature, hold greater practicality than optimistic predictions under extreme operating conditions. Pessimistic predictions will lead to more aggressive heat dissipation control to avoid permanent damage to the electric system due to high temperatures.

3.3 Comparisons with other Methods

To comprehensively validate the advantage of employing GRU to augment SSM, this paper also conducts experimental verification on other recurrent units. In the horizontal experiment, vanilla RNN and LSTM are employed to manage the state transition and control input of SSM, with comparisons made against GSSM. To ensure fairness in cross-sectional experiments, we conducted comparative experiments when the hidden state sizes were consistent and when the model parameter amounts were consistent (only the hidden size was modified), respectively.

This paper assesses the model performance across four dimensions: the infinite norm of error (ℓ_∞), mean absolute error (MAE), coefficient of determination (R^2), and inference time. The equation for the infinite norm is represented as

$$\ell_\infty(y, \hat{y}) = max(|y - \hat{y}|), \tag{8}$$

illustrating the upper limit of model prediction error. Mean absolute error portrays the average magnitude of model prediction error. The formula for the coefficient of determination R^2 is denoted as

$$R^2(y, \hat{y}) = 1 - \frac{\sum_{i=0}^{m} (y_i - \hat{y}_i)^2}{\sum_{i=0}^{m} (y_i - \bar{y})^2}, \tag{9}$$

where $\bar{y}$ is the mean of the measurement output, $\hat{y}$ is the prediction output, and y is the measurement output. The coefficient of determination indicates the extent to which the independent variable x elucidates variations in the dependent variable y. The value of R^2 closer to 1 indicates a superior fit of the model. Inference time reflects the complexity of the model.

The experimental results are presented in Table 2. The experimental results reveal that, with the same parameter count or the same hidden size, GSSM all exhibit superior prediction accuracy compared to

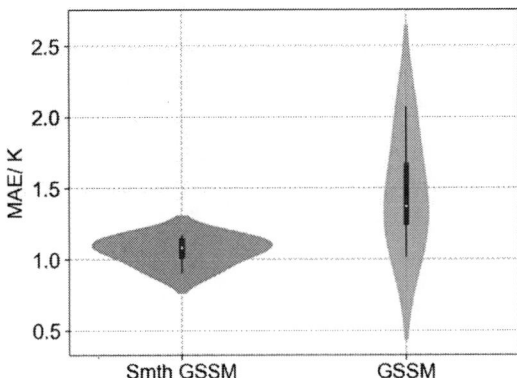

Fig. 6: The influence of smoothness constraints on the prediction accuracy of GSSM.

RNN-SSM and LSTM-SSM. Due to its simplistic architecture, vanilla RNN encounters challenges in effectively learning long-term dependencies, resulting in lower prediction accuracy albeit faster inference speeds. In response to vanilla RNN's struggle with learning long-term dependencies, LSTM introduces the input gate, output gate, and forget gate mechanisms, along with the cell state. Compared to GRU, LSTM incorporates an additional gate structure and cell state. However, due to its intricate architecture, LSTM is more susceptible to overfitting in industrial scenarios where data collection is costly.

3.4 Ablation Study

To substantiate the beneficial impact of the smooth regularization term on model performance, this paper conducts an ablation study focusing on this aspect. This paper evaluates the regularization effect of the smooth regularization term through the average MAE of five independent experiments. We compared the multiple prediction accuracies of GSSM with and without smoothness constraints while maintaining the other hyperparameters constant, and the experimental outcomes are shown in Fig. 6. The smooth regularization term exerts a beneficial influence on both the mean and variance of the prediction accuracy. Overall, incorporating the smoothly evolving properties of physical systems into data-driven models yields positive outcomes.

4 Conclusion

The proposed gated state-space model enhanced the accuracy and interpretability of the electric system temperature predictor. By harnessing the robust capabilities of GRU in time series modeling and incorporating the constraints of physical information, the GSSM successfully achieved a prediction with an average MAE of 1.042 K when applied to a real electric temperature dataset. Through comparison, this paper further substantiates the efficacy of GRU for enhancing SSM. Additionally, ablation experiments underscore the significance of regulating the similarity between the data-driven model and the physical system.

References

[1] W. Kirchgässner, O. Wallscheid, and J. Böcker, "Estimating electric motor temperatures with deep residual machine learning," *IEEE Trans. Power Electron.*, vol. 36, no. 7, pp. 7480–7488, 2020.

[2] J. Kim, J. Oh, and H. Lee, "Review on battery thermal management system for electric vehicles," *Appl. Therm. Eng.*, vol. 149, pp. 192–212, 2019.

[3] M. R. Amini, I. Kolmanovsky, and J. Sun, "Hierarchical mpc for robust eco-cooling of connected and automated vehicles and its application to electric vehicle battery thermal management," *IEEE Trans. Control Syst. Technol.*, vol. 29, no. 1, pp. 316–328, 2020.

[4] G. Feng, C. Lai, J. Tjong, and N. C. Kar, "Noninvasive kalman filter based permanent magnet temperature estimation for permanent magnet synchronous machines," *IEEE Trans. Power Electron.*, vol. 33, no. 12, pp. 10673–10682, 2018.

[5] Y. Xie, W. Li, X. Hu, X. Lin, Y. Zhang, D. Dan, F. Feng, B. Liu, and K. Li, "An enhanced online temperature estimation for lithium-ion batteries," *IEEE Trans. Transp. Electrif.*, vol. 6, no. 2, pp. 375–390, 2020.

[6] S. Zhao, F. Blaabjerg, and H. Wang, "An overview of artificial intelligence applications for power electronics," *IEEE Trans. Power Electron.*, vol. 36, no. 4, pp. 4633–4658, 2020.

[7] X. Liao, S. Chen, Y. Long, and S. Zhao, "Explainable neural dynamics models for electric motor temperature estimation," *Authorea Preprints*, 2023.

[8] O. Wallscheid, "Thermal monitoring of electric motors: State-of-the-art review and future challenges," *IEEE Open J. Industry Appl.*, vol. 2, pp. 204–223, 2021.

[9] X. Liao, S. Chen, P. Wen, and S. Zhao, "Remaining useful life with self-attention assisted physics-informed neural network," *Adv. Eng. Inf.*, vol. 58, p. 102195, 2023.

[10] P. Wen, Z.-S. Ye, Y. Li, S. Chen, P. Xie, and S. Zhao, "Physics-informed neural networks for prognostics and health management of lithium-ion batteries," *IEEE Trans. Intell. Veh.*, 2023.

[11] O. Wallscheid and J. Böcker, "Global identification of a low-order lumped-parameter thermal network for permanent magnet synchronous motors," *IEEE Trans. Energy Convers.*, vol. 31, no. 1, pp. 354–365, 2015.

[12] J. Drgoňa, A. R. Tuor, V. Chandan, and D. L. Vrabie, "Physics-constrained deep learning of multi-zone building thermal dynamics," *Energy Build.*, vol. 243, p. 110992, 2021.

[13] W. Kirchgässner, O. Wallscheid, and J. Böcker, "Thermal neural networks: Lumped-parameter thermal modeling with state-space machine learning," *Eng. Appl. Artif. Intell.*, vol. 117, p. 105537, 2023.

[14] K. Cho, B. Van Merriënboer, C. Gulcehre, D. Bahdanau, F. Bougares, H. Schwenk, and Y. Bengio, "Learning phrase representations using rnn encoder-decoder for statistical machine translation," *arXiv preprint arXiv:1406.1078*, 2014.

[15] A. Gu, K. Goel, and C. Re, "Efficiently modeling long sequences with structured state spaces," in *Int. Conf. Learn. Representations*, 2021.

[16] A. Gu and T. Dao, "Mamba: Linear-time sequence modeling with selective state spaces," *arXiv preprint arXiv:2312.00752*, 2023.

[17] J. Wang, T. Gangavarapu, J. N. Yan, and A. M. Rush, "Mambabyte: Token-free selective state space model," *arXiv preprint arXiv:2401.13660*, 2024.

[18] L. Zhu, B. Liao, Q. Zhang, X. Wang, W. Liu, and X. Wang, "Vision mamba: Efficient visual representation learning with bidirectional state space model," *arXiv preprint arXiv:2401.09417*, 2024.

[19] Y. Liu, Y. Tian, Y. Zhao, H. Yu, L. Xie, Y. Wang, Q. Ye, and Y. Liu, "Vmamba: Visual state space model," *arXiv preprint arXiv:2401.10166*, 2024.

[20] S. Hochreiter and J. Schmidhuber, "Long short-term memory," *Neural Comput.*, vol. 9, no. 8, pp. 1735–1780, 1997.

[21] D. P. Kingma and J. L. Ba, "Adam: A method for stochastic gradient descent," in *Int. Conf. Learn. Representations*, pp. 1–15, 2015.

[22] N. Srivastava, G. Hinton, A. Krizhevsky, I. Sutskever, and R. Salakhutdinov, "Dropout: a simple way to prevent neural networks from overfitting," *J. Mach. Learn. Res.*, vol. 15, no. 1, pp. 1929–1958, 2014.

PCIM Europe 2024, 11– 13 June 2024, Nuremberg   DOI: 10.30420/566262040

Novel Bidirectional Single-Stage Isolated 600-V GaN M-BDS-Based Single-/Three-Phase-Operable EV On-Board Charger

Sven Weihe [1], David Menzi [1], Jonas Huber [1], Daifei Zhang [1], Johann W. Kolar [1], Matthias Kasper [2], Kennith Kin Leong[2], Gerald Deboy[2]

[1] Power Electronic Systems Laboratory, ETH Zürich, Switzerland
[2] Infineon Technologies Austria AG, Austria

Corresponding author: Sven Weihe, weihe@lem.ee.ethz.ch
Speaker: Sven Weihe, weihe@lem.ee.ethz.ch

Abstract

Next generation On-Board Chargers (OBCs) should comprise only a single high-frequency-isolated ac-dc converter stage that realizes bidirectional power flow from not only a three-phase but also a single-phase grid to the Electric Vehicle (EV) battery and should cover a wide input-output voltage range. This paper introduces the novel isolated Y-Rectifier with series-resonant operation (iYR$_S$), which utilizes 600 V GaN Monolithic Bidirectional Switches (M-BDSs) and complies with the aforementioned requirements for next-generation OBCs. The operating principles and control methods for buck-boost operation from a three-phase and a single-phase mains with the same rated power are detailed and validated by circuit simulations. The voltage and current stresses of the main power components are derived, and a 6.6 kW design example indicates expected efficiencies of almost 97% for three-phase and 97.5% for single-phase operation.

1   Introduction

Electric Vehicle (EV) On-Board Chargers (OBCs) face very demanding requirements including galvanic isolation between the grid and the EV battery, a wide battery voltage range, nominal-power operation from three-phase and single-phase grids, high power density, and bidirectional power flow [1]. The state-of-the-art approach for OBCs is a two-stage system comprising of a power-factor-correction (PFC) rectifier front-end followed by an isolated dc-dc converter stage [2], which, however, is disadvantageous as the power is converted twice, and two converter stages need to be designed and built. High-frequency-(HF)-isolated single-stage ac-dc converters with either a Dual Active Bridge (DAB) [3], [4] or Series-Resonant (SR) operating mode [5] are thus an attractive alternative to meet OBC requirements. So far, however, the extensive research into single-stage isolated three-phase PFC ac-dc converters [3]–[22] has not yet considered single-phase operation, or only with a limited output power [23]. This shortcoming is addressed by

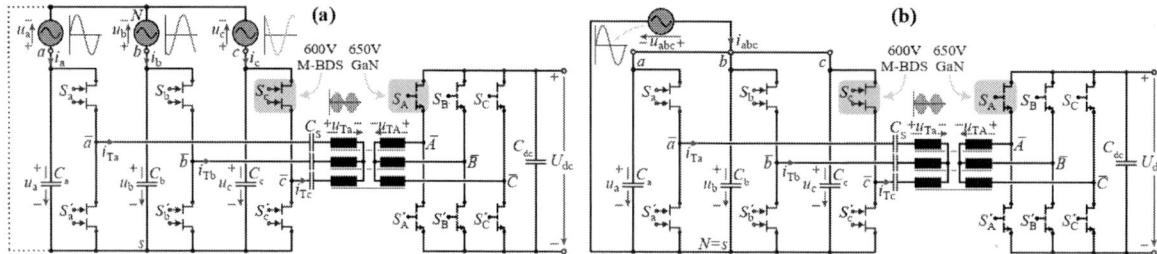

Fig. 1: The proposed topology of the isolated Y-Rectifier with series resonant operation (iYR$_S$) utilizing 600 V GaN M-BDSs in the ac-front-end to directly interface with the **(a)** three-phase (325 V$_{peak}$ line-to-neutral; 400 V line-to-line rms) and **(b)** single-phase (325 V$_{peak}$ line-to-neutral) grid. By employing a novel resonant modulation, bidirectional charging at nominal power is achieved with both configurations.

Tab. 1: Considered system specifications.

Parameter	Symbol	Value	Unit
Grid Voltage[1]	3ϕ, $u_{a,b,c}$	230	V_{rms}
	1ϕ, u_{abc}	230	V_{rms}
Grid Freq.	f_{ac}	50	Hz
Switching Freq.	f_{sw}	72	kHz
Rated Output Power	P_{dc}	6.6	kW
Output Voltage	U_{dc}	250-450	V

[1]line-to-neutral

the new bidirectional isolated Y-Rectifier topology[1] with a SR operating mode (iYR$_S$) as depicted in **Fig. 1**. The iYR$_S$ ac-front-end employs novel GaN-based Monolithic Bidirectional Switches (M-BDSs): with the ac-front-end M-BDSs blocking voltages defined by the grid peak line-to-neutral voltage, 600 V rated semiconductors can be used for interfacing the European 400 V (line-to-line rms) grid. The SR modulation enables full soft switching in the ac-front-end in both single- and three-phase operation for a significant share of the desired output voltage range $U_{dc} \in [250\,V, 450\,V]$.

In the following, first **Section 2** explains the operating method of the iYR$_S$ for three- and single-phase buck-boost operation via space vector (SV) diagrams and simulation results. Given the specifications in **Tab. 1**, **Section 3** then provides basic design considerations along with key component stresses and provides an efficiency estimate of an exemplary iYR$_S$ design. Finally, **Section 4** concludes the paper.

2 Operating Method

The method of operation of the iYR$_S$ topology is provided in this section for both, the three-phase and single-phase operation with the use of SV diagrams and circuit simulations. First, only boost operation is considered, and then the concept is extended to a buck-boost operation to enable a wide output voltage range.

2.1 Three-Phase

In the three-phase configuration (**Fig. 1a**), the ac-front-end semiconductors are switched synchronously with a duty cycle $d_{abc} = 50\%$ and translate the grid input voltages u_a, u_b, u_c into an

---

[1]Note that the main power circuit structure is identical to [22]. However, here a resonant operating mode is employed.

amplitude-modulated HF three-phase transformer voltage system u_{Ta}, u_{Tb}, u_{Tc} [22] as can be observed from the simulation results in (**Fig. 2a.i**). Here, the series capacitor C_S forms a resonant tank with the transformer inductance L_S in each phase, which is tuned to the switching frequency f_{sw} similar to a SR dc-dc converter [25].[2] In addition, the series capacitors C_S block half the instantaneous phase voltage and thus only the HF component varying between $\pm\frac{1}{2}u_a$, $\pm\frac{1}{2}u_b$, $\pm\frac{1}{2}u_c$ acts on the primary side of the resonant tank (a zoom-in on the waveforms is presented in **Fig. 2a.ii**). A SV representation of the primary-side transformer voltages ($\underline{u}_{Tabc}$) is presented in **Fig. 2b** and the two synchronous switching transitions of the primary-side stage, α and β (see u_{Ta} in **Fig. 2a.ii**), result in the primary-side SV $\underline{u}_{Tabc}$ in **Fig. 2b** toggling between $\pm\frac{1}{2}$ of the three-phase grid voltage SV $\underline{u}_{a,b,c}$. The dc-stage generates a secondary-side transformer voltage SV $\underline{u}_{TABC}$ (displayed in **Fig. 2b** for a unity transformer turns ratio) which is in phase with the primary-side voltage SV $\underline{u}_{Tabc}$ during the first ① and second ② half switching period $\frac{1}{2}T_{sw} = 1/(2f_{sw})$ and shows approximately equal amplitude,[3] resulting in naturally sinusoidal low-frequency (LF) grid currents $\bar{i}_a$, $\bar{i}_b$, $\bar{i}_c$. Note that the resonant tanks show approximately zero impedance at the switching frequency and thus the power flow can be adjusted by slightly adjusting the magnitude of the dc-stage voltage SV $|\underline{u}_{TABC}|$ similar to a SR dc-dc converter [25]: If, e.g., $|\underline{u}_{TABC}|$ is selected slightly smaller than $|\underline{u}_{Tabc}|$ the resulting voltage difference ΔU is applied to the resonant tanks which are energized, resulting in an increase in the three-phase power transfer and ultimately to an increase in the output voltage.

The operation of the dc-stage can be controlled by means of a simple Pulse Width Modulation (PWM) strategy with two duty cycles which are, e.g., for

---

[2]In contrast, in [22] the series capacitor C_S is sized to show negligible impedance compared to the transformer inductance L_S at the switching frequency to enable a DAB-type modulation.

[3]In the time domain this corresponds to the dc-stage generating HF differential-mode voltages u_{TA}, u_{TB}, u_{TC} recreating the voltage-time area of the primary-side transformer voltages u_{Ta}, u_{Tb}, u_{Tc} within each $\frac{1}{2}T_{sw}$ period as illustrated in **Fig. 2a.ii**.

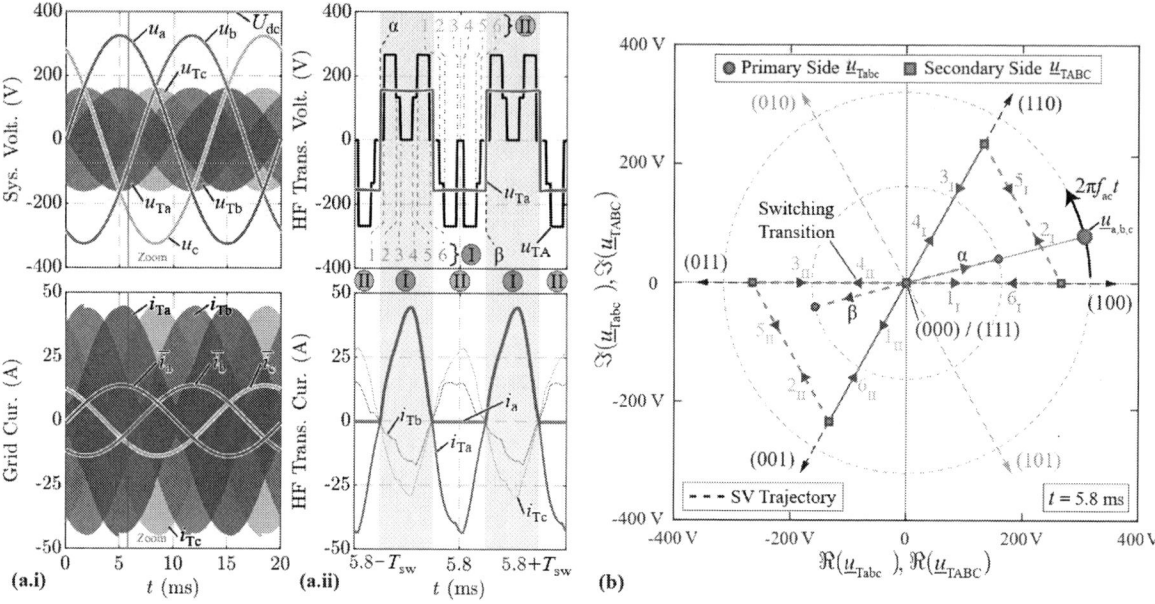

Fig. 2: (a) Simulated key voltage and current waveforms (generated using PLECS [24]) of the proposed iYR$_S$ topology shown in **Fig. 1a** for three-phase PFC operation over a mains period **(a.i)** and over two switching periods T_{sw} around $t = 5.8$ ms **(a.ii)** . **(b)** Space vector (SV) representation of the HF voltage waveforms in **(a.ii)**. The ac-front-end operation with $d_{abc} = 50\%$ results in a primary-side transformer voltage SV $\underline{u}_{Tabc}$ with half the magnitude of the grid voltage SV $\underline{u}_{a,b,c}$ and toggling between in-phase and anti-phase orientation in every half switching period (the primary-side switching transitions are labelled with α and β). The dc-stage generates a secondary-side voltage SV $\underline{u}_{TABC}$ recreating $\underline{u}_{Tabc}$ during the first ① and second ⑪ $\frac{1}{2}T_{sw}$ period (the dc-stage switching transitions are numbered with 1...6). Simulation parameters: $u_{a,b,c} = 230$ V (rms; line-neutral), $f_{sw} = 72$ kHz, $L_S = 30$ µH, $C_S = 163$ nF, $U_{dc} = 400$ V, $P_{dc} = 6.6$ kW.

phase a defined as

$$d_{A,I}(t) = \min\left(\frac{\Delta U + \hat{u}_{a,b,c}}{U_{dc}}\frac{N_2}{N_1}, 1\right)\frac{u_a(t)}{2\hat{u}_{a,b,c}} + \frac{1}{2} \quad (1)$$

for the first ① and $d_{A,II} = 1 - d_{A,I}$ for the second ⑪ $\frac{1}{2}T_{sw}$ period, with $d_{A,I}, d_{A,II} \in [0,1]$. Note that $\Delta U \ll \hat{u}_{a,b,c}$ allows to adjust the magnitude of the secondary-side voltage SV $|\underline{u}_{TABC}|$ to regulate the power flow. The switching signals are easily realised with a synchronised secondary-side carrier with $2f_{sw}$ (i.e., double the switching frequency of the ac-front-end) which is compared to $d_{A,I}$ and $d_{A,II}$ in alternation during the first and the second $\frac{1}{2}T_{sw}$ period, respectively. The resulting secondary-side voltage SV $\underline{u}_{TABC}$ trajectory[4] is presented in **Fig. 2b**

---

[4]Note that the transitions in **Fig. 2b** and **Fig. 3b** are highlighted just for illustration purposes: The voltage SVs $\underline{u}_{Tabc}$ and $\underline{u}_{TABC}$ are defined by corresponding discrete endpoints in the complex plane (corners and midpoint of a hexagon) and the transition from one space vector to the following happens instantaneously.

with the sequence of switching states

$$(000) \blacktriangleright_{1_I} (100) \blacktriangleright_{2_I} (110) \blacktriangleright_{3_I} (111) \blacktriangleright_{4_I} (110) \blacktriangleright_{5_I} (100) \blacktriangleright_{6_I} (000), \quad (2)$$

in the first $\frac{1}{2}T_{sw}$ period ①. Similarly, for the second $\frac{1}{2}T_{sw}$ period ⑪ the generated switching sequence is

$$(000) \blacktriangleright_{1_{II}} (001) \blacktriangleright_{2_{II}} (011) \blacktriangleright_{3_{II}} (111) \blacktriangleright_{4_{II}} (011) \blacktriangleright_{5_{II}} (001) \blacktriangleright_{6_{II}} (000). \quad (3)$$

The corresponding secondary-side time-domain voltage of phase a, u_{TA}, is shown in **Fig. 2a.ii**, which is in phase with the primary-side voltage u_{Ta} (i.e., in contrast to a DAB-type modulation, no HF phase shift is introduced in between u_{Ta} and u_{TA}) and shows an identical $\frac{1}{2}T_{sw}$ period average value, such that quasi-sinusoidal resonant tank currents i_{Ta}, i_{Tb}, i_{Tc} result.

To achieve symmetrical stresses of the dc-stage high- and low-side power transistors, the secondary-side carrier is flipped (i.e., phase shifted by 180°) for each 60° sector. Note that, alternatively, the dc-stage could be modulated with the SV modulation from [19] which enables to lower the (average) switching frequency from $2f_{sw}$ to $\frac{4}{3}f_{sw}$.

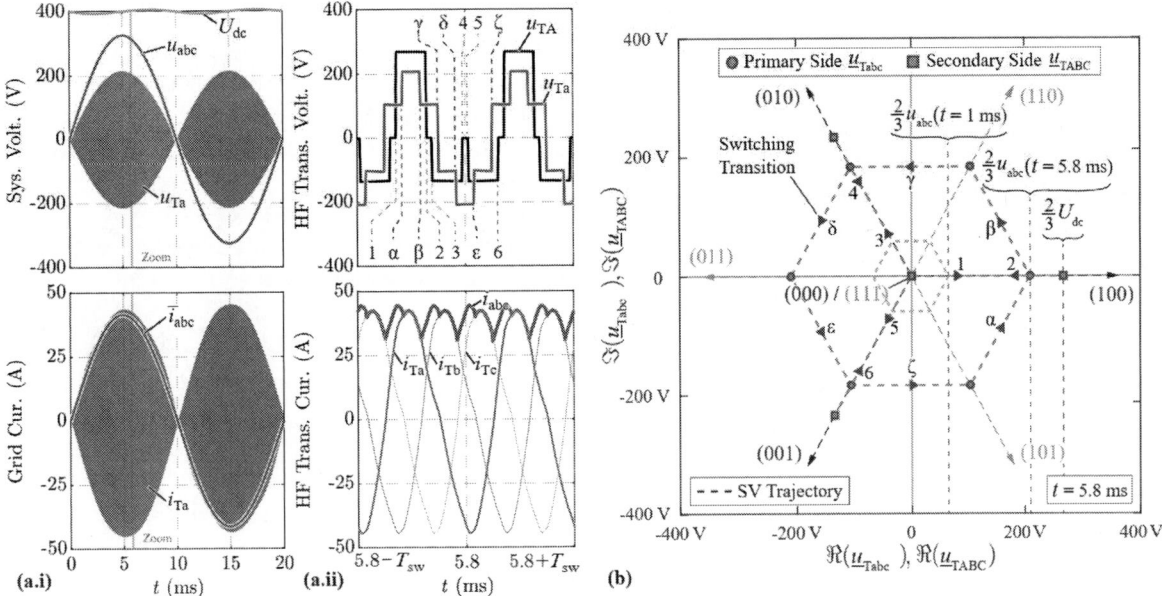

Fig. 3: (a) Simulated key waveforms (generated using PLECS [24]) of the proposed iYR$_S$ topology shown in **Fig. 1b** for single-phase PFC operation over a mains period **(a.i)** and over two switching periods T_{sw} around $t =$ 5.8 ms **(a.ii)**. **(b)** Space vector (SV) representation of the HF voltage waveforms in **(a.ii)**. In single-phase operation, both, the primary- and the secondary-side half-bridges operate with 120° PWM carrier phase shift, resulting in a hexagonal trajectory of the primary-side voltage SV $\underline{u}_{Tabc}$ in **(b)**. Note that the amplitude of the active primary-side voltage SV is proportional to the instantaneous grid voltage $u_{abc}(t)$ (i.e., the size of the hexagon changes with the grid voltage). The primary-side and secondary-side switching transitions are labelled with $\alpha \ldots \zeta$ and 1 ... 6, respectively. Simulation parameters: $u_{abc} = 230$ V$_{rms}$, $f_{sw} = 72$ kHz, $L_S = 30$ µH, $C_S = 163$ nF, $U_{dc} = 400$ V, $P_{dc} = 6.6$ kW.

2.2 Single-Phase

For single-phase operation (**Fig. 1b**), the ac-front-end half-bridges are parallel-connected to the grid line and neutral terminals and operate with $d_{abc} = 50\%$ and a 120° PWM carrier phase shift, thereby translating the single-phase grid voltage u_{abc} into a symmetric HF three-phase voltage system u_{Ta}, u_{Tb}, u_{Tc}. Thus, the operation resembles that of a multi-phase SR dc-dc converter, where the (bipolar) input voltage is varying with the grid input voltage u_{abc} as indicated in **Fig. 3a**. **Fig. 3b** presents the SV representation of the transformer voltages, and the aforementioned 120° PWM carrier phase shift of the ac-front half-bridges results in a hexagonal trajectory of the primary-side voltage SV $\underline{u}_{Tabc}$. Note that the instantaneous voltage SV magnitude $|\underline{u}_{Tabc}| = \frac{2}{3}u_{abc}$ is proportional to the instantaneous grid voltage value and thus the area of the primary-side voltage SV hexagon changes over time with twice the grid frequency $2f_{ac}$ and the time-domain transformer voltages u_{Ta}, u_{Tb}, u_{Tc} (see **Fig. 3a.i**) are amplitude modulated by the single-phase grid voltage u_{abc}. Note that in contrast to three-phase operation,

here the series capacitors C_S do not block any LF voltage components.

Similarly, the secondary side half-bridges operate at f_{sw} (i.e., the same switching frequency as the ac-front-end) with a 120° PWM carrier phase shift and identical duty cycles $d_A = d_B = d_C$ and

$$d_A = \frac{1}{\pi} \arcsin \left(\min \left(\frac{|u_{abc}(t)| + \Delta U}{U_{dc} \frac{N_1}{N_2}}, 1 \right) \right), \quad (4)$$

utilizing again the small voltage difference ΔU to regulate the power flow. During, e.g., the positive mains period, and for a duty cycle $d_A = d_B = d_C < 1/3$ (i.e., $|u_{abc}| < \frac{N_1}{N_2}U_{dc}$ for $\Delta U = 0$), the sequence of switching states is

$$(000)\underset{1}{\blacktriangleright}(100)\underset{2}{\blacktriangleright}(000)\underset{3}{\blacktriangleright}(010)\underset{4}{\blacktriangleright}(000)\underset{5}{\blacktriangleright}(001)\underset{6}{\blacktriangleright}(000), \quad (5)$$

as indicated for the voltage SV $\underline{u}_{TABC}$ in **Fig. 3b** (utilized dc-stage switching states are labelled in black; non-utilized switching states in gray). In contrast, for $1/3 < d_A = d_B = d_C < 1/2$ the secondary SV trajectory changes to a hexagonal shape with the

sequence of switching states

$$(101)\blacktriangleright(100)\blacktriangleright(110)\blacktriangleright(010)\blacktriangleright(011)\blacktriangleright(001)\blacktriangleright(101) \quad (6)$$

resulting in a counterclockwise sequence similar to the primary-side SV trajectory $\underline{u}_{\mathrm{Tabc}}$.

Note that, as the grid voltages becomes negative, the dc-stage switching signals generated according to **Eq. (5)** need to be inverted resulting, e.g., for phase a, in an effective conduction time of the high-side power transistor

$$d'_{\mathrm{A}} = \begin{cases} d_{\mathrm{A}}, & u_{\mathrm{abc}}(t) \geq 0 \\ 1 - d_{\mathrm{A}}, & u_{\mathrm{abc}}(t) < 0, \end{cases} \quad (7)$$

which advantageously facilitates equal power transistor current stresses of the top and bottom dc-stage power transistors.

2.3 Buck-Boost Operation

The presented three- and single-phase modulation concepts rely on controlling the power flow by adjusting the amplitude of the voltage SV $|\underline{u}_{\mathrm{TABC}}|$ generated by the dc-stage by a small ΔU. This, however, imposes a voltage limit $\frac{N_1}{N_2}U_{\mathrm{dc}} > \hat{u}_{\mathrm{a,b,c}}$ and

limits the iYR$_{\mathrm{S}}$ topology to boost operation, which is incompatible with the desired dc output ranges stated in **Tab. 1**. This limitation is resolved by additionally adjusting the ac-front-end modulation to maintain power flow control in buck operation with $\frac{N_1}{N_2}U_{\mathrm{dc}} < \hat{u}_{\mathrm{a,b,c}}$.

For three-phase operation, the ac-front-end duty cycle can then be defined as

$$d_{\mathrm{abc}} = \frac{1}{\pi} \arcsin\left(\min\left(\frac{U_{\mathrm{dc}}\frac{N_1}{N_2} - \Delta U}{\hat{u}_{\mathrm{a,b,c}}}, 1 \right) \right), \quad (8)$$

such that the ac-front-end takes over the power flow control in buck mode. Therefore, d_{abc} deviates from 50% only when $\frac{N_1}{N_2}U_{\mathrm{dc}} < \hat{u}_{\mathrm{a,b,c}}$ (for $\Delta U = 0$).

The ΔU-based power flow control defined with **Eq. (1)** and **Eq. (8)** automatically results in either the ac-front-end (boost) or the dc-stage (buck) operating in saturation with the maximum SV voltage amplitude, while the other stage assures power flow controllability. This enables the desired wide output voltage range operation and **Fig. 4a** shows simulation results where the iYR$_{\mathrm{S}}$ supplies an electronic dc load which sweeps the output

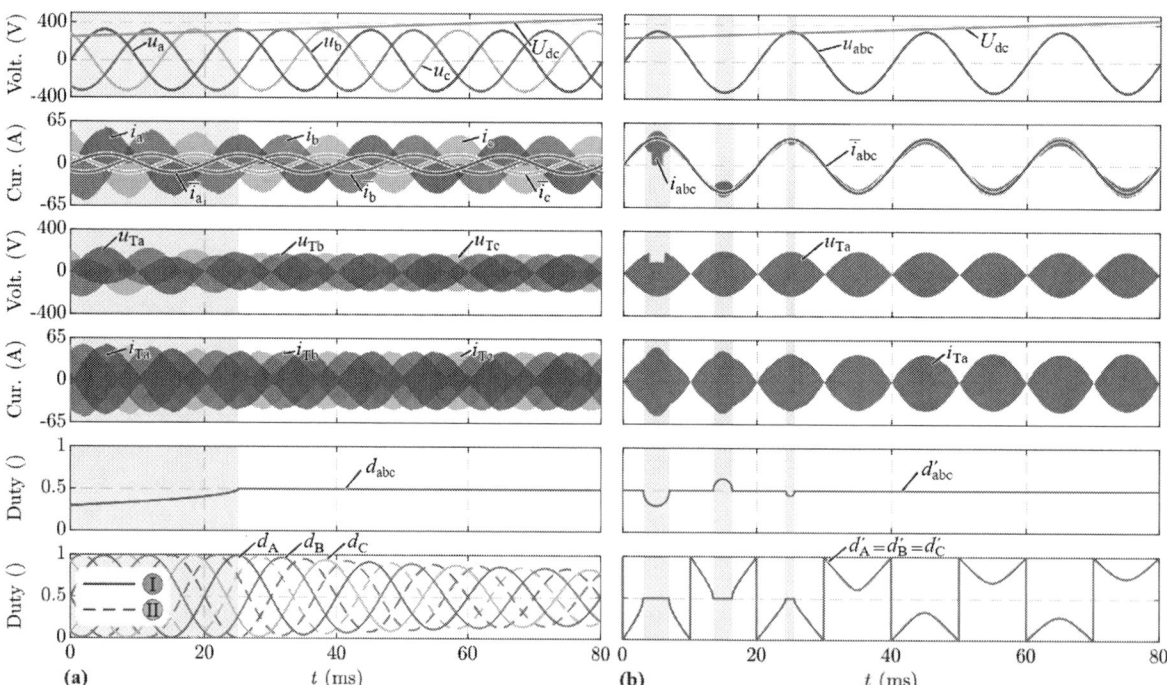

Fig. 4: Simulation waveforms (generated using PLECS [24]) showing the transition from buck (grey background) to boost operation of the iYR$_{\mathrm{S}}$ for a dc output voltage ramp from 250 V to 450 V for **(a)** three-phase and **(b)** single-phase operation. The secondary side is connected to an electronic load defining the dc voltage, and the iYR$_{\mathrm{S}}$ control maintains a constant power transfer of 6.6 kW.

voltage from 250 V to 450 V and the iYR$_S$ control ensures a continuous power transfer of $P_{dc} =$ 6.6 kW when transitioning from buck (highlighted in grey) to boost operation.

In contrast, for single-phase operation the modulation is adjusted based on the instantaneous absolute grid voltage value $|u_{abc}(t)|$ (instead of the grid voltage amplitude as in three-phase operation), i.e.,

$$d_{abc} = \frac{1}{\pi} \arcsin\left(\min\left(\frac{U_{dc}\frac{N_1}{N_2} - \Delta U}{|u_{abc}(t)|}, 1 \right) \right). \quad (9)$$

To realize equal conduction stresses for the high- and low-side M-BDSs over a mains period, the ac-front-end duty cycle is deviated symmetrically to values below and above 0.5 for the positive and negative grid interval, respectively, with

$$d'_{abc} = \begin{cases} d_{abc}, & u_{abc}(t) \geq 0 \\ 1 - d_{abc}, & u_{abc}(t) < 0. \end{cases} \quad (10)$$

Fig. 4b presents simulation results for single-phase operation, and a dynamic transition between buck (shaded grey areas) and boost operation can be observed. Note that the ac-front-end and dc-stage duty cycles deviate from 50% only in buck and boost operation, respectively.

3 Design and Performance Estimation

To provide a first performance evaluation of the iYR$_S$, a basic design example for a 6.6 kW converter according to **Tab. 1** is conducted (the resulting converter parameters are listed in **Tab. 2**) and the component stresses and losses are provided.

3.1 Component Selection

The dc-stage power transistors are subject to voltages up to $U_{dc} = 450$ V (not considering switching overvoltages) and can thus be realized with 650-V-rated GaN Systems GS-065-060-5-T-A which feature a low on-state resistance of 25 mΩ (typ.). The PLECS [24] thermal model (provided by the manufacturer) is used to assess the conduction and switching losses presented in **Tab. 3**, where further a thermal interface material (TIM) with a thermal impedance of 52 K mm²/W [26] is considered to interface the device case and a water-cooled cold plate with a (maximum) temperature of 60 ℃.
The M-BDSs of the ac-front-end are blocking voltages of up to $\hat{U}_{abc} = 325$ V, and can thus be realized

Tab. 2: Considered converter parameters.

Description	Identifier	Value	Unit
Input Cap.	C_a	2.5	μF
M-BDS[1]	S_a, S'_a	25	mΩ$_{typ}$
		650	V
Series Cap.	C_S	163	nF
Leakage Ind.	L_S	30	μH
Turns Ratio	$N_1 : N_2$	1:1	
DC Side Semi.[1]	S_A, S'_A	25	mΩ$_{typ}$
		650	V
DC Cap.	C_{dc}	4.2	mF

[1] GaN Systems GS-065-060-5-T-A (a virtual M-BDS version is assumed for the ac-front-end)

with 600 V/650 V GaN technology. The switching and conduction losses are estimated by assuming a virtual M-BDS version of the GS-065-060-5-T-A which has the same switching- and conduction-loss characteristics as the unipolar GaN device considered for the dc-stage.[5]
A transformer, with turns ratio $N_1 : N_2 = 1{:}1$ and a leakage inductance of $L_S = 30$ μH is considered. For brevity, the transformer losses (P_{T_a}) at rated power are obtained by assuming a typical efficiency of $\eta_T = 99.5$ %. The series capacitance $C_S = 163$ nF is selected such that the desired resonant tank frequency $f_0 = f_{sw}$ is achieved. The input capacitors $C_a = 2.5$ μF are selected such that a 2% reactive input current limit at nominal power operation is ensured. The dc output capacitance $C_{dc} = 4.2$ mF is sized for single-phase operation such that the maximum allowable peak-to-peak output voltage ripple $\Delta U_{dc} = 20$ V is not exceeded at the worst-case operating point with $U_{dc} = 250$ V and nominal power delivery $P_{dc} = 6.6$ kW. Note that a more compact system realization could be achieved by employing an active power pulsation buffer concept [27]–[29]. Assuming the use of capacitors with a high-quality dielectric material with a low dissipation factor, the filter, dc and series capacitor losses are neglected here.

---

[5]Note that such an M-BDS device does currently not exist. However, current R&D activities in the field of GaN M-BDS make it very likely that such a low on-resistance M-BDS device will be available in a next-generation product.

Tab. 3: Component stresses and loss evaluation for the converter in **Fig. 1** at a nominal output power of 6.6 kW and an output voltage of 400 V.

Parameter	3-Phase	1-Phase	Unit
$\hat{i}_{T_a}$	45.4	45.3	A
I_{T_a}	21.8	21.8	A$_{rms}$
$I_{S_a} = I_{S_a'}$[1]	15.4	15.4	A$_{rms}$
$I_{S_A} = I_{S_A'}$	15.4	15.6	A$_{rms}$
$P_{S_a,Cond} = P_{S_a',Cond}$[1]	9.0	9.0	W
$P_{S_a,Sw} = P_{S_a',Sw}$[1]	0.9	0.7	W
P_{T_a}	11	11	W
$P_{S_A,Cond} = P_{S_A',Cond}$	10.2	9.7	W
$P_{S_A,Sw} = P_{S_A',Sw}$	9.9	4.6	W
P_{Total}	213	177	W
η	96.9	97.4	%

[1]Note that unequal power transistor stresses result in three-phase buck operation.

3.2 Component Stresses and Performance

The component stresses, and primary loss sources of the converter design from **Section 3.1** are evaluated using PLECS [24] simulations considering nominal power operation with $U_{dc} = 400$ V for both, three and single-phase operation with the results listed in **Tab. 3**. The predominantly soft-switching operation of the ac-front-end M-BDSs results in low switching losses ($P_{S_a,Sw} = P_{S_a',Sw}$) for both, single- and three-phase operation, but is only possible for pure boost operating regions of the iYR$_S$.[6] It can be observed that the single-phase operation with a calculated efficiency of $\eta = 97.4$ % is superior to the three-phase operation with $\eta = 96.9$ %. The main reason for this performance deviation are the dc-side semiconductor hard-switching losses, which are elevated in three-phase operation due to the operation with twice the ac-front-end switching frequency $2 f_{sw}$ discussed earlier. As mentioned, the dc-stage could also be modulated according to [19] to lower the (average) switching frequency to $\frac{4}{3} f_{sw}$ which promises efficiency gains.

---

[6]Note that hard-switching transitions will occur when operating in buck mode; E.g. for nominal power operation with an output voltage $U_{dc} = 250$ V the calculated efficiency drops to $\eta = 95.8$ % in three-phase and $\eta = 97.2$ % in single-phase operation.

4 Conclusion

The requirements of next-generation On-Board Chargers (OBCs) demand compact and lightweight converter realizations that can operate under a broad range of operating conditions. This paper introduces a new isolated Y-rectifier with a series-resonant operation (iYR$_S$), which utilizes 600 V GaN M-BDSs in the ac-front-end to operate with nominal power in both, a 400 V (line-to-line rms) three-phase and a single-phase grid. The novel series-resonant modulation enables buck-boost operation and bidirectional power flow with fully sinusoidal grid currents and full soft switching of the ac-front-end transistors for a significant proportion of the desired output voltage range; thus, an exemplary 6.6 kW design achieves estimated efficiencies of 96.9% in three-phase and 97.4% in single-phase configuration.

References

[1] H. Wouters and W. Martinez, "Bidirectional on-board chargers for electric vehicles: State-of-the-art and future trends," *IEEE Trans. Power Electron.*, vol. 39, no. 1, pp. 693–716, Jan. 2024.

[2] C. Wei, H. Xie, Y. Liu, Z. Hu, and J. Shao, "A SiC based high efficiency 22kW bi-directional EV on-board charger," in *Proc. Power Convers. Intelligent Motion Conf. (PCIM)*, Nuremberg, Germany, 2021.

[3] D. Menzi, F. Krismer, T. Ohno, J. Huber, J. W. Kolar, and J. Everts, "Novel bidirectional single-stage isolated three-phase buck-boost PFC rectifier system," in *Proc. IEEE Appl. Power Electron. Conf. Expo. (APEC)*, Orlando, FL, USA, Mar. 2023.

[4] B. R. De Almeida, J. W. M. De Araujo, P. P. Praca, and D. De S. Oliveira, "A single-stage three-phase bidirectional AC/DC converter with high-frequency isolation and PFC," *IEEE Trans. Power Electron.*, vol. 33, no. 10, pp. 8298–8307, Oct. 2018.

[5] M. Zhang, H. Zou, Z. Chen, R. Yu, and A. Q. Huang, "A novel single-stage bidirectional isolated three-phase resonant mode AC-DC PFC converter," in *Proc. Energy Convers. Congr. Expo. (ECCE USA)*, Nashville, TN, USA, Oct. 2023, pp. 2222–2229.

[6] K. Ali, R. K. Surapaneni, P. Das, and S. K. Panda, "An SiC-MOSFET-based nine-switch single-stage three-phase AC–DC isolated converter," *IEEE Trans. Ind. Electron.*, vol. 64, no. 11, pp. 9083–9093, Nov. 2017.

[7] J. Lee, H. Jeong, T.-T. Le, and S. Choi, "Three-phase single-stage bidirectional CCM soft-switching AC–DC converter with minimum switch count," *IEEE Trans. Power Electron.*, vol. 38, no. 2, pp. 2052–2062, Feb. 2023.

[8] G. Li, J. Ruan, K. Wang, Y. Deng, X. He, and Y. Wang, "An interleaved three-phase PWM single-stage resonant rectifier with high-frequency isolation," *IEEE Trans. Ind. Electron.*, vol. 67, no. 8, pp. 6572–6582, Aug. 2020.

[9] M. J. Heller, F. Krismer, and J. W. Kolar, "EMI filter design for the integrated dual three-phase active bridge (D3AB) PFC rectifier," *IEEE Trans. Power Electron.*, vol. 37, no. 12, pp. 14527–14546, Dec. 2022.

[10] J. W. Kolar, U. Drofenik, and F. Zach, "Vienna Rectifier II—A novel single-stage high-frequency isolated three-phase PWM rectifier system," *IEEE Trans. Ind. Electron.*, vol. 46, no. 4, pp. 674–691, Aug. 1999.

[11] M. Silva, N. Hensgens, J. A. Oliver, P. Alou, Ó. García, and J. A. Cobos, "Isolated Swiss-forward three-phase rectifier with resonant reset," *IEEE Trans. Power Electron.*, vol. 31, no. 7, pp. 4795–4808, Jul. 2016.

[12] N. D. Weise, K. K. Mohapatra, and N. Mohan, "Universal utility interface for plug-in hybrid electric vehicles with vehicle-to-grid functionality," in *Proc. IEEE PES Gen. Meeting*, Minneapolis, MN, Jul. 2010, pp. 1–8.

[13] J. J. Sandoval, S. Essakiappan, and P. Enjeti, "A bidirectional series resonant matrix converter topology for electric vehicle DC fast charging," in *Proc. IEEE Appl. Power Electron. Conf. Expo. (APEC)*, Charlotte, NC, USA, Mar. 2015, pp. 3109–3116.

[14] M. J. Harrison, "A cyclo-converter and methods of operation," patent WO2008018802A2, Feb. 14, 2008.

[15] Y. Kosesoy, R. Bonten, H. Huisman, and J. Schellekens, "Control of a zero-voltage switching isolated series-resonant power circuit for direct 3-phase AC to DC conversion," in *Proc. Europ. Conf. Power Electron. Appl. (EPE/ECCE Europe)*, Hanover, Germany, Sep. 2022.

[16] Y. Xu, Z. Wang, Y. Shen, Z. Zou, and M. Liserre, "A VSC-based isolated matrix-type AC/DC converter without bidirectional power switches," *IEEE Trans. Ind. Electron.*, vol. 70, no. 12, pp. 12442–12452, Dec. 2023.

[17] P. Cortes, J. Huber, M. Silva, and J. W. Kolar, "New modulation and control scheme for phase-modular isolated matrix-type three-phase AC/DC converter," in *Proc. IEEE Ind. Electron. Soc. Conf. (IECON)*, Vienna, Austria, Nov. 2013, pp. 4899–4906.

[18] L. Gu and K. Peng, "A single-stage fault-tolerant three-phase bidirectional AC/DC converter with symmetric high-frequency Y-Delta connected transformers," *IEEE Trans. Power Electron.*, vol. 35, no. 9, pp. 9226–9237, Sep. 2020.

[19] R. Baranwal, K. V. Iyer, K. Basu, G. F. Castelino, and N. Mohan, "A reduced switch count single-stage three-phase bidirectional rectifier with high-frequency isolation," *IEEE Trans. Power Electron.*, vol. 33, no. 11, pp. 9520–9541, 2018.

[20] E. Asa, O. C. Onar, V. P. Galigekere, G.-J. Su, and B. Ozpineci, "A novel three-phase Oak Ridge AC/DC converter for wireless EV charger applications," in *Proc. IEEE Appl. Power Electron. Conf. Expo. (APEC)*, Phoenix, AZ, USA, Jun. 2021, pp. 437–443.

[21] J. E. Bosso, M. Llomplat, G. G. Oggier, and G. O. García, "Isolated bidirectional DC-to-three-phase AC converter for integration of renewable energy sources to electric grid," *IET Power Electronics*, vol. 12, no. 8, pp. 2058–2068, 2019.

[22] D. Menzi, J. W. Kolar, H. Sarango, Ó. Luciá, and J. Huber, "New 600 V GaN single-stage isolated bidirectional 400 V input three-phase PFC rectifier," in *Proc. Energy Convers. Congr. Expo. (ECCE USA)*, Nashville, TN, USA, Oct. 2023.

[23] H. Kim, J. Park, S. Kim, R. M. Hakim, H. Belkamel, and S. Choi, "A single-stage electrolytic capacitor-less EV charger with single- and three-phase compatibility," *IEEE Trans. Power Electron.*, vol. 37, no. 6, pp. 6780–6791, Jun. 2022.

[24] J. Allmeling and W. Hammer, "PLECS-piece-wise linear electrical circuit simulation for Simulink," in *Proc. IEEE Int. Conf. Power Electron. Drive Syst. (PEDS)*, Hong Kong, China, 1999, 355–360 vol.1.

[25] J. Jacobs, A. Averberg, and R. De Doncker, "Multiphase series resonant DC-to-DC converters: Stationary investigations," in *Proc. IEEE Power Electron. Specialists Conf. (PESC)*, Dresden, Germany, Jun. 2005, pp. 660–666.

[26] "TG-A1780," T-Global, Datasheet, www.tglobaltechnology.com.

[27] D. Neumayr, D. Bortis, and J. W. Kolar, "Ultra-compact power pulsation buffer for single-phase DC/AC converter systems," in *Proc. Energy Convers. Congr. Expo. (ECCE Asia)*, Hefei, China, May 2016, pp. 2732–2741.

[28] Z. Liao, N. C. Brooks, and R. C. Pilawa-Podgurski, "Design constraints for series-stacked energy decoupling buffers in single-phase converters," *IEEE Transactions on Power Electronics*, vol. 33, no. 9, pp. 7305–7308, Sep. 2018.

[29] D. Menzi, S. Weihe, J. A. Anderson, M. Kasper, J. Huber, and J. W. Kolar, "Novel S-Link enabling ultra-compact and ultra-efficient three-phase and single-phase operable on-board EV chargers," in *Proc. Control Modeling Power Electron. Workshop (COMPEL)*, ISSN: 2151-1004, Jun. 2023, pp. 1–7.

Application-specific investigation of inorganic potting material in drive trains

Sönke Fleck[1], Ulf Schümann[1]

[1] UAS Kiel

Corresponding author: Sönke Fleck, soenke.fleck@fh-kiel.de
Speaker: Sönke Fleck, soenke.fleck@fh-kiel.de

Abstract

The performance of electrical components is limited by their thermal behaviour. This paper focusses on the thermal behaviour of three different subassemblies with an inorganic potting, namely the DC link capacitor, the peripheral electronics and the windings of an electrical machine. It is shown in which applications the greatest performance advantages can be achieved by potting with the inorganic material. The thermal connection of the DC link capacitors cannot be improved compared to the current state of the art. However, there is great potential in the encapsulation of the peripheral electronics, as both the heat dissipation and the strength of the solder connections are improved. The direct connection of the winding head to the cooling jacket shows a considerable advantage for the motor windings. The current density can be significantly increased due to the improved heat dissipation and high temperature resistance of the cement.

1 Introduction

The next generation drive system requires complete integration of the electric machine and power electronics, preferably in a shared housing. The high power density results in an increasing focus on thermal management throughout the powertrain, since temperature is known to be one of the most important factors in service life considerations. At this point, the thermal conductivity of the potting materials plays a major role in several applications. Due to the increased basic thermal conductivity of cement compared to polymer (epoxy, silicone), these materials offer the potential to dissipate a large portion of the resulting power losses in high power density drive systems.

This paper examines the behaviour of three subassemblies, namely the motor winding of a hairpin motor, the driver board and the DC link capacitors.

The performance of electrical machines is limited by their thermal behaviour [1]. There are two ways to improve the thermal behaviour: Use technologies that generate less losses (e.g. hairpin windings [2]) or dissipate more heat from the system. The latter can be achieved, for example, by casting with highly thermally conductive material. In [3], a concrete example was used to show that increasing the thermal conductivity of the impregnating resin from $0,2\,W/(m\,K)$ to $5\,W/(m\,K)$ or $10\,W/(m\,K)$ can reduce the temperature increase in the winding head from 110°C to 80°C (-27,3%$_{rel}$,

-24,3%$_{abs}$) or 76°C (-30,9%$_{rel}$, -22,2%$_{abs}$) compared to the heatsink. In [4], the winding head is connected directly to the cooling jacket through the potting, as is also intended in this work. The measurement results show that the maximum temperature, depending on the operating point, can be reduced by approximately 10%-18% by potting.

As stated in [5] (Fig. 1), the causes of up to 96% of the failures of the peripheral electronics can be tackled by potting them. The encapsulation protects the board from humidity and increases the stiffness of the assembly. A higher stiffness leads, in general, to higher eigenfrequencies and lower warpage, though the added mass may counteract this effect.

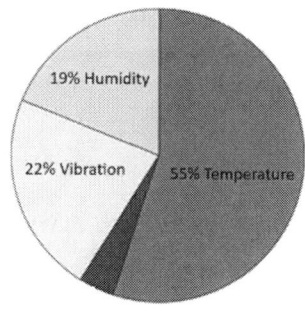

Fig. 1 Causes of Failure for Electronics [5]

PCBs usually have very small components and are cooled by conduction through the board and natural convection by the surrounding air. By efficiently spreading the heat, the potting can reduce the overall thermal resistance.

In addition to the power modules, the capacitors are critical components that have a crucial influence on the reliability and service life of the traction inverter. In addition to the applied voltage, the temperature is a decisive factor for ageing, as it has the greatest influence on electrochemical ageing. In this work, round-wound aluminium electrolytic capacitors are considered and the influence of the thermal connection to the heat sink is investigated by simulation.

2 Simulation Methods and Verification Measurements

2.1 Electrical Machine

To verify the results of the thermal simulation of the electrical machine, a reduced winding model of the motor is built, simulated and evaluated through an experiment. This reduced model consists of six windings mounted within an aluminium base (Fig. 2).

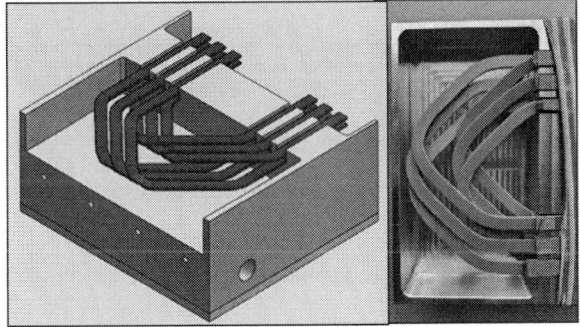

Fig. 2 Motorette Specimen for Verification

The test specimen is connected to a water cooling system to ensure a defined heat sink. The temperature is measured directly at the hairpins using thermocouples. The surface temperature is also monitored using an infrared camera.

The result of the measurement is compared with the simulation in Fig. 3.

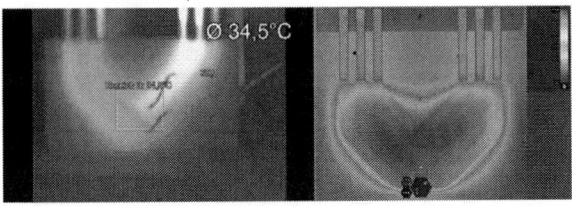

Fig. 3 Measurement and Simulation of Verification Measurement

The measurement with thermocouples results in a temperature of 32.3°C at the hairpins and thus show good agreement with the temperature of 31.9°C from the simulation. The relative deviation between simulation and measurement is -1.24%. This verifies the simulation. The thermographic IR measurement shows a temperature 34,5°C at the surface, corresponding to a relative deviation to the simulation of -7,54%. The difference between the measurement with thermocouples and IR measurement is indicating a systematic error in the IR measurement, most likely the emission coefficient was incorrectly selected.

The radiation on the right-hand side of the measurement (Fig. 3) is explained by the contact resistance at the interface to the current source.

The same simulation methods are then used to simulate a more complex model of a motor with hairpin windings and evaluate the influence of the cement potting on the temperature distribution of the motor by comparing it to a reference simulation with epoxy. The CAD-model of the unfilled complex simulation is shown in Fig. 4.

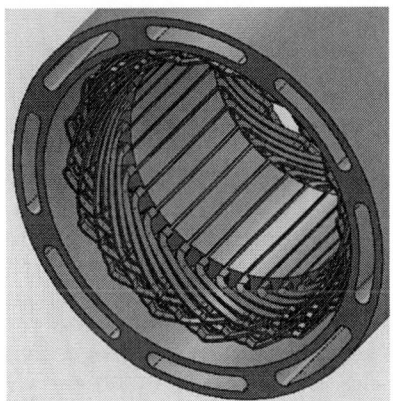

Fig. 4 CAD of complex motor model

A convection boundary condition is set on in the inside walls of the cooling channels as a heat sink. The heat transfer coefficient is calculated using the Nusselt correlation in Eq. 3.

$$d_{hyd} = 4 \cdot \frac{A}{U} \qquad \text{Eq. 1}$$

$$F = 1{,}615 \left(\frac{Pe \, d_{hyd}}{L} \right)^{\frac{1}{3}} \qquad \text{Eq. 2}$$

$$Nu = \left[49{,}37 + (F - 0{,}7)^3 \right]^{1/3} \qquad \text{Eq. 3}$$

Here A is the flow area, U the circumference, Pe the Peclet number and L the length of the cooling jacket. A total volume flow of 2 l/min is assumed, which is evenly distributed over all 8 channels, the volume flow for one channel is therefore 0.25 l/min. This results in a heat transfer coefficient of α = 168 W/(m²K) for the given geometry. The power loss is defined as Joule heat in the hairpins; current displacement effects and the temperature dependence of the electrical resistance are not taken into account here.

2.2 Peripheral Electronics

The circuit board under investigation is a driver for a SiC half bridge (Fig. 5). The DC-DC converters, the ICs and the gate resistors were identified as relevant heat sources. The comparative temperature measurement is carried out using thermocouples on the gate resistors and the ICs.

Fig. 5 Investigated PCB

The copper traces of the PCB are not explicitly taken into account in the simulation, but are implemented using a very rough trace mapping. This strong simplification allows a simple and fast calculation, the absolute values of the simulation deviate somewhat more from the measurement, but the temperature distribution shows a comparable pattern. The simulations are therefore evaluated comparatively without making absolute statements about the design. The variations simulated are the unpotted PCB as a reference, a fully potted variant and selective potting of the gate resistors.

The measurement results of the non-potted PCB are shown in Fig. 6. The gate resistors and ICs have the highest temperature at approximately 69°C and 62°C respectively.

After the PCB was potted on the top side (Fig. 5, right), the measurement was repeated. Tab. 1 shows the comparison of the potted and unpotted circuit board.

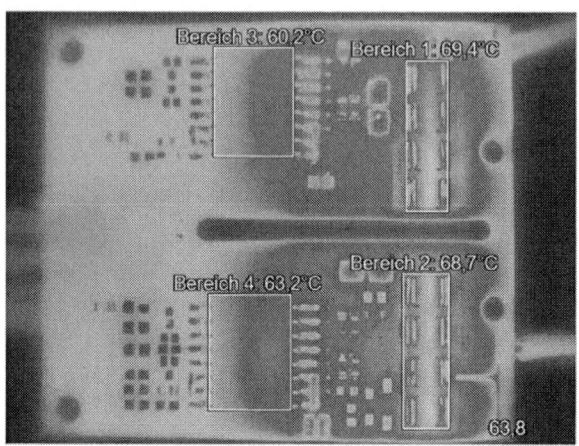

Fig. 6 Thermal Measurement of PCB

Tab. 1 PCB Measurement

	Reference	Potting
Gate 1	69,4°C	61,0°C
Gate 2	68,7°C	60,4°C
IC 1	60,2°C	64,8°C
IC 2	63,2°C	64,7°C

The potting therefore ensures a reduced maximum temperature due to good heat spreading.

In addition to the thermal performance, a harmonic vibration analysis of the 3 variants is carried out in order to be able to assess the influence of the potting on the mechanical behaviour of the circuit board. These simulations are also only of a comparative nature and are not verified by means of measurements.

2.3 DC-Link Capacitors

The casting of the aluminium electrolytic capacitors cannot be carried out with the cement at the current stage of development, as drying over 2 h at 150°C is mandatory. The capacitors cannot withstand this temperature undamaged, which is why a verification measurement is not possible. The simulative investigation of the capacitors is therefore only of a comparative nature.

The capacitor is modelled as a homogeneous body with orthotropic thermal conductivity. The heat sink is made of aluminium and is subjected to a heat transfer coefficient of 3000 W/(m²K) on the underside (Fig. 7).

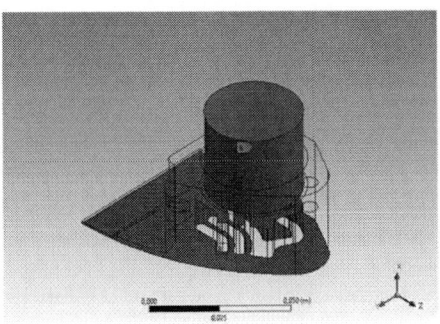

Fig. 7 Capacitor Model

3 Parametric Studies

3.1 Electrical Machine

In the parameter study, the influence of the thermal conductivity of the potting material on the motor with connection to the heat sink and a pure winding potting is first examined. The study shows that the influence of the thermal conductivity of the material decreases relatively quickly (Fig. 8)

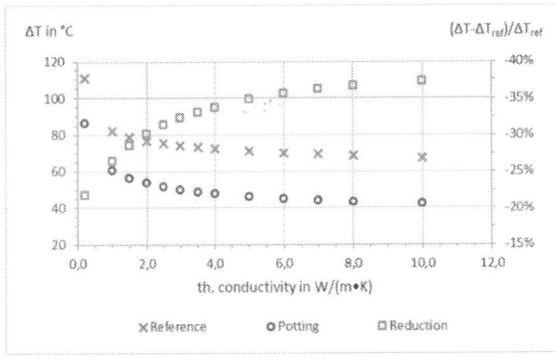

Fig. 8 Temperature over th. Conductivity of Potting

For material development, this means that an increase in thermal conductivity from a certain point (in this case approx. 3 W/(m K)) should become less important than, for example, flowability and fast processing times.

It can also be seen that the relative influence of thermal conductivity increases if the winding head is also connected directly to the cooling jacket.

The simulation of the more complex motor model shows a reduction of the maximum temperature from 177,6 °C ($\Delta T = 117,6$ K) to 112,3 °C ($\Delta T = 52,3$ K) for the chosen operating point for a thermal conductivity of 5 W/(m K) of the cement and 0,2 W/(m K) for the epoxy resin. By connecting the winding head directly to the cooling jacket, the maximum temperature can be reduced by

36,7% and the temperature increase compared to the heat sink is reduced by 55,5% (Fig. 9).

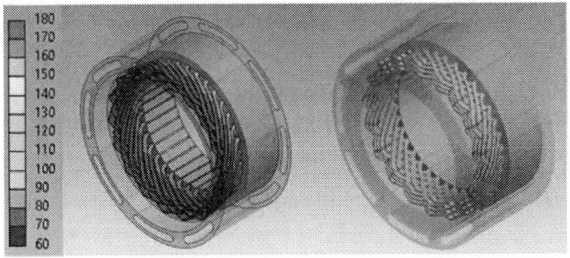

Fig. 9 Comparison of Simulation results for epoxy (left) and cement potting (right)

3.2 Peripheral Electronics

In agreement with the reference measurement, the simulations show that the maximum temperatures are reduced by the potting (Fig. 10). However, this influence is very small, as a large part of the heat spread is realized by the copper of the circuit board

Fig. 10 Simulation results of PCB

The modal analysis of the circuit board shows hardly any influence of the potting compound on the natural frequencies. The higher stiffness of the PCB is offset by the additional mass. The first natural frequency is around 2450 Hz for all variants.

3.3 DC-Link Capacitors

The parameter study of the capacitor simulation includes a variation of the thermal conductivity of the encapsulation material and the heat sink, the distances (radial and axial) between the capacitor and the heat sink, the power loss and the heat transfer coefficient. The result of the effect analysis is shown in Fig. 11.

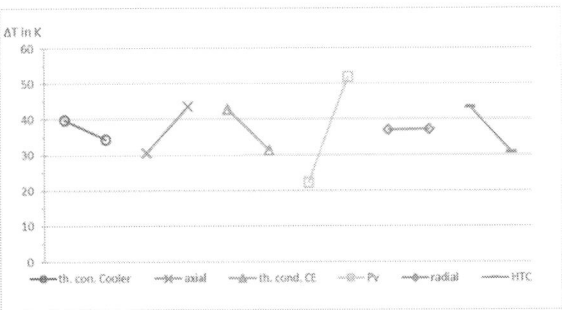

Fig. 11 Effect-Analysis for capacitor

The radial distance from the capacitor to the heat sink has a negligible influence, as the heat flow through the capacitor is mainly axial due to the wound structure. The heat transfer coefficient and the thermal conductivity of the heat sink and cement have comparable influences on the winding temperature. As the axial distance increases, the winding temperature also increases. This is another problem with encapsulating the capacitors with the cement, as the layer thickness of the cement is higher than with conventional TIM due to the potting process.

4 Conclusion

Cement encapsulation of DC link capacitors is not expedient with the material currently available. The high thermal conductivities and achievable layer thicknesses of the available TIM exceed the thermal performance of the cement and the high temperatures in the process currently make the application for electrolytic capacitors impossible.

Potting can have advantages for the driver board. Especially if the cement can be used to create a connection with a heat sink, potting offers a flexible option for thermal connection.

Potting offers enormous potential for the electrical machine. On the one hand, the maximum temperature can be massively reduced, and on the other, the cement offers excellent mechanical stability and a high continuous operating temperature. This allows the volumetric power density of the machine to be increased, as higher current densities can be used. As a result, either the output can be increased or material can be saved as the amount of copper required can be reduced. However, the added weight and the material costs need to be considered.

References

[1] M. Németh-Csóka, Thermisches Management elektrischer Maschinen, Wiesbaden: Springer Vieweg, 2018.

[2] J. Hagedorn, F. Sell-Le Blanc und J. Fleischer, Handbuch der Wickeltechnik für hocheffiziente Spulen und Motoren, Berlin-Heidelberg: Springer Vieweg, 2016.

[3] A. Huber, T. Nguyen-Xuan, N. Brossardt, F. Eckstein und M. Pfitzner, „Thermische Simulation eines hochdetaillierten Wickelkopfmodells einer elektrischen Antriebsmaschine," in ANSYS Conference & 32. CADFEM Users' Meeting, Nürnberg, 2014.

[4] Y. Yao, L. Gu, T. Fan, W. Sun und J. Luo, „Evaluation of Heat Transfer Characteristics of Aluminium Nitride (AlN) Potting Compound for the End Windings of Permanent Magnet Sychronous Machines," in 2011 International Conference on Electrical and Control Engineering, Yichang, China, 2011.

[5] T.-M. I. Băjenescu, Zuverlässige Bauelemente für elektronische Systeme, Wiesbaden: Springer Fachmedien Wiesbaden GmbH, 2020.

PCIM Europe 2024, 11– 13 June 2024, Nuremberg DOI: 10.30420/566262042

The Influence of the Glass Transition Temperature of Epoxy Mold Compounds on the Reliability of a Semiconductor Device

Stefan Schwab[1], Alexander Roth[1], Christoph Liebl[1], Timo Michael Müller[2]
[1] Infineon Technologies AG, Germany
[2] Infineon Technologies Asia Pacific Pte Ltd, Singapore

Corresponding author: Stefan Schwab, Stefan.schwab@infineon.com
Speaker: Stefan Schwab, Stefan.schwab@infineon.com

Abstract

Mold compounds used for chip encapsulation are an important part of discrete semiconductor packages. With increasing requirements such as high operational temperature, high voltage and increasing lifetime, the selection of the right mold compound is an important but challenging task. Thermomechanical properties of the mold compound, such as the storage modulus, the coefficient of thermal expansion (CTE) and the glass transition temperature (T_g), are key parameters in the selection. Various publications conclude that high T_g values (>>150 °C) are favorable or even required to ensure good device reliability. However, high T_g mold compounds can have several disadvantages such as lower temperature stability, higher water uptake and higher costs. For that reason, the influence of T_g on the overall package reliability is assessed in this publication by investigating 21 different materials with T_g values between 84 °C and 192 °C. Intrinsic material parameters such as weight loss at elevated temperature and humidity uptake have been measured. In addition, thermomechanical simulations were used to investigate the influence of T_g on the temperature dependent stress in the package. A package build up was carried out for some of the evaluated mold compounds to assess the performance at standard reliability conditions such as high temperature storage (HTS), temperature cycling (TC) and high temperature reverse bias tests (HTRB).

The results of this study show that while the T_g of the mold compound has a big influence on the properties, the use of high T_g materials do not necessarily increase the package performance and can even be a disadvantage. Considering that high T_g materials are typically more expensive than their low T_g alternatives, the use of well selected low T_g materials can come along with better cost competitiveness, without negative impact on the reliability of the device.

1 Introduction

Requirements for power semiconductor packages are getting more and more stringent. For instance, higher operational temperature (>200 °C), higher voltage capability and longer lifetime requirements are constantly requested by the market. At the same time, cost competitiveness remains a key target. Therefore, the selection of the right bill of material (BOM) for a discrete package is of high importance. The mold compound, which is used for the encapsulation, is one of the key elements of the semiconductor package. There are many requirements towards this material class, such as low contamination levels of ions, high thermal stability, and low cost. Furthermore, according to the literature, a high T_g mold compound is needed to achieve good reliability, especially for high temperature applications [1], [2], [3]. High T_g mold compounds are typically more expensive than their low

T_g alternatives. Therefore, the hypothesis of a need for high T_g materials for high temperature requirements needs to be evaluated.

2 Experimental

For the current study, 21 different mold compounds used for semiconductor packages from five different suppliers were investigated. They were selected in discussions with the suppliers focusing on high temperature reliability requirements and contain both, compounds in mass production and development grades. The compounds have T_g values between 84 °C and 192 °C. A list with some details about the EMCs used in the study is depicted in Table 1.

343

Table 1: Mold compounds used for evaluation. (MAR...multi aromatic; MF...multi-functional; BP...Biphenyl; DCPD...dicyclopentadiene; OCN...ortho-cresol novolac; BMI...bismaleimide; Sil...silicone)

EMC type	Resin type	Tg (TMA) (°C)	Filler content (wt.%)
A	MAR/MF/BP	135	87,5
B	DCPD/MAR	120	87
C	MAR/MAR	192	84,5
D	OCN/PN	153	79,5
E	MAR/MAR	105	88
F	BMI/PN	159	85
G	MAR/MF	171	84
H	MF/MF	188	84
I	MF/MF	171	84
J	MA/BP	109	90
K	MF/Sil	178	77
L	BP/MA	100	87
M	MAR/BP	90	84
N	OCN/PN	141	75
O	BP/OCN/MAR	108	89,5
P	BP/OCN/MAR	110	88,5
Q	BP/MAR	123	89,5
R	BP/MAR	123	89,5
S	MF/PN	129	84
T	OCN/PN	126	82
U	BP/MAR	84	87

To assess the influence of the T_g value on the properties of the mold compounds, various material characterizations were carried out.

All samples used for the evaluation were prepared at Infineon and measured in house to ensure comparability of the results. Test pieces were molded and post mold cured following the instructions from the suppliers. Typically, molding was carried out at 175 °C and a cure time of 90 seconds. Post mold curing was typically carried out at 175 °C for 4h. Some high T_g materials were molded and post mold cured at 200 °C instead, following the recommendations from the mold compound suppliers.

2.1 Weight Loss Study

To investigate the thermal stability of the mold compounds, a weight loss study at 200 °C was carried out. For this, four samples with a size of 35 x 35 x 1.5 mm³ were prepared for each material. The samples were pre-dried 24h at 125 °C to remove most of the humidity of the sample. Then the samples were stored at 200 °C for 1000 h. As there is still residual humidity in the sample at the beginning of the temperature treatment, the mold compounds were weighed after 24 h and this value was set as starting point of the weight loss. After 1000 h the samples were weighed again, and the weight difference calculated in wt.%. As the mold compounds have different amount of inorganic filler particles (which do not contribute to the weight loss), the weight loss was normalized to the polymer fraction of the mold compound for better comparison.

2.2 Humidity Uptake Study

To measure the humidity uptake of the mold compounds, six samples each of 60 x 10 x 1.5 mm³ were prepared and pre-dried at 125 °C for 24h. After that, the samples were weighed and put in a climate chamber at 85 °C and 85% rel. humidity for 240 h. The weight gain caused by humidity uptake was calculated in wt.%.

2.3 Dynamic Mechanical Analysis (DMA)

DMA measurements were carried out using 60 x 10 x 1.5 mm³ samples in a 3-point-bending setup with a Q800 dynamic mechanical analyzer from TA instruments. Samples were measured in a temperature range between -40 °C and 260 °C, using a 2 K/min ramp. A 10 µm amplitude was set with a force track of 125%. The frequency was set between 40 Hz and 1 Hz. Two heating cycles were carried out and data from the second heating cycle was used for comparison.

2.4 Thermomechanical Analysis (TMA)

TMA measurements were carried out using 4 x 4 x 4 mm³ cubes of mold compound using a Q400 TMA system from TA Instruments. The measurement was carried out between -40 °C and 260 °C using a heating rate of 5 K/min. Two heating cycles were carried out and the second cycle was used for comparison. The T_g value measured by TMA was used to compare the different mold compounds.

2.5 Adhesion Study

Button shear tests for adhesion measurements were carried out by molding 2 x 2 x 2 mm³ cubic buttons on a 0.5 mm thick lead frame. Cu K80 lead frames were used for the adhesion study. In addition, lead frames with the Infineon proprietary adhesion promotor technology, specifically developed for high reliability applications, were used as well in the study. 8 mold compound buttons per group and per stress condition were measured and the average shear force was used for comparison.

2.6 Simulation of Package Stress

Thermomechanical simulations were carried out to assess the interface stress in a semiconductor package using the finite element software Ansys. A simplified, schematic geometry of a semiconductor package (see Fig. 1) was used exemplarily for this investigation. Interface stress (shear and normal stress) was assessed at the interface between mold compound and leads of the package. To read out shear and normal stress values, the interface is modelled with bonded contact elements. Viscoelastic material models were set up based on DMA and TMA data of the mold compounds. During the simulation the mold compound and the rest of the model are connected at 175°C to simulate the molding process and then cooled down to -55°C . The interface stress was evaluated during a subsequent heating ramp with constant heating rate at steps of 10°C, the peak value of the heating ramp is 260°C. To minimize the impact of singularities the respective interface stress is averaged over 15% of the interface area with the highest stress level at each time step.

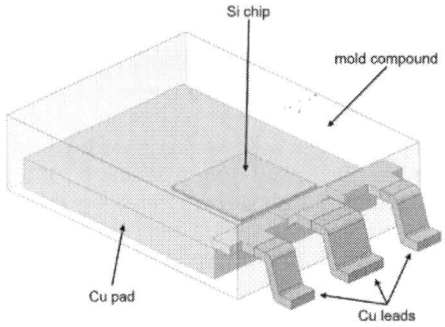

Fig. 1: Illustration of package model used for simulation.

2.7 Simulation of Power Cycling Performance

A finite element analysis (FEA) was carried out for different mold compounds to assess the power cycling (PC) performance. For that, a simplified model system was set up as shown in Fig. 2.

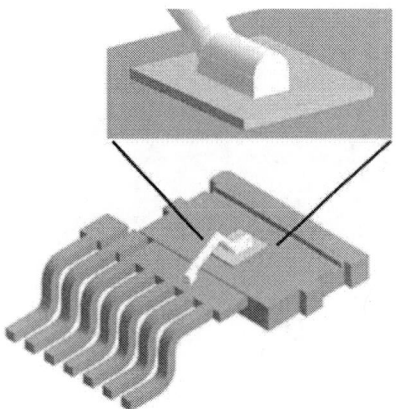

Fig. 2: CAD model for simulation (without mold compound)

The principal simulation flow (Fig. 3) consists of a transient thermal simulation where the transient temperature field is determined. A periodic, rectangular heat-flow load on the chip-surface and a convection boundary condition on the die pad bottom is applied. Those boundary conditions are varied to reach the required maximum temperatures. 4 t_{on}/t_{off} cycles were simulated to ensure a quasi-static thermal behavior.

1. Transient thermal simulation

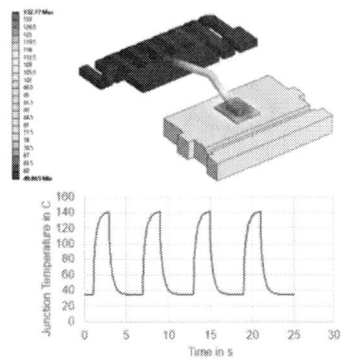

2. Static structural simulation

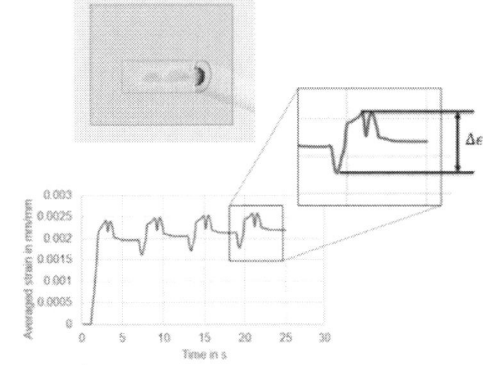

Fig. 3: Principal flow of PC simulation

3 Results & Discussion

3.1 Weight Loss and Humidity Uptake

The result of the weight loss study at 200°C are shown in Fig. 4 and Fig. 5.

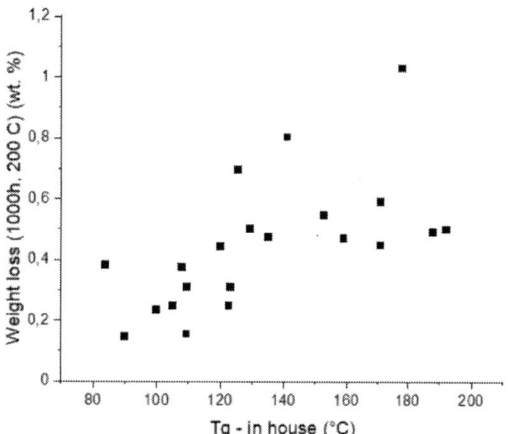

Fig. 4: *weight loss of different mold compounds at 200°C, 1000 h without filler calibration.*

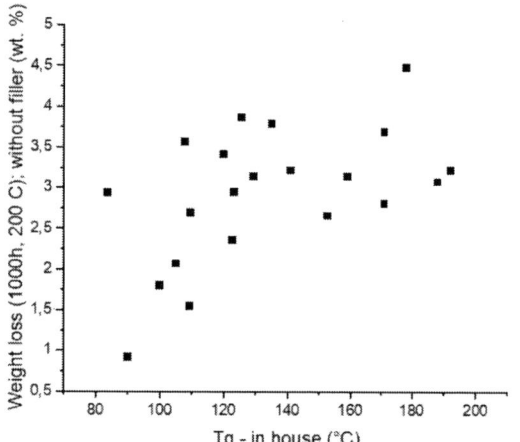

Fig. 5: *weight loss of different mold compounds at 200°C, 1000 h, calibrated by excluding the inorganic filler content.*

As shown, mold compounds with higher T_g values tend to have a higher weight loss compared to their low T_g alternatives. The water uptake of the different mold compounds is depicted in Fig. 6. High T_g materials show an increased water uptake compared to low T_g materials, which is a risk for humidity induced fails such as corrosion or delamination under humid conditions [4].

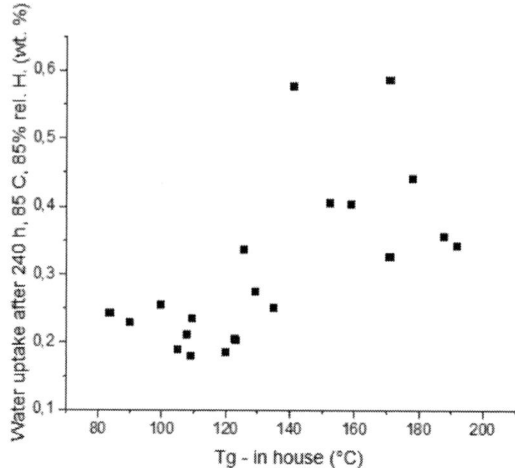

Fig. 6: *Humidity uptake of mold compounds with different T_g.*

The increased weight loss at high temperature as well as the higher water uptake of high T_g epoxy mold compounds can also be explained by the chemical composition of the polymer system.

There are several options to increase the T_g value of a polymer, as described in [5].

One option is by increasing the chain length of the individual polymer chains. This can be established by either using longer polymer chains in the un-molded mold compound (bi-stage) leading to a strongly increased viscosity during molding, or by reducing the catalyst concentration to generate fewer but longer polymer chains during molding. This would however severely increase the cure time of the mold compound.

Another option is to increase the cross linkage of the polymer, for instance by using multi-functional resins. This change typically has a lower negative impact on the molding process and is therefore the main development direction to increase the T_g in current epoxy mold compounds. However, by increasing the cross-linkage density, more functional groups are present in the final polymer system, which decreases the chemical stability and thus increasing the weight loss at higher temperature. In addition, the higher number of functional groups typically leads to an increased number of dipoles in the polymer system. This increases the hydrophilicity and thus the humidity uptake of the polymer. This is the main reason why high T_g mold compounds face higher weight loss at high temperature and higher moisture uptake.

3.2 Thermomechanical Stress

To assess the impact of the T_g on the thermomechanical stress in the package, simulations were carried out. As the thermomechanical stress is not only dependent on the T_g value, but also on the coefficient of thermal expansion (CTE) and the Youngs modulus, comparison of different mold compounds is challenging. Therefore, model systems are used to explore the impact of the T_g on the thermomechanical stress. Mold compound S was used as basis for this evaluation. Model systems with different T_g values are derived by shifting DMA and TMA measurement curves in the temperature axis, thus generating material models with the same properties such as CTE and modulus, but different T_g values. The shift of DMA and TMA data to generate the model systems is depicted in Fig. 7 and Fig. 8 respectively.

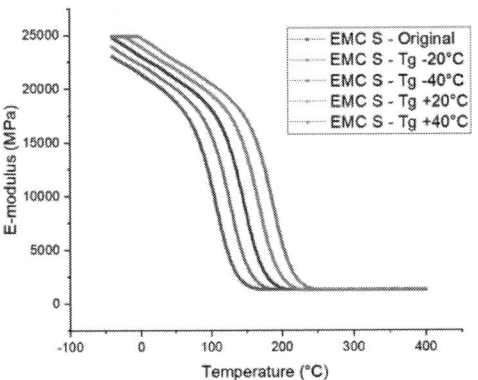

Fig. 7: Shifted DMA curves for T_g simulation of mold compound S. E-modulus capped at 25000 MPa

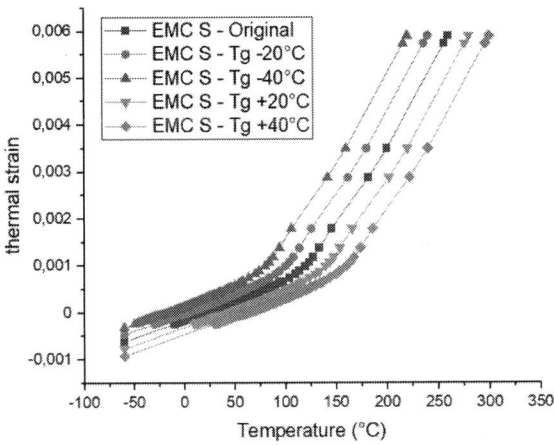

Fig. 8: Shifted DMA curves for T_g simulation of mold compound S.

The results of the stress simulation at the leads of the simplified package geometry (Fig. 1) are presented in Fig. 9. The stress level at the leads is shifted by changing the T_g value. A lower T_g value leads to a slight increase in the stress around the T_g value from around 8 to 10 MPa. However, at low temperatures around -50 °C, the stress of the low T_g mold compound is far lower (around 12 MPa) compared to the original compound (22 MPa). A higher T_g than the one from the original compound even leads to higher stress values at lower temperatures (>30 MPa). This shows that a higher T_g value can have negative impact on the stress level, which can lead to a higher risk of fails in temperature tests involving low temperatures (e.g. temperature cycling).

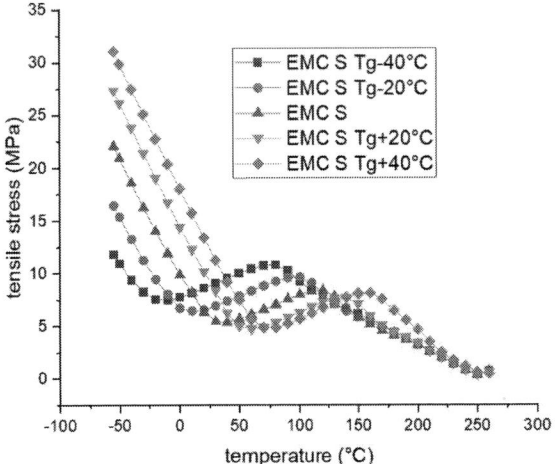

Fig. 9: Simulated thermomechanical stress at the lead to EMC interface with different T_g values based on EMC S

The thermomechanical stress in the package is influenced by many different parameters such as the CTE and modulus of the materials used. Therefore, the thermomechanical properties of the mold compound need to be carefully selected. However, the example above shows that a higher T_g value can have significant negative influence on the stress in the package.

3.3 Power Cycling (PC)

PC simulations with 9 different mold compounds (C, F, H, J, L, M, N, T, U) were carried out to investigate the influence of the T_g on the PC performance. Two different mean temperatures (T_{mean}) (80 °C and 140 °C) with three different ΔT values (45 K, 90 K and 120 K) were simulated to get a good overview of the over all PC performance. The results of the simulation are depicted in Fig. 10 and Fig. 11.

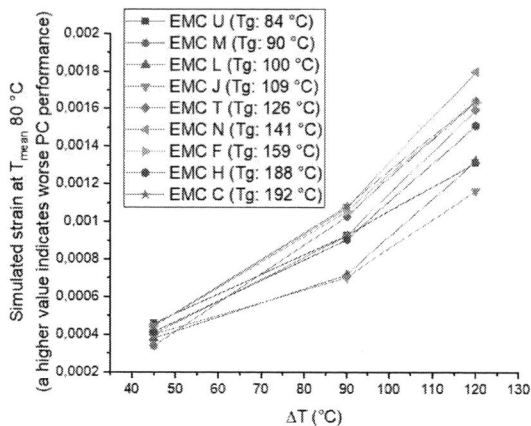

Fig. 10: Simulated PC strain for different mold compounds with T_{mean} 80 °C (sorted by T_g value)

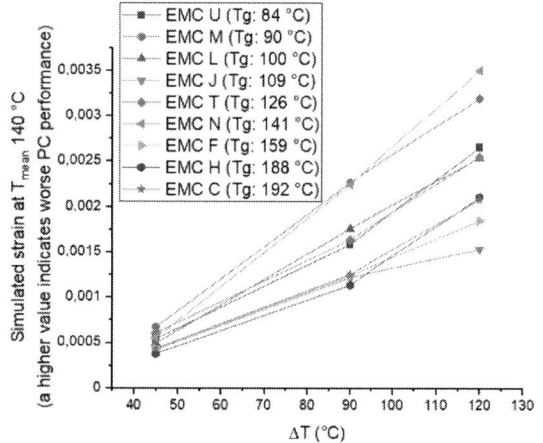

Fig. 11: Simulated PC strain for different mold compounds with T_{mean} 140 °C (sorted by T_g value)

The simulated strain at the chip/wire interface of the different mold compounds varies up to a factor of 3.

To investigate if there is any correlation between the T_g value and the simulated PC strain, the coefficient of determination (R^2) between T_g and the simulated strain was calculated for every stress condition and no correlation was observed (R^2 between 0.12 and 0.39).

3.4 Package Reliability

A package build-up with 8 of the mold compounds (D, J, L, O, P, S, T, U) was carried out in a QDAK package using a copper lead frame, a SiC MOSFET die and Al wires (500/75 μm). Infineon's internal adhesion promotor system was used to ensure good adhesion between the mold compound and the lead frame, wires and die respectively. Molding was carried out using the standard

parameters proposed by the mold compound suppliers. After initial electrical testing, samples were stressed by different stress conditions. After each stress condition, samples were tested electrically and scanning acoustic microscopy measurements (C-SAM) were carried out to look for delamination inside the package.

After high temperature storage at 200 °C and 2000 h, none of the samples showed electrical fails or delamination at any interface. This shows the excellent properties of the Infineon in house adhesion technology, which ensures excellent adhesion even after high stress conditions. To identify potential differences in the samples, the four groups P, J, L and T where further stressed for 1000 h at 225 °C. While all samples were electrical pass even after this increased stress level, delamination at leads at the bonding area was observed for sample P and T. Sample J and L showed no delamination at any interface, thus showing excellent high temperature capabilities. Both candidates have a low T_g (109 and 100 °C), indicating that low T_g materials show excellent stability at elevated temperature. This agrees with the results of the weight loss study described earlier.

Preconditioning tests (MSL1) we carried out with the 8 candidates as well. After preconditioning, all groups where electrically pass, but the two mold compounds with the highest T_g value of the package DOE (D, S) showed severe delamination at the die pad and leads. Mold compound U showed slight delamination at the leads. All other mold compounds did not show any delamination. Temperature cycling tests (TC) were carried out for the groups not facing delamination after preconditioning. Three different TC temperature ranges where used, with a temperature between -55/150 °C, -55/175 °C and -55/200 °C respectively. For the -55/150 °C TC, none of the samples showed electrical fails or delamination after 1000 cycles. However, starting die attach degradation was observed for mold compound T. For elevated TC testing of -55/175 °C, all mold compounds were electrically pass after 1000 cycles. Mold compound T showed die pad delamination. All mold compounds showed minor die attach degradation, with mold compounds O, J and L showing the lowest amount of degradation. At extreme TC conditions of -55/200 °C for 1000 cycles, all mold compounds except for mold compound O showed severe electrical fails due to die attach degradation (causing ΔV_{CE} issues). Mold compound O was electrically pass even at these extreme conditions. Mold compound O, J and L showed no delamination at any interface (despite the die attach degradation), while the rest of the groups showed delamination at leads and die pad. A summary table of the results is depicted in Table 2.

Table 2: Results of package DOE

#	T_g	HTS 200 °C 2000 h		HTS 225 °C 1000 h		Precon MSL1		Precon & TC -55/150 °C 1000x		Precon & TC -55/175 °C 1000x		Precon & TC -55/200 °C 1000x	
		el	delam	el	delam	el	delam	el	delam	el	delam	el	delam
O	108	ok	ok	n.A.	n.A.	ok	ok	ok	ok	ok	ok*	ok	ok*
P	110	ok	ok	ok	severe	ok	ok	ok	ok	ok	ok*	fail	severe
D	153	ok	ok	n.A.	n.A.	ok	severe	n.A.	n.A.	n.A.	n.A.	n.A.	n.A.
U	84	ok	ok	n.A.	n.A.	ok	minor	n.A.	n.A.	n.A.	n.A.	n.A.	n.A.
J	109	ok	ok	ok	ok	ok	ok	ok	ok	ok	ok*	fail	ok*
L	100	ok	ok	ok	ok	ok	ok	ok	ok	ok	ok*	fail	ok*
S	129	ok	ok	n.A.	n.A.	ok	severe	n.A.	n.A.	n.A.	n.A.	n.A.	n.A.
T	126	ok	ok	ok	severe	ok	ok	ok	ok*	ok	severe	fail	severe

* die attach degradation observed

In summary, mold compound O, J and L showed the best results in the package DOE, which is an indication, that low T_g materials perform well for high temperature applications.

In addition to the passive stress tests, high temperature reverse bias tests were carried out for mold compound J and L using a 1200 V SiC MOSFET device. HTRB was carried out at 200 °C for 1000 h with a reverse bias of 920 V (80%). No electrical fails or drifts were observed, showing that both compounds face no issues operating at a 200 °C junction temperature. The same samples passed a high voltage (920 V) / high humidity / high temperature stress tests (HV-H3TRB) without any issues.

4 Conclusion

The current study show that low T_g mold compounds tend to have lower weight loss at up to 200 °C as well as lower humidity uptake compared to high T_g materials. In addition, adhesion values on Cu and using Infineon's proprietary adhesion promotor technologies are not negatively influenced by a low T_g. Thermomechanical stress simulations showed that while the T_g value has an impact on the thermomechanical stress, high T_g values are not necessarily an advantage and selection of a mold compound with lower T_g can lead to a reduction in package stress.

For power cycling (PC), the result of the current study contrasts with those of a previous publication [4], concluding that using a mold compound above the T_g value is negatively influencing its PC performance. This can be true in some cases but should not be taken as a rule as other influences (coefficient of thermal expansion and modulus below and above T_g) seem to play an important role as well. Other than this hypothesis, the authors of the current study agree with the conclusion of [4], that lifetime modelling of molded packages is influenced by the change of properties below T_g vs above T_g must be considered to ensure accurate lifetime prediction for different mission profiles (T_{mean}, ΔT). Therefore, further investigations of the EMC influence on PC performance including package build up and PC testing on life samples are planned for the near future.

Package reliability data showed that excellent reliability at temperatures up to 200 °C can be achieved in passive and active stress tests using low T_g mold compounds together with Infineon's in house adhesion technologies.

Overall, this study showed that some low T_g molding compounds with T_g values around 100 °C did not show any issues in any of the relevant stress tests. When selecting the right compounds, advanced requirements such as good power cycling performance can be achieved with low T_g materials as well. Considering that low T_g mold compounds typically come at lower cost compared to their high T_g alternatives, low T_g materials lead to better cost competitiveness.

5 References

[1] B. M. Rok Šinkovec, "Effect of Organosilane Coupling Agents on Thermal, Rheological and Mechanical Properties of Silicate-Filled Epoxy Molding Compound," *Materials,* 2020.

[2] D. Ge, N. R. Subramanian, K. S. Yong, M. Y. Foo and S. L. Gan, "The impact of high glass transition temperature of molding compounds on power package warpage and stress performance," *2015 IEEE 17th Electronics Packaging and Technology Conference (EPTC),* 2015.

[3] A. Otto, R. Dudek, R. Döring and S. Rzepka, "Investigating the mold compounds influence on power cycling lifetime of discrete power devices," in *PCIM Europe 2019*, Nuernberg, 2019.

[4] A. Kwatra, D. Samet, V. T. Rambhatla and S. K. Sitaraman, "Effect of temperature and humidity conditioning on copper leadframe/mold compound interfacial delamination," *Microelectronics Reliability,* vol. 111, 2020.

[5] M. Gilbert, "Chapter 4 - Relation of Structure to Thermal and Mechanical Properties," in *Brydson's Plastics Materials (Eighth Edition)*, 2017, pp. 59-73.

[6] A. Teverovsky, "The Significance of Glass Transition Temperature of Molding Compounds for Screening and Reliability Qualification of COTS PEMs," *Engineering, Materials Science,* 2003.

PCIM Europe 2024, 11– 13 June 2024, Nuremberg   DOI: 10.30420/566262043

Corrosion Resistant Packaging for Power Semiconductor Modules – Modified Insulation Materials for Contaminated Environments

Michael Hanf[1], Andreas Brinkmann[2], Andrea Deißenberger[2], Volkmar Stenzel[2], Nando Kaminski[1]

[1] University of Bremen, Germany
[2] Fraunhofer IFAM, Bremen, Germany

Corresponding author and speaker: Michael Hanf, michael.hanf@uni-bremen.de

Abstract

The increasing electrification of high-power applications in various environments leads to more complex mission profiles and reliability issues for power semiconductor devices. A potential reason is corrosion mechanisms induced by contaminants like hydrogen-sulphide (H_2S) or similar species. Unfortunately, the mechanisms are not fully correlated to the surrounding environment and do not follow usual degradation models. While the improvement of degradation models and reproducible testing methods was part of recent investigations, increasing the robustness of the modules has not yet been. This study will show modified insulation materials to increase the robustness of IGBT-modules against H_2S-driven failure mechanisms.

1   Introduction

1.1   Environmental Reliability of IGBTs

The climatic reliability of IGBT power modules has been part of various investigations with respect to different stresses and applications. Especially climatic stress became more of a problem due to the increasing electrification in the mobility- and industry sector. In both applications, converter and power electronic circuits are typically not hermetically sealed and therefore, susceptible to the surrounding climate. In the past decade, moisture induced degradation was part of many research activities [1, 2] and improvements by the manufacturers [3, 4, 5]. But beyond the chip degradation, corrosion induced degradation of the packaging materials is a significant lifetime limiting process for silicone gel (Si-gel) potted multi-chip modules. If highly reactive species like Chlorine or Sulphur come along with the moist air, the corrosion mechanism changes completely and a significant amount of corrosion products are formed even at low temperatures [6, 7, 8]. Power modules, designed for industrial applications (≤ 1700 V) are often suffering from Sulphur-induced corrosion in specific applications, such as wastewater management, mining [8] and chemical industries. The

corrosion itself is commonly introduced by a corrosive gas, such as Hydrogen-Sulphide (H_2S) or Sulphur-Dioxide (SO_2). On metals like Copper (Cu) or Silver (Ag), both gases lead to the formation of corrosion products and depending on the surrounding climate, the particular corrosion mechanism is accelerated [7, 9].

1.2   Corrosion Mechanism

In power modules, Copper-corrosion is the predominant mechanism of interest and is mainly affecting the Direct Copper Bonding (DCB) ceramic substrates. Bare Copper, exposed to H_2S-containing moist air, forms Copper-Sulphide (Cu_2S / CuS) according to Eq (1 & 2) [10].

$$2Cu + H_2S \rightarrow Cu_2S + H_2 \qquad (1)$$
$$Cu + H_2S \rightarrow CuS + H_2 \qquad (2)$$

The reaction described in Eq. (1) is thermodynamically more beneficially compared to Eq. (2) [10] and therefore, Cu_2S can be defined as the primary corrosion product. Both reactions are not application relevant due to absence of water (H_2O) and the usage of bare Copper in the corrosion process. Copper is naturally forming an oxide-layer when

351

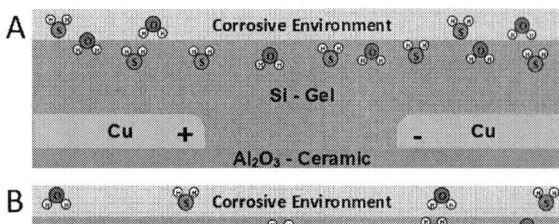

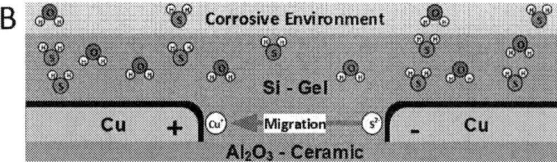

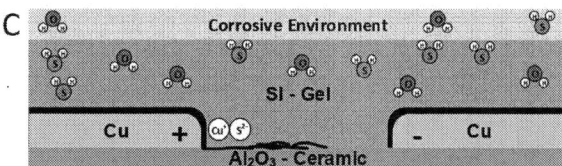

Figure 1: Corrosion mechanism in a Si-gel potted device with DCB-substrate in a moist, H_2S-containing environment.

exposed to air and therefore, the H_2S will mainly react with a Cu_2O layer on top of the Copper. At low relative humidity, Copper-oxide will not react with H_2S due to its thermodynamical stability [10]. But, the situation changes in the presence of high relative humidity (> 60 %) and the formation of multiple monolayers of water. H_2S will go into the aqueous phase after dissolving in the water layers and these layers are adsorbed by the Copper-oxide [10]. This effect is leading to the sulfidation of the upper oxide-layer and the formation of a brittle Cu_2S-surface layer [6, 11]. The corrosion process is self-limiting in a single gas atmosphere due to the passivating effect of the Cu_2S-layer and the impeded supply of Cu-ions through the oxide-layer.

In power modules, the dynamics of this corrosion process as well as the products are different due to the Si-gel potting. H_2O and H_2S can easily diffuse through the gel an react with the surface of the DCB-substrate. Furthermore, the concentration at the interface can be much higher compared to the surrounding climate [12, 13]. Alongside the Cu_2S-formation on the Cu-surface, a second, more critical mechanism takes place in the insulation trenches of the substrates [6, 7]: dendritic growth under the presence of an electrical field.

This effect can lead to shorts, permanent low-ohmic insulation distances and finally to failures of the complete device. The sequence of this particular mechanism is described in Fig. 1 with a simplified structure and the corresponding reactions in Eq. (3-5) [6, 7].

$$H_2S \rightleftarrows 2H^+ + S^{2-} \quad (3)$$
$$Cu \rightarrow Cu^+ + e^- \quad (4)$$
$$2Cu^+ + S^{2-} \rightarrow Cu_2S \quad (5)$$

First, H_2O and H_2S will diffuse into the Si-gel, with a time constant limited by the outer plastic packaging (Fig. 1, A). The H_2S will react at the cathode according to Eq. (3), forming Sulphide-ions ($p_H \gg 7$) which are able to migrate through the gel matrix [7] to the anode (Fig. 1, B), where positive charged Cu-ions are available (Eq. (4)). Simultaneously, the typical black, brittle tarnish is formed on top of the Cu-surface, independently of the electrical field (Fig. 1, B). Equation (5) describes the further reaction at the anode, forming Cu_2S-dendrites or dendritic structures (Fig. 1, C). In this case, the corrosion process is described as an anodic migration phenomenon [14] and is different from typical electro-chemical migration [15]. In [6], two different structures were described, an anodic corrosion product as described above, which is able to short the insulation distance, and a cathodic growth. These shorter dendrites are a side effect, linked to the H_2S-induced corrosion of the Cu-surface and are just slightly influenced by the electrical field. So far, they were just reported in accelerated testing [6, 8, 16] at high gas concentrations.

The formation of the primary, anodic growth can be accelerated by temperature, relative humidity, gas concentration and bias voltage [6, 7]. The influence of the relative humidity is predominant and leads to faster, more pronounced dendritic growth. This fact can be utilised for accelerated testing to separate the package-related corrosion from the degradation at chip-level [6].

Table 1: Overview of different device groups, modifications and test conditions

Device Types	Modification	Test Conditions
No. 1 (Module Group **1**)	Unmodified	85°C/85% rel.h./25ppm H_2S
No. 2 (Module Group **1**)	Conformal coating	85°C/85% rel.h./25ppm H_2S
No. 3 (Module Group **2**)	Unmodified	45°C/93% rel.h./50ppm H_2S
No. 4 (Module Group **2**)	H_2S trapping agent	45°C/93% rel.h./50ppm H_2S
No. 5 (Module Group **2**)	Corrosion inhibitor	45°C/93% rel.h./50ppm H_2S
No. 6 (Module Group **3**)	Si-gel sealing	45°C/93% rel.h./50ppm H_2S

1.3 Towards Robust Packaging

The understanding of the primary corrosion mechanism, inflicted by H_2S is key to develop corrosion resistant packaging solutions. A straight forward approach is the usage of hermetically sealed packages [17] as they are used for discrete devices and thyristors. The sealing would help to avoid any kind of contamination and the absence of moisture will prevent the corrosion. But, the significant higher pricing, makes this solution unattractive to high volume markets. This leaves only one option: the problem needs to be solved inside the package with different materials, designs and processes, just the same way it was done in the past with the moisture-induced degradation.

Starting with the substrate materials, the DCB-substrate with an Al_2O_3-ceramic was replaced by some manufactures by Insulated Metal Substrates (IMS) in recent years [18]. Typically, the conduction path is still made from Copper but, it allows the usage of Aluminium (Al), which is significantly more robust against corrosion. Another option is a surface treatment with Nickel (Ni)-plating, increasing the corrosion resistance significantly as known from PCBs [19]. Anyhow, the usage of different metals and surfaces are increasing the pricing, leads to more complex processing or comes with trade-offs in the thermomechanical performance. An easier approach is the growth of a thicker oxide-layer as discussed in [10], retarding the supply of Cu-ions to the surface to impede the reaction with H_2S.

Another straight forward approach also known from PCBs, is the usage of conformal coatings. These coatings will prevent the surface from high moisture levels as wells as contaminants and work as insulation materials for smaller distances. A main concern of these coatings is the adhesion and the cleanliness of the surface. On the one hand, if the adhesion to the surface is not intact, a closed water layer can build up, shorting the distance or lead to electro-chemical migration [15, 20]. On the other hand, anodic migration phenomena [14] can grow through the coating material, though with higher time constants. Anyhow, the application of a conformal coating needs the right material as well as a properly designed processing

to achieve a sufficient protection. Furthermore, an additional process step would be introduced into the production flow of the power module, leading to higher cost and time consumption.

As a significant part of the problem, the Si-gel was identified as the main diffusion path. Changing the gel to a hard mould [18] will solve the issue (H_2S-induced corrosion is not reported in hard mould), but is not commonly used and changes the thermo-mechanical behaviour. The gel might also be a chance of directly influencing the critical pollutants with modifications to inhibit the corrosion or trap and store the reactive species. The main advantage of this approach is changing just a single material, instead of the processing sequence of the module itself. But, modifications on the main insulation material for both package and chips is critical.

If the package cannot be sealed and the gel comes without modifications, another option is the sealing of the gel surface. As shown in [6], exposing the Si-gel surface directly to the environment, increases the corrosion rate significant. Sealing the surface with a flexible material can retard the ingress and extend the lifetime. For this, an additional, well defined material is required as well as a supplementary process step.

Overall a number of possibilities can be utilised to achieve a higher corrosion resistance against this particular mechanism. To prove the usability of each modification, accelerated corrosive gas tests under high voltage were executed on standard IGBT-modules. This excludes the option of modified substrate materials, but allows the usage of different insulation materials as well as investigations on the Cu-oxide-layer.

2 Devices and Test Procedure

2.1 Devices Under Test (DUT)

To test the different modifications, a standard IGBT-module package was used which allows optical analysis through the gel without permanently damaging the plastic housing. In [6, 16], a widely used package type with removable cover was utilised. Beside the advantage of the removable lit,

Table 2: DUTs for the investigations on conformal coatings

Split	Conformal Coating	Process Modification
Split 1	-	-
Split 2	Type 1	-
Split 3	Type 2	Process 1
Split 4	Type 2	Process 2

an insulation material can be potted directly on the chips and modifications like a conformal coating can be processed easily. For this experiment, three different device types were used as stated in Tab. 1, containing different chip sets as well as different substrate designs. Devices of group 1 (Tab. 1), were equipped with a modern, moisture robust 1700 V chip set in a half-bridge configuration. Regarding this test campaign, conformal coatings (No. 2) were processed on the metallic parts (chips, bond-wires, substrates) and potted with a standard Si-gel. The usability of a robust chip-set allowed a highly accelerated climate at 85°C and 85 % rel. humidity. All devices from group 2 and (Tab. 1) are half-bridge modules with an old generation of 1700 V IGBT- and Diode-chips, which are not moisture robust and tend to show a significant failure rate in a standard HV-H³TRB-test. Therefore, the test conditions were set to a milder climate with 45°C and 93 % rel. humidity. Basically, the modifications in group 2 were related to the Si-gel potting with a trapping agent (No. 4) and a corrosion inhibitor (No. 5) adaption. Furthermore, a third group (group 3) was introduced with a commercially available, modified 1700 V IGBT-module, using a sealing layer (No. 6) on top of the Si-gel surface. The moisture robustness of this chip-set is unknown and for comparison to group 2, the milder climate at 45°C and 93 % rel. humidity was chosen.

2.2 Test procedure

There is no standard test procedure available for power semiconductor modules at the time of this publication. In [7], a procedure at 85°C, 85 % rel. humidity and 50 ppm H_2S was proposed to build up on the knowledge of HV-H³TRB testing [1, 2]. But, as shown in [6], a milder climate with increased relative humidity can deliver a comparable result without causing moisture induced degradation at the chips. Furthermore, the concentration dependency is not fully clarified and experiments on bare Copper showed no significant difference between 25 and 50 ppm [7]. Group 1 was tested up to 1000 h at 25 ppm to achieve a longer ingress without causing shorts too early in the test. Devices of group 2 & 3 were exposed to 50 ppm for 720 h to test saturation and penetration effects due to the kind of modifications (saturation of trapping agent, penetration of sealing, etc.).

Anyway, the test was conducted as described in [6] with intermediate optical inspections after given time steps to monitor the actual growth over the test cycle. The chamber was designed according to IEC 60068-2-43, with a high voltage data acquisition system to observe the leakage currents of all devices. The bias voltage was set to 80 % of the nominal voltage (V_{nom}), as commonly used in the HV-H³TRB test.

3 Impact of Conformal Coatings

3.1 Concept of Modifications

Conformal coatings are widely used in electronic assemblies and highly used on PCBs, to extend the robustness against environmental effects such as moisture, pollutants and dust. Furthermore, these coatings can be used to increase the insulation capability of conduction paths, where high electrical fields are present. There are a number of coatings available to increase the robustness

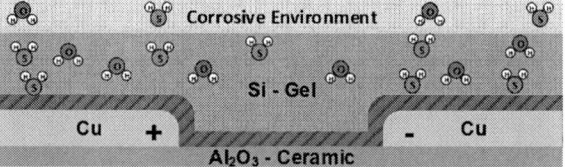

Figure 2: Application concept for a conformal coating in a Si-gel potted power module (red line)

against corrosion and moisture, but not all of them are able to stick on a DCB-substrate. Another critical aspect is the high temperature rating of power semiconductor modules of up to 175°C, which is significantly higher compared to most PCB applications. An option for these applications is Fluoro-polymer-coatings, which can handle the electrical field as well as the thermo-mechanical stress. These coatings are designed for moisture and corrosion protection [20] as well as a sufficient adhesion to a variety of materials.

The conformal coating is applied by the manufacturer of the IGBT-module in defined processes over the complete substrates (including chips). This leads to the situation at the corrosion interface as shown in Fig. 2. The applied coating acts as a barrier for H_2O and H_2S, avoiding any corrosion products on the Cu-surface as well as it supresses the ion migration. As long as the coating is not penetrated and the adhesion is intact, no signs of corrosion in critical areas should be found. Tab. 2 shows the corresponding device groups for this experiment with two different conformal coatings as well as a variation of the processes for one type of coating. Split 1 (Tab. 2) is the unmodified standard module, which is commercially available. Split 2 (Tab. 2) was modified with coating type 1, and both splits (1 and 2) were tested in the same run to get a direct comparison. Afterwards, splits 3 and 4 were tested together with a second type of coating and two different processes on the same module type.

Figure 3: Results after 1000 h of 85 °C, 85 % rel. humidity, 25 ppm H_2S and 80 % V_{nom} for the corresponding splits. The upper row represents area 1 around the auxiliary collector, the lower row area 2 around the chips of the high-side switch.

3.2 Results

Each module was compared in two different areas. Area 1 is around the auxiliary collector (Fig. 3, upper row) and area 2 is around the chips at the high-side switch (Fig. 3, lower row). In general, the corrosion products at the unmodified samples differ from the results in [6, 7, 8, 16] with a low number of isolated dendritic structures. The result here is more like a continues area of Cu_2S around the anode, without a significant growth from the cathode. Overall, the resulting corrosion state after 1000 h of harsh testing is acceptable without any shorted insulation trench. But this is not explainable for the devices from split 2 with the conformal coating of type 1. In Fig. 3, all samples from split 2 showed a high corrosion rate especially around the chips, with completely filled trenches. This is kind of a surprising result, due to the obviously extended corrosion state compared to the unmodified sample. As stated in Fig. 1, the Copper surface patina is a sufficient measure of the H_2S-molecules reaching the surface of the substrate. In comparison in Fig. 3, the patina of the unmodified sample is darker in area 1 and this fits to the comparable corrosion state. For area 2, this is not true with nearly the same colour around the Copper surface and a significant higher corrosion rate on the side of the coated samples. In conclusion, this type of corrosion protection is worse than the unmodified sample. Furthermore, the higher corrosion rate with different coloured structures look like an incomplete coating layer or a significant loss of adhesion.

With these first results, a second test with split 3 and conformal coating type 2 was executed, comparing not only the material but two different processes of applying the coating. Fig. 3 shows the DCB-substrates with a significant different result.

First of all, the patina looks much darker at the coated samples, which is still not fully explainable. The coating itself creates two new transitions between gel, coating and the substrates, leading to a more complex system in terms of diffusion. But nevertheless, the coating should work as a barrier to H_2S and H_2O but it does not over the full test time. Anyhow, coating type 2 shows a significant lower dendritic growth compared to type 1. Furthermore, the processing of the material is critical to the resulting corrosion state and process 2 reveals a more promising result.

To understand the impact of the processing of applying the conformal coating, it is worth to look at the intermediate optical analysis of splits 3 and 4 in Fig. 4. The pictures were taken after ~450 h of accumulated testing time with a clear indication of anodic growth in split 3. This analysis already indicates a significant difference in the time constant of the corrosion process and is attesting a better result to the combination applied to split 4.

3.3 Discussion

The test campaign on modified samples with conformal coating shows, that the coating can be part of the problem or might even lead to a higher number of corrosion products. But so far, if the correct material is utilised and the process applying the coating is optimised, at least a retardation of the corrosion can be achieved. Humidity wise, no impact of the coating on the reliability can be addressed and therefore, the qualification of a modified standard product is easier. On the other hand, the results in Fig. 3 indicate, that the processing of the material is critical and so is the cleanliness of the devices to achieve a sufficient adhesion, which will increase the effort of integrating this type of modification to a high-volume product.

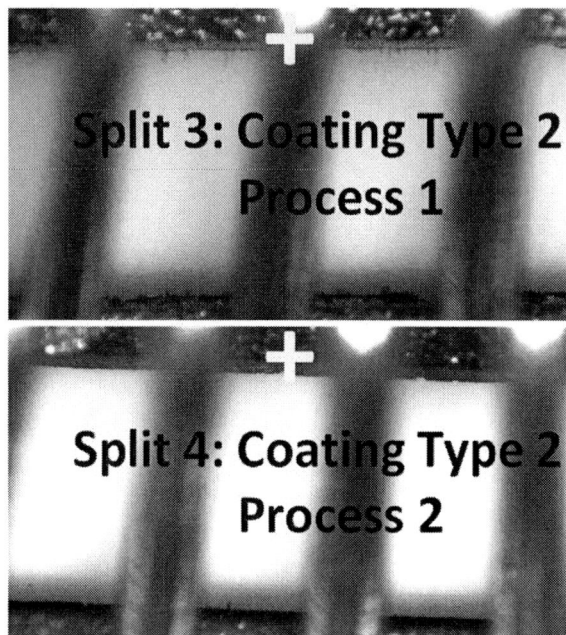

Figure 4: Comparison of split 3 and 4 after ~450 h of accumulated testing time. Left, anodic growth is already visible while the counterpart of split 4 shows no signs of anodic corrosion.

4 Modified Potting Materials

4.1 Basics

To prevent Sulphur induced corrosion effects, two methods for the minimisation of H_2S accumulation at the Copper surface were tried: a trapping agent with a high Sulphur affinity and an active corrosion inhibitor were added to the Si-gel, respectively. Before the modifications could be addressed, a suitable reference gel needed to be defined. As the composition of the commercial Si-gel, used in the reference modules was unknown, a similar material was formulated. A special focus was set on the homogeneity of the gel as well as a similar elasticity and viscosity. Furthermore, the self-developed formulation allowed a reduction of the components of the refence gel to a minimum, which decreases possible sources of error and undesired interactions with the modifying agents. Silver was selected as the trapping agent because it is known for its high affinity to Sulphur [21]. A disadvantage of Silver is its high conductivity, which was countered by using nano-scaled Silver for uniting a small concentration of the modification agent, while having a large active surface for the trapping reaction. In the bulk material, Silver reacts directly with H_2S according to Eq. (6).

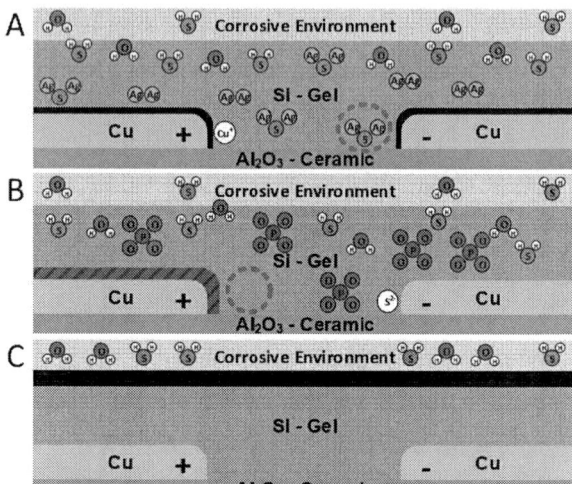

Figure 5: Modifications of the potting material with a trapping agent (A), an active corrosion inhibitor (B) and a sealing on top of the Si-gel surface

The basic principle of this modification is shown in Fig. 5 A with the corresponding reaction (Eq. (6)) in the corrosive gas test.

$$4Ag^0 + 2H_2S + O_2 \rightarrow 2Ag_2S + 2H_2O \qquad (6)$$

At the device surface, Copper and Silver compete for the Sulphide-ions that are formed and migrate through the gel (Fig. 1 B, Fig. 5 A). Hence, the triggered primary failure mechanism will be reduced, while the secondary mechanism, where Sulphide directly interacts at the surface is almost not affected. Furthermore, the Silver is able to react with the diffusing H_2S-moleculs in the gel to reduce the H_2S-concentration at the metallic interface.

For the second modification approach, Phosphates were used as an active corrosion inhibitor that is compatible with the Si-gel (Fig 5 B). The mode of action of the active corrosion inhibitor is based on several pillars. First is the passivation of the Copper surface, which is known from the corrosion protection of drinking water pipes [22].

$$3Cu_{2+} + 2PO_4^{3-} \rightarrow Cu_3(PO_4)_2 \qquad (7)$$

The surface Copper reacts with the Phosphate, forming superficial Copper-Phosphate, inhibiting the reaction with the corrosive species (Eq. (7)). Secondly, phosphate systems are used as buffering agents, keeping a neutral p_H whereas the Sulphide-formation proceeds in an alkaline medium (section 1.2). Furthermore, the counterions of the phosphates could act as trapping reagents comparable to the Silver, described above. By using Phosphates with differing solubilities, the durability of the inhibition process could be increased. The

Figure 6: Potting materials after curing: reference gel (A), Ag-trapping agent (B), Phosphate corrosion inhibitor (C), surface sealing (D).

principle is shown in Fig. 5 B with the formation of a $Cu_3(PO_4)_2$ layer at the anode, binding the Cu-ions, resulting of the electrical field.

The last modification was defined as a sealing layer between the environment and the Si-gel surface. Fig. 5 C shows the basic concept of the commercial module type. As stated before, the actual material of the layer is unknown and therefore, the concept itself is not fully clear. To reduce the reaction at the Copper-surface, a reduction of the H_2S-concentration is as sufficient as a reduction of the H_2O-concentration. If one, or both concentrations are affected, the actual corrosion at the interface is reduced.

All modifications were pictured after the curing process in Fig. 6, together with the reference gel (Fig. 6 A). An important feature of the proposed package is the optical analysis without removing the gel. But, the modifications are changing that benefit and need to be removed before presenting the information of possible growth.

4.2 Results of Modified Potting Materials

The modified samples as well as the references (group 2) and the devices from group 3 (sealed Si-gel surface) were tested in the same test run at 45°C, 93 % rel. humidity, 50 ppm H_2S and 80 % V_{nom}. It is important to note, that modifications at the main insulation material, the Si-gel, is very critical and the qualification process for such a material change is highly complex. It is known from previous testing, that the chips of this module are sensitive to moisture-induced corrosion and the impact of the gel modifications is not fully predictable. Therefore, a milder climate at 45°C and 93 % rel. humidity was chosen to trigger the correct failure mechanism [6], without over-stressing the semiconductors.

An interesting side-effect was investigated in terms of Cu-oxide thickness and its impact on the corrosion rate. Figure 7 shows the results of the two different kinds of references. In Fig. 7 A, the bare samples from the manufacturer, after the potting with a custom-made Si-gel is shown in comparison to the standard industry product in Fig. 7 B. Both samples were removed from the chamber after ~450 h of accumulated testing time with the results shown in Fig. 7 (lower row). As proposed in [10], the thicker oxide (defined by the colour of the surface) from Fig. 7 B retards the corrosion rate significantly, while the unprocessed sample in Fig. 7 A displays a high number of shorts already. Furthermore, the secondary effect from the cathode is much more pronounced. The impact of the Si-gel types remains unknown for this experiment but, the modified gel types were all applied on samples with the unprocessed oxide and the expectedly higher corrosion rate. In Fig. 6 B & C, both gel types are pictured and the samples are not fully optically inspectable through the gel. Therefore, the gel was removed and the results are shown in Fig. 8. In summary, the trapping agent and the corrosion inhibitor are lowering the corrosion rate significantly compared to the references in Fig. 7. Furthermore, the primary degradation mechanism is fully suppressed in Fig. 8 B, with the corrosion inhibitor. The cathodic growth is still present and occurs at all trenches, but no growth through migration can be detected. Fig. 8 A shows the situation for the trapping agent with a slightly different result. While the trenches at area 1 (around the gate pads) show no primary degradation mechanism, this is changing in the middle of the substrate (lower row, Fig 8 A). Here, a less pronounced anodic growth is found, and it can be stated, that the trapping agent is not able to store unlimited amounts of Sulphides. This

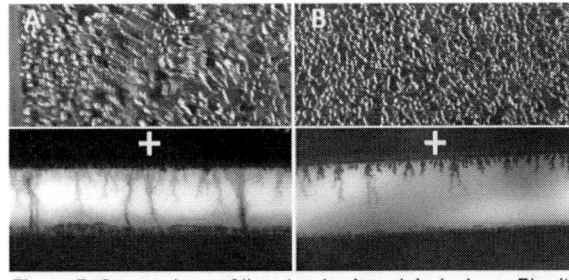

Figure 7: Comparison of the standard module (column B) with the experimental module (column A). The Cu-oxide layers are different, with a higher thickness at the standard module (B) and the resulting lower corrosion rate at the bottom pictures.

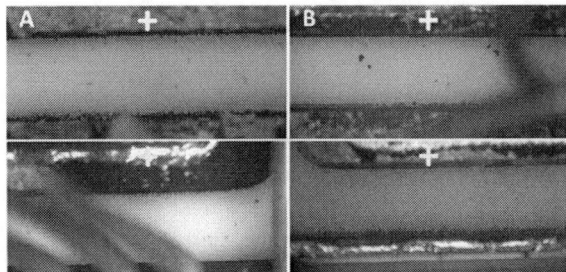

Figure 8: Resulting corrosion products at the DCB-substrates after removing the modified gels with: the trapping agent (column A) and the corrosion inhibitor (column B)

could be solved by increasing the trapping agent concentration, but is also impacting the curing of the gel as well as the electrical and mechanical properties. Nevertheless, the cathodic growth is also reduced in comparison to the corrosion inhibitor, which leads to the assumption, that the trapping agent is able to reduce the actual amount of Sulphides at the critical interface or at least the H_2S concentration. In conclusion, both gel modifications deliver an excellent result in terms of retarding the corrosion mechanism. After a total test time of 720 h, no shorts were found and in comparison to the references in Fig. 7, the growth rate is significantly reduced in terms of the trapping agent and not optically measurable for the corrosion inhibitor.

4.3 Results of the Si-Gel Sealing

The last modification, introduced in section 4.1 was the sealing of the Si-gel surface to impede the ingress of reactive species. As stated previously, the device concept was commercially available at the time of this investigation and tested under the same conditions like the modified gel types. The appearance of the sealing layer can be seen in Fig. 9 A, where a side-view of the module is shown with the corresponding sealing layer on top of the Si-gel surface. For a further analysis, the layer was removed and revealed a colour change over time. Fig. 9 C views the backside of the sealing after 720 h of testing at 50 ppm and this particular colour is similar to the initial state of the module. On the other hand, the exposed surface in Fig. 9 B is much darker and leads to the assumption, that this layer might be equipped with a trapping agent as well. Furthermore, the corrosion products in

Fig. 9 D are indicating a significant amount of corrosive species on the interface. But, in contrast to the modified gels, the surface patina of the Cu-pads is limited to particular areas with just little amounts of corrosion. Optically, the sealing was intact over the testing time, but still a significant number of dendrites were observed in the trenches of all samples, but no short was formed at any position. If the Si-gel under the sealing is a standard material and the sealing was not damaged, the number of possibilities on how this corrosion is possible are limited. One option would be a dispense of the reactive species over time, if the layer is not able to store more amounts of H_2S. In this case, the diffusion would continue with a lower concentration over time and the layer is just delaying the corrosion. But the overall concentration in the test is much higher compared to actual applications and therefore, this modification might increase the lifetime in the field significantly.

4.4 Discussion

The modification of the potting materials of a standard IGBT-module showed an overall high impact on the corrosion rate. A direct improvement of the Si-gel with a trapping agent is retarding the primary corrosion effect significantly, lowering the concentration of H_2S at the interface, and binds high amounts of Sulphides in the gel. But mixing Silver in an insulation material might cause problems in terms of other reliability tests. Furthermore, the electrical specifications of the gel will be different and can change the behaviour especially in active operation, where the insulation material is needed the most. Nano-Ag is a straight forward approach and if the trapping effect can be utilised

Figure 9: Pictures of the sealed device with the basic setup (A), colour of the sealing layer after 720h of test time (B), colour of the sealing on the backside (similar to the initial state, C) and the corrosion products after removing the gel (D)

with different elements, the electrical trade-offs might be reduced. A more promising result was achieved with the corrosion inhibitor. The main corrosion mechanism was fully suppressed and the modification with Phosphates are not influencing the electrical parameters of the gel as much as the Silver does. A sealing of the Si-gel surface was not performing as good as expected. The sealing is overall retarding the ingress into the gel but still, significant amounts of anodic dendrites were found after the test. In comparison to the gel modifications, this result is not satisfying and might cause problems in the applications based on the actual, not fully confirmed diffusion into the gel. If the layer was just saturated due to the high concentration in the test, then this concept can work in applications with significantly lower concentrations (which is expected). But if there is more to the diffusion process, the result might be similar in the applications.

5 Conclusion and Outlook

This investigation was a proof of concept of four different types of modified standard modules to enhance the robustness against Sulphur-induced corrosion. A reproducible testing is the key requirement for such investigations and results to previous test campaigns [6] were achieved.

The application of a conformal coating on the metallic surface is a straight forward approach but with a wide scattering in resulting corrosion states. Two different coatings were tested. While the first material was even worsening the result, the second product delivered a sufficiently improved corrosion resistance. But the processing of this coating is highly sensitive and needs further investigations with significantly larger statistics. In conclusion, it is possible to use such coatings in semiconductor modules to suppress Sulphur-induced corrosion.

In contrast to the conformal coatings, the modification of the Si-gel itself is more promising in terms of impeding the main anodic growth. While the trapping agent was just able to delay and limit the growth to very short anodic dendrites in particular areas, the corrosion inhibitor was able to prohibt the corrosion completely. Due to the lower number of trade-offs, the modification with Phosphates is the preferred way of further investigations and test campaigns. In this proof-of-concept-phase, the focus was completely set to Sulphur-induced corrosion and the only other parameters of interest were the blocking capability and the consistency of the gel. But for further development, the complete

specifications and the resulting qualification requirements need to be considered.

At the time of this test campaign, a commercially available product with a sealing layer was available on the market. But this concept did not work in the way it was expected to. The corrosion was not fully suppressed and significant dendritic growth was found in the areas of high electrical fields. Furthermore, the layer is supposed to act similarly to the trapping agent, used in the Si-gel modification. Anyway, the layer released the contaminant to the gel layer underneath and the corrosion started with only a time delay.

Overall, it is possible to modify standard power modules in a way, that corrosion due to contamination from the surrounding air is prevented. The presented modifications are just a selection of options. In terms of conformal coatings, different materials like Parylene are applied differently (gas phase) and were already proven to be able to insulate IGBT-modules [23]. Therefore, more materials and processes should be investigated to increase the robustness of such devices. The same holds true for the potting materials, with the options of different modifications depending on the application. If the diffusion through the material can be controlled, e.g. moisture should be part of further investigation despite the fact, that modern semiconductor chips tend to show a very high robustness already.

Acknowledgment

Parts of this work was supported by the ECSEL JU project iRel40 under the grant agreement No. 876659. The funding of the project comes from the Horizon 2020 research program and participating countries.

References

[1] C. Zorn, N. Kaminski, "Temperature-humidity-bias testing on insulated-gate bipolartransistor modules – failure modes and acceleration due to high voltage", IET Power Electronics: Special Issue on International Seminar on Power Semiconductors (ISPS'14), 2014

[2] J. Leppänen et al., "Aluminium corrosion in power semiconductor devices", Microelectronics Reliability vol. 137, 2022

[3] S. Honda, et al., "High Voltage Device Edge Termination for Wide Temperature Range plus Humidity with Surface Charge Control (SCC) Technology, 28th International Symposium on Power Semiconductor Devices and ICs (ISPSD'16), 2016

[4] C. Papadopoulos et al., "The influence of humidity on the high voltage blocking reliability of power IGBT modules and means of protection", Microelectronics Reliability vol. 88-90, 2018

[5] S. Kremp, O. Schilling, "Humidity robustness for high voltage power modules: Limiting mechanisms and improvement of lifetime", Microelectronics Reliability vol. 88-90, 2018

[6] M. Hanf et al., "Hydrogen sulphide (H_2S) single gas testing on power semiconductor modules under high voltage", Microelectronic Reliability vol. 138, 2022

[7] T. N. Wassermann et al., "A new high-voltage H_2S single noxious gas reliability test for power modules", Microelectronic Reliability vol. 100-101, 2019

[8] J. Rautio et al., "Cyclic temperature and humidity profile for mixed flowing gas tests of power semiconductor modules", 11th International Conference on Power Electronics and ECCE Asia (ICPE 2023 – ECCE Asia), 2023

[9] T. E. Graedel et al., "On the mechanism of silver and copper sulfidation by atmospheric H2S and OCS", Corrosion Science vol. 25 issue 12, 1985

[10] S. P. Sharma., "Reaction of Copper and Copper Oxide with H_2S", Electrochemical Society 127, 1980

[11] T. T. M. Tran et al., "The atmospheric corrosion of copper by hydrogen sulphide in underground conditions", Corrosion Science 45, 2003

[12] K. Hatori et al., "Humidity Absorption Behavior of Silicone Gel in HVIGBT Modules", 23rd European Conference on Power Electronics and Applications (EPE'21 ECCE Europe)

[13] J. Willner et al., "Examining differences in the uptake of corrosive gases in polymer films and its dependence on temperature and relative humidity using a novel procedure combining sample weathering and LA-ICP-MS analysis", SSRN Pre-print and available at: https://dx.doi.org/10.2139/ssrn.4672538

[14] M. Meier, H. Schweigart, "Corrosion in Power Electronics", 12th International Conference on Integrated Power Electronics Systems (CIPS'22), 2022

[15] E. L. Lee et al., "Review-Electrochemical Migration in Electronic Materials: Factors Affecting the Mechanism and Recent Strategies for Inhibition, J. Electrochem. Soc. 170, 2023

[16] M. Hanf, et al., "Sulphur related Corrosion in Power Modules and its Impact on the Switching Performance", International Seminar on Power Semiconductors (ISPS), 2014

[17] T. Bussarakons, "The Hermetic Surface Mount Device (SMD), Its Advantages and Solutions to Assembly Integration", White Paper International Rectifier, 2010

[18] S. Asada et al., "Resin Encapsulation Combined with Insulated Metal Baseplate for Improving Power Module Reliability", International Exhibition and Conference for Power Electronics, Intelligent Motion, Renewable Energy and Energy Management (PCIM'16 Europe), 2016

[19] E. Salahinejad et al., "Corrosion failure analysis of printed circuit boards exposed to H_2S-containing humid environments". Engineering Failure Analysis 79, 2017

[20] R. Ambat, K. Piotrowska, "Humiditiy and Electronics: Corrosion Reliability Issues and Preventive Measures", 1st Edition, 2021

[21] W. M. Haynes et al., "CRC Handbook of Chemistry and Physics", 95th Edition 2014-2015

[22] M. Edwards et al., "Phosphate inhibition of soluble copper corrosion by-product release", Corrosion Science 44, 2002

[23] S. Clausner et al., "Parylene as Coating for Power Semiconductor Devices", 15th International Seminar on Power Semiconductors (ISPS'21), 2021

PCIM Europe 2024, 11– 13 June 2024, Nuremberg   DOI: 10.30420/566262044

Investigation of Inorganic Encapsulation Materials in Power Electronic Systems for High Power Density Applications

Stefan Behrendt[1], Christophe Fery[2], Tamara Albert[2], Christiane Plikat[3], Rüdiger Knofe[4], Sönke Fleck[5], Ulf Schümann[5]

[1] Semikron Danfoss GmbH, Germany

[2] Heraeus Electronics GmbH & Co. KG, Germany

[3] Volkswagen AG, Germany

[4] Siemens AG, Germany

[5] University of Applied Sciences Kiel, Germany

Corresponding author: Stefan Behrendt, stefan.behrendt@semikron-danfoss.com
Speaker: Stefan Behrendt, stefan.behrendt@semikron-danfoss.com

Abstract

A technological limitation of power electronic systems is the encapsulation which is generally realized with organic polymers. This work discusses the use of inorganic ceramic encapsulation materials in power electronic systems. Body of the investigation is an automotive traction system. Besides the power module (PM) components like peripheral electronics (PE) as well as the motor are investigated. It can be shown that regarding the PM bonding wire temperatures at high current density levels can be reduced significantly. Additionally, on PE it can be observed that environmental testing (i.e. TST) is successfully passed. Also, the electrical performance of the electric motor can be improved.

1   Introduction

Further cost reduction of power electronic systems is required to support the adoption of electrification in our society, from renewable energies to mobility. To that extent, further miniaturization and improved reliability are needed, thus challenging the thermal management of such systems.

Polymeric encapsulation materials such as epoxies and silicone gels with a maximum operating temperature of up to 200°C are reaching their limit [1-3]. Furthermore, thermal conductivities of up to 3 W/(m·K) can only be achieved in epoxies at the cost of a challenging processability [4] [5]. Ceramic encapsulation materials (CE) provide very high thermal conductivities (TC), high mechanical stability and high maximum operating temperatures. Therefore, CE materials are promising candidates to overcome the shortcomings of polymer encapsulation materials.

In the context of an automotive traction system several subsystems can be defined which could benefit from alternative encapsulation materials like CE.

Regarding the power module of an automotive traction system new semiconductor technologies besides SiC are emerging. GaN as a wide bandgap (WBG) semiconductor material promises a low on-resistance and low gate charges. This results in a reduction of losses as well as enabling higher switching frequencies leading to more compact power converters with a higher power density [6].

Currently the potential of GaN devices is limited by their substrate technology (GaN-on-Si). New GaN-on-GaN substrates enable the manufacturing of vertical GaN devices which can provide current carrying capabilities of kA/cm² [7]. This translates to 90 A going through a semiconductor device with a size of 3 by 3 mm.

This kind of current density creates new challenges for the packaging of power electronic modules, especially for top side contacts of the semiconductor devices. Small devices like these do not provide enough contact area to mount a large amount of bonding wires for example. The effects of such a high power density on conventional bonding and joining technologies is addressed in this work.

Furthermore, the electrical motor as the center of the automotive traction system could also benefit from the application of CE materials. During operation the windings of the motor are heating up due to ohmic losses [8]. Higher temperatures lead to a decreased power factor of the motor and thus to a reduced efficiency [9].

Machines that are operated today usually do not have any active cooling applied to the windings of the stator. The encapsulation of these winding with a high TC material like CE could result in a lower operation temperature. Thus, leading to a higher efficiency which could be translated to a miniaturization of the motor as well as an extended reach for electrical vehicles (EV).

Other components of the automotive traction system like PCB based peripheral electronics could also gain an advantage from the encapsulation with CE. The combination of high TC and high mechanical stability could enhance the reliability of said components.

The goal of this investigation is to evaluate the application of CE materials inside an automotive power electronic system. To achieve this, the use of said materials in several subsystems is observed an analyzed.

2 Materials and methods

2.1 Inorganic encapsulation material

Heraeus is developing a novel cement-based ceramic potting compound with a temperature resistance up to 300°C and a thermal conductivity above 5 W/m·K. The material has no glass transition, thus ensuring stable performances over the operating temperature range.

In addition to that the material is non-flammable. It shows a good processability with a viscosity of ~4 Pa·s at room temperature as well as a good levelling and gap filling behavior (Tab. 1). It can be potted using standard vacuum casting equipment and shows a pot life of ~3 hours. After application, hardening is done at 85°C in a humid atmosphere and drying at 150°C.

Tab. 1: comparison of chosen key characteristics of ceramic encapsulation with typical silicone gel and epoxy

	CE	Silicone gel	Epoxy
Viscosity (pre curing)	4 Pa·s	5 Pa·s	n/a
Max. op. temp.	300°C	200°C	175°C
T_g	n/a	-50°C	185°C
TC	≥ 5 W/m·K	0,2 W/m·K	1,0 W/m·K

2.2 Power module – Thermal simulation

For evaluation of the thermal capabilities of an inorganic encapsulation material a simulation model representing a power module is constructed. It inherits one GaN semiconductor with an active area of 9 mm² and two 400 µm Al bonding wires. These components are located on an active metal brazed (AMB) ceramic substrate (Si_3N_4). The substrate is encapsulated by a CE material (5 W/m·K) and silicone gel (0,17 W/m·K) as a reference (Fig. 1).

Fig. 1: CAD model for thermal simulations regarding power module (encapsulation not shown)

Heat is injected into the system via defined power losses and joule heating. The top of the semiconductor and the small Cu island on the AMB are electrically contacted so that a controllable current can flow though the bonding wires. The current is varied from 10 A to 100 A.

A surface source is applied to the top of the chip. The power losses applied are based on a calculated on-resistance of the GaN device of 10,7 mΩ. This on-resistance represents a reasonable assumption for future vertical GaN devices based on [10][11].

Furthermore, a heat transfer coefficient (HTC) of 5000 W/m²·K to a coolant with a temperature of 17°C is applied to the bottom of the AMB substrate. This represents a turbulent direct water cooler like a Danfoss Shower Power®. All other surfaces of the model are applied with an HTC of 5 W/m²·K which is corresponding to natural convection to ambient (21°C).

The target criterion of this simulation is the resulting bonding wire and junction temperature. As critical limit a temperature of 185°C is defined. This critical temperature is based on the recrystallization of Aluminum which is accelerated significantly at such high temperatures [12].

Thus, leading to a reduction of tensile strength of the bonding wires which results in a significant reduction of module lifetime. These simulations were conducted via Simcenter FLOEFD by Siemens, Munich, Germany.

2.3 Peripheral electronics – Thermal Schock testing

To evaluate the use of inorganic encapsulation on and underneath typical electronic components of the system with regards to the solder joint reliability a standardized temperature shock test (TST) is performed.

Temperatures are set to -40°C as minimum and +125°C as maximum temperature with a testing procedure that is compliant to DIN IEC 60068-2-24 Na. As a testing vehicle a printed circuit board (PCB) with different passive components and typical packages is used (Fig. 2).

Fig. 2: Test PCB with different soldered components

This board is designed to test the usage of typical packages and components for peripheric electronics, assembled on a typical PCB (High T_g FR4, thickness: 1,6 mm; surface: ENIG). It also contains automatic 4-wire resistance measurement of most of the solder joints using daisy chain packages (BGA, QFP, QFN).

For the evaluation of the manufacturing quality of the solder joints an initial electrical measurement, automated optical inspection (AOI) and X-ray inspection is performed. The test group for the CE material was additional characterized after the material application.

In one test group the components are encapsulated with CE. Another group has no encapsulation as a reference. Cyclic electrical measurement and visual inspection (of the visual accessible solder joints and the surface of the inorganic material) is performed after 125, 250, 375, 500, 750, 1000, 1250, 1500, 1750 and 2000 cycles.

2.4 Electric motor – electrical and thermal characteristics

The inorganic encapsulation of the electric motor indicates numerous advantages regarding the characteristics of the electric stator with respect to improved mechanical strength and increased thermal conductivity. Furthermore, first electric test results show a positive impact on the partial discharge (PD) behavior which has been assessed by the institute elenia of the Technical University of

Fig. 3: Solely impregnated stator (left) and inorganic encapsulated stator (right)

Braunschweig.

As testing samples two stators of an electric machine for automotive application for a three-phase system in star configuration are taken out of the Volkswagen series production. Both stator windings are impregnated with resin. One specimen stays solely impregnated, the other is modified with regards to the cooling housing and encapsulated with CE (Fig. 3). The cooling housing then serves as potting form for the end windings.

To judge the impact of the modification on the electric insulation, PD measuring is realized for both stators according to IEC 60270 – High-voltage test techniques - Partial discharge measurements. Corresponding to this procedure, electric sinusoidal voltage of 50Hz is applied between windings and stator iron. The phases are connected inside the winding scheme (star configuration), hence, have the same potential difference towards the stator iron.

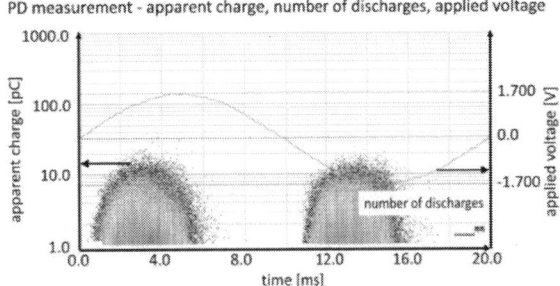

Fig. 4: Example of PD measurement diagram

Increasing the voltage until the partial discharges start defines the PD inception voltage (PDIV). This voltage level is kept at the inset point for one minute, the PD pulses are integrated and apparent charge displayed as shown in an exemplary result in Fig. 4. Then the voltage is further in-creased up to the RMS value of 1200V. In specific intervals at different voltage values the PD behavior is measured in an analogous way.

To detect the voltage value at which the partial discharges extinguish, the voltage is then again decreased incrementally and the PD extinction voltage (PDEV) is noted, similar to the PDIV measurement.

The later the discharges start, the higher the PDIV. A higher PDIV means less inhomogeneities in the insulation exist and the better its quality appears. According to that, a higher value for the PDEV points to a better insulation capability. Therefore, the values of the PDIV as well as the PDEV are important parameters to judge the insulation quality.

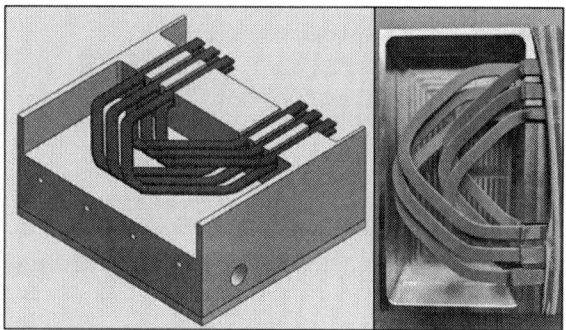

Fig. 6: Reduced demonstrator for the verification of the simulation model

Furthermore, the performance of electrical machines is limited by their thermal behavior. A thermal evaluation of the winding head under operation is also part of this work. To evaluate the thermal behavior of the electrical machine mentioned above is simulated with ANSYS by Ansys Inc, Canonsburg, PA, USA.

To verify the results of this simulation, a reduced winding model of the motor is built, simulated and evaluated via an Experiment. This reduced model consists of six windings mounted within an aluminum base (Fig. 6).

The stator windings are encapsulated with CE and connected to a current source. While a defined current is flowing through the copper, the surface temperature (T_s) of the CE is documented via thermographic IR and thermocouple measurement.

The same simulation methods are then used to simulate a more complex model of a motor with hairpin windings to evaluate the influence of the

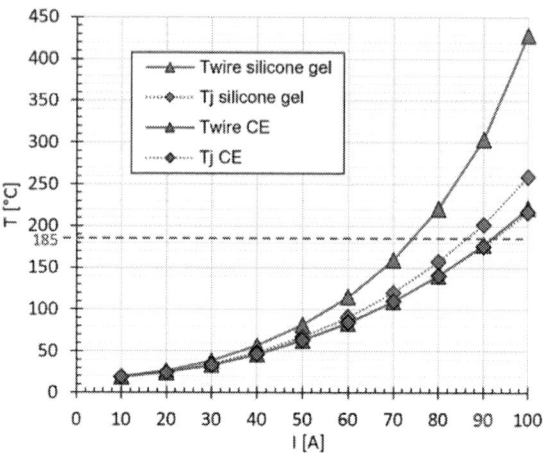

Fig. 5: Simulated temperatures of bonding wires (T_{wire}) and semiconductors (T_j) encapsulated in silicone gel and CE

CE encapsulation on the temperature distribution of the motor.

3 Results and discussion

3.1 Power module – Thermal simulation

The silicone gel encapsulated model shows a significant ΔT between the maximum bonding wire temperature (T_{wire}) and the semiconductors maximum junction temperature (T_j). Fig. 5 shows that an increased load current is leading to an increased ΔT between T_j and T_{wire}.

Although, the T_j at 80 A is still within a safe operating area (SOA) at 157°C, the bonding wires encapsulated by silicone gel reach a maximum tem-

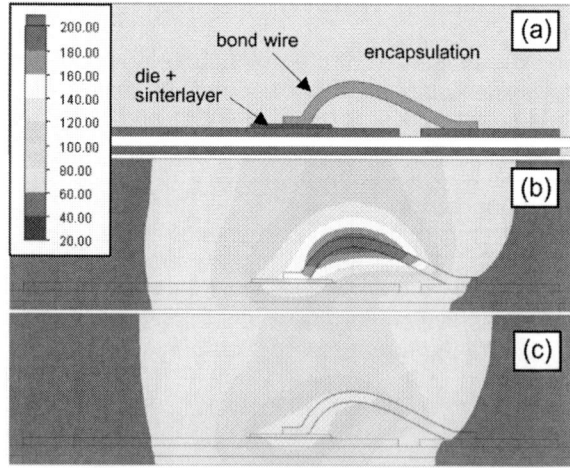

Fig. 7: Cross sectional view (a) with isotherms at 80 A; (b): silicone gel, (c): CE

perature of 221°C. Based on the assumed operation point (90 A) of upcoming vertical GaN devices in section 1, the potential of said devices can not be utilized when using silicone gel encapsulation. Furthermore, the resulting ΔT shows that the heat flux inside the system is flowing from the bonding wire to the semiconductor. As a result, the bonding wire is heating up the semiconductor with its losses. Fig. 7 shows how at 80 A the silicone encapsulated wire is the hottest component in the system (b). The current carrying capabilities of the wire are exceeded.

On the other hand, the CE encapsulated wire (c) still has nearly the same temperature as the chip underneath. Due to the enhanced heat dissipation the wire is still operating within its current carrying capabilities. This behavior can be observed throughout the whole applied current bandwidth. The average difference between T_j and T_{wire} in the range of 10 A to 100 A is 42,77K referring to the silicone gel. Within CE the average difference is only 0,28K.

Looking at the relative share of dissipated heat by the bonding wires, the thermal potential of CE materials is shown once more (Fig. 8). In silicone gel the wire is dissipating the heat to the chip and the AMB on which the second bond foot is mounted to almost equal shares.

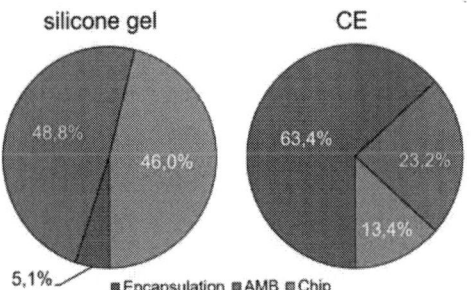

Fig. 8: Relative share of heat dissipation of bonding wires towards adjacent components

Due to the low TC of the silicone gel only 5,1% of the heat generated in the wires is dissipated to the encapsulation. The CE encapsulated wires are transferring 63,4% of the generated heat to the encapsulation. Thus, keeping the wire in the SOA that keeps the wire from exceeding its current carrying capabilities.

Therefore, it can be stated that CE materials enable the utilization of vertical GaN devices with regards to standard top contacting methods and referring to the configuration shown in this work.

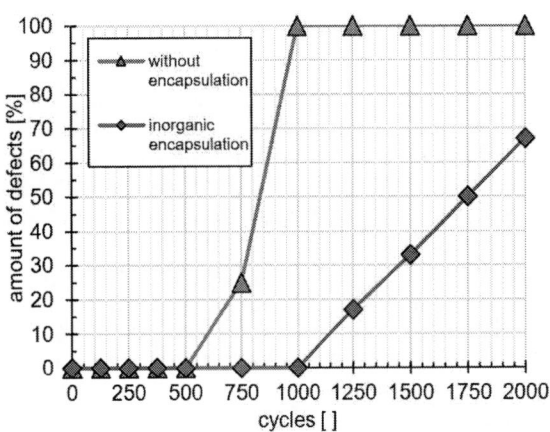

Fig. 10: Electrical defects of CR2512 during temperature shock test

3.2 Peripheral electronics – Thermal Schock testing

The only electrical interruptions (complete cracks) of solder joints were detected at the large ceramic resistors CR2512. The reliability of CR2512 is critical because of component geometry and material combination.

Specimen without encapsulation showed failures after 750 cycles whereas specimen with ceramic encapsulation showed no electrical failures until 1000 cycles (Fig. 10). The solder joints of the CR2512 without inorganic encapsulation show

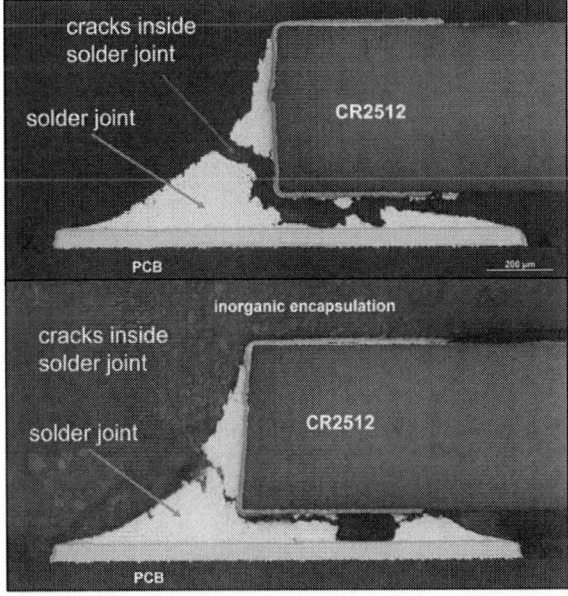

Fig. 9: Cross sections of solder joints after 1000 cycles; top: without encapsulation; bottom: with inorganic encapsulation

complete cracks earlier and the number of electrical interruptions was higher than the solder joints of the encapsulated components.

For a deeper evaluation of the possible cracks inside the CR2512 solder joints, cross sections of components with and without inorganic encapsulation were created (Fig. 9). The samples without encapsulation yield a complete crack inside the solder leading to an electrical failure.

Furthermore, the inorganic encapsulated specimen shows similar characteristics of solder cracking. In contrast to the non-encapsulated specimen there is only an increased resistance (9,4 Ω instead of 0,4 Ω at initial state) that can be measured instead of a failure.

Although, the soldering might be also affected by cracking, the electrical contact is maintained through the ceramic encapsulation fixating the parts in their position. Thus, keeping the electrical contact between the PCB and the component intact.

All other components didn't show any significant electrical changes of the electrical resistance or the capacitor characteristics. Beginning cracks inside the solder joints were also detectable on CR1206, CR0805 and CR0603 but without electrical interruption.

It can be stated that the use of inorganic encapsulation material yielded no negative effect on the solder joint reliability of the investigated components.

3.3 Electric motor – electrical and thermal characteristics

The ceramic encapsulation of the end winding seems to show numerous advantages regarding the characteristics of the electric stator. In addition to improved mechanical strength and increased thermal conductivity, first electric testing results show positive impacts on the PD behavior.

The PDIV compared to an identical but solely impregnated stator appears to be clearly increased. Furthermore, the high thermal conductivity of the inorganic potting material allows a good, direct connection of the winding head to the cooling jacket. The improved heat dissipation increases the continuous output of the machine with the same conductor cross-section.

The two testing objects only vary with respect to the cement potting, the measuring procedure has not been altered. Consequently, the obtained results can be regarded as directly referring to the potting influence.

In principle, temperature could affect the insulation behavior depending on probable inhomogeneities. Moreover, cement can be considered as material containing reaction and crystal water. Therefore,

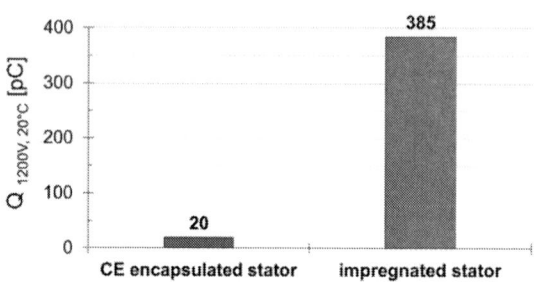

Fig. 11: Apparent charge of cement potted and solely impregnated stator at 20°C

the measurement procedure is carried out for both stators at 20°C and repeated after heating the stator to 180°C - to reduce potential water content - and after cooling it back down to 20°C (Fig. 12) to prove the possible dependency on water content.

As another parameter the value of the apparent charge allows to draw conclusions to the amount of charge which is affected by the PD. In Fig. 11 the measured values at 20°C and 1200V RMS are shown.

Due to limited availability the number of inspected samples is only one stator per insulation system. The interpretation of the results must be regarded as indicator for the potential and not as absolute value with specific confidence levels or failure tolerances. More test objects are needed to verify the described interpretation of measurement results.

Nevertheless, the visible tendency indicates promising material properties for the specific investigated application: The PD behavior of the CE encapsulated stator compared to an otherwise identical but solely impregnated stator appears to be clearly improved.

Thus, one can conclude, that the cement potting does improve the electric insulation system of the

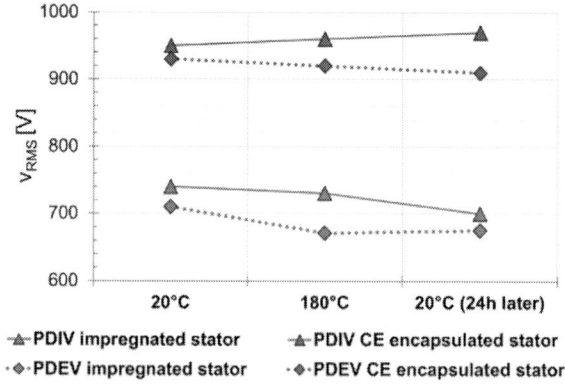

Fig. 12: PDIV and PDEV of cement potted and solely impregnated stator at different temperatures

regarded stator on behalf of the partial discharge behavior.

To verify the thermal simulation, a simple test setup is tested and measured. The result of the measurement is compared with the simulation in Fig. 13. The measurement with thermocouples results in a temperature of 32,3°C at the hairpins and thus a good agreement with the temperature of

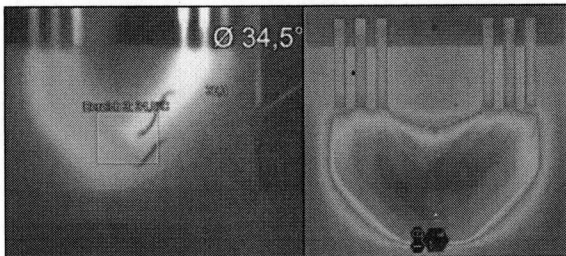

Fig. 13: Verification measurement (left) and simulation (right)

31,9°C from the simulation. The relative deviation between simulation and measurement is -1,24%. This verifies the simulation.

The thermographic IR measurement shows a temperature 34,5°C at the surface, corresponding to a relative deviation to the simulation of -7,54%. The difference between the measurement with thermocouples and IR measurement is indicating a systematic error in the IR measurement. Most likely the emission coefficient was incorrectly selected.

The radiation on the right-hand side of the measurement (Fig. 13, left) is explained by the contact resistance at the interface to the current source.

As shown in [13], for example, even increasing the thermal conductivity of the potting material results in a considerable reduction in temperature by up to 20%.

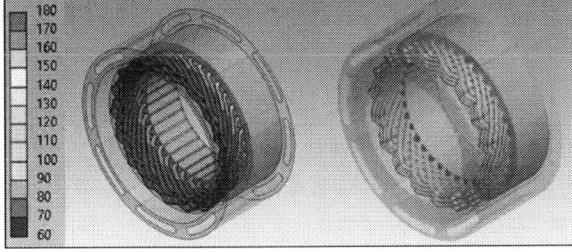

Fig. 14: Temperature distribution in the stator for a given operating point, left: epoxy right: CE

The simulation of the more complex motor model shows a reduction of the maximum temperature from 177,6 °C ($\Delta T = 117,6$ K) to 112,3 °C ($\Delta T = 52,3$ K) for the chosen operating point. By connecting the winding head directly to the cooling jacket, the maximum temperature can be reduced by up to 36,7% and the temperature increase compared to the heat sink is reduced by 55,5% (Fig. 14).

4 Conclusions

The investigations presented in this work show the enormous potential of inorganic encapsulation materials for automotive systems. Inside the power module CE material enable the operation at very high current densities (as expected with the introduction of vertical GaN devices) regarding bonding wires.

It was shown that the junction as well as the bonding wire temperature can be held inside a SOA when the components are encapsulated with CE. This remains true even at very high current levels and without any additional cooling efforts.

The wire temperature has not exceeded the chip temperature, which also could have an influence on reliability of such modules.

When the wire and the chip are at the same temperature, the thermal induced mechanical tensions between the components are only affected by the different coefficients of thermal expansion (CTE). Leading to a reduction of tensions in comparison to a scenario where the wire is significantly hotter that the chip. Thus, resulting in an enhanced reliability.

Peripheral electronics as well as the electrical motor can be cooled via thermal paths that can be achieved through the use of CE materials. The temperature cycling of CE encapsulated PCBs showed that there is no negative effect on the solder joint reliability.

It can even be discussed if the reliability is enhanced due to the mechanical fixation of the components. To confirm this phenomenon additional studies should be conducted.

Regarding the electrical motor it was shown that the thermal as well as the electrical characteristics can be enhanced using CE material. Experiments showed that the PDIV is reduced when the stator windings are encapsulated with CE. Therefore, leading to an improved insulation behavior.

Additionally, the thermal and mechanical coupling to the cooling jacket can enhance the cooling of the windings during operation. A reduction of more than 50K was shown in simulations. This leads to reduced losses inside the windings, resulting in an improved efficiency.

The results presented in this work show that power electronic systems can benefit from a wide range of use cases of inorganic encapsulation materials. Especially automotive systems with a very limited building space and increased reliability requirements are going to reach a level of power density in which coating and encapsulation materials used today cannot last. Ceramic encapsulation materials represent an alternative with very high potential for enabling the next generation of power electronic systems.

Acknowledgement

Acknowledgement to institute elenia of the Technical University of Braunschweig for supporting the PD measurements. The presented work received financial support by the Federal Ministry for Economic Affairs and Climate Action BMWK (Contract: 03ETE038)

References

[1] J. Muslim, O. Lesaint, R. Hanna, J. L. Reboud and N. I. Sinisuka, "Electrical Characterization of Dibenzyltoluene Liquid at High Temperatures up to 350°C," 2018 IEEE Conference on Electrical Insulation and Dielectric Phenomena (CEIDP), Cancun, Mexico, 2018, pp. 58-61

[2] P. Lall, Y. Zhang and J. Williamson, "Degradation Mechanisms of Epoxy Molding Compound Subjected to High Temperature Long Term Aging," 2021 20th IEEE Intersociety Conference on Thermal and Thermomechanical Phenomena in Electronic Systems (iTherm), San Diego, CA, USA, 2021, pp. 610-616

[3] A. Mavinkurve, L. Goumans and J. Martens, "Epoxy molding compounds for high temperature applications," 2013 Eurpoean Microelectronics Packaging Conference (EMPC), Grenoble, France, 2013, pp. 1-7

[4] A. Volke, M. Hornkamp, "IGBT modules Technologies, driver and application," 2. ed.,Infineon Technologies AG, 2012, p. 72 ff.

[5] M. Shibuya and L. Nguyen, "High Thermal Conductivity Mold Compounds for Advanced Packaging Applications," 2017 IEEE 67th Electronic Components and Technology Conference (ECTC), Orlando, FL, USA, 2017, pp. 1334-1339

[6] J. Everts, J. Das, J. van den Keybus, M. Germain, and J. Driesen, "GaN based power transistors for future power electronic converters," in Proc. IEEE Benelux Young Res. Symp., Leuven, Belgium, 2010

[7] Z. Hu, K. Nomoto, W. Li, Z. Zhang, N. Tanen, Q. T. Thieu, K. Sasaki, A. Kuramata, T. Nakamura, D. Jena, H. G. Xing, "Breakdown mechanism in 1 kA/cm2 and 960 V E-mode β-Ga2O3 vertical transistors," Appl. Phys. Lett. 17 September 2018; 113 (12): 122103

[8] S. Oechelsen, „Thermische Modellierung elektrischer Hochleistungsantriebe," Springer Vieweg, 2018, p.11

[9] S. Nategh, A. Krings, O. Wallmark and M. Leksell, "Evaluation of Impregnation Materials for Thermal Management of Liquid-Cooled Electric Machines," in IEEE Transactions on Industrial Electronics, vol. 61, no. 11, pp. 5956-5965, Nov. 2014

[10] S. Yang, S. Han, R. Li and K. Sheng, "1 kV/1.3 mΩ·cm2 vertical GaN-on-GaN Schottky barrier diodes with high switching performance," 2018 IEEE 30th International Symposium on Power Semiconductor Devices and ICs (ISPSD), Chicago, IL, USA, 2018, pp. 272-275

[11] S. Han, S. Yang and K. Sheng, "Conductivity Modulation in Vertical GaN PiN Diode: Evidence and Impact," in IEEE Electron Device Letters, vol. 42, no. 3, pp. 300-303

[12] V. T. Morgan, "Effect of elevated temperature operation on the tensile strength of overhead conductors," in IEEE Transactions on Power Delivery, vol. 11, no. 1, pp. 345-352, Jan. 1996

[13] A. Huber, T. Nguyen-Xuan, N. Brossardt, F. Eckstein, M. Pfitzner, „Thermische Simulation eines hochdetaillierten Wickelkopfmodells einer elektrischen Antriebsmaschine," 2014 ANSYS Conference & 32. CADFEM Users' Meeting, Nuremberg, Germany, 2014

PCIM Europe 2024, 11– 13 June 2024, Nuremberg   DOI: 10.30420/566262045

Characterization of Thermally Aged Silicone Gels for Power Semiconductor Modules

Elaheh Arjmand[1], Sonja Madloch[2], Thomas Spann[2], Philip Fletcher[3], Meghna De[1]

[1] Littelfuse, IXYS Westcode, UK
[2] Littelfuse, IXYS GmbH, Germany
[3] University of Bath, UK

Corresponding author: Elaheh Arjmand, earjmand@littelfuse.com
Speaker: Sonja Madloch, smadloch@littelfuse.com

Abstract

This paper presents a comprehensive study on the evaluation of silicone gels for power semiconductor module packages. Silicone gels are widely used as encapsulation materials serving many functions in the package such as preventing partial discharges or protecting inner module components from environmental impact. Five commercial gels were subjected to thermal aging at both extreme low and high temperature (-50°C and 200°C, respectively). Results of FTIR, DSC, TGA and as well as hardness measurements are presented. The results show the degradation and change of the material composition and properties through thermal load which supports selecting the most suitable silicone gel per application and the required working condition.

1   Introduction

Power semiconductor modules are part of a fast-growing market serving solutions for electric vehicles (EV), renewable energy, and industrial applications [1]. As a key to the next generation of products, innovative and robust power module packages must meet the requirements for higher power density, higher reliability, and lower costs [2]. Moreover, the use of wide band gap (WBG) semiconductor technologies such as SiC instead of established Si based semiconductor devices offer the potential for higher efficiency [3]. Simultaneously, this drives the adaptation of packaging technologies and materials to withstand the increased demand for operational conditions such as high temperatures and harsher environments. Silicone gels are considered as the main packaging and insulating materials for power semiconductor modules due to their dielectric properties, excellent sealing adhesion, and elasticity, especially in the presence of wire bonds [4,5]. During the operating condition of these packages the silicone gels will undergo thermal fluctuation, which in long term operations in a harsh environment lead to a certain change in the structure and properties of these gels.

In recent years, several studies have been published, focusing on the fundamental physical and chemical properties of the gels [6,7,8]. Studies were conducted to analyze the gels viscosity, adhesion [9, 10], and degradation behavior under high humidity [11,12]. The recent publications that addressed the analysis of dielectric properties of the gels [5, 13], focused on the influence of repetitive partial discharges on the silicone gels [14], the impact of high temperature on dielectric properties and internal structure of the gels [15,16,17]. For the qualification of next-generation power module packages, it is crucial to understand the influence of thermal aging on the material properties. There is currently a lack of comprehensive studies examining both low and high temperature aging, such as those encountered in automotive or rail applications with large temperature fluctuations.

This study presents the characterization results of commercial silicone gels which were subjected to thermal aging at high and low temperature.

2   Experimental plan

2.1   Gel Selection and Sample Preparation

Five commercially available gels were selected, which were labeled as gel A, B, C, D & E. With respect to operating temperature range, dielectric properties, volume resistivity, and curing conditions. Gels are either two-part agents or only one part which means no mixing is required. To protect business confidentiality, the main properties of the gels are not presented in this article. Table 1 shows the gel preparation methods of the five selected gels. Gels are cured according to datasheets that were provided by the manufacturers.

Preparation Requirement	Gel A	Gel B	Gel C	Gel D	Gel E
Vacuum time (mins)	20	20	20	20	20
Vacuum Pressure (mbar)	25	25	25	25	25
Curing Temp. (°C)	135	100	100	125	110
Curing Time (mins)	90	60	10	15	30

Table 1 Gel preparation details

For those gels that required mixing, it has been considered to mix the gels uniformly, for two minutes. The gels were then poured into labeled glass containers, were then kept under vacuum for 20 minutes to remove the trapped air. Finally, the gels were placed in an air circulated oven for curing according to the information given in Table 1.

2.2 Thermal Aging

The above-mentioned cured gel samples were divided into two groups. The first group were placed in the temperature-controlled oven at 200°C and the second group in a freezer at -50°C. The samples were randomly taken out after 500 hours , 1000 hours and 2000 hours for further analysis.

2.3 Characterization

2.3.1 Thermal Analysis

Thermogravimetric analysis (TGA) and differential scanning calorimetry (DSC) are the most widely used characterization methods for analysis of the thermal decomposition of materials such as silicone gels. The thermal stability of the gels was evaluated at various stages of the aging test using the equipment and parameters listed in Table 2.

Thermal analysis	System	Temp. range (°C)	Rate (°C/min)
TGA	Simultaneous Thermal Analyzer (STA) 6000	+30 to +990	20
DSC	Shimadzu DSC-60	-135 to -20	10

Table 2 Details of TGA and DSC system and parameters

2.3.2 Physio-Chemical Analysis

FTIR analysis is considered a fast acquisition method for spectroscopic information from polymer-based materials such as silicone gels. The change in the characteristic absorption of the key functional groups such as Si-O-Si, C-H, Si-CH$_3$ and Si-(CH$_3$)$_2$ bonds was monitored as-cured and through the thermal aging. The system used was a Nicolet™ iS50 FTIR Spectrometer in transmission mode, with a resolution of 4 cm^{-1} in scanning range of 400-4000 cm^{-1} with 30 scans. Fig. 1. shows the FTIR spectra of the five selected silicone gels in the as-cured condition.

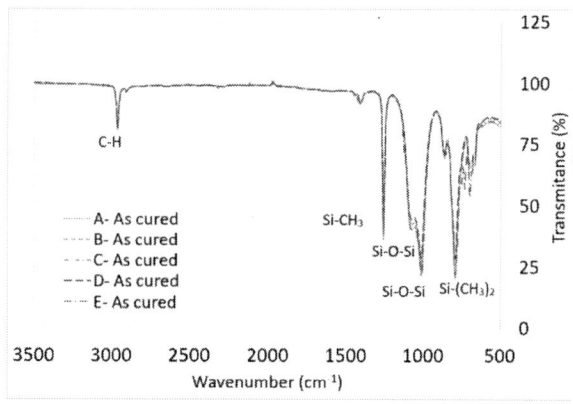

Fig. 1 FTIR spectra of the five selected gels, as-cured condition

2.3.3 Mechanical Analysis

Hardness measurements were conducted on the silicone gels using texture analyzer in compression mode. The mechanical resistance of the samples to stress was measured using a Stable Micro Systems TA.XTplus texture analyzer, based on industry standard for bloom strength, using the parameters shown on Table 3. Fig. 2 shows the texture analyzer with its typical graph results for gel A in the as-cured condition.

Test Parameters	Values
Probe diameter	10 mm
Test Speed	0.5 mm/s
Trigger force	4 g
Depth of measurement	10 mm

Table 3 Texture analyser test settings

PCIM Europe 2024, 11– 13 June 2024, Nuremberg   DOI: 10.30420/566262045

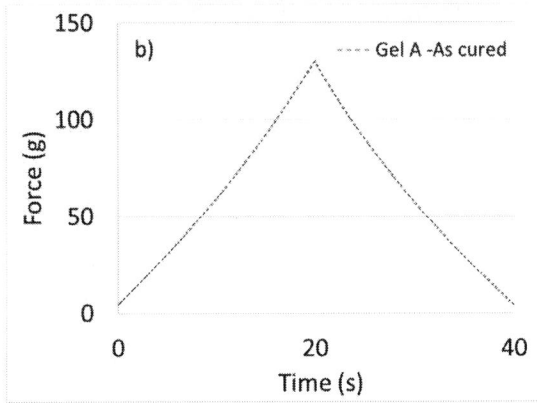

Fig. 2 a) Texture analyser and b) typical graph results – Gel A as cured.

2.3.4 Morphological Observation

The samples were closely observed during the aging tests to monitor changes in surface morphologies, crack initiation and delamination. A Keyence microscope VHX-6000 was used for visual observation and imaging. The details of the above-mentioned plan are presented in Table 4.

Methods	As cured	1000 hrs		2000 hrs	
		-50 °C	+ 200°C	-50 °C	+ 200°C
FTIR	✓	✓	✓	✓	✓
TGA	✓	✓	✓	✓	✓
DSC	✓	-	-	-	-
Hardness	✓	✓	✓	✓	✓

Table 4 Details of test characterization plan

3 Results and discussions

3.1 Thermal Analysis Results

3.1.1 TGA Results

In this section, the results of the TGA analysis were discussed. The TGA thermal curves were plotted for each single gel at different exposed temperatures. The key evaluation criteria given are based on: 1) thermal stability of the gels at operation temperature up to 250°C, 2) mass loss rate between 250°C-650°C and 3) mass loss at 650°C and beyond (see Fig. 3).

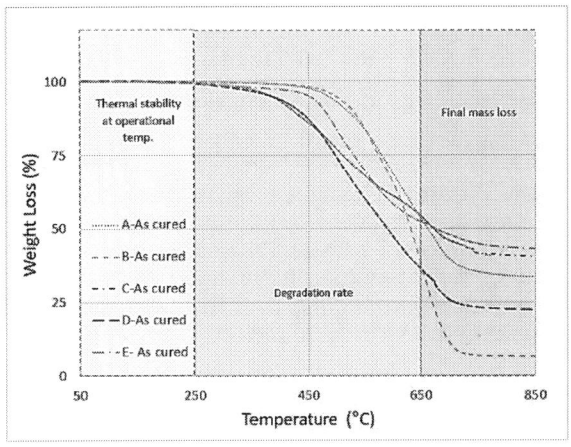

Fig. 3 TGA curve – the key characterization criteria – all the Si gels in as-cured condition

As shown in Fig. 4 all the gels remain stable as-cured and aged up to 130°C. The first thermal event occurred for gel E in the as-cured condition at 135°C, followed by gel D in the as-cured condition at 185°C. It appears that these two gels have the highest content of low temperature volatile materials. The remaining gels started degrading at -50°C, when aged for 1000 hours.

The thermal decomposition for the remaining gels started approximately at 250°C. It is observed as shown in Fig. 4 the thermal decomposition of most gels improved after high temperature aging including gel E. That might be the indication of possible cross-linking between molecular chains which results in a slower degradation rate [5, 7].

The onset temperature for each gel was extrapolated according to ISO 11358-1. The extrapolated data were plotted in the as-cured and after 1000 hours. at various aging temperatures (see Fig. 5)

371

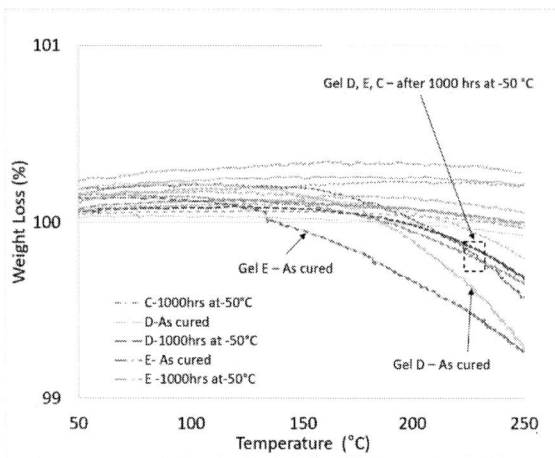

Fig. 4 Close-up TGA curves of gels – start of thermal decomposition temperature of some gels highlighted as-cured and after 1000 hours at various aging temperature.

Looking at the onset temperate results of the thermal endurance characterization similar trends were observed, most of the aged gels at high temperature showing higher onset temperature in compared to as-cured condition.

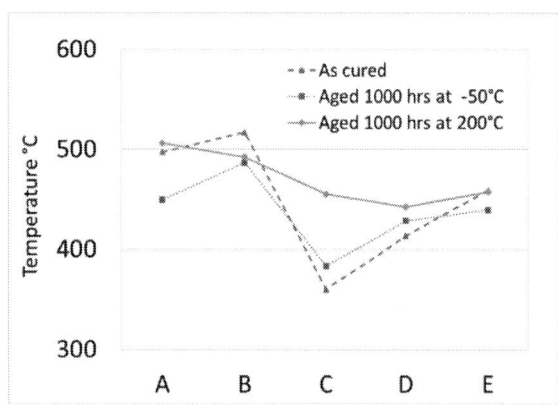

Fig. 5 TGA onset degradation temperatures, as-cured and after 1000 hours at various aging temperature.

The average rate of degradation is calculated as the percentage ratio of degraded mass at 5% mass loss and 60% mass loss to degradation time. The Fig. 6 shows the results of the average degradation rate. It appears that gel E among all the other gels, has the slowest degradation rates across all investigated conditions.

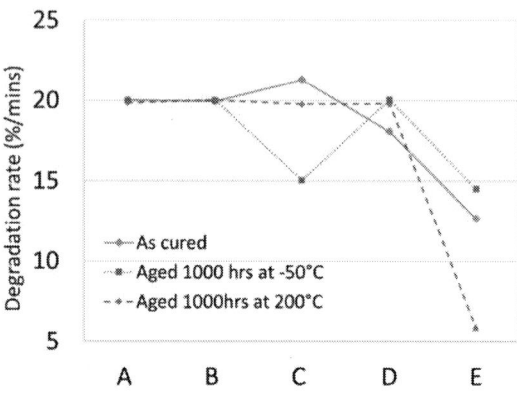

Fig. 6 TGA Average degradation range at 5% mass loss and 60% mass loss, as-cured and after 1000 hours at various aging temp.

Considering the overall mass loss at 650°C and beyond; as can be seen in Fig. 7, gel E shows the minimum mass loss followed by gel C.

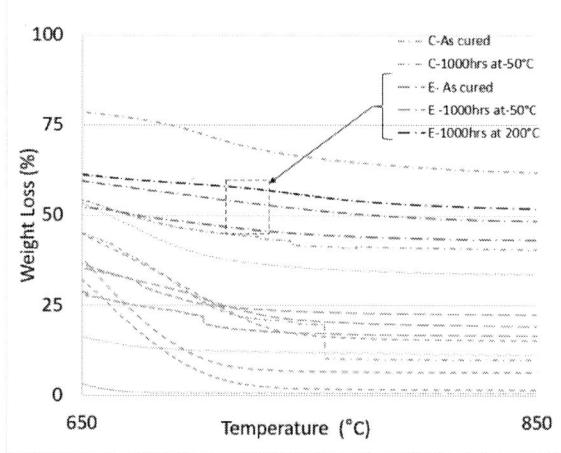

Fig. 7 Close-up TGA curves of all the selected gels (only those discussed highlighted), as-cured and after 1000 hours at various aging temperature, total mass loss at 650°C and beyond.

3.1.2 DSC Results

The thermal behavior of the selected gels in the as-cured condition was investigated using DSC. The thermograms of the selected gels were analyzed during cooling from -20°C to -150°C at a rate of 10°C/min. The results of two selected gels (C and D) were shown an exemplary in Fig. 8 and details of the observed thermal events are given in Table 5.

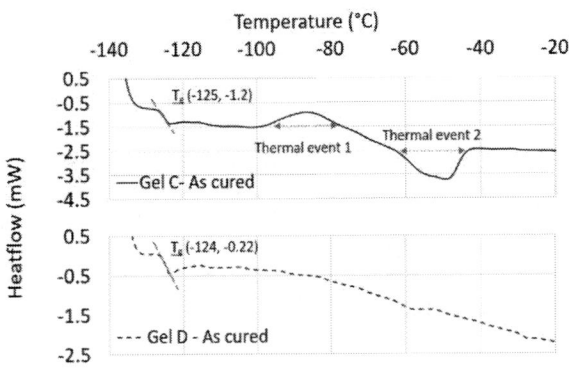

Fig. 8 DSC results of the gel C & D in as-cured condition

Gels	Glass state T_g onset (°C), (mW)	Thermal event 1 T_{on} to T_{end} (°C), (mW)	Thermal event 2 T_{on} to T_{end} (°C), (mW)
A	-127, -1.5	-85 to -104, -1.7	-32 to -52, -3.2
B	-*	-33 to -52, -3.16	-
C	-125, -1.2	-75 to -95, -1.5	-42.7 to -62, -2.5
D	-124, -0.22	-	-
E	-115, -0.53	-	-

* Glass transition temperature not detected within the tested temperature

Table 5 Details of thermal events observed for selected gels in as-cured condition based on DSC thermogram.

Overall, as detailed in Table 5, gels A and C have shown the most thermal events in below zero temperature. For gels A, B and C the thermal events start within the operational temperature range (typically down to -50 °C) of the majority of semiconductor modules. However, gels D and E have shown the best thermal stability, as no thermal events were observed within the studied temperature range.

3.2 FTIR Results

In this part, the results of the change in characteristic transmittance/absorption intensity of the key functional groups of the silicone gels are reported. Si-O-Si infrared bands were monitored between 1000-1010 and 1070-1090 wavenumber (cm^{-1}). C-H stretching was studied between 2955-2965 wavenumbers cm^{-1}, between 1255-1260 wavenumbers cm-1 for Si-CH3 and between 785-795 wavenumbers cm^{-1} for Si-(CH3)2, in the as-cured condition and after exposure to various aging temperatures.

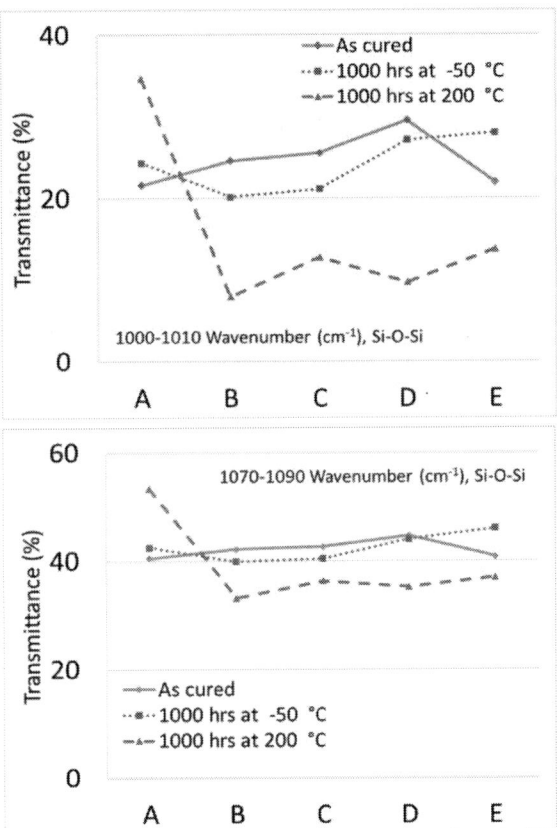

Fig. 9 Change in transmittance of the Si-O-Si bonds in as-cured and various thermal aging.

The transmittance peaks of the above-mentioned bonds were obtained from the FTIR data in as-cured conditions and at various aging temperatures. The data were plotted in Fig. 9 and Fig. 10. The gels A and E show less transmittance/ higher absorption density compared to gels B, C and D. This potentially indicate that the cross-linking degree of these A and E gels is higher after curing. Based on this observation, gels A & E behave more like an elastomer; therefore, less adhesion properties are expected [5,7].

As shown in Fig. 9 and Fig. 10, there are no apparent changes recorded in the transmittance intensity of the investigated functional groups after 1000 hours at -50°C, suggesting all gels have good stability at freezing temperatures.

The transmittance intensity of gel A stored at 200°C for 1000 hours has shown a significant increase in comparison to the other gels. The lower absorption in both functional groups indicates both chemical bonds were vulnerable and broken under the influence of high temperature. For the other gels, the absorption intensity increased at a comparable rate, suggesting the heat promoted an increase in the presence of the Si-O-Si and Si-CH3 bonds. Overall, the difference absorbance intensity of gels C and D in the as-cured and after thermal aging was low, which indicates good thermal stability within the investigated temperature range.

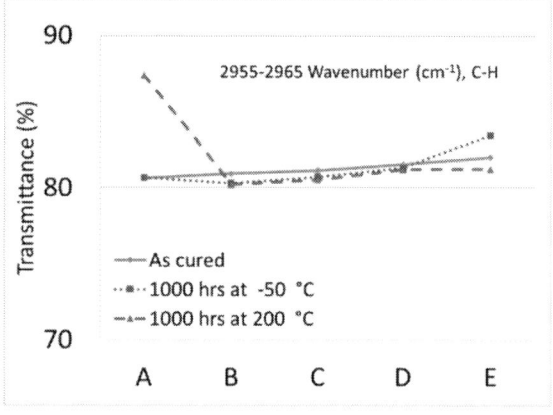

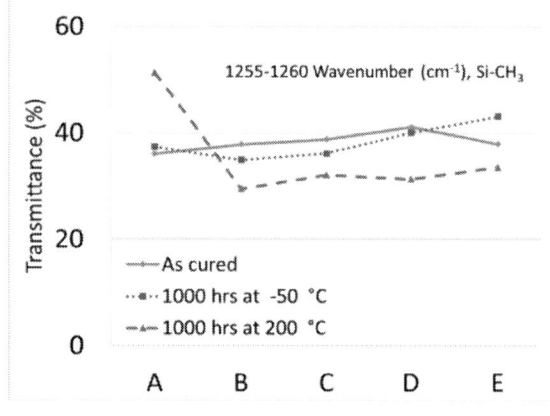

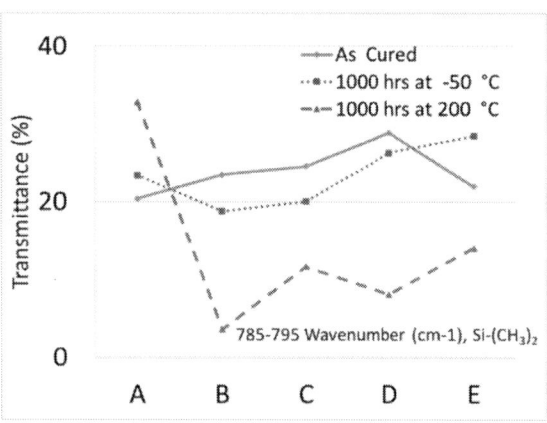

Fig. 10   Change in the transmittance of the C-H, Si-CH$_3$ and Si-(CH$_3$)$_2$ bonds in as-cured and various thermal aging.

3.3   Hardness Measurement Results

The results of the hardness testing of the gels are given in this section. The gel samples that were aged at -50°C, were taken from the freezer and allowed to thaw at room temperature. Then, the gels were kept in a controlled lab environment before texture analysis. As can be seen in Fig. 11, almost all the gels were stable in terms of hardness value compared to the as-cured value, after 1000 hours and 2000 hours. It indicates that the chemical degradation of the gels at low temperature is minimal [11].

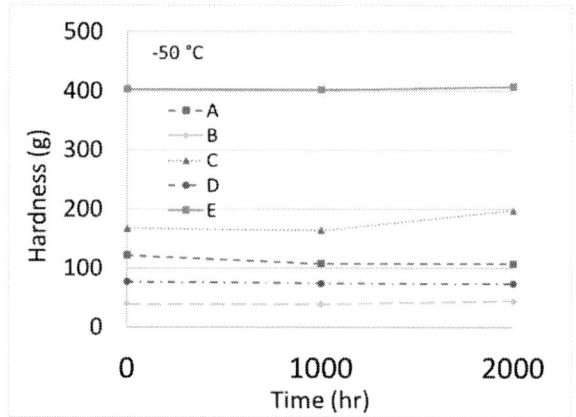

Fig. 11   Hardness graph of the selected gels exposed to -50°C

Fig. 12 presents the hardness values for the gels exposed to the 200°C temperature. As can be seen, all the gels have shown an enhanced degradation in physical properties over time. As already mentioned in the thermal analysis part, the new cross-links are formed in molecular chains of the gels resulting in an increase in the hardness of all the gels. As can be seen in Fig. 12, two measurements are missing after 2000 hours; first, gel B that became very hard, overloaded the tester using a 5kg load. Secondly, gel D became very brittle and formed several cracks and broke through during measurement.   Based on the measurement's values, gel C has shown the least change in hardness during the observed time period of aging at 200°C. However, overall, all the gels lost their mechanical properties such as tackiness, flexibility and   adhesion which are expected for the performance of the gel.   These observed physical changes are high likely to lead to poor performance of the power modules in applications that require higher operational temperature.

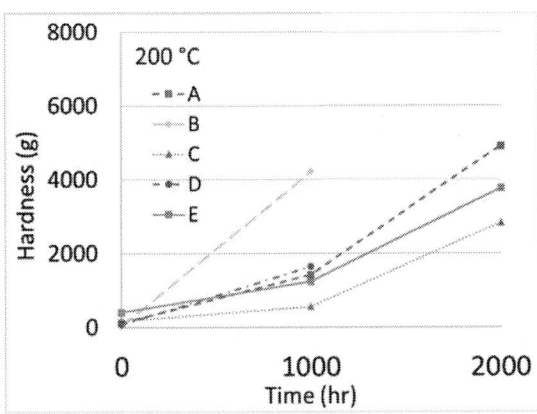

Fig. 12 Hardness graph of the selected gels exposed to 200°C

3.4 Observation Results

During the study, the gels in petri dishes and glass beakers were observed every 500 hours. The imaging was considered to determine if the change in material was visible. Two of the gels (A and B) exposed in the freezing temperature of -50°C started to show inner and surface cracking after 500 hours. Fig. 13 shows the surface condition of gel A after 500 hours.

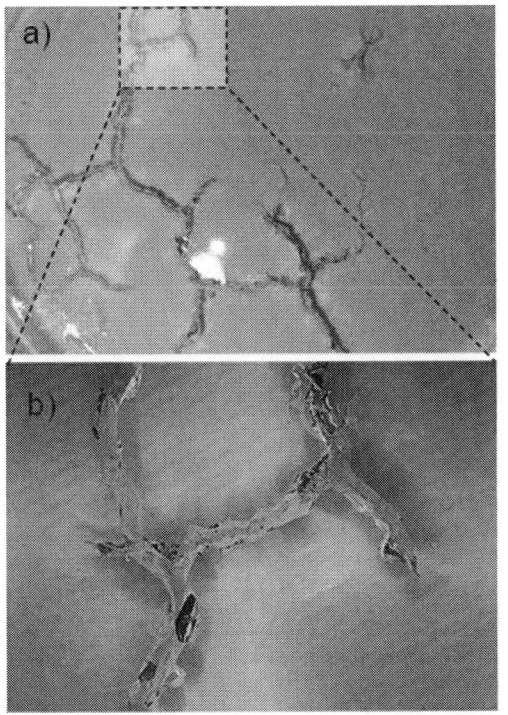

Fig. 13 a) Gel A surface condition, after 500 hours at -50°C b) close-up image of surface cracking.

Commenting on the visual observation and microscopic images, for the gels taken from the oven at 200°C, gels A and B started showing surface cracking after 500 hours. For gels C, D and E surface cracking was not observed. All the gels except gel E, developed discoloration from colorless/pale yellow to a darker shade of yellow/brown. Table 6. shows the observation with regard to the gel discoloration.

Gels	As cured	Post 1000 hrs	Post 2000 hrs
A			
B			
C			
D			
E			

Table 6 Close- up image of gels discoloration post high temperature aging

4 Summary of the Results

This section summarizes the investigation results of thermally aged silicone gels.

- As earlier mentioned, the gels E and D started showing slight weight loss within the operation conditions of the most available power semiconductor packages and the onset thermal decomposition temperature was also among the lowest. However, the amount of weight loss is less than 1%, meaning overall all the gels both as-cured and aged condition were showing stability within 30°C-250°C. Looking at the degradation rate beyond 250°C temperature, gel E degrades at a significant lower rate in comparison to the other gels and maintains its trend even when being exposed to both aging conditions. Based on the degradation rate results, gel E was selected as the most stable gel among the other gels.

- Overall performance of the gels in the as-cured condition based on the DSC results presented no major thermal event in freezing temperature for gels D and E. All the other gels have shown thermal events within the -50°C to -200°C temperature range, which is within the operational condition of the most semiconductor modules.

- Based on the FTIR results, the selected silicone gels show a broad overlap of infrared absorption bands which indicates a similar chemical composition (see Fig. 1). Looking at the adhesion capability of the gels in the as-cured condition, gels B, C and D showed better properties compared to the other gels. The FTIR results of the exposed gels under thermal aging indicate that, again the gels D and C have shown less variation in absorbance intensity, meaning thermal stabilities are superior compared to other gels investigated in this study.

- With respect to the gel strength measurement, the observation indicates all the gels maintained their strength after being exposed to -50°C. Measurement data post expose to 200°C indicate that gel C degrades at a slower rate than the other gels. Nevertheless, it is worth mentioning that, none of these gels, even those which are specified in their datasheets as high temperature gels, struggled to maintain their mechanical and physical properties through high temperature aging. The use of these gels for power semiconductor packages that require being exposed or tested at high temperature such as 200°C is highly questionable.

- In terms of visual observation exposed to extreme low and high temperature, both gel A and B are among the weakest performance gels. At the early stage of exposure hours, the micro surface cracking and inner cracking were observed for these gels. In packaging application those cracks have the potential to lead to the degradation of isolation properties.

Fig. 14 summarises the performance of the five selected gels from 1 (worst) to 5 (best) based on the investigation results obtained within this study.

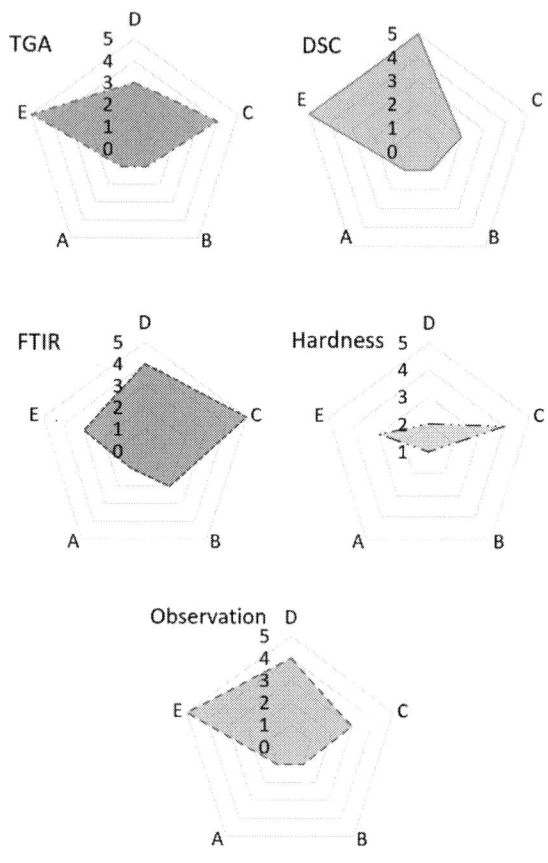

Fig. 14 The overall performance results of the five selected gels based on the analytical results

5 Conclusion

This paper presents the investigation results of different commercial silicone gels that were subjected to thermal aging at -50°C and 200°C. FTIR, TGA, DSC and as well as hardness measurements were conducted to evaluate the change in properties through temperature impact. The comprehensive results obtained from the different characterization methods allow a classification of the suitability of individual silicone gels for power semiconductor packages dependent on their application requirements.

Further investigation needs to be done to evaluate the interaction of the gels with other components within semiconductor packages. Furthermore, it is interesting to understand the effect of thermal aging on the dielectric properties of the gels.

6 Acknowledgment

The authors gratefully acknowledge the support of Mr. Jo Baker-Perrett from the Campden BRI for providing laboratory and technical support for this study.

References

[1] T. Do et al, "Reviewing of Using Wide-bandgap Power Semiconductor Devices in Electric Vehicle Systems: from Component to System", IEEE Vehicle Power and Propulsion Conference, 2020.

[2] M. Mars et al, "Power Electronics System Integration for Electric and Hybrid Vehicles", CIPS 2010.

[3] C. Chen et al., "A Review of SiC Power Module Packaging: Layout, Material System, and Integration".

[4] J. Hornberger, et al, "Silicon-carbide SiC Semiconductor Power Electronics for Extreme High Temperature Environments," in Proc. IEEE Aerosp. Conf.,2004

[5] B. Zhang et al, "Dielectric Properties Characterization and Evaluation of Commercial Silicone Gels for High-Voltage High-Power Power Electronics Module Packaging", IEEE, 2023.

[6] A. Morgan et al, "Characterization of Silicone Gel for High Temperature Encapsulation in High Voltage WBG Power Modules", International Symposium on Microelectronics, 2017

[7] L. Li et.al, "Analysis of Thermal Aging Characteristics of Silicone Gel for High-Voltage IGBT Packaging", IEEE 2023.

[8] X. Jianget al., "Comparative Analysis on Insulation Degradation Characteristics of Two Commercial Silicone Gels," 2022 IEEE, 2022.

[9] Carlos Montemayor, "Exploring the Performance of Silicone Gels at High and Low Temperature" April 2011.

[10] K. S. Siow et al., "Characterization of Silicone Gel Properties for High Power IGBT Modules and MEMS," 2015 IEEE Conference on Sustainable Utilization and Development In Engineering and Technology (CSUDET), Selangor, Malaysia, 2015.

[11] Kaixuan Li, et al. "Degradation Behaviors of Silicone Gel Encapsulation Material with Moisture Intrusion", Polymer Degradation and Stability, Volume 206, 2022.

[12] Tanaka, Nobuhiko et al. "Robust HVIGBT Module Design Against High Humidity." (2015).

[13] Suleimenova, Aliya, et al. "Insulation Evaluation in a SiC Power Module via Electric Field Simulation and Partial Discharge Measurement", 2020.

[14] X. Jiang, K. Li, Z. Yang, B. Zhang and X. Li, "Comparative Analysis on Insulation Degradation Characteristics of Two Commercial Silicone Gels," 2022 IEEE International Conference on High Voltage Engineering and Applications (ICHVE), Chongqing, China, 2022.

[15] Wang, Y. Gong, H. Ren, J. Wang and Q. Li, "High-Temperature Failure Mechanism and Lifetime Assessment of Silicone Gel Package Insulation for High-Power Electronic Devices Based on Pyrolysis Kinetics," in IEEE Transactions on Industry Applications, vol. 60, no. 1, pp. 1298-1309, Jan.-Feb. 2024.

[16] B. Zhang et al., "Electrical Properties of Silicone Gel for WBG-Based Power Module Packaging at High Temperatures," in IEEE Transactions on Dielectrics and Electrical Insulation, vol. 30, no. 2, pp. 852-861, April 2023

[17] Li, Qingfa & Wang, Wanjun & Wei, Zhe & Gong, Wenjie & Xu, Zhe & He, Dongxin. "Effect of Temperature Under Thermally Coupled Pulsed Electric Field on Silicone Electric Branches", 2023.

PCIM Europe 2024, 11– 13 June 2024, Nuremberg   DOI: 10.30420/566262046

A Coordinated Control of Hybrid Single-Phase AC/DC Microgrids Based on Natural Harmonic Injection Concept

Mehdi Baharizadeh [1] , Mohammad Sadegh Golsorkhi Esfahani [1] , Thomas Ebel [1]

[1] Centre for Industrial Electronics, Department of Mechanical and Electrical Engineering, University of Southern Denmark, Sønderborg, Denmark

Corresponding author: Mehdi Baharizadeh, smbaharizadeh@sdu.dk
Speaker: Mehdi Baharizadeh, smbaharizadeh@sdu.dk

Abstract

A hybrid single-phase AC/DC microgrid forms by connecting a single-phase AC microgrid to a DC one. In these systems, the DC microgrid experiences a natural injection of second harmonic current caused by the double-frequency component of the AC side instantaneous power. Based on the injected harmonic, a decentralized coordinated control is proposed to enable accurate power sharing among DC distributed energy resources (DERs) and among DC and AC DERs without imposing DC side voltage deviations. In this method, each DC DER adjusts its power based on the harmonic component frequency. Concurrently, the interlinking converter, which interconnects AC and DC microgrids, ensures the voltage level of the DC microgrid remains at the rated value.

1   Introduction

A microgrid is a part of a distribution system comprising distributed energy resources (DERs) and loads capable of autonomous operation as an island [1]. Depending on the distribution system's type, various forms of microgrids exist, including AC, DC, and Hybrid AC-DC configurations. In hybrid AC-DC microgrids, either a three-phase or a single-phase AC microgrid is connected to a DC one via an interlinking converter (IC) [2]-[4]. In the islanded mode of hybrid microgrids, as explored in this paper, a droop control method is commonly utilized to facilitate load sharing among DERs while maintaining voltage and frequency near the rated levels, all without the need for a communication network. In this method, to provide active power sharing in each microgrid, AC DERs employ active power-angular frequency (P-ω) droop, and DC DERs use power-voltage (P-V) droop. To expand power sharing to DERs throughout the hybrid microgrid, known as global power sharing (GPS), the IC exchanges an appropriate amount of active power determined based on the frequency drop in the AC microgrid and the voltage drop in the DC microgrid resulting from the implementation of droop characteristics [5],[6]. The droop control suffers from frequency and voltage deviations along with error in power sharing. In [7], a droop modification scheme is proposed to enable accurate

power sharing among DC DERs and accurate GPS. However, the voltage deviation in the DC microgrid persists. In [8], a distributed secondary control is presented to restore the frequency and voltage deviations and eliminate sharing errors. However, this scheme requires a communication network, which increases the cost and reduces the reliability. A harmonic injection scheme, presented in [9], eliminates errors in both power sharing among DC DERs and GPS without any communication link. The drawback of this method is injecting additional harmonics to the DC side which results in increased voltage ripples and flow of harmonic currents.

In hybrid single-phase AC/DC microgrids a double AC side frequency component naturally exists in IC instantaneous power. This component results in the natural injection of a second harmonic current to the DC microgrid. Based on the injected harmonic, this paper proposes a novel decentralized control method that provides both accurate power sharing among DC DERs and accurate GPS, without any imposed DC side voltage deviation. By measuring the frequency of the harmonic component, each DC DER regulates its power to realize accurate sharing. In addition, IC maintains the DC microgrid voltage level at the rated value. To realize the proposed method, DC DERs simultaneously control the harmonic voltage of the output filter capacitor and the DC current of the output filter

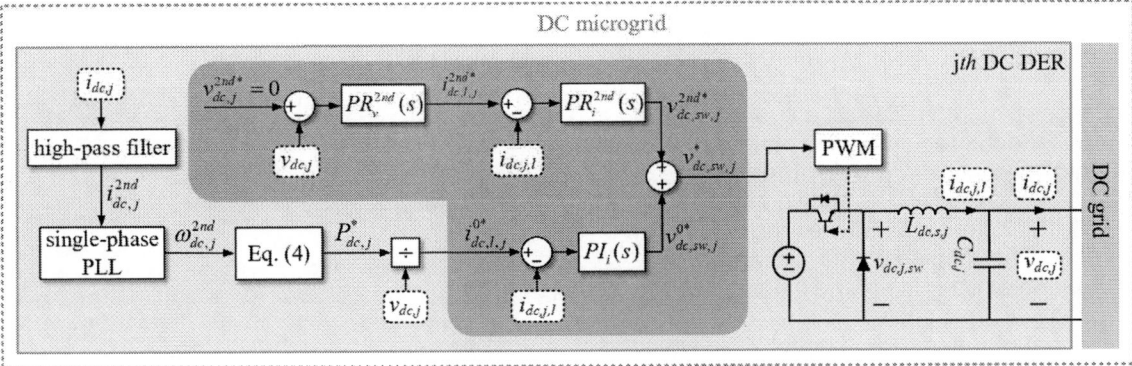

Fig. 1 Proposed control method of DC DERs

inductor. Compared to the existing solutions of [7]-[9] which focus on structures with a three-phase AC side, the proposed method presents a dedicated control scheme for single-phase AC/DC microgrids. In contrast with the secondary control of [8], the proposed method does not require any communication network. Finally, unlike the harmonic injection scheme of [9], which injects additional harmonics to the DC microgrid, the proposed method uses harmonic current that naturally exists in the DC side.

2 Proposed control method

2.1 Control Method of AC DERs

AC DERs utilize P-ω droop which adjusts the angular frequency of AC DERs according to the output active powers, as follows:

$$\omega_{ac,j}^* = \omega_{ac}^{nl} - m_{ac,j} \cdot P_{ac,j} \tag{1}$$

$$P_{ac,j}, m_{ac,j} = \frac{\Delta\omega_{ac}}{P_{ac,j}^{rated}} \tag{2}$$

where $\omega_{ac,j}^*$ and ω_{ac}^{nl} are reference value and no-load condition value of angular frequency, $P_{ac,j}$ and $P_{ac,j}^{rated}$ are low-pass filtered and rated active power, $\Delta\omega_{ac}$ is the intended angular frequency variation range, and $m_{ac,j}$ is the droop slope of the jth AC DER.

2.2 Proposed Control Method of DC DERs

To realize GPS, the relative output active power, i.e., the ratio of output active power to rated active power, should be the same for all AC and DC

DERs. Based on Eqs. (1) and (2), the relative output active power of each AC DER is calculated as:

$$P_{ac,j}^{rel} = \frac{\omega_{ac}^{nl} - \omega_{ac,j}}{\Delta\omega_{ac}} \tag{3}$$

Global property of frequency (steady-state frequency is the same at all AC buses) along with Eq. (3) implies the relative output active power of all AC DERs is equal. The angular frequency in the IC AC side could be applied to Eq. (3) to determine the relative output active power of AC DERs. On the other hand, the angular frequency of the second harmonic current injected to the DC microgrid is twice the angular frequency of the IC AC side. Nevertheless, by measuring this angular frequency, each DC DER can diagnose the unique relative output active power of AC DERs. Therefore, to realize accurate power sharing among DC DERs and accurate GPS, each DC DER should regulate its output power as:

$$P_{dc,j}^* = \frac{\left(\omega_{ac}^{nl} - 0.5\omega_{dc,j}^{2nd}\right) \cdot P_{dc,j}^{rated}}{\Delta\omega_{ac}} \tag{4}$$

where $P_{dc,j}^{rated}$ is rated power, and $\omega_{dc,j}^{2nd}$ is the measured second harmonic angular frequency of the jth DC DER. To realize this method in the jth DC DER, as shown in Fig. 1, first, by employing a high-pass filter and a single-phase phase-locked loop (PLL), angular frequency of the second harmonic component in the output current ($\omega_{dc,j}^{2nd}$) is extracted. Afterward, the measured angular frequency is applied to (4) for extracting the reference output power ($P_{dc,j}^*$). This reference value is realized by regulating dc component of the output filter inductor current ($i_{dc,l,i}^0$). Additionally, the second harmonic voltage at the output of the DER should be controlled. In this case, by utilizing a zero voltage reference, the harmonic output impedance of

DC DERs becomes negligible, which offers a diminished DC microgrid voltage ripple along with the harmonic current flowing toward the DC DERs. This harmonic current flow is an infrastructure for the proposed method. Hence, a special control for the simultaneous regulation of voltage (at second harmonic) and current (at zero frequency) is required. For this reason, as shown in the pink part of Fig. 1, the difference between the zero value of the second harmonic voltage reference ($v^{2nd*}_{dc,j}$) and the measured value of the output filter capacitor voltage ($v_{dc,i}$) is applied to a second harmonic proportional-resonant (PR) controller, in which its transfer function is as follows:

$$PR^{2nd}(s) = k_p^{2nd} + \frac{2k_r^{2nd} \cdot \omega_c^{2nd} \cdot s}{s^2 + 2\omega_c^{2nd} \cdot s + 4\omega_{ac}^{rated^2}} \quad (5)$$

Where k^{2nd}_p is the proportional gain, k^{2nd}_r is the resonant gain, and ω^{2nd}_c is the cut-off bandwidth of the second harmonic controller. This controller determines the second harmonic reference value of the output filter inductor current ($i^{2nd*}_{dc,l,j}$). Afterward, the difference between the second harmonic reference value and the measured value of the output filter inductor current is applied to another second harmonic PR controller. The controller extracts the reference voltage before the output filter ($v^{2nd*}_{dc,sw,j}$) required for realizing the second harmonic control. On the other hand, for regulating dc component of the output filter inductor current, its reference value ($i^{0*}_{dc,l,j}$) is compared with the measured current ($i_{dc,l,j}$) while the difference is applied to a PI controller (see Fig. 1). The PI controller extracts the reference voltage before output filter ($v^{0*}_{dc,sw,j}$) to realize dc component control. Finally, by summing DC and second harmonic reference voltages of before output filter, its total reference value ($v^*_{dc,sw,j}$) is extracted. $v^*_{dc,sw,j}$ is realized by PWM switching method.

2.3 Proposed Control Method of IC

Since the proposed control of DC DERs does not regulate the DC component of DC microgrid voltage, the IC becomes responsible for that. As shown in Fig. 2, the difference between the rated voltage (v^{rated}_{dc}) and the measured value of the IC DC side voltage ($v_{dc,ic}$) is applied to a PI controller which calculates the reference value of d-axis IC AC side current ($i^{d*}_{ac,ic}$). The reference current is tracked by the inner current loop. If the IC DC side voltage drops lower than the rated value, the voltage control loop decreases the d-axis current, which results in a decreased active power exchange from DC to AC side. Consequently, there will be an excess active power in the DC microgrid, which charges the DC side capacitor. This leads

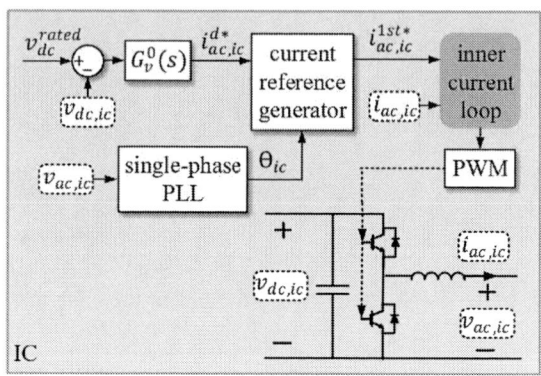

Fig. 2   Proposed Control Method of IC

to DC microgrid voltage increase till the IC DC side voltage reaches the desired level.

3   Simulation Results

Performance evaluation of the proposed method and its comparison with the conventional one, presented in [6], are provided based on time-domain simulations of a test hybrid single-phase AC/DC microgrid in PSIM. The test system shown in Fig. 3, includes 2 DC DERs, one AC DER, one IC, and both AC and DC side loads. The rated active power for the DC DERs is 5 kW and for the AC DER is 10 kW. The results are extracted by applying load changes in which ac side loads are decreased at t=2.5s and dc side load is increased at t=5.5s. Fig. 4(a) illustrates the DERs' active powers by employing the conventional control scheme. It is observed that the power sharing is inaccurate as the DC DERs have different active powers and the ratio of AC DER to DC DER output power is not equal to 2. On the other hand, results illustrated in Fig. 4(b) demonstrate, by employing the proposed method, even after load change, the active power of DC DERs remains the same and becomes equal to ½ of the AC DER. This verifies that the proposed method provides both accurate

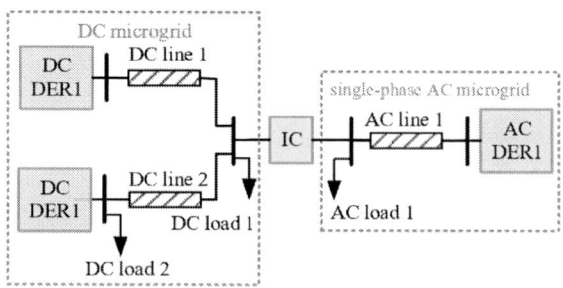

Fig. 3   Schematic of the test system

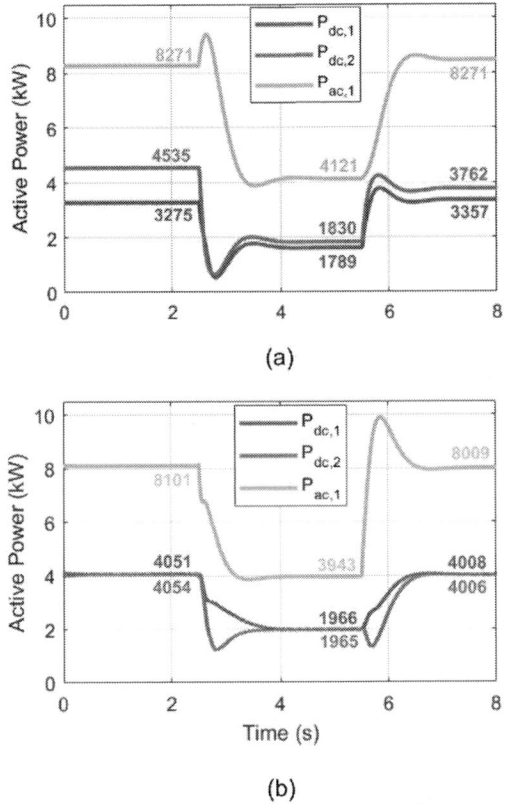

(a)

(b)

Fig. 4 Simulation results: active power generation of DERs by employing a) conventional control method, b) proposed control method.

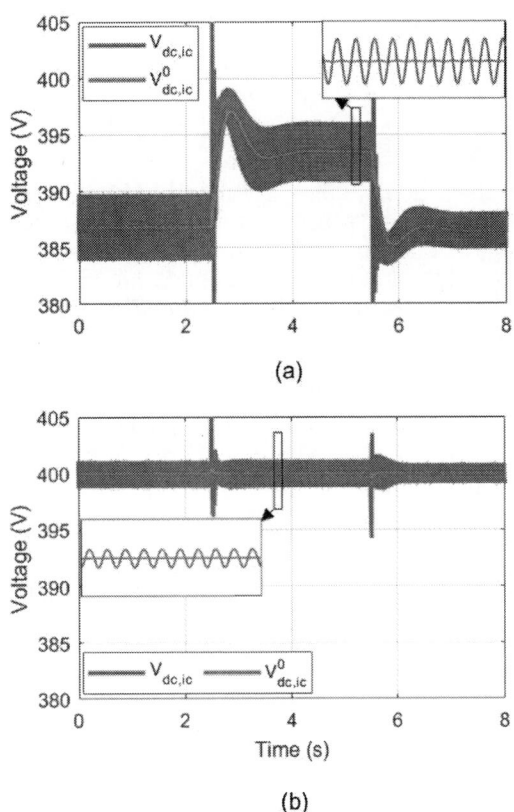

(a)

(b)

Fig. 5 Simulation results: voltage of IC dc side and its dc component by employing a) conventional control method, b) proposed control method.

power sharing among DC DERs and accurate GPS. Voltage of IC dc side by employing conventional and proposed methods are illustrated in Figs. 5(a) and 5(b) respectively. A comparison of these figures reveals the improved DC component voltage regulation and the reduced voltage ripple of the DC side by employing the proposed method. For the conventional scheme, the DC component of IC DC side voltage experiences a maximum deviation of 15 V from the rated value of 400 V, while this deviation is zero for the proposed method. The reason is in the proposed method IC is responsible for regulating the mentioned component exactly to the rated voltage which is not the case neither for IC nor for DC DERs for the conventional scheme. By employing the conventional method, the voltage ripple even reaches 1.5%, whereas this amount is 0.625% for the proposed scheme. The reason is reducing the second harmonic impedance in DC side which is a result of realizing second harmonic voltage control by DC DERs.

4 Conclusion

In this paper, a decentralized control method is proposed for hybrid single-phase AC-DC microgrids. The naturally injected second harmonic current to the DC microgrid is employed in the control loop of the DC DERs to enable accurate power sharing among DC DERs and accurate GPS without a requirement for communication network. A control scheme is presented for the DC DERs, which simultaneously regulates voltage (at second harmonic) and current (at zero frequency). The IC ensures the voltage level of the DC microgrid remains at the rated value. Comparative simulation results demonstrate that, unlike the conventional control scheme, the proposed method ensures accurate power sharing among all DC and AC DERs. Additionally, the DC component voltage regulation is improved, and voltage ripple is reduced in the DC microgrid by employing the proposed method.

References

[1] M. H. Saeed, W. Fangzong, B. A. Kalwar and S. Iqbal, "A Review on Microgrids' Challenges & Perspectives," in *IEEE Access*, vol. 9, pp. 166502-166517, 2021.

[2] Baharizadeh, M., Esfahani, M.S.G. and Kazemi, N., "Modified virtual frequency-voltage frame control scheme with zero sharing error for islanded AC microgrids," in *IET Generation, Transmission & Distribution*, vol. 17, no. 11, pp.2576-2586, 2023.

[3] Z. Li and M. Shahidehpour, "Small-Signal Modeling and Stability Analysis of Hybrid AC/DC Microgrids," in *IEEE Transactions on Smart Grid*, vol. 10, no. 2, pp. 2080-2095, March 2019.

[4] A. A. Hamad, M. E. Nassar, E. F. El-Saadany and M. M. A. Salama, "Optimal Configuration of Isolated Hybrid AC/DC Microgrids," in *IEEE Transactions on Smart Grid*, vol. 10, no. 3, pp. 2789-2798, May 2019.

[5] Nejabatkhah, F. and Li, Y.W., "Overview of power management strategies of hybrid AC/DC microgrid," in *IEEE Transactions on Power Electronics*, vol. 30, no. 12, pp.7072-7089, 2014.

[6] P. C. Loh, D. Li, Y. K. Chai, and F. Blaabjerg, "Autonomous Operation of Hybrid Microgrid With AC and DC Subgrids," in *IEEE Transactions on Power Electronics*, vol. 28, no. 5, pp. 2214-2223, May 2013.

[7] Baharizadeh, M., Karshenas, H.R. and Guerrero, J.M., "An improved power control strategy for hybrid AC-DC microgrids," in *International Journal of Electrical Power & Energy Systems*, vol. 95, pp. 364-373. 2018.

[8] E. Espina, R. Cárdenas-Dobson, J. W. Simpson-Porco, D. Sáez and M. Kazerani, "A Consensus-Based Secondary Control Strategy for Hybrid AC/DC Microgrids With Experimental Validation," in *IEEE Transactions on Power Electronics*, vol. 36, no. 5, pp. 5971-5984, May 2021.

[9] S. Peyghami, H. Mokhtari and F. Blaabjerg, "Autonomous Operation of a Hybrid AC/DC Microgrid With Multiple Interlinking Converters," in *IEEE Transactions on Smart Grid*, vol. 9, no. 6, pp. 6480-6488, Nov. 2018.

PCIM Europe 2024, 11– 13 June 2024, Nuremberg   DOI: 10.30420/566262047

A High Power-Density SiC-based TP PFC with a High Frequency Ripple Cancellation Leg

Ali Tausif [1], Ahmet Faruk Bakan [1], Serkan Dusmez [2]

[1] Yildiz Technical University, Istanbul, Turkey
[2] Huawei Technologies Duesseldorf GmbH, Nuremberg Research Center, Germany

Corresponding author: Serkan Dusmez, serkan.dusmez@huawei.com
Speaker: Serkan Dusmez, serkan.dusmez@huawei.com

Abstract

This paper introduces a novel ripple cancellation method by incorporating a high-frequency ripple cancellation leg comprising GaN FETs, connected in an interleaved manner with the SiC-based low-frequency half-bridge. The primary aim of the high-frequency leg is to generate an equal and opposite ripple current to counteract the low-frequency inductor current ripple, effectively canceling it out. This results in the elimination of low-frequency ripple at the input, leaving only the very high-frequency ripple current. Consequently, the differential-mode (DM) filter is tasked with filtering this high-frequency ripple current. The proposed approach streamlines the EMI filter to a single stage and offers the advantage of optimizing the design of the PFC inductor without the need to consider any consequences related to differential-mode noise; thus providing a very high-power density design for high power applications.

1 Introduction

Totem-pole (TP) power factor correction (PFC) converters have garnered widespread recognition for their simplicity, cost-effectiveness, efficiency, and high power densities. Notably, wide-band gap devices, with their zero reverse recovery current, have found a fitting application in TP PFCs [1–3]. Among these configurations, GaN-based hard-switched TP setups stand out, especially in the power range of 1.2 kW to 1.5 kW per leg, particularly when operated at high switching frequencies of up to 150 kHz [4, 5]. While these frequencies offer advantages at the system level, such as a reduction in boost inductor volume, challenges arise in reducing the EMI filter size, and high power demands necessitate interleaving legs due to the difficulty in dissipating heat from small GaN packages. To make a significant leap in the power density of a high-power converter, it becomes imperative to lower the amplitude of the input current ripple and push the current ripple frequency beyond 1.2 MHz [6]. Achieving such high switching frequencies while ensuring FETs remain within the thermal limits poses a notable challenge.
FETs with small die-size are considered to perform better at higher frequencies; however, their $R_{\text{ds,on}}$ values are usually high, leading to increased conduction losses in high-power applications, while their output capacitances(C_{oss}) are low, resulting in reduced switching losses. In order to use small sized GaNs and take the advantage of their low Q_{oss}, processing power of GaN must be kept low to avoid excessive losses and heat.
In the proposed innovative architecture, SiC devices take on the role of the low-frequency leg (operating at a few tens of kHz) to handle the primary power, while GaN FETs are strategically deployed for the high-frequency ripple cancellation leg. Since the high-frequency leg exclusively processes low-frequency ripple, small GaN FETs can be employed, effectively minimizing additional costs, volume, and power losses associated with this converter. This, in turn, allows for substantial reductions in both the main PFC inductor and EMI filter size and costs. The proposed method offers several key advantages over conventional designs:

- The elimination of the switching frequency ripple of the main inductor, thereby pushing the effective ripple frequency to a very high value, easing out DM noise filtering.

- The implementation of an EMI filter indepen-

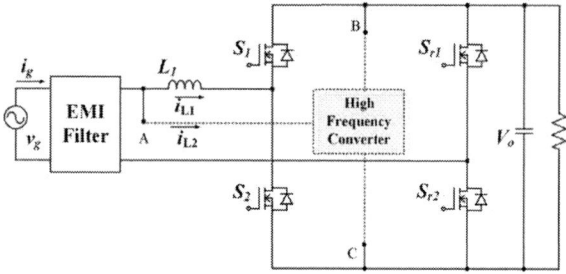

Fig. 1: Proposed converter architecture.

dent of the PFC inductor allows for the further optimization of the PFC inductor.

– The size of added auxiliary leg is considerably small in comparison to achieved EMI filter and boost inductor size reduction.

2 Proposed Method

2.1 Concept

The proposed converter comprises of a SiC-based half-bridge operating at 67 kHz to manage the primary power, paired with a GaN-based high frequency circuit connected as shown in Fig.1. Components L_1, S_1, and S_2 belong to the low frequency SiC leg. The purpose of GaN based high frequency converter is to generate equal and opposite ripple current to that of the first SiC-based half-bridge, ensuring efficient ripple cancellation in the system. The operation of the proposed converter mirrors that of a conventional totem-pole PFC, with the difference being the GaN-based high-frequency converter, which serves as a bidirectional buck-boost converter. Its objective is to produce a current that mirrors the ripple of the SiC leg but in the opposite polarity. This distinctive operation facilitates the cancellation of the low switching frequency fundamental component, i.e., 67 kHz, from the EMI spectrum. The illustration of the operational waveforms is shown in Fig. 2.

2.1.1 Design criteria of the high frequency leg

The main aim of the proposed converter is to enhance the power density by effectively reducing the size of the DM filter. It has been demonstrated that achieving this goal is feasible with a single-stage filter implementation, provided that the effective input ripple frequency is in the MHz range [6]. However, such a high frequency is unattainable with a hard-switched single-leg configuration, even when using GaN FETs at low power levels. Practically

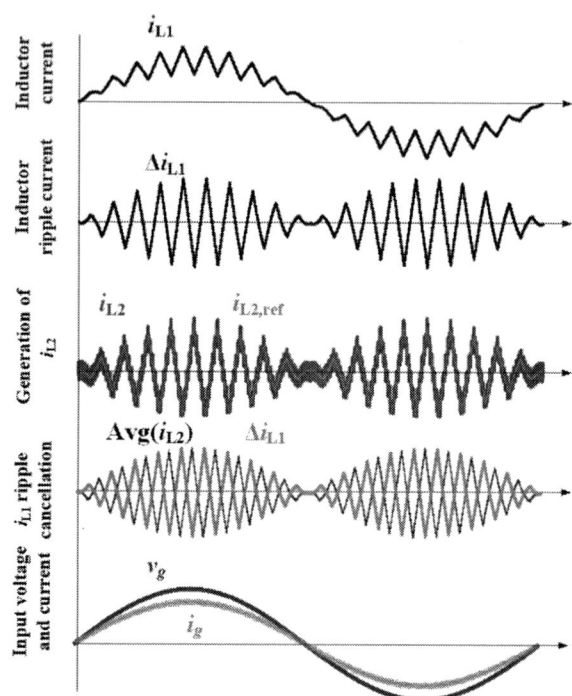

Fig. 2: Key waveforms of the proposed converter.

a single-leg 650V/150mΩ GaN-based half-bridge can operate at 500 kHz if the RMS current range falls between 2-4 A_{rms}. Nevertheless, this approach results in dissipation of 10-15 µJ during switching, necessitating the use of a heatsink. Although operating at 500 kHz reduces the size of the EMI filter, it does not confer a significant advantage, especially concerning a single-stage DM filter, which remains unfeasible. To achieve the target of increasing input ripple frequency beyond 1.2 MHz and keeping the current at minimum, various strategies have been explored. Interleaved converters were considered

Tab. 1: Design specifications

Output voltage (V_o)	400 V
Input voltage (V_g)	220 V_{rms}
Main inductor (L_1)	160 µH
Main switching freq. (f_{sw})	67 kHz
Aux switching freq. ($f_{sw,aux}$)	500 kHz
Auxiliary inductor (L_2)	10 µH (22x16x6mm)
GaN FETs	15mΩ / 200V x 6

PCIM Europe 2024, 11– 13 June 2024, Nuremberg   DOI: 10.30420/566262047

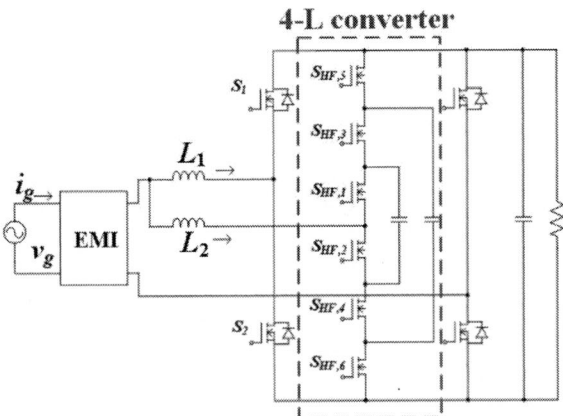

Fig. 3: Proposed high frequency 4-L converter.

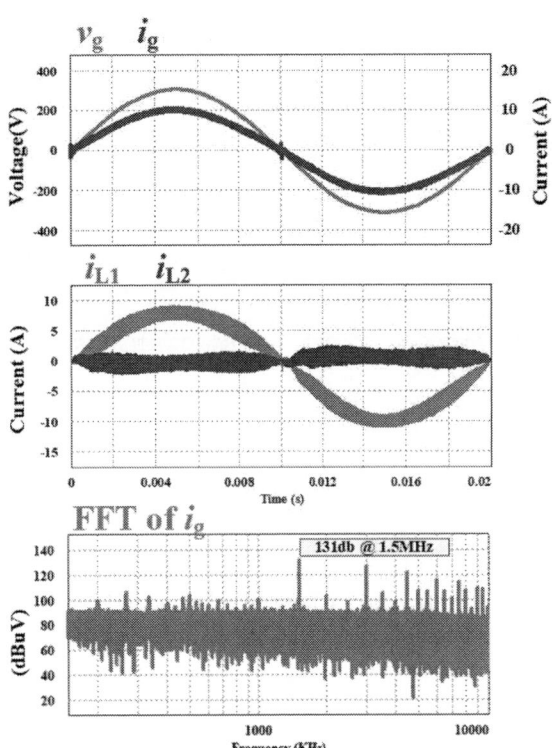

Fig. 4: Simulation results of grid voltage and current, inductor currents, and noise spectrum of input current.

but were found to produce excessive ripple per inductor, rendering them unsuitable for the low-power application.

A multilevel converter can be served as a promising solution for the high frequency leg. This not only increases the effective switching frequency but also reduces the ripple per inductor ($\Delta i_L/(n-1)$). The 4-level configuration, shown in Fig. 3, offers an advantageous compromise, utilizing available low voltage GaN FETs at 200 V with a high Figure-of-Merit (FOM) and keeping the current ripple close to 1.5 A_{rms} when the main inductance is set around 10 µH. The GaN FETs in this configuration can be operated at 500 kHz pushing the effective input current ripple frequency at 1.5 MHz. The design specifications of the proposed design are given in the Tab. 1.

2.1.2 Loss estimation of the high frequency leg

The losses linked with the high-frequency leg are estimated utilizing the loss models presented in [7,8]. It can be demonstrated that the most significant inductor loss on the planar inductor at a 1.5 MHz ripple is AC resistance losses resulting from proximity effect. This loss can be reduced to 1 W if the inductor is maintained around 10 µH with 1.5 A_{rms}. Similarly, under this condition, the DC losses are estimated to be 0.76 W. On the other hand, the losses associated with GaN FETs are estimated using the model presented in [8]. The switching loss and conduction loss per half-bridge amount to 0.65 W and 1.3 W respectively, resulting in a total loss of 5.85 W related to GaN FETs.

Therefore, for the given converter, employing a high-frequency converter using small-sized GaNs

(15mΩ/200V) and a planar inductor (22x16x6mm) enhances the power density of the converter by elevating the effective switching frequency to 1.5 MHz and enabling the realization of a single-stage DM filter. However, this improvement comes at the cost of approximately 7.5 W of additional power loss.

2.2 Simulation verification

The proposed concept has been successfully validated through simulations conducted in the PSIM environment. These simulations were executed with specific parameters, including an output power (P_o) of 1.5 kW, an output voltage (V_o) of 400 V, a grid voltage (V_g) of 220 V_{rms} at 50 Hz, and a f_{sw} of 67 kHz for the main leg and 500 kHz for the auxiliary 4-L high frequency leg. Additionally, the inductor values were set at 160 µH for L_1 and 10 µH for L_2. The simulation results, crucial for evaluating the concept's performance, have been meticulously analyzed, compared and visually presented in Fig. 4.

It is evident that the proposed converter effectively cancels out the 67 kHz ripple component in the input current, while introducing the 1.5 MHz compo-

PCIM Europe 2024, 11– 13 June 2024, Nuremberg   DOI: 10.30420/566262047

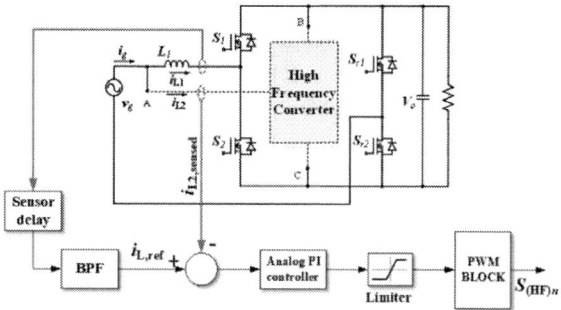

Fig. 5: Proposed converter with analog controller.

nents associated with the high-frequency converter.

3 Control Technique

Besides conventional cascaded controller for output voltage regulation, new independent controller is employed specifically for inductor ripple compensation, as shown in Fig. 5. The main inductor (L_1) current ripple is extracted and is fed as a reference of the auxiliary inductor (L_2) which is controlled using PI controller by generating duty of the high frequency converter's FETs.

3.1 High frequency controller's real-time implementation challenges

The reference for the high-frequency converter is derived from the sensed current of the SiC-based converter, filtering out only the ripple component using bandpass filter (BPF). This filtered signal is inversed and fed as an input to the controller of the high frequency converter. For a high-frequency converter to precisely track this signal, a controller with an extremely high bandwidth is essential. Considering the 67 kHz reference signal frequency, at least a minimum control loop bandwidth of 670 kHz is necessary for implementing the digital control.

Indeed, an analog controller would be a viable solution as shown in Fig. 5, although it presents its own set of challenges. Among these challenges, the most prevalent are the sensor delays and phase shifts induced by the BPF and analog controller, particularly at high frequencies.

These delays can lead to the generated current of auxiliary inductor follow the reference signal with a noticeable phase lag. This phase lag significantly distorts the performance of the proposed method, deviating it from the desired output.

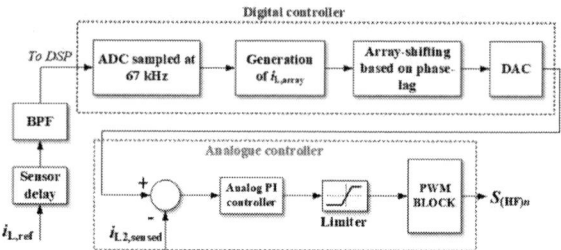

Fig. 6: Hybrid control block diagram.

3.2 Phase compensation and hybrid control solution

The solution to the aforementioned issues involves a hybrid approach, where the reference is initially processed within the digital controller, accounting for all potential delays, and then generated with a desired phase lead in order to compensate for the lag caused by the sensors and filters. This processed reference signal is subsequently fed to the analog controller via a digital-to-analog converter (DAC) of the MCU as shown in Fig. 6.

This approach enables the successful generation of high-frequency references, even up to a ripple frequency as high as 67 kHz. The illustration of this approach is shown in Fig. 7. It's important to note that the phase-lead in the generated reference is constrained by the resolution of the DAC.

The simulation results depicted in Fig. 8(a) reveal that prior to phase compensation, the resulting current waveform contains significant 67 kHz component. This occurrence stems from the phase difference between the reference and the auxiliary inductor current generated by the high-frequency converter, attributable to sensor and filter delays.

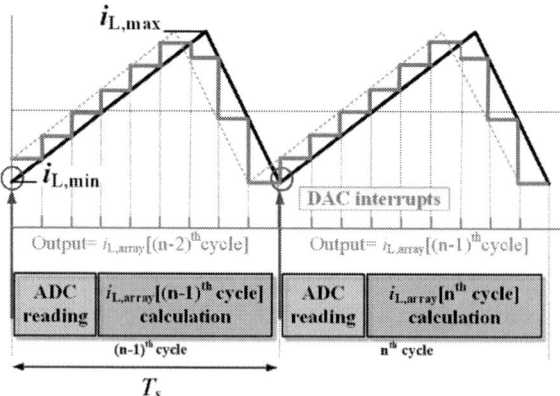

Fig. 7: Illustration of the generation of reference current using digital controller.

386

PCIM Europe 2024, 11– 13 June 2024, Nuremberg   DOI: 10.30420/566262047

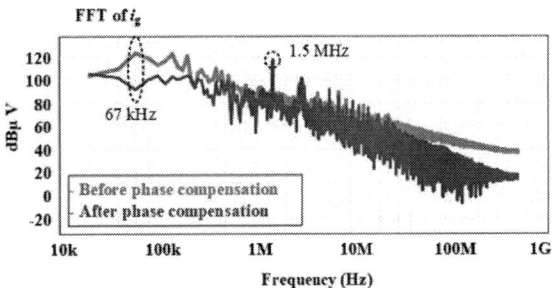

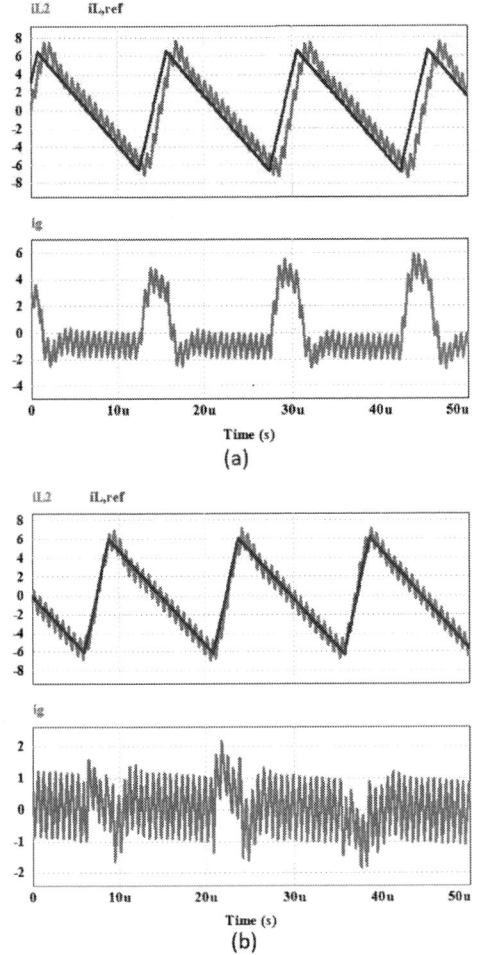

Fig. 9: FFT of the i_g before and after phase compensation.

Fig. 8: Simulation results showing currents (a) before phase compensation, and (b) after phase compensation.

Fig. 8(b) demonstrates that the phase difference is significantly reduced after implementing the proposed compensated approach. The resulting current waveform (i_g) indicates a negligible presence of the 67 kHz component after ripple cancellation, providing compelling evidence of the feasibility and effectiveness of this method at this particular frequency.

The FFT of i_g before and after phase compensation is presented in the Fig. 9 to facilitate a fair comparison. It is evident that prior to compensation, the significant presence of the high 67 kHz component necessitates a DM filter with a low corner frequency, resulting in minimal to no volume improvement despite the application of the proposed method. This emphasizes the crucial requirement

for precise alignment not only in magnitudes but also in phase between the reference and the generated current of the high-frequency converter in order for the proposed method to yield effectiveness.

4 Size comparison of the proposed converter

A comprehensive comparison of the proposed converter with single-leg TP PFC [3] and single-leg TP with GaN-based high-frequency half-bridge is provided in Fig. 10. The specifications of the main SiC converter are kept the same in all three topologies. A 650V 35 mΩ SiC FET operated at 67 kHz functions as the main converter in all considered topologies.

In the conventional TP PFC design, a bulky PFC inductor of 480 µH with dimensions of 61 x 61 x 40 mm is necessary, as the switching frequency is around 67 kHz, resulting in large input ripple and DM noise. To compensate for DM noise of 128 dBs at 201 kHz (i.e >150 kHz - EMI measurement range), a two-stage filter is required, the dimensions of which are provided in Fig. 10.

For the TP PFC with a single-leg high-frequency converter, the inclusion of the high-frequency leg reduces the requirement of the input PFC inductor to 160 µH, along with a decrease in size to 51x51x21mm. This simplification arises because the ripple generated at the input is eventually eliminated by the high-frequency leg. In this design, the high-frequency leg operates at a maximum of 500 kHz, primarily due to the thermal limits of the 150 mΩ FETs. Operating at this frequency generates significant loss and heat such that the use of a heat sink is necessary. The utilization of a 500 kHz high-frequency leg aids in reducing DM noise by pushing it to a higher frequency of 500 kHz, requir-

387

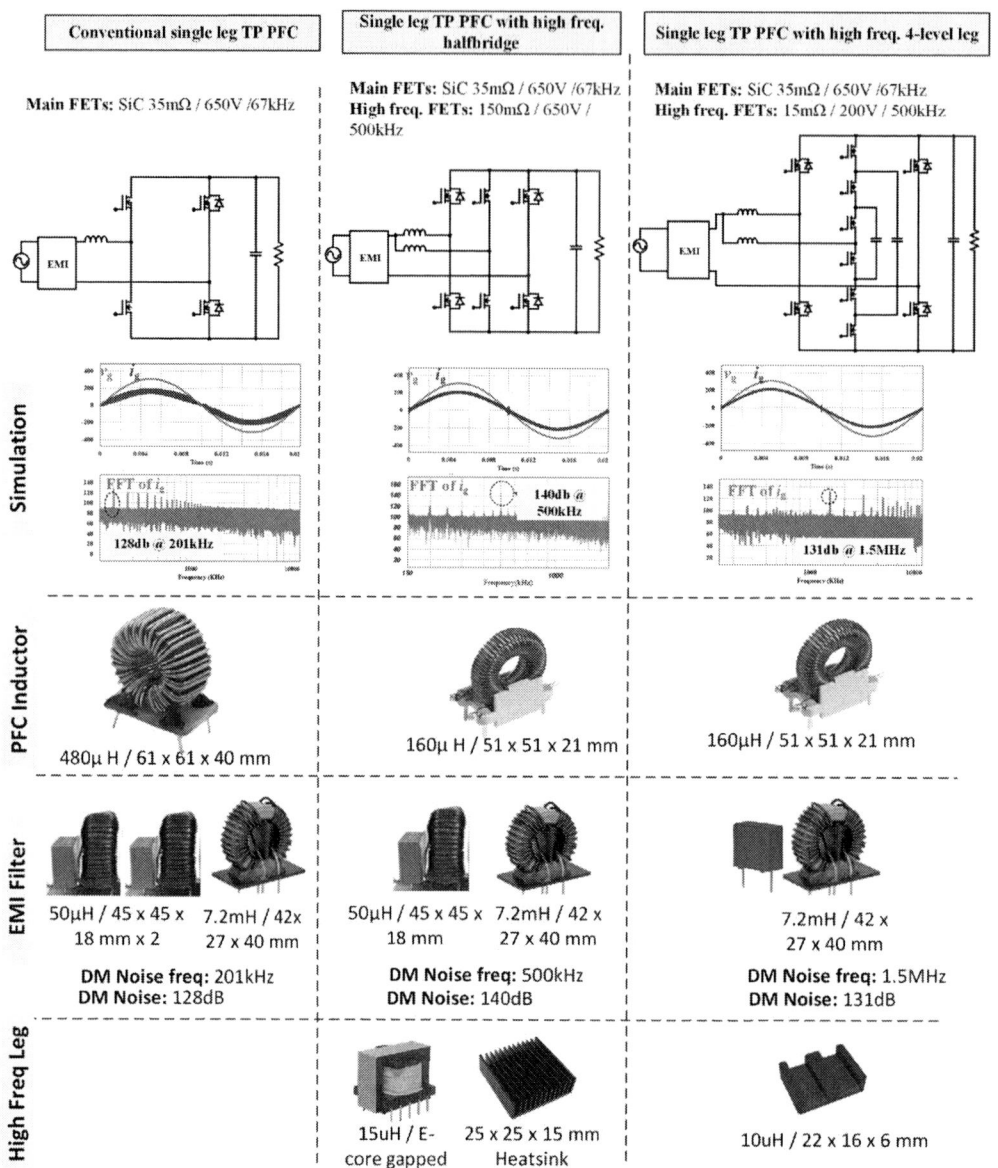

Fig. 10: Comparison of proposed design with conventional TP PFC and TP PFC with high frequency half-bridge for ripple cancellation circuits.

ing a comparatively smaller DM filter with a high corner frequency. However, the addition of a high frequency leg's inductor (i.e. 15 μH E-core gapped) and a heat sink (25 x 25 x 15 mm) for the high-frequency GaN switches contributes to the overall size of the converter.

In the case of the high-frequency 4-L converter, low-voltage small-die size 15 mΩ GaN FETs are utilized due to their superior performance when subjected to a high switching frequency of 500 kHz. The effective switching frequency is increased to

a staggering 1.5 MHz, and the ripple on the inductor is reduced likewise, necessitating a very small inductance of 10 μH, which can be implemented as a very low-profile planar inductor (22x16x6mm). Keeping the input PFC inductance the same as the previous case, the requirement for a DM filter is drastically reduced with the complete elimination of one filter stage, as illustrated in Fig. 10. In the present case, the DM noise observed is around 131 dB at 1.5 MHz, which can easily be eliminated using a single-stage filter.

The comparison reveals that the proposed design outperforms the other two TP PFC designs by significant margins in terms of overall volume. Therefore, it proves to be a viable solution for high-power-density converters.

5 Conclusion

This paper introduces a novel method to enhance the power density of TP PFCs by employing a high-frequency leg as an active filter, which cancels out the switching frequency ripple of the main PFC inductor. This approach eliminates the need for a bulky DM filter and makes it independent of the input PFC inductor design.

The current design demonstrates significant advantages in enhancing the overall volume of the single-leg SiC-based TP PFC by employing the high-frequency leg as a 4-Level converter. This approach not only provides the flexibility to elevate the effective input current ripple frequency up to 1.5 MHz, enabling the realization of a single-stage DM filter, but also reduces the magnitude of the ripple for its inductor, allowing for the use of planar inductors with negligible dimensions. Furthermore, the reduction of voltage stress on FETs permits the utilization of small-sized, low-voltage FETs, promising improved high-frequency performance. By introducing a high-frequency leg to a TP PFC, it becomes possible to improve the power density of the converter without increasing the switching frequency of the main converter. This addition incurs minimal extra losses, volume, and cost, thereby serving as a promising and cost-effective approach to enhance the overall power density of power supplies.

References

[1] B. Li, F. C. Lee and Q. Li, "High-Efficiency High-Density Critical Mode Rectifier/Inverter for WBG-Device-Based On-Board Charger," in IEEE Transactions on Industrial Electronics, vol. 64, no. 11, pp. 9114-9123, Nov. 2017, doi: 10.1109/TIE.2017.2716873

[2] K. Zhu, A. Bhalla and J. Dodge, "Enabling 99.3% Efficiency in 3.6 kW Totem-Pole PFC Using New 750 V Gen 4 SiC FETs," in IEEE Power Electronics Magazine, vol. 8, no. 4, pp. 30-37, Dec. 2021, doi: 10.1109/MPEL.2021.3123757.

[3] A. Salman, E. Ayerbe, J. Solovey, G. Moxey, S. Ryu and A. Barkley, "650 V Silicon Carbide MOSFETs in Totem-Pole Bridgeless PFC Design Achieves High Efficiency (80+ Titanium) without adding Complexity and Cost," PCIM Europe 2018; Nuremberg, Germany, 2018, pp. 1-8.

[4] R. Hou, Y. Shen, H. Zhao, H. Hu, J. Lu and T. Long, "Power Loss Characterization and Modeling for GaN-Based Hard-Switching Half-Bridges Considering Dynamic on-State Resistance," IEEE Trans. on Transportation Electrification, vol. 6, no. 2, pp. 540-553, June 2020

[5] B. Su, and Z. Lu, "An interleaved totem-pole boost bridgeless rectifier with reduced reverse-recovery problems for power factor correction," IEEE Trans. Power Electron., vol. 25, no. 6, pp. 1406–1415, Jun. 2010, doi: 10.1109/TPEL.2010.2040633.

[6] A. Tausif and S. Dusmez, "A Unified Differential Mode Noise Estimation Method and Filter Size Comparison in Single-Phase Multileg and Multilevel Totem-Pole PFC Converters,"IEEE Trans. on Power Electronics, vol. 38, no. 6, pp. 7197-7206, June 2023,

[7] A. Lordoglu, M. O. Gulbahce, D. A. Kocabas and S. Dusmez, "A New Optimization Method for Gapped and Distributed Core Magnetics in LLC Converter," in IEEE Access, vol. 11, pp. 14061-14072, 2023, doi: 10.1109/ACCESS.2023.3242869

[8] E. B. Bulut, D. A. Kocabas and S. Dusmez, "Optimization and Design Considerations of GaN-Based Multi-Level TP PFC Converters," in IEEE Access, vol. 11, pp. 47291-47303, 2023, doi: 10.1109/ACCESS.2023.3275751.

PCIM Europe 2024, 11– 13 June 2024, Nuremberg   DOI: 10.30420/566262048

High Frequency Active Filter for AC/DC High Power Converters

Sarah Sifoune[1,2] , Denis Labrousse[1,3] , Pierre-Etienne Lévy[1] , Cyrille Gautier[2] , Bertrand Revol[2]

[1] Université Paris-Saclay, ENS Paris-Saclay, CNRS, SATIE, France.
[2] SAFRAN Tech, SAFRAN Paris-Saclay, France.
[3] CNAM Paris, France.

Corresponding author: Sarah SIFOUNE, sarah.sifoune@ens-paris-saclay.fr
Speaker: Sarah SIFOUNE, sarah.sifoune@ens-paris-saclay.fr

Abstract

This paper proposes a high frequency switching active filter to reduce the volume and weight of conventional passive filter used to reduce electromagnetic interference. The selected solution is an AC/DC active rectifier (H-bridge) connected in parallel of an electronic polluting load. Based on harmonic current injection, the active filter provides current variations at the first harmonics of the switching frequency without affecting the fundamental. The control of the active filter is made through a feedback loop which is independent from the polluting load in order to obtain an adaptative filter design.

1   Introduction

To achieve carbon neutrality by 2050, a goal set by the European Union, the aeronautics sector is committed to aircraft electrification [1]. This involves massive use of high-power converters that makes Electromagnetic Interference (EMI) reduction more complicated, knowing that passive filters widely used are often too heavy and too voluminous. To reduce the volume and weight of these filters, many studies have been carried out. Among the proposed solutions, we find soft-switching converter with less switching harmonics or the addition of linear or nonlinear active filtering.

These non-linear active filters, widely used in low-frequency grid applications, were rarely studied for EMI applications. Nguyen et al. [2] aim to reduce the high frequency switching harmonics of a boost converter using a synchronous half bridge-based active EMI filter. It achieves a very high current attenuation of over 65 dB and 38 dB at the switching frequencies of 150 kHz and 1 MHz, respectively. Both converters, main boost and the half bridge used as active filter, share the same control which makes the filter totally dependent on the converter to which it is attached.

There are also a few applications in the literature for filtering harmonics due to switching on AC

loads with a switching frequency up to a few kHz [2-5]. Sato et al. [6] have proposed a hybrid PWM rectifier to avoid the high-frequency switching of the full current and so realizes lower EMI. The hybrid PWM rectifier consists of two bridge circuits, an independent active H-bridge switching filter to reduce switching harmonics in an active rectifier with a switching frequency of 1.2 kHz.

This paper proposes a fully independent high frequency active filter (HFAF) connected in parallel of a power converter considered as an electronic polluting load, whose principle is shown in Fig. 1. This structure of HFAF associated to a proper control are efficient regardless of the load with a switching frequency between 10 kHz and 40 kHz. In the second section, the design and the control strategy of the active filter are developed. Section 3 presents a filtering volume estimation for both cases passive filter and association of passive and active filter. The volume reduction with the use of HFAF is showed. Then some simulation results are presented and discussed. Finally, to confirm the effectiveness of the proposed HFAF on the reduction of the EMI, the in progress experimental developments are presented.

PCIM Europe 2024, 11– 13 June 2024, Nuremberg DOI: 10.30420/566262048

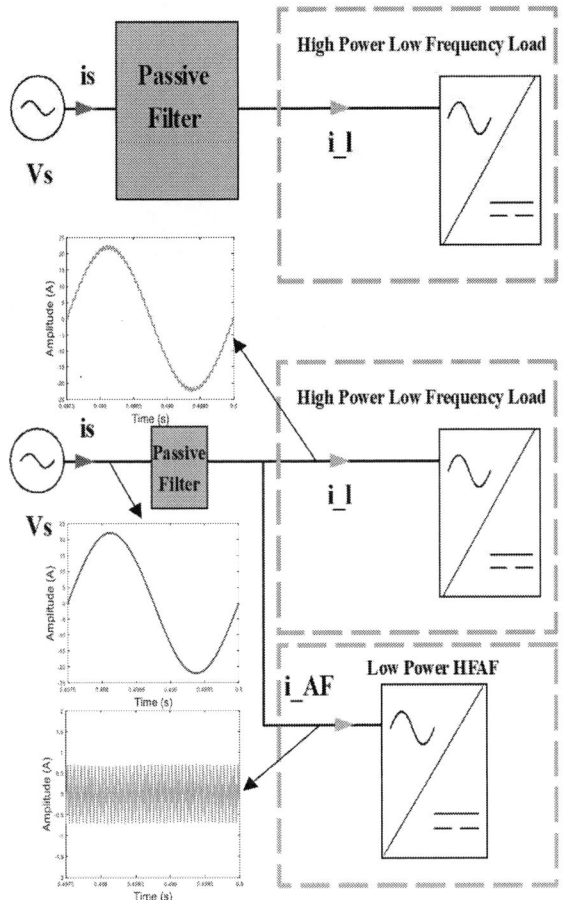

Fig. 1   HFAF principle with current waveforms.

2   Active filter design and control

The HFAF acts like a parallel current source able to provide current variations at switching frequency harmonics, so it must have a higher dynamic performance than the polluting converter. Thus, the switching frequency of the HFAF is at least around 500 kHz, what can be achieved through the use of wide gap components. It is constituted of three functions: measurement circuits, controlled converter and injection circuit. The measurement sensors must have a large frequency bandwidth. To control the current to be injected at high frequency, the filter is made up of an H-bridge voltage source inverter, based on power switches with high frequency and high voltage GaN HEMT transistors with a capacitive energy storage and an inductance for current control (Fig. 2).

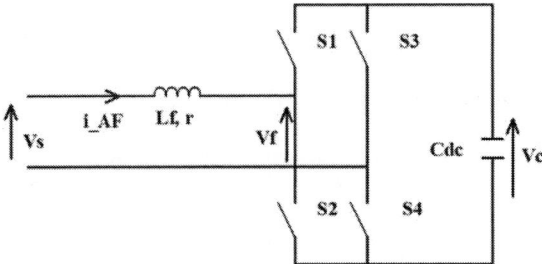

Fig. 2   H-bridge voltage source inverter.

The state model of the single-phase rectifier is described in (1).

$$\frac{d}{dt}\begin{pmatrix} i_{AF}(t) \\ V_c(t) \end{pmatrix} = \begin{bmatrix} \frac{-r}{L_f} & \frac{-f(t)}{L_f} \\ \frac{f(t)}{C_{dc}} & 0 \end{bmatrix}\begin{pmatrix} i_{AF}(t) \\ V_c(t) \end{pmatrix} + \begin{bmatrix} \frac{1}{L_f} & 0 \\ 0 & \frac{-1}{C_{dc}} \end{bmatrix}\begin{pmatrix} V_s(t) \\ 0 \end{pmatrix} \quad (1)$$

$f(t)$ is the modulation function.

In addition to its basic function as sources combining element. The inductance L_f must ensure current dynamics while limiting harmonic currents due to switching. The equation for the current variation of the active filter is given by (2).

$$L_f \frac{di_{AF}}{dt} = V_s - V_c \quad (2)$$

For an H-bridge full-wave control, the link between V_s and V_c is done with $f(t)$ in (3).

$$V_s = V_c.(2.f(t) - 1) \quad (3)$$

The current variation Δi_{AF} as a function of L_f is obtained by the association of (2) and (3) and is represented in (4).

$$\Delta i_{AF} = \frac{V_s - V_c}{L_f}\Delta t = \frac{V_s - V_c}{L_f}f(t)T_{dec}$$
$$\Delta i_{AF} = \frac{2f(t)(f(t) - 1)V_c}{L_f.f_{dec}} \quad (4)$$

The maximum variation on current is obtained for $f(t) = 1/2$. So L_f is chosen in order to limit this maximum current variation as showed in (5).

$$L_f = \frac{V_c}{2.\Delta i_{AF\_max}.f_{dec}} \quad (5)$$

Δi_{AF_max} represents the maximum value of the peak-to-peak current at the switching frequency. This current variation is limited as a ratio of the fundamental current. Most of the time, it doesn't exceed 40% of the fundamental.

The capacitor on the DC bus is designed to ensure that the steady-state voltage ripple is limited as much as possible. It must also be able to provide enough power under transient conditions. The following equation (6) gives the minimum value for C_{dc}:

$$C_{dc} = \frac{P_h}{\omega_h \cdot V_c \cdot \Delta V_c} \quad (6)$$

P_h is the maximum power to be provided in each half-period at the lowest harmonic, and ΔV_c is the peak-to-peak voltage ripple.

The active filter control strategy (Fig. 3) is based on the HFAF current (i_{AF}) parallel injection to compensate harmonics contained in the main current (i_L). The DC bus voltage (V_c) must be regulated and correspond to an active part of the current i_{AF}, that is in phase with the mains voltage (V_s), which requires the use of a PLL. The SOGI-PLL is used for its stability and simple implementation [7]. To identify harmonics, a second SOGI-PLL is used.

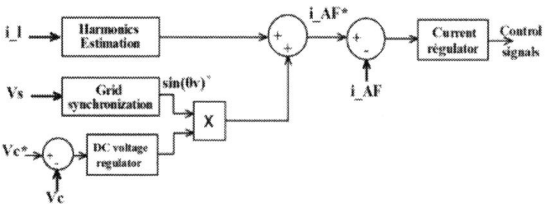

Fig. 3 Active filter control strategy.

The internal current control loop is a non-linear hysteresis controller. The error between the reference current and the inverter current i_{AF} is compared with the hysteresis band (Fig. 4). A control command is sent each time the error reaches one of the lower or upper boundary, so as to remain within it. Its main advantage is its ease of implementation.

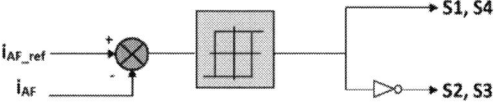

Fig. 4 Current hysteresis control loop.

In order to keep the DC bus voltage at a fixed value an external control loop is realized with a PI controller (Fig. 5).

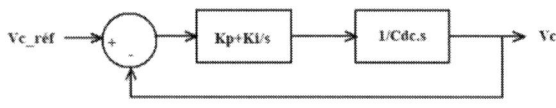

Fig. 5 DC voltage control loop.

To ensure synchronization with the grid, a phase-locked loop (SOGI-PLL) is used, based on a Second-Order Generalized Integrator (SOGI) to build

a quadrature signal generator [7]. The transfer functions of the SOGI block can be described according to Mason's formula (7).

$$G_\alpha(s) = \frac{V_\alpha(s)}{V_S(s)} = \frac{k\widehat{\omega}s}{s^2 + k\widehat{\omega}s + \widehat{\omega}^2}$$
$$G_\beta(s) = \frac{V_\beta(s)}{V_S(s)} = \frac{k\widehat{\omega}^2}{s^2 + k\widehat{\omega}s + \widehat{\omega}^2} \quad (7)$$

Then a Park transformation is applied to the quadrature signal. In equation (8), it's visible that V_d contains the information about the amplitude of the signal while V_q has the information about the phase angle.

$$\begin{bmatrix} V_d \\ V_q \end{bmatrix} = \begin{bmatrix} V_m \cos(\vartheta - \widehat{\vartheta}) \\ V_m \sin(\vartheta - \widehat{\vartheta}) \end{bmatrix} \approx \begin{bmatrix} V_m \\ V_m(\vartheta - \widehat{\vartheta}) \end{bmatrix} \quad (8)$$

The estimated pulse $\widehat{\omega}$ is then obtained by applying a PLL to V_q. The SOGI-PLL functional block is presented in Fig. 6.

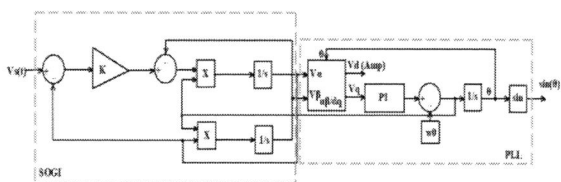

Fig. 6 SOGI-PLL functional block.

In order to estimate the harmonics due to the switching of the polluting load, without affecting the fundamental, it's important to identify them correctly. To do this, the SOGI-PLL presented above is used. The SOGI part gives the amplitude of the fundamental which enables the fundamental to be reconstructed. By subtracting it from the measured current we obtain the switching harmonics as presented in equation (9). The harmonic estimation method is described in Fig. 7.

$$i_{AF\_ref} = i_l - i_{l1} \quad (9)$$

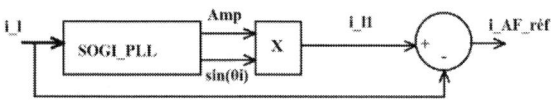

Fig. 7 Switching harmonics estimation method.

3 Filtering volume reduction with the use of HFAF

LC second-order passive filter is characterized by its cutoff frequency (f_c) and characteristic impedance (Z_c) represented by equations (10) and (11), respectively.

$$f_c = \frac{1}{2\pi\sqrt{L.C}} \qquad (10)$$

$$Z_C = \sqrt{\frac{L}{C}} \qquad (11)$$

To calculate the L and C elements of a 2nd order filter, we need to define the required attenuation for the harmonic at the switching frequency f_0. The cutoff frequency required to attenuate the line sufficiently at f_0 is given by the equation (12) [8].

$$f_c = f_0 . 10^{-\frac{Att_{requiredse}}{n.20}} \qquad (12)$$

(With n=2 for a second-order filter)

To calculate the required attenuation, we use an estimated value of the maximum current at the switching frequency. The current expressed in dBµA is subtracted from the DO160G standard. In this context, the literature often adds a value of 6 dBµA as a minimum safety margin. This value can be as high as 10 dBµA. The required attenuation is then described by equation (13).

$$Att_{required} = |I_h| - Limit(DO160G) + 6 \; dB\mu A \quad (13)$$

The DO160G standard is defined between 150 kHz and 152 MHz for conducted emission. The studied switching frequencies are between 10 kHz and 40 kHz. Thus, to limit efficiently the level of harmonics at these frequencies an expanded limit of the DO160G going to 10 kHz is proposed in Fig. 8.

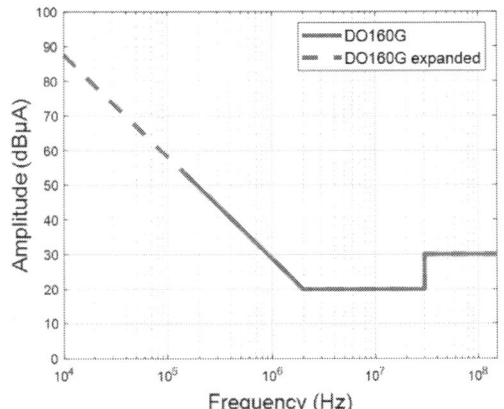

Fig. 8 DO160G expanded standard.

Based on the expression (14) for the product of areas in differential mode, it's possible to estimate the volume of the inductance of a second order passive filter.

$$A_e . S_B = \frac{L_f . I_{MD\_max} . I_{l\_eff} S_d}{J B_{max} K_b} \qquad (14)$$

The global volume of the inductance is then given by (15).

$$V_L = K_0 * (A_e . S_B)^{3/4} \qquad (15)$$

- A_e Effective magnetic core cross-section
- S_B Coiled surface
- S_d Copper cross-section
- K_b Coil filling coefficient
- B_{max} Induction at saturation
- K_0 Constant representative of the shape of the magnetic core
- V_L Global inductance volume

The analysis of the product of areas (14) shows that for the volume depends on the characteristics of the magnetic core, the current (so the power) and the value of the inductance which depends on the cutoff frequency (10) and the switching frequency f_0 (12).

For the same magnetic core characteristics, the volume of the inductance depends on the power and the switching frequency of the polluting load.

To estimate this volume, the maximum current ripple on the switching frequency is fixed at 40% of the nominal current. So, the required attenuation (equation (13)) is calculated based on the equation (16).

$$|I_h| = 120 + 20 \cdot \log(0.4 \cdot i_l) \; dB\mu A \qquad (16)$$

The cutoff frequency is calculated for different switching frequency between 10 kHz and 40 kHz, and for different power of the polluting load between 1 kW and 5 kW for a passive filter.

The cutoff frequency is also calculated for a passive filter used with the HFAF; in this case the switching frequency is fixed at the HFAF switching frequency 500 kHz. The required attenuation is calculated based on the maximum ripple current. It is limited at 40% of the active filter's nominal current i_{AF} which is at the maximum 40% of the load nominal current i_l. The results are presented in Fig. 9. It's clear that the association of the passive and the active filter makes the filtering design independent from the polluting load characteristics (switching frequency and current ripple). The cutoff frequency is higher for the HFAF in all cases, so the passive filter elements will be smaller.

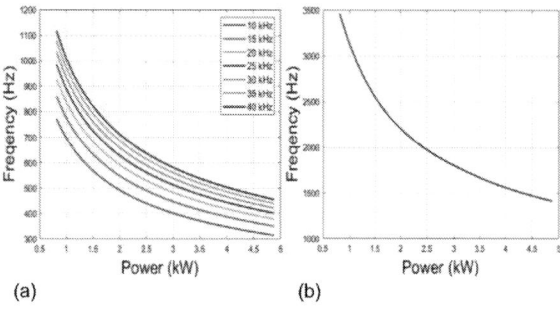

(a)   (b)

Fig. 9 Cutoff frequency f_c as a function of the polluting load power. (a) passive filter for different switching frequencies of the polluting load. (b) passive filter associated to the HFAF for a 500 kHz switching frequency of the HFAF.

To estimate the volume of the whole filter, equation (15) is used for the inductor volume with K_0 of a PM magnetic core. The capacitor is fixed at 10 µF since the most constraining for passive filter volume is the inductor and the voltage source that can influence a variation on the capacitor volume is fixed. The HFAF is a full bridge GaN based converter, GaNSystems proposes high power density (AC/DC) converters ($6W/cm^3$), according to this the HFAF volume is estimated according to its power. The global volume of the filter in both cases (passive and passive/active) is presented in Fig.10.

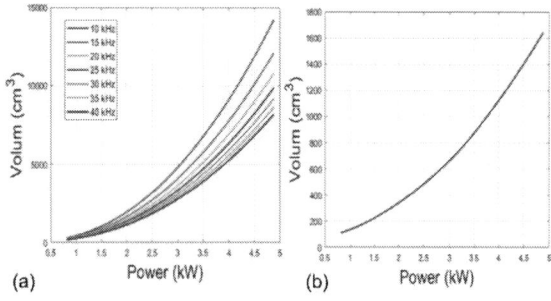

(a)   (b)

Fig. 10 Global filtering volume as a function of the polluting load power. (a) passive filter for different switching frequencies of the polluting load. (b) passive filter associated to the HFAF for a 500 kHz switching frequency of the HFAF.

The estimated filtering volume in Fig.10 shows that the association of HFAF and passive filter reduces the volume of the filtering significantly and that even more obvious for high power polluting load.

4 Simulation results

To evaluate the performance of the HFAF, simulations were run on MATLAB/Simulink. For an AC/DC converter as a polluting load in three cases: with a HFAF, with a HFAF associated to a passive filter and with a passive filter only. The values of the simulated system are summarized in Table 1.

Polluting load (AC/DC converter)	
Switching frequency f_{dec}	20 kHz
Nominal current i_L	22 A
Source voltage V_s	115 V / 400 Hz
High Frequency Active Filter (H-bridge inverter)	
Switching frequency f_{AF}	500 kHz – 2 MHz
Hysteresis bandwidth	± 0.1 A
DC Bus voltage	200 V
Passive filter	
Cutoff frequency f_{CPF}	656 Hz
L_{PF}	5.9 mH
C_{PF}	10 µF
Passive associated to active filter	
Cutoff frequency f_{CAF}	2.5 kHz
L_{AF}	0.4 mH
C_{AF}	10 µF

Table 1 Simulated system values.

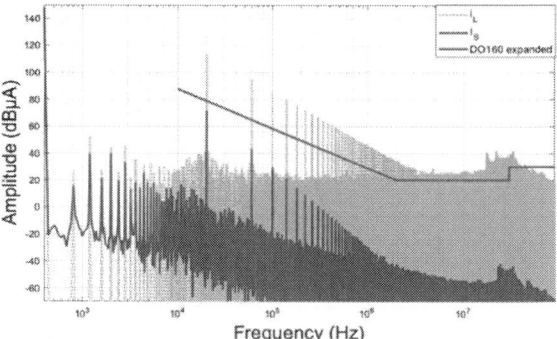

Fig. 11 Frequency domain currents from simulation results for a passive filter only.

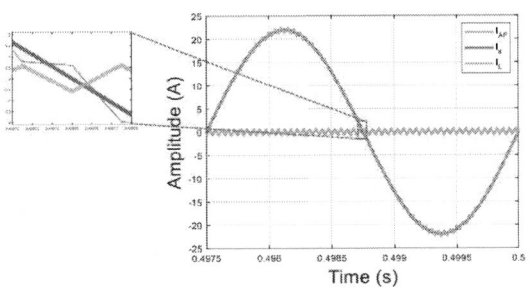

Fig. 12 Time domain currents from simulation results for a HFAF.

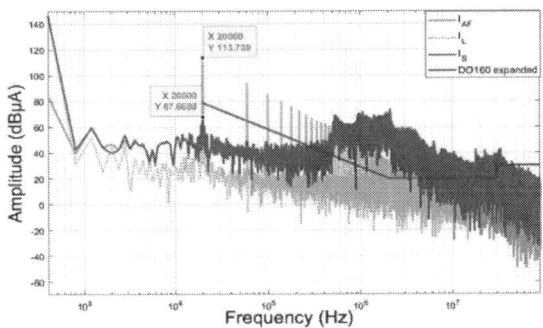

Fig. 13 Frequency domain currents from simulation results for a HFAF only.

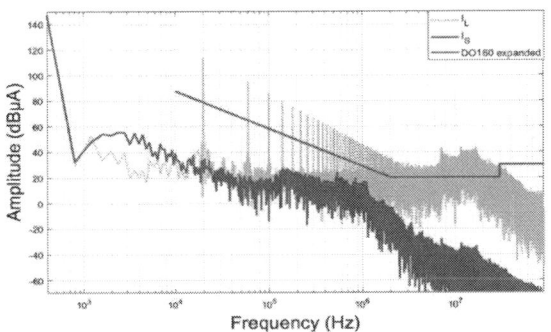

Fig. 14 Frequency domain currents from simulation results for a HFAF associated with passive filter.

The simulation results for the passive filter only are shown in Fig. 11. The volume of a passive filter used in this case is around 5000 cm^3 to be able to respect the expanded DO160G standard, and that without taking into account the parasitic interfering elements of the filter.

Fig. 12 illustrates how the current ripple due to the switching of the polluting load is provided by the HFAF and thus annulated in the source current i_s. Fig. 13 shows that the active filter is able to reduce the amplitude at the switching frequency (20 kHz) by 45 dB and reduces the level of the harmonics

to 50 dBμA up to the HFAF switching frequency (500 kHz). Therefore, the harmonics due to switching, initially contained in the i_s current, are pushed away from 20 kHz to 500 kHz with the active filter.

This allows the associated passive filter to be designed for an attenuation at 500 kHz, and thus reduces the volume of the passive elements, while still respecting the expanded DO160G standard (section 21) as shown in Fig.14. The estimated volume of all the filtering is around 1000 cm^3.

5 Experimental development

The HFAF prototype consists of a GaN HEMT transistors (GS66506T 650V Enhancement Mode GaN Transistor) voltage source inverter (Fig. 15), a measurement stage using WE-CST current sense transformer with a frequency range going up to 1 MHz, an injection stage (parallel plug) and a control implementation. For the high frequencies targeted, the current control is realized in an analog hysteresis while the rest of the control is implemented in the TMS320F28379D microcontroller. The H-bridge was developed based on an existing board designed by Safran Tech.

Fig. 15 H-bridge GaN based voltage source inverter.

A main board (Fig. 16) has also been designed to house all the functions required to operate the FAHF. This main board contains:

1. H-bridge interface and the passive elements required to operate as a voltage inverter;

2. Gird and pollutant load connections;

3. Passive filter connections if required;

4. Interface with the DSP for the digital part of the control (harmonic estimation, DC bus regulation and grid synchronization);

5. The second part of the control (i_{AF} current regulation with analog hysteresis and dead time);

6. The measurements and conditioning required for FAHF control (i_L current for harmonics estimation, i_{AF} current for injected current regulation, V_s voltage for grid synchronization and V_c voltage for DC bus voltage regulation);

7. Power supply.

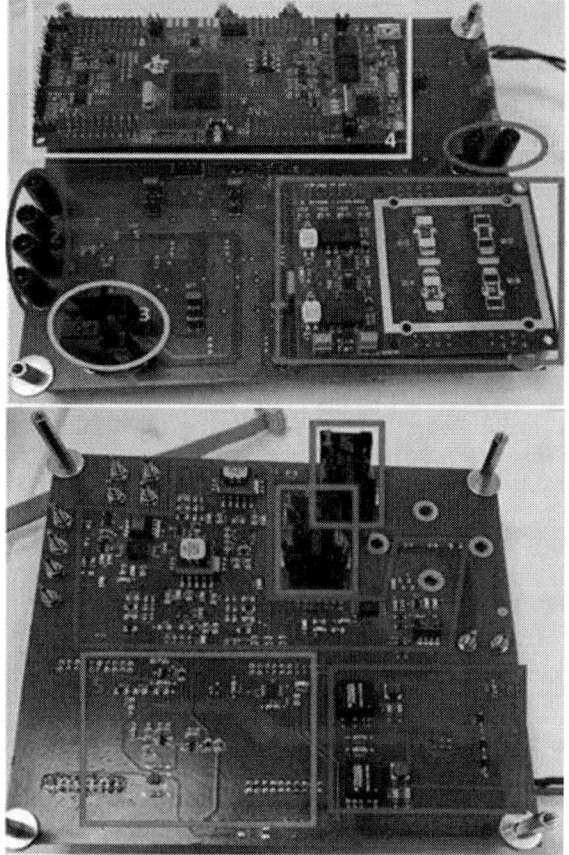

Fig. 16 Main board of the high frequency active filter.

6 Conclusion

A HFAF for differential mode EMI reduction with its design and the control strategy have been presented. The independence of the HFAF control from the polluting load control makes the HFAF reusable for different loads without having to design all over again.

The simulation results obtained are interesting, they demonstrate the interest of the HFAF by reporting the EMI constraint on the FAHF which not only reduces the passive filter volume and weight by pushing the harmonics up to 500 kHz. But also transfers the design of the passive filter to the HFAF for greater independence from the pollutant load.

The volume reduction due to the use of the association active and passive filters is demonstrated and this is more significant for high power polluting loads. The H-bridge GaN voltage source inverter and the main bord of the HFAF are realized and experimental tests are in progress.

References

[1] J. Amilhat, "Fly the Green Deal, Europe's Vision for Sustainable Aviation, Report of the Advisory Council for Aviation Research and Innovation in Europe (ACARE)".

[2] D. T. Nguyen, C. Deng, E. Macias, et A. J. Hanson, « Synchronously Switched Active EMI Filter », in *2022 IEEE Energy Conversion Congress and Exposition (ECCE)*, Detroit, MI, USA: IEEE, oct. 2022, p. 1-8. doi: 10.1109/ECCE50734.2022.9948006.

[3] S. Papadopoulos, M. Rashed, C. Klumpner, and P. Wheeler, "Investigations in the Modeling and Control of a Medium-Voltage Hybrid Inverter System That Uses a Low-Voltage/Low-Power Rated Auxiliary Current Source Inverter," *IEEE J. Emerg. Sel. Top. POWER Electron.*, vol. 4, no. 1, 2016.

[4] H. Bai, X. Wang, P. C. Loh, and F. Blaabjerg, "An Active Trap Filter for Switching Harmonic Attenuation of Low-Pulse-Ratio Inverters," *IEEE Trans. Power Electron.*, vol. 32, no. 12, pp. 9078–9092, Dec. 2017, doi: 10.1109/TPEL.2017.2657644.

[5] D. Bernet, L. Stefanski, and M. Hiller, "Integrating Voltage-Source Active Filters Into Grid-Connected Power Converters—Modeling, Control, and Experimental Verification," *IEEE Trans. Power Electron.*, vol. 36, no. 11, pp. 12218–12233, Nov. 2021, doi: 10.1109/TPEL.2021.3075068.

[6] Y. Sato, K. Kawamura, H. Morimoto, and K. Nezu, "Hybrid PWM rectifiers to reduce electromagnetic interference," in *Conference Record of the 2002 IEEE Industry Applications Conference. 37th IAS Annual Meeting (Cat. No.02CH37344)*, Pittsburgh, PA, USA: IEEE, 2002, pp. 2141–2146. doi: 10.1109/IAS.2002.1043827.

[7] J. Xu, H. Qian, Y. Hu, S. Bian, and S. Xie, "Overview of SOGI-Based Single-Phase Phase-Locked Loops for Grid Synchronization Under Complex Grid Conditions," *IEEE Access*, vol. 9, pp. 39275–39291, 2021, doi: 10.1109/ACCESS.2021.3063774.

[8] R. Chen, "Integrated EMI Filters for Switch Mode Power Supplies". Nov. 2004.

PCIM Europe 2024, 11– 13 June 2024, Nuremberg   DOI: 10.30420/566262049

Laboratory Setup for Accuracy Investigation of Electricity Meters and Monitors under Industry-Typical Operating Conditions

Xiaofei Guo[1], Matthias Schmidt[1], Michael Freiburg[2], Robin Abraham[1], Felix Hackeloeer[2], Lukas Christ[2], Christian Brenncke[3], Sascha Bierbach[4], Ralf Schellenberg[5]

[1] Physikalisch-Technische Bundesanstalt (PTB), Germany

[2] Faculty of Computer Science and Engineering Science, TH Köln, Germany

[3] ZERA GmbH, Germany

[4] Fischer und Kaufmann GmbH, Germany

[5] Alfred H. Schütte GmbH & CO. KG, Germany

Corresponding author: Xiaofei Guo, Xiaofei.Guo@ptb.de
Speaker: Matthias Schmidt, Matthias.Schmidt@ptb.de

Abstract

Electricity meters and power monitors are designed and calibrated for power frequency but increasingly exposed to harmonic distortion, harmonic active and reactive power components, or fast transitions. This leaves a gap in addressing real-world complexities in metrology. This study introduces a measurement setup created to investigate the impact of non-sinusoidal signals derived from industry-typical voltage and current signals on energy meter accuracy. Recordings of signals from actual industrial scenarios were used to synthesise test signals, to verify the test setup, and several energy meters were tested with the test signals in a laboratory environment.

1   Introduction

The energy transition is causing renewable energy to become an increasingly important source of energy [1]. This shift is driving the proliferation of semiconductors and converters, resulting in more complex waveforms [2]. In addition, industrial processes are increasingly equipped with inverter-based technologies, e.g. active drives, energy recuperation, etc. To sum up, industrial processes can be very nonlinear and dynamic, resulting in electrical signals different from sinusoidal. Efficiency is becoming more critical, and an increasing number of energy meters and power monitors are installed in public and industrial electricity grids. The complexity of these waveforms presents a challenge for meters, which must accurately measure energy consumption and/or supply. The energy in a certain time interval is generally calculated with formula (1). However, a significant question arises: Can electricity meters and power monitors measure energy consump-

tion/supply precisely in the face of complex waveforms? This needs to be addressed in the context of our evolving energy landscape.

$$E = \int_{t1}^{t2} v(t) \cdot i(t) dt \qquad (1)$$

Where E is the energy, v is the voltage, i is the current and t is the time.

The conventional certification process for energy meters focuses primarily on ensuring accurate measurements of pure sinusoidal signals [3] [4] [5] [6] [7], leaving an important gap in addressing the complexities of real-world signals. Researchers have investigated more complex signals [8], but the results lack real waveforms from actual industrial scenarios. This paper presents a new test setup and signal synthesis to be able to systematically study the impact of non-sinusoidal signals on energy meter precision, particularly in industrial contexts.

The meters are categorized based on the percentage error limit according to industrial standards. Class A, B, and C are defined by 50470-1 with error limits of 2.5%, 1.5%, and 1%, respectively, for a power factor of 1. Similarly, IEC 62053-21 and IEC 62053-22 define four accuracy classes with comparable accuracies. This paper does not discuss the categorization of meters as it is not relevant to the research project. A large range of different meters and power monitors have been tested. The meters, which shown an error more than 2 % in our tests, were deeper tested and discussed.

The research and the respective laboratory test setup on an NMI level aims to investigate whether the standard certification process, which guarantees accuracy in sinusoidal signals, is also applicable to the measurement of non-sinusoidal signals. MID-compliant [5] electrical energy meters for energy billing (MID-EM) and power monitors for monitoring energy consumption in manufacturing (PM) were tested.

2 Methods

2.1 Measurement of Industry-Typical Operating Conditions

To examine the impact of non-sinusoidal signals on accuracy of electricity meters and monitors used in industrial applications, voltage and current waveforms have been recorded. Recordings have been made from various authentic industrial settings. Due to dynamic production processes and the increased utilization of power electronics (e.g. electrical drives, active regenerative modules), industrial networks are increasingly operated under non-sinusoidal conditions.

Measurements of current and voltage waveforms have been conducted at different nodes of the industrial networks with a high-precision Power Quality and Network Analyzer (NEO PQA8000). As the industrial processes cannot be interrupted for measurement installation, Rogowski Coils have been used as primary sensors for the 3-phase current measurement whereas the voltage has been measured directly. To ensure high precision in the overall intended frequency range up to 10 kHz, the current sensors have been first calibrated and error-corrected, and the network analyzer has been initially compared against the PTB laboratory standard test setup. However, high accuracy would not be important in respect to the

goals of the study as the test signals are synthetized in the laboratory test setup and the main aim here is to quantify the frequency components.

Different industrial networks, machines or processes have been measured, e.g. hydraulic presses with active regeneration modules, galvanizing processes, manufacturing facilities with active drives, and large industrial machine tools. Exemplary measurements are shown in figure 1.

Fig.1: Exemplary pictures of measurements in industrial applications.

The current and voltage waveforms as well as classical Power Quality parameters have been measured and analysed to subsequently develop realistic test signals for the laboratory investigations.

Even if the signals and parameters show very individual pattern (cf. figure 3), common characteristics can be found. This allows us to generally categorize into periodic and non-periodic signals (in respect to the fundamental power frequency) and define further signal characteristics with direct implications for the test setup to be developed. The developed test signals based on the conducted site measurements are listed in table 5.

The captured signals are saved in CSV files as waveforms and loaded into the test computer. The periodic signals are then decomposed using Fourier transformation. To allow investigation of individual influence quantities, the test signals are synthetically reconstructed based on the decomposition in the laboratory environment. Meanwhile, non-periodic signals were directly replayed without further analysis or conversion. Non-periodic signals were synchronized to minimize leakage. Software low-pass filtering is also available if needed.

In addition to the recorded waveforms, several literature-based synthetic test signals with characteristic features not or only partly found in the field measurements have been examined to allow a systematic study of the errors of industrial energy meters and monitors.

2.2 Laboratory Environment for Examining Energy Meters by Using Test Signals

The recordings were reproduced in a laboratory environment to test energy meters and monitors. The experimental arrangement is demonstrated in figure 2. The measurement setup consists of a phantom power source and a reference measurement device, enabling an accuracy of up to 0.028% for 50 Hz periodic signals. The range of the measurand is restricted by the output of the phantom power source, which can supply up to 230 V and 14 A of DC power up to 150 kHz. [9]. The phantom power source consists of a computer control, digital-to-analog converters, and amplifiers. These components work together to accurately reproduce signals, resulting in the independent generation of current and voltage. The computer control and software will convert the periodic signal from the frequency domain to the time domain. The converted signal is synchronised to minimise leakage and can be repeated for as long as required. The DAC converts the digital time domain signals into analogue signals, which are then amplified by the current and voltage amplifier. The advantage of this configuration is its capacity to maintain low energy consumption while enabling the independent adoption of current and voltage waveforms in response to demand, without interfering between them [10]. The periodic signal can be freely synthetic produced as a form of Fourier series or other in the software predefined signals like rectangle signals or PWM signals. The signal generation is open-loop controlled.

Only transformer-connected meters were tested in this work to ensure maximum accuracy and uniform measurement conditions due to the use of phantom power sources. The sensor connections were directly connected to the current and voltage amplifiers, as illustrated in figure 2. To not endanger accuracy related losses, only one test object is tested at once.

The reference measurement device is a wideb and power analyser calibrated against a nation al power standard. The measurements were taken using a zero-flux current sensor, which provides a potential-free measurement and minimizes interference with the Device Under Test (DUT), allowing for wideband measuring up to 1 MHz.

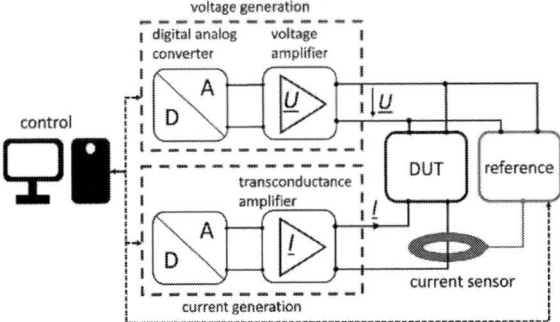

Fig. 2 Block diagram showing the recreation of the recorded waveforms. DUT stands for device under test.

2.3 Comparison of set and generated Test-signal Waveforms

Verification tests were carried out to verify that the regenerated signals (output of amplifiers) were in accordance with the set signals (input of amplifiers from control unit). Four synthetic signals were used to verify whether the generated signals aligned with the predefined signals. The test signals included a fundamental frequency of 50 Hz and up to four harmonics reaching 5000 Hz and are discussed in the following chapter.

2.4 Accuracy of Energy Meters and Power Monitors under Test Signal Conditions

Subsequently, the regenerated and synthesised test signals were utilized to investigate the accuracy of multiple energy meters and power monitors against the reference. Three of the waveforms in table 5 will be discussed in the chapter Results and Discussion. Tabel 1 lists the test devices

(DUTs). The DUTs were tested separately to eliminate any potential interference between them.

NUMBER	TYPE
1	MID-compliant electrical energy meters
2	power monitor
3	MID-compliant electrical energy meters
4	power monitor

Table 1 Devices under Test: Numbering and Types

3 Results and Discussion

3.1 Validation of the Measurement Setup

Table 2 shows the validation results for the power measurement of the reference instrument. The results represent the difference between the reference instrument in the setup and the Primary Power Standard of PTB. Figure 3 shows waveform of the PWM signal used in this validation, which was recorded at the output of a motor regulator with an 8 kHz switching frequency. The relative errors of the 50 Hz and PWM are under 0.05% for every quantity. However, the measurement with 50 Hz + 50 kHz shows a larger error, which can be explained by the total accuracy of 0.5% of reading + 0.002% of full scale for signals from 10 kHz to 100 kHz in the current sensor's data sheet.

Signal	Signal description	Rel. Error from P	Rel. Error from Q
50 HZ	50 Hz sinusoidal	-0.02%	-0.01%
50 Hz+ 50 kHz	50 Hz sinusoidal + 50 kHz sinusoidal	-0.41%	-0.39%
PWM	8 kHz switching frequency	-0.05%	0.03%

Table 2 Validation results for the power measurement of the reference instrument

Table presents the validation test results for the energy measurements. Figure 3 (b) shows the high dynamic waveform (hdw) recorded at the regulator of a hydraulic press. The relative errors of all three measurements were below 0.05% for each quantity.

Signal	Signal description	Rel. Error from P	Rel. Error from Q
hdw	hydraulic press	-0.02%	-0.01%
50 Hz	50 Hz sinusoidal	-0.02%	-0.01%
PWM	8 kHz switching frequency	-0.01%	-0.01%

Table 3 Validation results for the energy measurements of the reference instrument

Table 4 presents the validation results for the regenerated signals and their accordance with the set signals. The harmonics' amplitudes in the table were normalized to the amplitude of the fundamental frequency, which was either set at 50 Hz or measured at 50 Hz. The table compares the ratio between the harmonics and the fundamental frequency. The set and measured signals are sufficiently similar.

Signal	Freq. of Harm.	Ratio (set voltage)	Ratio (meas. voltage)	Ratio (set current)	Ratio (meas. current)
1	100	5,000	4,996	5,000	4,997
	150	5,000	4,999	5,000	4,997
2	250	5,000	5,000	5,000	5,000
	350	5,000	4,997	5,000	4,999
3	650	5,000	5,000	5,000	5,001
4	5000	5,000	4,992	5,000	5,073

Table 4 Harmonic Amplitude Validation in Mixed Frequency Outputs

3.2 Examination of Energy Meters by Use of Test Signals

The paper provides a brief overview of the results to demonstrate the test setup's capability, as complete information regarding the tests on the DUTs is content of the further investigations. Table 5 presents the results of the four meters, indicating the number of meters with a measurement error of more than 2%. All of the DUTs showed sufficient accuracy in measurements using PWM signals. The waveform generated by the hydraulic press has a high dynamic range. Mechanical machines, such as hydraulic presses, undergo a specific process cycle, such as lifting and pressing, and utilise active electrical energy regeneration units. This results in dynamic changes of electrical signals with transition times of less than 20ms, including variations in active power and power factor, as well as a rapid change in energy flow direction. Fig. 3(d) shows the changing direction of the energy flow. This caused a significant error in MID-compliant electrical energy meter number 3. The percentage error was up to -81.73%. The positive half-wave signal in fig. 3(c) was periodic and synthetic, and contained a DC component in the frequency domain. Three out of the four DUTs showed a very high percentage error ranging from -52.95% to -81.04%. If the DC component is removed from the half-wave, the percentage errors of all DUTs decrease to a sufficient level below 2%.

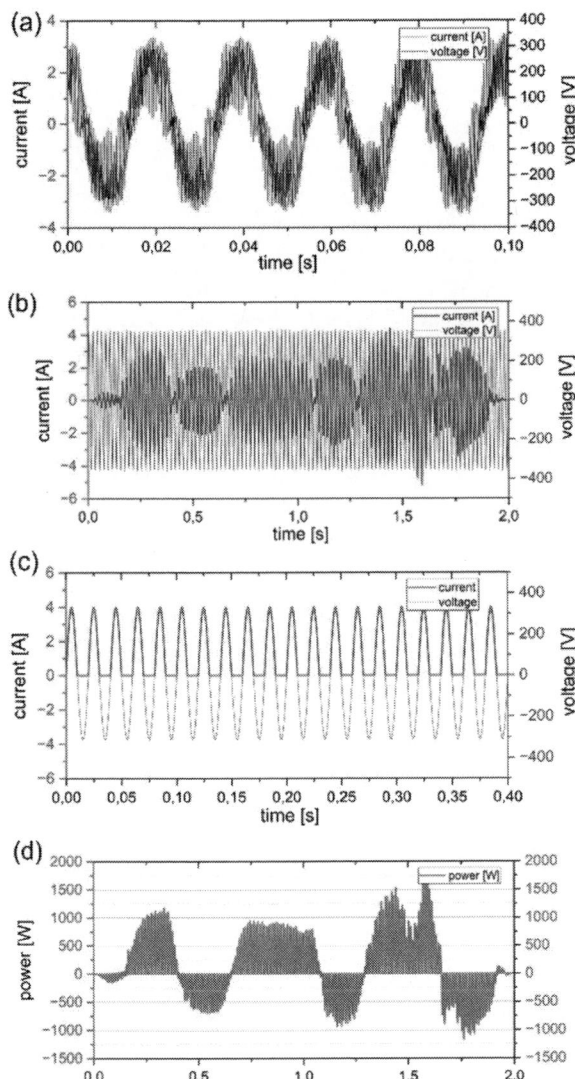

Fig. 3 (a) A PWM signal from an AC Motor controller, switching frequency 8 kHz (b) A signal produced by a hydraulic press as an example of non-periodic signals in respect to 50Hz power frequency.(c) A positive half-wave signal (d) The change in direction of the energy flow of the signal generated by a hydraulic press

Signal	Signal description	DUT with more than 2% error	accuracy
PWM	8 kHz switching frequency	0/4	0.16%
hdw	hydraulic press	1/4	0.20%
positive half-wave	periodic, synthetic	3/4	0.11%

Table 5 Test results of the meters under laboratory conditions.

4 Conclusion

This research describes a laboratory setup for testing the accuracy of electricity meters and monitors in controlled conditions different from pure sinusoidal waveforms. The setup allows for the verification of energy meters and monitors by capturing real waveforms from the field, reproducing and manipulating them in a laboratory environment with a high level of precision and flexibility. The laboratory setup can generate periodic and non-periodic signals, ensuring robust reproducibility. Possible limitations of the measurement setup include the lack of phase and amplitude correction in the energy measurement of the reference device. However, the relative error of the reference device by every waveform is under 0.2%, so the measurement result can still be considered sufficiently reliable.

This research presents an opportunity to discuss and evaluate the necessity of updating current standards. If real-life scenarios frequently involve signals that have the potential to cause errors, and these errors reach an unacceptable level, there is a case for arguing in favour of improving the existing standards towards more complex test signals. This could include adding test waveforms that more closely resemble realistic scenarios.

References

[1] IEA, "IEA (2022), Renewables 2022," 2022. [Online]. Available: https://www.iea.org/reports/renewables-2022/executive-summary. [Accessed 22 3 2024].

[2] A. Božiček, J. Kilter, T. Sarnet, I. Papič and B. Bl, "Harmonic Emissions of Power Electronic Devices Under Different Transmission Network Operating Conditions," *IEEE Transactions on Industry Applications,* vol. 54, no. 4, pp. 5216-5226, 2018. [3]   Reference in 11 pt type size and 12 pt line spacing.

[3] „DIN EN 50470-1," VDE, 2019.

[4] „DIN EN 50470-3," VDE, 2020

[5] European Parliament & Council, „Measuring Instruments Directive 2014/32/EU," 2014

[6] IEC, IEC 62053-22 Electricity Metering Equipment (A.C.)—Particular Requirements— Part 21, IEC, 2016

[7] IEC, IEC 62053-21: Electricity Metering Equipment (A.C.)—Particular Requirements— Part 21, IEC, 2016.

[8] L. Bartolomei, D. Cavaliere, A. Mingotti, L. Peretto und R. Tinarelli, „Testing of Electrical Energy Meters Subject to Realistic Distorted

Voltages and Currents," *Energies,* Bd. 13, 2020.

[9] M. Schmidt, F. Schilling, S. Bauer, X. Guo and B. Engel, "PTB'S New System for Calibrating Power Measurement Devices at Frequencies of up to 150 Khz [Manuscript submitted for publication]," in *XXIV IMEKO World Congress "Think Metrology"*, Hamburg, Germany, 2024.

[10] E. Mohns, G. Ramm und W. G. K. Ihlenfeld, „The PTB Primary Standard for Electrical AC Power," *Journal of Metrology Society of India,* Bd. 24, Nr. 1, pp. 15-19, 23.03.2009.

PCIM Europe 2024, 11– 13 June 2024, Nuremberg   DOI: 10.30420/566262050

Real-Time Evaluation of Weighting Factorless Predictive Control of LCL Filter Equipped Grid-Side Converters using Sorting Networks

Kristóf Bándy [1], Péter Stumpf [1], Zoltán Sütő [1]

[1] Department of Automation and Applied Informatics, Faculty of Electrical Engineering an dInformatics, Budapest University of Technology and Economics, Hungary.

Corresponding author: Kristóf Bándy, Bandy.Kristof@aut.bme.hu
Speaker: Kristóf Bándy, Bandy.Kristof@aut.bme.hu

Abstract

A novel multi-cost function model-based finite-set predictive control method is presented for grid-side converters equipped with LCL filters. This approach aims to overcome the tedious tuning of weighting factors. In the method, the cost functions are established based on various state variables and their references. Then the control options are ranked based on their performance in each of the cost functions. This ranking score determines the selection of the optimal control action. The control method is validated in a low-cost custom Hardware-in-the-Loop (HIL) environment using model-based rapid-prototyping tools of Matlab/Simulink. Simulation and measurement results show the viability of the method.

1   Introduction

Nowadays, Model Predictive Control (MPC) has become a promising technique in power electronics, such as controlling high performance electrical drives [1] or Grid Side Converters (GSC) in the utilization of renewable energies [2]. MPC schemes provide advantages like the possibility to use apriori insights about the system, formulation of multi-objective control laws by considering nonlinearities and constraints like voltage and current limits and controlling multiple state variables, all within a single loop. The typical technical challenges of MPC control techniques to be solved are the high computational load, limited modelling accuracy, selection and design of cost function(s) and calculation of proper weighting factors [3]. The aim of the current paper is to utilize a weighting factorless approach with low computational demand.

GSCs are connected to the utility via a line filter. Compared to a single inductive L filter as shown in [4], LCL filters can provide higher attenuation to higher order harmonics, which can be important in applications with limited switching frequencies [5]. Using an LCL filter between the grid and the GSC extends the number of state variables and thus increases the order of the system resulting in more

challenging implementation of MPC scheme [6]. In the paper the control objectives are defined to force the inverter output currents, filter capacitor voltages and grid currents to follow their reference signals. Although, these control objectives can be formulated into a single cost function by using weighting factors in the traditional MPC methods. However, the value of these weighting factors have a great impact on the performance of the system and their tuning is a nontrivial process. Paper [7] provides a review about weighting factor design approaches in the MPC scene. To overcome the issues caused by weighing factor selection, in the current paper the control objectives are defined by separate cost functions. Then, a sorting network is utilized to find the most appropriate control action. This multi cost function scheme was already established in [8] for the control of induction motor drives and this paper aims to extend the method to different applications and more extensive models.

The paper first introduces the state-space modeling of the LCL filter equipped grid-side converter, then the MPC control approach is presented, with the introduction of the sorting networks as optimal sorting solutions to the underlying optimization problem. The proposed control method is validated using a custom Hardware-in-the-Loop (HIL) approach, with experimental results showing the viability of the method in all GSC benchmark control maneuvers.

2 Mathematical Model

Figure 2 represents the LCL filter connected GSC that is the main focus of this paper. The three-phase two-level voltage source inverter (2L-VSI) can produce six active voltage vectors by connecting one or two output phase(s) to the positive DC rail of the inverter while connecting the rest of the phase(s) to the negative DC rail. These voltage vectors have an amplitude of $2/3V_{\mathrm{DC}}$, where V_{DC} is the dc-link voltage. The inverter can also produce the so-called zero voltage vector by connecting all phases to either the positive or the negative DC rail. These described voltage vectors can be observed in Fig. 1. It should be noted, that in the case of two-level GSC, only these active voltage vectors can be generated, while multilevel GSCs can generate other active output voltage vectors with smaller amplitudes and different angles as well.

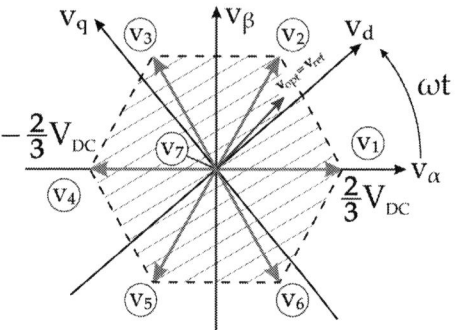

Fig. 1: Possible switch positions of a three-phase grid side converter, which can be realized two level inverters.

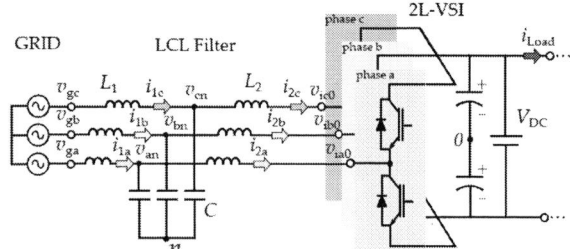

Fig. 2: Schematic topology of the system under discussion. The three-phase converter and the grid are connected via an LCL filter structure in all phases.

The dynamics of the GSC can be separated to the AC or grid-side and to the DC side, due to the nature of the converter. The AC-side of the GSC

the equivalent circuit can be observed in Fig. 2. The equations for the three-phase GSC can be established in the stationary coordinate system using space vector theory. The transformation of the quantities of the three-phase system to the $\alpha\beta$ stationary reference frame (SRF) can be established using the Clarke transformation.

$$\mathbf{x}_{\alpha\beta} = \begin{bmatrix} x_\alpha \\ x_\beta \end{bmatrix} = \frac{2}{3} \begin{bmatrix} 1 & -\frac{1}{2} & -\frac{1}{2} \\ 0 & \frac{\sqrt{3}}{2} & -\frac{\sqrt{3}}{2} \end{bmatrix} \begin{bmatrix} x_\mathrm{a} \\ x_\mathrm{b} \\ x_\mathrm{c} \end{bmatrix} \tag{1}$$

where $x_\mathrm{a}, x_\mathrm{b}$ and x_c are phase quantities. It is also important to introduce the rotating reference frame (RRF), that rotates about the same origin as the SRF, with an arbitrary ω angular speed (see Fig. 1). The transformation of the arbitrary $\mathbf{x}_{\alpha\beta}$ SRF quantity to RRF can be established via the Park transform described as:

$$\mathbf{x}_{\mathrm{dq}} = \begin{bmatrix} x_\mathrm{d} \\ x_\mathrm{q} \end{bmatrix} = \begin{bmatrix} \cos(\omega t) & \sin(\omega t) \\ -\sin(\omega t) & \cos(\omega t) \end{bmatrix} \begin{bmatrix} x_\alpha \\ x_\beta \end{bmatrix} \tag{2}$$

2.1 Continuous State-Space Model of the LCL filtered grid side of the converter

The system under review can be modeled with a linear state-space model of form in the rotating dq coordinate system that rotates about the origin with ω the electrical speed of the grid as

$$\frac{\mathrm{d}}{\mathrm{d}t}\mathbf{x}(t) = \mathbf{A}\mathbf{x}(t) + \mathbf{B}\mathbf{u}(t)$$
$$\mathbf{y}(t) = \mathbf{C}\mathbf{x}(t) + \mathbf{D}\mathbf{u}(t) \tag{3}$$

where, $\mathbf{x}(t)$ is the state-vector, $\mathbf{u}(t)$ is the input-vector, $\mathbf{y}(t)$ is the output vector, $\mathbf{A}$ is the state-matrix, $\mathbf{B}$ is the input-matrix, $\mathbf{C}$ is the output-matrix and $\mathbf{D}$ is the feedforward-matrix. The state vectors and matrices can be given respectively as:

$$\mathbf{x} = \begin{bmatrix} i_{1\mathrm{d}} & v_{\mathrm{fd}} & i_{2\mathrm{d}} & i_{1\mathrm{q}} & v_{\mathrm{fq}} & i_{2\mathrm{q}} \end{bmatrix}^\mathrm{T}$$
$$\mathbf{u} = \begin{bmatrix} v_{\mathrm{id}} & v_{\mathrm{iq}} & v_{\mathrm{gd}} & v_{\mathrm{gq}} \end{bmatrix}^\mathrm{T} \tag{4}$$

where, $i_{1\mathrm{d}}$ and $i_{1\mathrm{q}}$ are the grid side current, $i_{2\mathrm{d}}$ and $i_{2\mathrm{q}}$ are the converter side current, v_{fd} and v_{fq} are the filter capacitor voltage components in the state-vector and v_{id} and v_{iq} are the converter voltage, v_{gd} and v_{gq} are the grid voltage components in the input-vector.

404

$$\mathbf{A}(\omega) = \begin{bmatrix} -\frac{R_1}{L_1} & \frac{1}{L_1} & 0 & \omega & 0 & 0 \\ -\frac{1}{C_f} & 0 & \frac{1}{C_f} & 0 & \omega & 0 \\ 0 & -\frac{1}{L_2} & -\frac{R_2}{L_2} & 0 & 0 & \omega \\ -\omega & 0 & 0 & -\frac{R_1}{L_1} & \frac{1}{L_1} & 0 \\ 0 & -\omega & 0 & -\frac{1}{C_f} & 0 & \frac{1}{C_f} \\ 0 & 0 & -\omega & 0 & -\frac{1}{L_2} & -\frac{R_2}{L_2} \end{bmatrix}$$ (5)

$$\mathbf{B}(\omega) = \begin{bmatrix} 0 & 0 & -\frac{1}{L_1} & 0 \\ 0 & 0 & 0 & 0 \\ \frac{1}{L_2} & 0 & 0 & 0 \\ 0 & 0 & 0 & -\frac{1}{L_1} \\ 0 & 0 & 0 & 0 \\ 0 & \frac{1}{L_2} & 0 & 0 \end{bmatrix}$$

where, R_1 and R_2 are the resistance, L_1 and L_2 are the inductance of the grid and converter sides, respectively. C_f is the filter capacitor's capacitance. Furthermore, the output matrix $\mathbf{C}$ is the identity matrix of dimension six, and the feedforward-matrix $\mathbf{D}$ is the zero matrix, therefore the state and output variables are equal $\mathbf{y}(t) = \mathbf{x}(t)$.

2.2 Discretization

Using the established linear continuous state space model, the predictive control scheme can be derived using the discretized state-space model that allows the predictions of state variables for various input sequences. In the paper, only the one time-step horizon control is discussed. The discrete state-space model can be given as:

$$\mathbf{x}[k+1] = \mathbf{\Phi}(\omega)\mathbf{x}[k] + \mathbf{\Gamma}(\omega)\mathbf{u}[k]$$
$$\mathbf{y}[k] = \mathbf{H}\mathbf{x}[k] + \mathbf{F}\mathbf{u}[k]$$ (6)

where $\mathbf{\Phi}$, $\mathbf{\Gamma}$, $\mathbf{H}$ and $\mathbf{F}$ are the discretized equivalents of the $\mathbf{A}$ state, $\mathbf{B}$ input, $\mathbf{C}$ output, and $\mathbf{D}$ feedforward matrices respectively, with the bracket notation of $\mathbf{x}[k+1] = \mathbf{x}(t + T_s)$, with T_s denoting the sampling time.

In this paper, the discrete system model can be established using the exact discretization as $\mathbf{\Phi} = e^{\mathbf{A}T_s}$ and $\mathbf{\Gamma} = \mathbf{A}^{-1}(\mathbf{\Phi} - \mathbf{I})\mathbf{B}$. Since the parameters can be hypothesized as constant, the calculation of the discrete state-space matrices can be done offline.

2.3 Reference generation

In the formulation of the control law, it is necessary to establish the control objectives of all controlled variables. These references can be used to compare the various control options and allow the selection of an optimal action. The main objective in

this paper is to control the grid side current components, as these components directly influence the controlled system's performance. Using the v_{dc} voltage of the DC side capacitor a PI controller can set the i_{1d}^* reference that corresponds to the effective power of the system, while the i_{1q}^* can be set independently to achieve a required reactive power level.

In RRF and in steady state operation, the state variables can be assumed constant. Therefore the derivative side of the state-space model (3) can be assumed to be a zero vector. This way the filter voltage component references can be calculated using the state-matrix and the stator current references as:

$$v_{fd}^* = R_1 i_{1d}^* - \omega L_1 i_{1q}^* + v_{gd}$$
$$v_{fq}^* = R_1 i_{1q}^* + \omega L_1 i_{1d}^* + v_{gq}$$ (7)

Using the same reasoning, the filter current references can be obtained:

$$i_{2d}^* = i_{1d}^* - \omega C_f v_{fq}^*$$
$$i_{2q}^* = i_{1q}^* + \omega C_f v_{fd}^*$$ (8)

Therefore the desired reference vector can be constructed:

$$\mathbf{x}^* = \begin{bmatrix} i_{1d}^* & v_{fd}^* & i_{2d}^* & i_{1q}^* & v_{fq}^* & i_{2q}^* \end{bmatrix}^T$$ (9)

3 Proposed Model Predictive Control

In the finite-set model predictive control (FS-MPC) paradigm, the future state for all the available control options are established using the previously described discrete state-space model in (6) as

$$\mathbf{x}(\mathbf{u}_n)[k+1] = \mathbf{\Phi}x[k] + \mathbf{\Gamma}\mathbf{u}_n[k]$$ (10)

where the $\mathbf{u}_n[k]$ input vector refers to the combination of the nth possible state of the 2L-VSI and the actual state of the grid voltage as:

$$\mathbf{u}_n = \begin{bmatrix} v_{idn} & v_{iqn} & v_{gd} & v_{gq} \end{bmatrix}^T$$ (11)

The set of applicable control actions, denoted by $\mathbb{U}$, has a cardinality of 7 as the 2L-VSI can produce this many distinct voltage vectors. Further discussion will omit the bracket notation of the future time instant $[k+1]$, as all following functions require the effect caused by $\mathbf{u}_n[k]$ in the following time instant.

3.1 Multi Cost Function Method

In MPC schemes, the optimal control action is chosen using a cost function that evaluates the action's fitness to the set references or other criteria such as harmonic content and switching frequency in power electronic applications. In this paper the cost function is separated into three separate functions that establish the cost as the quadratic difference between the predicted state to the $\mathbf{u}_n$ control action and the calculated reference of the grid side current $J_1(\mathbf{u}_n)$, filter capacitor voltage $J_2(\mathbf{u}_n)$ and converter side current $J_3(\mathbf{u}_n)$, respectively.

$$
\begin{aligned}
J_1(\mathbf{u}_n) &= (i_{1\mathrm{d}}(\mathbf{u}_n) - i_{1\mathrm{d}}^*)^2 + (i_{1\mathrm{q}}(\mathbf{u}_n) - i_{1\mathrm{q}}^*)^2 \\
J_2(\mathbf{u}_n) &= (v_{\mathrm{fd}}(\mathbf{u}_n) - v_{\mathrm{fd}}^*)^2 + (v_{\mathrm{fq}}(\mathbf{u}_n) - v_{\mathrm{fq}}^*)^2 \quad (12) \\
J_3(\mathbf{u}_n) &= (i_{2\mathrm{d}}(\mathbf{u}_n) - i_{2\mathrm{d}}^*)^2 + (i_{2\mathrm{q}}(\mathbf{u}_n) - i_{2\mathrm{q}}^*)^2
\end{aligned}
$$

In order to arrive at a balanced approach between all the control objectives, which are the reference tracking of the three major electric quantities for the purposes of this paper, we propose a weighting factor-less method. This requires the establishment of ranking functions $R_i(\mathbf{u}_n)$ such that the received score reflects on the corresponding cost of the control option, such that for distinct control options $\mathbf{u}_n$ and $\mathbf{u}_m$, $R_i(\mathbf{u}_n) < R_i(\mathbf{u}_m)$ if $J_i(\mathbf{u}_n) < J_i(\mathbf{u}_m)$. Furthermore, the scoring function returns integer values such that the control option with the lowest cost value receives the value 1, the second lowest receives 2 and so on. This described process therefore requires the sorting of the cost functions. After the scores are established for each control option for all three cost functions the cumulative score $C(\mathbf{u}_n)$ of the control options can be calculated as:

$$
C(\mathbf{u}_n) = R_1(\mathbf{u}_n) + R_2(\mathbf{u}_n) + R_3(\mathbf{u}_n) \quad (13)
$$

Since lower scores were awarded for control actions which produced states closer to the reference, the $\mathbf{u}_{\mathrm{opt}}$ optimal control action is selected with the minimal cumulative score.

$$
\mathbf{u}_{\mathrm{opt}} = \arg\min_{\forall \mathbf{u}_n \in \mathbb{U}} C(\mathbf{u}_n) \quad (14)
$$

By using the presented approach, it is possible for multiple voltage vectors to have the same cumulative score $C(\mathbf{u}_n)$. In the paper, in this case the voltage vector resulting in least switching number is selected as optimal voltage vector.

It should be noted, even with this method we have the possibility to assign priorities to the control objectives by multiplying the ranking functions given in (13) by some gains. However, in this case these gains will have the similar role as weighting factors and the proposed scheme is not a fully weighting factor-less approach.

3.2 Sorting networks

In the previous description of the ranking function, it presents the problem of sorting through a fixed size array each iteration of the control algorithm. This fixed size property allows the use of sorting networks, that offer the least number of required comparisons to be made during execution. Thus, the execution time of the control algorithm can be minimized with this sorting solutions.

Figure 3 presents the comparator diagram of the sorting network used in the paper to sort the seven available cost function values. The horizontal lines represent the passing of the variables and the vertical connections stand for the comparators. The used sorting network has 16 comparators in total, which is proven optimal [9] for the sorting network of input size 7. It is important to note, that sorting networks are unidirectional, as shown on the figure. Furthermore, sorting networks allow parallelization, which allows further improvements in the execution of these sorting solutions.

3.3 Summary of the Proposed Control Method

The previously described proposed control method can be observed in Fig. 4. On the right hand side lies the LCL equipped converter. Measured are the DC capacitor voltage, the grid and converter side phase currents, the grid and the filter capacitors' line voltages. Using a phase locked loop algorithm, the electric angle of the grid is determined, that can be used to establish the RRF coordinate transformations described in (2). Using the externally set V_{DC}^* bus voltage reference and its actual value and the reactive power requirements, the reference vector can be established. Based on the current state, the predictions for all available voltage vectors can be established. Thus the respective cost functions can be calculated for all input actions, which need to be ranked as described above to arrive at the optimal control action. This described loop is repeated throughout the control process.

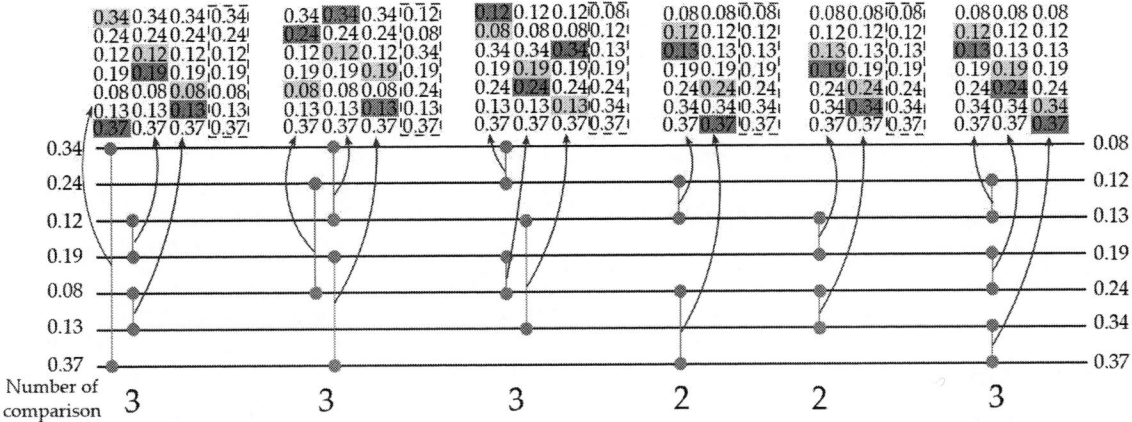

Fig. 3: The comparator diagram of the seven input sorting network, with an example random array of values fed to the left side of the network, that arrives sorted on the right side. All comparison operations are shown with the resulting array for each block of comparators. These blocks can be parallelized as there are no interlocking comparators.

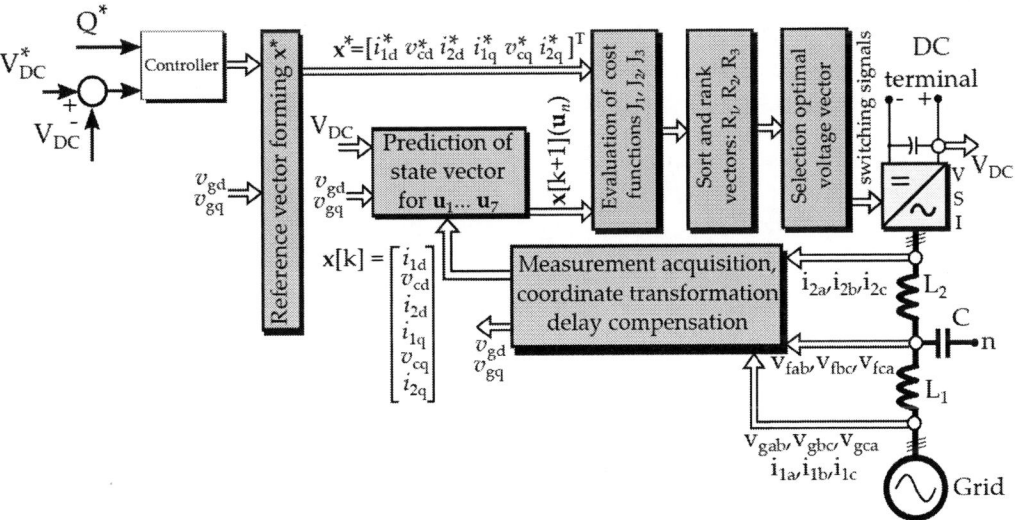

Fig. 4: The block diagram of the described FS-MPC algorithm controlling the LCL equipped grid-side converter.

4 Results

In this section the custom HIL simulation and the related testing environment is introduced, then the experimental results are shown for benchmark control maneuvers. The parameters of the system under discussion are described in Table 1.

4.1 Experimental Hardware

Recent times saw the increase of real-time HIL simulations to verify control concepts. The HIL concept allows the emulation of the high-power parts of the system via their replacement with powerful computerized models. Applying the controller to these hardware enables the replication of real world system behavior, thus control and prototyping functionalities can be validated without the risk of damaging high-power and high-cost equipment [10].

The HIL simulator used in this work is shown in Fig. 5. The core of the system is a Xilinx Zynq based commercial Zybo board from Digilent which contains a Field Programmable Gate Array (FPGA). FPGA makes it possible to run main circuit models with small time steps (in the presented case this is 40 ns). Since the FPGA has no analog outputs, $\Sigma - \Delta$ modulators with external low-pass filters offer a viable solution to create the analog sensor signals for the control board [11]. The HIL has been designed for a Texas Instruments Digital Signal Processor (DSP) based controller, but any control board can be connected to the HIL simulator by means of a simple signal level interface. The main circuit model is realized in Matlab/Simulink and optimized for code generation [12]. The concept allows for the development of more extensive circuit topologies as presented in this paper. The complete workflow from model development to deploying the bitstream to the device is created in Matlab/Simulink.

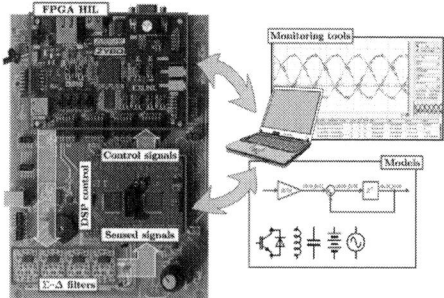

Fig. 5: Xilinx Zynq based low-cost HIL simulator.

Tab. 1: The default parameters of the LCL equipped grid-side converter

Symbol	Description	Value	
f	Grid Frequency	50	Hz
V_{rms}	Grid Line-to-Line voltage	400	V
V_{DC}	DC rail voltage	700	V
$i_{n,Load}$	Nominal DC load current	30	A
R_1	Grid-side resistance	0.5	Ω
L_1	Grid-side inductance	2.5	mH
C_f	Filter capacitance	60	μF
R_2	Converter-side resistance	0.5	Ω
L_2	Converter-side inductance	2.5	mH
C_{DC}	DC side capacitance	2000	μF

4.2 HIL Experiment Results

In order to present the potential of proposed ranking based FS-MPC approach for the control of the LCL filter equipped GSC the control algorithm was implemented on a low-cost dual-core TM320F28379D DSP running at 200 MHz clock frequency. The control algorithm including the measurement acquisition, delay compensation, predictions, sorting network and optimum finding was developed in C language using the IDE of Texas Instruments ™ for the DSP. In order to decrease the turnabout time, the computationally intensive calculations were established using the Control Law Accelerator (CLA)

of the DSP. The selected sampling frequency of the MPC algorithm $f_s = 1/T_s$=25 kHz, due to the measured turnabout time of the proposed control algorithm being 35 μs. The DSP reads from via its analog peripheral the line-to-line voltages of the grid (v_{gab}, v_{gbc} and v_{gca}), the DC bus voltage (V_{DC}) and the phase currents of the grid side inductor (i_{1a}, i_{1b} and i_{1c}) and similarly the converter side phase currents (i_{2a}, i_{2b} and i_{2c}). These signals were generated by the HIL enviroment using the $\Sigma - \Delta$ modulators. The DSP outputs the six control signals for the transistors of the 2L-VSI.

The external voltage control loop, which determines the i_{1d}^* reference value, ran at a smaller rate at 2.5 kHz. The actual electric angle used for the coordinate transformations (see Fig. 4) was established using a phase locked loop algorithm.

The signals were monitored and recorded by a custom-made software using serial communication between the HIL and a computer. The setup of the system can be seen on Fig. 5.

First and foremost the steady-state performance is presented for the nominal current load on the DC side under rectifier operation conditions in Fig. 6. The harmonic distortion was denoted for all three controlled phase variables. It can be deduced, that the filter effectively reduces the harmonic contents for the grid current i_{1a} compared to the converter side current i_{2a}. The average switching frequency was recorded to be 4692 Hz per switching element in the 2L-VSI.

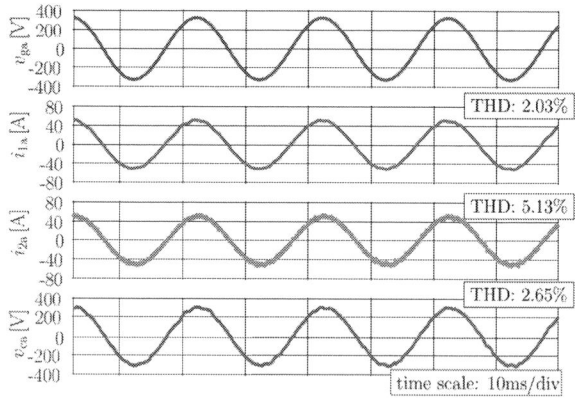

Fig. 6: Phase quantities of the system during steady-state performance under the nominal DC load current of 30 A.

Secondly, the operation mode was changed from rectifier to inverter maneuver is shown in Fig. 7, while the inverse operation is depicted in Fig. 8. In these cases the DC load current was changed in a step-wise fashion from positive to negative and vice-versa at nominal 30 A. As it can be seen the V_{DC} bus voltage has drop and increase in the former and in the latter case, due to the large modeled power compared to the relative small DC capacitor, and due to the response of the PI controller that sets the reference of the d axis directed grid current $i_{1\mathrm{d}}^*$ based on the bus voltage.

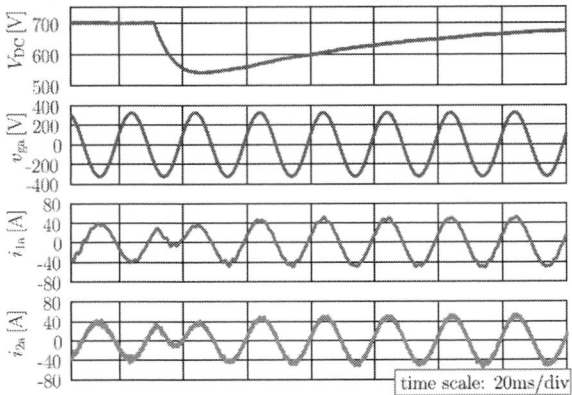

Fig. 7: Operation mode change from inverter to rectifier transient performance. (i_{Load} changes from -30 A to 30 A)

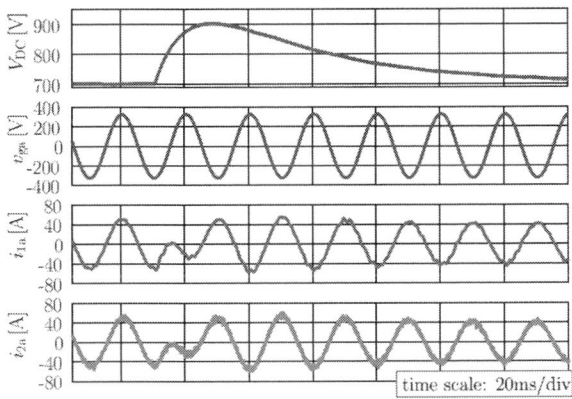

Fig. 8: Operation mode change from rectifier to inverter transient performance. (i_{Load} changes from 30 A to -30 A)

Usually unity power factor requirements are present in the requirements posed for grid side converters, which translates to the $i_{1\mathrm{q}}^*$ reference current to be set to zero. However, in order to achieve a more complete presentation to the capabilities of the proposed control scheme, a step-wise inductive current reference of $i_{1\mathrm{q}}^* = 30$ A was set in Fig. 9. Finally it is also important to observe the control performance with respect to the voltage changes in

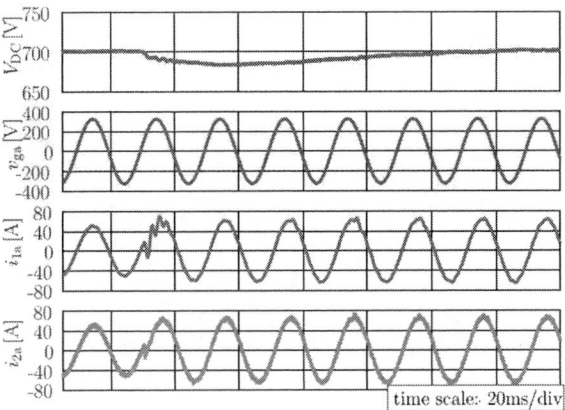

Fig. 9: Response wave forms to step-wise reactive (inductive) power requirement introduction to the control reference. ($i_{1\mathrm{q}}^*$ changes from 0 A to 30 A)

both the grid and the DC sides. The response to the sudden change in the grid voltage from 325 V to 225 V can be observed in Fig. 10. Whereas the step-wise control reference change from 700 V to 750 V is depicted in Fig. 11.

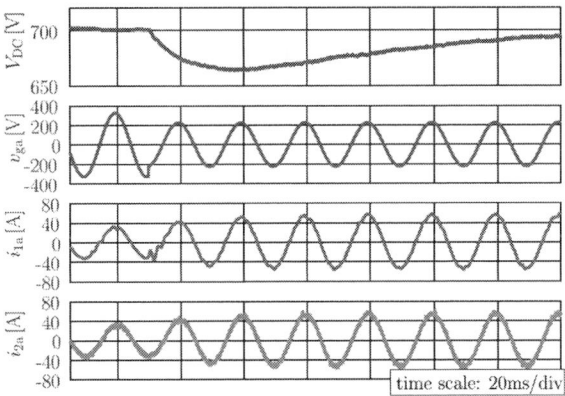

Fig. 10: Transient response to instantaneous grid voltage change. (v_{gd} changes from 325 V to 225 V in amplitude, in this case i_{Load} was 20 A)

5 Conclusions

In this paper, a novel multi cost function MPC approach was presented, with application for the LCL filter extended grid-side converter. The method allows the development of the optimal control law, without weighting factors and the associated tuning process. The algorithm efficiency is maximized with the use of sorting networks that allow optimal sorting performance for fixed size inputs. Simulation and experimental results using a custom HIL hardware demonstrates the performance of the pro-

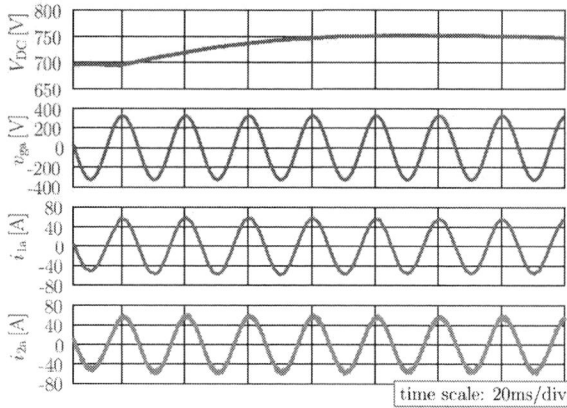

Fig. 11: Response to the step-wise increase in the DC bus voltage reference. (The reference of V_{DC} changes from 700 V to 750 V)

posed scheme. The paper also presents the operation of the HIL unit.

Acknowledgments

Supported by the ÚNKP-23-3-II-BME-60 New National Excellence Program of the Ministry for Culture and Innovation from the source of the National Research, Development and Innovation Fund. This work was supported by the National Research, Development, and Innovation Office under Grant FK 143429

References

[1] G. Mirzaeva and Y. Mo, "Model predictive control for industrial drive applications," *IEEE Transactions on Industry Applications*, vol. 59, no. 6, pp. 7897–7907, 2023. DOI: 10.1109/TIA.2023.3299887.

[2] M. E. Zarei, D. Ramirez, M. Prodanovic, and G. Venkataramanan, "Multivector model predictive power control for grid connected converters in renewable power plants," *IEEE Journal of Emerging and Selected Topics in Power Electronics*, vol. 10, no. 2, pp. 1466–1478, 2022. DOI: 10.1109/JESTPE.2021.3077953.

[3] I. Harbi, J. Rodriguez, E. Liegmann, H. Makhamreh, M. L. Heldwein, *et al.*, "Model-predictive control of multilevel inverters: Challenges, recent advances, and trends," *IEEE Transactions on Power Electronics*, vol. 38, no. 9, pp. 10845–10868, 2023. DOI: 10.1109/TPEL.2023.3288499.

[4] K. G. Bandy, P. P. Stumpf, M. M. R. Chowdhury, and Z. Suto, "Real-time validation of quadratic regression model-based predictive control of grid-side converters," in *PCIM Europe 2023; International Exhibition and Conference for Power Electronics, Intelligent Motion, Renewable Energy and Energy Management*, 2023, pp. 1–8. DOI: 10.30420/566091327.

[5] C. Xue, D. Zhou, and Y. Li, "Hybrid model predictive current and voltage control for lcl-filtered grid-connected inverter," *IEEE Journal of Emerging and Selected Topics in Power Electronics*, vol. 9, no. 5, pp. 5747–5760, 2021. DOI: 10.1109/JESTPE.2020.3049083.

[6] H. Zhang, Z. Ma, Z. Li, X. Zhang, Z. Liao, and G. Lin, "Multivariable sequential model predictive control of lcl-type grid connected inverter," in *2021 IEEE International Conference on Predictive Control of Electrical Drives and Power Electronics (PRECEDE)*, 2021, pp. 777–781. DOI: 10.1109/PRECEDE51386.2021.9680951.

[7] E. Zerdali, M. Rivera, and P. Wheeler, "A review on weighting factor design of finite control set model predictive control strategies for ac electric drives," *IEEE Transactions on Power Electronics*, pp. 1–16, 2024. DOI: 10.1109/TPEL.2024.3370550.

[8] K. Bándy and P. Stumpf, "Model predictive torque control for multilevel inverter fed induction machines using sorting networks," *IEEE Access*, vol. 9, pp. 13800–13813, 2021. DOI: 10.1109/ACCESS.2021.3052129.

[9] D. Knuth, *The Art Of Computer Programming, vol. 3: Sorting And Searching*. Addison-Wesley, 1973, pp. 391–392.

[10] A. Futo, T. Kokenyesi, I. Varjasi, Z. Suto, I. Vajk, *et al.*, "Real-time hil simulation of the discontinuous conduction mode in voltage source pwm power converters," *Journal of Power Electronics*, vol. 17, no. 6, pp. 1535–1544, Nov. 2017.

[11] T. Kokenyesi, M. Hegedus, S. Vereb, A. Balogh, Z. Suto, and I. Varjasi, "Fpga-driven dac with second order sliding mode control of filter model for hardware-in-the-loop simulators," in *2018 IEEE 18th International Power Electronics and Motion Control Conference (PEMC)*, 2018, pp. 824–829. DOI: 10.1109/EPEPEMC.2018.8521939.

[12] M. M. R. Chowdhury and Z. Sütő, "Validation of real-time simulation model of a three-phase active-front-end (afe) rectifier," in *2022 8th International Youth Conference on Energy (IYCE)*, 2022, pp. 1–5. DOI: 10.1109/IYCE54153.2022.9857546.

PCIM Europe 2024, 11– 13 June 2024, Nuremberg   DOI: 10.30420/566262051

Relaxed Robust Control with Pragmatic Shortage of Passivity for Wind, Storage and PV Power Converters

Mario Rizo[1], Sergio de Lopez[1], Pablo Moreno[1], Andres Agudo[1], Ana Rodriguez[1], Luis Diez[1]

[1] Gamesa Electric, Spain

Corresponding author: Sergio de Lopez, sergio.delopez@siemensgamesa.com
Speaker: Sergio de Lopez, sergio.delopez@siemensgamesa.com

Abstract

Passivity is gaining attention in the scenario of massive integration of inverter-based resources. Designing a power converter with passive input impedance is sufficient for stability, but not necessary. In fact, passive designs tend to conservatism providing degraded performance (high fault currents, slower settling times, etc). This paper analyzes the trade-off between passivity and performance. Taking advance of the inherent damping of transmission lines, a robust stability index, more relaxed than passivity, is used to tune the converter inner controller applying modern control optimization techniques. A good balance is hence obtained between performance and robust stability.

1 Introduction

The massive integration of Inverter-Based Resources (IBRs) plays a crucial role in modern power systems, facilitating the integration of renewable energy sources (RES) and enhancing grid flexibility and stability. These resources encompass a diverse array of technologies such as photovoltaic, wind turbines, and battery energy storage systems. In recent years, there has been a growing interest of the importance of incorporating passivity principles into the design of control strategies for IBRs taking into account the inherent nonlinear and time-varying characteristics. These systems are often subjected to uncertainties, varying grid conditions, and dynamic load profiles, which can lead to oscillations, voltage fluctuations, and stability issues if not appropriately addressed [1]. Among them, resonances between IBRs and grid passive elements (typically stray capacitors), called harmonic instability, are identified as a top priority stability concern by European TSOs [2]. Power filter, computation delay or inner controller condition the behavior of IBRs at target frequencies (from hundreds to some thousands of Hz). One of the main sources of uncertainity in order to design the inner control loops for IBRs is the equivalent grid impedance, so taking into account this uncertainty is a key topic in avoiding instability or undesired oscillations [3], [4].

In view of this fact, shaping a passive input-impedance is emerging as a common robust stability (RS) design goal for the controller of IBRs [3], [5]. According to the passivity theorem, the interconnection of passive sub-systems results in a stable complete system [6]. It is a powerful concept in terms of RS as it guarantees stability under a great uncertainty: the only assumption made about the grid is that it is passive. The design of passive controllers can be divided into two categories: (i) those approaches based on classic control, which consist of a heuristic process where analysis and re-design are iterative combined looking for the best possible passivity [7]–[9] and (ii) wholistic approaches relying on modern control, where passivity is an input objective within an optimization problem that synthetizes the controller [10]. Nevertheless, strict passivity can lead to conservative controllers with degraded performance [6]. Specifically, these alternatives have low bandwidth, which is ineffective to reject disturbances, leading to high faulty currents. Either, increasing the controller sampling frequency or redesigning the IBR power filter helps to solve this situation. However, this is not always possible due to computation and cost constrains.

This paper proposes a relaxed design criterion, where RS and performance are better balanced. A Pragmatic RS (PRS) index is created which, in contrast to passivity, alleviates uncertainty by adding general and conservative knowledge of transmission lines. The PRS index, frequency-based, configures the RS goal within a modern control design approach. The proposed method for harmonic stability is effectively applied to the current controller of an IBR, as demonstrated in Fig. 1. The figure presents a single-line simplified diagram of a three-phase grid-following converter equipped with an L-type filter and operating at a sampling frequency of $f_s = 5 \; kHz$. This configuration represents a challenging scenario in terms of harmonic stability.

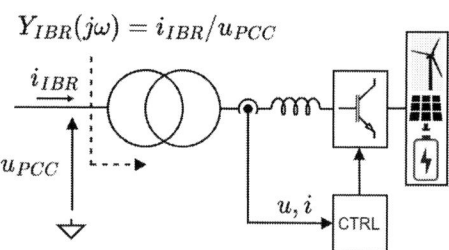

Fig. 1: Current-controlled IBR simplified diagram.

The term $Y_{IBR}(jw) = \frac{i_{IBR}}{v_{PCC}}$ defines the input admittance of the IBR, linking the relationship between the input current and the voltage in the Point of Common Coupling (PCC). The assessment of passivity consists of using the input-feedforward passivity (IFP) index, which matches with the conductance $G_{IBR} = Re\{Y_{IBR}\}$ for Single-Input Single-Output (SISO) systems.

The remainder of the paper is organized as follows. Section II offers a brief explanation of the theoretical basis for the proposed design, outlining the reasoning behind using a robust control approach within the framework of passivity control. Section III describes the plant model and the characterization of uncertainties. Following this, Section IV provides a comprehensive explanation of the methodology for creating a relaxed-passivity index, taking into account the inherent damping effects of transmission lines. Section V explores the trade-offs between performance and passivity, as well as alternative approaches for defining multiple control realizations. The conclusions drawn from the findings presented in this paper are outlined.

2 Theoretical Background

2.1 Control Problem definition

The control proposal is addressed within the context of the generalized control problem depicted in

Fig. 2 [10]. The plant $P(s)$ embodies a dynamic Multiple-Input Multiple-Output (MIMO) model, comprising the system to be controlled and interconnected weighting functions. These functions facilitate the translation of design objectives into a control design problem. The variables u and y define the input-output configuration of the controller. The relationship between w and z defines the performance goals.

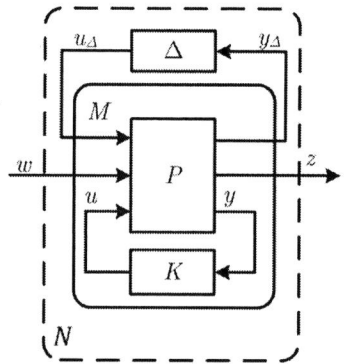

Fig. 2: Generalized structure for robust stability analysis.

In addition, the block Δ denotes the inherent uncertainty in the system to be controlled due to inadequacies in its modelling. A major source of uncertainty is the dynamic nature of grid conditions, which can be influenced by factors such as changes in load demand, integration of RES and grid disturbances. These dynamic variations are expressed, without loss of generality, as a feedback system from input u_Δ to output y_Δ.

The essence of the robust control strategy used in this paper lies in synthesizing a controller K that achieves specific goals for the closed-loop system $M = F_l(P, K)$, where F_l represents the lower linear fractional transformation (LFT). These goals are expressed as gain bounds between the input w and the output z. The controller is therefore the outcome of an optimization problem, where the norm to optimize is performance (soft requirement) under a passivity constraint (hard requirement). Moreover, it is crucial to ensure the stability of the complete system N taking into account the uncertainty of the grid.

2.2 Passivity-based Robust Control

The passivity concept is fundamental to tackle with the robust control problem definition. From an input-output point of view, passive systems dissipate an amount of energy that is less than or equal to the energy they receive. The relevance in control theory partly stems from the robust stability results associated with interconnected passive systems, leading to the Passivity Theorem [6]. If a particular system is designed to be passive, it ensures that when interconnected with various passive systems, instability concerns are mitigated.

Using indices, the notions of passivity excess and shortage provide a quantification of the degree of passivity of a system. In this paper is employed the IFP index as a metric for passivity design. For a stable linear system with transfer function $H(s)$ at frequency ω, the definition of the IFP index is as follows:

$$v_F(H(s), \omega) \triangleq \frac{1}{2}\lambda(H(j\omega) + H^*(j\omega)), \quad (1)$$

where λ represents the minimum eigenvalue. Indeed, the IFP index is defined as $v(H(s)) \triangleq min \; v_F(H(s), \omega)$ for the entire range of ω [6]. Therefore, a linear system is strictly passive if it is stable and $v > 0$. The stability requirement will depend on the damping of the grid. By modelling the damping of the transmission lines as an excess of passivity, the objective can be relaxed while maintaining robust stability.

3 Relaxed-Passivity Index

Mitigating uncertainty in the grid shifts the focus from strict passivity to more pragmatic criteria, resulting in improved performance. According to the Nyquist theorem [11], stability is given when the loci of $Z_G Y_{IBR}$ does not encircle -1, being Z_G the equivalent grid impedance. Analogously, a necessary condition for instability is a real axis intersection on the left of -1. This can be expressed as the left part of eqn. (2), which yields to the conditions on the right.

$$\begin{cases} Re\{Z_G Y_{IBR}\} < -1 \\ Im\{Z_G Y_{I}BR\} = 0 \end{cases} \Rightarrow \begin{cases} Re\{Y_{IBR}\} + Re\{Z_G^{-1}\} < 0 \\ Im\{Y_{IBR}\} = \frac{-Im\{Z_G\}Re\{Y_{IBR}\}}{Re\{Z_G\}} \end{cases} \quad (2)$$

Assuming the worst case scenario, where $Im\{Y_{IBR}\} = -Im\{Z_G\}Re\{Y_{IBR\}}/Re\{Z_G\}$ is always met, then the necessary instability condition in (2) yields to $G_{IBR} < -G_G$, being the grid conductance $G_G = Re\{Z_G^{-1}\}$. The opposite relation gives a sufficient stability condition, see eqn. 3. It reveals relaxation in contrast with the passivity requirement. This analysis also covers the interaction between inverters as Y_{IBR} can represent the aggregation of several IBRs and the short-circuit ratio (SCRs) of Z_G can be modified.

$$G_{IBR}(j\omega) \geq -G_G(j\omega) \quad (3)$$

Aimed to generalization, the PRS index μ is obtained using data from a large set, $\mathcal{Z}$, of grids: overhead, underground and submarine lines, multiple lengths, different voltage levels are considered.

A PI-section model (Fig. 3) is used to represent the transmission lines seen by the IBR plant. Although skin and proximity effects provide extra damping at target frequencies [12], they are not considered for the sake of conservativeness.

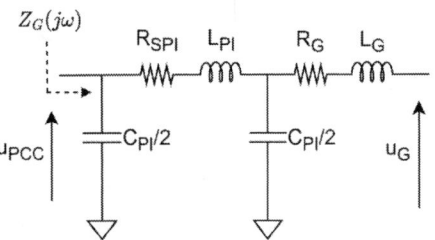

Fig. 3: Grid model with PI-section.

The approach entails performing parameter sweeps to characterize the grid impedance model, specifically in parameters such as R_{SPI}, SCRs, and different line lengths. Subsequently, it is defined a set $\mathcal{Z}$, where each element k corresponds to a frequency band of influence $\Delta\omega_k$ centered around the resonant frequency. Furthermore, the conductance values $G_G^k(j\Delta\omega_k)$ are recorded within this set. To obtain a frequency-dependent index, the distance of the line in PI is swept, yielding the values of $-G_G$ for the influence zones of each realization. For those cases where the zones of influence overlap, the maximum (most restrictive) value is retained. Finally, the PRS index is computed like an envelope of the stored values most restrictive for each frequency as described by (4).

$$\mu(j\omega) = \max_{k \; \omega\epsilon\Delta\omega_k} \left(-G_G^k(j\omega)\right),$$

$$G_G^k(j\omega) = Re\left\{\left(Z_G^k\right)^{-1}\right\}, \quad Z_G^k \epsilon \mathcal{Z} \quad (4)$$

Fig. 4.a displays the magnitude of three grid samples within $\mathcal{Z}$ and their frequency band of influence $\Delta\omega_k$. Moreover, Fig. 4.b shows their $-G_G^k(j\Delta\omega_k)$

and the resulting $\mu(j\omega)$ when all the elements $\mathcal{Z}$ in are used.

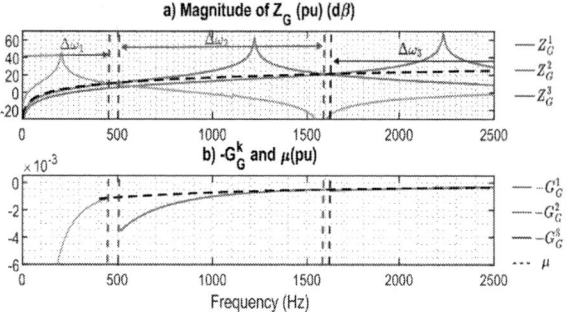

Fig. 4: Getting PRS index: a) Impedance magnitude of some line and frequency band of influence, b) obtaining μ following (4).

In Fig. 5, it is depicted the complete sweep of impedances to obtain a relaxed metric of passivity, taking into consideration the additional damping of the lines.

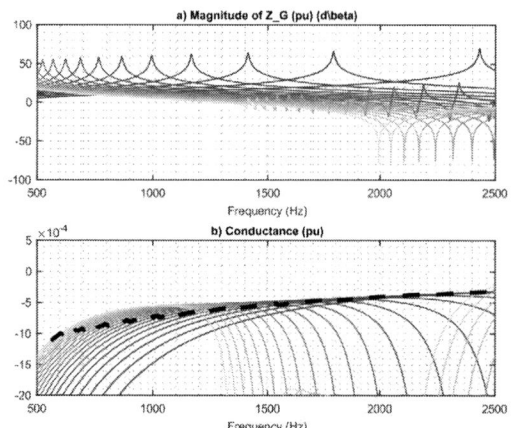

Fig. 5: Multiple realizations of the magnitude of Z_G and the negative of the conductance of Z_G.

In accordance with the robust control methodology, this newly relaxed passivity index needs to be incorporated into the generalized design framework based on weighting functions for uncertainty Δ. The approach adopted in this paper has involved finding a transfer function such that its real part aligns with the obtained conductance. To achieve the above, aiming for the index to have a physically basis, it is employed the impedance shown in Fig. 6 as the foundation for that transfer function. The structure of it (order and arrangement of elements) is such that it enables precise curve fitting of μ.

The transfer function of the pattern impedance

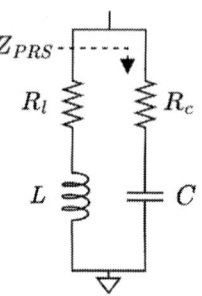

Fig. 6: Standard impedance Z_{PRS} such that $\mu(j\omega) = -G_{PRS}(j\omega)$.

Z_{PRS} and its corresponding admittance and conductance in the frequency domain are as follows:

$$Z_{PRS}(s) = \frac{s^2 LCR_c + s(L + CR_c R_l) + R_l}{s^2 LC + sC(R_c + R_l) + 1},$$

$$Y_{PRS}(j\omega) = \frac{(1 - \omega^2 LC) + j\omega C(R_c + R_l) + 1}{(R_l - \omega^2 LCR_c) + j\omega(L + CR_c R_l)},$$

$$G_{PRS}(j\omega) = \frac{\omega^4 (LC)^2 R_c + \omega^2 C^2 (R_c + R_l) R_c R_l + R_l}{\omega^4 (LCR_c)^2 + \omega^2 \left(L^2 + (CR_l R_c)^2\right) + R_l^2}$$

(5)

A curve fitting based on non-linear least squares is employed to obtain the values of L, C, R_c and R_l by comparing $G_{PRS}(j\omega)$ with the RS goal depicted in Fig. 7.

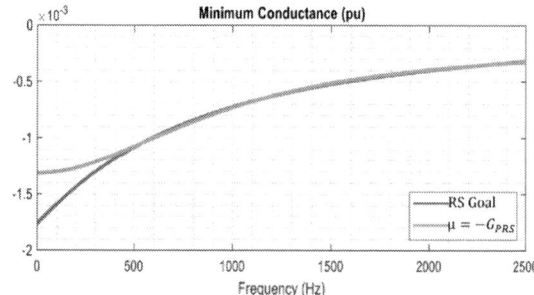

Fig. 7: Practical RS Index, which takes into account the minimum required conductance.

This criterion of relaxed passivity enables the implementation of control designs with a passivity shortage to establish a stable connection to the utility grid without negatively impacting the performance of the IBR. This will be demonstrated in subsequent sections.

4 Grid-Tied Current-Controlled VSCs

This paper analyzes the VSC control problem using the small-signal modelling that can be derived from symmetrical three-wire three-phase electrical systems by modeling them using space vectors [13]. By default, the space vector is in direct-quadrature (DQ) synchronous reference frame.

4.1 Plant model

The IBR is connected to the grid through a L filter topology (see Fig. 1) expressing the system dynamics with a continous state space representation as follows:

$$\frac{dx(t)}{dt} = A_{\sigma(t)}x(t) + B_{\sigma(t)}u(t)$$
$$y(t) = Cx(t) + Du(t), \tag{6}$$

The function $\sigma(t)$ represents the dynamic response of the IBR, influenced by operating conditions and accounting for nonlinear elements inherent in grid-following control systems, such as the Phase-Locked-Loop (PLL). Furthermore, the computational delay plus zero-order-hold (ZOH) and sampler model is defined as eqn. (7) and it is included in the overall plant definition by approximating each exponential term with a second order Pade.

$$H_d(s) = e^{-sT_s}\frac{1 - e^{-sT_s}}{sT_s}, \tag{7}$$

4.2 Current control structure

The design of the controller is guided by the specific control objectives. These objectives include tracking a current set-point with no steady-state error, achieving a damped response to step changes even in the absence of grid impedance, and ensuring robust stand-alone stability margins. In addition, the controller under consideration must meet passivity requirements to enhance the robustness of the system during grid interactions. The proposed configuration is illustrated in Fig. 8, with 4 degrees of freedom (DOF) represented by multiple complex constants integrated within the K_c block.

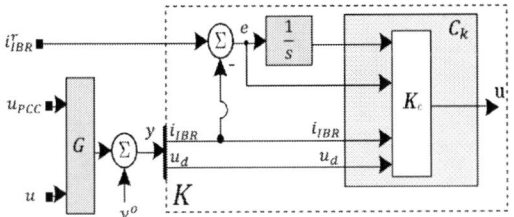

Fig. 8: Closed-loop model with detailed definition of the structure of K, including the tunable parts (Ck).

The dynamics of the output current of the IBR are the following (see Fig. 8):

$$\mathbf{i}_2 = \underbrace{S^{y^o \to \mathbf{i}_{IBR}}G^{\mathbf{u}\to y}K^{\mathbf{i}^r_{IBR}\to \mathbf{u}}}_{\mathbf{T}(s)}\mathbf{i}^r_{IBR} + \underbrace{S^{y^o \to \mathbf{i}_2}G^{\mathbf{u}_{pcc}\to y}}_{-\mathbf{Y}(s)}\mathbf{u}_{pcc} \tag{8}$$

where $S = [I - G^{\mathbf{u}\to y}K^{\mathbf{i}^r_{IBR}\to \mathbf{u}}]^{-1}$, the tracking dynamics denoted as $T(s)$ and the input admittance of the converter $Y(s)$.

4.3 PBRC synthesis problem

The problem formulation is particularized for the current controller design as depicted in Fig. 9, defining a multi-objective optimization problem [10], [11].

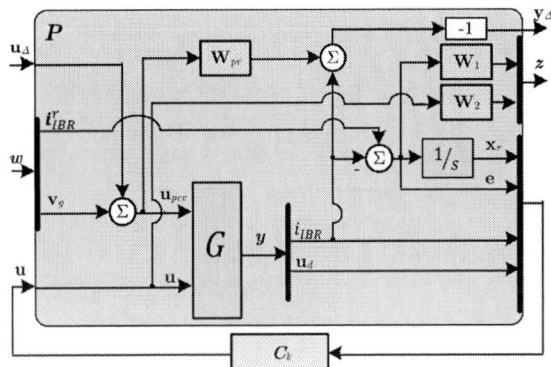

Fig. 9: Problem formulation for current controller design [10].

The function $W_1(s)$ affects the tracking behaviour, while $W_2(s)$ limits the gain to restrict the achievable control bandwidth and shape the boundaries of converter actuation. Furthermore, the weighting function W_{pr} addresses the passivity shortage.

The formulation of the transfer function $W_1(s)$ entails emphasizing high gain at the tracked frequency while ensuring that the tracking function T closely follows a first-order system to effectively dampen reference step changes. The expression for the generalized complex-valued weighting function in the continuous-time domain is provided in equation (9) [10].

$$W_1(s) = \frac{s + \omega_B^*}{s + \omega_B^* A_o} \tag{9}$$

The w_B^* is defined as the desired closed-loop bandwidth, denoting the frequency at which the straight-line approximation of the weight intersects 0 dB. Moreover, the $A_o \in (0, 1)$ is the steady state error and it can be selected arbitrarily small, but, avoiding a zero-value due to ill-conditioned optimization problem [14].

Concerning the function $W_2(s)$, it is established as a constant gain factor to mitigate excessive control effort resulting from constraints associated with the necessary DC voltage and the potential disturbances encountered by the IBR during operation.

Lastly, the W_{pr} weight is used to tackle with the passivity goal. It uses the relaxed-index based on knowledge of the damping of transmission lines described in previous section.

The synthesis and design of the IBR current control has been carried out in the Matlab environment, using the *systune* optimization tool to obtain the value of the C_k (see 9) controller constants. Algorithm 1 shows a simplified pseudo-code of the controller synthesis procedure.

Algorithm 1 Controller synthesis summary

    Input: $Model : G, T_s$ **and** Design objectives:
    $\omega_0, W_1, W_2, W_{pr}$
    Output: Controller gains: C_k

1: **Define plant and weighting functions:**
2:   G = ss(...,G, T_s);
3:   W_1 = ss(...,ω_0, ω_B^*);   ▷ % Performance weight
4:   W_2 = ss(...,k);   ▷ % Control effort weight
5:   W_{pr} = ss(...,μ);   ▷ % Passivity weight
6: **Compose generalized plant:**
7:   INT = ss($\frac{1}{s}$);   ▷ % Integrator
8:   P = connect(...,G,INT,W_1,W_2,W_{pr});
9:   $P \rightarrow P$ ▷ % Real-valued equivalent
10: **Define controller tunable parameters:**
11:   H_x = ss(...,realp(c_1),realp(p_1),...,realp(c_n), realp(p_n));
12:   C_k = ss(...,realp(k_i), realp(k_u));
13:   C_k = connect(...,C_k,H_x);
14: **Create closed-loop system:**
15:   M = lft(P,C_k);
16: **Define tuning requirements:**
17:   RSOFT = TuningGoal.Gain(w,z,1);
18:   RHARD = TuningGoal.Passivity(u_Δ,y_Δ,0);
19:   $[n, gH]$ = systune(M,RSOFT,RHARD,...);
20: **Goals evaluation:**
21: **if** $gH > 1$ **then**
22:     **goto Define plant and weights;**
23: **else if** $n > iter_{max}$ **then**
24:     **goto Define plant and weights;**
25: **else**
26:     **return** C_k
27: **end if**

---

5 Performance and Passivity Tradeoff

The proposed passivity index aims to illustrate performance improvement, while acknowledging a shortage of passivity in the current control design of the IBR. The design methodology considers the impact of DOF within the overall structure depicted in Fig. 8. Initially, passivity constraints are addressed by utilizing complex constants exclusively within the K_c subsystem. Subsequently, the introduction of additional DOF through transfer functions in feedback variables seeks to increase the flexibility and effectiveness of the RS design tool in identifying optimal solutions. Furthermore, various design iterations will explore the balance between performance and passivity, showcasing how the proposed index facilitates the development of resilient and consistent designs across diverse transmission lines.

Within this framework, the inputs from the optimization tool related to performance criteria, are used as soft objectives, namely, the weighting functions $W_1(s)$ and $W_2(s)$. Meanwhile, the passivity index is imposed through W_{pr} weighting function as a hard constraint.

The initial aspect explored in the control design analysis is the impact of the DOF in the *systune*-based approach using the strict passivity constraints ($\mathrm{Re}\{Y_{IBR}\} > 0$). This involves providing additional parameters to tune in order to improve performance whilst ensuring the strict passivity requirement. The method employed consists of incorporating transfer functions into the feedback signals, encompassing the error e, the current i_{IBR}, and the actuation feedback u_d. The form of the feedback filters is shown in the expression (10), replacing, therefore, the complex gains with these transfer functions that allows introducing zeros and poles to provide greater flexibility in the shaping of the admittance of the IBR.

$$H_x(s) = k_x \prod_{k=0}^{3} \frac{s - c_k^x}{s - \rho_k^x} \qquad (10)$$

$$c_k, \rho_k, k_x \in \mathbb{C}$$

Fig. 10 shows the comparison between both realizations.

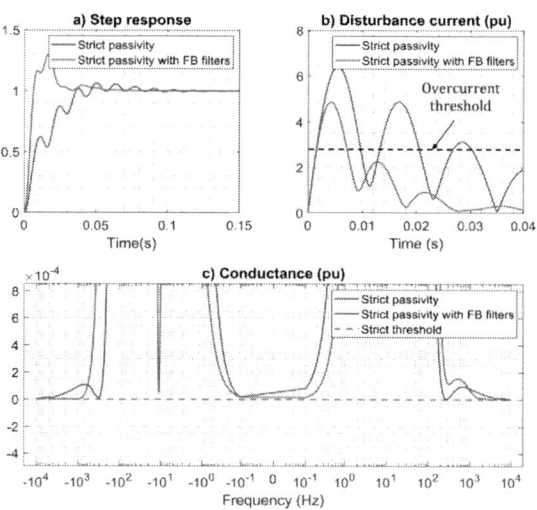

Fig. 10: Comparison using just complex gains and feedback transfer functions filters with the strict passivity criterion design, including (a) Time response, (b) The effect in disturbance current and (c) the conductance of the both realizations.

The tool achieves fully passive designs, as shown in Fig. 10.c, where the conductance remains above 0 across the entire frequency spectrum. However, imposing such strict passivity criteria can often lead to performance degradation within the system, resulting in larger settling times and excessive over-currents. Nevertheless, by incorporating transfer functions into the controller's feedback variables, significant improvements in design can be observed. These improvements are characterised by enhanced step responses and reduced over-currents resulting from disturbances.

Considering these results, the characteristics obtained do not lead to a finalised design. This is because the response to disturbances exceeds the hardware limits imposed on the IBR, as shown in Figure 10.b, and the settling time is still unsatisfactory.

The next approach is to exploit the pragmatic lack of passivity to achieve a more robust control design. In Fig. 11 is depicted the RS design again using just complex gains and using the feedback transfer functions to provide more DOF to tune.

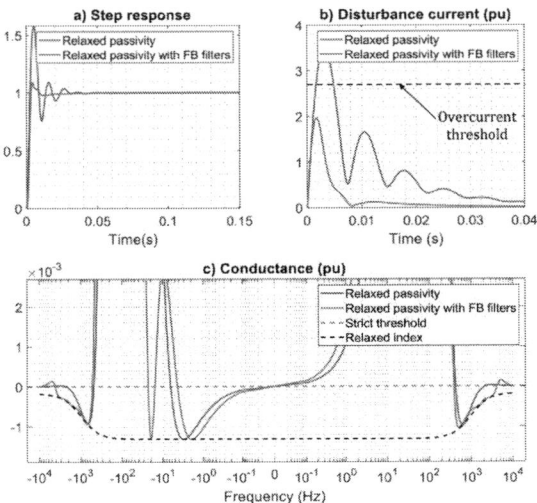

Fig. 11: Relaxed RS Index with complex gains against employing feedback filters, including (a) Time response, (b) The effect in disturbance current and (c) the conductance of the both realizations.

Again, it can be clearly seen in the conductance of both designs as the optimization tool provides solutions that meet the hard type requirement and tries to provide the best possible performance with the DOF available in the design. Moreover, the introduction of the pragmatic index for the design, allows the tool to achieve solutions closer to a real design demanded by the industry by significantly reducing the response to the disturbance and the response time before reference changes, when compared to the requirement to achieve a totally passive design.

The inclusion of additional tuning parameters yields a notable improvement in performance without compromising the passivity of the IBR. Moreover, this configuration effectively reduces disturbance currents, as demonstrated in Fig. 11.b, where the hardware limits are not exceeded.

To consolidate and clarify the diverse design iterations,

as well as highlight variances in the admittance characteristics of the IBR interfacing with the utility grid and its performance, all realizations are illustrated in Fig. 12 for ease of comparison.

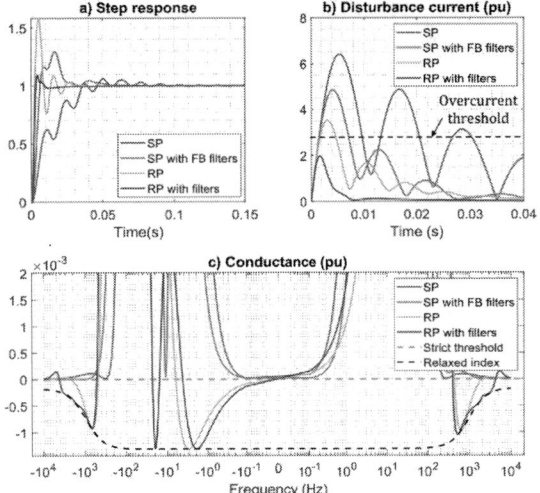

Fig. 12: Comparison between strict and relaxed passivity constraints, including (a) Time response, (b) The effect in disturbance current and (c) the conductance of the both realizations.

As can be observed, the requirement for strict passivity is noted to have a significant negative impact on overall system performance. By introducing the pragmatic passivity index into the optimization tool, enhanced flexibility is provided in achieving performance objectives set as soft constraints, resulting in improved controller designs. Furthermore, increasing the DOFs improves the robustness of the design, despite the resulting increase in complexity of the control structure and computational demands on the control board.

To confirm compliance with the Nyquist stability criterion, a sweep of the $Z_G^k Y_{IBR}$ transfer function was performed with varying grid impedances (see Fig. 13). It can be seen that in all cases, the critical stability point (-1 + 0j) is avoided, achieving a system with a pragmatic passivity requirement without degrading the performance of the equipment.

6 Conclusions

The increasing adoption of IBRs is a significant trend in the power system landscape, leading to a closer examination of the concept of passivity. Although the conventional wisdom suggests that designing power converters with passive input impedance is sufficient for ensuring stability, it is important to acknowledge that this is not necessary. While passive designs do guarantee stability, they often prioritize conservativeness, resulting in compromised performance.

This paper presents a pragmatic passivity index tailored to introduce prior knowledge of transmission lines linked

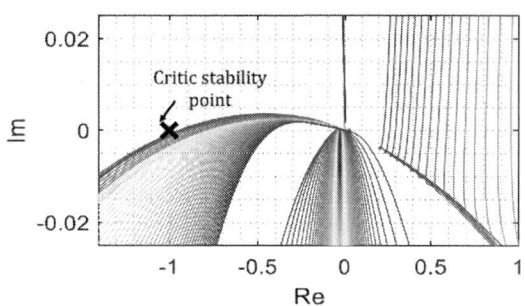

Fig. 13: Robust stability verification using Nyquist theorem and Nyquist diagram of $Z_G^k Y_{IBR}$

to the utility grid, facilitating the design of the innermost current control for tackling harmonic stability. To this end, the design is addressed as an optimization problem, incorporating specifications using weighting functions in the frequency domain. The trade-off between performance and passivity by enforcing both strict passivity requirements and our proposed more relaxed index is explored, evaluating their responses to reference steps and disturbances leading to over-currents. Furthermore, the impact of integrating additional DOFs into the design to enhance its overall effectiveness has been investigated.

Our analysis underscores the feasibility and benefits of a design approach that considers passivity, leverages minimal knowledge of connecting impedances, and harnesses inherent damping to optimize the performance of IBRs.

References

[1] N. Hatziargyriou, J. Milanovic, C. Rahmann, V. Ajjarapu, C. Canizares, *et al.*, "Definition and classification of power system stability – revisited and extended," *IEEE Transactions on Power Systems*, vol. 36, no. 4, pp. 3271–3281, 2021. DOI: 10.1109/TPWRS.2020.3041774.

[2] H. P. MIGRATE, *Report on systemic issues*, Available online at: https://ec.europa.eu/research/participants / documents / downloadPublic ? documentIds = 080166e5af08ecd7 & appId = PPGMS, last accessed on 12/03/2024, 2020.

[3] L. Harnefors, X. Wang, A. G. Yepes, and F. Blaabjerg, "Passivity-based stability assessment of grid-connected vscs—an overview," *IEEE Journal of Emerging and Selected Topics in Power Electronics*, vol. 4, no. 1, pp. 116–125, 2016. DOI: 10.1109/JESTPE.2015.2490549.

[4] L. Harnefors, A. G. Yepes, A. Vidal, and J. Doval-Gandoy, "Passivity-based controller design of grid-connected vscs for prevention of electrical resonance instability," *IEEE Transactions on Industrial*

Electronics, vol. 62, no. 2, pp. 702–710, 2015. DOI: 10.1109/TIE.2014.2336632.

[5] C. W. G. Document TB 754, *Ac side harmonics and appropriate harmonic limits for vsc hvdc*, 2019.

[6] B. E. Ydstie, "Process control: The passive systems approach (bao, k. and lee. p.l.; 2007)[book shelf]," *IEEE Control Systems Magazine*, vol. 30, no. 1, pp. 78–80, 2010. DOI: 10.1109/MCS.2009.935226.

[7] L. Harnefors, A. G. Yepes, A. Vidal, and J. Doval-Gandoy, "Passivity-based controller design of grid-connected vscs for prevention of electrical resonance instability," *IEEE Transactions on Industrial Electronics*, vol. 62, no. 2, pp. 702–710, 2015. DOI: 10.1109/TIE.2014.2336632.

[8] F. Hans, W. Schumacher, S.-F. Chou, and X. Wang, "Passivation of current-controlled grid-connected vscs using passivity indices," *IEEE Transactions on Industrial Electronics*, vol. 66, no. 11, pp. 8971–8980, 2019. DOI: 10.1109/TIE.2018.2883261.

[9] B. Hu, H. Nian, H. Li, L. Chen, S. Sahoo, and F. Blaabjerg, "Impedance reshaping band coupling and broadband passivity enhancement for dfig system," *IEEE Transactions on Power Electronics*, vol. 38, no. 8, pp. 9436–9447, 2023. DOI: 10.1109/TPEL.2023.3270364.

[10] J. Serrano-Delgado, S. Cobreces, M. Rizo, and E. J. Bueno, "Low-order passivity-based robust current control design for grid-tied vscs," *IEEE Transactions on Power Electronics*, vol. 36, no. 10, pp. 11 886–11 899, 2021. DOI: 10.1109/TPEL.2021.3068057.

[11] S. Skogestad and I. Postlethwaite, "Multivariable feedback control: Analysis and design," in Jan. 2005, vol. 2.

[12] Ł. H. Kocewiak, I. A. Aristi, B. Gustavsen, and A. Hołdyk, "Modelling of wind power plant transmission system for harmonic propagation and small-signal stability studies," *IET Renewable Power Generation*, vol. 13, no. 5, pp. 717–724, 2019. DOI: https://doi.org/10.1049/iet-rpg.2018.5077. eprint: https://ietresearch.onlinelibrary.wiley.com/doi/pdf/10.1049/iet-rpg.2018.5077.

[13] L. Harnefors, X. Wang, S.-F. Chou, M. Bongiorno, M. Hinkkanen, and M. Routimo, "Asymmetric complex-vector models with application to vsc–grid interaction," *IEEE Journal of Emerging and Selected Topics in Power Electronics*, vol. 8, no. 2, pp. 1911–1921, 2020. DOI: 10.1109/JESTPE.2020.2972070.

[14] Mathworks, *Control system toolbox*, 2020.

PCIM Europe 2024, 11– 13 June 2024, Nuremberg   DOI: 10.30420/566262052

An Effective DC Voltage Regulation of Active Front-End Rectifier through Model Predictive Control

Mobina Pouresmaeil [1], Jorma Kyyrä [1], Edris Pouresmaeil [1]

[1] Aalto university, Finland

Corresponding author: Mobina Pouresmaeil, Mobina.pouresmaeil@aalto.fi
Speaker: Mobina Pouresmaeil, Mobina.pouresmaeil@aalto.fi

Abstract

This paper proposes a novel control technique to improve the performance of the active front-end (AFE) rectifier. Employing a finite set model predictive control (FS-MPC) approach, this technique operates without any modulator. The proposed control method accurately and quickly regulates the DC-link voltage by tracking the reference active power, and it also tracks the reference reactive power on the AC side. The idea behind the control technique is to generate the required active power reference for the AC source to facilitate an effective DC voltage regulation without considering a direct control of DC-link voltage. The controller does not rely on awareness of the DC load and is robust to unknown load disturbances. Simulation results in MATLAB/Simulink verify the high and effective performance of the proposed control method.

1   Introduction

Electrolysis is a promising method for producing green hydrogen, i.e., hydrogen produced using renewable energy sources [1]. As electrolysis technology plays a crucial role in the transition to a sustainable energy future, it is necessary to ensure that its requirements are met. Precise control of DC output is one of the requirements for an efficient electrolysis process [2]. Voltage fluctuations can result in energy losses and diminish the overall efficiency of the electrolysis process. Therefore, providing a stable voltage is essential for optimizing the performance of the electrolyzer. Several power converter topologies are introduced for power-to-hydrogen electrolysis processes. An overview of the different power electronics converter topologies is presented in [3], with performance evaluation and comparison.

Active front-end (AFE) rectifier actively controls the rectification process and can be employed for this purpose. AFE rectifiers, with several advantages over diode rectifiers, find extensive application across various industrial sectors. A low harmonic distortion current on the AC side, and control of the power factor, are possible alongside with the main control objective of DC voltage regulation. AFE rectifiers can provide bidirectional power flow

and can be employed as an active filter for reducing total harmonic distortion [4]. The conventional control methods for AFE rectifiers can be classified into voltage-oriented control (VOC) [5] and the direct power control (DPC) [6], each consisting of two control loops. In VOC control of the AFE rectifiers, the inner control loop controls the current of the AC side to indirectly control the power, while in the DPC control concept the inner control loop directly controls the active and reactive power. Both control strategies have their advantages and limitations and can be employed depending on the requirements of the application [7].

Model predictive control (MPC) is increasingly recognized as a favorable control concept for power electronics devices[8]. MPC has primarily been employed for the inner control of the AFE rectifier, together with the PI controller for controlling the DC link voltage in the outer control loop [9]. However, this control method is not fast and accurate because of the PI controller's poor performance. Alternatively, different control strategies aimed at generating the required reference active power for the MPC control to compensate for the DC voltage error. [10] has proposed dynamic references for active power and the DC voltage. Despite the coupling between active power and the DC-link voltage, the cost function in the proposed control is designed

419

to simultaneously follow the reference active power and the reference DC voltage, in addition to tracking the reference reactive power. Besides, the control method is dependent on the knowledge of the DC load and is not robust to load disturbance. In [11] deriving the active power reference relies on the energy stored in the DC link capacitor. The measured and the calculated current of the DC side are employed to extract the required reference active power. Besides the interdependency and difficulty of the calculation, the issue is that the controller is not robust to the load change. Sensorless predictive control of AFE rectifier with voltage estimation is proposed in [12]. To address the challenges caused by the estimation-based control, the paper employs filters and parameter estimation through the Lyapunov technique, resulting in a complicated control method.

This paper proposes an MPC-based control technique for AFE rectifier in which the reference active power of the AC grid is obtained based on the output power and the power difference between the current and the next sampling instant. A simple estimation of the voltage is considered for the power error estimation. The control method facilitates the calculation while ensuring the desired DC link voltage regulation and reference reactive power tracking.

The remainder of the paper unfolds as follows: section 2 presents the AFE rectifier model and the fundamentals of the MPC-based DPC method. section 3 elaborates on the proposed reference generation method, while section 4 delves into the results and discussions. Lastly, section 5 concludes this study.

2 AFE Rectifier Modeling and the Proposed Control Technique

The model under study is an AFE rectifier, depicted in Fig. 1. The subject model comprises a fully controlled three-phase bridge, connected to the AC grid through filter L_f.

According to Fig. 1, dynamics of the system can be expressed by the following equations in $\alpha\beta$ frame, as:

$$v_s - L_f \frac{di_s}{dt} - R_f i_s = u_c \qquad (1)$$

where v_s, i_s and u_c are the space vectors in $\alpha\beta$ frame, representing the AC grid voltage and current, and the rectifier terminal voltage, respectively.

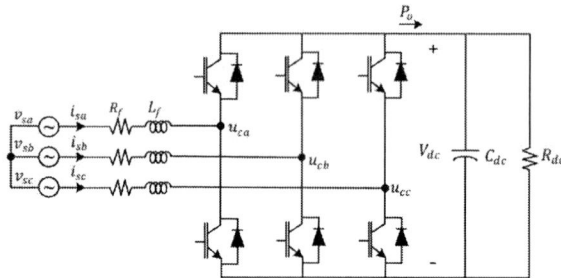

Fig. 1: The AFE rectifier diagram.

The proposed control scheme is illustrated in Fig. 2, which is an MPC-based control scheme. The MPC concept requires the discrete-time model of the system. Using the Euler approximation of the model represented by Eq. (1) and considering the sampling period of T_s, the discrete-time representation of the model can be described as:

$$i_s(k+1) = (1 - \frac{T_s R_f}{L_f})i_s(k) - \frac{T_s}{L_f}(v_s(k) - u_c(k))$$

$$(2)$$

where k and $k+1$ denote the current and subsequent sampling instants in the discrete model. Using the discrete model of the system, MPC predicts the future performance of the system in all the possible control actions and extracts the optimal control action to be applied. The eight possible switching states in the 2-level converter and the corresponding voltage vectors are listed in table 1. These voltage vectors are applied as the control action in the prediction process. Then the predicted values are utilized in the cost function defined for the predictive model. Finally, the optimal voltage vector can be derived from the minimization of the cost function.

Index	switching states	voltage vectors
s_1	$\{000\}$	0
s_2	$\{100\}$	$\frac{2V_{dc}}{3}$
s_3	$\{110\}$	$\frac{V_{dc}}{3} + j\frac{\sqrt{3}V_{dc}}{3}$
s_4	$\{010\}$	$\frac{-V_{dc}}{3} + j\frac{\sqrt{3}V_{dc}}{3}$
s_5	$\{011\}$	$\frac{-2V_{dc}}{3}$
s_6	$\{001\}$	$\frac{-V_{dc}}{3} - j\frac{\sqrt{3}V_{dc}}{3}$
s_7	$\{101\}$	$\frac{V_{dc}}{3} - j\frac{\sqrt{3}V_{dc}}{3}$
s_8	$\{111\}$	0

Tab. 1: The possible switching states and the corresponding voltage vectors.

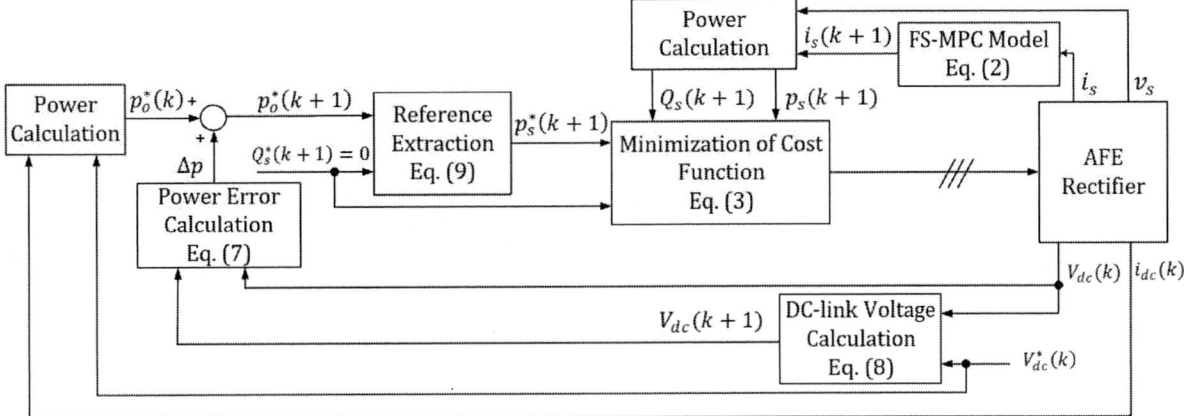

Fig. 2: The studied model and the overall control scheme.

In the AFE rectifier the main objective is to regulate the DC link voltage while transferring power from the AC source to the DC load. It is also important to control the power factor by controlling the active and reactive power on the AC side. Accordingly, the following cost function is specified which tracks the active and the reactive power references:

$$G = (p_s^*(k+1)-p_s(k+1))^2+(Q_s^*(k+1)-Q_s(k+1))^2 \tag{3}$$

where the instantaneous active and reactive power at instant $k+1$ can be calculated based on the measured grid voltage at time instant k and the predicted current at time instant $k+1$, as described in Eq. (4) and Eq. (5). It is presumed that the AC source voltage remains constant from the current to the next sampling time ($v_s(k+1) = v_s(k)$).

$$p_s(k+1) = v_{s\alpha}(k)i_{s\alpha}(k+1) + v_{s\beta}(k)i_{s\beta}(k+1) \tag{4}$$

$$Q_s(k+1) = v_{s\beta}(k)i_{s\alpha}(k+1) - v_{s\alpha}(k)i_{s\beta}(k+1) \tag{5}$$

3 The Proposed Reference Generation

DC-link voltage regulation can be realized indirectly by tracking the active power reference, given that the DC link voltage is coupled with the average active power of AC source. Therefore, in this section, the purpose is to generate the right active power reference for an accurate DC-link voltage regulation.

The active power reference of the DC side at time instance $k+1$ can be defined using the power reference at time instance k and the active power error (Δp) as:

$$p_o^*(k+1) = p_o^*(k) + \Delta p \tag{6}$$

The active power error can be calculated based on the transient energy stored of the DC link capacitor from the current to the next sampling instant, as follows:

$$\Delta p = \frac{C_{dc}}{2T_s}(v_{dc}(k+1)^2 - v_{dc}(k)^2) \tag{7}$$

In this equation, the DC-link voltage in the sampling time k can be measured and the DC-link voltage in the next sampling instant, $k+1$, can be defined. Fig. 3 illustrates how the capacitor voltage gradually changes to achieve its reference value in n steps.

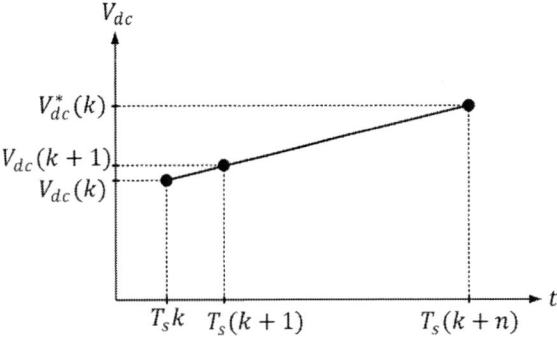

Fig. 3: The capacitor voltage, reaching its reference value in n steps.

With this figure, the DC-link voltage error can be estimated by dividing the voltage difference between the measured voltage at instant k and the DC-link reference voltage by n. Adding the estimated voltage error to the measured voltage gives the DC link voltage in the next sampling instant, defined as follows:

$$v_{dc}(k+1) = v_{dc}(k) + \frac{1}{n}(v_{dc}^*(k) - v_{dc}(k)) \quad (8)$$

Finally, the active power reference for the source can be computed through the output active power reference and the source reactive power reference as follows:

$$p_s^*(k+1) = \frac{3V_s}{4R_f}(V_s - \sqrt{a}) \quad (9)$$

$$a = V_s^2 - \frac{8R_f p_o^*(k+1)}{3} - \frac{16R_f^2 Q_s^{*2}(k+1)}{9V_s^2}$$

For a unity power factor operation system, the reference reactive power of $Q_s^*(k+1) = 0$ is considered in this study.

4 Results and Discussion

A comprehensive set of simulation results of the model shown in Fig. 2 are presented to validate the superior performance of the proposed model predictive-based control method. The parameters of the model and the proposed control method are detailed in table 2. In this study the DC voltage is considered as $V_{dc} = 630\,\text{V}$, associated to the active power of $P = 16\,\text{kW}$ in the studied model with DC load of $R_{dc} = 25\,\Omega$.

Parameters	Description	Values
f	Frequency	$50\,\text{Hz}$
v_s	AC side voltage	$400\,\text{V}$
V_{dc}	DC-link voltage	$630\,\text{V}$
C_{dc}	DC-link capacitor	$300\,\mu\text{F}$
L_f	Filter inductance	$4\,\text{mH}$
T_s	Sampling time	$10\,\mu\text{s}$
n	constant	100

Tab. 2: Parameters of the model and the proposed control method.

First, we investigate performance of the proposed MPC-based control strategy considering a change in reference DC voltage from $630\,\text{V}$ to $590\,\text{V}$ at $t = 0.2s$. Fig. 4 illustrates the simulation results in this case study. As can be seen, the proposed control strategy generates the desired power reference and tracks the reference appropriately, resulting in a regulated DC voltage. As shown in Fig. 4 (c) the unity power factor operation is achieved through the reference reactive power of $Q_s^* = 0$.

Next, we considered a DC load change from $25\,\Omega$ to $17\,\Omega$ to analyze the control performance in controlling DC voltage. Fig. 5 shows how the generated reference power changes to regulate the DC voltage at its reference value. The proposed control method maintains the DC voltage at its reference value through tracking the generated reference active power.

Finally, the controller performance in non-unity power factor is investigated. A step-up change of reference reactive power from $Q_s^* = 0$ to $Q_s^* = 2kVAR$ is carried out. Fig. 6 demonstrates high performance of the proposed controller in this condition, as it properly tracks the initial reference voltage and reference active power as well as the reference reactive power.

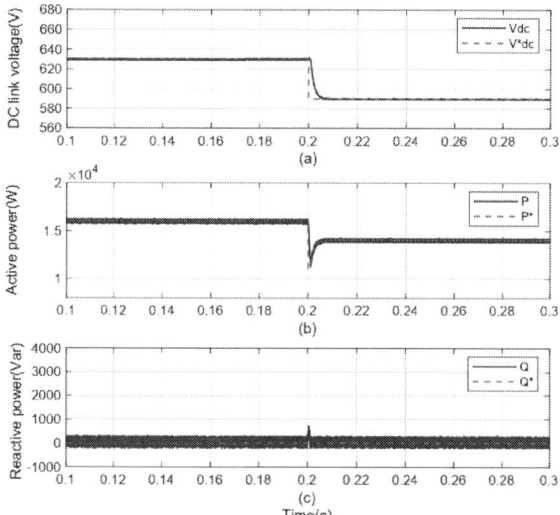

Fig. 4: Simulation results of change in reference DC-link voltage, (a) DC-link voltage, (b) Active power, and (c) Reactive power.

5 Conclusions

An FS-MPC-based direct power control method for AFE rectifiers is presented in this paper. The proposed control method provides several key benefits including the regulated DC voltage, controlled power factor on the AC side, and fast and accurate dynamic responses.

Simulation studies in MATLAB/Simulink have demonstrated the efficacy of the proposed control method in achieving different control objectives and providing efficient power conversion for applications such as the electrolysis process. Future research direction may include further optimization of the pro-

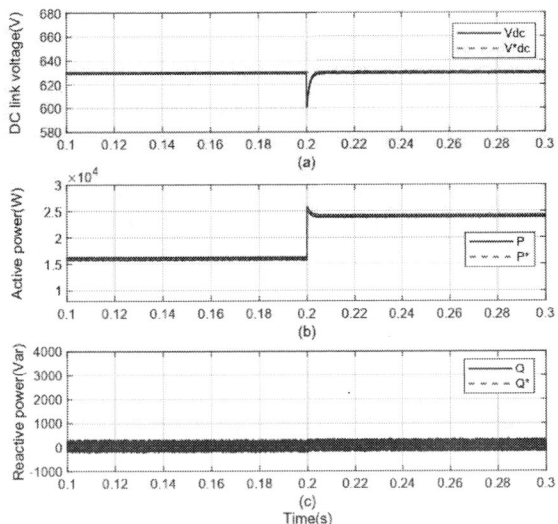

Fig. 5: Simulation results of change in DC load, (a) DC-link voltage, (b) Active power, and (c) Reactive power.

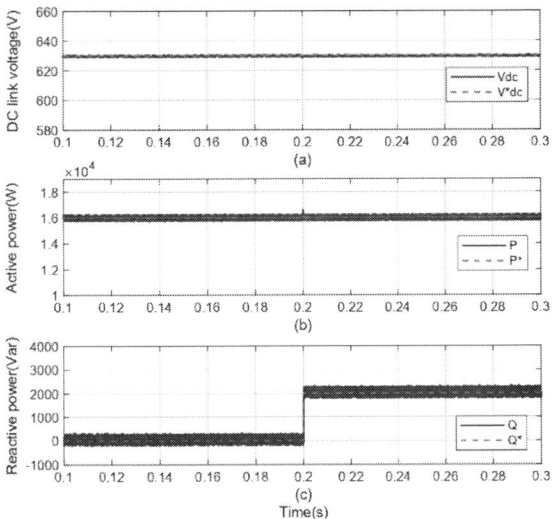

Fig. 6: Simulation results of change in reference reactive power, (a) DC-link voltage, (b) Active power, and (c) Reactive power.

posed control algorithm for obtaining higher power quality.

References

[1] Á. Iribarren, D. Elizondo, E. L. Barrios, H. Ibaiondo, A. Sanchez-Ruiz, *et al.*, "Dynamic modeling of a pressurized alkaline water electrolyzer: A multiphysics approach," *IEEE Transactions on Industry Applications*, 2023.

[2] B. Yodwong, D. Guilbert, M. Phattanasak, W. Kaewmanee, M. Hinaje, and G. Vitale, "Ac-dc converters for electrolyzer applications: State of the art and future challenges," *Electronics*, vol. 9, no. 6, p. 912, 2020.

[3] M. Chen, S.-F. Chou, F. Blaabjerg, and P. Davari, "Overview of power electronic converter topologies enabling large-scale hydrogen production via water electrolysis," *Applied Sciences*, vol. 12, no. 4, p. 1906, 2022.

[4] J. R. Rodríguez, J. W. Dixon, J. R. Espinoza, J. Pontt, and P. Lezana, "Pwm regenerative rectifiers: State of the art," *IEEE Transactions on Industrial Electronics*, vol. 52, no. 1, pp. 5–22, 2005.

[5] L. Tarisciotti, P. Zanchetta, A. Watson, J. C. Clare, M. Degano, and S. Bifaretti, "Modulated model predictive control for a three-phase active rectifier," *IEEE Transactions on Industry Applications*, vol. 51, no. 2, pp. 1610–1620, 2014.

[6] S. Vazquez, J. I. Leon, L. G. Franquelo, J. Rodriguez, H. A. Young, *et al.*, "Model predictive control: A review of its applications in power electronics," *IEEE industrial electronics magazine*, vol. 8, no. 1, pp. 16–31, 2014.

[7] M. Malinowski, M. P. Kazmierkowski, and A. M. Trzynadlowski, "A comparative study of control techniques for pwm rectifiers in ac adjustable speed drives," *IEEE Transactions on power electronics*, vol. 18, no. 6, pp. 1390–1396, 2003.

[8] M. Pouresmaeil, R. Sangrody, S. Taheri, D. Montesinos-Miracle, and E. Pouresmaeil, "Enhancing fault ride through capability of grid-forming virtual synchronous generators using model predictive control," *IEEE Journal of Emerging and Selected Topics in Industrial Electronics*, 2024.

[9] P. Cortes, J. Rodríguez, P. Antoniewicz, and M. Kazmierkowski, "Direct power control of an afe using predictive control," *IEEE Transactions on Power Electronics*, vol. 23, no. 5, pp. 2516–2523, 2008.

[10] D. E. Quevedo, R. P. Aguilera, M. A. Perez, P. Cortes, and R. Lizana, "Model predictive control of an afe rectifier with dynamic references," *IEEE transactions on power electronics*, vol. 27, no. 7, pp. 3128–3136, 2011.

[11] M. A. Pérez, R. L. Fuentes, and J. Rodriguez, "Predictive control of dc-link voltage in an active-front-end rectifier," in *2011 IEEE International Symposium on Industrial Electronics*, IEEE, 2011, pp. 1811–1816.

[12] M. Mehreganfar, M. H. Saeedinia, S. A. Davari, C. Garcia, and J. Rodriguez, "Sensorless predictive control of afe rectifier with robust adaptive inductance estimation," *IEEE Transactions on Industrial Informatics*, vol. 15, no. 6, pp. 3420–3431, 2018.

PCIM Europe 2024, 11– 13 June 2024, Nuremberg   DOI: 10.30420/566262053

Bi-directional 11-kW Multi-Level Active-Neutral-Point-Clamped AC-DC Converter Using 600 V / 750 V Si Super-Junction and SiC MOSFETs for High-Efficiency and High-Density Applications

Mengxing Chen[1], Manuel Escudero Rodriguez[1], Matteo-Alessandro Kutschak[1], David Meneses Herrera[2], Alex Rossi[1]

[1] Infineon Technologies Austria AG, Austria
[2] Infineon Technologies Nordic AB, Finland

Corresponding author: Mengxing Chen, Mengxing.Chen@infineon.com
Speaker: Mengxing Chen, Mengxing.Chen@infineon.com

Abstract

For bidirectional AC/DC power converters in electric vehicle (EV) charging station and onboard charger (OBC) applications, high efficiency and high power-density are two of the key success factors. This work presents a system solution of the 11-kW active-neutral-point-clamped (ANPC) AC/DC converter. Thanks to the combination of multi-level topology with 600 V Si super-junction (SJ) MOSFETs and 750 V SiC MOSFETs, a peak efficiency of 99.2% is reported in the work. Moreover, a superior power-density of 11.4 kW/L is achieved, thanks to the adoption of top-side-cooling packages. The operation principle, switching-cell design, system integration, thermal design and experimental results are presented.

1   Introduction

For electric vehicle (EV) charging station and onboard charger (OBC) applications, the 800 V_{DC} DC-bus finds more significance especially in higher power and three-phase feeding scenarios. Nowadays, as higher power delivery capability is required by the battery to enable faster charging, those power converters are typically rated at 11kW and above.

Such power converter requires a front-end power factor correction (PFC) stage to enable high waveform quality and superior power factor. Recently,

bi-directionality of the PFC stage sees increasing implementation in recent-released designs. This helps to realize advanced functionalities such as vehicle-to-grid and vehicle-to-load. Additionally, other performance metrics like efficiency, power density and cost are also under full consideration by design engineers.

Fig. 1 presents four possible topology candidates for the bi-directional PFC stage featuring 800 V_{DC} bus, i.e., in a single-leg representation. The 2L-HB is the well-established and simplest implementation, where the 1200V power semiconductors are

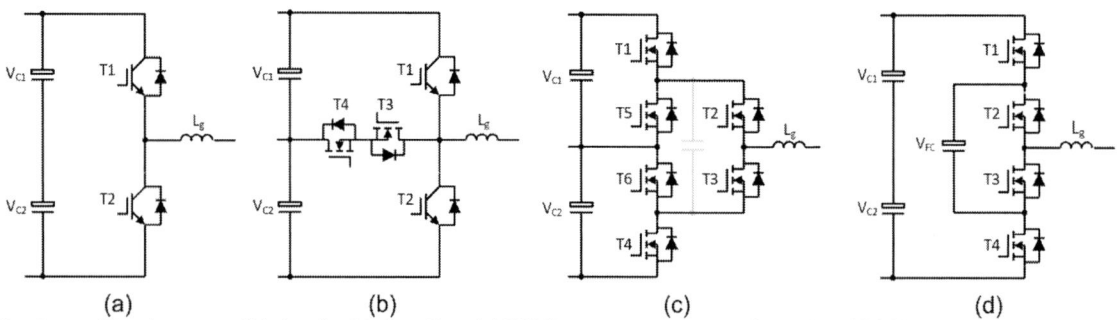

Fig. 1   Topology candidates for bi-directional AC/DC power converters featuring 800 V_{DC} bus in a single leg representation. (a) Two-level half bridge (2L-HB). (b) Three-level T-type neutral-point-clamped (3L-TNPC). (c) Three-level active-neutral-point-clamped (3L-ANPC). (d) Three-level flying-capacitor (3L-FC).

typically required. Nevertheless, the 2L-HB features high dv/dt and bulky PFC choke due to its intrinsic 2L-switching characteristic. An improvement of the 2L-HB is the 3L-TNPC, as shown in Fig. 1(b). Compared to 2L-HB, the 3L-TNPC enables half dv/dt switching characteristics and therefore offers smaller PFC choke, higher power density and less electro-magnetic interferences (EMI) [1]. Still, the 3L-TNPC requires 1200V power semiconductors, either Si IGBT or SiC MOSFETs, as the 2L-HB does to enable the 800 V_{DC} bus.

The 3L-ANPC converter [2], shown in Fig. 1(c), is an advancement of the classical 3L I-type NPC converter [3] by replacing two power diodes with power semiconductor switches at sockets of T_5 and T_6. Doing so, the 3L-ANPC features higher efficiency and better loss distributions [2], as well as more control flexibilities. Compared to 3L-TNPC, the 3L-ANPC is implemented with two additional power semiconductors with the benefit that all sockets can be implemented with 600 V – 750 V power semiconductors. Multiple modulation principles of 3L-ANPC converter can be found in literatures [2], [4-9]. The simplest pattern can be implemented with two fast-switching devices and four line-frequency switching, while the bi-directionality can still be maintained. More details of the 3L-ANPC modulation pattern will be discussed in Section II.

Last but not least, the 3L-FC converter [10] is a very attractive solution to enable the bi-directional PFC with 800 V_{DC} bus. For single phase-leg, four power semiconductor switches with 600 V – 750 V voltage rating in combination with a capacitor bank are required. By applying high-frequency phase-shift modulations, the apparent switching frequency the PFC choke sees is doubled, therefore making the designed PFC choke even smaller. Nevertheless, the flying capacitor voltage must be robustly maintained at half DC-bus voltage especially during start-up, surge, and dynamic load conditions. Therefore, the 3L-FC topology generally requires additional control complexity or additional circuitry to ensure its robustness.

This work focuses mainly on the 3L-ANPC converter for the PFC stage of EV charging and OBC applications with 800 V_{DC} bus. It is reported that the 3L-ANPC topology combining 600 V Si SJ MOSFETs and 750 V SiC MOSFETs in top-side-cooling (TSC) packages enables the design with high efficiency and high power-density.

2 Operation Principle

2.1 Modulation Scheme

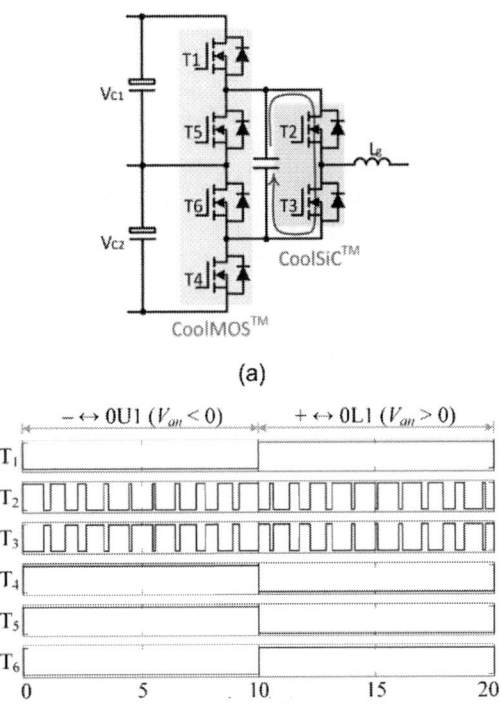

Fig. 2 The phase-leg configuration and modulation scheme adopted in this work. (a) Single-phase-leg representation of the 3L-ANPC topology with the combination of Si SJ-MOSFET and SiC MOSFET. (b) Modulation scheme of the 3L-ANPC phase leg, T_2 and T_3 at 65 kHz, and T_1, T_5, T_6 and T_4 at line-frequency switching. (V_{an}: grid voltage)

As it is mentioned in the introduction section, there are several modulation methods of 3L-ANPC converter in literatures [2], [4-9]. One of the significant differences among these modulation methods is the number of semiconductors under high-frequency and line-frequency switching. Those under line-frequency switching can be implemented with Si-based power semiconductors, e.g., Si IGBT and Si SJ MOSFET, while those under high-frequency switching shall be implemented with wide-band-gap (WBG) power semiconductors, e.g., SiC MOSFET and GaN HEMT.

The most straightforward modulation scheme utilizes only two power semiconductors in high-frequency switching, i.e., T_2 and T_3 in Fig. 2(a). With this scheme, the control and modulation complexity as well as the system cost of 3L-ANPC converter is minimized, compared to other schemes. On the other hand, the most advanced modulation scheme utilizes all six power semiconductors in high-frequency switching, so that the apparent switching frequency that the PFC choke sees is two times the actual switching frequency [4], [6]. Nevertheless, the system cost and complexity are much higher than the aforementioned scheme. Considering this, this work focuses on the first modulation scheme of the 3L-ANPC converter.

The phase-leg configuration of 3L-ANPC converter and the adopted modulation scheme are presented in Fig. 2(a) and 2(b) respectively. As mentioned before, the two devices T_2 and T_3 switch at 65 kHz, where WBG power semiconductors, e.g., SiC MOSFET and GaN HEMT, can be used. Moreover, the four devices T_1 and $T_4 - T_6$ commutate at the ac mains frequency and can be implemented with either Si SJ-MOSFETs or IGBTs. Since those devices bear only the conduction loss, it is flexible for the power converter designer to scale the power level by selecting the proper $R_{DS,ON}$. Thanks to the low $R_{DS,ON}$ achievable by the Si SJ-MOSFETs, the 3L-ANPC converter features much lower conduction loss compared to its counterparts. The $R_{DS,ON}$ selection of SiC MOSFETs T_2 and T_3 can be optimized between conduction losses and switching losses.

2.2 Switching-Loop Optimization

A shortcoming of many 3L-ANPC modulation schemes is the relatively long switching-loop of devices T_2 and T_3, no matter they are actively switching at high frequency or not [11-12]. This will cause significant overvoltage and oscillation on the drain-source voltages of T_2 and T_3. Therefore, the designers may need to slow down the switching speed suffering high switching losses.

A recent advancement of 3L-ANPC converter is adding an intermediate decoupling capacitor across the half-bridge housing T_2 and T_3 [7], [13], as it is depicted in Fig. 2(a). Then the resultant switching-loop with the de-coupling capacitor is significantly reduced.

3 System Demonstrator Design

3.1 System Introduction

In order to proof the concept of combining Si SJ-MOSFET and SiC MOSFET in a bi-directional 3L-ANPC converter, the system demonstrator is designed with high-efficiency and high-power-density targets.

The specifications of the system demonstrator are shown in Table 1. It is worth to mention that the system demonstrator targeting EV charging and OBC applications is able to handle both three-phase and single-phase input, delivering the output power of 11 kW and 7.3 kW, respectively. The output voltage can be maintained as 800 V_{DC} for both single- and three-phase input.

Parameter	Value
AC voltage (3-phase input)	318 – 480 V_{AC}
AC voltage (1-phase input)	180 – 275 V_{AC}
DC voltage	800 V_{DC}
Output power (3-phase input)	11 kW
Output power (1-phase input)	7.3 kW
Targeted peak efficiency	> 99.2 %
Targeted power density	> 10 kW/L

Table 1 System specification.

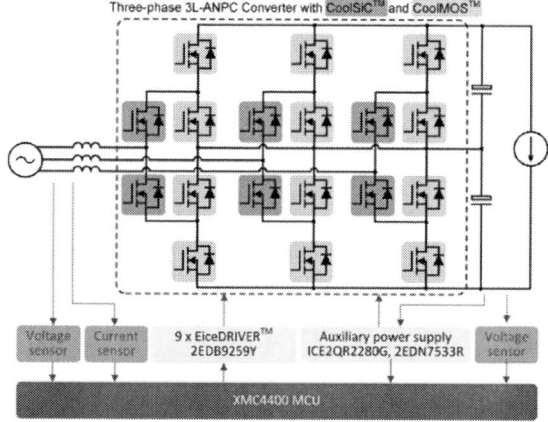

Fig. 3 System block diagram of the 11kW/800V 3L-ANPC demonstrator for OBC and EV charging applications.

A simplified system diagram is depicted in Fig. 3. The demonstrator is digitally controlled using Infineon micro-controller XMC4400 series. 2EDB9259Y dual-channel gate driver is selected and is capable to drive both Infineon CoolMOS™ and CoolSiC™. The auxiliary power supply unit is fed from the DC bus, i.e., both positive-half and negative-half DC rails. The quasi-resonant flyback controller ICE2QR2280 is used. Eleven channels of 18-V individual supply rails are generated out of the designed auxiliary power supply, which is enough to drive all of the power switches in this demonstrator.

The Si SJ-MOSFET and SiC MOSFET are offered in the top-side-cooling (TSC) package QDPAK in the prototype to demonstrate its capability to enable high-power-density designs. The estimated efficiency curves are presented in Fig. 4 with both 16 mΩ CoolSiC™ AIMDQ75R016M1H [14] and 40 mΩ CoolSiC™ AIMDQ75R040M1H [15]. The peak efficiency of both cases can reach 99.2 %, but

PCIM Europe 2024, 11– 13 June 2024, Nuremberg   DOI: 10.30420/566262053

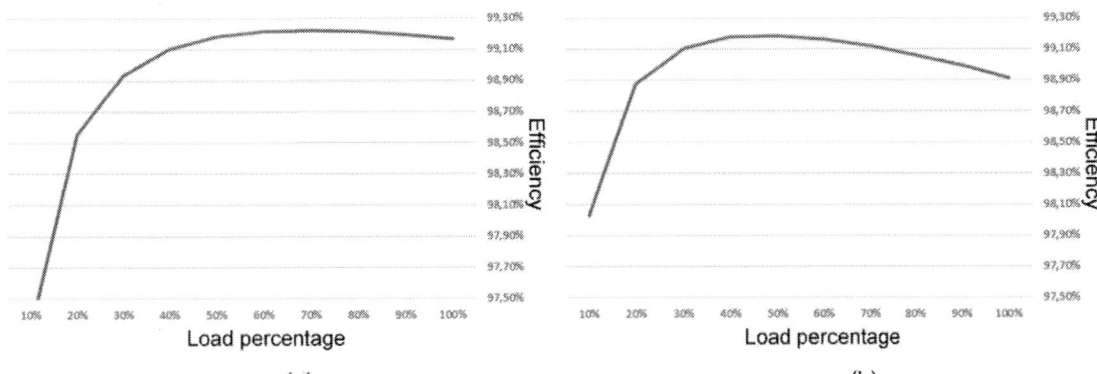

(a)   (b)

Fig. 4   Calculated system efficiency (including all auxiliary power). (a) With 16mΩ CoolSiC™ AIMDQ75R016M1H and 40 mΩ CoolMOS™ IPDQ60R040S7. (b) With 40 mΩ CoolSiC™ AIMDQ75R040M1H and 40 mΩ CoolMOS™ IPDQ60R040S7.

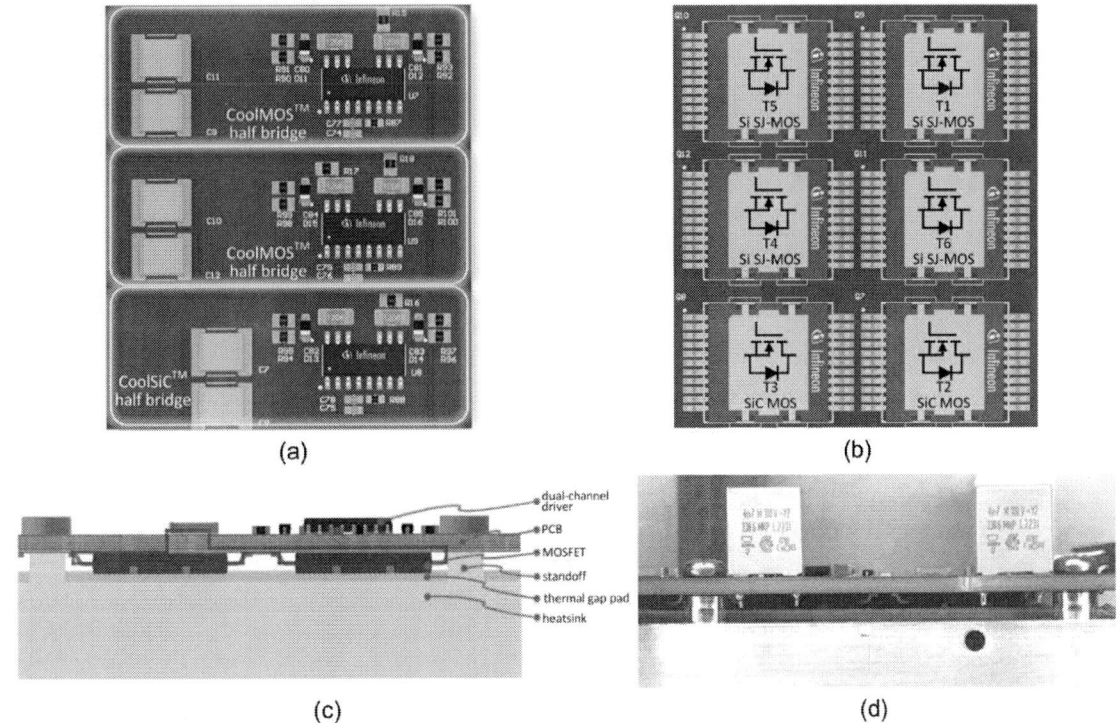

Fig. 5   3L-ANPC converter phase-leg and switching-cell design. (a) Phase-leg layout visualized from the top-side of PCB. (b) Phase-leg layout visualized from the bottom-side of PCB. (c) Cross-sectional view of the switching cell. (d) Side-view of the switching cell in the prototype.

higher full-load-efficiency can be achieved by using the 16 mΩ part. The 40 mΩ CoolMOS™ IPDQ60R040S7 [16] is used in the calculations. Additionally, the system efficiency can be further boosted by lower $R_{DS,ON}$ CoolMOS™ devices, as only conduction losses are generated in the low-frequency MOSFETs.

3.2   Phase-Leg and Switching-Cell Design

Following the defined specifications, the 11 kW / 800 V_{DC} 3L-ANPC demonstrator is designed and assembled. In this sub-section the phase-leg and switching cell design will be illustrated.

Figs. 5(a), 5(b), and 5(c) present the top-view, bottom-view, and the side-view of the designed 3L-ANPC phase leg, respectively. As it can be seen from Figs. 5(a) and 5(b), the 3L-ANPC phase-leg is organized as three half-bridges, with two Cool-MOS™ half bridges (i.e., switch pair T_1-T_5 and T_6-

427

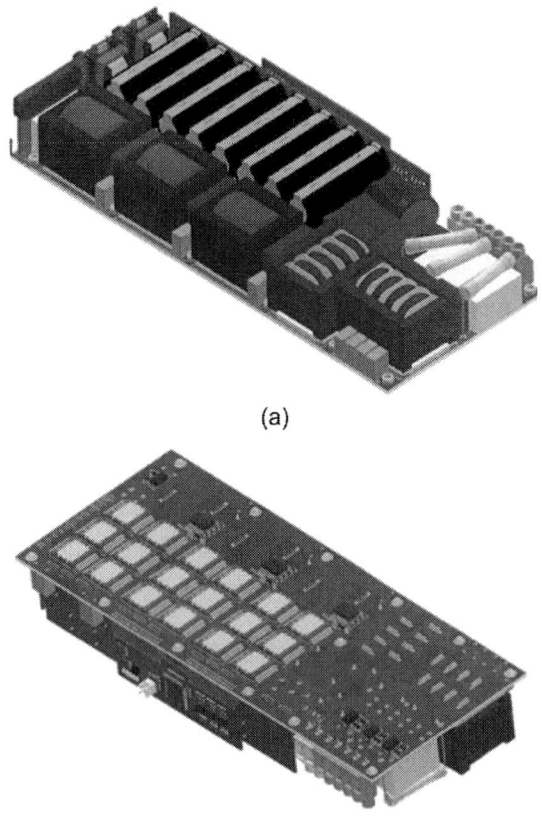

(a)

(b)

Fig. 6 CAD picture of the designed 11 kW / 800 V_{DC} 11.4 kW/L 3L-ANPC demonstrator in a form factor of: 243 mm in length, 99 mm in width, and 40 mm in height. (a) Top-side view. (b) Bottom-side view.

T_4) and one CoolSiC™ half bridge (i.e., switch pair T_2-T_3).

As it is mentioned, this work uses Infineon TSC package which allows the designer to layout decoupling capacitors, gate driver ICs and their peripheral circuits right above the power semiconductors, on top-side of the PCB. The TSC CoolMOS™ and CoolSiC™ are mounted right beneath their decoupling and driver circuitry. Therefore, very low gate-drive-path parasitic inductance can be achieved. The switching loop of fast SiC MOSFET leg is marked as the red solid line in Fig. 5(c), which is compact and fully utilizes the magnetic-field-cancellation technique (loop inductance estimated as < 7 nH).

Moreover, as shown in Figs. 5(b) and 5(c), the cooling pad of the device is facing downwards, enabling the direct heat-flow path towards the heatsink. More details about the cooling concept will be illustrated in section 4.

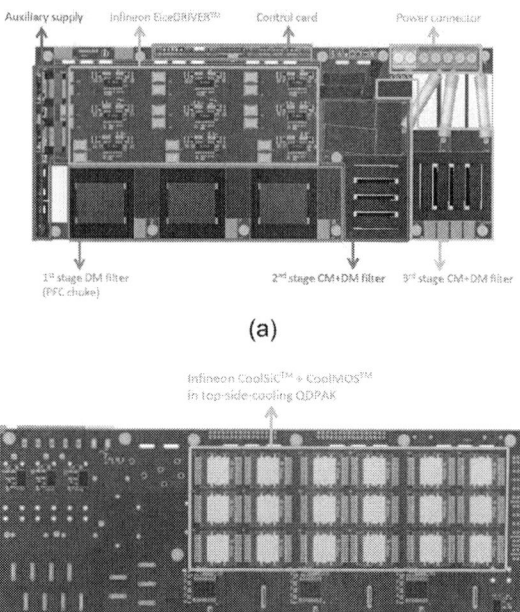

(a)

(c)

Fig. 7 CAD picture of the designed 11 kW / 800 V_{DC} 11.4 kW/L 3L-ANPC demonstrator. (a) Top-side view. (b) Bottom-side view.

Fig. 8 The designed high-density common-mode EMI filter with differential-model inductance integration [17]. Form factor: 43.8 mm in length, 36.0 mm in width, and 31 mm in height.

3.3 System Integration for High Power Density

The designed 11 kW / 800 V_{DC} 3L-ANPC demonstrator is presented in Fig. 6 and Fig. 7 from various viewing perspectives. The main power board is sitting in the bottom and houses the input/output power connector, the inrush protection circuit, the two-stage EMI filters, the PFC chokes, and the main 3L-ANPC power stage. All CoolSiC™ and CoolMOS™ parts in QDPAK (identical height profile can be guaranteed) are mounted on bottom side of main power board. Then, a flat and large-area cooling surface is achieved which thermally connects to the cooling system via the pre-defined thermal gap pad. Other SMD parts, i.e., EiceDRIVER™ and low-profile de-coupling capacitor, are mounted on top side of the PCB right above the QDPAKs. Three separate daughter cards are adopted in the design in order to achieve high power density and modularity, i.e., the auxiliary power supply card to the left-side, the capacitor card in the middle, and the control card to the top-side of the layout shown in Fig. 7(a). It's worth to mention that a special common-mode EMI filter is used with differential-mode inductance integration. The designed EMI filter is presented in Fig. 8, featuring the form factor of 43.8 mm by 36.0 mm by 31 mm. Magnetic shunts are inserted in between the common-mode windings to provide a higher-permeability path for the leakage flux. More information regarding this concept can be found in [17].

4 Cooling and Thermal Design Considerations

Reference [18] overviews the cooling concepts and their pros and cons for Infineon's TSC QDPAK, including gap pad, liquid gap filler, and phase-change materials. Specifically, the gap-pads are pre-cured silicone and ceramic powder materials. The commercially-available gap pads (soft or ultra-soft type) feature thermal-conductivity values ranging from approximately 2 to >10 W/(m·K). Normally the designers would need the high thermal-conductivity-grade gap pad to better dissipate the heat from the package. Nevertheless, there are also other performance metrics which the designer should pay attention to.

4.1 Designing Standoff Height

The gap pads generally feature higher operation temperature but also require permanent contact force to ensure the long-endurance of thermal performance. As it can be seen in Figs. 5(c) and 5(d), standoffs with the specified height are utilized in this work to define the PCB-to-heatsink distance $d_{standoff}$. Then, the thickness of thermal gap pad after contact $d_{gp,contact}$ is calculated as:

$$d_{gp,contact} = d_{standoff} - (d_{qdpak} \pm 0.1 \text{ mm}) \quad (1)$$

where, d_{qdpak} (= 2.4 mm) represents the median value of QDPAK thickness, with 0.1mm being the tolerance of package. In the application, it needs to be fulfilled that the maximum $d_{gp,contact}$ is always lower than the original gap-pad-thickness $d_{gp,origin}$

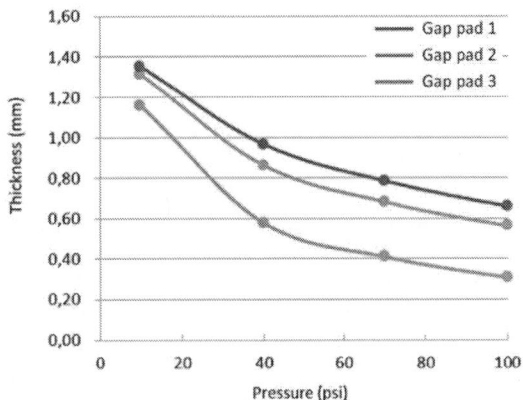

Fig. 9 Compressibility curves, i.e., thickness vs pressure, of three gap pads with an initial thickness of 1.5mm.

to ensure the permanent contact of package to gap pad, which gives:

$$d_{gp,contact} < d_{gp,origin} \quad (2)$$
$$d_{standoff} < d_{gp,origin} + (d_{qdpak} - 0.1 \text{ mm}) \quad (3)$$

Example: for thermal gap pad with 1.5 mm original thickness, the maximum allowable standoff height is calculated to be 3.8 mm. Theoretically, the thermal resistance for the gap pad material can be calculated by:

$$R_{th,gp} = d_{gp,contact} / (A_{gp} \cdot \lambda_{gp}) \quad (4)$$
$$= (d_{standoff} - (d_{qdpak} \pm 0.1 \text{ mm})) / (A_{gp} \cdot \lambda_{gp}) \quad (5)$$

where, A_{gp} and λ_{gp} are the effective heat transfer area and thermal conductivity of the applied gap pad, respectively. Generally, lowering the standoff height effectively reduces the thermal resistance. Nevertheless, the PCB bears more mechanical stress from the gap pad by lowering the standoff height. In this case, PCB warpage can happen [18], which may result in solder crack and component failure on PCB, especially the SMD ceramic capacitors.

4.2 Avoiding PCB Warpage

To prevent PCB warpage in the application, the first methodology is to increase the mechanically supportive level to PCB from the metallic chassis or cooling plate/heatsink. This can be done by placing more height-defining standoffs/spacers in the vicinity of TSC QDPAK, as it can be seen in Fig. 5(c). The second degree of freedom to alleviate PCB warpage is to use gap pad with higher compressibility. Fig. 9 presents the compressibility of three gap pads, i.e., the figure of thickness in relation to applied pressure, which is normally provided by gap-pad manufacturers. Gap pad 3 features the best compressibility, since it shows the most thickness variation as the identical amount of pressure is applied.

4.3 Insulation Requirements

Except for thermal conduction, the second critical functionality of gap pad is to provide galvanic insulation between the TSC QDPAK and the cooling plate/heatsink, which in many applications is electrically connected to the system protective earth. Depending on system specification, the power converter system generally requires an insulation level (from electrical potential to protective earth) in the range of several kilo-volts. To the authors' best knowledge, the gap pad manufacturers claim a dielectric breakdown strength ranging from 5 kV/mm to 10 kV/mm. Therefore, depending on the system insulation requirement and the selected gap pad's dielectric breakdown voltage, a minimum thickness of gap-pad layer (after pressure application) should be defined. In some cases, an additional insulation tape (double insulation) can be applied and attached to the cooling plate/heatsink to help strengthen the insulation level of the thermal setup. Considering all of the aforementioned aspects (tolerance, PCB warpage, insulation and etc.), in practice the contact (after compression) gap-pad thickness ranges from 0.5 mm to 1.5 mm.

4.4 Thermal Resistance Test

In order to validate the achievable thermal resistance accounting the above-mentioned design considerations, an experimental thermal-resistance characterization is conducted in our laboratory. Infineon CoolMOS™ with embedded temperature sensor IPDQ60T017S7 [19] is used for characterization, and high load current is injected into its body diode (up to 30 A) to generate significant conduction losses, which are precisely recorded. The temperature very close to the junction is measured via the embedded temperature-sensing diode. The 1.5 mm thick thermal gap pads from four different brands with multiple thermal-conductivity grade are tested. The device IPDQ60T017S7 is mounted on heatsink via a standoff with 3.68 mm height, which is smaller than the maximum allowable standoff height calculated in sub-section 4.1. The heatsink temperature is measured via a thermocouple. Then the thermal resistance can be precisely calculated by combining the measured conduction loss, embedded temperature-sensor (junction) and thermocouple (heatsink) values.

The measured junction-to-heatsink thermal-resistance R_{TH_JH} (K/W) results are presented in Fig.

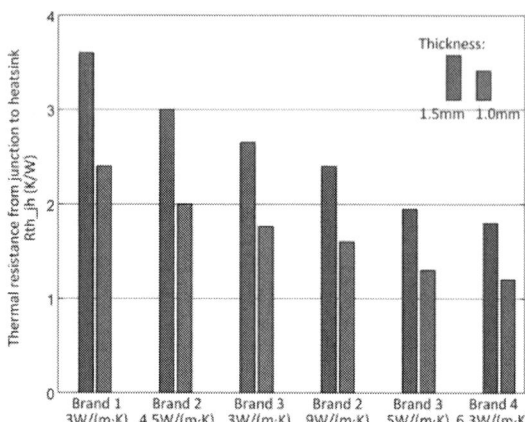

Fig. 10 Junction-to-heatsink thermal resistance $R_{TH,JH}$ (K/W) for CoolMOS™ IPDQ60T017S7 single device [19].

10. The blue bars represent the experimental results with 1.5 mm-original-thickness thermal gap pad. Red bars represent the R_{TH_JH} values using 1.0 mm-original-thickness thermal gap pad, which are calculated out of the 1.5 mm experimental measurement.

The thermal conductivities under testing ranges from 3 W/(m·K) to 9 W/(m·K), resulting in the thermal resistance values ranging from 3.6 K/W to 1.8 K/W, for the 1.5 mm-original-thickness gap pad. It should be noted that the thermal conductivity values listed in Fig. 10 are rated by manufacturers/brands. Different manufacturers may have different qualification standards for thermal conductivity, e.g., the 3 W/(m·K) gap pad from brand 3 features lower R_{TH_JH} than the 4.5 W/(m·K) gap pad from brand 2. Thermal gap pads with different thermal-conductivity grades also feature different level of cost and hardness. Generally speaking, the higher thermal-conductivity comes together with higher level of costs and hardness.

A significant reduction of R_{TH_JH} can be achieved by adopting a thinner thermal gap pad, e.g., with 1.0 mm original thickness. According to Fig. 10, with the 1.0mm gap pad, the achieved R_{TH_JH} values range from 2.4 K/W to 1.2 K/W. Nevertheless, the aforementioned aspects, i.e., tolerance, PCB warpage, insulation, shall be evaluated with more efforts.

5 Experimental Verification

The design system demonstrator is assembled and tested in our laboratory. Fig. 11 shows the fully assembled system demonstrator. Detailed explanations about the system integration and assembly can be found in section 3.3.

Fig. 11 The assembled 11 kW / 800 V_{DC} demonstrator with 11.4 kW/L power density achieved.

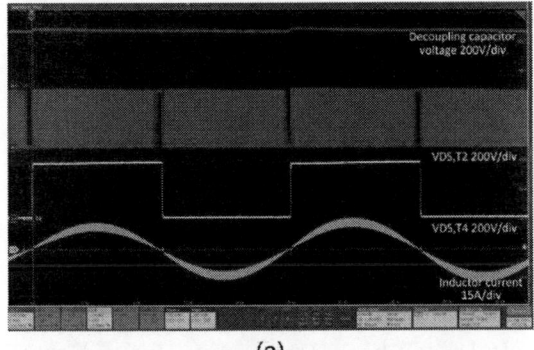

(a)

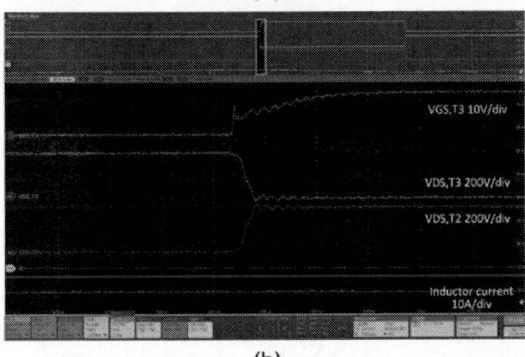

(b)

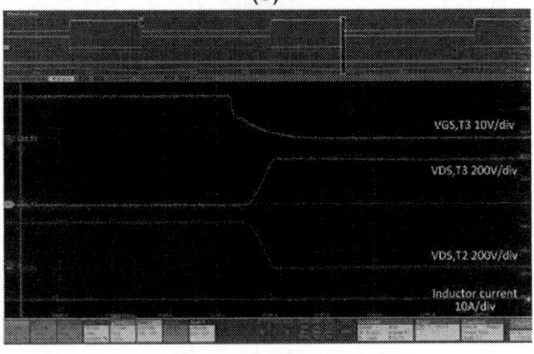

(c)

Fig. 12 The experimental waveforms under 800V bus voltage. (a) Double line-cycle view. (b) Turn-on transient of T_3 in negative line cycle. (c) Turn-off transient of T_3 in negative line cycle.

The functionalities of the auxiliary power supply card, the control card, and the capacitor card are verified first. Then the full functionality of the full system is tested. Fig. 12(a) presents the double line-cycle view, where a nice inductor current waveform quality can be seen out of the closed-loop control. The operation principle of the 3L-ANPC is successfully implemented, as CoolSiC™ $V_{DS,T2}$ is switching in high frequency while $V_{DS,T4}$ is switching in line frequency.

Figs. 12(b) and 12(c) present the turn-on and turn-off transients of CoolSiC™ devices T_2 and T_3 in a zoomed-in view. Thanks to the intermediate decoupling capacitor, the switching-loop inductance of the CoolSiC™ devices is optimized. Therefore, the drain-source voltage of T_2 and T_3 are relatively clean and no significant overshoot can be observed.

6 Conclusion

This work presents a new system solution for bidirectional AC/DC power converter in 800 V_{DC} DC bus EV charging and OBC application. High efficiency and high power-density are identified as two main motivations. The system solution combining 3L-ANPC topology, 600 V CoolMOS™ and 750V CoolSiC™ is presented in this work. A peak efficiency of 99.2 % and a power density of 11.4 kW/L are reported in this work in the power level of 11 kW. The newest TSC package QDPAK from Infineon is identified as the key enabler of high power-density and high efficiency designs. Then multiple design topics are covered in this paper, including the switching-cell design, system integration, thermal design considerations etc. Last but not least, the experimental results are shown to validate the feasibility of the proposal.

Acknowledgements

The authors would like to sincerely thank Dr. Martin Cheung, Mr. Rafael Garcia, and Mr. Nico Fontana from Infineon Technologies for technical discussion and supports. Special thanks to Dr. Francesco Pulsinelli for testing magnetic components. Thanks to Mr. Florian Wernegger and Ms. Ronja Krimm from Infineon Technologies for supporting mechanical and production topics, and thanks to Mr. Florian Zechner and Ms. Monika Ivanovic from Infineon Technologies for project management, communication, and marketing activities.

The authors would also like to thank our external partners and key component suppliers, i.e., Shinenergy Technology Co., Ltd. for providing customized magnetic components, Denka Co., Ltd., T-Global Technology Co., Ltd., Shenzhen

Aochuan Technology Co., Ltd., and Shin-Etsu Chemical Co., Ltd. for providing gap-pad samples.

References

[1] M. Schweizer and J. W. Kolar, "Design and Implementation of a Highly Efficient Three-Level T-Type Converter for Low-Voltage Applications," in *IEEE Transactions on Power Electronics*, vol. 28, no. 2, pp. 899-907, Feb. 2013.

[2] T. Bruckner, S. Bernet and H. Guldner, "The active NPC converter and its loss-balancing control," in *IEEE Transactions on Industrial Electronics*, vol. 52, no. 3, pp. 855-868, June 2005.

[3] A. Nabae, I. Takahashi and H. Akagi, "A New Neutral-Point-Clamped PWM Inverter," in *IEEE Transactions on Industry Applications*, vol. IA-17, no. 5, pp. 518-523, Sept. 1981.

[4] D. Floricau, E. Floricau and M. Dumitrescu, "Natural doubling of the apparent switching frequency using three-level ANPC converter," *2008 International School on Nonsinusoidal Currents and Compensation*, Lagow, Poland, 2008, pp. 1-6.

[5] Y. Jiao and F. C. Lee, "New Modulation Scheme for Three-Level Active Neutral-Point-Clamped Converter with Loss and Stress Reduction," in *IEEE Transactions on Industrial Electronics*, vol. 62, no. 9, pp. 5468-5479, Sept. 2015.

[6] Y. Deng, J. Li, K. H. Shin, T. Viitanen, M. Saeedifard and R. G. Harley, "Improved Modulation Scheme for Loss Balancing of Three-Level Active NPC Converters," in *IEEE Transactions on Power Electronics*, vol. 32, no. 4, pp. 2521-2532, April 2017.

[7] D. Zhang, J. He and S. Madhusoodhanan, "Three-Level Two-Stage Decoupled Active NPC Converter with Si IGBT and SiC MOSFET," in *IEEE Transactions on Industry Applications*, vol. 54, no. 6, pp. 6169-6178, Nov.-Dec. 2018.

[8] L. Zhang, X. Lou, C. Li, F. Wu, Y. Gu, G. Chen and D. Xu, "Evaluation of Different Si/SiC Hybrid Three-Level Active NPC Inverters for High Power Density," in *IEEE Transactions on Power Electronics*, vol. 35, no. 8, pp. 8224-8236, Aug. 2020.

[9] M. Chen, D. Pan, H. Wang, X. Wang and F. Blaabjerg, "Investigation of Switching Oscillations for Silicon Carbide MOSFETs in Three-Level Active Neutral-Point-Clamped Invert-

ers," in *IEEE Journal of Emerging and Selected Topics in Power Electronics*, vol. 9, no. 4, pp. 4839-4853, Aug. 2021.

[10] Jih-Sheng Lai and Fang Zheng Peng, "Multilevel converters-a new breed of power converters," in *IEEE Transactions on Industry Applications*, vol. 32, no. 3, pp. 509-517, May-June 1996.

[11] H. Gui et al., "Modeling and Mitigation of Multiloops Related Device Overvoltage in Three-Level Active Neutral Point Clamped Converter," in *IEEE Transactions on Power Electronics*, vol. 35, no. 8, pp. 7947-7959, Aug. 2020.

[12] B. Liu et al., "Effects of Junction Capacitances and Commutation Loops Associated with Line-Frequency Devices in Three-Level AC/DC Converters," in *IEEE Transactions on Power Electronics*, vol. 34, no. 7, pp. 6155-6170, July 2019.

[13] M. Najjar, A. Kouchaki, J. Nielsen, R. Dan Lazar and M. Nymand, "Design Procedure and Efficiency Analysis of a 99.3% Efficient 10 kW Three-Phase Three-Level Hybrid GaN/Si Active Neutral Point Clamped Converter," in *IEEE Transactions on Power Electronics*, vol. 37, no. 6, pp. 6698-6710, June 2022.

[14] Infineon Technologies, "AIMDQ75R016M1H – MOSFET – CoolSiC™ Automotive Power Device 750V G1," AIMDQ75R016M1H datasheet, Aug. 2023.

[15] Infineon Technologies, "AIMDQ75R040M1H – MOSFET – CoolSiC™ Automotive Power Device 750V G1," AIMDQ75R040M1H datasheet, Aug. 2023.

[16] Infineon Technologies, "IPDQ60R040S7 – MOSFET – 600V CoolMOS™ SJ S7 Power Device," IPDQ60R040S7 datasheet, Aug. 2021.

[17] M. Chen, M. Escudero, M.-A. Kutschak, "Differential-Mode Inductance Integration with Common-Mode EMI Filter," *in Proc. 2024 IEEE Applied Power Electronics Conference and Exposition (APEC)*, Long Beach, CA, USA, 2024, IS 12-4.

[18] "Innovative top-side cooled package solution for high-voltage applications – Application considerations for best performance," Appl. Note AN_2101_PL52_2103_112902, pp.1-30.

[19] Infineon Technologies, "IPDQ60T017S7 – MOSFET – 600V CoolMOS™ SJ S7 Power Device," IPDQ60T017S7 datasheet, Dec. 2023.

PCIM Europe 2024, 11– 13 June 2024, Nuremberg   DOI: 10.30420/566262054

A Study of Grid-Forming Inverter Control Strategy for Fault-Ride-Through Capability

Hirofumi Uemura[1] , Sachio Takano[1]
[1] Fuji Electric Co., Ltd., Japan

Corresponding author: Hirofumi Uemura, uemura-hirofumi@fujielectric.com
Speaker: Hirofumi Uemura, uemura-hirofumi@fujielectric.com

Abstract

This paper proposes a control strategy to give the GFM inverter high fault-ride-through capability. Supplemental control strategy for over current suppression and anti-out of synchronization with GFM inverter control are described. Proposed control strategy was implemented on a 2 kVA inverter hardware, and three different fault-ride-through tests were performed. As a result, the effectiveness of proposed control strategy was confirmed.

1   Introduction

In the 6th strategic energy plan[1], the Japanese government aims to reduce the country's greenhouse gas emissions by 46% by 2030 from the 2013 levels. In order to achieve that, there is an ambitious goal of 36-38% renewable energy resources integrated by 2030. Renewable energy includes not only photo voltaic (PV) and wind turbine (WT), but also hydropower (mainly pumped storage), geothermal, biomass, etc. PV and WT are connected to the grid by inverter, so called Inverter Based Resources (IBRs). With the ambitious goal, at least 20% of energy will be supplied by IBRs. While a large number of IBRs will be interconnected, the amount of interconnected synchronous generators (SGs) will be decreased, therefore, it is concerned that the frequency and voltage stabilities will be reduced due to the reduction of inertia and voltage source capability of SGs[2]. To address this issue, Virtual Synchronous Generator (VSG) control has been widely studied as a control strategy to provide virtual inertia or emulated inertia by IBR[3]. The purpose of these efforts is to provide a Grid-Forming (GFM) capability for IBRs, so called the GFM inverter.

One of the important capabilities of the GFM inverter is the Fault-Ride-Through (FRT). In literature [4], FRT capability test of the GFM inverter is conducted, however, it failed in some cases. For example, in Low-Voltage Ride-Through (LV-RT) test, over current that exceeding hardware rating

was supplied from the GFM inverter that was operating as a voltage source. As a result, the GFM inverter operation was terminated by overcurrent protection function that protects the equipment. It is required to add an Over Current Suppression (OCS) function in order to continue operation.

In addition, like SGs, there are cases where the GFM inverter (VSG) goes Out of Synchronization. In order to improve the synchronous stability of VSG, a Transient Damping Method (TDM) was proposed in literature [5]. It is also required to add an Anti-Out of Synchronization function in order to let IBRs stay synchronized with the grid even under OCS condition.

In this paper, supplemental control strategy for over current suppression and anti-out of synchronization with a GFM inverter control are proposed and described. Proposed control strategy is implemented on 2 kVA inverter hardware and experimental results on the effects of OCS and AOS functions for FRT capability are provided.

2   Control strategy for FRT with GFM inverter

Fig. 1 shows a system diagram including proposed control strategy for FRT capability with the GFM inverter. Using the three-phase instantaneous voltage and current measured at the Point of Measurement (POM), various measurement values that used in the control strategy are calculated through the measurement process.

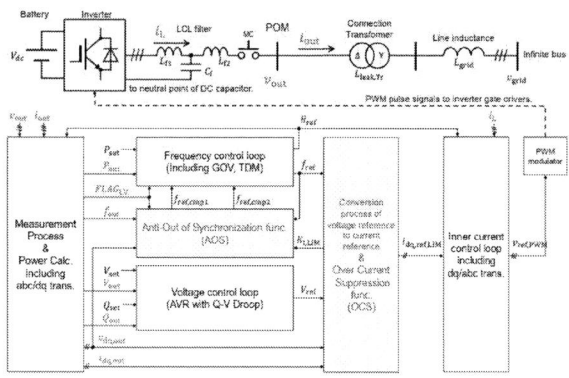

Fig. 1 System diagram of proposed control strategy.

2.1 Frequency control loop

Fig. 2 shows the contents of the frequency control loop. It includes the governor function (GOV) and TDM, and calculates the frequency reference value f_{ref} and the phase reference value θ_{ref} of the GFM inverter based on the SG's swing equation. Here, f_{n} is the nominal system frequency. Phase correction reference values, $f_{\text{ref,cmp1}}$ and $f_{\text{ref,cmp2}}$, calculated by the AOS function are shown here. Details of the AOS function is described in section 2.5.

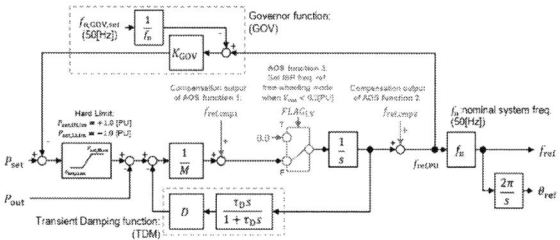

Fig. 2 Block diagram of frequency control loop.

2.2 Voltage control loop

Fig. 3 shows the contents of the voltage control loop. Automatic Voltage Regulator (AVR) with PI regulator and Q-V droop control are combined. It is also possible to operate the voltage control loop only with AVR (without Q-V droop). Here, the voltage magnitude reference value V_{ref} is calculated.

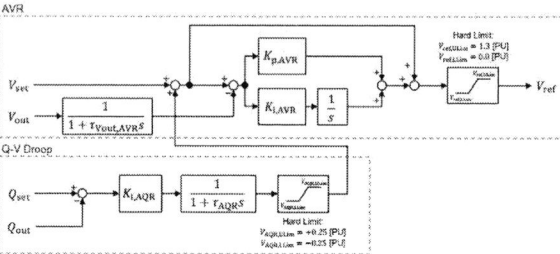

Fig. 3 Block diagram of voltage control loop.

2.3 Conversion process of voltage reference to current reference and OCS

Fig. 4 shows the contents of the conversion process of voltage reference to current reference. In this process, the basic current reference values $i_{\text{d,ref,0}}$ and $i_{\text{q,ref,0}}$ are obtained as follows.

$$i_{\text{d,ref,0}} = 0.95 \cdot i_{\text{d,out}} \tag{1}$$

$$i_{\text{q,ref,0}} = 0.95 \cdot i_{\text{q,out}} + \frac{K_{\text{Ci}} V_{\text{ref}} f_{\text{ref}}}{f_{\text{n}}} \tag{2}$$

Here, $i_{\text{d,out}}$ and $i_{\text{q,out}}$ are the output current measured at POM in the dq axis by dq conversion based on the phase reference θ_{ref}. In this process, the reactive current that consumed by output filter capacitor C_{f} is calculated and added in Eq. (2).

Furthermore, the voltage magnitude reference value of the GFM inverter V_{ref} is converted into the current reference value and added as follows.

$$i_{\text{d,ref}} = i_{\text{d,ref,0}} + \frac{V_{\text{ref}} - v_{\text{d,out}}}{K_{\text{p,ACR}}} \tag{3}$$

$$i_{\text{q,ref}} = i_{\text{q,ref,0}} - \frac{v_{\text{q,out}}}{K_{\text{p,ACR}}} \tag{4}$$

Here, $v_{\text{d,out}}$ and $v_{\text{q,out}}$ are the output voltage measured at POM in dq axis. $K_{\text{p,ACR}}$ is the proportional gain that used in the inner current control loop. The calculated current reference values $i_{\text{d,ref}}$ and $i_{\text{q,ref}}$ are passed to the OCS function.

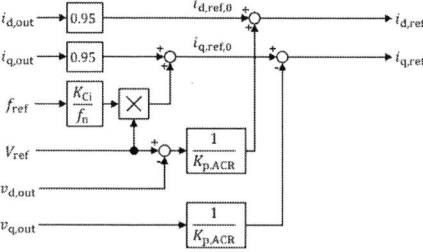

Fig. 4 Block diagram of conversion process of voltage reference to current reference.

Fig. 5 shows the contents of the OCS function. First, the current reference magnitude value $i_{s,ref}$ is calculated from $i_{d,ref}$ and $i_{q,ref}$. Then, the current reference suppression gain $K_{I,LIM}$ is calculated with the current reference limit value I_{LIM}. The current reference values are multiplied by $K_{I,LIM}$ in order to limit the current reference values within the settled limit value.

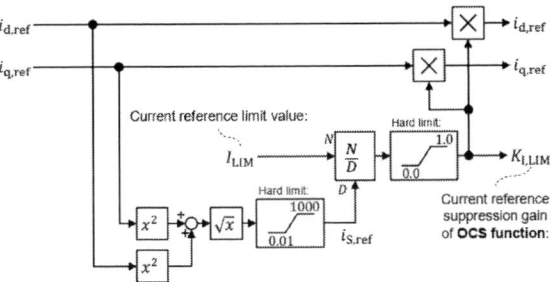

Fig. 5 Block diagram of OCS function.

Fig. 6 shows the relationship between the current reference magnitude value $i_{s,ref}$ and the current reference suppression gain $K_{I,LIM}$. If $i_{s,ref}$ is less than or equal to I_{LIM}, then $K_{I,LIM}$ is equal to one. Therefore, the current reference values are passed to the inner current control loop without being suppressed. When $i_{s,ref}$ exceeds I_{LIM}, then $K_{I,LIM}$ becomes less than one. As a result, the current reference values are suppressed.

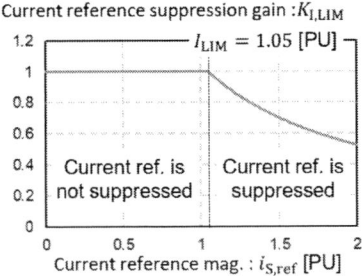

Fig. 6 Relationship between current reference

magnitude and suppression gain.

2.4 Inner current control loop

The inner current control loop is the same as the Automatic Current Regulator (ACR) used in general current control type inverters. Only propor-

tional control is used here. The instantaneous voltage reference value v_{pwm} for the PWM operation of the inverter unit is calculated. The calculation process of v_{pwm} in the dq axis are expressed as follows.

$$v_{d,pwm} = K_{p,ACR}(i_{d,ref,LIM} - i_{d,L}) + v_{d,out} \quad (5)$$

$$v_{q,pwm} = K_{p,ACR}(i_{q,ref,LIM} - i_{d,L}) + v_{q,out} \quad (6)$$

Here, $i_{d,L}$ and $i_{q,L}$ are the measured current of the main reactor L_{f1} of the LCL filter after dq conversion. When the current reference values $i_{d,ref,LIM}$ and $i_{q,ref,LIM}$ are not suppressed, Equation (5) and (6) become the following equations.

$$
\begin{aligned}
v_{d,pwm} &= K_{p,ACR}(i_{d,ref} - i_{d,L}) + v_{d,out} \\
&= K_{p,ACR}\left(i_{d,ref,0} + \frac{V_{ref} - v_{d,out}}{K_{p,ACR}} - i_{d,L}\right) + v_{d,out} \quad (7) \\
&= K_{p,ACR}(i_{d,ref,0} - i_{d,L}) + V_{ref}
\end{aligned}
$$

$$
\begin{aligned}
v_{q,pwm} &= K_{p,ACR}(i_{q,ref} - i_{d,L}) + v_{q,out} \\
&= K_{p,ACR}\left(i_{q,ref,0} - \frac{v_{q,out}}{K_{p,ACR}} - i_{d,L}\right) + v_{q,out} \quad (8) \\
&= K_{p,ACR}(i_{q,ref,0} - i_{d,L})
\end{aligned}
$$

When the current reference values are not suppressed, the voltage magnitude reference value V_{ref} appears in the instantaneous voltage reference value. Furthermore, the instantaneous voltage reference value v_{pwm} is obtained by inverse dq conversion from the values of equations (7) and (8) based on the phase reference value θ_{ref}. As a result, the output reference becomes the value of the voltage source. On the other hand, when the current reference values are suppressed, the instantaneous voltage reference value naturally becomes to the value of the current source, therefore, it is possible to seamlessly transition between the behavior of the voltage source and the behavior of the current source.

2.5 Anti-out of synchronization function

The AOS function prevents out-of-synchronization using the following three control strategies.

2.5.1 First strategy of AOS function

Fig. 7 shows the block diagram of first strategy of AOS function. Output signal $f_{ref,cmp1}$ consists of an element that follows the voltage phase at POM using only the proportional element of the Phase Locked Loop (PLL), and an element that follows the frequency at POM. $f_{ref,cmp1}$ is added into the freqeucny control loop shown in Fig. 2 in order to accelerates or decelerates the frequency reference of the GFM inverter. The aim is to follow the phase and frequency of the grid voltage to prevent out of synchronization. On the other hand, it is also a function to lose the voltage source behavior, such as inertial response and frequency response, therefore, it is designed to function only when OCS function is operating.

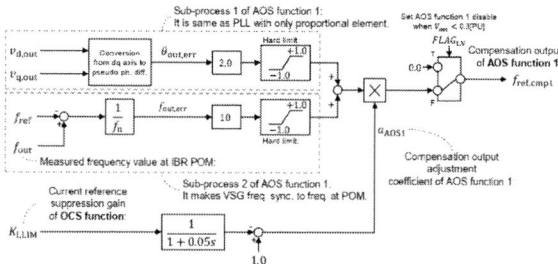

Fig. 7　Block diagram of first strategy of AOS function.

2.5.2　Second strategy of AOS function

Fig. 8 shows the block diagram of second strategy of AOS function. $f_{\text{ref,cmp2}}$ is calcluated by the phase difference between the measured output voltage at POM v_{out} and the reference voltage v_{pwm}. If the phase difference value is larger than the dead band, it compensates the frequency reference value. The phase difference detected here is the phase difference angle between the voltages v_{out} and v_{pwm} applied to both ends of the filter reactor ($L_{\text{f1}} + L_{\text{f2}}$). If this exceeds a certain value, it means that the reactor current i_{L} is also exceeds the rated value. In that case, it is likely to occur out of synchronization, therefore, second strategy of AOS function is applied in order to prevent that.

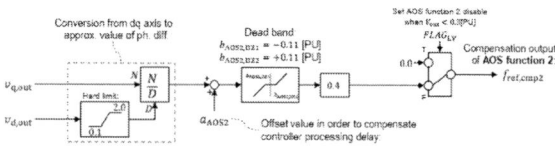

Fig. 8　Block diagram of second strategy of AOS function.

2.5.3　Third strategy of AOS function

The third strategy of AOS function is considered for the low grid voltage condition. If the measured voltage magnitude at POM V_{out} becomes less than 0.3 PU, it is determined that there is no other voltage source to synchronize with. In that case, the integrator that calculates the frequency reference value f_{ref} of the GFM inverter, is settled into the freewheeling mode by switching the input to zero as shown in Fig. 2. The frequency reference value becomes a constant value during the freewheeling mode.

3　Experimental results

3.1　Experimental set up

Fig. 9 shows the test circuit diagram. An AC voltage source that voltage and frequency can be controlled is used as the infinite bus. Between the inverter and the infinite bus, there are a filter reactor ($L_{\text{f1}} + L_{\text{f2}}$), leakage inductance $L_{\text{leak,Tr}}$ of the interconnection transformer, and line inductance L_{grid}. Summing up all of these, the reactance value X, which is unitized by the rated capacity and rated AC voltage of the inverter, is approximately 0.2 PU.

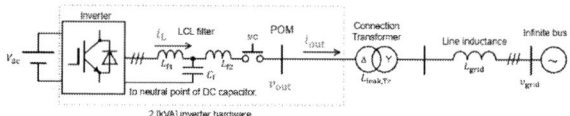

Fig. 9　Circuit diagram of experimental set up.

Proposed control strategy for GFM inverter is implemented on an inverter hardware with a rated AC voltage of 200 V and a rated capacity of 2 kVA. In order to confirm the FRT capability of proposed control strategy, Frequency Step Change RT (FSC-RT) test, Phase Step Change RT (PhSC-RT) test, and Low Voltage RT (LV-RT) test are conducted.

3.2　FSC-RT test result

In FSC-RT test, the frequency of the infinite bus is lowered stepwise from the nominal frequency of 50 Hz to 47 Hz, and returned to 50 Hz after 2.0 s. Here, very large values of the frequency fluctuation value and duration are experimentally tested. Table 1 shows the main control parameter settings of the inverter.

Output power setting	$P_{set} = 0.9$ [PU] $Q_{set} = 0.0$ [PU]
Frequency control	Inertia constant: $M = 4.7$ [s] TDM: $D = 15$ [PU], $\tau_D = 1.0$ [s] GOV: $K_{gov} = 25$ [PU]
Voltage control	Only AVR is in use $V_{set} = 1.0$ [PU]
OCS function	OCS is enabled $I_{LIM} = 1.05$ [PU]
AOS function	AOS functions are enabled
OVR/UVR	OVR: 1.15 [PU], 1.0 [s] UVR: 0.80 [PU], 1.2 [s]
OFR/UFR	OFR/UFR are disabled

Table 1 Control parameter settings for FSC-RT test.

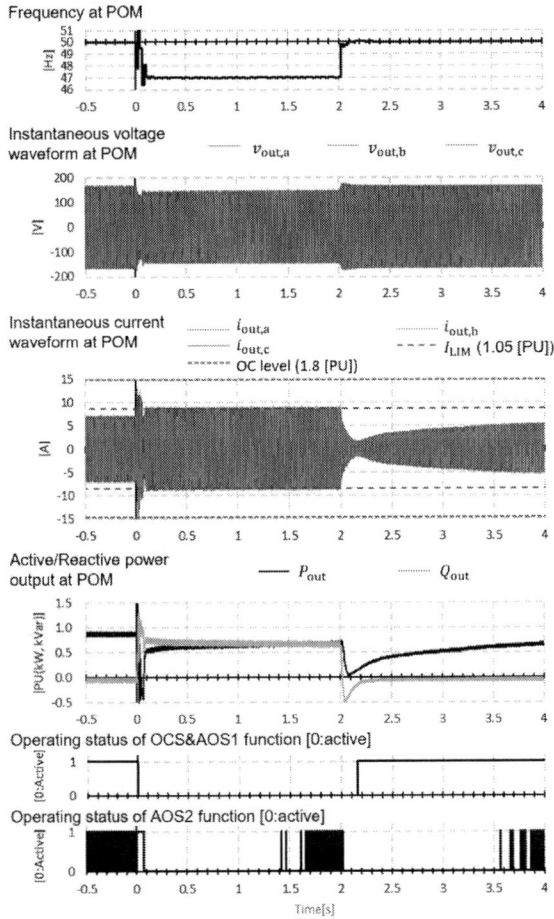

Fig. 10 FSC-RT test result.

Fig. 10 shows the FSC-RT test result. When the frequency drops to 47 Hz at Time = 0 s, the output current increases instantaneously due to the inertial response capability of the GFM inverter's voltage source behaviour. It exceeds the OCS function limit value of 1.05 PU, therefore, the OCS function is activated and inverter continues to operate at a constant output current condition. The AOS functions were also activated at the same time, and inverter operation could continue without becoming out of synchronization. On the other hand, as the reactive current output increased in the direction of lowering the voltage, the voltage at POM slightly decreased, however, the UVR was not activated here. When the frequency returns to 50 Hz at Time = 2.0 s, the OCS function is deactivated, and at the same time, the inertial response and frequency response capabilities of the GFM inverter are also restored. Though, the output current is accompanied by a transient change based on the swing equation, it eventually returns to the state before the frequency drop and becomes a steady state.

3.3 PhSC-RT test result

In PhSC-RT test, the phase of the infinite bus is suddenly changed by -60 degrees. Here, a very large amount of phase jump was experimentally tested. The main control parameters of the inverter are the same as shown in Table 1.

Fig. 11 shows the PhSC-RT test result. When the phase of the infinite bus suddenly changes by -60 degrees at Time = 0 s, the output current increases instantaneously due to the inertial response capability of the GFM inverter. At this time, the output current exceeds the instantaneous overcurrent level of the device protection function, therefore, the gate blocking function that implemented on the inverter hardware, was activated. After that, for approximately until 0.1 s, in addition to the gate blocking function, the OCS function and the AOS function are also activated. After 0.1 s to 0.4 s, only the OCS function and the AOS function are activated and the inverter continues to operate at the current limit value of 1.05 PU. Approximately after 0.4 s, the OCS function and the AOS function are deactivated and the operation state returns to the state before the phase jump and becomes a steady state.

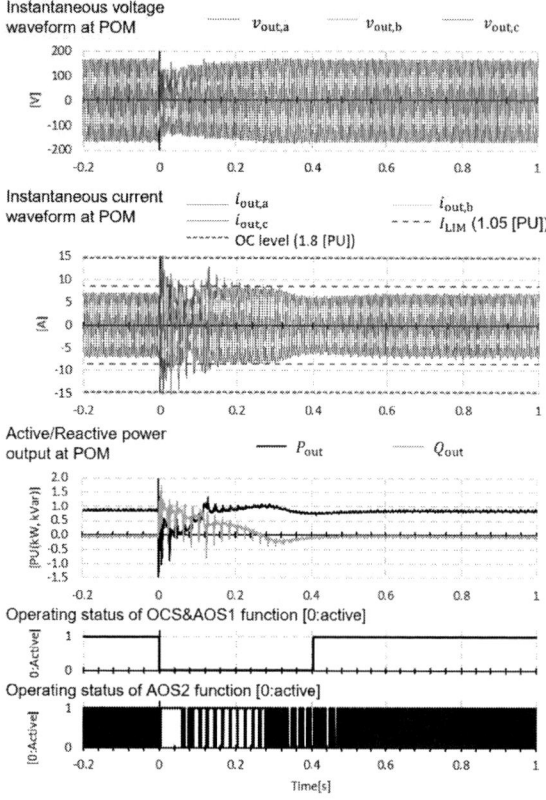

Fig. 11 PhSC-RT test result.

3.4 LV-RT test result

In the LV-RT test, the voltage magnitude of the infinite bus was decreased stepwise from 1.0 PU to 0 PU, and returned to 1.0 PU after 0.3 s. The main control parameters of the inverter are the same as in Table 1, however, the value of the active power output setting P_{set} was settled to 1.0 PU. Additionally, the second strategy of AOS function was not available for this test.

Fig.12 shows the LV-RT test results. When the voltage magnitude of the infinite bus decreases to 0 PU at Time = 0 s, the inverter output current rises instantaneously and exceeds the OCS function limit value of 1.05 PU, therefore, the OCS function is activated and the inverter continues to operate with a constant current output. At this time, the third strategy of AOS function is also activated at the same time and the frequency control of the inverter was operated in the freewheeling mode. Immediately after the voltage returns to 1.0 PU at Time = 0.3 s, the output active power and reactive power change greatly. This is due to the change in the phase difference between the grid voltage and

the inverter output voltage during the freewheeling mode. After Time = 0.6 s, the output active power returns to near the initial condition before the voltage fluctuation. At around Time = 0.93 s, the operation of the AOS function is deactivated, and the output state returns to the before the voltage drop, then becomes a steady state.

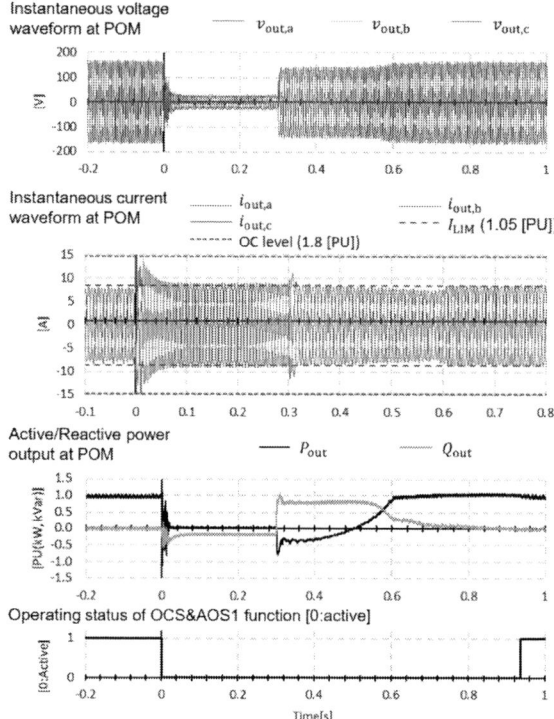

Fig. 12 LV-RT test result.

4 Conclusion

In this paper, we proposed supplemental control strategy for over current suppression and anti-out of synchronization in order to provide high FRT capability with GFM inverter. Proposed control strategy was implemented on 2 kVA inverter hardware and FSC-RT test, PhSC-RT test, and LV-RT test were performed. As results, it was confirmed that FRT capability for all test conditions. GFM inverters with high FRT capabilities are expected to contribute grid stabilization not only in normal operation condition, but also during and after abnormal condition.

References

[1] "The Sixth Strategic Energy Plan (Outline)", METI (Ministry of Economy, Trade and Industry, JAPAN), Updated on November 2021 (https://www.enecho.meti.go.jp/en/category/others/basic_plan/).

[2] A. Tayyebi, et. al., "Grid-Forming Converters – Inevitability, Control Strategies and Challenges in Future Grids Application", CIRED Workshop, Paper 0236, June 2018.

[3] A. Tayyebi, et. al., "Frequency Stability of Synchronous Machines and Grid-Forming Power Converters", IEEE Journal of Emerging and Selected Topics in Power Electronics, vol. 8, no. 2, pp.1004–1018, June 2020.

[4] H. Uemura, et. al., "Emulated Inertia Control of Grid-connected Inverter-based Power Supply Sources for Mass Integration of Renewable Energy Resources", IEEJ Journal of Industry Applications, Volume 12, Issue 3, Pages 368-375, May 2023.

[5] X. Xiong, et. al., "Transient Damping Method for Improving the Synchronization Stability of Virtual Synchronous Generators," in IEEE Transactions on Power Electronics, vol. 36, no. 7, pp. 7820-7831, July 2021.

PCIM Europe 2024, 11– 13 June 2024, Nuremberg   DOI: 10.30420/566262055

Film Capacitors for High Temperature AC-DC Inverter Applications

Adel Bastawros[1], Fumio Yu[2], Takeshi Horiguchi[3], Takashi Mori[3], and Kenichi Oshita[3]

[1] SABIC, USA

[2] SABIC, Japan

[3] Nichicon, Japan

Corresponding author and speaker: Adel Bastawros, Adel.Bastawros@SABIC.com

Abstract

High temperature commercial quality capacitors have been built and tested using newly introduced dielectric film that can operate at high temperatures reaching 150°C. At 900 volts and 150°C the capacitors passed 2000 hours of life testing and passed 3600 hours at 1000 volts and 130°C. Insulation resistance (IR), dielectric losses (tan δ) and equivalent series resistance (ESR) were stable over the test duration. A positive capacitance change (ΔC) was noted early in the test and remained acceptable at the end of the test.

Reaching 900V and 1000V performance was enabled by appropriate segmented metallization of the dielectric film and process optimization for building capacitor elements. Achieving 150°C performance was possible due to stable inherent dielectric characteristics (D_k and D_f) of the high heat polymer film. Operating at high temperature and high voltage is advantageous for efficiency improvement in AC-DC inverter modules, in particular those based on SiC technology. Additionally, the demand for active cooling can be reduced or eliminated giving rise to compact module design and bringing the capacitors closer to the SiC switch for lower induction losses. Compact designs may offer opportunity for weight reduction in electric vehicles.

1 Introduction

1.1 High Heat Performance, an Unmet Need

Societal demand for solutions to climate change is increasing. A key element for such solutions is switching the bulk of the internal combustion engine fleet to electric vehicles (EVs). For this big shift, perception of performance by consumers needs to change, particularly in range, acceleration and fast charging. EV companies want to move from conventional semiconductor technology, based on silicon, to wide band-gap (WBG) technology based on silicon carbide (SiC) and gallium nitride (GaN), to provide higher efficiency AC-DC inverter modules to deliver the desired performance improvements. These improvements are best realized when the WBG module operates under high voltages, frequencies and at higher temperatures. Components used in the module need to withstand the same operating conditions.

Film capacitors are a key passive component in the AC-DC inverter module. Incumbent capacitor technology relies typically on BOPP film (biaxially oriented polypropylene). BOPP is limited in its temperature handling ability to about 105°C along with implementation of active cooling of the electronic components. Active cooling systems consume valuable energy, add weight and require physical separation between components, which could lead to increase in induction losses. Other materials such as PET (polyethylene terephthalate) and PEN (polyethylene naphthalate) can survive higher temperatures. However, their use is limited to about 125°C due to excessive internal losses in the film at higher temperatures. Materials such as COC-PP blends (cyclic olefin copolymer - polypropylene) were recently introduced. However, these are still limited to about 125°C maximum operating temperature. Cooling systems or reducing the energy throughput is still necessary for capacitors made with PET, PEN and COC-PP films.

440

Adoption of Silicon Carbide (SiC) technology in AC-DC inverter modules in electric vehicles has signified the need for passive components, such as capacitors, that can operate at higher temperatures reaching 150°C. Operating at higher temperatures beyond the limits of incumbent BOPP film has been an ongoing industry and academic challenge [1,2,3].

1.2 High Heat Material Solution

SABIC has been developing new materials for use in ultra-thin film dielectric applications with operating temperatures upwards of 150°C [4]. The current work addresses this need and offers a solution of an ultra-thin high heat dielectric film for capacitors operating at high temperatures and high voltages. A high-heat engineering thermoplastic material was developed by SABIC to balance the often conflicting electrical and thermal requirements of the application and the challenge of processing into very thin film and optimizing performance in demanding downstream processes (e.g., metallization, slitting, capacitor winding, and heat treatment).

SABIC's ELCRES™ HTV150A film was engineered to provide a new generation of ultra-thin dielectric films and offer a different solution to meet the need for high temperature capacitor films [5]. The base resin in HTV150A film is an amorphous engineering thermoplastic resin with high glass transition temperature (T_g) of 205°C and relatively stable loss factor (D_f) at temperatures up to 150°C, avoiding the issues seen in some crystalline resins where the loss factor can change significantly at higher temperatures, or at temperatures near the glass transition temperature of the material. The polymer's aromatic carbon/aliphatic carbon ratio and net polarizability are balanced to deliver a higher dielectric constant than conventional BOPP while maintaining adequate self-clearing for many high voltage applications [6].

Reliability life-testing of HTV150A film has been previously demonstrated at conservative voltage of 100V/μm at 150°C for 2000 hours [5]. This corresponds to 500V for 5μm-based capacitors and 300V/μm for 3μm-based capacitors. The main objective of the current work is to demonstrate stable capacitor performance at a higher operating voltage up to 1000V, or 200V/μm for 5μm-based capacitors. Preliminary performance results for 3μm-based capacitors operating at 600V, or 200V/μm are also discussed.

1.3 Optimized Capacitor Design

Nichicon implemented high heat HTV150A films into capacitor designs to demonstrate performance at high temperatures and high voltages to offer candidate components to AC-DC module manufacturers.

Based on the dielectric, thermal and mechanical characteristics of the film, appropriate processing parameters were tuned to fit the thermo-mechanical characteristics of the HTV150A film. The film was metallized prior to capacitor building in standard metallization schemes. Segmentation (patterning) of the metallization was employed to maximize the operating voltage. Segmented metallization offers a mechanism to eliminate (electrically isolate) regions (i.e., segments) of the metallized surface where an electrical breakdown can occur. Similar to blowing a fuse, the affected segment is disconnected from the rest of the metallized surface whenever an imperfection within the segment causes local current increase. An optimum segmentation design must have appropriate responsivity to local current spikes, but not be overly sensitive to avoid excessive premature isolation of segments and large reductions in capacitance.

2 Dielectric Film Characteristics

In the current work, HTV150A dielectric film made with an advanced high-heat engineering thermoplastic material is selected for building high heat temperature capacitors. The film has desirable permittivity and dissipation losses [5]. Fig.1 shows D_k at room temperature and at 150°C as a function of frequency with (a) at room temperature and (b) at 150°C, respectively. HTV150A has a D_k of 2.9 that is very stable over the temperature range. By comparison, PEN has similar D_k performance whereas BOPP is 30% lower.

At 150°C, HTV150A and PEN retain their D_k performance. BOPP has no D_k value since the material will degrade at 150°C.

Fig. 1 (c and d) shows dissipation losses, represented by D_f, at room temperature and 150°C, respectively. Dissipation losses are lower for HTV150A film vs PEN at room temperature. At 150°C HTV150A film maintains stable performance, whereas losses in PEN increase. The increased losses limit use of PEN films beyond 125°C. BOPP has low losses at low temperatures but cannot survive higher temperature levels.

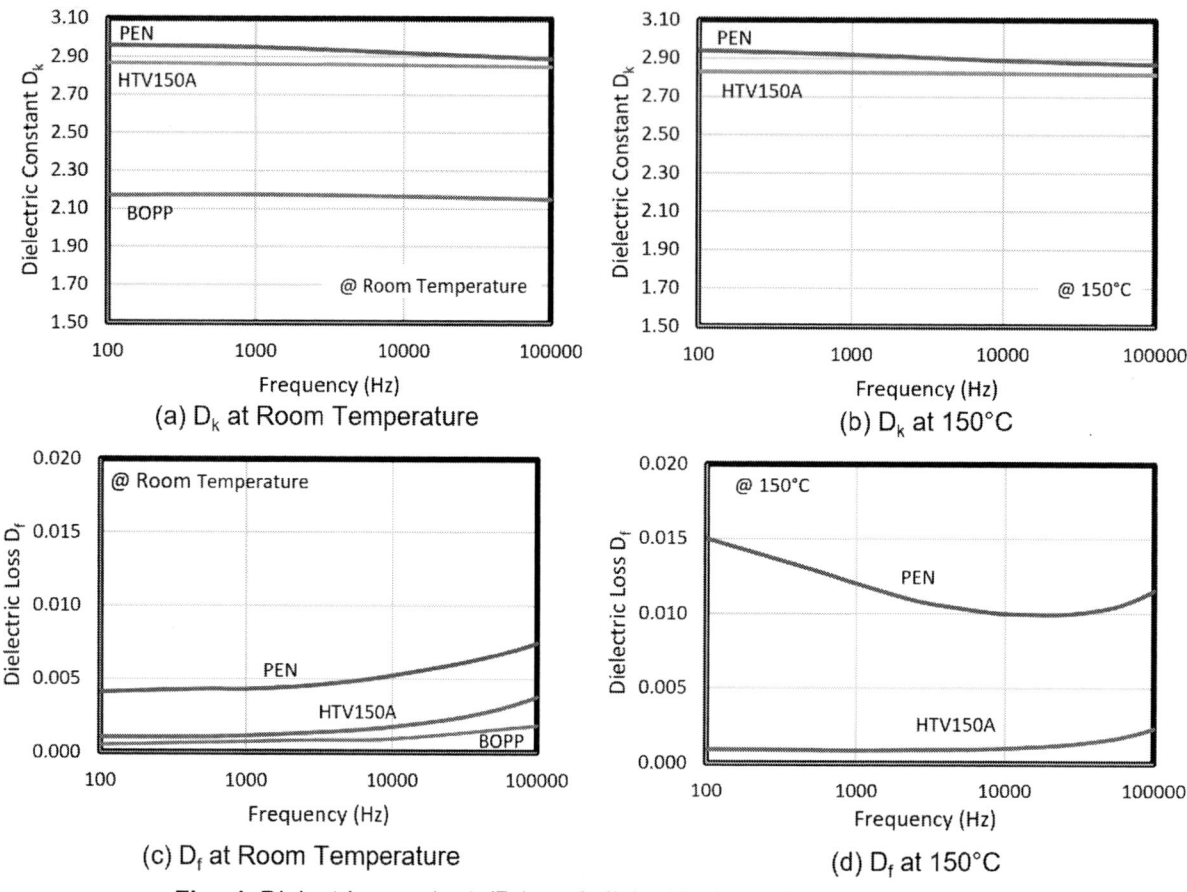

Fig. 1 Dielectric constant (D_k) and dielectric loss (D_f) for HTV150A film

Breakdown voltage (BDV) is measured according to ASTM D149 on unmetallized film. Fig. 2 shows HTV150A film having a highest BDV of about 850 V/µm at room temperature and 720 V/µm at 150°C. PEN and BOPP have lower values; noting that no value exists for BOPP at 150°C and the test for PEN is performed up to 125°C since losses limit use of PEN at higher temperatures.

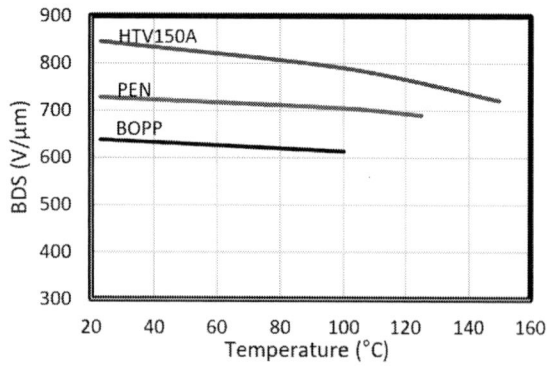

Fig. 2 Breakdown strength (V/µm)

3 Capacitor Building and Testing

Metallized 5µm and 3µm films were used to build test elements. Metallized film has 20Ω/sq active electrode, 5Ω/sq heavy edge and constant width margin. Round elements were made from the film using standard capacitor winding equipment following process parameters suited for HTV150A film characteristics. The round elements were flattened to an oval shape, followed by heat conditioning to stabilize the geometry and tighten the elements. Thermal end-spray and soldered terminals were applied to provide electrical connectivity. High temperature epoxy potting was applied to all elements to create finished capacitors for testing. Reference characteristics for each capacitor were recorded, including initial capacitance, internal losses (tan δ), insulation resistance (IR) and equivalent series resistance (ESR).

4 Capacitor Performance

To determine appropriate operating voltage, voltage stress tests were performed at temperatures of interest. Reliability life-testing is then performed at the operating voltage and temperature.

4.1 Voltage Stress Testing

A group of six similar 5µm-based capacitors was used to determine the operating voltage for each of 130°C and 150°C continuous temperature levels. Voltage was applied at 100V increments over 60 second steps during which change in capacitance was monitored. Maximum voltage for the capacitors is determined from the response charts as the voltage corresponding to 10% capacitance drop. Fig.3 (a and b) shows the voltage stress test results, averaged for the six capacitors, at 130°C and 150°C, respectively. The maximum voltage is 1650V at 130°C and 1550V at 150°C. Based on the stress test results operating voltages of 900V and 1000V were selected for the 5µm-based capacitors for 150°C and 130°C temperatures, respectively. Similarly, for the 3µm-based capacitors 600V operating voltage was determined for 150°C.

4.2 Reliability Testing (Life-Testing)

Reliability testing was performed on two groups of finished capacitors (5µm) subjected to constant voltage of 900V at 150°C and 1000V at 130°C. Typically the test is run for 2000 hours. In the current work the test was extended to 3600 hours for 130°C. Preliminary testing on a group of 3µm-based capacitors was done at 150°C and 600V for 1500 hours.

Throughout the test, change in capacitance (ΔC), insulation resistance (IR), dielectric losses (represented by tan δ) and equivalent series resistance (ESR) were monitored over the test duration.

For 130°C and 1000V, capacitance value over time is visualized in Fig. 4(a) for the 5µm-based capacitors. Performance is represented by the average value for the group of 10 capacitors; standard deviation is also shown. After the first 50 hours there is a slight increase (~5%) in capacitance. This is likely due to additional tightening of the wound element when subjected to initial heat; a known condition that can be mitigated by adjusting the winding and heat conditioning steps during capacitor building.

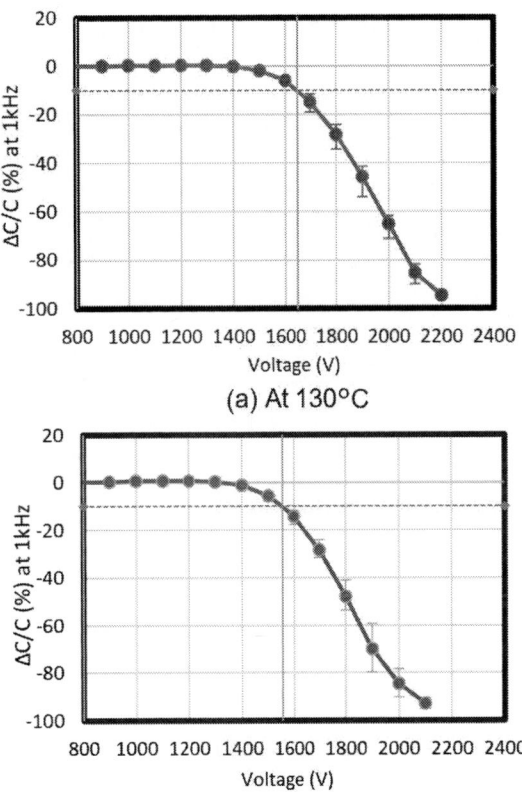

(a) At 130°C

(b) At 150°C

Fig. 3 Voltage Stress test on Capacitors made with HTV150A film

The change in capacitance over time, ΔC/C%, is relatively stable and remains positive (gain) as shown in Fig. 4(b). The change is within 10% threshold of the initial value. Excluding the initial capacitance increase, capacitance change during the remainder of the test (i.e., 3550 hours) is actually less than 5%.

Loss factor, tan δ, is shown in Fig. 4(c). Expectation is that tan δ remains less than twice the initial value over the duration of the test. Tan δ had very little change, from an average value of 0.07% at the beginning of the test to 0.09% after 3600 hours. Insulation resistance (IR) and equivalent series resistance (ESR) are shown in Fig. 4 (d and e). IR and ESR remained stable over the 3600 hours of testing.

Similarly, at 150°C and 900V over 2000 hours of life-testing, capacitance level had an initial increase of 8~10% and remained positive (gain) throughout the test (see Fig. 5 (a and b)). This initial increase is slightly higher than at 130°C,

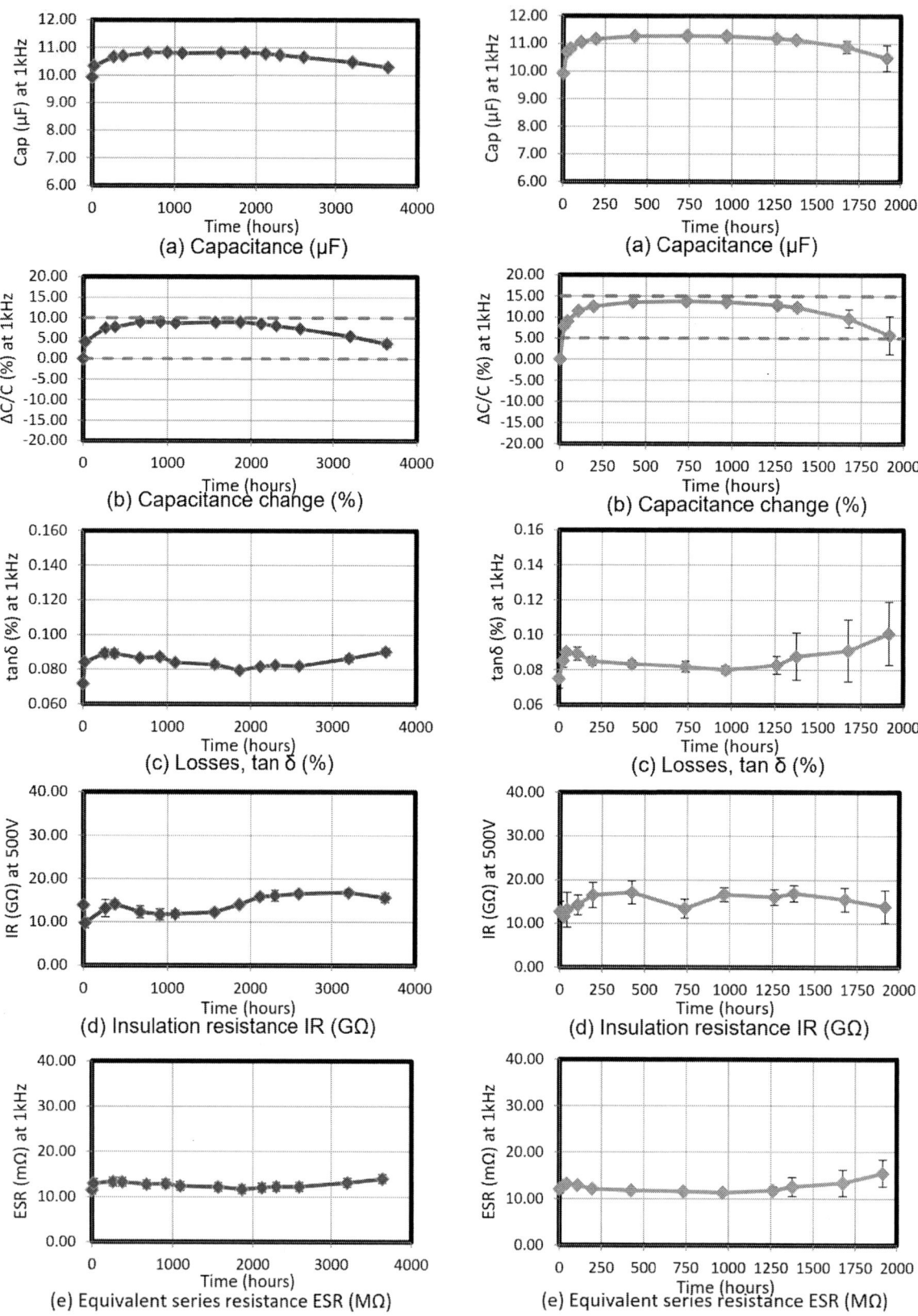

Fig. 4 Reliabilty life-testing at 130°C and 1000V for 5μm-based capacitors

Fig. 5 Reliabilty life-testing at 150°C and 900V for 5μm-based capacitors

likely due to more tightening of the wound element at a higher temperature. Excluding this initial increase, capacitance over the test duration was within 5% (gain).

Loss factor, tan δ, insulation resistance (IR) and equivalent series resistance (ESR), shown in Fig. 5 (c, d and e), remained stable within acceptable limits over the 2000-hour test duration.

For the 3µm-based capacitors the trends and stability over time are similar to 5µm-based capacitors. Fig. 6 (a) shows capacitance value over time at 150°C and 600V. Capacitance change (%) is shown in Fig. 6 (b). An initial gain in capacitance of ~15% is noted within the first 50 hours and remained stable thereafter. This initial increase is likely due to tightening of the wound element when heat is applied and can be mitigated through further tuning of the winding tension as well as heat conditioning after winding. Longer testing (>1500 hours) will be repeated on the modified capacitors. Excluding this initial increase, capacitance over the test duration was within 5% (gain). Loss factor, tan δ and equivalent series resistance (ESR), shown in Fig. 5 (c and d), remained stable within acceptable limits over the 1500-hour test duration.

Passing reliability testing offers new opportunities for the module designer to consider efficient module architectures whereby active cooling may be reduced or eliminated, components may be brought closer together for reduced induction losses and better thermal management or combining components together into integrated drives for efficiency improvements and possibilities for weight reduction.

5   Conclusions

High-heat ELCRES™ HTV150A dielectric films have been used to build high temperature capacitors. Under 1000V applied voltage, the 5µm-based capacitors passed accelerated reliability life testing at 130°C for 3600 hours and passed 2000 hours at 150°C under 900V. Early results for 3µm-based capacitors showed similar stable trend over 1500 hours of lie-testing under 600V at 150°C.

Excluding an initial increase in capacitance, likely caused by tightening of the film winding upon application of heat, the capacitance change ΔC remained within 5% throughout the test duration. Dissipation losses, tan δ, remained lower than twice the starting value; insulation resistance IR

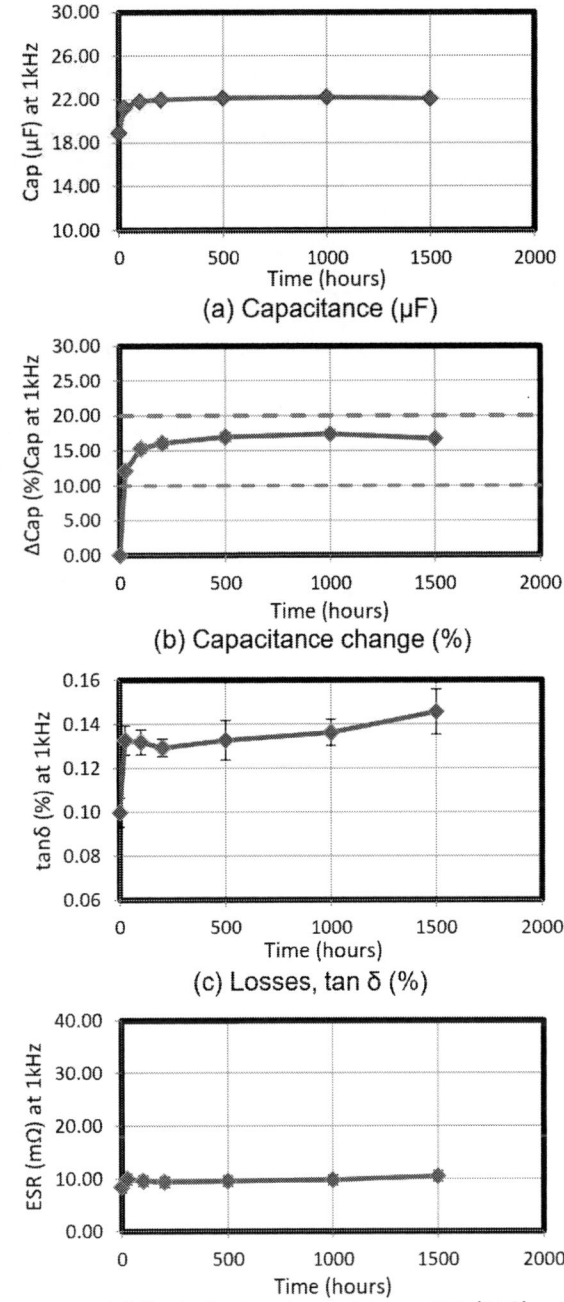

Fig. 6 Reliability life-testing at 150°C and 600V for 3µm-based capacitors

and equivalent series resistance ESR remained stable.

Capacitors made with HTV150A film are well positioned to help realizing full benefits of SiC and GaN MOSFETs when used in AC-DC inverters for EV applications.

6 Acknowledgements

The authors are indebted to Nichicon's and SABIC's Global Technology Teams for valuable insights on capacitor building and testing and on material and film characterizations. The examplary concerted efforts were key for the successful outcome of this work.

7 References

[1] L. Caliari, P. Bettacchi, E. Boni, D. Montanari, A. Gamberini, L. Barbieri, and F. Bergamaschi, "KEMET film capacitors for high temperature, high voltage and high current". Capacitor and Resistor Technology Symposium (CARTS) International Proc. ECA (Electronics Components, Assemblies & Materials Association), pp. 1-15, 2013.

[2] I. W. Clelland and R. A. Price, "Polymer Film Capacitors", APEC 2011, Special Session 1.3.4, pp. 2-4, 2011.

[3] D. Tan, L. Zhang, Q. Chen, et al, "High-Temperature Capacitor Polymer Films". Journal of Electronic Materials, Vol. 43, 4569–4575, 2014.

[4] N. Pfeiffenberger, F. Milandou, and M. Niemeyer, "High Temperature Polyetherimide Film Development," IEEE Trans. Dielectrics and Electrical Insulation Vol. 25, No. 1, pp 120-126, 2018.

[5] A. Bastawros, A. Pingitore, J. Mahood, M. Niemeyer, F. Yu and T. Sugawara, "New Generation Capacitor Films for 150°C High Voltage AC-DC Inverter Applications," PCIM Europe 2022;, pp. 1517-1523, 2022.

[6] A. Bastawros, A. Pingitore, C. Grabowski, P. Flanagan, and M. Buratto, "High Temperature Capacitor Films with Reduced Dissipation Losses for High Voltage AC-DC Inverters," proc. PCIM Europe 2023, pp. 1314-1317, 2023.

SABIC and brands marked with ™ are trademarks of SABIC. Any brands, products or services of other companies referenced in this document are the trademarks, service marks and/or trade names of their respective holders.

PCIM Europe 2024, 11– 13 June 2024, Nuremberg   DOI: 10.30420/566262056

Loss Reduction by Laminating Ferrite E Cores

Lukas Reißenweber[1], Julia Rogner[1], Alexander Stadler[1]

[1] Coburg University of Applied Sciences and Arts, Germany

Corresponding author: Lukas Reißenweber, lukas.reissenweber@hs-coburg.de

Abstract

Due to the increasing switching frequencies in power electronics, macroscopic eddy currents in ferrites and the associated losses have to be increasingly considered. Especially for cores with larger cross sections, the effect occurs already at lower frequencies. The reduction of eddy current losses in SiFe cores by lamination is already well known. This work applies the principle of lamination to ferrite cores of the sizes E70, E42 and E36 of the material MF106 (comparable to N97). The effect of lamination is investigated in the small-signal range using the loss tangent (tanδ) and at higher flux densities up to 500 kHz at 25 °C and 100 °C using rectangular voltage excitation.

1 Introduction

Due to the increasing switching frequencies in power electronics, passive components can be dimensioned smaller. This reduces the consumption of materials and thus also costs. However, it also poses new challenges for the engineers of magnetic components in order to get a design that is as compact and efficient as possible. Ferrite is usually used as the core material at higher frequencies, since it combines relatively good magnetic properties with low losses and is also cost efficient. Precisely because of the increasing frequencies, effects such as eddy currents, dimensional resonances and structural defects in ferrites are playing an increasingly important role [1]–[10]. Especially for large cores with large cross sections these effects have to be considered. The structure of isolated ferrite grains [11] suppresses macroscopic eddy currents. But the electric and dielectric properties of the grain isolation depend on frequency and temperature [3], [8], [12], [13]. The higher the frequency and the temperature the higher the conductivity. Due to the increased conductivity, macroscopic eddy currents occur more intensely throughout the core cross section and cause losses. The resulting fields are superimposed and lead to a kind of magnetic skin effect. At the same time, the high permeability together with the high permittivity lead to dimensional resonance effects. The resulting flux distribution in the core depends on the fields of the eddy currents and the dimensional resonance effects [1]–[9]. Which in turn depend on the ge-

ometry of the core and its frequency, temperature, field strength dependent magnetic, electric and dielectric properties. One possibility to reduce the macroscopic eddy currents and its losses is the lamination of ferrite cores as it is well known from SiFe cores and already from ferrites in filter applications [3], [4], [10], [14]. The effects of laminating ferrite E cores of sizes E70, E42 and E36 of the material Manifer 106 (MF106) are investigated in this work. The material MF106 is comparable to e.g. N97.

2 Fundamentals

2.1 Principle of Lamination

Figure 1 a) shows the magnitude of the complex impedance normalized to the value at $f = 10\,\text{kHz}$ and b) the argument of different sized E cores made of MF106 at $\vartheta = 25\,°\text{C}$. It can be seen that due to the reasons mentioned before, the largest core already loses impedance at lower frequencies. Also, the argument of the larger core first falls to zero and below.

A flux distribution in an E70 core at $f_s = 300\,\text{kHz}$ and $\vartheta_s = 100\,°\text{C}$ considering the fields of the eddy currents and the dimensional resonance effects is illustrated in fig. 2 a) for a bulk core with a thickness of 32 mm and in b) for a laminated core made of 8 layers. The layer thickness is 4 mm and the distance between them is 0.5 mm. It shows the distribution of the magnitude of magnetic flux density $|\underline{B}|$ in one core half normalized to the highest occurring flux density $|\underline{B}_{\text{bulk}}(y = 0mm)|$ of the cross section of the bulk core. The values

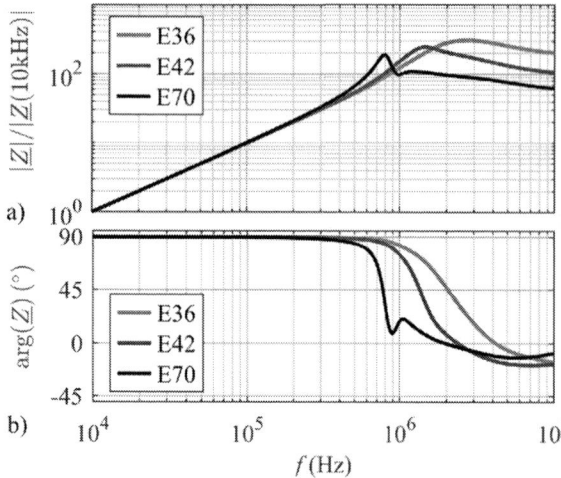

Fig. 1: a) Normalized magnitude of complex impedance and b) argument of E cores of the sizes E36, E42 and E70 made of MF106 at $\vartheta = 25\,^\circ\mathrm{C}$.

of the real $\mu_\mathrm{r}'(fs,\vartheta_\mathrm{s}) = 2845$ and the imaginary part $\mu_\mathrm{r}''(fs,\vartheta_\mathrm{s}) = 111$ of the relative permeability, the frequency dependent electrical conductivity $\sigma + \omega\varepsilon''(fs,\vartheta_\mathrm{s}) = 1.2\,\mathrm{A/(Vm)}$ and the real part of the relative permittivity $\varepsilon_\mathrm{r}'(fs,\vartheta_\mathrm{s}) = 101 \cdot 10^3$ are used. The values are based on measurements on toroidal cores made of MF106 ($d_\mathrm{a} = 24.9\,\mathrm{mm}, d_\mathrm{i} = 14.6\,\mathrm{mm}, h = 9.9\,\mathrm{mm}$) with measurement set-ups according to [8], [13], [14] and subsequent correction of the material data regarding field displacement effects within the toroidal cores according to [8]. It can be seen, that in the middle limb of the bulk core the flux density is significant increased in the center and decreased in the outer region. In the laminated core the distribution is more homogeneous. Such distributions are experimentally verified on toroidal cores by measurements in [3], [4]. The effect is even more significant for cores with larger cross sections [10], [14]. In many applications, this means that the overall flux density in the core cannot be particularly high in order to keep losses low. Therefore, the core is not well utilized in terms of power density. In fig. 3 the associated flux density, is depicted for the x- and z-direction at $y = 0\,\mathrm{mm}$ normalized to the maximum flux density of the cross section of the bulk core. For the x-direction of the laminated core, the cut line was placed through the center of one of the inner layers $z = 2.25\,\mathrm{mm}$. It can be seen, that in the bulk core the flux density in the outer regions of the middle limb is lower. In the inner area it is increased. Due to the smaller cross-sectional

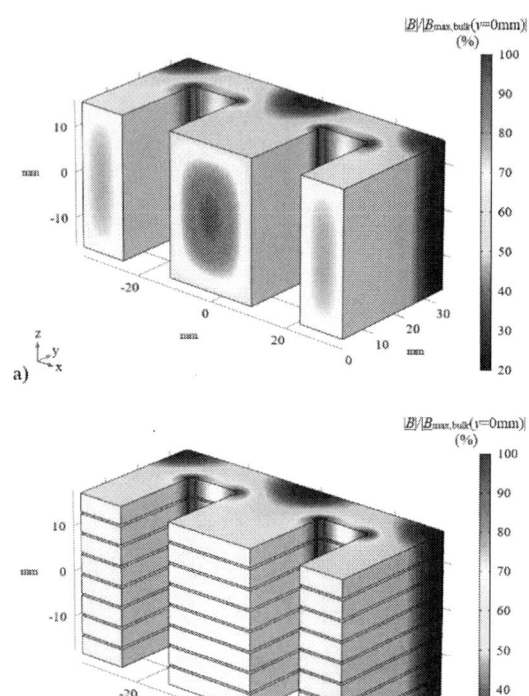

Fig. 2: Distribution of normalized magnitude of flux density in one core half of an ungapped a) bulk and b) laminated E70 core at $f_\mathrm{s} = 300\,\mathrm{kHz}$ and $\vartheta_\mathrm{s} = 100\,^\circ\mathrm{C}$.

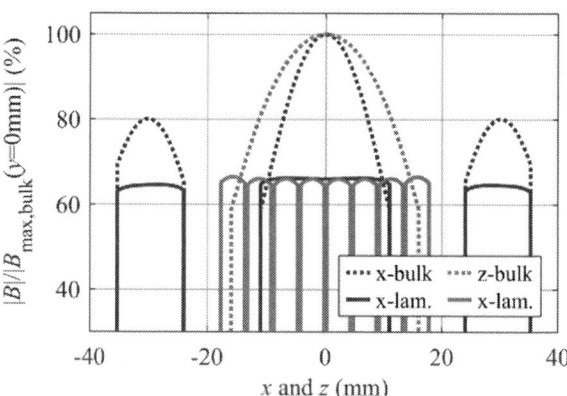

Fig. 3: Distribution of the normalized magnetic flux density in the cross section in x- and z-direction of a bulk and a laminated E70 core at $f_\mathrm{s} = 300\,\mathrm{kHz}$ and $\vartheta_\mathrm{s} = 100\,^\circ\mathrm{C}$.

areas, this effect is significantly reduced in the laminated core. This is shown by the more uniformly distributed flux density. This makes it possible to increase the total flux density and thus reduce the core cross section and volume. The smaller circumference also reduces the winding length and thus the losses and costs. Calculations of core loss and structural effects are provided in [6]–[8], [11], [15], [16]. Due to the reduction of eddy currents and the associated fields, laminated cores have a higher impedance at higher frequencies. This is particularly interesting for filtering EMI. Temperature and frequency increase the effect of macroscopic eddy currents. Figure 4 depicts the frequency depended conductivity $\sigma + \omega\varepsilon''$ of the material MF106 at different temperatures calculated from measured initial values [8]. This shows that the eddy currents are more pronounced at the temperatures at which the ferrite is used in the application.

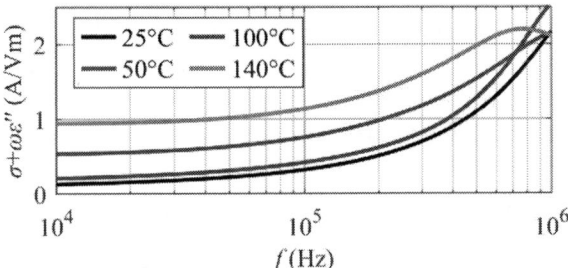

Fig. 4: Frequency dependent conductivity $\sigma + \omega\varepsilon''$ of the material MF106 at different temperatures.

2.2 Laminated E Cores

For the investigation E cores of the sizes E36, E42 and E70 of the material MF106 are used. The layers for the laminated cores were produced by grinding down bulk cores symmetrical on both sides. Also the chamfers have been removed from the original (bulk) E70 core. The machining of the cores and also the position from which the layer is produced has an influence on the magnetic material properties [14]. For the insulation of the layers from each other 0.5 mm thick plates of FR4 are inserted between them. Figure 5 shows an overview of the cores of the respective sizes E36, E42 and E70 and layer thicknesses. Table 1 provides an overview of all investigated core samples and their number. $N_{\text{lam.}}$ is the number of layers and $h_{\text{lam.}}$ the layer thickness. The parameter l_{e} stands for the effective core length and A_{e} for the effective cross section of the core. The values were taken from the data sheets and the effective cross sections were scaled according to the thickness of the cores. Figure 6 shows a laminated E70 core with FR4 elements for positioning in a clamping device. The

Fig. 5: Bulk and laminated E36, E42 and E70 cores of material MF106.

inner FR4 elements have a thickness of 0.5 mm. Since the material values of ferrites are dependent on mechanical pressure, heat-resistant plastic compression springs ensure equal contact pressure for all samples. For the small-signal measurements, the cores were measured without an air gap. For the large-signal measurements a 75 μm thick foil was inserted into the outer legs.

Fig. 6: Laminated E70 core with FR4 elements for positioning the layers and compression springs for equal contact pressure in a clamping device.

2.3 Small-Signal Set-Up

The measurements for the comparison of the losses in the small-signal range have been made with the impedance analyzer E4990A from Keysight and the test fixture 16047E. Figure 7 a) shows a DUT in front of the impedance analyzer. Short/Open/Load compensation was performed on the impedance analyzer with the test fixture. For the impedance of the DUT $\underline{Z}_{\text{DUT}}$, the measured impedance $\underline{Z}_{\text{xm}}$ was additionally corrected by an open $\underline{Z}_{\text{o}}$ and a short $\underline{Z}_{\text{s}}$ measurement [17]:

$$\underline{Z}_{\text{DUT}} = \frac{\underline{Z}_{\text{s}} - \underline{Z}_{\text{xm}}}{\underline{Z}_{\text{xm}} - \underline{Z}_{\text{o}}} \underline{Z}_{\text{o}} \tag{1}$$

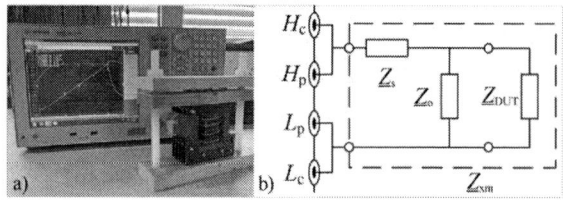

Fig. 7: a) DUT at impedance analyzer and b) ECM of residuals of the test fixture [17].

The equivalent circuit model for the residuals of the test fixture is depicted in fig. 7 b). The impedance of the measurement winding $\underline{Z}_{\mathrm{wire}}$ was subtracted from the measurement of the core with winding $\underline{Z}_{\mathrm{core+wire}}$ to obtain the impedance of the core $\underline{Z}_{\mathrm{core}}$:

$$\underline{Z}_{\mathrm{core}} = \underline{Z}_{\mathrm{core+wire}} - \underline{Z}_{\mathrm{wire}} \qquad (2)$$

As a result of the processing, the laminated cores can show a reduced real part of the complex permeability compared to the bulk cores [14]. A slight decrease was observed for cores E36 and E42. The comparison is therefore based on the loss tangent $\tan\delta$:

$$\tan\delta = \frac{\mathrm{Re}\{\underline{Z}_{\mathrm{core}}\}}{\mathrm{Im}\{\underline{Z}_{\mathrm{core}}\}} \qquad (3)$$

For the E70 cores there was no change in the real part of the permeability noticeable.

Core	$N_{\mathrm{lam.}}$	$h_{\mathrm{lam.}}$ (mm)	Nbr. of samples	l_e (mm)	A_e (mm²)
E70	1	30 (bulk)			640.3
	2	16		150	
	4	8			683
	8	4			
E42	1	15 (bulk)			
	2	7.5			
	3	5	3	97.4	178
	5	3			
	10	1.5			
E36	1	11 (bulk)			120
	2	7.5			
	3	5		81	
	5	3			163.6
	10	1.5			

Tab. 1: Overview of investigated E core samples.

2.4 Loss Measurement at Higher Flux Densities

For the loss measurement at higher flux densities the two winding method is used. The measurements are carried out with a HDO6054B oscilloscope from Teledyne LeCroy. The primary current is measured with a Pearson current monitor 2877 and the secondary voltage with a PP024 voltage probe. The runtime compensation of the current monitor is determined by comparing a small and large-signal measurement of an air coil [8]. The circuit used is a SiC half bridge configuration that provides a symmetrical square wave voltage. The voltage steepness is reduced by adding capacitors $C_{\mathrm{par}} = 0.5\,\mathrm{nF}$ in parallel to the semiconductors and adjusting the dead time. An equivalent circuit diagram shows fig. 8 a) and b) the set-up (without oven). For the measurements at $\vartheta = 100\,^{\circ}\mathrm{C}$, the test samples were placed in an oven. The cables with BNC connection for the voltage measurement were not changed. They were positioned so that the probe sits outside the oven.

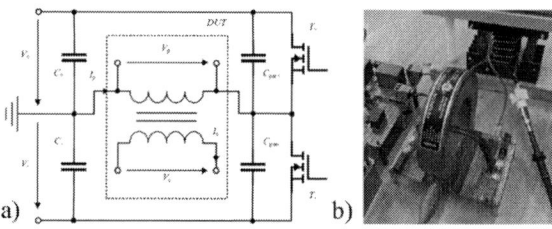

Fig. 8: a) ECM of used half bridge and b) set-up with DUT.

The magnetic field strength $H(t)$ is given by [8], [18], [19]:

$$H(t) = \frac{N_{\mathrm{p}}}{l_e}\left[i_{\mathrm{p}}(t) - \frac{N_{\mathrm{s}}}{N_{\mathrm{p}}}i_{\mathrm{s}}(t)\right] \qquad (4)$$

N_{p} is the number of turns of the primary winding and $i_{\mathrm{p}}(t)$ the primary current. N_{s} is the number of turns of the secondary winding. The winding ratio is $N_{\mathrm{s}}/N_{\mathrm{p}} = 2/4$. The current $i_{\mathrm{s}}(t)$ charges the capacitance of the probe $C_{\mathrm{probe}} = 12\,\mathrm{pF}$. Measurements have shown that the secondary current can be neglected. The magnetic flux density $B(t)$ depends on the secondary voltage $v_{\mathrm{s}}(t)$ [8], [18], [19]:

$$B(t) = \frac{1}{N_{\mathrm{s}}A_e}\int_0^T v_{\mathrm{s}}(t)\mathrm{d}t \qquad (5)$$

With the magnetic field strength and the magnetic flux density the mean core losses per volume $\bar{p}_{\mathrm{v}}$

can be calculated [8], [18]:

$$\bar{p}_v = \frac{1}{T} \int_0^T H(t) \cdot \frac{\delta B(t)}{\delta t} dt \tag{6}$$

To validate the measurements, all cores without an air gap were additionally measured with sinusoidal signals using a B-H Analyzer SY-8219 from Iwatsu at $\vartheta_s = 25\,°C$.

3 Measurement Results

3.1 Small-Signal Measurements

In fig. 9 a) the magnitude normalized to the 10 kHz value and in b) the argument of the bulk and the laminated E42 cores without an air gap are depicted. The magnitude of the impedance is maintained with decreasing film thickness up to higher frequencies. The argument shows that this is the inductive part of the component. This is interesting for filtering high-frequency noise. Since here at higher frequencies more reactive content contributes to the attenuation [3], [4], [10], [14]. The E36 cores show a comparable behavior. Here the effects of the lamination occur at even higher frequencies. In fig. 10 a) the

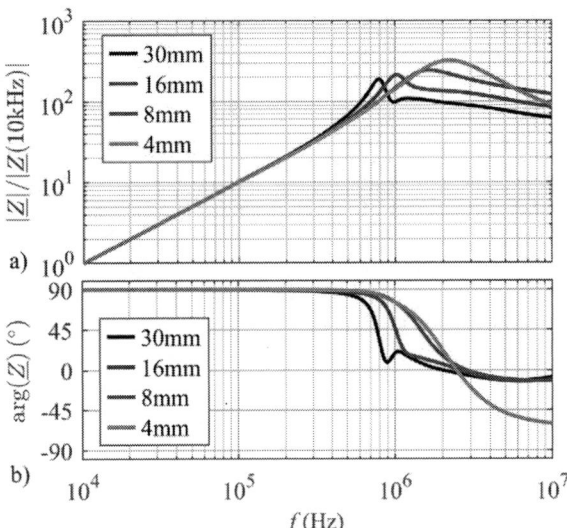

Fig. 10: Normalized magnitude of complex impedance and b) argument of E cores of the size E70 made of MF106 at $\vartheta = 25\,°C$.

magnitude and the argument decrease more than for the other cores and also below that.

For ferrites in power applications, the reduction of losses is of interest. Therefore, in the following the measurements of the cores are compared by the loss tangent from eq. 3. Figure 11 shows the loss tangent of the E36 and E42 cores with different layer thickness in the lower frequency range at $\vartheta_s = 25\,°C$. For the E36 cores, the reduction of the film thickness has no noticeable effect on the loss factor. The graphs of the E42 core show that halving the layer thickness from 15 mm to 7.5 mm leads to a slight reduction. The reduction to 5 mm halves the loss factor compared to the bulk core and brings it into line with the E36 core. Figure 12 depicts the loss factor for the E36 and E42 cores at higher frequencies at $\vartheta_s = 25\,°C$. Here it can be seen that the loss tangent is significant smaller as the layer thickness decreases. The loss tangent of E42 laminates is higher than that of E36 laminates with the same thickness. In the case of the E70 core a reduction takes place in the frequency range in which it is used. In fig. 13 the curves of the loss tangent are shown. In the range up to 700 kHz, this is reduced further and further as the layer thickness decreases. The values of 8 mm are close to those of 4 mm. Above 700 kHz, the 4 mm are slightly higher than those of 8 mm. Tab. 2 shows a comparison of the values at certain frequencies and the reduction of the loss tangent compared to the bulk core.

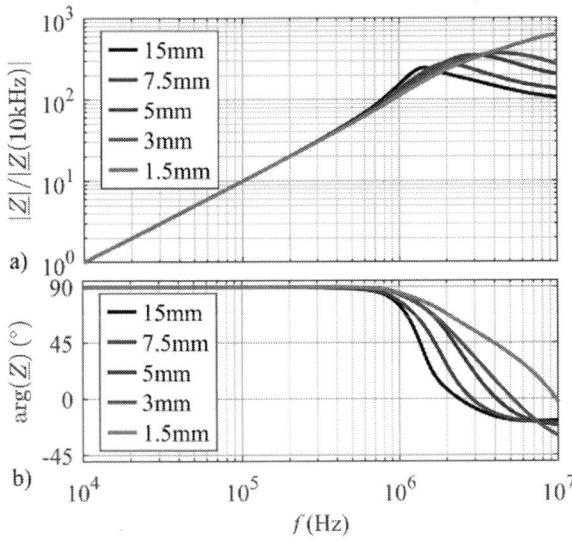

Fig. 9: a) Normalized magnitude of complex impedance and b) argument of E cores made of MF106 of the size E42 at $\vartheta = 25\,°C$.

magnitude normalized to the 10 kHz value and in b) the argument of the bulk and the laminated E70 cores without an air gap are depicted. Also, here the magnitude of the impedance is maintained with decreasing film thickness up to higher frequencies. For the core with a layer thickness of 4 mm, the

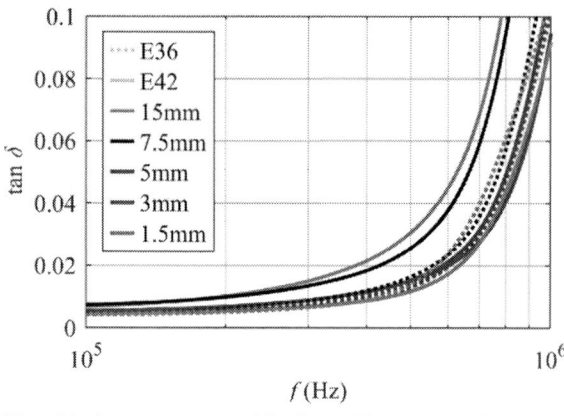

Fig. 11: Loss tangent of bulk and laminated E36 and E42 cores made of MF106 in the lower frequency range at $\vartheta = 25\,°C$.

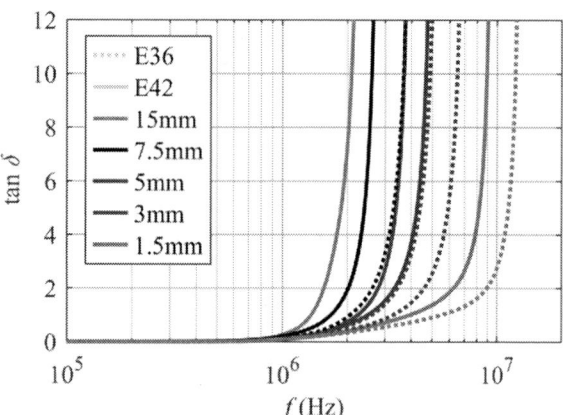

Fig. 12: Loss tangent of bulk and laminated E36 and E42 cores made of MF106 in the higher frequency range at $\vartheta = 25\,°C$.

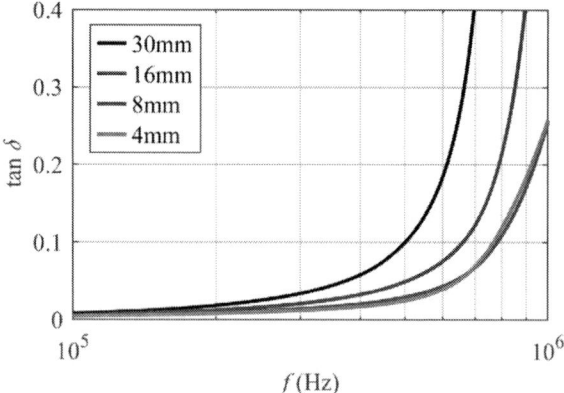

Fig. 13: Loss tangent of bulk and laminated E70 cores made of MF106 at $\vartheta = 25\,°C$.

3.2 Core Losses at Higher Flux Densities

The small-signal measurements are an indication of loss reduction using lamination. However, the

$h_{lam.}$ (mm)	30 (bulk)	16	8	4	16	8	4
f (kHz)	tanδ				tanδ reduction against bulk (%)		
100	0.008	0.006	0.005	0.005	25.0	37.5	37.5
200	0.018	0.012	0.009	0.008	33.3	50.0	55.6
300	0.034	0.021	0.014	0.012	38.2	58.8	64.7
400	0.058	0.032	0.020	0.017	44.8	65.5	70.7
500	0.099	0.049	0.029	0.024	50.5	70.7	75.8

Tab. 2: Loss tangent and reduction of it for E70 cores.

losses at higher flux densities are more important for the application. Figure 14 shows the measured primary current and the measured secondary voltage for the $\hat{B} = 50\,mT$ value of an E70 core with 8 layers at $f = 500\,kHz$ and $\vartheta = 25\,°C$. The slope of the primary voltage was adjusted so that there is minimum current drop (not visible in the figure) due to the parasitic capacitance at the switching moment. Overall, it has been shown that a thinner layer thickness is also associated with a lower current drop. The influence of charging and discharging the parasitics of the winding and the high permittivity core on the flux density and losses are discussed in [20]. The flux density associated with the

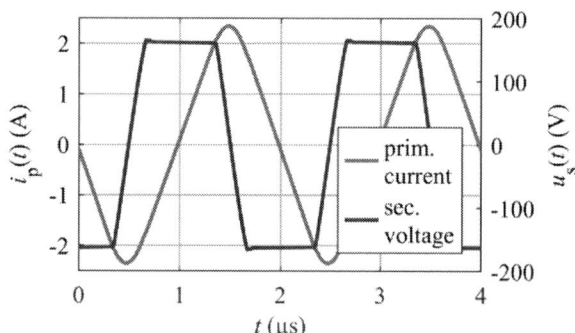

Fig. 14: Measured primary current and secondary voltage at E70 core with 8 layers at $\hat{B} = 50\,mT$, $f = 500\,kHz$ and $\vartheta = 25\,°C$.

voltage of fig. 14, calculated from Eq. 5 is shown in Fig. 15. Figure 16 compares BH-curves of selected test specimens of all layer thicknesses $f = 500\,kHz$ and $\vartheta = 25\,°C$. The bulging is reduced with decreasing layer thickness.

Figure 17 shows the measured power loss density per volume at $\vartheta = 25\,°C$. It can be seen that the power loss density decreases with decreasing layer thickness. More at high frequencies than at low

PCIM Europe 2024, 11– 13 June 2024, Nuremberg   DOI: 10.30420/566262056

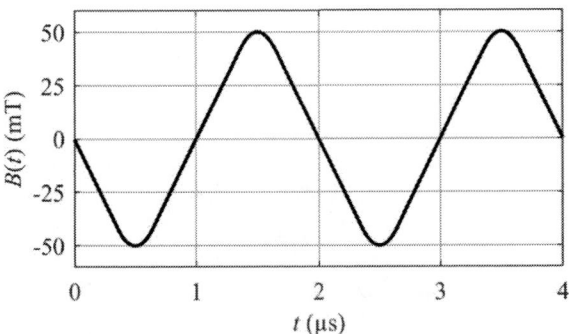

Fig. 15: Calculated flux density of E70 core with layer thickness 4 mm at $\hat{B} = 50\,\text{mT}$, $f = 500\,\text{kHz}$ and $\vartheta = 25\,°\text{C}$.

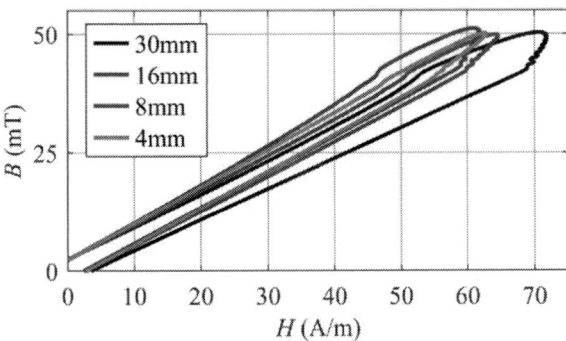

Fig. 16: BH-curves of bulk and laminated E70 cores at $\hat{B} = 50\,\text{mT}$, $f = 500\,\text{kHz}$ and $\vartheta = 25\,°\text{C}$.

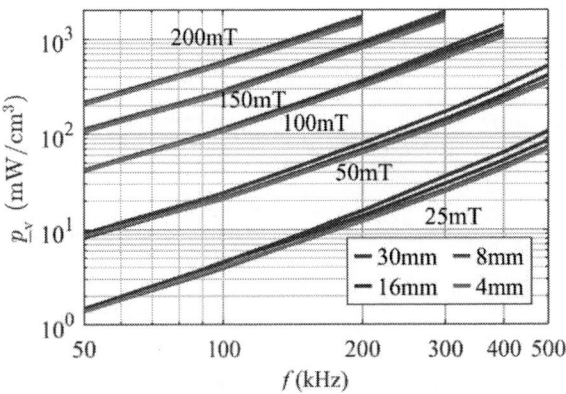

Fig. 17: Power loss density of bulk and laminated E70 cores made of MF106 at $\vartheta = 25\,°\text{C}$.

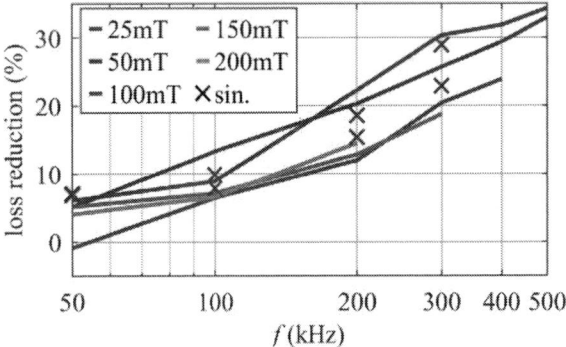

Fig. 18: Reduction in power loss density of the E70 core with 4 mm layers compared to the bulk core at $\vartheta = 25\,°\text{C}$.

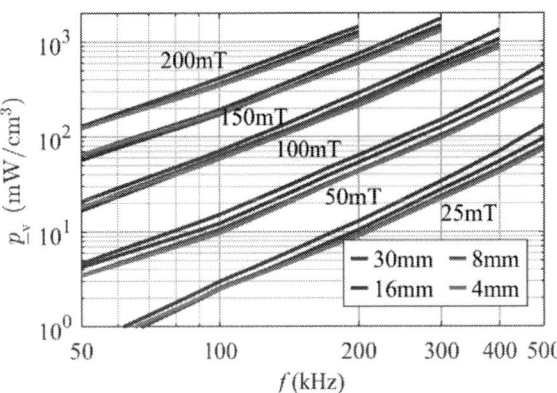

Fig. 19: Power loss density of bulk and laminated E70 cores made of MF106 at $\vartheta = 100\,°\text{C}$.

frequencies. The curves narrow towards the low frequencies. The calculated reduction of the loss density of the cores with 4 mm layers compared to the bulk cores is shown in fig. 18. It can also be seen here that the reduction increases with increasing frequency. This is somewhat more pronounced at low flux densities than at high flux densities. The results at $\hat{B} = 25\,mT$ and $\hat{B} = 50\,mT$ with sinusoidal signals are also shown. These are slightly lower but show a comparable trend.   In fig. 19 the measured power loss density per volume at $\vartheta = 100\,°\text{C}$ is shown. Also here the power loss density decreases with decreasing layer thickness. More at high frequencies than at low frequencies. Towards the low frequencies the curves narrow less than with the curves of $\vartheta = 25\,°\text{C}$. The calculated reduction in losses of the cores with 4 mm layers compared to the bulk core is shown in fig. 20. It behaves like the measurements at $\vartheta = 25\,°\text{C}$, but the reduction is even higher.

The reduction in power losses density for the different layer thicknesses at $f = 300\,\text{kHz}$ for $\vartheta = 25\,°\text{C}$ and $\vartheta = 100\,°\text{C}$ is shown in fig. 21. There is an almost linear relationship with the layer thickness.

The reduction is higher at $\vartheta = 100\,°\text{C}$ and decreases with increasing flux density at both temperatures.

Figure 22 shows the reduction of the power loss density for the different layer thickness at $\hat{B} = 50\,\text{mT}$ for $\vartheta = 25\,°\text{C}$ and $\vartheta = 100\,°\text{C}$. Figure 23 shows it for $\hat{B} = 150\,\text{mT}$. These confirm once

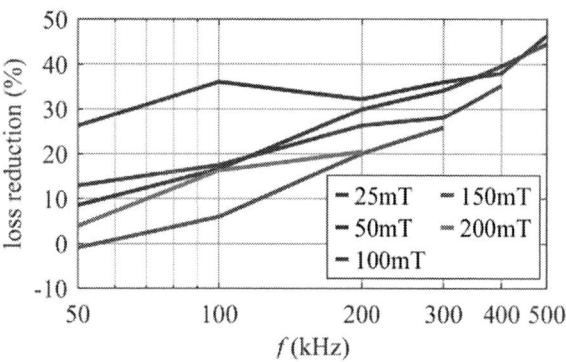

Fig. 20: Reduction in power loss density of the E70 core with 4 mm layers compared to the bulk core at $\vartheta = 100\,°C$.

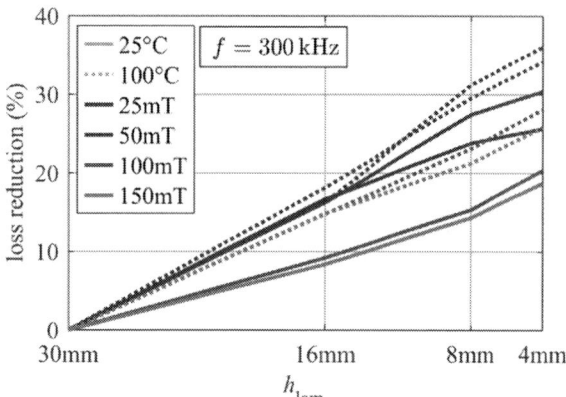

Fig. 21: Reduction in power loss density of E70 cores with different layer thickness for different flux densities at $f = 300\,kHz$ for $\vartheta = 25\,°C$ and $\vartheta = 100\,°C$.

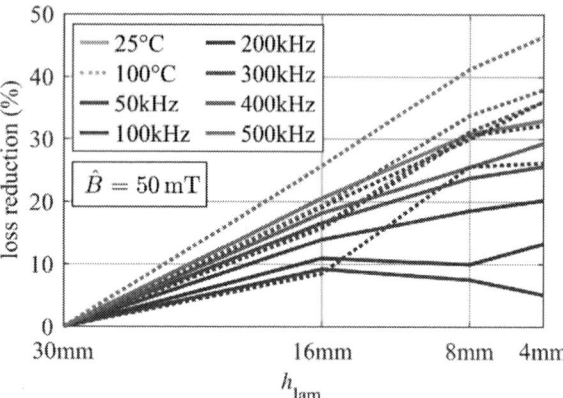

Fig. 22: Reduction in power loss density of E70 cores with different layer thickness for different frequencies at $\hat{B} = 50\,mT$ for $\vartheta = 25\,°C$ and $\vartheta = 100\,°C$.

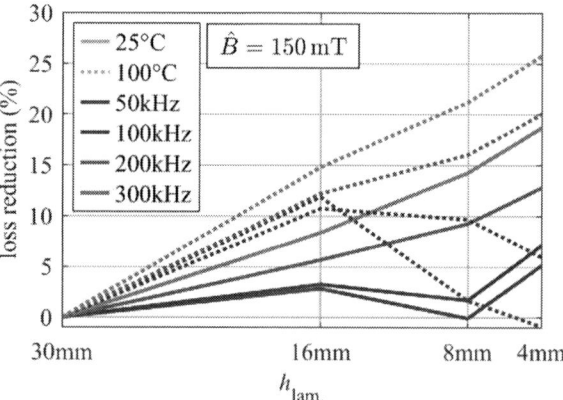

Fig. 23: Reduction in power loss density of E70 cores with different layer thickness for different frequencies at $\hat{B} = 150\,mT$ for $\vartheta = 25\,°C$ and $\vartheta = 100\,°C$.

again that the reduction increases with reduced layer thickness and increasing frequency and temperature. At the lower flux density in fig. 22 more than at the higher flux density in fig. 23.

3.3 Conclusion

Macroscopic eddy currents in ferrites are playing an increasingly important role with the increasing frequencies in power electronics. On E36, E42 and E70 cores of the material MF106, the effects of lamination on the impedance and especially on the loss factor were investigated in the small-signal range at room temperature. For all cores, a higher impedance at high frequencies could be realized with lower layer thicknesses. This means that high-frequency EMI can be better attenuated. In the case of the loss factor, the smaller E36 and E42 cores only show a significant reduction at higher frequencies. For the E70 cores, the loss factor

is already noticeable reduced in the usual operating range. The E70 cores were measured in terms of their power loss density at different frequencies, flux densities and temperatures. Overall, the losses of the E70 cores can be reduced further and further with decreasing layer thickness. At high frequencies and high temperatures more than at low frequencies. The reduction in losses could be confirmed for low flux densities and at room temperature using measurements with sine waves. Laminating ferrite cores can reduces the power loss and increases the impedance at high frequencies.

References

[1] G. Skutt, "High-Frequency Dimensional Effects in Ferrite-Core Magnetic Devices," *PHD Thesis, Virginia Polytechnic Institute and State University*, 1996.

[2] G. Skutt and F. Lee, "Characterization of dimensional effects in ferrite-core magnetic devices," in *PESC Record. 27th Annual IEEE Power Electronics Specialists Conference*, vol. 2, 1996, 1435–1440 vol.2. DOI: 10.1109/PESC.1996.548770.

[3] M. Kacki, "Investigation of the high-frequency effects in Mn-Zn ferrites for EMI filter applications," *PHD Thesis, University College Cork*, 2022.

[4] M. Kacki, M. S. Rylko, J. G. Hayes, and C. R. Sullivan, "A Study of Flux Distribution and Impedance in Solid and Laminar Ferrite Cores," in *2019 IEEE Applied Power Electronics Conference and Exposition (APEC)*, Anaheim, CA, USA: IEEE, Mar. 2019, pp. 2681–2687. DOI: 10.1109/APEC.2019. 8722252.

[5] M. Kacki, M. S. Rylko, J. G. Hayes, and C. R. Sullivan, "Magnetic core dimensional effects - flux propagation in ferrites," in *PSMA Workshop during IEEE Applied Power Electronics Conference*, 2018.

[6] W. Hauser, "Modellbildung für strukturabhängige Effekte in Ferritkernen," *PHD Thesis, Friedrich-Alexander-Universität Erlangen-Nürnberg*, 2018.

[7] W. Hauser and M. Albach, "Analytic model of structural effects in toroid cores with rectangular cross section," in *2016 6th International Electric Drives Production Conference (EDPC)*, Nuremberg, Germany: IEEE, Nov. 2016, pp. 60–66. DOI: 10.1109/EDPC.2016.7851315.

[8] A. Stadler, "Messtechnische Bestimmung und Simulation der Kernverluste in weichmagnetischen Materialien," *PHD Thesis, Friedrich-Alexander-Universität Erlangen-Nürnberg*, 2009.

[9] A. Stadler, M. Albach, and A. Bucher, "Calculation of Core Losses in Toroids with Rectangular Cross Section," in *2006 12th International Power Electronics and Motion Control Conference*, Portoroz: IEEE, Aug. 2006, pp. 828–833. DOI: 10.1109/EPEPEMC.2006.4778502.

[10] S. Takahashi, "Experimental investigation of the dimensional effect on small-signal characteristics of common-mode inductors," *IEEE Access*, vol. 10, pp. 123 068–123 079, 2022. DOI: 10.1109/ACCESS.2022.3223436.

[11] T. Dimier and J. Biela, "Eddy Current Loss Model for Ferrite Ring Cores Based on a Meta-Material Model of the Core Properties," *IEEE Transactions on Magnetics*, vol. 58, no. 2, pp. 1–5, Feb. 2022. DOI: 10.1109/TMAG.2021.3084812.

[12] M. Kacki, M. S. Rylko, J. G. Hayes, and C. R. Sullivan, "A Practical Method to Define High Frequency Electrical Properties of MnZn Ferrites," in *2020 IEEE Applied Power Electronics Conference and Exposition (APEC)*, New Orleans, LA, USA: IEEE, Mar. 2020, pp. 216–222. DOI: 10.1109/APEC39645.2020.9124101.

[13] A. Stadler, M. Albach, and A. Lindner, "A Practical Method to Measure Electrical AC Conductivity of MnZn Ferrites Using Conventional Toroids," *IEEE Transactions on Magnetics*, vol. 46, no. 2, pp. 678–681, Feb. 2010. DOI: 10.1109/TMAG. 2009.2030157.

[14] L. Reißenweber, F. Wohlrath, and A. Stadler, "Substitution of nanocrystalline toroid by laminated ferrite toroid in the application of a common-mode choke," in *2022 24th European Conference on Power Electronics and Applications (EPE'22 ECCE Europe)*, 2022, pp. 1–9.

[15] B. Wunsch, T. Christen, S. Skibin, and V. Forsstrom, "Broadband circuit model of a ferrite core, including dimensional resonance, saturation, and hysteresis," *IEEE Transactions on Magnetics*, vol. 55, no. 7, pp. 1–5, 2019. DOI: 10.1109/TMAG. 2019.2901216.

[16] M. Baumann, C. Drexler, J. Pfeiffer, J. Schueltzke, E. Lorenz, and M. Schmidhuber, "Investigation of core-loss mechanisms in large-scale ferrite cores for high-frequency applications," in *2022 24th European Conference on Power Electronics and Applications (EPE'22 ECCE Europe)*, 2022, pp. 1–10.

[17] Keysight, "Impedance measurement handbook 6th edition," 2020.

[18] E. C. Snelling, "Soft ferrites, properties and application," London: Iliffe Books Ltd., 1969.

[19] E. Stenglein, B. Kohlhepp, D. Kübrich, M. Albach, and T. Dürbaum, "Gan-half-bridge for core loss measurements under rectangular ac voltage and dc bias of the magnetic flux density," *IEEE Transactions on Instrumentation and Measurement*, vol. 69, no. 9, pp. 6312–6321, 2020. DOI: 10.1109/TIM.2020.2972140.

[20] D. Serrano, H. Li, S. Wang, T. Guillod, M. Luo, *et al.*, "Why magnet: Quantifying the complexity of modeling power magnetic material characteristics," *IEEE Transactions on Power Electronics*, vol. 38, no. 11, pp. 14 292–14 316, 2023. DOI: 10.1109/TPEL.2023.3291084.

PCIM Europe 2024, 11– 13 June 2024, Nuremberg   DOI: 10.30420/566262057

Multigap Toroidal Transformer and Inductors for Overcoming Fringing Losses in High Frequency Converters

Pau Colomer[1], Marc Maneja[2], David Prados[3]

[1] Prax, Spain
[2] Prax, Spain
[3] Prax, Spain

Corresponding author: Pau Colomer, pau.colomer@prax-power.com
Speaker: Pau Colomer, pau.colomer@prax-power.com

Abstract

This paper offers an overview of achieving a reduction in fringing losses in high-frequency magnetics for resonant converter applications. Through the utilization of ferrite toroidal cores with split gaps into several smaller gaps, resulting in a multi-gap ferrite toroidal core, substantial reductions in fringing losses can be achieved, leading to higher efficiency or size reduction of magnetics. The study presents practical examples of multi-gap toroidal transformer and resonant inductor designs, including optimization of the number of gaps and gap sizes using finite element analysis. Additionally, the paper discusses wire selection strategies to minimize losses and provides comparative results against designs in E or PQ format. Overall, this study demonstrates the effectiveness of the multi-gap toroidal approach in mitigating fringing losses and improving the performance of high-frequency magnetics in resonant converter applications.

1   Introduction

1.1   Main Effects Generating Losses in Windings

Over the past few years, there has been an increase in the use of resonant topologies in the design of high-frequency converters, such as LLC, DAB, CLLC, and others. This trend is particularly noticeable in the power range of thousands of watts.

In these resonant topologies, magnetic components, especially the main transformer and resonant inductor, play a crucial role in achieving the desired power efficiency and compact size without generating excessive heat. Proper calculation of magnetic losses is essential to achieve efficiency and size goals for the converter.

Losses in magnetic components are typically divided into two main groups: core losses and winding losses. This paper focuses on winding losses, specifically addressing how to mitigate the excess losses in windings caused by the fringing effect.

Winding losses in magnetic components are typically divided into three main effects: the skin effect, proximity effect, and fringing effect.

The skin effect occurs when a high-frequency current concentrates on the outer side of the conductor, increasing its effective resistance, known as Rac.

Similarly, the proximity effect results in an increase in the winding's effective resistance due to the influence of the several layers of the same winding.

The fringing effect is the most influential and perhaps the most challenging to simulate and quantify. It arises due to the presence of the magnetic field circulation in the core, which extends into the gap area and generates additional losses in the windings. These losses manifest as an increase in the winding's effective resistance and are directly related to the gap size, operating frequency, and distance to the windings (refer to Fig. 1).

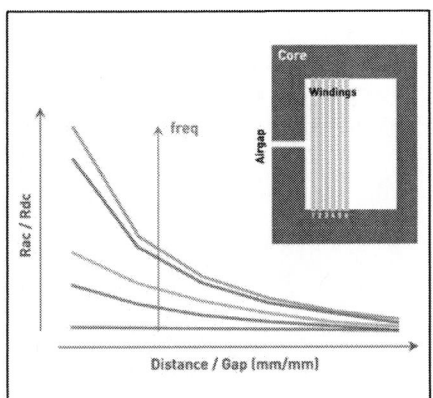

Fig. 1 Winding R_{ac} due fringing effect versus distance to air-gap and frequency

Resonant topologies require the use of large air gaps for both resonant inductor and main transformer, particularly when employing ferrite cores. Ferrites cores are the preferred option due to their low core losses, which becomes crucial with increasing converter switching frequencies.

The presence of such air gaps, along with the bending of the magnetic flux in that area, leads to additional losses in the windings and subsequent temperature or heat rise. This heat accumulation, depicted by the red area in Fig. 2, is notably concentrated near the core's center leg gap, posing challenges in heat management.

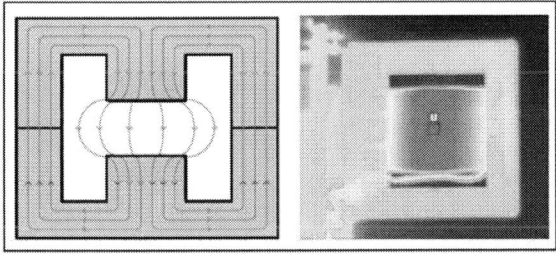

Fig. 2 Representation of the magnetic flux bent in the gap area and the associated heating.

1.2 Traditional Fringing Losses Mitigation Strategies

Currently, the fringing effect is mitigated in widely used formats such as PQ, PM, or ETD through various methods, with one common approach being the division or splitting of the length of the gap into several smaller gaps. This distributed-gap approach is typically implemented in the core center leg of the ferrite (as shown in Fig. 3). By dividing the gap into smaller segments, the bent magnetic flux is reduced, consequently decreasing the winding losses caused by the fringing effect.

While this approach effectively mitigates the fringing effect, it has its drawbacks. Implementing smaller gaps often necessitates customized cores and expensive tooling. Furthermore, dimensioning the gap size becomes an iterative process, posing challenges, especially when multiple iterations are required. Additionally, despite mitigating the fringing effect, the hot-spot winding temperature tends to remain concentrated in the inner layers, posing cooling challenges associated with heat dissipation.

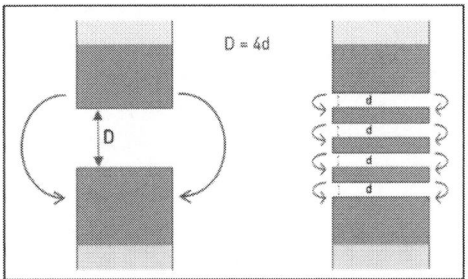

Fig. 3 Representation of a multi-gap approach on PQ, PM, ETD ferrite core formats

2 The Toroidal Multi-Gap Approach

2.1 Introduction

In this paper, we propose an innovative approach to improve current methods for mitigating the fringing effect in magnetic components. Our proposal involves utilizing a toroidal shape made of low-loss ferrite material, incorporating multi-gap technology.

To implement this approach, we suggest developing a simple tool for creating ferrite toroid segments. These segments can then be assembled together to form a complete toroidal shape. By splitting the total gap into smaller gaps across these segments, we can achieve a multi-gap toroidal configuration.

The primary objective of this approach is to reduce the AC resistance of the magnetic component by increasing the number of gaps. As more gaps are introduced within the toroidal structure, the magnetic flux is better distributed, thereby minimizing the fringing effect and reducing winding losses.

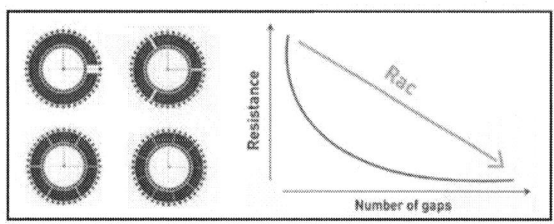

Fig. 4 Representation of the reduction of the R_{ac} by a toroidal multi-gap ferrite core.

2.2 Finite Elements Analysis Simulation

To validate and quantify the reduction in Rac, we conducted finite element analysis simulations to investigate the influence of the number of gaps on the AC resistance of the windings. Initially, we simulated a single large gap of 6mm in a toroidal structure subjected to a sinusoidal current at 200kHz. Subsequently, we divided this large gap into multiple smaller gaps, starting with 3, then 6, 9, and so forth.

The direct impact of the fringing effect on the gap length can be observed in Fig. 5. This graph illustrates how increasing the number of gaps leads to a reduction in the AC resistance of the windings, demonstrating the effectiveness of our proposed multi-gap toroidal configuration in mitigating winding losses caused by the fringing effect.

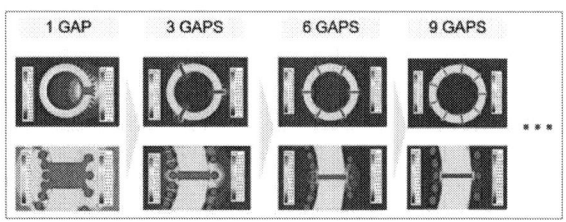

Fig. 5 FEM simulation of a ferrite toroid with 6mm air-gap at 200kHz

Finite element analysis quantifies the value of Rac in each case, allowing us to plot the results in a graph. We simulated up to 24 gaps and obtained the corresponding AC resistance values.

As shown in Table 1 and Fig. 6, distributing the gap into 12-15 smaller gaps, resulting in an individual gap length of 0.5mm, proves to be the most optimized approach. Increasing the number of gaps beyond this range continues to reduce the effective resistance, but the additional benefits may not justify the extra manufacturing effort.

It's important to note that these results are based on simulations using a 200kHz waveform. Generally, the value of Rac will vary with the operating frequency, exhibiting a direct relationship with it.

SIM	GAP (mm)	# GAPS	L/GAP (mm)	# TURNS	Rac (Ω)
1	6	1	6,00	40	184,40
2	6	3	2,00	40	65,40
3	6	6	1,00	40	34,60
4	6	9	0,67	40	21,00
5	6	12	0,50	40	13,23
6	6	15	0,40	40	9,99
7	6	18	0,33	40	8,28
8	6	24	0,25	40	6,23

Table 1 FEM R_{ac} results over different number of gaps.

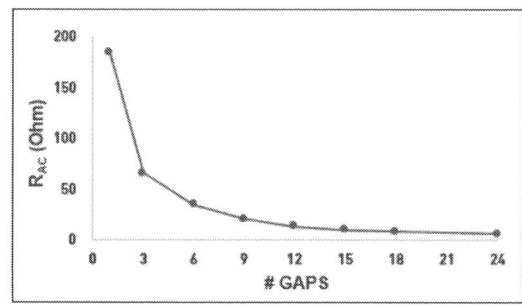

Fig. 5 Graphic representation of R_{AC} versus number of gaps.

2.3 Manufacturing Process

The manufacturing process is defined in a straightforward manner. For rapid sample production, the individual segments are placed over a simple sample tooling. FR4 laminates are then inserted between each ferrite segment. Subsequently, glue and tape are applied to form a toroidal multi-gap core, as depicted in Fig. 6.

Fig. 6 Samples construction: Core segment and FR4 boards for gap definition

For mass production, we can consider designing plastic covers with the gaps already fixed, along with implementing automatic taping, as illustrated in Fig. 7.

Fig. 7 Production construction: Plastic cover including gaps for fast assembly.

2.4 Inductor Design and Test Results

To validate our simulations, we constructed a multi-gap ferrite toroid inductor with an inductance value of 100uH. Performance evaluation was conducted by applying a sinusoidal waveform at 150kHz with 7 Amps RMS.

The design features a multi-gap ferrite toroid with a diameter of 45mm and a height of 27mm. We used thin strand HF litz wire for winding to ensure low winding Rac. The total gap measures 5.5mm, distributed over 8 segments, resulting in each individual gap length of approximately 0.7mm.

Utilizing the Dowell equation to calculate winding effective resistance and the Steinmetz equation for core losses, we determined the copper and core losses, along with the corresponding final temperature of the component at a 25°C ambient (refer to Table 2).

It's important to note that this simulation does not account for any additional losses due to the fringing effect.

Core Size	Toroid 45x27mm
Gap Definition	G5.5mm (x8 segments)
Turns & Wire	37 T; Litz wire 200x0.07mm
Copper Loss	6.73 W
Core Loss	2.89 W
Final Temp.	78°C

Table 2 Inductor simulation results L = 100 uH; I_{RMS} = 7 Amps; f = 150 kHz; Sinusoidal

In Figure 8 it is shown a picture of the manufactured sample.

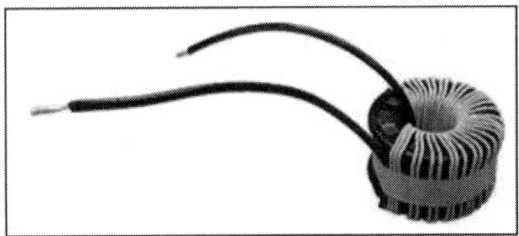

Fig. 8 100uH inductor with multi-gap ferrite toroid core.

We evaluated the performance of the sample by applying the defined waveform. Upon measuring the temperature, we observed that the estimated and tested values were nearly identical, with a difference of less than 2°C. There were not many additional losses generated by the fringing effect, which indicating configuration minimizes fringing losses.

Component	Multi-Gap Inductor
Measured L	94.9 uH
Applied Waveform	I_{pk} = 10.4 A; I_{rms} = 6.9 A
Final Temp.	79.8°C

Table 3 Hardware test results

Despite copper losses being more than double those of core losses, there were no hot spots in the windings, and the temperature difference compared to the core was less than 10°C. This indicates a relatively good thermal link between the windings and the core.

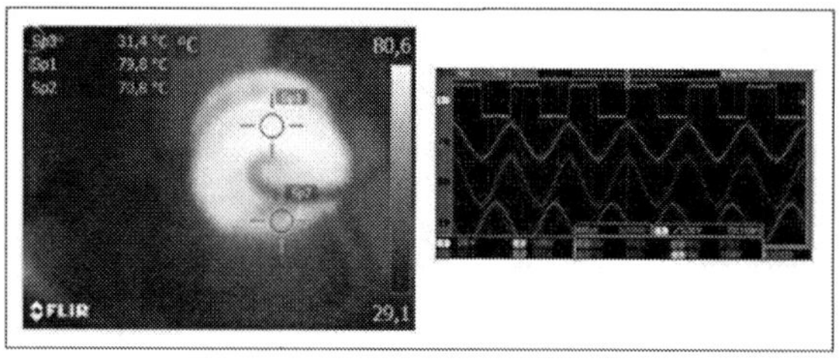

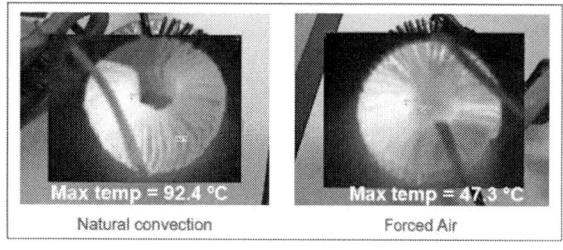

Fig. 9 Measured temperature and applied waveforms over multi-gap inductor.

An additional benefit of the toroidal shape is its heat dissipation capabilities. With both windings and core in contact with each other, the component typically experiences uniform heating, resulting in minimal temperature differences between the core and windings.

Furthermore, the toroidal shape demonstrates excellent heat management with any type of cooling method. Figure 10 illustrates the temperature of the component under the same load conditions with natural air cooling compared to forced air cooling. It shows a reduction of 45°C with forced air cooling. Therefore, it is relatively straightforward to cool down both the winding and core, whether with air cooling or by attaching a heat sink.

Fig. 10 Measured temperature with natural and forced air over a multi-gap toroid inductor.

3 Design Evaluation

3.1 Wolfspeed 20kW LLC Converter

For the final section of the paper, we evaluated a reference design from Wolfspeed. This design comprises a 20 kW LLC Resonant converter, featuring an input voltage of 800 V and an output voltage reaching up to 550 V. The operating frequency ranges from 150 kHz to 400 kHz.

The transformer in the Wolfspeed reference design utilizes 2 x PQ6560 cores, while the resonant inductor is constructed using 2 x PQ3535 cores, as depicted in Fig. 11.

Fig. 11 Wolfspeed 20kW reference design board showing transformer and resonant inductor.

The objective of this study was to redesign both the transformer and inductor using traditional PQ format and multi-gap toroidal technology to find the optimal solution. We then compared these designs with the current magnetics definition in the Wolfspeed reference design.

This study was conducted in collaboration with Frenetic, an online magnetic design software company.

It's worth noting that the Wolfspeed resonant inductor design exhibits a current derating due to excessive heat. Therefore, we incorporated this power derating factor into our design comparisons.

3.2 Resonant Inductor Design

In Figure 12, a summary of the simulated results for the resonant inductor is presented. The first row displays the design generated by Frenetic using software simulation tools. The second row represents the multi-gap toroid approach, while the third row depicts the analysis of the inductor in the Wolfspeed reference design.

The toroidal multi-gap approach exhibits superior performance in terms of losses, temperature, and power density. Although utilizing the PQ65/54 shape can achieve a similar component size, cooling becomes challenging, suggesting the necessity of a larger core.

As previously mentioned, the power density of the Wolfspeed design is adequate for its derating definition. Therefore, its power density is the lowest despite having a smaller total component size.

Core Format	Component Size	Core Loss	Copper Loss	Total Loss	Final Temp	Power Density
PQ65/54	68x64x63 mm	16,61 W	33,90 W	50,51 W	175°C (2 m/s)	74 mW/mm³
Multi-Gap Toroid	70x70x51 mm	12,04 W	24,18 W	36,22 W	120°C (2 m/s)	80 mW/mm³
PQ35/35 (x2)	83x39x39 mm	9,84 W	45,60 W	55,44 W	180°C (2 m/s)	51 mW/mm³

Fig. 12   Simulated results for the resonant inductor

Core Format	Power Density	Total Loss	Max Temp	Relative cost
1x (PQ65/54+PQ84/54)	120 mW/mm³	89,22 W	175°C	-20%
2x (Multi-Gap Toroid)	142 mW/mm³	77,67 W	120°C	**-44%**
2x (PQ35/35+PQ65/60)	89 mW/mm³	129,48 W	180°C	-

Fig. 14   Summary of simulated results and cost-analysis comparison

3.3 Transformer Design

The same approach was taken with the main transformer, aiming to design with the PQ format while keeping losses as low as possible and ensuring a proper temperature rise, regardless of the size.

In Figure 13, the difference between the toroidal multi-gap design and the PQ shape in terms of cooling capabilities can be observed. Both PQ designs require a larger size to ensure proper temperature control compared to the multi-gap toroid approach, resulting in a lower power density.

It's important to note that the multi-gap toroid approach exhibits a similar temperature rise with 50% higher losses compared to the PQ approach. This allows for a reduction in size and therefore improves the converter power density.

Core Format	Component Size	Core Loss	Copper Loss	Total Loss	Final Temp	Power Density
PQ84/54	87x78x63mm	12,20 W	26,51 W	38,71 W	110 °C (2 m/s)	46 mW/mm³
Multi-Gap Toroid	70x70x66mm	13,33 W	28,12 W	41,45 W	120 °C (2 m/s)	62 mW/mm³
PQ65/60 (x2)	132x62x64mm	33,22 W	40,82 W	74,04 W	125 °C (2 m/s)	38 mW/mm³

Fig. 13   Simulated results for the main transformer

3.4 Summary

The final analysis summarizes the results for both the transformer and resonant inductor. The multi-gap toroid approach exhibits similar losses to the most optimal design using a PQ format, but with improved power density and significantly better temperature performance.

Additionally, relative cost comparisons versus the Wolfspeed reference design approach are included. There is an improvement in cost compared to the PQ format, with a reduction of over 40% from the original cost. It's important to note that this cost reduction figure should be taken as tentative, but it serves as an indication that the multi-gap toroid approach is a cost-effective component solution.

4 Market Designs Examples

In the final section of the paper, we present examples of multi-gap ferrite toroid inductor and transformer designs that have already been introduced in the market.

One such example is an LLC transformer and resonant inductor set developed in 2018 for an on-board charger in a full electric vehicle application. This design utilizes toroidal multi-gap ferrite for both the transformer and resonant inductor. The windings are made with triple insulated litz wire to ensure low effective resistance and meet high isolation and working voltage requirements.

The first generation of this design entered mass production in 2020. It consists of two separate components interconnected using the same wire.

During 2022, we developed the evolution of this design by integrating the resonant inductor into the transformer leakage inductance. By decoupling or decreasing the coupling factor between primary and secondary windings, we achieved controlled leakage inductance within a tolerance of +/-10%. This second generation design significantly reduced size and cost compared to the two-component set of the first generation, while maintaining similar electrical, thermal, and safety performance.

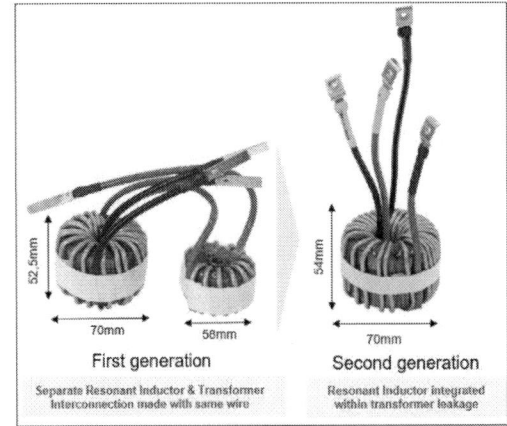

Fig. 15   First and second generation transformer and resonant inductor for an 11kW OBC for an EV car

5 Conclusion

In summary and conclusion, the toroidal multi-gap approach effectively addresses several challenges in magnetic component design. It utilizes distributed gap technology to minimize fringing losses, while also increasing the winding area by eliminating the need to place windings far away from the gap.

Additionally, this approach considers the use of low-loss ferrite materials, enabling the handling of high currents and operating frequencies. With windings located on the external part of the component, it exhibits excellent heat management with any type of cooling method.

Furthermore, the toroidal multi-gap approach proves to be a cost-effective solution when compared to traditional formats as EE, PQ and PM, making it a promising choice for various applications in power electronics.

References

[1] A. Van den Bossche, V. C. Valchev. *"Inductors and Transformers for Power Electronics"*. CRC press, 2005.

[2] W. G. Hurley, W. H.Wölfle. *"Transformers and Inductors for Power Electronics. Theory, Design and Applications"*. John Wiley & Sons Ltd, 2013.

[3] R. W. Erickson, D. Maksimovic. *"Fundamentals of Power Electronics"*. Kluwer Academic Publishers, 2004.

PCIM Europe 2024, 11– 13 June 2024, Nuremberg   DOI: 10.30420/566262058

Study on Sample Geometries for Ferrite Characterisation in the MHz Range

Till Piepenbrock[1], Lukas Keuck[2], Sebastian Schachten[1], Joachim Böcker[1], Frank Schafmeister[1]

[1] Paderborn University, Germany
[2] HELLA GmbH & Co. KGaA, Germany

Corresponding author: Till Piepenbrock, piepenbrock@lea.upb.de

The Power Point Presentation will be available after the conference.

Abstract

Ferrite material modelling becomes more difficult with high frequencies due to eddy current losses and dimensional resonance effects. This paper discusses the characterisation of core material properties for hysteresis and eddy current losses separately using customised thin walled ferrite samples. In a simulation study, it is shown that, with standard characterisation procedures, electric losses are incorrectly assigned to magnetic losses and vice versa for high frequencies. Further, the dimensional resonance leads to an inhomogeneous magnetic flux density over the core cross-section and complicates the correct allocation of a single large-signal magnetic flux density to specific measurement results. In an experimental test setup it is shown that for both, permeability and permittivity measurements, thin core probes show the least parasitic high-frequency effects and are thus suited best for accurate high-frequency characterisation.

1   Introduction

Empirical formulae (e.g. based on Steinmetz Equation) are commonly used to estimate the core losses of soft ferrite components. Parameters are usually derived from core loss data published by manufacturers measured with toroidal cores (e.g. T34x20x12). Especially at higher frequencies and for larger core types, applying these empirical approaches leads to massive measurement errors. This can be explained by three main reasons: First, eddy current losses in the core become dominant with larger core types. Although these eddy current losses are recorded in the measurements, they are incorrectly assigned to magnetic hysteresis losses. This would not be a problem for the calculation of overall losses, if this loss type could be scaled with the core volume in the same manner as hysteresis losses. Unfortunately, this is not the case. Second, the hysteresis losses strongly depend on the flux density which varies within the cross-section. Third, dimensional resonance occurs in large core designs already at medium frequencies which causes a drastic loss increase and a changed inductance. The term "dimensional res-

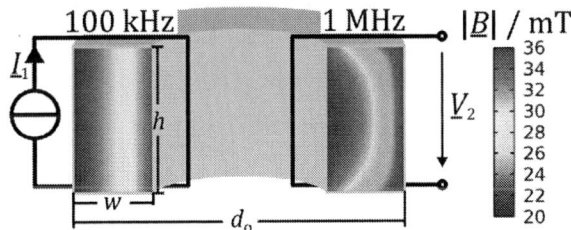

Fig. 1: Two-winding toroid (T34x20x12) measurement setup with magnetic flux density FEM simulation solution for $100\,\mathrm{kHz}$ and $1\,\mathrm{MHz}$.

onance" means that wave propagation must be taken into account at higher frequencies, which is negligible at lower frequencies [1]–[5].

To yield accurate models for geometric optimisations of high-frequency magnetic cores, accurate measurements of magnetic permeability and dielectric permittivity are crucial. For both properties it is desired to avoid parasitic high-frequency effects during the measurements and obtain pure material properties independent of the probe geometry. Only such data is suited as material data for e.g. FEM simulations in a multi-objective optimisation process.

The characterisation of the permeability in

463

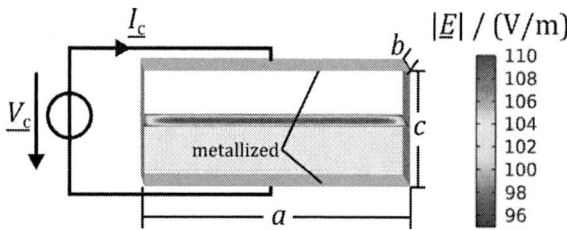

Fig. 2: Cuboid capacitor setup for dielectric hysteresis characterisation at $1\,\mathrm{MHz}$ with $b = 4\,\mathrm{mm}$.

datasheets is carried out for toroidal cores under sinusoidal magnetic flux excitation. The existing standard IEC-60404-6 is valid up to $100\,\mathrm{kHz}$ and needs to be revised for higher frequencies [6].

Although at high frequencies the dielectric hysteresis (in terms of eddy currents) strongly affects the field distribution in the core and thus the losses, permittivity values are usually missing in datasheets. Typically, the permittivity is measured with a capacitor probe (cf. Fig. 2) [5], [7].

This paper reveals the parasitic effects for both types of measurements by an FEM simulation study. According to this analysis, the systematic errors are explained and it is discussed which geometrical properties must be fulfilled for the measurement probes to minimise them.

Finally, practical measurements in the MHz frequency range show the effect of different core geometries on the measurement results.

2 Theoretical Background

Assuming sinusoidal magnetic flux density and the absence of DC flux bias, the magnetic and dielectric hysteresis can be described by nearly elliptical trajectories [3], [5], [8]. Consequently, the magnetic and dielectric flux densities and field strengths can be denoted as RMS phasors

$$\underline{B} = \underline{\mu}\,\underline{H}, \text{ and} \tag{1}$$

$$\underline{\tilde{D}} = \underline{\tilde{\varepsilon}}\,\underline{E} \tag{2}$$

with the complex permeability and permittivity being defined according to

$$\underline{\mu} = |\underline{\mu}|\,\mathrm{e}^{-\mathrm{j}\,\zeta_\mu}, \text{ and} \tag{3}$$

$$\underline{\tilde{\varepsilon}} = \underline{\varepsilon} - \mathrm{j}\frac{\kappa}{\omega} = |\underline{\tilde{\varepsilon}}|\,\mathrm{e}^{-\mathrm{j}\,\zeta_{\tilde{\varepsilon}}}. \tag{4}$$

The definition of the complex equivalent permittivity $\underline{\tilde{\varepsilon}}$ in Eq. 4 includes the bulk conductivity κ.

Assuming a linear, homogeneous and isotropic material, the Maxwell equations (Eq. 5 - 8) can be denoted with

$$\nabla \cdot \underline{E} = 0, \tag{5}$$

$$\nabla \cdot \underline{H} = 0, \tag{6}$$

$$\nabla \times \underline{E} = -\mathrm{j}\omega\underline{\mu}\,\underline{H}, \text{ and} \tag{7}$$

$$\nabla \times \underline{H} = \mathrm{j}\omega\underline{\tilde{\varepsilon}}\,\underline{E}. \tag{8}$$

Consequently, Poynting's theorem can be derived according to

$$\oint_{\delta V} (\underline{E} \times \underline{H}^*) \cdot \mathrm{d}\boldsymbol{S}$$
$$= \int_V (\underbrace{\omega\,\mathrm{Im}(\underline{\mu})\,|\underline{H}|^2}_{p_{\mathrm{mag.loss}}} + \underbrace{\omega\,\mathrm{Im}(\underline{\tilde{\varepsilon}})\,|\underline{E}|^2}_{p_{\mathrm{el.loss}}})\,\mathrm{d}V$$
$$+ \mathrm{j}\omega \underbrace{\int_V (\mathrm{Re}(\underline{\mu})\,|\underline{H}|^2 - \mathrm{Re}(\underline{\tilde{\varepsilon}})\,|\underline{E}|^2)\,\mathrm{d}V}_{\text{reactive power}}. \tag{9}$$

The key advantage of using Eq. 9 over most empirical loss formulae is to allow a physical interpretation and to separate between magnetic $p_{\mathrm{mag.loss}}$ and electric loss density $p_{\mathrm{el.loss}}$ [3]–[5], [9].

2.1 Theory of Toroid Core Measurement

The toroidal core measurements are standardised for frequencies below $100\,\mathrm{kHz}$ in IEC-60404-6 [6]. For the magnetic hysteresis characterisation, the primary current $\underline{I}_1$ excites a magnetic flux in the core and the secondary voltage $\underline{V}_2$ is measured with a high-impedance probe (cf. Fig. 1). For N_1 primary and N_2 secondary turns, both quantities define the complex mutual impedance

$$\underline{Z}_\mathrm{m} = \mathrm{j}\omega\underline{L}_\mathrm{m} = \frac{N_1}{N_2} \cdot \frac{\underline{V}_2}{\underline{I}_1}. \tag{10}$$

Assuming no air flux through the secondary winding, the mutual impedance equals the magnetising impedance. To obtain the permeability from the magnetising impedance, the geometry of the core has to be considered. Instead of using the geometrical cross-section, length, and volume, for toroidal core probes, the effective core parameters A_{eff}, l_{eff}, and V_{eff} are used.

$$l_{\mathrm{eff}} = 2\pi \frac{ab}{b-a} \ln\frac{b}{a} \tag{11}$$

$$A_{\mathrm{eff}} = h \frac{ab}{b-a} \ln^2\frac{b}{a} \tag{12}$$

$$V_{\mathrm{eff}} = l_{\mathrm{eff}}\,A_{\mathrm{eff}} \tag{13}$$

However, the effective parameters match with the low-frequency field solution [7].

Assuming the absence of dimensional resonance, the relative complex permeability $\underline{\mu}_r$ can be calculated by employing the effective cross-section and magnetic path length (with the parameters from Fig. 1) [7].

$$\underline{\mu}_r = \frac{2\pi}{j\omega\mu_0\,N_1^2\,h\,\ln\left(\frac{1}{1-\frac{2w}{d_o}}\right)} \cdot \underline{Z}_m \qquad (14)$$

2.2 Theory of Cuboid Core Measurement

Analogously, the complex permittivity can be calculated assuming an ideal plate capacitor model by measuring the complex impedance of the cuboid core, which results from the exciting capacitor voltage $\underline{V}_c$ and the measured current $\underline{I}_c$ (cf. Fig. 2).

Assuming the absence of dimensional resonance, the complex relative permittivity follows with

$$\underline{\tilde{\varepsilon}}_r = \frac{c}{j\omega\varepsilon_0\,a\,b} \cdot \frac{\underline{I}_c}{\underline{V}_c}. \qquad (15)$$

3 Simulation Studies of the Material Measurements

In two simulation studies, the magnetic and dielectric characterisation measurements at high frequencies are investigated for systematic deviations that occur even with ideally assumed measurements. Therefore, two FEM models (cf. Fig. 1 and Fig. 2) are set up with an idealised material (homogeneous, isotropic, linear, and independent of the frequency). For the FEM simulation, the relative complex permeability is chosen to

$$\underline{\mu}_r^{\mathrm{FEM}} = 1750 \cdot e^{-j\cdot 10^\circ} \qquad (16)$$

and the relative complex equivalent permittivity to

$$\underline{\tilde{\varepsilon}}_r^{\mathrm{FEM}} = 60000 \cdot e^{-j\cdot 20^\circ}. \qquad (17)$$

The assumption of constant complex permeability and permittivity is not realistic, but it is chosen on purpose, to simplify the analysis of the simulation results and avoid mixing different causes of deviations.

For both measurement setups, the frequency and core widths are varied exemplary and the effect

on the systematic measurement deviations is analysed. Therefore, the resulting voltage and current from the simulations are inserted in the idealized formulae Eq. 14 and 15 neglecting the dimensional resonance to obtain the geometry- and frequency-dependent estimation of the permeability and permittivity. These calculated core properties are compared to the true material definition.

3.1 Simulation Study: Permeability Characterisation

For the permeability characterisation, it is desired to measure only magnetic hysteresis losses and no electric losses. In a toroidal core, the magnetic flux density varies naturally with the core radius. This low-frequency field solution is already taken into account in Eq. 14. However, for real non-linear materials, this variation can lead to measurement inaccuracies. For high frequencies and larger toroid thickness w, the magnetic flux distribution in the toroid cross-section varies more (cf. Fig. 3).

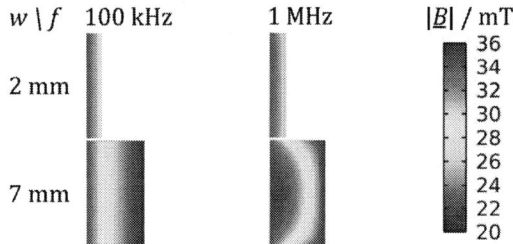

Fig. 3: Magnetic flux density distribution in the right cross-sections of two different toroids at $100\,\mathrm{kHz}$ and $1\,\mathrm{MHz}$. The position of the cross-section refers to the right side of Fig. 1.

For high frequencies, the breakdown of loss mechanisms identifies electric losses (eddy currents) adding a significant part to the losses of the magnetic hysteresis measurement. For the probe of $7\,\mathrm{mm}$ thickness, at $1\,\mathrm{MHz}$ over $20\,\%$ of the measured losses are contributed from electric losses. In a measurement, these parasitic losses are incorrectly assigned to magnetic hysteresis losses (cf. Fig 4). This leads to an overestimation of the measured magnetic loss angle ζ_μ (cf. Fig. 5). By reducing the toroid thickness w to $2\,\mathrm{mm}$ the wrongly assigned electric losses can be reduced to less than $3\,\%$. As a further effect, due to the dimensional resonance, the magnetising inductance of the measured toroid increases and consequently the amplitude of the complex permeability $|\underline{\mu}_r|$ is

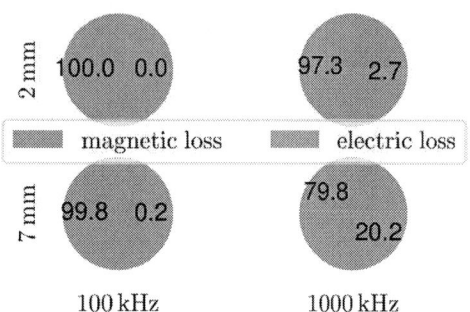

Fig. 4: Breakdown of loss mechanisms for different frequencies f and toroid thicknesses w.

overestimated (cf. Fig 5). Again, the overestimation is drastically reduced with the choice of a smaller toroid thickness.

3.1.1 Large-signal Allocation Problem

The above-mentioned two problems occur even for linear materials and are practically present in small-signal measurements. Since the permeability naturally varies non-linearly with the amplitude of the magnetic flux density, in datasheets usually large-signal measurements are provided at certain magnetic flux densities. Due to the dimensional resonance, the magnetic flux density varies greatly in the cross-section, but in a measurement it is assigned to a single effective peak flux density b_{eff}. For sinusoidal excitation, the effective peak magnetic flux density can be calculated with the law of induction and the effective cross-section for the absence of dimensional resonance using the RMS secondary voltage V_2 according to

$$b_{\text{eff}} = \frac{V_2}{2\pi f A_{\text{eff}} N_2}. \tag{18}$$

To compare the geometry- and frequency-dependent deviation of the effective magnetic flux density, for each exemplary toroid the effective magnetic flux density is normalised on its value at $100\,\text{kHz}$ as a low-frequency reference. The discrepancy of the effective flux density is again pronounced for larger core thicknesses w (cf. Fig. 5). This deviation indicates the variation of the field solution compared to the low-frequency solution within the core cross-section. For the linearly assumed material, different flux density amplitudes of the different core probes simply lead to an overestimation of the loss angle and the amplitude according to the geometric resonance. As the permeability

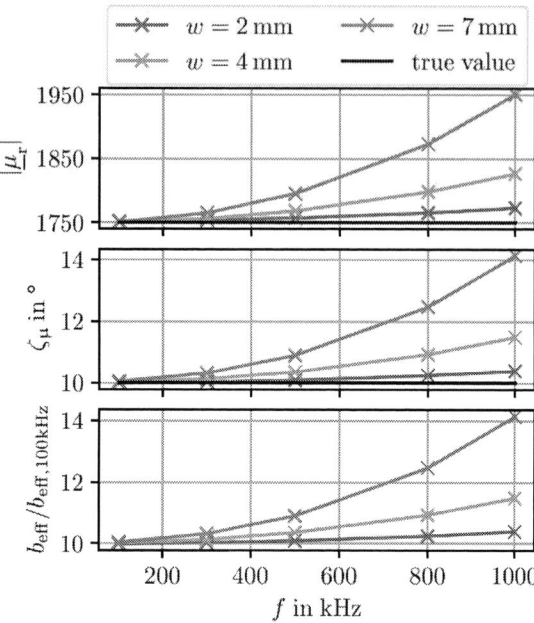

Fig. 5: Determined geometry and frequency permeability according to Eq. 14 for different toroid thicknesses w.

practically shows a non-linear dependency on the magnetic flux density, with respect to the amplitude and the loss angle, the situation for too thick core samples is more complicated.

3.2 Simulation Study: Permittivity Characterisation

Analogous to the characterisation of permeability, dimensional resonance is also observed when characterising permittivity at high frequencies. For the permittivity characterisation, it is desired to measure only dielectric hysteresis losses and no magnetic hysteresis losses. For high frequencies and larger cuboid thickness b, the electric field distribution in the cuboid cross-section varies more (cf. Fig. 6).

For high frequencies, the breakdown of loss mechanisms reveals magnetic losses add a significant part to the losses of the dielectric hysteresis characterisation (cf. Fig. 7).

This leads to an overestimation of the measured dielectric loss angle $\zeta_{\tilde{\varepsilon}}$ (cf. Fig. 8). Further, due to the dimensional resonance, the capacitance of the measured cuboid increases and consequently the amplitude of the complex permittivity is overestimated as well (cf. Fig. 8).

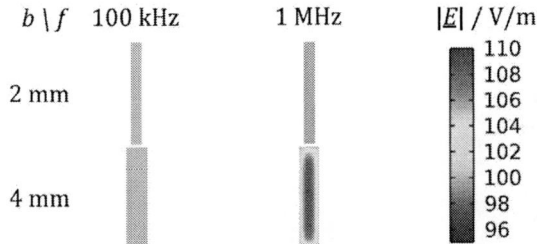

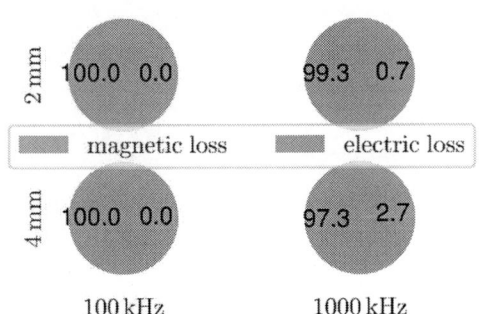

Fig. 6: Electric field distribution in the cross-sections of two different cuboids at $100\,\mathrm{kHz}$ and $1\,\mathrm{MHz}$. The position of the cross-sections refers to Fig. 2.

Fig. 7: Breakdown of loss mechanisms for different frequencies f and cuboid thicknesses b.

As the permittivity does not vary significantly with the large-signal excitation, the situation is less complicated compared to the permeability measurement in terms of the correct allocation of an effective electric field strength e_{eff} despite the field inhomogeneity.

4 Experimental Results

To verify the theoretical findings from the FEM simulations, experimental measurements of differently sized ferrite toroids and cuboids are analysed. To allow a fair comparison and avoid batch differences, two toroid cores and two cuboid cores are CNC-ground from a single block of the high-frequency MnZn power ferrite 3F46, which is recommended for $1\,\mathrm{MHz}$ to $3\,\mathrm{MHz}$ [10].

The small-signal measurements are performed with the impedance analyser WayneKerr 6515B [11].

4.1 Permeability Characterisation

For the permeability characterisation two customised toroids "T36x32x16" and "T24x18x15" with the dimensions in millimetre are produced.

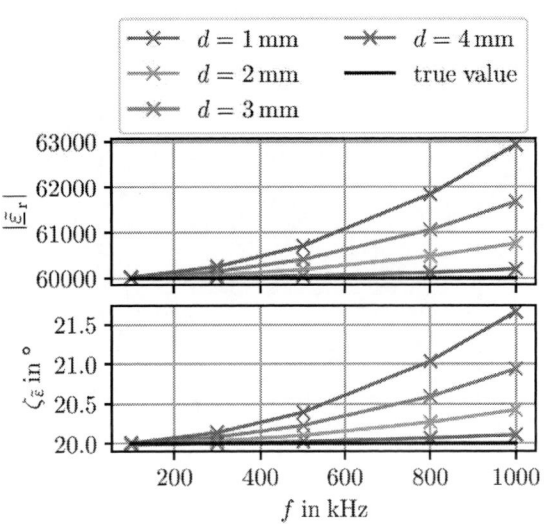

Fig. 8: Determined permittivity according to Eq. 15 for different cuboid thicknesses b.

4.1.1 Small-signal Toroid Measurement

As the small-signal permeability characterisation is carried out with an impedance analyser, it is not possible to apply the two-winding method directly. Instead of measuring $\underline{I}_1$ and $\underline{V}_2$ individually, three impedance measurements are performed to extract the transformer impedance ECD parameters (cf. Fig. 9).

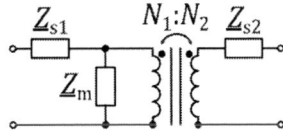

Fig. 9: Impedance ECD of a two-winding transformer.

From two open loop and one closed loop measurements, the self impedances $\underline{Z}_{11}$ and $\underline{Z}_{22}$ as well as the short-circuit impedance $\underline{Z}_s$ can be obtained. Choosing the transformer ratio n as the turns ratio $\frac{N_1}{N_2}$ yields

$$\underline{Z}_{11} = \underline{Z}_{\mathrm{m}} + \underline{Z}_{s1}, \tag{19}$$

$$\underline{Z}_{22} = \frac{\underline{Z}_{\mathrm{m}}}{n^2} + \underline{Z}_{s2}, \text{ and} \tag{20}$$

$$\underline{Z}_{s} = \underline{Z}_{s1} + (\underline{Z}_{s2}\, n^2)||\underline{Z}_{\mathrm{m}}. \tag{21}$$

Consequently, the mutual impedance which is used for the permeability calculation (cf. Eq. 14) can be

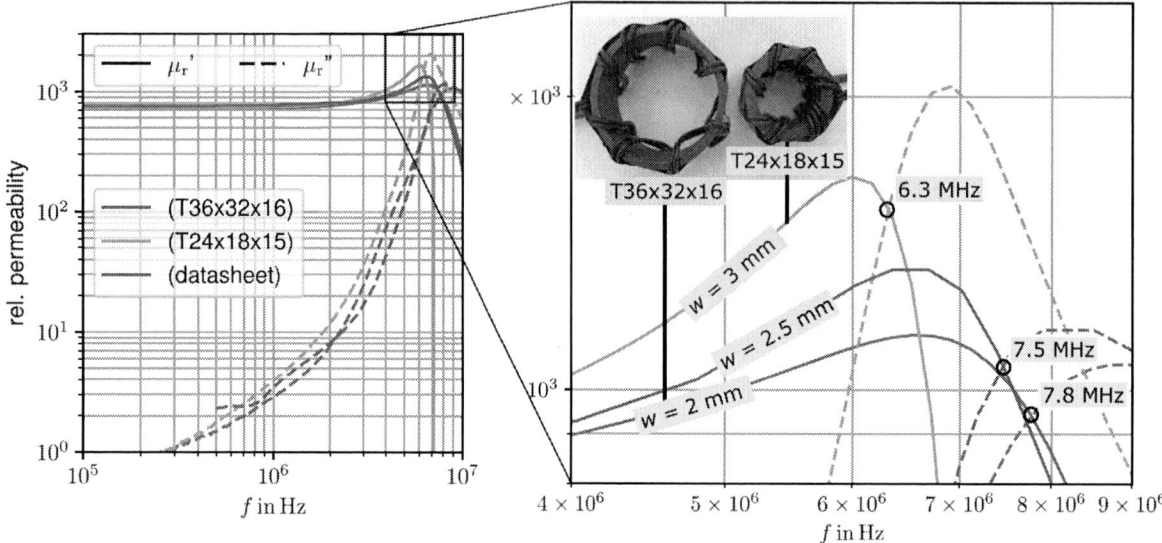

Fig. 10: Small-signal permeability (real μ_r' and imaginary part μ_r'') of 3F46 measured on different toroid probes vs. datasheet. In the right part of the figure with marked points $\zeta_\mu = 45°$.

expressed by

$$\underline{Z}_m = n \sqrt{\underline{Z}_{22}} \sqrt{\underline{Z}_{11} - \underline{Z}_s}. \qquad (22)$$

The non-physical solution of Eq. 22 with a negative real part can be neglected.

A typical datasheet figure is the small-signal real and imaginary part over the frequency (cf. Fig. 10). For the permeability characterisation, a good indicator for comparing the influence of parasitic eddy currents is the crossing of real and imaginary parts ($\zeta_\mu = 45°$) of differently sized toroid probes. The higher the frequency is, where this crossing occurs, the less parasitic high-frequency effects are recorded and the better is the accuracy of the permeability measurement. As Fig. 10 shows, the frequency at which $\zeta_\mu = 45°$ is drastically increased with smaller wall thickness w. Even though the overall dimensions of the probe "T36x32x16" are greater compared to "T24x18x15" as well as the one used in the datasheet characterisation ("T14x9x5"), it turns out that the flattest high-frequency behaviour is achieved with the customised core having a wall thickness of $w = 2\,\text{mm}$.

4.1.2 Large-signal Toroid Measurement

For the large-signal permeability measurement, the magnetic flux density adds a very non-linear dependency to the frequency and temperature. A typical large-signal datasheet plot is the relation of the loss density p_v over the frequency for constant magnetic

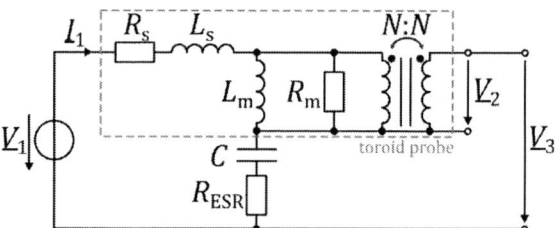

Fig. 11: Circuit diagram of the compensated measurement method with number of turns $N_1 = N_2 = N$ [12].

flux densities. According to Eq. 18, a single effective peak magnetic flux density b_{eff} calculated from the secondary voltage $\underline{V}_2$ is therefore allocated to one measurement result. As shown in the FEM simulations, for high frequencies this allocation is not trivial, due to the inhomogeneous magnetic flux in the core.

The two-winding toroidal measurement method is the standard procedure for large-signal measurements (cf. Sec. 2.1). At high frequencies, the large reactive power component in the direct power calculation with $\underline{V}_2$ and $\underline{I}_1$ can lead to large errors in the active power due to different delay times in current and voltage measurement.

To overcome this problem, a capacitor is connected in series to the primary winding and its capacitance C is chosen in resonance with the mutual inductance L_m. The dissipated power $P_{\text{comp.}}$ of

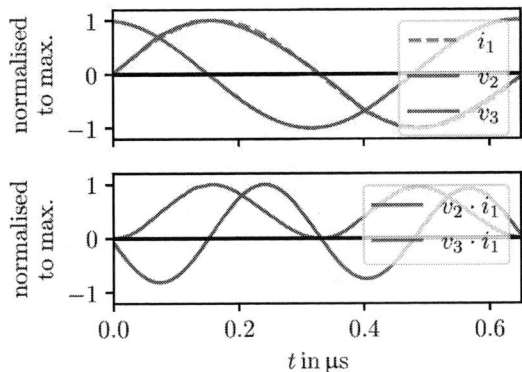

Fig. 12: Exemplary capacitive-compensated measurement at $1.52\,\mathrm{MHz}$ with measurement signals i_1, v_2, v_3, and the corresponding power normalised to their maxima.

both components can be calculated with $\underline{V}_3$ and $\underline{I}_1$ which are in phase for a perfectly tuned resonant circuit. In that case, the loss calculation is simplified with the primary RMS current I_1 and RMS voltage V_3 according to

$$P_{\text{comp.}} = \mathrm{Re}\left(\underline{V}_3\,\underline{I}_1^*\right) \overset{\text{in-phase}}{=} V_3\,I_1. \qquad (23)$$

An exemplary measurement shows that the normalised signals i_1 and v_3 overlap almost perfectly. Even if the compensation is not perfect, the capacitive-compensated measurement method allows to measure the magnetic flux density more reliably at high frequencies [12].

Correcting the measured power by the capacitor's equivalent series resistance R_{ESR}, the dissipated loss density p_{v} of the ferrite sample follows with the toroid volume V_{eff} as

$$p_{\text{v}} = \frac{P_{\text{comp.}} - R_{\text{ESR}}\,I_1^2}{V_{\text{eff}}}. \qquad (24)$$

The amplitude of the measured permeability $|\underline{\mu}|$ can simply be calculated from Eq. 14 by employing the measured secondary voltage $\underline{V}_2$. To obtain the loss angle ζ_μ from the capacitive-compensated measurement method, the definition of the magnetic hysteresis losses $p_{\text{mag. loss}}$ (cf. Ponyting theorem, Eq. 9) can be used resulting in

$$\zeta_\mu = \arcsin\left(\frac{|\underline{\mu}|\,p_{\text{v}}}{\pi\,f\,b_{\text{eff}}^2}\right). \qquad (25)$$

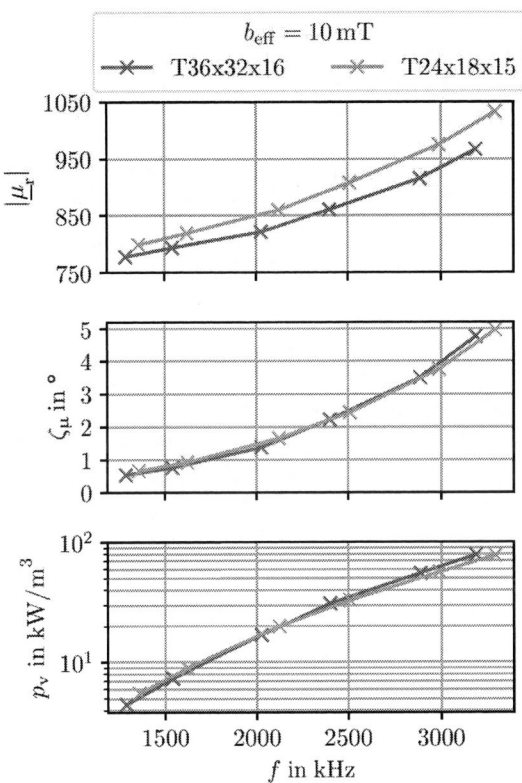

Fig. 13: Complex permeability and specific core losses measured on "T36x32x16" and "T24x18x15" 3F46 toroid cores for $10\,\mathrm{mT}$.

In Eq. 25 it is assumed, that the measured loss density p_{v} can completely be assigned to the magnetic hysteresis loss density $p_{\text{mag. loss}}$.

The measurement results show an overestimation of the relative magnetic permeability amplitude $|\underline{\mu}_{\text{r}}|$ for the sample probe with a toroid thickness of $4\,\mathrm{mm}$ compared to the probe with $2\,\mathrm{mm}$ of over $5\,\%$ at $3\,\mathrm{MHz}$ (cf. Fig. 13). As the permeability of both probes is nearly the same for the lowest measured frequency of $1\,\mathrm{MHz}$ this overestimation can be explained by the dimensional resonance.

In an experimental setup the permeability amplitude and the loss density are measured for a magnetic flux density of $10\,\mathrm{mT}$. Different to the small-signal measurements and the FEM simulation with the simplified amplitude-independent permeability, the measured losses of the $4\,\mathrm{mm}$ sample are not clearly exceeding the $2\,\mathrm{mm}$ sample over the full frequency range. A possible explanation for this result is that increased eddy current losses are compensated by the overestimated and wrongly allocated effective flux density b_{eff} as discussed in Sec. 3.1.1.

4.2 Permittivity Characterisation on Cuboid Cores

To exemplary investigate the geometry dependency of the permittivity characterisation, two thin cuboid cores "C25x2x16" and "C25x4x16" with the dimensions in millimetre are CNC-ground (cf. Fig. 14).

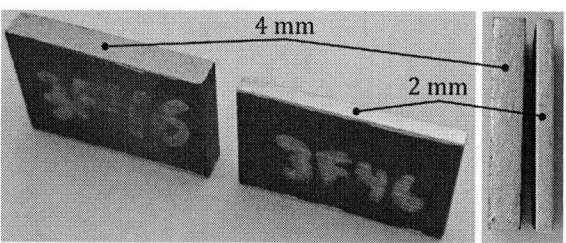

Fig. 14: Silver-coated cuboid samples with the dimensions "C25x2x16" and "C25x4x16".

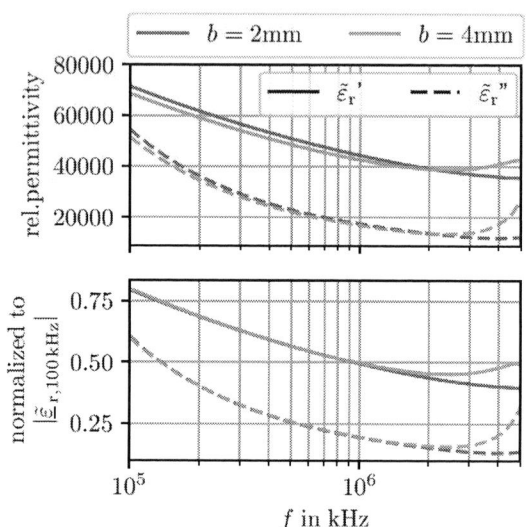

Fig. 15: Small-signal permittivity measurements of 3F46 on two different cuboid probes with the dimensions "C25x2x16" and "C25x4x16".

According to recent publications [5], [7], the permittivity is nearly independent of the field amplitude. Consequently, only a small-signal characterisation is enrolled in this study.

The complex equivalent permittivity is measured for a frequency range of $100\,\mathrm{kHz}$ to $5\,\mathrm{MHz}$. For low frequencies, the permittivity of both probes differs slightly from each other. A possible reason for this can be found in mechanical stress due to the pressure of the fixture or the grinding process [13], [14].

To allow a better comparison of the high-frequency behaviour, the measurement curves of real and imaginary parts are normalised to the amplitude at $100\,\mathrm{kHz}$. It can be clearly seen that the sample of $2\,\mathrm{mm}$ thickness shows a flat frequency behaviour compared to the sample with $4\,\mathrm{mm}$ up to several MHz.

5 Conclusion

The appropriate choice of sample dimensions for the characterisation of high-frequency ferrite core materials is essential. With an FEM-based simulation assuming linear material behaviour the phenomenon of dimensional resonance is identified as a systematic reason for inaccurate measurements of material properties. For the magnetic hysteresis characterisation, parasitic electric (eddy current)

losses are measured and for the permittivity characterisation, the magnetic hysteresis losses are wrongly assigned to the electric losses. It is practically shown that not the overall probe size but the wall thickness is the decision-making dimension for toroid and cuboid cores. The large-signal permeability characterisation reveals another source of error, as the allocation of the magnetic flux density to a certain measurement is non-trivial for thicker wall thicknesses.

6 Acknowledgements

This research work was funded by the Deutsche Forschungsgemeinschaft (DFG, German Research Foundation – BO 2535/22-1). The ferrite block used to produce the core samples was kindly provided by Ferroxcube.

References

[1] F. G. Brockman, P. H. Dowling, and W. G. Steneck, "Dimensional Effects Resulting from a High Dielectric Constant Found in a Ferromagnetic Ferrite," *Physical Review*, vol. 77, no. 1, pp. 85–93, 1950. DOI: 10.1103/PhysRev.77.85.

[2] M. Kacki, M. S. Rylko, J. G. Hayes, and C. R. Sullivan, "A Study of Flux Distribution and Impedance in Solid and Laminar Ferrite Cores," in *2019 IEEE Applied Power Electronics Conference and Exposition (APEC)*, IEEE, 2019, pp. 2681–2687. DOI: 10.1109/APEC.2019.8722252.

[3] G. R. Skutt, "High-Frequency Dimensional Effects in Ferrite-Core Magnetic Devices (PhD thesis)," Ph.D. dissertation, Virginia Polytechnic Institute and State University, Blacksburg, Virginia, 1996.

[4] J. Böcker, "Analysis of the Magnetic Skin Effekt in Motors and Inductors," in *SPEEDAM 2020 proceedings*, Sorrento, Italy: IEEE, 2020, pp. 103–107. DOI: 10.1109/SPEEDAM48782.2020.9161895.

[5] L. Keuck, *Entwurf eines einstufigen Ladewandlers auf Basis eines LLC-Resonanzwandlers*. Paderborn, 2023.

[6] IEC, *DIN EN IEC 60404-6 (VDE 0354-6): 2022-05 Magnetische Werkstoffe - Teil 6: Verfahren zur Messung der magnetischen Eigenschaften weichmagnetischer metallischer und pulverförmiger Werkstoffe bei Frequenzen im Bereich 20 Hz bis 100 kHz mittels Ringproben*, VDE, 5-2022.

[7] Alexander Stadler, "Messtechnische Bestimmung und Simulation der Kernverluste in weichmagnetischen Materialien," doctoral thesis, Friedrich-Alexander-Universität Erlangen-Nürnberg (FAU), 2010.

[8] H. Li, D. Serrano, S. Wang, and M. Chen, "MagNet-AI: Neural Network as Datasheet for Magnetics Modeling and Material Recommendation," *IEEE Transactions on Power Electronics*, vol. 38, no. 12, pp. 15854–15869, 2023. DOI: 10.1109/TPEL.2023.3309233.

[9] J. Böcker, "Concept Study of an LLC Converter with Magnetically Resonant Inductor," in *2022 International Symposium on Power Electronics, Electrical Drives, Automation and Motion (SPEEDAM)*, IEEE, 2022, pp. 166–170. DOI: 10.1109/SPEEDAM53979.2022.9842047.

[10] Ferroxcube. "3F46: Material specification: data sheet." (2016), [Online]. Available: https://www.ferroxcube.com/upload/media/product/file/MDS/3f46.pdf (visited on 01/30/2024).

[11] Wayne Kerr Electronics. "Precision Impedance Analyzers: 6500B Series." (), [Online]. Available: http://www.waynekerrtest.com/datasheet/instruments/wk6500b.pdf (visited on 03/20/2024).

[12] M. Mu, Q. Li, D. J. Gilham, F. C. Lee, and K. D. T. Ngo, "New Core Loss Measurement Method for High-Frequency Magnetic Materials," *IEEE Transactions on Power Electronics*, vol. 29, no. 8, pp. 4374–4381, 2014. DOI: 10.1109/TPEL.2013.2286830.

[13] M. Kacki, M. S. Rylko, J. G. Hayes, and C. R. Sullivan, "Measurement Methods for High-Frequency Characterizations of Permeability, Permittivity, and Core Loss of Mn-Zn Ferrite Cores," *IEEE Transactions on Power Electronics*, vol. 37, no. 12, pp. 15152–15162, 2022. DOI: 10.1109/TPEL.2022.3189671.

[14] A. van Groenou, "Grinding of ferrites, some mechanical and magnetic aspects," *IEEE Transactions on Magnetics*, vol. 11, no. 5, pp. 1446–1451, 1975. DOI: 10.1109/TMAG.1975.1058839.

PCIM Europe 2024, 11– 13 June 2024, Nuremberg   DOI: 10.30420/566262059

FEM-Supported and Non-Destructive Magnetic Characterization Method for Non-Laminated Steel

Stefan Tobler[1], Simon Nigsch[1]

[1] OST - Eastern Switzerland University of Applied Sciences, Switzerland

Corresponding author: Stefan Tobler, stefan.tobler@ost.ch
Speaker: Stefan Tobler, stefan.tobler@ost.ch

Abstract

For magnetic characterization where the influence of eddy current effects cannot be prevented by given measurement conditions, like insufficient lamination, other approaches are needed to eliminate their impact. This paper presents a Finite Element Method (FEM)-based method for B-H curve determination for non-laminated steel where non-significant eddy current effects influence the results. In addition to the conventional hysteresis measurement, the sample is simulated in FEM to generate a compensation for achieving the most accurate characterization. An example material characterization is performed and its result is used to simulate the magnetic force on a test device. These simulation results were then compared with measurements for verification. As these experimental verification demonstrates, simulations with compensated material characteristic achieve a more accurate result than uncompensated. The method is not limited to a specific sample shape and can therefore be used for different materials, core shapes and applications.

1   Introduction

For proper designs of magnetic components or electrical machinery, the magnetic properties of selected materials, such as B-H curve or iron losses, must be known or well estimated. While electrical main components of magnetic actuators are commonly made of specific rated steels and other metallic alloys, structural parts are built with materials selected mainly for their mechanical properties. Some of these parts are also exposed to various magnetic fields, if they are placed within a leakage flux path or even part of the magnetization path. The magnetic material specifications are usually provided by manufacturers. Mechanical steels often lack the description of magnetic properties. Thus, magnetic characterization must be performed. Methods for this are known and can be found in publications and corresponding standards [1], [2]. Several measuring configurations with different sample shapes, such as Epstein frame, toroidal core or single sheet setups can be used and compared with each other [3]. In all of these approaches, the influences of eddy current effects are not considered. Either because the core sample is a stack of insulated material sheets, or the excitation frequency results in a non-problematic penetration depth in relation to the sample thickness.

Sheet dimension or measurement frequency have an influence on the different types of losses that occur, such as hysteresis and core losses [4]. If these influences cannot be neglected and the eddy current effects are significant, the characterization process needs to be adapted appropriately or a DC measurement must be performed. For DC measurement, the magnetic flux is measured by inserting a probe in the magnetic path. This implies cutting the test object and therefore destroying it. This paper presents a method where the proposed approaches are adapted, which allows a higher frequency to be used to characterize materials of test objects without lamination or destruction.

2   Method

To reduce the influence of eddy current or skin effect, a more accurate characterization and extended measurement method of the non-laminated test object is presented here. Figure 1 shows the workflow of B-H curve determination with additional post-processing and a Finite Element Method (FEM)-supported compensation.

PCIM Europe 2024, 11– 13 June 2024, Nuremberg   DOI: 10.30420/566262059

Measurements   **Post processing**

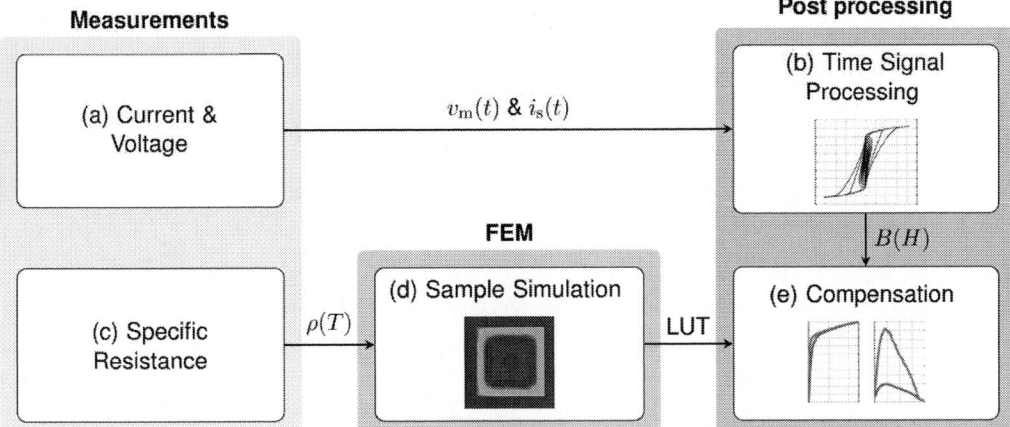

Fig. 1: Workflow of the Finite Element Method (FEM)-compensated characterization method. The result of the adjusted B-H curve determination from the measured time signals is compensated with an LUT. This is obtained with a FEM simulation with the test object geometry and measured specific resistance.

The magnetization curve is determined using an AC sinusoidal current injection at low frequency [2]. In this method, an excitation with a different amplitude is performed for each sample point on the B-H curve (a). As shown in Fig. 2, a current source as well as a voltage and current meter are connected to the windings of the specimen. The magnetic field strength H and flux density B are obtained from the measured time signals $v_\mathrm{m}(t)$ and $i_\mathrm{s}(t)$ using Eqs. (1) and (2). N is the number of turns, l_m the magnetic core length and A_m the core cross-sectional area.

$$H(t) = \frac{N i_\mathrm{s}(t)}{l_\mathrm{m}} \tag{1}$$

$$B(t) = \frac{1}{N A_\mathrm{m}} \int v_\mathrm{m}(t) dt \tag{2}$$

The hysteresis at the individual excitation levels can then be plotted and the initial B-H curve can be obtained (b) and is visualized in Fig. 5. However, the field strength determination is already subject to errors caused by eddy current effects, which must be considered and are discussed in the following section 2.1. Additionally, an occurring skin effect also has an influence on the calculated flux density. For a further improvement of the results, the field attenuation in the core is compensated with FEM data. The look up tables (LUT) used for this purpose are generated by FEM simulations of the measurement setup (d). Hence, the specific resistance ρ of the test object material must be known or measured (c). The FEM compensation (e) is addressed in section 2.2.

2.1 Excitation current measurement

While the magnetic flux density can be correctly determined via an additional measuring winding to measure v_m, the measured source current i_s also includes the current of the eddy current losses in the core. These losses can be modeled with a parallel resistor to the excitation winding. Figure 2 shows the measurement circuit including the equivalent resistor R_el of the eddy current losses.

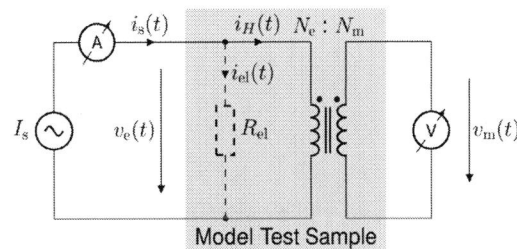

Fig. 2: Measurement circuit with equivalent resistor R_el for eddy current loss modeling.

To model the equivalent resistor R_el, a toroidal core with a circular cross section area with radius R and the magnetic path l_m is considered. As shown in Fig. 3, a winding with N turns carrying the sinusoidal current $i = \sqrt{2}\, I \sin(\omega t)$ generates a magnetic flux density B. Assuming no skin effect (and therefore a skin depth $\delta \gg R$) and a infinite large toroidal core, the magnetic flux density B in the core is homogeneous.

When considering the flux Φ through the area A defined by the radius r in the Fig. 3, the induced electrical field strength E in the circumference of

473

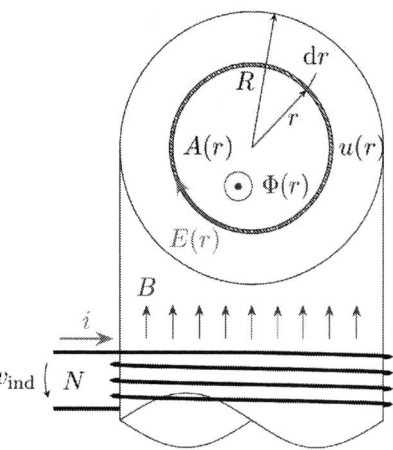

Fig. 3: Toroidal core drawing to identify equivalent resistor R_{el} for eddy current losses with homogeneous field in the core.

this area can be determined by Eq. (3).

$$E(r) = -\frac{1}{u(r)}\frac{d\Phi(r)}{dt} \quad (3)$$

With this induced electrical field strength E and the specific core material resistance ρ the induced losses P_{el} in the full core follow with Eq. (4).

$$P_{el} = \int_V \frac{E^2(r)}{\rho}\,dV = \int_0^R \frac{E^2(r)}{\rho}\,u(r)\,l_m\,dr \quad (4)$$

With the induced voltage in the winding

$$V_{ind} = I\,\omega L = I\,\omega N^2\frac{\mu R^2\pi}{L_m} \quad (5)$$

and the eddy current losses from Eq. (4) the equivalent resistor R_{el} can be described by Eq. (6).

$$R_{el} = \frac{V_{ind}^2}{P_{el}} = \frac{8\pi N^2\rho}{l_m} \quad (6)$$

The equivalent resistor R_{el} is purely ohmic for these conditions where no skin effect is present. Therefore, as Fig. 4 shows, the current through the equivalent resistor i_{el} is in phase with the induced voltage v_m and is leading by 90° the true excitation current of the field strength i_H. At the zero crossing of the measured voltage v_m or peak of the calculated magnetic flux density B_{max}, the current through the equivalent resistor i_{el} is zero. At this point the measured source current i_s is equal to the relevant excitation current i_H and is used to determine the magnetic field strength H. As shown in Fig. 5, for

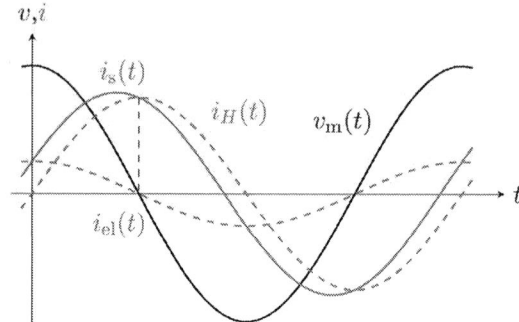

Fig. 4: Corresponding time signals of the measurement circuit. For a purely ohmic equivalent resistor, the current i_{el} is zero at the maximum of the magnetic flux determined by v_m. At this point the measured source current i_s is equal to the excitation current of the magnetic field strength i_H and is used to determine the magnetic field strength.

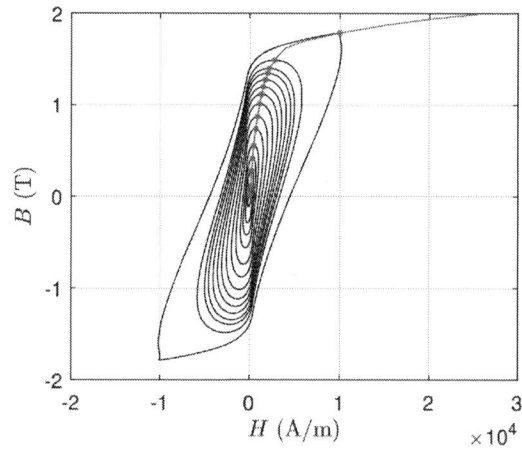

Fig. 5: Initial B-H curve determined with B_{max} from the time signals and display of some corresponding measured hysteresis.

the determination of the initial B-H curve, the maximum flux density B_{max} of a hysteresis loop and the corresponding magnetic field strength are used for each sample points.

The more pronounced the skin effect is, the less accurate the determination of the excitation current becomes. The equivalent resistor R_{el} must then be modeled as an impedance Z_{el}. For more pronounced skin effect the behavior of this impedance becomes more inductive.

Considering again the toroidal core with a circular cross section with radius R and the magnetic path l_m, the winding with N turns carries again the sinusoidal current $i = \sqrt{2}\,I\sin(\omega t)$. Assuming

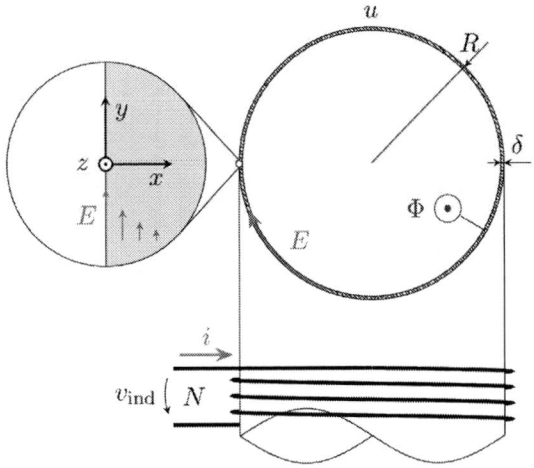

Fig. 6: Toroidal core drawing to identify equivalent impedance Z_{el} for eddy current losses with most pronounced skin effect. A coordinate system is introduced on the surface of the toroidal core.

now a most pronounced skin effect and an infinite large toroidal core, a coordinate system on the surface of the core is defined, as shown in Fig. 6. Its x-coordinate is the distance to the surface, the y-direction is along the circumference of the cross section and the z-direction lies parallel to the magnetic path and also along the surface.

With the condition that the magnetic field strength $\vec{H}$ in the core points in the direction of z ($H_x = 0$, $H_y = 0$) and the remaining component is constant in the direction of y ($\frac{\partial H_z}{\partial y} = 0$), Ampere's law can be simplified to Eq. (7).

$$-\frac{\partial H_z}{\partial x} = \frac{1}{\rho} E_y \tag{7}$$

Similar can be applied to the electrical field strength $\vec{E}$ and Farady's law, which is reduced to Eq. (8). The components of the electrical field strength in the directions of x and y are set to zero ($E_x = 0$, $E_z = 0$) and the remaining y-component is constant along z ($\frac{\partial E_y}{\partial z} = 0$).

$$\frac{\partial E_y}{\partial x} = -j\omega\mu H_z \tag{8}$$

By combining Eqs. (7) and (8), the differential equation Eq. (9) is obtained with the solution Eq. (10), which is the electrical field strength in the core.

$$\frac{\partial^2 E_y}{\partial x^2} = j\frac{\omega\mu}{\rho} E_y \tag{9}$$

$$E_y = E_0 \, e^{-\frac{x}{\delta}(1+j)} \tag{10}$$

Inserting the solution Eq. (10) into the simplified Farady's law from Eq. (8) leads to the megnetic field strength in Eq. (11) in the toroidal core.

$$H_z = H_0 \, e^{-\frac{x}{\delta}(1+j)} \tag{11}$$

The total flux Φ in the core cross section follows with the integral over the area in Eq. (12).

$$\Phi = \int_A B_z \, dA = u \int_0^\infty \mu H_z \, dx \tag{12}$$

Rearranging the expression of the induced voltage in the winding v_{ind} and inserting the derivative of the magnetic flux Φ from Eq. (12) gives the searched equivalent impedance Z_{el} for the eddy current losses.

$$Z_{el} = \frac{v_{ind}}{i} = \frac{2\sqrt{2}\pi N^2 R}{l_m} \frac{\rho}{\delta} \, e^{j\frac{\pi}{4}} \tag{13}$$

For most pronounced skin effect, the impedance has a phase angle of 45°. With this equivalent impedance, the current of the eddy current losses is not zero at the peak of the flux density and the determination of the excitation current becomes less accurate.

2.2 FEM compensation

In addition to eddy current losses, the skin effect also influences the magnetic property determination. Due to the attenuation of the alternating fields within the core, the effective magnetic cross-sectional area is reduced. Consequently, the magnetic flux density B and thus the permeability are determined incorrectly. True permeability $\mu_{R,true}$ is the actual value of the material property and entered into the simulation. Effective permeability $\mu_{R,eff}$ is the value determined by measurement or simulation with skin effect influence. Figure 7 visualizes the deviation between true permeability $\mu_{R,true}$ and effective permeability $\mu_{R,eff}$ of the test object material at 20 °C. For simplification, the permeability is assumed to be independent of the flux density in the simulation. For a relative permeability of $\mu_R > 100$, the effective permeability at an excitation frequency of 15 Hz is not matching the results in DC. To compensate for this effect, the measured permeability is corrected with this data as LUT. This table is generated with FEM simulations of the test object and different permeability values. Accordingly, the resistivity of the test object material must be known or determined.

Furthermore, it should be mentioned that at low

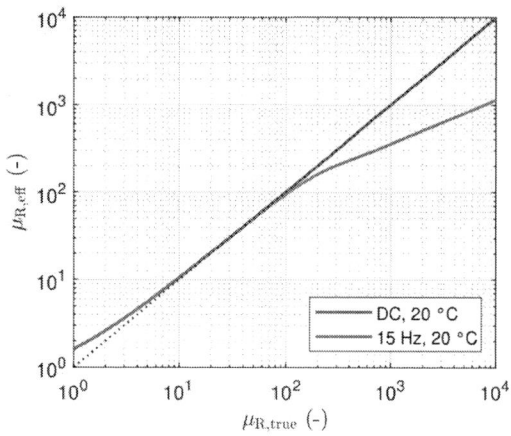

Fig. 7: Effective permeability $\mu_{R,eff}$ versus the true applied permeability $\mu_{R,true}$ for DC and at 15 Hz at 20 °C. The curves were generated with FEM, the sample geometry and the measured resistivity and serve as LUT in the compensation.

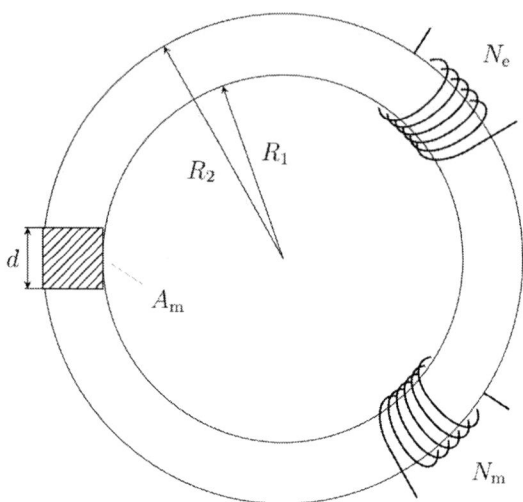

Fig. 8: Toroidal sample core with a square cross section and equipped excitation and measurement windings.

permeability ($\mu_R < 10$), both the AC and DC measurements provide a permeability that is too high. In this range, the magnetic flux outside the core must be considered as well, since the measured core reluctance is in the range of the leakage reluctance.

3 Experimental Verification

3.1 Characterization of Sample Core

The given test object to characterize is an toroidal core with a square cross section, as shown in Fig. 8. Its dimensions are given in Tab. 1. To characterize the sample core the specific resistance must also be measured.

Tab. 1: Sample toroidal core specification

Designation	Description	Value
R_1	Inner radius	40 mm
R_2	Outer radius	50 mm
d	Core depth	10 mm
A_m	Core cross section	100 mm²

3.1.1 Specific Resistance

The specific resistance was measured using DC current injection [2]. With various injected currents in the range of 1 A to 2.5 A the appearing voltages were measured at multiple positions on the surface of the toroidal core. The results from each measurement point pair were averaged to estimate the specific resistance of the sample material.

3.1.2 Magnetic Characterization

The measurement circuit for the B-H curve determination is shown in Fig. 2. The specification of the measurement setup is given in Tab. 2. Excitation is accomplished with a sinusoidal current with a frequency of 15 Hz and a peak amplitude of up to 35 A fed into the excitation winding with N_e turns to reach saturation.

Tab. 2: Excitement current measurement setup

Designation	Description	Value
N_e	Excit. winding turns	350
I_e	Peak excit. current	≤ 35 A
f	Excit. frequency	15 Hz
N_m	Meas. winding turns	350

The magnetic field strength H is calculated by sampling and processing the injected current $i_s(t)$. To determine the magnetic flux density B, the voltage $v_m(t)$ is measured at a second winding with N_m turns. This winding is placed below the excitation winding to be positioned as close as possible to the core.

All measurements were performed at defined and controlled temperatures inside a climate chamber.

PCIM Europe 2024, 11– 13 June 2024, Nuremberg   DOI: 10.30420/566262059

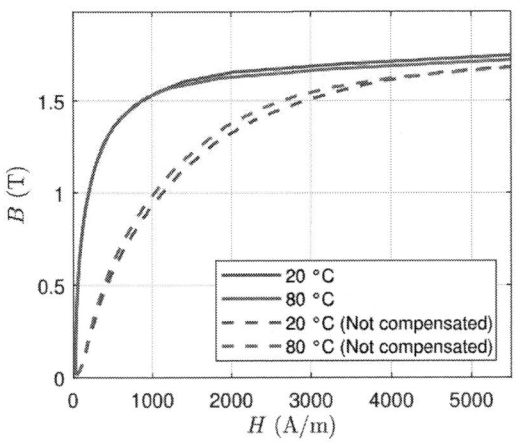

Fig. 9: Comparison of resulting initial B-H curve before and after compensation at the temperatures 20 °C and 80 °C.

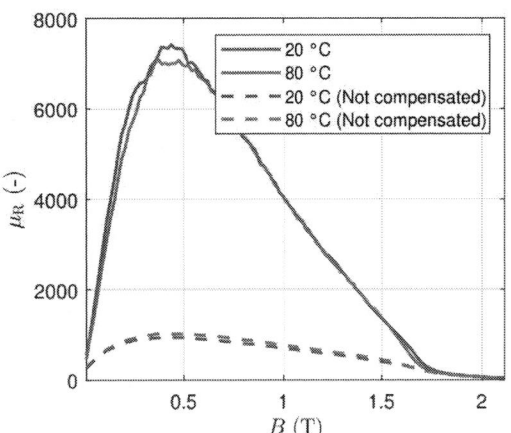

Fig. 10: Comparison of resulting relative permeability μ_R depending on flux density B before and after compensation at the temperatures 20 °C and 80 °C.

3.1.3 Results and Discussion

Table 3 presents the measured specific resistance ρ of the ring core sample at 20 °C and 80 °C. The results were utilized in the core simulation with FEM to generate the LUT for compensating the eddy current effect.

Tab. 3: Results of specific resistance determination of the ring sample core

Temperature T in °C	Specific Resistance ρ in $\mu\Omega m$
20	0.178
80	0.217

Figures 9 and 10 show the resulting initial B-H curve and relative permeability μ_R in relation to the flux density B before and after compensation for temperatures of 20 °C and 80 °C. The effective magnetic cross-section is reduced due to the attenuation of the alternating fields within the core. The measured flux density B was found to be too low when calculated using the physical cross-sectional area of the sample and the induced voltage v_m. As a result, the relative permeability was corrected to a higher value and the initial B-H curve was adjusted to a steeper slope using the LUT obtained from the FEM simulation. Figure 10 shows that the peak relative permeability increases from around 1000 to approximately 7000 when considering the relationship between relative permeability μ_R and flux density B. Due to the lower temperature, which results in a higher specific resistance ρ and a lower

skin depth δ, the permeability at 20 °C is more compensated than at 80 °C.

3.2 Application Measurement

To validate the characterization method, the compensated and uncompensated characterization results are used in simulations and compared with measurements. The material properties obtained at 20 °C will be further used for this validation. A test device with magnetic main components manufactured from the same material as the sample ring core is excited with DC currents of 466 mA and 744 mA. The current is applied to the embedded winding in the lower magnetic body. Figure 11 shows a simplified drawing of the application measurement setup. The force generated in z-direction

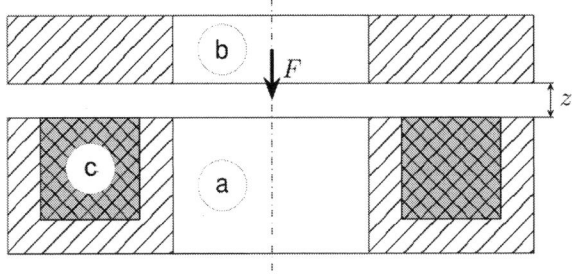

Fig. 11: Simplified drawing of the test device. The generated force between magnetic body (a) and armature (b) is measured as a function of the gap z. The current source is connected to the embedded winding (c) in the lower magnetic body.

477

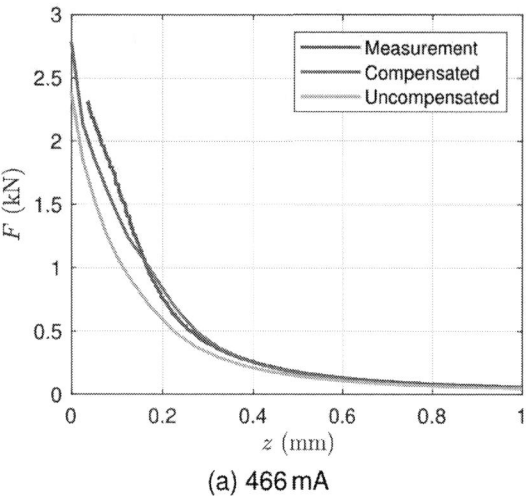

(a) 466 mA

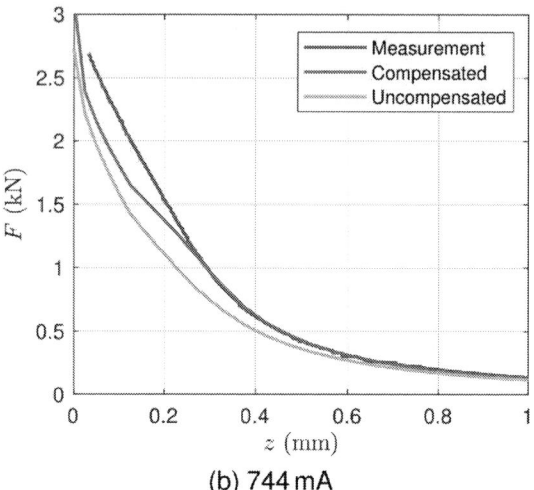

(b) 744 mA

Fig. 12: Comparison of simulated and measured magnetic force F of the test device as a function of the gap z for excitement currents 466 mA and 744 mA. The simulations were performed with the determined compensated and uncompensated material characteristics.

is measured as a function of the gap z between the magnetic body and armature. The setup is also simulated in FEM with the material characteristics before and after compensation for comparison.

3.2.1 Results and Discussion

Figure 12 shows the final results of the experimental verification as plot of the generated force as a function of gap z. Tables 4 and 5 list the percentage errors for selected gaps. When comparing simulation and measurement, the simulation with compensated material properties achieves a more accurate result than the uncompensated. The largest deviation occurs at a small air gap. For small gaps the material reluctance dominates the magnetic path. Due to the high magnetization at small gaps, the saturation flux density has a major influence at high excitation currents. Therefore the deviation between measurement and simulation is larger for a small gap and higher current. This is also observed in Figs. 12a and 12b.

4 Summary and Conclusion

This paper presents a non-destructive and FEM-based method for determining the initial B-H curve of a non-laminated steel test object where eddy current effects significantly affect the measurements. In addition to the hysteresis determination using voltage and current measurements, the resistivity of the test object material must be measured, and the test object modeled and simulated in FEM. The simulation results are used to compensate for the

Tab. 4: Percentage errors of the generated force comparison for 466 mA and different gaps

Gap z	Error in %	
in mm	Compensated	Uncompensated
0.1	−14.1	−34.7
0.2	9.4	−22.9
0.3	6.2	−18.1
0.5	−5.8	−19.3
1.0	−17.9	−25.2

Tab. 5: Percentage errors of the generated force comparison for 744 mA and different gaps

Gap z	Error in %	
in mm	Compensated	Uncompensated
0.1	−17.2	−27.2
0.2	−10.4	−27.8
0.3	−0.5	−23.4
0.5	−1.3	−17.0
1.0	−7.0	−15.2

influence of eddy current effects in the solid material of the test object to achieve the most accurate material characterization. Depending on the particular application, it is possible to measure the initial B-H curve of a test object, which would otherwise be impossible without cutting the test object because of the hardware limitations of equipment in many laboratories. As the experimental verification shows, more accurate results can be obtained with the compensated material characteristic.

However, the method is limited by the occurrence of significant eddy current effects caused by the combination of material properties (permeability and resistance), the cross section size and excitation frequency. This leads to an error in the determination of the magnetic field strength, derived from the measured source current, due to the assumption of a purely ohmic equivalent impedance for the eddy current losses, and in the simulation due to the simplification of a linear permeability of the test material.

Furthermore, the compensation is very sensitive to the true relative permeability $\mu_{R,true}$, as can be seen in Fig. 7. Thus the quality of the compensation is highly dependent on the accuracy of the measurements and FEM simulation.

The method is not limited to a specific test object shape and can be applied to all types of magnetic components. However, core shapes with homo-geneous fields, such as the toroidal core, are better suited for the proposed method. Otherwise, in addition to the eddy current effects, geometrical influences such as corners or a changing cross-sectional area would lead to further deviations due to the simplification of a linear permeability in the FEM simulation. Better results could be obtained with a smaller cross-sectional area or thickness of the test core as well as with a lower excitation frequency.

References

[1] A. Krings and J. Soulard, "Experimental characterization of magnetic materials for electrical machine applications," in *2015 IEEE Workshop on Electrical Machines Design, Control and Diagnosis (WEMDCD)*, 2015, pp. 85–89. DOI: 10.1109/WEMDCD.2015.7194514.

[2] M. Muhit, "Magnetic and electric characterization of materials for electrical machines," Jan. 2011.

[3] S. Lee, J.-T. Park, and S.-J. Kim, "Examination of magnetic properties of nonoriented electrical steels using ring-type specimens," *Journal of Magnetism and Magnetic Materials*, vol. 557, p. 169 471, 2022. DOI: https://doi.org/10.1016/j.jmmm.2022.169471.

[4] A. Krings and J. Soulard, "Overview and comparison of iron loss models for electrical machines," *Journal of Electrical Engineering*, vol. 10, pp. 162–169, May 2010.

PCIM Europe 2024, 11– 13 June 2024, Nuremberg   DOI: 10.30420/566262060

Highly-Compact Bearingless Axial-Flux Motor for a Pediatric Implantable Fontan Blood Pump

A. Horat [1], R. V. Giuffrida [1], S. Miric [2], M. Granegger [3], M. Hübler [4], J. Huber [1], J. W. Kolar [1]

[1] Power Electronic Systems Laboratory, ETH Zurich, Zurich, Switzerland
[2] Innsbruck Drive and Energy Systems Laboratory, University of Innsbruck, Innsbruck, Austria
[3] Department of Cardiac Surgery, Medical University of Vienna, Vienna, Austria
[4] Universitätsklinikum Hamburg-Eppendorf (UKE), Hamburg, Germany

Corresponding author: Andreas Horat, horat@lem.ee.ethz.ch
Speaker: Andreas Horat, horat@lem.ee.ethz.ch

Abstract

A pediatric implantable rotary blood pump (RBP) is under development in a research collaboration between the ETH Zurich, the University of Innsbruck, the University Medical Center Hamburg-Eppendorf and the Medical University of Vienna, in order to assist the Fontan circulation in newborns with a single ventricle. The RBP is driven by a small bearingless double-stator axial-flux permanent magnet synchronous machine, providing 2.2 mNm of torque at a rotational speed of 5500 rpm. The paper tackles the crucial challenge of defining an axial/angular position measurement concept with adequate resolution and bandwidth to enable magnetic levitation. As the sensors are integrated close the motor's winding, the measurement signals are significantly disturbed by stray fields generated by the phase currents. Such disturbances are compensated to obtain a usable signal for closed-loop position control. The experimental results show that, with the proposed compensation, the measurement errors are reduced to only 45 µm and 2.2°.

1 Introduction

One out of thousand children is born with a single ventricle [1], and the only treatment option is to undergo a series of three surgeries to create the Fontan circulation. Unfortunately, this is just a palliative solution and the patients' life quality and expectancy remains significantly limited. In fact, in the Fontan circulation, the single ventricle is responsible for sustaining both the systemic and pulmonary circulations, which are surgically connected in series through the total cavopulmonary connection (TCPC). This can lead to a progressive decline of the hemodynamics, ventricular failure, and even premature death [2]. These adverse consequences can be mitigated by using a cavopulmonary assist device (CPAD), i.e., an implantable blood pump. In the last decades, a large variety of rotary blood pumps (RBPs) has been developed and a few found its way to the clinical practice, but mainly for adult patients [3]. Only few VADs, mostly based on RBPs, meet the spatial requirements for long-term pedi-

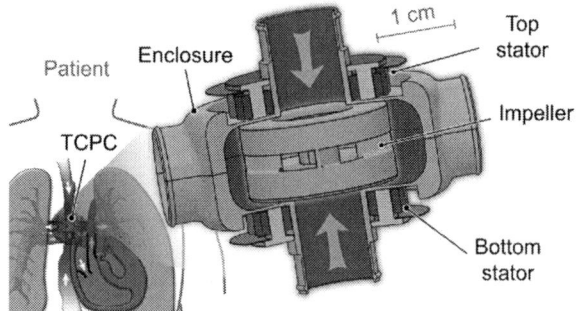

Fig. 1: The pediatric implantable cavopulmonary support device. The small blood pump is located at the total cavopulmonary connection (TCPC).

atric implantation, and could be re-purposed in the course of pilot studies [4]. In contrast, the development of (natively) pediatric implantable VADs has lagged behind, with the result that the only device approved is the Berlin EXCOR, a paracorporeal blood pump to support the left or right ventricle [5].

Recently, the company also introduced a cannula to support the cavopulmonary circulation of Fontan patients [6]. The *PediaFlow* [7], [8] and the *Jarvik 2015* [9] are two promising examples of natively pediatric VADs, where unfortunately only the first one is still under development. However, they are not tailored for cavopulmonary support in Fontan patients. Noticeably, a RBP based on a Bearingless Dual-Stator Axial-Flux Motor has been proposed [10], but it is not specifically designed for cavopulmonary support in Fontan patients, with a ≈ 30 % larger volume with respect to the proposed RBP. Furthermore, the chosen position sensors used to levitate the impeller are not integrated within the motor's volume. In this context, a long-term assistive cavopulmonary device, specifically designed to support the Fontan circulation, has been proposed in [11]. As illustrated in **Fig. 1(a)**, the small rotary blood pump is located at the TCPC, and propels the blood entering from its two axial inlets centrifugally to the two radial outlets. A first version of the pump with mechanical bearings has been designed, realized, and experimentally verified [11], [12]. The next step in the development is to exploit the key advantages of a magnetically levitated impeller, as in state-of-the-art third-generation rotary blood pumps [13]. This paper presents the electric motor required to drive the impeller and enable magnetic levitation, a Bearingless Dual-Stator Axial-Flux Motor (BDSAFM), i.e., with simultaneous bearing force and torque generation. Integrating magnetic bearings in the existing motor design is challenging. Of crucial importance is the choice of the appropriate displacement sensors, which is the focus of the paper. The sensors are required to measure and control the axial and angular positions of the impeller with sufficient accuracy and bandwidth. In particular, the number of sensors and their location in the system has to be defined properly, to guarantee satisfactory measurement accuracy. Furthermore, due to the targeted pediatric use, the device needs to be extremely compact, which significantly restricts the design options.

The paper is organized as follows. **Sec. II** describes the BDSAFM, explaining how torque and bearing force are generated. **Sec. III** discusses the position measurement concept with details on the sensors chosen, their location in the motor, and the required compensation of the measurement disturbances due to the phase currents in the winding. **Sec. IV** presents the realized motor prototype and

Tab. 1: Geometric parameters of the BDSAFM.

Name	Symbol	Value	Unit
Inner diameter	d_{in}	11.5	mm
Outer diameter	d_{out}	20.5	mm
Stator height	h_{st}	3.8	mm
Coil height	h_{c}	2.8	mm
PM thickness	d_{PM}	1.6	mm
Nominal air gap	d_{ag}	1.3	mm

the test bench needed to commission it. **Sec. V** reports the experimental measurements demonstrating successful correction of the measurement disturbances. Finally, **Sec. VI** concludes the paper.

2 Bearingless Dual-Stator Axial-Flux Motor

The proposed BDSAFM that drives and magnetically levitates the impeller is shown in **Fig. 2(a)**.

The chosen PMSM topology is based on the first version of the pump with mechanical bearings [11], [12]. The geometric parameters of the motor are reported in **Table 1**. Each of the two (top and bottom) stators has $N_{\mathrm{s}} = 9$ slots and a multi-layer concentrated winding with $N_{\mathrm{c}} = 9$ coils having $N_{\mathrm{t}} = 106$ turns. Embedded in the impeller, there are two rotors, each interacting with one stator. Each rotor consists of a set of axially magnetized permanent magnets, creating $N_{\mathrm{p}} = 3$ rotor pole pairs. Each set is backed by a thin disc of ferromagnetic material, which closes the magnetic circuit. Simultaneous torque and bearing force generation with the proposed BDSAFM can be explained considering the one-third sector view in **Fig. 1(b)**. Using field-oriented control with the electrical angle $\varphi_{\mathrm{el}} = N_{\mathrm{p}}\,\varphi_{\mathrm{mech}}$, tangential force (and

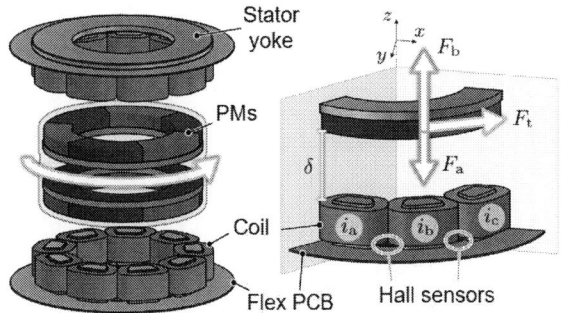

Fig. 2: The proposed bearingless, dual-stator, axial-flux motor (BDSAFM), together with a one-third section view to visualize its operating principle.

hence torque) generation is standard, i.e., it is obtained with the quadrature-component of the stator currents $i_{\mathrm{q},\{\mathrm{top,bot}\}}$ in the rotor-oriented frame:

$$M_{\mathrm{d}} = k_{\mathrm{m}}\left(i_{\mathrm{q,top}} + i_{\mathrm{q,bot}}\right). \tag{1}$$

In a BDSAFM, the bearing force is instead generated by a combination of field weakening with one stator and field strengthening with the other, using the direct-component of the stator currents $i_{\mathrm{d},\{\mathrm{top,bot}\}}$ in the rotor-oriented frame, hence

$$F_{\mathrm{b}} = k_{\mathrm{b}}\left(i_{\mathrm{d,top}} - i_{\mathrm{d,bot}}\right). \tag{2}$$

The generated bearing force F_{b} is controlled to counteract the magnetic attraction force F_{a} and maintain the rotor at the nominal distance $\delta_{\{\mathrm{top,bot}\}}$ from the corresponding stator. F_{a} is the resultant of the reluctance forces existing between each rotor and its corresponding stator, and can be linearly approximated as

$$F_{\mathrm{a}} = \left(F_{\mathrm{a,top}} - F_{\mathrm{a,bot}}\right) \approx k_{\mathrm{a}}\, z_{\mathrm{sec}}, \tag{3}$$

where $z_{\mathrm{sec}} = \delta_{\mathrm{top}} - \delta_{\mathrm{bot}}$ thus defined is the axial position of the impeller (at the considered sector).

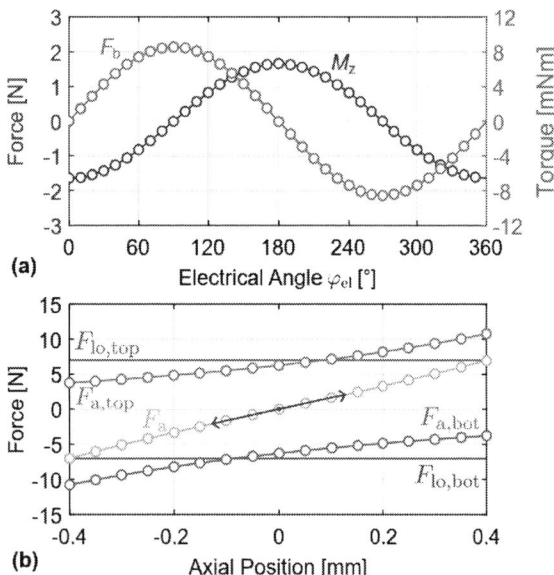

Fig. 3: (a) Simulated axial force and torque versus φ_{el} for 1 A of current amplitude. The peaks of the two curves are shifted by 90° electrical, indicating that force and torque generation are decoupled and correspond to the d- and q-components. **(b)** Simulated attraction forces for the top and bottom stators $F_{\mathrm{a},\{\mathrm{top,bot}\}}$ and their difference, i.e. the net attraction force F_{a}, with a negative slope (stiffness).

This force has a destabilizing effect, as it points towards the stator with the smaller air gap length, and it would be zero only in the ideal condition of a perfectly axially centered impeller. Finally, to achieve full 6-DoF controllability of the impeller's position and orientation, i.e, including tilting about the x and y axes, at least three distinct bearing forces are required. Therefore, each sector needs to be controlled individually, which requires a total of 9 currents per stator, i.e., 18 phase currents for the full motor. The achievable torque and bearing machine constants are predicted using Finite Element Method (FEM) simulations. The results reported in **Fig. 3(a)** clearly show decoupled generation of the drive torque M_{z} and the bearing force F_{b} as indicated by **Eq. (1)** and **Eq. (2)**, with the machine constants $k_{\mathrm{m}} = 8.6\,\mathrm{mN\,m/A}$ and $k_{\mathrm{b}} = 1.7\,\mathrm{N/A}$. Furthermore, an attraction constant (negative stiffness) of $k_{\mathrm{a}} = 20\,\mathrm{N/mm}$ is predicted (cf. **Fig. 3(b)**). Importantly, the simulated force at $z = \pm 0.4\,\mathrm{mm}$ represents the liftoff force $F_{\mathrm{lo}} = 8\,\mathrm{N}$ that needs to be provided to the impeller at the startup.

3 Position Measurement Concept

To control the axial and angular positions of the impeller, accurate measurements are required. This section presents the proposed position measurement concept, with details on the chosen type of sensors, their number and location in the motor, and the compensation of the undesired measurement disturbances due to the phase currents.

3.1 Type/Number of Sensors and Post-Processing

Measuring the position of a sealed impeller requires contactless position sensors such as eddy-current or Hall sensors. Eddy-current sensors typically offer superior signal quality due to their high-frequency operating principle. [14] In contrast, Hall sensors are less robust against low frequency disturbances. Nevertheless, due to the extremely tight spatial constrains of the considered application, Hall sensors are selected. They are simple to use and available in miniature packaging formats, which allows for close integration within the volume of the motor. As illustrated schematically in **Fig. 4**, the proposed measurement concept requires six Hall sensors per sector, i.e., three per side, electrically displaced by 120° (corresponding to 40° mechanically with $N_{\mathrm{p}} = 3$). This way, the measured magnetic flux densities $h_{\{1,2,3\}}$ and $h_{\{4,5,6\}}$ are ideally three 120°-phase-shifted sinusoidal signals for both top and bottom

PCIM Europe 2024, 11– 13 June 2024, Nuremberg DOI: 10.30420/566262060

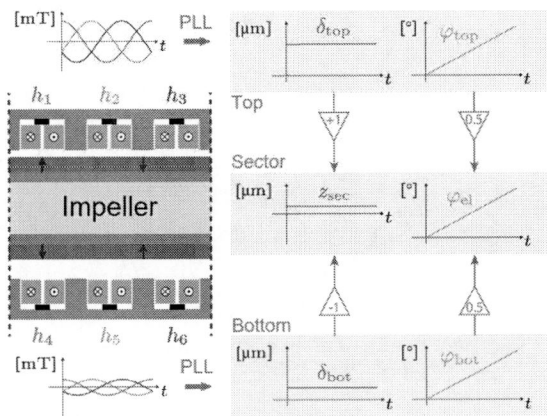

Fig. 4: Proposed measurement concept, requiring six Hall sensors per sector (three per side top/bottom). The measured flux densities are post-processed with PLLs to obtain an axial and angular position measurement per side $\delta_{\{top,bot\}}$ and $\varphi_{\{top,bot\}}$. The two are subtracted to obtain the axial position measurement of the sector z_{sec}, and averaged for φ_{el}.

sides. From these, two PLLs are used to provide two measurements of the angular position of the impeller $\varphi_{\{top,bot\}}$. The angle $\varphi_{el} = 0.5 (\varphi_{top} + \varphi_{bot})$ is then used both for field-oriented control and to calculate the angular speed ω_z. The amplitude information returned by the PLLs is instead used to determine two measurements $\delta_{\{top,bot\}}$ of the axial position of the impeller. These are then subtracted to obtain the final differential measurement z_{sec} for each sector, finally used for magnetic levitation control.

3.2 Sensor Location and Sensitivity

In order to ensure that the signals $h_{\{1,...,6\}}$ meet the desired ideal conditions as far as possible, the most favorable location of the sensors in the motor is studied using 3D FEM simulations. This is an essential step, as it guides the selection of an appropriate Hall sensor offering sufficient axial position sensitivity and bandwidth. Due to the tight spatial constraints in the considered stator geometry, there is very few options. The outer dimensions of a single sensor (including connecting pins) are limited to a maximum of 2-by-2 mm, in order to fit within a stator slot. It is only possible to place a thin flex-PCB hosting the Hall sensors either above the coils (top of the teeth, **Case 1**) or below the coils (bottom of the teeth, **Case 2**), as visible in **Fig. 5(a)**. Moreover, the total thickness of the sensor and the flex-PCB should be minimized as far as possible, to avoid reducing the height (i.e., number of turns) of

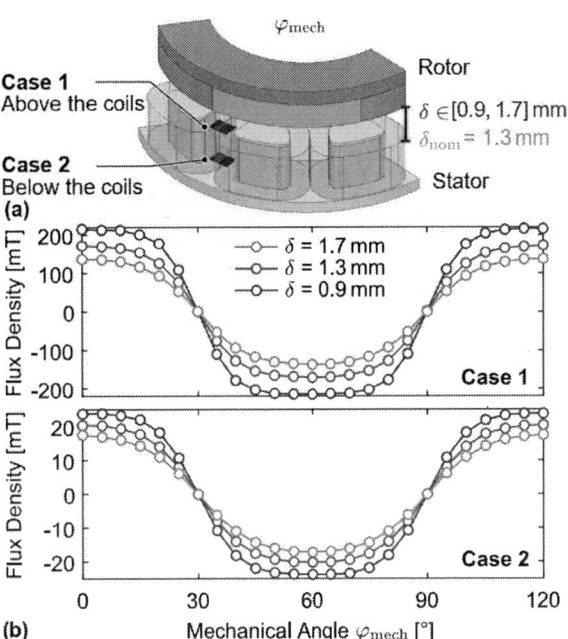

Fig. 5: **(a)** Sector view with two possible locations for the Hall sensors, i.e., above or below the coils. **(b)** Corresponding 3D-FEM results for the two considered locations and different values of the air gap δ. The signals exhibit a third harmonic distortion.

the coils excessively. The FEM results in **Fig. 5(b)** show the average magnetic flux density B_{hall}, calculated over the sensor planes indicated in **Fig. 5(a)**, versus the rotary position φ_{mech}. With a peak of 180 mT, the signal amplitude for **Case 1** is too large, with the risk of saturating the Hall element of the sensor, especially for the case $\delta_{min} = 0.9$ mm (z_{sec} = –0.4 mm, touchdown). An off-the-shelf sensor of suitable dimensions with such measurement range could in fact not be found. Therefore, as already visible in **Fig. 2**, the selected location is at the bottom of the stator's teeth, i.e., **Case 2**. With an amplitude of 17.5 mT at $\delta_{max} = 1.7$ mm (furthest), 20 mT at $\delta_{nom} = d_{ag} = 1.3$ mm (nominal), and 24 mT at δ_{min} = 0.9 mm (closest), the predicted axial position sensitivity is of about 8.1 mT/mm. The chosen sensor is the *ams AS5510*, which offers a resolution of 97 µT and a sampling frequency of 20 kHz. Therefore, the achievable axial position resolution is of 12 µm, which is adequately small compared to the considered 0.8 mm range of axial motion. Noticeably, for both cases there is a certain (spatial) third harmonic distortion on the measured flux density. As visible further in **Fig. 8**, this translates into a tolerable magnitude ripple of 50 µm.

483

3.3 Measurement Disturbance due to the Phase Currents

A crucial problem to consider is that, by being placed very closely to the motor's winding, the Hall sensors are inevitably disturbed by the phase currents. The entity of such disturbance can be studied in detail with the aid of FEM simulations. To first visualize and describe it qualitatively, consider the magnetic field distribution in **Fig. 6**, obtained with FEM simulations for a simplified 2D geometry. To solely include the magnetic field generated by the currents circulating in the coils, the PMs are removed from the model at this stage. It is hence assumed that the field component generated by the phase currents and the one generated by the PMs can be superimposed, and that the relative permeability of the PMs is $\mu_{r,PM} \approx \mu_0$. For this analysis, an ideal ferromagnetic linear material ($\mu_{r,Fe}$ = 4000) is chosen for the magnetic cores. Consider now the chosen sensor location at the bottom of a stator slot. When the two adjacent coils are energized, in this example with the same current $I_{left} = I_{right} = I_{test} = 1$ A, there is some stray field through the sensor at the selected sensor location. With an average value of 20 mT, this disturbance is as large as the measurement signal. Therefore, it is necessary to consider it explicitly, characterize it in detail, and finally compensate it. It is also a further reason to place the Hall sensors away from the rotor, as the current disturbance could potentially saturate the sensors.

3.3.1 Characterization on a Single Sensor

The detailed characterization of the disturbance introduced by the phase currents is conducted on the complete 3D FEM model of the motor (i.e., with nonlinear materials and PMs included). The

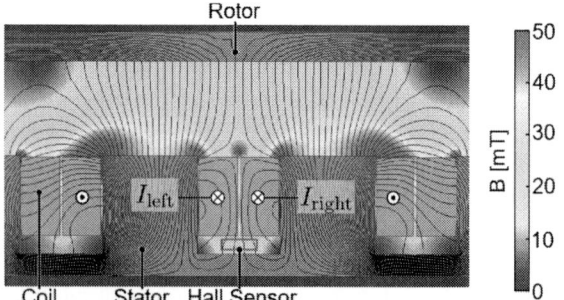

Fig. 6: Simplified 2D-FEM simulation to visualize the magnetic field distribution generated by the currents in the coils, causing a stray field that disturbs the sensor's measurement signal.

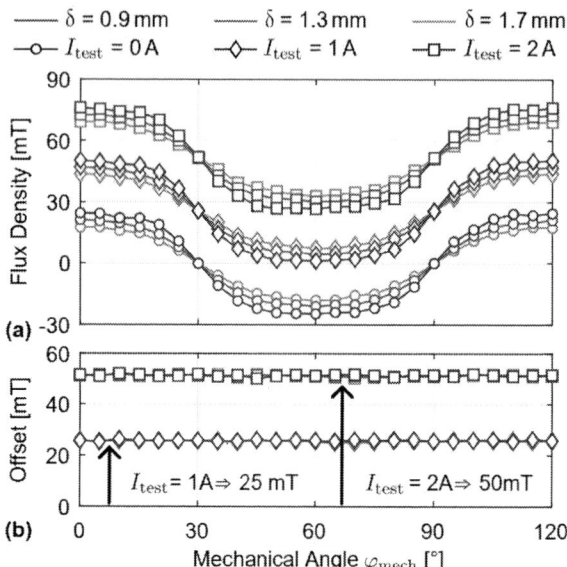

Fig. 7: FEM characterization of the measurement disturbance on a single sensor. **(a)** Average flux densities at the sensor plane for different axial positions and currents. **(b)** Calculated offsets for the cases I_{test} = 1 A and 2 A by subtracting the corresponding values obtained for I_{test} = 0 A. The offset scales linearly with current and does nearly not depend on the rotor's axial/angular position.

goal is to assess whether the amount of stray field through the sensor depends on the axial and angular positions of the rotor, and to verify its linearity with respect to the assigned current excitation in the adjacent coils. Therefore, the simulations are conducted for different angular and axial positions (again $\varphi_{mech} = \{0,5,...120\}$° and δ = $\{0.9,1.3,1.7\}$mm) and currents ($I_{test} = \{0,1,2\}$A), with the same current I_{test} in the two adjacent coils), obtaining the results of **Fig. 7(a)**. As can be observed, the flux density curves are the same as in **Case 2** of **Fig. 5(b)** for I_{test} = 0 A, and they are shifted up by a certain offset for the other two cases $I_{test} = \{1,2\}$A. This can be better visualized in **Fig. 7(b)**, showing the difference between the signals for $I_{test} = \{1,2\}$A and the corresponding ones for I_{test} = 0 A. With an offset of 25 mT for I_{test} = 1 A and 50 mT for I_{test} = 2 A, the linearity of the disturbance is verified, with 25 mT/A. Furthermore and importantly, the disturbance offset does not exhibit any relevant dependency on the axial or angular positions of the rotor. This result is reasonable and can be understood by observing how the stray field closes within the stator itself in **Fig. 6**.

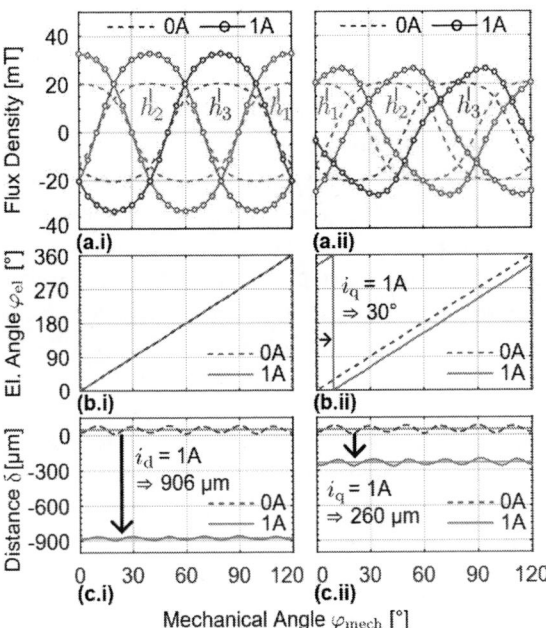

Fig. 8: FEM characterization of the measurement disturbance for typical operation with **(.i)** i_d = 1 A and **(.ii)** i_q = 1 A. **(a)** Average flux densities $h_{\{1,2,3\}}$ at the three sensor planes of a sector side (top or bottom). **(b)** Calculated angle using the PLL. **(c)** Calculated amplitude using the PLL.

3.3.2 Characterization for Typical Operation

With these intermediate simulation results, the analysis can be brought a step further. Assigning the same current to both the adjacent coils is useful to study and characterize the disturbance on a single sensor, but is a case that never occurs in practice. Two realistic cases for typical operation are when d- and q-currents are assigned to the motor's winding for field-oriented control. As seen, the former is used to generate the bearing force, whereas the latter for the drive torque. For the FEM results of **Fig. 8**, the nominal axial position δ = 1.3 mm (z_{sec} = 0 mm) and 1 A of d- or q-current are considered. The analysis is extended to one side of a sector, i.e., considering three sensor signals $h_{\{1,2,3\}}$. For the d-case, $h_{\{1,2,3\}}$ are amplified by a factor 1.6, which is due to field strengthening. This does not influence the electrical angle φ_{el} detected by the PLL, but of course strongly influences the amplitude and hence the detected axial position δ, with a large error of 906 µm. Conversely, for the q-case, the signals $h_{\{1,2,3\}}$ are rather distorted. This affects both the detected electrical angle φ_{el}, with about 30° of error, and the amplitude δ, with a less pronounced but still relevant 260 µm error. Therefore,

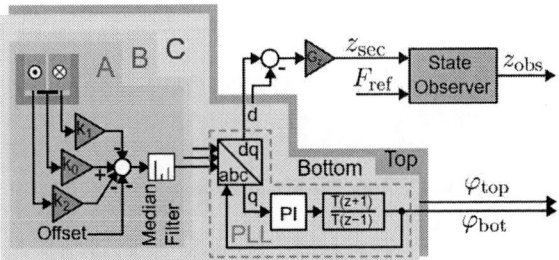

Fig. 9: Diagram of the post-processing of the Hall signals, with the proposed disturbance compensation (one sector).

as expected, it is essential to compensate the studied disturbance, i.e. to derive a signal suitable for controlling the impeller's position.

3.4 Proposed Disturbance Compensation

The complete diagram of the Hall sensor signals post-processing is shown in **Fig. 9**. The disturbance due to the phase currents is compensated individually for each sensor before feeding the signals to the PLLs. This is because, in practice, the manufacturing tolerances will influence the final position of each sensor in the motor. Due to the verified linearity, the disturbance can be modeled by a simple weighted sum and can be compensated as

$$h_{comp} = k_0 \cdot h_{raw} - \hat{k}_1 \cdot I_1 - \hat{k}_2 \cdot I_2, \qquad (4)$$

where $k_0 \approx 1$ is an adjustment factor to account for sensor and manufacturing tolerances. The estimated factors $\hat{k}_1$ and $\hat{k}_2$ depend on the exact coil geometry and sensor position and should hence be specifically calibrated. As explained further in **Sec. 5.1**, due to the limited sampling frequency of the Hall sensor and its inherent transmission delay, a median filter is used to further improve the signals when an incorrect compensation occurs. The further processing of the compensated signals is straightforward and follows the same steps already illustrated in **Fig. 4**.

4 Hardware Prototype and Test Bench

The hardware prototype of the BDSAFM is shown in **Fig. 10**. The stator yoke and the rotor back iron are realized with special ferromagnetic materials, namely the *Valbruna MG2* and the *VACOFLUX50*, which offer high magnetic saturation (2 T and 2.3 T), and are hence particularly suited for a compact realization. The nine concentrated coils have N_t =

106 turns and are realized with a 0.125 G1B enameled copper wire. Finally, the permanent magnets are custom made with thickness $d_{\mathrm{PM}} = 1.6\,\mathrm{mm}$ and N48 magnetization grade. The motor is fed by a custom 18-phase inverter, based on the *MP6536* (*Monolithic Power Supply*) three-phase inverter module. The controller is implemented on a *AMD Xilinx Zynq 7020* module. Due to the number of current controllers to be executed simultaneously, they are implemented in the FPGA. They are executed with an update rate of 500 kHz allowing a bandwidth of $f_{\mathrm{cc}} = 10\,\mathrm{kHz}$. To commission the proposed BDSAFM, the test bench in **Fig. 10** is realized. The impeller assembly (containing the two rotors) is connected to a 6-axis force/torque sensor (*BOTA systems*) mounted to a top fixture. This allows measuring not only the generated torque, the axial attraction force, and the bearing force, but also the tilting torques applied to the impeller. This can be done for different relative positions of the impeller and the stator assembly, which is mounted on two precision vertical (*standa 7VT174-5*) and rotary (*standa 7R150*) positioning stages, fixed to the baseplate. The stator assembly features an interconnection PCB and a round stainless steel mount replacing the pump's enclosure.

5 Experimental Results

This section presents the experimental measurements conducted on the presented testbench. These include the verification of the machine constants and of the measured sensor signals, with

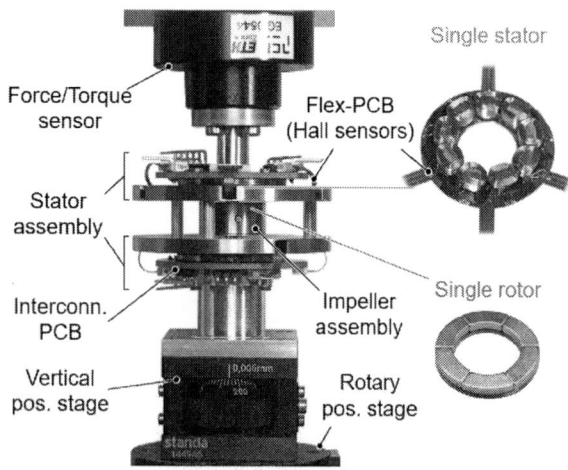

Fig. 10: Experimental test bench for the proposed BD-SAFM. The stator and rotor assemblies are shown in detail.

particular focus on the proposed disturbance compensation.

5.1 Machine Constants and Sensor Signals Verification

The measurements in **Fig. 11(a)** demonstrate successful and well-decoupled torque and bearing force generation, with the torque constant $k_{\mathrm{m}} = 9.1\,\mathrm{mN\,m/A}$ and the bearing constant $k_{\mathrm{b}} = 1.6\,\mathrm{N/A}$ per stator (top and bottom). The measured attraction constant is $k_a = 15\,\mathrm{N/mm}$. With the maximum z-position of $z_{\mathrm{max}} = 0.4\,\mathrm{mm}$, it results to a maximum required liftoff force of $F_{\mathrm{lo}} = 6.0\,\mathrm{N}$. The required liftoff current is hence $i_{\mathrm{d,lo}} = 1.88\,\mathrm{A}$. Finally, the measured Hall sensor signals are shown in **Fig. 11(b)**. The signal amplitude of 25 mT corresponds to the predicted one from FEM, and also the expected third harmonic distortion is visible.

5.2 Disturbance Compensation: Calibration

The individual position of each sensor introduces a gain factor, which is measured by normalizing the sensor amplitude when rotating the mover at a fixed vertical and radial position. It is crucial that the impeller is centered vertically, such that the difference of the amplitudes is zero. The sensor zero value offset is specified as maximum 0.45 mT, translating to

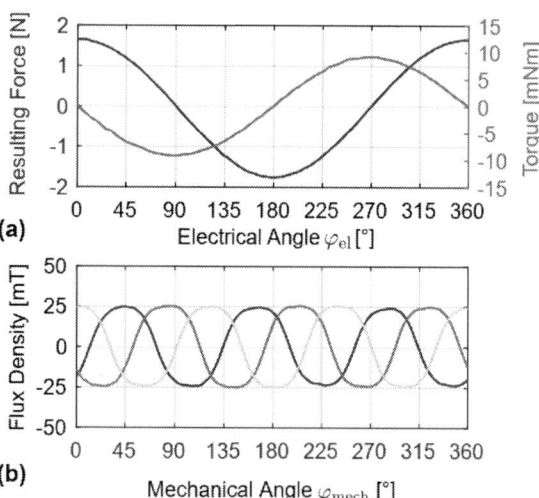

Fig. 11: (a) Measured force and torque generated for a fixed, horizontal rotor position and a rotating stator field with $\hat{I} = 1\,\mathrm{A}$. The torque- and bearing force-generating components are clearly decoupled. (b) Measured sensor signals of a sector side for a complete revolution of the impeller at the nominal (centered) axial position.

a maximum offset of 30 bit. This offset is measured without a rotor present and zero phase currents and then subtracted from the measured signal. The disturbance compensation factors ($\hat{k}_1$ and $\hat{k}_2$, for each sensor) are calibrated at the nominal (centered) axial position $z = 0\,\mu m$ with the rotor mechanically fixed and positioned using the test bench. To determine the individual compensation factors for each sensor, the electrical angle is assigned such that only one of the two adjacent coils at a time carries a current. As it is not possible to thermally sustain a continuous DC current in the winding for the whole duration of the measurements, it is decided to apply four short (20 ms duration) current steps with amplitudes from 0.5 A to 2 A in steps of 0.5 A. As shown in **Fig. 12** (red signal – uncompensated), the linearity of the disturbance is verified, at least at the considered axial position, with 10 mT/A. However, for the 2 A case, a slight saturation of the rotor core leads to an overcompensation. All in all, the proposed compensation proves to be effective for currents till about 1.5 A and low frequency, with a maximum residual signal disturbance of only 2 mT. Importantly, the magnified signal edge shows that the compensation results incorrect for the first two samples after a current step. This is due to the

low possible readout frequency of the Hall sensor, i.e. 20 kS/s, compared with the high current sampling rate of 500 kHz and controller bandwidth of 10 kHz. For the chosen Hall sensor, the frequency behavior is not entirely specified, and it is hence investigated experimentally. No low pass behavior could be observed up to 2 kHz. However, the sensor exhibits a zero-order hold characteristic with 50 μs hold time. This can only be compensated accurately if the sampling frequency is well above the signal bandwidth or the sampling instant is exactly known. Neither is the case for fast-changing current steps with a current controller bandwidth of 10 kHz. This leads to an over- or under-compensation for one to two measurement samples. These disturbances are estimated to be above 100 μm, which is not acceptable for magnetic levitation. For this reason, a median filter with window three is used to successfully suppress the uncompensated spikes with minimal additional phase lag.

5.3 Disturbance-Compensated Measurements for Typical Operation

Once the compensation factors for each sensor are calibrated, the effectiveness of the proposed compensation for the realistic case of d- and q-current excitations can be assessed experimentally. Also in this case the rotor is fixed in the test bench and its position adjusted with the positioning stages. Two short d- and q-current steps are commanded. The amplitude of the d-current step is inverted to -1 A, to emulate the current step necessary for liftoff in the bottom stator. The measurements are conducted for three axial positions ($z = \{-0.4, 0, 0.4\}$ mm) and three angular positions ($\varphi_{el} = \{0, 30, 60\}°$). An exemplary repetition is reported in **Fig. 13**. The three Hall sensors of one sector are recorded (cf. **Fig. 13 (a)**) and used to derive the axial and rotational positions. The d-current step leads to an axial position error of 600 μm, whereas the rotational angle is only disturbed marginally, with an error of about 2°. Also in this case, if uncompensated, the disturbance in the axial position is larger than the axial motion range. The disturbance on the rotational angle is non-zero due to machine imperfections. For the q-current step, the expected angle error of 24° can be measured, as well as an error in the axial position measurement of 200 μm. The proposed compensation reduces the influence of the stray magnetic field created by the phase currents by a factor of about 10 to a value below $\pm 45\,\mu$m.

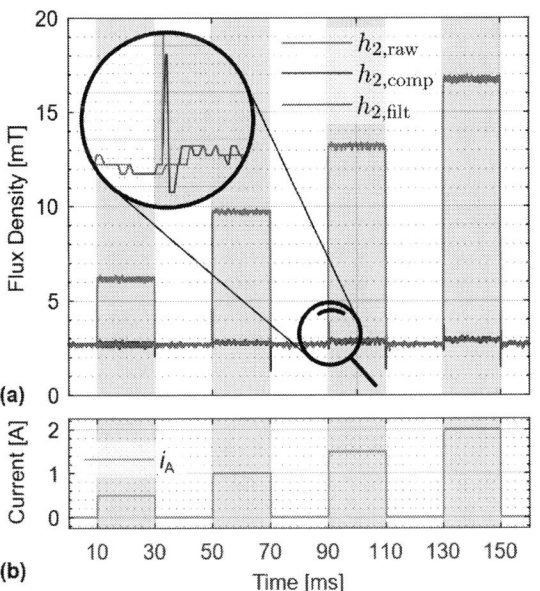

Fig. 12: Investigation of the measurement disturbance on a single sensor without and with compensation. The zoomed view shows the effectiveness of the median filter in suppressing the signal spikes occurring due to incorrect compensation at the edge of the current steps.

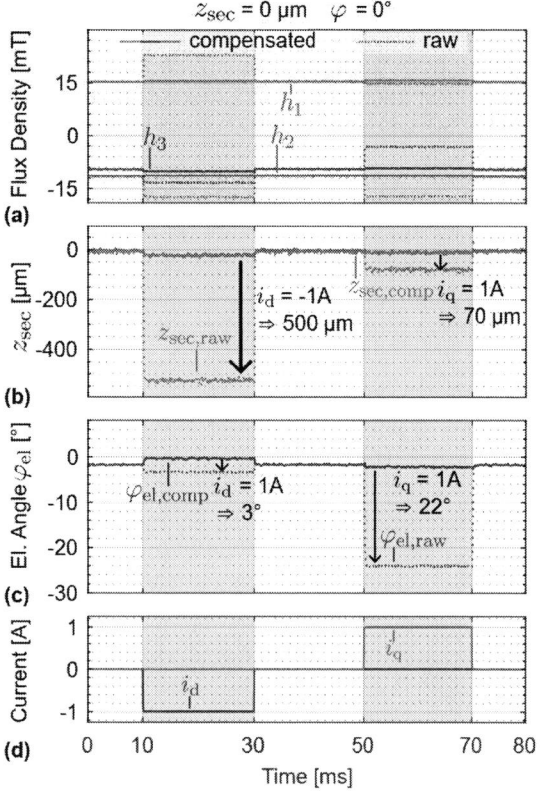

Fig. 13: Final measurement of the calculated z_{sec} and φ_{el} in one sector shown as measured (raw) values and with the proposed compensation over time. **(a)** Flux density $h_{1,2,3}$ of the three sensors. **(b)** measured z_{sec} disturbance. **(c)** electrical angular disturbance φ_{el}. **(d)** current pulses in d- and q-axis.

z_{sec}	$-0.4\,\text{mm}$		$0\,\text{mm}$		$0.4\,\text{mm}$	
φ_{el}	i_d	i_q	i_d	i_q	i_d	i_q
$0°$	14.4	14.0	23.9	31.7	45.0	18.0
$30°$	17.0	6.9	29.6	33.2	39.2	14.2
$60°$	18.1	8.1	33.7	25.0	42.2	10.9

Tab. 2: Maximum absolute z-position error in micrometer for all six sectors at maximum and minimum height and three electrical angles when disturbed with $i_d = -1\,\text{A}$ or $i_q = 1\,\text{A}$.

z_{sec}	$-0.4\,\text{mm}$		$0\,\text{mm}$		$0.4\,\text{mm}$	
φ_{el}	i_d	i_q	i_d	i_q	i_d	i_q
$0°$	1.53	0.58	1.33	1.44	2.02	1.89
$30°$	1.46	1.17	1.19	0.62	1.97	0.78
$60°$	1.90	2.03	1.47	1.35	1.76	0.27

Tab. 3: Maximum absolute angular position error in degrees for all six sectors at maximum and minimum height and three electrical angles when disturbed with $i_d = -1\,\text{A}$ or $i_q = 1\,\text{A}$.

6 Conclusion

Pediatric implantable rotary blood pumps (RBPs), specifically designed for cavopulmonary support in Fontan circulation and currently under development in a research collaboration between ETH Zürich, the University of Innsbruck, the University Medical Center Hamburg-Eppendorf and the Medical University of Vienna shall exploit the key advantages of magnetic bearings for improved durability and hemocompatibility. However, the design and integration of extremely compact bearingless motors poses numerous challenges, especially for the position measurement of the levitated impeller. In this paper, an axial/angular position measurement concept for a compact bearingless double-stator axial-flux motor (BLDSAFM) has been proposed. Importantly, this requires a compensation of the measurement disturbances caused by the currents in the motor windings. The experimental results demonstrate that the compensation works and that clean and usable position measurement signals with disturbances below 45 µm and 2.2° can be achieved. Therefore, in future work, the proposed sensing concept can be used for closed-loop position control of the BLDSAFM.

Acknowledgments: The authors are very much indebted to the "EVER Foundation", which generously supports the research on rotary blood pumps for cavopulmonary support in Fontan patients at the Power Electronic Systems Laboratory of ETH Zurich.

References

[1] P. W. O'Leary, "Prevalence, Clinical Presentation and Natural History of Patients with Single Ventricle," *Progress in Pediatric Cardiology*, vol. 16, no. 1, pp. 31–38, 2002.

[2] M. Gewillig, "The Fontan Circulation," *Heart*, vol. 91, no. 6, pp. 839–846, 2005. DOI: 10.1136/HRT.2004.051789.

[3] Ş. M. Moisă, A. Burlacu, C. Brinza, E. Cinteză, L. I. Butnariu, *et al.*, "An Up-to-Date Literature Review on Ventricular Assist Devices Experience in Pediatric Hearts," *Life*, vol. 12, no. 12, 2022. DOI: 10.3390/life12122001.

[4] I. Adachi, S. Burki, and C. D. Fraser, "Current Status of Pediatric Ventricular Assist Device Support," *Seminars in Thoracic and Cardiovascular Surgery: Pediatric Cardiac Surgery Annual*, vol. 20, pp. 2–8, 2017. DOI: 10.1053/j.pcsu.2016.09.010.

[5] C. S. Almond, D. L. Morales, E. H. Blackstone, M. W. Turrentine, M. Imamura, *et al.*, "Berlin Heart EXCOR Pediatric Ventricular Assist Device for Bridge to Heart Transplantation in US Children," *Circulation*, vol. 127, no. 16, pp. 1702–1711, 2013. DOI: 10.1161/CIRCULATIONAHA.112.000685.

[6] B. Karner, E. Urganci, J. Schlein, E. Base, S. Greil, *et al.*, "First-in-man use of the EXCOR Venous Cannula for Combined Cavopulmonary and Systemic Ventricular Support in Fontan Circulation Failure," *The Journal of Heart and Lung Transplantation*, vol. 41, no. 10, pp. 1533–1536, 2022. DOI: 10.1016/j.healun.2022.06.009.

[7] H. S. Borovetz, S. E. Olia, and J. F. Antaki, "Toward the Development of the PediaFlow™ Pediatric Ventricular Assist Device: Past, Present, Future," *Applications in Engineering Science*, vol. 11, pp. 100–113, 2022. DOI: 10.1016/j.apples.2022.100113.

[8] M. D. Noh, J. F. Antaki, M. Ricci, J. Gardiner, D. Paden, *et al.*, "Magnetic Design for the PediaFlow Ventricular Assist Device," *Artificial Organs*, vol. 32, no. 2, pp. 127–135, 2008. DOI: 10.1111/j.1525-1594.2007.00501.x.

[9] I. Adachi, S. Burki, F. Zafar, and D. L. S. Morales, "Pediatric Ventricular Assist Devices," *Journal of Thoracic Disease*, vol. 7, no. 12, pp. 2194–2202, 2015. DOI: 10.3978/j.issn.2072-1439.2015.12.61.

[10] M. Osa, T. Masuzawa, K. Yamaguchi, and E. Tatsumi, "Double Stator Axial Gap Type Ultra-Compact 5-DOF Controlled Self-Bearing Motor for Rotary Pediatric Ventricular Assist Device," *IEEE Transactions on Industry Applications*, vol. 57, no. 6, pp. 6744–6753, 2021.

[11] A. Escher, C. Strauch, E. J. Hubmann, M. Hübler, D. Bortis, *et al.*, "A Cavopulmonary Assist Device for Long-Term Therapy of Fontan Patients," in *Seminars in Thoracic and Cardiovascular Surgery*, Elsevier, vol. 34, 2022, pp. 238–248.

[12] E. J. Hubmann, D. Bortis, M. Flankl, J. W. Kolar, M. Granegger, *et al.*, "Optimization and Calorimetric Analysis of Axial Flux Permanent Magnet Motor for Implantable Blood Pump Assisting the Fontan Circulation," in *Proc. of the IEEE 22nd Intern. Conf. on Electr. Machines and Systems (ICEMS)*, IEEE, 2019, pp. 1–8.

[13] H. Hoshi, T. Shinshi, and S. Takatani, "Third-Generation Blood Pumps with Mechanical Noncontact Magnetic Bearings," *Artificial Organs*, vol. 30, no. 5, pp. 324–338, 2006.

[14] A. J. Fleming, "A review of nanometer resolution position sensors: Operation and performance," *Sensors and Actuators A: Physical*, vol. 190, pp. 106–126, 2013. DOI: 10.1016/j.sna.2012.10.016.

PCIM Europe 2024, 11– 13 June 2024, Nuremberg   DOI: 10.30420/566262061

A Novel Permanent Magnet Synchronous Motor Drive for Reaction Wheels in Satellites

Barış ÇOLAK[1], Süleyman ÇETİNKAYA[1]

[1] TUBITAK UZAY (Space Technologies Research Institute), TÜRKİYE

Corresponding author: Barış Çolak, baris.colak@tubitak.gov.tr
Speaker: Barış Çolak, baris.colak@tubitak.gov.tr

Abstract

TUBITAK UZAY's Reaction Wheel offers zero-crossing and micro-vibration qualities, as well as satisfactory performance in terms of torque, momentum storage, and wheel speed measuring. A reliable, cost-effective, and flexible motor drive that uses only inexpensive, low-resolution hall sensors to achieve field-oriented control is proposed. An anti-fuse FPGA is employed as a digital controller, and the 3-phase inverter uses 50 kHz PWM. These help to enhance the transient response of the torque control loop and reduce the torque ripple of the motor compared to conventional drives employing a trapezoidal control system.

1   Introduction

For the last decade, TUBITAK UZAY (Space Technologies Research Institute) has been developing equipment for indigenously built satellite projects. One of the equipment developed by this initiative is the Reaction Wheel (RW), whose technological knowledge has been qualified and incorporated into TUBITAK UZAY's expertise. This equipment has been built for IMECE, TURKSAT 6-A, and AYAP-1 missions, which are Low Earth Orbit (LEO) earth observation satellite, Geosynchronous Orbit (GEO) telecommunication satellite, and Lunar Research Project Phase-1 satellite, respectively [1].

RW's are momentum-exchanging components, which are comprised of a high inertia rotor (flywheel) coupled to an electric motor and are used for precision control of the attitude and stabilization of satellites [2] [3]. Typically, four RW's (4th for redundancy considerations) are required to control the attitude of the satellite in 3-axis [4] [5]. They are commonly employed in small and medium-sized spacecraft due to their wide torque range, low power consumption, and mechanical simplicity, resulting in good reliability, simple utilization, and low manufacturing costs.

However, the RW design must meet numerous restrictions to ensure precise and effective satellite control, including supporting launch requirements and space environment risks. RW's have to operate nearly continuously during the lifetime of the spacecraft. Moreover, to stabilize the satellite's attitude, the torque management of the wheels needs to be highly accurate.

Conventionally, reaction wheels with heritage employ a brushless DC motor and a driver with trapezoidal control [6] [7]. The limited capacity of the space components market (due to harsh environmental challenges) and reliability specifications have restricted the performance and variety of the motor drives used in space applications.

The utilization of commercial off-the-shelf components and recent advancements in space component technology have resulted in enhancements to motor drives. This paper employs these advances to present a novel field-oriented controlled PWM drive for a Permanent Magnet Synchronous Motor (PMSM).

This paper proposes a reliable, cost-effective, and flexible motor drive. Instead of complex, less reliable, and fragile speed/position encoders, only inexpensive, low-resolution hall sensors achieve field-oriented control. After processing the signals from the hall sensors, the controller estimates the rotor's location and speed to minimize torque ripple and provide smooth motor control.

An anti-fuse FPGA is employed as a digital controller, and the 3-phase inverter uses 50 kHz PWM. These help to enhance the torque control loop's transient response and reduce the motor's torque ripple. The ripple generated by the new motor drive is %50 reduced compared to the conventional control methods.

2 Overall Architecture & Features

Reaction Wheel (Fig. 1) developed by TUBITAK UZAY comprises two units:

1) Electrical Unit consists of the controller (FPGA) and interface to the Attitude Control Subsystem (ACSS) task through On Board Computer

2) Mechanical Unit, consists of the motor, flywheel part, and the 3-phase inverter

Fig. 1 RW Mechanical (L) and Electrical Unit (R)

Feature	Value
Angular momentum capacity @4000 rpm	15 Nms
Output torque capacity	>200 mNm
Operating voltage	23-35 V
Operating temperature	-20 °C / 50 °C
Total Mass	8.2 kg
Dynamic imbalance	<20 gcm2
Static imbalance	< 3 gcm
Dimensions for Mechanical Unit	330∅x150 mm
Highest Power	<150 Watt
"Zero net torque at maximum speed" power	<25 Watt
"Zero torque-zero speed" power (Stand by)	<10 Watt
Typical Loss @ 3500 rpm	19 mNm
TM/TC Interface / Speed	CAN / 20 Hz

Table 1 Reaction Wheel Characteristics

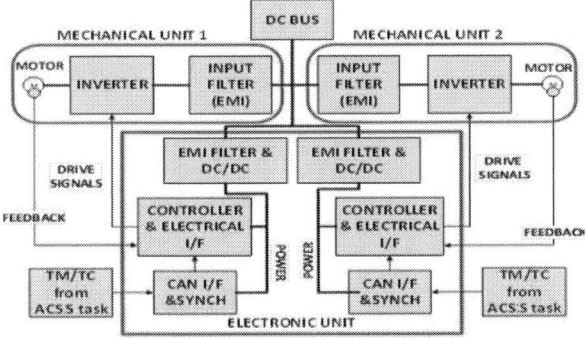

Fig. 2 Electrical Functional Diagram of RW

The electrical functional diagram is given in Fig. 2. Electrical Unit receives telecommands (torque) and sends telemetry signals (speed) to control the reaction wheel's speed under the ACSS task. It also observes the electrical motor (position and current feedback) to establish the Mechanical Unit inverter's drive signals (Pulse Width Modulation (PWM)). Electrical Unit is capable of controlling two independent Mechanical Units.

Mechanical Unit comprises several components, including the pre-tensioning mechanism, flywheel, hub, ball bearing, shaft, permanent magnet synchronous motor (PMSM), printed circuit board, and connection elements that make the top cover and the base (Fig. 3). The air pressure inside of the Mechanical Unit is decreased to approximately 1/100 of the nominal atmospheric pressure by a vacuum pump to minimize friction losses of the flywheel and motor (rotating parts). Nevertheless, Mechanical Unit still has very low air pressure after being sealed to distribute the heat inside the assembly and prevent hot spots.

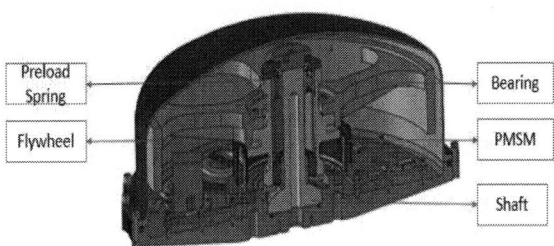

Fig. 3 Cross-Sectional View of Mechanical Unit

3 The Electric Motor

Among the existing motor technologies, Permanent Magnet Alternating Current (PMAC) motors are the most suitable motor types to be used in space technologies due to their advantages such as long lifetime (no brushes), highly linear torque to current ratio, high power density, high efficiency (especially compared to induction motors), high reliability, and high acceleration rate [6] [7] [8] [9] [10]. PMAC motors can be classified according to their back-EMF profiles: Brushless Direct Current (BLDC) Motor and The Permanent Magnet Synchronous Motor (PMSM).

Being an electronically commutating permanent magnet DC motor [9], BLDC motors have found broad interest in space applications owing to their advantages, such as simple controllability and the requirement for relatively low-cost sensors.

Parameter	Value
Continuous Torque @ Max Speed	0.99 Nm
Max Continuous Speed	4000 RPM
Max Continuous Power	413.2 W
Voltage Constant (l-l)	5.4 V_{pkl-l}/ kRPM
Torque Constant	0.063 Nm /A_{rms}
Motor Constant	0.138 N-m/$\sqrt{W}$
Resistance @ Max Temp	0.196 Ω
Inductance	6.75 µH
Number of Magnetic Poles	32
Outer Diameter	110 mm
Through Hole Diameter	89.5 mm
Axial Height	21.2 mm
Mass	0.304 kg

Table 2 Motor Parameters ThinGAP TO-110

However, they have certain disadvantages compared to PMSM. They produce torque ripple at commutations (due to non-idealities), create more harmonic content on the bus current, which could result in thermal issues, and are not suitable for drives using voltage modulation techniques to enhance the performance of the motor. Therefore, the authors aimed to exploit the advantages of PMSM while keeping the hardware as simple as possible without compromising the performance.

There are other classifications when comparing the different PMSM types [6]. Due to the following advantages, a coreless surface mount PMSM (SPMSM) has been preferred among the other candidates:

• SPMSM employs very low rotor losses, which is difficult to handle in space. In addition, this provides an additional cooling benefit that extends the life of magnets and other heat-sensitive components.

• SPMSM is better when comparing windage losses, which may affect high-speed losses.

• The stiffness of the rotor is much better for PMSM, which enables high-speed operations.

• Torque to inertia is essential for fast response, and PMSM has the best characteristic.

• The main advantage of coreless SPMSM is its low torque ripple or pulsation, which may create stability problems for spacecraft or blurred optical images for earth observation satellites.

• The core-magnet interaction (cogging torque) is very low or absent in coreless SPMSM; thus, it limits the high-speed losses.

ThinGap's TO-110 series motor has been used in Mechanical Unit, and its parameters are given in Table 2. This motor has been selected due to its low axial height, which conforms to the strict volume and height restrictions.

4 Electronic Hardware

4.1 The Control Method

Before proceeding with the hardware design, one must decide on the control method to define the components' characteristics. There are several alternatives for the control method regarding the sensors:

1. Sensorless control methods: These methods do not need speed or position observers. However, they lack performance stability in all speed regions, which is unacceptable for space applications. [11] [12].

2. High-resolution sensors: Since they are sensitive electronic devices, they cannot be both reliable and cost-effective. Also, they usually need custom designs based on the type of application. Therefore, they are rarely utilized in space applications.

3. Low-resolution sensors: Hall sensors are the most preferred types due to their low radiation sensitivity and high reliability. They are best used with analog controllers for BLDC motors. However, they need an observer for high-precision applications.

Since Hall sensors are the optimal choice for space applications, an observer should be implemented to enhance the performance of the existing technology. In order to achieve this goal, a digital controller is implemented with a Field Programmable Gate Array (FPGA) instead of an analog controller implemented with discrete components because building hardware with discrete components involves a significant amount of time and effort.

The digital controller is a high-speed anti-fuse FPGA (Microchip AX2000) with triple module redundancy. FPGA is responsible for all communication, control, and observation functions. The field-oriented control (FOC) algorithm is implemented in the controller to achieve the highest performance of the PMSM. The FOC algorithm is well-known for terrestrial applications and is given in Fig. 4 [13] [14].

The FOC initially transforms the 3-phase stationary coordinate system (ABC) into a 2-phase orthogonal coordinate system (α-β) using the Clarke transformation. The Park transformation converts the α-β coordinate system to a 2-phase orthogonal coordinate system (d-q), which rotates in sync with the rotor. The position of the rotor is sensed using HALL sensors and fed into the FPGA's position estimator block. As a result, the flux and torque equations are decoupled and linearly managed by changing stator current vectors on the d and q axes.

In the ACSS task, i_{qref} refers to the intended torque obtained from the speed controller via telecommand. The FPGA includes a simple speed controller for independent testing, but this controller is bypassed in flying configuration. i_{dref} is connected to magnetic flux in the rotor and is set to zero because the magnetic field is generated by magnets placed on the rotor of a PMSM unless the motor is in the field weakening region. This closed-loop control strategy keeps the flux and torque vectors orthogonally aligned for optimal efficiency and minimal torque ripple.

The proportional-integral (PI) controller calculates the deviation of the current vectors from the reference values and generates a voltage output value (V_{inv_q} & V_{inv_d}) to be applied to the inverter. These values are transformed into (α-β) and (ABC) stationary reference frames through inverse-Park and inverse-Clarke transformations. V_{inv_abc} voltages are applied to the inverter through the Pulse Width Modulation (PWM) block that operates at 50 kHz.

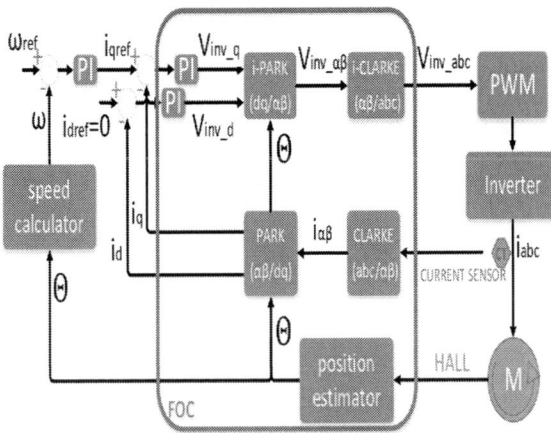

Fig. 4   Diagram of FOC with Hall Sensors

4.2 Power Inverter

Since the self-inductance of the coreless SPMSM is very small compared to other types of PMSM, the time constant of the R-L(phase resistance and inductance) circuit is very low, which is very difficult to handle with moderate switching frequencies (20 -100 kHz). In order to reduce the current ripple in the motor windings, inductors in the range of a few hundred µH, depending on the mission, are added series to the motor phase inductances. The 3-phase inverter is depicted in Fig. 5.

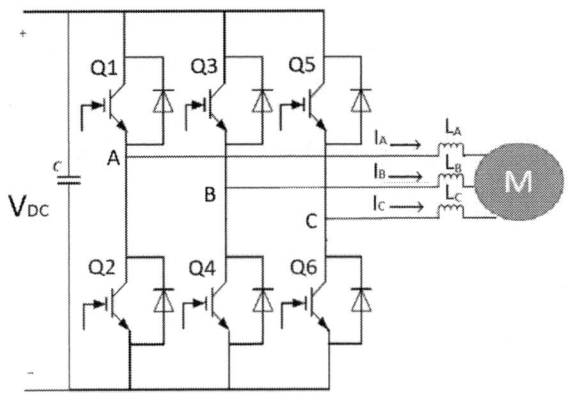

Fig. 5   Three phase inverter

Since the bus voltage (V_{DC}) differs for various missions, the power components are selected accordingly. For 28 V missions (IMECE and AYAP-1), STRH40N6 from ST and for 100 V missions (TURKSAT 6A), BUY25CS12J from Infineon are used as MOSFETs.

4.3 FOC with Hall Sensors

Since low-resolution switch-type Hall position sensors are inexpensive and can offer specific speed and rotor position information, they have emerged as the best option for balancing performance and cost. Several studies discussed switch-type Hall position sensors, explained how low-resolution sensors operate, and explained error-causing factors [11] [16].

If an overview is provided about the estimation techniques for rotor position and speed, one can see three main methods:

1. Interpolation method: Average methods are not adequate for high-speed dynamic performance; moreover, the least squares estimation method requires intensive calculations for especially the 32-pole motor used here.
2. Filter method: This method also needs intensive processing power, which is usually not possible in space applications. Moreover, it is dependent on motor parameters and challenging to find an optimal filter parameter, which makes it unsuitable for this application.

3. Observer method: This method relies on observing back-EMF signals, which are more than trivial to measure. Hence, during observation, the performance could be limited. Some other studies improve this method, but they also require intensive mathematical calculation for this application [11].

The Hall sensor signals are analyzed with respect to back-EMF signals to understand the cause of deviations when the inverter is not operating. Figure 6 shows that motor poles are asymmetrical since only one Hall sensor signal is shown, and the deviation goes as high as 7.5°. Therefore, the position estimation method needs to correct the magnet poles' asymmetries and the hall sensors' misplacement.

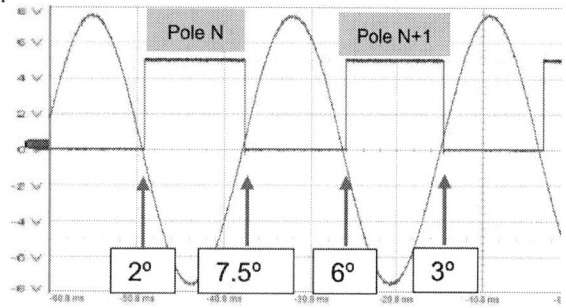

Fig. 6 Hall sensor signals vs back-EMF signal

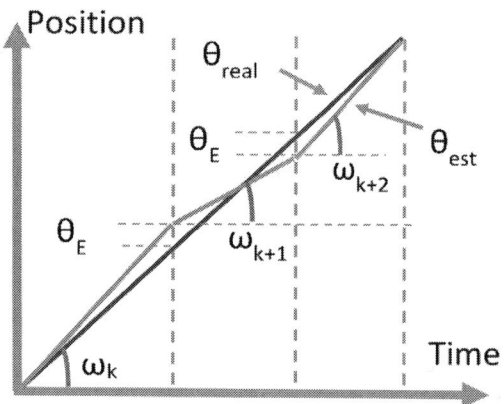

Fig. 7 Hall Sensor Position Estimation

The method developed in this study does not adjust the position directly when the Hall sensor state changes, as in [17]. Instead, the speed (derivative of the position, ω_k) is updated in every state change according to the calculation for the previous sector.

As shown in Fig. 7, θ_E shows the error between the actual position (θ_{real}) and the estimated position (θ_{est}). Typically, the position estimation methods use the Hall sensor duration of 60° (electrical angle). In contrast, the method in this paper uses 360° durations (the rising or falling edge of the same Hall signal) to calculate the speed of the rotor. However, the calculation is repeated every 60° using the previous 360° window.

This calculation, however, still needs to be corrected as seen in Fig. 6. The rising edge of (N)th pole lags 2°, but that of (N+1)th pole lags 6°, which are not equal. To overcome this error the considered compensation is as follows:

• The rising edge of a Hall sensor is triggered, and the duration between the previous and the current rising edge of the same Hall sensor is calculated (ΔT).

• The speed, ω_{k+1}, is calculated for that 60° sector (k+1) as in Eq. (1).

$$\omega_{k+1} = 360° / \Delta T \qquad (1)$$

• The acceleration for sector (k+1), $\Delta\omega_{k+1}$, is also calculated according to Eq. (2) [18], where T_m is motor output torque (proportional to i_{qref}), b is the friction constant, c is the static (Coulomb) friction, ω_r is the mechanical speed of the rotor, and J_{rot} is the inertia of rotating parts.

$$\Delta\omega_{k+1} = [(60°)*(T_m - b*\omega_r - c*sgn(\omega_r)] / J_{rot} \qquad (2)$$

• If $\omega_{k+1} > \omega_k$, $(\Delta\omega_{k+1})/2$ is subtracted from ω_{k+1}, otherwise $(\Delta\omega_{k+1})/2$ is added to ω_{k+1}.

Therefore, in every 60°, if the speed increases or decreases abnormally, the method limits the deviation to acceptable levels. Since this method relies on the system parameters, these should be known precisely to apply this method. Thanks to the nature of the reaction wheel, the loading or frictional characteristics are almost constant, and the changes could be neglected. Moreover, the advantage of using a digital controller is that these parameters can also be updated if any deviation is observed during the spacecraft's mission.

5 Results

During startup, since the speed is not known as in Fig. 7, the position is kept constant during the next 60° like a six-step operation. After reaching a certain speed, the Hall sensors' FOC algorithm starts. The startup sequence is shown in Fig. 8 using the FOC algorithm. It can be observed that the transition is very smooth. The RMS value of the currents is constant before and after the transition. Therefore, no torque pulsation occurs during startup as required.

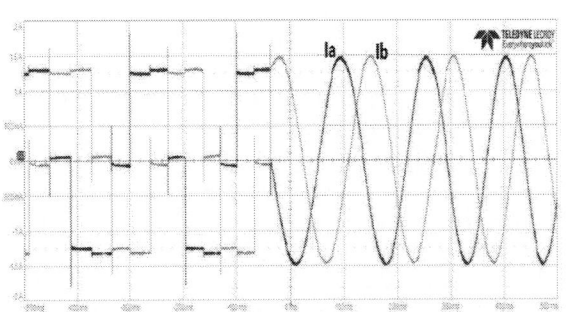

Fig. 8 Startup transition to FOC algorithm

As it is well known, the back-EMF signal and line currents should be in phase to achieve maximum output torque. Fig. 9 depicts the phase difference between the back-EMF signal between phases A - B (V_{AB}) and I_A-I_B (calculated using a digital oscilloscope).

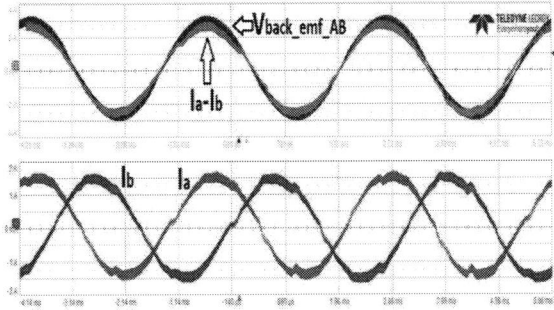

Fig. 9 Phase currents and back-EMF signals

Fig. 10 shows that the measured torque output during the current reference (i_{qref}) command changes when the Hall sensor estimation is not used. The peak-to-peak torque ripple is approximately 55 mNm.

Fig. 11 shows that the measured torque output during the current reference (i_{qref}) command changes when the Hall sensor estimation is active. The peak-to-peak torque ripple is reduced to approximately 32 mNm.

The torque ripple enhancement for various speed and loading conditions is given in Table 3. For low speeds and low torque regions, the ripple is reduced by 50%.

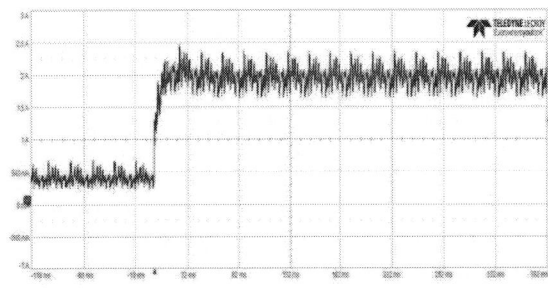

Fig. 10 Torque response when no method is used

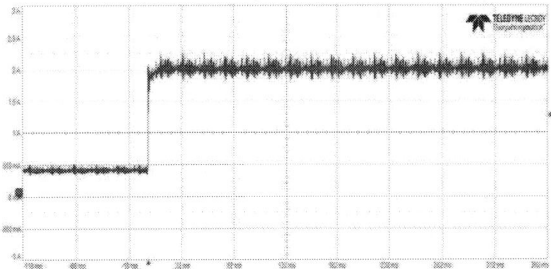

Fig. 11 Torque response when estimation is active

Speed (RPM)	Torque command (mNm)	Torque Ripple with method (mNm)	Torque Ripple without method (mNm)
1000	50	21	10
2000	50	28	15
3000	50	37	22
1000	100	28	13
2000	100	34	20
3000	100	45	29
1000	200	42	21
2000	200	51	28
3000	200	72	40

Table 3 Torque ripple enhancements

6 Conclusion

The performance of the developed Reaction Wheel has been verified in terms of torque output, torque losses, torque stability, and micro-vibration criteria. Instead of using a brushless DC motor and a driver with trapezoidal control, a PMSM and a digital controller are used to enhance the performance of the reaction wheel in terms of torque pulsations, efficiency, EMC performance, and transient response. In order to achieve this improvement, low-cost Hall sensors and controllers are employed in conjunction with a position estimator and field-oriented control algorithm. Results are satisfactory when compared to conventional motor drives used for space applications.

References

[1] B. Çolak, A. Özdemir, Ş. Ötenkaya, B. Selamlar, M.E. Öncüler, and S. Çetinkaya, "First Reaction Wheel Qualified For Space in Türkiye", Recent Advances in Space Technologies'23 (RAST 2023), June 2023.

[2] A. Shirazi and M. Mirshams, "Design and Performance Simulation of a Satellite Momentum Exchange Actuator", Australian Journal of Mechanical Engineering, vol. 14, no. 1, pp. 1–9, 2016.

[3] W. J. Larson and J. R. Wertz, "Space Mission Analysis and Design", Space Technology Library. El Segundo, CA. Microcosm Press, 2005.

[4] S. Nicosia and P. Tomei, "Nonlinear Observer and Output Feedback Attitude Control of Spacecraft," IEEE Transactions on Aerospace and Electronic Systems, vol. 28, no. 4, pp. 970–977, Oct. 1992, DOI: 10.1109/7.165360.

[5] B. H. Kenny, R. Jansen, P. Kascak, T. Dever, and W. Santiago, "Integrated Power and Attitude Control with Two Flywheels," IEEE Transactions on Aerospace and Electronic Systems, vol. 41, no. 4, pp. 1431–1449, Oct. 2005, DOI: 10.1109/TAES.2005.1561894.

[6] E. D. Ganev, "High-Performance Electric Drives for Aerospace: More Electric Architectures Part I – Electric Machines," in 2007 IEEE Power Engineering Society General Meeting, Jun. 2007, pp. 1–8. DOI: 10.1109/PES.2007.385463.

[7] J. Fang, X. Zhou and G. Liu, "Precise Accelerated Torque Control for Small Inductance Brushless DC Motor", IEEE Transactions on power Electronics, vol. 3, no. 28, March 2013.

[8] C. Qi, Z. Wang, and S. Su, "A Torque Control Method of Reaction Wheel Driven by Coreless PMSM," International Journal of Control and Automation, vol. 10, no. 1, pp. 379–386, Jan. 2017, DOI: 10.14257/ijca.2017.10.1.34.

[9] R. Krishnan, "Electric Motor Drives Modeling, Analysis and Control," Prentice Hall 2001.

[10] K. Sumitra, M.K.Giridharan, "Comparison between Field Oriented Control of BLDCM and PMSM Using SVPWM in Minimizing Torque Ripples in Satellite Application," IJSRD - International Journal for Scientific Research & Development, Vol. 3, Issue 09, 2015

[11] J. Wang, Q. Jiang, and D. Xiong, "Review of Rotor Position and Speed Estimation Method of PMSM with Hall Sensor," 2021 IEEE 16th Conference on Industrial Electronics and Applications (ICIEA), pp. 1832–1837, 2021

[12] M. Ebadpour, M.B.B. Sharifian, and M.R. Feyzi, "A Cost-Effective Position Sensorless Control for Four-Switch Three-Phase Brushless DC Motor Drives Using Single Current Sensor," International Review of Automatic Control (I.RE.A.CO.), Vol. 4, N. 3, May 2011

[13] D. W. Novotny and T. A. Lipo, "Vector Control and Dynamics of AC Drives", 1st edition. Oxford : New York: Clarendon Press, 1996.

[14] G. S. Buja and M. P. Kazmierkowski, "Direct Torque Control of PWM Inverter-Fed AC Motors - A Survey," IEEE Transactions on Industrial Electronics, vol. 51, no. 4, pp. 744–757, Aug. 2004, DOI: 10.1109/TIE.2004.831717.

[15] M. Akrami, E. Jamshigpour, and V. Frick, "Application of Hall Position Sensor in Control and Position Estimation of PMSM- A Review," IEEE International Conference on Environment and Electrical Engineering, 2023, DOI: 10.1109/EEEIC/ ICPSEUROPE57605.2023.10194763

[16] M. Ebadpour, M.B.B. Sharifian, and M.R. Feyzi, "A Cost-Effective Position Sensorless Control for Four-Switch Three-Phase Brushless DC Motor Drives Using Single Current Sensor," International Review of Automatic Control (I.RE.A.CO.), Vol. 4, N. 3, May 2011

[17] X. Zhang and W Zhang, "An Improved Rotor Position Estimation in PMSM with Low-Resolution Hall-Effect Sensors," 17th International Conference on Electrical Machines and Systems (ICEMS), 2014.

[18] V. Carrara and H.K. Kuga, "Torque and Speed Control Loops of a Reaction Wheel" 11th International Conference on Vibration Problems, 2013.

PCIM Europe 2024, 11– 13 June 2024, Nuremberg   DOI: 10.30420/566262062

Exploring High Frequency Operation of Motor Drives: Practical Insights on Efficiency and Loss

Asantha Kempitiya[1], Hrach Amirkhanian[1], Steve Oknaian[1] and Jannik Gade[2]

[1] Infineon Technologies, USA
[2] Infineon Technologies, Austria

Corresponding author and speaker: Asantha Kempitiya, asantha.kempitiya@infineon.com

Abstract

This work provides a comprehensive investigation of high switching frequency operation in motor drive systems. Two motor drive systems: with and without a stator core are developed and controlled via inverters utilizing Infineon's 100V 4mΩ OptiMOS silicon technology. The study shows that higher power factor of the coreless system increases inverter efficiency by more than 2 times at maximum load. Motor efficiency increases for both systems due to the reduction of motor losses driven primarily by harmonic loss reduction as evident from the coreless system results. The performance of the coreless system improves significantly with switching frequency, confirming that wideband technologies would be the ideal choice for inverters. Surprisingly, the comparison of motor and system power losses indicate that once the switching frequency for a specific application is determined, motors with lower L/R, such as coreless motors, perform better at lower motor speeds. In contrast, motors with cores that have higher L/R provide better performance at higher motor speeds due to the reduction of harmonic losses.

1 Introduction

The advances in modern silicon and wide bandgap technologies allow power converters such as inverters to operate efficiently up to several hundreds of kHz and even MHz with considerably low dynamic losses [1]. Furthermore, when considering only the electric motor, higher switching frequency can result in increased motor efficiency, lower torque ripple and faster control response [2]. However, the overall impact of operating at increased switching frequency on a motor drive system, must include the intricate interaction between the inverter and the electric machine. As such, in this work two motor drive systems that consist of motors with and without a stator core are exercised under various degrees of freedom using an automated test procedure to investigate a motor drive system's entire region of operation to provide a comparative analysis with interesting insights.

2 Methodology

2.1.1 Motor Generator Test Platform

To experimentally investigate the value proposition of high switching frequency operation in motor drives, amongst various other motor and inverter parameters, a test setup is developed as shown in Fig. 1. The first motor generator platform comprises of a Makita motor from a power drill while the second platform consist of a coreless procured from CLS. The specifications for the motor provided in the following table.

Parameter	Makita Motor (with stator core)	CLS Motor (without stator core)	Units
Resistance	24.5	40	mΩ
Inductance	37.5	6.25	µH

Table 1. Comparison of motor specifications.

Both motor-generator test setups are driven by a three-phase inverter utilizing Infineon's 100V 4mΩ OptiMOS silicon technology. Sensored field-oriented control (FOC) is implemented by two XMC 4400 drive cards that control motor and generator inverters under speed and torque control modes, respectively.

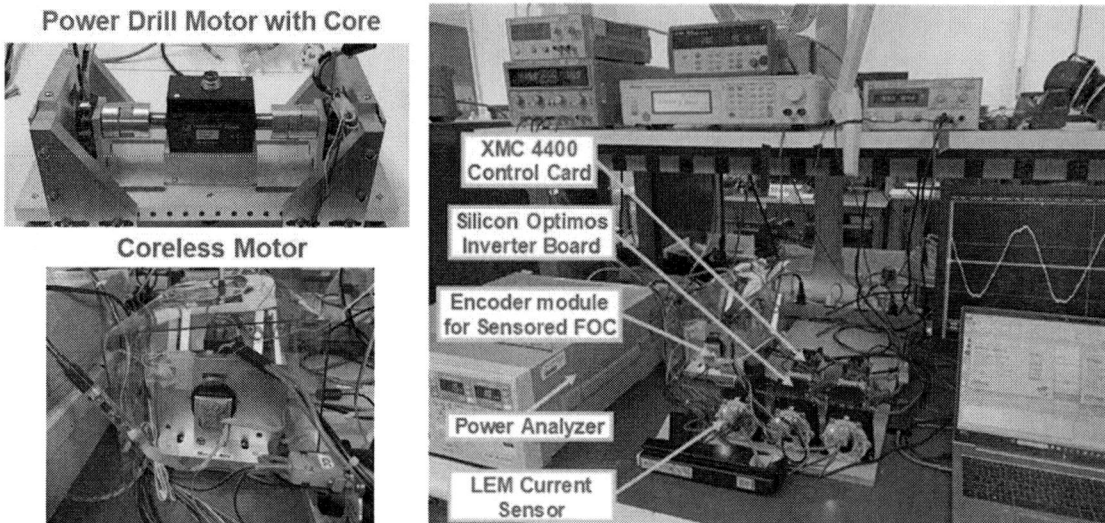

Fig. 1: 48V Motor generator test setups of motors with and without stator cores.

2.1.2 Automated Test Sequence

The automated test procedure implements a 4-dimensional parameter sweep, where motor speed and torque along with inverter switching frequency and deadtime are varied for a comprehensive analysis. As demonstrated in the example waveforms of Fig. 2, for a motor speed of 2000kRPM, two inverter switching frequencies of 20kHz and 100kHz at two dead times of 25ns and 100ns respectively are applied for each torque profile step

to facilitate the evaluation of the motor drive under increasing inverter and motor power levels.

In order to maintain similar motor winding temperatures throughout the test, as shown in the last plot of Fig. 2, a rest time is employed after the application each torque profile step. During the entire test sequence, inverter, motor and system efficiency is measured over motor speed, torque, switching frequency and post processed to evaluate motor drive system behavior.

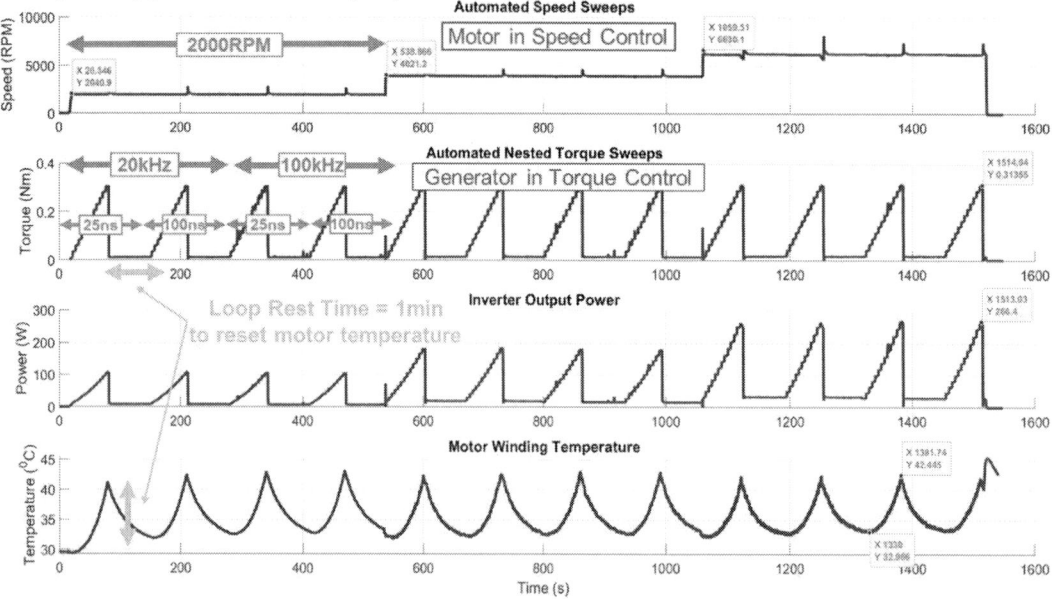

Fig. 2: Example waveforms demonstrating automated test procedure with Makita motor.

PCIM Europe 2024, 11– 13 June 2024, Nuremberg   DOI: 10.30420/566262062

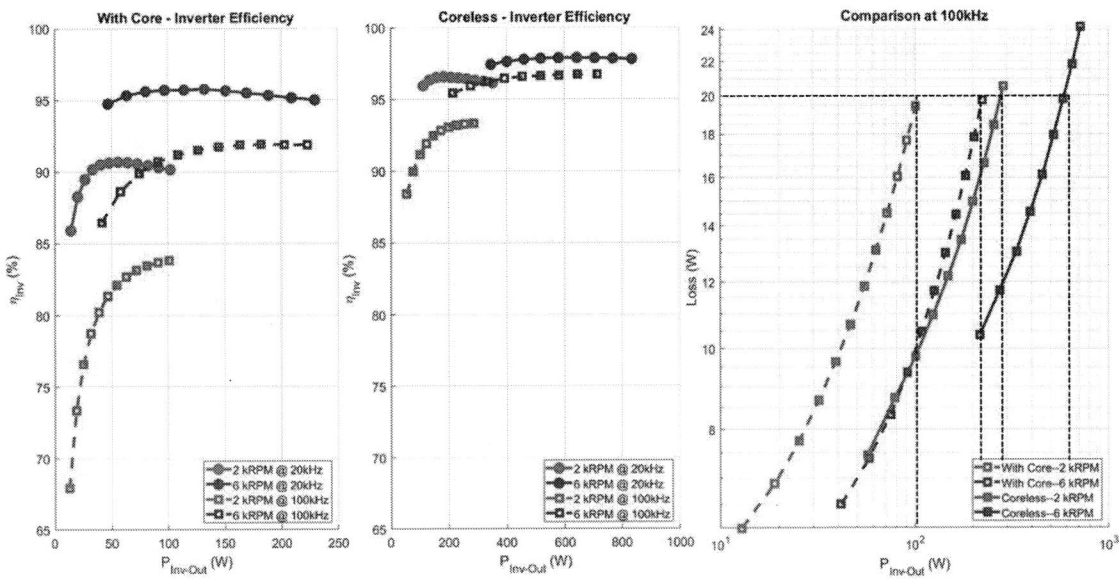

Fig. 3: Measured inverter efficiency curves for with core and coreless motors and their comparison.

3 Measurement Results

3.1 Inverter Efficiency

The first two plots of Fig. 3 show inverter efficiency curves for both test platforms. As expected, both systems exhibit higher switching frequency due to higher switching losses that reduce inverter efficiency. The power factor of each system increases with increasing motor speed due to the impedance increase of the motor windings that result in increased inverter efficiency with motor speed.

Comparing both systems, the coreless system shows increased inverter efficiency due to higher effective power factor. The third plot of Fig. 3 shows for the same total inverter power loss of

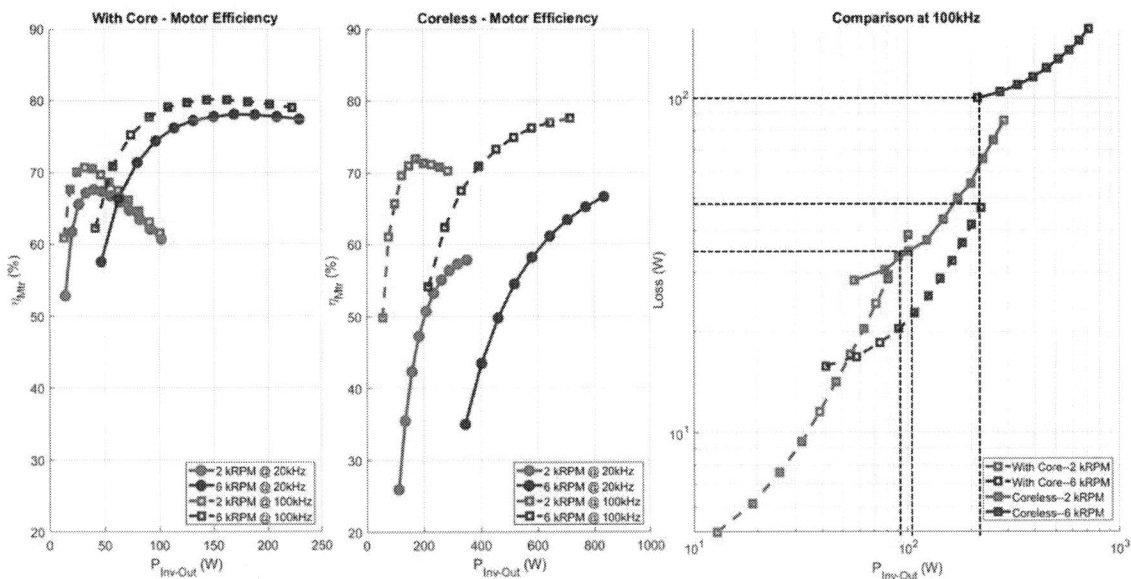

Fig. 4: Measured motor efficiency curves for with core and coreless motors and their comparison.

499

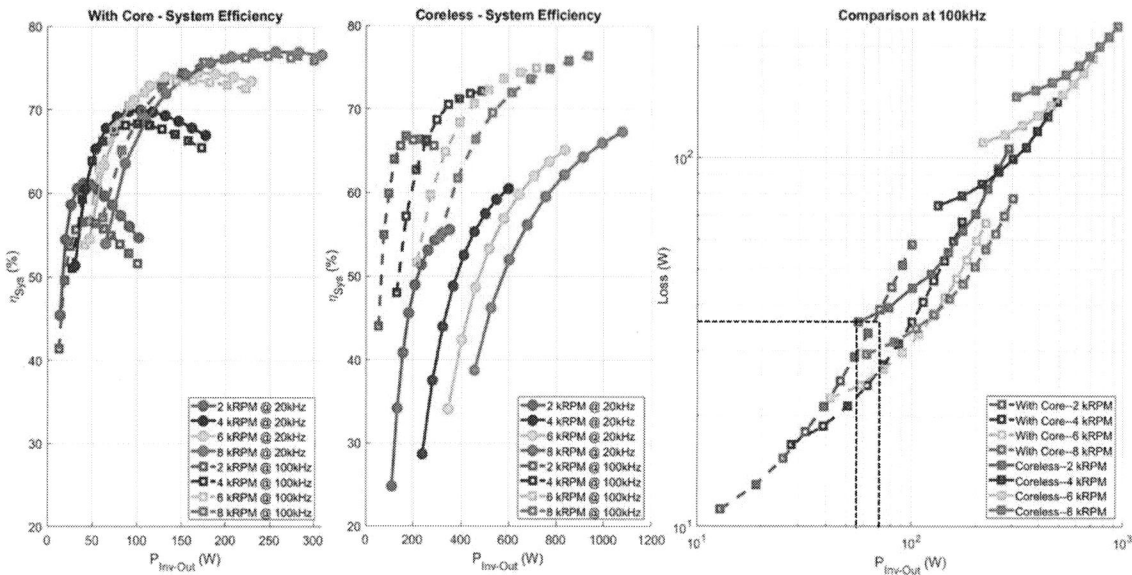

Fig. 5: Measured system efficiency curves for with core and coreless motors and their comparison.

~20W occurring at high load conditions, inverter efficiency increase allows the coreless system to produce ~2.3 and 2.6 times more inverter output power at 2kRPM and 6kRPM, respectively.

3.2 Motor Efficiency

As shown in Fig. 4, higher switching frequency operation results in motor efficiency increase for both systems. This phenomenon can be attributed to the reduction of hysteresis [3] and harmonic [4] losses of the motor when operating at higher switching frequency. As hysteretic losses are absent in the coreless system that shows considerable gain in motor efficiency, harmonic losses appear to be the dominant factor contributing to this phenomenon. The third plot in Fig. 4, shows that even though higher efficiency increase is observed in the coreless system, the motor winding time constant (determined by L/R) of approximately 10 times higher in the coreless system produces more harmonic-related motor losses.

Only at 2kRPM, does the coreless system show a 1.2 times increase in output power at 38W of motor loss. For example, at a motor speed of 2kRPM the coreless motor produces 2.1 times more motor losses for an equivalent inverter output power of ~220W. The trend indicates that motor efficiency of the coreless system improves significantly with switching frequency.

3.3 System Efficiency

System efficiency curves shown in Fig. 5 indicate that unlike GaN with OptiMOS [5] there is a higher reduction in inverter efficiency with increasing switching frequency. Even though, motor efficiency increases at all speeds, its magnitude fails to compensate the reduction in inverter efficiency at most operating points for the system with the core. The only exception to this case is at lower power levels for a motor speed of 6kRPM where system efficiency increases for higher switching frequency. For the coreless system, system efficiency increases at all speeds providing at least an approximate ~10% increase when operated at higher switching frequency. Similar to motor efficiency and power loss curves, the third plot of Fig.5 shows that system power loss is higher at all speeds, with the exception of 2kRPM motor speed where ~1.1 times more inverter output power can be generated for an equivalent system power loss of ~39W.

4 Conclusion

This work evaluates the interdependency between inverter and electric machine for two motor drive systems: with and without a stator core. The motors controlled via inverters utilizing Infineon's 100V 4mΩ OptiMOS silicon technology are exercised under various degrees of freedom using an automated test procedure. The study shows that higher effective power factor of the coreless system facilitates more than 2 times increase in the inverter efficiency for a total inverter power loss of ~20W at maximum load. Furthermore, regardless of silicon technology, motor efficiency increases for both core and coreless systems due to the reduction of motor losses.

This phenomenon is primarily driven by harmonic loss reduction that occurs at higher switching frequency as evident from the coreless system results. Since motor and system efficiency of the coreless system shows larger gain with switching frequency, wideband gap devices would be the ideal technology of choice. Interestingly, the comparison of motor and system power losses indicate that a motor with lower L/R time constant is typically suited for a low motor speed for a specific switching frequency. This contradicts the widely accepted notion that coreless motors with inherently lower inductance and higher resistance perform better when operated at higher motor speeds for a fixed switching frequency. In summary, this work provides a comprehensive investigation of motor drive system performance highlighting the impact of high switching frequency operation in motor drives, including the effects of other key motor and inverter parameters.

References

[1] B. J. Baliga. "Power semiconductor device figure of merit for high frequency applica-tions," IEEE electron Device Lett., Vol. 10, p. 455, Oct. 1989.

[2] H. Järvisalo, J. Korhonen, J. Honkanen and P. Silventoinen, "Considerations for a high-speed PMSM drive featuring a GaN-ANPC inverter," 2017 19th European Conference on Power Electronics and Applications (EPE'17 ECCE Europe), pp. P.1-P.6., 2017.

[3] B. Eriksson and E. Rudqvist, "Mapping of IPM machine efficiency and Noise Vibration Harshness", 2022.

[4] J. Miller and J. Soulard, "Energy-Efficiency of Electrical Machine and Drive with SiC Transistors", 2015.

[5] A. Kempitiya, H. Amirkhanian, S. Yerra and K. Kelkar, "100V GaN for Highly Efficient 1kW Motor Drive Applications," 2022 IEEE 9th Workshop on Wide Bandgap Power De-vices & Applications (WiPDA), Redondo Beach, CA, USA, 2022, pp. 238-241.

PCIM Europe 2024, 11– 13 June 2024, Nuremberg   DOI: 10.30420/566262063

High Power Density System Design for GaN-based LV Motor Drives

Marco Cannone[1], Edward A. Jones[1], Martin Wattenberg[1]

[1] Infineon Technologies Austria AG, Austria

Corresponding author:	Marco Cannone, marco.cannone@infineon.com
Speaker:	Marco Cannone, marco.cannone@infineon.com

Abstract

This work presents a complete motor drive system design that leverages Gallium Nitride (GaN) transistors to improve power density, especially in battery-powered applications. These applications such as drones, cobots, and e-mobility require small and powerful motor control boards, which are often integrated within the motor enclosure where the thermal budget is limited. The design presented here reaches a maximum power of almost 2 kW with a supply voltage of 48 V, corresponding to a power density of 3.2 kW/in³ (200 W/cm³). It includes the power stages for each phase leg, as well the control, sensing/protection, and auxiliary supply circuits to form a complete system. The paper presents the methodology of the PCB and thermal design used to integrate all the necessary components into one small board (29 x 51 x 9mm), as well as experimental verification of its performance.

1 Introduction

Designing a motor drive with GaN transistors enables high switching frequencies ($\geq$100 kHz), without significant loss penalty. The consequence of higher frequency on the hardware design is the reduction of the size of the DC-link capacitors, as shown in [1], with a subsequent increase in power density. The inverter presented here uses a switching frequency of 100 kHz to reduce the bulk capacitance requirement and input voltage ripple. This allows the DC-link to use only ceramic capacitors and avoid electrolytics to obtain a low-profile design, as presented also in [2], which greatly improves power density and permits more flexibility on the design of the heatsink. The small size of the GaN transistors also helps to achieve the highest power density, due to the lower specific on-resistance of GaN HEMTs compared to Si MOSFETs. To be able to achieve the target of 2 kW in the a small form factor of 29 x 51 x 9mm, the concept of transistors in parallel is used to scale up the power, as also shown in [3].

Prior work has focused on the design of high-density power stage design for GaN motor drives. This paper incorporates these design principles into a complete system architecture, with all the additional circuits necessary for a practical motor drive system. The design must therefore address new topics, such as shielding between the controller and nearby power stages in the presence of high dV/dt, as well as adapting the thermal design to fit the final form factor of the system.

2 System design approach

2.1 Schematic design

The schematic in **Fig. 1** reports all the fundamental circuit blocks implemented in the final board. The power block is composed of a three-phase inverter with two GaN transistors in parallel for each position in the phase leg. Each pair of paralleled HEMTs is driven by a single-channel gate driver without any gate resistors. The mounted driver (1EDN7126U) is available in four different drive strengths from 2 down to 0.5 A with respective part numbers from 1EDN7116U up to 1END7146U. Removing the gate resistors ensures a short, uninterrupted path for the connection with the GaN transistor, thus minimizing gate loop parasitics. Despite the lack of gate resistors, it is possible to maintain design flexibility to slow down or speed up the switching transitions by changing the driver to any of the four footprint-compatible options in this family.

The low-side drivers are supplied by a step-down DC/DC converter from the bus voltage (48 V from the battery) to 5 V, and the high-side is supplied through regulated bootstrapping from the low-side driver. The phase current is sensed by an in-phase Hall sensor, the TLI4971, which ensures galvanic

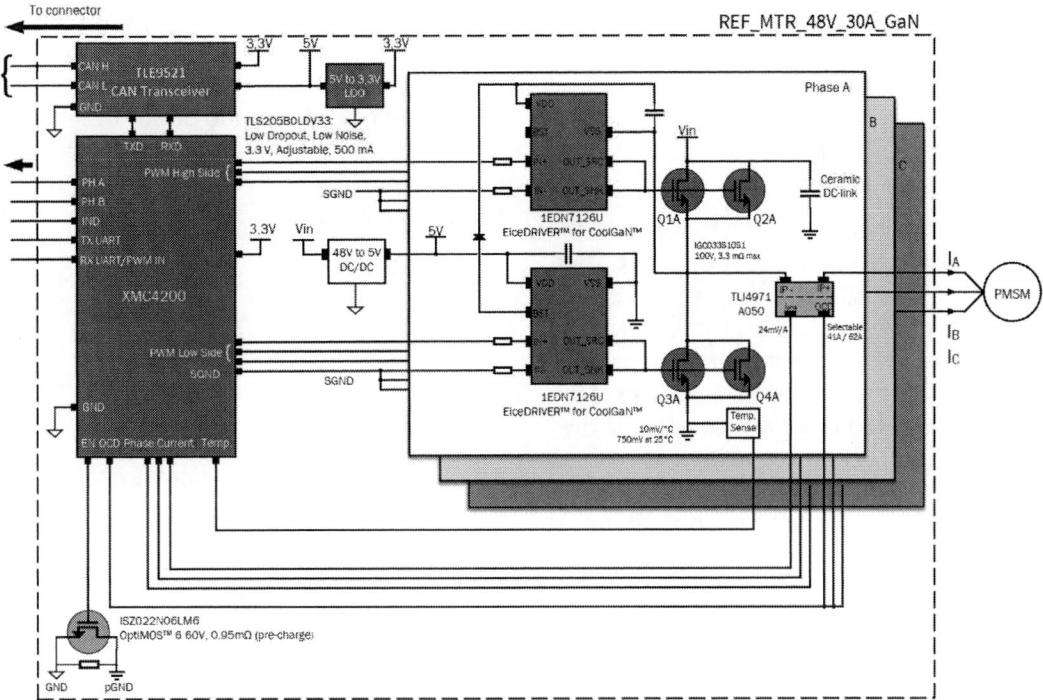

Fig. 1: Schematic of the proposed motor drive.

isolation between the switching phase voltage and the ground-referenced controller. The board temperature is also measured near the power block, providing a continuous indication to the controller to avoid over-temperature failure.

The control block is composed of an XMC4200 microcontroller and all relevant interfaces necessary to communicate with the nearby motherboard. Such peripherals are the Controller Area Network (CAN) and the Universal Asynchronous Receiver Transmitter (UART), where the CAN has a dedicated transceiver to combine a single-ended received and transmitted data (RXD and TXD) to the differential output (CANL and CANH). The controller circuit and current sensors are supplied by a small Low Drop Out (LDO) regulator, able to step down the auxiliary voltage to 3.3 V. This approach has the added benefit of providing additional power supply rejection ration (PSRR) to the analog sensors. Furthermore, the system has a precharge functionality, driven by a small logic-level transistor (ISZ022N06LM6) directly controlled by the microcontroller, which can avoid overvoltage and overcurrent caused by inrush to the bulk capacitors when the battery is plugged in.

2.2 DC-link sizing

A key advantage of GaN HEMTs is their higher switching frequency attainable with the same power-loss budget. As discussed in [1] the DC-link size is inversely proportional to the switching frequency f_{sw}.

The total DC-link capacitance C_{DC} can be reduced due to the high switching frequency. The lower capacitance requirement permits using several multi-layer ceramic capacitors (MLCC) in parallel, distributed across the PCB, instead of one or two bulky electrolytic capacitors.

This design choice optimizes the capacitance per unit of volume ($\mu F/mm^3$), as reported in [4], and delocalized the ripple currents to improve EMI across the PCB. Choosing ceramic capacitors over electrolytics also lowers the effective equivalent series resistance and inductance (ESR / ESL) with respect to other capacitor technologies, as highlighted in [5]. The low ESR reduces the power dissipation ($P_{cap.}$) caused by the capacitor RMS current ($I_{C_{DC}}$), in Eq. (1) and Eq. (2) from [6].

$$I_{C_{DC}} \approx 0.65 \cdot I_{phs} \tag{1}$$

$$P_{C_{DC}} = ESR \cdot I_{C_{DC}}^2 \tag{2}$$

The resulting voltage ripple with a given capacitance and frequency can be calculated at a particular phase current (ΔV_{pp}) according to

$$\Delta V_{pp} \approx \frac{I_{phs} \cdot \sqrt{2}}{4 \cdot f_{sw} \cdot C_{DC}}. \tag{3}$$

It is important to note that MLCCs experience a DC bias effect, which reduces their capacitance when a DC voltage is applied as illustrated in [7]. The resulting $C_{DC,eff}$ must therefore be used to calculate the voltage ripple, with a different effective capacitance depending on the DC operating voltage of the system. The calculated voltage ripples are shown in **table 1** for different DC input voltage conditions and switching frequencies, to demonstrate the capability of this system to achieve the desired ripple depending on the requirements of the application.

Tab. 1: Voltage ripple at 30 A_{RMS} for different dc voltage and f_{sw} conditions

V_{DC}	36 V	48 V	60 V
$C_{DC,eff}$	45 µF	28 µF	23 µF
50 kHz	4.7 V_{pp}	7.5 V_{pp}	9.4 V_{pp}
100 kHz	2.4 V_{pp}	3.8 V_{pp}	4.7 V_{pp}
150 kHz	1.6 V_{pp}	2.5 V_{pp}	3.1 V_{pp}
200 kHz	1.2 V_{pp}	1.9 V_{pp}	2.4 V_{pp}

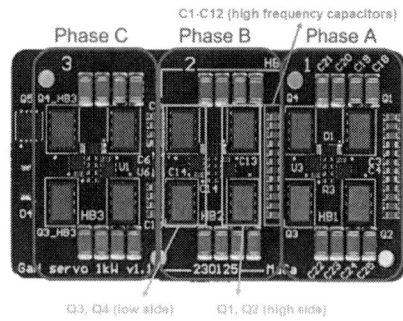

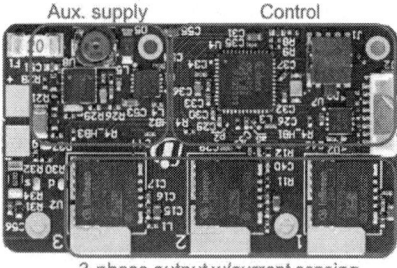

Fig. 2: Top and bottom view of the motor drive system.

2.3 PCB architecture and layout

In order to fit the entire system in just 29 x 51 x 9 mm, it is necessary to implement the two-sided design shown in **Fig. 2**, where the power block is located on the top side and the control block on the bottom side of the PCB. The structure of the stack-up is divided into layer pairs made by 70 µm of copper, as shown in **Fig. 3**, where the copper polygons are strategically placed to minimize the parasitics of the power loop and gate loop, using an optimal loop method as reported in [8].

The GaN commutation loops are closed between the top layer and the 1st inner layer in **Fig. 4**, which guarantees a magnetic field cancellation and reduces the stray inductance as previously shown in [9].

The presence of the gate driver and the paralleled GaN transistors on the top side of the PCB does not permit routing of the phase node in this area. For this reason, it becomes necessary to locate the phase on inner layers 2 and 3 as shown in **Fig. 5**. This phase node copper connects to all the transistors for each half-bridge (Q1, Q2, Q3, Q4) on the top layer using vias, and brings the current down to the Hall sensors located on the bottom layer.

Because the analog and digital signal traces of such a design are particularly vulnerable to noise introduced by the switching events, careful attention must be paid to the layout and stack-up to avoid coupling between power and signal domains. Fast dV/dt during the switching transitions of the GaN HEMTs can inject current into adjacent layers through the capacitive coupling as explained in [10]. Any overlap between DC and switching potentials

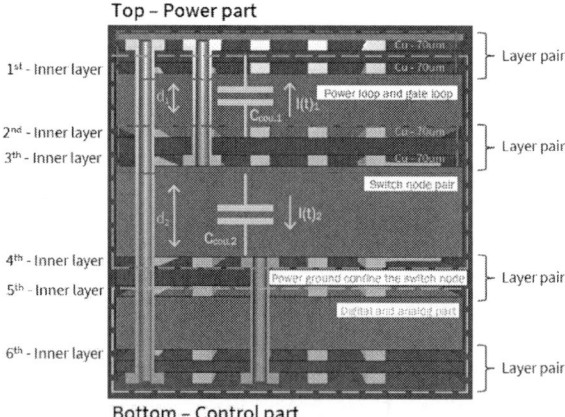

Fig. 3: Stack-up of the PCB.

PCIM Europe 2024, 11– 13 June 2024, Nuremberg   DOI: 10.30420/566262063

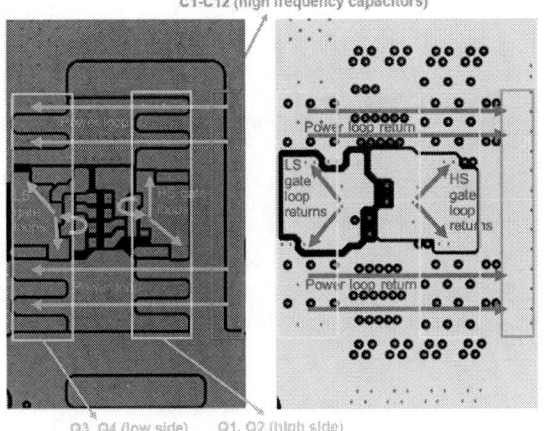

Fig. 4: Top-layer with GaN HEMTs and gate driver (left); 1st inner layer to close loops (right)

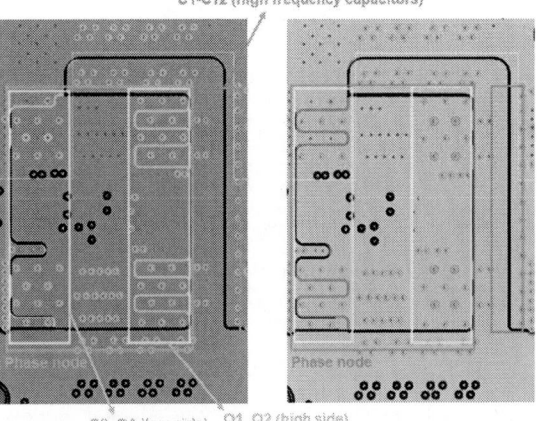

Fig. 5: 2nd and 3rd inner layers with phase node copper

leads to coupling capacitance $C_{coupling}$ which can be approximated using the area A and respective distance d as

$$C_{coupling} = \epsilon_0 \cdot \epsilon_r \cdot \frac{A}{d}. \qquad (4)$$

The coupling can be reduced by increasing the distance d between the layers, as reported in Eq. (4). This distance is based on the prepreg thickness from the switch node pair to the nearest layer. The stack-up alternates thin core, where the critical commutation loops are located, with thicker prepreg (3 or 4 times the core thickness) between each layer pair.

By shielding the switch node copper on inner layers 2 and 3 with a ground plane on inner layer 4, unwanted noise can be kept away from sensitive analog and digital circuits. A very thick prepreg layer in the center of the board limits the stray capacitance between inner layers 3 and 4. This concept mitigates the cross-talk problem and leaves the possibility to operate the system at high dV/dt without functional problems.

The via configuration is also an important aspect of this PCB architecture, with three types of vias in the board as pictured in **Fig. 3**. Through-hole vias are primarily used to connect the control signals to the power block, as well as connecting the phase node from inner layers to the in-phase current sensors and output connectors on the bottom side of the board. However, the critical loops for the power block are connected using blind vias on the top four layers only. Meanwhile, the sensitive control sig-

nals are routed using blind vias on the bottom four layers only. This reinforces the two-sided design approach, with the high dV/dt and di/dt nets remaining on the top half of the board and the control signals safely shielded in the quieter bottom half.

2.4 Thermal design

The high density of components on both sides of the design, as shown in **Fig. 2**, challenges the extraction of heat. The amount of copper in the proposed stack-up helps to spread heat on the bottom side, but this effect is limited by the very small size of the PCB. The low-profile concept of this work permits high flexibility on the top side, making it easier to design the heat-spreading plate.

The cross-section and the simple thermal circuit in **Fig. 6** emphasize where most of the heat is dissipated. In particular, the heat-spreader on top of the GaN device is the lowest thermal conductivity path. To improve the top-side cooling, an ultra-soft Thermal Interface Material (TIM) pad, with a conductivity of 12.5 W/mK, is used to fill the air gap and provide galvanic isolation.

Fig. 7 demonstrates the placement of the low profile design on a custom heat-spreading plate, which also serves as an enclosure to protect the electronics components and contain any radiated EMI.

3 Experimental results

3.1 Initial testing as buck converter

Operation of the system in buck mode uses only one half-bridge of the inverter, to verify the behavior gate and drain waveforms. This testing validates the correct turn-on and turn-off of the power stage

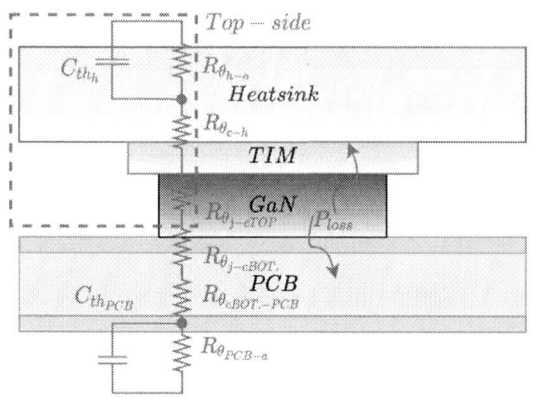

Fig. 6: Cross section of the board with a simplified thermal equivalent circuit.

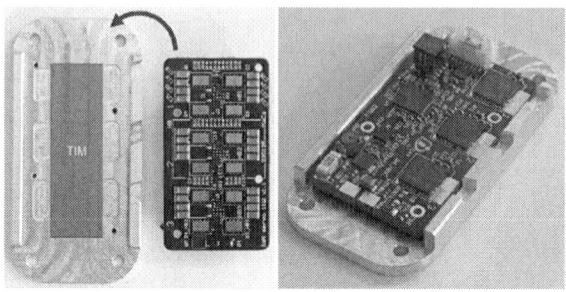

Fig. 7: Placement of the system board on a heat-spreading plate/enclosure.

with high current and steady-state thermal stress, while monitoring voltage ringing and optimizing the dead time.

Fig. 8 reports the phase-node waveforms with four different gate driver variants, from 2 A down to 0.5 A as explained in section 2.1. The consequences of stronger gate driving is a shorter switching transition and higher dV/dt, leading to lower switching loss and potentially causing EMI issues as described in [11]. **Fig. 8** shows that the strongest gate driver (1EDN7116U) causes a significant ringing near the maximum rating of the GaN device, and a slope of approximately 70 V/ns, as evaluated between the 10% and the 90% of the min/max values in the waveform. The other drivers reduced the amount of ringing and the slope, down to 13 V/ns for the weakest driver (1EDN7146U).

However, it is important to note that this board cannot accommodate optimal probing points in such a small form factor, and the waveforms may be slightly exaggerated in peak values and dv/dt by the twisted-pair connections at each probe tip. Nevertheless, the gate driver 1EDN7126U was selected for the final design based on the measured overshoot voltage and ringing captured here.

3.2 Testing as a three-phase inverter

The system was next operated as a three-phase inverter to determine the thermally limited maximum operating power of the board. In this testing, each phase of the Device Under Test (DUT) in **Fig. 9** was modulated by a V/f algorithm as describned in [12]. In this test configuration, a fixed value of phase frequency (f_{ph} = 300 Hz) is set, and the modulation index of all the output phases A, B, and C

are varied linearly, with 120° phase shift between each phase.

The load in this setup was a resistive-inductive load rather than a motor, as shown in **Fig. 9**. This choice permits a more repeatable setup, compared to a motor setup with a mechanical brake.

The measurement environment is composed of three resistive-inductive elements connected in star configuration and a 3-channel power analyzer (WT5000) measuring all three current phases by LEM 100-S sensor and the output voltage through a Norma star point adapter (by Fluke). The adapter is necessary to sense symmetrically all the phases, usually used whenever the star point of the motor or RL load is not accessible. The input voltage is acquired with a datalogger and the current is evaluated across a shunt resistor. The temperature is captured by a Flir thermal camera viewing the top of the GaN devices. A fan was used to provide forced air cooling of 2 m/s. A capacitor with more than 10 mF was placed between the power supply unit (PSU) and the DUT decouples the input of the board from the high inductance of the cable and the PSU, providing more stable input current for the shunt resistor measurement.

The power dissipated from the board is evaluated

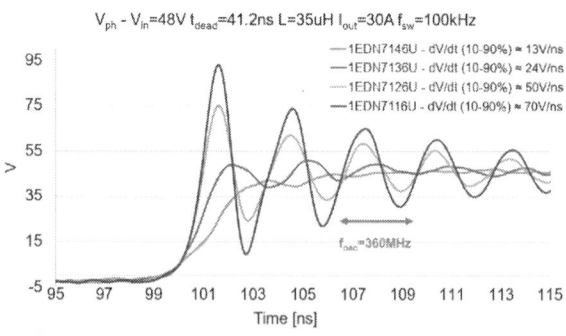

Fig. 8: Phase node voltage during hard turn-on.

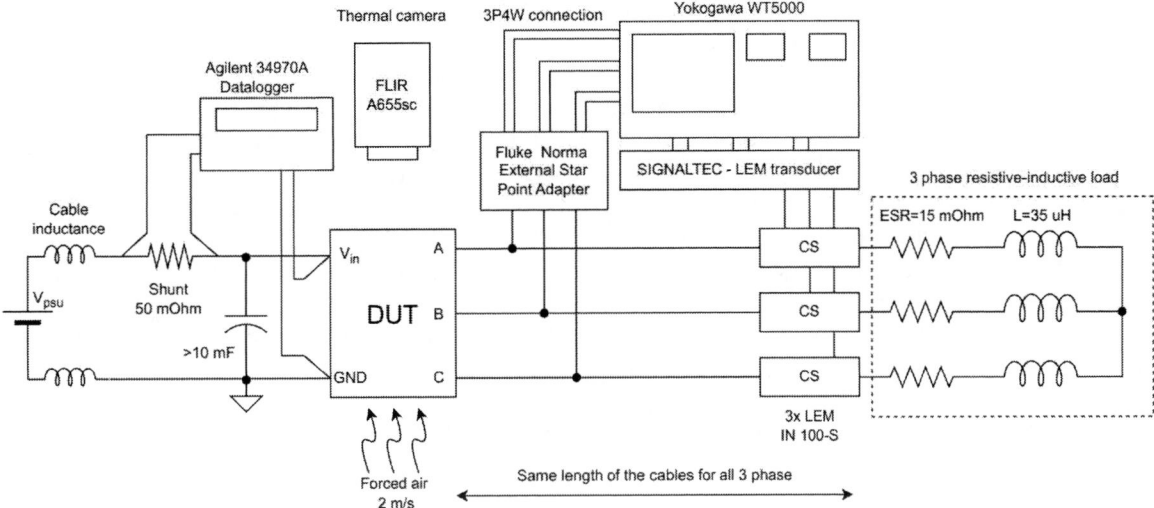

Fig. 9: Three phase inverter measurement setup.

in Eq. (5). The output power (P_{out}) is defined by the losses of the resistive part of the load, the active power (P_A, P_B and P_C).

$$P_{loss} = P_{in} - P_{out} = (V_{in} \cdot I_{in}) - (P_A + P_B + P_C) \quad (5)$$

The complete system's power consumption is considered here, including the auxiliary power consumed by the controller IC, gate drivers, and sensors, as well as the losses of the dc/dc housekeeping supply used to generate the 5 V and 3.3 V rails from the main input voltage. These auxiliary circuit losses account for approximately 1.2 W, which impacts the overall thermal budget of the system. The graph in **Fig. 10** shows the overall system power losses with two different conditions of cooling and two different gate drivers. The cooling conditions were evaluated first with natural convection and then with a forced air of 2m/s, both with

no heatsink and an ambient temperature of 25 °C. Different gate drivers were mounted to evaluate the different switching losses with a driver of 2 A (1EDN7116U) down to a 0.5 A (1EDN7146U) in sink and source. It is visible in **Fig. 10** that the impact of gate driver on power loss becomes more evident at the high current in both thermal conditions. At low current, the switching loss is dominated by the charge-related component rather than the driver-related I-V overlap component. The difference at the highest measured current conditions was 1 W with natural convection and 3 W with forced air. The temperature on the hottest point of the board is shown in **Fig. 11**, corresponding to the GaN device top case temperatures. The junction temperature was limited to 120°C maximum during testing, based on the specified maximum junction temperature of 150°C for this device. Due to the very

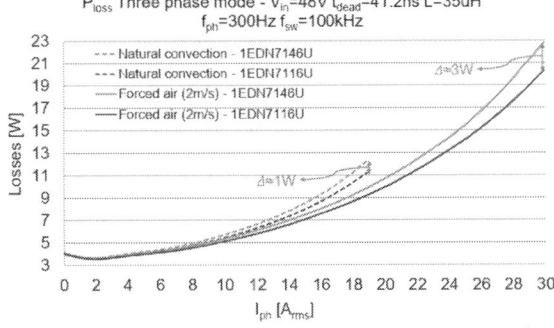

Fig. 10: Losses of the motor drive system with different gate driver strength and cooling conditions.

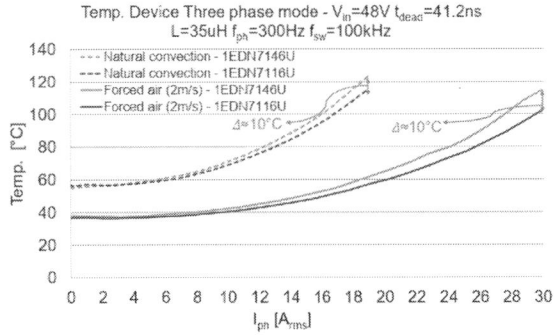

Fig. 11: Temperature of the GaN devices for different gate driver strengths and cooling conditions.

small thermal resistance from the junction to the top-side case of 0.5 K/W, it is reasonable to assume that the top-side case temperature is approximately equivalent to the junction temperature of the device. **Fig. 11** indicates that the stronger gate driver reduces junction temperature by 10°C at the highest current measured with forced air or natural convection, owing to the shorter overlap during the faster switching transition, as previously shown in **Fig. 8**. Throughout the experimental evaluation of the three-phase inverter, the auxiliary circuits were also functionally validated, including the control circuit, current sensing, encoder input, and communication circuits. Despite some noise observed on the encoder signal, the system performed without issue even at the highest dV/dt tested. Current sensing signals were accurately read by the control IC's ADC, even though the V/f control used for experimental evaluation did not require these current signals to run the inverter. This indicates that the PCB architecture and design techniques presented in this paper are able to effectively integrate a fast-switching GaN inverter power stage into a complete motor drive system.

4 Conclusion

This work presents design considerations as well as experimental results for a very compact three-phase GaN-based inverter. The PCB was carefully defined with an architecture that partitions the power circuits and control circuits between the top and bottom half of the stack up, with shielding ground copper in between. The switching commutations are confined to the first inner layer to enable a high dV/dt without compromising the other functions of the board.

The operation at 100 kHz or above allows the design to use a smaller MLCC-based DC-link while maintaining a low input voltage ripple. The total system size is 29 x 51 x 9 mm and includes all auxiliary components such as controller and housekeeping supplies. Such design can be easily integrated within the enclosure of brushless DC motors, eliminating long cables and improving overall system power density. The tests show the highest performance with a strong driver and high dV/dt. However, the loss and temperature penalty of the weakest driver and lowest dV/dt may be acceptable in very noise-sensitive environments or in combination with longer cables.

In future experimental investigations, the system will be tested in a field oriented control (FOC) scheme, including sensored and sensorless FOC, where the encoder and motor position sensors can be omitted from the design. Subsequent system designs may be improved by using new digital control ICs and current sensing technologies, to reduce the overall size and power consumption of these circuits. In the future, a GaN integrated power stage (IPS) with integrated current and temperature sensing may offer even further improvement in solution size, while also simplifying the PCB layout of each phase leg.

References

[1] M. Vujacic, M. Hammami, M. Srndovic, and G. Grandi, "Analysis of dc-link voltage switching ripple in three-phase pwm inverters," *Energies*, vol. 11, no. 2, p. 471, 2018.

[2] M. Wattenberg, E. A. Jones, and J. Sanchez, "A low-profile gan-based integrated motor drive for 48v foc applications," in *PCIM Europe digital days 2021; International Exhibition and Conference for Power Electronics, Intelligent Motion, Renewable Energy and Energy Management*, pp. 1–8, VDE, 2021.

[3] M. Wattenberg, O. G. Lorenz, and J. Sanchez, "A multi-kilowatt low-profile gan inverter for light electric vehicles and high-power tools," in *2023 IEEE Applied Power Electronics Conference and Exposition (APEC)*, pp. 1443–1450, IEEE, 2023.

[4] A. Hopkins, B. Hopfensperger, and P. Mellor, "Dc-link capacitor reduction in low voltage and high power integrated modular motor drives," in *2019 IEEE Energy Conversion Congress and Exposition (ECCE)*, pp. 3208–3214, IEEE, 2019.

[5] J. Cain, "Comparison of multilayer ceramic and tantalum capacitors," *AVX Technical Publication, AVX Corporation*.

[6] J. W. Kolar, T. M. Wolbank, and M. Schrodl, "Analytical calculation of the rms current stress on the dc link capacitor of voltage dc link pwm converter systems," 1999.

[7] S. Cen, "Dc bias characteristics of ceramic capacitors," *KYOCERA AVX Components Corporation*.

[8] D. Reusch and J. Strydom, "Understanding the effect of pcb layout on circuit performance in a high-frequency gallium-nitride-based point of load converter," *IEEE Transactions on Power Electronics*, vol. 29, no. 4, pp. 2008–2015, 2013.

[9] B. Sun, Z. Zhang, and M. A. Andersen, "Research of pcb parasitic inductance in the gan transistor power loop," in *2019 IEEE Workshop on Wide Bandgap Power Devices and Applications in Asia (WiPDA Asia)*, pp. 1–5, IEEE, 2019.

[10] D. A. Elena, R. Roxana, and M. O. Popescu, "Capacitive coupling interference phenomena between three conductors on the surface of pcb," in *2013 4th International Symposium on Electrical and Electronics Engineering (ISEEE)*, pp. 1–4, IEEE, 2013.

[11] Y. Xu, X. Yuan, F. Ye, Z. Wang, Y. Zhang, M. Diab, and W. Zhou, "Impact of high switching speed and high switching frequency of wide-bandgap motor drives on electric machines," *IEEE Access*, vol. 9, pp. 82866–82880, 2021.

[12] "Electric motor control," in *Electric Motor Control* (S. H. Kim, ed.), Elsevier, 2017.

PCIM Europe 2024, 11– 13 June 2024, Nuremberg　　DOI: 10.30420/566262064

Design of GaN Transistor based Variable Speed Drive Inverter with Output Voltage Filtering

Kaspars Kroics[1] (iD), Ugis Sirmelis[2], Valerijs Maricevs[1]
[1] Riga Technical University, Latvia
[2] Energotronix Ltd, Latvia

Corresponding author:　Kaspars Kroičs, Kaspars.Kroics@rtu.lv
Speaker:　　　　　　　Kaspars Kroičs, Kaspars.Kroics@rtu.lv

Abstract

GaN transistors allow to achieve higher switching frequency with high efficiency. However, the rapid voltage rise at the output is the main challenge to utilize full potential of GaN based variable frequency drives for three phase motors. The paper provides experimentally based designs of the inverters with limited slew rate of the switching actions at the output to utilize higher switching frequency and improve performance of the motor drive.

1 Introduction

GaN material provide significant advantages over conventional silicon one as it can be seen in Fig. 1. High electron mobility of GaN allows reduction of on-resistance of the transistors at smaller die and this leads to minimization of input and output capacitances. Therefore GaN transistors can be used at a higher switching frequencies with a small switching losses [1].

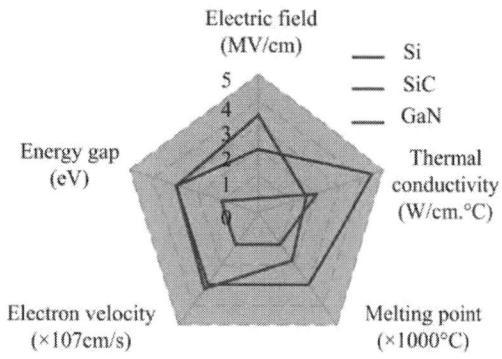

Fig. 1 Comparison of different types of transistor materials [2].

Three phase permanent magnet synchronous machines (PMSM) are used extensively in various applications due to high efficiency. PMSMs typically are fed by pulse width modulated inverter drives. To increase efficiency further, in the inverter design GaN semiconductor devices can be integrated. However, fast switching transition of GaN will generate high *dv/dt* that will cause additional losses into motor, EMI, and bearing current. These are the reasons why there is just minor efficiency improvement [3]. Rapid voltage rise in a short time (high *dv/dt*) will lead to undesired oscillations and overvoltage in cable connecting motor and inverter and in the motor itself [4]. Therefore, the voltage rise and fall times should be limited. One option to prevent this is to reduce the voltage rise time of the transistor by using different gate resistance values or by implementing closed-loop gate control [5], or with increasing of the gate-to-drain capacitance of the GaN transistor with an external additional capacitance [6]. These solutions are going to slow down the dv/dt transient, but at the same time will increase the switching losses.

Another approach is to use output filter. These filters have usually been bulky, but the high switching frequency permits use of small filters with a high resonant frequency. Such filters have potential to increase the system efficiency because the motor is fed with filtered voltages [7]. Application of different filters have been analysed in scientific papers [8], [9], [10] and others. The filter may influence the control system of the motor. Maintaining the switching frequency high there is a wider frequency range to place resonance frequency of the filter, but anyway some damping resistor could be necessary.

This paper analysis practical aspects of the design of GaN transistor based three phase inverter and experimental research on implementation of high *dv/dt* limiting solutions and implementation aspects into the motor inverter.

510

2 GaN transistor based three phase inverter

Size of the GaN transistors is much smaller than of Si one. This allows design of the converter with more compact size if proper cooling solutions are applied. Often GaN transistors or integrated circuits are bottom cooled devices. In that case most of the components should be placed on one side of the PCB, but other side is intended for the heat sink.

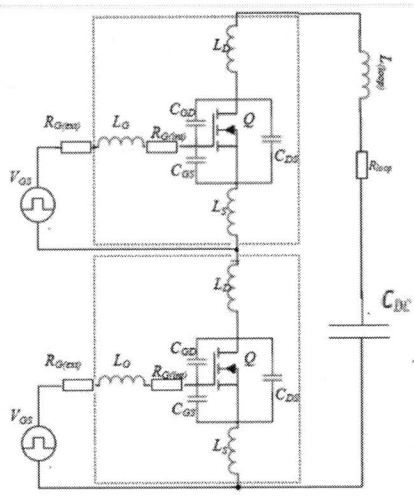

Fig. 2 Parasitic components of transistors connected in the half bridge circuit.

Fig. 2 shows schematics of half bridge connection of the transistors, which is also used in the three-phase inverter design. In the figure there are shown main parasitic elements that influence output waveforms. The parasitic elements describe package of the transistor, transistor structure itself and traces of the PCB. Three of half bridge circuits are needed to create three phase PMSM motor inverter. During turn-on transient the transistor drain to source capacitance C_{DS} of one transistor is bypassed but C_{DS} of other transistor is charged. During this time the parasitic inductances L_{loop}, L_S and C_{DS} create resonant loop [11] with a resonant frequency ω_{ON}:

$$\omega_{ON} = \frac{1}{\sqrt{(L_{loop} + L_D + L_S) \cdot C_{DS}}} \quad . \quad (1)$$

During turn-off, capacitances C_{DS} and C_{GD} of the transistor are resonating with the resonating frequency of ω_{OFF}:

$$\omega_{ON} = \frac{1}{\sqrt{(L_{loop} + L_D + L_S) \cdot (C_{GD} + C_{CS})}} \quad . \quad (2)$$

The long PCB traces between DC bus capacitor and both of half bridge transistors can create significant large loop inductance which produces large turn on and turn off oscillations in both V_{GS} and V_{DS}. Therefor this loop should be optimized to minimize this inductance as much as possible. On the market there is available integrated circuits with GaN transistors half bridge and driver integrated inside the chip. This simplifies the design of the inverter PCB, but anyway the DC link capacitor or at least part of it should be placed as close as possible to this chip.

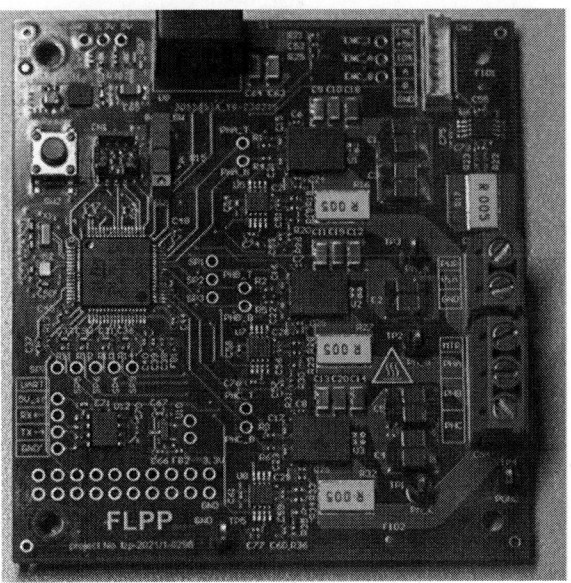

Fig. 3 Developed 50 V three-phase GaN transistor-based inverter.

50V GaN transistor-based three phase inverter PCB board was developed with three Texas instruments LMG5200MOFT integrated circuits which consists of GaN transistors in half bridge connection and integrated bootstrap gate driver. The experimental prototype can be seen in Fig. 3. The DC link capacitors are distributed into two parts – larger ceramic capacitors are placed a bit apart of transistors, but smaller capacitors are placed in direct proximity to the GaN transistor-based IC. On the other side of PCB there is only few components to save space for the heat sink.

For driving the transistor gates, 5 V voltage, which was recommended in the datasheet, was used. For high side switching the IC has integrated diode for bootstrapping purpose and has dedicated outputs for bootstrapping capacitor. Because of the integrated driver, the control of the gates is made transistor–transistor logic (TTL) compatible with voltage level up to 12 V, but the threshold for gate activation is 2 V and therefore is compatible with

STM32G4 3.3 V logic. The switching frequency for the inverter can reach 500 kHz within safe temperature range. For the control of the converter STM32G4 microcontroller was used. This microcontroller offers features like filter math accelerator (FMAC), high resolution timer (HRTIM), ADC and processor working at frequency up to 170MHz [12] and is available for relatively low cost.

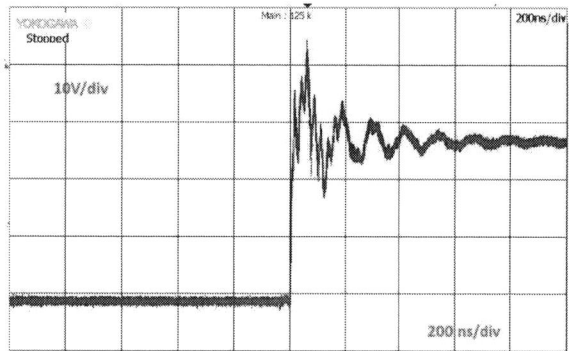

Fig. 4 Voltage on the low-side transistor during transients.

In Fig. 4 is shown transient process of the switching action of the transistor. As can be seen there is around 20 percent overshoot, but since maximum voltage of transistors is 80 V, but maximum DC bus voltage is going to be 50 V this is acceptable.

The gate control of GaN transistors is significantly different from typical silicon Si MOSFETs. One of the main differences in terms of gate control is that the gate threshold voltage $V_{GS(th)}$, the gate plateau and the maximum gate voltage $V_{GS(max)}$ are lower. To turn off the GaN semiconductor, the gate voltage should be maintained below the minimum gate threshold voltage, which is around 1 V. This can be a challenging in applications where the GaN transistor drain is exposed to a high dv/dt. In such situation the GaN transistor can be turned on from the EMI voltage spike on the gate pin. The e-mode GaN transistors have specific behaviour: they do not have internal body diode but when the gate is turned off and drain is negative with respect to source, the GaN transistor conducts current from source to drain, but the voltage drop between source and drain during dead-time conduction is higher than in the case of Silicon transistor. This voltage drop decreases efficiency of the inverter, therefore optimal dead-timing should be used. Driver must have high common-mode transient immunity. The printed circuit board of the driver circuit should be placed near to the transistor gate to limit stray inductances and capacitances. High dv/dt causes high common mode currents, this is

a reason why it is important to minimize coupling capacitances of isolated power supply and use a gate driver with high common mode transient immunity.

Fig. 5 GaN transistor half bridge with the driver solution.

Considering previous information for 200 V inverter PCB board design isolated driver circuits have been selected. To isolate gate driving voltages low power switched mode power supplies with high isolation voltage and low capacitance between primary and secondary side have been selected. The driver solution and half bridge of GS-065-004-1-L GaN transistors can be seen in Fig. 5. This solution is much more expensive than bootstrap driver one, but for research it is safer to operate converter with isolated driver solution.

Since GaN transistors are fast switching devices, the main limiting factor for dead time reduction is delay mismatch of the driver circuit. The new driver circuits have been released recently so the dead time in GaN transistor inverter can be small. Small dead time results in increased efficiency [16] since the current will flow in reverse conduction region through transistor only for short time. By optimization of the value of the dead time it is possible improve waveform of the inverter sinusoidal signal. In this case dead-time equal to 50 ns was used. Figure 6 shows waveforms of the output current of the inverter. The inverter is tested in the open loop condition without any feedback. To generate the sinusoidal pulse with modulation (SPWM) the table of 600 values were recorded into memory of microcontroller. As can be seen in Fig.6 there is almost no influence of the dead time on current waveform. This allows to avoid the need for dead time compensation techniques [17]. High switching frequency reduces current ripples and as can be seen in Fig. 6 the current is close to sinusoidal.

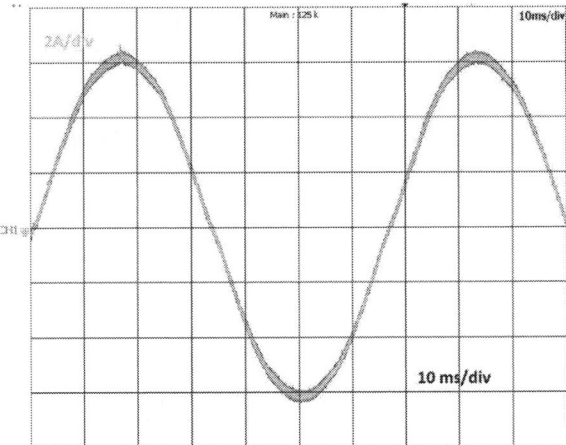

Fig. 6 Output current of the inverter.

3 Inverter with sinusoidal output voltage filter

For *dvt/dt* limitation at the output of the inverter can be limited by using passive filter. The most widely used type of filter is LC filter. It is used in this case to filter output voltage of the inverter thus avoiding high dv/dt and losses into the motor. LC filters influence the current delay in phase, and it makes more difficult to develop stable controller of the motor. The resonant frequency of the filter should not be close to the switching frequency of the converter to avoid ringing of the output voltage. The cut off frequency of the LC filter can be calculated by the well-known equation:

$$f_{cut} = \frac{1}{2\pi\sqrt{(L_1 \cdot C_1)}} \quad . \tag{3}$$

In order to obtain an almost sinusoidal motor voltage, the resonance frequency of the filter should be below the lowest harmonic frequency of the inverter voltage caused by pulse width modulation [10]. The resonance frequency has also to be above the fundamental frequency of the motor voltage. By applying GaN transistors into motor inverter it is possible to increase switching frequency

therefore there is wider frequency range where to select cut-off frequency.

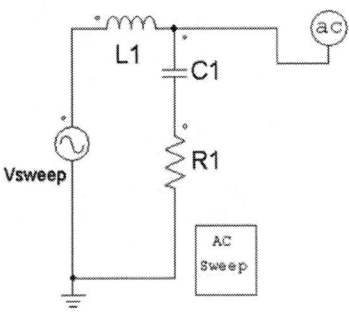

Fig. 7 LC filter with a damping resistor.

To investigate filter response behaviour AC sweep model was created in PSIM software shown in Fig. 7. The switching frequency has been selected equal to 500 kHz. The cut-off frequency of the filter can be selected at 6-10 times lower frequency. The fundamental frequency is expected lower that 100 Hz therefore the resonant frequency 10 times higher can be considered as the lowest frequency cut-off frequency. Based on these considerations it is possible to select values of inductor and capacitor. The LC filter with L_1=0.01 mH, C_1=20 μF and damping resistor 0.5 Ω has been simulated. The result can be seen in Fig. 9. As can be seen the amplification at 100 kHz is at around -60 dB which is acceptable. It can be expected that the output voltage will be close to sinusoidal. The cut-off frequency is around 10 kHz so there is space for fast controller implementation.

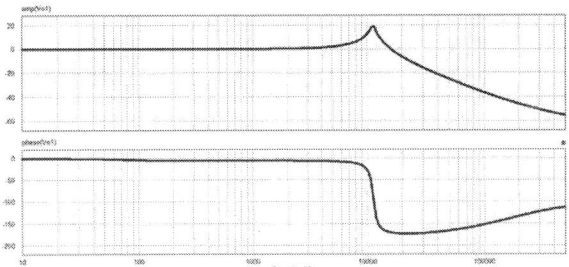

Fig. 9 LC filter gain and phase with a damping resistor.

To test the output waveform of the filter LC filter has been created with a similar parameter as in simulations. The inductor was made from windings of Litz wire. EELP 22/6/16 ferrite core made from PC200 material was used. Four 4.5 μF capacitors was connected in parallel with a damping resistor of 0.5 Ω. Experimental setup can be seen in Fig. 10. Power supply (1) provides power, GaN transistor-based inverter (2) creates the high frequency pulses that go to the filter (4) and supplies power to the motor (3). STM32 microcontroller was used to generate sinusoidal PWM signal with 500 kHz PWM frequency. Teknic M-2310P-LN-04K PMSM servomotor (3) was used for experiments. The output voltage can be seen in figure 11. The temperature of the motor have been reduced, but there are additional losses in the filter and transistors due to increased frequency. Working at high frequency the size of the filter can be small, but voltage drop and influence on control system should be considered. Overall as it has been shown in the literature efficiency can be increased but the size of the inverter is increased due to filter.

Fig. 10 Experimental setup for inverter with sinusoidal filter testing.

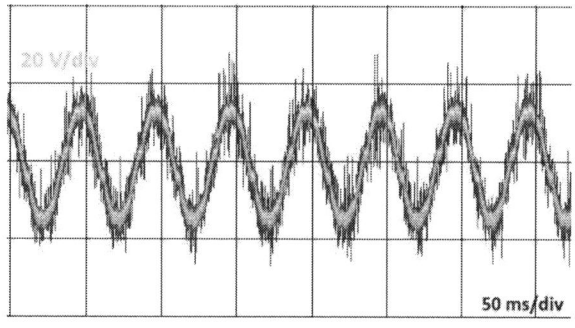

Fig. 11 Experimental waveform of the output voltage of inverter with sinusoidal filter.

4 Gate driver based dv/dt reduction

For high power application the size of the filter will be significant and it ads also additional costs therefore other solutions can be preferable. The simplest way to reduce dv/dt is to slow down the transient process by influencing gate voltage of the transistor. As one of the best solution can be con-

sidered artificial increasing of C_{GD} [18]. The manufactures has provided the Spice model of GS-065-004-1-L GaN transistor. In the model it is possible to add additional C_{GD} capacitance and change gate resistor obtaining desired dv/dt. LTSpice model with transistor half bridge has been created and C_{GD} value changed obtaining lower dv/dt.

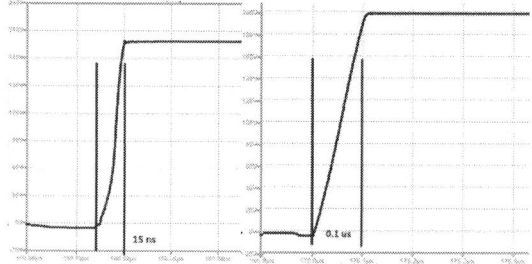

Fig. 12 Voltage rise time without and with increased C_{GD}.

Fig. 13 Experimental PCB of GaN half bridge with added additional C_{GD}.

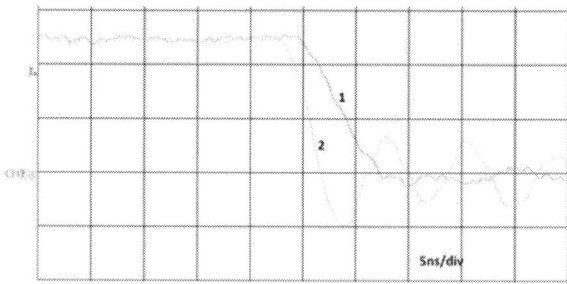

Fig. 14 Experimental fall time without (2) and with increased C_{GD} (1).

Figure 13 show transistor half bridge with added 50 pF capacitor. As can be seen in Fig. 14 this approach allows reduction of dv/dt. The results are close to simulation model in case parasitic inductances of PCB is considered. This is very simple and effective method to limit dv/dt. Unfortunately, slower transient process is going to increase switching losses thus destroying overall efficiency.

GaN transistors will need larger heat sink to cool down these losses.

5 Passive filter for dv/dt reduction with active control

In [19] it is proposed to use undamped LC filter for dv/dt reduction with additional charge pulse in addition to the SPWM signal. There was proposed to implement this control signal by using FPGA. Here microcontroller will be used to implement this signal pattern. This additional control signal allows to achieve a voltage transition without any overshoot. Moreover, in contrast to the passive filter, the energy stored in the filter capacitor is not dissipated but can be recovered. As a drawback can be mentioned two additional switching actions that is going to increase switching losses.

The inverter pulse pattern edges should be broken down into narrower pulses, which control the filter LC circuit. This results into an output voltage without overshoot typically seen in common LC circuits. To implement this filter there is need to generate PWM with added specific filter charge and discharge signal. In this case STM32G4 microcontroller will be used to generate such specific signal. The precharge signal length can be calculated as follows [19]:

$$t_{rise} = 2.094 \cdot \sqrt{LC}. \tag{4}$$

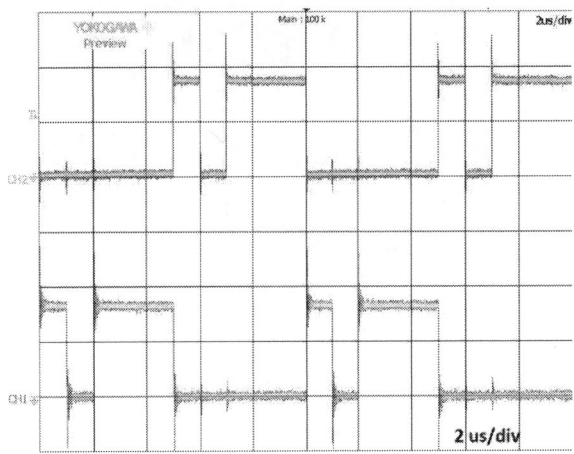

Fig. 15 Timer settings to obtain desired PWM signal with filter precharge signal.

The explanation of generating of such PWM signal is explained in Fig. 15. There is used two timers for each inverter phase and they are synchronized

together. One timer is used for high side transistor driving, but other for low side transistor driving. There are used 3 compare registers and timer overflow to change output to the opposite state. Figure 15 explains the generation of desired pulse sequence. In Figure 15 there are two timers, but in the figure they are shown as one just with different compare registers CMP and CMPa. Practically obtained waveform can be seen in Fig. 16.

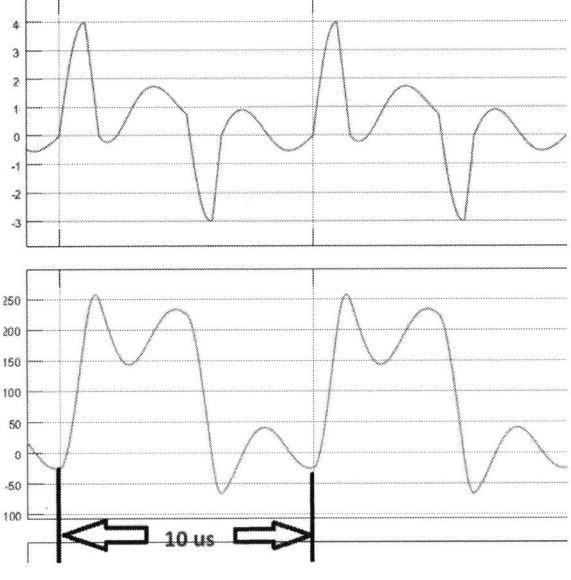

Fig. 16 PWM signal with filter precharge signal.

Fig. 17 Simulation of inverter with LC filter with active control.

Simulation model was created in Matlab Simulink of inverter, LC filter and control signal generator. Fig. 17 shows simulation results, the filter is charged and discharged with these pulses. There can be seen some oscillations which can be explained with load influence since in [19] it is mentioned that some additional pulse should be generated to compensate load influence. Generation of such pulse, which depends on the load current, with microcontroller could be challenging [20].

Fig. 18 One half bridge of the inverter with implemented LC filter with active control.

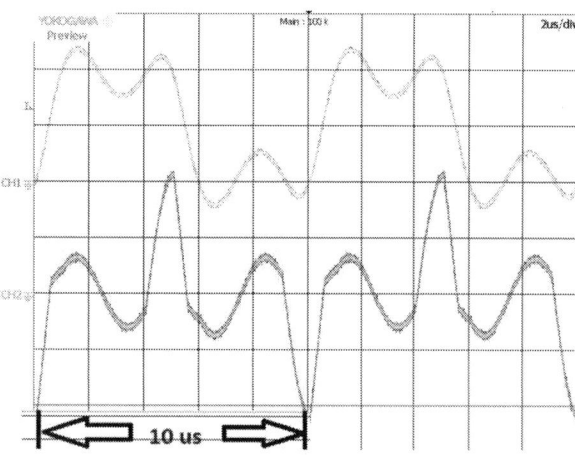

Fig. 19 Experimental waveforms of inverter with LC filter with active control – output voltage and current.

The filter was implemented into inverter design shown in Fig. 18. Inductance was equal to 3.3 µH and capacitance 100 nF. The filter charge pulse calculated by Eq. 4 equal to 1 µs was generated. Experimental waveforms can be seen in Fig. 19. Although it is possible to generate such pulses with a microcontroller it makes more difficult to generate sinusoidal PWM therefore more advanced controllers are preferable. Moreover, there are additional switching actions of GaN transistors and therefore switching losses are increased.

6 Conclusions

Different dv/dt reduction methods for GaN transistor-based inverters were analyzed from practical implementation point of view. Application of GaN transistors in motor drives boosts the efficiency and improves the waveform of the current. Fast switching actions of these transistors create high dv/dt that should be limited in motor drive application. Gate voltage influencing by changing gate resistor value and adding additional capacitor between gate and drain is the simplest method for implementation and capacitor is small, but losses into transistor will increase thus leading to larger heatsink and power density reduction.

Sinusoidal output filter at high switching frequency can be developed compact in size. Such solution contributes to EMI reduction, motor efficiency improvement and dv/dt reduction. In comparison to other solutions the power density is not the best and it is difficult to avoid current spikes and ringing and therefore passive damping is required which decrease efficiency. The filter will influence transfer function and can limit the control bandwidth.

Additional signal pulse can be introduced into control signal to charge undamped dv/dt LC filter thus controlling current spikes. Such control can be implemented with conventional microcontroller, but it makes more difficult generation of sinusoidal PWM. The filter is small, but anyway the loss into transistors is increased since switching actions are increased twice. This is a promising way to reduce dv/dt, but should be developed improved solutions for more simple control without increasing losses into transistors.

Acknowledgment

This research work has been supported by the Latvian Council of Science, project "Fast transient response and high efficiency GaN based BLDC motor converter with a dual power supply", project No. lzp-2021/1- 0298.

References

[1] E. A. Jones, F. F. Wang, and D. Costinett, "Review of Commercial GaN Power Devices and GaN-Based Converter Design Challenges," *IEEE Journal of Emerging and Selected Topics in Power Electronics*, vol. 4, no. 3, pp. 707–719, Sep. 2016.

[2] J. Chen, X. Du, Q. Luo, X. Zhang, P. Sun, and L. Zhou, "A Review of Switching Oscillations of Wide Bandgap Semiconductor Devices," *IEEE Transactions on Power Electronics*, vol. 35, no. 12, pp. 13182–13199, Dec. 2020.

[3] J. Wang, Y. Li, and Y. Han, "Integrated Modular Motor Drive Design With GaN Power FETs," *IEEE Transactions on Industry Applications*, vol. 51, no. 4, pp. 3198–3207, Jul. 2015.

[4] K. Choksi, Y. Wu, Mustafeez-ul-Hassan, and F. Luo, "Evaluation of Factors Impacting Reflected Wave Phenomenon in WBG Based Motor Drives," in *2022 International Power Electronics Conference (IPEC-Himeji 2022-ECCE Asia)*, May 2022, pp. 736–740.

[5] H. Kim, A. Anurag, S. Acharya, and S. Bhattacharya, "Analytical Study of SiC MOSFET Based Inverter Output dv/dt Mitigation and Loss Comparison With a Passive dv/dt Filter for High Frequency Motor Drive Applications," *IEEE Access*, vol. 9, pp. 15228–15238, 2021.

[6] N. G. M. Thao, K. Naruse, and K. Fujisaki, "Reduction of Harmonics and Inverter Temperature in Experimental GaN-based Motor Drive System at High Frequencies Using LC Filter," in *2022 IEEE Ninth International Conference on Communications and Electronics (ICCE)*, Jul. 2022, pp. 507–512.

[7] F. Maislinger, H. Ertl, G. Stojcic, and L. Siplika, "Performance of a Two-Stage Actively Damped LC Filter for GaN/SiC Motor Inverters," in *2018 IEEE 18th International Power Electronics and Motion Control Conference (PEMC)*, Aug. 2018, pp. 177–183.

[8] M. Haider, M. Guacci, D. Bortis, J. W. Kolar, and Y. Ono, "Analysis and Evaluation of Active/Hybrid/Passive dv/dt-Filter Concepts for Next Generation SiC-Based Variable Speed Drive Inverter Systems," in *2020 IEEE Energy Conversion Congress and Exposition (ECCE)*, Oct. 2020, pp. 4923–4930.

[9] X. Chen, D. Xu, F. Liu, and J. Zhang, "A Novel Inverter-Output Passive Filter for Reducing Both Differential- and Common-Mode dv/dt at the Motor Terminals in PWM Drive Systems," *IEEE Transactions on Industrial Electronics*, vol. 54, no. 1, pp. 419–426, Feb. 2007.

[10] P. Mishra and R. Maheshwari, "Design, Analysis, and Impacts of Sinusoidal LC Filter on Pulsewidth Modulated Inverter Fed-Induction Motor Drive," *IEEE Transactions on Industrial Electronics*, vol. 67, no. 4, pp. 2678–2688, Apr. 2020.

[11] K. Kroičs, "Considirations for Development of High Speed Response BLDC Motor Drive with GaN Semiconductors," in *2022 IEEE 7th International Energy Conference (ENERGYCON)*, May 2022, pp. 1–5.

[12] "RM0440 Reference manual - Google Search." Accessed: Jan. 05, 2024. [Online]. Available: https://www.st.com/resource/en/reference_manual/rm0440-stm32g4-series-advanced-armbased-32bit-mcus-stmicroelectronics.pdf

[13] J. Strydom and D. Reusch, "Dead-time optimization for maximum efficiency," *Efficient Power Conversion (EPC) whitepaper: WP012*, 2013.

[14] D. Han and B. Sarlioglu, "Deadtime Effect on GaN-Based Synchronous Boost Converter and Analytical Model for Optimal Deadtime Selection," *IEEE Transactions on Power Electronics*, vol. 31, no. 1, pp. 601–612, Jan. 2016.

[15] P. M. Roschatt, R. A. McMahon, and S. Pickering, "Investigation of dead-time behaviour in GaN DC-DC buck converter with a negative gate voltage," in *2015 9th International Conference on Power Electronics and ECCE Asia (ICPE-ECCE Asia)*, Jun. 2015, pp. 1047–1052.

[16] L. Hoffmann, C. Gautier, S. Lefebvre, and F. Costa, "Optimization of the Driver of GaN Power Transistors Through Measurement of Their Thermal Behavior," *IEEE Transactions on Power Electronics*, vol. 29, no. 5, pp. 2359–2366, May 2014.

[17] F. Chierchie, E. E. Paolini, and L. Stefanazzi, "Dead-Time Distortion Shaping," *IEEE Transactions on Power Electronics*, vol. 34, no. 1, pp. 53–63, Jan. 2019.

[18] "Comparative Evaluation of Gate Driver and LC-Filter Based dv/dt-Limitation for SiC-Based Motor-Integrated Variable Speed Drive Inverters | IEEE Journals & Magazine | IEEE Xplore." Accessed: Apr. 09, 2024. [Online]. Available: https://ieeexplore.ieee.org/document/10144373

[19] J.-P. Strom, J. Korhonen, J. Tyster, and P. Silventoinen, "Active du/dt—New Output-Filtering Approach for Inverter-Fed Electric Drives," *IEEE Transactions on Industrial Electronics*, vol. 58, no. 9, pp. 3840–3847, Sep. 2011.

[20] Kroičs, K.; Būmanis, A. BLDC Motor Speed Control with Digital Adaptive PID-Fuzzy Controller and Reduced Harmonic Content. *Energies* 2024, 17, 1311.

PCIM Europe 2024, 11– 13 June 2024, Nuremberg   DOI: 10.30420/566262065

The 8th Generation LV100 IGBT Module with Higher Current Rating

Daichi Otori[1], Masaomi Miyazawa[1], Stumpf Eugen[2], Koichi Masuda[2]

[1] Power Device Works, Mitsubishi Electric Corp., Japan

[2] Mitsubishi Electric Europe B.V., Germany

Speaker: Daichi Otori, Otori.Daichi@ap.MitsubishiElectric.co.jp
Corresponding author: Daichi Otori, Otori.Daichi@ap.MitsubishiElectric.co.jp

Abstract

This paper presents 8th generation 1800A/1200V IGBT power module designed for industrial applications. In this power module, cutting-edge 8th generation IGBTs and diodes are mounted. Significantly lower power losses can be achieved compared to conventional power module by adopting Split-Dummy-Active (SDA) gate structure and reducing chip thickness which can be realized by adopting Controlling charge carrier Plasma Layer (CPL) structure. In particular, the SDA structure reduces E_{on} approximately 60% with the same recovery dv/dt as the conventional module. By significant cutting of power losses, this module can increase power density. By adopting these technologies and expanding the chip areas, the 8th generation 1200V IGBT power module achieves a rated current of 1800A, which is 1.5 times higher than the conventional 1200V IGBT power module, in the same Mitsubishi Electric LV100-package.

Key words: 8th generation, LV100-type half bride package, High current density, SDA gate structure, CPL structure

1   Introduction

In recent years, the renewable energy markets such as PV and ESS are steadily growing as countermeasures against global warming. In these markets, demands of power semiconductors which are the core of power conversion systems are significantly increasing. In particular, 1200V-class IGBT power modules are widely integrated into the systems, and the IGBT modules with higher output power operation are required to improve system performance and cut system costs. Engineers are designing systems with high power capabilities within limited space. Therefore, the IGBT modules are required to increase output power while maintaining in a conventional package size. There are two approaches to increase the output power. The first one is to raise output voltage, the second one is to increase output current. Opting to raise the output voltage requires users to significantly redesign because many devices around the IGBT module also need to increase their rated voltage. On the other hand, choosing to increase current provides a way to boost output power with less user's design changes than raising voltage. Increasing the

power density of IGBT modules is important to increase the output current over the conventional 1200V-class IGBT power modules. "Reduce loss" and "Increase heat dissipation" are preferable ways to increase the power density.

Figure 1 shows the normalized ratio of calculated DC and AC power losses under 3-level A-NPC topology conditions for the 7th generation 1200A/1200V rated product in an available LV100-package shown in Figure 2 [1]. It can be noticed that DC power loss of IGBT and diode have the high rate of power losses among outer, inner, neutral conditions. It can be also noticed that the turn-on switching power loss ratio of the outer condition is large. Therefore, reductions of IGBT DC, diode DC and turn-on switching power losses are highly effective in decreasing of total power losses. Expansion of chip area is possible way to increase heat dissipation.   The 8th generation chips are optimized for the above power losses reduction. In addition, the 8th generation chips are improved DC electrical characteristics and heat dissipation by expanding the chip areas. The LV100-package is an industrial package designed to have low Ls (parasitic internal package inductances) by adopting a parallel plate electrode structure [2]. In addition, the LV100 can provide excellent current balance for internal chips and can be used in parallel, so it has been used for a variety of purposes in recent years.

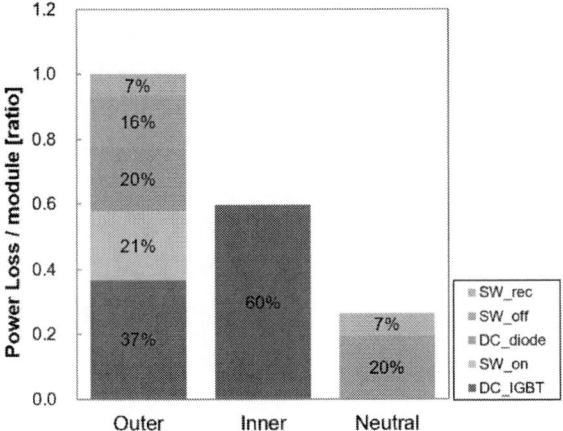

Condition : T_{vj}=150 °C, V_{cc}=750 V, M=0.75, PF=1, f_c=3 kHz, f_o=50 Hz, 3-level A-NPC topology

Fig. 1 The rate of power loss of conventional 1200V-class module.

Fig. 2 Industrial LV100-package appearance.

2 The 8th generation chip technology

The 8th generation chips have mainly adopted Split-Dummy-Active (SDA) gate structure and Controlling charge carrier Plasma Layer (CPL) structure, and these technologies are explained in this section.

Turn-on switching power loss can be reduced by high-speed switching operation of IGBT module. However, high-speed switching operation provides high recovery dv/dt. It is known as an EMI irradiation noise source and stresses user's motor insulation. So the recovery dv/dt should be limited

below a certain value by increasing gate resistance (R_G). However, large R_G slows down the switching speed of IGBT module and causes higher turn-on switching power loss. Therefore, it is necessary to reduce the recovery dv/dt without increasing of R_G. In Figures 3 and 4, schematic cross-sectional views of the 7th generation and the 8th generation Carrier-Stored-Trench-Gate-Bipolar-Transistor (CSTBT™) of IGBT are showed [3]. In the 7th generation CSTBT™, the active trenches are connected to the gate and the dummy trenches are connected to the emitter are placed alternately. In contrast, in the 8th generation CSTBT™, the dummy trenches are replaced with SDA trenches. In this configuration, the electrode inside the trench is divided into two stages, the upper electrode of the SDA trench connects to the emitter, while the lower one connects to the gate. In addition, the CPL is applied to the backside buffer of the 8th generation CSTBT™. Figures 5-(a) and (b) show the chip characteristics of the 7th generation and the SDA structures, showing the emitter current (I_E) dependence of recovery dv/dt. The horizontal axis in Figure 5-(a) shows the area at below 100% of the rated I_E, and the horizontal axis in Figure 5-(b) shows the area at below 2% of the rated I_E. The SDA structure increases the gate-collector capacitance (C_{GC}) without increasing gate-emitter capacitance (C_{GE}). As shown in Figure 5-(a) and (b), increasing C_{GC} can reduce the recovery dv/dt under low currents without affecting the recovery dv/dt under high currents [4]. This effect of the SDA structure is important because the recovery dv/dt is generally highest under low currents.

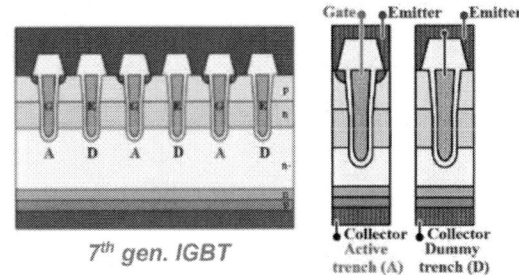

Fig. 3 Schematic view of the 7th gen. CSTBT™.

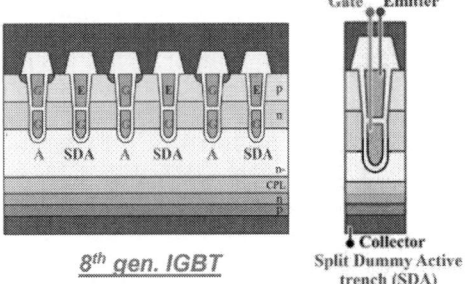

Fig. 4 Schematic view of the 8th gen. CSTBT™.

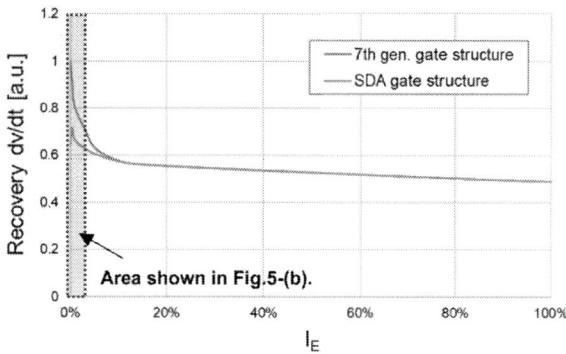

Condition : T_{vj}=25 °C, V_{cc}=750 V, V_{GE}=15 V, R_G=0 Ω

Fig. 5-(a) Chip characteristics (0-100% area) : Emitter current dependence of recovery dv/dt.

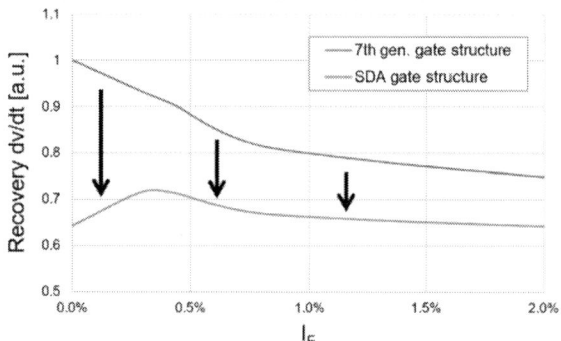

Condition : T_{vj}=25 °C, V_{cc}=750 V, V_{GE}=15 V, R_G=0 Ω

Fig. 5-(b) Chip characteristics (0-2% area) : Emitter current dependence of recovery dv/dt.

DC and switching power losses can generally be reduced by cutting the chip thickness. The chip thickness should be optimized considering a breakdown voltage. During turn-off operation with high di/dt, a turn-off surge voltage is generated. If this turn-off surge voltage exceeds the breakdown voltage of the IGBT, the IGBT module would be destroyed. Hence, it is important to suppress the turn-off surge voltage in order to reduce the IGBT thickness and enable turn-off operation at high di/dt.

The 8th generation IGBT has adjusted the backside buffer design [5]. The CPL structure provides turn-off softness. The turn-off waveforms of the IGBTs with and without CPL are shown in Figure 6. The IGBT without CPL shows high and sharp turn-off surge voltage. The surge voltage peak is exceeding the rated blocking voltage of 1200V. In contrast, the IGBT with CPL shows suppression of the turn-off surge voltage and oscillation compared to the IGBT without CPL. The surge voltage peak is kept below the rated

blocking voltage of 1200V. They are compared under the same conditions because the occurrence levels of surge voltage and oscillation are generally affected by external Ls and operating conditions, etc. Both chip thicknesses are matched to the 8th generation IGBT. They are compared by changing only the backside buffer design. Therefore, the 8th generation IGBT with CPL has enabled turn-off operation under higher di/dt and reduced the chip thickness. Consequently, the 8th generation IGBT could be cut power losses.

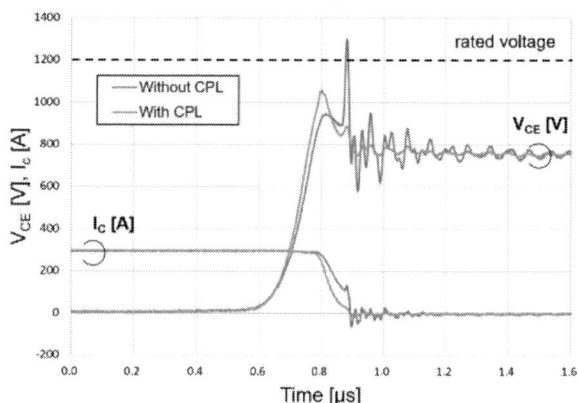

Condition : T_{vj}=150 °C, V_{cc}=750 V, V_{GE}=15 V, R_G=1.6 Ω, I_C=rated current

Fig. 6 Turn-off waveforms of the IGBTs with and without CPL.

The 8th generation chips have applied technologies such as SDA gate structure and CPL structure to increase power density.

3 Evaluation results of the 8th generation IGBT module

The 8th generation 1800A/1200V rated IGBT module equipped with the 8th generation chips described in section 2 in the LV100-package and the conventional CM1200DW-24T (1200A/1200V rated) equipped with the 7th generation chips are evaluated.

Figure 7 shows the chip areas and a normalized comparison of junction-case thermal resistance ($R_{th(j-c)}$). The chip areas of 8th generation 1200V-class chipsets are optimized for the chip mounting areas in the LV100-package. The 8th generation IGBT can reduce $R_{th(j-c)}$ by adopting the 39% larger chip area than the 7th generation IGBT area.

Expanded the IGBT area can also reduce DC power loss of IGBT.

The 8th generation diode has adjusted loss trade-off and optimized chip thickness to bring out the performance of the 8th generation IGBT modules. Furthermore, the diode can reduce $R_{th(j-c)}$ by adopting the 18% larger chip area than the 7th generation diode area. Expanded the diode area can also reduce DC power loss of diode. In addition, the 8th generation IGBT module has expanded the area available for chip mounting by optimizing the design inside the package.

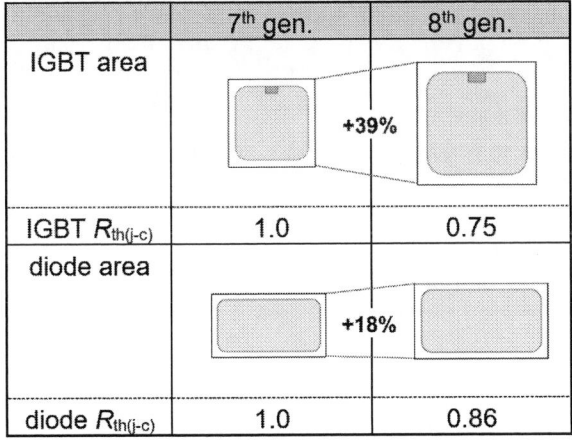

	7th gen.	8th gen.
IGBT area		**+39%**
IGBT $R_{th(j-c)}$	1.0	0.75
diode area		**+18%**
diode $R_{th(j-c)}$	1.0	0.86

Fig. 7 1200V-class chip area and $R_{th(j-c)}$.

Figure 8-(a) and (b) show the I_E dependence of recovery dv/dt when evaluated under describing conditions. The horizontal axis in Figure 8-(a) shows the area at below 100% of the rated I_E, and the horizontal axis in Figure 8-(b) shows the area at below 2% of the rated I_E. Similar to chip characteristics, both the 8th generation and the conventional module have higher recovery dv/dt under the low currents than under the high currents. In particular, the recovery dv/dt of conventional module is significantly higher under the area at below 0.5% of rated I_E. In contrast, the recovery dv/dt of the 8th generation module is reduced in all areas at below the rated I_E compared to the conventional module. In particular, the recovery dv/dt of the 8th generation module under the low currents at below 0.5% of rated I_E is effectively reduced by having been applied the SDA structure. The recovery dv/dt in high current areas is reduced by expanding the chip areas and trade-off tuning.

Figure 9 shows the Emitter-Collector voltage (V_{EC}) waveforms during the recovery operation at the maximum recovery dv/dt of the 8th generation and the conventional modules (shown in Figure 8-(b), the conventional module's condition: I_E=0%,

the 8th generation module's condition: I_E=0.1%). These waveforms also confirm that the 8th generation module's dv/dt is lower than the conventional module's dv/dt.

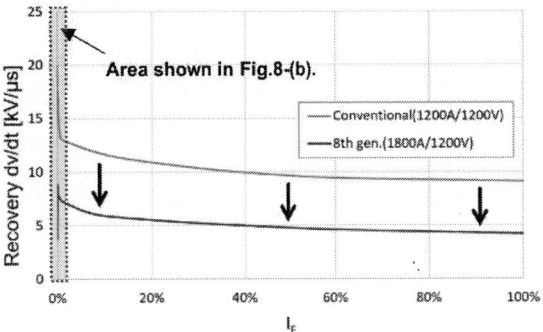

Condition : T_{vj}=25 °C, V_{cc}=750 V, V_{GE}=15 V, R_G=1.6 Ω

Fig. 8-(a) Module characteristics (0-100% area) : Emitter current dependence of recovery dv/dt.

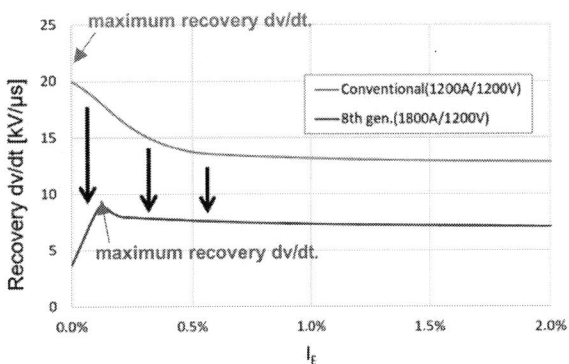

Condition : T_{vj}=25 °C, V_{cc}=750 V, V_{GE}=15 V, R_G=1.6 Ω

Fig. 8-(b) Module characteristics (0-2% area) : Emitter current dependence of recovery dv/dt.

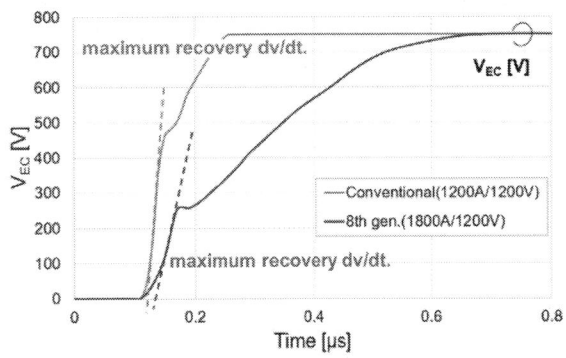

Condition : T_{vj}=25 °C, V_{cc}=750 V, V_{GE}=15 V, R_G=1.6 Ω, I_E= the current at maximum recovery dv/dt

Fig. 9 Recovery waveforms at maximum recovery dv/dt.

Figure 10 shows the I_E dependence of recovery dv/dt. The recovery dv/dt of the 8th generation is matched with 20kV/µs, which is the maximum value of the conventional module. The 8th generation module's R_G is adjusted for matching in the recovery dv/dt of the conventional module. The 8th generation module can use the smaller R_G than the conventional module because the 8th generation module has a low recovery dv/dt. The reduced R_G provide high-speed switching operation for the 8th generation module. This can reduce Turn-on switching energy per pulse (E_{on}).

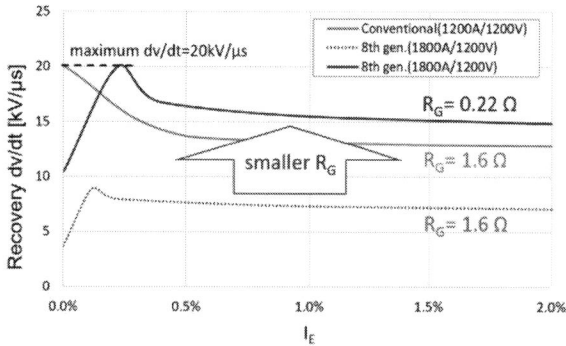

Condition : T_{vj}=25 °C, V_{cc}=750 V, V_{GE}=15 V

Fig. 10 Module characteristics (0-2% area) : Recovery dv/dt dependence on emitter current and R_G.

Figure 11 shows relationship between E_{on} and recovery dv/dt of the 8th generation and the conventional IGBT modules. The 8th generation module has reduced E_{on} under recovery dv/dt=20kV/µs by approximately 60% compared to the conventional module.

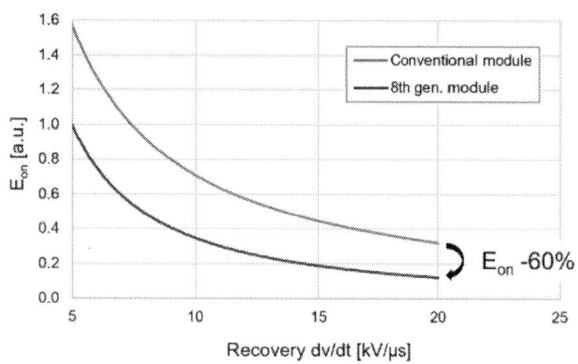

Condition : T_{vj}=150 °C, V_{cc}=750 V, V_{GE}=15 V, I_C=1800 A, dv/dt=20 kV/µs

Fig. 11 Recovery dv/dt dependence of E_{on}.

4 Electrical characteristics and output power of the 8th generation IGBT module

Figure 12 shows the electrical characteristics of the conventional CM1200DW-24T (1200A/1200V rated) and the 8th generation 1800A/1200V rated LV100 IGBT power module. The conventional module uses available minimum R_G condition, and the recovery dv/dt at that time is 20kV/µs. The 8th generation module's R_G has adjusted to match the recovery dv/dt of the conventional module. The loss results of the 8th generation module compared to the conventional module are described below.

(a). V_{CEsat}

V_{CEsat} is reduced by expanding the chip area and reducing the chip thickness.

(b). V_{EC}

V_{EC} is reduced by expanding the chip area and optimizing the chip thickness and trade-off tuning.

(c). E_{on}

E_{on} is reduced by applying the SDA structure and reducing the chip thickness.

(d). E_{off}

E_{off} is increased by expanding the chip area, but it is reduced by cutting the chip thickness. As a result, E_{off} is adjusted to loss curve similar to the conventional module.

(e). E_{rr}

E_{rr} is increased by expanding the chip area and optimizing trade-off adjusting.

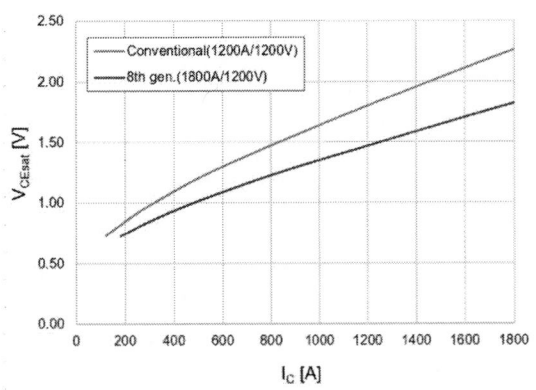

(a) I_C dependence of V_{CEsat}.

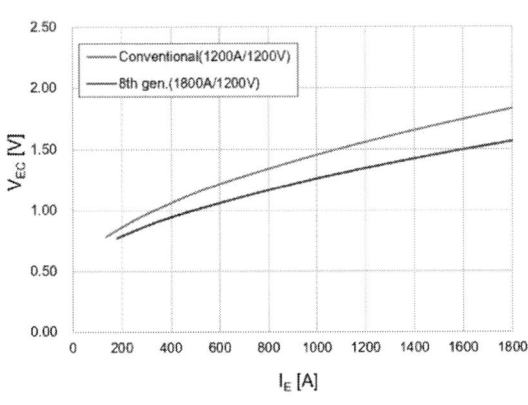

(b) I_E dependence of V_{EC}.

Condition : T_{vj}=150 °C, V_{GE}=15 V, I_C,I_E=~1800 A

Fig. 12-(a),(b) Electrical characteristics (DC) of the conventional and the 8th generation IGBT modules.

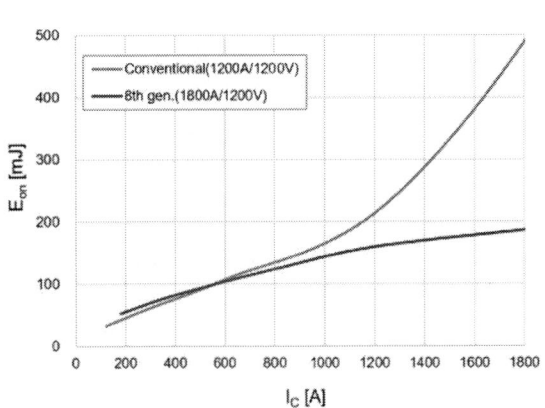

(c) I_C dependence of E_{on}.

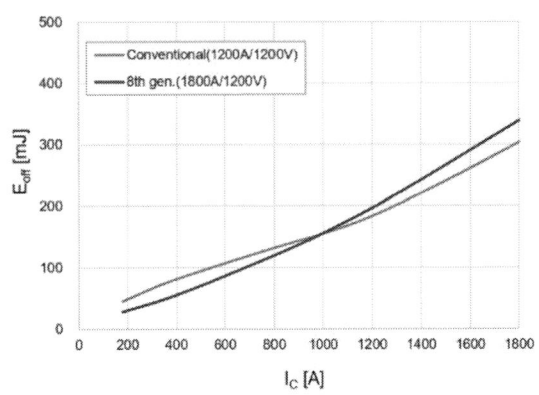

(d) I_C dependence of E_{off}.

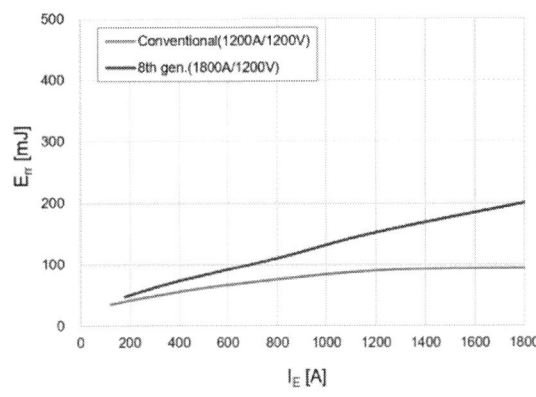

(e) I_E dependence of E_{rr}.

Condition : T_{vj}=150 °C, V_{cc}=750 V, V_{GE}=15 V, I_C,I_E=~1800 A, dv/dt=20 kV/µs

Fig. 12-(c),(d),(e) Electrical characteristics (AC) of the conventional and the 8th generation IGBT modules.

Figure 13 shows relationship between carrier frequency (f_c) and output current (I_{out}) of the IGBT modules. The horizontal axis shows the f_c condition, and the vertical axis shows the running value of the I_{out}. This is calculated by the electrical characteristics in Figure 12. This result shows that the 8th generation IGBT module can achieve approximately 25% more output power that compared to the conventional module. I_{out} can also be further increased by optimizing cooling condition and other factors.

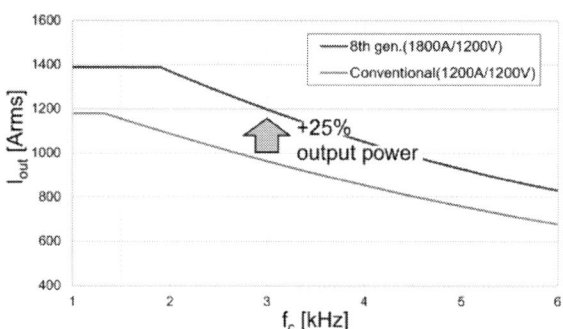

Condition : T_{vj}=150 °C, V_{cc}=750 V, M=0.75, PF=1, f_o=50 Hz, forced air cooling, T_a=40 °C, 3-level A-NPC topology

Fig. 13 Output power comparison.

5 Conclusion

The 8th generation 1200V LV100 IGBT power module with higher rated current is designed for industrial applications. The 8th generation IGBT module demonstrates superior power losses by adopting the SDA gate structure and the CPL structure. In particular, the SDA gate structure in IGBT reduces E_{on} by approximately 60% compared to the conventional module with the same recovery dv/dt. In addition, the 8th generation chip areas have been expanded to enhance heat dissipation and cut DC losses. By cutting losses, this module has increased power density. Hence, the 8th generation 1200V IGBT module achieves a rated current of 1800A with the same package as the conventional LV100 as shown in Figure 14.

6 Reference

[1] "7th Generation T-series Industrial LV100-type Application Note", Mitsubishi Electric Power module application note, March 2023, https://www.mitsubishielectric.co.jp/semicond uctors/powerdevices/application_notes/indust rial_lv100_e.pdf

[2] M. Miyazawa, M. Tabata, H. Muraoka, T. Hieda, T. Radke, "7th Generation IGBT Module for Industrial Apprications" PCIM 2014.

[3] H. Takahashi et al. "Carrier Stored Trench-Gate Bipolar Transistor (CSTBT) -A Novel Power Device for High Voltage Application-", ISPSD 1996.

[4] K. Konishi, K. Nishi, K. Sako, A. Furukawa, "Split-Dummy-Active CSTBTTM for Improving Recovery dV/dt and Turn-on Switching Loss Tradeoff" ISPSD 2022.

[5] K. Suzuki, K. Nishi, M. Kaneda, A. Furukawa, "N-buffer design optimization for Short Circuit SOA ruggedness in 1200V class IGBT" ISPSD 2018.

[6] K. Konishi, R. Kamibaba, M. Umeyama, A. Narazaki, T. Takahashi, A. Furukawa and M. Tarutani, "Experimental Demonstration of the Active Trench Layout Tuned 1200V CSTBT™ for Lower dV/dt Surge and Turn-on Switching Loss", Proc. ISPSD 2016.

[7] S. Machida, T. Sugiyama, M. Ishiko, S. Yasuda, J. Satio and K. Hamada, "Investication of Correlation between Device Structures and Switching Losses of IGBTs", Proc. ISPSD 2009.

[8] S. Machida, K. Ito, and Y. Yamashita, "Approaching the Limit of Switching Loss Reduction in Si-IGBTs", Proc. ISPSD 2014.

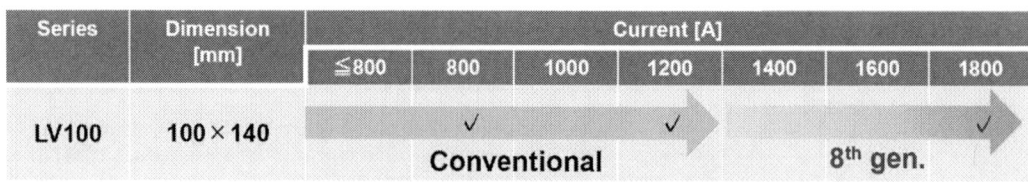

Fig. 14 Product lineup of 1200V rated LV100.

PCIM Europe 2024, 11– 13 June 2024, Nuremberg   DOI: 10.30420/566262065

Loss Reduction by Laminating Ferrite E Cores

Lukas Reißenweber[1], Julia Rogner[1], Alexander Stadler[1]

[1] Coburg University of Applied Sciences and Arts, Germany

Corresponding author: Lukas Reißenweber, lukas.reissenweber@hs-coburg.de

Abstract

Due to the increasing switching frequencies in power electronics, macroscopic eddy currents in ferrites and the associated losses have to be increasingly considered. Especially for cores with larger cross sections, the effect occurs already at lower frequencies. The reduction of eddy current losses in SiFe cores by lamination is already well known. This work applies the principle of lamination to ferrite cores of the sizes E70, E42 and E36 of the material MF106 (comparable to N97). The effect of lamination is investigated in the small-signal range using the loss tangent ($\tan\delta$) and at higher flux densities up to 500 kHz at 25 °C and 100 °C using rectangular voltage excitation.

1 Introduction

Due to the increasing switching frequencies in power electronics, passive components can be dimensioned smaller. This reduces the consumption of materials and thus also costs. However, it also poses new challenges for the engineers of magnetic components in order to get a design that is as compact and efficient as possible. Ferrite is usually used as the core material at higher frequencies, since it combines relatively good magnetic properties with low losses and is also cost efficient. Precisely because of the increasing frequencies, effects such as eddy currents, dimensional resonances and structural defects in ferrites are playing an increasingly important role [1]–[10]. Especially for large cores with large cross sections these effects have to be considered. The structure of isolated ferrite grains [11] suppresses macroscopic eddy currents. But the electric and dielectric properties of the grain isolation depend on frequency and temperature [3], [8], [12], [13]. The higher the frequency and the temperature the higher the conductivity. Due to the increased conductivity, macroscopic eddy currents occur more intensely throughout the core cross section and cause losses. The resulting fields are superimposed and lead to a kind of magnetic skin effect. At the same time, the high permeability together with the high permittivity lead to dimensional resonance effects. The resulting flux distribution in the core depends on the fields of the eddy currents and the dimensional resonance effects [1]–[9]. Which in turn depend on the geometry of the core and its frequency, temperature, field strength dependent magnetic, electric and dielectric properties. One possibility to reduce the macroscopic eddy currents and its losses is the lamination of ferrite cores as it is well known from SiFe cores and already from ferrites in filter applications [3], [4], [10], [14]. The effects of laminating ferrite E cores of sizes E70, E42 and E36 of the material Manifer 106 (MF106) are investigated in this work. The material MF106 is comparable to e.g. N97.

2 Fundamentals

2.1 Principle of Lamination

Figure 1 a) shows the magnitude of the complex impedance normalized to the value at $f = 10\,\text{kHz}$ and b) the argument of different sized E cores made of MF106 at $\vartheta = 25\,°\text{C}$. It can be seen that due to the reasons mentioned before, the largest core already loses impedance at lower frequencies. Also, the argument of the larger core first falls to zero and below.

A flux distribution in an E70 core at $f_\text{s} = 300\,\text{kHz}$ and $\vartheta_\text{s} = 100\,°\text{C}$ considering the fields of the eddy currents and the dimensional resonance effects is illustrated in fig. 2 a) for a bulk core with a thickness of 32 mm and in b) for a laminated core made of 8 layers. The layer thickness is 4 mm and the distance between them is 0.5 mm. It shows the distribution of the magnitude of magnetic flux density $|\underline{B}|$ in one core half normalized to the highest occurring flux density $|\underline{B}_\text{bulk}(y = 0mm)|$ of

525

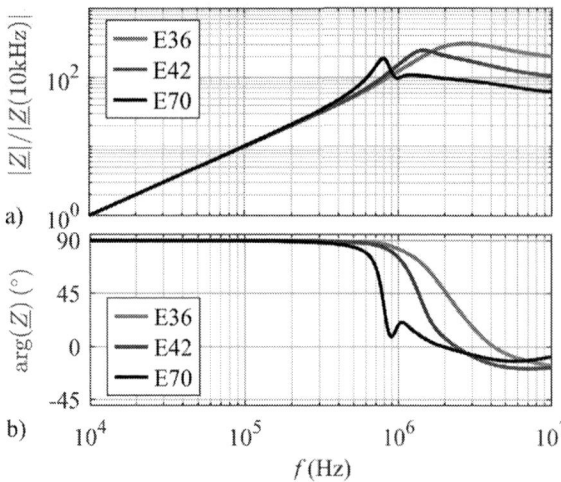

Fig. 1: a) Normalized magnitude of complex impedance and b) argument of E cores of the sizes E36, E42 and E70 made of MF106 at $\vartheta = 25\,°\mathrm{C}$.

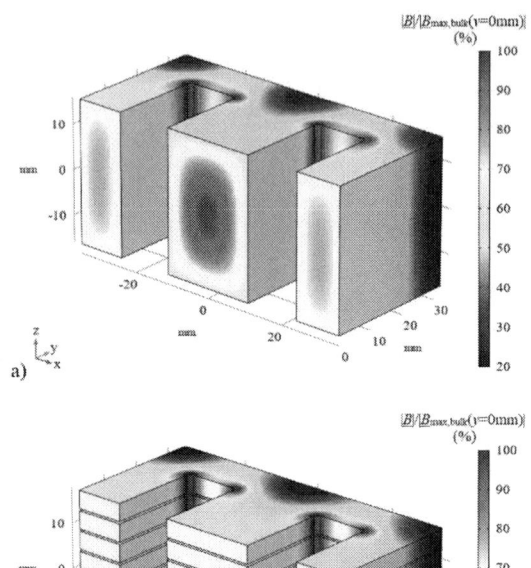

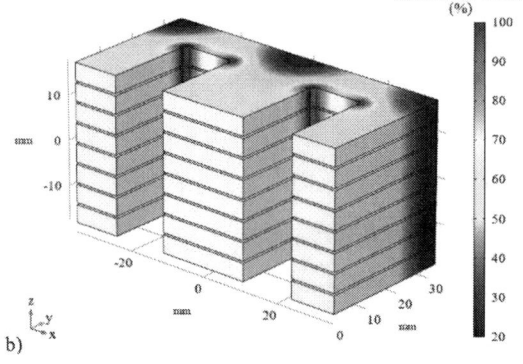

Fig. 2: Distribution of normalized magnitude of flux density in one core half of an ungapped a) bulk and b) laminated E70 core at $f_\mathrm{s} = 300\,\mathrm{kHz}$ and $\vartheta_\mathrm{s} = 100\,°\mathrm{C}$.

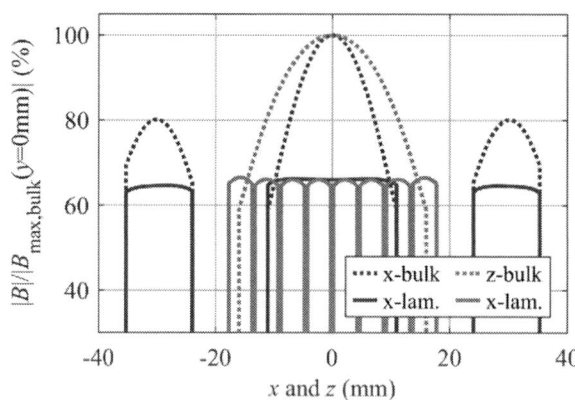

Fig. 3: Distribution of the normalized magnetic flux density in the cross section in x- and z-direction of a bulk and a laminated E70 core at $f_\mathrm{s} = 300\,\mathrm{kHz}$ and $\vartheta_\mathrm{s} = 100\,°\mathrm{C}$.

the cross section of the bulk core. The values of the real $\mu_\mathrm{r}'(fs, \vartheta_\mathrm{s}) = 2845$ and the imaginary part $\mu_\mathrm{r}''(fs, \vartheta_\mathrm{s}) = 111$ of the relative permeability, the frequency dependent electrical conductivity $\sigma + \omega\varepsilon''(fs, \vartheta_\mathrm{s}) = 1.2\,\mathrm{A/(Vm)}$ and the real part of the relative permittivity $\varepsilon_\mathrm{r}'(fs, \vartheta_\mathrm{s}) = 101 \cdot 10^3$ are used. The values are based on measurements on toroidal cores made of MF106 ($d_\mathrm{a} = 24.9\,\mathrm{mm}$, $d_\mathrm{i} = 14.6\,\mathrm{mm}$, $h = 9.9\,\mathrm{mm}$) with measurement set-ups according to [8], [13], [14] and subsequent correction of the material data regarding field displacement effects within the toroidal cores according to [8]. It can be seen, that in the middle limb of the bulk core the flux density is significant increased in the center and decreased in the outer region. In the laminated core the distribution is more homogeneous. Such distributions are experimentally verified on toroidal cores by measurements in [3], [4]. The effect is even more significant for cores with larger cross sections [10], [14]. In many applications, this means that the overall flux density in the core cannot be particularly high in order to keep losses low. Therefore, the core is not well utilized in terms of power density. In fig. 3 the associated flux density, is depicted for the x- and z-direction at $y = 0\,\mathrm{mm}$ normalized to the maximum flux density of the cross section of the bulk core. For the x-direction of the laminated core, the cut line was placed through the center of one of the inner layers $z = 2.25\,\mathrm{mm}$. It can be seen, that in the bulk core the flux density in the outer regions of the middle limb is lower. In the inner area it is

increased. The same applies to the outer legs. Due to the smaller cross-sectional areas, this effect is significantly reduced in the laminated core. This is shown by the more uniformly distributed flux density. This makes it possible to increase the total flux density and thus reduce the core cross section and volume. The smaller circumference also reduces the winding length and thus the losses and costs. Calculations of core loss and structural effects are provided in [6]–[8], [11], [15], [16]. Due to the reduction of eddy currents and the associated fields, laminated cores have a higher impedance at higher frequencies. This is particularly interesting for filtering EMI. Temperature and frequency increase the effect of macroscopic eddy currents. Figure 4 depicts the frequency depended conductivity $\sigma + \omega\varepsilon''$ of the material MF106 at different temperatures calculated from measured initial values [8]. This shows that the eddy currents are more pronounced at the temperatures at which the ferrite is used in the application.

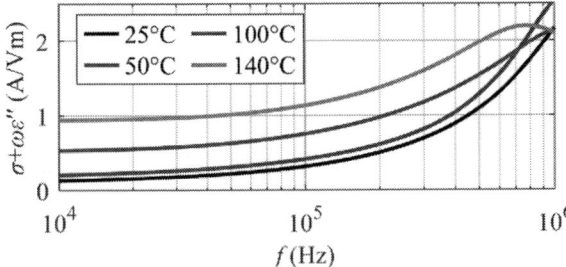

Fig. 4: Frequency dependent conductivity $\sigma + \omega\varepsilon''$ of the material MF106 at different temperatures.

2.2 Laminated E Cores

For the investigation E cores of the sizes E36, E42 and E70 of the material MF106 are used. The layers for the laminated cores were produced by grinding down bulk cores symmetrical on both sides. Also the chamfers have been removed from the original (bulk) E70 core. The machining of the cores and also the position from which the layer is produced has an influence on the magnetic material properties [14]. For the insulation of the layers from each other 0.5 mm thick plates of FR4 are inserted between them. Figure 5 shows an overview of the cores of the respective sizes E36, E42 and E70 and layer thicknesses. Table 1 provides an overview of all investigated core samples and their number. $N_{\mathrm{lam.}}$ is the number of layers and $h_{\mathrm{lam.}}$ the layer thickness. The parameter l_e stands for the effective core length and A_e for the effective cross section of the core. The values were taken from the data sheets and the effective cross sections were scaled according to the thickness of the cores.

Fig. 5: Bulk and laminated E36, E42 and E70 cores of material MF106.

Figure 6 shows a laminated E70 core with FR4 elements for positioning in a clamping device. The inner FR4 elements have a thickness of 0.5 mm. Since the material values of ferrites are dependent on mechanical pressure, heat-resistant plastic compression springs ensure equal contact pressure for all samples. For the small-signal measurements, the cores were measured without an air gap. For the large-signal measurements a 75 μm thick foil was inserted into the outer legs.

Fig. 6: Laminated E70 core with FR4 elements for positioning the layers and compression springs for equal contact pressure in a clamping device.

2.3 Small-Signal Set-Up

The measurements for the comparison of the losses in the small-signal range have been made with the impedance analyzer E4990A from Keysight and the test fixture 16047E. Figure 7 a) shows a DUT in front of the impedance analyzer. Short/Open/Load compensation was performed on the impedance analyzer with the test fixture. For the impedance of the DUT $\underline{Z}_{\mathrm{DUT}}$, the measured impedance $\underline{Z}_{\mathrm{xm}}$ was additionally corrected by an open $\underline{Z}_o$ and a short $\underline{Z}_s$ measurement [17]:

$$\underline{Z}_{\mathrm{DUT}} = \frac{\underline{Z}_s - \underline{Z}_{\mathrm{xm}}}{\underline{Z}_{\mathrm{xm}} - \underline{Z}_o} \underline{Z}_o \qquad (1)$$

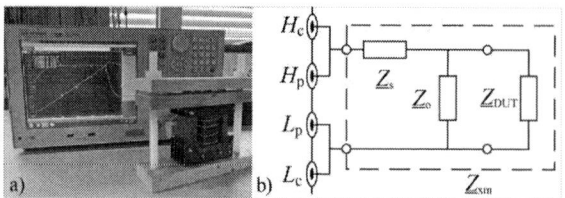

Fig. 7: a) DUT at impedance analyzer and b) ECM of residuals of the test fixture [17].

The equivalent circuit model for the residuals of the test fixture is depicted in fig. 7 b). The impedance of the measurement winding $\underline{Z}_{\text{wire}}$ was subtracted from the measurement of the core with winding $\underline{Z}_{\text{core+wire}}$ to obtain the impedance of the core $\underline{Z}_{\text{core}}$:

$$\underline{Z}_{\text{core}} = \underline{Z}_{\text{core+wire}} - \underline{Z}_{\text{wire}} \tag{2}$$

As a result of the processing, the laminated cores can show a reduced real part of the complex permeability compared to the bulk cores [14]. A slight decrease was observed for cores E36 and E42. The comparison is therefore based on the loss tangent $\tan\delta$:

$$\tan\delta = \frac{\text{Re}\{\underline{Z}_{\text{core}}\}}{\text{Im}\{\underline{Z}_{\text{core}}\}} \tag{3}$$

For the E70 cores there was no change in the real part of the permeability noticeable.

Core	$N_{\text{lam.}}$	$h_{\text{lam.}}$ (mm)	Nbr. of samples	l_e (mm)	A_e (mm²)
E70	1	30 (bulk)			640.3
	2	16		150	
	4	8			683
	8	4			
E42	1	15 (bulk)			
	2	7.5			
	3	5	3	97.4	178
	5	3			
	10	1.5			
E36	1	11 (bulk)			120
	2	7.5			
	3	5		81	
	5	3			163.6
	10	1.5			

Tab. 1: Overview of investigated E core samples.

2.4 Loss Measurement at Higher Flux Densities

For the loss measurement at higher flux densities the two winding method is used. The measurements are carried out with a HDO6054B oscilloscope from Teledyne LeCroy. The primary current is measured with a Pearson current monitor 2877 and the secondary voltage with a PP024 voltage probe. The runtime compensation of the current monitor is determined by comparing a small and large-signal measurement of an air coil [8]. The circuit used is a SiC half bridge configuration that provides a symmetrical square wave voltage. The voltage steepness is reduced by adding capacitors $C_{\text{par}} = 0.5\,\text{nF}$ in parallel to the semiconductors and adjusting the dead time. An equivalent circuit diagram shows fig. 8 a) and b) the set-up (without oven). For the measurements at $\vartheta = 100\,°\text{C}$, the test samples were placed in an oven. The cables with BNC connection for the voltage measurement were not changed. They were positioned so that the probe sits outside the oven.

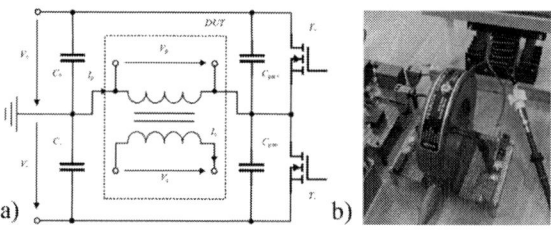

Fig. 8: a) ECM of used half bridge and b) set-up with DUT.

The magnetic field strength $H(t)$ is given by [8], [18], [19]:

$$H(t) = \frac{N_{\text{p}}}{l_e}\left[i_{\text{p}}(t) - \frac{N_{\text{s}}}{N_{\text{p}}}i_{\text{s}}(t)\right] \tag{4}$$

N_{p} is the number of turns of the primary winding and $i_{\text{p}}(t)$ the primary current. N_{s} is the number of turns of the secondary winding. The winding ratio is $N_{\text{s}}/N_{\text{p}} = 2/4$. The current $i_{\text{s}}(t)$ charges the capacitance of the probe $C_{\text{probe}} = 12\,\text{pF}$. Measurements have shown that the secondary current can be neglected. The magnetic flux density $B(t)$ depends on the secondary voltage $v_{\text{s}}(t)$ [8], [18], [19]:

$$B(t) = \frac{1}{N_{\text{s}}A_e}\int_0^T v_{\text{s}}(t)\text{d}t \tag{5}$$

With the magnetic field strength and the magnetic flux density the mean core losses per volume $\overline{p}_{\text{v}}$

can be calculated [8], [18]:

$$\bar{p}_{\mathrm v} = \frac{1}{T}\int_0^T H(t)\cdot\frac{\delta B(t)}{\delta t}\mathrm{d}t \qquad (6)$$

To validate the measurements, all cores without an air gap were additionally measured with sinusoidal signals using a B-H Analyzer SY-8219 from Iwatsu at $\vartheta_{\mathrm s} = 25\,°\mathrm{C}$.

3 Measurement Results

3.1 Small-Signal Measurements

In fig. 9 a) the magnitude normalized to the 10 kHz value and in b) the argument of the bulk and the laminated E42 cores without an air gap are depicted. The magnitude of the impedance is maintained with decreasing layer thickness up to higher frequencies. The argument shows that this is the inductive part of the component. This is interesting for filtering high-frequency noise. Since here at higher frequencies more reactive content contributes to the attenuation [3], [4], [10], [14]. The E36 cores show a comparable behavior. Here the effects of the lamination occur at even higher frequencies. In fig. 10 a)

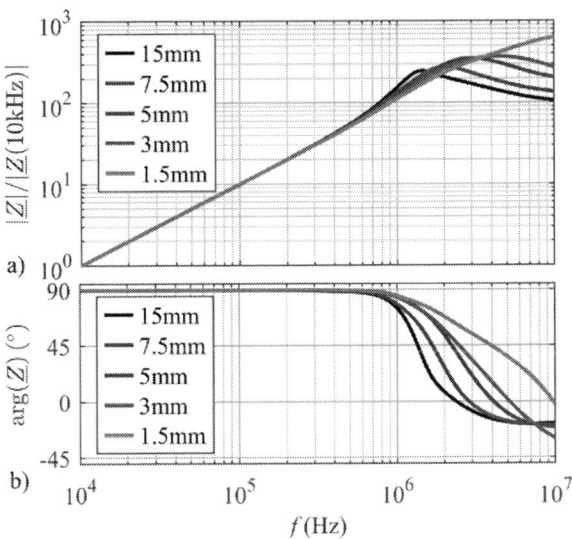

Fig. 9: a) Normalized magnitude of complex impedance and b) argument of E cores made of MF106 of the size E42 at $\vartheta = 25\,°\mathrm{C}$.

the magnitude normalized to the 10 kHz value and in b) the argument of the bulk and the laminated E70 cores without an air gap are depicted. Also, here the magnitude of the impedance is maintained with decreasing layer thickness up to higher frequencies. For the core with a layer thickness of 4 mm, the magnitude and the argument decrease

more from 2.3 MHz than the other cores and also fall below.

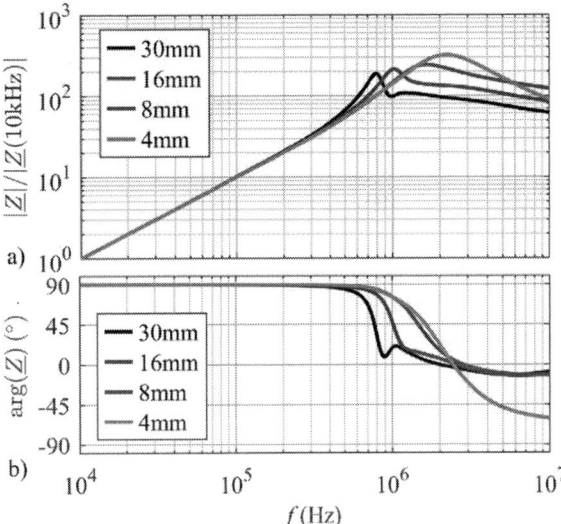

Fig. 10: Normalized magnitude of complex impedance and b) argument of E cores of the size E70 made of MF106 at $\vartheta = 25\,°\mathrm{C}$.

For ferrites in power applications, the reduction of losses is of interest. Therefore, in the following the measurements of the cores are compared by the loss tangent from eq. 3. Figure 11 shows the loss tangent of the E36 and E42 cores with different layer thickness in the lower frequency range at $\vartheta_{\mathrm s} = 25\,°\mathrm{C}$. For the E36 cores, the reduction of the layer thickness has no noticeable effect on the loss factor. The graphs of the E42 core show that halving the layer thickness from 15 mm to 7.5 mm leads to a slight reduction. The reduction to 5 mm halves the loss factor compared to the bulk core and brings it into line with the E36 core. Figure 12 depicts the loss factor for the E36 and E42 cores at higher frequencies at $\vartheta_{\mathrm s} = 25\,°\mathrm{C}$. Here it can be seen that the loss tangent is significant smaller as the layer thickness decreases. The loss tangent of E42 laminates is higher than that of E36 laminates with the same thickness. In the case of the E70 core a reduction takes place in the frequency range in which it is used. In fig. 13 the curves of the loss tangent are shown. In the range up to 700 kHz, this is reduced further and further as the layer thickness decreases. The values of 8 mm are close to those of 4 mm. Above 700 kHz, the 4 mm are slightly higher than those of 8 mm. Table 2 shows a comparison of the values at certain frequencies and the reduction of the loss tangent compared to

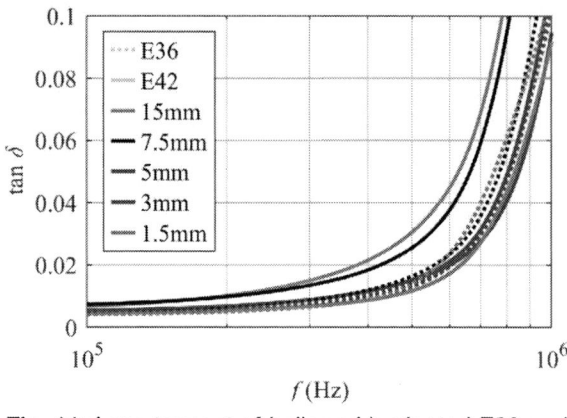

Fig. 11: Loss tangent of bulk and laminated E36 and E42 cores made of MF106 in the lower frequency range at $\vartheta = 25\,°\mathrm{C}$.

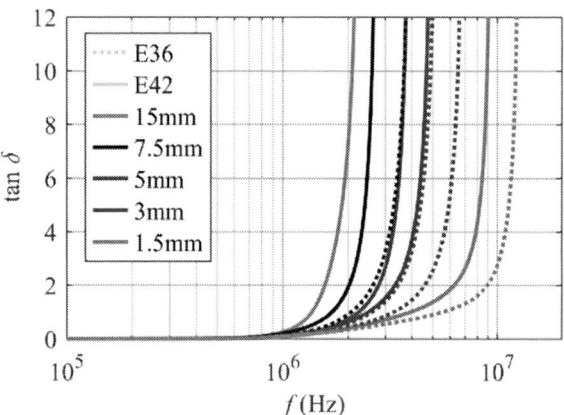

Fig. 12: Loss tangent of bulk and laminated E36 and E42 cores made of MF106 in the higher frequency range at $\vartheta = 25\,°\mathrm{C}$.

the bulk core.

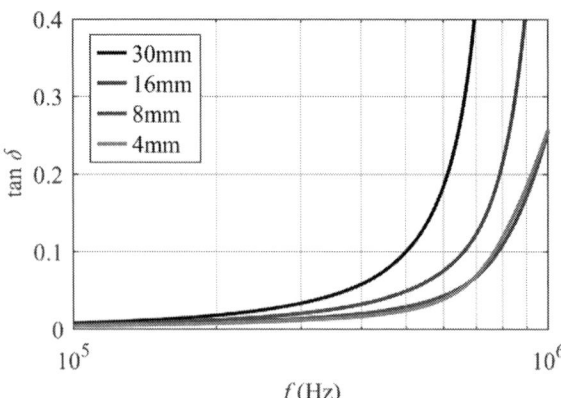

Fig. 13: Loss tangent of bulk and laminated E70 cores made of MF106 at $\vartheta = 25\,°\mathrm{C}$.

$h_{lam.}$ (mm)	30 (bulk)	16	8	4	16	8	4
f (kHz)	tanδ				tanδ reduction against bulk (%)		
100	0.008	0.006	0.005	0.005	25.0	37.5	37.5
200	0.018	0.012	0.009	0.008	33.3	50.0	55.6
300	0.034	0.021	0.014	0.012	38.2	58.8	64.7
400	0.058	0.032	0.020	0.017	44.8	65.5	70.7
500	0.099	0.049	0.029	0.024	50.5	70.7	75.8

Tab. 2: Loss tangent and reduction of it for E70 cores.

3.2 Core Losses at Higher Flux Densities

The small-signal measurements are an indication of loss reduction using lamination. However, the losses at higher flux densities are more important for the application. Figure 14 shows the measured primary current and the measured secondary voltage for the $\hat{B} = 50\,\mathrm{mT}$ value of an E70 core with 8 layers at $f = 500\,\mathrm{kHz}$ and $\vartheta = 25\,°\mathrm{C}$. The slope of the primary voltage was adjusted so that there is minimum current drop (not visible in the figure) due to the parasitic capacitance at the switching moment. Overall, it has been shown that a thinner layer thickness is also associated with a lower current drop. The influence of charging and discharging the parasitics of the winding and the high permittivity core on the flux density and losses are discussed in [20]. The flux density associated with the

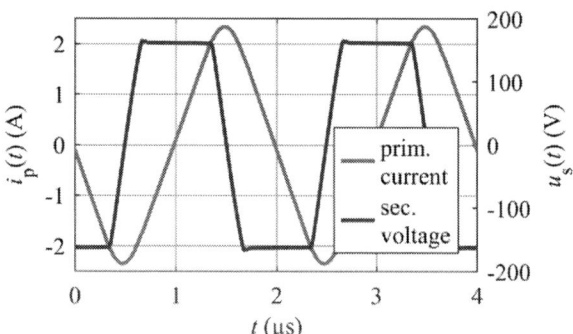

Fig. 14: Measured primary current and secondary voltage at E70 core with 8 layers at $\hat{B} = 50\,\mathrm{mT}$, $f = 500\,\mathrm{kHz}$ and $\vartheta = 25\,°\mathrm{C}$.

voltage of fig. 14, calculated from eq. 5 is shown in fig. 15. Figure 16 compares BH-curves of selected test specimens of all layer thicknesses $f = 500\,\mathrm{kHz}$ and $\vartheta = 25\,°\mathrm{C}$. The bulging is reduced with decreasing layer thickness.

Figure 17 shows the measured power loss density

PCIM Europe 2024, 11– 13 June 2024, Nuremberg　　　DOI: 10.30420/566262065

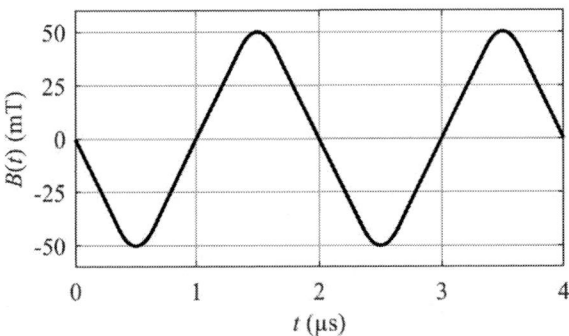

Fig. 15: Calculated flux density of E70 core with layer thickness 4 mm at $\hat{B} = 50\,\text{mT}$, $f = 500\,\text{kHz}$ and $\vartheta = 25\,°\text{C}$.

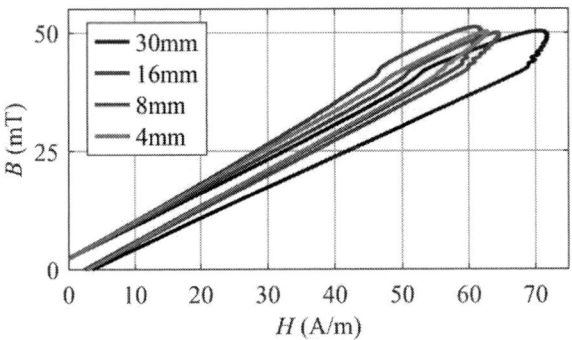

Fig. 16: BH-curves of bulk and laminated E70 cores at $\hat{B} = 50\,\text{mT}$, $f = 500\,\text{kHz}$ and $\vartheta = 25\,°\text{C}$.

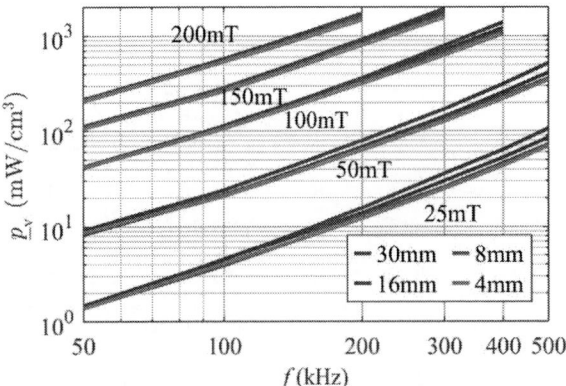

Fig. 17: Power loss density of bulk and laminated E70 cores made of MF106 at $\vartheta = 25\,°\text{C}$.

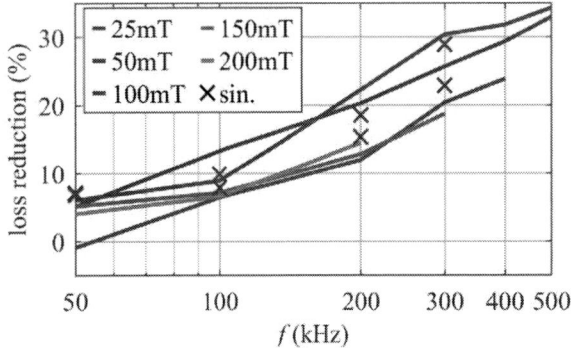

Fig. 18: Reduction in power loss density of the E70 core with 4 mm layers compared to the bulk core at $\vartheta = 25\,°\text{C}$.

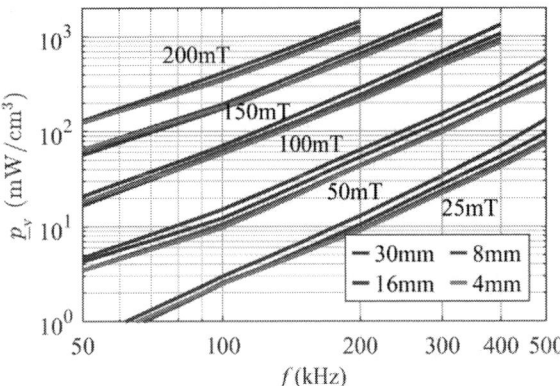

Fig. 19: Power loss density of bulk and laminated E70 cores made of MF106 at $\vartheta = 100\,°\text{C}$.

per volume at $\vartheta = 25\,°\text{C}$. It can be seen that the power loss density decreases with decreasing layer thickness. More at high frequencies than at low frequencies. The curves narrow towards the low frequencies. The calculated reduction of the loss density of the cores with 4 mm layers compared to the bulk cores is shown in fig. 18. It can also be seen here that the reduction increases with increasing frequency. This is somewhat more pronounced at low flux densities than at high flux densities. The results at $\hat{B} = 25\,\text{mT}$ and $\hat{B} = 50\,\text{mT}$ with sinusoidal signals are also shown. These are slightly lower but show a comparable trend. In fig. 19 the measured power loss density per volume at $\vartheta = 100\,°\text{C}$ is shown. Also here the power loss density decreases with decreasing layer thickness. More at high frequencies than at low frequencies. Towards the low frequencies the curves narrow less than with the curves of $\vartheta = 25\,°\text{C}$. The calculated reduction in losses of the cores with 4 mm layers compared to the bulk core is shown in fig. 20. It behaves like the measurements at $\vartheta = 25\,°\text{C}$, but the reduction is even higher.

The reduction in power loss density for the different layer thicknesses for $f = 300\,\text{kHz}$ at $\vartheta = 25\,°\text{C}$ and $\vartheta = 100\,°\text{C}$ is shown in fig. 21. There is an almost linear relationship with the layer thickness. The reduction is higher at $\vartheta = 100\,°\text{C}$ and decreases with increasing flux density at both temperatures.

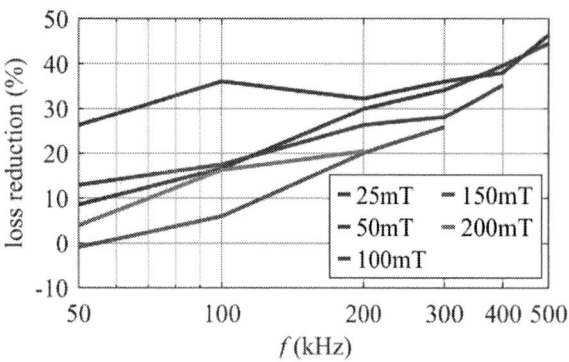

Fig. 20: Reduction in power loss density of the E70 core with 4 mm layers compared to the bulk core at $\vartheta = 100\,°\text{C}$.

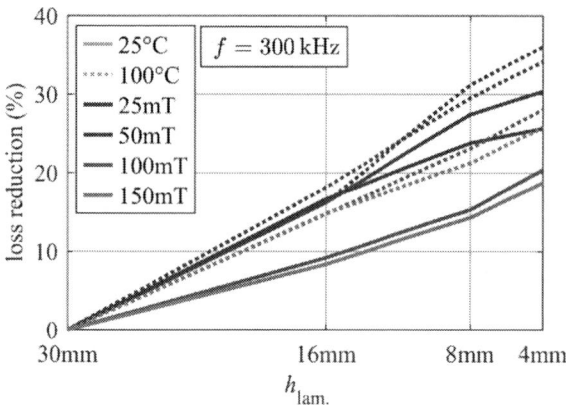

Fig. 21: Reduction in power loss density of E70 cores with different layer thickness for different flux densities for $f = 300\,\text{kHz}$ at $\vartheta = 25\,°\text{C}$ and $\vartheta = 100\,°\text{C}$.

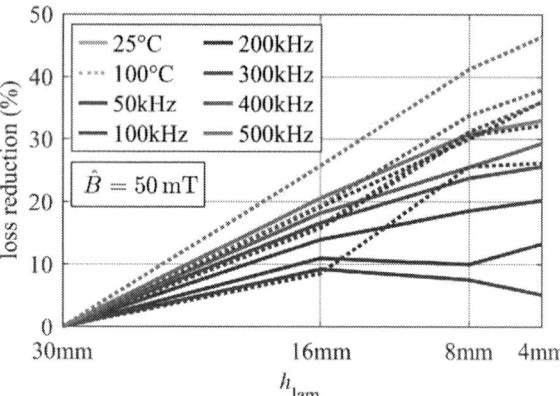

Fig. 22: Reduction in power loss density of E70 cores with different layer thickness for different frequencies for $\hat{B} = 50\,\text{mT}$ at $\vartheta = 25\,°\text{C}$ and $\vartheta = 100\,°\text{C}$.

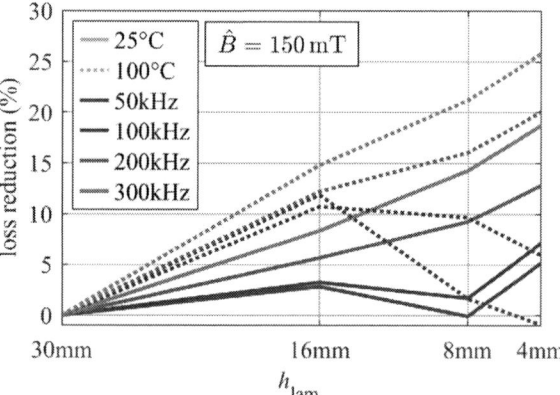

Fig. 23: Reduction in power loss density of E70 cores with different layer thickness for different frequencies at $\hat{B} = 150\,\text{mT}$ for $\vartheta = 25\,°\text{C}$ and $\vartheta = 100\,°\text{C}$.

Figure 22 shows the reduction of the power loss density for the different layer thickness for $\hat{B} = 50\,\text{mT}$ at $\vartheta = 25\,°\text{C}$ and $\vartheta = 100\,°\text{C}$. Figure 23 shows it for $\hat{B} = 150\,\text{mT}$. These confirm once again that the reduction increases with reduced layer thickness and increasing frequency and temperature. At the lower flux density in fig. 22 this is more than at the higher flux density in fig. 23.

3.3 Conclusion

Macroscopic eddy currents in ferrites are playing an increasingly important role with the increasing frequencies in power electronics. On E36, E42 and E70 cores of the material MF106, the effects of lamination on the impedance and especially on the loss factor were investigated in the small-signal range at room temperature. For all cores, a higher impedance at high frequencies could be realized with lower layer thicknesses. This means that high-frequency EMI can be better attenuated. In the case of the loss factor, the smaller E36 and E42 cores only show a significant reduction at higher frequencies. For the E70 cores, the loss factor is already noticeable reduced in the usual operating range. The E70 cores were measured in terms of their power loss density at different frequencies, flux densities and temperatures. Overall, the losses of the E70 cores can be reduced further and further with decreasing layer thickness. At high frequencies and high temperatures more than at low frequencies. The reduction in losses could be confirmed for low flux densities and at room temperature using measurements with sine waves. Laminating ferrite cores can reduce the power loss and increases the impedance at high frequencies.

References

[1] G. Skutt, "High-Frequency Dimensional Effects in Ferrite-Core Magnetic Devices," *PHD Thesis, Virginia Polytechnic Institute and State University*, 1996.

[2] G. Skutt and F. Lee, "Characterization of dimensional effects in ferrite-core magnetic devices," in *PESC Record. 27th Annual IEEE Power Electronics Specialists Conference*, vol. 2, 1996, 1435–1440 vol.2. DOI: 10.1109/PESC.1996.548770.

[3] M. Kacki, "Investigation of the high-frequency effects in Mn-Zn ferrites for EMI filter applications," *PHD Thesis, University College Cork*, 2022.

[4] M. Kacki, M. S. Rylko, J. G. Hayes, and C. R. Sullivan, "A Study of Flux Distribution and Impedance in Solid and Laminar Ferrite Cores," in *2019 IEEE Applied Power Electronics Conference and Exposition (APEC)*, Anaheim, CA, USA: IEEE, Mar. 2019, pp. 2681–2687. DOI: 10.1109/APEC.2019.8722252.

[5] M. Kacki, M. S. Rylko, J. G. Hayes, and C. R. Sullivan, "Magnetic core dimensional effects - flux propagation in ferrites," in *PSMA Workshop during IEEE Applied Power Electronics Conference*, 2018.

[6] W. Hauser, "Modellbildung für strukturabhängige Effekte in Ferritkernen," *PHD Thesis, Friedrich-Alexander-Universität Erlangen-Nürnberg*, 2018.

[7] W. Hauser and M. Albach, "Analytic model of structural effects in toroid cores with rectangular cross section," in *2016 6th International Electric Drives Production Conference (EDPC)*, Nuremberg, Germany: IEEE, Nov. 2016, pp. 60–66. DOI: 10.1109/EDPC.2016.7851315.

[8] A. Stadler, "Messtechnische Bestimmung und Simulation der Kernverluste in weichmagnetischen Materialien," *PHD Thesis, Friedrich-Alexander-Universität Erlangen-Nürnberg*, 2009.

[9] A. Stadler, M. Albach, and A. Bucher, "Calculation of Core Losses in Toroids with Rectangular Cross Section," in *2006 12th International Power Electronics and Motion Control Conference*, Portoroz: IEEE, Aug. 2006, pp. 828–833. DOI: 10.1109/EPEPEMC.2006.4778502.

[10] S. Takahashi, "Experimental investigation of the dimensional effect on small-signal characteristics of common-mode inductors," *IEEE Access*, vol. 10, pp. 123 068–123 079, 2022. DOI: 10.1109/ACCESS.2022.3223436.

[11] T. Dimier and J. Biela, "Eddy Current Loss Model for Ferrite Ring Cores Based on a Meta-Material Model of the Core Properties," *IEEE Transactions on Magnetics*, vol. 58, no. 2, pp. 1–5, Feb. 2022. DOI: 10.1109/TMAG.2021.3084812.

[12] M. Kacki, M. S. Rylko, J. G. Hayes, and C. R. Sullivan, "A Practical Method to Define High Frequency Electrical Properties of MnZn Ferrites," in *2020 IEEE Applied Power Electronics Conference and Exposition (APEC)*, New Orleans, LA, USA: IEEE, Mar. 2020, pp. 216–222. DOI: 10.1109/APEC39645.2020.9124101.

[13] A. Stadler, M. Albach, and A. Lindner, "A Practical Method to Measure Electrical AC Conductivity of MnZn Ferrites Using Conventional Toroids," *IEEE Transactions on Magnetics*, vol. 46, no. 2, pp. 678–681, Feb. 2010. DOI: 10.1109/TMAG.2009.2030157.

[14] L. Reißenweber, F. Wohlrath, and A. Stadler, "Substitution of nanocrystalline toroid by laminated ferrite toroid in the application of a common-mode choke," in *2022 24th European Conference on Power Electronics and Applications (EPE'22 ECCE Europe)*, 2022, pp. 1–9.

[15] B. Wunsch, T. Christen, S. Skibin, and V. Forsstrom, "Broadband circuit model of a ferrite core, including dimensional resonance, saturation, and hysteresis," *IEEE Transactions on Magnetics*, vol. 55, no. 7, pp. 1–5, 2019. DOI: 10.1109/TMAG.2019.2901216.

[16] M. Baumann, C. Drexler, J. Pfeiffer, J. Schueltzke, E. Lorenz, and M. Schmidhuber, "Investigation of core-loss mechanisms in large-scale ferrite cores for high-frequency applications," in *2022 24th European Conference on Power Electronics and Applications (EPE'22 ECCE Europe)*, 2022, pp. 1–10.

[17] Keysight, "Impedance measurement handbook 6th edition," 2020.

[18] E. C. Snelling, "Soft ferrites, properties and application," London: Iliffe Books Ltd., 1969.

[19] E. Stenglein, B. Kohlhepp, D. Kübrich, M. Albach, and T. Dürbaum, "Gan-half-bridge for core loss measurements under rectangular ac voltage and dc bias of the magnetic flux density," *IEEE Transactions on Instrumentation and Measurement*, vol. 69, no. 9, pp. 6312–6321, 2020. DOI: 10.1109/TIM.2020.2972140.

[20] D. Serrano, H. Li, S. Wang, T. Guillod, M. Luo, et al., "Why magnet: Quantifying the complexity of modeling power magnetic material characteristics," *IEEE Transactions on Power Electronics*, vol. 38, no. 11, pp. 14 292–14 316, 2023. DOI: 10.1109/TPEL.2023.3291084.

PCIM Europe 2024, 11– 13 June 2024, Nuremberg   DOI: 10.30420/566262066

New Planar 4.5 kV Split-gate (SG) Si-IGBT Device for Improved Switching Characteristics and High Frequency Operation

Gaurav Gupta[1], Jeremy Jones[1], Boni Boksteen[1], Babak Nikberg[1], Luca De-Michielis[1], Gontran Pâques[2]

[1] Hitachi Energy, Fabrikstrasse 3, 5600 Lenzburg, Switzerland
[2] Hitachi Energy Research, Segelhofstrasse 1 A, 5405 Baden-Daettwil, Switzerland

Corresponding author: Gaurav Gupta, gaurav.gupta1@hitachienergy.com

Abstract

This paper presents the new split-gate (SG)-IGBT design concept implemented in Hitachi Energy's 4.5 kV enhanced-planar Si-IGBT technology platform. The proposed SG-IGBT design, shows lower gate capacitances that lead to improved switching characteristics as observed from both TCAD simulations and experimental investigations. Our new design shows about a 33% reduction in the duration of Miller plateau in turn-off switching, a 38% reduction in the measured gate-charge and about 60% reduction in the gate-leakage current. The lower gate capacitances in SG-IGBT design also lead to an improved trade-off between switching speed (dI/dt) and IGBT turn-on losses (E_{on}), leading to about a 15% reduction in the combined IGBT and diode losses at the same dI/dt. Furthermore, our new SG-IGBT design is accompanied by additional cell protection features which prevents any reduction in blocking capability and other static characteristics typically observed in conventional SG devices.

1 Introduction

Optimizing the device parasitic capacitances is a key aspect in IGBT design for improved switching characteristics as required especially for high frequency applications [1], [2]. The parasitic gate capacitances namely, gate-emitter capacitance, C_{ge} and gate-collector capacitance, C_{gc} also known as Miller capacitance play crucial role in determining switching behavior of any MOS (metal-oxide-semiconductor) based device concept in general. Well-optimized gate capacitances can potentially improve trade-off relation between IGBT turn-on switching losses, E_{on}, and its switching speed, dI/dt, leading to overall lower turn-on switching losses and improved resistance to electromagnetic interference failures (EMI) [3], [4]. Turn-off switching will also benefit from device with lower gate-capacitances in terms of switching speed and turn-off losses, E_{off}. Additionally, lower gate capacitances would result in improved gate charge, Q_g, characteristics leading to lower gate drive current and higher gate power efficiency facilitating high frequency applications [5], [6]. In this direction, split-

gate (SG) design is an interesting concept and has been widely reported before for trench gate vertical double-diffused MOS (VDMOS) power MOSFETs [7], [8]. More recently, SG concept has also been investigated for trench IGBTs [9]–[12]. Other than trench gate design for Si devices, there are some reports where SG designs were also investigated for planar gate SiC MOSFETs [13], [14]. However, there has been no reports of experimental investigation of SG for planar Si IGBT device especially for high voltage (>3.3 kV) application. SG designs, especially for planar device, tends to degrade device static characteristics due to presence of high electric field at the edges of gate electrode leading to premature breakdown [14].

In this work, we report for the first time the experimental as well as TCAD investigation of new SG concept as applied to our high voltage (4.5 kV) enhanced planar Si-IGBT device [15], [16]. In contrast to previous reports, our SG design is accompanied with cell protection features to (1) not compromise the device reliability at high electric fields and (2) keep static characteristics largely unaffected while improving its dynamic performance.

PCIM Europe 2024, 11– 13 June 2024, Nuremberg   DOI: 10.30420/566262066

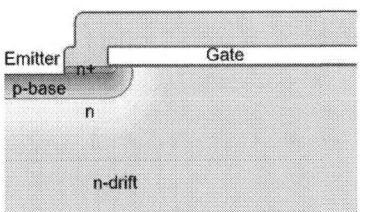 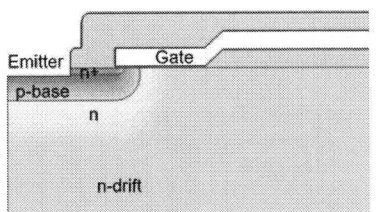 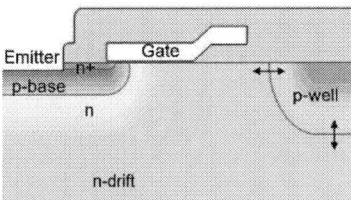

Fig. 1: Schematic cross-section of (a) a conventional planar cell, (b) planar cell with stepped (gate)-oxide used as a reference in this work and (c) new proposed split-gate (SG) planar IGBT

The outline of this paper is as follows. In section 2, we will first introduce our new device structure followed by analysis of TCAD simulated results. In section 3, we will present static and dynamic characterization of our experimentally realized SG-IGBT device and its comparison to the reference device. Finally, in section 4, we will conclude our findings.

2   Device structure and TCAD simulation results

The schematic cross-sections in Figure 1 show the evolution of different planar IGBT cell concepts as optimized for lower gate capacitances. Figure 1(a) shows the cross-section of typical planar cell where gate electrode runs continuously over thin gate-oxide in between the two active cells. Stepping the thin gate-oxide to thicker oxide (also known as terraced oxide) as shown in Fig. 1(b) is a reasonably good solution to reduce gate capacitances without affecting the device on-state characteristics. However, remaining gate electrode over the thick oxide still contributes to the parasitic capacitances leaving room for further optimization.

Figure 1(c) shows the schematic cross-section of our proposed design where the significant part of the gate electrode in between active cells was removed to realize the split gate structure. Unlike the typical SG structures reported in literature, we terminate our gate electrode on the top of a thicker oxide. This feature reduces the electric field at the edges of the SG electrodes. Additionally, our design features deep p-well placed in between MOS-cells to further shield the SG edges from a high electric field. The depth and lateral extent of such deep p-well is engineered for maximum benefit in the electric field reduction without affecting the device on-state characteristics. The incorporation of such well-optimized cell protection features pre-

vents any premature breakdown at the edges of the gate poly-silicon due to presence of high electric field.

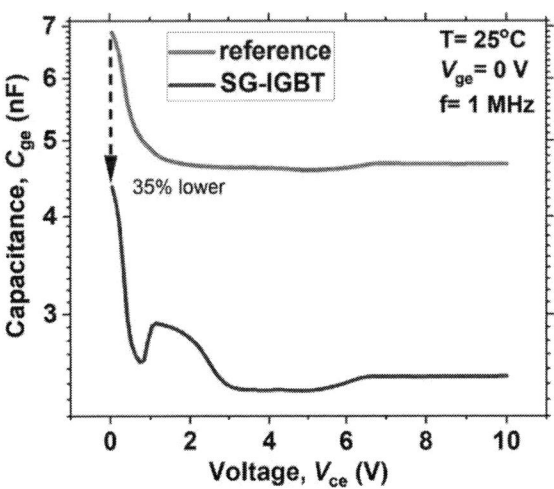

Fig. 2: TCAD simulated $C - V$ characteristics showing C_{ge} for reference design and the proposed SG-IGBT device.

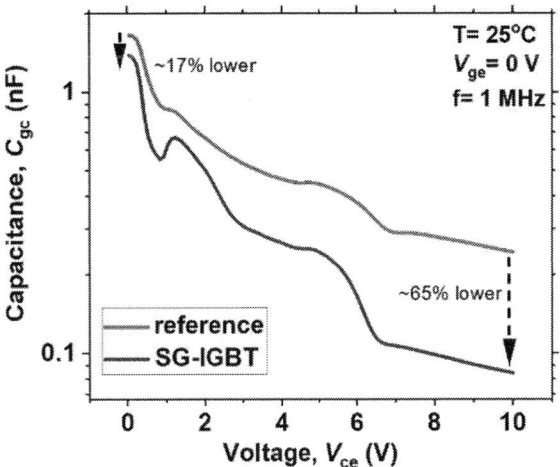

Fig. 3: TCAD simulated $C - V$ characteristics showing C_{gc} comparison for the two different IGBT designs.

535

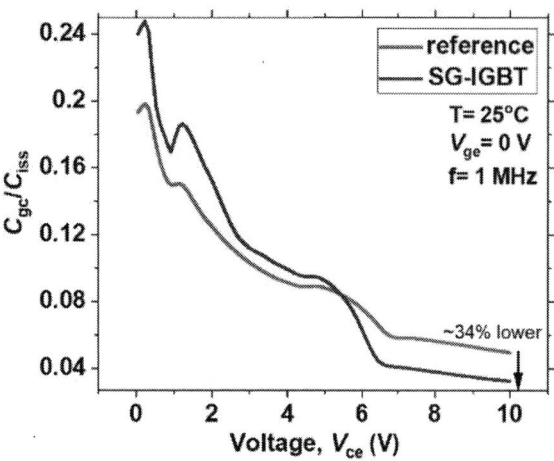

Fig. 4: TCAD simulated $C-V$ characteristics comparing the ratio C_{gc}/C_{iss} for the two different IGBT designs.

Next, we will discuss the TCAD simulation study of the proposed device (Fig. 1(c)) and its comparison to the reference design (Fig. 1(b)) as implemented for a planar 4.5 kV Si-IGBT. Figure 2, 3 show the simulated individual gate capacitances, C_{ge} and C_{gc}, of SG-IGBT device and their comparison to that of the reference device. As can be noted from Fig. 2, SG-IGBT device showed ~35% reduction in the C_{ge} at low V_{ce} values. The implementation of SG concept also led to ~17% reduction in the C_{gc} at low V_{ce} values and ~65% reduction at high V_{ce} values as shown in Fig. 3. This improvement in device parasitic gate capacitances is expected to yield benefits in its switching characteristics in terms of speed and losses.

Further, in [2], it has been shown that the ratio, C_{gc}/C_{iss}, where $C_{iss}= (C_{ge}+C_{gc})$ is the total input capacitance, is a reasonable index to study the trade-off between IGBT switching losses versus its switching speed, dI/dt. The authors reasoned that the total IGBT switching losses can be largely described as:

$$(E_{on} + E_{off}) \propto C_{iss}R_g(1 + C_{gc}/C_{iss}) \qquad (1)$$

Here, its been assumed that the static parameters such as on-state voltage drop (V_{cesat}), threshold voltage (V_{th}) and transconductance (g_m) are same for devices under comparison. Now, for a fixed dI/dt ($\propto [C_{iss}R_g]^{-1}$) as limited by EMI requirements [3], [4], a lower value of C_{gc}/C_{iss} ratio is desired in order to reduce the IGBT total switching losses. Figure 4 shows the comparison of capacitance ra-

tio, C_{gc}/C_{iss}, between our proposed SG-IGBT device and the reference design. It can be noted that the SG-IGBT shows a favorable C_{gc}/C_{iss} index (34% lower) at higher V_{ce} values. Therefore, our SG-IGBT device is expected to show an improved switching losses versus dI/dt trade-off. At lower V_{ce} values, however the SG-IGBT device showed an unfavorable C_{gc}/C_{iss} index. This is due to smaller reduction in C_{gc} value compare to drop in C_{ge} value in this region (see Fig. 2, 3). For further improved trade-off characteristics i.e. lower C_{gc}/C_{iss} index, the reduction in the C_{gc} in SG design should be complemented with some increment in C_{ge}.

Note that, in the above analysis, it has been assumed that EMI noise is mainly governed by dI/dt. In [10], the authors have also addressed the case where EMI noise is mainly determined by dV/dt.

Figure 5 shows the simulated blocking characteristics of our SG-IGBT device in comparison to the reference device. SG-IGBT device without any cell protection features tends to show a reduced blocking capability compared to the reference device. With the implementation of our proposed cell protection features, the high electric field at the edges of gate polysilicon in SG design is relieved and blocking capability is restored as also shown in Fig. 5. Further, other static characteristics such as on-states losses (see Fig. 6) and transfer characteristics (V_{th} and g_m) remain largely unaffected with the implementation of SG concept.

TCAD simulated turn-off characteristics are shown

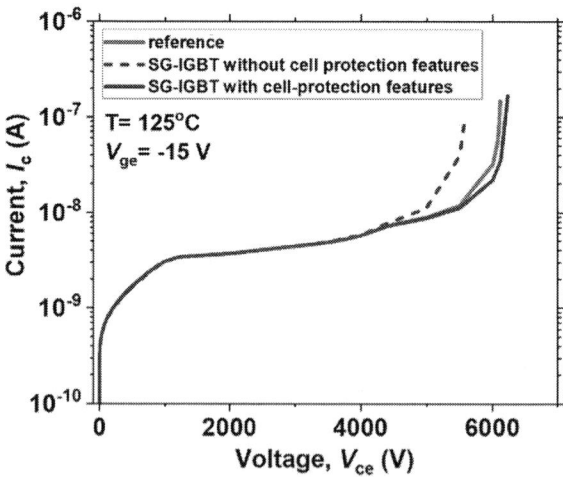

Fig. 5: TCAD simulated blocking characteristics comparing the 3 different design options.

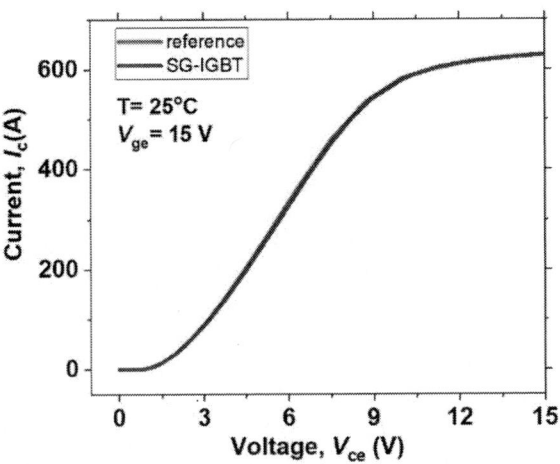

Fig. 6: TCAD simulated output characteristics of two different IGBTs designs.

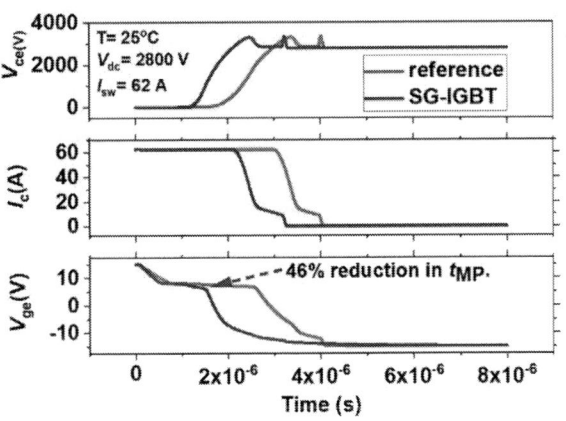

Fig. 7: TCAD simulated turn-off switching waveforms of the two different IGBT designs.

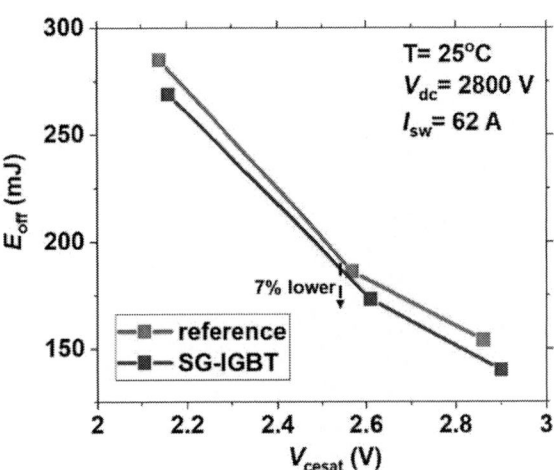

Fig. 8: TCAD simulated technology curve showing the trade-off between on-state losses, V_{cesat} and turn-off switching losses, E_{off} for SG-IGBT and its comparison to the reference device.

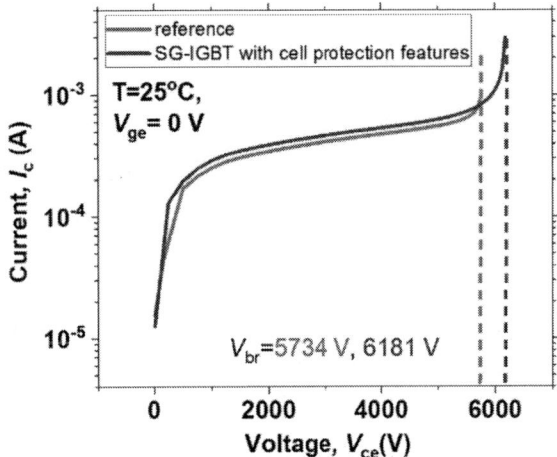

Fig. 9: Measured blocking characteristics of different IGBTs designs at T=25°C.

in Fig. 7. SG-IGBT shows faster gate switching characteristics attributed to its overall lower input capacitance C_{iss} as observed in $C - V$ characteristics before. Lower C_{gc} in our SG design resulted in about 46% reduction in Miller plateau duration, t_{MP} ($\propto C_{gc} R_g$), along with similar reduction in the simulated gate-collector charge, Q_{gc} as compared to the reference device making this design potentially interesting for higher frequency applications. SG design also benefits from a smaller accumulation layer near the surface due to reduced gate coverage which in turn also facilitates faster turn-off. Turn-off switching losses, E_{off}, which in general are weak function of switching speed [12], [17], showed about 7% reduction with some marginal improvement in the technology trade-off as shown in Fig. 8 for SG-IGBT design.

3 Experimental results

SG-IGBT design as proposed in the previous section was also experimentally realized along with the reference device. In this section, we will discuss the electrical measurement results of SG-IGBT device and its comparison to the reference design. We will also try to correlate the results of our experimental investigation to the trends observed in the simulation study where the strong agreement has been observed.

Figure 9 shows the measured blocking characteristics of realized SG-IGBT design and its compari-

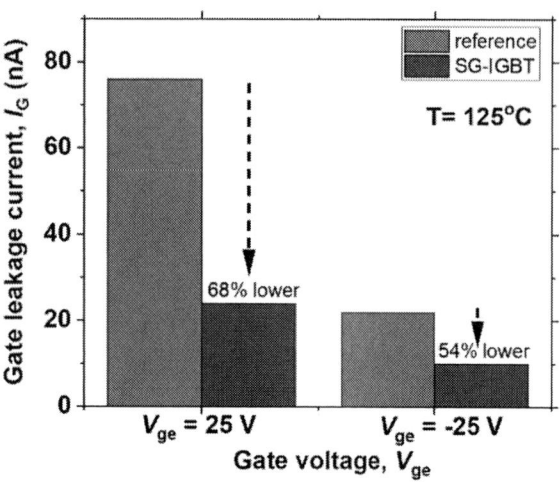

Fig. 10: Measured gate-leakages of different IGBT designs at T=125°C.

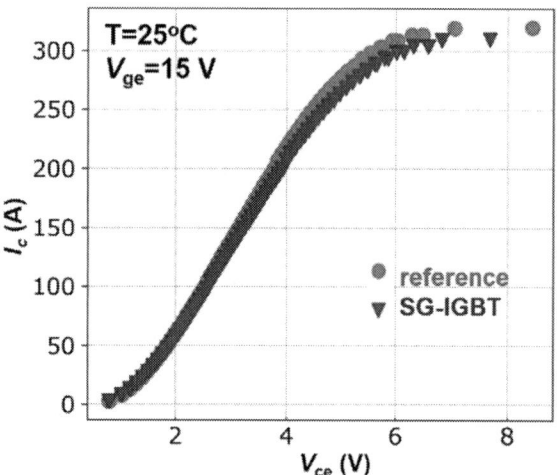

Fig. 11: Measured output characteristics of SG-IGBT device and its comparison to the reference design at T=25°C.

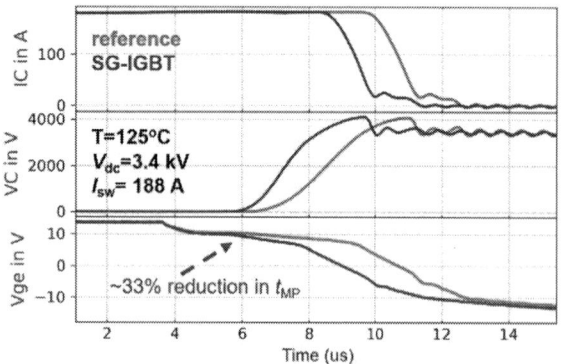

Fig. 12: Measured turn-off switching characteristics of different IGBT designs at SOA conditions.

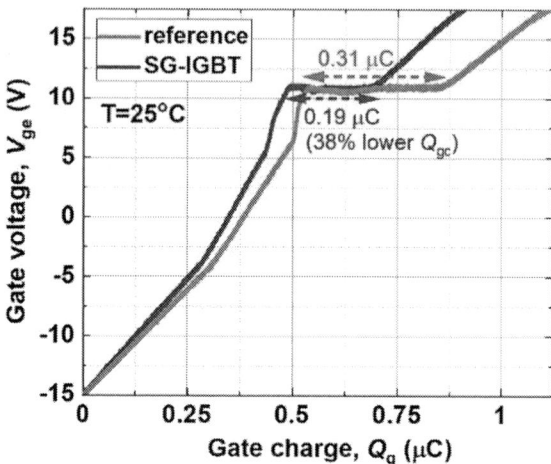

Fig. 13: Measured gate-charge characteristics of different IGBT designs at T=25°C.

son to the reference. SG-IGBT with cell protection features shows improved blocking with respect to the reference possibly due to suppression of high electric field near SG edges as also expected from TCAD simulations. The measured SG-IGBT design also shows ~60% reduction in the gate-leakage (see Fig. 10) as compared to the reference due to overall smaller gate-polysilicon coverage in the active area of the MOS-cell. Further, the implementation of SG design has no adverse effect on device on-state characteristics (see Fig. 11) as desired and is in agreement with our simulation results. Other static characteristics such as high temperature collector-emitter leakage current and threshold voltage also remained unchanged for the two devices.

Figure 12 shows the measured turn-off switching characteristics of our SG-IGBT design and its comparison to the reference device at SOA conditions (V_{ce}=3.4 kV, I_c=2xI_{nom}). The SG-IGBT device shows similar reduction in Miller plateau duration, t_{MP}, as expected from TCAD simulations due to smaller C_{gc}. Marginal improvements in the E_{off} was also noted as also observed in TCAD simulations before for SG-IGBT device.

The reduction in C_{gc} for a SG design is also reflected in its measured gate-charge characteristics as shown in Fig. 13 where it shows about 38% lower gate-collector charge, Q_{gc}, in comparison to the reference device. This suggests that a smaller gate-current is required to switch the gate voltage from *on* to *off* state or vice-versa in a given

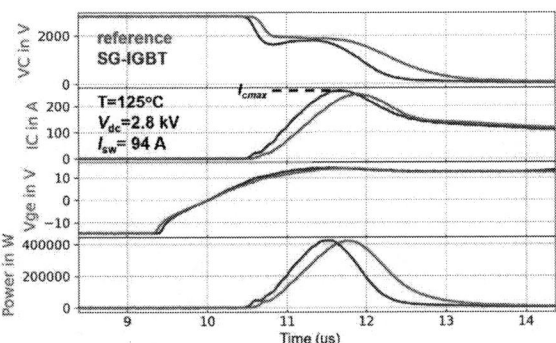

Fig. 14: Measured turn-on switching characteristics of SG-IGBT device and its comparison to that of the reference device at nominal conditions ($I_c=I_{nom}$).

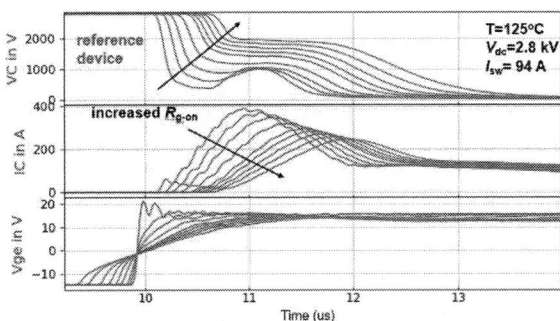

Fig. 15: Measured turn-on switching waveforms of reference device for different values of R_{g-on} plotted at nominal conditions ($I_c=I_{nom}$).

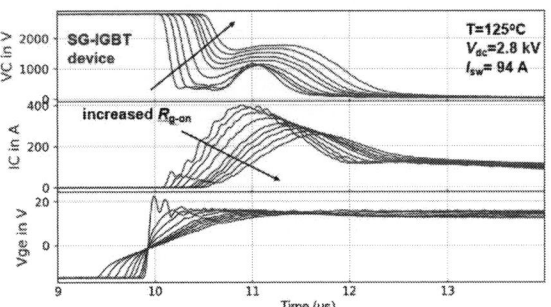

Fig. 16: Measured turn-on switching waveforms of SG-IGBT for different values of R_{g-on} plotted at nominal conditions ($I_c=I_{nom}$).

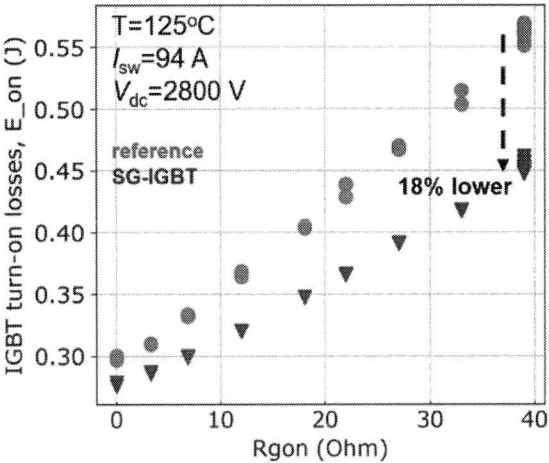

Fig. 17: Plot showing IGBT turn-on losses, E_{on} for varying R_{gon} as measured for different IGBT designs at T=125°C.

time which could simplify the gate unit design [6]. Further, lower Q_{gc} would also facilitates higher frequency operation limiting gate power dissipation ($P_{gate}=Q_g V_{ge} f$) where f is the operating switching frequency.

Figure 14 shows the measured turn-on switching characteristics of SG-IGBT device and its comparison to that of the reference device at nominal switching conditions. As expected from its overall lower C_{iss}, SG-IGBT shows faster switching with higher dI/dt resulting in lower ($\sim$ 18%) turn-on switching losses, E_{on}. SG-IGBT with lower C_{gc} also showed a steeper V_{ce} tail at the end of the turn-on cycle [9].

We also characterized these devices at different dI/dt conditions (by varying gate resistance, R_{g-on}) as shown in Fig. 15 and Fig. 16 where good turn-on controllability can be observed for both devices. SG-IGBT consistently showed lower E_{on} for varying R_{g-on} as shown in Fig. 17. It can be noted that the improvement resulting from reduced gate capacitances are more significant at higher values of R_{g-on} [10] where possibly the RC product is more influenced by a capacitance value than at the lower values of R_{g-on}. Further, the observed hump in V_{ce} at low R_{g-on} values (see Fig 15 and 16) for both devices where they possibly go into desaturation ($I_{cmax} > I_{sat}$) also adds to additional turn-on losses.

In addition, the SG design also showed improved trade-off between measured peak collector current, I_{cmax}, ($\propto$ dI/dt [4]) versus E_{on} as shown in Fig. 18, 19 for two different switching currents, I_{sw}. This essentially implies that for the same switching speed

(limited by EMI considerations [3]), the SG-IGBT showed improved E_{on} which is also expected from its lower C_{gc}/C_{iss} index (refer to Fig. 4) as observed via TCAD simulations in the previous section. It

PCIM Europe 2024, 11– 13 June 2024, Nuremberg DOI: 10.30420/566262066

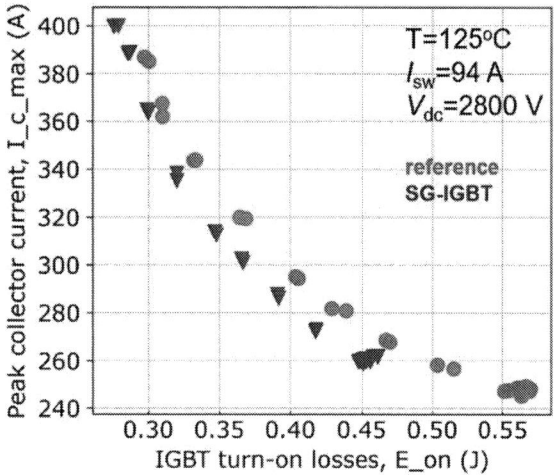

Fig. 18: Plot showing turn-on I_{cmax} vs E_{on} trade-off as measured for different IGBT designs at T=125°C at I_{sw}= 94 A.

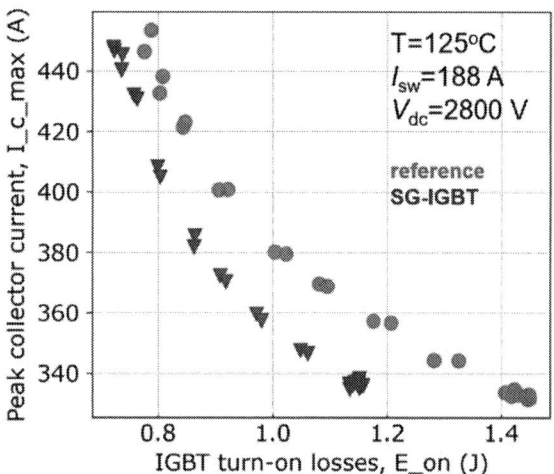

Fig. 19: Plot showing turn-on I_{cmax} vs E_{on} trade-off as measured for different IGBT designs at T=125°C at I_{sw}= 188 A (2xI_{nom}).

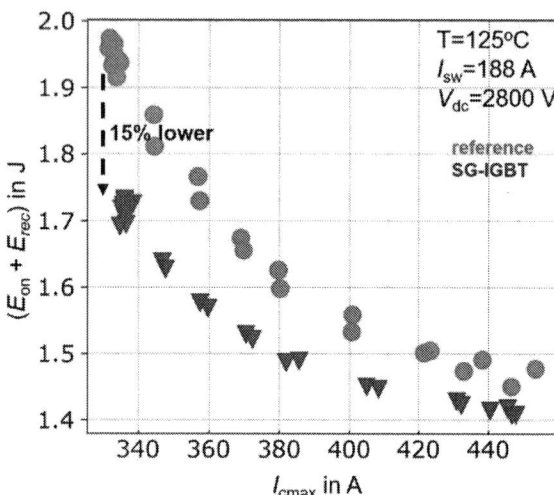

Fig. 20: Plot showing the variation of total turn-on losses, E_{tot}, for varying $I_{cmax}(\propto dI/dt)$ as measured for different IGBT designs at T=125°C at I_{sw}= 188 A (2xI_{nom}).

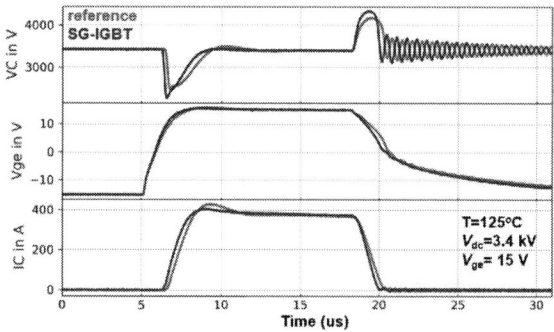

Fig. 21: Measured short circuit characteristics of our SG-IGBT device and its comparison to that of the reference device at T=125°C.

can also be noted that the I_{cmax} versus E_{on} trade-off showed greater improvement at higher (2xI_{nom}) switching current, I_{sw}, which again could possibly be explained based on the C_{gc}/C_{iss} index as shown in Fig. 4, where it becomes more favorable at higher V_{ce} values.

Furthermore, the improved trade-off also results in overall reduction in total turn-on losses, E_{tot} (E_{on} + E_{rec}) where E_{rec} is the diode reverse recovery loss as shown in Fig. 20. This is mainly originating from the improvement in E_{on} while E_{rec} remains largely unchanged for the same dI/dt.

It is also worthwhile to mention that no degradation in the SOA capability was observed for a SG device as compared to the reference design. Both devices showed high reverse-bias (RB)-SOA capability ($\sim$ 5x I_{nom}) when measured at T=125°C and V_{ce}=3.4 kV. SG-IGBT device also showed similar short-circuit (SC) characteristics as that of the reference device (see Fig. 21) with no adverse effect in its SC capability ($> 12\ \mu$s). The small differences in the V_{ce} overshoot between two devices can be attributed to faster switching (higher dI/dt) characteristics of SG-IGBT due to its lower gate capacitances.

540

4 Conclusion

In this work, we have presented Hitachi Energy's new SG-IGBT design concept as implemented in its latest 4.5 kV enhanced planar cell technology platform and its comparison to our previous generation device. The proposed design was extensively investigated using TCAD simulations and electrical measurements where strong agreement was observed. The new design with lower parasitic gate capacitances showed promising results in terms of improved turn-off switching and gate charge characteristics making it suitable for higher frequency applications. SG-IGBT also showed improved trade-off between IGBT turn-on switching losses and its switching speed (dI/dt) while keeping its SOA capability largely unchanged with respect to the reference design. Moreover, incorporation of additional cell protection features in our design prevents any deterioration of other static parameters unlike observed in conventional SG devices. Such design concept can be implemented on device level across varied voltage classes benefiting Hitachi Energy's different product ranges.

5 Acknowledgment

The authors acknowledge the support of Hitachi Energy wafer fabrication facility for manufacturing these devices. The authors would also like to thank BIMOS R&D chip group and Product evaluation and qualification (PEQ) group members at Hitachi Energy for all the discussion and help in this work.

References

[1] J. Lutz, H. Schlangenotto, U. Scheuermann, and R. De Doncker, *Semiconductor Power Devices:Physics, characteristics, reliability.* Springer, 2018.

[2] S. Machida, T. Sugiyama, M. Ishiko, S. Yasuda, J. Saito, and K. Hamada, "Investigation of correlation between device structures and switching losses of IGBTs," in *2009 21st International Symposium on Power Semiconductor Devices & IC's*, IEEE, 2009, pp. 136–139.

[3] S Momota, M Otsuki, K Ishii, H Takubo, and Y Seki, "Analysis on the low current turn-on behavior of IGBT module," in *12th International Symposium on Power Semiconductor Devices & ICs. Proceedings (Cat. No. 00CH37094)*, IEEE, 2000, pp. 359–362.

[4] K. Nishi, T. Takahashi, and A. Narazaki, "Analysis the complex tradeoff among Eon-Vcesat-SCSOA and EMI noise through the single chip evaluation method," in *2019 31st International Symposium on Power Semiconductor Devices and ICs (ISPSD)*, IEEE, 2019, pp. 475–478.

[5] B. J. Baliga, *Fundamentals of power semiconductor devices.* Springer Science & Business Media, 2010.

[6] *Use Gate Charge to Design the Gate Drive Circuit for Power MOSFETs and IGBTs*, International Rectifier, Application note AN-944.

[7] B. J. Baliga, *Vertical MOSFETs having trench-based gate electrodes within deeper trench-based source electrodes*, US Patent 6,621,121, 2003.

[8] R. K. Williams, M. N. Darwish, R. A. Blanchard, R. Siemieniec, P. Rutter, and Y. Kawaguchi, "The trench power MOSFET: Part I—History, technology, and prospects," *IEEE Transactions on Electron Devices*, vol. 64, no. 3, pp. 674–691, 2017.

[9] K Ohi, Y Ikura, A Yoshimoto, K Sugimura, Y Onozawa, *et al.*, "Ultra low miller capacitance trench-gate IGBT with the split gate structure," in *2015 IEEE 27th International Symposium on Power Semiconductor Devices & IC's (ISPSD)*, IEEE, 2015, pp. 25–28.

[10] K. Nishi and A. Narazaki, "CSTBT™ based split-gate RC-IGBT with low loss and EMI noise," in *2020 32nd International Symposium on Power Semiconductor Devices and ICs (ISPSD)*, IEEE, 2020, pp. 138–141.

[11] Y. He, H. Luo, R. Qin, X. Luo, Y. Yao, *et al.*, "A split-gate trench IGBT with low Miller capacitance and d V/dt noise," *Journal of Computational Electronics*, vol. 20, no. 1, pp. 568–574, 2021.

[12] K. Konishi, K. Nishi, K. Sako, and A. Furukawa, "Split-Dummy-Active CSTBT™ for Improving Recovery dV/dt and Turn-on Switching Loss Tradeoff," in *2022 IEEE 34th International Symposium on Power Semiconductor Devices and ICs (ISPSD)*, IEEE, 2022, pp. 273–276.

[13] A. Agarwal, K. Han, and B. J. Baliga, "2.3 kV 4H-SiC accumulation-channel split-gate planar power MOSFETs with reduced gate charge," *IEEE Journal of the Electron Devices Society*, vol. 8, pp. 499–504, 2020.

[14] J. Yoon and K. Kim, "A 3.3 kV 4H-SiC split gate MOSFET with a central implant region for superior trade-off between static and switching performance," *Journal of Semiconductors*, vol. 42, no. 6, p. 062 803, 2021.

[15] M. Andenna, B. Boksteen, D. Prindle, L. De-Michielis, V. Botan, *et al.*, "Rugged 4500V HiPak Module with 1500A Current Rating and 150° C Capability for Traction Application," in *PCIM Europe digital days 2020; International Exhibition and Conference for Power Electronics, Intelligent Motion, Renewable Energy and Energy Management*, VDE, 2020, pp. 1–8.

[16] M Rahimo, A Kopta, and S Linder, "Novel enhanced-planar IGBT technology rated up to 6.5 kV for lower losses and higher SOA capability," in *2006 IEEE International Symposium on Power Semiconductor Devices and IC's*, IEEE, 2006, pp. 1–4.

[17] S. Machida, K. Ito, and Y. Yamashita, "Approaching the limit of switching loss reduction in Si-IGBTs," in *2014 IEEE 26th International Symposium on Power Semiconductor Devices & IC's (ISPSD)*, IEEE, 2014, pp. 107–110.

PCIM Europe 2024, 11– 13 June 2024, Nuremberg   DOI: 10.30420/566262067

4.5 kV Double-Gate Reverse-Conducting Press-Pack IEGT

Satoshi Yoshida[1], Tatsunori Sakano[1], Ryohei Gejo[2], Takahiro Kato[2], Atsushi Yamaoka[1], Tomoaki Inokuchi[1], and Kazuto takao[1]

[1] Toshiba corporation, Japan

[2] Toshiba Electronic Devices & Storage Corporation, Japan

Corresponding author: Satoshi Yoshida, satoshi25.yoshida@toshiba.co.jp
Speaker: Satoshi Yoshida, satoshi25.yoshida@toshiba.co.jp

Abstract

A 4.5 kV double-gate (DG) reverse-conducting (RC) press-pack (PP) injection-enhanced gate transistor (IEGT) was developed. The DG-RC-PPI (PP-IEGT is abbreviated as PPI) was designed to control two kinds of gates in a DG-RC-IEGT device without chip-to-chip imbalance or interference between the gates. The fabricated DG-RC-PPI was examined under a switching condition of V_{CE}=2800 V, I_C=2100 A. By applying gate control, turn-off switching loss and reverse-recovery loss were reduced by 30% and 18%, respectively. As a result, we succeeded in reducing total switching loss by 16% compared with a conventional single-gate PPI.

1 Introduction

A press-pack (PP) injection-enhanced gate transistor (IEGT) (hereafter, PP-IEGT is abbreviated as PPI) is a key component in high-power applications such as high voltage direct current (HVDC) transmission systems and industrial motor drives, owing to its high-power capability, low thermal resistance, rupture resistance and high reliability [1,2]. In such applications, further improvement of device performance (i.e., reduction of device losses) is required to achieve higher power density and reduce energy consumption.

Much effort has been devoted to developing device structures and process technologies aimed at improving insulated-gate bipolar transistors (IGBTs). In contrast to those conventional approaches, gate control techniques have the potential to realize breakthroughs in device performance [3-10]. In our previous works [3-7], we proposed a double-gate (DG) reverse-conducting (RC) IEGT capable of reducing both turn-off loss and reverse-recovery loss by applying gate control techniques.

However, those studies were limited to demonstrations on a single chip, and no demonstration on PPI has been reported. A PPI requires a careful gate path design to drive parallelized IEGT chips uniformly, and thus simple substitution of conventional single-gate chips with double-gate chips is not possible.

In this work, we first discuss gate path design considerations for a DG-RC-PPI, and then present a newly developed 4.5 kV DG-RC-PPI and demonstrate switching loss reduction by using gate control techniques.

2 Device structure and gate control technique of DG-RC-IEGT

Figure 1 shows a schematic structure of a DG-RC-IEGT implemented in a DG-RC-PPI, which is the same structure in our previous report [3]. The DG-RC-IEGT has two kinds of gates: the main gate (MG) and the control gate (CG). The MG is the same structure as a conventional single-gate IEGT device, whereas the CG has no N$^+$ emitter layer. Therefore, the CG can control carrier density in N$^-$ drift region by controlling the hole current without increasing channel density.

The gate control sequence of the DG-RC-IEGT is shown in Fig.2 in the example of the double-pulse test circuit. The low-side and high-side device operates in IEGT mode and diode (reverse-conducting) mode, respectively. In the turn-off switching of the IEGT-mode device, the CG is turned off earlier than the MG by DT$_{off}$, which is called double-gate control. By applying a negative voltage (-15 V) to the CG, the carrier density stored in the N$^-$ drift region is extracted by the hole current through the p-type inversion layer around the bottom of the CG trenches, thereby reducing turn-off loss. In the reverse-recovery of the diode-mode device, by applying a positive gate voltage (+15 V) during DT_{rr} before the reverse-recovery, the carrier density in the N⁻ drift region is extracted by the electron current through the n-type inversion layer in the P-base region, thereby reducing reverse-recovery loss. This type of gate control is called desaturation control.

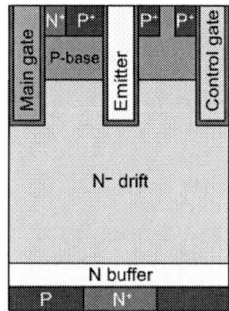

Fig. 1   Schematic structure of the DG-RC-IEGT.

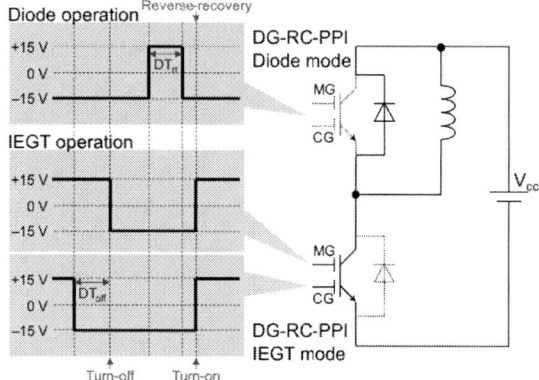

Fig. 2   Gate control sequence of the DG-RC-IEGT.

3   Gate path design for DG-RC-PPI

Figure 3 shows a schematic structure of the DG-RC-PPI. Gate signals are input to gate terminals and subsequently distributed to 42 parallelized chips via a gate printed circuit board (PCB). In parallel operation of double-gate devices, as shown in Fig. 4, the self-inductance of the gate circuit (L_{MGi} and L_{CGi}) and the mutual inductance between the MG and CG ($M_{MGi\text{-}CGi}$) are key parameters. This is because self-inductance leads to unbalanced switching delays, resulting in device breakdown due to current concentration in a specific chip. In addition, MG–CG mutual inductance causes interference between the MGs and the CGs.

As shown in Fig. 5, the developed gate PCB consists of three layers: the MG, the CG, and the emitter-sense (ES) layer. By inserting the ES layer between the MG layer and the CG layer, the MG current and the ES current become anti-parallel, thereby reducing the self-inductance of the MGs (L_{MGi}). This is also true for the CG current–ES current pair, that is, self-inductance of the CGs (L_{CGi}). Furthermore, the electro-magnetic coupling between the MG current and the CG current is shielded by the ES layer, which reduces MG–CG mutual inductance.

Figure 6 shows the simulated self-inductance of each chip with and without the ES layer. By inserting the ES layer, both the self-inductance values and their variance are reduced, which is at the same level as those of a single-gate PPI [11].

Figure 7 shows the simulated MG–CG mutual inductance. The ES layer suppresses the mutual inductance values from several tens of nH to only a few nH and reduces the variance by about 70%.

A fabricated gate PCB for the DG-RC-PPI was examined using test devices with capacitors equivalent to gate capacitances, as shown in Fig. 8. The MG and CG voltages at each chip were measured when the gate signal was input from the gate drivers to the gate terminals. Figure 9(a) shows the measured MG and CG voltage waveforms. In this case, the CG is turned off 20 μsec earlier than the MG. The waveforms of chip numbers 1–42 completely overlap. To discuss the imbalance of switching delays, we define the switching delay time τ_{delay} as the time taken for the gate voltage to decrease from +14.5 V (the beginning of voltage drop) to +6 V (around the gate threshold). Figure 9(b) and (c) show the normalized switching delay time τ_{delay} of the MG and CG, respectively. The difference in the switching delay time τ_{delay} between chips is less than the capacitor tolerance (±5%), indicating the absence of chip-to-chip imbalance. Furthermore, there is no destructive interference between the MG and CG voltage waveforms because the MG–CG mutual inductance is successfully reduced.

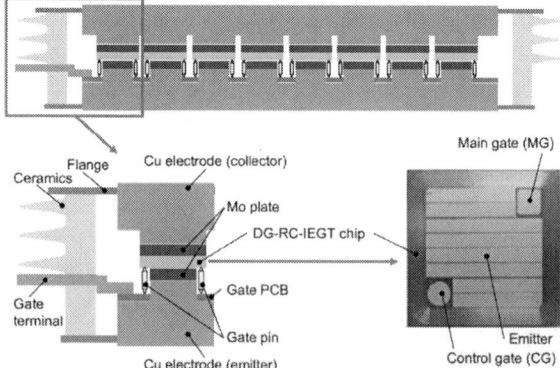

Fig. 3   Schematic structure of the DG-RC-PPI.

PCIM Europe 2024, 11– 13 June 2024, Nuremberg DOI: 10.30420/566262067

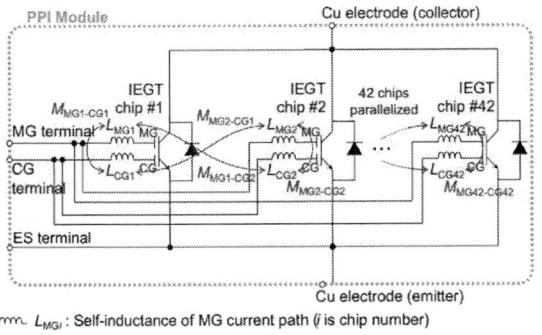

Fig. 4 Parasitic gate inductance of the DG-RC-PPI.

- ⏦ L_{MGi} : Self-inductance of MG current path (i is chip number)
- ⏦ L_{CGi} : Self-inductance of CG current path (i is chip number)
- ◀┄▶ $M_{MGi\text{-}CGj}$: Mutual inductance between MG and CG (i and j are chip numbers)

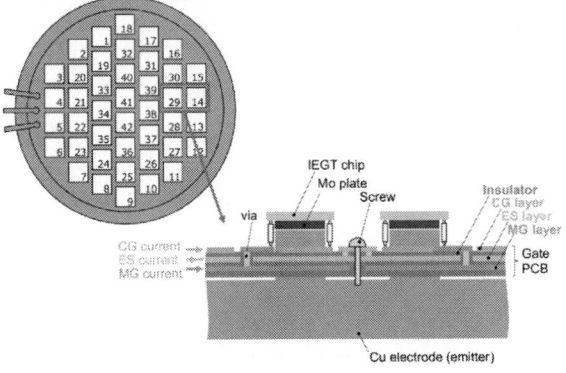

Fig. 5 Schematic structure of the gate PCB.

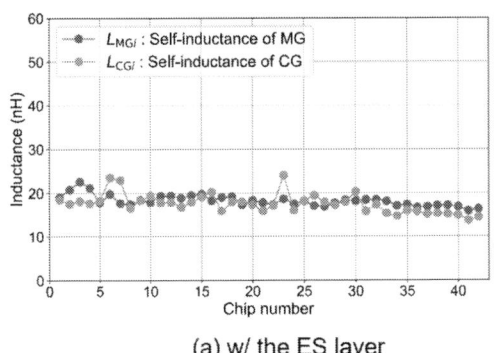

(a) w/ the ES layer

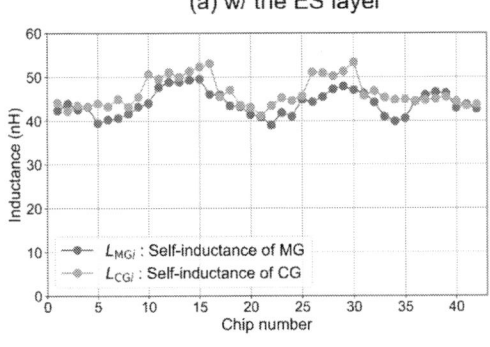

(b) w/o the ES layer

Fig. 6 Simulation results of self-inductance. The chip numbers are shown in Fig. 5.

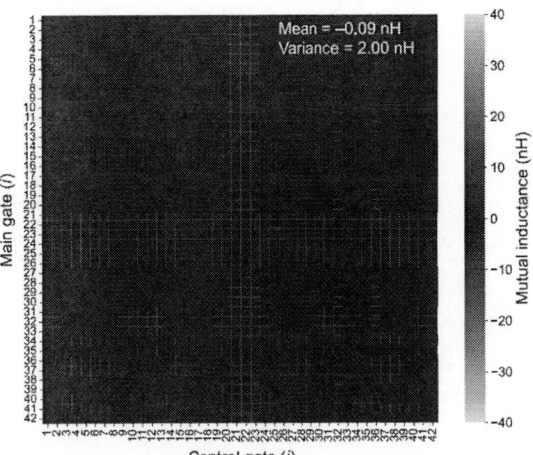

(a) w/ the ES layer

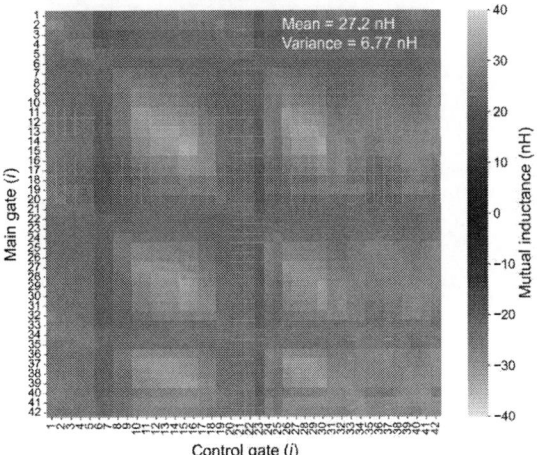

(b) w/o the ES layer

Fig. 7 Simulation results of MG–CG mutual inductance. The mutual inductance between the MG of chip i and the CG of the chip j is mapped.

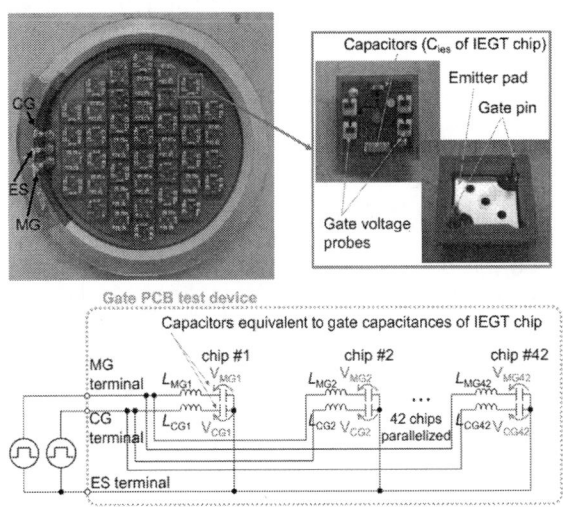

Fig. 8 Gate PCB test device and equivalent circuit.

545

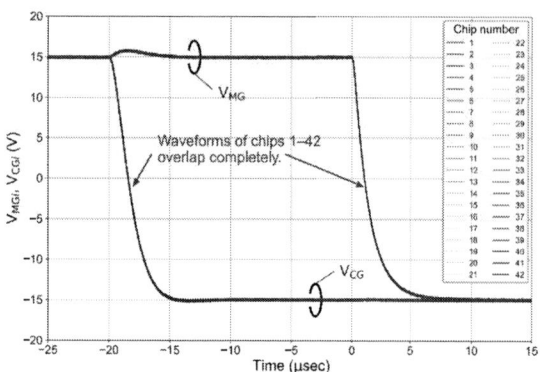

(a) Voltage waveforms of the gate PCB test device

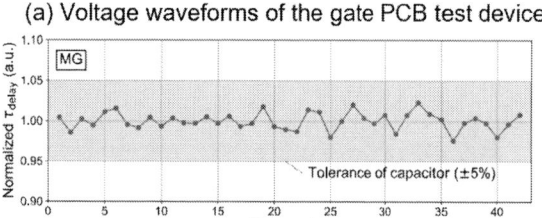

(b) Normalized switching delay time τ_{delay} of MGs

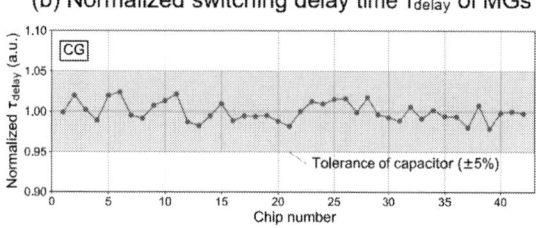

(c) Normalized switching delay time τ_{delay} of CGs

Fig. 9 Verification results of the fabricated gate PCB.

4 Demonstration of the switching loss reduction on DG-RC-PPI

Figure 10 shows the fabricated DG-RC-PPI. Switching characteristics were measured using the setup shown in Fig. 2 at V_{CE}=2800 V and I_C=2100 A.

Figure 11 shows the turn-off waveforms. By applying double-gate control, the transition of V_{CE} becomes faster and the tail current is reduced, thereby reducing turn-off loss by 30%. Figure 12 shows the E_{off}–dV_{CE}/dt trade-off measured by varying the gate resistance value. As the figure shows, the trade-off is improved over the measured gate resistance range by applying double-gate control.

Figure 13 shows the reverse-recovery waveforms. By applying desaturation control, the reverse-recovery current decreases, and the reverse-recovery loss is reduced by 18%. In addition, because of the reduction of the reverse-recovery current, the turn-on loss of the IEGT-mode device in the opposite arm is also reduced by 10%, as shown in

Fig. 14. Figure 15 shows the E_{rr}+E_{on}–dI_C/dt trade-off measured by varying the turn-on gate resistance value. It was shown that the trade-off on reverse-recovery can be also improved by applying desaturation control.

As shown in Fig. 16, applying gate control reduced total switching loss by 16%. Thus, the switching loss reduction of DG-RC-PPI was successfully demonstrated.

Finally, we discuss the timing parameter (DT_{off} and DT_{rr}). Figure 17 shows the DT_{off} dependence of the E_{off} at I_C=1000 A and 2100 A. E_{off} was saturated with DT_{off} longer than 80 µsec for both I_C = 1000 A and 2100 A, indicating that the optimal DT_{off} timing is almost independent of the load current. Similarly, Fig. 18 shows the DT_{rr} dependence of the E_{on}+E_{rr} at I_C=1000 A and 2100 A. The trend that the E_{on}+E_{rr} saturates for DT_{rr} longer than 80 µsec is also seen in both I_C=1000 A and 2100 A. Accordingly, the DT_{off} and DT_{rr} can be set to optimal fixed values even if the load current changes over time, and thus no complicated load current feedback mechanisms are necessary, which makes the DG-RC-PPI easy to use.

Fig. 10 Photo of the fabricated DG-RC-PPI.

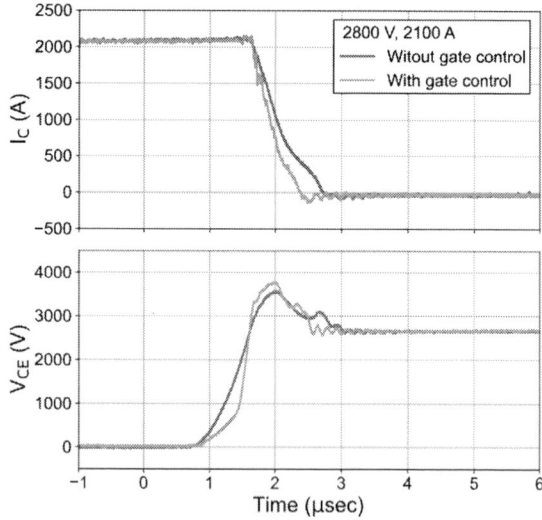

Fig. 11 Turn-off switching waveforms.

PCIM Europe 2024, 11– 13 June 2024, Nuremberg   DOI: 10.30420/566262067

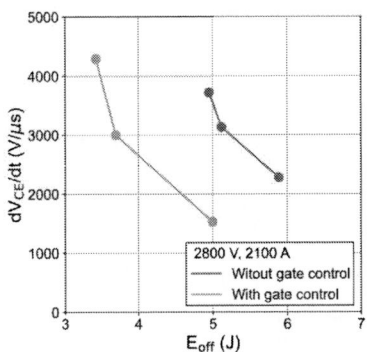

Fig. 12 Trade-off between the E_{off} and dV_{CE}/dt on the turn-off switching.

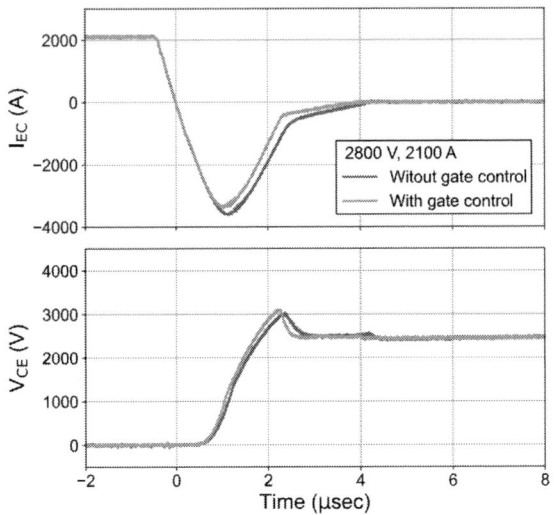

Fig. 13 Reverse-recovery waveforms of the diode-mode DG-RC-PPI.

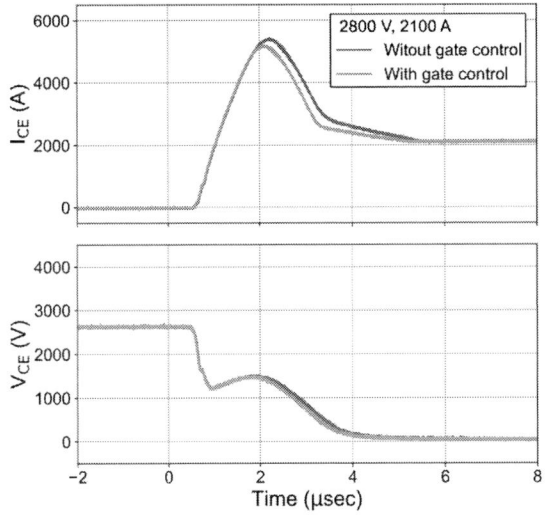

Fig. 14 Turn-on switching waveforms of the IEGT-mode DG-RC-PPI.

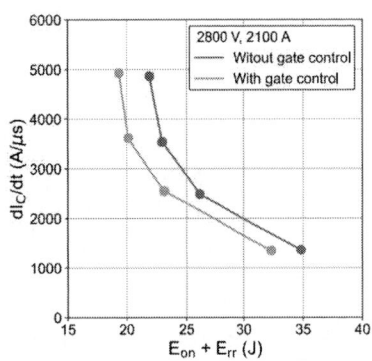

Fig. 15 Trade-off between the $E_{on} + E_{rr}$ and dI_C/dt.

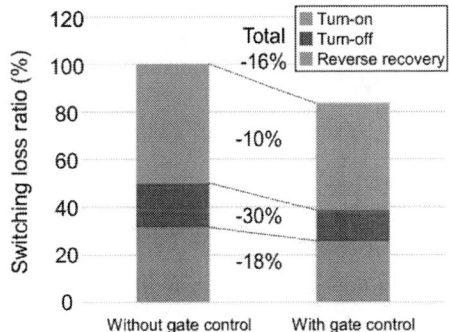

Fig. 16 Comparison of measured switching loss ratios with and without the gate control.

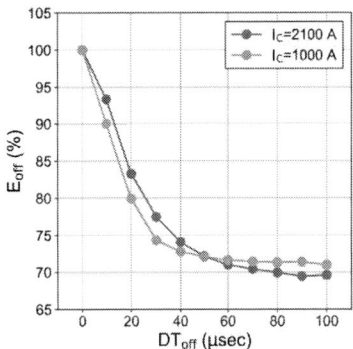

Fig. 17 Measured DT_{off} dependence on E_{off}.

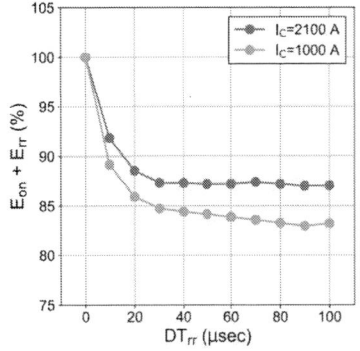

Fig. 18 Measured DT_{rr} dependence on $E_{on} + E_{rr}$.

547

5 Conclusion

In this paper, a newly developed 4.5 kV DG-RC-PPI was presented. In the development of the DG-RC-PPI, the gate path was designed to suppress self- and mutual inductance in order to avoid the imbalance of switching delays and the interference between gates, respectively. Inserting an ES layer in the middle layer of the gate PCB was shown to be effective in the inductance simulation, and uniform gate drive was verified for 42 chips on a fabricated gate PCB. The switching characteristics of the DG-RC-PPI were investigated under the switching condition of V_{CE}=2800 V, I_C=2100 A. By applying double gate control, turn-off switching loss was reduced by 30%. In addition, by applying desaturation control, reverse-recovery loss and turn-on loss were reduced by 18% and 10%, respectively. As a result, total switching loss was reduced by 16% compared with a conventional single-gate PPI. Moreover, it was shown that the gate control parameters are easy to tune. Thus, it was successfully demonstrated that the DG-RC-PPI has great potential for improving the performance of power electronics applications.

Acknowledgement

The authors would like to thank Mr. S. Hayase, Mr. K. Yosida, Mr. N. Tsukamoto, Mr. N. Aikou, Mr. T. Ide and Mr. H. Kitazawa for supporting this study.

References

[1] I. Omura, T. Domon, E. Miyake, Y. Sakiyama, T. Ogura, M. Hiyoshi, N. Yamano, and H. Ohashi, "Electrical and Mechanical Package Design for 4.5kV Ultra High power IEGT with 6kA Turn-off Capability", Proc. ISPSD'03, pp. 114-117 (2003).

[2] R. Kotani, T. Nitta, N. Tsukamoto, H. Kitazawa, M. Kitagawa, T. Kawano, and G. Tchouangue, "4.5kV rupture resistant Press Pack IEGT", Proc. PCIM'18, pp. 619-622 (2018).

[3] R. Gejo, T. Sakano, A. Kawakami, T. Kato, S. Hayase, T. Inokuchi, and K. Takao, "4.5 kV Double-gate RC-IEGT with Hole Control Gate", Proc. ISPSD'22, pp. 277-280 (2022).

[4] T. Sakano, K, Takao, Y. Iwakaji, and T. Matsudai, "Ultra-Low Switching Loss Triple-Gate controlled IGBT" Proc. ISPSD'21, pp. 363-366 (2021).

[5] T. Sakano, K. Adachi, T. inokuchi, K. Takao, Y. Iwakaji, R. gejo, T. Matsudai, "Three-level Gate Drive Technique for Enhancing Switching Loss Reduction in Triple-Gate IGBTs", Proc. ISPSD'22, pp. 117-120 (2022).

[6] Y. Iwakaji, T. matsudai, T. Sakano, and K. Takao, "Analysis of dependence of dVCE/dt on turn-off characteristics with a 1200 V double-gate insulated gate bipolar transistor", Jpn. J. Appl. Phys., vol. 60, pp. SBBD02 (2021).

[7] Y. Kobayashi, M. Fukui, T. Matsudai, T. Saraya, K. Itou, T. Takakura, S. Suzuki, R. Gejo, T. Sakano, T. kato, T. inokuchi, K. takao, T. Hiramoto, "Single-Back and Double-Front Gate-Controlled IGBT for Achieving Low Turn-Off Loss", Proc. ISPSD'23, pp. 207-210 (2023).

[8] M. Sumitomo, H. Sakane, K. Arakawa, Y. Higuchi, and M. Matsui, "Injection Control Technique for High Speed Switching with a double gate PNM-IGBT", Proc. ISPSD'13, pp. 33-36 (2013).

[9] M. Mori, T. Miyoshi, T. Furukawa, Y. takeuchi, Y. Hotta, and M. Shiraishi, "An Innovative Silicon Power Device (i-Si) through Time and Space Control of a Stored Carrier (TASC)", Proc. ISPSD'18, pp. 520-523 (2018).

[10] D. Domes, "Control Method for a Reverse Conducting IGBT", Proc. PCIM'15, pp. 147-154 (2015).

[11] H. Y. Long, M. R. Sweet, E. M. S. Narayanan, and G. Li, "Reliability study and modelling of IGBT press-pack power modules", Proc. APEC'17, pp.2711-2717 (2017).

PCIM Europe 2024, 11– 13 June 2024, Nuremberg   DOI: 10.30420/566262068

Evaluation of a 3 kV Polarization Superjunction GaN HEMT

Alireza Sheikhan[1] ⓘ, E. M. Sankara Narayanan[1], Hiroji Kawai[2], Shuichi Yagi[2], Hironobu Narui[2]

[1] The University of Sheffield, UK

[2] Powdec K.K., Japan

Corresponding author: Alireza Sheikhan, asheikhan1@sheffield.ac.uk

Speaker: Alireza Sheikhan, asheikhan1@sheffield.ac.uk

Abstract

Gallium Nitride (GaN) offers unique material properties making it more suitable for high frequency, high voltage, power dense applications. Our GaN devices benefit from high density polarization charges of 2DEG and 2DHG, which coexist in respective heterojunctions of a double heterostructure to form a charge balanced, polarization based super junction (PSJ). This capability enables design of area efficient, scalable, high performance transistors and diodes. This paper demonstrates a large area, 3 kV PSJ GaN high electron mobility transistor (HEMT) fabricated on sapphire. The device characteristics, working principle and switching performance are experimentally investigated. The device shows a low on-state resistance of 210 mΩ at 25°C. At elevated operating temperatures, it can be observed that the turn-off switching losses are not affected while the turn-on losses show a small increase. Also, a wide range of dV/dt controllability can be achieved through intelligent control of gate without significant increase in losses to meet various application requirements.

1   Introduction

In recent years, GaN based devices have gained significant popularity and market acceptance due to their superior material characteristics compared to conventional silicon (Si). GaN offers a wide bandgap of 3.4 eV, a high critical electric field of 3.3 MV cm^{-1} and high electron mobility which facilitates construction of high voltage, low resistance diodes and transistors capable of operating at high frequencies and temperatures [1,2]. Such properties are ideal for power dense efficient power electronics of the future [3,4]. The high frequency characteristic of GaN HEMTs is key to achieve high power density power conversions as the size of passive components shrink at high frequency.[1][5-7]. GaN has already established its commercial success in lighting, power supplies and radio frequency applications. However, the development of high voltage devices has been hindered by a few factors [1]; One of the key challenges in fabrication of high voltage GaN power transistors and diodes is the management of electric fields under off-state conditions. Conventional lateral GaN HEMTs utilize several field plates and sophisticated processing to manage field crowding, which adds to manufacturing overheads [1][8,9]. Furthermore, these devices use cheaply available Si as substrate material for epitaxial growth of GaN layers, which limits the performances. Moreover, thick GaN transition and buffer layers are required

for high voltage devices to account for lattice mismatch and prevent vertical breakdown [1]. Sapphire substrates, on the other hand, offer insulating and cost-effective thin GaN layer solutions to achieve high voltage capability, particularly in the form of PSJ technology [2]. PSJ devices work on a similar manner to conventional superjunction (SJ) which is based on the precise control of charges through impurity doping methods. However, in the PSJ concept, the charge balance is achieved due to the coexistence of positive and negative polarization charges in a double heterojunction formed by GaN/AlGaN/GaN layers [8]. Therefore, a uniform distribution of electric field is realized in PSJ devices with minimum drift region length in such a way that these devices can offer performance beyond the one-dimensional material limits [10]. GaN HEMTs often show very high dV/dt switching transitions which is beneficial in reducing switching losses [3]. However, they typically offer limited controllability which can limit their use in applications such as motor drives where dV/dt needs to be limited to prevent damage and adhere to EMI requirements [11-15]. Also, the sharp switching edges of GaN can result in unavoidable oscillatory behavior due to presence of stray inductances in circuit layout and packing [7]. Therefore, a comprehensive understanding of their switching behavior is essential to fully take advantage of their performance. In this paper, a large area, 3 kV GaN PSJ HEMT fabricated on sapphire is demonstrated for the first time. The device characteristics, working principle and switching behavior are experimentally investigated and the results are discussed in detail. The effect of temperature on switching characteristics and power losses are also analyzed.

2 Operating Physics and Characteristics

2.1 Polarization Superjunction

SJ in Si was the key to realize high voltage, low resistance devices which works on the basis of charge balance within the drift region to minimize its width. PSJ devices work on a similar principle to conventional SJ [16]. The main difference is that the charge balance is originated from polarization based charges in the two dimensional electron gas (2DEG) and two dimensional hole gas (2DHG) which are formed at their respective interface within a double heterojunction (GaN/AlGaN/GaN) [5,17].

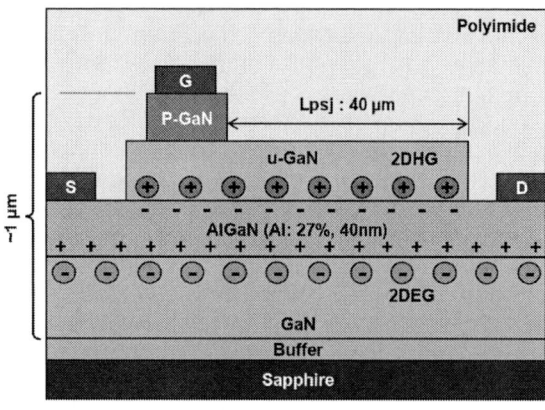

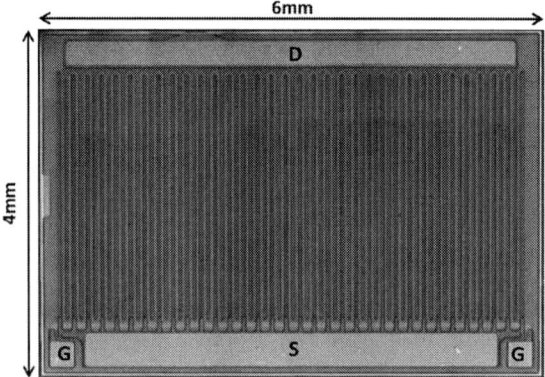

Fig. 1 Cross-sectional view of the device and top view of a 3 kV GaN PSJ HEMT chip. Gate, drain, and source are marked as G, D and S respectively.

In previous works, it has been demonstrated that breakdown voltage can be enhanced with a small increase in on-state resistance by extending the gate-drain distance (L_{PSJ}) [1]. Figure 1 illustrates a typical cross-sectional diagram of a depletion mode (d-mode) 3 kV GaN PSJ HEMT along with a top view of a chip.

The number of fingers connected to the drain, and source terminals are 33 and 34 respectively. The presence of the 2DEG provides a channel for current flow in the absence of gate bias ($V_{GS} = 0$ V) resulting in a normally on d-mode operation. The threshold voltage depends upon the thickness of the AlGaN layer. Once a negative gate bias larger than the threshold voltage of -4.9 V is applied, the device enters its blocking state. The p-GaN gate provides an ohmic contact to the 2DHG which facilitates injection or extraction of holes for collapse free operation. A positive gate bias can be applied to reduce the on-state resistance by attracting additional electrons beneath the 2DHG. Application of positive gate bias is restricted by two p-n junction diodes that are formed between (1) gate-drain (D_{GD}) and (2) gate-source (D_{GS}) which limits the positive gate drive voltage to around ~3.5 V. Exceeding this limit leads to excessive gate current injection that can lead to degradation.

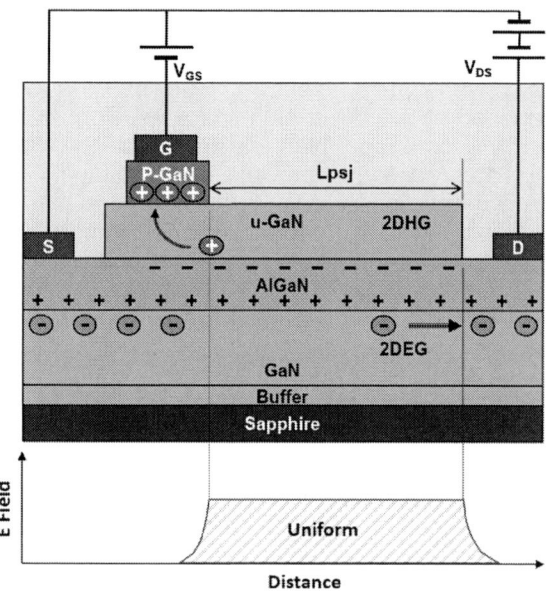

Fig. 2 Cross-sectional view of the GaN PSJ HEMT under off-state condition.

In the off state, the 2DHG and 2DEG are depleted through the gate and drain terminals respectively as depicted in Fig. 2. Consequently, a charge balance state is achieved which results in a uniform distribution of electric field over the drift region length (L_{PSJ}). This mechanism has enabled the design of high voltage GaN devices with a prospective of up to 10 kV blocking capability [18]. The device does not have a body diode and the reverse conduction can be facilitated through the 2DEG or an external GaN diode [4].

2.2 Current Voltage Characteristics

The measured output I-V characteristics of the 3 kV GaN PSJ HEMT at 25°C are presented in Fig. 3. The measurements were carried out under pulse mode to avoid self-heating.

The on-state resistance of the device at room temperature corresponds to 210 mΩ and 305 mΩ at the gate voltage of 3 V and 0 V respectively. The on-state resistance of the device was extracted from the measured I-V characteristics at different junction temperatures as shown Fig. 4.

The device exhibits a positive temperature coefficient of on-state resistance which facilitates even distribution of current sharing between paralleled devices without the risk of thermal runaway. The transfer characteristics were measured at drain voltage of 10 V and at different temperatures as shown in Fig. 5.

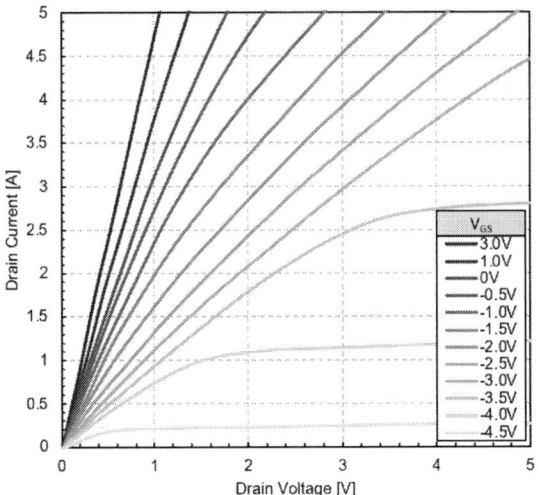

Fig. 3 Measured output characteristics of the 3 kV GaN PSJ HEMT at 25°C and pulse width of 250 μs.

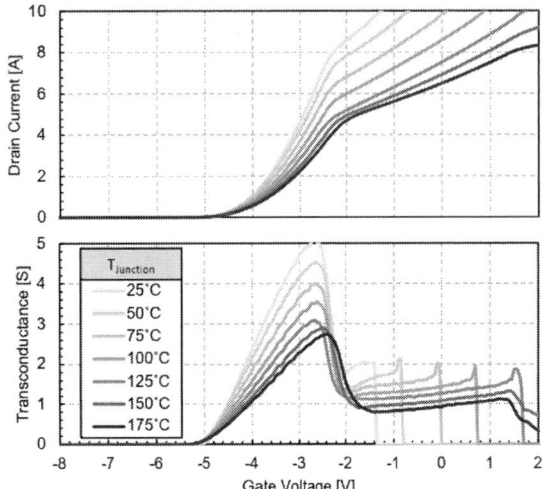

Fig. 5 Measured transfer characteristics of the 3 kV GaN PSJ HEMT at 25°C. The drain voltage is at 10 V, and the compliance current is set at 10 A.

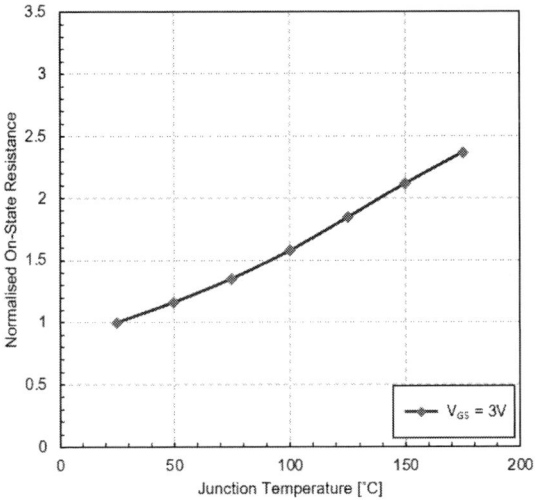

Fig. 4 Normalized on-state resistance versus temperature.

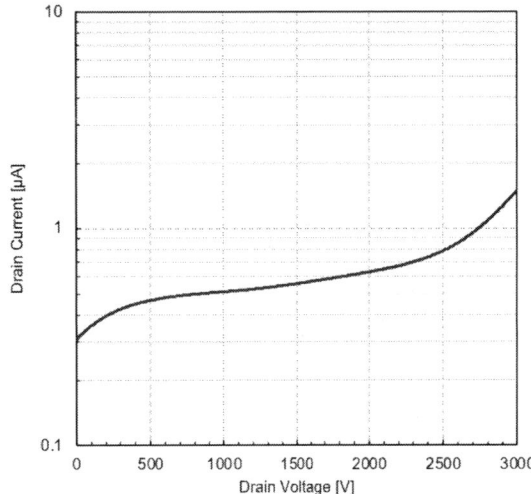

Fig. 6 Measured off-state I-V characteristics. The gate voltage is -10 V. The maximum drain voltage is limited by the instrument.

The slope of the transfer curves indicates the transconductance g_m which can be seen to decrease at higher temperatures. A threshold voltage of -4.9 V was extracted at drain voltage of 1 V and current of 1mA at room temperature. The threshold voltage rises by only ~6 % at 175°C which indicates high switching stability.

The off-state leakage current was measured at RT as illustrated in Fig. 6. The drift region to support 3 kV is 40 µm as shown previously in Fig. 1. The leakage current is 1 µA at 3 kV at 25°C. This is particularly important in high voltage devices to have low leakage current.

2.3 Dynamic On-State Resistance

One of the challenges in the development of power GaN devices, is the presence of dynamic on-state resistance (dynamic R_{dson}) which occurs due to current collapse phenomena [6]. The increase in resistance is attributed to electrons that are trapped near the channel which deplete the 2DEG causing its resistance to increase [6]. This is particularly problematic if high electric field peaks are present [19]. In GaN PSJ devices, current collapse is suppressed by the effective use of the p-GaN to inject or extract holes into the 2DHG which results in a uniform distribution of electric field. Figure 7 illustrates variation in the dynamic resistance as a function of applied voltage.

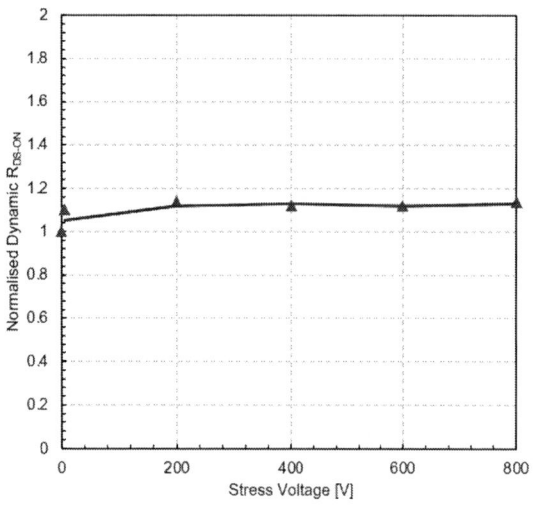

Fig. 7 Typical dynamic on-state resistance as function of applied stress voltage. The stress period is 10 s.

It can be observed that the on-state resistance does not change significantly with the applied voltage. This is because the electric field spreads evenly across the drift region.

2.4 Capacitance Characteristics

The input (C_{ISS}), output (C_{OSS}) and reverse transfer (C_{RSS}) capacitances of the device were measured using an Agilent B1505 power device analyzer at 1 MHz as shown in Fig. 8.

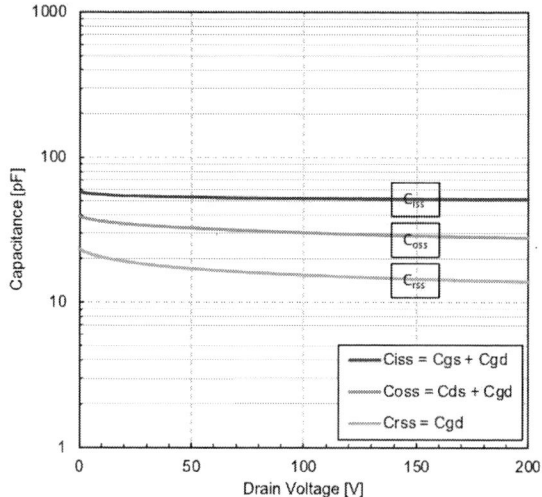

Fig. 8 Measured capacitance versus drain voltage measured at 1 MHz. The gate voltage is -20 V.

Amongst these capacitances, the gate-drain capacitance ($C_{GD} = C_{RSS}$) determines the switching speed as given by Eq. (1)

$$\frac{dV_{DS}}{dt} = \frac{I_G}{C_{GD}} = \frac{V_{GS}}{R_G . C_{GD}} \tag{1}$$

In GaN PSJ HEMTs, C_{GD} originates from the overlap of the 2DEG and 2DHG which is optimized for dV/dt controllability [3].

3 Experiment

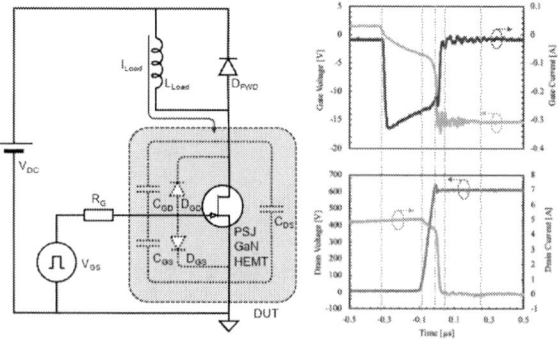

Fig. 9 Experimental setup and typical turn-off switching waveforms [3].

The switching performance was evaluated by a clamped inductive switching test (commonly referred to as double pulse test) set up as shown in Fig. 9. The switching voltage is limited by the measurement setup and safety considerations.

A 1.18 mH inductor is used as the load with a freewheeling SiC Schottky diode (D_{FWD}). The turn-off sequence can be described as follows; Initially, the gate-source capacitance (C_{GS}) is charged via the external gate resistor (R_G) by the gate driver (V_{GS}). In this period, the device is still on and carrying the full load current (I_D). The gate voltage is then maintained to keep the drain current flowing while the gate-drain capacitance (C_{GD}) begins to charge by the gate current (I_G). During this period, the drain voltage (V_{DS}) begins to climb to its final value set by the supply (V_{DC}) depleting the 2DEG and 2DHG. The slew rate is governed by the V_{GS} and R_G as given by Eq. (1). It should be noted that the C_{GD} decreases as the depletion expands. Once the V_{DS} is at supply voltage, the current starts to decrease and diverts to the D_{FWD}. Meanwhile, the V_{GS} continues to ascend to its final value set by the gate driver. An overshoot voltage occurs due to stray inductances as given by Eq. (2).

$$V_{Overshoot} = L_{Stray} \cdot \frac{dI_{Drain}}{dt} \qquad (2)$$

It is critical to reduce stray inductances to minimise the surge voltage. The turn-on sequence is essentially the same as the turn-off in backward sequence. During the switching period, energy is dissipated in the form of heat due to overlap of current and voltage as given by Eq. (3).

$$E_{SWT} = E_{On} + E_{Off} = \int V_{Drain} \cdot I_{Drain}\, dt \qquad (3)$$

The integration is performed over the turn-on and turn-off transitions.

4 Switching Characteristics

4.1 Turn-Off dV/dt Controllability

The turn-off dV/dt is evaluated by analyzing the effect of gate drive parameters on switching slew rate as shown in Fig. 10. In this experiment, the switching voltage and current are set at 600 V and 5 A respectively. The rise time and dV/dt were extracted from the measured waveforms as shown in Fig. 11.

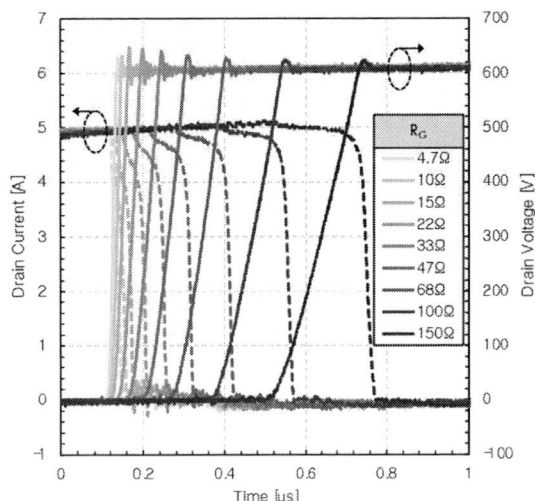

Fig. 10 Measured turn-off switching waveforms at different gate resistances and at 25°C.

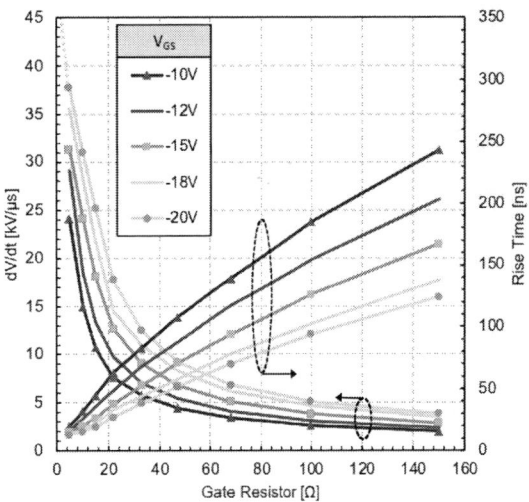

Fig. 11 Measured turn-off dV/dt and rise time at different gate resistances and gate voltages and at 25°C.

It can be observed that the device can operate at a wide range of dV/dt values which can be adjusted by the gate voltage and gate resistance based on the application requirement. The turn-off switching energy losses were calculated as shown in Fig. 12.

The dissipated energy increases at higher gate resistances and lower gate voltages due to the lower slew rates and longer switching time. It should be noted that the dissipated energy is at least an order of magnitude lower than Si IGBTs counterparts.

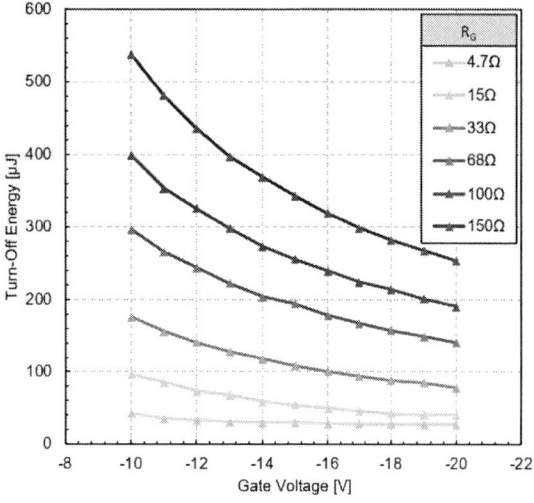

Fig. 12 Measured turn-off energy at different gate resistances and gate voltages and at 25°C.

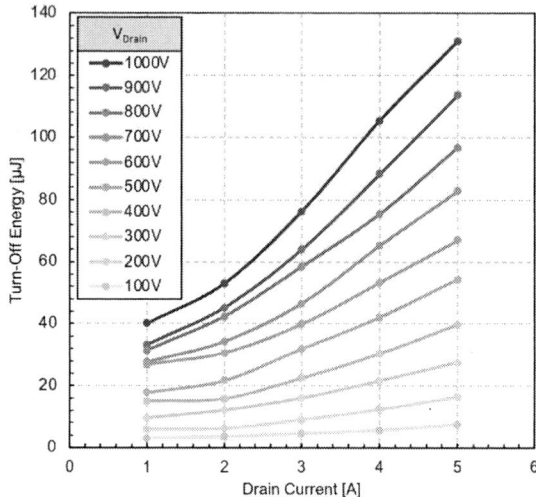

Fig. 14 Measured turn-off dissipated energy at various load currents and voltages and at 25°C. R_G = 15 Ω and V_G = +2 V/-15 V.

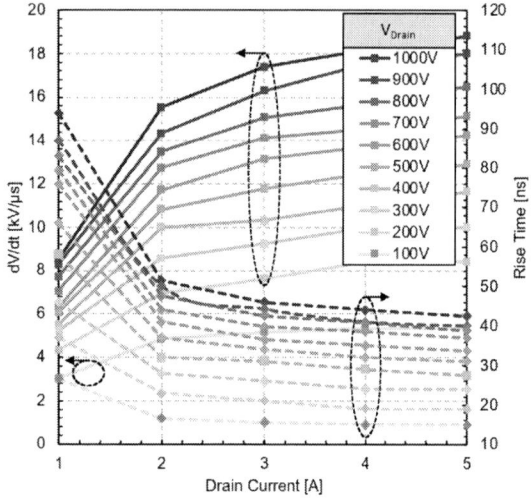

Fig. 13 Measured dV/dt and rise time as function of load current and voltage and at 25°C. R_G = 15 Ω and V_G = +2 V/-15 V.

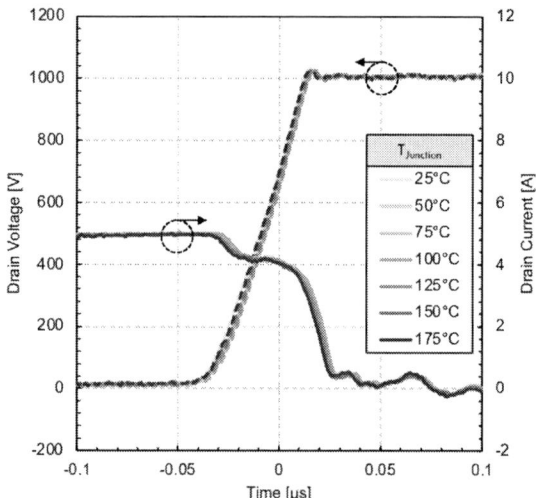

Fig. 15 Measured turn-off switching waveforms at different junction temperatures. R_G = 15 Ω and V_G = +2 V/-15 V.

4.2 Influence of Load current and Voltage on Turn-Off dV/dt

Figure 13 shows the influence of switching voltage and load current on dV/dt and rise time. The gate resistance is 15 Ω. While the switching voltage has a direct impact on switching speed, the incremental increase in dV/dt gets smaller at high voltage.

The corresponding energy losses at different voltages and currents are shown in Fig. 14.

4.3 Influence of Temperature and Load Current

Power semiconductor devices often are required to operate at elevated ambient temperatures. In this section, the influence of temperature on switching characteristics is evaluated. The turn-off and turn-on switching waveforms at different temperatures are presented in Fig. 15 and Fig. 16 respectively.

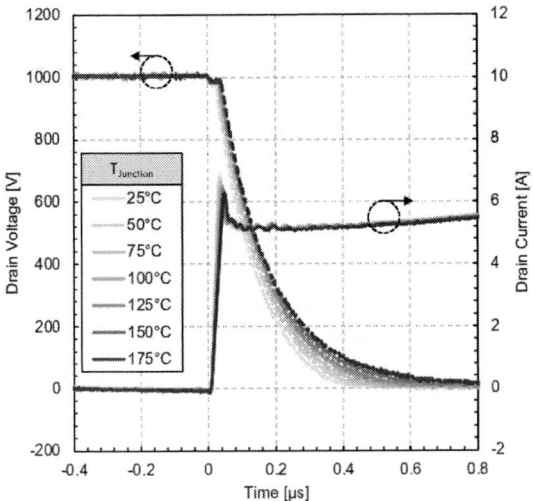

Fig. 16 Measured turn-on switching waveforms at different junction temperatures. $R_G = 15\,\Omega$ and $V_G = +2\,V/-15\,V$.

It can be observed that the temperature has no significant effect on the turn-off switching performance. The turn-on, however, is affected by the temperature due to the changes in the transconductance which decreases at elevated temperatures as previously illustrated in Fig. 5. Figure 17 illustrates the voltage rise and fall time for turn-off and turn-on transients respectively at various temperatures. The corresponding switching energy losses were calculated separately for turn-on and turn-off as shown in Fig. 18.

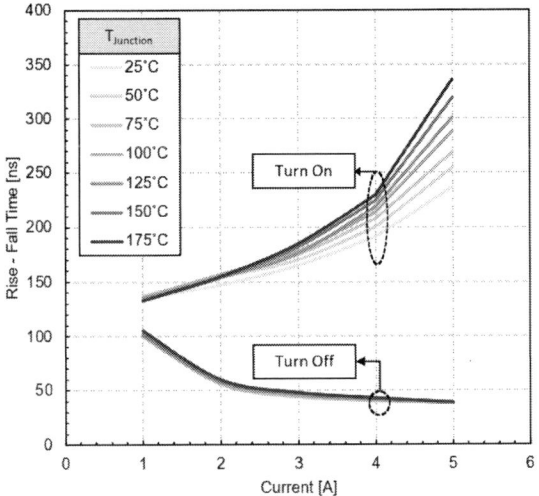

Fig. 17 Measured voltage rise (turn-off) and fall (turn-on) time at different current levels and junction temperatures.

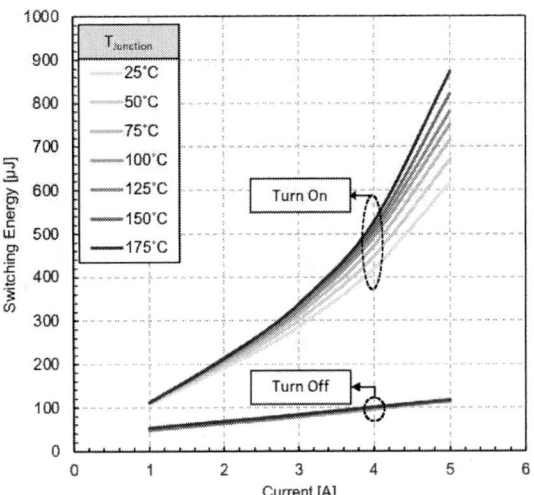

Fig. 18 Measured switching energy as a function of load current at different junction temperatures.

Due to slower transition of fall time at higher junction temperatures, the turn-on energy losses increase. The turn-off switching energy is almost constant regardless of the operating temperature. Overall, the results show potential benefits of GaN for high temperature applications.

5 Conclusion

In this paper, a large area, 3 kV PSJ GaN HEMT is demonstrated for the first time. The device characteristics, working mechanism and switching performance were investigated in detail. The device demonstrates a wide range of slew rates which can be adjusted through intelligent control of the gate resistance and gate voltage to fit various application requirements while maintaining a low power loss profile. The high temperature switching performance shows that the turn-off is effectively unaffected at elevated temperatures. The turn-on shows an increase in switching losses at higher operating temperature. Nevertheless, the losses are much lower than Si counterparts. The results are highly promising for the application of PSJ GaN technology in high voltage power electronic applications.

References

[1] H. Kawai, S. Yagi, S. Hirata, F. Nakamura, T. Saito, Y. Kamiyama, M. Yamamoto, H. Amano, V. Unni and E. M. S. Narayanan, "Low cost high voltage GaN polarization superjunction field effect transistors,"

Physica Status Solidi (a), vol. 214, no. 8, p. 1600834, 2017.

[2] "The 2018 GaN power electronics roadmap," *Journal of Physics D: Applied Physics,* vol. 51, no. 16, p. 163001, 2018.

[3] A. Sheikhan, E. M. S. Narayanan, H. Kawai, S. Yagi and H. Narui, "Evaluation of turn-off dV/dt controllability and switching characteristics of 1.2 kV GaN polarisation superjunction heterostructure field-effect transistors," *Japanese Journal of Applied Physics,* vol. 62, p. 064502, 2023.

[4] A. Sheikhan, G. Narayanankutty, E. M. S. Narayanan, H. Kawai, S. Yagi and H. Narui, "Analysis of 1.2 kV GaN polarisation superjunction diode surge current capability," *Japanese Journal of Applied Physics,* vol. 62, no. 1, p. 014501, 2023.

[5] A. Nakajima, Y. Sumida, M. H. Dhyani, H. Kawai and E. M. S. Narayanan, "High Density Two-Dimensional Hole Gas Induced by Negative Polarization at GaN/AlGaN Heterointerface," *Applied Physics Express,* vol. 3, no. 12, p. 121004, 2010.

[6] W. Saito, T. Nitta, Y. Kakiuchi, Y. Saito, K. Tsuda, I. Omura and M. Yamaguchi, "Suppression of Dynamic On-Resistance Increase and Gate Charge Measurements in High-Voltage GaN-HEMTs With Optimized Field-Plate Structure," *IEEE Transactions on Electron Devices,* vol. 54, no. 8, pp. 1825-1830, 2007.

[7] E. A. Jones, F. F. Wang and D. Costinett, "Review of Commercial GaN Power Devices and GaN-Based Converter Design Challenges," *IEEE Journal of Emerging and Selected Topics in Power Electronics,* vol. 4, no. 3, pp. 707 - 719, 2016.

[8] A. Nakajima, V. Unni, K. G. Menon, M. H. Dhyani, E. M. S. Narayanan, Y. Sumida and H. Kawai, "GaN-based Bidirectional Super HFETs Using Polarization Junction Concept on Insulator Substrate," in *International Symposium on Power Semiconductor Devices and ICs,* Bruges, 2012.

[9] V. Unni, H. Y. Long, H. Yan, A. Nakajima, H. Kawai and E. M. S. Narayanan, "Analysis of drain current saturation behaviour in GaN polarisation super junction HFETs," *IET Power Electronics,* vol. 11, no. 14, pp. 2198-2203, 2018.

[10] C. Hu, "Optimum doping profile for minimum ohmic resistance and high-breakdown voltage," *IEEE Transactions on Electron Devices ,* vol. 26, no. 3, pp. 243-244, 1979.

[11] H. Akagi, "Influence of high dv/dt switching on a motor drive system: a practical solution to EMI issues," in *International Symposium on Power Semiconductor Devices and ICs,* Kitakyushu, 2004.

[12] A. V. Jouanne and P. N. Enjeti, "Design Considerations for an Inverter Output Filter to Mitigate the Effects of Long Motor Leads in ASD Applications," in *Applied Power Electronics Conference,* San Jose, 1996.

[13] E. Persson, "Transient effects in application of PWM inverters to induction motors," *IEEE Transactions on Industry Applications,* vol. 28, no. 5, pp. 1095-1101, 1992.

[14] A. V. Jouanne and P. N. Enjeti, "Design considerations for an inverter output filter to mitigate the effects of long motor leads in ASD applications," *IEEE Transactions on Industry Applications,* vol. 33, no. 5, pp. 1138-1145, 1997.

[15] A. V. Jouanne, P. N. Enjeti and W. Gray, "Application issues for PWM adjustable speed AC motor drives," *IEEE Industry Applications Magazine,* vol. 2, no. 5, pp. 10-18, 1996.

[16] A. Nakajima, M. H. Dhyani, E. M. S. Narayanan, Y. Sumida and H. Kawai, "GaN based Super HFETs over 700V using the polarization junction concept," in *23rd International Symposium on Power Semiconductor Devices and ICs,* San Diego, 2011.

[17] A. Nakajima, K. Adachi, M. Shimizu and H. Okumura, "Improvement of unipolar power device performance using a polarization junction," *Applied Physics Letters,* vol. 89, no. 19, p. 193501, 2006.

[18] Powdec K.K., [Online]. Available: www.powdec.co.jp. [Accessed 1 May 2024].

[19] W. Saito, M. Kuraguchi, Y. Takada, K. Tsuda, T. Domon, I. Omura and M. Yamaguchi, "380v/1.9A GaN power-HEMT: current collapse phenomena under high applied voltage and demonstration of 27.1 MHz class-E amplifier," in *IEEE International Electron Devices Meeting,* Washington, 2005.

PCIM Europe 2024, 11– 13 June 2024, Nuremberg   DOI: 10.30420/566262069

More than 1200 V Breakdown and Low Area-Specific On-State Resistances by Progress in Lateral GaN-on-Si and GaN-on-Insulator Technologies

Richard Reiner[1], Stefan Mönch[1,2], Stefan Müller[1], Patrick Waltereit[1], Fouad Benkhelifa[1], Michael Basler[1], Michael Mikulla[1], and Rüdiger Quay[1,3].

[1] Fraunhofer IAF, Tullastrasse 72, 79108 Freiburg, Germany

[2] University of Stuttgart, IEW, Pfaffenwaldring 47, 70569 Stuttgart, Germany

[3] University of Freiburg, INATECH, Emmy-Noether-Str. 2, 79110 Freiburg, Germany

Corresponding author and speaker: Richard Reiner, richard.reiner@iaf.fraunhofer.de

Abstract

This work focuses on the recent developments of lateral GaN-HEMTs for the 1200 V class. Results of GaN-on-Si and GaN-on-SiC technology with static off-state voltages of over 1200 V and low area-specific on-state resistances are presented. The switching performance is demonstrated by a GaN-on-Si power device in a double-pulse setup with a voltage up to 1100 V and a static specific resistance of $R_{ON} \cdot A = 2.5 \ m\Omega \cdot cm^2$. The results are compared with state-of-the-art devices and technologies. In addition, economic aspects of GaN technologies on Si and highly insulating substrates are discussed.

1 Introduction

GaN-on-Si HEMTs have been commercially available for several years and demonstrate excellent switching performance. Compared to SiC-MOSFETs, the technology is relatively cost-effective because GaN-HEMTs are typically processed on inexpensive large-area Si wafers. However, the heteroepitaxy of thin GaN-HEMT layers on conductive Si (111) substrates leads to limitations in terms of the maximum achievable breakdown voltage. Apart from a few exceptions, commercial GaN-HEMT's are available up to a voltage class of 650 V (see Fig. 1). As a result of this constraint, todays available GaN-HEMTs are limited to lower voltage (≤ 650 V) and therefore lower power classes compared to predominant vertical technologies Si IGBTs, and SiC MOSFET technologies. However, there is a high demand for wide bandgap performance, cost-effective power transistors in many upcoming applications of the energy transition, such as charging technology and electric drive technology for electromobility, with 800 V battery voltages.

Recent developments show that GaN technology can also address higher voltage classes (>650 V). There are two developments in this regard. Firstly, the GaN epitaxy on Si wafers is continuously improving. Secondly, there is a development

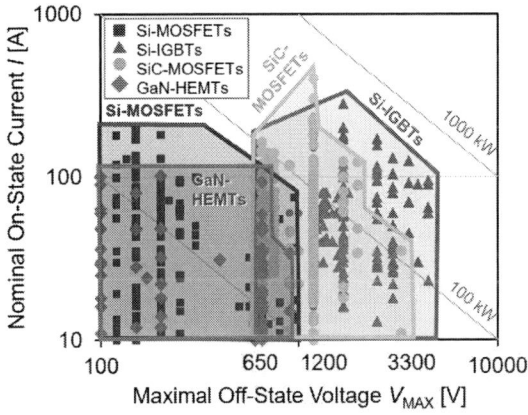

Fig. 1: Nominal current as a function of blocking voltage for various commercial power semi-conductors. Values are taken from distributors online information.

towards alternative, insulating carrier substrates. This work discusses the topic by focusing on recent achievements for GaN-on-Si and GaN on isolated substrates technologies (GaN-on-X_{ISO}, X_{ISO} = Sapphire, SiC, QST™, GaN). Furthermore, we show our own results of GaN-on-Si HEMTs and GaN-on-SiC HEMTs with off-state voltages of over 1200 V and low area-specific on-state resistances.

557

2 Progresses Towards Lateral 1200 V GaN-Technologies

Current developments in material growth and process technology are aiming for high-voltage devices beyond 650 V by improved GaN epitaxy on conductive Si, as well as different isolating substrate materials. In the following we will shortly review some examples of current achievements towards lateral 1200 V GaN HEMTs. GaNPower Inc. has demonstrated the power switching capability at 800 V and around 8 A for a 1200 V commercial GaN-on-Si HEMTs [1]. Power Integrations Inc. launches 900 V and recently 1250 V devices using a GaN-on-Sapphire technology [2,3]. A product-related 1200 V GaN-on-Sapphire technology is presented in [4]. An inductive switching waveform in a half bridge is demonstrated at 720 V/28 A. A GaN-on-Sapphire device with active passivation is presented with a breakdown of 2230 V in static measurements [5]. A monolithic integrated half-bridge in a 1200 V GaN-on-Sapphire technology is switched at 800 V/100 MHz an published in [6]. A static breakdown of 1.4 kV and a specific on-resistance of $R_{ON} \cdot A = 6.73$ m$\Omega \cdot$cm² is published in [7]. A GaN buffer layer qualified for 1200 V applications on 200 mm QST™ substrates, with a hard breakdown exceeding 1800 V is promised in [8]. Furthermore, static breakdown voltages above 1200 V and 950 V switching results are achieved with GaN-HEMTs using an un-doped AlN-buffer and SiC-substrate material [9].

3 In-House Fabricated Devices

3.1 GaN-on-Si Technology

In the following, we present the characteristics of the GaN-on-Si devices based on our own technology (Fraunhofer IAF). Fig. 2 shows the results of a parameter variation of the gate-drain distance L_{GD} on small GaN-on-Si test devices.

The details and the characteristics of the Superlattice epitaxial buffer structure were already presented in [10]. The buffer structure was grown by metal organic chemical vapor deposition (MOCVD) reactor on an 800 µm-thick 4" highly conductive Si (111) substrate. A 120 nm-thick AlN nucleation layer and 80 periods of AlN/GaN were first grown, followed by a thick GaN layer. Part of the GaN buffer layer was doped with carbon up to $3 \cdot 10^{-19}$ cm³ to ensure high vertical and lateral isolation. Finally, a thin AlGaN barrier and GaN cap layer were grown on the buffer structure to form a 2DEG (two-dimensional electron gas) for

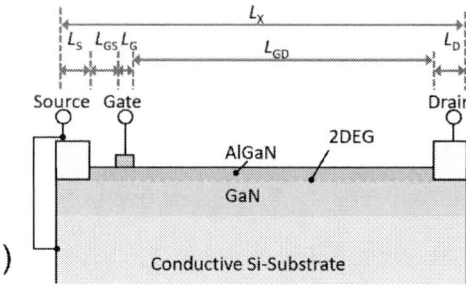

a)

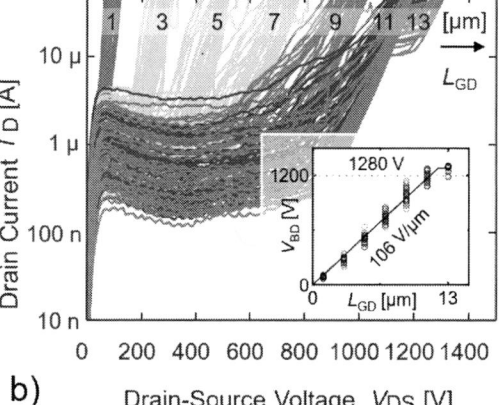

b)

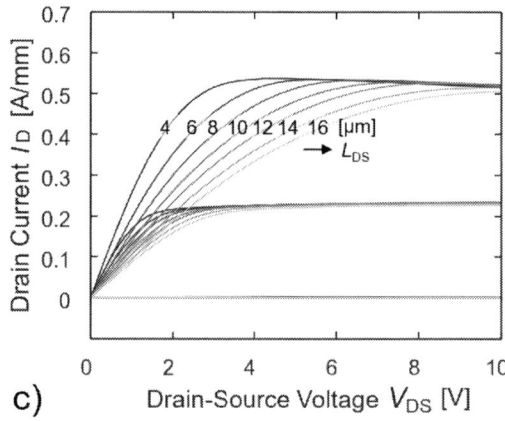

c)

Fig. 2: Gate-drain distance L_{GD} variation on small gate width test transistor (W = 50 µm, L_{GS} = 2 µm, L_G = 1 µm, L_{GD} = var. param.)
a) Cross-section of a GaN-on-Si HEMT without field plated, as use as test device in these measurements.
b) Off-state measurements for different gate-drain distances (measurements of all 37-wafer cells are overlayed).
c) On-state measurements for different drain-source distances (averaged currents of all 37-wafer cells).

the HEMT devices. The Al concentration of the barrier layer is 22%, with 24.5 nm thickness. The sheet resistance of the 2DEG is around $R_{SH} = 616\ \Omega\square$. The growth conditions of the structure were optimized to improve the layers quality, the vertical isolation and to reduce the internal strain build up during the growth and the cooling down process.

The process technology [11,12] used in this work, features a D-mode Schottky gate-metallization, an ohmic-contact metallization, a few hundred nanometer thick interconnect metal, a few micrometers thick top metallization realized by electroplating, two different interlevel dielectric layers with the corresponding openings, and a final passivation.

3.2 GaN-on-Si Measurement Results

The dimensions of the test devices are $L_{GS} = 2\ \mu m$, $L_G = 1\ \mu m$, $L_D = L_S = 5\ \mu m$, $W = 50\ \mu m$, $A = L_x \cdot W$. These test devices are designed without field plates. The explanation of dimensioning is illustrated in Fig. 2 a and breakdown V_{BD} vs. depletion length L_{DS} is shown in Fig. 2 b. The measurements were conducted on all cells on the wafer. The GaN-on-Si technology achieves a lateral isolation of $V_{BD}/L_{GD} = 106\ V/\mu m$, maximal breakdown of $V_{BD,MAX} = 1280\ V$ (limited by the vertical buffer breakdown) Additionally, the on-state resistances of the structures were characterized by on-wafer mappings. Fig. 2 c displays the output characteristics for all cells, corresponding to the parameterized drift lengths.

The dynamic on-state resistance was measured with an AURIGA™ test measurement setup with off-state stress voltages $V_{DS,\ OFF}$ up to 1000 V and shown in Fig. 3. The measurements were conducted on-wafer using test HEMTs with a gate width $W = 1\ mm$, and intrinsic dimensions $L_{GD}/L_G/L_{GS} = 15/1/2\ \mu m$ (see Fig. 3 b). The measurement setup initially applies an off-state stress voltage. Then, the device is brought into the on-state, and the on-resistance is measured (see Fig. 3 a). The measurement results are depicted in Fig. 3 c, illustrating the increase in dynamic on-state resistance dyn. R_{ON} as a function of the off-state stress voltage $V_{DS,\ OFF}$. The on-state resistance increases from an initial value of about 10 Ω to approximately 25 Ω at 1000 V stress voltage.

In the following we will present the performance of a large area GaN device (see Fig. 4 a) with a gate width of $W = 87\ mm$, intrinsic dimensions $L_{GD}/L_G/L_{GS} = 12.5/1/2\ \mu m$ a chip area of $A_{CHIP} = 2 \times 2\ mm^2$, and an active device area of $A = A_{ACTIVE} = 1.85 \times 1.2\ mm^2$. This is the chip area

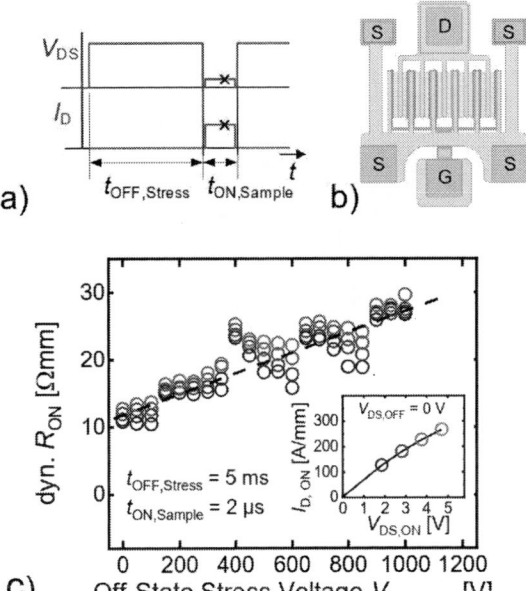

Fig. 3: Dynamic on-state characterization.
a) Pulse pattern of the dynamics on-state measurement using AURIGA™ pulse setup.
b) Layout of the test transistor $W = 1\ mm$.
c) Measurement results of the on-state resistance dyn. R_{ON} after an off-state stress voltage $V_{DS,OFF}$.

without bond-pad area, but including the area of the finger metallization. The on-state resistance was measured to be $R_{ON} = 110\ m\Omega$ (see Fig. 4 b) and a corresponding area specific on-state resistance of $R_{ON} \cdot A = 2.5\ m\Omega \cdot cm^2$. The device was tested in a double pulse measurement setup as shown in Fig. 4 c. The coil has an inductance of $L = 860\ mH$ and uses a SiC freewheeling diode. The pulse pattern is generated by a microcontroller (MSP430) and a gate driver (TC4452). The off-state voltage on the device has been continuously increased from $V_{DS,OFF} = 500\ V$ up to 1100 V with 100 V steps, as shown in Fig. 4 d. The on-state current chosen to be $I_{D,ON} = 0.5\ A$.

3.3 Characterization on In-House Fabricated GaN-on-SiC Devices

Within the next, we present the characteristics of the GaN-on-SiC devices based on our highly isolated GaN buffer technology. The GaN-on-SiC technology achieves a lateral isolation of V_{BD}/L_{GD} = 98 V/μm, a sheet resistance around 600 Ω□, and maximal on-state currents of 800 mA/mm. No maximum voltage limit has been measured, because the SiC substrate is highly isolating and thus the devices are not affected by a vertical

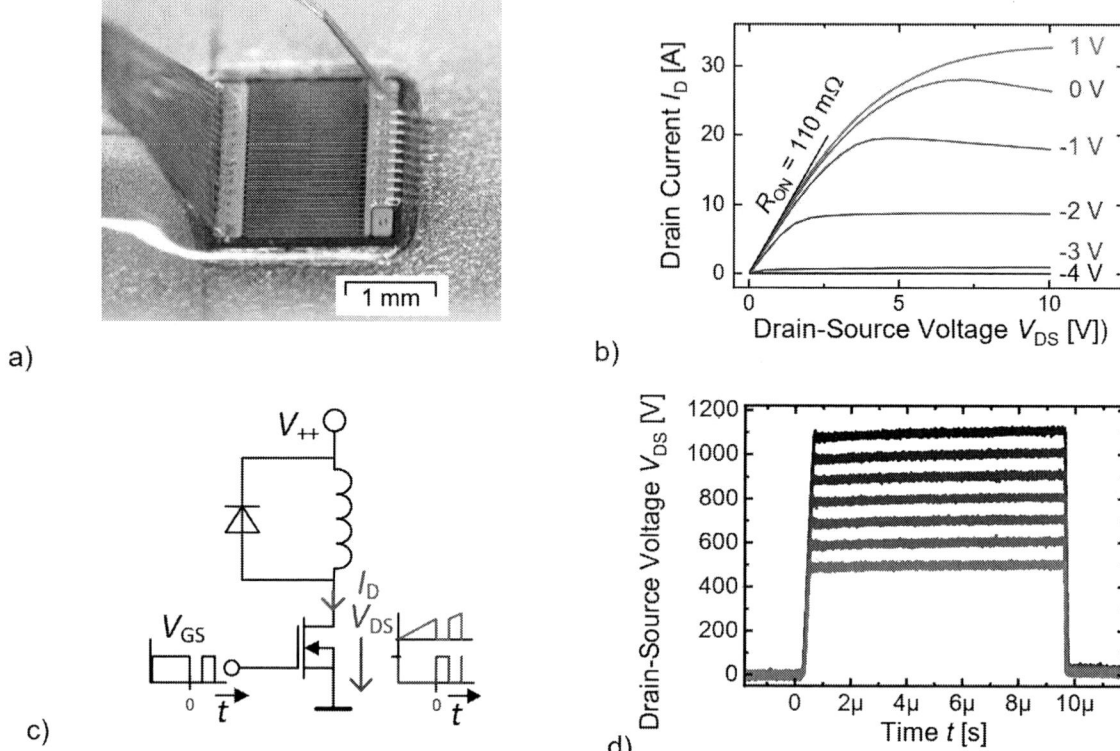

a)

b)

c)

d)

Fig. 4: Performance of a large area GaN-on-Si device with W = 87 mm, $L_{GD}/L_G/L_{GS}$ = 12.5/1/1.5 µm
a) Microscope image of the device in a TO220 engineering package with bonding wires.
b) Output characteristics of the packaged device. The pulse time was set to be t_{PLS} = 100 µs.
c) Schematics of the double pulse measurement setup. d) Drain source voltage during the switch-off pulse.

breakdown. However, this wafer had a faulty ohmic contact. The transistors therefore had an unwanted forward voltage of approx. 1.3 V. Nevertheless, the wafer shows the potential of GaN-on-SiC technology. GaN on highly insulating SiC substrates can achieve similar performance to GaN-on-Si technology. Compared to the conductive Si substrate, there is no vertical breakdown with the insulating SiC. This means that with GaN-on-SiC there is no voltage limitation due to vertical breakdown.

4 Discussion

Lateral GaN technologies aim for higher blocking voltages than 650 V. The approach of GaN on insulating sapphire substrates is particularly promising, with 1250 V devices already entering the market today [3]. However, the ongoing improvements in GaN-on-Si technology toward higher voltages are also motivating. Furthermore, there is already a commercially available product promising 1200 V. [1]. In addition to the increase of the blocking voltage, there is also potential for improvements in the on-state behavior. Today's

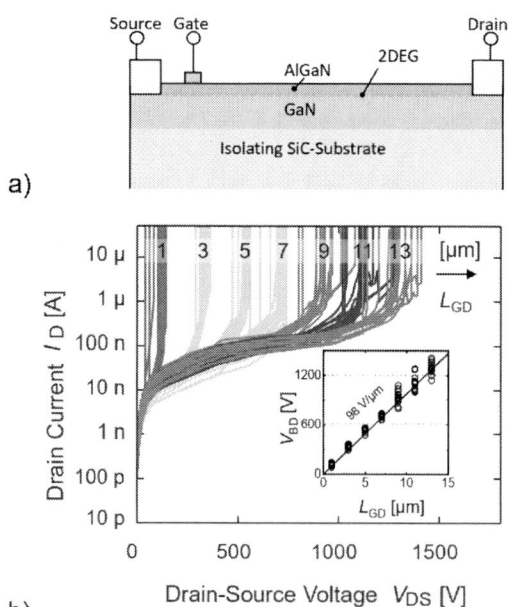

a)

b)

Fig. 5: Gate-drain distance L_{GD} variation on small gate width HEMTs: a) Cross-section of a GaN-on-SiC. b) Off-state measurements cells.

GaN transistors are often oversized (regarding depletion length L_{GD}). With further development in the fields of design, epitaxy and process technology and increasing knowledge in the field of reliability, GaN HEMTs are becoming more mature. This will reduce the area-specific on-state resistance $R_{ON} \cdot A$ and improve the overall performance of lateral GaN technologies. So far, the performance (figure-of-merit: $R_{ON} \cdot A$ vs. V_{BD}, see Fig. 6 a) GaN technologies are still away from the theoretical limit for lateral devices predicted by W. Saito [13]. However, continuous progress can be observed, and concepts such as active passivation [5,14] (similar to REduced SURface Field = RESURF concept [15] as lateral super junction concept) could even exceed W. Saito's limit. This work shows the potential of a state-of-the-art GaN-on-Si technology with small-

gate width test transistors (Fig. 2) with static breakdown around 1280 V as well as a large area device demonstrator switching up to 1100 V (Fig. 4). The experimental results from Fig. 2 are shown as green areas in Fig. 6. We expect that commercial GaN-on-Si devices will also exploit this performance. Furthermore, literature (chapter 2) and results of this work (chapter 3.3) show that GaN on insulating substrates (X_{ISO}) is not limited by vertical breakdowns and can reach voltages of 1200 V and more. The sheet resistance of the two-dimensional electron gas of the HEMT is not influenced by the carrier substrate, so the performance is expected to be similar to that of GaN-on-Si. The comparable performance of GaN on isolating SiC to GaN-on-Si has been confirmed in our own work with a lateral isolation of $V_{BD}/L_{GD} \approx 100$ V/µm and a sheet resistance around $R_{SH} \approx 600$ Ω□. In Fig. 6 a the green area has been extended by an ochre-green area for insulating substrates (GaN-on-X_{ISO}).

Commercial GaN-on-Si HEMTs are grown on inexpensive, large diameter substrates (≥ 8 Inch), and fabricated in CMOS-compatible factories. Therefore, the chip area of GaN-on-Si is already significantly cheaper than that of SiC (see Fig. 6 b), although the market entry of GaN-on-Si was several years later than SiC. But GaN-on-Sapphire also has the potential to become a cost-effective, high-performance technology. Sapphire substrates are already used for LED applications in large quantities and are available with large diameters, high material quality and at low cost. GaN-on-Sapphire wafers are already significantly cheaper than SiC wafers and almost as cheap as GaN-on-Si. This means that 1200 V class GaN-on-Si and GaN-on-Sapphire have the potential to take over many markets that are currently dominated by SiC.

However, GaN-on-Sapphire still has some open research questions that need to be investigated: For example, how are the thermal properties compared to GaN-on-Si? Sapphire has lower thermal conductivity. However, the substrate can be thinned, and often the GaN epitaxial layers are the thermal bottleneck, rather than the carrier substrate. New investigations are therefore necessary here. New epitaxial concepts for GaN-on-Sapphire have to be developed and investigated. Sapphire is highly insulating, so thinner buffer layers may be sufficient for very high voltage (≥ 1200V) compared to GaN-on-Si. This could also mean a further cost reduction for epitaxy. However, little is known in the literature about reliability and trapping effects and the associated dynamic on-state resistance. However, it is encouraging that components are already

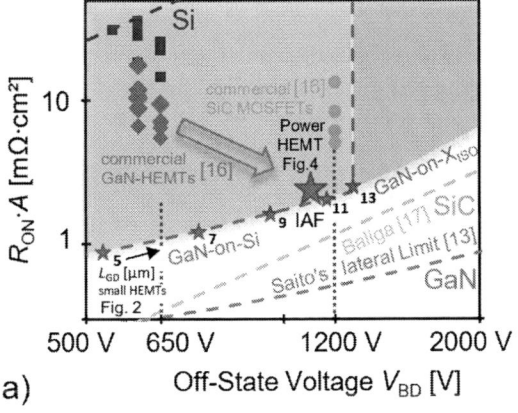

a)

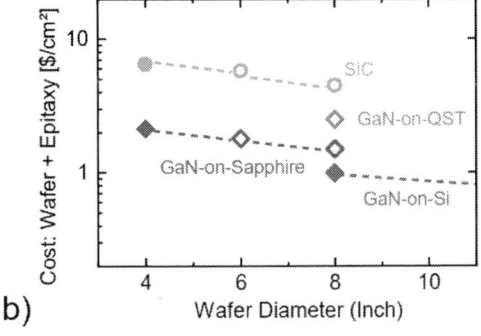

b)

Fig. 6: a) area specific on-state resistance as a function of the off-state voltage. The devices of Fig. 2 and Fig. 4. are marked as small and larger star and compared to theoretical limits [13,17] and values of some commercial devices [16].
b) Estimation of wafer costs and corresponding size, data collected from [18-20], and QST added and estimated by IAF.

available on the market at 900 V [2] and 1250 V [3].

Another interesting aspect of research could be, that GaN on insulating substrates are advantageous for monolithically integrated GaN power circuits (GaN-Power ICs). Many components such as gate drivers, sensors and control circuits can be integrated into lateral GaN-on-Si technology (low side integration) [21]. However, if entire topologies such as half-bridges (single-phase, multi-phase, or multilevel) are to be integrated [22] (high-side integration), this leads to problems such as back-gating, trapping, and crosstalk [23].

5 Conclusion

Lateral GaN HEMT technologies can reach the 1200 V voltage class and can serve many markets that are currently occupied by SiC-MOSFETs and Si-IGBTs. GaN-on-Si is voltage-limited by vertical buffer breakdowns. However, the epitaxy is continuously being developed to make it feasible for use at 1200 V. GaN on insulating substrates (GaN-on-X_{ISO}, X_{ISO} = Sapphire, SiC, QST™, GaN) suppress the vertical breakdown mechanism, which enables GaN-on-X_{ISO} the potential for higher voltages ≥ 1200 V. Both GaN-on-Si and GaN-on-Sapphire wafers are already more cost-effective than SiC wafers. This work presents static measurements on GaN-on-Si HEMTs up to 1280 V and double-pulse measurements up to 1100 V. The performance of the devices demonstrates the potential of GaN-on-Si technology but also its voltage limitations. On the other hand, GaN-on-X_{ISO}, with similar lateral isolation and on-state performance, is not voltage-limited. This work presents static results on GaN on high-insulating SiC substrates with a comparable sheet resistance around 600 $\Omega_\square$ and lateral breakdown voltage of $V_{BD}/L_{GD} \approx 100$ V/μm. GaN-on-X_{ISO} has further potential as an improved technology for monolithic integration (GaN Power ICs), because back-gating and cross talk is eliminated by the isolated Substrate. However, these new GaN-on-X_{ISO} technologies open research questions regarding reliability, dynamic behavior, and thermal behavior that need to be investigated. The Fraunhofer IAF is dedicated to addressing these issues in the GaN4EmoBiL project [24].

Acknowledgements

In the GaN4EmoBiL project [24], the Fraunhofer IAF (lead), the University of Stuttgart, Robert Bosch GmbH, and Ambibox GmbH are researching cost-effective bidirectional charging technologies for battery voltages of 800 V. In the Fraunhofer IAF subproject, cost-effective 1200 V GaN-on-X_{ISO} devices are being developed.

This work was supported by the German BMWK - Federal Ministry of Economics and Climate Protection (BMWK) within the projects GaN4EmoBiL (FKZ: 01MV23003A)

The authors thank their colleagues from the Fraunhofer IAF's epitaxy, technology and microelectronic departments for their contributions.

References

[1] GaNPower International Inc, "GaNPower Demonstrates Industry's First 1200 V Single-Die E-Mode GaN Power Devices" [Online-Download, 06.10.2023] https://iganpower.com/ganpower-demonstrates-industrys-first-1200-v-single-die-e-mode-gan-power-devices

[2] Power Integrations Inc, "Power Integrations Launches 900 V GaN Flyback Switcher ICs", March 20, 2023, [Online-Download, 06.10.2023] https://s27.q4cdn.com/802031818/files/doc_news/Power-Integrations-Launches-900-V-GaN-Flyback-Switcher-ICs-2023.pdf

[3] Power Integrations Inc, "Power Integrations Releases Ground-Breaking 1250-Volt GaN Switcher IC", Oct. 30, 2023, [Online-Download, 04.03.2024] https://investors.power.com/news/news-details/2023/Power-Integrations-Releases-Ground-Breaking-1250-Volt-GaN-Switcher-IC/default.aspx

[4] G. Gupta *et al.*, "1200 V GaN Switches on Sapphire Substrate," *2022 IEEE 34th International Symposium on Power Semiconductor Devices and ICs (ISPSD)*, Vancouver, BC, Canada, 2022, pp. 349-352, doi: 10.1109/ISPSD49238.2022.9813640.

[5] J. Cui *et al.*, "Demonstration of 1200-V E-Mode GaN-on-Sapphire Power Transistor with Low Dynamic ON-Resistance Based on Active Passivation Technique," in *IEEE Electron Device Letters*, vol. 45, no. 2, pp. 220-223, Feb. 2024, doi: 10.1109/LED.2023.3341413.

[6] S. Li et al., "1200V E-mode GaN Monolithic Integration Platform on Sapphire with Ultra-thin Buffer Technology," *2023 International Electron Devices Meeting (IEDM),* San Francisco, CA, USA, 2023, pp. 1-4, doi: 10.1109/IEDM45741.2023.10413753.

[7] J. Cui et al., "Method to Study Dynamic Depletion Behaviors in High-Voltage *(BV=1.4 kV)* p-GaN Gate HEMT on Sapphire Substrate," *2023 35th International Symposium on Power Semiconductor Devices and ICs (ISPSD),* Hong Kong, 2023, pp. 127-130, doi: 10.1109/ISPSD57135.2023.10147490.

[8] Press release of Imec and AIXTRON: "Imec and AIXTRON Demonstrate 200 mm GaN Epitaxy on AIX G5+ C for 1200V Applications with Breakdown in Excess of 1800V" LEUVEN (Belgium), APRIL 29, 2021.

[9] O. Hilt et al., "10 A/950 V switching of GaN-channel HFETs with non-doped AIN buffer," *2023 35th International Symposium on Power Semiconductor Devices and ICs (ISPSD),* Hong Kong, 2023, pp. 374-377, doi: 10.1109/ISPSD57135.2023.10147681.

[10] S. Moench et al., "Monolithic Integrated AlGaN/ GaN Power Converter Topologies on High-Voltage AIN/GaN Superlattice Buffer." physica status solidi (a) 218.3 (2021): 2000404.

[11] P. Waltereit, R. Reiner, H. Czap, et al.,"GaN-based high voltage transistors for efficient power switching." *Physica status solidi c* 10.5, pp. 831-834, 2013.

[12] F. Benkhelifa, D. Krausse, S. Müller, et al., "AlGaN/GaN HEMTs for high voltage applications." *Proc. 5th Space Agency-MOD (ESAMOD) Round Table Workshop GaN Component Technol.* 2010.

[13] W. Saito, I. Omura, T. Ogura and H. Ohashi, „Theoretical limit estimation of lateral wide band gap semiconductor power switching device," *in Solid State Electronics* vol. 48.9, pp.: 1555-1562, 2004.

[14] J. Yang et al., "Enhanced robustness against hot-electron-induced degradation in active-passivation p-GaN gate HEMT." Applied Physics Letters 124.10 (2024).

[15] J. A. Appels and H. M. J. Vaes, "High voltage thin layer devices (RESURF devices)," *1979 International Electron Devices Meeting,* Washington, DC, USA, 1979, pp. 238-241, doi: 10.1109/IEDM.1979.189589.

[16] R. Reiner et al., "Lateral GaN Power Devices and Integrated GaN Power Circuits: Status and Recent Progress," *Components of Power Electronics and their Applications 2023; ETG Symposium,* Bad Nauheim, Germany, 2023, pp. 89-96.

[17] B. J. Baliga, "Power semiconductor device figure of merit for high-frequency applications", *IEEE Electron Device Letters,* Vol.: 10 Issue:10, pp.: 455 – 457, Oct. 1989.

[18] T. Ayari, P. Chiu, Yole-Development, "Power GaN 2022", "*Market and Technology Report 2022*", 2022.

[19] Y. Zhang, A. Dadgar, and T. Palacios, "Gallium nitride vertical power devices on foreign substrates: a review and outlook." *Journal of Physics D: Applied Physics* 51(27), 273001, 2018.

[20] Y. Zhang, "GaNPower Devices: Current Status, Challenges, and Emerging Technologies." *tutorial presentations slides, at the WiPDA2022,* 2022.

[21] M. Basler et al., "Building Blocks for GaN Power Integration," in *IEEE Access,* vol. 9, pp. 163122-163137, 2021, doi: 10.1109/ACCESS.2021.3132667.

[22] S. Mönch et al., " GaN power converter and high-side IC substrate issues on Si, p-n junction, or SOI." *e-Prime - Advances in Electrical Engineering, Electronics and Energy,.* Vol. 4, 2023,100171, doi: 10.1016/j.prime.2023.100171.

[23] B. Weiss et al., "Substrate biasing effects in a high-voltage, monolithically-integrated half-bridge GaN-Chip," *2017 IEEE 5th Workshop on Wide Bandgap Power Devices and Applications (WiPDA),* Albuquerque, NM, USA, 2017, pp. 265-272, doi: 10.1109/WiPDA.2017.8170558.

[24] Fraunhofer IAF, "Batteries on wheels — New charging technology to make e-cars suitable for mass use as mobile power storage units,". press release – GaN4EmoBiL -BMWK - project-start, August 01, 2023.

PCIM Europe 2024, 11– 13 June 2024, Nuremberg   DOI: 10.30420/566262070

Novel 200 V MOSFET Technology Pushes Motor Drive Inverter Efficiency to an Unprecedented Level

Mark Thomas[1], Ralf Siemieniec[1], Elvir Kahrimanovic[1], Laszlo Juhasz[1], Michael Hutzler[1], Kapil Kelkar[2]

[1] Infineon Technologies Austria AG, Austria
[2] Infineon Technologies Americas Corp., USA

Corresponding author: Mark Thomas, mark.thomas@infineon.com
Speaker: Mark Thomas, mark.thomas@infineon.com

Abstract

This work introduces the latest 200 V trench MOSFET technology released to the market. Based on the advantages of 3D charge compensation, the new cell design combines the benefits of low conduction and switching losses with good ruggedness, excellent body diode properties and an extremely tight threshold voltage spread. These features result in a well-balanced all-round performer that will bring significant improvements to a broad range of target applications. They also make the devices an ideal fit especially for high power motor-drive applications, which require an easy paralleling of many devices. The presented results focus on the use of the new devices in such applications.

1   Introduction

New power MOSFET devices dedicated to motor-drive applications are required to provide improvements across a wide range of device parameters. Device losses in the application are primarily associated with on-state resistance (conduction losses), although charges are also significant (switching losses) as well as the body diode forward voltage drop (conduction losses during dead time). In addition, it is advantageous if the MOSFET shows a small reverse-recovery charge and an improved linearity of the output and Miller capacitances. This reduces unwanted oscillations and excessive voltage overshoots ensuring compliance with EMI limits. Furthermore, a small variation of the threshold voltage between different devices eases the paralleling of devices to serve a wide range of output current needs. A small reverse-recovery charge is also beneficial to

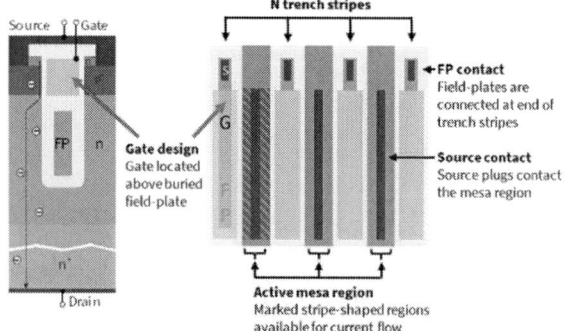
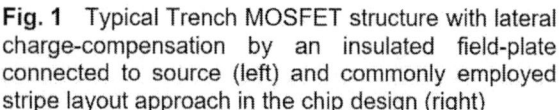

Fig. 1 Typical Trench MOSFET structure with lateral charge-compensation by an insulated field-plate connected to source (left) and commonly employed stripe layout approach in the chip design (right)

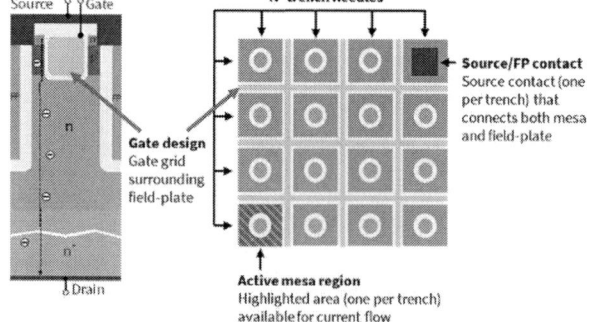

Fig. 2 Trench MOSFET structure with lateral charge-compensation by an insulated field-plate and separated gate trench (left) and the new grid-like layout approach in the improved chip design (right)

ensure a high commutation ruggedness. Finally, the MOSFET also needs to offer sufficient robustness to survive critical operation conditions that may occur occasionally.

2 New device approach

2.1 Advanced cell design

New MOSFET devices are required to provide improvements across all figures of merit. To meet these requirements, a novel cell-design approach has been developed which uses true three-dimensional charge compensation. Today's state-of-the-art MOSFET technologies use an insulated deep field plate underneath and separated from the gate electrode and employ a stripe layout as depicted in Fig. 1 [1]. This new generation separates the field plate trench, which is now formed with a needle-like structure, from a grid-like gate trench that surrounds the needles [2] as shown in Fig. 2. This increases the silicon area available for current conduction allowing for a further reduction in the overall on-resistance [2].

To reduce the two figure-of-merits (FOM) that are essential to achieve good switching properties, $FOM_G = R_{DS(on)} \times Q_G$ and $FOM_{GD} = R_{DS(on)} \times Q_{GD}$, the gate trench underwent a complete redesign to minimize its lateral extension.

The use of a gate grid additionally yields a far more even distribution of the gate resistance across the chip supporting faster switching of the device. This strongly improved homogeneity is also advantageous for device robustness, for example avalanche ruggedness, by reducing the probability that a part of the chip is affected by gate signal delays [3] or parasitic turn-on. In earlier transistor generations, both gate signal delay and parasitic turn-on degrade the device ruggedness as power dissipation is limited to just a part of the chip. In addition, the direct connection of the field-plate in the needle to the source metal supports a high avalanche ruggedness. In a striped layout, the local field-plate potential varies along a trench stripe and consequently may alter the local breakdown voltage [4]. This can lead to an inhomogeneous power dissipation over the chip area degrading device ruggedness.

2.2 Device properties

Fig. 3 summarizes the realized parameter improvements for the new OptiMOS™ 6 200 V technology [5] based on the new grid-like layout

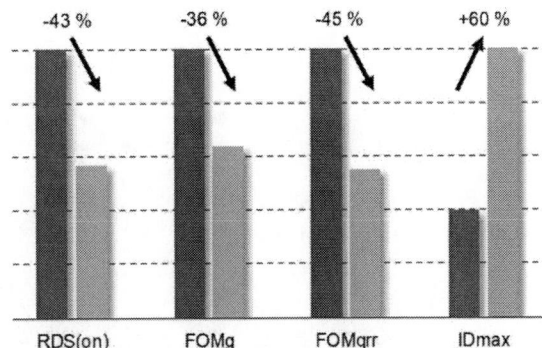

Fig. 3 Improvement in device performance for best-in-class 200 V devices in TO-263 package

with trench needles. All key parameters show a significant improvement over the earlier OptiMOS™ 3 technology [6].

As the new 200 V devices mainly target battery power and motor drive applications such as light-electric vehicles (LEVs) and forklifts, it was especially important to further reduce the reverse-recovery charge with respect to the previous generation with a fast diode [6]. This not only enables a further reduction of switching losses, but also improves the EMI behavior and ensures a high commutation ruggedness [7]. Thanks to the new advanced cell design, the on-resistance of the device is considerably reduced allowing the device to increase the drain current capability by a remarkable 60 % in the same package footprint.

3 Single switch in 3-phase motor-drives inverter test platform

3.1 Introduction of the test platform

This platform serves to compare the performance of the new OptiMOS™ 6 technology to its predecessor OptiMOS™ 3 using best-in-class single switches in a standard TO-263 package. This test platform represents a three-phase motor drive inverter and consists of three modules; gate drive board, power board and DC capacitor bank.

The power PCB employs a single layer insulated metal substrate (IMS) board with aluminum core to ensure an excellent thermal performance. The switching frequency of the inverter is 10 kHz with sinusoidal SVM modulation. The platform delivers a single phase current of 33 Arms at a DC bus voltage of 144 V. The dead time between high and low side switches is set to 600 ns. Fig. 4 shows the test platform.

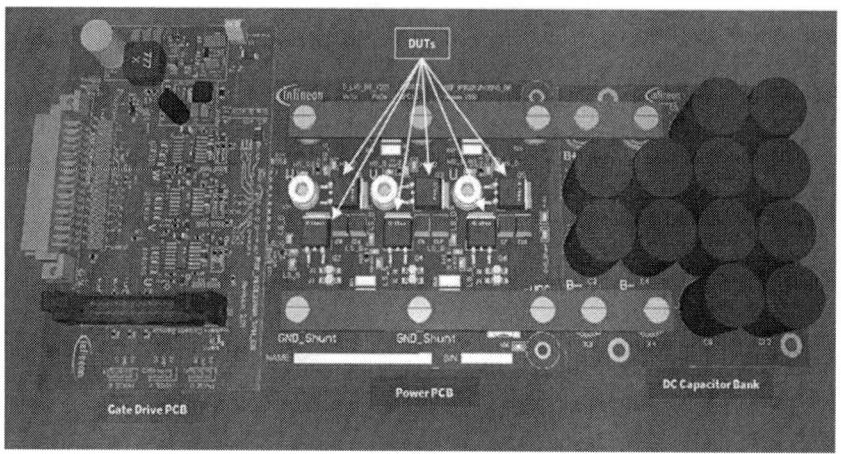

Fig. 4 3 phase motor-drives inverter test platform

3.2 Device performance comparison

Fig. 5 compares the switching behavior of devices from the previous and new generations of technology . The new OptiMOS™ 6 waveforms reveal a cleaner and more linear turn-on and turn-off, translating into lower switching losses. The total power losses are reduced by 45 %, with 39 % less switching losses and 49 % less conduction losses. Consequently the new devices run much cooler with a maximum temperature of 63.6 °C, compared to 95.8 °C in the case of the previous generation as indicated by the thermal images shown in Fig. 6. This enables a significant increase of the output power. For the same device temperature, the current per phase can be increased by an impressive 38 %.

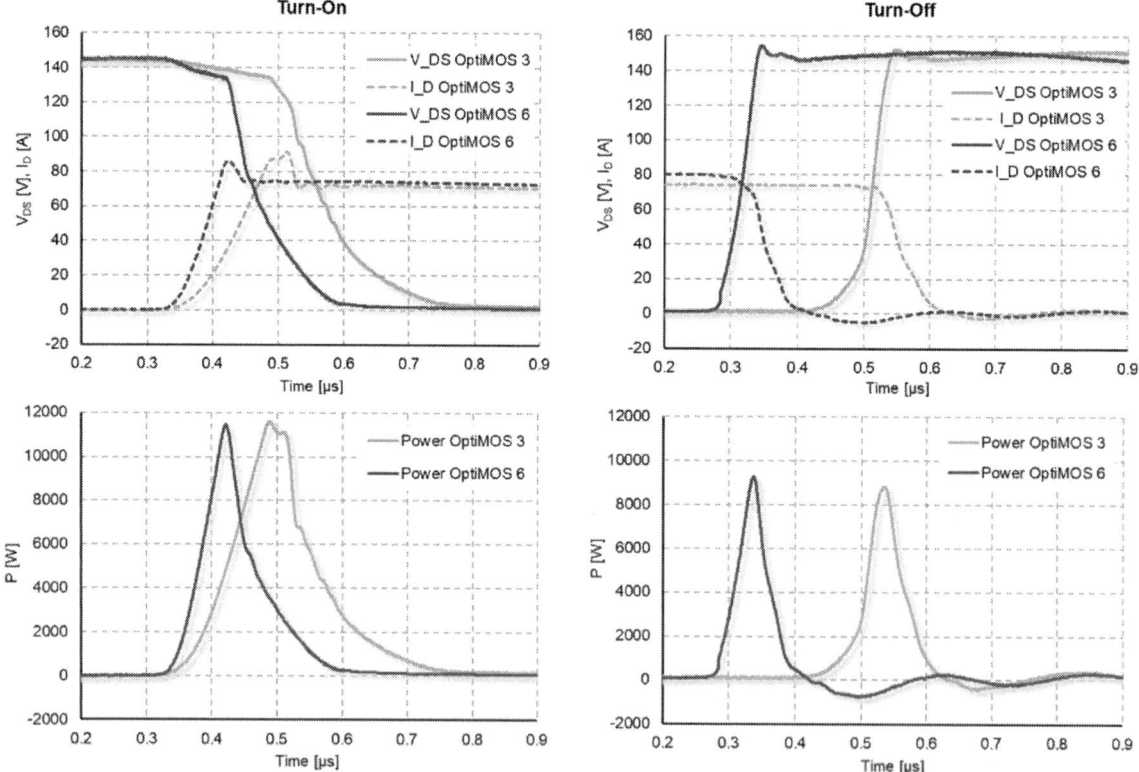

Fig. 5 Comparison of switching waveforms (top) and instantaneous power (bottom) for the new OptiMOS™ 6 and the predecessor OptiMOS™ 3 technology (I_D = 75 A)

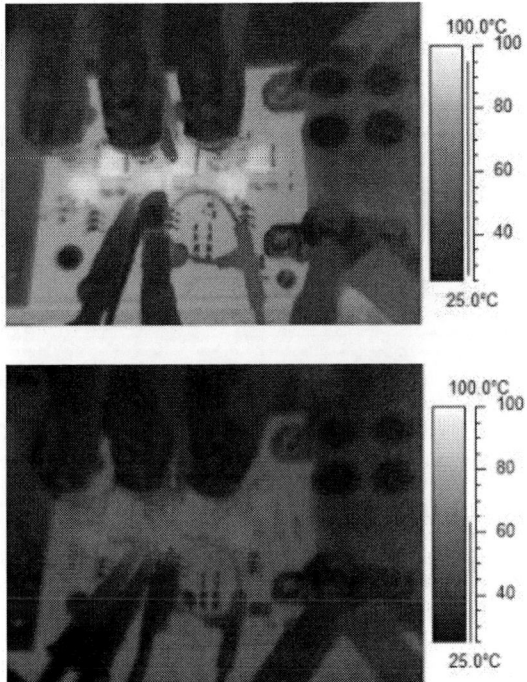

Fig. 6 Comparison of device temperature at identical power output between OptiMOS™ 3 (top) with 95.8°C and the new OptiMOS™ 6 (bottom) with 63.6°C

4 Device paralleling in a modified commercial 3-phase inverter

4.1 Introduction of the test platform

This application compares the performance of the latest 200 V technology with that of its predecessor under hard switching conditions in a motor drives application. The modified commercially available inverter employs a common B6 topology as depicted in Fig. 7, with a nominal input voltage of 144 V, an average current output of 135 Arms and a 1-minute phase RMS output current of 500 Arms. The inverter is sized to drive a 65 kW AC induction motor. In total the power board contains 96 MOSFETs, with 16 devices paralleled in each leg. The use of an insulated metal substrate power base provides superior heat transfer for increased reliability and performance. All tests are performed at a switching frequency of 10 kHz, with a dead time of approximately 1 µs. Fig. 8 illustrates the general functional block diagram of the test environment.

The inverter uses devices in standard TO-263-3 packages, making it relatively easy to swap between different generations. The investigation presented in this work focusses on the comparison of the new OptiMOS™ 6 200 V devices with the preceeding technology. The comparison uses best-in-class devices, where the new OptiMOS™ 6 devices have an on-resistance of 6.8 mΩ whist the previous OptiMOS™ 3 devices have an on-resistance of 11.7 mΩ.

In the presented measurements, the motor was running in a load condition with a phase current of 160 Arms. To enable the loss calculations, the measured values, among others, include low-side MOSFET current, high- and low-side drain-to-source and gate-to-source voltages and the phase current. All measurements extended over one complete electrical period of the motor.

4.2 Test results in the motor drives inverter

The first investigation determined the mean losses per MOSFET. The results are shown in Fig. 9. The comparison includes the overall mean losses per MOSFET as well as the separate conduction, turn-on and turn-off losses. The overall loss reduction accounts for a remarkable 36%. It is also worth mentioning that the new OptiMOS™ 6 200 V devices achieve a reduction in all of the loss contributors.

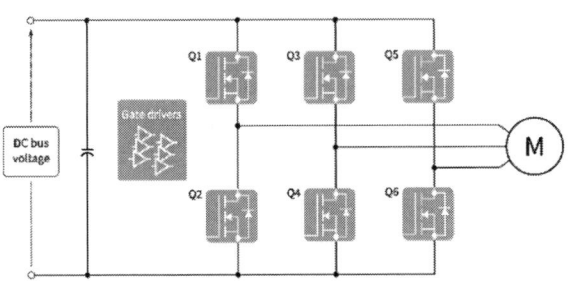

Fig. 7 Basic schematic of the B6 inverter

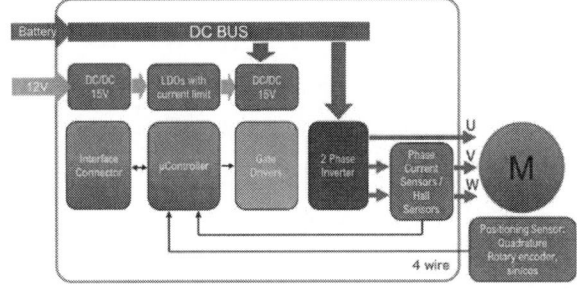

Fig. 8 General functional block diagram

Beside a reduction of the losses, it is important that the devices provide a clean switching behavior. Switching waveforms are measured for a single MOSFET. Fig. 10 gives an overview of the investigated device parameters, and at which positions in the circuit these values are measured. All gate-to-source and drain-to-source voltages are calculated from two separate measurements, taken from the respective electrode to ground. The current through the MOSFET is measured by a Rogowski coil at the source of the low-side device.

Fig. 11 shows the switching waveforms of the new generation of devices when the high-side switch is turned-on. Fig. 12 depicts the transients for the turn-off of the high-side switch, with the low-side MOSFET operating in synchronous rectification mode. In both cases, the use of the OptiMOS™ 6 devices result in clean waveforms. There is no visible ringing, and the slew rates are rather linear which is advantageous for the EMI behavior.

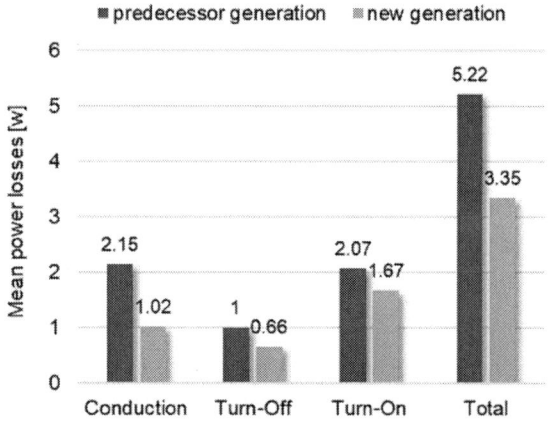

Fig. 9 Comparison of mean power losses per MOSFET

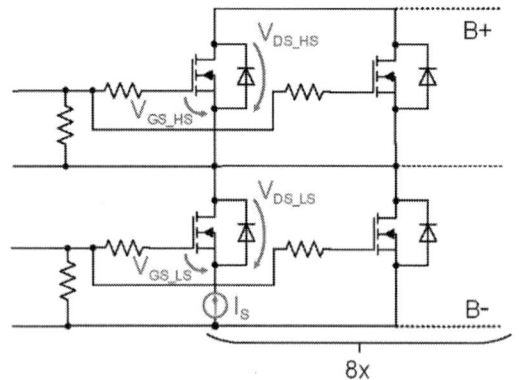

Fig. 10 Indication of test points for the waveform measurements

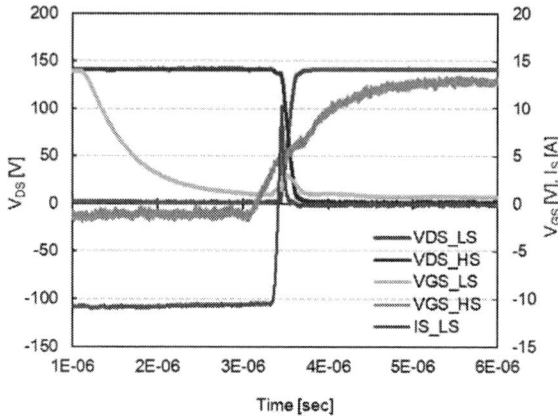

Fig. 11 New device switching waveforms at turn-on of the high-side switch

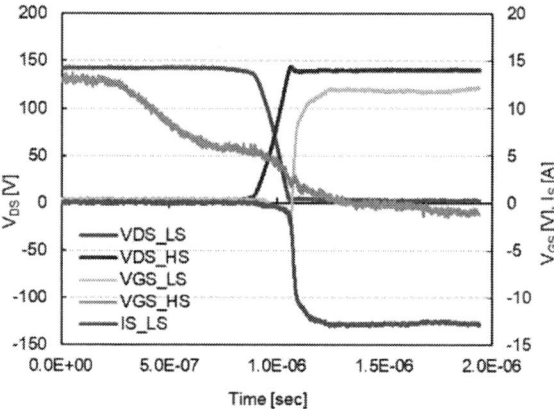

Fig. 12 New device switching waveforms at turn-off of the high-side switch

This implies that the significantly improved device performance does not degrade the EMI behavior. Indeed, this is confirmed by a comparison of the radiated emission of the two technology generations as shown in Fig. 13 and Fig. 14. The radiated emission measurements were done in accordance with the applicable standard EN 12895.

5 Conclusion

This work introduces our latest 200V power MOSFET technology that delivers improvements in all important device parameters and combines the benefits of low on-state resistance with a superior switching performance.

The remarkable progress in the overall device performance is enabled by substantial improvements at the device technology level. This has culminated in a unique device structure, which is the first to employ three-dimensional charge

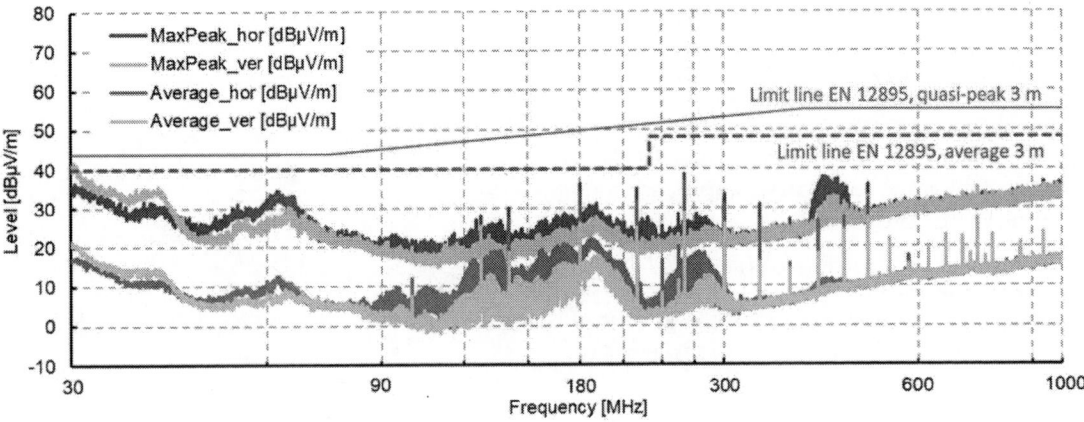

Fig. 13 Measurement of radiated emission in the application using the predecessor device generation

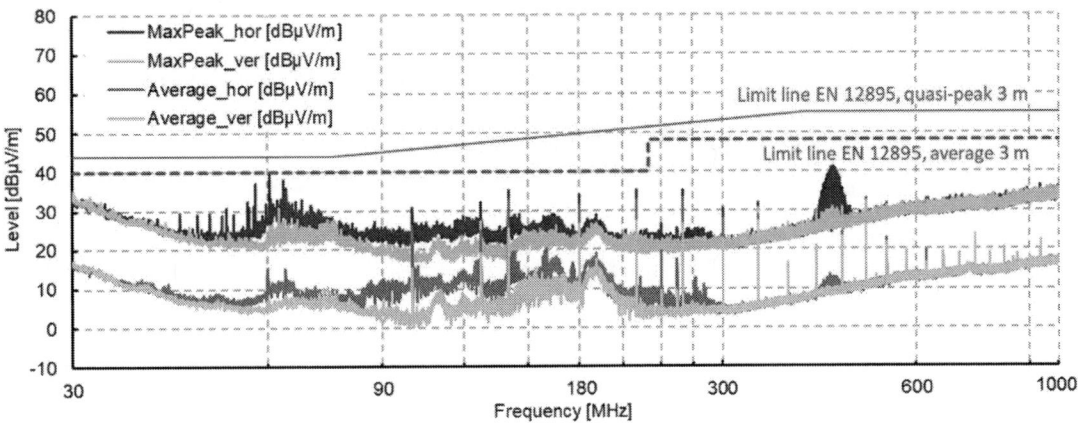

Fig. 14 Measurement of radiated emission in the application using the new device generation

compensation combined with a gate grid. The new design provides a so-far unmatched homogeneity of the gate and field plate resistance across the chip. The reduction achieved in the on-resistance, together with a low output charge and the improved switching homogeneity across the device area, enhance the system efficiency in the tested applications across all load conditions. The new device structure is also beneficial for the internal body diode of the MOSFET. Because the silicon area conducting current is increased, the body diode current density is decreased, which for the same current level means a decreased reverse recovery charge.

Motor drive applications benefit immensely from the new technology, as both conduction and switching losses reduce. The devices can easily be massively paralleled achieving clean switching waveforms. The good switching properties are also confirmed by radiated emission measurements which stay well within the required limits.

The significantly improved device performance even allows a reduction of up to half the number of paralleled devices required (depending on the application), or alternatively the use of smaller footprints, without having a negative impact on the temperature of the devices. This not only provides an advantage in terms of bill-of-materials (BOM) costs, but also the opportunity to reduce the area on the PCB.

This opens the door for a further optimization at the system design level, which is expected to further boost efficiency, reduce the converter or inverter size and increase the power density.

6 Acknowledgements

We thank Jannik Gade, Volodymyr Yakobniuk and Nikola Ivic for the measurement support on the motor drives inverter.

We also want to thank Adrian Finney for carefully editing this article.

7 References

[1] A. Schlögl, F. Hirler, J. Ropohl, U. Hiller, M. Rösch, N. Soufi-Amlashi and R. Siemieniec, "A new robust power MOSFET family in the voltage range 80 V – 150 V with superior low RDSon, excellent switching properties and improved body diode", Proc. EPE, Dresden, Germany, 2005

[2] R. Siemieniec, M. Hutzler, C. Braz, T. Naeve, E. Pree, H. Hofer, I. Neumann, D. Laforet, "A new power MOSFET technology achieves a further milestone in efficiency", Proc. EPE, Hannover, Germany, 2022

[3] I. Pawel, R. Siemieniec, and M. Rösch, "Multi-Cell Effects during Unclamped Inductive Switching of Power MOSFETs", Proc. MIEL, Niš, Serbia, 2008

[4] I. Pawel, R. Siemieniec, and M. Born, "Theoretical Evaluation of Maximum Doping Concentration, Breakdown Voltage and On-state Resistance of Field-Plate Compensated Devices", Proc. ISPS, Prague, Czech Republic, 2008

[5] Infineon Technologies AG, "OptiMOS™ 6 Power MOSFET 200V", Datasheet, 2024

[6] Infineon Technologies AG, "OptiMOS™ Fast Diode 200V IPB117N20NFD", Datasheet, 2014

[7] R. Siemieniec, O. Blank, M. Hutzler, L.J. Yip and J. Sanchez, „Robustness of MOSFET devices under hard commutation of the body diode", Proc. EPE, Lille, France, 2013

PCIM Europe 2024, 11– 13 June 2024, Nuremberg   DOI: 10.30420/566262071

Moisture Robust Chip Design – Improved Edge-Terminations for High Lifetime under High Humid Conditions

Arnost Kopta[1], Michael Hanf[3], Yanrui Ju[2], Raffael Schnell[2], Nando Kaminski[3]

[1]SwissSEM Technologies AG, Switzerland (until January 2023)

[2]SwissSEM Technologies AG, Switzerland

[3]University of Bremen, Germany

Corresponding author: Michael Hanf michael.hanf@uni-bremen.de
Speaker: Michael Hanf michael.hanf@uni-bremen.de

Abstract

In more and more applications and operation sites, power semiconductor modules are facing harsh environmental conditions. Especially moisture induced degradation is limiting the service life of Silicon-IGBT (Si-IGBT) modules and is the topic of a number of research activities. As moisture mainly stresses areas of high electrical fields, the focus is on the edge-termination. This work will contribute to the reliability enhancement under accelerated humidity testing by means of an improved edge-termination. Different designs are presented, which significantly increase the lifetime under HV-H³TRB conditions.

1 Introduction

Si-IGBT modules are not hermetically sealed and moisture can ingress through the plastic package and the Silicone-gel (Si-gel) potting. With a low time-constant, moisture will reach the chip interface [1] and can lead to corrosion of the metallic structures. A number of investigations [2, 3, 4, 5, 6, 7] had succeeded to quantify humidity induced degradation in power semiconductor devices, which led to a significant improvement in the overall reliability [8].

A fundamental basis for the verification of humidity induced degradation was the improvement in accelerated testing [2]. A test procedure with an applied bias voltage close to the nominal voltage (V_{nom}) enabled fast and application relevant testing. From these tests, typical failure mechanisms like aluminium corrosion [2, 7], dendritic growth [2] and mobile ions movement [9] were identified. While dendrites in power semiconductor modules were assumed to be linked to the process quality and cleanliness, the aluminium corrosion at the edge-termination of the chips and the mobile ion accumulation were much more complex problems to solve. Especially early failures in the accelerated tests were caused by mobile ion

accumulating on top of the edge-termination [9]. A decrease in the blocking capability occurred after surface charges accumulated on the guard rings, leading to much higher electrical field peaks on the surface [9]. Furthermore, typical edge-terminations with guard rings are insulated by a layer of SiO_2 between the silicon surface and the passivation [6]. At this interface, an electrical field can generate carriers, which disturb the electrical field distribution [6]. To prevent an accumulation at this critical spot, a semi-insulating layer was introduced in [6]. This enables a conduction path for the carriers and together with an improved passivation [9], early failures in high relative humidity can be avoided. In longer terms, the reliability of silicon IGBTs is highly affected by aluminium corrosion [2, 7]. Due to the presence of water, a galvanic cell is formed at the edge-termination, leading to electro-chemical migration and the resulting damage at the corresponding metal parts [2, 7]. A straight forward approach was used in [5] with an additional inorganic oxide layer on top of the aluminium field plates. This method allowed the usage of an already standardised edge-termination with field plates by adding a layer on top. But the usage of aluminium field plates remains a risk in humid environments and should be avoided. A different approach was published by [4], combining [9] and [5] by adding a silicon-nitride (Si_3N_4) layer on top of the field plate and passivating it afterwards with polyimide. While silicon-nitride can also corrode in presence of water [11], the time constant is higher compared to aluminium and can increase the lifetime in harsh environments. Furthermore, the failure analysis in [4] revealed a delamination of the polyimide layer and degradation of the semi-insulating layer (SIPOS). An improvement of both materials was sufficient to pass the standard test criterion. A similar improvement was shown in [10], with a floating field ring edge-termination without field plates. In an HV-H³TRB test, a delamination of the polyimide occurred, due to Si_3N_4 corrosion and the resulting outgassing. The solution was another inorganic layer between the polyimide and the silicon-nitride to improve the adhesion. In summary, the edge-termination design is critical in terms of accelerated humidity testing and especially under high voltage, a number of failure mechanisms have been documented in this area. To overcome common issues the following design features were verified to be effective:

- A semi-insulating layer between silicon and insulator to overcome mobile ions
- An additional passivation layer between insulator and organic passivation for improved adhesion
- Avoiding aluminium field plates in regions of high electrical fields

Overall, these technologies come with a significant processing effort and increase the complexity of the chips. Furthermore, some of these improvements can lead to a change in the dynamic and static behaviour of the semiconductors. Therefore, a reduction of processing efforts as well as an optimised design will reduce the possible failure mechanism as well as it will increase the competitiveness of the chips.

2 Testing Procedure

Latest results, published in [8], suggest an overall high lifetime of modern IGBT-modules in the standard HV-H³TRB-test. But the detailed structure of the edge-termination designs, which can handle this kind of stress, remains unpublished. Therefore, a test campaign was carried out with a novel optimised design to verify the performance under highly humid climate.

The HV-H³TRB test is the widely accepted, accelerated test to verify the humidity robustness of power electronic devices. At 85 °C, 85 % relative humidity (rel. h.) and up to 80 % of the nominal voltage (V_{nom}) the test will require at least 1000 h of testing time without significant degradation in terms of leakage current and blocking capability. The degradation behaviour is documented e.g. in [2, 3] and can be evaluated by the leakage current monitoring or even better, by intermediate measurements at dry conditions. The units tested in this experiment were 1200V ED-Type modules (EconoDual footprint) equipped with both IGBTs and diodes. Flux free solder is used and no wet cleaning with a potentially harmful cleaning agent or redistribution of residues is needed. It is worth to mention that the diodes have the same edge-termination concept as the IGBTs, except for differences like the ring doping levels and consequently slightly adjusted ring spacings. As the IGBTs and diodes are connected in parallel, the diodes were also tested, underlining the general strength of this termination design philosophy. In the test, just the high-side switches, containing three IGBT and three diode chips, were connected to high voltage. All samples (Tab. 1) were tested up to 2000 h, with blocking curve measurements at room temperature performed before the test, at an intermediate step after 1000 h testing time and at the end of the test.

3 Edge-Termination Design

3.1 Design Targets and Principles

The junction termination of a commercial device has to fulfil several conflicting requirements, consisting of a mixture of technical and financial aspects. The different requirements can be weighted differently depending on the target application. Nevertheless, they have to be addressed in one way or another. Below the main

Table 1: Devices under test with the corresponding edge-termination design

Split	Quantity	Edge-Termination Design
Split 1	4	P+ rings with reference silicon rich Si_xN_4
Split 2	4	P+ rings with metal plugs and reference silicon rich Si_xN_4
Split 3	4	P+ rings with metal plugs and high conductive silicon rich Si_xN_4
Split 4	4	P+ rings with metal plugs and stoichiometric Si_3N_4 passivation

Figure 1: ED-type 1200V/750A half-bridge IGBT module which was used for this test campaign

design goals and the corresponding commercial aspects are summarised:

1. The termination needs to ensure the specified blocking voltage requirements with margin. This is primarily to guarantee a good yield all the way from the first check on wafer level until the final test prior to customer delivery. The blocking voltage of a device is a combination of the capability of the junction termination and the design of the N-Base, buffer and the anode. A poor termination performance can partly be compensated by a thicker N-Base, which then however leads to higher electrical losses.

2. It must be robust against production process variation. This is to ensure the desired blocking voltage is reached with a high yield. The termination must be robust against variations in the starting material as well as photo lithography tolerances and process induced charges.

3. Be as compact as possible to allow for the largest possible active area. This in turn gives the lowest possible losses or allows for smaller and thus, less costly devices. This is especially important for small IGBT and diode chips.

4. The termination should require as few manufacturing steps as possible to ensure a low cost of the final product and overall good manufacturing efficiency.

5. Low leakage currents: The device needs to be capable of switching at $T_j = 175°C$ which puts limits on the maximum allowed leakage currents to prevent any thermal runaway effects. In IGBTs, the leakage current is mainly determined by the buffer/anode design but also the passivation layer (here Si_xN_4) can contribute if not composed adequately.

6. Be robust against environmental impacts during the packaging process and especially in the field – The main topic of this paper.

The size of the termination is involved in several of the points above making it an important trade-off and optimisation parameter. It is less important for large devices, but in practice nevertheless significant as a new termination will likely be used in a variety of different chip sizes over time. Given the very competitive price and performance levels of today's IGBTs, even small improvements can make a difference.

A too narrow termination design might limit the blocking voltage of the device and in addition be more sensitive to process variations. This can impact the production yield but can also lead to reliability issues. This can be fixed by using a thicker/higher resistivity N-base, which however increases the electrical losses or deteriorates the switching performance.

The total width of the presented IGBT termination measured starting from the transition zone 0 to the beginning of the sawing street is about 360 μm. This value is very similar to the termination size of the latest commercially available generation of IGBTs from two major competitors. These competitors have their homebases on two different continents and are assumed to be a good reference in this field. One of the competitors utilises a design based on the concept of Variation of Lateral Doping (VLD), whereas the other one uses a special adaption of the floating field ring design.

The here investigated termination structure profits from a highly integrated process flow, requiring the fewest possible amount of lithography steps.

3.2 Development Approach

The termination was developed using a combination of device simulations and experimental trials, with a clear focus on the latter. Simulations were deployed to generate the starting distribution of the floating field rings. A limited number of simulations with possible perturbations like oxide charges and a dielectric layer with a high permittivity to emulate the influence of a humidity-soaked polyimide layer were also done in order to check the general sensitivity of the basic structure. Beyond this point, it is however difficult to efficiently use simulations mainly due to the simple fact that the Si_xN_4 is hard to simulate due to the absence of adequate models. The Si_xN_4 layer has a very high, yet not infinite resistivity, which influences the potential distribution of the P^+ rings it is connected to. This is the main feature of this termination design, but makes it difficult to simulate precisely. While environmental impacts are anyway not easy to simulate, the easiest and fastest way is to switch to experiments as soon as possible in the development process. In an early stage, the ring distribution was optimised based on a simple variation of the ring spacings starting from the simulated ideal case. Devices with both, smaller and bigger gaps were produced and the optimum design based on the sensitivity of the blocking voltage and initial H^3TRB results were selected and later used as a base design in the experiments presented in this paper. In this way it is believed that the optimum design with a low sensitivity towards external influence parameters as well as possible weaknesses can be achieved.

One important aspect of the design considerations is the location where the avalanching starts when a voltage is applied. In order to reach the highest possible breakdown voltage for a given number of rings, all of them should ideally start avalanching at the same time. From a reliability point of view, it is however preferable that the avalanche breakdown happens mainly at the main junction. In tendency, the peak electric fields here are further away from the silicon surface and the passivation layers are therefore less stressed than when the highest fields occur towards the outer rings. To control the location of the breakdown point, it is important that the simulation models reproduce this at least with some minimum accuracy having the influence of the Si_xN_4 layer in mind. During the experimental phase, the breakdown simulation results were therefore checked with measurements using Emission Microscopy (EMMI), which is a good method to visualise avalanche breakdowns in power devices. Figure 2 shows the result of such an investigation.

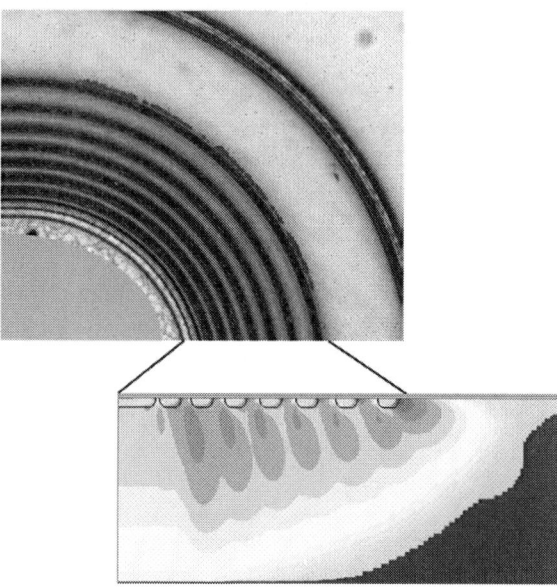

Figure 2: Top: Avalanche breakdown visualisation using Emission Microscopy (EMMI). Bottom: Simulated avalanche breakdown (impact ionisation).

The top part displays the avalanche location (blue) of an intermediate design, which in this case happens at the last termination ring. The bottom part of the figure shows the corresponding simulation results with the maximum impact ionisation rate also located at the last ring. As mentioned above, this is not the ideal location for the main breakdown point in terms of reliability. The final devices were adjusted to have this point at the main junction.

3.3 IGBT Termination

Figure 3 shows a schematic view of the investigated IGBT edge-termination. The corresponding top view can be seen in Fig 8 (left column). The termination consists of several floating P^+-rings, in the vertical direction protected by a material stack consisting of SiO_2 (both thermally grown and deposited), a passivation layer made of silicon rich nitride (Si_xN_4) and finally a thick layer of Polyimide. In the horizontal direction, the edge-termination consists of the following functional zones: **AA** is the last trench cell (stripe) of the active area and is followed by zone **0**, a transition area where the emitter metal ends and the passivation stack begins. Zone 0 has a P^+ diffusion, which is well connected to the emitter potential to prevent any high fields in this sensitive part of the device. This is followed by **R1**, the first termination ring. **R2...n** denotes the remaining rings. The rings are spaced by increasing distances from each other, which have been optimised to reduce the electric field peaks

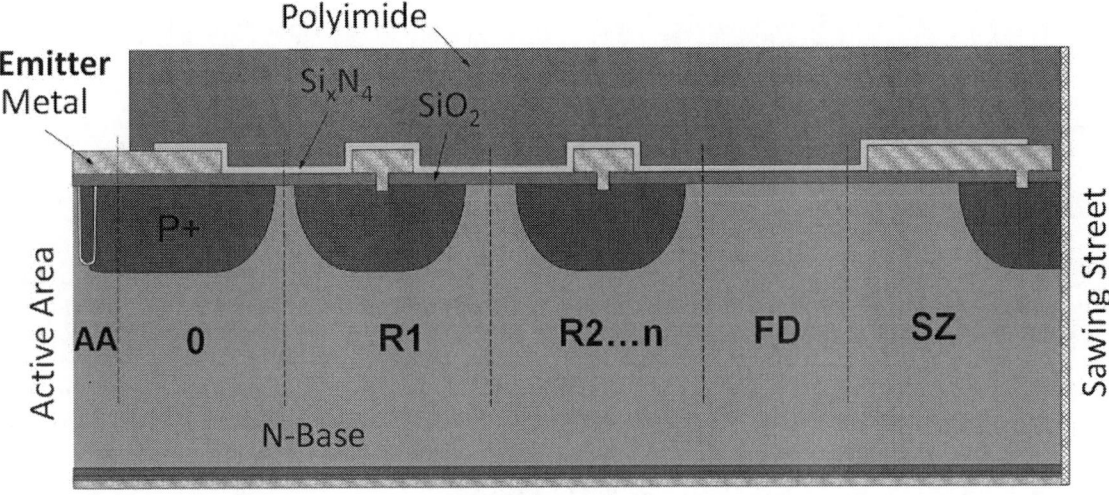

Figure 3: IGBT edge-termination with the corresponding internal structures as well as the different passivation layers

and distribute the potential evenly over all the rings. The field rings do not have any field-plates as these are considered to be problematic for reliability under humid conditions. This is mainly due to the high electric fields they inherently induce in the passivation materials leading to the failure modes described in the introduction section above. The reference design utilises small metal-plugs in the rings to enhance the electrical contact between the Si_xN_4 layer and the P^+ diffusion of the ring. After the last ring, the Field Decay **FD** zone follows. The width of this zone was designed to allow the electrics field to decrease to a low value preventing it from reaching the sawing street. Finally, there is a Safety Zone **SZ** just next to the sawing street to take care of any remaining fields, which might have found their way through the field decay zone FD. This can happen due to surface charges introduced by environmental impacts like moisture and ion migration during the lifetime of the device. The safety zone consists of a P^+ diffusion, which has the task to provide a well-defined potential (the collector potential) to a metallic field plate, the only one in the design. This field plate has the task to bend any remaining electric fields away from the sawing street. Thanks to the FD Zone, there will only be small residual potential differences between the edge of the field plate and the underlaying N-base. As a consequence, the electric field at the edge of the field-plate will only be small and insignificant compared to electric fields encountered in termination designs using field-plates in their main termination rings. The field plate in the Safety Zone is therefore considered as uncritical in terms of reliability and is assumed not to play any

important role in the investigations done in this work.

The Si_xN_4 layer has an important role as it electrically connects the emitter metal with all the rings and the sawing street. It has been tuned to have a very high, but still finite resistivity, which is achieved by having a higher silicon content than in stoichiometric Si_3N_4. The hypothesis is that any undesired disturbances from mobile ions outside of the silicon, should be balanced to some extent by its ability to conduct and in this way to provide corresponding mirror charges. The main goal of the experiments in this work was to check this hypothesis using different passivation layer compositions and hence resistivities as well as two different approaches of how to contact the field rings to this layer.

In the middle of each ring, there is a metal plug, which has the main role to establish a good connection between the P^+ diffusion and the Si_xN_4 passivation (Fig. 4 left). The metal plug does not extend over the P^+ ring periphery (even at full depletion during blocking) and has therefore no other electrical function. Based on simulations, there were no electric field peaks at the plug corners, as it would be the case for structure elements like field plates. This is an important design criterion to ensure a high reliability, especially in HV-H³TRB tests. To investigate the actual role of the metal plug, a second version without it was manufactured. In this case, (Fig. 4 right), the passivation contacts the P^+ ring directly, most probably with a higher contact resistance.

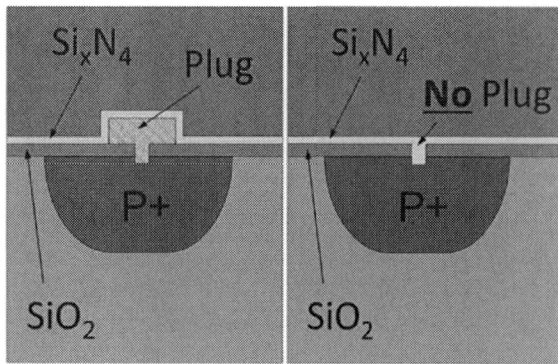

Figure 4: Versions of the termination design with metal plug (left picture) and floating ring (right picture)

3.4 Diode Termination

The diode termination was designed based on the same principals as the IGBT. The biggest difference was the doping concentration of the floating rings and the design of the transition zone 0. The schematic cross-section of the diode termination is similar to the IGBT in Fig. 3 and the corresponding top view in Fig 7 (right column).

The diode has been designed using a low anode emitter-efficiency to control its reverse recovery characteristics. To allow for highly streamlined processing with the fewest possible lithography steps of the emitter-controlled diode, the guard rings have a significantly lower doping concentration compared to those of the IGBT. Careful adjustment of the ring width and spacings is required to match the blocking performance. The diode termination used in the experiment in this work had the reference design with metal plugs contacting each field ring.

In diodes, the transition zone 0 needs to be designed with much more care compared to the case of IGBTs. The reason is that this part determines the reverse recovery robustness of the entire diode. The zone 0 in the investigated diodes was made up of several rings which partly were cross-diffused into each other. This structure aims at separating high current regions from any parts with high electric fields during the diode reverse recovery. Other than that, zone 0 has only a limited impact on the blocking characteristics and thus, on the reliability of the diode termination.

4 Test results

4.1 Results of HV-H³TRB Testing

As described before, standard test parameters were applied and all samples from Tab. 1 were tested for 2000 h. Until the end of the test time, no device failed with an overcurrent event and therefore, all devices made it to the final measurement. But there are significant differences in the degradation behaviour over time for the respective splits. Figure 5 shows an example of the leakage current, displaying one sample of each split with a typical current evolution. While the devices of split 1 and split 2 show no signs of degradation, the leakage currents of split 3 and 4 develop significant oscillations and increased current levels. In [2], this phase of high oscillations occurred after 2000 h (at 65 % V_{nom}) and was the first indicator of a degraded blocking capability. With this in mind, the blocking voltage curves in Fig. 6 are highly consistent with the monitoring. While split 1 and 2 remain at their initial curve, the other two splits show a different behaviour. Split 3 is more or less reproducing the results, known from Aluminium corrosion, with a constant initial leakage level and a reduced avalanche inception voltage [2]. Split 4 on the other hand, reaches the initial blocking voltage but the leakage current level is significantly higher. It is even increasing over time to a final level of more than 10 times the initial value. In terms of a standard failure criterion, split 3 and split 4 do not even pass the 1000 h test. The situation is completely different for both splits with the standard Silicon-rich Nitride (1 and 2) and both splits passed the test without significant signs of degradation.

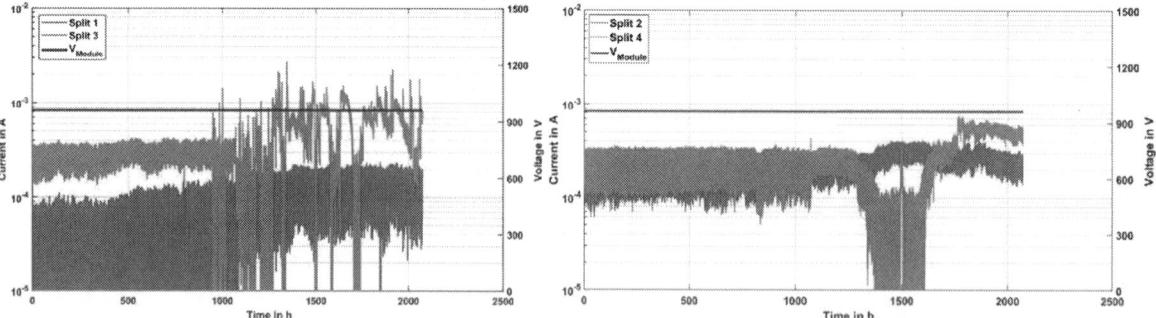

Figure 5: Leakage current evolution over 2000 h of testing at 85°C, 85 % rel. h. and 80 % V_{nom} for one sample of each split. On the left, split 1 and 3, on the right, split 2 and 4 are depicted.

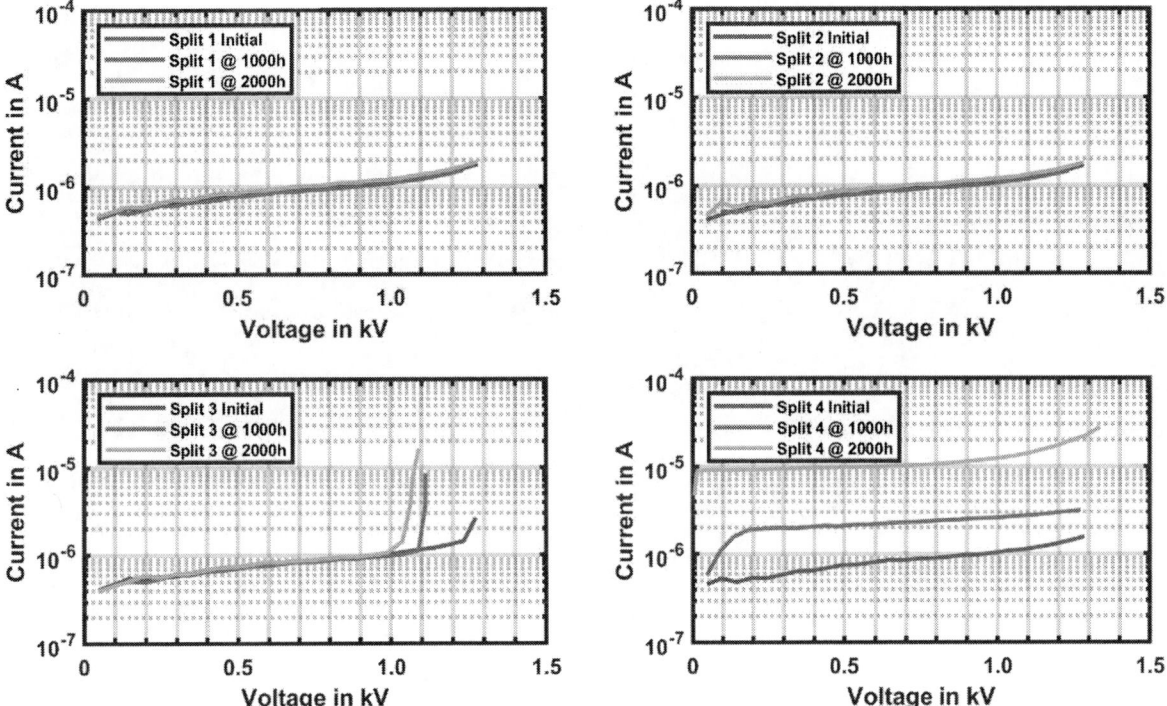

Figure 6: Blocking voltage measurement after 1000 h and 2000 h in comparison to the initial value at room temperature. The leakage current is normalised on the initial temperature.

In conclusion:

- Devices with the standard Si_xN_4 passed
- Devices with stochiometric and high conductive passivations are not able to pass
- No impact of the contact type to the passivation (metal plug) can be assigned

Just looking at the bare test results, Silicon-rich Nitride shows the overall best performance.

4.2 Further Analysis of Failed Devices

An important information in terms of the correlation with the applications and field returns, would be a sufficient failure analysis. Devices, which suffered from Aluminium corrosion [2, 5, 7] were significantly easier to analyse due to the situation, that the critical areas were mostly covered by just Polyimide. Therefore, after the removal of the Si-gel and the Polyimide, usually the edge-termination corrosion was visible already. But using a Silicon-Nitride-passivation does not allow an easy optical analysis. A removal is not possible and even after removing the Polyimide, a picture like Fig. 7 is revealed, with no signs of corrosion. In [4, 7, 10], the failure analysis was supported by a focused ion beam (FIB) and cuttings into interesting areas. Both tools are not available for this analysis and therefore, a different approach was chosen. In [12] an experiment with moisture saturated modules was utilised to reveal the impact of a certain degradation mechanism on package level, different from this study. It is expected, that the Silicon-Nitride passivation is oxidised and delaminated as found in [4, 10]. Therefore, moisture saturated devices should accumulate water in cavities, formed by the corrosion and delamination process. The setup is similar to [12] with a storage under 100 % rel. h. for at least 24 h at 50°C. Each device was then removed from the cabinet and measured at 50°C in a curve tracer to reveal the blocking curve under moist conditions. In Fig. 8, the results of one sample per split are shown with a clear result for the splits 1 and 2. In both cases, no negative effects under higher temperature and higher moisture content is visible. In contrast, this is not the case for both degraded splits 3 and 4 with a significant impact of moisture on the blocking curve. The avalanche inception voltage in Fig. 9 is shifted to voltages below 1 kV and therefore, the loss of blocking capability is even worse. Furthermore, the effect is reversible and after an additional drying process at 50°C, the blocking capability is shifted back to higher voltages. In conclusion, both degraded splits 3 and 4 show a significant moisture influence and it can be assumed that delamination and/or corrosion in the edge-termination region is present. A control experiment with the corresponding low-side switch

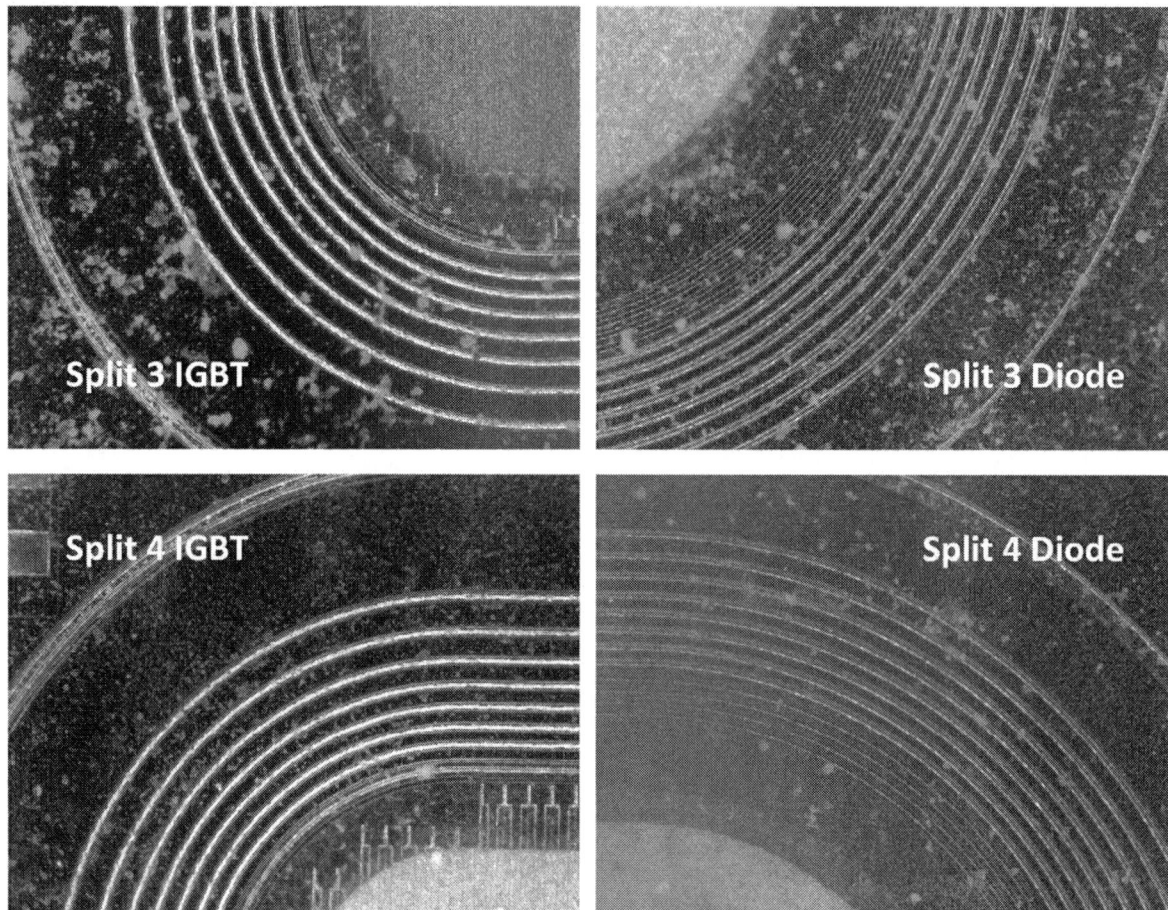

Figure 7: Edge-termination regions of split 3 and 4 with a tested IGBT (left column) and the parallel diodes (right column). The Polyimide was partly removed and the crystalline particles are residues from the sample preparation process

(same test time but no bias voltage) reproduced the initial blocking capability with no signs of degradation. This measurement is able to detect chip related degradation without complex analysis methods. But for process or design improvements a root cause needs to be identified and consequently, this method is just suitable for a "quick & dirty" comparison of different test groups.

4.3 Discussion

The four different edge-termination designs and passivation materials showed a significant impact on the reliability in the HV-H³TRB test. Figures 6 and 8 indicate a high robustness of the approach with a Silicon-rich Nitride (Si_xN_4), which is outperforming the other materials. A modification with higher conductivity (split 3) reproduced a degradation behaviour as it is known from the past [2, 7] with a decreasing blocking capability at dry conditions. On the other hand, the stoichiometric Si_3N_4-passivation shows a significant increase of the leakage current, while preserving its initial blocking capability. In both

cases, the failure criteria according to AQG324 are reached or at least approached. The underlying degradation mechanism remains unknown, but the impact on the blocking capability is highly influenced by moisture and therefore, corrosion and/or delamination of the passivation layer(s) can be assumed.

5 Conclusion

In this investigation, four different edge-termination designs were tested under HV-H³TRB conditions for 2000 h. The designs differ in the passivation layers as well as in the connection of these layers to the P^+ rings. Both designs with the standard Silicon-rich Nitride (Si_xN_4, splits 1 and 2) revealed an outstanding performance with no signs of degradation within in the test time. Furthermore, devices of these splits were also out-performing the other two splits 3 and 4 under high humid conditions in the blocking curve measurements. Split 1 utilised a semi-floating field ring structure with the same performance as split 2

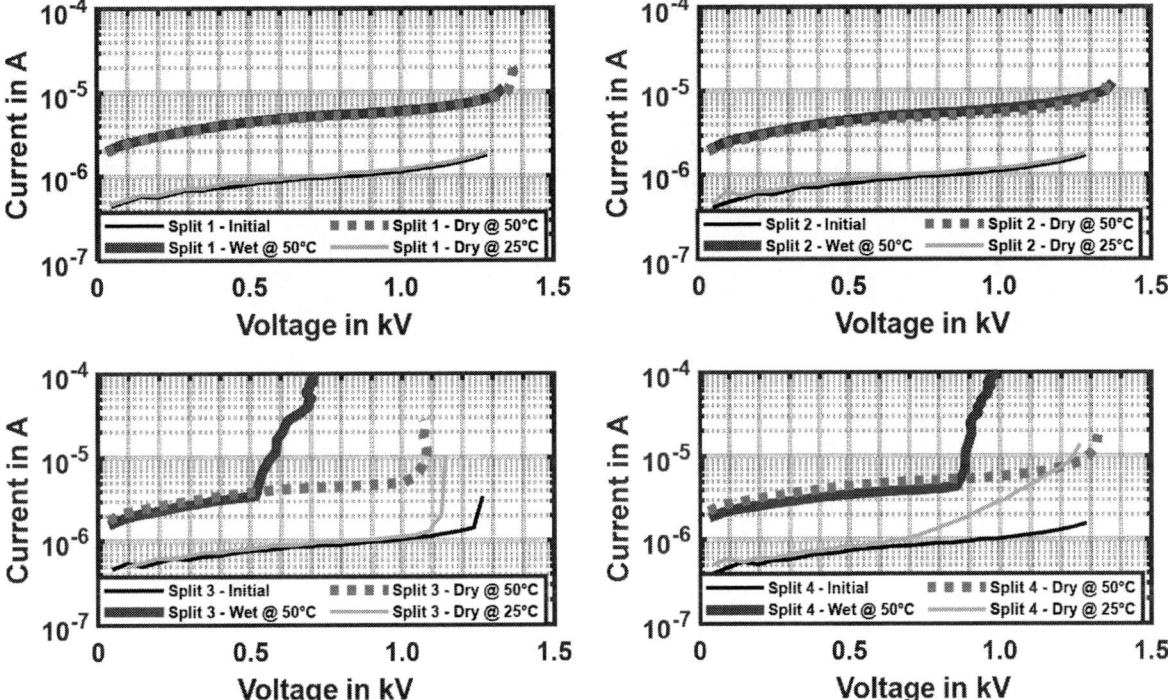

Figure 8: Blocking curves of the corresponding split under moist and dry conditions at 50°C in comparison to the initial curve as well as the final curve after 2000 h of testing at dry conditions and 25°C.

that has a metal plug connected passivation. Therefore, the connection type of the P+ region is not significantly changing the reliability in the presented time frame. Splits 3 and 4 showed an inferior performance in the HV-H³TRB test, with significant changes in the blocking behaviour as well as noticeable oscillations in the leakage monitoring. Under high humid conditions, the blocking capability of those splits was substantially reduced and the effect was reversible through a drying procedure. This result matches the assumption of delamination and corrosion at the passivation layer, which is reducing the performance and depends on the material composition as well as the processing.

In conclusion, the presented edge-termination design with an optimised the passivation stack on top of the field rings manufactured with a minimum number of process steps proves to be a very competitive solution. Combined with the outstanding performance in the HV-H³TRB tests, this newly developed termination is one of the leading-edge designs in this voltage class.

References

[1] K. Hatori, K. Nakamura, W. Noboru, N. Soltau, E. Wiesner, "Humidity Absorption Behavior of Silicone Gel in HVIGBT Modules", 23rd European Conference on Power Electronics and Applications (EPE'21 ECCE Europe)

[2] C. Zorn, N. Kaminski, "Temperature-humidity-bias testing on insulated-gate bipolartransistor modules – failure modes and acceleration due to high voltage", IET Power Electronics: Special Issue on International Seminar on Power Semiconductors (ISPS'14), 2014

[3] J.-H. Peters, M. Hanf, S. Clausner, C. Zorn, N. Kaminski, "Improved HV-H³TRB robustness of a 1700V IGBT Chip set in standard power modules", Microelectronics Reliability Vol. 126, 2021

[4] C. Papadopoulos, C. Corvasce, A. Kopta, D. Schneider, G. Pâques, M. Rahimo, "The influence of humidity on the high voltage blocking reliability of power IGBT modules and means of protection", Microelectronics Reliability Vol. 88-90, 2018

[5] S. Kremp, O. Schilling, "Humidity robustness for high voltage power modules: limiting mechanisms and improvement of lifetime", Microelectronics Reliability Vol. 88-90, 2018

[6] S. Honda T. Harada, A. Nishii, Z. Chen, K. Shimizu, "High Voltage Device Edge Termination for Wide Temperature Range plus Humidity with Surface Charge Control (SCC) Technology", 28th International Symposium on Power Semiconductor Devices and ICs (ISPSD'16), 2016

[7] J. Leppänen, J. Ingman, J.-H. Peters, M. Hanf, R. Ross, G. Koopmans, J. Jormanainen, A. Forsström, G. Ross, N. Kaminski, V. Vuorinen, "Aluminium corrosion in power semiconductor devices", Microelectronics Reliability vol. 137, 2022

[8] J.-H. Peters, M. Hanf, S. Clausner, N. Kaminski, „Step-Change in HV-H³TRB Performance of Latest Silicon IGBTs and Advanced Humidity Testing Methods", 13th International Conference on Integrated Power Electronic Systems (CIPS'24), 2024

[9] N. Tanaka, K. Ota, S. Iura, Y. Kusakabe, K. Nakamura, E. Wiesner, E. Thal, "Robust HVIGBT module design against high humidity", International Exhibition and Conference for Power Electronics, Intelligent Motion, Renewable Energy and Energy Management (PCIM'15 Europe), 2015

[10] C. Papadopoulos, B. Boksteen, G. Pâques, C. Corvasce, "Humidity Robustness of IGBT Guard Ring Termination", International Exhibition and Conference for Power Electronics, Intelligent Motion, Renewable Energy and Energy Management (PCIM'19 Europe), 2019

[11] J. W. Osenbach, "Water-Induced Corrosion of Materials Used for Semiconductor Passivation", j. Electrochem. Soc. Vol 140, No. 12, 1993

[12] M. Hanf, R. Schnell, S. Matthias, N. Kaminski, "Sulphur related Corrosion in Power Modules and its Impact on the Switching Performance", International Seminar on Power Semiconductors (ISPS), 2023

PCIM Europe 2024, 11– 13 June 2024, Nuremberg   DOI: 10.30420/566262072

Method for Measuring the Initial State of a Solder Joint Delamination in a 3D PCB Integration Assembly of SiC

Souhila Bouzerd[1*], Mickaël Petit[2], Céline Combettes[3], Vincent Bley[3], Laurent Dupont[1]

[1] Univ. Gustave Eiffel, Univ. Paris-Saclay, ENS Paris-Saclay, CNRS, SATIE, Versailles, France

[2] Le CNAM, Univ. Paris-Saclay, ENS Paris-Saclay, CNRS, SATIE, Paris, France, HESAM Université

[3] LAPLACE, Université de Toulouse, CNRS, INPT, UPS, Toulouse, France

*Corresponding author: Souhila BOUZERD souhila.bouzerd@univ-eiffel.fr

Speaker: Souhila BOUZERD souhila.bouzerd@univ-eiffel.fr

Abstract

This article presents a method for evaluating the initiation of delamination at a corner of a solder joint in a 3D power electronics assembly of SiC MOSFETs on a PCB (printed circuit board) substrate. This development is part of an effort to assess the robustness of technological choices for a new model of wide bandgap component assembly. These assemblies consist of a PCB substrate (layers of copper, epoxy, and glass fibers) and a soldered copper block corresponding to the heat-sink base-plate. However, conventional methods do not effectively meet the need for detecting the initiation of solder delamination for the assembly design. The method relies on potential measurements that exhibit greater sensitivity to detect the onset of solder delamination than conventional method. Finite element simulations are carried out to evaluate of the method's sensitivity and discriminating factors such as geometry and materials involved. Based on the numerical results, dedicated prototypes of the assembly are manufactured with control of delamination initiated at a corner of the solder joint. Confrontation of experimental results and numerical studies offers good perspective as a complementary method to detect the delamination initiation propagation of joining technologies.

1   Introduction

Power electronics is demanded to improve energy transfer efficiency while reducing system size. The reduction in converter size and efficiency has been notably enhanced by the use of wide bandgap semiconductor components. However, the packaging of these semiconductors requires particular attention. Thus, a reduction in the size of the switching cell is necessary to eliminate the negative impact of parasitic elements in the switching cell design (over voltages and oscillations) amplified by the high performance of wide bandgap semiconductors. Additionally, a 3D packaging with double-sided water cooling [1] is proposed to ensure better cooling, but this type of cooling makes the system heavier, more complex, more expensive, and overall, less efficient. Therefore, it seems useful to improve module cooling while keeping it lighter to meet the needs of electric mobility where weight is a major constraint [2]. Air cooling is proposed, but to maintain sufficiently high thermal performance, the packaging of the semiconductors needs to be adapted to this type of cooling. A 3D integration technology with a minimal number of thermal interfaces is proposed in the TAPIR technology (Compact and Modular Power Modules with Integrated Cooling) [3]. This technology involves integrating wide bandgap transistors into a suitable PCB substrate and in a heat exchange configuration using air coolers placed on both sides of the technological structure for double-sided cooling, as illustrated in Figure 1. The heat-sinks used in this technology are brazed on both sides of the PCB substrate. Each heat-sink is placed at the same electrical potential as the metallization connected to each side of the active part embedded in the PCB substrate. The use of air cooling requires a significant extent of thermal exchange surface due to the lower heat transfer coefficient compared to that obtained by water cooling. This condition implies significant dimensions for the heat exchanger, but these heat-sinks are brazed onto 70 µm copper layers placed on a PCB substrate (epoxy + glass fibers).

PCIM Europe 2024, 11– 13 June 2024, Nuremberg   DOI: 10.30420/566262072

a) Prototype technology

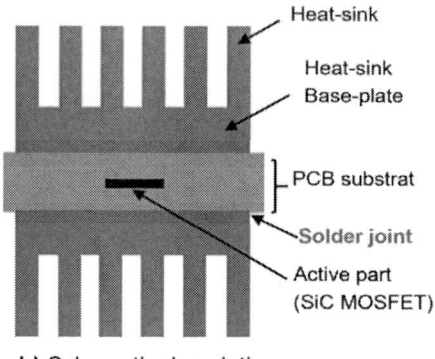

b) Schematic description

Figure 1: Presentation of the 3D integration model of TAPIR [3].

However, there are differences in the coefficients of thermal expansion (CTE) and in the behaviors of the materials used. Additionally, one of the common imperfections of brazing technology is the variation in thickness between the center and the corner of the solder joint. This implies that one of the corners of the solder joint bears more stress, leading to the initiation of delamination at one of the corners [4]. The main solder damage is a delamination resulting from the crack propagation in the solder joint induced by repetitive thermal cycles [5, 6]. An illustration of the delamination is depicted in Figure 2 with a comparison of the solder joint state after 100 (a) and 300 (b) thermal cycles. The identification of delamination is essential for recognizing variations in behavior among various technologies. Therefore, it is necessary to propose indicators that can be used to qualify, or even quantify, the initiation of delamination on a corner of the solder joint by a non-destructive method.

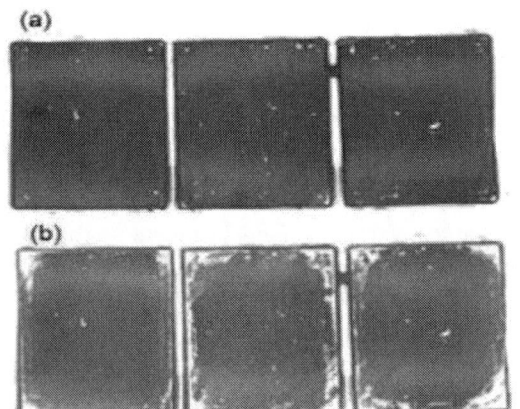

Figure 2 : Acoustic microscopy scan image of SAC solder joint after thermal cycling tests (a) 100 cycles (b) 300 cycles [6].

In the literature, exist several conventional methods for detecting delamination of power module solder joints. These include methods based on thermal performance measurements ,[7] electrical capacitance measurement [8], lock-in thermography [9], X-ray tomography [10], and acoustic microscopy [8].

However, these methods are not directly applicable in our perspectives of thermal cycling aging tests used to evaluate the robustness of the technological choices of the power electronic assembly. They are either complicated to implement or insufficiently sensitive to detect delamination initiation to operate a rigorous comparison of the technological proposals. Moreover, the area is inaccessible to a dense thermal flux for characterization by a thermal method typically used for damage assessment during accelerated aging campaigns.

This study presents a solution for detecting delamination initiation in a solder joint between two metallic blocs. Sensitivity to delamination on the order of one percent has been achieved using electrical measurements, enabling a detailed comparison of different technological choices. Firstly, this article provides a description of the numerical model associated with an experimental study on specifically developed test prototypes. Lastly, a discussion elaborates on the results to draw conclusions and provide future perspectives.

2 Numerical analysis

2.1 Numerical model

A finite element simulation was carried out using COMSOL Multiphysics software. The numerical study helps to identify the parameters which affect the performances of the proposed method used to detect the initialization of the solder delamination. The studied finite element model (FEM) is a 3D geometry under steady-state electrical study conditions, without considering thermal effects. The model studied is composed of three parts. These three parts are parallelepipeds corresponding to the top metallization of a PCB substrate (35 mm × 35 mm) using copper with a thickness of 70 µm and 105 µm, the heat-sink base-plate (17×17 mm) using nickeled copper with a thickness of 3 mm. The copper electrical resistivity is $1.7×10^{-8}$ Ω.m [11]. The third part is the solder joint (17×17 mm) using Sn-3.0Ag-0.5Cu (SAC305) with a thickness of 100 µm and electrical resistivity of $1.04×10^{-6}$ Ω.m [12] (Figure3). The materials properties presented are the theoretical values. Resistivity measurement of the PCB substrate metallization copper is done. The experimental value is $1.40×10^{-8}$ Ω.m. The numerical model is set up respecting the experimental value measured.

A steady state current injection of 100 A is imposed on the corner of the metallization of the PCB substrate on the same side as the corner A of the solder joint. The injection ground (Mi) is imposed on the heat-sink base-plate. Then electrical potential evaluations (Vm) are done on the PCB substrate metallization near the corner A of the solder joint while the measurements ground (Mm) is imposed on the heat-sink base-plate as illustrated in the figure 3. Since the dimensions of the PCB substrate metallization and the solder joint are reduced, this makes potential electrical measurements very sensitive to their position.

Simulations were performed for models with uniform thickness solder joint and without degradation. Additionally, simulations were performed for models with delaminated solder joint. The delamination is represented by the red cercle in the figure 3. It is imposed on the corner A of the solder joint with different rates. An appropriate mesh has been controlled for this study to refine the simulation results. In particular, an adaptive triangular mesh is established on the edges of the solder joint to refine simulation results in the delamination zone. Since the simulation values are influenced by the concentrations of current lines that induce the measured potentials. This influence, computation times and the error values of measurement are used to fix the limits of the controlled the mesh.

2.2 Delamination indicator

The potential electrical measurements are analyzed to provide an indicator of the delamination rate imposed on the weld joint corner. Also, to evaluate the sensibility of the method to the various rates of the delamination imposed on the solder joint corner. The delamination indicator is calculated using the following equation:

$$\Delta VA = \frac{VA_D - VA_{ND}}{VA_D} \times 100 \qquad (1)$$

Where, ΔVA, delamination indicator, VA_D, electrical potential measured on delaminated prototype, VA_{ND}, electrical potential measured on non-delaminated prototype.

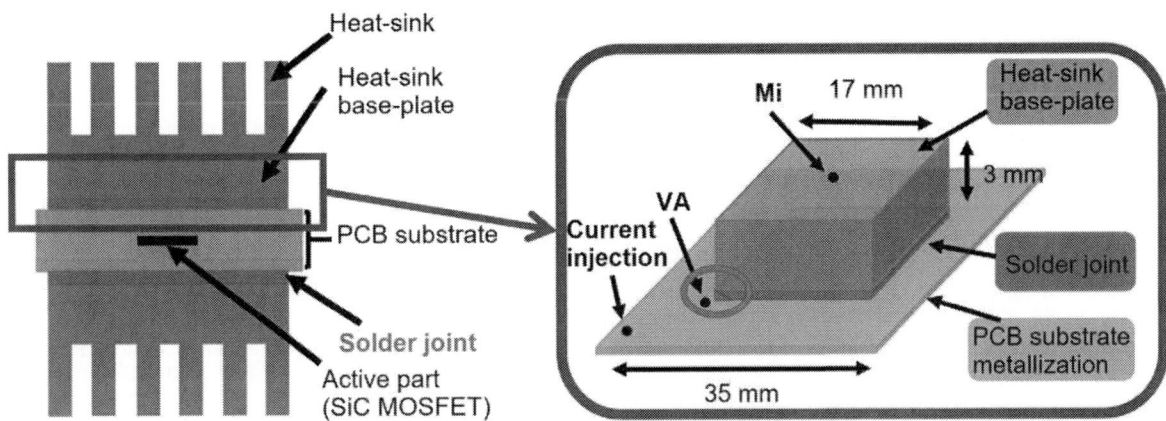

Figure 3: Geometries and boundary conditions of the FEM showing current injection and electrical potential measurements positions.

2.3 Simulations results

The results of realized simulations show a sensitivity to the materials used for the heat-sink base-plate. Copper has higher electrical conductivity 5.96×10^7 S.m^{-1} compared to aluminum 3.50×10^7 S.m^{-1}. This allows for more sensitivity in measurements on prototypes with a heat-sink base-plate with copper than those with aluminum. Since current lines concentration is greater in copper than in aluminum, in the vicinity of the soldering defect. Figure 4 illustrates the sensitivity of measurements to the materials of the heat-sink base-plate, the model used is a prototype with PCB substrate metallization thickness of 70 μm.

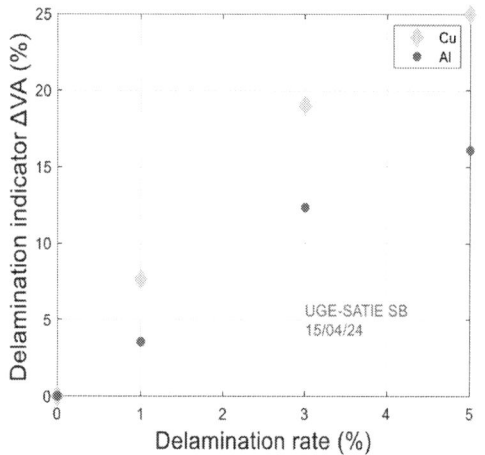

Figure 4: Comparison method sensitivity between Copper and Aluminum materials.

More simulations are done in order to study the effect of the delamination rate over the delamination indicator ΔVA. 1%, 3% and 5% delamination rates are imposed in the corner A of the solder join where the evaluated electric potential VA without imposed delamination is 5 mV. The figure 5 illustrates that the delamination indicator in proportional to the delamination rates.

In the context of studying module aging, a sensitivity study of the method to the solder joint condition was conducted. Firstly, we examined the method's sensitivity to voids that may exist in the solder joint during the manufacturing process. These voids are represented in simulations by a lack of material in the solder.

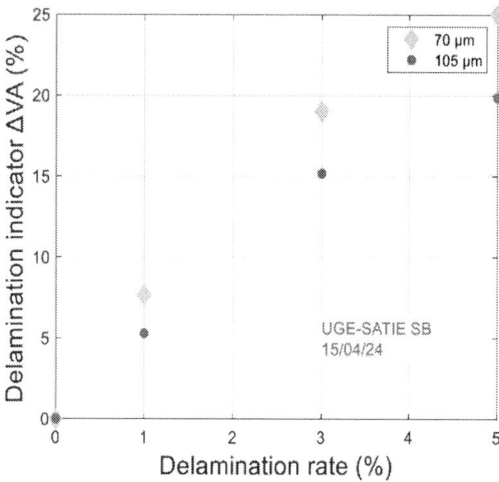

Figure 5: Effect of the delamination rates over the delamination indicator.

Figure 6 presents a comparison between two models with the same specifications, but voids was imposed on one of them in its solder joint.

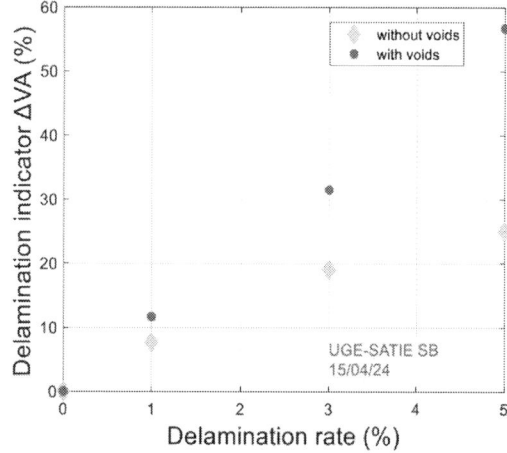

Figure 6: Effect of the voids in solder joint over the delamination indicator.

We concluded that the method is sensitive to voids. For an aging study, it is imperative to perform a measurement before starting the aging cycles in order to distinguish the effect of voids from the effect of delamination. Secondly, we examined the method's sensitivity to the solder joint resistivity ρ. The resistivity of the solder joint can potentially increase due to several factors, including temperature variations induced by thermal cycles. Simulations with different values of the solder joint resistivity are conducted. The values used are the resistivity multiplied by two and by four. Figure 7 illustrates the effect of the solder joint resistivity on the delamination indicator.

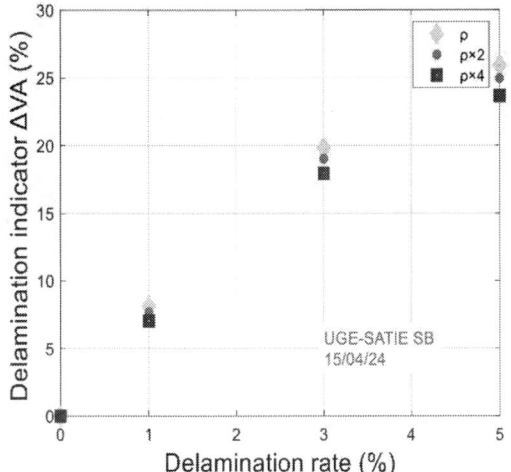

Figure 7: Effect of the solder joint resistivity over the delamination indicator.

We concluded that the method is robust to changes in the solder joint resistivity and ageing of the copper substrate metallization.

3 Definition of a test prototype

Simplified prototypes were created for this study to validate the encouraging simulation results obtained. These prototypes consist of a PCB substrate (copper layers, epoxy, and glass fibers) (35 mm × 35 mm) with copper layers thickness of 70 μm and other prototypes with copper layers thickness of 105 μm, a copper block representing the heat-sink base-plate (17×17 mm) with a thickness of 3 mm, and with Sn-3.0Ag-0.5Cu solder joint (SAC305) with a thickness of 100 μm was used. The PCB substrate metallization is designed with a gold finish to increase the wettability and limit voids in the solder joint. Also, the PCB substrate metallization is performed without an insulating coating to enable current injections and electrical potential measurements. The copper block corresponding to the heat-sink base-plate has been nickel-plated to prevent copper oxidation before soldering onto PCB substrates.

Controlled solder delamination is achieved by using a solder-stop in the desired corner of the solder joint with a desired sharp to design the delamination rate. Even if there is a solder on the solder-stop, there is no electrical contact between the PCB substrate metallization and the heat-sink base-plate. Specifically, at the corner where de-

lamination is imposed. So, it provides good robustness. The figure 8 illustrates the protypes parts before and after soldering.

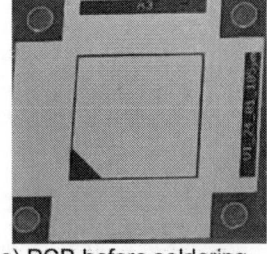 

a) PCB before soldering heat-sink base-plate.

b) After heat-sink base-plate SAC305 soldering

Figure 8: Dedicated PCB substrate before and after heat-sink base-plate soldering.

An assessment of the solder joint condition is conducted using Scanning Acoustic Microscopy (SAM) (Figure 9).

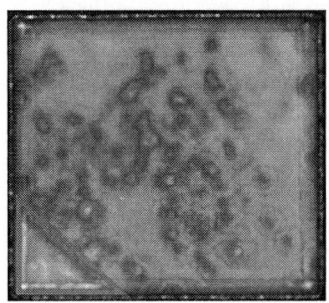

Figure 9: Scanning Acoustic Microscopy of the prototype with 5% imposed delamination (delimited by the red triangle).

Additionally, an evaluation of the zones where imposed delamination is performed. This analysis of the solder joint condition is primarily performed to adjust the measured delamination values with those applied in the numerical model and expected to be observed in the prototypes. The values of the delamination surfaces desired and measured on the prototypes calculated with respect to the solder joint surface (17×17 mm^2) are summarized in Table 2.

Table 2: Desired and measured delamination rates.

PCB substrate metallization thickness (μm)	Delamination percentage (%)	Surface calculated (mm^2)	Surface measured (mm^2)
70	1	2.89	2.27
	3	8.67	10.49
	5	14.45	12.66
105	1	2.89	0.92
	3	8.67	3.25
	5	14.45	11.91

Figure 10 : Panoramic view of the experimental test bench.

4 Experimental study

After obtaining the results from the simulations, an experimental study is implemented to verify these numerical findings. A test bench is set up for current injection and electrical potential measurements. Then, to evaluate the sensitivity of the electrical method, the decision was made to define an injection of a high level of current to increase the signal-to-noise ratio. Due to the small dimensions of the prototypes, the measurement of electrical potential is highly sensitive to the measurement location. Therefore, probes are used to achieve greater measurement accuracy. These probes are limited to a current level less than 3 A in DC mode. In order to be able to inject 100 A in DC mode and avoid being influenced by self-heating, the test bench allows a pulsed current injection of approximately 100 A for 300 µs. The electrical measurements are performed using HBM GEN3i acquisition unit that offers 2MEch/s frequency and 18 bits ADC resolution. A panoramic view of the test bench is shown in Figure 10.

Before proceeding with measurements, an experimental protocol is established. Firstly, the repeatability of measurements is verified to account for constraints affecting measurements (vibrations, nature of measurement tools, etc.). Finally, electrical measurements are conducted using a measurement probe to precisely place measurements and current injection.

4.1 Repeatability of the measurements

The first step is to verify the repeatability of the measurements from two aspects. The first is the positioning of the measurements in the same prototype. The other side is to verify the measurements over different prototypes with the same specifications. We take the example of prototypes

with PCB substrate metallization 70 µm, and without any imposed delamination in the corner of the solder joint. The objective of this measures is to check the repeatability of the measurements in order to validate the measurement procedure and the reliability of the obtained results. The repeatability verification protocol involved removing the probes and taking out the prototype, then putting it back and reinstalling the measurement and injection probes. The purpose of this is to verify the repeatability of the probe positioning to ensure that all measurements were taken at the same point. This ensures that the comparison of measurements between different prototypes will be meaningful. La figure 11 presents the curve of the current injection and the electric potential measurement VA.

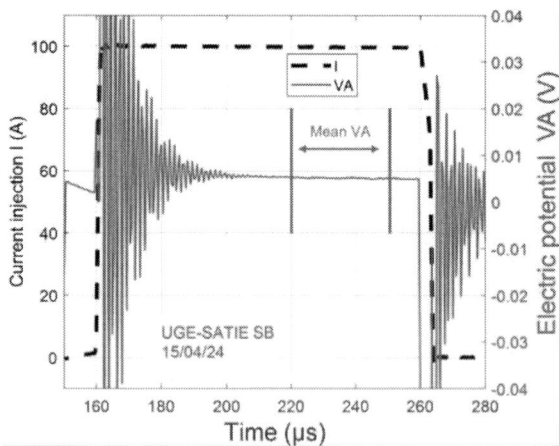

Figure 11: Current injection and electric potential measurement.

The results presented in the figure 11 allow us to calculate the mean value of the electric potential

VA between the two time triggers from 220 μs to 250 μs.

Figure 12 presents a comparison between a series of 5 measurements on the same prototype without any delamination rate imposed on the corner of the solder joint and with PCB substrate metallization thickness of 70 μm. The standard deviation of the realized measurements is 0.03 % for the first prototype and 0.01 % for the second one.

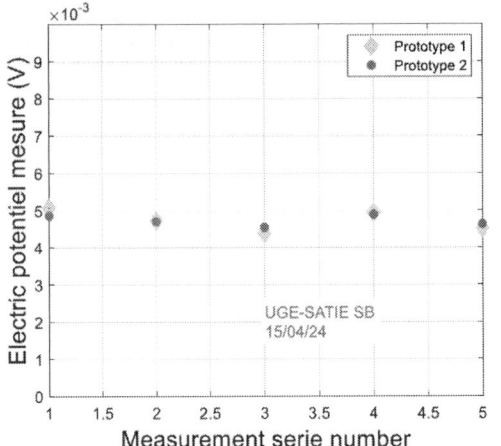

Figure 12 : Comparison of a series of measurements.

Also, the figure 12 presents a comparison between the measurements of the electrical potential on the PCB substrate metallization realized on two different prototypes but with the same specifications (solder joint without any imposed delamination, thickness of the PCB substrate metallization).

4.2 Sensitivity of measurements to delamination rates

A series of electrical potential measurements is conducted on all prototypes. These prototypes include those without delamination, as well as those with delamination imposed at corner A of the solder joint. The imposed delamination rates are 1%, 3%, and 5% at corner A of the solder joint. The prototypes have two thicknesses of metallization on the PCB substrate, 70 μm and 105 μm. Based on the measurements taken, the delamination indicator is calculated using Equation 1. Figure 13 shows the variation of the delamination indicator as a function of the imposed delamination rate at corner A.

We determined that the delamination indicator is proportional to the imposed delamination rate and inversely proportional to the thickness of the PCB substrate metallization. The experimental results demonstrate that the method is sensitive to very small imposed delamination rates at corner A of the solder joint. Thus, the electrical method allows detection of delamination initiated at a corner A of the solder joint. This method is complementary to other existing methods used to evaluate the integrity of an assembly such as thermal performance measurement.

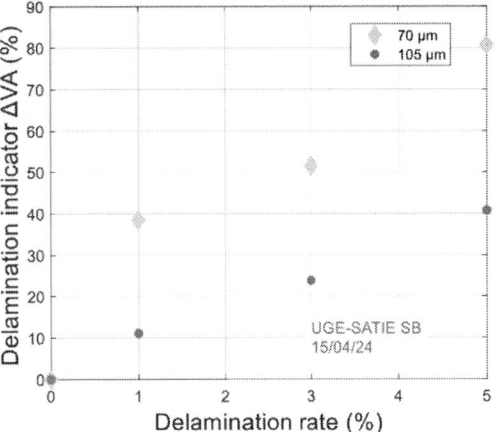

Figure 13: Variation of the delamination indicator depending on the delamination rate.

5 Conclusion

In the case of integrating SiC MOSFET components on a PCB model, defining air cooling requires reducing the number of thermal interfaces by brazing large heat-sinks directly onto the substrate metallization. To assess the relevance of the technological choices in the assembly, it's necessary to detect the initiation of delamination in this solder joint. While conventional methods exist for this purpose, they are not easily applicable to the studied assembly model. The method based on potential electrical measurements allows for the detection of delamination initiation and also monitoring the aging of the assembly composing the power module. The numerical and experimental results have proven that the electrical method exhibits sensitivity to the delamination rates imposed on corner A of the solder joint. Thus, the numerical study has demonstrated that the measurement is not sensitive to voids. The method is specifically sensitive to low delamination rates on the order of 1% of the solder joint surface. The delamination indicator evaluated for these low delamination rates, through numerical and experimental studies, enables the initiation delamination detection. This method proves to be complementary to other existing methods. The next steps will involve improving the sensitivity evaluation on corner A as well as on the other two corners. Additionally, evaluating the robustness of the proposed solution (different rates and locations of delamination, optimization of geometries for "design to sensors"), and generalizing the method to other power electronic assembly technologies (sintered joint or intermetallic diffusion, electrical or mechanical connections). Also, high-frequency alternating current injection will be used to take advantage of the skin effect to increase the sensitivity of the method.

Acknowledgment

This work was supported by the ANR agency (ANR - DESTINI - ANR-21-CE05-0037-01) for funding this research.

ORCID

Souhila BOUZERD https://orcid.org/0009-0007-6816-8810

Laurent Dupont https://orcid.org/0000-0001-7597-0317

Mickaël Petit https://orcid.org/0000-0002-4886-4727

Vincent Bley https://orcid.org/0000-0002-1264-0764

References

[1] Bryan C. Charboneau, Fei Wang, Jacobus Daniel van Wyk, Dushan Boroyevich, Zhenxian Liang and al., « Double-Sided Liquid Cooling for Power Semiconductor Devices Using Embedded Power Packaging », IEEE Trans. Ind. Appl., vol. 44, no 5, p. 1645-1655, sept. 2008.

[2] Justin Broughton, Vanessa Smet, Rao R. Tummala and Yogendra K. Joshi « Review of Thermal Packaging Technologies for Automotive Power Electronics for Traction Purposes », Journal of Electronic Packaging, 1 décembre 2018.

[3] W. F. Bikinga, Jean-Luc Schanen, Bachir Mezrag, Eric Vagnon, Vincent Bley and al., « TAPIR (compacT and modulAr Power modules with IntegRated cooling) Technology: Goals and Challenges », in 2021 Third International Symposium on 3D Power Electronics Integration and Manufacturing (3D-PEIM), Osaka, Japan: IEEE, juin 2021.

[4] M. Ciappa, « Selected failure mechanisms of modern power modules », Microelectron. Reliab., vol. 42, no 4, p. 653-667, avr. 2002.

[5] M. Bouarroudj, Z. Khatir, S. Lefebvre and L. Dupont, « Thermo-mechanical investigations on the effects of the solder meniscus design in solder joint lifetime for power electronic devices », International Conference on Thermal,

Mechanical and Multi-Physics Simulation Experiments in Microelectronics and Micro-Systems. EuroSime London, 2007.

[6] S.-W. Guo, Y.-W. Huang and H.-K. Liao, « High reliability lead free solder evaluations in power module application », 12th International Microsystems, Packaging, Assembly and Circuits Technology Conference (IMPACT), Taipei: IEEE, oct. 2017.

[7] F. Arabi, L. Theolier, D. Martineau, J.-Y. Deletage, M. Medina and E. Woirgard, « Power electronic assemblies: Thermo-mechanical degradations of gold-tin solder for attaching devices », Microelectron. Reliab., vol. 64, p. 409-414, sept. 2016.

[8] L. Dupont, Z. Khatir, S. Lefebvre, R. Meuret, B. Parmentier and S. Bontemps, « Electrical characterizations and evaluation of thermo-mechanical stresses of a power module dedicated to high temperature applications », in 2005 European Conference on Power Electronics and Applications, p. 11 pp.-P.11, Dresden, Germany: IEEE, 2005.

[9] W. Qiu, B. Zee, K. L. Pey and N. Raghavan, « Thermal simulations of lock-in-thermography for failure analysis of integrated circuits », in 2023 IEEE International Symposium on the Physical and Failure Analysis of Integrated Circuits (IPFA), p. 1-6, Pulau Pinang, Malaysia: IEEE, juill. 2023.

[10] W. Sabbah, R. Riva, S. Hascoet, C. Buttay, S. Azzopardi and al., « Evaluation of silver-sintering die attach », Conference on Integrated Power Systems (CIPS), Germany, 2012.

[11] N. Ismaïl, A. Jalar, A. Afdzaluddin and M. A. Bakar, « Electrical resistivity of Sn–3.0Ag–0.5Cu solder joint with the incorporation of carbon nanotubes », NanoMITe Annual Symposium & Nanotechnology Malaysia Annual Symposium 2019.

[12] K. Guth, D. Siepe, J. Gorlich, H. Torwesten, R. Roth and al, « New assembly and interconnects beyond sintering methods », Conference on Integrated Power Systems (CIPS), Germany, 2012.

PCIM Europe 2024, 11– 13 June 2024, Nuremberg   DOI: 10.30420/566262073

Generic Lifetime Model for Wire Bonds Degradation in IGBT Modules Based on a Fracture Mechanics Parameter

Merouane Ouhab[1], Nicolas Degrenne[1], Yusaku Ito[2], Masaki Taya[2]

[1] Mitsubishi Electric R&D Centre Europe (MERCE), France

[2] Mitsubishi Electric Corporation (MELCO), Japan

Corresponding author: Merouane Ouhab, m.ouhab@fr.merce.mee.com
Speaker: Merouane Ouhab, m.ouhab@fr.merce.mee.com

Abstract

Junction temperature variation (ΔTj), minimum junction temperature (Tjmin) and heating time (ton) are pointed out as the most influencing load parameters limiting the lifetime of standard power modules (PMs). In this work, a large series of DC power cycling tests (PCTs) are conducted on IGBT PMs targeting a considerable range of variation for each load parameter. The tested PMs are based on heavy aluminum wires (400um diameter) used as top electrical interconnections. Post-analysis of tested devices and acquired measurements during the PCTs revealed wire bond lift-off as the predominant observed failure mechanism. A generic Physics-of-Failure (PoF) model is proposed to establish different trends and dependencies according to each parameter ΔTj, Tjmin and ton. It relies on nonlinear fracture mechanics, and more precisely a crack growth model describing crack propagation at the wire/metallization interface.

1   Introduction

To emulate lifetime consumption under operating conditions, accelerated power cycling [1] represents an efficient testing tool standardized to evaluate reliability of power electronics devices. In the case of PMs with a standard packaging structure, wire bonds and die-solders are considered as the weakest interconnections, which are the most prone to failure by thermomechanical fatigue. From several years ago, extensive PCTs were performed under DC and PWM conditions to understand the occurring failure mechanisms to model the effect of their main driving force parameters, i. e. ΔTj, Tjmin and ton [2]. As a first approach, many empirical lifetime models were proposed to consider the effect of each load parameter separately. As relevant references from literature we can cite LESIT, CIPS and SkiM63 lifetime models. Since the empirical approach suffers from uncertainty particularly when extrapolating the results towards non-tested ranges, development of physics-based models became necessary. Based on a review of the existing physics-based lifetime models, no generic solution was provided to account reasonably for the three main load parameters, i. e. ΔTj, Tjmin and ton. Within this scope, we carried out in this study several DC PCTs (26 tests) on a standard PM, and we targeted various loading conditions. As a continuity to our proposed physics-based model in [3], we generalize it in this paper to account for additional parameters (ΔTj and Tjmin).

2   Experiment: PCTs and failure analysis

26 DC PCTs were carried out on an IGBT PM of a standard packaging technology, rated 1200V/150A. The device has a 3-phase inverter topology where only the middle leg is tested as it is illustrated in **Fig. 1**. During the heating stage, the top and the bottom IGBTs, T1 and T2 respectively, are conducting the load current. During the cooling stage, a sensing current of 100mA is injected through T1 and T2 in order to estimate the junction temperature of each device using collector-emitter (V_{CE}) voltage drop as a Temperature Sensitive Electrical Parameter (TSEP). The tested range regarding each load parameter is [30°C, 130°C] for

589

ΔTj, [25°C, 95°C] for Th and [0.05s, 32s] for ton. Detailed power cycling conditions are listed in **Table 1**. In the following, PCT # 7 is used as a reference test, hence all the lifetime data is normalized to its lifetime calculated using the B63 characteristic from Weibull distribution.

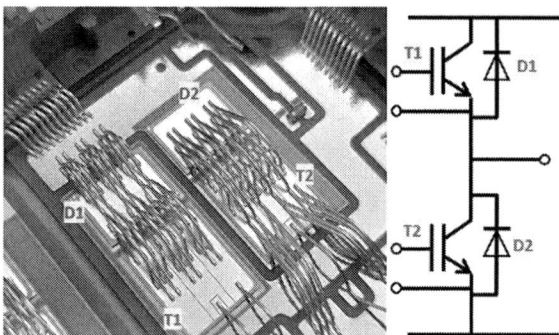

Fig. 1   Middle leg of the studied PM

To monitor degradation of wire bonds and solders during PCTs, the on-state voltage at load current (V_{CE}) and the junction-to-heatsink thermal resistance (Rthj-h) were measured respectively. Threshold values of +10% in V_{CE} and +20% in Rthj-h were set as an end-of-life criteria for the calculation of the number of cycles to failure (Nf).

PCT #	ΔTj (°C)	ton (s)	Th (°C)	I (A)	Samples #
1	30	0.05	45	165	6
2	30	0.05	65	165	6
3	30	0.05	95	165	6
4	40	0.5	45	119	6
5	50	0.26	35	150	3
6	50	1	45	119	6
7	60	1	45	131	3
8	60	3	45	110	3
9	70	0.2	35	150	6
10	70	3	35	120	6
11	70	3	45	125	6
12	90	1	35	120	3
13	90	1	45	171	3
14	90	3	35	135	6
15	90	3	45	153	3
16	90	8	35	120	3
17	90	20	35	120	6
18	90	22	45	113	6
19	90	32	35	135	3
20	110	3	25	173	3
21	110	3	35	150	6
22	110	20	28	130	6
23	110	22	45	143	6
24	130	3	25	180	3
25	130	8	25	170	6
26	130	22	25	157	6

Table 1 PCTs conditions

In all the experiments, the failure indicators (V_{CE} & Rthj-h) revealed that wire bonds lift-off was the predominant failure mechanism, with minor or no visible solders degradation. For illustration, we can see in **Fig. 2** typical V_{CE} and Rthj-h curves as function of the number of cycles normalized to the lifetime of the reference PCT # 7. The demonstrated measurements were acquired during PCTs # 6, 7 & 11 conducted at ΔTj=50°C, 60°C & 70°C respectively.

(a)

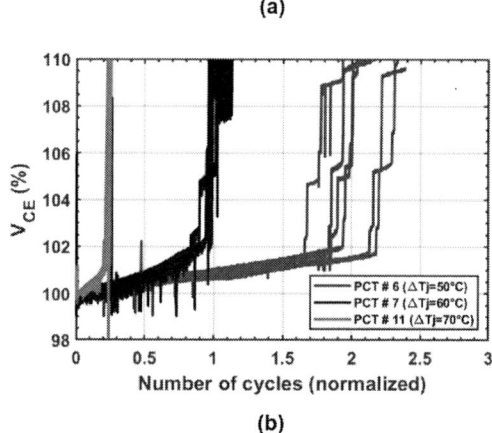

(b)

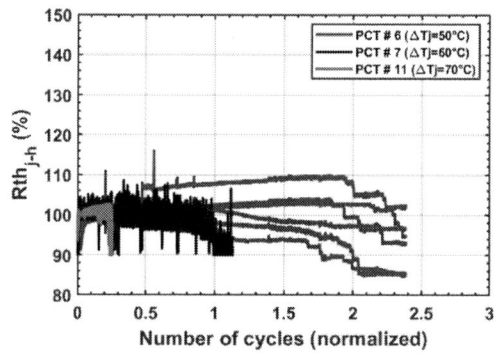

Fig. 2 Damage indicators during PCT at ΔTj=50°C, 60°C & 70°C, (a) on-state voltage, (b) junction-to-heatsink thermal resistance

As shown in **Fig. 3**, selected SAM (Scanning Acoustic Microscopy) images made at different critical joints (wires contact, die- & DCB-solders) were analyzed after PCTs # 20 & 26. The images confirm furthermore no critical damage inside both die- and DBC-solders even if the related PCTs conditions are supposed to be highly stressing for the solder joints. However, the images emphasize again that wire-bonds contact (black spots) disappeared partially or completely from the top surface of IGBT (T1) in particular.

PCIM Europe 2024, 11– 13 June 2024, Nuremberg   DOI: 10.30420/566262073

Fig. 3 SAM images after selected PCTs, (a) ΔTj=110°C, ton=3s (PCT # 20), (b) ΔTj=130°C, ton=22s (PCT # 26)

Measured lifetime data was processed statistically using Weibull distribution to extract the B63 characteristic lifetime for each PCT and also to measure the data quality. For demonstration, in **Fig. 4** the effect of junction temperature variation and heating time are highlighted based on PCTs # 18 & 23 and PCTs # 14 & 17 respectively. As expected, when ΔTj or ton increase the lifetime decreases. The estimated shape parameter for these PCTs (higher than 15) clearly indicates an ageing process.

Based on the B63 lifetime calculated for all the PCTs data, an empirical model was built as a function of different load parameters, where the effect of ΔTj and ton is correlated with a power law (Coffin-Manson-type) and the effect of Tjmin with an Arrhenius term. Finally, the model expression is given by the following equation:

$$N_f^{Emp} = A \times \Delta T_j^{C_1} \times t_{on}^{C_2} \times e^{\left(\frac{E_a}{k_B \cdot (T_{jmin}+273K)}\right)} \quad (1)$$

Where: A, C_1 & C_2 represent materials constants, E_a the activation energy and k_B the Boltzmann constant.

A multi-linear regression is used to identify these constants to achieve the best fitting between the empirical model and the lifetime data. This is reduced to minimizing the following cost-function:

$$\sum_{i=1}^{i=25} \left(ln\left(N_{f,i}^{Emp}\right) - ln\left(N_{f,i}^{PCT}\right) \right)^2 = min \quad (2)$$

(a)

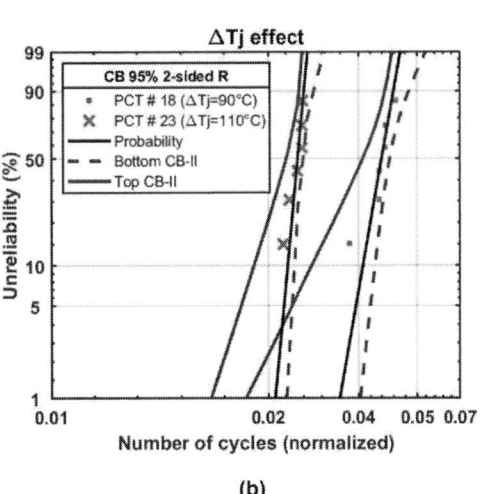

(b)

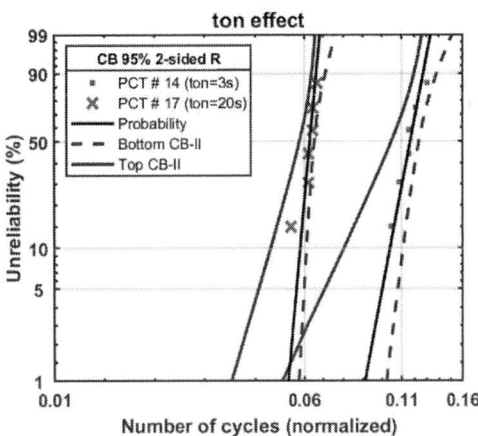

Fig. 4 Weibull distribution of selected PCTs data, (a) ΔTj-dependency (b) ton-dependency

Solving the previous equation led to the constants given in **Table 2**, where the constant A is normalized to the B63 lifetime of the reference PCT.

A (°C⁻ᵅ.sᵝ)	C_1 (1)	C_2 (1)	E_a (eV)
1.2968×10⁷	-4.2322	-0.45	0.0309

Table 2 Empirical model constants

591

For analysis purpose, lifetime data together with the empirical model estimations are normalized and plotted in **Fig. 5 (a-d)** to highlight their dependency to ΔTj, ton and Tjmin respectively. For better visualization and readability, three cyclic colors (black, red & blue) are used in the legend, where each color refers to groups of tests with common conditions. As demonstrated in **Fig. 5 (a) & (b)**, ΔTj is separated into high (>70°C) and low (<70°C) ranges respectively. At high ΔTj, the empirical model demonstrates a relatively good agreement compared to PCTs data with a remarkable mismatch at very high ΔTj values. As it can be observed, the slope of measured lifetime data decreases as ΔTj increases, unfortunately the empirical model fails to follow this behavior, thus several linear regressions might be necessary to identify adequate parameters for each ΔTj interval to reduce the inherent deviation. At low ΔTj, the empirical model provides generally acceptable estimations except at ΔTj=30°C and high heatsink temperatures, i. e. Th=65°C & 95°C.

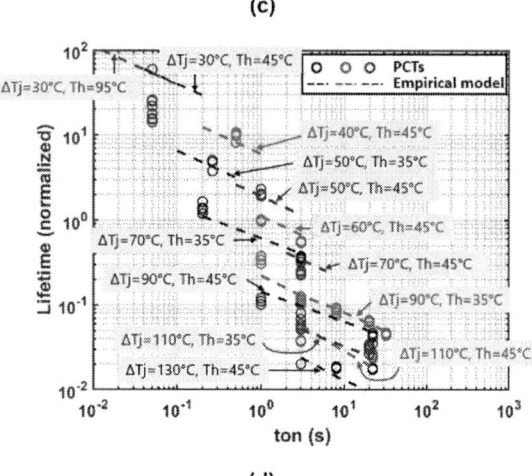

(c)

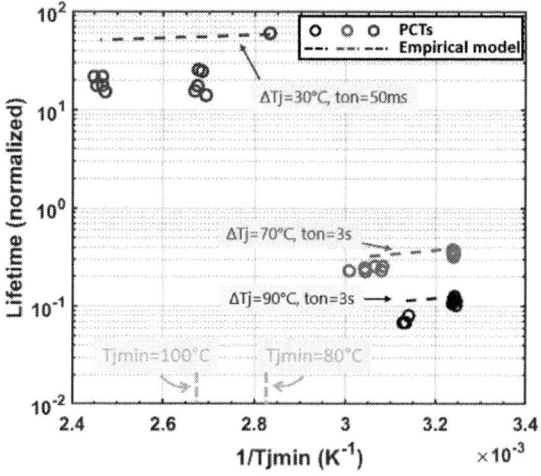

(d)

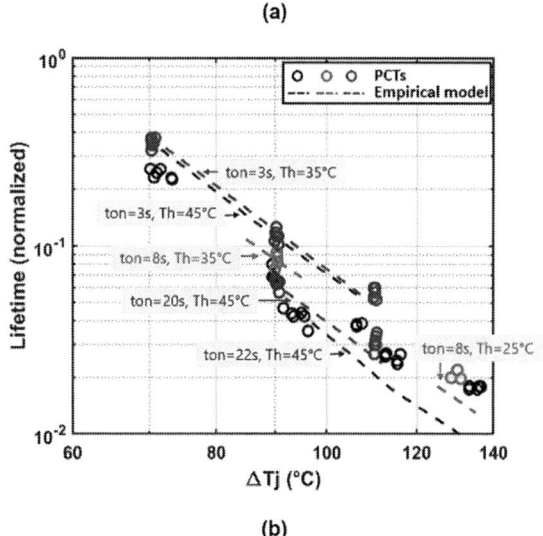

(a)

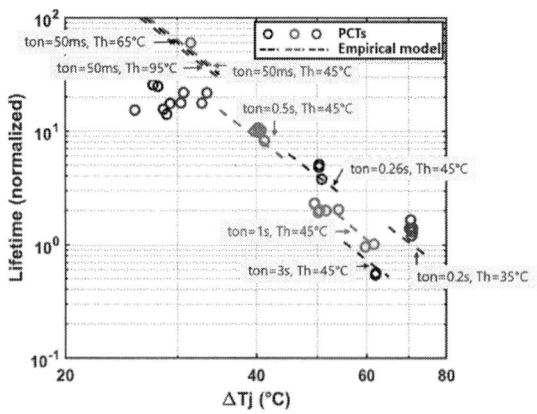

(b)

Fig. 5 PCTs data vs empirical lifetime model, (a) High ΔTj-dependency, (b) Low ΔTj-dependency, (c) ton-dependency, (d) Tjmin-dependency

As shown in **Fig. 5 (c)**, the heating time demonstrates a relative decreasing impact on the lifetime as ton increases revealing an attenuation effect. Similar results were reported in [4] and [5] under both DC and PWM (Pulse Width Modulation) PCTs conditions, this behavior was assumed to be related to the maximum plastic deformation achieved at high values of pulse duration leading to a stationary situation. In the same figure, the empirical model is plotted showing a good fitting with the lifetime data but only over the tested intervals. Obviously, using a power law, the attenuation effect due to ton increase is not predictable. Hence, results extrapolation becomes subject to high uncertainty, and more testing effort will be required.

In **Fig. 5 (d)**, the impact of Tjmin is illustrated for three PCTs groups. Based on the measured lifetime data, three Tjmin intervals can be highlighted:

Tjmin<80°C, 80°C≤Tjmin<100°C and Tjmin≥100°C. In the first interval, a low activation energy (Ea=0.0309eV given in **Table 1**) is measured regarding the limited impact of Tjmin on the lifetime, see black and red circles. Using this value, lifetime extrapolation to Tjmin≥100°C using the empirical model induces a non-negligeable error, see the blue dashed line. In the second interval, a higher activation energy is measured (Ea=0.5598eV) due to the significant influence on the lifetime. The change in Ea value has been observed in the literature [6] and was related to a shift in the degradation mechanism from wire bonds to die-solders, which is not aligned with the findings of this work. In the third interval, a lifetime saturation is observed. This phenomenon has been also investigated in [7] & [8], which revealed that during thermal cycling at high Tjmax, softening and sub-grains coarsening occur in wires aluminum material, this microstructural modification was correlated with an annealing effect causing a slower fatigue rate. In summary, the empirical model with an Arrhenius term to account for Tjmin impact is not sufficient to cover the above-mentioned non-linearities due to the sudden high change in Ea value, and also to the observed lifetime saturation. Moreover, the activation energy is highly related to the tested device and to the packaging materials. As it can be seen in **Table 3**, Ea value needs to be determined for each specific device variant, which means that using universal empirical constants cannot be fulfilled.

	Eₐ (eV)	Tjmin range (°C)	Degradation zone
This work	0.0309, 0.5598	[25, 80], [80, 100]	Wire bond
LESIT [9]	0.8	[25, 85]	Wire bond
CIPS 08 [10]	0.111	[20, 120]	Wire bond, die-solder
Schilling [11]	0.080	[45, 95]	Wire bond
Schmidt [6]	0.069, 0.159	[-20, 66], [-20, 90]	Wire bond, die-solder
Scheuermann [12]	0.066	[-25, 85]	Wire bond

Table 3 Measured activation energy from several references.

Based on the previous analysis, development of a generic physics-based lifetime model is very desired and should be helpful to:

- Reduce the gap in terms of physical comprehension
- Increase confidence in lifetime extrapolation

- Reduce the number of required experiments

3 PoF modelling and comparison with experiment

Within the scope of modelling wire bond wear-out, a simplified thermomechanical FEM is developed to evaluate simultaneously the impact of ΔTj, Tjmin & ton on the stress/strain fields surrounding a crack-tip introduced at the wire/metallization interface. The thermo-mechanical properties used in the simulation model for different components: silicon (Si) die, aluminum (Al) metallization and aluminum wire are reported in [13-15]. The silicon chip is considered as a linear elastic material. The aluminum material assigned to the wire and the metallization is modelled by an elastic-plastic-creep behavior. The main properties of Si and Al materials are given in **Table 3**.

Material	Young's Modulus (Gpa)	Poisson ratio	CTE
Al	Fig. 6 (a)	0.33	Fig. 6 (b)
Si	165.7	0.22	2.6

Table 3 Mechanical properties of aluminum and silicon [13-14]

Young's modulus and Coefficient of Thermal Expansion (CTE) are temperature-dependent for the aluminum material as illustrated in the figures below.

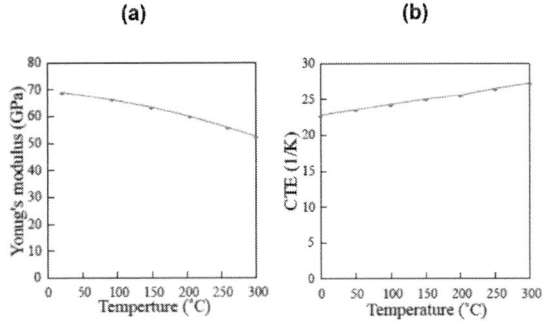

Fig. 6 (a) Young's modulus and (b) CTE of Aluminum material [13-14]

The plastic-creep properties of the aluminum material are modeled based on measured data reported in [15]. This data was acquired from several isothermal tensile tests performed on thick aluminum wires of 400um diameter. Typical mechanical responses can be seen in **Fig. 7**.

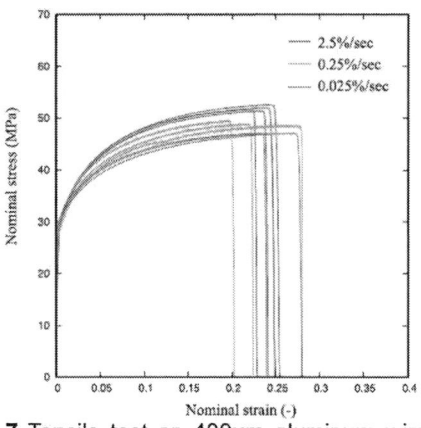

Fig. 7 Tensile test on 400um aluminum wires at a temperature of 20°C and different strain rates [15]

Based on the collected data at different temperatures ranging from 20°C to 250°C, and different strain rates ranging from 2.5%/s to 0.025%/s, the plastic behavior was modeled using Ludwik constitutive equation given by:

$$\sigma = \sigma_0 + k_1 . \varepsilon_p^{k_2} \tag{3}$$

Where σ_0, k_1 & k_2 are material constants. σ & ε_p are the stress and the plastic strain respectively. To account for the creep strain, a transient creep law is used based on Norton-Bailey constitutive equation:

$$\varepsilon_{cr} = A\sigma^n t^m \tag{4}$$

Where A, n and m are constant material parameters. σ and t represent the true stress and the holding time during the tensile test respectively.

Fitting iteratively the two equations (3) & (4) on the experimental data allowed to identify different constants as shown in **Table 4 & 5**.

T (°C)	σ_0 (MPa)	k_1 (MPa)	k_2 (1)
20	19.94	84.97	0.3932
120	20.99	61.86	0.3810
220	13.16	40.12	0.3779
250	12.02	40.05	0.4282

Table 4 Plastic model parameters [15]

T (°C)	A (MPa^{-n}.s^{-m})	n (1)	m (1)
20	2.826×10^{-6}	1.964	0.3680
120	6.877×10^{-6}	2.310	0.3232
220	2.725×10^{-5}	2.031	0.3798

250	3.533×10^{-5}	1.999	0.4989

Table 5 Creep model parameters [15]

Since the thermo-mechanical properties are time- & temperature-dependent, and the PCTs cover a wide range regarding each load parameter, we propose in this study a PoF lifetime model based on nonlinear fracture mechanics. Within this scope, to characterize time- & temperature-dependent stress/strain fields near the crack-tip at the interface between the wire and the metallization, creep J integral (J*) is selected as a driving force parameter of the crack growth. Using J* parameter, the time crack growth rate can be expressed as follow:

$$\frac{da}{dt} = \alpha[J^*]^\beta \tag{5}$$

Therefore, the cycle crack growth rate can be estimated using an average integration of the time crack growth over the heating time period:

$$\frac{da}{dN} = \int_0^{t_{on}} \frac{da}{dt} dt \tag{6}$$

As J* is dependent on the power cycling load conditions, we finally find:

$$\frac{da}{dN} = \int_0^{t_{on}} \alpha\big[J^*\big(\Delta T_j, t_{on}, T_{jmin}\big)\big]^\beta dt$$
$$= \alpha \Delta J^*\big(\Delta T_j, t_{on}, T_{jmin}, \beta\big) \tag{7}$$

Where: a represents the crack length at the wire/metallization interface, α & β are a scale parameter and an exponent, respectively, which depend on the loading type, i. e. for time-dependency $\alpha = \alpha_1$ and $\beta = \beta_1$, and for temperature-dependency $\alpha = \alpha_2$ and $\beta = \beta_2$. Thus, to calculate the lifetime ratio between ton1 and ton2 at constant ΔTj and Tjmin values, we can simply use the following expression:

$$R_{\frac{t_{on1}}{t_{on2}}} = \frac{N_f(t_{on1})}{N_f(t_{on2})} \tag{8}$$

where:

$$N_f(t_{on}) = \int_{a_0}^{a_f} (\alpha_1)^{-1}\big[\Delta J^*(t_{on}, \beta_1)\big]^{-1} da \tag{9}$$

This yields the following ratio:

$$R_{\frac{t_{on1}}{t_{on2}}} = \frac{\Delta J^*(t_{on2}, \beta_1)}{\Delta J^*(t_{on1}, \beta_1)} \tag{10}$$

In the same way, to calculate the lifetime ratio between $\Delta Tj1$ and $\Delta Tj2$ at constant ton and Tjmin values, equation (10) becomes:

$$R_{\frac{\Delta T_{j1}}{\Delta T_{j2}}} = \frac{\Delta J^*(\Delta T_{j2}, \beta_2)}{\Delta J^*(\Delta T_{j1}, \beta_2)} \tag{11}$$

Similarly to ΔTj effect, to calculate the lifetime ratio between Tjmin1 and Tjmin2, we have:

$$R_{\frac{T_{jmin1}}{T_{jmin2}}} = \frac{\Delta J^*(T_{jmin2}, \beta_2)}{\Delta J^*(T_{jmin1}, \beta_2)} \tag{12}$$

As it can be noted, using the lifetime ratio definition allows to reduce the lifetime model dependency to only β parameter. This means for a given wire material with no prior-art, three PCTs are sufficient to establish fully the proposed model in this study. The first PCT represents the reference test, and the two remaining tests will be used to identify β_1 and β_2 for time- and temperature-dependencies. In a previous work [3], β_1 has been already defined using a value of $\beta_1 = 0.66$ that was reported in [16]. However, β_2 is to be calculated based on two PCTs performed at two different ΔTj or Tjmin values. Herein, using the reference PCT (# 7) together with PCT # 13, we calculate β_2 as follow:

$$R_{\frac{\Delta T_j=90°C}{\Delta T_j=60°C}} = \frac{N_f(\Delta T_j = 90°C)}{N_f(\Delta T_j = 60°C)} \tag{13}$$

This implies the following equation:

$$\frac{\Delta J^*(60°C, \beta_2)}{\Delta J^*(90°C, \beta_2)} = 0.11 \tag{14}$$

Solving the previous equation gives $\beta_2=2.1$. Physically, β_1 and β_2 are related to the exponents m and n in equation (4) respectively, where the temperature stressors (ΔTj & Tjmin) are correlated with the stress (σ), and the heating time (ton) is correlated with the holding time (t). Development of these relations is out of the scope of this work.

The fracture parameter J* is calculated on an evaluation contour surrounding a crack-tip in the $x_1 x_2$ plane, as illustrated in **Fig. 8**. Its mathematical expression is given by:

$$J^* = \int_\Gamma \left(W^* dx_2 - T_i * \left(\frac{\partial \dot{u}_i}{\partial x_1}\right) * ds \right) \tag{15}$$

Where:

$$W^* = \int_0^{\dot{\varepsilon}_{ij}} \sigma_{ij} \, d\dot{\varepsilon}_{ij} \tag{16}$$

And:

$$T_i = \sigma_{ij} n_j \tag{17}$$

With: $i = 1, 2$ & $j = 1, 2$.

Here, Γ represents a contour integral surrounding the crack-tip, u the displacement field vector, ds is an element of arc length along Γ. W^* the rate of strain energy density associated with the stress point σ_{ij} and strain rate $\dot{\varepsilon}_{ij}$, T the outward directed traction vector, n the normal vector to the contour path Γ, a_0 & a_f are the initial and the final crack lengths respectively. To be mentioned, the ratio of cyclic J* (ΔJ^*) at two different ton, ΔTj or Tjmin values is independent of the crack length.

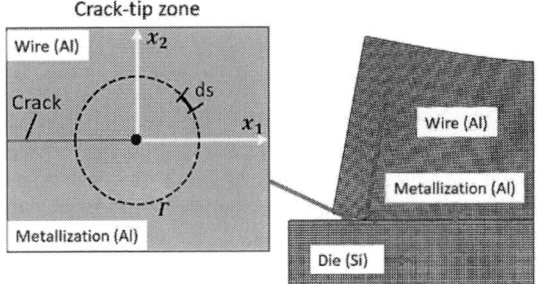

Fig. 8 Crack-tip zone at wire-metallization interface

As the thermomechanical stresses are mainly induced at wire-metallization interface due to CTE mismatch in the stack, a half structure of the wire contact zone is studied, with a symmetry line applied vertically in the middle of the bond interface. The structure is constrained at 2 points to reproduce wire contact mechanical environment, see **Fig. 9**. Aiming to reduce numerical deficiencies near the crack-tip, 3 paths/contours with circular shapes have been defined to calculate an average value of J*.

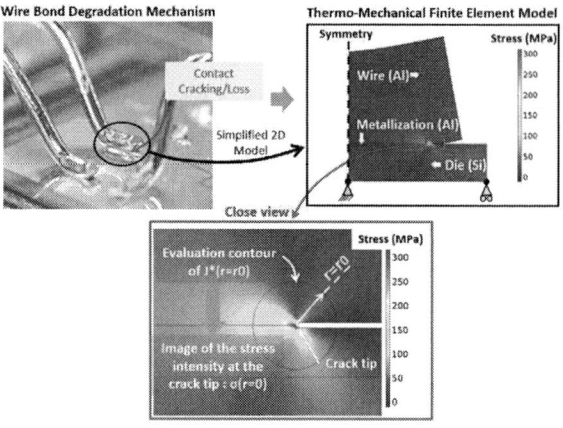

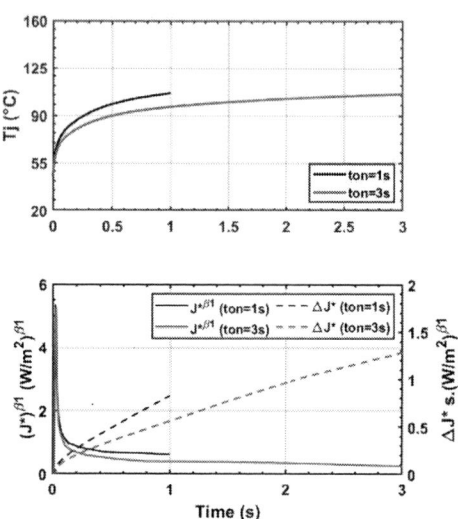

Fig. 9 Thermo-mechanical FEM of wire bond

To simulate PCTs conditions, J* response was evaluated using three steady-state thermal cycles considering each specific PCT conditions. This aims to reach thermomechanical steady-state by the 3rd cycle. The simulation results are then processed to calculate $(J^*)^\beta$ and ΔJ^* responses. Typical curves are plotted in **Fig. 10 (a) & (b)** to highlight ΔTj and ton effects respectively.

At constant ton and Tjmin, the difference in ΔJ^* between high and low ΔTj is mostly due to the transient period (highest variation in Tj) this is related mainly to the elastic-plastic behavior of the Al material. However, at constant ΔTj and Tjmin, the difference in ΔJ^* between short and long heating times increases over time which indicates creep behavior taking place.

(a)

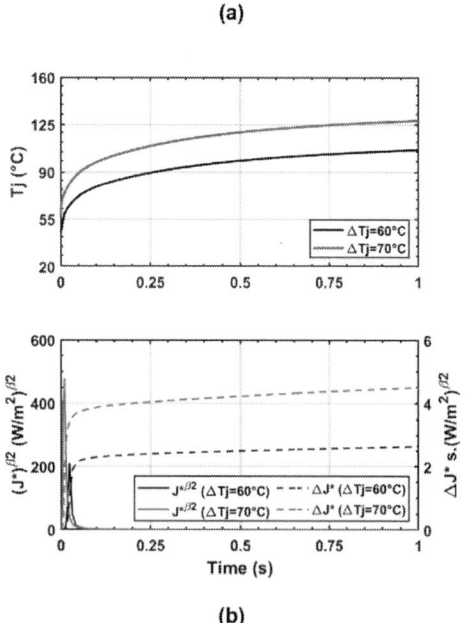

(b)

Fig. 10: Simulation responses of $J^{*\beta}$ and ΔJ^* to typical junction temperature profiles, (a) ΔTj-dependency: ΔTj=60°C, 70°C & constant ton=1s, (b) ton-dependency: ton=1s, 3s & constant ΔTj =60°C

Finally, using the simulation results as input to the proposed model, lifetime can be estimated for the whole set of the PCTs. As shown in **Fig. 11 (a-d)**, comparison between experiment and theory is illustrated with respect to ΔTj, ton and Tjmin dependencies.

In the high ΔTj range (**Fig. 11 (a)**), the PoF model shows a high fidelity to the lifetime data. Under high thermal stress conditions, large inelastic strain (plastic and creep) occurs dominating elastic strain, thus the rate of energy density (W*) in J* expression can be reduced to the rate of dissipated energy density. At constant heating time and zero traction vector, ΔJ^* can be correlated to the cyclic dissipated energy, which is a well-accepted parameter to describe low cycle fatigue using Coffin-Manson approximation. In the low ΔTj range (**Fig. 11 (b)**), the PoF model fits adequately with the lifetime data. Under very low thermal stress conditions, elastic strain dominates inelastic strain. Hence, the rate of energy density (W*) becomes equivalent to the rate of stored (elastic) energy density. At constant heating time and zero traction vector, ΔJ^* can be correlated to the cyclic stored energy which is commonly used in the high cycle fatigue region using Basquin approximation.

In **Fig. 11 (c)**, the PoF model provides acceptable estimations along the full tested heating time interval. At high ton values (ton≥1s), the creep strain is the main parameter governing the energy density. The higher is the heating time, the higher is the creep strain and the energy density. At relatively low ton values (ton<1s), creep strain and temperature rate (dT/dt) are competing parameters to govern the energy density. For ton>100s, the PoF model estimates a lifetime saturation which is due to stress relaxation (or creep) in the aluminum material. Relaxation is characterized by a decreasing thermal stress and a decreasing strain rate ($\dot{\varepsilon}$) approaching zero. This leads the rate of energy density (W*) to approach zero. Thus, any longer heating time would not cause any change in ΔJ^* nor in the lifetime.

As shown in **Fig. 11 (d)**, the PoF model follows the PCTs data over the three highlighted Tjmin intervals characterized by: low Ea value, high Ea value and lifetime saturation (Ea~0). The increase of the activation energy is related to the high increase of creep strain exponent (n) measured between 20 and 120°C, see **Table 5**. As the temperature increases, the creep strain is more activated causing more dissipated energy density. Above 120°C, the creep strain exponent (n) decreases slightly which suggests a lifetime increase, but as the yield stress (σ_0) decreases significantly at temperatures above 120°C (see **Table 4**), the amount of the energy density is maintained constant leading to a constant lifetime.

Finally, the proposed model in this work delivers reasonable results over all the tested ranges of ΔTj, ton and Tjmin with a standard error 2.3 times less than the empirical model. In addition, a physics-based extrapolation of the model allows to learn more about lifetime behavior over non-tested intervals, as it helps to reduce the testing efforts particularly at low ΔTj, low and high ton.

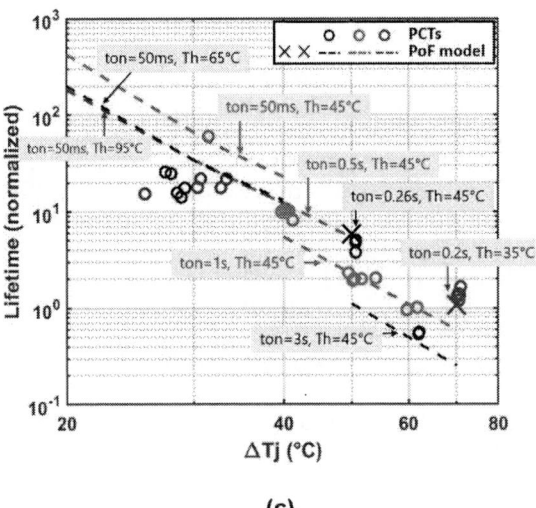

(c)

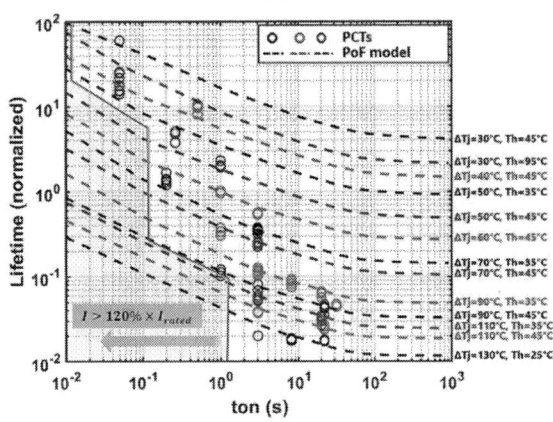

(d)

(a)

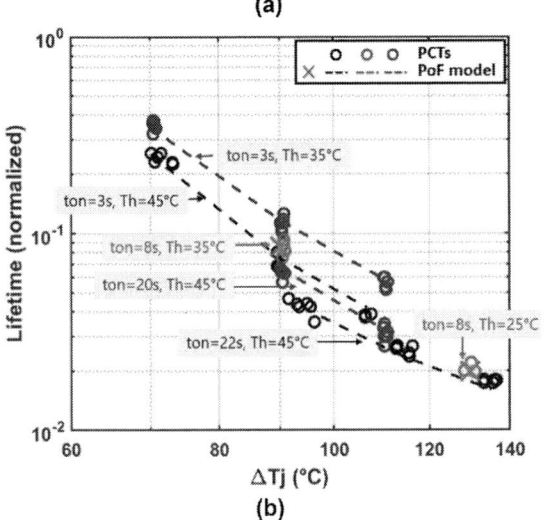

(b)

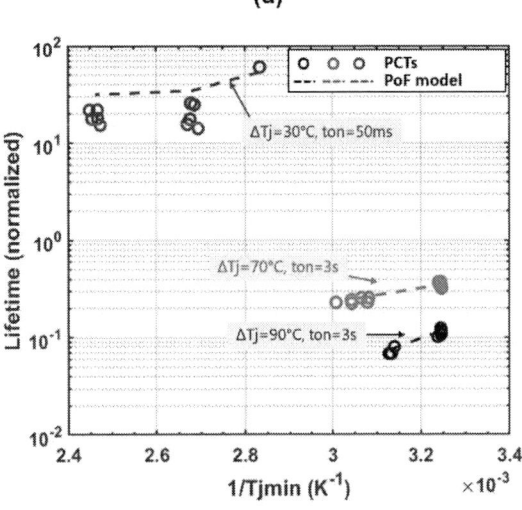

Fig. 11: PCTs data vs PoF model, (a) High ΔTj-dependency, (b) Low ΔTj-dependency, (c) ton-dependency, (d) Tjmin-dependency

4 Conclusion

Various PCT conditions were carried out on a standard IGBT PM. The tests targeted the following ranges: ΔTj within [30°C, 130°C], Tjmin within [25°C, 95°C] and ton within [0.05s, 32s]. As expected, a lifetime decrease was observed as ΔTj, Tjmin or ton increases. Post-analysis of failed DUTs confirmed that wire bonds degradation was the main failure mechanisms. Using nonlinear fracture mechanics, a creep crack growth model dedicated to the degradation of wire-metallization interface provided reasonable lifetime estimations compared to the experiment, with only 2 PCT results used to identify fully the model parameters. The proposed lifetime model revealed interesting features to be an efficient tool to extrapolate with confidence the PCTs results towards high effort or unfeasible testing conditions, such as low ΔTj, very long and very short ton values.

5 References

[1] T. Harder, "Qualification of Power Modules for Use in Power Electronics Converter Units in Motor Vehicles", ECPE Guideline AQG 324, May 2019.

[2] N. Dornic et al., "Stress-Based Model for Lifetime Estimation of Bond Wire Contacts Using Power Cycling Tests and Finite-Element Modeling", in IEEE Journal of Emerging and Selected Topics in Power Electronics, vol. 7, no. 3, pp. 1659-1667, Sept. 2019, doi: 10.1109/JESTPE.2019.2918941.

[3] M. Ouhab, N. Degrenne, Y. Ito and S. Izuo, "Physics-of-Failure Model to Explain the Heating-Time Effect on IGBT Power Modules Lifetime," PCIM Europe 2023, Nuremberg, Germany, 2023, pp. 1-6, doi: 10.30420/566091007.

[4] U. Choi, F. Blaabjerg and S. Jørgensen, "Study on Effect of Junction Temperature Swing Duration on Lifetime of Transfer Molded Power IGBT Modules", in IEEE Transactions on Power Electronics, vol. 32, no. 8, pp. 6434-6443, Aug. 2017, doi: 10.1109/TPEL.2016.2618917.

[5] U. Scheuermann, R. Schmidt and P. Newman, "Power cycling testing with different load pulse durations", 7th IET International Conference on Power Electronics, Machines and Drives (PEMD 2014), Manchester, 2014, pp. 1-6, doi: 10.1049/cp.2014.0475.

[6] R. Schmidt, F. Zeyss and U. Scheuermann, „Impact of absolute junction temperature on power cycling lifetime," in Proc. of the EPE'13 ECCE Europe, 2013.

[7] Y. Yamada, Y. Takaku, Y. Yagi, I. Nakagawa, T. Atsumi, M. Shirai, and I. Ohnuma, "Reliability of wire-bonding and solder joint for high temperature operation of power semiconductor device," Microelectron. Reliab., vol. 47, no. 12, pp. 2147–2151, Dec. 2007.

[8] P. A. Agyakwa, M. R. Corfield, L. Yang, J. F. Li, V. M. F. Marques, and C. M. Johnson, "Microstructural evolution of ultrasonically bonded high purity Al wire during extended range thermal cycling," Microelectron. Reliab., vol. 51, no. 2, pp. 406–415, Feb. 2011.

[9] M. Held, P. Jacob, G. Nicoletti, P. Scacco and M. H. Poech, "Fast power cycling test of IGBT modules in traction application," in Proc. of the 2nd PEDS, pp. 425-430, 1997.

[10] R. Bayerer, T. Herrmann, T. Licht, J. Lutz and M. Feller, "Model for power cycling lifetime of IGBT modules - various factors influencing lifetime," in Proc. of the 5th CIPS, pp. 1-6, 2008.

[11] O. Schilling, M. Schäfer, K. Mainka, M. Thoben and F. Sauerland, „Power cycling testing and FE modelling focussed on Al wire bond fatigue in high power IGBT modules," Microelectronics Reliability, vol. 52, pp. 2347-2352, 2012.

[12] U. Scheuermann and R. Schmidt, "Impact of load pulse duration on power cycling lifetime of Al wire bonds," Microelectronics Reliability, vol. 53, pp. 1687-1691, 2013.

[13] Y. Setoguchi, N. Shishido, M. Koganemaru, T. Ikeda, Y. Hayama and N. Miyazaki, "Thermal fatigue analysis for aluminium wire bonding of power modules considering transient creep", Proc. of The Computational Mechanics Conference, 2018.

[14] Y. Setoguchi, N. Shishido, M. Koganemaru, T. Ikeda, Y. Hayama, N. Miyazaki, "Thermal elasto-plastic creep analysis for Al wirebond used in power module", Proc. of The Computational Mechanics Conference, 2017.

[15] Nobuyuki Shishido, Yoshiki Setoguchi, Yuto Kumagai, Masaaki Koganemaru, Toru Ikeda, Yutaka Hayama, Noriyuki Miyazaki, "Characterization of plastic and creep behavior in thick aluminum wire for power modules, Microelectronics Reliability", Volume 123, 2021, 114185, ISSN 0026-2714.

[16] Hall, David and McDowell, and Saxena, Ashok, (1998), CRACK TIP PARAMETERS FOR CREEP-BRITTLE CRACK GROWTH. Fatigue & Fracture of Engineering Materials & Structures, 21: 387-401.

PCIM Europe 2024, 11– 13 June 2024, Nuremberg   DOI: 10.30420/566262074

Modular Coaxial Power Converter for High-Density Integration into Medium-Voltage Cables

Mark Cairnie [1], Aakash Kamalapur [1], Rajaie Nassar [1], Jack Knoll [1], Qingrui Yuchi[1], Jung-Soo Bae[1], Christina DiMarino [1], Khai Ngo [1], Guo-Quan Lu[1], Qiang Li[1], Dushan Boroyevich[1], Jierui Zhou[2], Yang Cao[2], Douglas DeVoto[3], Bidzina Kekelia[3]

[1] Center for Power Electronic Systems (CPES), Virginia Polytechnic and State University, USA
[2] Electrical Insulation Research Center (EIRC), The University of Connecticut, USA
[3] National Renewable Energy Laboratory, USA

Corresponding author: Mark Cairnie, mcairnie@vt.edu
Speaker: Mark Cairnie, mcairnie@vt.edu

Abstract

This work proposes to combine the functionality benefits of power electronics with the power density benefits of medium-voltage cables to create a streamlined, high-density power electronics solution that seamlessly integrates with medium-voltage cables. Located at the ends of a medium- or high-voltage line, the proposed converter uses a cascade of coaxial power conversion cells to gradually step down the voltage, and excels in high step-down applications. By mimicking the coaxial geometry of medium-voltage cables, the converter preserves the axisymmetric electric field of the cable which, when combined with a solid insulating dielectric, provides a voltage scaling advantage over conventional planar and PCB-based converter solutions. Similar to medium voltage cables, the converter is fully passively cooled. A passive cooling strategy allows for combined installation with existing medium voltage cable systems without the added cost, maintenance needs, infrastructure, and reliability concerns associated with active cooling systems. The scalability of the modular structure in combination with the integration benefits provide a flexible power electronics system that can adapt to the evolving demands of the grid.

1 Introduction

The goal of this work is to demonstrate a new concept that will enable a compact, flexible, scalable, and adaptable medium-voltage (MV) distribution network that can better adapt to the growing and changing sources, demands, and usage patterns [1]. To achieve this, this work proposes to combine the benefits of power electronics and MV cables to create a cohesive structure that can replace bulky substation components while enhancing functionality. Located at the ends of an MV line, the system (shown in Fig. 1) uses a cascade of coaxial power conversion cells to gradually step down the voltage to the levels required by the loads. By mimicking the coaxial geometry of HV cables, the modules are able to inherit some of the beneficial properties of cables such as the uniform electric fields while seamlessly integrating with the line [2].

The proposed concept could revolutionize distribu-

tion systems by enabling significant improvements in power density while achieving high efficiency and scalability to higher voltages and power. This innovation could pave the way for active line terminations with built-in features, such as bidirectional power flow control and STATCOM functionality, and could enable immense flexibility (e.g., the ability to interface lines with different frequencies and voltages) [3]. This flexibility and functionality could allow seamless integration of distributed renewables, energy storage, and EV fast charging, contributing to the reduction of foreign energy imports, greenhouse gas emissions, and energy loss.

The objective of this work is to build and demonstrate a prototype coaxial power conversion cell that can be utilized in a ±5 kV DC to ±400 V DC installation with a combined power rating of 300 kW. The enabling technological innovations include: modular high-step-down isolated-stacked ćuk (iSĆuk) topology, a variant of the ćuk converter [4] with low passive requirements; high-frequency coaxial in-

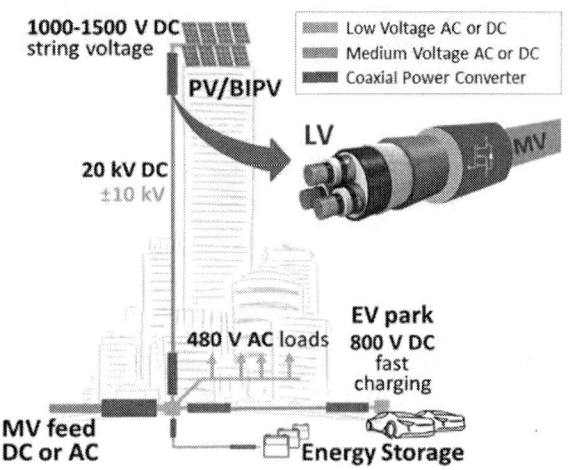

Fig. 1: Coaxial power conversion cells integrated directly into MV cables providing voltage step-up, step-down, AC-DC and DC-AC conversion in a future grid with distributed energy storage and renewable resources.

Fig. 2: Underground cable vault housing three MV cable splices that are accessible via a manhole for routine inspection, maintenance, and possible replacement. Source [6].

ductor; high-energy-density structural capacitors; high-temperature, high-dielectric-strength insulation; MV silicon carbide (SiC) power modules; and a passive, self-contained cooling system.

2 Thermal Considerations

Similar to MV cables, the proposed coaxial power converter is designed to be fully passively cooled, allowing the cells to be deployed without the need for additional cooling infrastructure (e.g., pumps, radiators, fans, etc.). This has the potential to significantly increase overall system reliability. A study in [5] suggests that thermal related failures make up as much as 16% of transformer failures in MV substations. Unfortunately, the performance of passive cooling is low relative to active cooling strategies, requiring significant surface area, in addition to being a strong function of the environment.

To further design the passive cooling system, the environment was narrowed down to underground cable vaults, as they are an optimal installation site for early adaptors of this technology. Underground vaults are commonly used to house cable splices and transformers in urban environments where space above-ground is not always available [7]. The vault provides access to the splices and equipment via a manhole cover or grate, allowing routine inspection, maintenance, and replacement when necessary. Fig. 2 shows an example of an underground cable vault which houses three MV cable splices.

Since underground vaults are also used to house transformers, the cooling performance that can be achieved in this environment has been studied in the literature over the past decades, and the techniques for extracting heat vary widely [8]–[10]. Typically, the vaults are vented to allow natural convection currents to form, extracting the heat while drawing in cool air. In some cases, the vaults are allowed to flood naturally and can even be intentionally flooded in order to enhance the cooling capability via conduction into the water.

Due to the potential for flooding, the case temperature of the system needs to be maintained below 100°C, otherwise boiling can occur, causing fouling of the case. In addition, despite the vault ventilation, ambient temperature in the vault can reach up to 50°C, which is a function of both the vent size and shape as well as the weather and geographic location, as discussed in [11].

Using the existing literature on cooling in underground vaults along with available standards, such as IEC-60287 [11], a radial fin structure was designed to increase the surface area of the converter, increasing the total heat rejection from free convection. The fin structure (Fig. 3) was based on the design of commercial fin tubes used in liquid-to-air heat exchangers in chemical processing applications [12]. The design, performance, and manufacturing of these structures is well understood and can easily be scaled further to increase the total heat rejection [13], [14].

The prototype shown in Fig. 3 experimentally

PCIM Europe 2024, 11– 13 June 2024, Nuremberg DOI: 10.30420/566262074

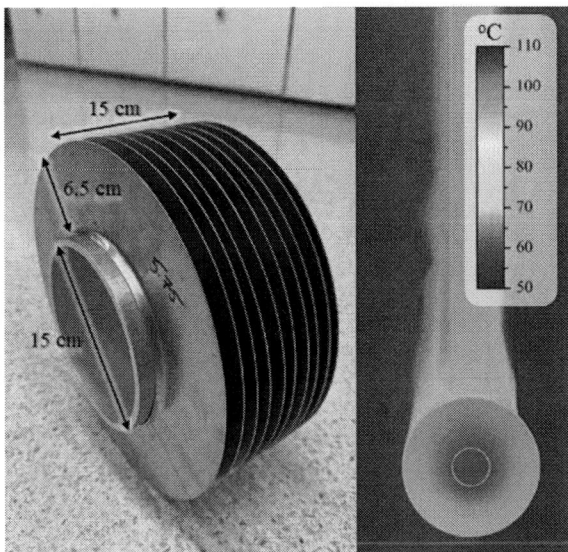

Fig. 3: Left: prototype radial fin structure with stamped, 20-gauge aluminum fins measuring 28 cm OD, 15 cm ID and spaced 1.2 cm apart. Right: CFD simulation showing the natural convection currents forming in the air around the fin structure.

demonstrated a dissipation of 280 W without exceeding a case temperature of 100°C, translating to 1400 W per meter of length. Typical underground vaults can be as large as 9 m, allowing a theoretical maximum heat dissipation of 12.6 kW per cable in a typical vault [15]. This loss figure could be further improved via extending the fins beyond 6.5 cm. Fig. 4 shows a theoretical rendering of three parallel installations of three, cascaded 50 kW coaxial power conversion cells, (9 cells total or 450 kW) in a 3 x 1.5 x 1.5 m vented underground cable vault.

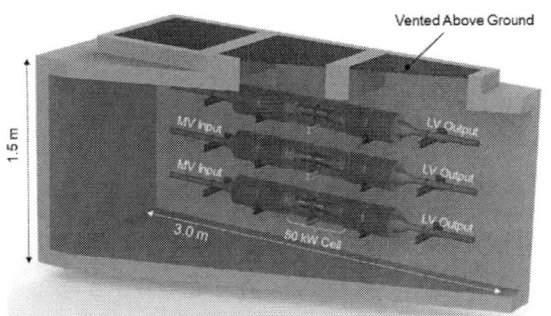

Fig. 4: Render of three parallel installations of three, cascaded 50-kW coaxial converters with radial fin structures, in a vented underground cable vault. Nine total cells with a theoretical combined output power of 450 kW can be passively cooled.

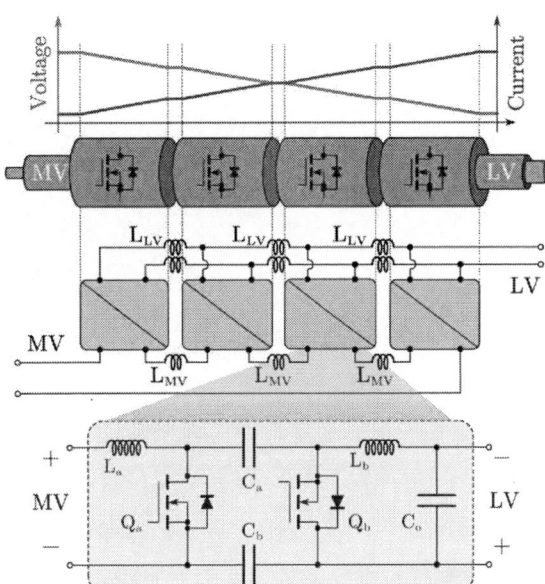

Fig. 5: Cascaded coaxial power conversion cells based on the iSĆuk topology are arranged in a series-input parallel-output (ISOP) configuration to gradually step-down the voltage while increasing current along the length of the cable.

3 Electrical Considerations

The architecture in Fig. 5, depicts the fundamental principle of stacking multiple dc-to-dc converter cells in series to process the input voltage. Each cell processes a fraction of the input voltage, denoted as MV/n, where n represents the number of stacked cells. This architecture integrates four pivotal features. Firstly, isolation between each cell's input and output ports is achieved through capacitors, which enhances the energy density compared to the use of magnetic components [16]. Secondly, a unified single-phase duty cycle control scheme is implemented across all cells. This ensures uniform steady-state time-evolution and distributes stress equally among the passive components within each cell. Thirdly, the output ports of all cells are connected in parallel, which allows each cell to contribute an equal fraction of the total output power, expressed as P_{out}/n, to the load. Finally, the fourth feature, distinct to the structural design of the system, involves passive cooling of all cells. The elongated design of the power components within each cell facilitates enhanced passive heat dissipation. However, this configuration introduces additional parasitic elements (depicted in Fig. 5) between the inter-cell terminations of input and output ports,

PCIM Europe 2024, 11– 13 June 2024, Nuremberg   DOI: 10.30420/566262074

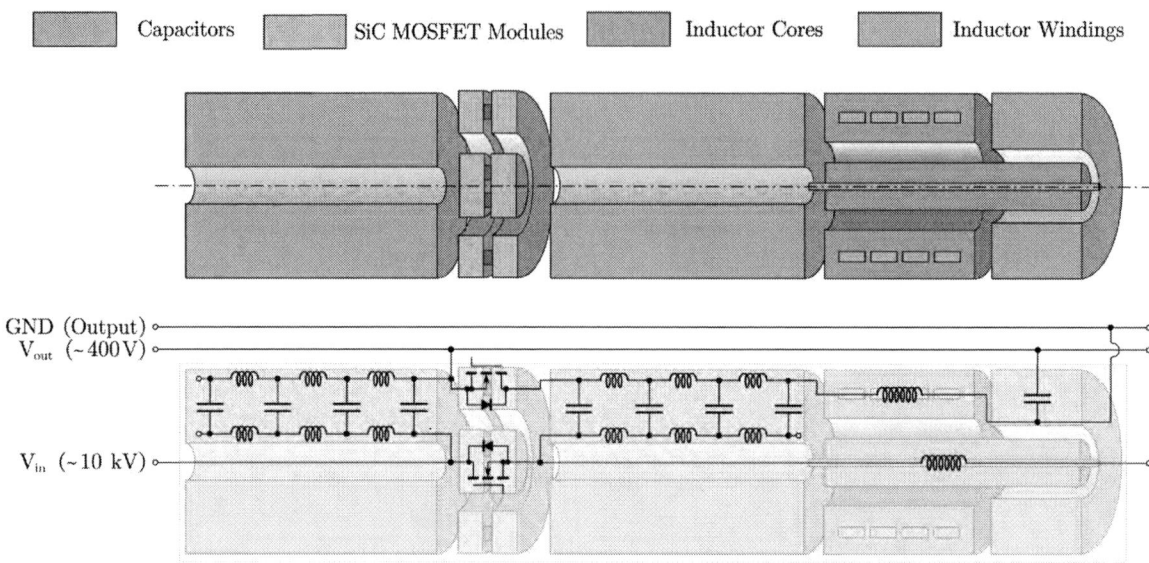

Fig. 6: Coaxial component integration of the iSĆuk converter showing the SiC MOSFET modules, inductor and capacitor mimicking the coaxial structure of the cable to minimize disturbances to the electric field distribution.

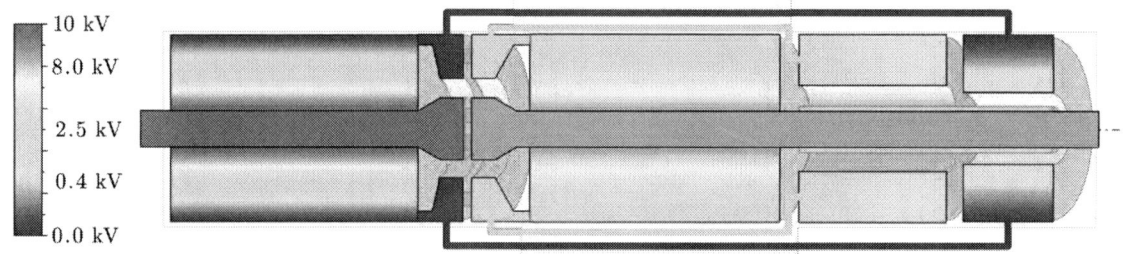

Fig. 7: Coaxial potential pattern inside the coaxial power conversion cell. Voltage is graded radially, with the highest voltage components at the center of the structure and the low voltage components around the circumference.

L_{MV} and L_{LV}. These parasitics impact the dynamic performance of the system and increase the stresses in the power components.

To implement the architecture, the iSĆuk converter (shown in Fig. 5) is selected due to its effective management of parasitic components. The converter naturally absorbs input-port parasitics L_{MV} into the input inductor L_a, decouples output-port parasitics L_{LV} through the output capacitor C_o, and confines the commutation loop parasitics to within each cell. Additionally, to further reduce voltage spikes in the switches, the iSĆuk converter is operated to allow negative inductor currents, which facilitate zero-voltage switching (ZVS) turn-on of the active switch Q_a.

To best integrate the power conversion cells into the MV cable, each component is designed to mimic the coaxial structure of the cable, which in theory,

minimizes disruptions to the electric field pattern and thus minimizes the need for additional field grading and electrical insulation structures. Fig. 6 shows the integration strategy for the iSĆuk topology. Note the potential pattern of the integrated cell in Fig. 7 shows the highest voltage components at the center of the cell and the lowest voltage components at the outside of the cell. The voltage is graded radially, similar to a MV cable. The iSĆuk topology was developed specifically for its ability to be integrated coaxially without overlapping traces that break the axial symmetry of the structure.

4 Coaxial Component Design

Fig. 8 shows the custom, coaxial SiC modules, capacitors, and inductors assembled into a 50 kW power conversion cell measuring 53 cm long and 13 cm in diameter. To achieve its full power rating, the cell must be potted into a radial fin structure,

602

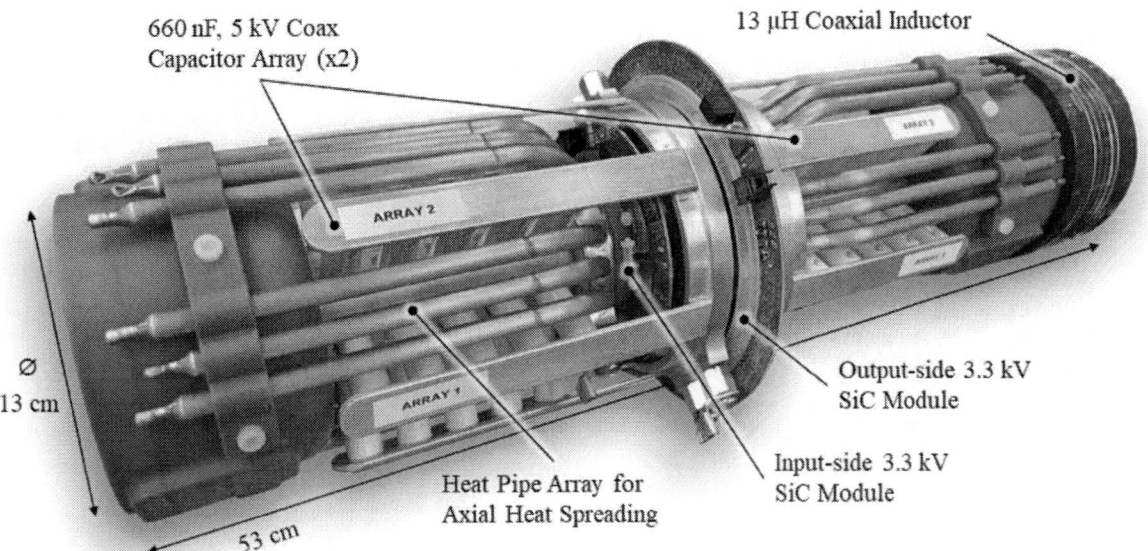

Fig. 8: Integrated coaxial capacitors, SiC switch modules, and inductors. Completed prototype measures 53 cm long with a 13 cm diameter and is rated for 50 kW (with radial fin structure installed, not shown) and a maximum input voltage of 5 kV.

such as that shown in Fig. 3. The cell is potted with a high temperature potting that is both electrically insulating and has a high thermal conductivity. The heat pipes shown in the cell assist with heat transfer by conducting heat away from the SiC modules, distributing it axially to achieve a uniform case temperature. The following sections will discuss additional details and considerations for each of the three main components: SiC MOSFET modules, axial flux inductor, and film capacitor.

4.1 Coaxial SiC MOSFET Modules

Two SiC MOSFET modules are needed, each containing a single switch position comprised of paralleled 3.3 kV, 25 mΩ SiC MOSFETs from Microchip (PN: MSC027SMA330D/S). The modules are designed to nest within one another, forming a coaxial structure (Fig. 9) with the MV, input-side switch in the center and the low-voltage, output-side switch on the outer circumference.

Similar to press-pack technology [17], the modules do not contain a conventional substrate and instead sandwich the die between two Au-plated, Cu contacts. Fig. 10 shows the internal structure of the inner modules, which is scaled up to form the outer module. The dies are attached to a Mo stage to assist with CTE grading. Both the die and stage are affixed using nano-Ag preform from NBE-tech. The dies are wirebonded with 5-mil Al wedge bonds to a silver-plated AlN spacer which makes contact

to the Cu source terminal (not shown in Fig. 10). The Cu terminals then serve as both the electrical contacts for drain and source as well as the cooling surfaces through which heat is extracted.

4.2 Coaxial Capacitors

The coaxial capacitor array is shown in Fig. 11. The array is comprised of 30 discrete, cylindrical PET film capacitors, arranged such that the common MV

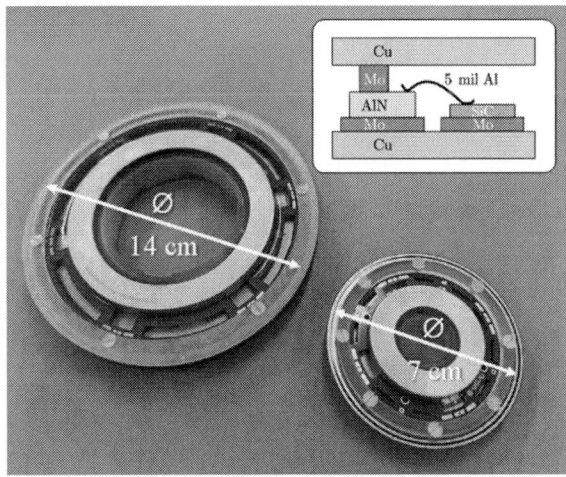

Fig. 9: Coaxial SiC modules with Au-plated Cu contacts and PCB slip-rings for the gate and kelvin source connections. The MV input-side module (right) nests inside of the low-voltage, output-side module (left).

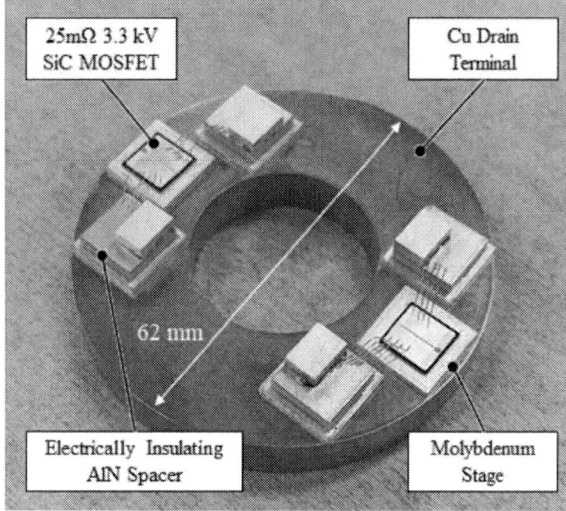

Fig. 10: Coaxial SiC MOSFET module with SiC die located between two Au-plated Cu contacts which serve as both the electrical terminals and the cooling surfaces.

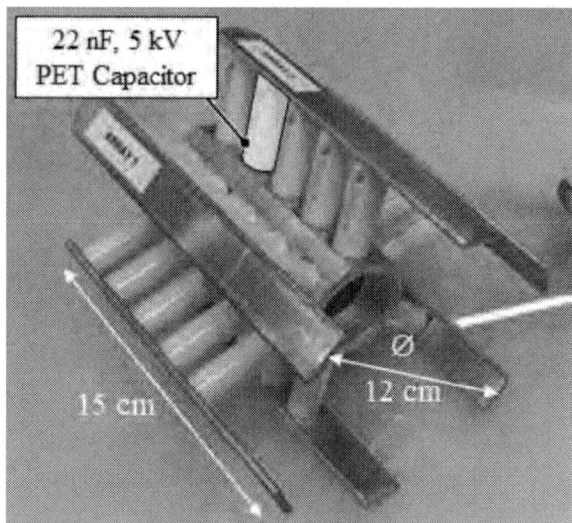

Fig. 11: Coaxial film capacitor array comprised of 30 discrete cylindrical capacitor rolls with a cumulative capacitance of 660 nF, and a voltage rating of 5 kV.

terminal is at the center of the array while the low-voltage busbar terminals are arranged around the outer circumference. The capacitors are bonded to the Sn-plated terminals via Nanofoil® preform from Indium Corp. The preform bonds the exposed Sn-Zn endspray metallization of the capacitors directly to the Sn-plated busbars without the need for solder and without overheating the capacitor winding.

Due to the coaxial structure, magnetic fields caused by opposing current in the inner and outer conductor can cancel, resulting in very low ESL at high frequency relative to the size of the capacitor. Fig. 12 shows the measured impedance of the array. Note that due to the distributed capacitance, multiple resonant points are present, forming a resonant band from 1.5 MHz up to 13 MHz. Below this band, the impedance is purely capacitive and above this band, it is purely inductive. Multiple resonant points in the impedance plot indicate that there may be larger circulating energy in the capacitors that needs to be considered from a topology design standpoint. Further investigation on this topic is required.

4.3 Coaxial Inductor

The output inductor structure and assembly process is shown in Fig. 13. It consists of a solenoidal winding enclosed by low-loss ferrite cores that are shaped to conform to the system's coaxial geometry. On either side of the winding, inner and outer ferrite rings (5 mm thick) utilize a quasi-distributed

gap (3.7 mm total on each side) to control the inductance while reducing fringing field losses in the winding [18]. The gaps are made from aluminum-nitride sheets to enhance the thermal conductivity of the inductor. A 12 mm-thick cap at either end of the structure completes the magnetic path around the helical winding. The magnetic field travels axially within the inner and outer segments of the core, and radially in each of the end caps.

This structure is similar to the one discussed in [19] with noticeable differences in geometry, frequency and application. For this inductor, a hole in the center of the structure is required for system integration, limiting the available cross-sectional

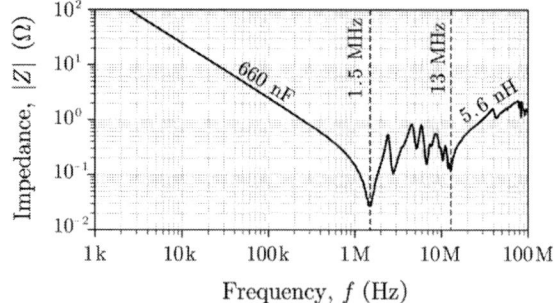

Fig. 12: Frequency response of 5 kV, 660 nF coaxial capacitor array showing the resulting resonant band from 1.5 MHz to 13 MHz and a high frequency ESL of 5.6 nH.

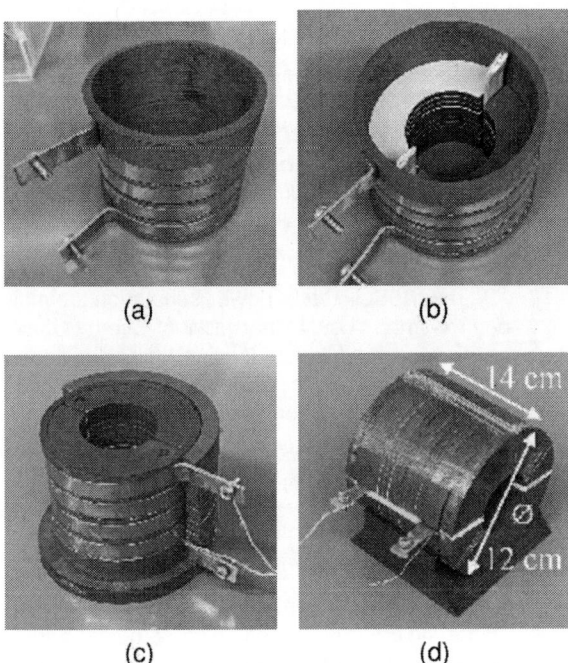

Fig. 13: Assembly of a 13 µH coaxial inductor. a) rectangular winding with bobbin; b) inner core stack with ceramic spacers in the airgaps; c) installation of outer core stack; and d) completed assembly measuring 12 cm OD by 14 cm long.

area available for the core and winding window. Following guidelines from [18] for the turn-to-gap spacing, the resulting winding window is narrow and tall, requiring the long and thin rectangular turns seen in Fig. 13a to better utilize the available winding area. Due to the turn's long aspect ratio, accurately controlling the winding position within the window becomes important. This is achieved using a customized 3D-printed bobbin, made form an electrically-nonconducting, thermally conducting filament [20]. Another consideration is the 2 mm thick copper turns, which is five times larger than the typical choice of two skin depths at 100 kHz for high-frequency inductors. This lowers the winding's dc resistance since the output inductor carries the full load current.

As for the core, while the axial gaps control the inductance, additional circumferential gaps are needed in all the ferrite pieces. In ungapped outer rings, a toroidal flux component is unintentionally induced due to the axial propagation of the current through these rings. Circumferential gaps are thus utilized in the outer rings to avoid increased core loss and the risk of saturation. Circumferential

gaps in the inner rings and end caps diminish the parasitic coupling with other system components running along the central axis through the inductor due to similar toroidal flux components in these pieces.

While the designed inductor theoretically meets the system's loss budget, experimental loss verification is still needed. Further work is also required to ensure proper thermal management and EMI compliance.

5 Results

To validate the coaxial converter, the system in Fig. 8 is tested at 1 kV-to-200 V, 10 kW. The simulation results for this operating point shown in Fig. 14 are in complete accordance with the experimental drain-source voltages of the switches shown in Fig. 15. To reach the rated power of 50 kW, the radial fin structure from Fig. 3 needs to be installed and potted into place. The converter demonstrates zero-voltage switching during the turn-on transition of the active switches. Furthermore, it also experiences resonances while inverting and rectifying power.

6 Summary and Conclusions

In this work, a 50 kW coaxial power converter cell was developed which draws inspiration from MV power cables. Like MV cables, the converter utilizes a coaxial form factor and solid insulation to manage electric fields and is fully passively cooled, enabling seamless integration with the cable. The converter is based on a variant of the ćuk topology which is capacitively isolated, allowing the converter cells to be cascaded to achieve higher power levels and step-down ratios. Custom coaxial components were developed, including SiC MOSFET modules, as well as coaxial inductors and capacitors.

A cooling strategy was developed which allows for passive cooling at a rate of up to 1400 W/m in underground cable vaults. Underground vaults are a prime installation location for early technology adaptors as it provides opportunities for retrofits as well as access for inspection and maintenance. As the technology matures, the seamless integration with the cable, in conjunction with the ability to passively cool, may enable coaxial converter cells to be installed anywhere there are cables. This has the potential to revolutionize distribution systems by providing a new level of flexibility and adaptability to meet the demands of future grids.

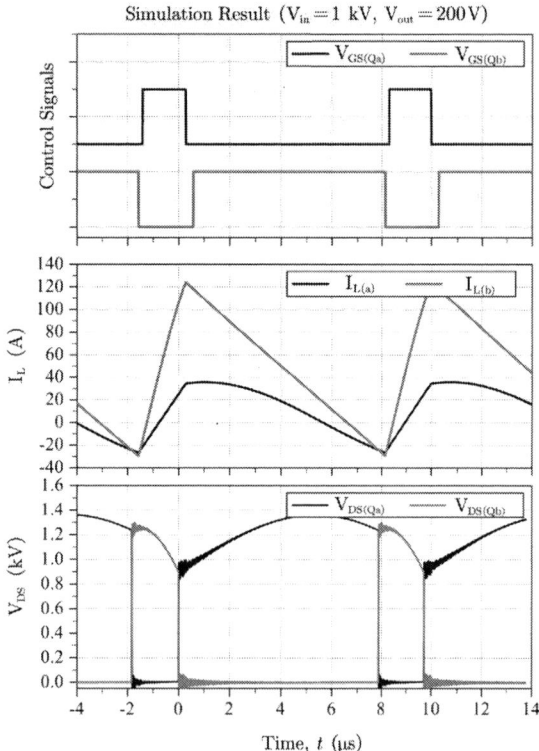

Fig. 14: Simulated inductor currents and switch voltages at V_{in} =1 kV and V_{out} = 200V

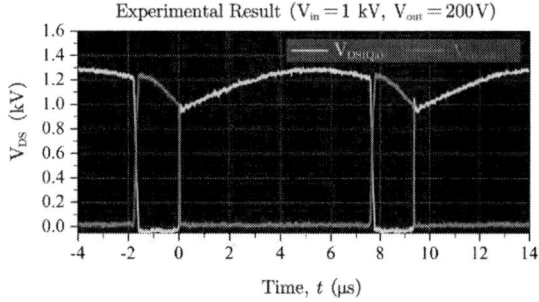

Fig. 15: Experimental drain-source voltages for 1 kV DC to 200 V DC test. Peak dV/dt is 40 V/ns during Q_a turn-off and 33 V/ns during Q_b turn-on.

7 Acknowledgment

The authors acknowledge the funding support from the U.S. Department of Energy, Advanced Research Project Agency – Energy (ARPA-E) through award: DE-AR0001568.

The authors greatly appreciate in-kind donations from NBE Tech LLC for the nano-silver sintering paste and preform as well as from Microchip Technology Inc. who donated the 3.3 kV SiC MOSFET dies (PN: MSC027SMA330D/S). The authors would also like to thank Electronic Concepts Inc., whose expertise in capacitor design and fabrication was a key enabler of the coaxial capacitor structure. Lastly, the authors would like to acknowledge the financial support provided by the Bradley Department of Electrical and Computer Engineering via the Bradley Graduate Fellowship at Virginia Tech.

References

[1] "Oe report: Solid state power substation technology roadmap," U.S. Department of Energy Office of Electricity. (Jul. 2022), [Online]. Available: https://www.energy.gov/oe/articles/oe-report-solid-state-power-substation-technology-roadmap.

[2] "Bs 6622 xlpe pvc 12.7/22kv cable," Eland Cables. (), [Online]. Available: https://www.elandcables.com/handlers/downloadpdf.ashx?url=%5C%2Fmedia%5C%2Fcl0l350j%5C%2Fbs-6622-xlpe-pvc-22kv-cable.pdf.

[3] "Chapter 3: Enabling modernization of the electric power system, technology assessments: Transmission and distribution components," *U.S. Department of Energy, Quadrennial Technology Review*, 2015.

[4] S. Cuk and R. Middlebrook, "A new optimum topology switching dc-to-dc converter," in *1977 IEEE Power Electronics Specialists Conference*, 1977, pp. 160–179. DOI: 10.1109/PESC.1977.7070814.

[5] F. Vahidi and S. Tenbohlen, "Statistical failure analysis of european substation transformers," Nov. 2014.

[6] D. Cisilino. "Brugg cables," Thorne & Derrick International. (Feb. 2024), [Online]. Available: https://www.powerandcables.com/tag/underground-cable-vault/.

[7] T. D. Bracken, R. F. Rankin, R. S. Senior, R. Kavet, and L. G. Geissinger, "Magnetic-field exposures of cable splicers in electrical network distribution vaults," *Applied Occupational and Environmental Hygiene*, vol. 16, no. 3, pp. 369–379, 2001. DOI: 10.1080/10473220118938.

[8] W. A. Stewart, "Cooling of distribution transformers in vented underground vaults," *IEEE Transactions on Power Apparatus and Systems*, vol. PAS-88, no. 6, pp. 843–853, 1969. DOI: 10.1109/TPAS.1969.292401.

[9] J. P. A. Sandraz, F. de León, and J. Cultrera, "Validated transient heat-transfer model for underground transformer in rectangular vault," *IEEE Transactions on Power Delivery*, vol. 28, no. 3, pp. 1770–1778, 2013. DOI: 10.1109/TPWRD.2013.2260183.

[10] P. Bedge, J. Spaulding, D. Zimmerle, and G. Duggan, "Modeling the winding hot-spot temperature and aging of enclosed vault transformers using a physics-based heat transfer model," in *2020 IEEE/PES Transmission and Distribution Conference and Exposition*, 2020, pp. 1–5. DOI: 10.1109/TD39804.2020.9299993.

[11] "Electric cables – Calculation of the current rating," *IEC 60287*, 2014.

[12] R. L. WEBB, "Air-side heat transfer in finned tube heat exchangers," *Heat Transfer Engineering*, vol. 1, no. 3, pp. 33–49, 1980. DOI: 10.1080/01457638008939561.

[13] H. Shokouhmand, S. Mahjoub, and M. R. Salimpour, "Constructal design of finned tubes used in air-cooled heat exchangers," *Journal of Mechanical Science and Technology*, vol. 28, pp. 2385–2391, 2014. DOI: 10.1007/s12206-014-0145-z.

[14] R. Kocurek and J. Adamiec, "Manufacturing technologies of finned tubes," *Advances in Materials Sciences*, vol. 13, Oct. 2013. DOI: 10.2478/adms-2013-0009.

[15] M. E. B. Scott E. Newland, "Construction Challenges of 345kV Underground," *Electrical Transmission Line and Substation Structures*, pp. 317–325, 2000. DOI: 10.1061/40790(218)29.

[16] M. D. Seeman, V. W. Ng, H.-P. Le, M. John, E. Alon, and S. R. Sanders, "A comparative analysis of Switched-Capacitor and inductor-based DC-DC conversion technologies," en, in *2010 IEEE 12th Workshop on Control and Modeling for Power Electronics (COMPEL)*, Boulder, CO, USA: IEEE, Jun. 2010, pp. 1–7. DOI: 10.1109/COMPEL.2010.5562407.

[17] S. Kaufmann, T. Lang, and R. Chokhawala, "Innovative press pack modules for high power igbts," in *Proceedings of the 13th International Symposium on Power Semiconductor Devices & ICs. IPSD '01 (IEEE Cat. No.01CH37216)*, 2001, pp. 59–62. DOI: 10.1109/ISPSD.2001.934559.

[18] J. Hu and C. Sullivan, "Ac resistance of planar power inductors and the quasidistributed gap technique," *IEEE Transactions on Power Electronics*, vol. 16, no. 4, pp. 558–567, 2001. DOI: 10.1109/63.931082.

[19] R. S. Yang, A. J. Hanson, B. A. Reese, C. R. Sullivan, and D. J. Perreault, "A low-loss inductor structure and design guidelines for high-frequency applications," *IEEE Transactions on Power Electronics*, vol. 34, no. 10, pp. 9993–10 005, 2019. DOI: 10.1109/TPEL.2019.2892397.

[20] "Ice9 nylon - tcpoly - advanced 3d printer materials," TCPoly. (2022), [Online]. Available: https://tcpoly.com/rigid-ice9/.

PCIM Europe 2024, 11– 13 June 2024, Nuremberg   DOI: 10.30420/566262075

Controlled Inductor Based BCM Buck Converters

Shmuel (Sam) Ben-Yaakov, Ziv Gellman, Iris Eting, and Yivgeni Semidotskih

Department of Electrical and Computer Engineering, Ben Gurion University of the Negev, Israel

Corresponding author: Shmuel (Sam) Ben-Yaakov, sby@bgu.ac.il
Speaker: Ziv Gellman, zivgell@post.bgu.ac.il

Abstract

The study presents and demonstrates an efficiency enhancement of Buck converters by a controlled inductor that achieves BCM mode at fixed frequency. The paper includes analysis, simulation, and experimental verification. Laboratory measurements show an improvement of up to 4% in efficiency compared to a conventionally designed Buck converter. The intended future work includes neural network (NN) training for three parameters control: duty cycle, inductance and switching frequency.

1   Introduction

PWM converters conventionally apply fixed inductors as intermediate energy storage elements. Numerous studies have shown that the performance, as well as the efficiency of PWM converters and resonant converters, can be improved by applying a controlled inductor (CI) [1-8]. This study investigated the use of a CI in a BCM Buck converter.

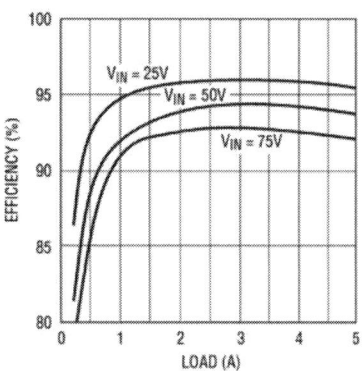

Fig. 1: Typical efficiency vs load current of a step-down switching converter as given in datasheet of a DC-DC controller (LTC3703, Analog Devices).

Fig.1 is an example of the efficiency of a Buck converter as given by the manufacturer (Analog Devices). Typical of many switch mode converters, there is a significant deterioration of the efficiency at lower power levels. This is due

to the increase in the ratio of switching losses to the power level. Many a times, the lower power range is in fact the typical operating point, especially when redundancy is used.

Buck converters can be operated in three modes: Continuous Current Mode (CCM), Discontinuous Current Mode (DCM) and Border line Current Mode (BCM) with inductor current waveforms as shown in Fig. 2.

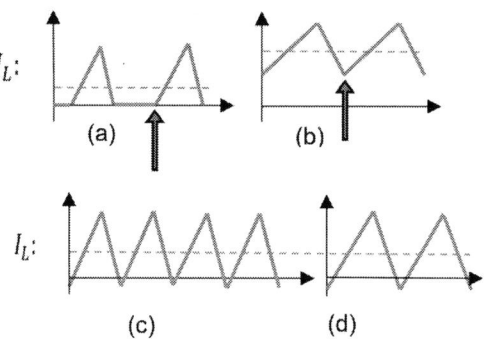

Fig. 2: Inductor current in different modes. Arrows indicate instances of hard switching. (a) DCM. (b) CCM. (c) BCM obtained by a frequency change at higher input voltages. (d) BCM obtained by a controlled inductor.

Since BCM mode achieves zero voltage switching (ZVS) as the high side switch turns on, it eliminates the harmful reverse recovery spikes and losses as well as Coss losses and can thus potentially improve the efficiency at moderate power levels. To ensure self-commutation of the midpoint at BCM, the inductor current needs to reach negative values to store sufficient energy in the inductor needed to charge the capacitances. At high power levels, the improvement by BCM may be hampered by increased conduction losses. However, even at moderate power levels, the potential improvement in efficiency in BCM, that is obtained by frequency change, is offset by the losses due to the increased switching frequency as the input voltage increases (Fig. 2c). This is because the inductor ripple current is dependent on the product $(L \cdot f_s)$.

$$\Delta I_L = \frac{I_{av}}{2} = V_{out} * \left(1 - \frac{V_{out}}{V_{in}}\right) * \frac{1}{L \cdot f_s} \qquad (1)$$

where:

ΔI_L = inductor's current ripple

V_{out} = output voltage

V_{in} = input voltage

L = inductance of inductor

f_s = switching frequency

That is, for a constant L, there is a need to increase the frequency when V_{in} increases, to sustain BCM while keep the ripple constant for same average current. Alternatively, as Eq. (1) implies, the frequency can be kept constant while increasing the inductance of the inductor and thus avoiding the penalty of a switching frequency increase (Fig. 2d).

The objective of this study was to explore the latter approach using a controlled inductor (CI).

2 Design concept

This study applied the CI originally suggested in [9]. It is built around an E core (Fig. 3) with a gap in the mid arm on which the inductor winding is placed. The side arms windings are for DC current bias that controls the inductance. This is achieved by pushing the ferromagnetic material of the outer frame toward saturation, as depicted intuitively in Fig. 4.

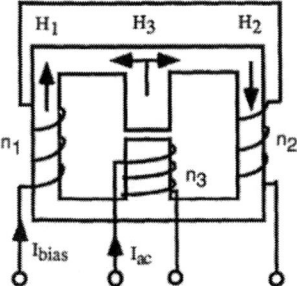

Fig. 3: The variable inductor concept.

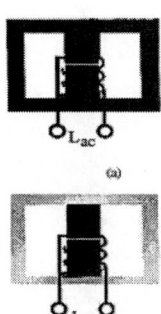

Fig. 4: The effect of the DC bias current on the incremental permeability of CI outer frame. Darker shade signifies higher permeability.

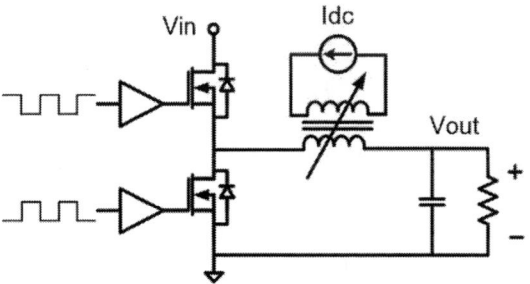

Fig. 5: The investigated controlled inductor Buck.

As the DC bias current (Idc) increases the magnetic incremental permeability of the ferromagnetic material of the outer frame decreases. Since this is also in the flux path of the inductor section, the relative permeability seen by the inductor windings decreases. This permeability change is illustrated in Fig. 4. The change causes a reduction in the relative permeability of the core path seen by the inductor and the change in the permeability translates into a decrease in the inductance according to Eq. (2).

$$L = \frac{\mu_0 \cdot \mu_r(I_{dc}) \cdot N^2 \cdot A_e}{l_e} \qquad (2)$$

where:

μ_0 = permeability of free space

μ_r = relative permeability

N = number of turns in the inductance coil

A_e = effective area of the core

l_e = effective length of the core

The studied Buck topology consisted of a half bridge, the CI, and the output section (Fig. 5). This structure was used to explore the efficiency improvement that can be obtained when operating the converter at a fixed frequency while it is kept in the BCM by the CI.

3 Simulations

The simulations schematics were based on the design of a Buck converter evaluation board (DC501A, Analog Devices), which was then used in the experimental part (Fig. 6).

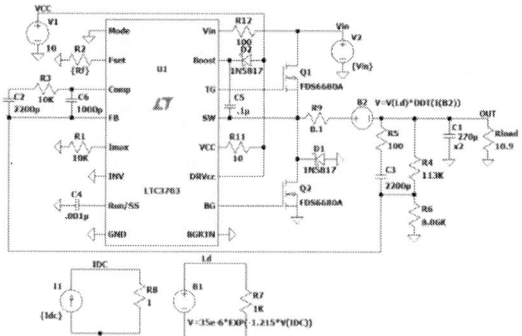

Fig. 6: LTspice simulation circuit of the DC501A evaluation board based on the LTC3703 controller (Analog Devices).

Figure 6 is the simulation schematics of the buck converter utilizing the LTC3703 controller. Q1 and Q2 are external power MOSFETS in a half-bridge configuration. The controller can operate in one of two modes. When the Mode input is connected to VCC, it operates in pulse-skip mode which inhibits negative inductor currents. Connection of Mode to GND, as done in this study, keeps the lower MOSFET (Q2) on during the OFF duration and hence allows negative inductor current which is needed for the CI control operation – to ensure self-commutation. Rf of the Fset terminal of the controller was set for 260kHz in the case of original operation, for 110kHz in the CI controlled operation and was adjusted as required in the variable F operation.

The nonlinear inductor was emulated by the behavioral dependent source (B2) which realizes the state equation of an inductor:

$$V = L_d \frac{dI}{dt} \qquad (3)$$

where:

L_d = the incremental inductance of the CI

I = the CI AC current

The source I1, (Fig. 6) was the command of the DC winding current (V(I_{DC})) while the inductance Ld was emulated by B1 to produce (V(L_d)) which

approximates the experimental CI inductance by an exponential function.

$$V = 35 \cdot 10^{-6} \cdot e^{(-1.215 \cdot V(I_{DC}))} \qquad (4)$$

The simulations covered three cases:

1. The original evaluation board design
2. Constant inductance with variable frequency to maintain BCM
3. Variable inductance with fixed frequency

In each case the input and output powers (including the power loss of the DC winding) were measured, and the power losses and efficiency were calculated.

System parameters of the simulations and experimental stages are summarized in Table 1.

Parameter	Value
$V_{out}[V]$	12
$I_{out}[A]$	1.1
$V_{in}[V]$	20-40
$P_{out}[W]$	13.24

Table 1 Parameters values used in the simulation.

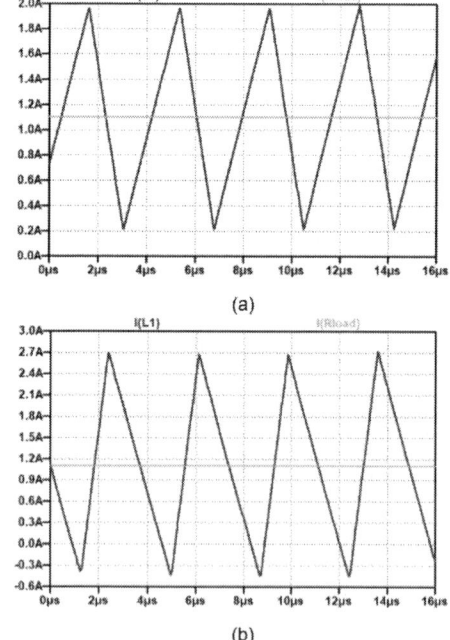

(a)

(b)

Fig. 7: Inductor current in original design operation, 20V input (a), 40V input (b).

In the first simulation run, the performance of the original evaluation board design was tested at a constant 260kHz switching frequency, and the loss and efficiency registered. In this case the inductor current was found to be positive at 20V and slightly negative at 40V as shown in Fig. 7a,b.

Two additional cases were then simulated. In the first one, the inductance was kept constant at the original value of 10μH while the frequency was varied to obtain the highest efficiency at each input voltage. In the second case, the frequency was fixed to 110KHz while the inductance was altered to reach maximum efficiency at BCM.

The results of the simulation runs are summarized in Figs. 8-10. As seen in Fig. 8 and Fig. 9, to reach maximum efficiency, the frequency and the inductance need to be increased as the input voltage rises. And as Fig. 10 reveals, the power loss with the CI was found to be significantly lower than the losses in the variable frequency case. while both were forced to be at BCM.

Figure 11 is a comparison of the three cases: original design with fixed inductance with a fixed F, constant inductance and variable frequency (F controlled), and constant frequency with a variable inductance (L controlled). It is evident that the F controlled method increases the losses at higher input voltages, while the L controlled method improves the efficiency for all the simulated operating points. This trend is seen at higher input voltage up to 72V, Fig. 12, beyond the experimental range of 20V-40V.

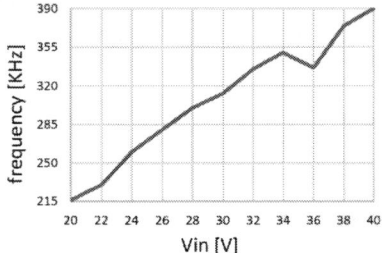

Fig. 8: Frequency changes due to Vin increase as required to keep the operation at BCM.

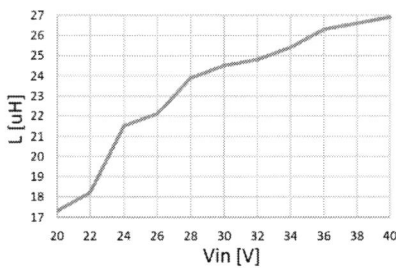

Fig. 9: inductance changes due to Vin as required to keep the operation at BCM.

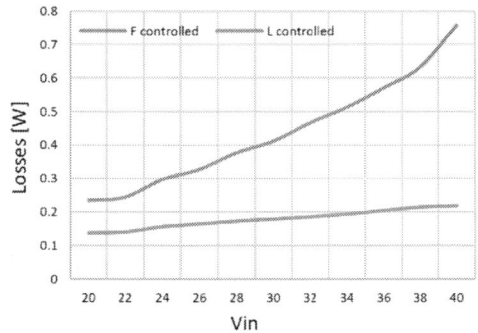

Fig. 10: Power losses vs input voltage.

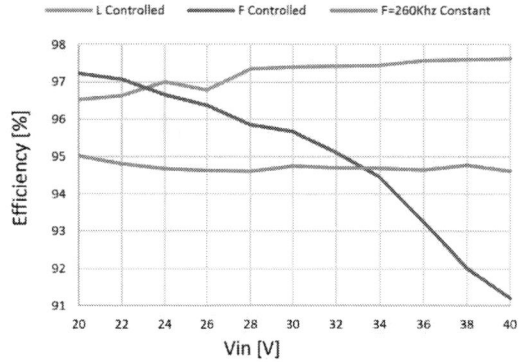

Fig. 11: Simulation efficiency comparison between constant inductor constant frequency, constant inductor F control and constant frequency L control .

Parameter	Value
$V_{out}[V]$	12
$I_{out}[A]$	2.5
$V_{in}[V]$	40-72
$V_{out}[W]$	30

Table 2 Parameters Values for extended voltage switching losses simulation.

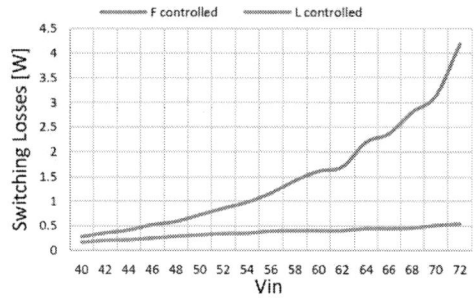

Fig. 12: switching losses for high input voltages.

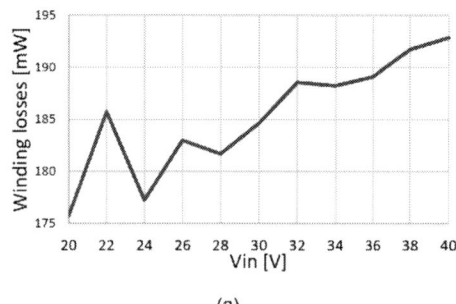

(a)

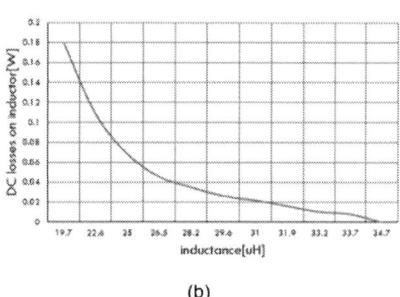

(b)

Fig. 13: Losses in CI method, winding losses vs input voltage (a), DC losses vs inductance value (b).

Figure 13 is the simulated value of the DC winding losses. As expected, the losses are relatively high at low CI inductance which is obtained by a higher DC current.

The simulation runs were also used to estimate the inductor energy needed at BCM to ensure self-commutation of the midpoint (Fig. 14). This is based on the need of stored energy to charge the transistor capacitances. simulation was made to calculate the amount of energy as shown in Fig. 14.

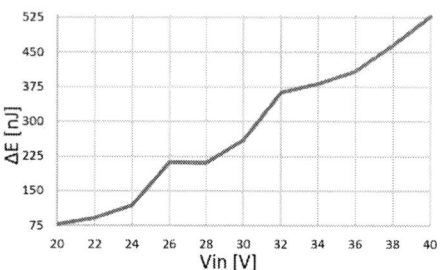

Fig. 14: Stored energy (ΔE) required to charge load the transistors outer capacitance.

Another approach tested by simulation is a combination of the frequency adjustments method and the CI method. This was tested for an input voltage of 30V and three frequencies. The optimum inductances and frequencies are given in Table 3 and summarized in Fig. 15.

Frequency [KHz]	L[µH]
80	31
110	24.5
200	12.6

Table 3 Inductance value for maximum efficiency in different frequencies.

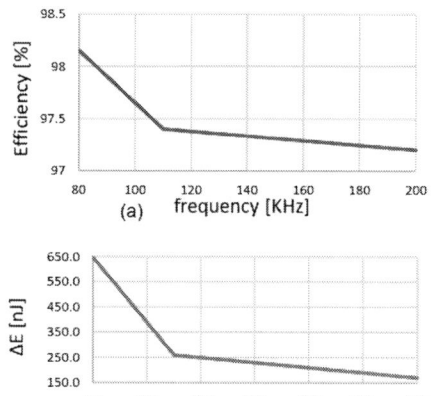

Fig. 15: Results of controlled inductor simulation at three different frequencies per Table 3. Efficiency vs frequency (a), stored energy as function of frequency (b).

The main conclusions out of the series of the conducted simulation, is that lower frequencies improve efficiency in CI operation with fixed frequency, and that the optimal solution may lie in combining the two methods: L and F control.

4 Experimental and results

Adjustments were made to the evaluation board (Fig. 16) allowing to switch between the regular inductor and the variable inductor. The loss and efficiency values calculations accounted for all power sources connected to the board, VCC, V_{in}, and V_{DC} (CI DC winding loss). All measurements were made for a constant output voltage of 12V and a constant power of 13.24W. A digital multimeter (DMM) was used to capture the necessary voltage and current data.

The CI was obtained from Hulda Transformers, Israel (cat. # 100037042010) with a range of 11µH to 35µH for DC bias current range of 0 to 1A (Fig. 17).

The lab experiment procedure was according to the flow chart of Fig. 18.

Fig. 16: DC501A Evaluation board after adjustments.

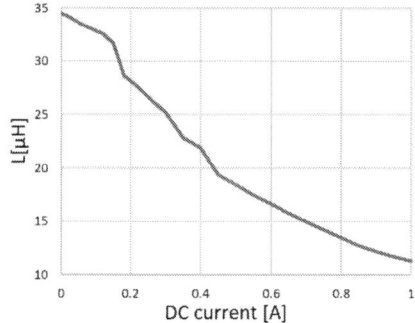

Fig. 17. The L= f(Idc) response of the experimental controlled inductor (Hulda transformers cat. #100037042010).

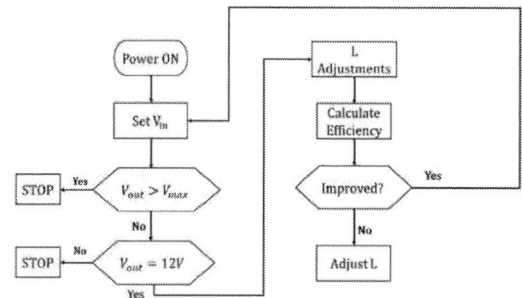

Fig. 18: Experiment flow chart.

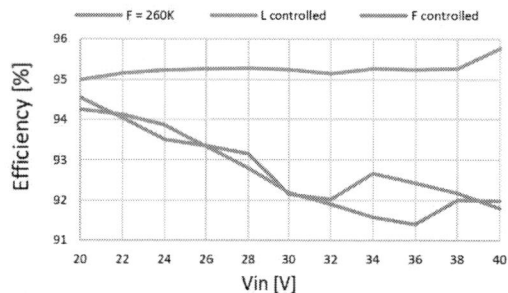

Fig. 19: Measured efficiency as various input voltages of the original design, F controlled and L controlled options.

As evident from the results shown in Fig. 19, applying the CI method resulted in higher efficiency than the board's original efficiency (F=260Khz and L=10µH), and higher than the alternative method of BCM with frequency variation when input voltage is changed. Efficiency improvement was found to be up to 4%. Similar to the simulations part (Fig. 11), the CI method is the most efficient. The frequency adjustments with a fixed L method becomes less efficient as the input voltage rises. For both methods, the simulations and the experiments had the same behavior. However, the simulations of the original board design exhibit almost constant efficiency while in the lab the efficiency decreases when the input voltage was increased.

5 Future work

The results of this study support the conclusion of numerous previous studies by others, that a CI can improve the efficiency of PWM converters. This calls for a two-dimensional control in DC-DC converters, CI control plus the conventional duty cycle control. This study reveals that a three-dimensional control D, L and frequency can further improve the efficiency. One way to implement such a control is a neural network (NN) that can be trained to optimize the operation over the operational envelop. Similar to the proposal of [10], a NN can be set up for this case as described by the generic block diagram of Fig. 20. The training could be done by simulation while perhaps final tuning on a sample converter. It is planned to pursue this approach in the next phase of the study.

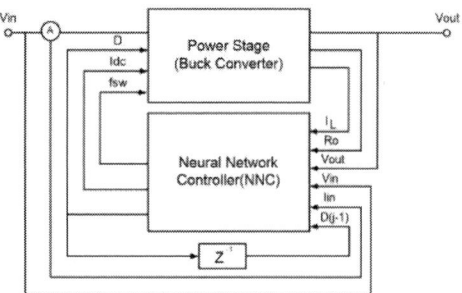

Fig. 20 Suggest a NN setup for a three-dimensional control (D, L, F) training of a PWM converter.

6   Conclusions

Analysis, simulation, and laboratory measurements have verified that applying a CI for fixed frequency BCM Buck converters can enhance efficiency. Compared to CCM BUCK with fixed inductance and frequency, and to BCM with constant inductance and variable frequency, an efficiency improvement of up to 4% was demonstrated, achieving an efficiency of 95.2% over a large input voltage variation at moderate power level. Although the experimental CI was considerably larger in size than the original inductor, it is estimated that an optimum CI design will not increase the overall size considerably while still improving the efficiency.

7   References

[1] R. L. Brandt, "DC-DC converter having magnetic feedback "United States, US 20060097711A1, May. 11, 2006.

[2] Y. Li *et al.*, "Extension of ZVS Region of Series–Series WPT Systems by an Auxiliary Variable Inductor for Improving Efficiency," in *IEEE Transactions on Power Electronics*, vol. 36, no. 7, pp. 7513-7525, July 2021.

[3] M. W. Beraki, J. P. F. Trovão, M. S. Perdigão and M. R. Dubois, "Variable Inductor Based Bidirectional DC–DC Converter for Electric Vehicles," in *IEEE Transactions on Vehicular Technology*, vol. 66, no. 10, pp. 8764-8772, Oct. 2017

[4] T. Yan, T. Chen, A. Huang, W. Chen, T. Cao, "Cuk PFC Converter Based on Variable Inductor", Electronics 2023, 12(10), 2245.

[5] A. P. Mendes, B. Baptista, M. S. Perdigão and A. M. S. Mendes, "Experimental analysis of a DC current-controlled variable inductor in a DC-DC converter," *2019 IEEE International Conference on Industrial Technology (ICIT)*, Melbourne, VIC, Australia, 2019, pp. 440-445.

[6] M. W. Beraki, J. P. F. Trovão, M. S. Perdigão and M. R. Dubois, "Variable Inductor Based Bidirectional DC–DC Converter for Electric Vehicles," in *IEEE Transactions on Vehicular Technology*, vol. 66, no. 10, pp. 8764-8772, Oct. 2017.

[7] Beraki, Mebrahtom W.; Trovão, João Pedro F.; Perdigão, Marina S. Wiley "Performance enhancement of powertrain DC–DC converter using variable inductor" IET electrical systems in transportation, 2021, Vol.11 (2), p.161-170.

[8] Omar Abu Mohareb, "Efficiency Enhanced DC-DC Converter Using Dynamic Inductor Control" Wiesbaden: Springer Fachmedien Wiesbaden : Imprint: Springer Vieweg -- 2019 1st ed. 2019.

[9] D. Medini and S. Ben-Yaakov, "A current-controlled variable-inductor for high frequency resonant power circuits," *Proceedings of 1994 IEEE Applied Power Electronics Conference and Exposition - ASPEC'94*, Orlando, FL, USA, 1994, pp. 219-225.

[10] N. Kodner, D. Adar and S. Ben-Yaakov, "Neural Network Controllers for Switch Mode Systems: Off Line Training by an "Ideal Controller" Data," *Eighteenth Convention of Electrical and Electronics Engineers in Israel*, Tel Aviv, Israel, 1995.

PCIM Europe 2024, 11– 13 June 2024, Nuremberg   DOI: 10.30420/566262076

Influence of varying Common Mode Choke Sizes on the Performance and Stability of an Active EMI Filter

Patrick Körner[1,2], Philip Brockerhoff[1], Felix Müller[1], Marco Jung[2]

[1] Vitesco Technologies, Germany
[2] Bonn-Rhein-Sieg University of Applied Science, Germany

Corresponding author: Patrick Körner, patrick.koerner@vitesco.com
Speaker: Patrick Körner, patrick.koerner@vitesco.com

Abstract

Power electronic converters generate conducted and radiated Electromagnetic Interference (EMI). An Active EMI Filter (AEF) is smaller in size, volume, weight, and cost than a similarly performing Passive EMI Filter (PEF). This paper proposes a Voltage Sense Current Inject (VSCI) Feedback (FB)-type AEF for Common Mode (CM) attenuation which can be installed at the AC-input of an 11 kW On-Board Charger (OBC). The simulation model of a 4-phase (4ph) Common Mode Choke (CMC) is presented, which is used to model the behavior of three different CMCs. Three AEF designs are compared to each other in terms of their performance, while a stability assessment for the AEF FB-loop is described which is based on the Nyquist stability criterion. It is shown that an AEF can reduce the CMC impedance by approximately 58 % in comparison to a baseline PEF. This is verified by both simulation and experimental validation with a Vector Network Analyzer (VNA). The results of a transient Electromagnetic Compatibility (EMC) simulation for the Power Factor Correction (PFC)-stage underline the performance of the proposed AEF.

1 Introduction

Modern power converters utilize high switching frequencies to reduce the size of passive components, which increases the converter's power density.

On system level, further volume and weight reductions are possible by integrating different functions within one compartment. In the case of an OBC for Electric Vehicles (EVs), the charging function is often combined with a HV/LV-DC/DC converter in the same housing. [1]

Higher integration levels and switching with higher frequencies can lead to increased levels of conducted and radiated EMI. As for OBCs, automotive standards for grid-connected charging (e.g. ECE R10 [2]), restrict the conducted emissions towards a certain limit in the frequency range from 150 kHz to 30 MHz. EMI filters are used to attenuate the conducted emissions. Since the penetration of power electronic equipment in the grid is increasing, worldwide standard committees are already recommending new EMC constraints in the frequency range from 2 kHz to 150 kHz, also referred to as supraharmonics [3]. High-power applications, like OBCs, already have PEFs that are large in volume, weight, and cost thus the filter is only designed to achieve proper EMI attenuation starting at 150 kHz. When the filter needs to attenuate emissions also in the lower frequency band, its volume would increase even further because larger passive components would be needed.

A solution to decrease the size of PEFs, is the usage of AEFs [4], [5]. Previously conducted research shows that an AEF can reduce the filter volume by 40 % [6], while magnetic components like CMCs can be reduced by 50 % in volume [7]. AEFs can be implemented to attenuate CM or Differential Mode (DM) EMI. A further categorization is possible, based on the used sensing and cancellation method [4]. AEFs have been presented for various applications like DC/DC converters, AC/DC converters, and inverters [8], [9].

This paper proposes a CM VSCI FB-type AEF, which is used at the AC-input of an 11 kW OBC. As a baseline design, a PEF is used as a performance indicator for the filter's CM Insertion Loss (IL). The AEF is implemented for three different CMC impedance sizes. This gives a qualitative example what CMC impedance reduction is feasible until the AEF achieves a similar CM performance as the baseline PEF, which uses the CMC with the largest impedance. In section 2 a model for a 4ph CMC is presented, which is verified by a comparison between simulation and measurement results. This is done for three different CMCs. Section 3 describes the used stability assessment for the AEF FB-loop and investigates the behavior for two different termination cases. The achieved simulation and experimental results for the different EMI filter implementations are presented in section 4. Measurement results for the CM IL are shown. Furthermore, an EMC simulation of the PFC-stage proves the AEF performance, when it is installed in the OBC.

2 Common Mode Choke

The total amount of Y-capacitance, which can be installed between the power lines and protective earth is limited. The reason for this are safety requirements, which limit the maximum allowed earth leakage currents. [10] Therefore, CMCs are a key component in the EMI filter to provide high enough CM attenuation to deal with generated CM currents. A 4ph CMC consists of four windings, which are wound in the same direction so that a magnetic flux, generated by DM currents, is canceled. CM currents lead to a residual magnetic flux in the magnetic core and therefore, the CMC represents an inductance. The magnetic core typically consists of a high-performance magnetic material, like ferrite, nanocrystalline, or amorphous materials [10]. In this paper, a nanocrystalline magnetic core with four sectional windings is considered.

2.1 High-Frequency Equivalent Model

In the literature, lumped [10]–[16] and behavioral models [17], [18] are presented to accurately depict the high-frequency behavior of CMCs. To model the CMC's impedance behavior with reasonable accuracy and modeling effort, the in Fig. 1 shown equivalent circuit is proposed as an adapted model from [10].

The inductive behavior of the CMC is represented by the inductances L_1 and L_2, both having the magnetic coupling factors k_1 and k_2 towards the other sectional windings. The parasitic Z_{dm}, which is caused by magnetic stray fields, determines the coupling factors $k_1 < 1$ and $k_2 < 1$, while R_{ser} represents the ohmic resistance of the copper wire. Nanocrystalline CMC cores have an imaginary part

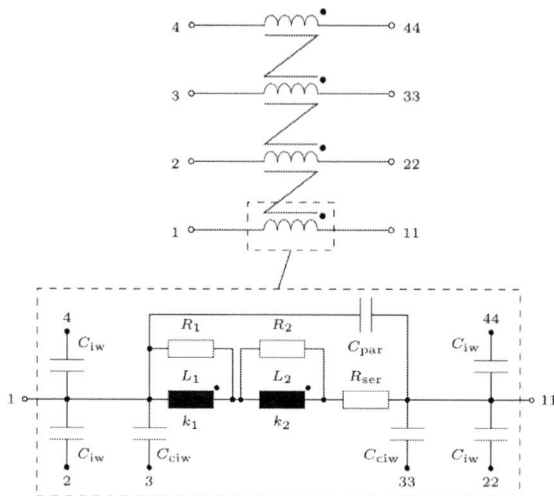

Fig. 1: High frequency equivalent circuit of a 4ph CMC.

of the complex permeability that is frequency-dependent [18]. Therefore, a decrease in the total Z_{cm} for a rising frequency can be observed. This is modeled by both L_1 and L_2, having differing values and parallel resistances.

The parasitic interwinding capacitance C_{iw} represents the coupling from one sectional winding towards the next sectional winding. As a 4ph CMC is investigated, a cross-interwinding capacitance C_{ciw} is introduced which represents the coupling towards the winding, which is placed on the other side of the magnetic core. The cross-interwinding capacitance is smaller than the interwinding capacitance. The parasitic capacitance C_{par} is caused by the parallel wound turns of each section. For simplification and because C_{iw} and C_{ciw} are much smaller than the parasitic parallel capacitance C_{par}, both C_{iw} and C_{ciw} are distributed equally between the input and output of each sectional winding. The most significant influence on the parallel resonance point of Z_{cm} has C_{par}.

As the conducting ground layer of the PCB, onto which the CMC is installed also influences the High Frequency (HF) behavior, the authors refer to [11] for further model enhancement.

2.2 Model Verification

To verify the developed model and to enable an accurate AEF simulation, the used CMCs are characterized with a VNA and are modeled in a circuit simulation software. The comparison between measurement and simulation for Z_{cm} is shown in Fig. 2.

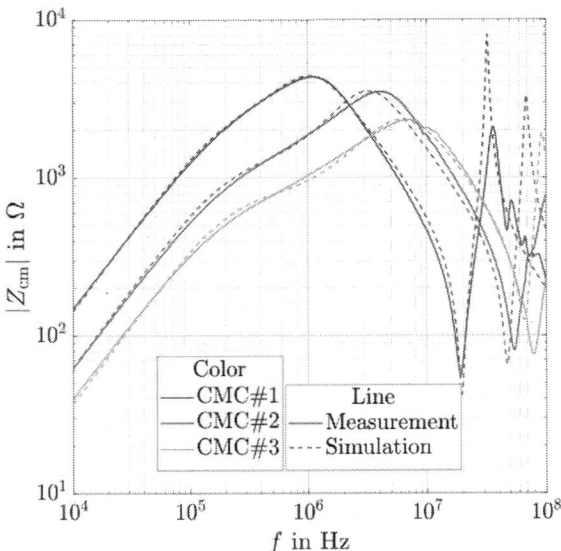

Fig. 2: Comparison between the measured and simulated Z_{cm} for the three different CMCs.

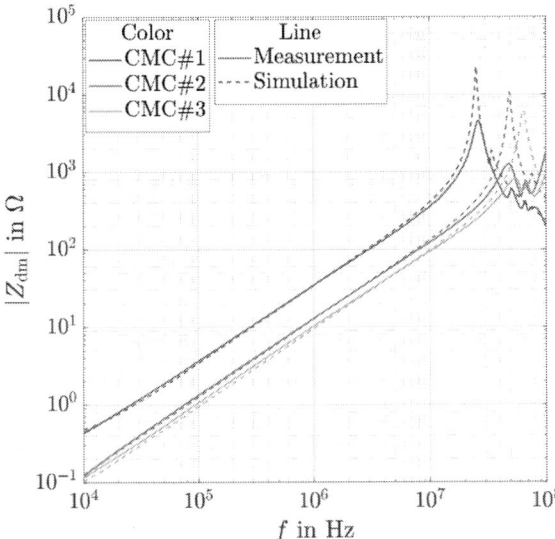

Fig. 3: Comparison between the measured and simulated Z_{dm} for the three different CMCs.

CMC#1 is used in the baseline PEF implementation. The CMCs CMC#2 and CMC#3 are realized by reducing the number of turns which are present on each one of the sectional windings. The size of the magnetic core stays the same for all CMCs. When CMC#1 has N_1 turns, the reduction of Z_{cm} at low frequencies (e.g. 10 kHz) for CMC#2 (N_2 turns) and CMC#3 (N_3 turns) can be calculated with:

$$x \approx 1 - \left(\frac{N_i}{N_1}\right)^2, i \in \{1; 2; 3\} \tag{1}$$

When compared to CMC#1, the impedance of CMC#2 and CMC#3 is reduced by 58 % and 73 %, respectively. This impedance reduction and the fact that fewer turns per section lead to a decrease in parasitic C_{par}, shifts the parallel resonance point of the CMC towards higher frequencies (cf. Fig. 2). In terms of filter performance, this is a positive effect, since the smaller CMCs provide higher Z_{cm} for frequencies beyond 2 MHz in comparison to CMC#1. Figure 3 shows the parasitic Z_{dm} for the three CMCs. Since the magnetic coupling factor stays nearly the same for the three CMCs, the decrease in parasitic Z_{dm} can also be calculated with Eq. (1).

3 Active EMI Filter Design

The proposed CM AEF is a VSCI FB-type topology. A simplified schematic of the complete EMI filter is shown in Fig. 4.

VSCI AEFs are applicable when the sensing and injection stage can be decoupled with a high impedance from the EMI source and the EMI sink. In the case of an OBC, the PFC stage acts as an EMI source, while the connected power grid represents the EMI sink. The VSCI topology is especially beneficial for high-power applications since only capacitive coupling towards the power lines is needed [4]. To ensure that a reasonably high decoupling impedance is present, the VSCI AEF is placed between two CMCs.

The AEF itself replaces a stage of passive Y-capacitors, which is present between both CMCs for a PEF implementation. CM disturbances are sensed via the C_{Y2} capacitors, which are part of an input high-pass filter H_{in}. As mentioned in [19], [20], this high-pass filter needs to properly attenuate the 50 Hz mains component. In a balanced three-phase system, this is not a major concern, as the 50 Hz component is canceled by summation of all three phases and the neutral. Due to the fact, that the shown AEF from Fig. 4 will also be used for

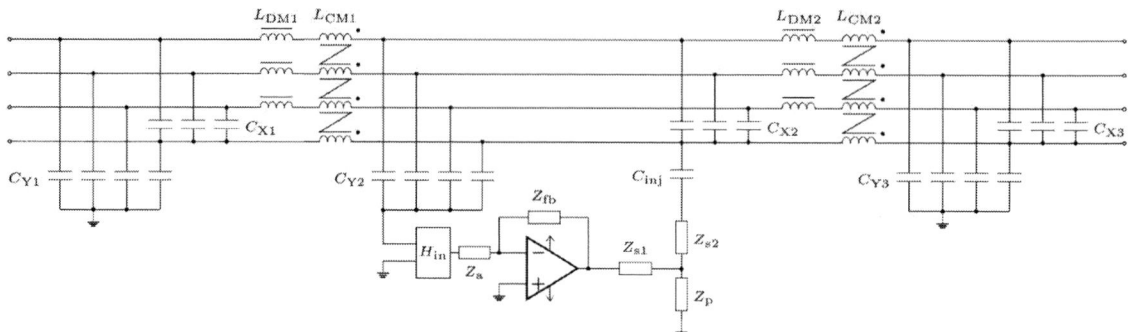

Fig. 4: Investigated VSCI CM AEF, which is used at the AC-input of an 11 kW OBC [4].

single-phase charging, the now visible 50 Hz CM component must be damped to prevent an overload of the the amplifier's output, which is limited by its supply voltage. An inverting Operational Amplifier (OpAmp), generates the needed high gain for the FB-loop. The injection stage consists of a single Y-capacitor C_{inj} and an impedance branch, which provides damping to ensure a stable FB-loop. An injected cancellation current is distributed between the lines via C_{X2} as it is large enough in value. This allows the usage of only one single C_{inj} that is connected to the X-capacitor star point at the neutral. The value of C_{inj} should not exceed the sum of the replaced Y-capacitors of a PEF implementation to ensure that the same leakage current level can be maintained.

3.1 Equivalent Circuit Model for CM EMI

To enable an analytical stability investigation of the feedback loop, a CM equivalent circuit model of the system from Fig. 4 is built. The derived model is shown in Fig. 5. It includes equivalent CM impedance representations of the EMI source Z_{source} and the EMI sink Z_{sink} which determine the filter performance and are important for stability investigation. The EMI sink impedance Z_{sink} is the CM equivalent of the Artificial Mains Network (AMN) that is placed between the power grid and the AC-input of the OBC and is used to measure the conducted emissions in accordance with [2]. The needed CM EMI source impedance Z_{source} can be approximated as the sum of the parasitic capacitances of the power stage and the DC-link towards ground potential [21].

3.2 Stability

Straightforward approaches for studying the stability of the AEF feedback system like bode diagrams are widely used in the literature [8], [20]. The au-

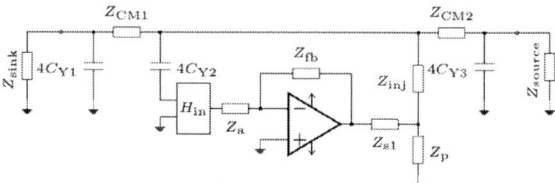

Fig. 5: CM equivalent circuit diagram of the shown AEF from Fig. 4, which is suitable for analytical loop gain analysis.

thors chose the Nyquist stability criterion, while also using the root locus diagram to check the system's stability similarly as it was done in [19], [22], [23].

3.2.1 Considered Termination Cases

This paper proposes a stability assessment for two cases, which are important for practical AEF application and can lead to either a stable or an unstable system. The following cases are considered in this paper:

1. AEF is installed in an OBC and is connected to the AMN. This means that the CM EMI source impedance Z_{source} and the AMN CM equivalent impedance Z_{sink} terminate the EMI filter.

2. A VNA is used to measure the CM IL of the AEF. Therefore, the EMI filter is terminated with 50 Ω on both the input and output side.

3.2.2 Design of the Feedback Loop Gain

The feedback network of the OpAmp is scaled to have a mid-gain band of approximately 45 dB for the frequency range, where EMI attenuation is required ($f \geq 150\,\mathrm{kHz}$). This was found to be a good trade-off between AEF performance, while still offering the possibility to ensure system stability.
It is found that a major concern for the stability of the feedback loop is the lower frequency range around 10 kHz, in which the phase of the system gets close

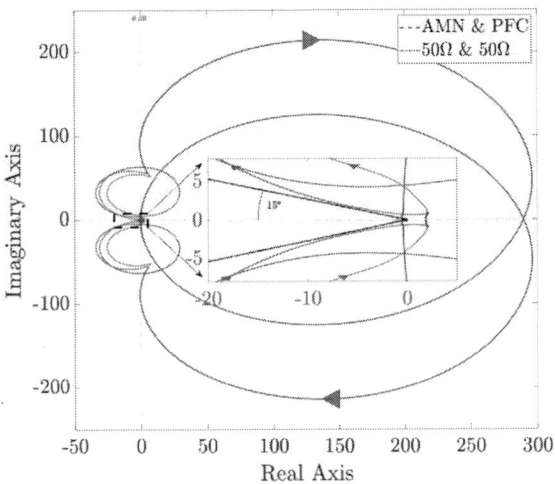

Fig. 6: Nyquist plot of the AEF open-loop gain, with investigation of the $(-1+j0)$ point for CMC#1. The system is closed loop stable.

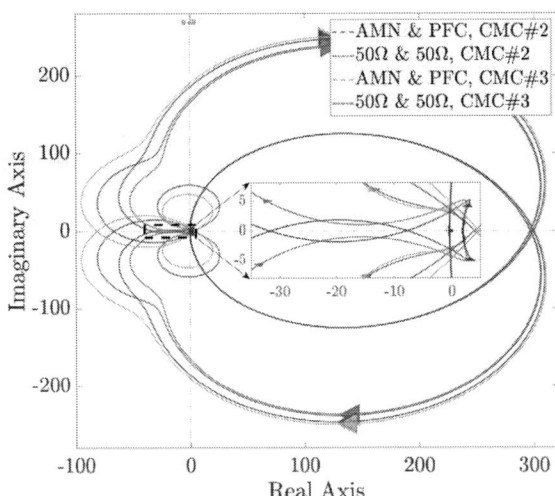

Fig. 7: Stability investigation for a variation in the CMC size, without rescaling of the damping network. Nyquist plots for CMC#2 and CMC#3 are shown for the two termination cases. The AEF is instable when installed in the OBC for both CMCs.

to $-180°$. This can be observed in Fig. 6, where a desired margin of $> 15°$ towards $-180°$ is set as a design requirement for the feedback loop gain. The critical $(-1+j0)$ point in the Nyquist diagram is not encircled. This means the designed system is closed-loop stable. Without a proper damping network scaling, the system tends to be unstable. The voltage divider that is placed at the output of the amplifier was proposed in [24] and serves the purpose of ensuring system stability well. It is necessary to use components in the damping network that are robust in terms of temperature dependency and aging to ensure that the system remains stable for the considered operating boundaries.

3.2.3 Influence of different CMC Impedances on the Stability

When the CMC is exchanged with one that has a smaller impedance (cf. Fig. 2), as it is done exemplary in Fig. 7, it can be observed that the system gets unstable now for case 1, visible by the encirclement of the critical $(-1+j0)$ point. Nevertheless, both systems with CMC#2 and CMC#3 are stable when terminated with $50\,\Omega$ on both sides. This is an interesting finding, as one may observe a stable AEF during a CM IL measurement with a VNA, thus the system will oscillate after it is placed in the OBC.

Generally, it is found that a change in CMC impedance causes a pair of complex conjugated zeros to be shifted into the right half plane. For a properly designed system, as it is visible in Fig. 8, all zeros are located in the left half-plain, thus not

causing the problem of potential instability due to the introduced phase-shift. As for Fig. 9, which is calculated for CMC#2, the damping network must be rescaled. Same applies, when CMC#3 is used.

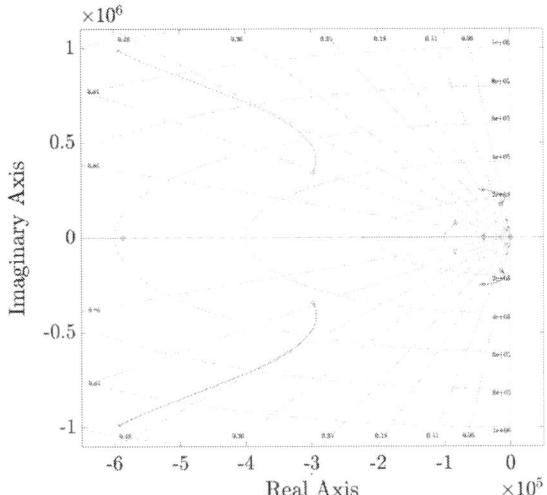

Fig. 8: Particular of the root locus for CMC#1, evaluated for the use-case OBC. The damping network is properly scaled, all zeros and poles are located in the left half plane.

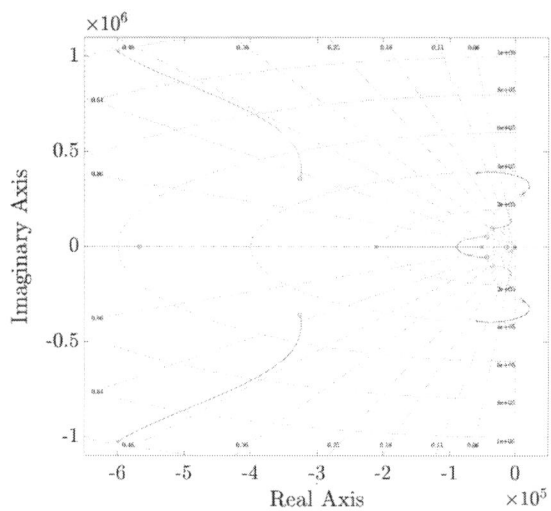

Fig. 9: Particular of the root locus for CMC#2, evaluated for the use-case OBC. The damping network is not properly re-scaled, a pair of complex conjugated zeros is in the right half plane.

4 Simulation and Experimental Results

To determine what CMC impedance reduction is feasible with an AEF, three different designs are investigated, utilizing the different CMCs from Fig. 2. For a comparable result, the damping network of the AEF is rescaled for each CMC impedance individually with the same stability assessment, as described in subsection 3.2.

4.1 Measurement Setup for the CM IL

The developed AEF is experimentally validated by measurements with a VNA to check the achievable CM IL in a 50 Ω system. Furthermore, the measurement is compared to simulations which are used to design the AEF. Overall the comparison between simulation and measurement shows a satisfactory match (cf. Fig. 10).

To conduct the measurement, an enclosed metallic box is used to provide the necessary ground reference to measure the CM IL of the EMI filter. Due to this setup, parasitic inductances and resistances in the ground path, limit the measurable S_{21} towards a certain value. This plays a crucial role for higher frequencies than 1 MHz, since the influence of the ground path parasitics gets more significant. The simulation is carried out with an ideal ground setup, which causes the difference in measurement and simulation at higher frequencies.

4.2 Results for the CM IL

The resulting CM IL curves are depicted in Fig. 10. Two identical CMCs of size CMC#1 are used as a baseline PEF (cf. Fig. 4, AEF structure is replaced by four passive Y-capacitors). It is already known that the corresponding PEF provides high enough CM attenuation so that the OBC can pass conducted emission tests according to the AC-charging limits, specified in ECE R10 [2].

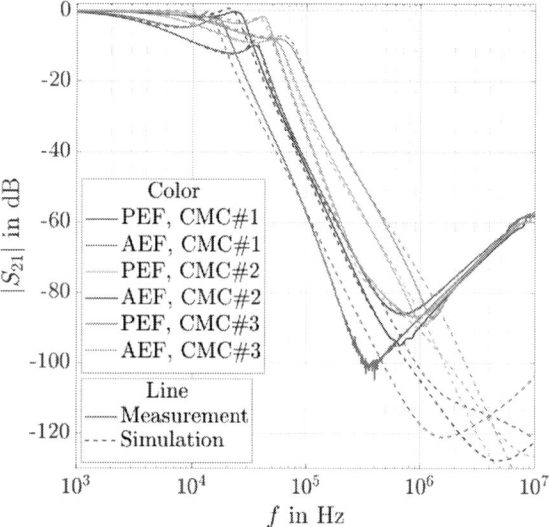

Fig. 10: Measured and simulated CM IL for three different CMC impedances. For each CMC, a PEF implementation and the corresponding AEF is shown.

An AEF implementation for CMC#1 is able to increase the achievable CM IL of the filter by approximately 18 dB at 150 kHz. This additional IL is now used to reduce the CMC impedance.

For comparison, the CM IL for the PEFs built with CMC#2 and CMC#3 is also provided. As expected, the effective cutoff frequency of the PEFs shifts towards higher frequencies due to the smaller CMC impedances in comparison to the baseline PEF.

The AEF, which is designed for CMC#2, shows a similar CM performance as the baseline PEF for the frequency range starting from 150 kHz. This means that the impedance of the CMC can be reduced by approximately 58 % thus a similar CM performance can be expected.

As for the AEF with CMC#3, the performance nearly reaches the desired CM IL of the baseline design. About 5 dB less CM attenuation at 150 kHz is achieved with the AEF when comparing it to the baseline PEF.

4.3 EMC Simulation of the PFC

As explained in subsection 3.2, it is important to validate the AEF performance and functionality also in the actual OBC to ensure that the feedback loop is correctly designed. This is done with a transient EMC simulation of the PFC stage running at 11 kW. The considered PFC is a Vienna-Rectifier working with a switching frequency of 70 kHz. In [4], a brief description is given of how the simulation setup can look like and what power stage parasitics must be included to also properly model the CM emissions. The simulation setup is suitable to accurately predict the conducted emissions up to a frequency of 1 MHz. This serves the purpose of AEF validation well.

Fig. 11 shows the simulated emission spectra of the OBC. The switching harmonics are present at multiples of 70 kHz. As it can be seen, the PEF in Fig. 11a is suitable to offer high enough attenuation, so that the emission level stays within the given limits according to ECE R10 [2].

The shown emission spectrum consists of DM and CM emissions. Since this paper only targets CM emissions with the proposed AEF, a reduction in emission level is caused by a reduction of CM noise level. Notably is the fact that the authors expect a slight increase in DM emissions when the AEF is implemented with a smaller CMC impedance size. The reason for this is the decrease in leakage inductance of the CMC, which represents a DM impedance (cf. Fig. 3). This can be observed in the simulation when comparing the 70 kHz and 140 kHz peaks from Fig. 11a to the peaks in Fig. 11c. These are the first and second switching harmonics, whereas a significant amount of DM noise is present. Due to the decreased DM inductance of CMC#2, the peak is higher (cf. Fig. 11c) than for the larger CMC (cf. Figs. 11a and 11b).

When comparing Fig. 11b to the baseline PEF for frequencies of 210 kHz and above, a decrease in emission level can be observed. To fully utilize this performance increase, the CMC impedance is reduced for the simulation depicted in Fig. 11c. As it can be seen, the emissions stay within the given limits although it is possible to reduce the CMC impedance by 58 %. As a main design criterion, the third harmonic at 210 kHz must be taken into account as this frequency is closest to the limits. At higher frequencies, still the AEF shows better performance than the baseline PEF.

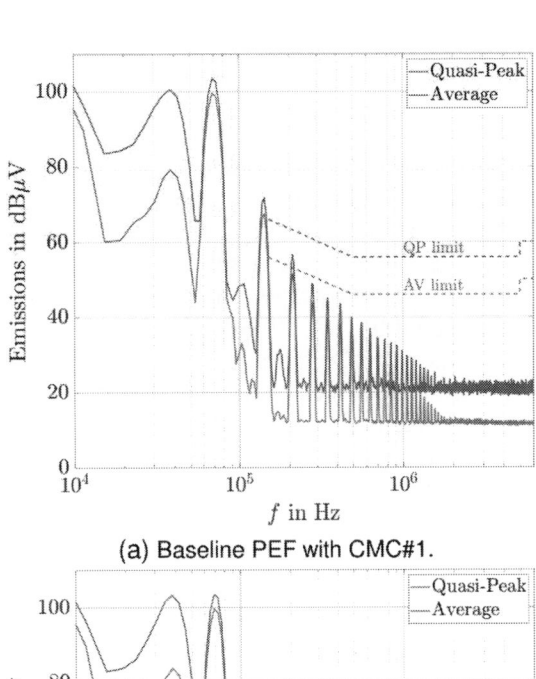

(a) Baseline PEF with CMC#1.

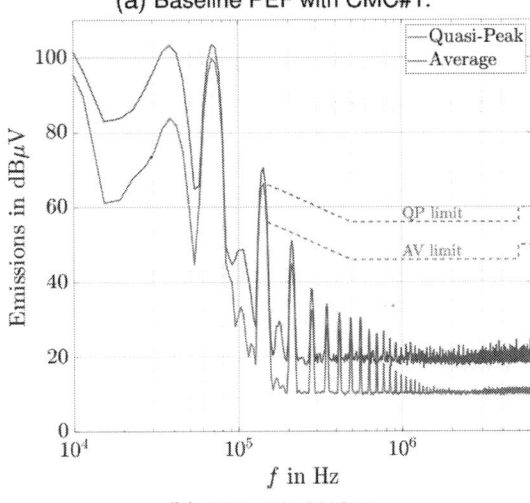

(b) AEF with CMC#1.

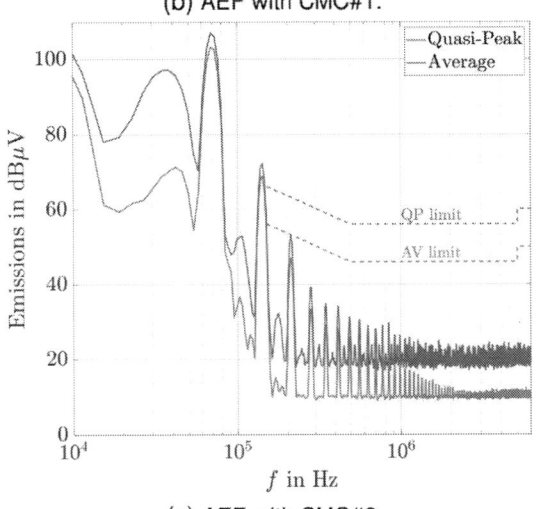

(c) AEF with CMC#2.

Fig. 11: Simulated emission spectra at the AMN for the baseline PEF implementation and two AEF implementations.

4.4 Discussion

It can be found that a CMC impedance reduction of approximately 58 % is feasible with the proposed AEF. The authors come to this conclusion since it was shown with the presented results in Fig. 10, that an AEF which uses CMC#2 can achieve a similar performance as the baseline PEF, even outperforming it for frequencies higher than 200 kHz. Also, the AEF build with the CMC#3, achieves a similar performance as the baseline PEF, although it appears to have slightly less CM IL in the 150 kHz range (cf. Fig. 10).

A positive effect of the size reduction is that CMCs with fewer turns have a smaller R_{ser} (cf. Fig. 1), which also leads to decreased power losses in the EMI filter. When it is possible to reduce the necessary number of turns on the magnetic core by half, the copper losses can be effectively reduced by 50 %. Horizontal CMCs are typically cooled by spreading the heat from its top surface to a heatsink. When less heat must be dissipated, the cooling concept can be simplified and pure air cooling might be sufficient.

The investigation in this paper is limited to the restriction that the CMC impedance is reduced by a turn-reduction for each sectional winding while the magnetic core stays fixed in size and volume. This has the positive effect, that the parasitic parallel capacitance C_{par} is reduced, thus the parallel resonance point of the CMC is shifted to higher frequencies and offers better performance. Since the size of the magnetic core is the main cost driver for CMCs, in a second step the magnetic core size will be decreased. Therefore, further investigations are necessary as it must be ensured that the magnetic core is large enough to prevent it from saturating. This evaluation is left for future work.

5 Conclusion

EMI Filters are an important part of power electronic converters to attenuate conducted and radiated emissions. A model for a 4ph CMC is shown, which can depict the HF behavior for the CM and DM impedance. The simulation model is verified by a measurement of the CMC impedance behavior of three different CMCs.

This paper proposes a VSCI FB-type AEF for CM attenuation, which is implemented at the AC-input of an 11 kW OBC. A stability evaluation principle for the system is shown, which uses the Nyquist stability criterion. The authors evaluate two impor-

tant cases that show a different behavior stability-wise for the three considered CMC impedances. It was found that an AEF can be in a stable operating state, when a VNA terminates the input and output of the AEF with 50 Ω, although instability can be observed when it is installed in an OBC.

In terms of performance, the CM IL is measured with a VNA for three different AEF implementations that are designed for three different CMCs. The AEFs are compared to a baseline PEF. With the VNA measurement, it was found that the AEF can reduce the size of the CMC impedance by more than 58 %. This result is also verified with an EMC simulation of the PFC stage of the OBC.

Acknowledgements

The authors would like to thank the German Federal Ministry for Economic Affairs and Climate Action as well as the Project Coordinator DLR for funding the research project CombiPower leading to this paper. Only the authors are responsible for the content of this publication.

References

[1] S. Schmalzl, P. Brockerhoff, B. Nyiredi, C. Bottke, W. Heimann, et al., "Functional Integrated Electronics for HV Architectures (Second Report)," JSAE Annual Congress, pp. 1–6, May 2023.

[2] United Nations Economic Commission for Europe, ECE R10: Uniform provisions concerning the approval of vehicles with regard to electromagnetic compatibility, Feb. 2017.

[3] A. Ganjavi, D. Kumar, F. Zare, and P. Davari, "Analyzing the Effect of Choke Placement on Differential-Mode Supra-Harmonics in Variable Frequency Drives: New Standardization," in IEEE 3rd International Conference on Sustainable Energy and Future Electric Transportation (SEFET), Bhubaneswar, India: IEEE, Aug. 2023, pp. 1–6. DOI: 10.1109/SeFeT57834.2023.10245535.

[4] P. Körner, P. Brockerhoff, and F. Müller, "Analysis of Passive and Active EMI Filters for On-Board Chargers in Electric Vehicles," in 25th European Conference on Power Electronics and Applications (EPE'23 ECCE Europe), Sep. 2023, pp. 01–08.

[5] B. Narayanasamy and F. Luo, "A Survey of Active EMI Filters for Conducted EMI Noise Reduction in Power Electronic Converters," IEEE Transactions on Electromagnetic Compatibility, vol. 61, no. 6, pp. 2040–2049, Dec. 2019. DOI: 10.1109/TEMC.2019.2953055.

[6] J. Biela, A. Wirthmueller, R. Waespe, M. L. Heldwein, K. Raggl, and J. W. Kolar, "Passive and Active Hybrid Integrated EMI Filters," *IEEE Transactions on Power Electronics*, vol. 24, no. 5, pp. 1340–1349, May 2009. DOI: 10.1109/TPEL.2009.2012404.

[7] M. C. Di Piazza, A. Ragusa, and G. Vitale, "Design of Grid-Side Electromagnetic Interference Filters in AC Motor Drives With Motor-Side Common Mode Active Compensation," *IEEE Transactions on Electromagnetic Compatibility*, vol. 51, no. 3, pp. 673–682, Aug. 2009. DOI: 10.1109/TEMC.2009.2025595.

[8] D. Shin, S. Jeong, and J. Kim, "Quantified Design Guidelines of a Compact Transformerless Active EMI Filter for Performance, Stability, and High Voltage Immunity," *IEEE Transactions on Power Electronics*, vol. 33, no. 8, pp. 6723–6737, Aug. 2018. DOI: 10.1109/TPEL.2017.2763972.

[9] A. Bendicks, M. Gerten, and S. Frei, "Active Cancellation of Periodic CM EMI at the Input of a Motor Inverter by Injecting Synthesized and Synchronized Signals (S3-AEF)," *IEEE Transactions on Power Electronics*, vol. 37, no. 10, pp. 11951–11961, Oct. 2022. DOI: 10.1109/TPEL.2022.3172205.

[10] M. L. Heldwein, L. Dalessandro, and J. W. Kolar, "The Three-Phase Common-Mode Inductor: Modeling and Design Issues," *IEEE Transactions on Industrial Electronics*, vol. 58, no. 8, pp. 3264–3274, Aug. 2011. DOI: 10.1109/TIE.2010.2089949.

[11] C. Domínguez-Palacios, P. Gonzalez-Vizuete, and J. Bernal Escuela, "Effect of Conducting Surfaces on the Performance of Common Mode Chokes," in *IEEE International Symposium on Electromagnetic Compatibility and IEEE Asia-Pacific Symposium on Electromagnetic Compatibility (EMC/APEMC)*, May 2018, pp. 363–368. DOI: 10.1109/ISEMC.2018.8393799.

[12] C. Domínguez-Palacios, P. González-Vizuete, M. A. Martín-Prats, and J. B. Mendez, "Smart Shielding Techniques for Common Mode Chokes in EMI Filters," *IEEE Transactions on Electromagnetic Compatibility*, vol. 61, no. 4, pp. 1329–1336, Aug. 2019. DOI: 10.1109/TEMC.2019.2918863.

[13] M. Kumar and K. Jayaraman, "Common Mode Impedance Shaping Choke to Attenuate the Conducted EMI in Three Phase Drive," in *IEEE International Conference on Power Electronics, Drives and Energy Systems (PEDES)*, Dec. 2020, pp. 1–5. DOI: 10.1109/PEDES49360.2020.9379346.

[14] A. Ojeda-Rodríguez, C. Dominguez-Palacíos, J. Bernal-Méndez, and M. Martín-Prats, "Simple and Accurate Characterization of Nanocrystalline Common Mode Chokes," in *IEEE International*

Symposium on Electromagnetic Compatibility & Signal/Power Integrity (EMCSI), Aug. 2022, pp. 472–477. DOI: 10.1109/EMCSI39492.2022.9889371.

[15] A. Ojeda-Rodríguez, J. Bernal-Méndez, and M. A. Martín-Prats, "Modal Theory and Approach for Accurate Characterization of Common Mode Chokes," *IEEE Transactions on Power Electronics*, pp. 1–15, 2023. DOI: 10.1109/TPEL.2023.3286007.

[16] W. Tan, C. Cuellar, X. Margueron, and N. Idir, "A High Frequency Equivalent Circuit and Parameter Extraction Procedure for Common Mode Choke in the EMI Filter," *IEEE Trans. on Power Electronics*, vol. 28, no. 3, pp. 1157–1166, Mar. 2013. DOI: 10.1109/TPEL.2012.2209206.

[17] I. Stevanovic, S. Skibin, M. Masti, and M. Laitinen, "Behavioral Modeling of Chokes for EMI Simulations in Power Electronics," *IEEE Transactions on Power Electronics*, vol. 28, pp. 695–705, Feb. 2013. DOI: 10.1109/TPEL.2012.2203319.

[18] M. Illia, L. Koleff, and G. Griepentrog, "Non-Ideal Model of the Common Mode Choke for EMI Filters," in *IEEE Applied Power Electronics Conference and Exposition (APEC)*, Tampa, FL, USA: IEEE, Mar. 2017, pp. 938–944. DOI: 10.1109/APEC.2017.7930809.

[19] M. L. Heldwein, "EMC Filtering of Three-Phase PWM Converters," Ph.D. dissertation, ETH Zurich, 2008. DOI: 10.3929/ethz-a-005635188.

[20] A. Kumar, Y. Hou, Y. Ramadass, T. Merkin, T. Hegarty, and A. Obidat, "An Active EMI Filter for High-Power Off-Line Applications," in *IEEE Applied Power Electronics Conference and Exposition (APEC)*, Mar. 2023, pp. 2063–2067. DOI: 10.1109/APEC43580.2023.10131427.

[21] H. Zhang, L. Yang, S. Wang, and J. Puukko, "Common-Mode EMI Noise Modeling and Reduction With Balance Technique for Three-Level Neutral Point Clamped Topology," *IEEE Transactions on Industrial Electronics*, vol. 64, no. 9, pp. 7563–7573, Sep. 2017. DOI: 10.1109/TIE.2017.2677344.

[22] M. C. Di Piazza, A. Ragusa, and G. Vitale, "An Optimized Feedback Common Mode Active Filter for Vehicular Induction Motor Drives," *IEEE Transactions on Power Electronics*, vol. 26, no. 11, pp. 3153–3162, Nov. 2011. DOI: 10.1109/TPEL.2011.2147801.

[23] G. F. Franklin, J. D. Powell, and A. Emami-Naeini, *Feedback Control of Dynamic Systems*, 6th ed. Upper Saddle River [N.J.]: Pearson, 2010.

[24] T. Hegarty, A. Kumar, R. Blattner, and A. Obidat, "An Active EMI Filter for Common-Mode EMI Mitigation in High- Power AC Systems," in *PCIM Europe: International Exhibition and Conference for Power Electronics, Intelligent Motion, Renewable Energy and Energy Management*, May 2023, pp. 1–6. DOI: 10.30420/566091339.

PCIM Europe 2024, 11–13 June 2024, Nuremberg   DOI: 10.30420/566262077

A High Efficiency Battery Charger with Maximum Power Point Tracking for Magnetic Energy Harvesters

Antonio-Miguel Muñoz-Gómez[1] , Javier Ballestín-Fuertes[1] , José-Francisco Sanz-Osorio [2]

[1] CIRCE Foundation, Zaragoza, Spain

[2] Instituto Universitario de Investigación Mixto CIRCE Universidad de Zaragoza-Fundación CIRCE, Zaragoza, Spain

Corresponding author & speaker: Antonio-Miguel Muñoz-Gómez, amimunoz@fcirce.es

Abstract

Energy harvesting technologies are increasingly common as sources of energy for Internet of Things (IoT) devices. Magnetic energy harvesting from power cables is a particularly promising technology for smart grid, infrastructure and environmental monitoring, and aerial robotic applications. This paper presents a high-efficiency two-stage battery charger that utilizes an active bridge rectifier and a synchronous buck converter to reduce conversion losses. Efficient energy harvesting is achieved by tracking the maximum power point. Experimental results suggest that charging efficiency of up to 90% can be achieved while providing energy for high-energy demanding IoT applications.

1   Introduction

With billions of devices connected, the growth of the Internet of Things (IoT) is a worldwide phenomenon that is revolutionizing our society. The IoT is establishing a broad network of devices that can interconnect and exchange data with each other, encompassing smartphones, smart home devices, industrial sensors, and healthcare equipment. This expansion is driven by technological developments, including upgraded connectivity features enabling advanced network topologies, powerful processors equipping edge computing capabilities, and the evolution of sophisticated software and algorithms to analyze and compute large volumes of data. As a result, the IoT is fostering new opportunities to increase efficiency, improve productivity, and create disruptive business models capable of transforming industries, urban areas, and our daily lives.

Traditional methods of powering devices, like batteries, are impractical for many IoT applications due to the large number of devices and difficult-to-access locations. This makes energy harvesting a key component of IoT expansion. Energy harvesting technologies enable the capture and conversion of ambient energy, including solar [1], [2], [3], thermal [4], wind [5], mechanical vibrations[6], radio frequency [7], [8] or electric [9] and magnetic[10], [11], [12] fields, into electrical energy. This harvested energy can power IoT devices reducing the need for conventional power sources and enabling deployment of IoT devices in remote locations and for longer periods. Thus, energy harvesting will play a crucial role in supporting and enabling IoT expansion, offering a sustainable and practical solution for powering the billions of connected devices that comprise the IoT ecosystem.

The energy usage of an IoT device may differ based on factors such as the number and quality of sensors, frequency of data refresh, complexity of processing algorithms, data transmission and reception, and activation of power-saving features. Energy harvester technology is a feasible alternative that in some cases can entirely serve as an energy source while enhancing sensor lifespan and minimizing maintenance downtime and operational expenses. In general, an IoT device with energy harvesting components comprises of several essential parts:

- Transducers: convert harvested energy into electrical energy.
- Power converters: electronic stages that extract energy from the transducers and convert the voltage into a suitable form of electricity for electronic devices.
- Energy storage: stores harvested energy and decouples the source from the load.
- Sensors: for data collection and quality assurance.
- Information management: involves edge computing, connectivity, data storage and user interfaces.

Efficient converters and energy storage are essential in energy harvesting systems due to the intermittency of energy sources. Power converters transform the electricity from the energy harvester into a suitable form for electronic devices or energy storage, such as supercapacitors [13] or batteries [1], [2], [10]. A large number of single [1] and dual stage [2], [6], [8], [11] power electronic circuits have been documented in the academic literature. Single-stage circuits are used for AC-DC or DC-DC conversion, while two-stage circuits usually combine both AC-DC and DC-DC converters. A variety of topologies, including boost [2], [8], [13], buck [1], [3], buck-boost [6], [10], [11] have been suggested to achieve high output voltage and power, overcoming the aforementioned difficulties. The proposed circuits are characterized by the possibility of achieved high energy density in the power electronics circuit, thereby reducing the size and cost of the device while maintaining optimum efficiency.

The voltage and power levels in energy harvesters differ greatly depending on the source and are constantly changing due to external variables. In order to achieve optimal energy harvesting and to reach the Maximum Power Point (MPP), several methods have been proposed to optimize impedance matching and to track peak power. The MPP is located on the characteristic I-V curve and defines the maximum power produced by a particular device at a specific time. Maximum Power Point Tracking (MPPT) methods establish control strategies to track peak power and can be categorized into four groups depending on their tracking methods [14]: classical [2], intelligent [3], optimization [15] and hybrid [16]. Although conventional methods typically perform well in regular conditions, their ability to accurately track MPP in complex situations is limited. Intelligent, hybrid, and optimized techniques require intricate design structures but, they are capable of tracking the overall maximum power point under challenging circumstances and they demonstrate greater tracking speed and stability. Despite their effectiveness, advanced technologies are more expensive than conventional methods, so [14] highlight that the traditional Perturb and Observe (P&O) method remains the industry's most widely used approach to designing MPPT controllers.

A promising technique for harvesting energy is to capture it from current-carrying components of the power grid, such as overhead and underground power cables. Magnetic Energy Harvesters (MEHs) can act as transducers that capture the magnetic field surrounding a power line due to the electrical current flowing through the cable. The MEH harvests energy from magnetic fields (H-field) by using a magnetic core that concentrates the magnetic flux and induces the secondary coil from the primary winding to transfer it to the load, essentially acting as a transformer, hence MEH is also called inductive energy harvesting. The harvested energy can be converted into electrical energy and subsequently used to power IoT electronic devices.

MEHs have the capability to produce tens or hundreds of watts, making them a suitable choice for powering advanced IoT devices with edge computing, energy-intensive sensors and high-bandwidth communications. MEHs have received considerable attention in literature for their viability in several applications, including:

- Smart grids, for real-time monitoring of power lines [9], [17], [18],
- Infrastructure and environment monitoring [17], [19],
- Robot and drone charging for autonomous operation in remote regions [10], [11].

This innovative technology is particularly advantageous for enabling operational intelligence in smart grid applications where sensors are often located in remote or inaccessible places. Figure 1 highlights the main IoT applications proposed by the authors for deployment via MEHs in power lines. This practical solution enables independent wireless sensors by eliminating the need for frequent battery replacement and ultimately minimizing maintenance costs.

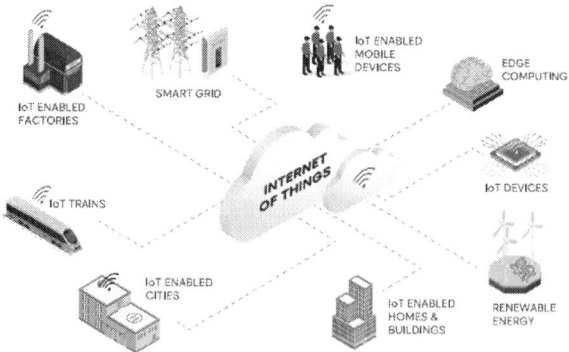

Fig. 1   IoT applications proposed for deployment using magnetic energy harvesters

In this work, a MEH is used as the energy source and a three-cell lithium-polymer battery as the energy storage device. The unregulated alternating voltage from the energy harvester is rectified and adapted to charge the battery using a two-stage battery charger, with the objective of improving the efficiency of the stages from previous work [10]. The first stage involves an active bridge which enables AC to DC full-wave rectification with reduced power losses. Then, a synchronous buck converter with a MPPT control method based on P&O is proposed in the second stage for adaptive limited current with pulse-width modulation control. Furthermore, the constant voltage mode is employed to protect the battery when it is fully charged. The proposed design incorporates ultra-low power integrated circuits for energy harvesting applications, ensuring high efficiency conversion for charging batteries. The energy source is considered to be power lines carrying currents of 10 to hundreds of amperes. These current levels are commonly available in low-voltage electrical distribution and medium to high-voltage underground and overhead cables for distribution and transmission lines.

The document is structured as follows. Section 1 introduces energy harvesting as an energy source for deploying IoT and offers an overview of converter topologies and control strategies. Section 2 defines the boundary conditions for the power conversion stage, established by the MEH and the battery, at the input and output, respectively. Section 3 describes the proposed solution for the power stage, responsible of the charge of the battery, and Section 4 presents the results and a comparative analysis of the converter stages. Finally, Section 5 discusses the findings, and Section 6 concludes the paper.

2 Harvesting method

This paper proposes the use of a MEH to extract energy from the oscillating magnetic field surrounding a power line, generated by the current flowing through the conductor. The transducer converts magnetic field into electromotive force for charging a battery through a power converter, as illustrated in Fig. 2. The MEH unit comprises a ferromagnetic core that enhances the magnetic flux pathway, the primary winding which is essentially the power line, powered by the AC current, and the secondary winding, which induces voltage from the primary to deliver it to the load. High permeability materials are required to provide high magnetic induction in the ferromagnetic core. Furthermore, it should have low hysteresis and eddy current losses to prevent overheating and minimize energy wastage. When cost is a significant factor, cores made from silicon steel are a suitable option. On the other hand, for high performance designs, nickel and nanocrystalline materials are proving to be the optimum choice.

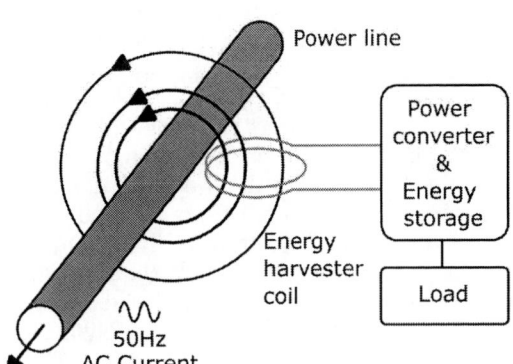

Fig. 2 Conceptual diagram of magnetic field in a power line with transducer, power converter, energy storage and load

In order to convert the electromotive force from the energy harvester into an appropriate form for electronics, a two-stage power converter is proposed. The first stage converts AC to DC, and the second stage provides adequate current and voltage to store energy in a battery. Once the energy has been harvested and stored, it will be available for powering sensors, microcontrollers, motors, actuators and wireless communication modules that have high power requirements. The conduction losses can be a significant factor in high-efficiency systems. Losses occur when current flows through a device with finite resistance, such as a transistor or diode. A low current transducer, power converter and battery are preferable to minimize conduction losses. Additionally, it is essential to limit the voltage range between power converter stages for increased energy conversion efficiency.

The development of a power system using MEH technology that can improve efficiency over a broad range of operating conditions is the main objective of this research. This proposal exploits the I-V curve characteristic of the MEH transducer whereby increases in current through the power cable minimally impact the voltage at the maximum power point (Vmp). The authors present an approach to improve efficiency by designing the transducer with Vmp only slightly above the maximum battery voltage for most operating points. Employing this approach will boost power converter efficiency thanks to the slight difference between input and output voltage.

2.1 Lithium-polymer battery

In order to provide the appropriate voltage for electronic devices such as sensors, cameras and robotic effectors, it was considered to use a 3-cell lithium-polymer battery as the energy storage device. This battery chemistry was chosen for its robust and versatile characteristics, which provide high energy density and long life. This battery pack can provide a range of up to 12.6 V when fully charged and a minimum of 9 V when discharged. The nominal voltage of the battery is 11.1 V and has a capacity of 3300 mAh. The maximum charging rate for lithium-polymer cells is 1C, and the discharging rate is 30C. Therefore, the charger for this battery configuration can provide a maximum charging current of 3.3 A. Table 1 illustrates the specifications of the selected lithium-polymer battery pack.

Parameter	Value	Unit
Chemic	Lithium-polymer	
Cells / configuration	3 serial	
Capacity	3300	mAh
Minimum volt.	9	V
Nominal voltage	11.1	V
Maximum volt.	12.6	V
Charge rate	1C	
Discharge rate	30C	
Weight	252	g
Batt. pack size	131*42.5*19.5	mm

Table 1 Battery pack specification

2.2 Magnetic energy harvester

Magnetic energy harvesters have a wide range of output voltages depending on mechanical design, the core material B-H curve, primary current and load. On the other hand, the relationship between voltage and current output can be modified by choosing the number of turns of the secondary coil. A split ferromagnetic core made of grain-oriented silicon steel was selected in this proposal for MEH due to its efficiency, permeability, affordability and manufacturing adaptability. To facilitate the placement of high-diameter power lines, such as medium-voltage underground cables, and secondary winding, a split core with a large window area was selected.

The secondary coil has been wound to guarantee that the voltage delivered in DC after rectification is 10% greater than the maximum battery voltage once the primary current reaches 100 A. This allows for accommodating fluctuations in the magnetic circuit under the conditions of possible gaps or core misalignment and provides sufficient voltage for the conversion stages. Figure 3 and Table 2 present the MEH specifications. This MEH is referenced in this article as the transducer for the design of the power converter.

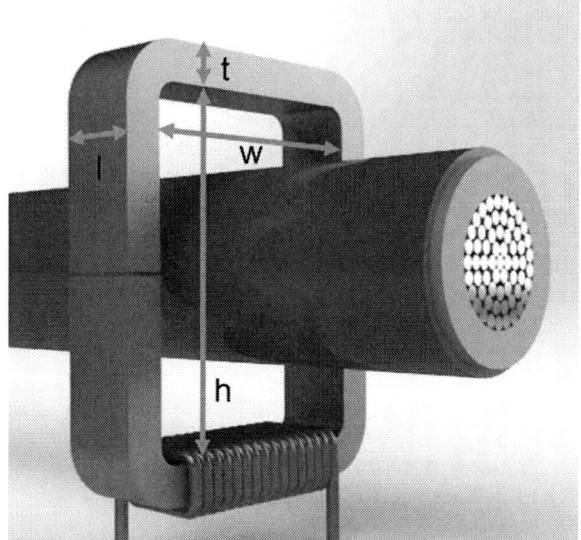

Fig. 3 Magnetic energy harvester with split core and secondary coil in an underground power line

Power Line	Value	Unit
Current	0-600	A
Frequency	50/60	Hz
Core	**Value**	**Unit**
Weight	550	g
Material	Grain-oriented silicon steel	
Width (w)	68	mm
Height (h)	90	mm
Length (l)	21	mm
Thickness (t)	9.5	mm
Coil	**Value**	**Unit**
Turns second.	185	
Material	Enamelled copper	
Wire diameter	1.5	mm

Table 2 Magnetic energy harvester specification

The characterization of the MEH reveals a direct correlation between the MPP and primary current, as illustrated in Fig. 4a. This correlation is observed in the range of primary current ranging from tens to hundreds of amperes. In contrast, energy harvested with a primary current lower than 50 A is negligible. This paper suggests optimizing the power converter's working range to the interval in which the highest amount of energy can be harvested. This strategy penalizes low primary current cases, which may not be optimal for power lines with low current. However, energy storage can benefit medium and high-current power lines by providing an energy buffer. When the primary has a current of 100 A, 5 W can be harvested. This relationship grows linearly from 100 A to 600 A, the maximum current considered in this paper, where

40 W can be extracted. In spite of this, the battery charger is able to operate at lower currents; however, it is not able to reach the MPP.

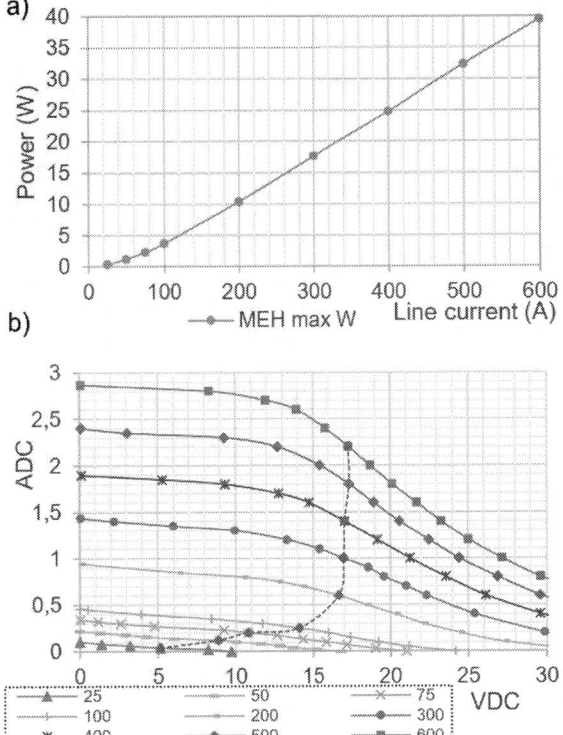

Fig. 4 Parameters of the magnetic energy harvester as a function of a 50 Hz primary current in the range of 0-600 A: a) Maximum power point for MEH b) V-I curve in DC after rectification stage

Figure 4b illustrates the V-I curve for the MEH transducer after rectification stage. This curve shows the relationship between the output current and the voltage in DC, where the active output power in MEH is the product $P = V \times I \times cos(\theta)$, being θ the phase difference between the current and the voltage. For a given current in the primary cable, the output power is dependent on the voltage as the current is almost constant between the short-circuit point and the knee of the curve. The I-V curve comprises three significant points:

- Short-circuit output current: This occurs when the output terminals are shorted, resulting in a 0 voltage output.
- Open-circuit voltage: This occurs when the output is open, resulting in maximum voltage but 0 A current. Some saturation may be observed in the MEH core for high current in primary in this point.
- Maximum Power Point : This is where maximum output power is obtained. Beyond this point, the curve drops sharply and the output power drops.

A comprehensive grasp of the features of MEH is fundamental in enhancing the performance of the power converter and maximizing its power output. It is worth noting that although the power cable current ranges from 100 A to 600 A, the output voltage of MPP is only marginally impacted. Nevertheless, the current and resulting power decreases as the power cable current goes down. Despite the wide voltage range, it is mainly in a narrow range that the maximum power is obtained. Therefore, battery charger takes advantage of this feature to simplify design and reduce costs.

3 Proposed battery charger

A lithium-ion/polymer battery charger is proposed as a conditioning circuit to provide regulated energy from an MEH to the energy storage element. Figure 5 shows the block diagram of the proposed two-stage battery charger including the following:

- An active bridge that enables AC to DC full-wave rectification;
- A synchronous buck converter equipped with a MPPT controller;
- An overvoltage protection circuitry.

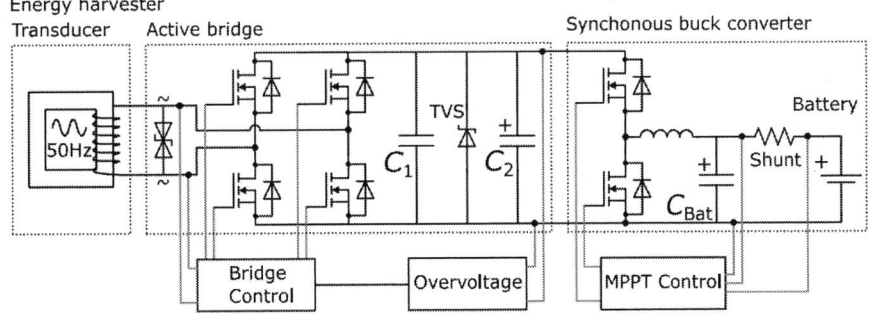

Fig. 5 Battery charger for magnetic energy harvester diagram

The proposed design was implemented incorporating ultra-low power integrated circuits suitable for energy harvesting applications, ensuring high efficiency conversion processes for charging batteries. The proposal utilized demonstration circuits for the converter stages and a custom printed circuit board for overvoltage protection. Technical specifications of the power stages are shown in Table 3.

Active MOSFET bridge rectifier			
Parameter	Min	Max	Unit
Input Voltage Range	9	40	V
Load Current		4	A
Synchronous buck converter MPPT			
Input Voltage Range	5	35	V
Battery Voltage Range	2,9	35	V
Charge current		3,2	A

Table 3 Electric specifications of stages in battery charger for energy harvester

3.1 Active MOSFET bridge rectifier

AC induced voltage is generated by the MEH through electromagnetic induction from the power line. This voltage is synchronized with the grid frequency, which is generally 50 Hz or 60 Hz worldwide. The amplitude of the secondary voltage and current varies with the primary current and is dependent on the load and the core B-H curve. A rectification stage is then proposed to meet the DC requirements. To perform full-wave rectification similar to a diode bridge but with much lower power losses, an active bridge with four N-channel MOSFETs is proposed. This arrangement simplifies thermal design, eliminates expensive heat sinks, and can reduce PCB size while increasing available output voltage. A bypass ceramic capacitor C_1 for noise suppression and a smoothing capacitor C_2 to suppress voltage ripples are employed. Additionally, a unidirectional TVS diode can be incorporated to safeguard the second stage from transient overvoltage events. The demonstration circuit 1823B from Analog Devices [20] was selected for the active rectifier bridge. This device uses an LT4320 controller and is suitable for applications that require high current AC to DC full-wave rectification.

3.2 Buck converter with a MPPT control

A synchronous buck converter is proposed to step down voltage from the active rectifier to the battery. Step-down converters are the preferred option for electronic DC-DC conversion due to their inherent high efficiency. Buck converters are the most efficient architecture because current is directed to the output during the entire cycle. The high-side MOSFET is connected between the rectified voltage and the inductor. When this MOSFET is turned on, it allows current to flow from the source, through the inductor and load, and back to the source. Low-side MOSFET is connected between the inductor and ground. When the high-side MOSFET is turned off, the low-side MOSFET turns on. This allows the inductor to discharge its stored energy into the load. Both MOSFETs work synchronized to provide more efficient conversion than traditional (asynchronous) buck converter that rely on diodes due to less resistance and voltage drop.

The lithium-ion/polymer battery charger integrates adaptive limited current and constant voltage feedback control loops in order to prevent overcharging. The adaptive limited current (ALC) mode starts the battery charge until the battery reaches its target voltage. An MPPT algorithm is used within the ALC mode to optimize power extraction from the MEH. Due to variations in power cable current, the MPPT control produces variable output power. To track the highest energy output of an energy harvester, it is necessary to operate at the MPP on the I-V curve. This determines the maximum power produced by the specific harvester at any given time and requires ongoing tracking via the MPPT algorithm. The ALC control loop transitioned to the constant voltage (CV) control loop when the battery voltage reaches the programmed voltage limit. The Multi-Cell Step-Down battery charger with MPP tracking algorithm is provided by the Analog Devices demonstration circuit DC2038A-A [21], which is based on the LTC4162-LADM.

3.3 Overvoltage protection

Overvoltage protection is implemented to avoid damaging the components of the system due to a low level of energy demand or due to voltage spikes. Additionally, it is necessary to devise strategies in MEH to prevent saturation of the core under large current and protect the system from overvoltage caused by power line short-circuits. An overvoltage protection circuit can be integrated into the AC-DC converter. When the smoothing capacitor reaches a predetermined voltage limit, the low-side MOSFETs are activated while the high-side MOSFETs are disabled. The secondary coil of the energy harvester is short-circuited using the back-to-back MOSFET configuration to prevent any additional energy flow to capacitor C_2. A hysteresis is programmed to release the short circuit when the voltage in C_2 drops to a suitable level, allowing the device to drain energy from MEH. In harsh electrical conditions, the device is protected from transient overvoltage events by including bidirectional TVS diodes on the AC side and unidirectional diodes on the DC side.

4 Results

As depicted in Fig. 6, a prototype transducer based on an MEH, power stages including an active bridge rectifier and a synchronous buck converter with MPPT control, overvoltage protection and a lithium-polymer battery was assembled.

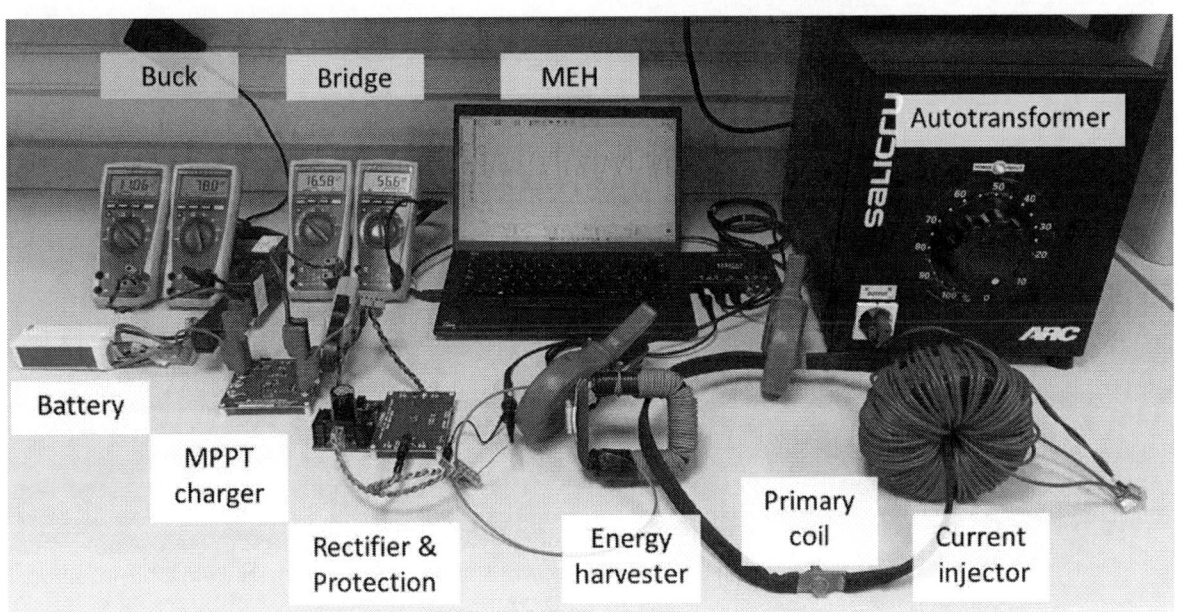

Fig. 6 Laboratory test setup for charging a li-po battery using a MEH with power converters

Line current (A)		51,55	76,78	106,8	208,4	303,2	399	514,1	582
Magnetic Energy Harvester	V	9,866	10,440	10,960	14,720	15,320	15,350	15,210	15,290
	I	0,139	0,273	0,429	0,826	1,276	1,775	2,392	2,727
	W	1,149	2,431	4,089	10,724	17,614	24,787	33,286	38,232
Bridge	V	11,940	11,950	11,950	16,720	17,040	16,960	17,060	17,240
	I	0,088	0,191	0,328	0,620	0,999	1,406	1,863	2,144
	W	1,051	2,282	3,920	10,366	17,023	23,846	31,783	36,963
Buck	V	10,530	10,530	10,530	11,030	10,870	10,780	10,700	10,700
	I	0,090	0,204	0,356	0,901	1,480	2,085	2,802	3,215
	W	0,948	2,148	3,749	9,938	16,088	22,476	29,981	34,401
Overall efficiency		82,5	88,4	91,7	92,7	91,3	90,7	90,1	90,0
Efficiency Bridge		91,5	93,9	95,9	96,7	96,6	96,2	95,5	96,7
Efficiency Buck		90,2	94,1	95,6	95,9	94,5	94,3	94,3	93,1

Table 4 Average voltage (V), current (I) and power (W) in stages of the system at different line current

Afterward, relevant environment tests were carried out to assess functionality over a wide range of operating conditions. To simulate a power cable, an alternating magnetic field is generated by a primary coil fitted with a current injector and applied to the MEH. The induced current in the primary coil is controlled by adjusting the output voltage provided by an autotransformer. As shown in Fig. 6, voltage, current and active power are measured in each stage, first the AC parameters provided for the MEH and then the DC of the bridge and buck outputs. The transducer harvests energy from the magnetic field, and the active rectifier converts the induced AC to DC. The synchronous buck converter equipped with MPPT was used to supply energy to the battery. A series of 50 Hz AC primary current points were emulated to assess the amount of energy that could be obtained from the power unit and the efficiency of the power electronics. The charging test was performed during the ALC phase when the charger current was not limited by battery capacity. Table 4 shows the results of each power stage for the test performed. Fig. 7 shows the relationship between the energy measured at each stage of the battery charger, starting at the MEH, then at the output of the active bridge, and finally the amount of energy transferred to the battery by the buck converter in a series of line currents.

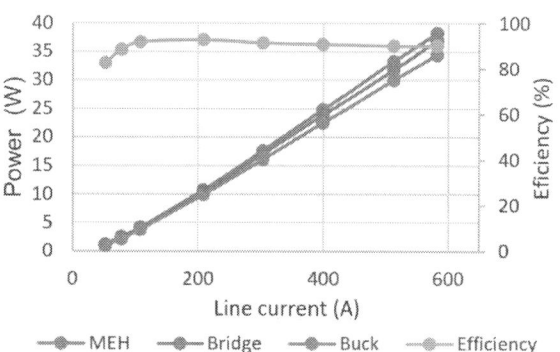

Fig. 7 Active energy in power stages and total power conversion efficiency (based on Table 4)

Efficiency
From 100 A to 600 A the average reported efficiency for the active bridge rectifier is 96.3%. On the other hand, the synchronous buck converter achieves an average efficiency of η = 94.6%. Taking both stages into account, the battery charger achieves a charging efficiency η greater than 90%, with an average efficiency η = 91.1% and a standard deviation σ = 0.88.

Maximum power point tracking
The effectiveness of the LTC4162's MPPT algorithm was evaluated by comparing the maximum power points achieved experimentally with a configurable electronic load to those achieved with the battery charger. Figure 8 provides a visual representation showing successful tracking from 100 to 600 A.

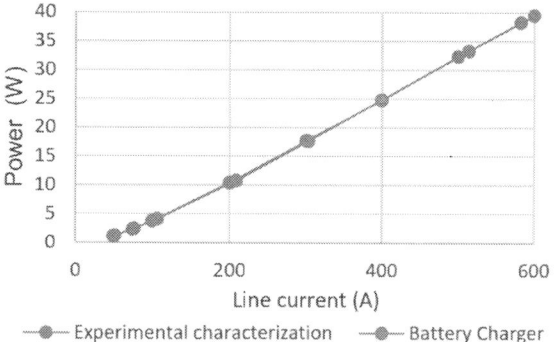

Fig. 8 Experimental MPP harvested with an electronic load and the LTC4162 using MPP tracking

Voltage at the maximum power point
Fig. 9 shows a comparison of the Vmp obtained from the electronic load characterization and that obtained with the LTC4162 tracker. The graph shows that the Vmp obtained from both methods are consistent over the 100 to 600 A range.

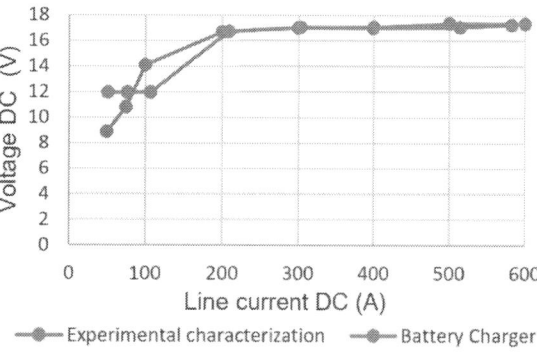

Fig. 9 Voltage after rectification stage at MPP of energy harvested experimentally with an electronic load and the LTC4162 using MPP tracking

5 Discussion

An improvement in efficiency was observed across a wide range of operational points when comparing to previous work [10]. Specifically, the efficiency peak reported in the previous work was η = 79.7%, remaining almost constant between 150 and 600 A. In contrast, the present study achieved η = 91% on average within the same range as Table 5 and Fig. 10 shows. This strategy penalizes low primary current cases, which may not be optimal for low current power lines. However, utilizing the linearity of Vmp in the transducer for power converter design and adapting to battery voltage presents an advantage in medium and high-current power lines. This approach presents an innovative strategy for efficient power conversion in MEH. Despite the technique focused on improving the energy harvesting from 100 A to 600 A, the battery charger is able to work with lower currents as shown in Table 4. The charger also works with 50 and 75 A in line current, but since the voltage of the battery is higher than the MPP voltage, the charger works in a point with a voltage higher of the battery voltage as shown in Fig. 9, shifting slightly from the MPP.

Line (A)	100	200	300	400	500	600	Aver.
[10]	72,6	78,7	79,7	79,3	78,7	78,0	77,8
Proposed	91,7	92,7	91,3	90,7	90,1	90,0	91,1

Table 5 Efficiency comparison between previous (Aerial Core) and current research

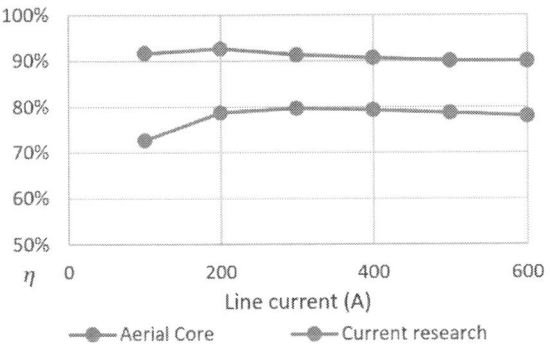

Fig. 10 Diagram based on table 5

Battery charge
The LTC4162's MPPT algorithm primarily supplies current to the battery when the magnetic field in the line induces enough voltage to charge the main capacitor C_2. The battery is then charged with a 100 Hz frequency component due to the 50 Hz grid frequency. Recent studies have shown that very high current can be applied to the cells during very short pulses ($\geq$100 Hz) with negligible capacity loss, even for relatively long periods in LIFEPO4 chemistries [22]. Another study found that for Li-ion cells, a ripple of 100 and 1000 Hz not only did not negatively affect the LFP lifetime, but also slightly improved the cell lifetime performance compared to DC cases [23]. Based on cited

studies, premature battery ageing due to control technology is not expected.

6 Conclusion

A high-efficiency two-stage battery charger with maximum power point tracking for effective energy harvesting was proposed and studied in this paper. An active MOSFET bridge rectifier and a synchronous buck converter was designed for the efficient charging of lithium-ion/polymer batteries from watt-scale magnetic harvesting sources. The battery charger proposed achieves highest charging efficiency by identifying an optimal conversion point that leverages the narrow voltage band of the maximum power point for primary currents ranging from 100 to 600 A, all while maintaining an affordable design. Experimental results suggest that charging efficiencies of up 90% can be achieved, representing a significant improvement over the conventional method. This technology has been demonstrated to be a suitable option for powering high-energy-consuming IoT devices, reducing the size and cost of the device while maintaining optimum efficiency. Future work includes testing this technology under harsh line conditions to validate successful MPP tracking and overvoltage protection.

Acknowledgments

The research has been carried out in the framework of the REEFLEX (https://reeflexhe.eu/) project funded by the European Union under the Horizon Europe programme (HORIZON-CL5-2022-D3-01).

References

[1] T. J. Lee, P. K. Su, and C. C. Wang, "20V HV Energy Harvesting Circuit with ACC/CV Mode and MPPT Control for a 5 W Solar Panel," *2020 2nd International Conference on Smart Power and Internet Energy Systems, SPIES 2020*, pp. 205–208, 2020, doi: 10.1109/SPIES48661.2020.9243091.

[2] R. Sabzehgar, R. Ghali, and P. Fajri, "A Novel Combined Control Strategy for a Two-Stage Parallel Full-Wave ZCS Quasi Resonant Boost Converter for PV-Based Battery Charging Systems with Maximum Power Point Tracking," *Electricity*, vol. 3, no. 1, pp. 145–161, 2022, doi: 10.3390/electricity3010009.

[3] J. L. Seguel and S. I. Seleme, "Robust digital control strategy based on fuzzy logic for a solar charger of VRLA batteries," *Energies (Basel)*, vol. 14, no. 4, 2021, doi: 10.3390/en14041001.

[4] W. Dipon, B. Gamboa, M. Estrada, W. P. Flynn, R. Guo, and A. Bhalla, "Self-Sustainable IoT-Based Remote Sensing Powered by Energy Harvesting Using Stacked Piezoelectric Transducer and Thermoelectric Generator," *Micromachines (Basel)*, vol. 14, no. 7, 2023, doi: 10.3390/mi14071428.

[5] A. Jushi, A. Pegatoquet, and T. N. Le, "Wind Energy Harvesting for Autonomous Wireless Sensor Networks," *Proceedings - 19th Euromicro Conference on Digital System Design, DSD 2016*, pp. 301–308, 2016, doi: 10.1109/DSD.2016.43.

[6] M. Kamran, M. Edla, A. M. Thabet, D. Mikio, and V. Bui, "A Self-Powered FBRJT AC–DC Conversion Circuit for Piezoelectric Energy Harvesting Systems," *Designs (Basel)*, vol. 7, no. 4, 2023, doi: 10.3390/designs7040094.

[7] M. Caselli and A. Boni, "Modeling and design of 3-D MPPT for ultra low power RF energy harvesters," *Integration*, vol. 72, no. February, pp. 21–28, 2020, doi: 10.1016/j.vlsi.2020.02.008.

[8] J. Kim and I. Kwon, "Design of a High-Efficiency DC-DC Boost Converter for RF Energy Harvesting IoT Sensors," *Sensors*, vol. 22, no. 24, 2022, doi: 10.3390/s222410007.

[9] J. Bendík, M. Cenký, and O. Hromkovič, "Energy Harvesting Device for Smart Monitoring of MV Overhead Power Lines—Theoretical Concept and Experimental Construction," *Sensors*, vol. 23, no. 17, 2023, doi: 10.3390/s23177538.

[10] A.-M. Muñoz-Gómez, J.-M. Marredo-Píriz, J. Ballestín-Fuertes, and J.-F. Sanz-Osorio, "A Novel Charging Station on Overhead Power Lines for Autonomous Unmanned Drones," *Applied Sciences*, vol. 13, no. 18, p. 10175, 2023, doi: 10.3390/app131810175.

[11] V. D. Hoang and E. Ebeid, "Advanced Magnetic Energy Harvester for Charging Drones from Overhead Powerlines," in *2023 25th European Conference on Power Electronics and Applications, EPE 2023 ECCE Europe*, EPE Association, 2023, pp. 1–9. doi: 10.23919/EPE23ECCEEurope58414.2023.10264597.

[12] M. S. Noohi and M. Habibi, "An energy-efficient CMOS interface circuit with maximum power point tracking and power management capabilities for self-powered sensor node applications using 50/60 Hz transmission line magnetic field harvesters," *Electrical Engineering*, vol. 105, no. 3, pp. 1413–1430, 2023, doi: 10.1007/s00202-023-01740-7.

[13] S. Kim, J. H. Lam, J. Kim, and P. H. Chou, "Burst-transfer boost charger for supercapacitors from subwatt-scale harvesting sources," *J Power Sources*, vol. 520, no. July 2021, p. 230745, 2022, doi: 10.1016/j.jpowsour.2021.230745.

[14] M. L. Katche, A. B. Makokha, S. O. Zachary, and M. S. Adaramola, "A Comprehensive Review of Maximum Power Point Tracking (MPPT) Techniques Used in Solar PV Systems," *Energies (Basel)*, vol. 16, no. 5, 2023, doi: 10.3390/en16052206.

[15] G. Al-Muthanna *et al.*, "A High Speed MPPT Control Utilizing a Hybrid PSO-PID Controller under Partially Shaded Photovoltaic Battery Chargers," *Sustainability (Switzerland)*, vol. 15, no. 4, 2023, doi: 10.3390/su15043578.

[16] S. A. Ibrahim, A. Nasr, and M. A. Enany, "Maximum Power Point Tracking Using ANFIS for a Reconfigurable PV-Based Battery Charger under Non-Uniform Operating Conditions," *IEEE Access*, vol. 9, pp. 114457–114467, 2021, doi: 10.1109/ACCESS.2021.3103039.

[17] F. Yang, L. Du, H. Yu, and P. Huang, "Magnetic and electric energy harvesting technologies in power grids: A review," *Sensors (Switzerland)*, vol. 20, no. 5. MDPI AG, Mar. 01, 2020. doi: 10.3390/s20051496.

[18] A.-M. Muñoz-Gómez, J. Granado-Fornas, J. Muñoz-Cruzado Alba, and J.-F. Sanz-Osorio, "Wireless self-powered monitoring system for underground cable joints: a real use-case," in *27 th International Conference on Electricity Distribution*, 2023, pp. 12–15. doi: 10.1049/icp.2023.1189.

[19] A.-M. Muñoz-Gómez, J. Muñoz-Cruzado Alba, and J.-F. Sanz-Osorio, "A Novel Application in Overhead Power Lines of Wireless Self- Powered Monitoring System for Early Detection of Forest Fires .," in *PCIM Europe 2023*, 2023, pp. 9–11. doi: 10.30420/566091176.

[20] Analog Devices, "dc1823b." Accessed: Feb. 22, 2024. [Online]. Available: https://www.analog.com/en/resources/evaluation-hardware-and-software/evaluation-boards-kits/dc1823b.html

[21] Analog Devices, "dc2038a." Accessed: Feb. 22, 2024. [Online]. Available: https://www.analog.com/en/resources/evaluation-hardware-and-software/evaluation-boards-kits/dc2038a.html

[22] A. Ghassemi, A. F. Hollenkamp, P. Chakraborty Banerjee, and B. Bahrani, "Impact of high-amplitude alternating current on LiFePO4 battery life performance: Investigation of AC-preheating and microcycling effects," *Appl Energy*, vol. 314, no. March, p. 118940, 2022, doi: 10.1016/j.apenergy.2022.118940.

[23] A. Ghassemi, P. Chakraborty Banerjee, A. F. Hollenkamp, Z. Zhang, and B. Bahrani, "Effects of alternating current on Li-ion battery performance: Monitoring degradative processes with in-situ characterization techniques," *Appl Energy*, vol. 284, no. December 2020, p. 116192, 2021, doi: 10.1016/j.apenergy.2020.116192.

PCIM Europe 2024, 11– 13 June 2024, Nuremberg   DOI: 10.30420/566262078

Symmetric Flying-capacitor Boost Converter for Medium-voltage Photovoltaic Applications

Luis Gabriel Alves Rodrigues[1] ⓘ, Guillaume Piquet-Boisson[1] ⓘ, Anthony Bier[1] ⓘ, Arnaud Revel[1], Stéphane Catellani[1]

[1] French Alternative Energies and Atomic Energy Commission – CEA/INES, France

Corresponding author: Luis Gabriel Alves Rodrigues, luisgabriel.alvesrodrigues@cea.fr
Speaker: Luis Gabriel Alves Rodrigues, luisgabriel.alvesrodrigues@cea.fr

Abstract

Photovoltaic (PV) energy is seen today as an essential energy source to achieve a carbon-neutral society. With that purpose, great pressure exists in terms of energy costs (€/MWh). In utility-scale PV systems, the increase of system voltages beyond low-voltage standards are being studied. In this paper, a novel medium-voltage DC/DC boost converter is proposed, capable of operating with 3 kV PV strings. The investigated topology is the 5-level symmetric flying-capacitor equipped with 1.2 kV/16 mΩ SiC MOSFETs. Experimental switching characterization results are provided and the main design aspects of the 80 kW SiC-based prototype are discussed. The estimated peak and European efficiencies are 99.5 and 99.33%, respectively.

1 Introduction

In the current energy scenario towards a carbon-neutral society, photovoltaics (PV) plays a crucial role, having reached 1.2 TW$_p$ of cumulative installed power in 2023. Among all the renewable energy sources, solar power concentrates, for several years in a row, the largest share of new investments worldwide. [1] However, further deployment of PV technology still relies on cost reduction. In utility-scale PV power plants, an adopted and lately proven strategy to cap costs has been the increase of system voltage levels on both PV and grid sides. Through the implementation of higher voltages, cable cross-section and the number of combining boxes can be reduced, as well as installation and maintenance costs [2]. This trend has led to system voltages at, or close to, the limit of low-voltage (LV) standards, i.e., 1.5 kV$_{DC}$ and 1 kV$_{AC}$. For that reason, recent R&D activities have been focused on medium-voltage (MV) AC and DC PV architectures [3], [4], [5], [6], [7]. Figure 1-a) and b) show, respectively, examples of grid-connected MVAC and MVDC PV systems.

A key enabler of MV PV systems concerns the insulation of PV panels [2]. In the literature, a few solutions addressing this subject start to emerge. In [8], tailor-made aluminum frame PV panels forming a 3 kV string are fabricated and tested, showing a high insulation resistance of 4 GΩm² (33 MΩ/string) at 14 kV. The authors of [5]

propose the utilization of dual-glass PV panels in order to achieve enough insulation resistance to be compatible with MV applications. From a power electronics perspective, MV PV conversion systems can be implemented with high-voltage semiconductors and/or multilevel topologies. In [8], with the aim of tracking the maximum power

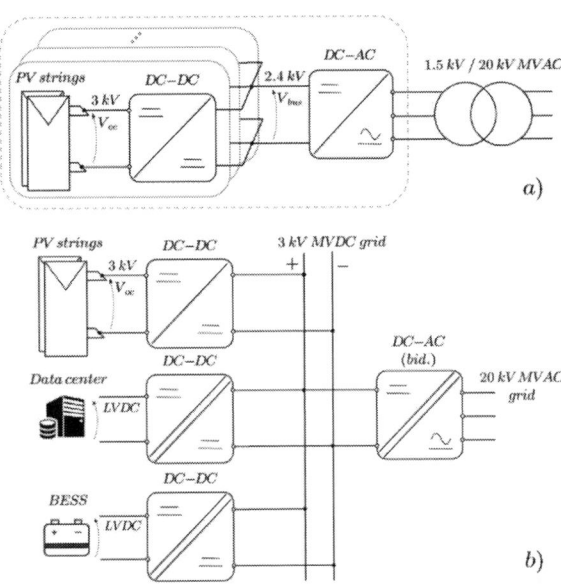

Fig. 1 – a) Two-stage MV PV string inverter proposed in [5] and b) MVDC grid in backbone configuration, as discussed in [3], including a 3 kV PV DC/DC converter (or string MPPT optimizer).

PCIM Europe 2024, 11– 13 June 2024, Nuremberg   DOI: 10.30420/566262078

point (MPPT) from the 3 kV$_{OC}$ PV string, an 18 kW two-level DC/DC converter based on Si 6.5 kV IGBTs is proposed. In [5], a two-stage MV PV string inverter is discussed, where the MPPT function is provided by a 3.3 kV SiC-based parallel-interleaved boost converter. Both aforementioned MV converter solutions make use of high-voltage semiconductors, which present poor switching behavior and relatively high costs, when compared to their LV counterparts (e.g. 1.2 kV). These drawbacks could limit the attractiveness of such MV PV systems. To cope with that, this paper presents a multilevel 3 kV DC/DC boost converter using widespread 1.2 kV SiC devices (cf. Fig. 2). The proposed topology is the 5-level symmetric flying-capacitor (5LSFC) converter. The 5LSFC is capable of performing the MPPT functionality on a 3 kV$_{OC}$ PV string in a two-stage MV string inverter configuration, as shown in Fig. 1-a), or as a standalone DC/DC converter (commercially known as string MPPT optimizer) connected to a 3 kV MVDC grid, as depicted in Fig. 1-b).

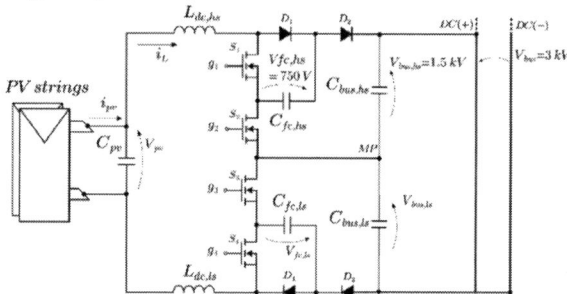

Fig. 2 – *5-level symmetric flying-capacitor boost converter.*

The aim of this paper is to discuss the most important operating principles and design aspects

of an 80 kW 5LSFC prototype. Table 1 provides the main 5LSFC converter characteristics.

Nom. MVDC grid voltage (V$_{MVDC}$)	3 kV
Nominal Power (P$_n$)	80 kW
Number of PV strings in parallel	2
PV string open-circuit voltage (V$_{OC}$)	3 kV
MPP range	1.9–2.8 kV
Min. PV voltage for operation (V$_{pv,min}$)	800 V
Max. RMS PV current (I$_{PV,max}$)	42 A
Peak and European Efficiencies (%)	99.4 & 99.2

Table 1   Specifications of the 5LSFC converter.

2 5-level symmetric flying-capacitor converter

2.1 Operating Principle

The 5LSFC topology (cf. Fig. 2) has its origin from the association of two 3-level flying-capacitor (FC) structures. Thus, four controllable switches (S$_1$...S$_4$) and four diodes (D$_1$...D$_4$) are necessary. In this arrangement, two switching cells (SC) arise in the high-side part (HS) of the topology, namely SC1 and SC2. The switching path of SC1 is formed by S$_1$, D$_1$ and flying capacitor C$_{fc,hs}$. The switching path of SC2 is formed by S$_2$, D$_2$, flying-capacitor C$_{fc,hs}$ and half-bus capacitor C$_{bus,hs}$. In a symmetric manner, another two switching cells exist for the low-side (LS) part of the topology. In terms of semiconductor blocking voltages, all switches must withstand V$_{bus}$/4 in nominal operation, i.e., 750 V. That is valid as far as the

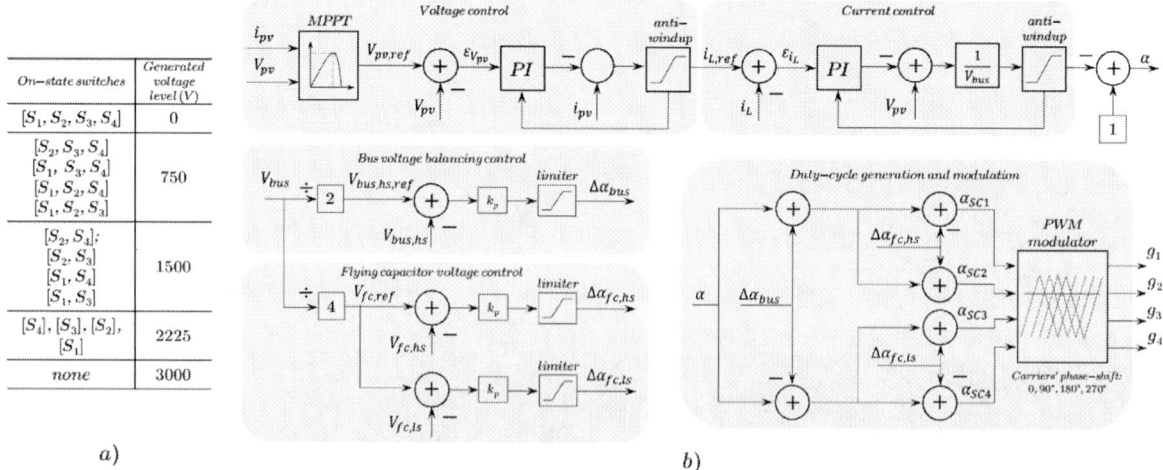

Fig. 3 – *a) Possible switching states and their correspondent generated voltage level, and b) adopted control strategy for the 5LSFC grid-connected converter.*

flying-capacitor and half-bus capacitor voltages are controlled to 750 V and 1.5 kV, respectively.

In order to properly generate the gate signals, pulse-width modulation is implemented with a phase-shift of 90° between switches, with the S_1-S_3-S_2-S_4 sequence. That allows interleaving both HS and LS 3-level FC structures. In doing so, the proposed converter is capable of generating five voltage levels at the input inductor, depending on the switching states. Figure 3-a) depicts all possible switching states and their respective generated voltage levels. [9] As a result of the interleaving technique, the current ripple frequency in the input inductors ($L_{dc,hs,ls}$) is four times the switching frequency (F_{sw}). In continuous-conduction mode (CCM), the maximum current ripple is expressed as

$$\Delta i_{L,max} = \frac{V_{bus}}{4 \cdot (n-1)^2 \cdot F_{sw} \cdot L_{dc,eq}} \qquad (1)$$

where n is the number of levels (n=5) and $L_{dc,eq}$ is the sum of the input inductance values ($L_{dc,hs}$ and $L_{dc,ls}$). The 5LSFC voltage gain is given by Eq. (2), where α is the duty ratio.

$$\frac{V_{bus}}{V_{pv}} = \frac{1}{1-\alpha} \qquad (2)$$

2.2 Control strategy

Figure 3-b) shows the control strategy used in the 5LSFC topology. Here, two voltage and current control loops are cascaded, as usual, in order to perform the MPPT from the PV strings and operate as a grid-tied converter. For a given flying-capacitor, its voltage is regulated to $V_{bus}/4$ through a proportional controller, which provides a slight duty-ratio difference ($\Delta\alpha_{fc}$) between the two SCs connected to it. For the HS part, one has $\alpha_{SC1} = \alpha - \Delta\alpha_{fc,hs}$ and $\alpha_{SC2} = \alpha + \Delta\alpha_{fc,hs}$, where α is the general duty-ratio to operate the converter at a given V_{pv} and i_{pv}, and α_{SC} is the actual duty-ratio applied to the semiconductors.

On the other hand, the voltage between the two half-buses also need to be regulated to $V_{bus}/2$, implying the modification of the aforementioned α_{SC}. In a similar strategy as before, a proportional controller provides a duty-ratio difference ($\Delta\alpha_{bus}$) between the two 3-level FC converters. Considering the SC1, it yields $\alpha_{SC1} = \alpha - \Delta\alpha_{fc,hs} + \Delta\alpha_{bus}$.

2.3 Switching Characterization

In a preliminary design phase, two device combinations have been taken into account for the 5LSFC converter: SiC MOSFETs in synchronous

rectification (SR) mode and SiC MOSFETs and SiC Schottky diodes. Thanks to the possibility of reducing conduction losses, the solution based on SiC MOSFETs in SR mode has been chosen. The implemented devices are 1.2 kV 16 mΩ SiC MOSFETs (C3M0016120K) in a TO 247-4 packaging.

In order to test the selected semiconductors under the application conditions while taking into account hardware constraints, Double-Pulse Test (DPT) is carried out. The DPT allows us to *i*) properly define the external gate resistor value (R_g), *ii*) obtain accurate data for switching losses (E_{sw}), *iii*) define dead-time (T_{dead}) for the SR switch, *iv*) test gate-driver circuitry and *v*) inspect potential harmful drain-to-source (V_{ds}) or gate-to-source (V_{gs}) overvoltages and oscillations. Figure 4 presents the employed DPT circuit with its corresponding prototype, where the diode function is performed by the MOSFET S_{D2} in SR mode.

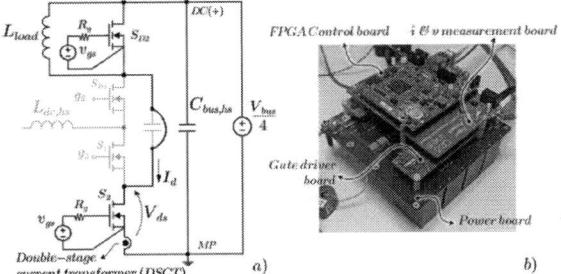

Fig. 4 – a) DPT circuit to assess the switching behavior and b) DPT prototype presenting the same layout as in the 5LSFC converter. Gray parts in a) are not assembled onto the prototype. The TO 247-4 devices are mounted vertically under the power board.

The implemented DPT circuit is basically the HS part of the 5LSFC. It is worth noting that the converter's PCB layout is employed in the DPT prototype. The parasitic inductances of both switching cells SC1 and SC2 are assessed with a precision impedance analyzer at 20 MHz, presenting, respectively, 7.55 nH and 13.89 nH. These values account for the parasitics of PCB layout and decoupling ceramic capacitors. As a worst-case, the switches forming SC2 are mounted on the prototype in order to proceed to the switching characterization under different R_g and junction temperatures (T_j), which is controlled by a hotplate.

In the developed DPT prototype, gate-driver, FPGA controller (sbRIO-9607) and power board are located on separate PCBs. On the gate-driver board, it is possible to set different R_g values for the turn-on ($R_{g,on}$) and turn-off ($R_{g,off}$) transitions. All MOSFETs are gated with V_{gs} = -4/+15 V. The

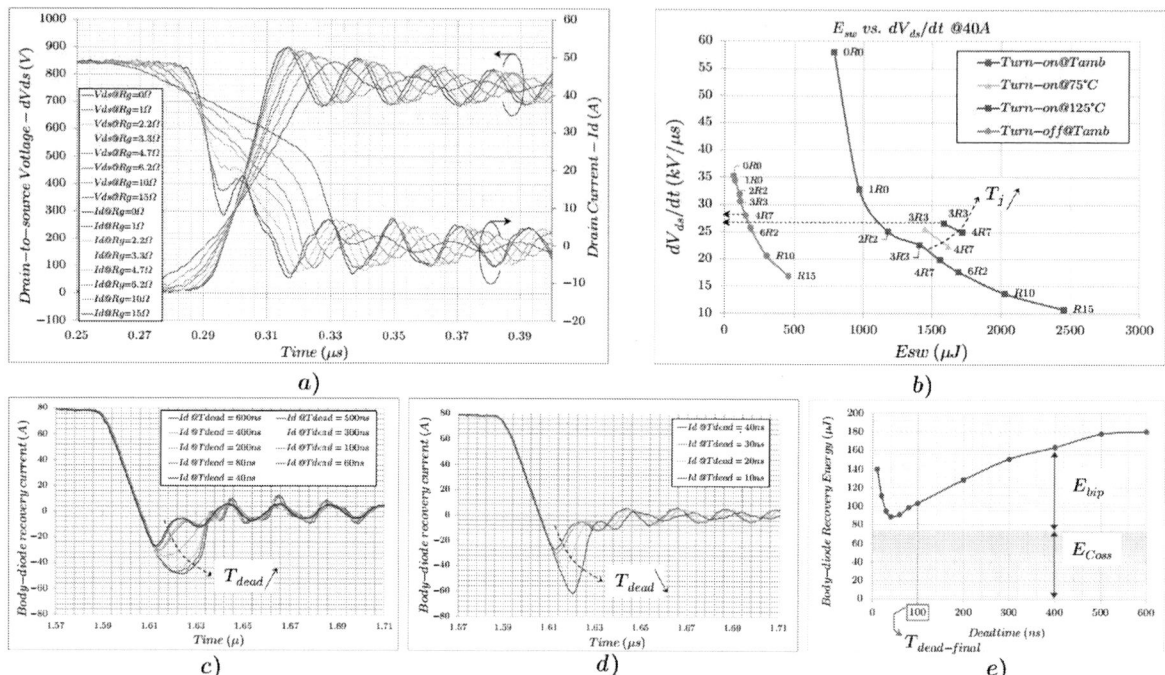

a)

b)

c)

d)

e)

Fig. 5 – *a) Typical I_d and V_{ds} switching waveforms at turn-off (V_{bus} = 750V, I_{load} = 50 A and T_{amb}) for different R_g values, b) dV_{ds}/dt vs. switching energies at V_{ds} = 750V and T_{amb} for several R_g values, c) and d) body-diode recovery current at T_j = 150°C and different T_{dead} values and e) body-diode recovery energy vs. T_{dead} at T_j = 150°C.*

drain current (I_d) measurement technique is a double-stage current transformer, as described in [10], with the first stage (10 mm NiZn HF70 toroidal ferrite with 7 turns) being mounted directly on the TO 247-4 power source lead. The drain-to-source voltage (V_{ds}) is measured with a 1:100 passive probe (PPE 2 kV).

Typical turn-off switching waveforms at V_{bus} = 750 V, I_{load} = 50 A and room temperature (T_{amb} = 20°C) are depicted in Fig. 5-a). Here, the SR MOSFET (S_{D2} in Fig. 4-a)) is permanently blocked at V_{gs} = -4 V.

The main goal in this first test phase is to experimentally determine the $R_{g,on,off}$ for the controlled device S_2, considering the trade-off between E_{sw} and dV_{ds}/dt at the converter nominal current ($I_d \approx 40$ A). As discussed in [11], switching speed is related to high-frequency EMI generation potential. Furthermore, this trade-off E_{sw} vs. dV_{ds}/dt is improved if the turn-on and off transition times are matched [12]. Figure 5-b) depicts the dV_{ds}/dt (calculated with 30-70% of V_{ds}) as a function of E_{sw} for several R_g values. As it can be seen, there is a strong non-linear behavior between switching speed and losses. Also, the turn-on dV_{ds}/dt and $E_{sw,on}$ of the device S_2 increase with T_j – due to the recovery effect of S_{D2} MOSFET body-diode. At turn-off event, dV_{ds}/dt and $E_{sw,off}$ are not affected by the T_j, only the charge/discharge of devices' parasitic capacitances come into play

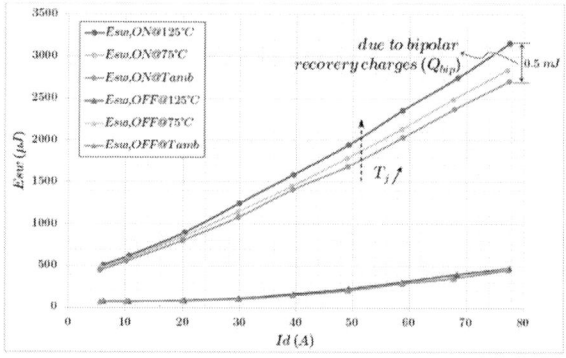

Fig. 6 – *E_{sw} vs. I_d with $R_{g,on}$ = 3.3 Ω and $R_{g,off}$ = 4.7 Ω.*

(not shown in Fig. 5-b)). Taking all that into account, the selected R_g values for the S_2 MOSFET are as follows: $R_{g,on}$ = 3.3 Ω and $R_{g,off}$ = 4.7 Ω, limiting the $dV_{ds,on,off}/dt$ to 27 kV/µs while still capping switching losses. Figure 6 presents the E_{sw} vs. I_d for the selected R_g values at different T_j.

Having defined the external gate resistors for the controllable MOSFET, the SR dead-time influence on switching losses is investigated. For this purpose, the DPT circuit of Fig 4-a) is modified: L_{load} is connected in parallel to S_2, S_{D2} becomes the controllable device and S_2 operates now in SR mode. That enables to evaluate the body-diode switching waveforms without replacing the current and voltage probes. To efficiently perform synchronous rectification, the R_g for the MOSFET in SR mode is set to a minimum value, i.e., $R_{g,SR,on} = R_{g,SR,off} = 1\ \Omega$, enabling to control its channel as quickly as possible while avoiding hazardous gate-to-source overvoltages and oscillations. Fig. 4 c-d) show the impact of varying T_{dead} on body-diode recovery current. The total recovered energy during body-diode turn-off is defined as $E_{rec} = E_{Coss} + E_{bip}$ [13], where E_{Coss} is the necessary energy to create the space-charge region (i.e., charging process of MOSFET output capacitance C_{oss}), and E_{bip} corresponds to the energy required to remove the bipolar carriers from the device's drift region. In E_{rec}, only the term E_{bip} accounts for semiconductor losses during body-diode turn-off. E_{Coss} can be experimentally assessed when turning off the body-diode at room temperature and $I_d \approx 0$. In that conditions, $E_{rec} = E_{Coss}$. Figure 4-e) displays the dependency of E_{rec} vs. T_{dead} at $T_j = 150°C$. At $T_{dead} = 40$ ns, the MOSFET body-diode E_{rec} reaches its minimum value – the majority of charges coming from the $E_{Coss} = 75\ \mu J$ (at that same point E_{bip} accounts for 15 μJ). With a safety margin of 60 ns for jitter and delay tolerances, T_{dead} is finally set to 100 ns.

With the DPT configured as in Fig. 4-a), Fig. 7 shows the results for E_{sw} vs. I_d with the selected R_g and T_{dead}. From that, one concludes that the MOSFET body-diode now behaves virtually as a unipolar device, the decrease of $E_{sw,on}$ being related to the negative temperature coefficient of $V_{gs,th}$ (device's gate-to-source threshold voltage).

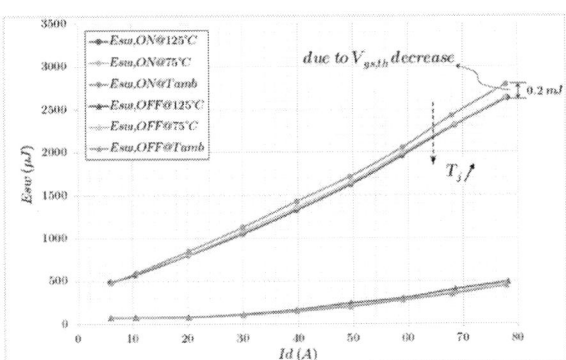

Fig. 7 – E_{sw} vs. I_d with $R_{g,on} = 3.3\ \Omega$, $R_{g,off} = 4.7\ \Omega$, $R_{g,SR,on,off} = 1\ \Omega$ and $T_{dead} = 100$ ns at $V_{ds} = 750$ V.

2.4 Input Inductor – L_{dc}

Both input inductors $L_{dc,hs,ls}$ are identical and share the same design constraints, yielding to $L_{dc,eq} = L_{dc,hs} + L_{dc,ls} = 2 \cdot L_{dc}$. The L_{dc} value depends on the current ripple (Δi_L) and can be calculated from Eq. (3). L_{dc} shall be designed to operate in the worst DC-bias condition, i.e., at nominal power

and minimum PV MPP voltage. In addition to this, Δi_L should also be taken into account.

$$\Delta i_L = \begin{vmatrix} \dfrac{V_{bus}\left(\frac{1}{4} - \alpha\right)\alpha}{F_{sw} \cdot L_{dc,eq}} & if\ 0 < \alpha < 0.25 \\[2em] \dfrac{V_{bus}\left(\frac{1}{2} - \alpha\right)\left(\alpha - \frac{1}{4}\right)}{F_{sw} \cdot L_{dc,eq}} & if\ 0.25 < \alpha < 0.5 \\[2em] \dfrac{V_{bus}\left(\frac{3}{4} - \alpha\right)\left(\alpha - \frac{1}{2}\right)}{F_{sw} \cdot L_{dc,eq}} & if\ 0.5 < \alpha < 0.75 \\[2em] \dfrac{V_{bus}(1 - \alpha)\left(\alpha - \frac{3}{4}\right)}{F_{sw} \cdot L_{dc,eq}} & if\ 0.75 < \alpha < 1 \end{vmatrix} \quad (3)$$

Figure 8 shows the normalized values for the input inductor current ripple and the maximum allowed RMS PV current (I_{PV}) (see Table 1) vs. α.

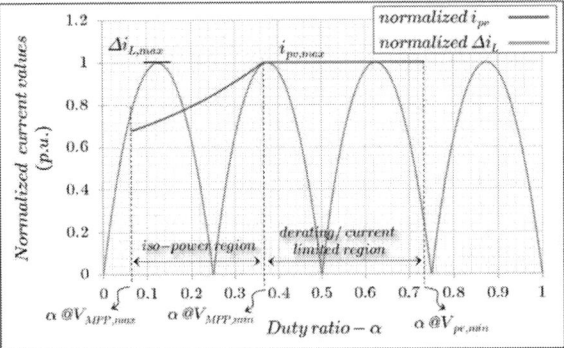

Fig. 8 – Normalized values for the inductor current ripple and the maximum allowed PV current as a function of the duty ratio.

From there, one concludes that the maximum current ripple coincides with the maximum PV current (42 A). That represents the operating point for which the inductor should be designed.

Powder core materials are considered for L_{dc}, thanks to their superior DC-bias and soft-saturation capabilities in comparison to ferrites [14], which are suitable for the application. In terms of core geometry, only toroidal shapes are taken in account. Litz wire (80 strands, Ø = 0.28 mm) is used in the coil with 2 wires in parallel. The design variables are listed as follows: i) powder core material – Edge, KoolMµ and KoolMµ Max, from Magnetics, ii) relative permeability – µr (ranging from 14 to 125), iii) core size (ext. diam. ranging from 50.8 to 132 mm), iv) number of stacked cores – n_{core} (up to 4) and current ripple – Δi_L (ranging from 5 to 20% of I_{PV}). From the aforementioned set of variables, the design space might present thousands of available solutions. In order to fairly compare the performance of different materials and fully explore the design space, the inductor design routine of Fig. 9 is implemented.

PCIM Europe 2024, 11– 13 June 2024, Nuremberg   DOI: 10.30420/566262078

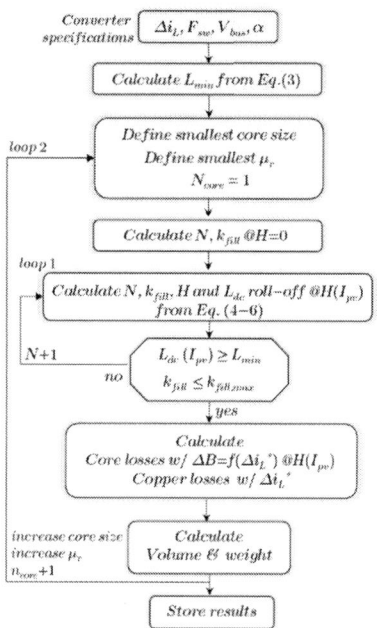

Fig. 9 – *Flowchart procedure for the inductor design.*

Contrary to what is often found in the literature [15], here the L_{dc} is modeled as a current-dependent non-linear inductor. The calculation of the roll-off inductance is taken into account considering the roll-off function ($\mu_{roll} = f(H)$) provided by the core's manufacturer. Within the design procedure, the design constraints are the maximum filling factor ($k_{fill,max}$), maximum number of stacked cores ($n_{core,max}$) and the required minimum inductance value from Eq. (3). Furthermore, an arbitrary limit value for μ_{roll} is not specified. In doing so, design solutions presenting a strong non-linearity are also investigated. The inductor design is not considered to be thermally constrained, since forced-convection is employed and the designs of interest are those presenting low-losses.

Equations (4-6) describe the relationships between number of turns (N), magnetizing field (H) and biased inductance taking into account the roll-off function ($L_{dc}=f(\mu_{roll})$).

$$N = \sqrt{L_{dc} \cdot 10^9 / A_l} \quad (4)$$

$$H = \frac{N \cdot i_L}{l_e} \quad (5)$$

$$L_{dc} = N^2 \cdot A_l \cdot \mu_{roll}(H) \quad (6)$$

where A_l and l_e stand for core's permeance (usually referred to as inductance factor) and equivalent magnetic path length, respectively.

For highly-biased current-dependent inductors, constant current slopes during the switching period do not occur. This non-linearity is described by

$$\Delta i_L(L_{dc}) = \frac{v_L}{L_{dc}(i_L(t))} \cdot t_{ramp} \quad (7)$$

where v_L is the applied voltage across L_{dc} within the switching period (T_{sw}) and t_{ramp} is the corresponding charging-time. In that case, the actual current ripple, noted here as Δi_L^*, shall be greater than the one predicted by Eq. (3). In order to better estimate core losses, the inductor current is computed solving Eq. 7 by a linear piece-wise approximation [14]. Note that both v_L and t_{ramp} must be described in the same manner as Δi_L in Eq. (3) – for the sake of brevity, it is not discussed here. Then, t_{ramp} is divided in a high-enough number of intervals (500 in our case) and Eq. (4-7) are solved for each of them. Thus, the actual inductor ripple Δi_L^* can be found, serving to calculate the H-field and lately the B-field swing in the core (the relation B = f(H) are provided by the core's manufacturer for each material and μ_r). Core losses are calculated based on Steinmetz coefficients also given in the manufacturer's datasheet. Copper losses account for DC and AC resistances (calculated for odd current harmonics 1 to 7) due to skin-effect based on Levasseur's approximation with a fixed winding temperature of 95°C.

Figure 10 shows the design results considering $F_{ripple} = 4*F_{sw} = 140$ kHz for the Edge, KoolMµ and

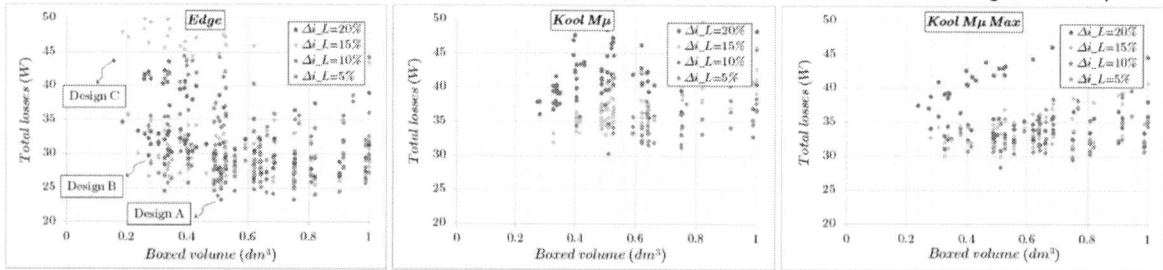

Fig. 10 – *Summary of inductor design results for materials Edge, Kool Mµ and Kool Mµ Max. Parameters: $k_{fill,max}=0.5$, $V_{mpp}=1900V$, $V_{bus}=3kV$, $I_{PV}=42A$, $F_{ripple}=140kHz$, $n_{core,max}=4$, $T_{winding}=95°C$, 2 Litz wires in parallel.*

640

KoolMµ Max materials. The choice of F_{sw} is further discussed in §3.

For all evaluated materials, minimum losses are achieved with a 10% current ripple. Further ripple reduction to 5% (increased L_{dc} value), leads to higher overall losses or non-feasible designs, mainly due to the $n_{core,max}$ constraint. Compared to Kool Mµ, Kool Mµ Max -based inductors present a slightly better trade-off between volume and losses, which might not compensate the related material cost increase – from [16], €$_{koolMµ-Max}$=2*€$_{koolMµ}$. On the other hand, the Edge material provides the best solutions both in terms of volume and losses. Table 2 summarizes the main characteristics of three Edge selected designs, namely A, B and C (see Fig. 10).

	Design A Δi_{Ldc}=10%	Design B Δi_{Ldc}=15%	Design C Δi_{Ldc}=20%
Core Ref.	59777	59876	59871
$L_{dc@42A}$ (µH)	160.3	108.6	80.5
N	20	29	33
$\mu_{roll@42A}$ (%)	97.7	95	87
n_{core}	2	2	1
OD x HT (mm)	77.8 x 26	77.8 x 13	77.8 x 13
μ_r	60	60	75
P_{core} (W)	9.5	15.8	32
P_{Cu} (W)	13.7	13.04	11.6
Vol (dm³)	0.51	0.28	0.158

Table 2 Comparison of selected inductor designs.

From the aforementioned results, design B is the final adopted L_{dc} solution, presenting the best volume/losses trade-off and a suitable form factor.

2.5 DC-link and Flying Capacitors – C_{bus} & C_{fc}

In the 5LSFC topology, the capacitors' design constraints are the same as in a typical 3-level FC converter. The capacitors are sized in order to limit their associated voltage ripple and self-heating, which affects the capacitor useful lifetime. For both C_{bus} and C_{fc}, the maximum RMS current stress ($I_{RMS,max}$) occurs at P_n and $I_{PV,max}$. Analytical expressions for $I_{RMS,max}$ are derived from [17]. In our case, $I_{RMS,max}$ values are 36 A and 15 A for C_{fc} and C_{bus}, respectively. With F_{sw} = 35 kHz, the minimum required capacitances to limit the voltage ripple to 5% of V_{bus}/4 are $C_{fc,min}$ = 12 µF and $C_{bus,min}$ = 4 µF.

In the design routine, a database with polypropylene metallized film capacitors (MKP) is

created taking into account four different component families: C4AU, C4AQ, C4AQ-M and C4AQ-P, from Kemet. All capacitors showing a voltage rating (V_{rating}) greater than 750 V are taken into account, i.e., 326 capacitor references in total. C_{bus} is implemented as a series association of two capacitors. Then, for a given component reference, the minimum number of capacitors in parallel is calculated to limit their maximum internal temperature to 105°C (or 125°C for C4AQ-P) and to satisfy the minimum required capacitance value. Capacitor losses are not a design criteria here, since they are usually limited to a few watts per capacitor bank. However, PCB surface and capacitor bank volume are of interest. Figure 11 depicts possible solutions for C_{fc} with regard to the required PCB surface.

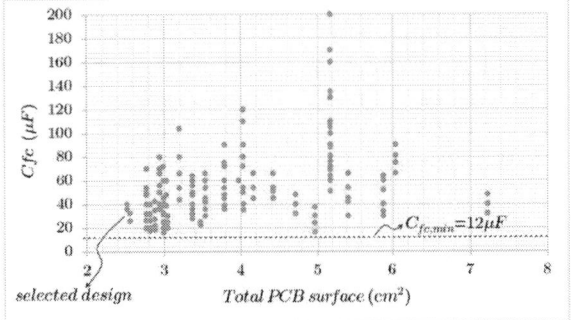

Fig. 11 – Possible solutions for C_{fc} capacitor bank.

From there, the choice of C_{fc} is made based on the minimum capacitance value and required PCB surface. Also, component availability must be taken into account. Finally, the selected reference is C4AQIBU4650M18J (800 V/6.5 µF) with 5 capacitors in parallel, i.e., C_{fc} = 32.5 µF. The resultant capacitor useful lifetime is checked, leading to 78.000 hours at 85°C (hot-spot temperature). C_{bus} is designed with a similar approach, leading to the series association of C4AUIBW5300M3LJ (800 V/30 µF) with two components in parallel. The corresponding lifetime is 80.000 hours at 85°.

Due to the severe capacitor current stress in the FC topologies, the final values of C_{fc} and C_{bus} are much greater than the ones defined simply by the voltage ripple limitation constraint. As a result, the increase of F_{sw} will not have a major impact on capacitors' overall volume, since $I_{RMS,max}$ does not depend on F_{sw} [17]. The increase of power density would require, for example, the implementation of ceramic capacitors, which present superior current capabilities.

3 Efficiency Estimation

3.1 Modelling of Semiconductors Losses

For both controllable and synchronous rectified switches (S and S_D), MOSFET on-state resistance is modelled as $R_{ds,on} = f(T_j, I_d)$. Based on Fig. 7, switching losses are modelled by a second order polynomial equation as

$$E_{sw} = aI_{pv}^2 + bI_{pv} + c \tag{8}$$

where a, b and c are fitting coefficients. The semiconductor RMS current stresses are described as follows

$$I_{S_{RMS}} = I_{pv} \cdot \sqrt{\alpha} \tag{9}$$

$$I_{S_{D_{RMS}}} = I_{pv} \cdot \sqrt{1-\alpha} \tag{10}$$

Then, conduction and switching losses become

$$P_{S_{cond}} = R_{ds,on}(T_j, I_{pv}) \cdot I_{S_{RMS}}^2 \tag{11}$$

$$P_{sw} = F_{sw} \cdot E_{sw}(I_{pv}) \tag{12}$$

The total 5LSFC semiconductor losses is

$$P_{tot} = 4 \cdot (P_{S_{cond}} + P_{S_{D_{cond}}} + P_{sw}) \tag{13}$$

3.2 5LSFC Converter Thermal Limit

To study the thermal limit of semiconductors, the maximum output power is calculated for a large range of switching frequency values. In order to account for the MOSFET $R_{ds,on}$ electro-thermal coupling, a thermal model is implemented considering the following thermal resistances: $R_{th,j-c} = 0.27$ K/W, $R_{th,c-hs} = 0.4$ K/W and $R_{th,hs-amb} = 0.4$ K/W (actual R_{th} values corresponding to the converter prototype of Fig. 14). Figure 12 shows the 5LSFC thermal limit for the criteria $T_{j,max} = 150°C$.

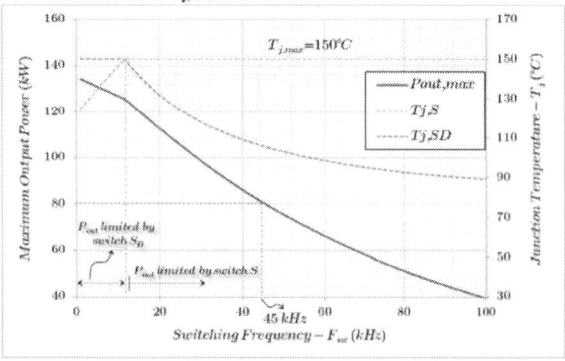

Fig. 12 – Thermal limit of semiconductors with $V_{pv} = 1900$ V and $T_{amb} = 45°C$.

As it can be seen, the converter output power is capped by the switches in SR mode, if $F_{sw} < 11$ kHz. For higher switching frequencies, the controllable switches become the limiting factor. At 80 kW, the maximum allowed F_{sw} is 45 kHz, leading to a current ripple at 180 kHz.

3.3 Choice of Switching Frequency – F_{sw}

In order to comply with existing EMC PV standards, e.g. EN 62920, EMI filters should be implemented in the converter [18]. A well-known way to minimize the volume and losses of those filters is to set the ripple frequency to be smaller than and as close as possible to 150 kHz (lowest frequency for conducted emission measurements). Furthermore, F_{sw} should be limited to meet the targeted efficiency requirements given in Table 1. Considering all those aspects, F_{sw} is defined as 35 kHz, leading to a current ripple at 140 kHz. Figure 13-a) shows the predicted semiconductor efficiency as a function of the PV voltage and input PV power.

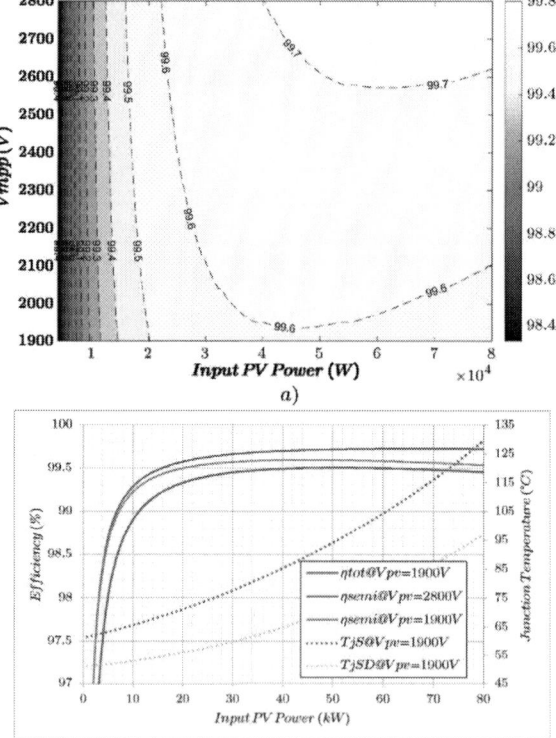

Fig. 13 – a) Semiconductor efficiency as a function of the PV voltage (V_{MPP}) and input power and b) Converter efficiency and T_j vs. input power. η_{tot} corresponds to the overall converter efficiency including the passives L_{dc}, C_{fc} and C_{bus}. For a) and b) $F_{sw} = 35$ kHz and $T_{amb} = 45°C$

Figure 13-b) depicts the junction temperatures and overall converter efficiency including passive losses. From that, one concludes that the targeted peak and European efficiencies are met, with η_{peak} = 99.5% and n_{euro} = 99.33%, leaving a certain margin for non-modelled losses such as auxiliary supply, PCB and cabling Joule losses, etc.

The 3D-view of the 5LSFC converter is presented in Fig. 14.

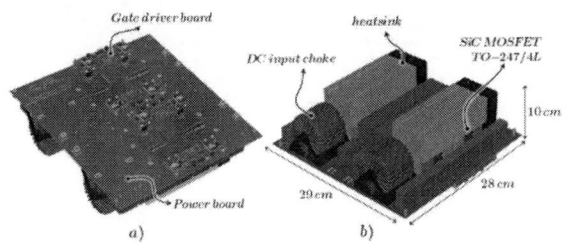

Fig. 14 – *3D-view of the 5LSFC converter.*

4 Conclusion and Perspectives

In this paper, a comprehensive design approach of a 3 kV 80 kW step-up symmetric flying-capacitor converter is presented, targeting MV PV utility-scale power plants. Main design aspects concerning control, modulation, semiconductors and passive components are discussed. A methodology for switching characterization in synchronous rectification mode is proposed, together with extensive experimental data. For the design of passives, especial attention is given to the choice of inductor powder core materials and current ripple, taking into account losses and volume trade-offs. Finally, based on the developed models, converter efficiency estimations are carried out, showing a peak and European efficiency of 99.5% and 99.33%, respectively. The future work includes a hardware realization of the proposed 80 kW 5LSFC converter to validate the calculated performances.

Acknowledgements

This work was performed in the framework of the TIGON project (Towards Intelligent DC-based hybrid Grids Optimizing the Network performance). The TIGON project has received funding from the European Union's Horizon 2020 research and innovation programme under grant agreement N°957769.

References

[1] "RENEWABLES 2023 GLOBAL STATUS REPORT." Accessed: Apr. 18, 2024. [Online]. Available: https://www.ren21.net/gsr-2023/

[2] L. Scarpa, G. Chicco, F. Spertino, P. M. Tumino, and M. Nunnari, "Technical Solutions and Standards Upgrade for Photovoltaic Systems Operated Over 1500 Vdc," in *2018 IEEE 4th International Forum on Research and Technology for Society and Industry (RTSI)*, Sep. 2018, pp. 1–6. doi: 10.1109/RTSI.2018.8548360.

[3] J. K. Steinke, P. Maibach, G. Ortiz, F. Canales, and P. Steimer, "MVDC Applications and Technology," 2019.

[4] J. Munoz-Cruzado Alba *et al.*, "Solid-State Transformers for DC-AC Hybrid Grids: a Case Study of TIGON Project," in *PCIM Europe 2023; International Exhibition and Conference for Power Electronics, Intelligent Motion, Renewable Energy and Energy Management*, May 2023, pp. 1–9. doi: 10.30420/566091253.

[5] C. Drexler, J. Pfeiffer, M. Schmidhuber, D. Derix, M. Geiss, and J. Thoma, "Magnetic Component Design for Medium Voltage Photovoltaic Application," in *PCIM Europe 2022; International Exhibition and Conference for Power Electronics, Intelligent Motion, Renewable Energy and Energy Management*, May 2022, pp. 1–8. doi: 10.30420/565822252.

[6] H. Wang, J. Lu, X. Huang, Y. Wang, and H. Xu, "Design of PV MVDC Converter with Wide Output Voltage Range for Series DC System," *Energies*, vol. 14, no. 6, p. 1617, Jan. 2021, doi: 10.3390/en14061617.

[7] M. N. Ngo, P. Ladoux, J. Martin, S. Sanchez, and A. Bier, "Performance Evaluation of SiC MOSFETs for Isolated DC-DC Conversion in Medium Voltage Photovoltaic Power Plants," in *PCIM Europe 2022; International Exhibition and Conference for Power Electronics, Intelligent Motion, Renewable Energy and Energy Management*, May 2022, pp. 1–9. doi: 10.30420/565822043.

[8] A. Bier, O. Wiss, and P. Messaoudi, "A 3 kV, 18 kW medium-voltage PV plant demonstrator," 2017.

[9] M. Stojadinović, "Bidirectional DC-DC Converters for MVDC Applications," Doctoral Thesis, ETH Zurich, 2020. doi: 10.3929/ethz-b-000439302.

[10] L. G. A. Rodrigues, J. Martin, S. Catellani, and J.-P. Ferrieux, "Characterization of 1.7 kV SiC MOSFET Modules for Medium/High Power Current Source Inverter in Photovoltaic Applications," in *PCIM Europe 2017; International*

Exhibition and Conference for Power Electronics, Intelligent Motion, Renewable Energy and Energy Management; Proceedings of, VDE, 2017, pp. 1–8.

[11] N. Oswald, P. Anthony, N. McNeill, and B. H. Stark, "An Experimental Investigation of the Tradeoff between Switching Losses and EMI Generation With Hard-Switched All-Si, Si-SiC, and All-SiC Device Combinations," *IEEE Transactions on Power Electronics*, vol. 29, no. 5, pp. 2393–2407, May 2014, doi: 10.1109/TPEL.2013.2278919.

[12] N. Oswald, B. H. Stark, D. Holliday, C. Hargis, and B. Drury, "Analysis of Shaped Pulse Transitions in Power Electronic Switching Waveforms for Reduced EMI Generation," *IEEE Transactions on Industry Applications*, vol. 47, no. 5, pp. 2154–2165, Sep. 2011, doi: 10.1109/TIA.2011.2161971.

[13] P. Sochor, A. Huerner, M. Hell, and R. Elpelt, "Understanding the Turn-off Behavior of SiC MOSFET Body Diodes in Fast Switching Applications," 2021.

[14] J. Imaoka, K. Okamoto, M. Shoyama, Y. Ishikura, M. Noah, and M. Yamamoto, "Modeling, Magnetic Design, Simulation Methods, and Experimental Evaluation of Various Powder Cores Used in Power Converters Considering Their DC Superimposition Characteristics," *IEEE Transactions on Power Electronics*, vol. 34, no. 9, pp. 9033–9051, Sep. 2019, doi: 10.1109/TPEL.2018.2886044.

[15] R. M. Burkart, H. Uemura, and J. W. Kolar, "Optimal inductor design for 3-phase voltage-source PWM converters considering different magnetic materials and a wide switching frequency range," in *2014 International Power Electronics Conference (IPEC-Hiroshima 2014 - ECCE ASIA)*, Hiroshima, Japan: IEEE, May 2014, pp. 891–898. doi: 10.1109/IPEC.2014.6869693.

[16] "Magnetics - New 2024 Powder Cores Catalog." Accessed: Apr. 23, 2024. [Online]. Available: https://www.mag-inc.com/New-2024-Powder-Cores-Catalog

[17] F. Hamma, T. A. Meynard, F. Tourkhani, and P. Viarouge, "Characteristics and design of multilevel choppers," in *Proceedings of PESC '95 - Power Electronics Specialist Conference*, Jun. 1995, pp. 1208–1214 vol.2. doi: 10.1109/PESC.1995.474968.

[18] M. Tadbiri-Nooshabadi, J.-L. Schanen, H. Iman-Eini, and L. G. A. Rodrigues, "Frequency Model for EMC Study of Multi-Level Flying Capacitor Boost Converter," in *2023 IEEE Energy Conversion Congress and Exposition (ECCE)*, Oct. 2023, pp. 3677–3683. doi: 10.1109/ECCE53617.2023.10362484.

PCIM Europe 2024, 11– 13 June 2024, Nuremberg   DOI: 10.30420/566262079

Comparison of Si IGBT, SiC MOSFET and Adjustable Hybrid Switch PV Inverters for Different Geographical Locations

Tanya Thekemuriyil[1], Dario Schneider[1], Jaspera Dominique Rohner[1], Munaf T.A. Rahimo[2], Vipluv Aga[3], Silvia Mastellone[1], Renato Amaral Minamisawa[1]

[1] University of Applied Sciences and Arts Northwestern Switzerland, Switzerland
[2] MTAL GmbH, Switzerland
[3] SOLEXTRON AG, Switzerland

Corresponding author: Tanya Thekemuriyil, tanya.thekemuriyil@fhnw.ch
Speaker: Tanya Thekemuriyil, tanya.thekemuriyil@fhnw.ch

Abstract

This paper compares a three-level three-phase SiC MOSFET and Adjustable Hybrid Switch (AHS) photovoltaic (PV) inverter to a commercially available Si IGBT PV inverter. The comparison extends beyond mere peak efficiency values to include the profitability aspect of adopting novel semiconductor technologies over conventional Si IGBTs. The study analyzes the performance of different PV inverters based on solar conditions and local electricity prices in a region. This research helps choose the right transistor configuration for specific application sites while highlighting a tradeoff between energy and profit gains. It demonstrates that solar irradiation profile, upfront costs and operating income are crucial factors for the widespread adoption of SiC technology.

1   Introduction

Researchers are actively exploring methods to enhance the performance of power electronic hardware in renewable energy systems. One significant trend is the adoption of Silicon Carbide Metal-Oxide-Semiconductor Field-Effect Transistors (SiC MOSFETs), gradually replacing traditional Silicon Insulated-Gate Bipolar Transistors (Si IGBTs) in applications where the latter once dominated. This transition is motivated by the dual objectives of improving overall efficiency and advancing decarbonization efforts [1].

SiC MOSFETs are attracting attention in photovoltaic (PV) inverters, primarily due to their reduced losses, which can enhance the overall efficiency of PV systems [2]. Consequently, numerous studies [1]-[4] are focusing on the efficiency and reliability of SiC MOSFETs in PV inverters. However, SiC MOSFETs also present several challenges in PV converters, including increased costs, design complexity, and reliability concerns [5]. To address these challenges, several studies [5]-[7] have developed PV inverters with hybrid switches combining silicon and silicon carbide, aiming to improve

performance while leveraging the cost-effectiveness of Si technology. Many research papers delve into optimizing the switching behavior of SiC MOSFETs and Si IGBTs within these hybrid configurations to further enhance performance [5], including the development of the Adjustable Hybrid Switch (AHS) converter [8]-[9].

Interestingly, despite considerable research efforts, there is limited information available on the total energy yield of converters over their operational life resulting from replacing Si IGBTs with SiC MOSFETs or hybrid switches. This analysis is highly dependent on the solar profile of the region, particularly the operating power points and the frequency at which these power points occur. Existing studies have predominantly limited their focus to mere peak efficiency for novel transistor technologies and datasheet metrics for off-the-shelf converters [3], [10]-[12]. To bridge this gap, this study undertakes a comprehensive analysis to estimate energy profits, considering the operation of inverters utilizing real solar profiles. By focusing on real-world data, we aim to provide more accurate insights into the potential energy gains achievable with different transistor technologies in PV converters.

645

It is widely accepted that SiC MOSFETs exhibit superior efficiency compared to their counterparts, followed by hybrid switches and the least for Si IGBTs. Despite their promising performance, novel technologies like SiC MOSFETs face hurdles in widespread market adoption, particularly within the PV inverter domain. It is worth addressing here the conflicting interests of the key stakeholders for PV inverter systems in the adoption of a particular transistor technology [3]. The increased upfront costs associated with SiC MOSFET converters hinder PV inverter manufacturers from readily adopting SiC technology. Selling inverters at higher prices in competitive markets becomes challenging due to these elevated costs. Currently, SiC usage in PV inverters is primarily limited to diodes within converters. From an installer perspective, PV installers often prioritize cost-effectiveness over performance improvements. Installing solar plants at a better price takes precedence for them, rather than investing in cutting-edge technologies like SiC MOSFETs. Additionally, there exists limited understanding among end users regarding the added benefits of SiC MOSFETs. To address this, the cost-saving potential of novel transistor technologies needs to be communicated through specific parameters, such as in electric vehicles (EV) where SiC MOSFET based inverters can extend the range of EVs. Policymakers strive for technology adoption, but they require additional data beyond datasheet values (European efficiency) [13]. A better understanding of the long-term inverter performance under real operating conditions would aid their decision-making process.

The need for a comprehensive financial analysis of various inverter technologies, in addition to performance studies, is evident. This investigation relies on unique location-based solar profiles and local electricity tariffs. Solar profiles unique to each location determine the operating power point of the inverter on the efficiency curve, thereby impacting the overall efficiency. The same inverter operating in two different locations exhibits varied performance over its lifetime. Figure 1 illustrates two histograms comparing the annual solar profile for a 10-kW plant in Berlin and Sevilla. Berlin, with more operations in the low-power region, struggles to consistently achieve targeted datasheet efficiency values compared to Sevilla. The operating points of Berlin demand inverters with different transistor technology that have higher efficiency in the low-power region. Additionally, assessing whether operational cost gains (by feeding in more energy to the grid and consuming more solar energy directly) can cover the expense of replacing Si IGBTs with SiC MOSFETs or hybrid switches is influenced by location-specific solar and consumption profiles, along with local electricity tariffs. Therefore, the application site of an inverter significantly influences the choice of semiconductor technology.

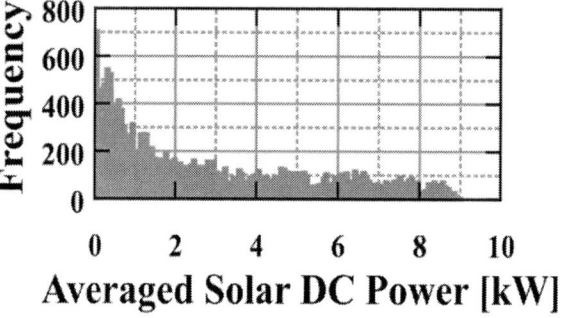

(a)

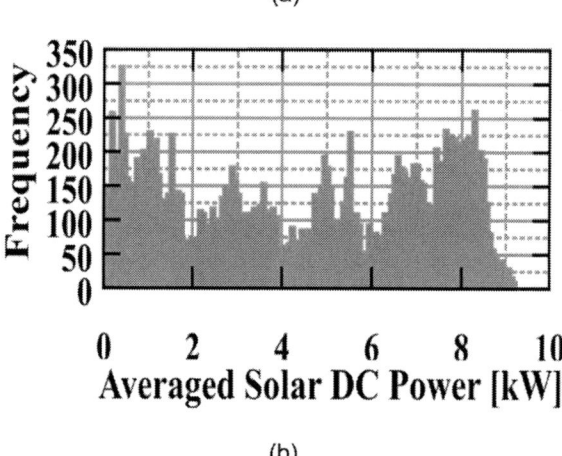

(b)

Fig. 1 Histograms of PV generated DC power in kilowatts in (a) Berlin (b) Sevilla.

The primary focus of this investigation is to compare the energy and cost-saving potential of SiC MOSFETs and AHS converters against the off-the-shelf Si IGBT converter across various locations. The outcomes of this study aim to provide valuable insights for end users, plant installers, inverter manufacturers, and policymakers to standardize one transistor technology in specific regions. The methodology is outlined as follows: (i) characterizing and comparing the efficiency of a 10 kW SiC MOSFET converter and AHS converter with a commercially available 10 kW Si IGBT PV inverter (ii) tracking actual solar irradiance, weather conditions, and residential consumption profiles in cities across three European countries using commercial software to evaluate the energy yield for the PV plant over its 10-year lifespan (iii) analyzing self-consumption profiles and local electricity tariffs to assess whether cost savings

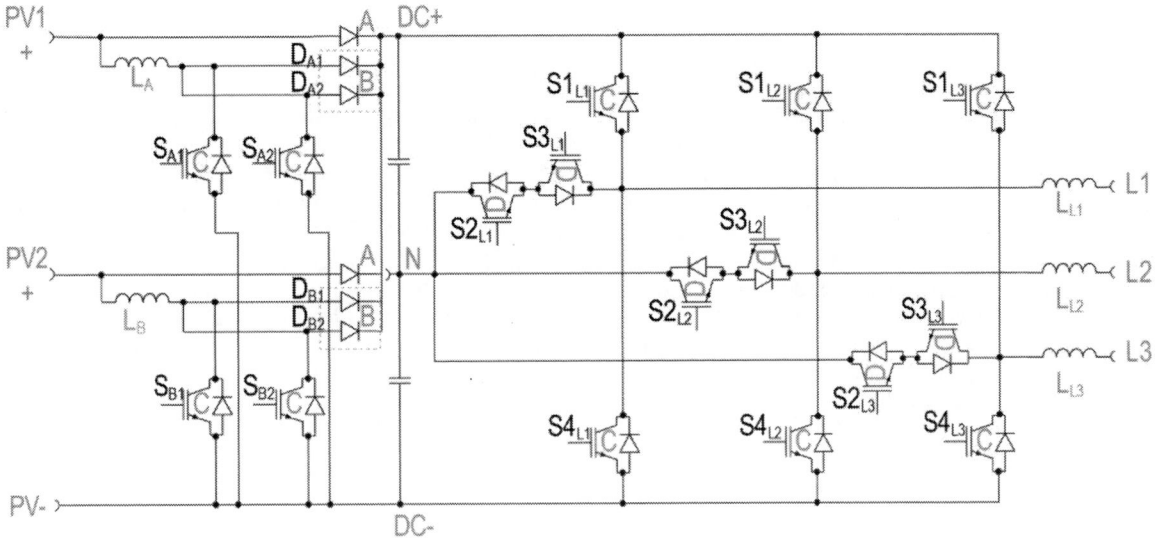

Fig. 2 Topology of the commercial 10 kW PV Inverter.

from SiC MOSFETs or AHS converters justify their upfront costs compared to conventional Si IGBT inverters.

2 Modelling of PV Inverters

2.1 Commercial Si PV Inverter

The study utilizes a three-phase 10 kW off-the-shelf converter as a benchmark, which incorporates two maximum power point trackers (MPPTs), each with one input. Operating within a voltage range of 140 V to 980 V, and with a maximum input current of 11 A per MPPT, the inverter can deliver a maximum output current of 16.9 A when connected to the grid. The study scrutinizes the inverter to examine its fundamental design parameters, particularly the topology and semiconductor components employed. The identified topology, consisting of two boost converters in the first stage and a T-type inverter in the second stage, is illustrated in Fig. 2. Semiconductor specifications for each switch are outlined in Table 1, noting that all the devices are housed in discrete TO-247 packages. Additionally, while silicon carbide diodes are utilized for the boost converter stage, no switches in the inverter stage employ silicon carbides.

The inverter is simulated in PLECS, considering individual semiconductor loss models based solely on available transistor datasheet characteristics. A DC-link voltage of 800 V is chosen, with DC link capacitors sized according to reference standards.

Ref.	Device Model	Rating	$/unit
A	WND45P16W	1.6 kV, 45 A	2.93
B	FFSH10120ADN-F155	1.2 kV 10 A	5.54
C	NGTB40N120FL3WG	1.2 kV 40 A	7.80
D	STGWA40H65DFB	650 V 40 A	3.47

Table 1 Semiconductor models in the commercial 10 kW PV Inverter.

Conventional practices are used to size the filters of the converter [14]. Maximum power dissipation in the switches is calibrated using semiconductor datasheet information to determine thermal resistance of heatsinks and thermal interface materials (TIM) [15]. The control system comprises an MPPT controller in the boost stage to maximize power extraction from PV strings, utilizing voltage and current controllers to regulate PV voltage by managing injected current. The T-type converter's controller manages DC-link voltage and current transferred to the grid. Space vector pulse width modulation (SVPWM) is applied in the inverter control [16]. Many design parameters, including DC-link, LCL, and boost filter, as well as thermal models of heatsink and TIM, and control algorithms, are derived from conventional practices and references. It is acknowledged that the simulation model may not perfectly align with the manufacturer's inverter model, but the study's primary focus is to compare different transistor technologies for performance. Authors believe that an approximated model cannot be inaccurate for the intended study. Additionally, reference [3] indicates that passive component losses remain consistent regardless of transistor technology.

The investigation included characterizing efficiency under different switching frequencies ranging from 8 to 20 kHz, maintaining a constant input voltage of 470 V to align with efficiency curves provided in the manufacturer's datasheets. To ensure consistency, a reference voltage of 470 V was utilized in the simulation model instead of the reference voltage generated by the MPPT controller. The results of the characterization are shown in Fig. 3. and is compared to the datasheet characteristics. The least mean absolute error is attained for a switching frequency of 16 kHz.

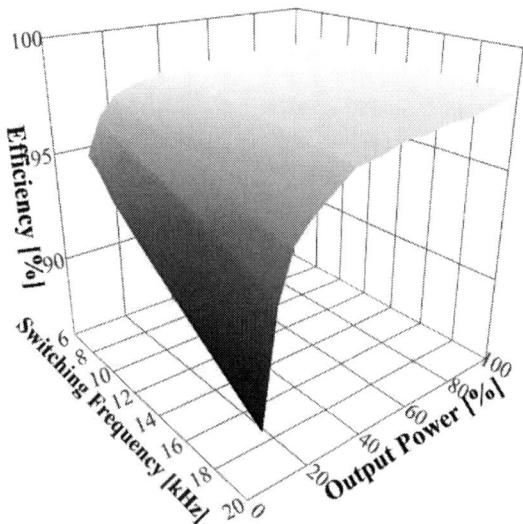

Fig. 3 Efficiency characterisation of the commercial PV inverter for varying switching frequencies and constant voltage of 470 V.

2.2 SiC MOSFET PV Inverter

The design parameters of the developed Si IGBT inverter model serve as the basis for all simulation models, maintaining consistency across comparisons. Even though SiC MOSFETs have the capability to switch at higher frequencies [2], the switching frequency (16 kHz) remains unchanged to ensure a fair comparison with Si IGBT converters. Consequently, only the Si IGBTs from the established model are substituted with novel transistor technologies to construct new models. The SiC MOSFETs replacing the Si IGBTs are outlined in Table 2, where the selected switches possess comparable voltage and current ratings to their Si IGBT counterparts in real hardware. This approach ensures that the comparison accurately reflects the potential performance improvements offered by the novel transistor technologies.

Ref.	Device Model	Rating	$/unit
C	E3M0040120K	1.2 kV 41 A	23.19
D	C3M0045065K	650 V 35 A	19.13

Table 2 SiC MOSFET semiconductor models to replace Si IGBTs.

2.3 AHS PV Inverter

The hybrid Si/SiC switch (HyS) represents a compromise solution, integrating both SiC MOSFET and Si IGBT in parallel [17]. This configuration holds the promise of enhanced performance at a reduced cost by leveraging the advantages of both SiC MOSFET and Si IGBT technologies. Additionally, there's a technique called the adjustable hybrid switch (AHS), where SiC MOSFETs and Si IGBTs are switched separately based on the operating power requirements. This approach has been demonstrated to enhance efficiency in electric vehicles significantly [8]. The AHS operates based on an optimization algorithm, activating the transistor in the hybrid configuration with the highest efficiency for the given operating power. One critical consideration in this process is ensuring that the current ratings of the semiconductors meet the required operating power constraints. The AHS mechanism focuses on extracting maximum efficiency from the converter throughout its operational lifespan, rather than just achieving peak efficiency at specific instances [9]. This consideration for efficiency over the converter's operating life is the primary reason why the authors of this study chose to utilize the AHS mechanism instead of opting for the HyS configuration alone.

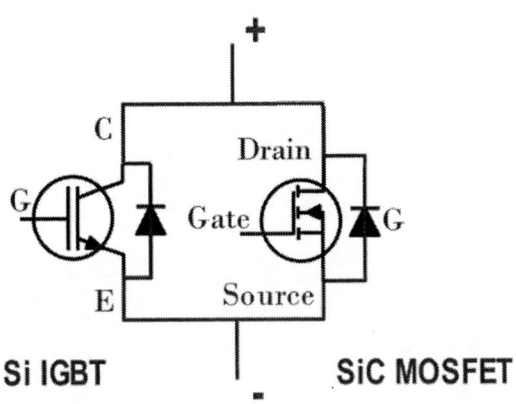

Fig. 4 Hybrid Switch (HyS).

In the simulation model, a single Si IGBT is substituted with a hybrid switch comprising a SiC MOSFET and Si IGBT in parallel, as depicted in Fig. 4. The characteristics of these newly chosen switches are detailed in Table 3. While the voltage ratings remain consistent with those of the replaced devices, the current rating of the Si IGBT in the hybrid switch represents approximately 62.5% of the current capacity of a fully Si IGBT solution. The remaining demanded current is met with SiC MOSFETs. This selection was made based on device availability and cost considerations. Although assigning a higher current percentage to the SiC MOSFET would have been advantageous, it is constrained by cost considerations.

Ref.	Device Model	Rating	$/unit
C	NGTB25N120FL3WG	1.2 kV, 25 A	6.13
	E3M0160120K	1.2 kV 16 A	11.52
D	STGWA20H65DFB2	650 V 20 A	2.94
	C3M0120065K	650 V 13.5 A	7.17

Table 3 Hybrid Switch (HyS) semiconductor models to replace Si IGBTs.

Research [9] suggests that it's optimal to operate the SiC MOSFET of the hybrid switch independently in the low-power region. Operating beyond the current capacity of the SiC MOSFET, which is approximately 37.5% of the rated output current of the converter, may lead to device damage. This necessitates the activation of the Si IGBT in conjunction with the SiC MOSFET or the deactivation of the SiC MOSFET while activating the Si IGBT. As HyS outperform Si IGBTs at every operating current beyond the 37.5% threshold of the rated converter current, a straightforward AHS algorithm is integrated into the simulation model. Accordingly, a cutoff point of 6.3 A of rms phase current is selected. Below this threshold, the SiC MOSFET operates independently, while the additional Si IGBT of the hybrid is activated beyond this cutoff level. This approach ensures optimal performance throughout the operation of the converter.

3 Characterisation of Si, SiC and AHS PV Inverter

To accurately assess the performance of both stages of the inverter, including MPPT, the simulation no longer operates under constant voltages. Instead, a Trina vertex S TSM-420DE09R.08 (420 Wp) PV panel is selected and modeled in PLECS as a non-linear current source based on the panel's IV characteristics, accounting for variable insolation.

To size a 10 kW PV plant, a total of 24 of the selected panels are required. These panels are configured into two strings for efficient operation. Specifically, 12 panels are connected in series to form a single string, which then serves as the input for each MPPT. This configuration ensures that the MPPT trackers receive adequate input for optimizing power generation from the solar panels.

The three simulation models, each employing different transistor technologies, are then characterized for efficiency across the entire power range by varying the incident solar radiation. Power is evenly distributed between the two inputs of the inverter, ensuring that each input receives an equal share of the generated power from the solar panels. The results of the characterization of three simulation models are depicted in Fig. 5. SiC MOSFET based converter attains the maximum efficiency, whereas Si IGBT converter exhibits least performance. AHS featuring intermediate efficiency drops its performance around the cut-off level followed by a regain in efficiency. This converter distinguishes itself from the conventional efficiency curves. AHS maintains a better performance than Si IGBT converter however at a reduced cost than SiC MOSFET converter.

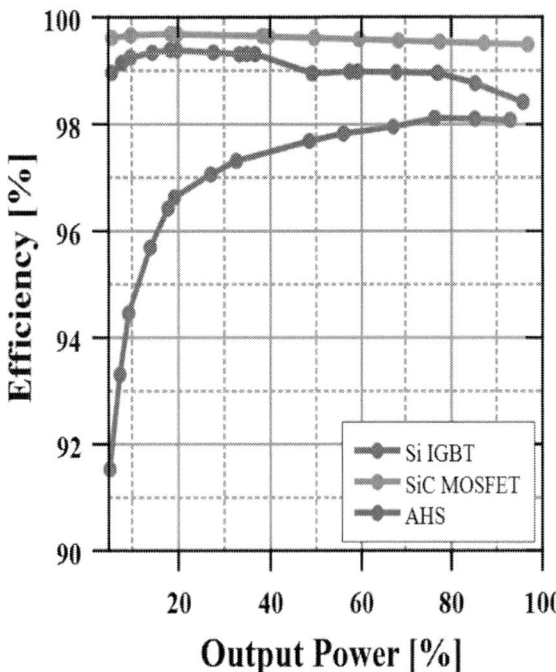

Fig. 5 Efficiency Characteristics of different Inverter configurations.

4 Energy Analysis of PV Inverters

To evaluate the performance of different PV inverters across diverse geographical locations, three distinct cities in Germany, Switzerland, and Spain have been selected for analysis. Using the commercial software tool "SOLEXTRON," the DC energy generation from PV panels is simulated based on the unique solar irradiation and temperature profiles of each region. Various design parameters of the PV plants are inputted into the software to facilitate simulation, with these parameters remaining consistent regardless of the geographical location under consideration. These parameters include those for a standard 10 kW residential installation, featuring two PV strings, each comprising 12 panels oriented at a 30-degree inclination in south orientation, as depicted in Fig.6.

The software generates profiles detailing solar irradiation and the corresponding DC energy generated at 15-minute intervals throughout the span of one year. To conduct a comparative assessment of energy production among three distinct inverter configurations, the efficiency curves obtained in section 3 are re-plotted to illustrate efficiency versus the input power of the inverter. From these efficiency versus input power graphs, we interpolate the efficiency for each generated DC power at every instance to estimate the output AC power. This process is repeated for the three inverter models across nine different regions, enabling a detailed assessment of their performance under various solar and climatic conditions.

Fig. 6 3-D simulated model of the 10-kW residential plant.

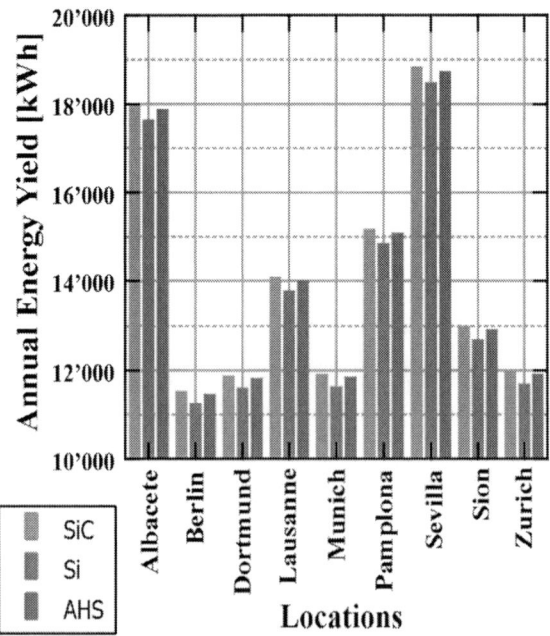

Fig. 7 Annual energy yield of inverters in different locations.

650

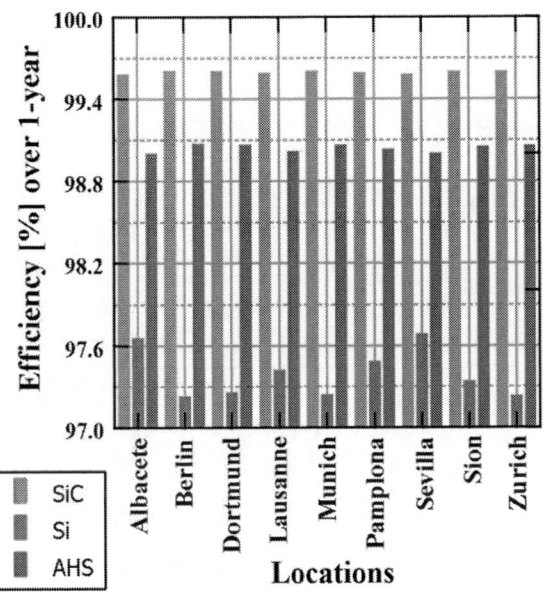

Fig. 8 Efficiency based on energy yield of different inverters over one year in different locations.

The annual energy yield of the PV plant is determined by integrating the resulting AC power output over the duration of one year, and the results are illustrated in Fig. 7. The production is improved utilizing novel transistor technologies. Additionally, assuming a consistent yearly production profile over the entire 10-year lifespan of the system (according to manufacturer offered warranty period), the total energy yield over its operational lifetime is evaluated. This analysis offers valuable insights into how various geographical locations and different transistor technologies can impact energy production over the lifespan of an inverter.

Figure 8 shows the efficiency of different inverter configurations over the course of one year in various geographical locations. Surprisingly, the selected commercial inverter fails to provide the promised European efficiency of 98.1% or peak efficiency of 98.6%.

5 Cost Analysis of PV Inverters

The cost-saving potential of a SiC MOSFET converter and an AHS converter compared to a Si IGBT inverter is dependent on two key factors. Firstly, it involves considering the additional expenditure incurred in upfront costs when replacing Si IGBTs with SiC MOSFETs or AHS. Secondly, it evaluates the operational cost savings resulting from enhanced energy production achieved by employing novel transistor technologies. This indicates that the financial analysis must focus only on

the profit and loss associated with replacing Si IGBTs from a commercial Si IGBT inverter with SiC MOSFETs or AHS. Therefore, all other aspects of the PV plant, including panel cost, installation prices, maintenance costs, etc., are not considered into the analysis and are assumed to remain consistent across all transistor technologies in every location. The analysis aims to provide a clear assessment of the economic implications of adopting SiC MOSFETs or AHS inverters without the influence of other variables within the PV system.

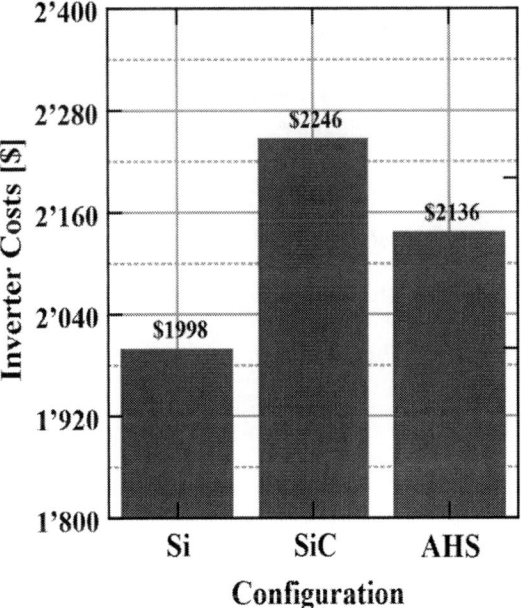

Fig. 9 Inverter Upfront Costs.

The upfront costs of the chosen commercial inverter are analyzed within the current market to establish a baseline. By deducting the retail prices of the semiconductors present in the inverter and adding the costs of SiC MOSFET switches or hybrid switches, the costs of SiC and AHS inverters are estimated. The upfront costs of different inverters are shown in Fig. 9.

Understanding the electricity energy consumption patterns of households within different regions supports cost analysis of the different transistor technologies over their operating life. The individual residential load profiles are also generated in the "SOLEXTRON" tool. The profitability of the operation of converters is explored with two forms of financial gains. Firstly, the less energy consumed from the grid by directly consuming generated solar energy is monetized through local electricity prices for 10 years as in (1),

$$Cost\ Saving = 10 \times Electricity\ tariff$$

$$\times \sum_{j=1}^{365} \sum_{i=1}^{24} E_{i,j} \qquad (1)$$

where,

$$E_{i,j} = \begin{cases} g_{i,j}, if\ g_{i,j} < c_{i,j} \\ c_{i,j}, else \end{cases} \qquad (2)$$

and $g_{i,j}$ and $c_{i,j}$ are the generated PV energy and demanded energy by the household at every i^{th} hour of jth day. Secondly, the surplus energy generated from the PV plant fed to the grid adds to the financial profit based on regional feed-in-tariff as in (3),

$$Cost\ Saving = 10 \times Feed\ in\ tariff$$

$$\times \sum_{j=1}^{365} \sum_{i=1}^{24} E_{i,j} \qquad (3)$$

where,

$$E_{i,j} = \begin{cases} g_{i,j} - c_{i,j}, if\ g_{i,j} > c_{i,j} \\ 0, else \end{cases} \qquad (4)$$

The local electricity tariffs and regional feed-in-tariff by country as of 2023 are summarized in Table 4. The attained retail prices are based on a thorough web search in the present market.

Country	Electricity Price ($/kWh)	Feed-in-tariff ($/kWh)
Spain	0.2	0.065
Germany	0.45	0.089
Switzerland	0.2912	0.14

Table 4 Regional list of local electricity and feed-in-tariff in $/kWh.

The total operational profits over the 10-year lifespan of the converters are shown in Fig 10. Fig.11 demonstrates how much savings can be made using SiC and AHS technology in comparison to the profits of the Si Inverter as zero baseline in each location.

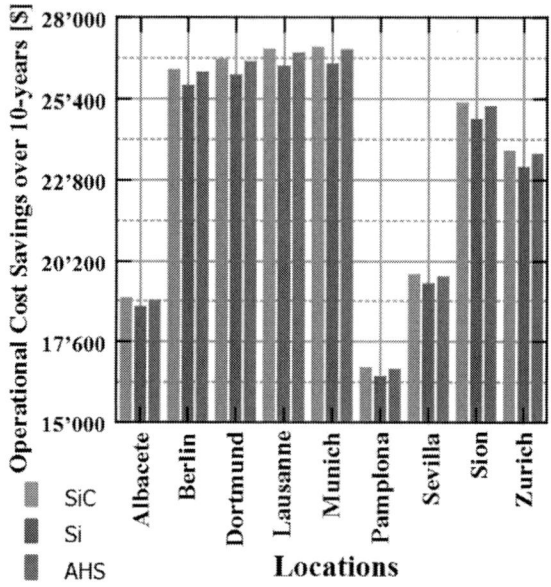

Fig. 10 Operational cost savings of different inverters over 10-year lifespan due to reduced grid import and higher grid export.

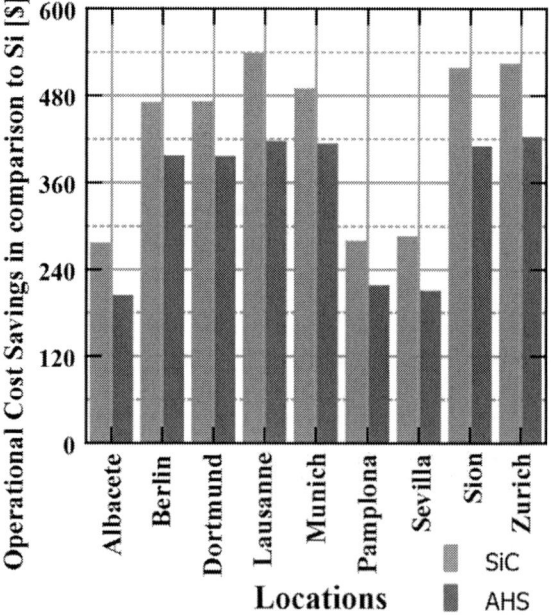

Fig. 11 Operational cost savings over 10 years due to reduced grid import and higher grid export with SiC and AHS inverter compared to Si IGBT inverter in different locations.

The degree to which an inverter technology is profitable in a specific region is determined by the question if the upfront costs are covered by operational profit. Fig 12. depicts how profitable it is to replace Si IGBTs with SiC MOSFETs and AHS considering both operational and upfront costs. The profit is clearly based on the regional solar profile and the current market of local electricity prices and inverter costs. Adopting the SiC technology in the selected inverter is mainly beneficial for locations with relatively low solar irradiation and higher electricity tariffs. Although the benefits are less in other regions, the end consumers can be convinced that there are no personal losses in investing in novel technologies.

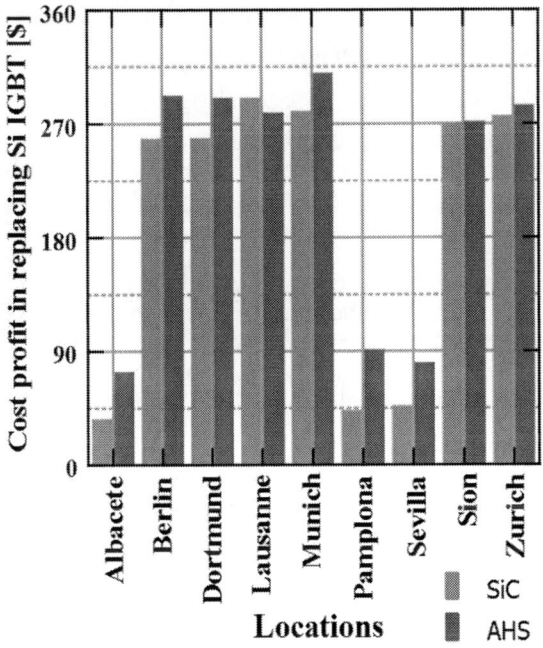

Fig. 12 Total cost savings due to replacing Si IGBTs with SiC and AHS inverter in different locations.

It is to be noted that the inverter under investigation here is comparatively inexpensive in competition to other PV inverters and uses cheaper discrete devices. Replacing Si IGBT modules in other expensive inverters with expensive SiC modules change the outcomes. One promising way for enhancing the cost benefit of SiC (silicon carbide) technology lies in optimizing photovoltaic (PV) plant design according to inverter performance. With SiC technology facilitating higher energy production, there exists an opportunity to implement design alterations that could reduce the initial investment required for PV plants. This way of optimizing the design of PV plant is identified as the future scope of the study by the authors.

6 Conclusion

This paper investigates the performance of commercial PV inverters to assess the energy yield during its operational life in different regions. A comparative analysis is conducted by replacing the Si IGBT from the PV inverters model with SiC MOSFET and hybrid solution. On a different dimension, the study considers the cost expenditure and resulting profit gains of Si, SiC and hybrid inverters in selected locations. The approach allows for a detailed understanding of the economic viability of different inverters in each location offering a standardized solution for residential solar power generation. In this conference, we will present a systematic study on the potential of each inverter technology.

References

[1] M. Buffolo. D. Favero, A. Marcuzzi , C. De Santi, G. Meneghesso et al., "Review and Outlook on GaN and SiC Power Devices: Industrial State-of-the-Art, Applications, and Perspectives," in IEEE Transactions on Electron Devices, vol. 71, no. 3, pp. 1344-1355, March 2024, doi: 10.1109/TED.2023.3346369.

[2] W. Choi, D. Kim, D. Son and S. Kim, "Kelvin Source Package to Maximize 1200V SiC MOSFET Performance in Solar Inverter Applications," 2023 11th International Conference on Power Electronics and ECCE Asia (ICPE 2023 - ECCE Asia), Jeju Island, Korea, Republic of, 2023, pp. 1959-1965, doi: 10.23919/ICPE2023-EC-CEAsia54778.2023.10213545.

[3] T. Eskilson, A. Jehle, P. Schmidt, M. Makoschitz and F. Baumgartner, "Identifying the potential of SiC technology for PV inverters," 2023 25th European Conference on Power Electronics and Applications (EPE'23 ECCE Europe), Aalborg, Denmark, 2023, pp. 1-7, doi: 10.23919/EPE23ECCEEurope58414.2023.10264500.

[4] Villanueva, I.;Vázquez, N. Vaquero, J. Hernández, C. López-Tapia, H. et al., "Photovoltaic Inverter Reliability Study through SiC Switches Redundant Structures," Technologies 2023, 11, 59. https://doi.org/10.3390/technologies11020059

[5] M. Stecca, C. Tan, J. Xu, T. B. Soeiro, P. Bauer and P. Palensky, "Hybrid Si/SiC Switch Modulation With Minimum SiC MOSFET Conduction in Grid Connected Voltage Source Converters," in IEEE Journal of Emerging and

Selected Topics in Power Electronics, vol. 10, no. 4, pp. 4275-4289, Aug. 2022, doi: 10.1109/JESTPE.2022.3146581.

[6] H. Liu, T. Zhao and X. Wu, "Performance Evaluation of Si/SiC Hybrid Switch-Based Three-Level Active Neutral-Point-Clamped Inverter," in IEEE Open Journal of Industry Applications, vol. 3, pp. 90-103, 2022, doi: 10.1109/OJIA.2022.3179225.

[7] Kshatri, S.S.; Dhillon, J.; Mishra, S.; Haghighi, A.T.; Hunt, J.D.; Patro, E.R. Comparative Reliability Assessment of Hybrid Si/SiC and Conventional Si Power Module Based PV Inverter Considering Mission Profile of India and Denmark Locations

[8] M. Rahimo, R. A. Minamisawa, S. Mastellone, T.Koottungal, J. Spoendlin et al., "An Advanced Adjustable Switch Hybrid (ASH) Concept for High Power Automotive Converters," PCIM Europe digital days 2021; International Exhibition and Conference for Power Electronics, Intelligent Motion, Renewable Energy and Energy Management, Online, 2021, pp. 1-8.

[9] T. Thekemuriyil, M. T. A. Rahimo, R. A. Minamisawa and S. Mastellone, "Performance Assessment of the Adjustable Hybrid Switch Converter for E-mobility Applications," 2023 25th European Conference on Power Electronics and Applications (EPE'23 ECCE Europe), Aalborg, Denmark, 2023, pp. 1-8, doi: 10.23919/EPE23ECCEEurope58414.2023.10264282.

[10] J. Wu, Yuying Wu, Ning He, Wenxing Zhong, Seiki Igarashi et al., "Impact of SiC MOSFET on PV Inverter," 2018 IEEE Energy Conversion Congress and Exposition (ECCE), Portland, OR, USA, 2018, pp. 1853-1860, doi: 10.1109/ECCE.2018.8558284.

[11] A. Anthon, Z. Zhang, M. A. E. Andersen, D. G. Holmes, B. McGrath and C. A. Teixeira, "The Benefits of SiC mosfets in a T-Type Inverter for Grid-Tie Applications," in IEEE Transactions on Power Electronics, vol. 32, no. 4, pp. 2808-2821, April 2017, doi: 10.1109/TPEL.2016.2582344.

[12] M. H. Ahmed, M. Wang, M. A. S. Hassan and I. Ullah, "Power Loss Model and Efficiency Analysis of Three-Phase Inverter Based on SiC MOSFETs for PV Applications," in IEEE Access, vol. 7, pp. 75768-75781, 2019, doi: 10.1109/ACCESS.2019.2922741.

[13] B. S. Hansen, "Policy measures to drive WBG for end use equipment," 2023 25th European Conference on Power Electronics and Applications (EPE'23 ECCE Europe), Aalborg, Denmark, 2023, pp. 1-7, doi: 10.23919/EPE23ECCEEurope58414.2023.10264489.

[14] Texas Instruments. "10-kW, Bidirectional Three-Phase Three-Level (T-type) Inverter and PFC Reference Design". March 2018.

[15] P. Bruyere, G. P. Boisson and G. Perez, "Design of a serial impingement cooling heatsink for a 30 kW PV string inverter," 2022 24th European Conference on Power Electronics and Applications (EPE'22 ECCE Europe), Hanover, Germany, 2022, pp. 1-11.

[16] M. Fatima, A. S. Siddiqui and S. K. Sinha, "Implementation of Three-Phase two Stage Solar PV Inverter for Grid Connection," 2022 8th International Conference on Advanced Computing and Communication Systems (ICACCS), Coimbatore, India, 2022, pp. 1325-1329, doi: 10.1109/ICACCS54159.2022.9785351.

[17] M. Rahimo, Francisco Canales, Renato Amaral Minamisawa, Umamaheswara Vemulapati, Masatoshi Aketa et al., "The Cross Switch "XS" Silicon and Silicon Carbide Hybrid Concept," Proceedings of PCIM Europe 2015; International Exhibition and Conference for Power Electronics, Intelligent Motion, Renewable Energy and Energy Management, Nuremberg, Germany, 2015, pp. 1-8.

PCIM Europe 2024, 11– 13 June 2024, Nuremberg   DOI: 10.30420/566262080

Optimising a Power Module for Electrical and Thermal Performance and Symmetry Using EDA Tools

Wilfried Wessel[1], Roland Bátai[2], Gergő Juhász[2], Florian Bauer[3], Dr. Andreas Schwarzbacher[1], Prof. Dr. Christian Jakob[4]

[1] TU Dublin, Irland

[2] Vincotech Hungária Kft., Hungary

[3] Siemens EDA GmbH, Germany

[4] University of Applied Sciences Darmstadt, Germany

Corresponding author: Wilfried Wessel, D22128972@mytudublin.ie
Speaker: Wilfried Wessel, D22128972@mytudublin.ie

Abstract

The paper describes a comprehensive automation of a power module design and verification process. State-of-the-art design and verification tools are combined in a novel way to achieve full automatic optimisation of a real industry use case. The paper presents significant electrical and thermal performance benefits. Device temperature differences are reduced by 35%, and inductance differences for single devices are reduced by 99%. The results represent a significant step in the field of power module development.

1   Introduction

In the next three years, the market for power modules will grow with a CAGR of 10.5% [1]. This high growth rate demands novel development methods. This is because it is not possible to employ engineers at the same rate as the CAGR. On top of that, growing highly competent employees proportional to the market growth rate is not sustainable. There are multiple reasons for this. Firstly, these employees are not available in the market. Secondly, adding more engineers to a project does not linearly increase productivity. The pressure for those engineers would still be high, which leads experience-wise to long working hours and stress. The method described in this paper accommodates the bottleneck of experts. Using novel optimisation technologies, as described in this paper, increases the performance and quality of power modules but also has a significant effect on the environment [4]. The current work method of manufacturing a design idea and validating the physical prototype takes material resources and labour hours. Even if other state-of-the-art methods, such as [5], are available, they are not widely used. This is because fragmented simulation tools need to be glued together. In such a workflow, a huge amount of engineering resources is required to synchronise between the thermal, mechanical and electrical domains [5]. Another major disadvantage is

that the literature confused optimisation with manual design adjustment and simulation. This manual process is repeated until a sufficient design has been found. The novel approach in this paper proves that full automatic optimisation is possible, and the manual steps of design adjustments and synchronising domains can be eliminated from the workflow. Paper [6] shows a typical optimisation of a power module with a high count of parallel SiCFETs. The paper concludes with a good improvement in the overall module performance. This clearly illustrates that the outcome is highly dependent on engineering expertise. The human in the process is the limiting factor in processing simulated data and making an optimal design decision. Of course, the design will most likely fulfil the specifications, but it is unclear if it is the best possible design.

Revolving around the utilisation of the Siemens EDA toolchain, the novel core of the paper is to fully integrate, connect and automate power module design and simulation. This ensures that all requirements, including manufacturing parameters, electrical, thermal, and mechanical parameters, can be verified and fulfilled. This paper describes a full automatic multi-objective optimisation to overcome the disadvantages, including electrical and thermal parameters. The novel approach to enhancing power modules' electrical and thermal performance uses an enhanced analytics search algorithm called SHERPA[2]. The method is applied to a power module used in solar inverters of the Vincotech power module. Comprising multiple SiC MOSFETs in parallel, this design serves as a representative platform for demonstrating the effectiveness of the proposed methodology. The main reason for parallelising is to extend the power module's power range. Therefore, equal current sharing is crucial among all devices. Because the current imbalance has significant effects on reliability and product lifetime. Paper [13] describes the effect of higher temperatures on SiC MOSFET devices. It concludes that higher temperatures led to faster switching and decreased gate-source voltage for the same drain current. This behaviour can cause a negative feedback loop if electrical and thermal parameters are not optimised for all parallel devices. The used method is capable of performing this kind of design optimisation. The time needed for this task is two days compared to the four weeks usually required.

The paper's first section provides an overview of the design before optimisation, helping identify the parameters to optimise.

The paper's second section discusses the first innovative method of fully automatically changing copper structures.

Section three picks up the results of section two to run a full automatic placement optimisation. The optimisation considers the electrical, thermal, and manufacturing domains. This is the second outstanding part of this paper.

Section four describes the thermal and electrical characterisation and the functional testing results completely.

The paper concludes with a 35% improvement in thermal performance. This represents a significant step toward more efficient and reliable power module designs. It proves that a complete automatic geometrical optimisation is possible and discloses additional parameters not considered in the first run.

2 Identifying Optimisation Parameters

The Vincotech design shown in Fig. 1 represents the critical part of the power module. The red rectangle highlights the arrangement of parallelised semiconductors. In addition to the advanced neutral point clamped (ANPC) topology, it contains features such as gate resistors, a Kelvin emitter to improve switching, and a temperature sensor. The main design challenge is the placement and connection of the five high-speed SiC-MOSFETs marked red in Fig. 1. The tight form factor (58 mm x 34 mm), the increased bill of material, compared to a simple two transistor topology, to implement the ANPC topology and the given pin locations force the SiC-MOSFETs to be placed in the current direction. Placing the five SiC-MOSFETs in the current direction results in a non-ideal design with huge disadvantages. Optimising these designs and reducing the disadvantages in existing non-ideal parallel designs in a state-of-the-art design tool differentiates this work from others, e.g. [7]. The weak points of a parallel circuit structure implemented as series placement in the current direction, as shown in Fig. 1, are a common drain-source path and a common gate-source path. The design below has a single-gate driver for all five SiC-MOSFETs. This makes it worse since each device sees a different gate-source voltage. In the measurement, this leads to a high-temperature difference between the leftmost device, called T13_a, and the rightmost device in Fig. 2, called T13_e, of 7.72 K as reference see Table 2.

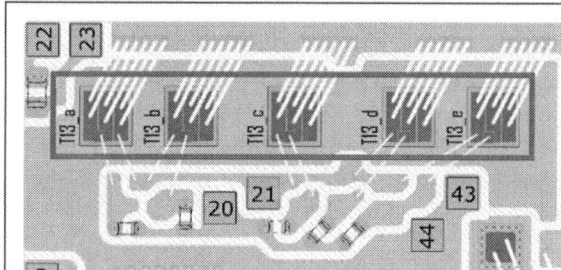

Fig. 1   Initial Substrate Design

The power module is analysed electrically and thermally using the Siemens EDA toolchain to see if the same effect can be reproduced by simulation. HyperLynx Advanced Solvers-Fast 3D is used for the electrical simulation. The solver is a quasi-static solver. Other industry papers [14] show that such solvers have high accuracy and a major performance advantage on power module simulation compared to the full 3D solution. This is also crucial for the later optimisation run. The equivalent circuit used to analyse the unsymmetric power loop is represented in Fig. 2. The circuit shows that the high side IGBT T11 and one out of five SiC-MOSFETs are shorted to extract the resistance and inductance between the ports P_{G1} and P_{G2}. All other SiC-MOSFETs are open and not conducting.

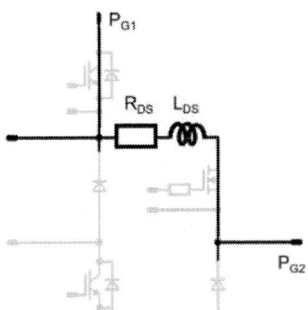

Fig. 2   Equivalent Circuit for Parasitic Extraction with HyperLynx Advanced Solver - Fast 3D

Table 1 summarises the simulated results. Considering T13_a and T13_e as extreme cases, the temperature difference under AC load conditions can be explained easily. T13_e has 1.52 nH less inductance and 7.46 mΩ less resistance than T13_a. Assuming no differences in R_{DSON} and the gate signals, T13_e drives most of the current. The described case is true for dynamical switching.

Device	Power Loop Inductance @ 100 MHz	Power Loop Resistance @ 100 MHz
T13_a	12.61 nH	35.83 mΩ
T13_b	12.37 nH	34.62 mΩ
T13_c	11.96 nH	32.61 mΩ
T13_d	11.50 nH	30.44 mΩ
T13_e	11.09 nH	28.37 mΩ

Table 1   Electrical Characteristics of the Power Loop Prior Optimisation

A conductive heat transfer simulation with Siemens Simcenter FLOEFD, a CAD-integrated CFD software, was performed for the thermal simulation. The Siemens toolchain natively supports the data exchange and maintaining data integrity. For the simulation, each of the five devices gets a unit load of 1 watts. The bottom surface of the power module is assigned to a wall boundary with a fixed temperature of 300 K. Default materials are used for Cu, Al_2O_3, Al and SiC. This is a suitable method for the simulation and optimisation of value difference compared to absolute values. The thermal simulation results in Table 2 show a slight difference in thermal resistance between all devices. The deviation between the highest (T13_a) and lowest (T13_c) thermal resistance is 0.042 K / W. These are excellent values. Table 2 also shows that the thermal design of the power module is not the reason for the high-temperature differences of 7.72 K. The reason must be the high difference in inductance or resistance.

Device	Simulated R_{TH}	Measured Avg. Temperature
T13_a	0.836 K / W	58.89 °C
T13_b	0.824 K / W	59.91 °C
T13_c	0.794 K / W	NA
T13_d	0.817 K /W	63.30 °C
T13_e	0.796 K /W	66.61 °C

Table 2   Thermal Characteristics Prior Optimisation

Concluding the first section, an optimisation of the electrical parameters is required to lower the differences in device temperature. Because the characteristics of the device affecting the current sharing are temperature-dependent. The paper focuses on optimising the power loop differences of 1.52 nH in inductance and 7.46 mΩ in resistance. The next sections are a detailed descriptions of the methods used for optimising these values.

3   Automatic Copper Optimisation

Possible input parameters are all the power module design criteria, such as the definition of copper structures, bond wire placement, component placement, and the adjustment of technology parameters such as substrate thickness or bond wire diameter. The use cases can even be extended to mechanical design. This makes the optimisation extremely complex. The main advantage of splitting the task is reducing time to result. Starting with the copper optimisation, does not fit to all designs. In other cases, it might be better to optimise the placement first or consider a redesign for a better starting point. In this case, a redesign is not possible to maintain compatibility with the application's printed circuit board (PCB). Fig. 3 shows the basic concept of introducing a parallel structure in the existing design. The copper plane between the SiC-MOSFETs T13_b and T13_c is divided into two separate current paths. This method is most promising for optimisation. Using a T-shaped obstruct allows the parasitics on both sides of the T-shape to be adjusted individually. Individual parameters are illustrated in Fig. 3 as well. The picture shows that this task can no longer be sufficiently solved by using trial and error. The reason for this is that it is a multi-objective optimisation. As an optimisation objective, the reduction of the difference in resistance and inductance of T13_a and T13_e are selected.

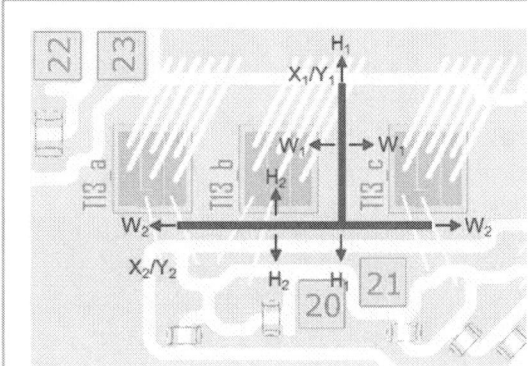

Fig. 3   Plane Obstruct to Force Electrical Current

Siemens Simcenter HEEDS must be introduced as a missing link for optimisation. HEEDS is a powerful design space exploration and optimisation software. The solution can interfere with almost all CAD and CAE tools available in the market. For this paper, three custom portals to control Xpedition, HyperLynx Advanced Solvers (HLAS), and FloEFD were developed. The flow used in this section is represented in Fig. 4. It contains the electrical and manufacturing domains. In the flow diagram Fig. 4, all parameters defined in Fig. 3 are used as variable inputs for HEEDS. After the initial design parameters are set, HEEDS executes the portals. A portal is, for example, a Python script for Xpedition and HyperLynx or a VB.NET application for FLOEFD. The portals then call the native tool functionality using the tools' API. The Xpedition portal in HEEDS adjusts the X_1/Y_1, H_1/W_1, X_2/Y_2 and H_2/W_2 directly in the power module design software. This also allows checking for 2D and 3D design rules (DRC) to ensure manufacturability. A new Python script for Xpedition is created by starting an optimisation with the adjusted values for X_1/Y_1, H_1/W_1, X_2/Y_2 and H_2/W_2. In the next step, the adjusted Python script is executed. If no DRC violations occur, a CCE file is exported from Xpedition. The CCE file format is the standard exchange

format between Xpedtion and HyperLynx Advanced Solvers. It includes all layout and technology information. If a DRC occurs, a new set of X_1/Y_1, H_1/W_1, X_2/Y_2 and H_2/W_2 is tested. In the next step, HEEDS will launch the HyperLynx Advanced Solvers (HLAS) portal. No additional adjustments are required in this step. As a result, the values R_S and L_S, as shown in Fig. 2 for T13_a and T13_e, are exported in a text file. Based on the difference in the response of R_S and L_S for T13_a and T13_a, a new set of parameters for X_1/Y_1, H_1/W_1, X_2/Y_2 and H_2/W_2 is generated and the process starts again. The difference in magnitude of R_S and L_S required additional weighting on the inductance difference. In the use case described in the paper, the difference in inductance is multiplied by 10^6. Regarding Table 1, this factor gives both values approximately the same weight.

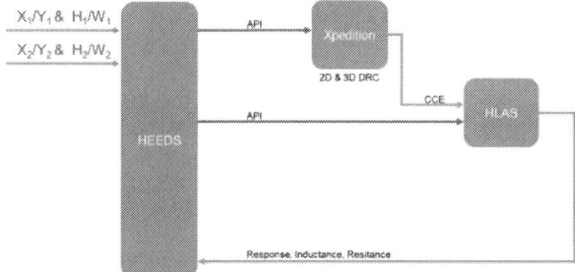

Fig. 4   Copper Optimisation Flow using Siemens
HEEDS and HLAS

As an optimisation method for this project, SHERPA was used [2] for the parameter optimisation. SHERPA is described by Lars Hartel as "SHERPA is a hyperparameter optimisation library for machine learning models. It is specifically designed for problems with computationally expensive, iterative function evaluations, such as the hyperparameter tuning of deep neural networks." [2]. For this reason, it is perfect for power module optimisation. In other literature, such as [8], SHERPA is used for thermal characterisation. Optimising a complete power module is a novel way of using this hybrid parameter optimisation. Fig. 5 shows an overlap of all responses. The plot shows all design values, from left to right, T13_a resistance, T13_a inductance, T13_e resistance, T13_e inductance and ends with the difference of both values called dRAE and dLAE. The bold black line represents the baseline design, and the yellow line is the best design by definition of the objectives. Light blue are all values of the other iterations. The baseline design includes the initial parameters for X_1/Y_1, H_1/W_1, X_2/Y_2, and H_2/W_2, and is not the original design. The plot helps in such complex cases to find other solutions. Even if manufacturing constraints are maintained automatically, there is still a trade-off between an excellent electrical, thermal and mechanical design and the most suitable design for manufacturing. In other words, a slightly worse design could have advantages for manufacturing. Another reason to look for a different design is that not all requirements were implemented in the cost function. This can be used to keep things simple and to search for a solution graphically with the existing data. In the case described here, the best design 15 is also the engineer's choice. Fig. 5 shows that the resistance and inductance for T13_a were increased to minimise the parasitic differences. The yellow line is above the black line. In the case of T13_e, the resistance and inductance were decreased. The best solution was already found after 23 iterations. The design set was limited to 25 designs. The overall Simulation time was 2 hours and 45 minutes (Intel Xeon Gold 5220R, 256 GB RAM). This is usually the time to set up a single design and simulation by hand.

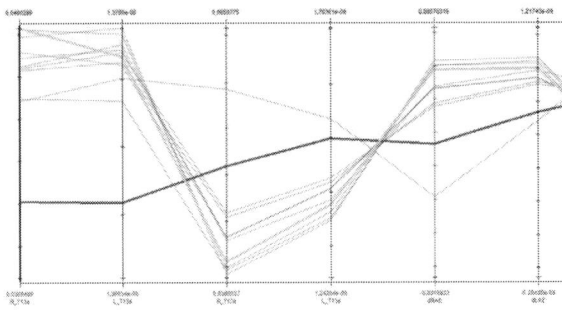

Fig. 5 Overlapping Best and Initial Case

Table 3 summarises all results compared to the original design, which gives a much better understanding of the introduced changes.

Variable	Best Value	Original Value
R_T13_a	40.03 mΩ	35.83 mΩ
L_T13_a	13.76 nH	12.61 nH
R_T13_e	40.58 mΩ	28.37 mΩ
L_T13_e	13.77 nH	11.09 nH
dRAE	0.55 mΩ	7.46 mΩ
dLAE	0.01 nH	1.52 nH

Table 3 Result Comparison After First Optimisation

The table shows a general increase of resistance and inductance of the best design of 4.2 mΩ and 1.15 nH for T13_a and 12.21 mΩ and 2.68 nH for T13_e. These values are compensated with the impressive optimisation results of 0.55 mΩ difference in resistance compared to 7.46 mΩ and 0.01 nH difference in inductance compared to 1.52 nH.

4 Automatic Placement and Bond Wire Optimisation

This section further improves the design by matching the parasitic inductance and resistance of T13_b, T13_c, and T13_d with those of the other two devices. As a foundation for this section, the best design of the previous section is used. Cascading different optimisation results is one of the best practices to reduce interactions and speed up the time needed to achieve the results the paper wants to show. Due to the copper changes done in the second section and the component placement between T13_a and T13_e, the thermal domain was added to the analysis. Based on the initial placement and the surrounding copper, a new set of input parameters X_b/Y_b, X_c/Y_c and X_d/Y_d can be derived. All six input variables were limited to a range of +/- 1 mm with a resolution of 10um to avoid false design variants. For example, the initial value X_b in the original design is 13.5 mm, regarding the board origin in the bottom left corner in Fig. 1. Using the described settings, the SiC-MOSFET can move between 12.5 mm and 14.5 mm in the horizontal direction. The benefits of using an integrated design environment are that the design rule check (DRC) filters all designs unsuitable for manufacturing, and the bond wire location changes according to the placement of the transistors.

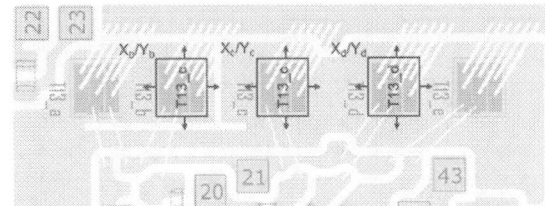

Fig. 6 X/Y Parameters of Components T13_b to

T13_d

Fig. 7 shows the updated optimisation flow. Compared to Fig. 4, the flow now includes the thermal analysis using FloEFD. It is also visible that the inputs to HEEDS changed to X_b/Y_b, X_c/Y_c and X_d/Y_d. The same Xpedition portal can be reused. Like the HyperLynx Advanced Solvers portal, the FLOEFD portal uses the initial simulation template created in section 1 to analyse the initial design behaviour.

This means all simulation settings, such as heat sources, material assignment, and goals assignment, are the same. This ensures comparability.

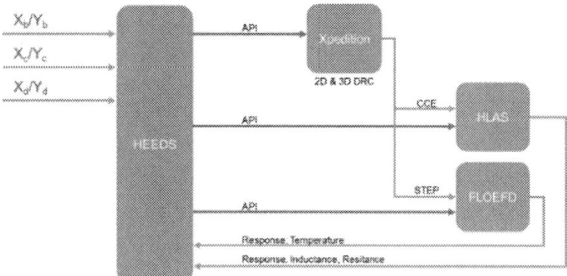

Fig. 7 Placement Optimisation Flow using Siemens HEEDS, HLAS and FLOEFD

To keep the excellent matching of T13_a and T13_e, the positions of these devices were not changed. For the optimisation of the other devices two groups were formed. The first group contains T13_a and T13_b and the second group T13_c, T13_d and T13_e. The group were formed by the thermal dependencies. For example, changing the position of T13_b in X_b/Y_b has a much higher impact due to thermal cross-coupling in T13_a than in T13_c. The reason is the copper cut-out introduced in section 3. Reconsidering the optimisation objectives, it shows that to match electrical parasitics between T13_a and T13_b, both devices should be placed as close as possible. To match thermal characteristics, both devices should be placed as far away from each other as possible. The optimisation can be applied to find the sweet spot in the middle. The cost function Eq. 1 is the summed absolute difference of the parasitic resistance and inductance of T13_a and T13_b. In this case, the weighting of the inductance becomes even more important than the weighting in section 3.

$$f_{cost} = |\Delta R_{T13}| + |\Delta L_{T13} \cdot 10^6| \quad (1)$$

Another change compared to section 3 is that the number of designs was increased to 101. This was required since, in this paper, only one device was moved at the time. The overlapped diagram in Fig. 8 of all objectives is used to identify the best design. For better visibility, the diagram does not show all 101 design iterations. It only shows the baseline design in black and the best design with design ID 57 in yellow. The baseline design represents the initial design condition for this optimisation. This is the optimisation result from section 3, not the original design. In the left part of Fig. 8, it can be seen that all parasitics were further decreased compared to the baseline. Looking at the cost functions, a slight increase from 0.00218 to

0.00248 in the difference between T13_a and T13_b can be seen. The second group shows a significant improvement from 0.00862 to 0.00762. Compared to all other design variants, the difference in thermal resistance between all devices improved significantly from 0.359 K / W to 0.247 K / W. There are design variants available with significant improvements in a single objective. Still, the design with the ID 57 is the best trade-off overall objectives. The optimisation over 101 designs took 8 hours and 10 minutes (Intel Xeon Gold 5220R, 256 GB RAM). Including the thermal domain, this excellent performance result matches time-to-result expectations. This means the optimisation can run overnight.

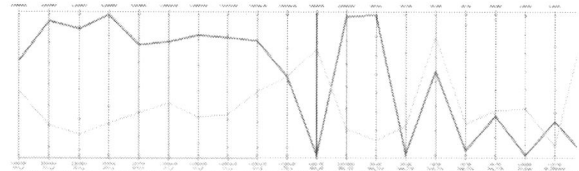

Fig. 8 Optimisation Result of the Second Run. Comparison of Design ID 57 with the Baseline

Table 4 compares the optimised design with the original design in all relevant parameters for this project. The parasitic inductance and resistance of T13_a and T13_e still match well. In conclusion, the placement of the other three devices did not significantly interfere with the optimisation done in section 2. The table shows that the copper cutout and the placement adjustment led to an increase in the electrical and thermal resistance, e.g. for T13_c, the electrical resistance increased from 32.61 mΩ to 41.70 mΩ and the thermal resistance from 0.794 K / W to 1.042 K / W. The table also shows a significant improvement for the maximum difference in inductance of 1.05 nH and the maximum difference in resistance of 7.20 mΩ. For the inductance, this is a reduction of approximately 30 %; for the resistance, it is 4.5 %. The maximum difference in thermal resistance noticeably increased. Looking at Table 4, T13_c has the worst thermal and electrical designs. Measurements in the next section will show that these effects compensate each other.

Variable	Value Design 57	Original Value
R_T13_b	34.50 mΩ	34.62 mΩ
L_T13_b	11.45 nH	12.37 nH
R_T13_c	41.70 mΩ	32.61 mΩ
L_T13_c	12.96 nH	11.96 nH
R_T13_d	39.30 mΩ	30.44 mΩ
L_T13_d	12.46 nH	11.50 nH
R_{TH} T13_a	0.912 K / W	0.836 K / W
R_{TH} T13_b	0,912 K / W	0.824 K / W
R_{TH} T13_c	1.042 K / W	0.794 K / W
R_{TH} T13_d	0.918 K / W	0.817 K /W
R_{TH} T13_e	0.811 K / W	0.796 K /W
Max dL	1.05 nH	1.52 nH
Max dR	7.20 mΩ	7.46 mΩ
Max dR_{TH}	0.23 K / W	0.043 K / W

Table 4 Final Comparison After Second Optimisation

The key takeaway from this optimisation is that equalising the device temperatures has multiple parameters that must be considered. This paper considered only two of them. Firstly, the electrical parameter is improved or worsened, and secondly, the thermal parameter is improved or worsened. When working with existing designs of parallel SiC-MOSFET circuits without major redesign, it is most likely that the electrical and thermal parameters must be worsened to achieve the goal of equal device temperatures. This means the design under normal load conditions performs much better than the non-optimised design with lower parasitics per single device. Design 57, shown in Fig. 9, was manufactured without any redrawing. This is a severe time saver and ensures data integrity.

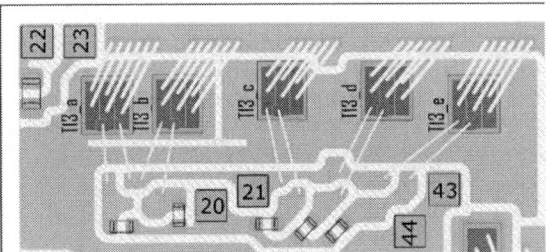

Fig. 9 Final Design 57 for Manufacturing

5 Analysis of Measurement Results

With the standard characterisation methods such as double pulse test or thermal characterisation [9][10], it was not possible to show if the optimisation had a positive effect on the power module behaviour under load conditions.

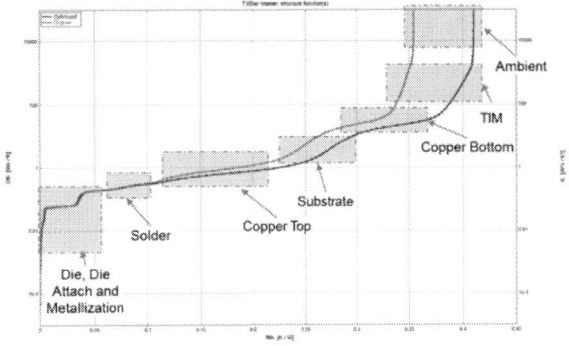

Fig. 10 Structure Function of the Original Design (green) and the Optimised Design (blue)

For example, the thermal characterisation in Fig. 10 shows the structure-function of the original and optimised design. The results show differences in the "Copper Top" and "Substrate" regions. The differences in "Copper Top" are caused by the design changes in the previous sections. Differences in the substrate material cause the difference in the "Substrate" region. Both changes will cause a higher module temperature, but which device is the hottest cannot be differentiated. For this reason, a function test of the original solar inverter system was required. The core of the whole test is the fast stationary infrared camera of INFRATEC [11]. The high image acquisition rate of up to 105 kHz guarantees reliable measurement, including heating and cooling phases during operation. IN-FRATEC says: „The ImageIR® 5300 has been designed specifically for capturing and recording extremely fast running thermal processes." [11]. The inverter works with a frequency of 50 kHz. This shows that the camera is an excellent choice for measuring the thermal behaviour. The main challenge to analysing the power module was that it was mounted under the main inverter board. A second challenge was that the power module was completely encapsulated. For this reason, a section of the board and the module enclosure was cut out. Modifying the board takes up to three weeks. For this reason, such measurements are not feasible for an efficient development process. Instead, simulation must be used. The result of the cutout is shown in Fig. 11. The cutout itself does not influence the power module or system behaviour. Another major disadvantage is the press-fit pins used to connect the power module to the main inverter board. The pins are only capable of up to five use cycles. This makes testing of a larger series of power modules impossible. The effect of component or manufacturing tolerances, e.g., placement with 50 µm, cannot be analysed.

The functional test considered two load cases: 60 A DC current and 57 A AC. These tests are essential to check whether all five SiC-MOSFETS T13_a to T13_e can withstand the nominal forward current without issues, e.g., exceeding the T_{jmax} threshold. For the safe test in the laboratory, only one-third of the nominal current of 180 A was used.

Optimised
Power Module

Fig. 11 Detailed View of the Inverter Application

Showing a Board Cutout

The measurement of the AC load case is the most important factor in qualifying the effects of the optimisation. For this reason, power circulation [12] was used as an accurate measurement method for AC power losses. The load in both cases was set at 57 A and between two times 500 V. Fig. 12 and Fig. 13 show the result of the AC measurements. A fourth measurement point, P2, was added to the original design's measurement. Compared to C1, C2, and C3, this is a single-point measurement. Based on the position of T13_e, temperatures can

vary. P1 was set to represent the average similar to C1, C2, and C3. It can be identified that both figures show a temperature increase in the current direction T13_a to T13_e.

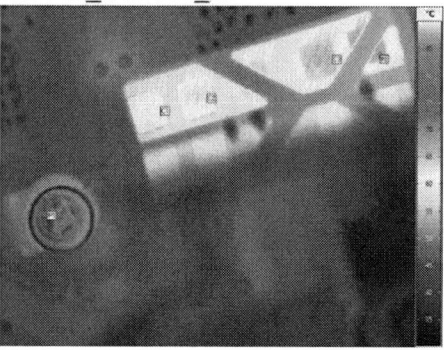

Fig. 12 Thermal Measurement Under AC Load

Conditions of 57 A 2 x 500V of the Original Design

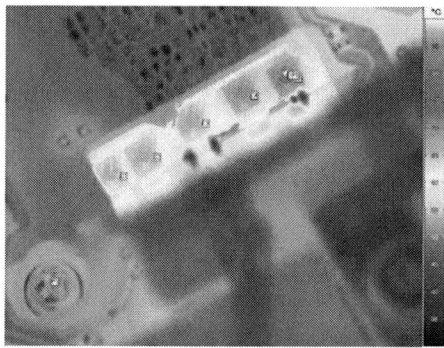

Fig. 13 Thermal Measurement Under AC Load

Conditions of 57 A 2 x 500V of the Optimised Design

Table 5 is used for a detailed analysis. The table shows a significant improvement in temperature differences comparing the extreme cases T13_a and T13_e. The difference in the original design is 7.72 K, and in the optimised design, 4.97 K. This is an improvement of 35%. For the first optimisation, this result is excellent.

Device	Avg. Temperatur Original	Avg. Temperatur Optimised
T13_a	58.89 °C	62.58 °C
T13_b	59.91 °C	63.69 °C
T13_c	NA	63.82 °C
T13_d	63.30 °C	65.40 °C
T13_e	66.61 °C	67.86 °C

Table 5 Final Comparison After Second Optimisation

6 Conclusion

The paper introduced a novel method to fully automate the power module design, verification and optimisation process. By splitting up the optimisation process into two independent steps and cascading the single results, significant improvements in equalising device temperatures of 35% or 2.75 K were achieved. Additionally, the overall runtime for both processes was 2 hours and 45 minutes for the first run and 8 hours and 10 minutes for the second run. These are excellent results for an optimisation combining electrical and thermal effects. The paper also shows the potential for further optimisation time improvements by moving all components simultaneously. The paper's contribution to enhancing the environment cannot be estimated directly [4]. The novel optimisation techniques proved the capability to make a power module more reliable. Lower temperature differences mean lower thermal and mechanical stress on single devices. This enhances the lifetime of a power module or even reduces the number of redesigns, which directly decreases the CO_2 emission. Another key aspect of why low-temperature differences are important is described in [13]. The paper shows that the switching behaviour of parallel SiC-MOSFETs can create a negative feedback loop. This can lead to critical problems. The holistic approach in the paper proves that the methods help to tackle these problems and the growing demand for engineering resources. A single engineer can configure multiple optimisations and let them run automatically compared to manual design and simulation setup adjustments. The paper also proves that the Siemens EDA toolchain is suitable for driving these use cases. Using the Siemens EDA toolchain is also a disadvantage since it is limited in accessibility for organisations and engineers. With the shortage of engineering resources in mind, the user-friendly environment and the high out-of-the-box range of functions compensate for this disadvantage. The paper showed a multi-object optimisation, and the potential of SHERPA was validated. Future projects must analyse how to further optimise the 4.97 K temperature difference to, for example, a 0.50 K difference. The paper results show that such an optimisation is only possible when the simulation of dynamic switching characteristics is included to extract real dynamic and static losses. This would improve the analysis and optimisation of electrical and a thermal network only for simultaneous electrical and thermal analysis. The results also show that simple equivalent circuits, as used in this paper, are not suitable when common current paths are used. For this reason, non-reduced and fully coupled parasitic models must be used in future. Especially if layout structures are unknown, a fully coupled model is required. Besides the improvements and advantages of the paper, the main finding is that significant differences in the source path cannot be fully compensated by changing the drain path of parallel SiC-MOSFET layouts. This also highlights that future projects must include the power and the gate-loop for optimisation. For this, simulations of full switching waveforms, including all parasitic substrate effects, are required. Additional comparisons between SHERPA and other algorithms, such as genetic algorithms can also extend the paper. The main focus of this comparison should be time to result. Overall, the paper shows significant advantages to measurements, as shown in section 5. For example, large measurement series were not realisable due to the limitations of the press-fit pins. No such limitations are available using the proposed method. This shows that design approaches based on trial and error are no longer feasible, and full automatic geometrical optimisation, including all domains, is required.

References

[1] Dr. S. Agarwal, 'Status of the Power Module Packaging Industry 2021',2021, www.yole.fr, pp. 17.

[2] L. Hertel, J. Collado, P. Sadowski, J. Ott, and P. Baldi, 'Sherpa: Robust hyperparameter optimization for machine learning', SoftwareX, vol. 12, p. 100591, Jul. 2020, doi: 10.1016/j.softx.2020.100591.

[4] S. Wilke, 'Treibhausgasminderungsziele Deutschlands', Umweltbundesamt. Accessed: Apr. 01, 2024. [Online]. Available: https://www.umweltbundesamt.de/daten/klima/treibhausgasminderungsziele-deutschlands

[5] Y. Yang, Y. Ge, Z. J. Wang, and Y. Kang, 'An Automated Electro-Thermal-Mechanical Co-Simulation Methodology Based on PSpice-MATLAB-COMSOL for SiC Power Module Design', in 2021 IEEE Workshop on Wide Bandgap Power Devices and Applications in Asia (WiPDA Asia), Aug. 2021, pp. 499–503. doi: 10.1109/WiPDAAsia51810.2021.9656074

[6] W. Li, S. Mao, H. Liu, J. Fan, and K. Zeng, 'Current Uniformity Optimization of Multi-Chip SiC Module for High-Power Applications', presented at the 2022 IEEE International Power Electronics and Application Conference and Exposition (PEAC), Guangzhou,Guangdong, China: IEEE, Nov. 2022, pp. 545–550. doi: 10.1109/PEAC56338.2022.9959680.

[7] T. M. Evans et al., 'PowerSynth: A Power Module Layout Generation Tool', IEEE Trans. Power Electron., vol. 34, no. 6, pp. 5063–5078, Jun. 2019, doi: 10.1109/TPEL.2018.2870346.

[8] J. K. Kim, 'Thermal characterization of automotive power module with SHERPA', in 2021 IEEE 23rd Electronics Packaging Technology Conference (EPTC), Singapore, Singapore: IEEE, Dec. 2021, pp. 479–482. doi: 10.1109/EPTC53413.2021.9663866.

[9] B. Robin, V.-V. Andras, B. Byron, W. Gang, and H. W. Voon, 'Full-circuit 3D electro-thermal modeling'. Siemens DI SW, 2019.

[10] Simcenter T3STER', Siemens Digital Industries Software. Accessed: Mar. 31, 2024. [Online]. Available: https://plm.sw.siemens.com/en-US/simcenter/physical-testing/t3ster/

[11] 'ImageIR®5300 - High-speed Thermography Camera with Large Pitch'. InfraTec GmbH, 2021. [Online]. Available: https://media.infratec.eu/infratec-imageir-5300-h-en-mail.pdf?mp_enc=bXBfZGlyPTY1MTY3Jm1wX2lkPTE2NTk2MDczMzg=

[12] E. Temesi, 'Applying Power Circulation Theory to More Accurately Measure Power Loss and Efficiency'. Vincotech, 2015.

[13] A. J. Lelis, D. B. Habersat, R. Green, and N. Goldsman, 'Temperature-Dependence of SiC MOSFET Threshold-Voltage Instability', MSF, vol. 600–603, pp. 807–810, Sep. 2008, doi: 10.4028/www.scientific.net/MSF.600-603.807.

PCIM Europe 2024, 11– 13 June 2024, Nuremberg   DOI: 10.30420/566262081

Conductor-Based Modeling of Voltage Distribution along a Single-Tooth Winding of Electrical Machines

Hujun Peng ©[1], Yue Yu[1], Svetomir Stevic ©[1], Niklas Driendl ©[1], Kay Hameyer ©[1]

[1] Institute of Electrical Machines (IEM), RWTH Aachen University, Germany

Corresponding author: Hujun Peng, hujun.peng@iem.rwth-aachen.de
Speaker: Hujun Peng, hujun.peng@iem.rwth-aachen.de

Abstract

This work developed a conductor-based model to determine the distribution of the electrical potential along the conductors of a single-tooth winding of a permanent magnet synchronous machine. More importantly, this work considers the proximity effect between various conductors belonging to the single-tooth winding, besides the skin effect in each conductor. The model is parameterized by using impedance measurement results. Then, the model accuracy is validated in the frequency domain using the impedance measurement results and in the time domain applying the voltage signal measurements. Thereby, an averaged relative error of 4.47 % is observed between the simulated and measured voltage peaks along the single-tooth winding on the test bench.

1   Introduction

Single-tooth windings are increasingly gaining attention in electrical machines thanks to their capacity for achieving significantly elevated copper filling factors and compact end windings. This design preference enhances operational efficiency and amplifies the power density of electrical machines [1], [2]. Conversely, the application of wide band-gap semiconductors like SiC and GaN is on the rise in electric vehicle powertrains. Their merits, including heightened switching frequency and enhanced efficiency compared to the Si-based IGBT, are evident. However, these advancements come hand in hand with challenges, such as steep voltage slew rates due to their abbreviated switching times. The resulting swift voltage fluctuations foster oscillatory behavior, causing an uneven voltage distribution and escalating the likelihood of insulation system failures due to partial discharge. This oscillation phenomenon becomes particularly conspicuous when employing single-tooth windings [3]. Consequently, the intricate interplay between single-tooth windings and these wide band-gap semiconductors introduces formidable challenges when integrated into electric powertrains. To tackle this issue, researchers are underway to discern the voltage distribution within electrical machines. Based on this insight, the insulation system can be more judiciously devised. Numerous researchers are actively

delving into this subject. For instance, [4] optimizes the insulation layer thickness of the hairpin winding by leveraging voltage distribution based on a white box model. Research in [5] introduces white box models for stator coils of electrical machines. However, these models need to consider proximity effects in the high-frequency range.

In a departure from existing works, our study in this paper strives to pinpoint the voltage distribution along conductors, adopting a resolute grey box model instead of the conventional white box model [6], grounded in finite element analysis. Our approach meticulously accounts for the frequency-dependent proximity effect between neighboring conductors of a single-tooth winding. The so-called conductor-based grey box model amalgamates the computational efficiency of the classical grey box method with the meticulous resolution of the traditional white box method.

The paper's structure unfolds as follows: Section 2 provides a comprehensive panorama of impedance measurement, encompassing the measurement configuration and results. Section 3 introduces the conductor-based grey box modeling method. Section 4 rigorously validates the model in the frequency domain, meticulously scrutinizing impedance spectra of various topologies through a harmonious interplay of modeling and measurement. Subsequently, Section 5 unveils the vali-

665

PCIM Europe 2024, 11– 13 June 2024, Nuremberg   DOI: 10.30420/566262081

dation results of the grey box model in the time domain, meticulously examined against test bench measurements. Lastly, Section 6 encapsulates the discerning conclusions drawn from this exhaustive study.

2 Measurement configuration and results

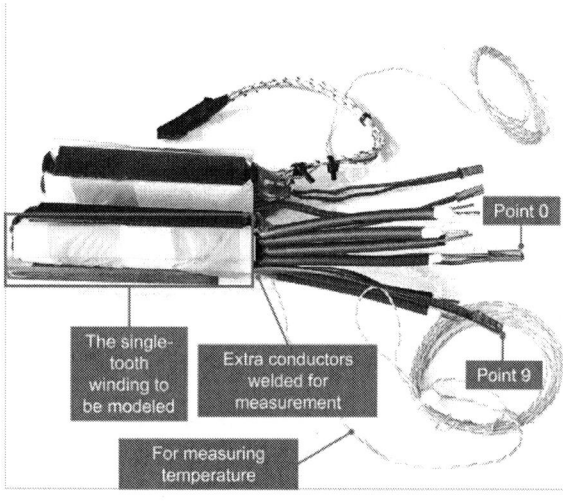

(a)

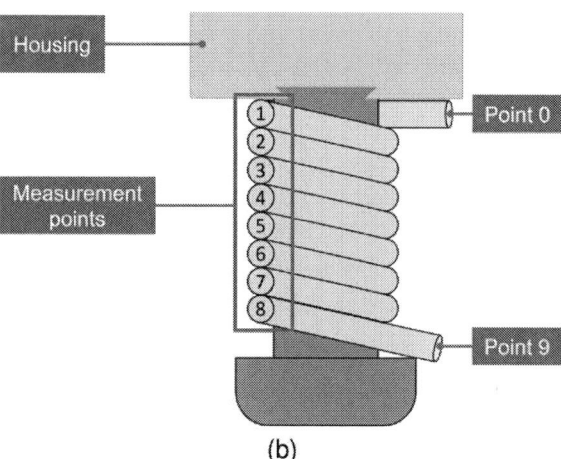

(b)

Fig. 1: The structure of the single-tooth winding: (a) The single-tooth winding, (b) The schematic presentation of the single-tooth winding.

Fig. 1a presents the single-tooth winding slated for modeling, while Fig. 1b unveils the intricate structure of the modeled tooth winding. Rigorous precision in modeling demands thorough measurements at points between the single-tooth winding terminals and the housing. The impedance spectra of the single-tooth winding are meticulously acquired

through the following setup: The impedance analyzer, the 'Keysight E4990A,' takes center stage in measuring impedance spectra, covering a frequency spectrum from 100 Hz to 10 MHz. In order to facilitate in-depth modeling analysis and subsequent parameter identification for the conductor-based model, each intermediate point between turns of the single-tooth winding is systematically assigned a numerical identifier ranging from 0 to 9. This sequential numbering aligns in ascending order from the initiation to the termination of the single-tooth winding, as portrayed in Fig. 1b. The impedance measurements encompass the evaluation of 'turn-to-turn' impedance, denoting the impedance between distinct pacing points labeled as 'i-to-j.' Additionally, a measurement point affixed to the housing enables the comprehensive assessment of 'turn-to-gnd' impedance at various positions along the single-tooth winding, capturing the impedance between each turn and the housing. Fig. 2 meticulously illustrates the outcomes of these measurements, accentuating the conspicuous similarities between various turns, a reflection of the inherent structural attributes of a concentrated single-tooth winding. In the frequency domain below 1 MHz, the 'turn-to-turn' impedance predominantly embraces inductive characteristics, intensifying with ascending frequency. As the frequency escalates, the substantial influence of parasitic capacitance comes to the forefront, with the phase angle of the 'turn-to-turn' impedance nearing -90 ° at frequencies exceeding 5 MHz. Contrastingly, the 'turn-to-gnd' impedance chiefly mirrors the traits of the insulation layer at lower frequencies. With the frequency surge, a distinctive peak-shaped protrusion emerges in the phase angle of the 'turn-to-gnd' impedance above 3 MHz, signifying inductive behavior.

The following section will introduce a grey box model utilizing the measured impedance spectra of the single-tooth winding.

3 Grey box modeling method

3.1 Overview of the model based a Π-shaped unit

The conductor-based model of the single-tooth winding is developed with the following considerations:

- Capacitive coupling between turns and the housing,

666

PCIM Europe 2024, 11– 13 June 2024, Nuremberg   DOI: 10.30420/566262081

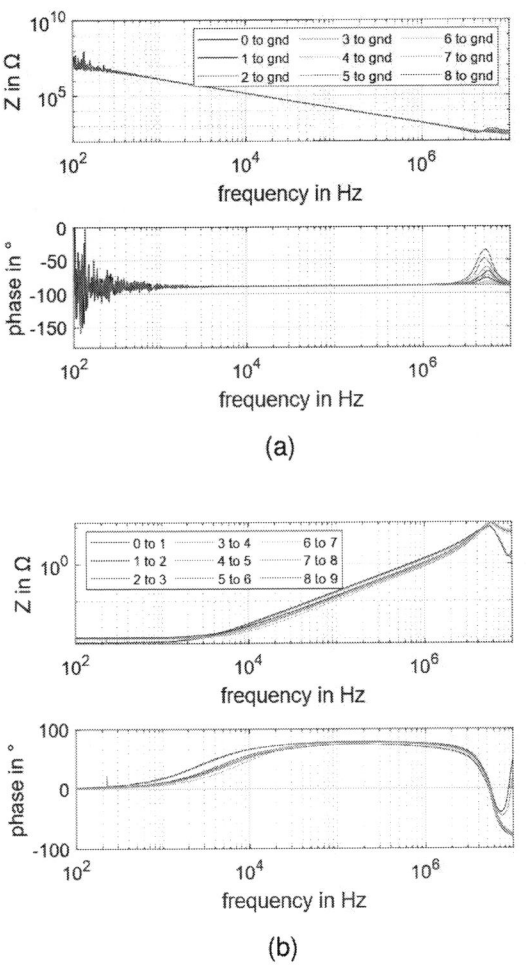

(a)

(b)

Fig. 2: Impedance spectra measured at different points along the single-tooth winding: (a) Topologies 'turn-to-gnd', (b) Topologies 'turn-to-turn'.

– Capacitive coupling between adjacent turns,

– Frequency-dependent skin effects observed in each individual turn,

– Frequency-dependent proximity effects between turns.

At its core, the model delineates the structure of a single turn, spanning between pacing points i and $i+1$, establishing the foundational framework for the single-tooth winding. Capitalizing on the impedance intricacies revealed within the single-tooth winding, as discerned from the outcomes presented in Fig. 2, the entire winding architecture unfolds systematically through the replication of this elemental unit. This pivotal Π-shaped unit, laid out in Fig. 3, serves as the cornerstone. Subsequent

sections meticulously introduce each sub-structure, intricately modeled with a corresponding equivalent electrical circuit. It is imperative to underscore that integrating inductive coupling among diverse turns is an indispensable element within the model.

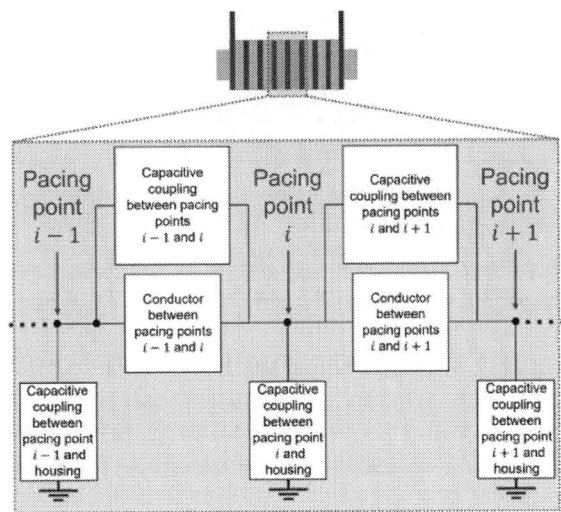

Fig. 3: Basic model structure featuring Π-shaped units bridging between pacing points.

3.2 Incorporation of capacitive coupling modeling between turns and the housing

As portrayed in Fig. 4, the model structure meticulously outlines the capacitive coupling between measurement points i and the motor housing. This configuration is purposefully crafted to encapsulate the capacitive interactions between the winding and the housing. For the emulation of the insulation layer, an initial selection involves a single resistor with significant resistance, effectively approximating the insulation behavior at low frequencies, particularly its quasi-DC characteristics. The importance of parasitic capacitance between the winding and housing becomes pronounced when scrutinizing the insulation layer's capacitive performance. In order to address this, a capacitor is introduced in parallel with the insulation resistor in the model, portraying the capacitive coupling between turns (or winding) and the housing. Additionally, an extra parallel branch featuring a resistor and a capacitor connected in series is incorporated to offer a simplified representation of lossy capacity.

667

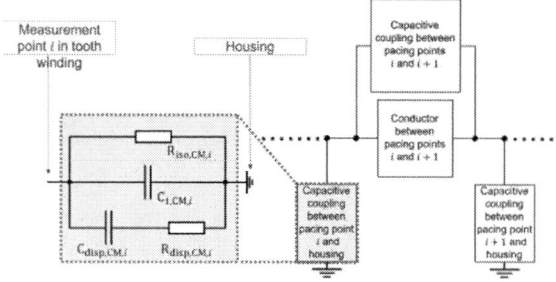

Fig. 4: Modeling capacitive coupling between each turn conductor and the winding housing.

3.3 Capturing the capacitive coupling between turns

Elevating the importance of capturing the capacitive coupling between neighboring turns in the winding, the model adopts a strategy akin to the coupling model between turns and the housing. Here, two capacitors are introduced in parallel— one with and the other without a resistor. This deliberate design seeks to encompass the influence of both parasitic capacitance and dielectric dissipation, elucidated in Fig. 5. These interconnected neighboring capacitive couplings indirectly address the capacitive coupling between conductors that are not in immediate proximity. These interconnected neighboring couplings are arranged in series within the model, effectively portraying the behavior of parasitic capacitance in scenarios involving higher frequencies.

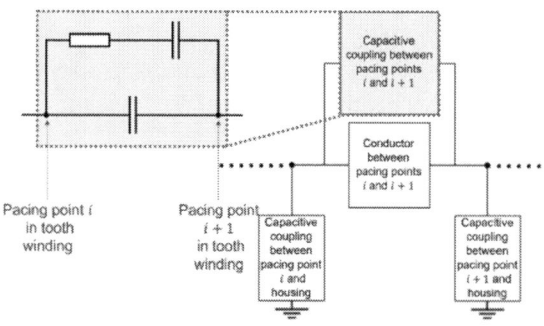

Fig. 5: Modeling capacitive coupling between two adjacent turns.

3.4 Inductive coupling within and between the conductors of the single-tooth winding

In order to tackle the frequency-dependent characteristics of self-inductance and resistance within each turn, the conductor-based model incorporates a sophisticated RL-Ladder structure, showcased as

'Part-A' in Fig. 6. However, it is essential to underscore that the skin effect primarily characterizes the behavior of an individual conductor. Equally crucial is the consideration of the proximity effect between conductors within the single-tooth winding. Given these nuances, a more intricate conductor-based model for the single-tooth winding has been meticulously formulated, as presented in Fig. 6. Each conductor now integrates both 'Part A and 'Part B.' Specifically, 'Part A features a sophisticated four-

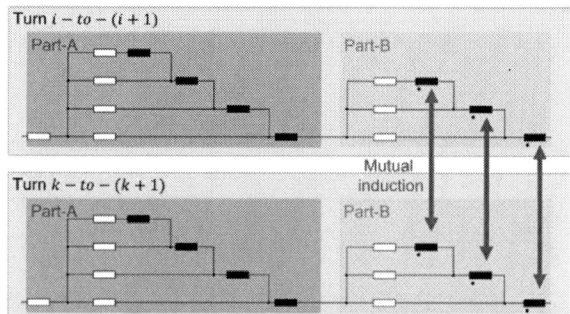

Fig. 6: Modeling differential mode impedance of each turn conductor.

layer RL-Ladder structure aimed at mimicking the skin effect within each conductor. The 'Part-B' section seamlessly integrates a three-layer RL-Ladder within the model. This particular segment is meticulously crafted to delineate the nuances of the proximity effect. Moreover, the proximity effect encapsulates inductive coupling across all layers between distinct turns, vividly illustrated by bidirectional arrows in Fig. 6. Each instance of mutual inductance is precisely articulated through the utilization of a matrix, outlined as follows:

$$\begin{bmatrix} L_{11,k} & M_{12,k} & \cdots & M_{19,k} \\ M_{21,k} & L_{22,k} & \cdots & M_{29,k} \\ \vdots & \vdots & \ddots & \vdots \\ M_{91,k} & M_{92,k} & \cdots & L_{99,k} \end{bmatrix}$$

with

$$M_{ij,k} = M_{ji,k}, \tag{1}$$

whereby the symbol k represents the index of the layer number in the RL-Ladder associated with the corresponding inductance in 'Part-B,' ranging from 1 to 3. The numerical value 9 denotes the presence of nine conductors within the single-tooth winding. The inductive coupling factors for each mutual inductance must adhere to the range of 0 to 1.

4 Validation of the model in the frequency domain

The parameters are determined through optimization algorithms that aim to minimize the disparity between the modeled impedance spectra and the actual measurements. In this context, a genetic algorithm is utilized. Different measurement topologies are considered to establish the cost function the genetic algorithm requires.

After optimizing the parameters, the validation of the proposed high-frequency model commences in the frequency domain. The model generates impedance spectra for different terminal configurations, including 'turn-to-turn' and 'turn-to-gnd.' For instance, Fig. 7a displays the fitting performance of the impedance spectra for the '8-to-9' topology, while Fig. 7b illustrates the performance for the '0-to-gnd' topology. Fig. 8 consolidates the statistical outcomes of fitting impedance spectra, incorporating the mean average percentage error (MAPE), defined as:

$$MAPE = \frac{100\%}{N} \sum_{i=1}^{N} |\frac{|Z_i| - |\hat{Z}_i|}{|Z_i|}|, \qquad (2)$$

whereby the Z_i and $\hat{Z}_i$ represent the measured and modeled impedance at the i-th sampled frequency point, respectively. The parameter N signifies the number of sampling points within the frequency domain, spanning from 10 kHz to 10 MHz. Within this frequency spectrum, the model attains a validation average impedance error of 6.09 % for 'turn-to-turn' topologies and 11.71 % for 'turn-to-gnd' topologies. The overall average error across all topologies aggregates to 9.37 %.

5 Validation of the model on the test bench

5.1 Measurement setup for the time-domain validation

A high-voltage impulse generator equipped with SiC power modules and varying frequencies is employed during the measurement. This generator produces a bipolar voltage signal featuring a 50 % duty ratio. The produced impulse signal is utilized as the input voltage applied across the two terminals of the winding. Precisely for this validation instance, measurement point 0 and the winding housing (GND) are deliberately selected as the input points. The resulting voltage signals between

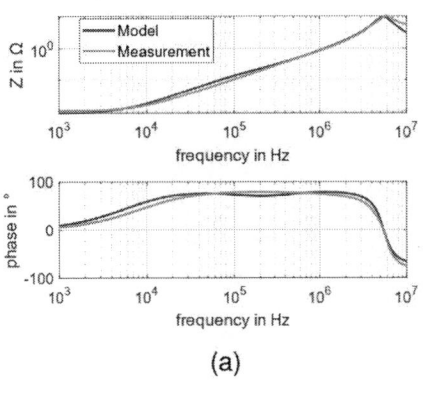

(a)

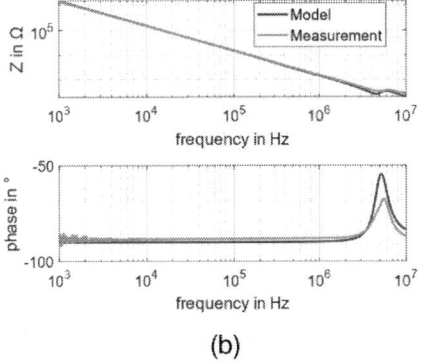

(b)

Fig. 7: Verification of the impedance modeling in the frequency domain through illustrative examples: (a) Topology '8-to-9', (b) Topology '0-to-gnd'.

other measurement points are then meticulously recorded as output signals, as visually depicted in Fig. 9b. After this, the measured signals undergo a thorough validation process by being compared with their corresponding simulated voltage signals. The DC-link voltage is 100 V, the switching frequency is 10 kHz, and the voltage slew rate is about 6 V/ns.

5.2 Validation results in the time-domain

In Fig. 10, the voltage dynamics between the turn and housing are showcased for various outputs, presenting measured and modeled signals. A distinct phase error comes to light in Fig. 10b, where the simulated voltage signal displays an unusual decrease followed by an increase, contrary to the straightforward upward transition observed in the corresponding measured signal. This phase error magnifies along the single-tooth winding, evident when comparing Fig. 10c to Fig. 10b. As a consequence of this phase error, the absolute maximum of the simulated voltage signal is relatively dimin-

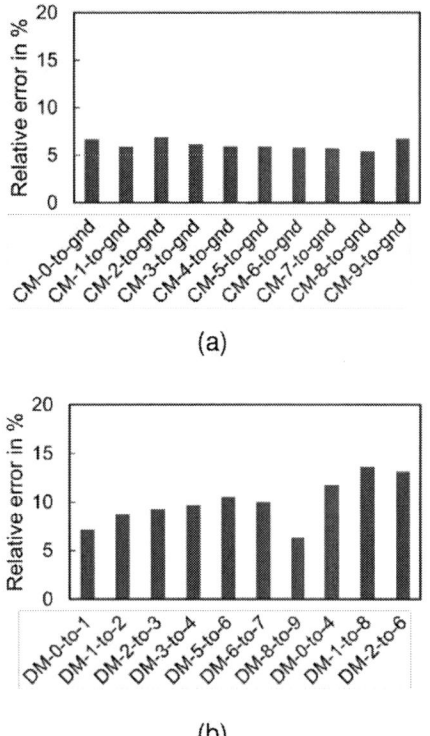

(a)

(b)

Fig. 8: Discrepancies in the impedance modeling within the frequency domain for various measurement topologies: (a) Topologies 'turn-to-gnd', (b) Topologies 'turn-to-turn'.

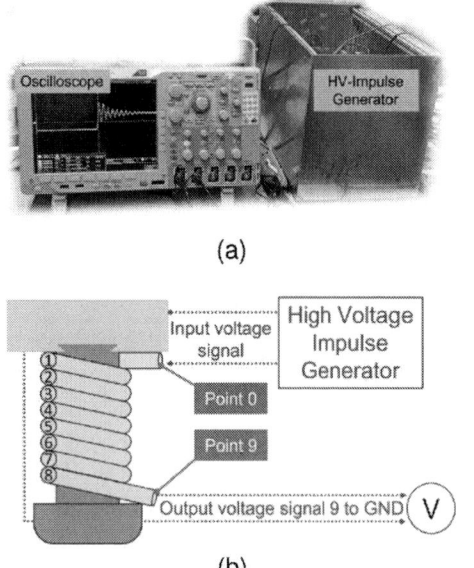

(a)

(b)

Fig. 9: Measurement setup for the time-domain validation:(a) High voltage impulse generator and oscilloscope, (b) Measurement terminals.

ished compared to the corresponding measured maximum in Fig. 10c. In order to rectify this issue, the evaluation will now rely on the step amplitude in the simulated voltage signal rather than the absolute maximum, as illustrated in Fig. 11. This adjustment effectively compensates for the phase error. Fig. 11 visually illustrates the computation of the voltage relative error $\Delta_{relative}$ with the compensation considered. Additionally, the oscillation

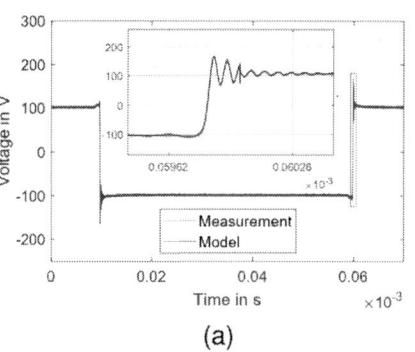

(a)

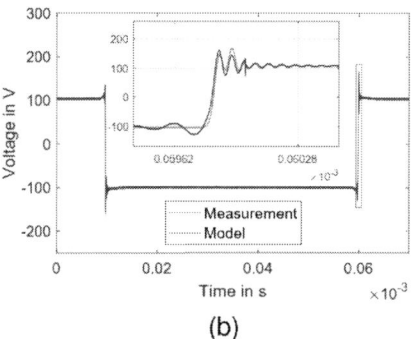

(b)

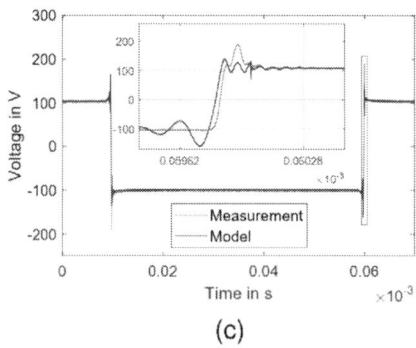

(c)

Fig. 10: Model validation using the turn-to-housing voltage signals in the time domain with a slew rate of $6\,V\,ns^{-1}$, PWM-frequency of $10\,kHz$ and DC-link voltage of $100\,V$: (a) 1-to-gnd, (b) 4-to-gnd, (c) 9-to-gnd.

frequency values during the switching process are

670

PCIM Europe 2024, 11– 13 June 2024, Nuremberg DOI: 10.30420/566262081

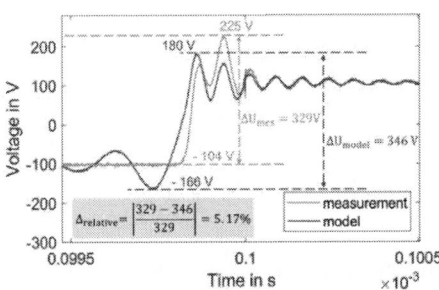

Fig. 11: Comparison between the measured and modeled voltage signals for the topology '4-to-gnd' and the simulated voltage step value with compensation of the phase error considered.

specifically chosen to facilitate a comparison between the modeled and measured voltage signals. Determining the oscillation frequency involves analyzing the signal segment that follows the maximum point of the voltage signal.

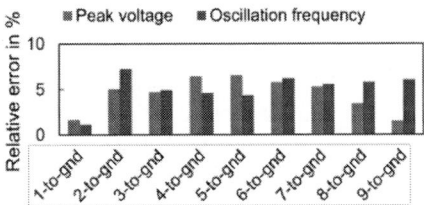

Fig. 12: Relative deviations during validating the turn-to-housing voltage signals using the impulse generator with a slew rate 6 V ns^{-1}, DC-link voltage 100 V and PWM-frequency 10 kHz.

Fig. 12 presents the comprehensive statistical results from the validation process, encompassing all turn-to-housing voltage signals at a switching frequency of 10 kHz and a slew rate of 6 V ns^{-1}. The average relative error for the peak voltage stands at 4.47 %, while the relative error for the oscillation frequency is 5.09 %.

6 Conclusions

This paper conducts high-frequency modeling of a single-tooth winding using the grey box modeling method, considering frequency-dependent parasitic effects. These effects encompass capacitive and inductive coupling among different turn conductors and capacitive coupling between the conductors and the winding housing. The validation of this high-frequency model for the single-tooth winding is first carried out in the frequency domain, achieving an impressive average relative error of less than 10 % across the frequency spectrum from 10 kHz to 10 MHz. In addition, a validation process is executed in the time domain based on test bench measurements. The observed average relative error in predicting the maximum voltage is 4.47 %. Moreover, the model exhibits notable accuracy in estimating the oscillation frequency of the voltage during the switching process, with a relative error of less than 5.09 %. Consequently, this model demonstrates high fidelity, making it a valuable tool for accurately predicting the maximum voltage stress along the single-tooth winding. Such precision holds substantial potential for mitigating the need for overdesigning insulation systems in the next generation of electrical machines.

References

[1] A. M. El-Refaie, "Fractional-slot concentrated-windings synchronous permanent magnet machines: Opportunities and challenges," *IEEE Transactions on industrial Electronics*, vol. 57, no. 1, pp. 107–121, 2009.

[2] I. Petrov, M. Polikarpova, P. Ponomarev, P. Lindh, and J. Pyrhönen, "Investigation of additional ac losses in tooth-coil winding pmsm with high electrical frequency," in *2016 XXII International Conference on Electrical Machines (ICEM)*, IEEE, 2016, pp. 1841–1846.

[3] F. Chiang, "Effects of high frequency voltage stress on air insulation and solid insulation," in *2010 IEEE Symposium on Product Compliance Engineering Proceedings*, IEEE, 2010, pp. 1–10.

[4] J. Dittmann, M. England, and B. Ponick, "Design of hairpin windings considering the transient potential distribution," *e & i Elektrotechnik und Informationstechnik*, vol. 140, no. 2, pp. 271–280, 2023.

[5] A. Boglietti and E. Carpaneto, "Induction motor high frequency model," in *Conference Record of the 1999 IEEE Industry Applications Conference. Thirty-Fourth IAS Annual Meeting (Cat. No. 99CH36370)*, IEEE, vol. 3, 1999, pp. 1551–1558.

[6] D. Zheng, G. Lu, and P. Zhang, "An improved online stator insulation monitoring method based on common-mode impedance spectrum considering the effect of aging position," *IEEE Transactions on Industry Applications*, vol. 58, no. 3, pp. 3558–3566, 2022.

PCIM Europe 2024, 11– 13 June 2024, Nuremberg   DOI: 10.30420/566262082

Reduction of PWM Harmonics with Carrier Phase Shifting in a Dual-Stator PMSM with Magnetic Coupled Windings

Bünyamin Tekir[1], Robert Zipprich[1], Jan Winter[1], Marcus Ziegler[1]

[1] University of Kassel, Germany

Corresponding author: Bünyamin Tekir, b.tekir@uni-kassel.de
Speaker: Bünyamin Tekir, b.tekir@uni-kassel.de

Abstract

The compensation of PWM harmonics by a carrier phase shift-SVPWM (CPS-SVPWM) in the dual-stator PMSM is analyzed. For this purpose, the carriers of a SVPWM were phase-shifted by 180°. To verify the analytical methods and to create approximately real conditions, a co-simulation with MATLAB-Simulink and Ansys-Maxwell was used. It was shown that the torque ripple of the DS-PMSM can be reduced, while keeping the same signal quality and lower switching frequencies of the converters. Again, it was also shown that the magnetic losses can be reduced, while the switching frequencies of the inverters remain the same.

1   Introduction

The aviation industry has set new standards with the electrification of aircraft powertrains [1]. The concepts of the more-electric-aircraft (MEA) and the all-electric-aircraft (AEA) are being used to research new approaches to electric drive systems. This results in higher safety requirements for the electric drive train. In addition to the power electronic components, the revision of the electric drive machine is an important research approach.

Important boundary conditions here are an optimum power-to-weight ratio and a lower fault tolerance, which means that the drive unit must also have redundancy for operation in aeroplanes. Permanent magnet dual-stator synchronous machines (DS-PMSM) have established themselves in aviation applications both in terms of redundancy and power-to-weight ratio [2][3].

The two-layer tooth coil winding allows a higher power factor than any other winding, as the concentrated winding results in a much smaller winding head and thus better use of the inductances [4].

Redundancy is provided by the operation of two separate winding systems, which are driven by independent two-level voltage converters, as shown in Fig. 1. This topology of two separate voltage converters and the independent control and supply of the winding systems guarantees 50% of the power in the event of a drive path failure. Due to the use of the SVPWM to control the two-level voltage converter, this leads to a current ripple in the DS-PMSM, which has an effect on the torque ripple [5][6]. This factor is amplified by the magnetic coupling of both winding systems [7]. However, the torque ripple of the DS-PMSM with coupled winding systems can be reduced by carrying out a carrier signal phase shift of the SVPWM.

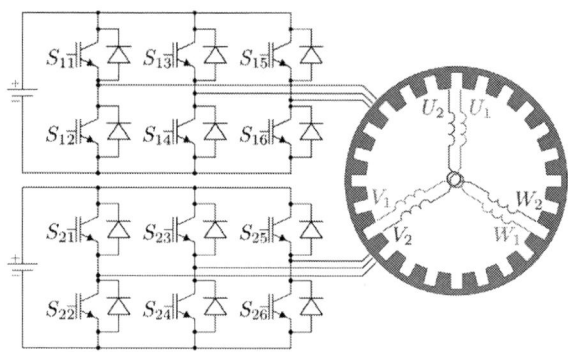

Fig. 1: Topology of the dual three phase synchronous machine with separate star points.

2   Dual-Stator PMSM with coupled windings

2.1   Two winding system

A dual-stator PMSM consists of two separate winding systems with two separate star points. This means that both winding systems are electrically isolated from each other. Figure 2 shows a DS-

PCIM Europe 2024, 11– 13 June 2024, Nuremberg · DOI: 10.30420/566262082

PMSM with 24 slots and 10 pole pairs (20 embedded magnets) and a two-layer tooth coil winding. For this purpose, the winding systems were wound across the stator according to the symmetry conditions [8]. The two winding systems are marked in colour.

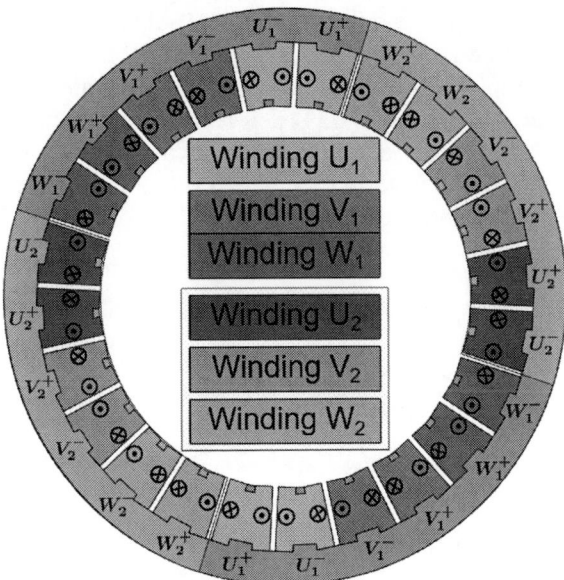

Fig. 2: Dual-stator PMSM with initial winding and weak magnetic coupling named as C_{min}.

It can be seen that each winding system has 12 slots. If the windings of both winding systems U1, V1, W1 and U2, V2, W2 are halved so that each phase system does not have 40 windings per slot, but $40/2 = 20$ windings, both phase systems can be distributed over the entire stator without violating the symmetry conditions [9]. For an even distribution, both phase systems are arranged alternately for each half of the initial winding [9]. This results in a winding system for a dual-stator synchronous machine as shown in Fig. 3.

As both winding systems share a slot, this results in a higher magnetic coupling. The magnetic coupling is determined according to Eq. (1).

$$C = \frac{M_{x_1,x_2}}{\sqrt{L_{x_1} \cdot L_{x_2}}} \quad (1)$$

Figure 4 illustrates the coupling factor between the two winding arrangements. Based on this, the winding system with the lower coupling factor is called C_{min} and the winding system with the high coupling factor is called $C_{max,IO}$. The IO in the index stands for In-Out of the alternating arrangement.

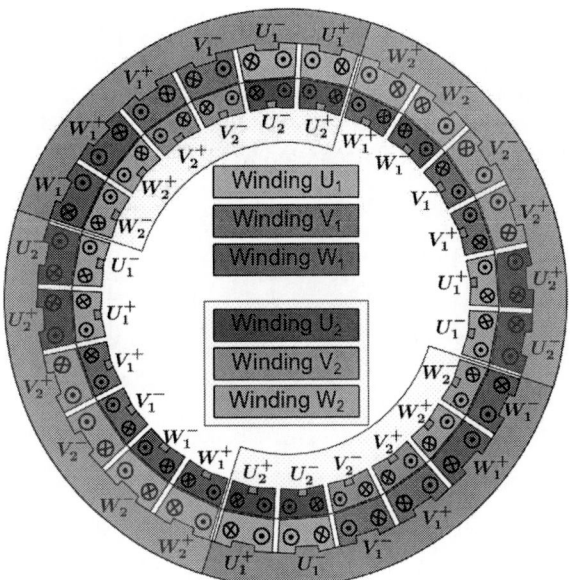

Fig. 3: Dual-stator PMSM with modified winding and strong magnetic coupling named as $C_{max,IO}$.

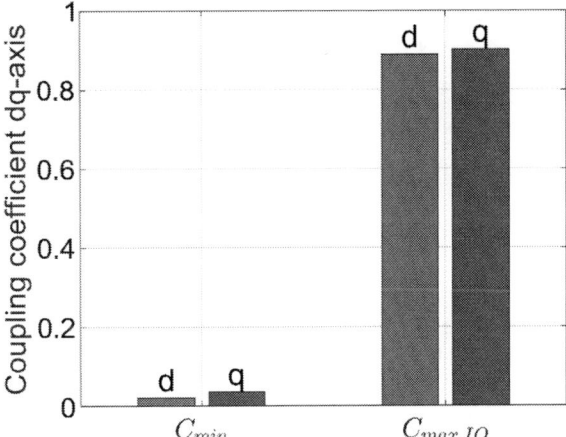

Fig. 4: Coupling factor in dq-coordinates between the two winding systems for both winding concepts.

2.2 Analytical torque generation of DS-PMSM

To take a closer look at the DS-PMSM, the torque generation of the electric machine should first be considered separately for the winding systems. The torque of a PMSM with embedded magnets can be described according to Eq. (2).

$$T_e = \frac{3}{2} \cdot p \cdot [\Psi_{PM} \cdot i_q + (L_d - L_q) \cdot i_d \cdot i_q] \quad (2)$$

In Eq. (2), the parameters p for the number of pole pairs of the machine, Ψ_{PM} for the flux linkage of

673

the magnets in the rotor and the dq currents $i_{d,q}$ can be defined. To describe the torque of a DS-PMSM, the dq equations of the winding systems are considered. The flux linkage results from the dq transformation according to Eq. (3).

$$
\begin{bmatrix} \psi_{d_1} \\ \psi_{q_1} \\ \psi_{d_2} \\ \psi_{q_2} \end{bmatrix} = \begin{bmatrix} L_{d_1} & M_{d_1 q_1} & M_{d_1 d_2} & M_{d_1 q_2} \\ M_{q_1 d_1} & L_{q_1} & M_{q_1 d_2} & M_{q_1 q_2} \\ M_{d_2 d_1} & M_{d_2 q_1} & L_{d_2} & M_{d_2 q_2} \\ M_{q_2 d_1} & M_{q_2 q_1} & M_{q_2 d_2} & L_{q_2} \end{bmatrix} \begin{bmatrix} i_{d_1} \\ i_{q_1} \\ i_{d_2} \\ i_{q_2} \end{bmatrix}
$$
$$
+ \begin{bmatrix} \Psi_{PM} \\ 0 \\ \Psi_{PM} \\ 0 \end{bmatrix}
$$
(3)

Despite the electrically isolated winding systems with separate neutral points, there is still a magnetic coupling, which is to be considered as mutual inductance M_{x_1,x_2} in Eq. (3). For the mutual inductance, $M_{x_1,x_2} = M_{x_2,x_1}$ and the inductive coupling in the same phase system between the d- and q-axis as well as the cross-coupling inductances are negligible, which is why they can be set to zero [10]. This allows Eq. (3) to be simplified to Eq. (4)-Eq. (7).

$$\psi_{d_1} = L_{d_1} \cdot i_{d_1} + M_{d_1,d_2} \cdot i_{d_2} + \Psi_{PM} \qquad (4)$$

$$\psi_{q_1} = L_{q_1} \cdot i_{q_1} + M_{q_1,q_2} \cdot i_{q_2} \qquad (5)$$

$$\psi_{d_2} = L_{d_2} \cdot i_{d_2} + M_{d_1,d_2} \cdot i_{d_1} + \Psi_{PM} \qquad (6)$$

$$\psi_{q_2} = L_{q_2} \cdot i_{q_2} + M_{q_1,q_2} \cdot i_{q_1} \qquad (7)$$

The dq equations of the DS-PMSM determined above can now be used to calculate the torque in Eq. (8). The matrix $\mathbf{K}$ results from the dq transformation of Eq. (8).

$$T_e = \frac{3}{2} \cdot p \cdot \vec{i}_{dq}^T \cdot \mathbf{K} \cdot \vec{\psi}_{dq} \qquad (8)$$

The result is the analytical torque equation for a permanently magnetised dual-stator synchronous machine in Eq. (9).

$$
\begin{aligned}
T_{e,DS} = &\frac{3}{2} p (\Psi_{PM} [i_{q_1} + i_{q_2}] + [L_{d_1} - L_{q_1}] i_{d_1} i_{q_1} + \\
&[L_{d_2} - L_{q_2}] \cdot i_{d_2} i_{q_2} + \\
&[M_{d_1,d_2} - M_{q_1,q_2}] \cdot [i_{d_2} i_{q_1} + i_{d_1} i_{q_2}])
\end{aligned}
$$
(9)

Here it becomes clear that both winding systems influence each other when generating torque due to the magnetic coupling.

3 Analysis of PWM-harmonics of a DS-PMSM

The space vector pulse width modulation (SVPWM) method generates an unavoidable current ripple in the DS-PMSM due to its discrete switching operations. For a more precise analytical view, the current that is fed into the DS-PMSM can be considered as a Fourier series, which can be used to determine the torque. For the exact derivation, a reference is made to [11]. It is important to note that the carrier signal is a triangular function. In Eq. (10), the voltage for a phase is represented as a Fourier series.

$$
u_x(t) = \frac{4U_{DC}}{\pi} \sum_{\substack{m=0,\, if\ n=1 \\ or \\ m>0,\, if\ n=\infty}}^{\infty} \sum_{\substack{n=1 \\ or \\ n=\infty}}^{\infty} C_{mn} \cdot \ldots
$$
$$
\cos\left(m\left[\omega_c t + \varphi_c\right] + n\left[\omega_0 t + \varphi_0\right]\right)
$$
(10)

$$\Rightarrow C_{mn} = \frac{1}{q} J_n \left(q\frac{\pi}{2}M\right) \cdot \sin\left([q+n]\frac{\pi}{2}\right)$$

Equation (10) is made up of the variables M for the modulation factor, J_n for the Bessel-function, ω_c for the carrier signal angular frequency and ω_0 for the fundamental frequency of the phase voltage. φ_c represents the initial angle of the carrier signal, where φ_0 represents the initial angle of the phase voltage. The variables m and n are the harmonic index variables.

Now the phase current $i_x(t)$ can be calculated on the basis of the voltage $u_x(t)$ according to Eq. (11) [12].

$$i_x(t) = \frac{u_x(t)}{\sqrt{R_S^2 + L_x^2}} \qquad (11)$$

The phase currents of both winding systems can be converted into equal quantities using the dq transformation, as can be seen in Eq. (12) and Eq. (13).

d_x – system:

$$i_{d_x} = i_{d_x,DC} + i_{d_x,ripple}(m,n) \quad (12)$$

$$i_{d_x,ripple}(m,n) = C_{mn}^d \cdot \ldots$$
$$\cos\left(\left[n\omega_0 + m\omega_c\right]t + n\varphi_0 + m\varphi_c\right)$$

$$C_{mn}^d = C_{mn} \cdot \frac{2}{3} \begin{bmatrix} \cos(\theta) \\ \cos(\theta - 120°) \\ \cos(\theta + 120°) \end{bmatrix}^T \begin{bmatrix} i_{U_{1,2}}(t) \\ i_{V_{1,2}}(t) \\ i_{W_{1,2}}(t) \end{bmatrix}$$

q_x – system:

$$i_{q_x} = i_{q_x,DC} + i_{q_x,ripple}(m,n) \quad (13)$$

$$i_{q_x,ripple}(m,n) = C_{mn}^q \cdot \ldots$$
$$\cos\left(\left[n\omega_0 + m\omega_c\right]t + n\varphi_0 + m\varphi_c\right)$$

$$C_{mn}^q = C_{mn} \cdot \frac{2}{3} \begin{bmatrix} -\sin(\theta) \\ -\sin(\theta - 120°) \\ -\sin(\theta + 120°) \end{bmatrix}^T \begin{bmatrix} i_{U_{1,2}}(t) \\ i_{V_{1,2}}(t) \\ i_{W_{1,2}}(t) \end{bmatrix}$$

This makes it possible to represent the torque of the DS-PMSM as an equation (see Eq. (12) and Eq. (13)) from the DC values of the currents (see Eq. (14)). Here it can be seen that the torque is separated into a DC component DC and a wavy component $ripple$. The DC component corresponds to the torque, which has already been derived in Eq. (9).

$$T_{e,DS} = T_{DC_{12}} + T_{ripple_{12}}(m,n) \quad (14)$$

The ripple component of the torque is now made up of the harmonics generated by the SVPWM. As the DS-PMSM contains two phase systems and is fed via two independent inverters (see Fig. 7), the total torque is divided into two components. System "1" for winding system 1 and system "2" for winding system 2. If Eq. (12) and Eq. (13) are used to determine the total torque ripple, this results in Eq. (15) with the assumption that $i_{U_1,V_1,W_1} = i_{U_2,V_2,W_2}$ applies to the impressed current.

Equation (15) shows that a difference is formed between the phase angle of the carrier signal φ_c and the angle δ_c. The variable δ_c is the required carrier signal phase shift. This process is shown in Fig. 5.

$$T_{ripple_{12}}(m,n) = \frac{3}{2}p(\ldots$$
$$\Psi_{PM}\left[C_{mn}^q \left\{ \begin{array}{l} \cos\left(\omega_{nm}t + n\varphi_0 + m\varphi_c\right) + \\ \cos\left(\omega_{nm}t + n\varphi_0 + m(\varphi_c - \delta_c)\right) \end{array} \right\} \right]$$
$$+ \left[L_{d_1} - L_{q_1}\right] C_{mn}^d C_{mn}^q \left\{ \begin{array}{l} \cos\left(\omega_{nm}t + n\varphi_0 + m\varphi_c\right) \cdot \ldots \\ \cos\left(\omega_{nm}t + n\varphi_0 + m\varphi_c\right) \end{array} \right\}$$
$$+ \left[L_{d_2} - L_{q_2}\right] C_{mn}^d C_{mn}^q \left\{ \begin{array}{l} \cos\left(\omega_{nm}t + n\varphi_0 + m(\varphi_c - \delta_c)\right) \cdot \ldots \\ \cos\left(\omega_{nm}t + n\varphi_0 + m(\varphi_c - \delta_c)\right) \end{array} \right\}$$
$$+ \left[M_{d_1,d_2} - M_{q_1,q_2}\right] C_{mn}^d C_{mn}^q \cdot \ldots$$
$$\left(\left\{ \begin{array}{l} \cos\left(\omega_{nm}t + n\varphi_0 + m(\varphi_c - \delta_c)\right) \cdot \ldots \\ \cos\left(\omega_{nm}t + n\varphi_0 + m\varphi_c\right) \end{array} \right\} \right.$$
$$\left. + \left\{ \begin{array}{l} \cos\left(\omega_{nm}t + n\varphi_0 + m\varphi_c\right) \cdot \ldots \\ \cos\left(\omega_{nm}t + n\varphi_0 + m(\varphi_c - \delta_c)\right) \end{array} \right\} \right).$$
$$\Rightarrow \omega_{nm} = n\omega_0 + m\omega_c \quad (15)$$

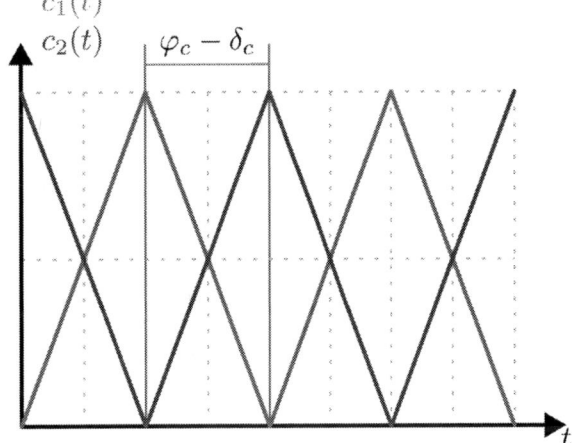

Fig. 5: Illustration of the carrier signal phase shifting process.

4 System description and simulation

In this section, the torque ripple of the new alternating winding arrangement $C_{max,IO}$ is analysed with a carrier signal phase shift. In addition to the torque, the magnetic losses are also considered, as these are an key factor in the efficiency of a permanently excited synchronous machine. The dual-stator PMSM is analysed using a co-simulation environment with MatLab/Simulink, Ansys Simplorer and Ansys Maxwell. The advantage of this simulation environment is a comprehensive analytical and numerical calculation, which means that more detailed simulation results can be expected. Based on co-simulation (Co-Sim), the non-linear behaviour of the electrical machine can be taken into account in dynamic and stationary states. The simulation results are presented following the system description.

4.1 Simulation environment

A coupled transient simulation includes several software interfaces [13]. In this case, there are three interfaces. Figure 6 shows the structure of the co-sim used as a block diagram.

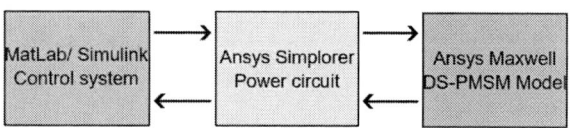

Fig. 6: Structural design of the co-simulation.

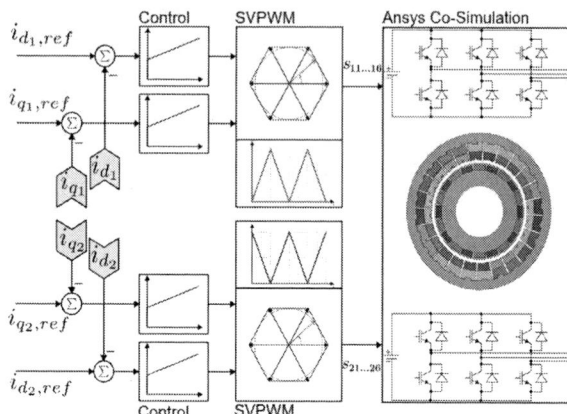

Fig. 7: Control system for a DS-PMSM with the simulation environment.

The control system for controlling the dual-stator PMSM is set up in Simulink. This is a parallel current control system where the dq currents are each controlled with two PI controllers. Both control branches are independent of each other. The control parameters for the PI controllers were determined according to the absolute value optimum. The control structure can also be extended to include speed control, which was not considered here. In addition, the dq axes are decoupled from each other and are then modulated via space vector modulation (SVPWM), whereby six active and two passive switching states can be switched according to the switching matrix of the inverter. With SVPWM, the carrier signals now differ. A carrier signal phase shift of $\varphi_c - \delta_c = \pi$ is defined, as can be seen in Fig. 7. The defined switching states from the SVPWM are then passed on via the interface between Simulink and Ansys Simplorer. There are two parallel two-level voltage converters here (see Fig. 7).

IGBT semiconductors with freewheeling diodes are used as power switches. The switched voltages are then passed between the Ansys Simplorer interface to Ansys Maxwell, where a finite element analysis of the DS-PMSM takes place. The simulation results are then forwarded to Simulink via Ansys Simplorer. The time step t_s of the co-simulation is defined as a fixed time step. It is important to mention again that the model of the electrical machine is not a reduced-order model.

4.2 Steady-state operating case

In the case of steady-state operation, a fully stabilised state applies. Switching frequencies of $f_{sw} = 5\,\mathrm{kHz}$, $f_{sw} = 10\,\mathrm{kHz}$ and $f_{sw} = 20\,\mathrm{kHz}$ are compared with each other. For comparison, the torques between the winding arrangement C_{min} and the winding arrangement $C_{max,IO}$ are shown in Fig. 8.

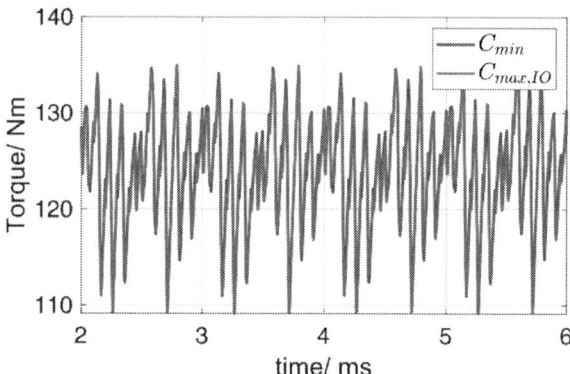

Fig. 8: Generated torque comparison of the winding concepts C_{min} and $C_{max,IO}$.

It can be seen that both winding concepts have the same torque curve and the same torque ripple. It can therefore be stated that the winding arrangement $C_{max,IO}$ has no different behaviour to the default winding arrangement C_{min}. On this basis, the carrier signal phase shift SVPWM (CPS-SVPWM) is analysed below for the winding arrangement $C_{max,IO}$. Figure 9 shows all the resulting torques with the switching frequencies f_{sw} to be analysed.

PCIM Europe 2024, 11– 13 June 2024, Nuremberg   DOI: 10.30420/566262082

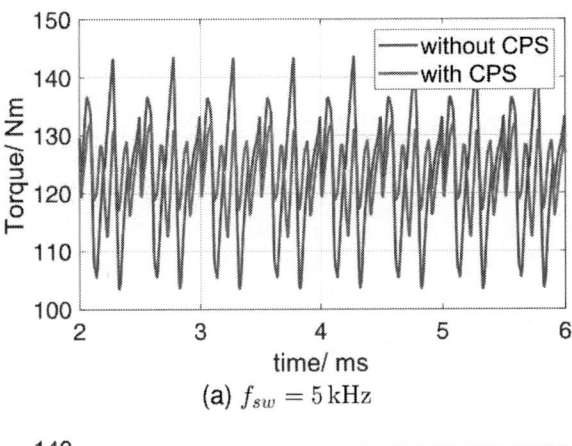

(a) $f_{sw} = 5\,\text{kHz}$

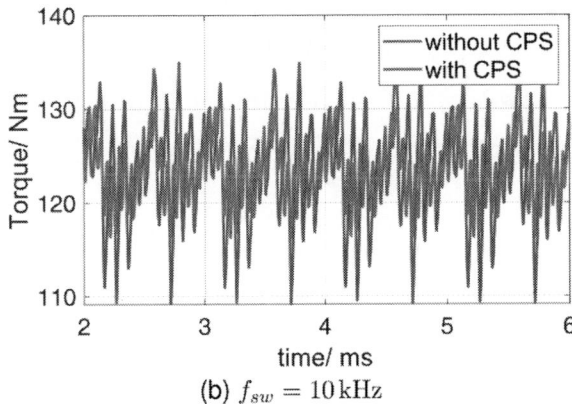

(b) $f_{sw} = 10\,\text{kHz}$

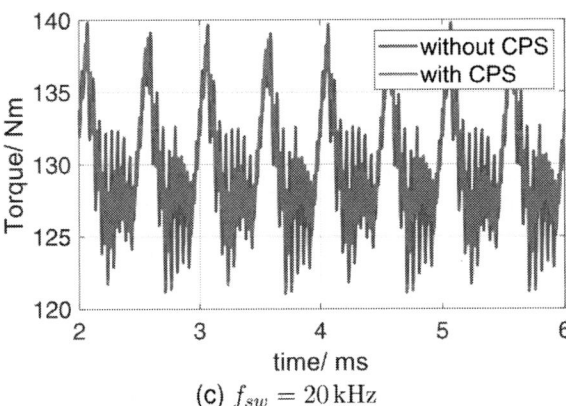

(c) $f_{sw} = 20\,\text{kHz}$

Fig. 9: Comparison of torques at different switching frequency with and without a CPS-SVPWM

f_{SW}	Torque ripple-factor	
	without CPS	with CPS
5 kHz	32.27%	15.84%
10 kHz	20.94%	11.47%
20 kHz	14.5%	11.63%

Tab. 1: Comparison of the torque ripple-factor at different switching frequencies.

If the results from Tab. 1 are analysed, it can be seen that the ripple factor can be almost halved at a switching frequency of $f_{sw} = 5\,\text{kHz}$ and $f_{sw} = 10\,\text{kHz}$. At the switching frequency $f_{sw} = 20\,\text{kHz}$ it can be seen that the ripple factor only has a difference of 3%, which leads to the statement that the advantages of a carrier signal phase shift are reduced with a higher frequency. By halving the torque ripple, the switching frequencies can also be reduced with the same quality of torque ripple.

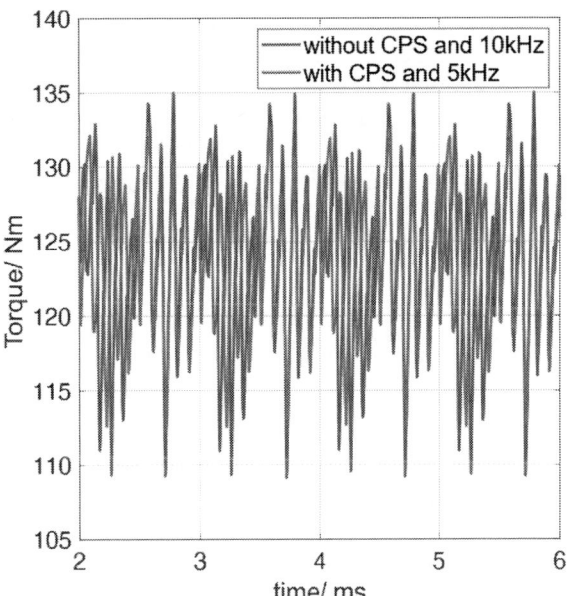

Fig. 10: Equal torque ripple at half the switching frequency.

Figure 9 show the stationary torques without and with carrier signal phase shift. It can be seen for all three torques that the torques with the CPS-SVPWM have a lower ripple than the torques without the CPS-SVPWM. For a quantified statement, the ripple factor of the torques was calculated with the respective switching frequencies (see Tab. 1).

Figure 10 shows that the torque curve of the DS-PMSM with a CPS-SVPWM and a switching frequency of $f_{sw} = 5\,\text{kHz}$ approximates the torque curve of the DS-PMSM without a CPS-SVPWM and a switching frequency of $f_{sw} = 10\,\text{kHz}$ very well. By reducing the switching frequency with the same quality of the torque ripple, switching losses can be reduced in both inverters.

677

PCIM Europe 2024, 11– 13 June 2024, Nuremberg   DOI: 10.30420/566262082

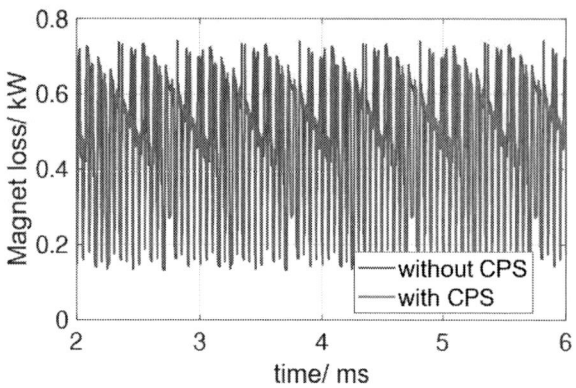

Fig. 11: Visualisation of magnetic losses.

A lower torque ripple indicates a lower power loss, which plays an important role in terms of efficiency and efficiency optimisation. Figure 11 shows the magnetic losses of the dual-stator PMSM. This clearly shows that an improvement in magnetic losses can be achieved with the CPS. The quantified values have been determined again in Tab. 2 for the three switching frequencies. A reduction in magnetic losses can be seen at all switching frequencies. Other losses, such as copper or core losses, are not considered.

	Magnetic losses DS-PMSM	
f_{SW}	without CPS	with CPS
5 kHz	743.74 W	675.85 W
10 kHz	730.18 W	670.88 W
20 kHz	726.75 W	682.08 W

Tab. 2: Comparison of the magnetic losses at different switching frequencies.

4.3 Dynamic operating case

For the dynamic investigation of the effect of the carrier signal phase shift, the step response of the q current is considered. The d current is constantly regulated to $0\,A$. Figure 12a shows the setpoint step without carrier signal phase shift. The currents in both winding systems are controlled in exactly the same way and are exactly one above the other. Figure 12b, on the other hand, shows the current steps with the carrier signal phase shift. A difference between the q currents of both winding systems can be seen here. The difference arises from the carrier signal phase shift (CPS). Due to the dynamic current control, certain deviations occur because the PI controllers work dynamically against the deviation.

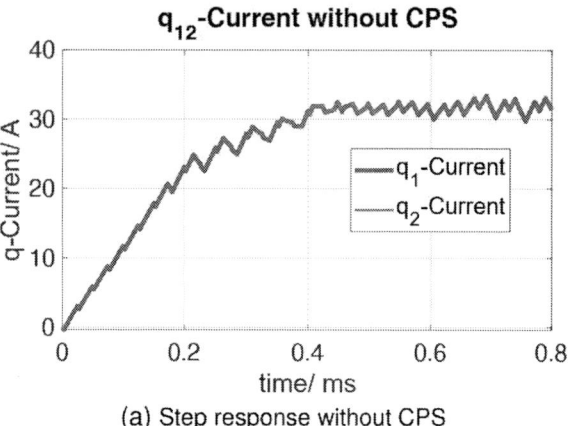

(a) Step response without CPS

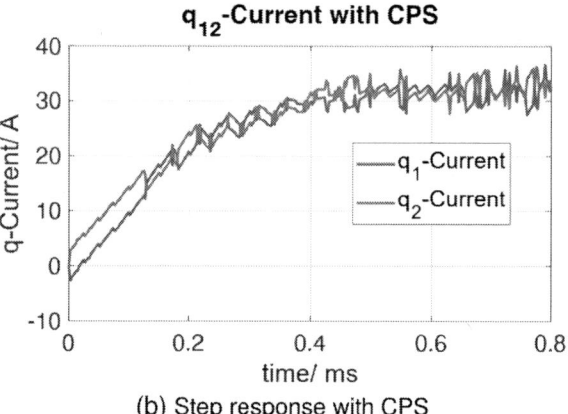

(b) Step response with CPS

Fig. 12: Current step response for current control with and without CPS-SVPWM .

Figure 13 shows the step response of the torques. Despite the different current curves (see Fig. 12), the dynamic torque curve is the same. It should be noted that there is a certain dynamic influence on the dynamic behaviour due to the current control, which was parameterised in the same way for both dual-stator machines.

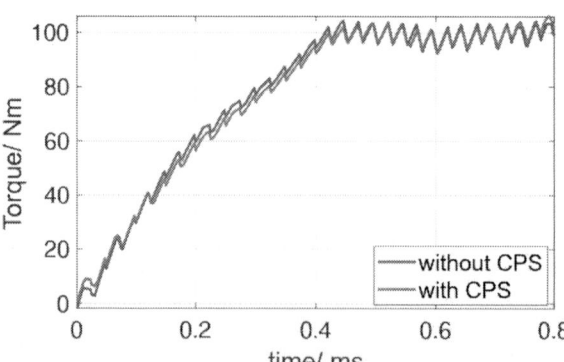

Fig. 13: Torque step response.

5 Discussion

The investigations show that a symmetrically adapted winding arrangement with alternating distribution $C_{max,IO}$ of the dual-stator PMSM has no disadvantages due to the mutual influence of the magnetic coupling. Nevertheless, different fluxes are to be expected in the machine. This effect can be used to advantage with the carrier signal phase shift SVPWM to optimise the behaviour of the dual-stator PMSM. This allows the PWM harmonics to be reduced and the torque ripple to be minimised. Three typical switching frequencies from drive technology were used as comparison parameters. It was also shown that the CPS-SVPWM can achieve the same torque quality with half the switching frequency, which in turn leads to a reduction in switching losses. Furthermore, the CPS-SVPWM not only reduces the torque ripple, but also the magnetic losses of the machine, as the complementary currents, due to the CPS, cancel out the harmonics. In the dynamic case, it was analysed that the dq currents have a different course due to the carrier signal phase shift, whereas in the dynamic case without CPS, the torque-forming q currents lie exactly on top of each other. Despite the different currents, the rise times are the same.

6 Conclusion

The targeted use of the carrier signal phase shift SVPWM (CPS-SVPWM) can reduce PWM harmonics, which results in a reduction in torque ripple, switching frequencies and magnetic losses in a dual-stator PMSM with an alternating winding arrangement $C_{max,IO}$. Both the dynamics and the control behaviour show no differences to a DS-PMSM with and without a CPS-SVPWM. The combination of a magnetically coupled winding arrangement and a CPS-SVPWM can be used to advantage. In addition, the use of FPGAs for faster control algorithms and fast-switching power semiconductors harbours further research potential. This will make it possible to realise drive trains with increased efficiency in the future.

References

[1] R. Alexander, D. Meyer, and J. Wang, "A comparison of electric vehicle power systems to predict architectures, voltage levels, power requirements, and load characteristics of the future all-electric aircraft," in *2018 IEEE Transportation Electrification Conference and Expo (ITEC)*, 2018, pp. 194–200. DOI: 10.1109/ITEC.2018.8450240.

[2] H. Hu, M. Zha, L. Chang, and W. Kong, "Design of double redundancy controller for electric actuator based on dual three-phase pmsm," in *2021 IEEE 2nd China International Youth Conference on Electrical Engineering (CIYCEE)*, 2021, pp. 1–6. DOI: 10.1109/CIYCEE53554.2021.9676922.

[3] F. Barrero and M. J. Duran, "Recent advances in the design, modeling, and control of multiphase machines—part i," *IEEE Transactions on Industrial Electronics*, vol. 63, no. 1, pp. 449–458, 2016. DOI: 10.1109/TIE.2015.2447733.

[4] A. E. Hoffer, I. Petrov, J. J. Pyrhönen, J. A. Tapia, and G. Bramerdorfer, "Analysis of a tooth-coil winding permanent-magnet synchronous machine with an unequal teeth width," *IEEE Access*, vol. 8, pp. 71 512–71 524, 2020. DOI: 10.1109/ACCESS.2020.2987872.

[5] N. K. Bajjuri and A. K. Jain, "Torque ripple reduction in double-inverter fed wound rotor induction machine drives using pwm techniques," *IEEE Transactions on Industrial Electronics*, vol. 66, no. 6, pp. 4250–4261, 2019. DOI: 10.1109/TIE.2018.2866110.

[6] J. Prieto, E. Levi, F. Barrero, and S. Toral, "Output current ripple analysis for asymmetrical six-phase drives using double zero-sequence injection pwm," in *IECON 2011 - 37th Annual Conference of the IEEE Industrial Electronics Society*, 2011, pp. 3692–3697. DOI: 10.1109/IECON.2011.6119909.

[7] Y. Jia, J. Mo, K. Chen, and J. F. Pan, "Torque ripple suppression method for dual-stator motors based on cross-coupling control," in *2022 IEEE 9th International Conference on Power Electronics Systems and Applications (PESA)*, 2022, pp. 1–5. DOI: 10.1109/PESA55501.2022.10038344.

[8] A. Munteanu, "Six-phase fractional slot concentrated winding permanent magnet synchronous motor with reduced torque ripple," in *2022 International Conference and Exposition on Electrical And Power Engineering (EPE)*, 2022, pp. 527–530. DOI: 10.1109/EPE56121.2022.9959854.

[9] J. Winter, R. Zipprich, B. Tekir, and M. Ziegler, "Investigation of the influence caused by coupling the winding systems in a dual-stator pmsm using matlab/maxwell co-sim," in *PCIM Europe 2023; International Exhibition and Conference for Power Electronics, Intelligent Motion, Renewable Energy and Energy Management*, 2023, pp. 1–8. DOI: 10.30420/566091300.

[10] J. Karttunen, S. Kallio, P. Peltoniemi, P. Silventoinen, and O. Pyrhönen, "Dual three-phase permanent magnet synchronous machine supplied by two independent voltage source inverters," in *International Symposium on Power Electronics*

Power Electronics, Electrical Drives, Automation and Motion, 2012, pp. 741–747. DOI: 10.1109/SPEEDAM.2012.6264448.

[11] D. G. Holmes and T. A. Lipo, *Pulse width modulation for power converters: principles and practice*. John Wiley & Sons, 2003, vol. 18.

[12] F. Wang, Y. Zhang, X. Zeng, W. Kong, and R. Qu, "Torque harmonic elimination for dual three-phase pmsms based on cps-svpwm," in *2021 IEEE 4th International Electrical and Energy Conference (CIEEC)*, 2021, pp. 1–5. DOI: 10.1109/CIEEC50170.2021.9510761.

[13] A. Rihar, P. Zajec, and D. Vončina, "Cosimulation of ansys simplerer and matlab/simulink," in *2017 19th International Conference on Electrical Drives and Power Electronics (EDPE)*, 2017, pp. 313–317. DOI: 10.1109/EDPE.2017.8123222.

AUTHOR INDEX

Abbas, Khizra 764
Ackermann, Martin....................................1336
Aiello, Giuseppe 1217
Akbari, Saeed..2094
Akturk, Akin 739
Alauzet, Louis...2811
Albert, Tianlong..1759
Alfonso, Irene Maria Torres.....................2503
Alfonzetti, Emanuela 1844
Allioua, Abdelmoumin..............................2128
Ammar, Ahmed 1087
Appleby, Matthew......................................3276
Arai, Nobuhide 298
Araujo, Lucas..1673
Arnaudov, Dimitar 2268
Askan, Kenan..1545
Aspalter, Paul..2258
Augustin, Tim 3086
Aunon, Fernando 1467
Ausseresse, Pierrick 1082
Austrup, Isabel 2956
Babaki, Amir 1227
Bagheribavaryani, Mohammadreza 1418
Baharizadeh, Mehdi...................................378
Bai, Yeriel 1804
Baker, Nick 1923
Bándy, Kristóf...............................403, 2566
Barcelos, Renan Pillon...............................264
Barón, Kevin Muñoz 1978
Barth, Henry 2838
Basso, Christophe 3096
Bastawros, Adel 440, 1951
Batista, Emmanuel 2394
Baudais, Briac...3187
Behrendt, Stefan 361
Beiranvand, Hamzeh..................................1105
Beyerle, Raphael..958
Bhatia, Tamanna 1259
Bicer, Ekin Alp 40
Bimmel, Luc..3206
Blechinger, Christoph 1717
Block, Marius 2217
Bockholt, Yannick.....................................3334
Böhning, Lukas 2208
Boldyrjew-Mast, Roman............................723
Bosnjic, Zlatko..1788
Boutry, Arthur 1878
Bouzerd, Souhila.......................................581

Branas, Christian 2286
Brandl, Anja Katerina................................1613
Breidenstein, Daniel..................................1634
Bürger, Matthias 863
Cairnie, Mark 599
Calmels, Alain 3305
Cammarata, Federica 1289
Campos, Adriana 2663
Cannone, Marco 502
Capobianco, Thomas Anthony 1168
Çay, Yunus..3247
Cepin, Simon...1051
Chaisakdanugull, Chanuch.........................3067
Chatroux, Daniel 2278
Chatterjee, Bhaskar 774
Chen, Mengxing 424
Cherief, Wahid..1910
Cho, Wonjin Dylan 1046
Choo, Vin Loong 1775
Chorfi, Ilias 2175
Cinik, Sadik..2453
Colak, Baris 490
Colomer, Pau 456
Conilh, Christophe.....................................2227
Corbitt, Anna.......................135, 1123, 1821
Croston, Jose Andres Aguilar.....................3150
Curbow, Austin..1475
Cusumano, Andrea.....................................1627
Czerwenka, Philipp.....................1139, 3034
Daire, Baptiste...3110
Dasch, Michael..1907
Davoodi, Hossein 1013
Debbadi, Karthik 2963
Deboy, Gerald 15
Dedew, Mohamed Lemine 34
Delaforge, Timothé....................................1797
Denk, Marco..1192
Despesse, Ghislain.....................................797
Diz, Sergio De Lopez 411
Do, Nguyen Nghia 1428
Dresel, Lars 2737
Du, Xinyuan 1987
Duijsen, Peter Van..........1658, 2248, 2657, 3213
Dumollard, Yannick 1751
Dupont, Max 93
Dusmez, Serkan 383, 2334, 3060
Eichler, Felix...3020
Eyama, Takaaki 56

Fabian, Benjamin	190
Fenske, Florian	3390
Fey, Justin	1902
Fleck, Soenke	338
Förster, Nikolas	3237
Fotteler, Oleg	3328
Fräger, Lukas	926
Frank, Michael	754
Frank, Wolfgang	1770
Frei, Steffen	2478, 3007
Fuchs-Gade, Jannik	2632
Fuhrmann, Jan	1315
Gackowski, Bartosz	1504
Gandluru, Veera Bharath Chandra Reddy	2167
Gavin, Serge	1101
Gebhard, Thomas	1128
Gebhardt, Mathias	2769
Gellman, Ziv	608
Gendrin, Martin	909
Ghanbari, Alireza Ramezan	3175
Ghosh, Priyanka	1523
Gick, Sebastian	1264
Gioda, Alexis	3400
Girgin, Mehmet Oguz	3353
Giuffrida, Simone	248
Giuffrida, Vittorio	1065
Gleissner, Michael	2803
Goff, Gregoire Le	3160
Gomez, Antonio Miguel Munoz	625
Gottardo, Davide	2461
Gragger, Johannes	2104
Graham, Robert	1410
Groon, Fabian	3380
Groos, Gerhard	986
Guan, Jiajia	2591, 3395
Gudala, Bhavana	2524
Guiot, Eric	1604
Gunes, Ekrem R.	3221
Gupta, Gaurav	534
Gürlek, Yavuz	745
Haake, Daniel	2538
Haas, Tobias	2326, 3017
Haehre, Karsten	214
Haensel, Stefan	230
Hanf, Michael	351, 571
Harmand, Thomas	2138
Hasegawa, Kazunori	3002
Hauenschild, Philipp	1969
Hegarty, Timothy	1092
Hegde, Niranjan	1374
Heimler, Patrick	1955
Hellinger, Rolf	1

Hepp, Maximilian	3045
Herrera, Adolfo	1057
Herrmann, Clemens	731
Hertline, Joseph	1886
Herzog, Fabian	3136
Hirao, Takashi	1007
Hironaka, Yoichi	699
Hoffmann, Lennart	3264
Horat, Andreas	480
Hornbuckle, Malachi	2724
Hosseinzadehlish, Mana	1402, 1610
Hu, Jhih-Cheng	791
Huber, Jonas	254
Huerner, Andreas	681
Huselstein, Jean-Jacques	2547
Husev, Oleksandr	893
Igartuburu, Daniel San Laureano	2303
Imai, Ayano	180
Ippisch, Matthias	2638
Irifune, Hiroyuki	2028
Jahn, Simon	883
Jamal, Adeel	1346
Jappe, Tiago	2843
Jegal, Junhyeok	1590
Jha, Kunal	2930
Jia, Minli	2730
Jo, David	1732
Jones, Jeremy	1031
Kaiser, Jeremias	1538
Kampert, Erik	2342
Kanatzar, Paul	1361
Kangjia, He	62
Karout, Mohammed Amer	1835
Kasko, Igor	1991
Kato, Koji	1368
Kaufmann-Bühler, Marius	2400
Kawabata, Junya	2049
Keilmann, Robert	2972
Kempitiya, Asantha	497
Klever, Severin	1561
Knappstein, Lukas	1745
Knecht, Martin	3142
Koch, Jan-Niklas	2240
Koczy, Dawid	1651
Kohlhepp, Benedikt	2316
Koi, Kenichi	67
Kono, Hiroshi	2022
Kopischke, Ruben	2796
Körner, Patrick	615
Kragl, Robert	1385
Kreppel, Thomas	2416
Krigar, Tim	174

Kroics, Kaspars 510
Kugener, Jeff 3315, 3318
Kurukuru, Varaha Satya Bharath 875
Kuzmanoska, Sara 2745
Ladentin, Kevin 1964
Lambert, Adrien 1574
Langfermann, Sascha 1516
Lavery, Melanie 1485
Lee, Chih Hui 1152
Lee, Jongmu 1712
Lee, Kihyun 1724, 1737
Lemaitre, Damien 2596
Lenz, Travis 1352
Lenzen, Patrick 903
Leung, Wing Tai 74
Liao, Xinyuan 322
Lim, Alex 2937
Lindner, Lars 2370
Lippold, Florian 2981
Liu, Baihan 1072
Liu, Iris 1222
Liu, Yusi 3181
Lottis, Christian 3347, 3358
Lotz, Marc René 1457
Lu, Juncheng 19, 837
Lucia, Oscar 2448, 2513
Lutzen, Hauke 976
Lv, Jianwei 1872
Ma, Kwokwai 2778
Machtinger, Katharina 2119
Madloch, Sonja 369
Maheshwari, Ramkrishan 3118
Mai, Annette 284
Maier, Jannik 1642
Mandrioli, Riccardo 2576
Mannen, Tomoyuki 831
Mari, Jorge 843
Marie, Alexandre 2819
Martano, Emanuele 2874
Martínez, Alfonso 2359
Masuda, Akiyoshi 1018
Mauromicale, Giuseppe 2751
Mazzer, Simone 2162, 2532
McRae, Tim 2190, 2364, 3042
Medina-Garcia, Alfredo 1207
Meligy, Ahmed 1495
Menzel, Steffen 933
Merrouche, Abdennour 1113
Minamisawa, Renato Amaral 2036
Mirkovic, Nikola 2488
Mo, Xianghao 2386
Mochizuki, Yo 870

Mönch, Stefan 167
Mueller, Lukas 2425
Mühlfeld, Christian 3340
Muralikrishna, Ajay Krishna Voppu 2886
Nachete, Idriss 2408
Nakako, Hideo 49
Nawaz, Muhammad 2013
Nehmer, Dominik 24
Neira, Sebastian 2700
Neuner, Matthias 3296
Nikiforidis, Ioannis 916, 2718
Nkembi, Armel Asongu 2469
Oberdieck, Karl 1828
O'Keeffe, Rosemary 1249
Olalla, David 1814
Ong, Shu Ee 2942
Orlando, Stefano 1434, 2673
Otori, Daichi 518
Otte, Raphael 2234
Ouhab, Merouane 589, 2948
Owzareck, Michael 3371
Palma, Marco 1568
Panchal, Pranav 315
Paradkar, Sachin Shridhar 2627
Patterson, Andrew 1330
Paul, Indrajit 1133
Peng, Hujun 665
Petzold, Tom 1158
Pham, Thanh-Toan 2786
Philippe, Antoine 1441, 1449
Phung, Thanh Hai 1555, 2612
Piccioni, Andrea 2680
Piepenbrock, Till 463
Poller, Tilo 2914
Porpora, Francesco 222
Pouresmaeil, Mobina 419
Prince, Aswathy M. 2850
Rabay, Battist 2082
Radix, Bryan 2112
Radomsky, Lukas 3286
Randerath, Joschka 3256
Raßmann, Rando 2350
Rauh, Michael 690
Rebenklau, Lars 2088
Reddy, Niranjan Suravarapu 2273
Rehlaender, Philipp 2686
Reimann, René 2377
Reiner, Richard 557
Reißenweber, Lukas 447, 525
Reitz, Niclas 1393
Ren, Linhao 1917
Ren, Xufu 803

Rendek, Karol	2831
Reymond-Laruina, Frédéric	103
Rezaeizadeh, Amin	1686
Ribarich, Tom	1254
Ribeiro, Kelly	2294
Rillo, Oriol Subirats	290
Ringelmann, Tim	2708
Rodrigues, Luis Alves	635
Rodriguez, Manuel Escudero	812
Rodruigez, Manuel Escudero	3077
Rosensaft, Boris	2620
Rudzki, Jacek	1942
Ruoff, Dominik	1277
Ruppert, Lukas	1703
Sakai, Junya	2006
Salomez, Florentin	2431
Samura, Koki	2764
Sankari, Rasched	197
Sawada, Takashi	161
Schindler, Stefanie	1147
Schindler, Tobias	140
Schmidhuber, Michael	2438
Schmidt, Matthias	397
Schmidt, Paul	1999
Schmitz, Laurids	2995
Schnell, Raffael	855
Schnitzler, Ruben	1851
Schulte, Felix	3364
Schulz, Martin	1077
Schwab, Stefan	343
Schwarz, Niklas	1200
Scuto, Alfio	2921
Seber, Elizabeth	888
Sekar, Ajith Kumar	2041
Sen, Gokhan	784
Seo, Hansol	1896
Sheikhan, Alireza	549, 2606
Shi, Sanbao	1212, 3029
Sifoune, Sarah	390
Singer, Mehyeddine	2309
Solomakha, Oleksandr	1174
Somarin, Hasan Mousavi	2494
Sos, Carlos Costas	205
Sousa, Gean	2557
Srikrishna, N. H	3269
Steenbock, Liska	3169
Steiner, Felix	1891
Stone, David A.	949
Subotic, Stefan	274
Sugie, Hisashi	1765
Sun, Qing	2791
Suzuki, Keita	2053

Syed, Hadiuzzaman	2758
Talits, Kevin	997
Tan, John Emmanuel	150
Tanikawa, Kohei	1272
Tarmoom, Ehab	942
Tekir, Bünyamin	672
Tengvall, Sebastian	1380
Thamm, Merlin	1532
Thekemuriyil, Tanya	645, 2986
Thirukoluri, Rajani Kumar	1865
Thomas, Mark	564
Thönnessen, André	1584
Tigira, Sandu	3130
To, Pham Ha Trieu	707
Tobler, Stefan	472
Tokorozuki, Takeshi	849
Torrisi, Marco	822
Tranchero, Maurizio	1322
Troudi, Rami	1185
Tuncay, Sebnem	1858
Uemura, Hirofumi	433
Ueno, Masaki	2909
Ugur, Abdulkerim	3229
Uhlemann, Andre	1283
Urbaneck, Daniel	2152
Varadarajan, Kamal	1598
Vemulapati, Umamaheswara Reddy	1025
Vinciguerra, Vincenzo	1039
Vobecky, Jan	1002
Vogelsberger, Markus	1307
Vogt, Michael	2866
Vuletic, Radovan	305
Walter, Michael	1297
Wang, Hamlin	2185
Wang, Hao	2693
Wang, Lei	1242
Wang, Lisheng	2060
Wang, Qilei	113
Wang, Rui	84
Wang, Yushi	966
Watanabe, Hiroki	3125
Weckbrodt, Julien	121
Wei, Frank	2146
Wei, Suhang	2074
Weihe, Sven	330
Wen, Jin	2182, 3103
Wessel, Wilfried	655
Wietschel, Martin	7
Wille, Christopher	128
Winkler, Paul	1511
Xie, Dong	2880
Xie, Luhong	1930

Yadav, Sachin..2583
Yan, Xingda 1680
Yan, Yiyang..1809
Ye, Yijun 2826
Ye, Zhong 1621, 2646
Yoshida, Satoshi..543
Yoshioka, Kentaro..2067
Yu, Renze..717
Yu, Sean 1180, 2518
Yu, Sheng-Yang 2200
Zeng, Chenhang 1693
Zhang, Chi 1781
Zhang, Hongpeng 238
Zhang, Huaiyuan...2901
Zhang, Yi 1936
Zhao, Yue 3197
Zheng, Zexiang 2860
Zhu, Shiwu..2893
Zipperstein, David 2651
Zipprich, Robert...1664
Zocher, Markus..1233

VDE VERLAG GMBH
Bismarckstr. 33
P.O.B. 12 01 43
10625 Berlin, Germany

ISBN 978-1-7138-9966-2